Jane's

ELECTRO-OPTIC SYSTEMS

Edited by Michael J Gething

Ninth Edition

2003-2004

Total number of entries 1,641 New and updated entries 712
Total number of images 1,566 New images 70

Visit jeos.janes.com and view the list of latest updates that have been added to the online version of *Jane's Electro-Optic Systems* subsequent to this print edition.

ISBN 0 7106 2543 X

In the USA and its dependencies
Jane's Information Group Inc, 110 N Royal Street, Suite 200, Alexandria, Virginia 22314, USA

EDITORIAL AND ADMINISTRATION

Director: Ian Kay, e-mail: Ian.Kay@janes.com

Managing Editors: David Shipton, e-mail: David.Shipton@janes.com
Simon Michell, e-mail: Simon.Michell@janes.com and

Content Services Director: Anita Slade, e-mail: Anita.Slade@janes.com

Content Systems Manager: Jo Agius, e-mail: Jo.Agius@janes.com

Pre-Press Manager: Christopher Morris, e-mail: Christopher.Morris@janes.com

Team Leader: Melanie Rovery, e-mail: Melanie.Rovery@janes.com

Content Editor: Melanie Rovery, e-mail: Melanie.Rovery@janes.com

Production Controller: Victoria Powell, e-mail: Victoria.Powell@janes.com

Content Update: Jacqui Beard, Information Collection Co-Ordinator
Tel: (+44 20) 87 00 38 08 Fax: (+44 20) 87 00 39 59
e-mail: yearbook@janes.com

Jane's Information Group Limited, Sentinel House, 163 Brighton Road, Coulsdon, Surrey CR5 2YH, UK
Tel: (+44 20) 87 00 37 00 Fax: (+44 20) 87 00 37 88
e-mail: jeos@janes.com

SALES OFFICE

Send Europe, Middle East and Africa enquiries to: *Mike Gwynn – Head of Information Sales*
Jane's Information Group Limited, Sentinel House, 163 Brighton Road, Coulsdon, Surrey CR5 2YH, UK
Tel: (+44 20) 87 00 37 00 Fax: (+44 20) 87 63 10 06
e-mail: info@janes.com

Send USA enquiries to: *Robert Loughman – Sales Director*
Jane's Information Group Inc, 110 N Royal Street, Suite 200, Alexandria 22314, USA
Tel: (+1 703) 683 37 00 Fax: (+1 703) 836 02 97 Telex: 6819193
Tel: (+1 800) 824 07 68 Fax: (+1 800) 836 02 97
e-mail: info@janes.com

Send Asia enquiries to: *David Fisher – Group Business Manager*
Jane's Information Group Asia, 78 Shenton Way , #10-02, Singapore 079120
Tel: (+65) 63 25 08 66 Fax: (+65) 62 26 11 85
e-mail: info@janes.com.sg

Send Australia/New Zealand enquiries to: *Pauline Roberts – Business Manager*
Jane's Information Group, PO Box 3502, Rozelle Delivery Centre, New South Wales 2039, Australia
Tel: (+61 2) 85 87 79 00 Fax: (+61 2) 85 87 79 01
e-mail: info@janes.thomson.com.au

Send Middle East enquiries to: *Ali Abdellatif Siali*
Jane's Information Group, PO Box 502138, Dubai, United Arab Emirates
Tel (+971 4) 390 23 35 Fax (+971 4) 390 88 48
e-mail: mideast@janes.com

Send Japan enquiries to
Jane's Information Group, Palaceside Building, 5F, 1-1-1, Hitotsubashi, Chiyoda-ku, Tokyo 100-0003, Japan
Tel: (+81 3) 52 18 76 82 Fax (+81 3) 52 22 12 80
e-mail: norihisa.fukuyama@janes.jp

ADVERTISEMENT SALES OFFICES

(Head Office)
Jane's Information Group
Sentinel House, 163 Brighton Road,
Coulsdon, Surrey CR5 2YH, UK
Tel: (+44 20) 87 00 37 00
Fax: (+44 20) 87 00 38 59/37 44
e-mail: defadsales@janes.com

Richard West, Senior Key Accounts Manager
Tel: (+44 1892) 72 55 80 Fax: (+44 1892) 72 55 81
e-mail: richard.west@janes.com

Joni Beeden, Advertising Sales Executive
Tel: (+44 20) 87 00 39 63 Fax: (+44 20) 87 00 38 59/37 44
e-mail: joni.beeden@janes.com

Nicky Eakins, Advertising Sales Executive
Tel: (+44 20) 87 00 38 53 Fax: (+44 20) 87 00 38 59/37 44
e-mail: nicky.eakins@janes.com

(USA/Canada office)
Jane's Information Group
110 N Royal Street, Suite 200,
Alexandria, Virginia 22314, USA
Tel: (+1 703) 683 37 00
Fax: (+1 703) 836 55 37
e-mail: defadsales@janes.com

USA and Canada
Katie Taplett, US Advertising Sales Director
Tel: (+1 703) 683 37 00 Fax: (+1 703) 836 55 37
e-mail: katie.taplett@janes.com

Northern USA and Eastern Canada
Harry Carter, Northeast Region Advertising Sales Manager
Tel: (+1 703) 683 37 00 Fax: (+1 703) 836 55 37
e-mail: harry.carter@janes.com

Southeastern USA
Kristin D Schulze, Advertising Sales Manager
PO Box 270190, Tampa, Florida 33688-0190
Tel: (+1 813) 961 81 32 Fax: (+1 813) 961 96 42
e-mail: kristin@intnet.net

Western USA and Western Canada
Richard L Ayer
127 Avenida Del Mar, Suite 2A, San Clemente, California 92672
Tel: (+1 949) 366 84 55 Fax: (+1 949) 366 92 89
e-mail: ayercomm@earthlink.com

Australia: *Richard West* (see UK Head Office)

Benelux: *Nicky Eakins* (see UK Head Office)

Brazil: *Katie Taplett* (see USA address)

Corporate Accounts: Simon Kay
33 St John's Street, Crowthorne, Berkshire RG45 7NQ, UK
Tel: (+44 1344) 77 71 23 Mobile: (+44 7702) 54 96 84
Fax: (+44 1344) 77 58 85
e-mail: simon.kay@btclick.com

Eastern Europe (excl. Poland) : MCW Media & Consulting Wehrstedt
Dr Uwe H Wehrstedt
Hagenbreite 9, D-06463 Ermsleben, Germany
Tel: (+49) 0700/WEHRSTEDT / (+49) 03 47 43/620 90
Fax: (+49) 03 47 43/620 91
e-mail: info@Wehrstedt.org

France: Patrice Février
BP 418, 35 avenue MacMahon,
F-75824 Paris Cedex 17, France
Tel: (+33 1) 45 72 33 11 Fax: (+33 1) 45 72 17 95
e-mail: patrice.fevrier@wanadoo.fr

Germany and Austria: *MCW Media & Consulting Wehrstedt* (see Eastern Europe)

Greece: *Nicky Eakins* (see UK Head Office)

Hong Kong: *Joni Beeden* (see UK Head Office)

India: *Joni Beeden* (see UK Head Office)

Israel: Oreet – International Media
15 Kinneret Street, IL-51201 Bene Berak, Israel
Tel: (+972 3) 570 65 27 Fax: (+972 3) 570 65 27
e-mail: admin@oreet-marcom.com
Defence: Liat Shaham
e-mail: liat_s@oreet-marcom.com

Italy and Switzerland: Ediconsult Internazionale Srl
Piazza Fontane Marose 3, I-16123 Genoa, Italy
Tel: (+39 010) 58 36 84 Fax: (+39 010) 56 65 78
e-mail: genova@ediconsult.com

Middle East: *Joni Beeden* (see UK Head Office)

Pakistan: *Joni Beeden* (see UK Head Office)

Poland: *Nicky Eakins* (see UK Head Office)

Russia: Vladimir N Usov, PO Box 98, Nizhniy Tagil,
Sverdlovsk Region, 622018, Russia
Tel/Fax: (+7 3435) 23 02 68
e-mail: uvn125@uraltelecom.ru

Scandinavia: The Falsten Partnership
PO Box 27, Brighton BN41 2AX, UK
Tel: (+44 1273) 41 33 44 Fax: (+ 44 1273) 41 33 55
e-mail: sales@falsten.com

Singapore: *Richard West/Joni Beeden* (see UK Head Office)

South Africa: *Richard West* (see UK Head Office)

South Korea: Infonet Group Inc
Sanbu Rennaissance Tower 902, 456 Gongdukdong, Mapogu, Seoul, South Korea
Contact: Mr Jongseog Lee
Tel: (+82 2) 716 99 22
Fax: (+82 2) 716 95 31
e-mail: jslee@infonetgroup.co.kr

Spain: Via Exclusivas SL
Contact: Julio de Andres
Viriato 69SC, E-28010 Madrid, Spain
Tel: (+34 91) 448 76 22 Fax: (+34 91) 446 02 14
e-mail: j.a.deandres@viaexclusivas.com

Turkey: *Richard West* (see UK Head Office)

ADVERTISING COPY
Linda Letori (Jane's UK Head Office)
Tel: (+44 20) 87 00 37 42 Fax: (+44 20) 87 00 38 59/37 44
e-mail: linda.letori@janes.com

For North America, South America and Caribbean only:
Lia Johns (Jane's USA/Canada Office)
Tel: (+1 703) 683 37 00 Fax: (+1 703) 836 55 37
e-mail: lia.johns@janes.com

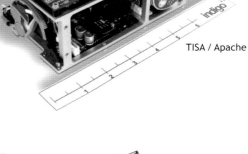

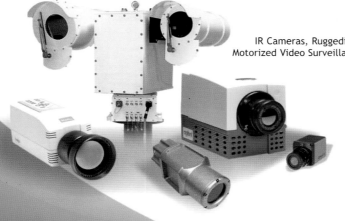

Contents

Jane's Electro-Optic Systems website: jeos.janes.com

How to use *Jane's Electro-Optic Systems*

Jane's Electro-Optic Systems provides information on naval, land and airborne military electro-optic systems or systems with electro-optic elements. These electro-optic elements include either infra-red, thermal imaging, image intensifying or laser technology. Systems included are either in production, in service or under development.

Jane's Electro-Optic Systems is generally organised in terms of the main application of the equipment and is divided into four main sections. The application sections cover naval, land and airborne systems. Each of the application sections is further divided into subsections, broadly on missiles, countermeasures, weapon control and observation and surveillance. This, like any classification system, has its limitations. Some systems have more than one application and could be put in several different subsections. Certain thermal imagers, for example, may be used on a variety of platforms. Where this is the case, they have normally been assigned to the land section, since this is considered the main area of use. Some systems do not conveniently fall into any category and rather than have a 'Miscellaneous' section, they have been placed in the section with the closest correspondence in terms of application. The aim has been to keep the classification as simple as possible while being meaningful in terms of operational use.

At the beginning of the application and technology sections is a summary providing a brief explanation of the types of systems to be included in each subsection. Entries are structured as follows:

Title

Development/Description: contains information on the development history, application and technical details of the system.

Operational status: contains information on the development/production/sales/service status of the system.

Specifications: lists the main technical and performance parameters.

Contractor: lists the prime contractors.

In addition to the main text, *Jane's Electro-Optic Systems* includes two indexes: an alphabetical systems index and a manufacturers' index that correlates entries to manufacturers, listed alphabetically by manufacturer. There is also a list of contractors including address, telephone, fax numbers and e-mail addresses.

To help users evaluate the data of this edition, the following identifiers have been used:

● ***VERIFIED*** The editor has made a detailed examination of the entry's content and checked its relevance and accuracy for publication in the new edition to the best of his ability.

● ***UPDATED*** During the verification process, significant changes to content have been made to reflect the latest position known to Jane's at the time of publication.

● ***NEW ENTRY*** Information on new equipment and/or systems appearing for the first time in the title.

● ***NEW*** New images are identified as ***NEW***. Some are followed by a seven-digit number for ease of identification by our image library.

Any update to the content of this product will appear online as it occurs (see jeos.janes.com for the additional benefits of an online subscription and details of our free online trial) and will be incorporated annually into future print editions of *Jane's Electro-Optic Systems.*

Total number of entries 1,641 New and updated entries 712
Total number of images 1,566 New images 70

Visit jeos.janes.com and view the list of latest updates that have been added to the online version of *Jane's Electro-Optic Systems* subsequent to this print edition.

British Library Cataloguing-in-Publication Data.
A catalogue record for this book is available from the British Library.

Printed and bound in Great Britain by Biddles Ltd, Guildford and Kings Lynn

Quality Policy

Jane's Information Group is the world's leading unclassified information integrator for military, government and commercial organisations worldwide. To maintain this position, the Company will strive to meet and exceed customers' expectations in the design, production and fulfilment of goods and services.

Information published by Jane's is renowned for its accuracy, authority and impartiality, and the Company is committed to seeking ongoing improvement in both products and processes.

Jane's will at all times endeavour to respond directly to market demands and will also ensure that customer satisfaction is measured and employees are encouraged to question and suggest improvements to working practices.

Jane's will continue to invest in its people through training and development, to meet the Investor in People standards and changing customer requirements.

Jane's

Glossary

This glossary deals with technical terms only and not standard SI or other units, names of organisations, or of specific programmes; the latter will be found in the general index.

A few words of explanation are provided where appropriate. For further technical detail, an excellent reference text is *"The Infra-Red and Electro-Optical Systems Handbook"*, edited by J S Accetta and D L Shumaker, published by SPIE/ERIM (1993); Volume 5 of this set is particularly relevant.

Because of the potential for confusion between different video standards and different measures of image resolution, some further notes on these topics are provided at the end of this glossary.

AA	Anti-Aircraft
AAM	Air-to-Air Missile
AAW	Anti-Air Warfare
ABC	Automatic Brightness Control (for image intensifiers)
Absorption coefficient	Fraction of energy absorbed per unit pathlength
AC	Alternating Current
ACLOS	Automatic Command to Line of Sight (guidance mode of a missile)
AEW	Airborne Early Warning (aircraft)
AFV	Armoured Fighting Vehicle
AGC	Automatic Gain Control
AGL	Above Ground Level (height of an aircraft)
Angle of elevation	The angle between the line of sight and the horizontal plane
AP	Armour-Piercing (ammunition)
APDS	Armour-Piercing Discarding Sabot (ammunition)
APFSDS	Armour-Piercing Fin-Stabilised Discarding Sabot (ammunition)
APC	Armoured Personnel Carrier
APD	Avalanche Photodiode (provides higher gain than PIN diode detector; often used in LRF receivers)
Anti-reflection coating	A thin film of material applied to an optical surface to reduce the reflectivity and increase the transmission of radiation through the surface
ASuW	Anti-Surface Warfare
ASW	Anti-Submarine Warfare
ATBM	Anti-Tactical Ballistic Missile
ATGM	Anti-Tank Guided Missile (almost synonymous with ATGW)
ATGW	Anti-Tank Guided Weapon
BIT	Built-In Test
BITE	Built-In Test Equipment
C^2	Command and Control
C^3I	Command, Control, Communications and Intelligence
CAS	Close Air Support
CCD	Charge Coupled Device (solid-state TV imaging detector chip)
CCIR	see note below on video standards
CCTV	Closed-Circuit TV
CEP	Circular Error Probability (a measure of the accuracy of bomb or missile targeting)
CIC	Command and Information Centre (on a ship)
CIWS	Close-In Weapons System
CLGP	Cannon-Launched Guided Projectile
CLOS	Command to Line of Sight (guidance mode of a missile)
CMOS	Complementary Metal Oxide Silicon
CMT	Cadmium Mercury Telluride (commonly used IR detector material). Also known as MCT. Made in PV or PC variants qv
COTS	Commercial Off-The-Shelf
CRT	Cathode Ray Tube (display)
CW	Continuous Wave
DAS	Defensive Aids System
DC	Direct Current
DF	Direction Finding
DFoV	Dual FoV (Field of View)
DIRCM	Directed/Directional IR Countermeasure
Divergence	The bending of light beams away from each other, for example by a lens
ECCM	Electronic Counter Countermeasure (capability to resist ECM)
ECM	Electronic Countermeasure
EFL	Effective Focal Length
EFP	Explosively Formed Projectile (type of missile warhead)
EMC	Electromagnetic Compatibility
EMI	Electromagnetic Immunity
E-O	Electro-Optic(al)
E-O detector	A component that detects radiation by the effect of light in generating an electrical signal
EOCCM	EO Counter Countermeasure (capability to resist EOCM)
EOCM	EO Countermeasure
EOD	Explosive Ordnance Disposal
ERA	Explosive Reactive Armour
ESM	Electronic Support Measures
EW	Electronic Warfare
F (or f) number	The ratio of the focal length of a lens to its diameter
FAC	Forward Air Controller
FAC	Fast Attack Craft
FCS	Fire-Control System
FDDI	Fibre-Distributed Data Interface
FFT	Fast Fourier Transform
FLIR	Forward Looking Infra-Red (typically a fixed-direction narrow-FOV system, with a display for the user)
FM	Frequency Modulation
FO	Fibre Optic (sometimes used in the form of a twister to invert an image, or as a taper to couple an image intensifier to a CCD camera)
FOV	Field of View
FPA	Focal Plane Array (as opposed to a scanned array)
Gen (or GEN) I, II, III	The generations of image intensifiers used in NVG. Earliest electrostatically focused Gen I tubes had low gain. Gen II introduced MCP for much higher gain; Gen III introduced improved III-V (GaAs) photocathodes. A confusing variety of proprietary names are also used such as SuperGen and GEN II Super
GPS	Global Positioning System
H	Horizontal (referring to FOV)
HE	High-Energy (warhead explosive)
HEAT	High-Energy Anti-Tank (ammunition)
HEL	High-Energy Laser
HESH	High-Energy Squash Head (ammunition)
HF	High-Frequency
HMD	Helmet-Mounted Display
HOE	Holographic Optical Element
HUD	Head-Up Display
HVM	High- (or Hyper) Velocity Missile
ICCD	Intensified CCD (CCD TV camera with image intensifying stage)
ICV	Infantry Combat Vehicle
IDCA	Integrated Detector/Cooler Assembly
IFF	Interrogation Friend or Foe
IFV	Infantry Fighting Vehicle
II or I^2	Image intensifier(d)
IIR, I^2R	Imaging IR (as distinct from earlier generation scanned IR systems)
INS	Inertial Navigation System
IR	Infra-Red
IRCCD	Infra-Red CCD

IRCCM	IR Counter Countermeasure (capability to resist IRCM)
IRCM	IR Countermeasure
IRFPA	IR Focal Plane Array
IRLS	IR Line Scan
IRST	IR Search and Track (differs from FLIR in that the FOV is mechanically steerable in the direction of choice, the primary destination of the image information is a computer rather than a display screen and autotracking functions are built in).
JT	Joule Thomson (cooler for IR detector). A cooling technique which uses the expansion of high-pressure gas. By forcing the gas, usually nitrogen or argon, through a narrow nozzle, the gas expands and absorbs heat causing its surroundings to cool
KE	Kinetic Energy (weapon)
KTP	Potassium Titanate Phosphate (non-linear crystal used for laser frequency or wavelength shifting)
LAN	Local Area Network
Laser designator	An instrument for weapon delivery applications, the laser illuminates the target with a coded signal. The attacking missile launched from a platform which can be some distance from the designator, has a laser sensor which detects the reflected code signal from the target and provides the homing signal to guide the missile to the target
Laser range-finder	An instrument to measure the range of a target
LAV	Light Armoured Vehicle
LCD	Liquid Crystal Display
LED	Light Emitting Diode
LLADS	Low-Level Air Defence System
LLTV, LLLTV	Low-Light Level TV
LOAL	Lock-On After Launch
LOBL	Lock-On Before Launch
LOROP	Long Range Oblique Photographic
LOS	Line Of Sight
LPE	Liquid Phase Epitaxy (method of manufacturing IR detectors)
LRF	Laser Range-finder
LRU	Line Replaceable Unit
LST	Laser Spot Tracker
LTD	Laser Target Designator
LWIR	Long-Wave IR (the 8 to 12 μm band)
LWR/LWS	Laser Warning Receiver/System
MANPADS	Man-Portable Air Defence System
MBT	Main Battle Tank
MCLOS	Manual CLOS (guidance mode of a missile)
MCM	Mine Countermeasures (ship)
MCP	Microchannel Plate
MCT	Mercury Cadmium Telluride (see CMT)
MLU	Mid-Life Update
MMI	Man/Machine Interface
MOVPE	Metal Organic Vapour Phase Deposition (method of manufacturing IR detectors)
MRTD (or MRT)	Minimum Resolvable Temperature Difference (a subjective measure of the thermal contrast sensitivity of an IR system including its display, usually quoted in °C or K at a given image resolution expressed in lp/mrad).
MTBF	Mean Time Between Failures
MTI	Moving Target Indication
MTTR	Mean Time To Repair
MWIR	Mid-Wave IR (the 3 to 5 μm band)
NBC	Nuclear, Biological and Chemical
NDI	Non-Developmental Item
NEI	Noise Equivalent Irradiance
NETD (or NET)	Noise Equivalent Temperature Difference (differs from MRTD, in that it is a measure of contrast sensitivity defined as equivalent to the electronic noise level of the receiver)
NFOV	Narrow Field Of View (for system having more than one FOV)
NGS	Naval Gunfire Support
NVB	Night Vision Binocular
NVG	Night Vision Goggles
OEM	Original Equipment Manufacturer
OPO	Optical Parametric Oscillator (non-linear crystal, for example KTP, used for shifting laser wavelength)
OTA	Overfly Top-Attack (anti-armour missile attack mode)
PC	Personal Computer
PC	Photoconductive (mode of operation of a photodetector)
PGM	Precision Guided Munition (often SAL guided)
PIN	Positive-Intrinsic-Negative (type of semiconductor photodiode structure)
PN	Proportional Navigation (guidance mode of a missile)
PPI	Plan Position Indicator (radar display)
PRF	Pulse Repetition Frequency
PST	Lead Scandium Tantalate (a thermal detector material)
PV	Photovoltaic (mode of operation of a photodetector)
QWIP	Quantum Well Infra-red Photodetector
RAM	Radar-Absorbing Material
Raman effect	When light is scattered through a transparent material, part of the light is scattered in all directions. The frequency of much of the scattered light is identical to the frequency of the incident beam. A part of the scattered light has frequencies different from the frequency of the incident beam by values related to the emission or absorption energies of the atoms or molecules of the scattering material. This part is called Raman scattering. If the frequency ν of the incident light is varied, then the frequencies of the Raman scattered photons maintain constant frequency differences from ν
RCS	Radar Cross Section
RF	Radar Frequency
RHA	Rolled Homogeneous Armour
RMS	Root Mean Square
RPV	Remotely Piloted Vehicle (see also UAV)
RWR	Radar Warning Receiver
SACLOS	Semi-Automatic CLOS (guidance mode of a missile)
SAL	Semi-Active Laser (missile guidance using laser designation)
SAM	Surface-to-Air Missile
SAR	Synthetic Aperture Radar
SFW	Sensor Fused Weapon
SLR	Single Lens Reflex (camera)
SP	Self-Propelled
SPRITE	Signal Processing In The Element (a proprietary technique performing on-chip signal integration in a scanned IR detector)
SRU	Shop Replaceable Unit
SSBN	Submarine (ballistic missile, nuclear powered)
SSKP	Single-Shot Kill Probability
SSN	Submarine (nuclear powered)
TBM	Tactical Ballistic Missile
TDI	Time Delay and Integration
TEL	Transporter-Erector-Launcher (for TBM)
Thermo-electric cooling	A cooling technique which exploits the 'Peltier Effect' by which current flowing across a junction between two dissimilar materials causing one material to heat while the other cools
TI	Thermal Imager/Imaging
TICM	Thermal Imaging Common Module
TIS	Thermal Imaging System
TOF	Time of Flight
TV	Television
TVL	TV Lines (a measure of image resolution)
TWT	Travelling Wave Tube
UAV	Unmanned Aerial Vehicle
UHF	Ultra-High Frequency
UTM	Universal Transverse Mercator
UV	Ultra-violet (wavelengths shorter than 400 nm)
V	Vertical (referring to FOV)
VCR	Video Cassette Recorder
VHF	Very High Frequency
VLSI	Very Large Scale Integration (of electronic circuits)
WFOV	Wide Field of View (for system having more than one FOV)
WRA	Weapon Replaceable Assembly (US term for LRU qv)

UNITS

Angle

Fields of view and resolutions of EO systems may be expressed in a variety of different units. Angle may be denominated in:

- degrees (°)
- mrad (milliradians, that is, one thousandth of a radian – 1 mrad being approximately 0.0573°)
- mil (1 mil is 1/6400 of a circle, that is, 0.05625°, almost equal to 1 mrad)
- grad (1 grad is 1/100 of a right angle, that is, 0.9°)

Image resolutions may be expressed as:

- TV lines (per picture height)
- lp/mm (line pairs per mm) or cyc/mm (cycles per mm), referred to the linear image size
- lp/mrad or cyc/mrad, in angular terms

Linear and angular scales are related to each other through the focal length of the system.

Laser beamwidths are often expressed in mrad, but the definition may be stated in terms of the width at half maximum (that is, 50 per cent amplitude), the width at 90 per cent points, 1/e points (37 per cent), or $1/e^2$ points (13.5 per cent).

Wavelength

Wavelengths in the visible region are usually expressed in nm (nanometres (10^{-9} m), and range from approximately 400 nm (or 0.4 µm) in the blue to 700 nm (0.7 µm) in the red. Infra-red wavelengths tend to be denominated in µm (micrometres (10^{-6} m) or microns; 1 µm = 1000 nm). Ultra-violet wavelengths are less than 400 nm.

Illumination/luminance

Lux – the S.I. unit of illumination. Typical ambient light levels range through:
10^{-4} – Overcast (starlit) sky
10^{-2} – Starlight
10^{-1} – Full moon
10 – Twilight
10^4 – Overcast day
10^5 – Bright sunlight
foot-Lambert (fL) – unit of luminance of a source used in the US. Elsewhere Candela/m^2 (approximately 3.43 fL) is generally used.

Materials

Some of the common materials are mentioned in the preceding glossary listing, but a separate summary here is thought helpful.

Beryllium oxide	A dielectric ceramic semiconductor material with high electrical resistivity and high thermal conductivity
Cadmium Mercury Telluride (CdHgTe)	A material which is sensitive to IR radiation and which generates an electrical output when stimulated. The most common IR detector material
Germanium (Ge)	A shiny semi-conductor material used for windows and lenses in infra-red imaging systems
Indium Antimonide (InSb)	A semiconductor material used as an infra-red detector for radiation of wavelengths of 1 to 6 µm (near to mid-wave IR)
Lead Scandium Tantalate (PST)	A thermal detector material
Lead Selenide (PbSe)	A photoconductive detector material, sensitive to the infra-red portion of the spectrum covering wavelengths of 1 to 7 µm
Lithium Fluoride (LiF)	A crystalline material used for windows and other components in the ultra-violet, visible and infra-red. It has very high transmittance from 140 nm (in the UV) to its infra-red absorption edge at 1.8 µm
Lithium Niobate ($LiNbO_3$)	A crystalline ferroelectric material with very high electro-optic and piezoelectric coefficients. Lithium Niobate is used as a pyroelectric material in pyroelectric infra-red detectors
Lithium Tantalate ($LiTaO_3$)	A pyroelectric material used for pyroelectric infra-red detectors
Nd:Glass	Neodymium:glass
Neodymium: YAG	Neodymium Yttrium Aluminium Garnet. Yttrium aluminium garnet doped with neodymium is the lasing medium of the Nd:YAG laser. The laser wavelength is 1.064 µm. Uses include laser range-finding and laser radar
Potassium Titanate Phosphate (KTP)	A non-linear crystal used for laser frequency or wavelength shifting
Zinc Sulphide (ZnS)	A polycrystalline material which transmits in the infra-red

VIDEO STANDARDS

Video standards are set by bodies such as the CCIR and EIA.

- CCIR (Comité Consultatif des Radio Communication/International Radio Consultative Committee)
- EIA (Electronic Industries Association (US). Produce RS (recommended standards)

Standards in common usage include:

- CCIR. Set of CCTV standards, used outside US and Japan (625 line, 50 Hz)
- NTSC. Broadcast standard in US and Japan. Equivalent to RS-170A
- PAL. European broadcast standard (625 lines, 50 fields/s, 2:1 interlace). Equivalent to CCIR System 1
- RS-170. Monochrome video (525 lines, 60 fields/s, 2:1 interlace)
- RS-170A. Colour, comparable to RS-170
- RS-330. Similar to RS 170 (525 lines, 60 fields/s, 2:1 interlace)
- RS-343A. High-resolution monochrome CCTV (875 lines, 50 or 60 fields/s)

Alphabetical list of advertisers

A

Aselsan
PO Box 101, Yenimahalle 06172, Ankara,
Turkey .... *Facing inside front cover*

D

Diversified Optical Products
282 Main Street, Salem, USA .... [7]

E

El-Op
Advanced Technology Park, Kiryat Weizmann, PO Box 1165,
IL-76111 Rehovot, Israel .... *Outside back cover*

F

FLIR SYSTEMS
Hill Avenue, Kings Hill, West Malling,
Kent ME19 4AQ .... *Outside front cover*

I

Indigo Systems
50 Castilian Drive, Goleta, California,
USA 93117 .... *Facing pages* [2] *and* [3]

R

Radamec Defence Systems
Bridge Road, Chertsey, Surrey KT16 8LJ, UK .... *Bookmark*

S

Sagem
Le Ponant de Paris, 27 rue Leblanc, F-75512 Paris Cedex,
France .... *Opposite title page*

V

Vectronix AG
Max-Schmidheiny-Strasse, CH-9435 Heerbrugg,
Switzerland .... *Inside front cover*

Users' Charter

This publication is brought to you by Jane's Information Group, a global company with more than 100 years of innovation and unrivalled reputation for impartiality, accuracy and authority.

Our collection and output of information and images is not dictated by any political or commercial affiliation. Our reportage is undertaken without fear of, or favour from, any government, alliance, state or corporation.

We publish information that is collected overtly from unclassified sources, although much could be regarded as extremely sensitive or not publicly accessible.

Our validation and analysis aims to eradicate misinformation or disinformation as well as factual errors; our objective is always to produce the most accurate and authoritative data.

In the event of any significant inaccuracies, we undertake to draw these to the readers' attention to preserve the highly valued relationship of trust and credibility with our customers worldwide.

If you believe that these policies have been breached by this title, you are invited to contact the editor.

A copy of Jane's Information Group's Code of Conduct for its editorial teams is available from the publisher.

INVESTOR IN PEOPLE

Executive Overview

To take over a Yearbook four-fifths the way through its production cycle is no easy task and so I begin the Executive Overview to the ninth edition of *Jane's Electro-Optic Systems* by acknowledging the great debt of gratitude I owe for the work undertaken by my predecessor, Keith Atkin. On his output, I have begun to build my own style, based on 30 years as an aerospace/defence writer and editor, 10 of which (come December 2003) have been with Jane's.

In this age of instant electronic communication, it is impossible to state categorically that everything contained within this tome is 100 per cent up-to-date and accurate. It is a fact of life that, 30 minutes after signing off the last page of a publication and the presses have begun to roll, the probability of a major announcement that will effect some element of the content is high. With four weeks between writing these words and seeing bound copies of the book, I ruefully expect at least one such announcement. However this may effect the content of *Jane's Electro-Optic Systems*, I crave the reader's indulgence. The product remains a valuable 'snapshot in time'. The online service will of course be updated as soon as new information is received.

Yet, the 'instant erudition' provided by the World Wide Web and e-communication can work both ways. Like all Jane's products, *Jane's Electro-Optic Systems* is now updated on a monthly basis for electronic and Online customers. For an editor, this theoretically spreads the workload more evenly. Of course, it can also mean some entries receiving multiple updates over the course of 12 months.

Meanwhile, the company is constantly seeking to improve delivery times and means and, to this end, 2004 will see the introduction of a dedicated microsite for the book.

This Executive Overview continues the previous practice of selecting a number of topical subjects for highlighting and comment. The result should, I trust, continue to make a relevant and interesting introduction to the book.

In addition to a battlefield radar, this LAV-25 Coyote (8 × 8) reconnaissance vehicle of the Canadian army features an electro-optic package. A long-range TV camera gives day/night recognition/detection out to 18 km; a high-performance FLIR offers all-weather recognition/identification of targets out to 12 km; and an eye-safe laser range-finder with a 10 km range offers accuracy to ±5 m. The latest generation of reconnaissance vehicles should offer improved capabilities (General Dynamics Land Systems – Canada) ***NEW***/0093524

MARKET ANALYSIS

With electro-optic (E-O) systems taking an increasing role in sensor suites within military weapons systems, one cannot ignore their impact on the conduct of the Second Gulf War.

From the gathering of intelligence by varying reconnaissance assets through the targeting and delivery of precision-guided munitions (of all types) to the use of night-vision devices in close combat, E-O systems have been in the forefront of the run-up to, and the war itself. The challenges to E-O technology thrown up by the large spectrum of climatic conditions encountered in Iraq have been vast.

Experience in combat invariably highlights shortcomings in either the equipment itself or a missing capability. One isolated example might be the June 2003 award of contracts to Raytheon and FLIR Systems by the US Special Forces for improved thermal imaging systems for several types of Special Forces helicopters.

Elsewhere, the US Coast Guard is making much of E-O sensors to equip the various seaborne and airborne elements of its major 'Deepwater' modernisation programme. If this programme crosses to the paramilitary market, one innovative civil use of E-O has been in the rapid introduction at some airports of thermal imaging to monitor airline passengers for potential SARS (Severe Acute Respiratory Syndrome) cases. Also the use of thermal imaging to detect survivors in crash/earthquake situations continues.

THE INDUSTRY

While one might argue that 'rationalisation' within the defence industry has declined, mergers and acquisitions continue to be announced. Northrop Grumman completed acquisition of TRW; L-3 Communications has acquired Canada's Wescam, while (too late to be reflected in product entries) Elbit subsidiary, El-Op, has acquired OIP of Belgium.

Meanwhile, the digestion of merged and acquired companies by bigger players has seen further corporate re-organisations emerge – BAE Systems being a particular case in point but Thales is another example. Sagem's Defence and Security division has been re-titled 'Aerospace and Defence', while the names 'Leica' and 'Leica-Vectronix' disappear as the parent company re-brands as Vectronix AG.

As recently as 15 July, Northrop Grumman announced two major teaming agreements in the E-O field. The first involves the company's Defensive Systems Division and Israel Aircraft Industries' TAMAM Division combining their expertise in the design and production of E-O payloads for intelligence, surveillance, reconnaissance and targeting missions.

The second also involves the company's Defensive Systems Division, which has signed a teaming agreement with the Sonoma Design Group, a technical engineering-based company located in Santa Rosa, California. This agreement involves the design and production of a high-performance E-O and Infra-Red (IR) system, called Night Hunter II, again, for intelligence, surveillance, reconnaissance and targeting roles.

MISSILE DEFENCE

Following the unilateral suspension of the Anti-Ballistic Missile Treaty by the United States, to which Russia appears to have acceded de facto (referred to in last year's Foreword), work on high-powered lasers continues.

The US Airborne Laser (ABL) programme is progressing although the in-service date has moved to around 2008. Following the first flight of the Boeing 747-400 modified as the platform for the system on 18 July 2002, the US Missile Defense Agency (MDA) is moving towards a demonstration against a threat-representative target by early 2005.

Meanwhile, the MDA is soliciting industry contributions towards a future ABL in six major areas: a strategic illuminator laser; an advanced inertial reference unit; advanced detectors (ranging from 64 × 64 elements to 256 × 256); small laser transmitters for Ladar; an electromechanical regenerative chemical oxygen iodine laser; and a hydrogen fluoride overtone laser demonstration.

Notably, MDA funding for the Space Tracking and Surveillance System (formerly the Space-based Infrared Systems Low satellite system) has been scaled back with the on-orbit experiment of two satellites now in the 2007 timeframe.

On 12 June 2003, the US Department of Defense and the UK Ministry of Defence signed a framework Memorandum of Understanding to develop bilateral co-operation on missile defence. No doubt this will spawn collaborative deals between British and American companies in the future.

NAVAL SYSTEMS

Sweden leads the field in stealth warship developments, with the second 'Visby' class frigate being launched on 27 June 2003. Work continues in Singapore, the UK and the US, with the US Navy now firmly wedded to the new DD (X) family of stealth warships (including the Littoral Combat Ship), with the so-called Northrop Grumman/Raytheon 'Gold Team' designated 'lead design agent'. Already the US Navy is seriously considering the use of high-power lasers for future armament as well as passive E-O sensors.

Passivity remains vital for such vessels to remain stealthy and the use of Infra-Red Search and Track (IRST) sensors will become more widespread, particularly for detection of air- and land-launched missiles. Detection of hostile vessels lurking in inlets and amongst islands which radar would not be able to discriminate from the background clutter is another important facility that IRSTs bring to the stealthy warship. Although an expensive asset, IRSTs are being acknowledged as a necessary defensive aid for major assets.

Meanwhile the application of thermal imaging to submarine periscope systems continues with the US Navy planning to equip its submarine fleet with an omnidirectional (360° panoramic) imaging system. This involves an uncooled thermal imager and ultra-high-resolution visible-light camera mounted in the antenna radome at the top of the periscope mast.

This thermal weapon sight developed by Thales for the UK Future Infantry Soldier Technology (FIST) programme is one of the applications for the STAIRS A 256 × 128 pixel PST array. Electro-optic systems move ever-closer to close combat (H Keeris) ***NEW***/0083439

LAND SYSTEMS

- **Defensive Aids**

As Operation 'Iraqi Freedom' demonstrated, E-O targeting systems have matured considerably. While weather or obscurants can still mask targets, the success rates of penetration have vastly improved since the 1991 Gulf and Kosovo wars. Countermeasures continue to become more sophisticated with improved computing and image processing. Major land systems such as main battle tanks (MBTs) and other armoured fighting vehicles (AFVs) get no cheaper to procure and require both passive and active protection. E-O systems make their contribution with laser-detection sensors and obscurant dischargers are becoming common. Innovative design of obscurants to counter specific E-O sensors continues.

- **Infantry Weapon Systems**

Various programmes to identify and equip the infantryman of the future with advanced weaponry and sensors move forward. In the US, it is Land Warrior, the UK has FIST (Future Integrated Soldier Technology) and in France it is known as ECAD (Equipement de Combattant Debarque). Similar

Following its operations in Sierra Leone (2000) and Afghanistan (2001) with Chinook HC.2 helicopters, the UK Royal Air Force initiated a Night Enhancement Package (NEP), which included a BAE Systems nose-mounted FLIR turret and improved night-vision goggle displays. Initially, 10 helicopters were upgraded with the NEP, the first of which became operational in August 2002. This photo shows an NEP-equipped Chinook HC.2 with the BAE Systems FLIR turret inset (Patrick Allen/Jane's) ***NEW***/0554945

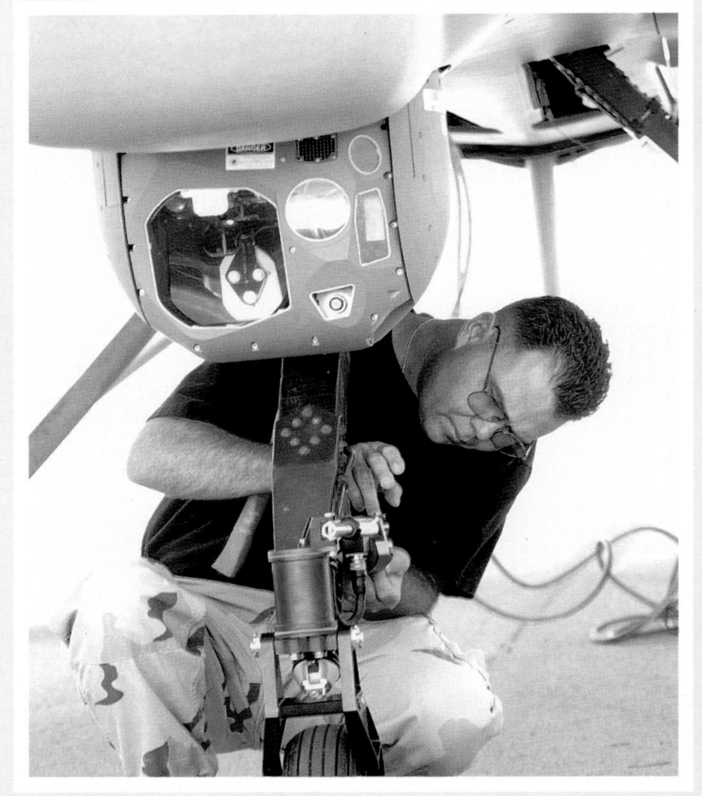

A Raytheon AN/AAS-52 Multispectral Targeting System (MTS) protrudes from the nose of a General Atomics MQ-11 Predator unmanned aerial vehicle, as the nose landing gear is inspected after an operational flight from Tallil air base in Iraq on 21 July 2003. The AN/AAS-52 MTS – a combined electro-optic, infra-red and laser ranging system – is used to target the AGM-114 Hellfire missiles which arm the MQ-1 version of the Predator (US Air Force/2nd Lt. Gerardo Gonzalez) ***NEW***/0554756

programmes are underway in Australia, the Netherlands and elsewhere. Almost certainly the activities of Special Forces in Afghanistan and Iraq by the US, UK, Australia and others will have added to the database to tailor E-O sensors to the user's requirements.

Meanwhile, the acquisition of both thermal and image intensifying sights for rifles, machine guns and other crew-served support weapons was boosted by the impending and actual conflict in Iraq.

• Image Intensifying Systems

Echoing the 2002 Foreword, for provision of night vision for infantrymen 'on the hoof', image intensifying sights and surveillance devices remain popular. Weight and affordability remain the deciding criteria on procurement.

The debate over the merits of Gen II, SuperGen, Gen III, so-called Gen IV and other tubes continues. Much publicity continues to be generated by competing manufacturers and the side-by-side independent comparison of all available products suggested in 2002 would be very interesting.

• Armoured Fighting Vehicles

The round of integrating thermal sighting systems onto MBTs used by the major powers has seen further upgrading of older vehicles. At last, the Czech T-72 upgrade programme (which includes the Galileo Avionica TURMS-T fire-control system) is moving forward, although in nowhere near the quantity originally suggested. Australia has adopted the Thales Optronics Catherine thermal imager as the core of the thermal sight for its Leopard 1 MBTs.

The upgrade of Infantry Fighting/Combat Vehicles (IFVs/ICVs) with thermal imagers continues in the US and UK. The M2A3 and M3A3 upgrades for the Bradley family of IFVs include thermal (IR) sensors; while production elements of the UK's Battle Group Thermal Imaging (BGTI) programme are undergoing user trials on Warrior ICVs (as well as Scimitar reconnaissance vehicles) as these words are written. The UK's STAIRS-C module has already been adopted for the thermal sight for the Stormer-mounted Starstreak HVM system.

The introduction of equivalent systems on new build projects such as the US Army's Stryker and the UK's Future Rapid Effects System (FRES) confirms the comments made last year that ''the cascade effect resulting from affordability and experience will continue'' as the costs, weight, size and complexity of such sensors continues to reduce. It will be interesting to see whether the less-costly uncooled thermal sensors will eventually match the higher performance of the more-expensive cooled scanning systems.

Passive electro-optic systems become more important as warships become more stealthy. Illustrated is the second 'Visby' class stealth corvette for the Swedish Navy – HMS Helsingborg *– at its launch at Kockums' Karlskrona shipyard on 27 June 2003. It is due to enter operational service in mid-2005. First-of-class* HMS Visby *has been undergoing extensive trials since its delivery to Sweden's Defence Materiel Administration in 2002 and is scheduled to enter service in January 2005* (Kockums) ***NEW***/0531197

The prototype Boeing YAL-1A Airborne Laser aircraft, coming into land at Edwards Air Force Base, California, on 19 December 2002, from where the development testing is proceeding (US Air Force/Edwards AFB) ***NEW***/0554837

AIRBORNE SYSTEMS

• Unmanned Aerial Vehicles (UAVs)

The comments made last year that ''UAVs are now a potent force'' have proved true, starting in Afghanistan and then moving across to Operation 'Iraqi Freedom'. US Predator and Hunter UAVs have evolved into Unmanned Combat Aerial Vehicles (UCAVs) by the addition of E-O sensors and precision-guided air-to-surface weaponry. The addition of the BAT (Brilliant Anti-Tank) submunition to the Hunter UAV was made in record time, while further development of the BAT seeker for use in what is now referred to as 'urban CAS' (Close Air Support) continues apace.

With the trend of increased use of UAVs and UCAVs firmly established, manufacturers of such craft and traditional electro-optic systems suppliers are forging urgent new links. The Northrop Grumman activity mentioned earlier is but one example. Whether programmes are for large-, medium-, small- or micro-UAVs, E-O sensors will be included.

• Defensive Aids Systems (DAS)

The employment of IRST as an 'offensive sensor' on fast-jet combat aircraft – initially by Russian MiG-29s and planned for Typhoon, Rafale, F/A-22 Raptor and the F-35 (Joint Strike Fighter) – has evolved to defensive aids as well. The use of such systems on board transport aircraft (cargo and VIP) and helicopters has increased, notably with the Anglo-US Nemesis/DIRCM system being deployed on C-17A Globemaster III airlifters. The 'offensive' systems for the fighters mentioned above are well on the way to be given 'defensive' use, which might well overtake the original concept.

• Precision-Guided Munitions (PGMs)

While the application of E-O sensors to various types of missile and bomb is not new, the extent to which PGMs (which bracket now includes those with Global Positioning System (GPS) guidance) were used operationally has mushroomed.

According to statistics revealed by Group Captain (Gp Capt) Greg Bagwell, Gp Capt – Offensive Operations at Headquarters, Royal Air Force (RAF) Strike Command on 10 June 2003, of the 29,155 air-to-surface weapons dropped by allied aircraft during Operation 'Iraqi Freedom', 68 per cent were PGMs. Of those 19,904 PGMs, 58 per cent were laser-guided bombs (LGBs), with the remaining 42 per cent being GPS-guided weapons.

Gp Capt Bagwell revealed that during Operation 'Telic' (as the UK element of Operation 'Iraqi Freedom' was known) the RAF dropped 27 MBDA Storm Shadow cruise missiles (with terminal guidance from an imaging infrared seeker); 360 Raytheon Enhanced Paveway II/III weapons (a GPS/inertial navigation-guided version of the standard Paveway II/III LGB); 255 Raytheon Paveway II/III LGBs; and 30 Raytheon AGM-65G2 Maverick air-to-surface missiles (with an imaging infra-red seeker).

As can be seen statistically, and confirmed to the editor in conversation, the Enhanced Paveway was the RAF's "weapon of choice". It was no surprise therefore, that a few days later, Raytheon's Paveway IV was selected to meet the RAF's new precision-guided bomb requirement.

The increased use (indeed, availability) of PGMs, apart from the obvious need to accurately hit enemy targets, was to reduce civilian casualties – the so-called 'collateral damage factor'. As a consequence, for some urban CAS missions, inert Paveway training rounds (with concrete rather than explosive warheads) were used, relying on kinetic energy and impact to achieve the destructive power.

This example of electro-optic imagery, showing the US Naval Air Station at China Lake, is from the Integrated Sensor Suite on board an RQ-4 Global Hawk

NEW/0054856

ACKNOWLEDGMENTS

My predecessor, Keith Atkin, joins with me to thank those manufacturers, armed forces, research and development establishments and expert individuals who have contributed to *Jane's Electro-Optic Systems*, particularly those who were in receipt of urgent requests for clarification of 'this or that' point after my arrival on the scene. No doubt over the next year, they will be ensuring that I remain 'up-to-speed' on the particular developments and products on which they focus.

Meanwhile at Jane's Coulsdon HQ, my thanks go to Simon Michell and David Shipton for guiding me through the minefield which the transition from a 'news' to a 'reference' environment has entailed. One ignores at his peril the gallant ladies and gentlemen of our content editing/production teams – Melanié Rovery, Emma Donald, Nicole Smith, Janet Seymour, Jack Brenchley and others – who have taken my input, be it electronic or manuscript, and created the product you now hold. To them go my sincere thanks. Thanks too, to Belinda Dodman and Jacqui Beard who, as the 'information focal point' in the fight to gather updated data (against what, at times, must have seemed an ever-increasing headwind), have shown robust determination in the face of encroaching deadlines.

I am also indebted to my colleagues at Jane's, whether in 'news' or 'reference' side of the house, who have answered specific or general questions. Particular thanks go to Christopher F Foss, E R (Ted) Hooton, Doug Richardson, Richard Scott and Martin Streetly.

If something has slipped through, please let me know ... there is always the next electronic update on the horizon.

Michael J Gething — July 2003

Michael J Gething, *AMRAeS, MCIJ*

Michael J Gething has been an aviation/defence journalist and editor since 1973, when he joined the staff of the Royal Aeronautical Society's publication *Aerospace*. In October 1976, he moved to *DEFENCE magazine* where he spent 17 years, eight of them as Editor, before joining Jane's Information Group in December 1993 to edit *Jane's Defence Systems Modernisation*. In 1997, this evolved into *Jane's Defence Upgrades*. With the incorporation of *JDU* in *International Defence Review* in June 2003, he became IDR's Upgrades Editor and began work on *Jane's Electro-Optic Systems*.

Between 1972 and 1979, Michael produced the aircraft modelling and aviation interest pages for *Air Cadet News*, newspaper of the Air Training Corps, in which he served as a Flying Officer in the Training Branch of the Royal Air Force Volunteer Reserve (1972-1986). He was also the last editor of the *Airfix Magazine* in 1993. Together with Günter Endres, he produced the 2002 edition of *Jane's Aircraft Recognition Guide*, and among his other solo published works are *Sky Guardians – the Air Defence of Great Britain, Air Power 2000* and *F-15 Eagle*.

An Associate Member of the Royal Aeronautical Society and a Member of the Chartered Institute of Journalists, Michael also belongs to Air-Britain and the Air Power Association. He is married with a son (in the RAF) and a daughter (at college) and lives in deepest Sussex.

NAVAL SYSTEMS

Submarine weapon control systems
Optronic masts
Periscopes

Ship-launched missiles
Surface-to-surface missiles
Surface-to-air missiles

Ship close-in weapon systems
Surface-to-air missiles/guns
Guns

Ship countermeasure systems
Laser dazzle systems
Laser warning systems

Ship weapon control systems
Fire control
Infra-red search and track
Surveillance
Thermal imagers

NAVAL SYSTEMS – SECTION SUMMARY

This section includes electro-optic systems reported as deployed on naval vessels or developed for naval applications. Systems are grouped in the following subsections according to their type:

Submarine weapon control systems

Optronic masts
Non-hull penetrating submarine masts.

Periscopes
Search and attack periscopes, but excluding periscopes that do not contain a thermal imager, image intensifier or laser range-finder.

Ship-launched missiles

Surface-to-surface missiles
Ship-launched surface-to-surface missiles with a laser seeker, or a scanning or imaging infra-red seeker in at least one variant of the missile class.

Surface-to-air missiles
Ship-launched surface-to-air missiles with a scanning or imaging infra-red seeker in at least one variant of the missile class or with an electro-optic fire-control system or an optional electro-optic adjunct to a radar fire-control system.

Ship close-in weapon systems

Surface-to-air missiles/guns
Close-in weapon systems with an electro-optic element and which combine a gun with a high rate of fire and a short range missile system.

Guns
Close-in weapon systems with an inbuilt electro-optic sight and a gun with a high rate of fire.

Ship countermeasure systems

Laser dazzle systems
Shipborne active laser countermeasures systems.

Laser warning systems
Shipborne laser warners

Ship weapon control system

Fire control
Shipborne fire-control systems for guns and/or missiles that include a laser range-finder, thermal imager or image intensifying camera as either part of the fire-control system or an optional adjunct to the fire-control system (see also surveillance systems in this section for closely related systems).

Infra-red search and track
Shipborne scanning infra-red detection systems, primarily deployed for protection against sea-skimming missile threats.

Surveillance
Shipborne electro-optic sensors used primarily for observation and surveillance and not specifically associated with weapon control systems. Some systems may, however, be used for limited control of light guns.

Thermal imagers
Thermal imagers that have been qualified for naval applications or are known to be used on shipborne systems.

SUBMARINE WEAPON CONTROL SYSTEMS

OPTRONIC MASTS

Kollmorgen Model 86 optronic mast series

Type
Submarine mast.

Description
Kollmorgen was awarded a contract in 1988 by the US Defense Advanced Research Agency (DARPA) to develop the Model 86 non-hull penetrating optronic mast which was operationally tested aboard the USS *Memphis* SSN submarine.

Model 86 includes a sensor unit, a hydraulically operated mast which is streamlined and connects by an external cable to an electronic interface unit and a control/display console internal to the hull. Sensor information can be processed and displayed in a dedicated operating console, or incorporated into the main combat consoles.

Features of the optronic mast include:
(a) 3-5 or 8-12 μm thermal imaging sensor
(b) high-definition monochrome and colour CCD TV cameras for daylight, low-light level and 'quick-look' viewing
(c) three-axis line of sight stabilisation to eliminate ship's motion and mast vibrations
(d) ESM warning to detect radar threats
(e) rotating sensor package (sealed statically) with quick response and low power consumption
(f) manual or automatic mast control with a 'quick-look' mode.

Operational status
In production. The Model 86 has been installed in US Navy 'Los Angeles' class submarines.

Specifications
Field of view:
Visual: 16 × 12°, 4 × 3°
IR: 9 × 6.75°, 4 × 3°
Monochrome camera: 1,035 × 1,940 pixels
3 Chip colour camera: 480 × 640 pixels
IR sensor: 8-12 μm FLIR or 3-5 μm FPA
Elevation
Visual: −10 to +60°
IR: −10 to +55°
Additional optional configurations include: mission critical camera, RAS/RAM, laser range-finder

Contractor
Kollmorgen Electro-Optical.

VERIFIED

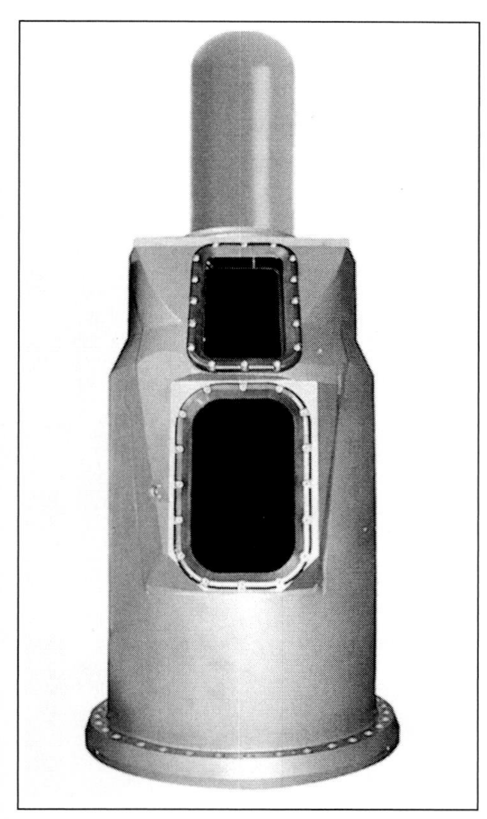

Masthead of Model 86 optronic mast
0088167

Kollmorgen, Calzoni Universal Modular Mast

Type
Submarine mast.

Description
The Universal Modular Mast consists of a cartridge assembly including a structural module, a mast fairing subassembly, a hoist cylinder and a closed-door mechanism. The use of standard interface allows integration of a variety of above-water sensors and communications antennas to be fitted to the mast system. The modularity of the design enables easy and quick installation and de-installation from the platform and the possibility, with minor changes, of using common subassemblies for different sensor payloads. The use of a cartridge concept in which the faired mast, bearings and hydraulic actuation are incorporated into a single unit, ameliorates alignment problems.

The two-stage design allows for higher sensor positioning and height adjustments.

The sensors and communication systems in the mast include electro-optic imaging, electronic support measures, radio and satellite communications and radar. Integration is achieved via standard interface modules. Calzoni has proposed its modular submarine bridge fin as an extension of the mast structure.

Operational status
In production for the US Navy's 'Virginia' class New Attack Submarine.

Contractors
Calzoni S.r.L.
Kollmorgen Electro-Optical.

UPDATED

The Calzoni non-penetrating hoist mast for submarines

Kollmorgen AN/BUS-1 Photonics Mast Program (PMP)

Type
Submarine mast.

Description
The electro-optical sensor system for the programme includes an eye-safe laser range-finder, two high-definition televisions (colour and monochrome) and a mid-wave staring infra-red sensor in a single multispectral head window. The sensor system includes an eye-safe laser range-finder, ESM, direction-finding and communications antenna. The AN/BUS-1 (PMP) is non-hull penetrating.

Features of the Photonics Mast Program include:
(a) Colour television
(b) Monochrome HDTV
(c) Thermal imaging
(d) Eyesafe Laser Range-finder
(e) ESM - Omni-directional and DF monopulse

Operational status
In production for the US Navy's 'Virginia' class New Attack Submarine.

Contractor
Kollmorgen Electro-Optical.

UPDATED

SAGEM Infra-red Mast (IMS)

Type
Submarine mast.

Description
The Infra-red Mast (IMS) combines the capabilities of SAGEM's non-hull penetrating masts while including a single infra-red channel, which uses SAGEM's IRIS 8 to 12 μm thermal camera. The reduced dimensions of the above-water component and of the radar cross-section area, achieved by careful design and by covering the exposed part of the periscope (head) with Radar Absorbent Material (RAM), have both improved submarines' capacity for covert operation while providing a day and night capability.

The main characteristics of IMS are a 210 mm diameter head which includes two-axis gyrostabilised line of sight and a dual field of view IRCCD thermal imaging system. An antenna module is integrated on top of the head to provide ESM warning and GPS. Passive range-finding is carried out on the controller's screen using the stadiametric technique. As well as the direct view there are panoramic surveillance and 'look around' modes of operation. The system is designed to be fitted on any type of non-hull penetrating hoisting device.

Operational status
In production for several export customers. IMS has been in service aboard two of the Royal Danish Navy's Type 207 'Narhvalen' class submarines since 1994.

Contractor
SAGEM SA Aerospace & Defence Division.

VERIFIED

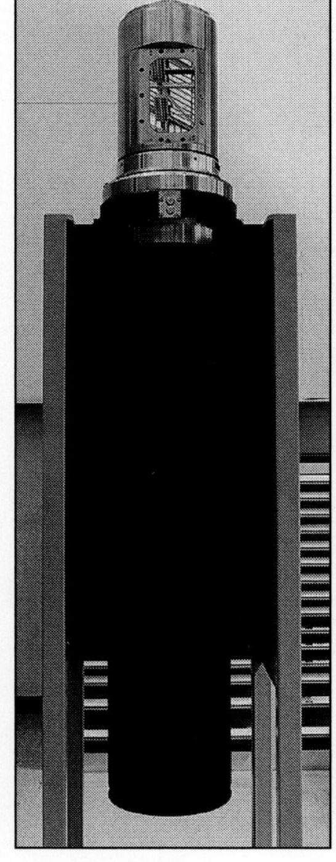

SAGEM Infra-red Mast of the Danish Navy's 'Nahrvalen' class submarines

SAGEM Optoradar Mast (OMS)

Type
Submarine mast.

Description
SAGEM's Optoradar Mast (OMS) combines the capabilities of SAGEM's optronic mast with the integration of a navigation radar. A single mast for these dual functions reduces the possibility of detection. The 360 mm diameter head includes X-band navigation radar; a dual field of view IRCCD thermal imaging system; a high-definition TV system with two magnifications; and one-axis gyrostabilised line of sight. Azimuth stabilised surveillance can be presented on one of four range scales from 4 to 32 km. Up to five targets can be tracked and automatic target acquisition is provided. An antenna module is integrated on top of the head to provide ESM warning and GPS. Passive rangefinding is carried out on the controller's screen using the stadiametric technique. As well as the direct view there are panoramic surveillance and 'look around' modes of operation. The radar cross-section area has been reduced by covering the head with Radar Absorbent Material (RAM). The system is designed to be fitted on any type of non-hull penetrating hoisting device.

Operational status
Operational on board French SSBN *'Le Triomphant'* class submarines.

Contractor
SAGEM SA Aerospace & Defence Division.

VERIFIED

The SAGEM OMS, optoradar mast for new-generation SSBN

SAGEM Search Mast System (SMS)

Type
Submarine mast.

Development
A prototype of SAGEM's Search Mast System (SMS) was trialled on board a French Navy 'Daphne' class submarine in 1992. The system has also been trialled on Swedish Navy *Vastergotland* in 1993, Royal Norwegian Navy 'Type 207' in 1994-95 and a South Korean 'Type 209' in 1995.

Description
The optronic mast combines the advantages of SAGEM's search optronic periscopes with the increased safety of non-penetrating masts. The 320 mm

SAGEM Search Mast System (SMS)

diameter head includes a high-definition TV system with four fields of view/ magnifications (×1.5, ×3, ×6 and ×12) and a dual field of view IRCCD thermal imaging system, SAGEM's IRIS, along with two-axis gyrostabilised line of sight. An antenna module may be integrated on top of the head to provide ESM warning and GPS. Reduction of the radar cross-section area has been achieved by covering the exposed part of the mast (head) with Radar Absorbent Material (RAM).

As well as the direct view there are panoramic surveillance and 'look around' modes of operation. Passive range-finding is carried out on the controller's screen using the stadiametric technique. The system is designed to be fitted on any type of non-hull penetrating hoisting device.

Operational status

In production for several export customers.

Contractor

SAGEM SA Aerospace & Defence Division.

VERIFIED

Thales Optronics CM010 Optronic mast

Type

Submarine mast.

Description

All optronic masts in the CM010 family are non-hull penetrating. They offer a wide choice in sensor technology, including either a DRS (formerly Boeing) 3 to 5 μm or 8 to 12 μm thermal imager using either Thales' HDTI or Synergi (Thales, Zeiss Optronik). Image Intensification, high-definition monochrome television and colour television sensors, as well as support for high-sensitivity, broadband ESM, communications and GPS sensors are optional for all systems. Images captured by the system are complemented by advanced image manipulation and image processing capabilities which further enhance the operational advantages of the system.

Variants include:

CM010 – 3-5 μm thermal imager, monochrome and colour TV
CM011 – image intensifier, monochrome and colour TV
CM012 – 8-12 μm thermal imager, monochrome and colour TV
CM013 – 3-5 μm thermal imager with additional 8-12 μm thermal imager and image intensifier optional; monochrome and colour TV
CM014 – 8-12 μm thermal imager with additional 3-5 μm thermal imager optional

The optronic mast system is controlled and operated from a dedicated remote-control console, equipped with a high-resolution monitor display, allowing the command team to gain a complete above-water picture. Alternatively, the system can be controlled and operated from suitably equipped multifunction consoles. Additionally, the common mast raising equipment facilitates integration with other payloads such as dedicated ESM, radar, satcom and communications packages. Stealth features reduce acoustic, visual, radar and thermal signatures.

Programmable modes of operation include quick look round, continuous view and snapshot. Real-time image processing is combined with target analysis on live/ recorded images.

Operational status

In production and deployed at sea.

Contractor

Thales Optronics.

UPDATED

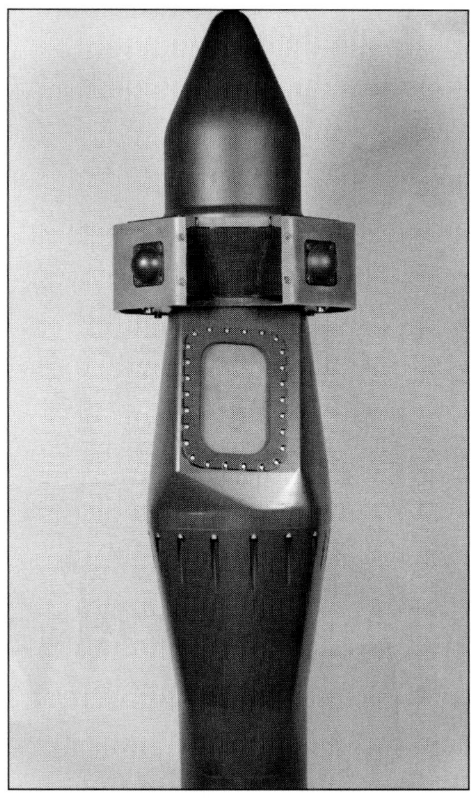

The CM010 sensor head unit
0055091

SUBMARINE WEAPON CONTROL SYSTEMS

PERISCOPES

Eloptro submarine periscope upgrades

Type
Submarine periscope (upgrades).

Description
Eloptro is engaged in the upgrade of search and attack periscopes including Daphne and Type 209 class submarines.

Upgrading of both Search and Attack Periscopes typically covers the following:
- Improvement of existing optical characteristics, particularly field curvature, chromatic aberration and transmittance by redesigning the optical layout and by using modern optical design software, modern glass materials and thin film technology.
- Improvement in the transmittance of the periscope which is achieved by using state-of-the-art anti-reflection coating and using a minimum number of components necessary for each subsystem. This results in the approximate doubling of the transmittance of the periscope system. The attack periscope's exit pupil diameter is typically increased from 4 to 5 mm, thus increasing the luminous flux transmitted by 56 per cent.
- A Passive Range-Finder (PRF), based on the split image principle, is integrated into the periscopes. The accuracy exceeds that of the active sonar range-finder of the submarine.
- Binocular eyepieces, with a capability to switch to monocular vision.
- A television (TV) capability consisting of Day TV (DPI) and Night Vision TV (NTV). The direct view optics and the TV are mutually inclusive, which means that the visual image seen through the eyepiece can be displayed simultaneously on a TV monitor. For both the DPI and the TV, the image is displayed on a TV monitor situated elsewhere in the system, as well as on a video display unit situated on the ocular box of the periscope. In addition, remote periscope control at a multifunction console (providing 'penetrant' optronic periscope capability) can be fitted.
- Recording of the TV images by means of a digital video recorder.
- The Night Vision TV (NTV) is achieved by means of an IIT and CCD camera low light level television.
- PRF capability for the TV, achieved electronically.
- Attachment of a 35 mm still camera to the eyepiece, with improved resolution due to a reduction in field curvature and axial chromatic aberration. In addition, Eloptro also has the technical capability to redesign periscopes to include laser range-finding capabilities.

Improvement to reliability aspects of the periscopes includes the following:
- Minimising the number of moving assembles and subassemblies inside the periscope tube.
- Accommodating two image intensifier tubes inside the periscope tube for redundancy.
- The image intensifier tube assembly is situated outside the path of the direct view optics should a failure occur.
- Should the passive range-finder's electronics fail, mechanical backup returns the image split to the zero position to provide unobstructed direct vision.
- Electronics interchangeability between periscopes.

Operational status
Available.

Contractor
Eloptro (Pty) Ltd.

VERIFIED

Kollmorgen Model 76

Type
Submarine periscope.

Description
The Kollmorgen Model 76 is a modular periscope system with common components for the attack and search versions. The basic difference is that the attack periscopes have smaller heads, while the search periscopes' larger heads act as multipurpose reconnaissance platforms.

The system consists of a mast unit with optical train, a display and control unit including a split-beam binocular eyepiece, a 35 mm camera and training handles. In addition to the mast unit there is a hoisting yoke, a control unit and a junction box unit.

The display and control unit includes a control panel, system focus, mode select, stadiameter control and microphone. The attack periscope includes a broadband antenna and crystal video receiver ESM system, together with a display and control panel on the control unit.

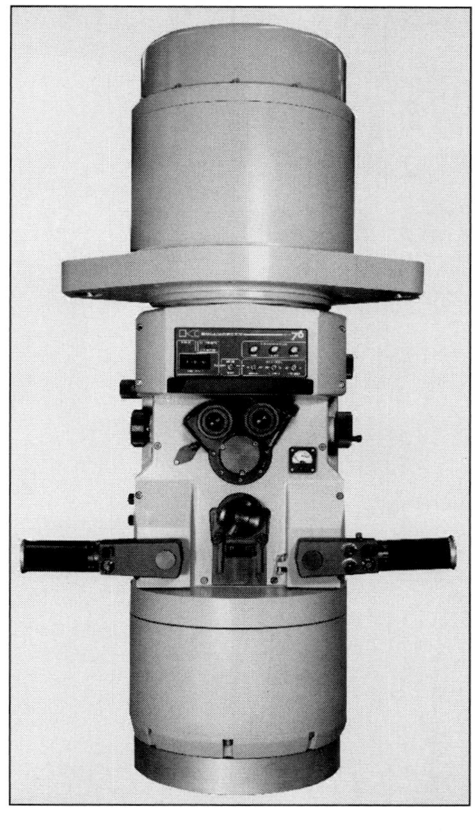

Model 76 periscope

The basic periscope systems are: stabilised line of sight; integral torque drive motor with auto-torque assist; ×1.5, ×6 and ×12 magnification; mechanical bearing dials; eyepiece data display - range, relative, true and elevation; binocular viewing eyepiece; heated head window; digital interfaces; photocamera - 35 mm; optical stadiameter; high-optical light transmission; fail-safe elevation stabilisation line of sight; image intensification (night vision); LLLTV camera or CCD camera - integral; ESM early warning; remote-control operator console; laser range-finder - attack (optional); RAM (optional); videotape recorder; infra-red capability - 3 to 5 μm (optional).

Operational status
In production. The Model 76 is fitted in a number of submarines including Brazilian and Turkish Type 209 submarines and the improved 'Sauro' class of the Italian Navy. Also operational in Swedish 'Nacken' and Israeli 'Dolphin' class and others.

Specifications
Diameter: 190.42 mm
Elevation
Attack: –10 to +74° (+90° detection)
Search: –10 to +60° (+76° detection)
Magnification: ×1.5, ×6, ×12
Field of view: 4°, 8°, 32°

Contractor
Kollmorgen Electro-Optical.

VERIFIED

Kollmorgen Model 90 optronic periscope system

Type
Submarine periscope.

Description
The Model 90 optronic periscope system completed sea trials in 1992 with delivery of production systems beginning in 1995. It has been developed to allow the operator to search the sea surface during day and night utilising a thermal imaging subsystem and, at the same time, to supply a direct viewing visual channel.

The periscope system combines a wide range of sensors in one periscope: a thermal imaging camera, CCD TV camera, 35 mm photographic camera, laser rangefinder as well as passive TV and visual stadiameter, omni radar early warning

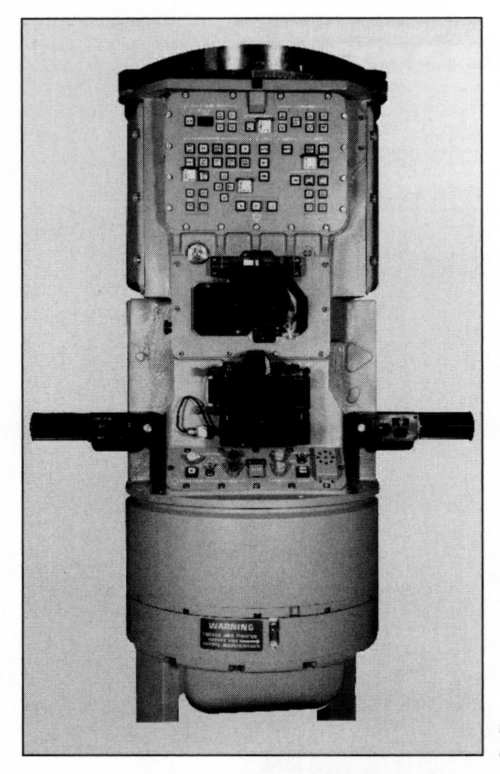
Model 90 electro-optic mast eyepiece unit

Model 90 optronic masthead unit

antenna, a radar direction-finding antenna and GPS. The periscope provides high performance by utilising accurate line of sight stabilisation to compensate for induced vibrations and platform motion to the visual and thermal lines of sight. Additionally, in combination with this function, the operator is provided with a periscope rotation and line of sight elevation rate control which allows fast direction and target tracking. The operator has a direct view of the scene in addition to a video display and eyepiece data display of target range, target bearing and line of sight elevation angle.

A remote-control station is supplied as part of the Model 90 optronic periscope system, in addition to a complete control datalink to the submarine fire-control system.

Operational status
In production. The mast is operational and in service with an undisclosed country.

Specifications
Periscope tube diameter: 190.42 mm
Line of sight elevation
Visual and TV: −10 to +74°
Thermal imaging: −10 to +55°
Fields of view:
Visual: 2.5 ,4, 8 and 32°
Infrared: 4.4, 10°
Magnification: ×1.5, ×6, ×12, ×18

Contractor
Kollmorgen Electro-Optical.

UPDATED

LOMO PLC Classical periscope

Type
Submarine periscope.

Development
LOMO PLC of Russia have been developing the Classical periscope for the Kilo and Amur classes of submarine. This equipment is also intended for export.

Description
The full Classical periscope system includes features to enable surface observation, day/night target acquistion and classification, range and bearing measurement, celestial sight taking, satellite navigation, preliminary acquisition of radio signals and video recording. For observation there is an optical visual channel, a TV day and low-light channel and a thermal imaging channel. The range of capabilities can be tailored to customer requirements.

Operational status
In development.

Specifications
Persicope tube diameter: 180 or 260 mm (dependent on mast head type)
Entrance pupil to eyepiece distance: 7-12 m
Azimuth aiming range: ±210°
Azimuth angle measurement error: 2-10 min
Celestial reference elevation error: 2-3 min
Max traverse rate: 20°/s
Mass: <2,000 kg
Visual channel
Magnification: ×2 (option ×4) and ×8
Fields of view: 40° (option 20°) and 10°
Elevation aiming range: −10 to +60°
TV day and low-light channel
Field of view: 18°
Elevation aiming range: −10 to +30°
Thermal imager channel
Field of view: 10°
Laser range-finder
Wavelength: 1.54 μm (eyesafe) or 1.06 μm (optional)
Measurement range: 60 m to 18.5 km
Accuracy: 5-10 m
Antenna module reception: GPS, Glonass and radio

Contractor
LOMO PLC.

LOMO PLC non-retractable periscope

Type
Submarine periscope.

Description
The optional capabilities of the LOMO non-retractable periscope system include surface observation, day and night target acquisition and classification, range and bearing measurement, eyesafe laser range-finding, celestial sight taking, satellite navigation, preliminary acquisition of radio signals and video recording. The basic system is a single tube optical periscope. A twin tube system is optional. Two axis stabilisation is provided.

Operational status
In development.

Specifications
Periscope tube diameter: 260 mm
Azimuth aiming range: ±210°
Azimuth angle measurement error: 2-10 min
Celestial reference elevation error: 2-3 min
Max traverse rate: 20°/s
Two axis stabilisation error: 30 s
Mass: 3,000-4,000 kg
Visual channel
Magnification: ×2 (option ×4) and ×8
Fields of view: 40° (option 20°) and 10°
Elevation aiming range: −10 to +60°

Contractor
LOMO PLC.

Raytheon NESSIE GEN II programme

Type
Submarine periscope sensor.

Development
In April 1996, Raytheon Systems Company was awarded a US$7.9 million US Navy development contract for an advanced electro-optical system for submarines. The engineering and manufacturing development contract for the programme, NESSIE GEN II, was awarded by the Naval Undersea Warfare Center, New London, Connecticut. The NESSIE module is intended to be mounted on top of the Type 22 search periscope used aboard 'Los Angeles' class submarines.

Description
The optronic system will incorporate a third-generation infra-red sensor and a commercial-off-the-shelf low-light level television. The system will be housed in a modified periscope furnished by the government. Kollmorgen Corp, as the major subcontractor to Raytheon, will be responsible for the periscope and submarine related activities, including system integration and test. The infra-red sensor features a mid-wavelength staring focal planar-array. The manufacturer claims that this third-generation technology has demonstrated unprecedented image quality and range performance. The reflective optics enable the system simultaneously to image both the visible and infra-red spectral regions.

Operational status
Under development.

Contractor
Raytheon Company, (El Segundo).

UPDATED

SAGEM attack periscope (APS)

Type
Submarine periscope.

Description
SAGEM has incorporated several improvements into its attack periscope. The head size has been reduced to a minimum because of a requirement for maximum discretion, without significantly degrading the periscope's attack-phase

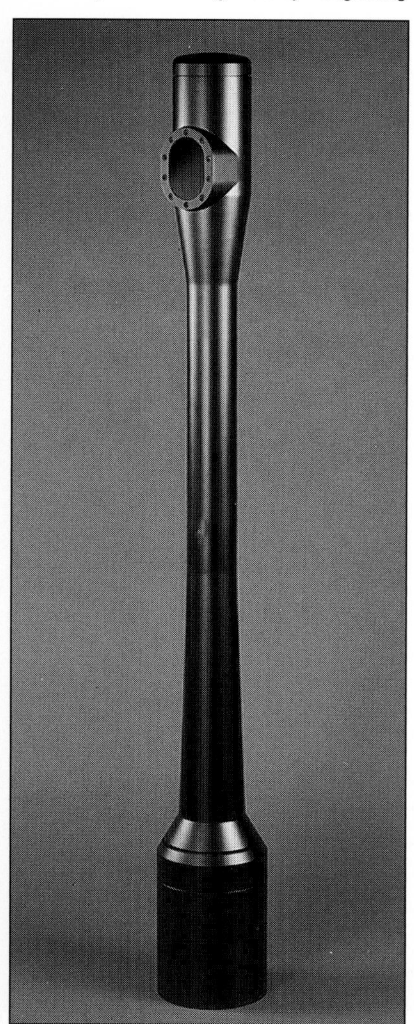

Head of the SAGEM APS attack periscope
0002019

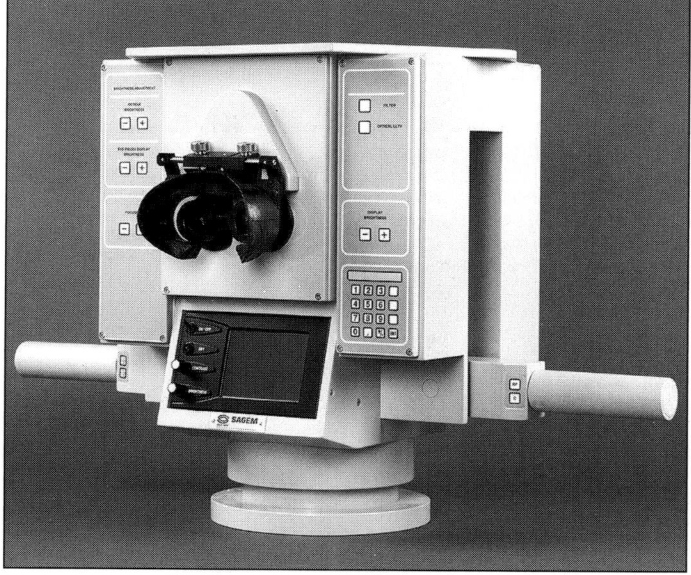

Ocular box of the SAGEM APS attack periscope 0002020

performance. The radar cross-section area has been minimised by careful design and by covering the exposed part of the periscope (head) with Radar Absorbent Material (RAM).

The 140 mm head includes single axis stabilised line of sight; a single optical channel with four fields of view (×1.5, ×3, ×6, and ×12); Low-Light TV channel (LLTV); ocular box with a colour TV camera. A third-generation IR camera may be integrated in place of the LLTV.

An antenna module may be integrated on top of the head to provide ESM warning and GPS. The periscope can be remotely controlled from a multifunction common console.

Operational status
Developed for French submarines and for foreign navies.

Specifications
Optical fields of view: 30, 15, 7.5 and 3.75°

Contractor
SAGEM SA, Aerospace & Defence Division.

UPDATED

SAGEM search periscope (SPS) (PIVAIR)

Type
Submarine periscope.

Description
The SAGEM search periscope (SPS) is the latest version of the PIVAIR family (incorporating a high-accuracy sextant mode). It is operational on board 'Le

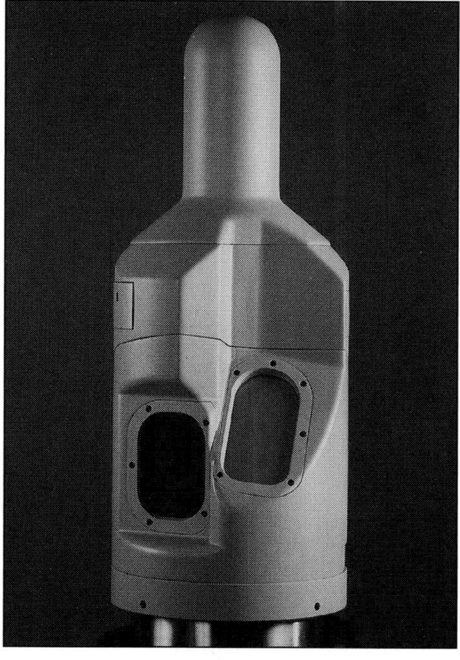

SAGEM SPS search periscope

Triomphant', the French Navy's newest class of SSBN. Various electro-optic sensors (high-definition TV and thermal imaging systems) are integrated in the periscope, permitting day and night vision and better detection/identification capability in all weathers. There is dual-axis stabilisation of the Line Of Sight (LOS) for all channels, direct optical and infra-red, improving image quality. Rapid search is available through the use of the infra-red panoramic surveillance mode, scanning the horizon over 360° and an automatic 'look around' mode which minimises above water exposure time. An antenna module is integrated into the top of the periscope head for communications, ESM warning and GPS. The radar cross-section area has been reduced by covering the exposed part of the periscope (head and upper fairing) with Radar Absorbent Material (RAM). The new fairing design has also reduced wake and head vibration through vortex shedding.

Operational status

Operational on board all French nuclear submarines (SSBN-SSN).

Contractor

SAGEM SA, Aerospace & Defence Division.

UPDATED

SAGEM ST 5 periscopes

Type

Attack periscope.

Description

SAGEM produces advanced attack and surveillance periscopes.

The ST 5 attack periscope head is stabilised by a rate gyroscope with an image intensified TV microcamera for night vision. The design is so compact, that it is fitted in the tiny ST 5 periscope head, which itself has been specially shaped and covered by Radiation Absorbent Material (RAM) to reduce its radar cross-section. It uses a fixed eyepiece with magnifications of ×1.5, ×6 and ×12 giving fields of view of 36° and 7° over an elevation arc of –10 to +30°.

Operational status

The ST 5 periscopes are in service on board the French Navy SSNs and foreign SSK submarines.

Contractor

SAGEM SA, Aerospace & Defence Division.

UPDATED

SAGEM ST 5 attack periscope

Thales Optronics CK038 search and CH088 attack periscopes

Type

Submarine periscope.

Description

The CK038 is a fully electronic search periscope intended for SSK submarines in the 600 to 1,800 ton range. The CK038 is optimised for low susceptibility to visual counter-detection and has an optical system designed for maximum light gathering to accommodate watch-keeping, night viewing and use in poor visibility. For use as a stand-alone system or as part of an optronic mast and periscope visual system,

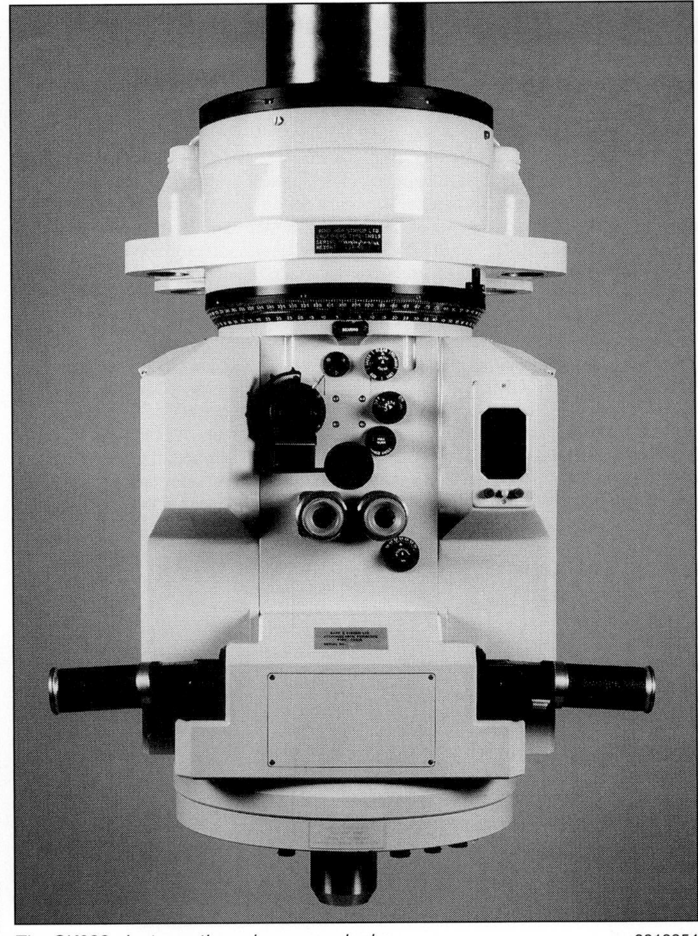

The CK038 electro-optic periscope ocular box 0010854

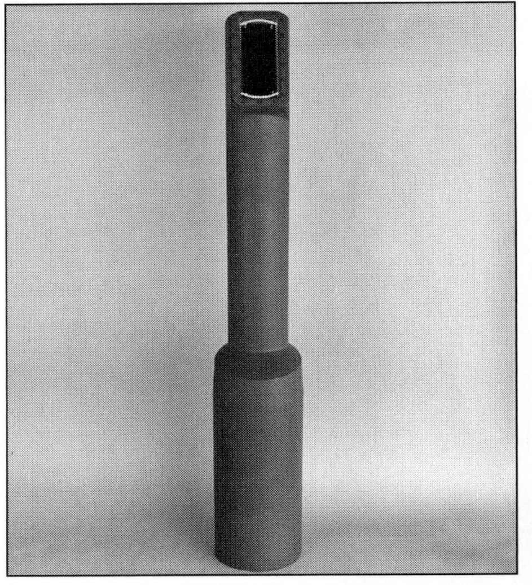

The CK038 search periscope head 0010855

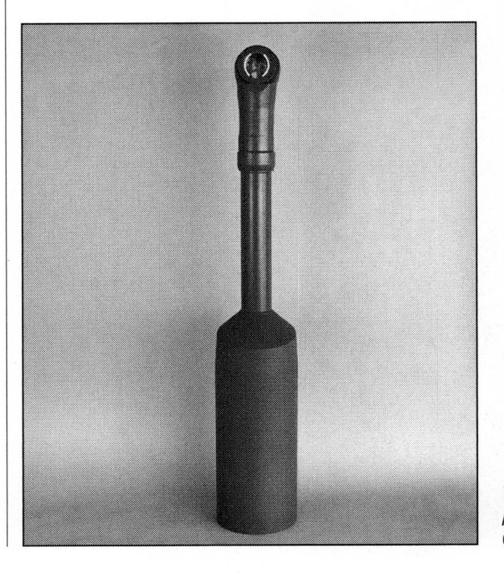

The CH088 attack periscope head 0010856

the CK038 is fitted with an image intensifier, a low-light level TV and optional thermal imaging sensor. Other standard features include a weapons system interface, 35 mm camera, heated top window, BITE system, stabilisation, stadiametric range-finding and power drive azimuth rotation. Data and images from either the CK038 or CH088 can be relayed to the CM010 advanced optronics mast. Optional features include GPS and ESM sensors.

The CH088 attack variant is available with the same features without the ESM/GPS facility.

Operational status

CK038 is in service.

Specifications

Tube diameter: 190 mm
Mechanical length: 10,990 mm (CK038), 10,929 mm (CH088)
Optical length: 10,400 mm
Weight: 800 kg
Magnification: ×1.5, ×6 (with ×3, ×12 electronic optional)
Elevation (line of sight): −10 to +60° (CK038), −10 to +30° (CH088)
Elevation (edge of field): −26 to +76° (CK038), −26 to +46° (CH088)

Contractor

Thales Optronics.

VERIFIED

Thales Optronics CK043, CH093

Type

Submarine periscope

Description

The CK043 search and CH093 attack periscopes are suitable for submarines of 1,800 tons and greater. Both periscopes have four fields of view. In addition to the optical path, CK043 has thermal imaging and low-light television sensors, while CH093 has image intensification and low-light television sensors. These can be viewed directly or at a multifunction, remote- control and viewing console. Key tactical data can be viewed either on the consoles or by eyepiece injection. Other key operational features include line of sight stabilisation, optical and electro-optic range-finding, position fixing (sextant), still photography, internal communications (intercom) and support for a comprehensive ESM sensor suite. Through full integration with the submarine command and control system the periscope performs a primary role in the detection and identification of surface vessels and aircraft, intelligence gathering and weapon system support.

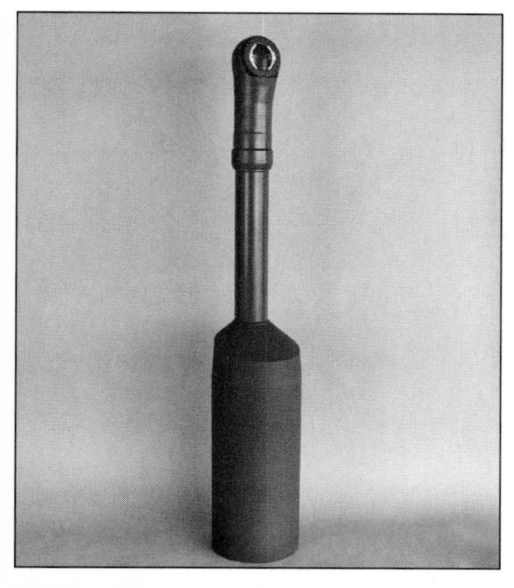

CH093 attack periscope head
0010858

CK043 search periscope head
0010857

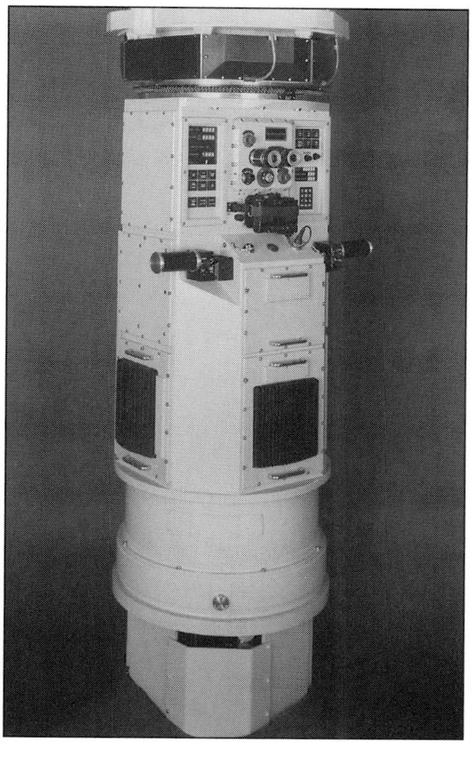

The CK043/CH093 optronic periscope ocular box

Operational status

In service on the Royal Australian Navy 'Collins' class submarines.

Specifications

Tube diameter: 254 mm
Length: 14,300 mm
Weight: 1,500-1,900 kg
Magnification: ×1.5, ×3, ×6, ×12
Elevation: −15 to +60°

Contractor

Thales Optronics.

VERIFIED

Thales Optronics compact periscopes (CK032, CK037, CK039, CK041, CK044, CK044S, CK060)

Type

Submarine periscope.

Description

Thales (formerly Pilkington) Optronics compact periscopes are a family of periscopes tailored for small submarines between 50 and 400 tons. These instruments offer a range of standard features including image intensification, stabilisation, stadiametric range-finding, elevation of line of sight (−15 to +60°), still photography, heated window and weapon system interface. Optional features include a TV camera, ESM, communications sensor and GPS sensors.

The compact periscope mast head
0010860

Operational status

The CK032 to CK041 are all in service; CK044 and CK060 are in development.

Specifications

Tube diameter: 127 mm
Length: 3,500-5,400 mm
Weight: 150-250 kg
Magnification: ×1.5, ×6
Elevation: –15 to +60°

Contractor

Thales Optronics.

VERIFIED

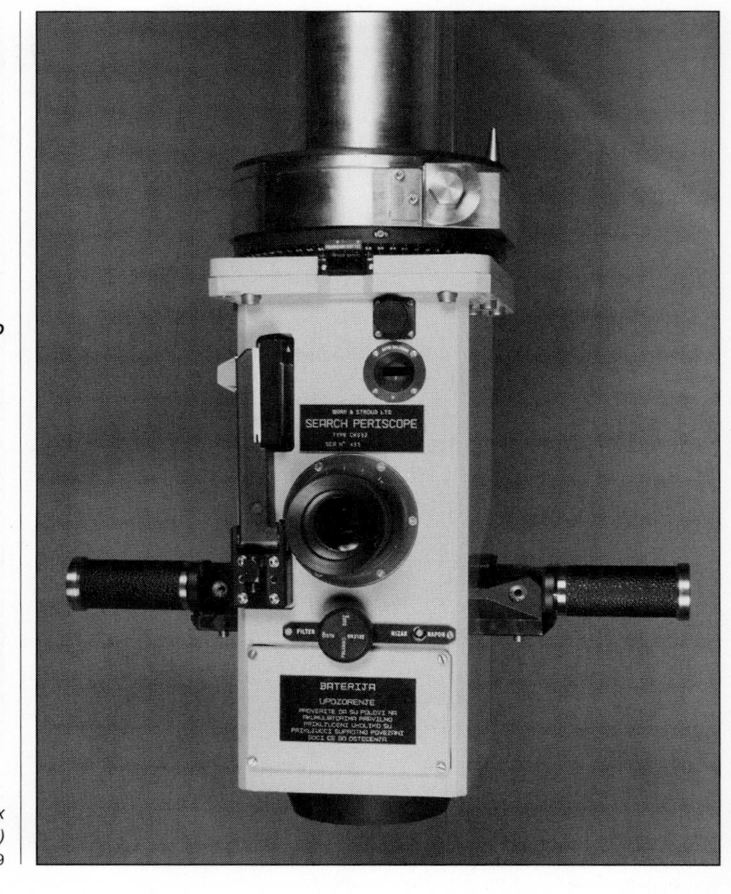

Compact periscope ocular box (with bracket for 35 mm camera)
0010859

SHIP-LAUNCHED MISSILES

SURFACE-TO-SURFACE MISSILES

Boeing Sea SLAM

Type
Ship-launched missile - surface-to-surface.

Development
Summary: SLAM uses the Maverick imaging infra-red seeker from Raytheon.

To provide a conventionally tipped, precision-attack, air-launched weapon, Boeing (formerly McDonnell Douglas) developed SLAM (Standoff Land Attack Missile) during the late 1980s from the successful Harpoon missile. The first production missile AGM- 84E, was delivered to the US Navy in November 1988. A version of this missile for shipborne use was test fired from the 'Ticonderoga' (CG 47) class cruiser USS *Lake Champlain* (CG 57) in 1990. The firing was to demonstrate the capability as part of a US Navy test and valuation programme to decide whether or not a ship-launched SLAM would be of operational value.

In February 1995 Boeing received a US$3 million contract for another two ship-launched SLAM demonstrations, in support of the Surface Fire Support programme. The demonstration involves two missiles controlled by an SH-60B helicopter and an F/A-18 strike aircraft.

Sea SLAM was successfully test launched in August 1996. An extended range version of the SLAM, known as SLAM Expanded Response (SLAM-ER) or AGM-84H, began development in 1994. It was a private venture proposal for retrofitting into existing SLAMs, but there are no plans at the moment for a ship-launched SLAM-ER.

Description
SLAM is a conventional Harpoon missile, a slim cylinder with a pointed nose and cruciform wing-fin configuration. Both the wings and fins are of cropped-delta planform with the former being broad while the latter are narrow. The radar seeker is replaced by a Raytheon imaging infra-red seeker, a datalink and a multichannel Rockwell-Collins Global Positioning System (GPS). The seeker is identical to that used in the Maverick air-to-surface missile, while the datalink is the same as that used in the Walleye 'smart' bomb system. The GPS system includes a navigation processor and a Mk 82 interface, so the missile remains compatible with the SWG-1A weapon control system. The missile's mid-course guidance unit has minor hardware and software modifications. The missile is 65 cm longer and 90.72 kg heavier than a conventional Harpoon.

During the cruising phase the missile autopilot is updated through the GPS receiver/processor to ensure the seeker is pointed directly towards the target, which is especially useful as the missile is also capable of multiple waypoints. The seeker is activated during the last 60 seconds of the terminal phase and sends a video image through the datalink and a specific aiming point on the target is selected and the seeker is locked on to it. After lock on the missile autonomously tracks the target.

Operational status
Ship-launched Sea SLAM has been successfully tested.

Contractor
Boeing Integrated Defense Systems.

UPDATED

A ship-launched Sea SLAM is fired from the USS Lake Champlian

CPMIEC, CATIC CSSC-2 'Silkworm' (SY-1/HY-1)/ CSSC-3 'Seersucker' (HY-2/FL-1/FL-3A)

Type
Ship-launched missile - surface-to-surface.

Development
Summary: the HY-2A version has an infra-red seeker.

In 1959, the former Soviet Union supplied China with examples of the SS-N-2A 'Styx' P15 which China decided to manufacture under licence as the SY-1. Development of the first systems began in March 1960 and was not completed until November 1966 when they entered service with both fast attack craft and coast defence units. This missile later received the US designation CCS-N-2 (later CSSC-2 'Silkworm').

In 1976, it was decided to combine the features of the coast defence and shipborne missiles to produce one which could be used for either role. From 1978 production of the new missile, simply designated HY-1, superseded that of the HY-1 Coastal.

In 1983 the Nanchang Aircraft Factory began to improve the HY-1. The objectives were to improve the range, ECCM and sea-skimming performance. In 1987 the missile, designated HY-1A, was accepted for service, with the first being delivered the following year.

In 1970, development of an improved version, HY-2 ('Seersucker'), had begun. Apparently, this incorporates technology of the SS-N-2C 'Styx' or P-21, although this version of the missile does not appear to have been supplied to China by the former Soviet Union.

A derivative of the SY-1/HY-1 has also been produced by the Nanchang Factory as FL-1 and is marketed by CATIC as an improved version FL-3A. North Korea is known to produce Chinese 'Styx' missiles.

In January 1995, it was reported that Iran was developing its own version of 'Silkworm', initially for coast defence purposes.

Description
The CSSC-2/CSSC-3 type missiles have the same general external appearance being aircraft shaped with rounded nose, fixed delta planform wings and triple tail surfaces at 120° angles along the bottom of the missile.

Targets are located with a Type 254 radar (based on the Russian 'Square Tie') and the firing sequence is operated from a dedicated console with its own analogue computer. Once the target is confirmed as hostile, the operator manually sets a range gate around it and feeds in the location, speed and ballistic data. The computer then feeds the firing solution to the missile autopilot.

After launch it climbs to one of five preset altitudes (between 100 and 350 m) which depend upon the target ranges. The guidance package includes a radio altimeter. The upper altitude for short-range (20 km) targets is 150 m while for longer distances it will be up to 350 m, the distance being dictated by the need to track the missile with the Square Tie radar. For short ranges a glide angle terminal phase is used, for longer ranges a high-angle attack is preferred.

The HY-1 is essentially the same as the SY-1 but the fuselage has been strengthened while both the active radar seeker and autopilot have been improved. In the deep sea vessels the Type 254 radar is used for missile fire control and is supplemented by a dedicated surface search radar.

These features have all been included in the HY-2 which offers a selection of seekers. In addition to the I-band system of HY-1, the missile is offered with an infra-red seeker (HY-2A) and a monopulse active radar seeker (HY-2G). During the cruise phase HY-2 flies at an altitude of 30 m, but for the terminal phase it descends to about 8 m.

Operational status
This missile is in service with the Chinese Navy where it is used by surface vessels and the coastal defence forces. The Chinese continue to use a variety of 'Styx' missiles and their derivatives, some of which have been exported.

Specifications
HY-2
Length: 7.36 m
Diameter: 76 cm
Wing span: 2.40 m
Launch weight: 2,500 kg
Speed: M0.9
Range: 51 n miles/(95 km)
Guidance: All use autopilot and active radar. HY-2A has an alternative IR seeker instead of radar designated HY-2.

Contractors
China National Precision Machinery Import & Export Corporation.
China National Aero-Technology Import and Export Corporation (CATIC).

VERIFIED

EADS – LFK Polyphem/Triton

Type
Short-range anti-ship missile.

Development
In 1984 Aerospatiale and Deutsch Aerospace (now part of EADS) began a fibre optic guided missile technology demonstrator programme and four flights were made using a modified SS 12. The success of this programme led to the beginning of a development programme in 1992 by Euromissile (of which Aerospatiale and EADS are partners) to study both army and naval applications using extensive captive flight tests and this was joined in 1994 by Italy. The first prototype was flown in 1995 and in July 1996 Aerospatiale revealed the missile, now named Polyphem, had flown 32 n miles (60 km). By October 1996 Euromissile stated the naval missile, Polyphem-S, had been selected for the German and Italian Navy's Type 212 submarines and for the German Navy's planned new K130 class of corvettes. It would appear that the missiles have been selected in principle to equip the Italian Type 212 submarines but there remains considerable opposition to them in the German Navy. Development of the missile has since been taken over by DASA's Lenkflugkorpersysteme (LFK) now EADS-LFK.

Nevertheless, the German Navy is to fund a demonstrator programme which may attract foreign investment. Naval applications for the missile include precision anti-ship and surface-to-surface attacks as well as a submarine-launched anti-helicopter missile. In the Spring of 1996 Euromissile teamed with Northrop Grumman's Huntsville Engineering Center to market Polyphem within the United States with Polyphem-S offered as a special operations, precision strike weapon from nuclear-powered attack submarines but this project did not progress.

By the Autumn of 1998, declining interest in the anti-armour version of Polyphem by the German Army led to increased importance for the naval version of the missile for both ship (K130) and submarine (U212) applications. DASA officials stated that government support for this project, certainly for a development contract, would depend upon the weapon successfully completing flight tests and a qualification programme.

At Euronaval, in October 1998, DASA revealed a redesigned submarine-launched version as Sub-Polyphem being developed in association with submarine manufacturers Howaldtswerke-Deutsche Werft AG (missile launcher-container) and Kongsberg Defence & Aerospace (weapon-control system) and that the German Defence Ministry had placed a contract for an experimental study which was completed with live firings by the end of 2000. By the Spring of 1999, this version had been redesignated Triton and by the early Autumn of 1999 initial tests had been completed of the booster for this missile. A Polyphem missile was successfully test-flown at a German Army range on 5 September 2000 to demonstrate unspooling of the fibre-optic cable, the robustness of the command guidance link and the redesign of the wing deployment mechanism. Validation/demonstration of Triton began in March 2001 with launching tests from a Type 206A submarine a year later and qualification was anticipated in mid-2003 and a production contract anticipated in June 2004.

In the Summer of 2000, it was reported that German Army interest in Polyphem had declined and that it wished to withdraw from the programme when system engineering and flight tests are completed in 2001. Doubts had been expressed about the German Navy's ability to fund the naval programme if the army was to withdraw, but the weapon has been selected for the German Navy's K-130 corvette programme, the first of which is scheduled to enter service in 2005. Ship test sets will be delivered by December 2003. LFK hinted early in 1999 that Triton had been sold to an unidentified customer which was believed to be Israel for use in the 'Dolphin' class submarines. Test firings of Triton were scheduled for the autumn of 2001 with the first systems available for customers from 2005.

Description
Polyphem/Triton is a fibre optic guided weapon guided remotely with the aid of an electro-optic sensor whose images are displayed at the control console through the fibre optic cable capable of providing data at more than 200 Mbytes/s. The following description is of the ship-launched version for land-attack (Polyphem/Triton) and anti-helicopter (Triton) roles, the Triton having 60 per cent commonality with Polyphem.

Externally the missile is a slim cylinder with rounded nose, two air inlets are on the side in front of the four thin-aspect wings, which are just aft of the middle of the missile and arranged in cruciform configuration, and stub fins near the rear behind

An illustration of the Polyphem surface-ship export launcher 0083580

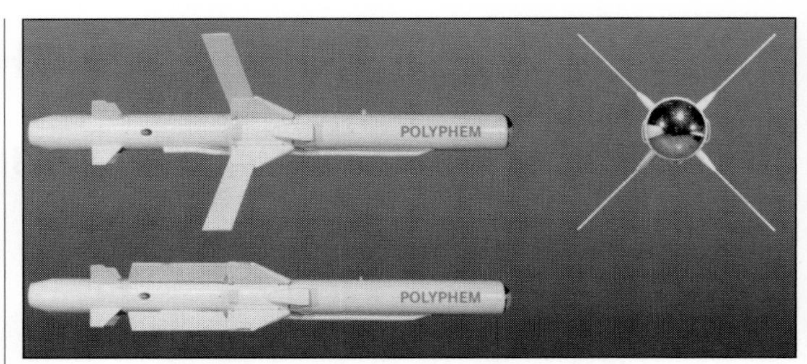

The Polyphem ship-launched missile with wings folded and extended 0017994

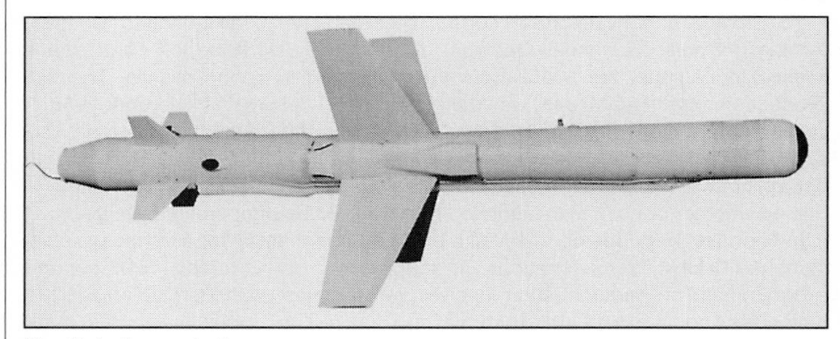
The Polyphem missile

two side-mounted rocket nozzles. The wings fold along the body before launching and there are shaped hinges where the aerodynamic surface meets the body. Triton will feature a booster from Bayern-Chemie.

Internally, the nose has a 3.4 to 5.5 μm infra-red imager with focal planar array and platinum-silicon detectors on a triple-axis, gyrostabilised platform, with a field of view of 10 × 7.5° and a pan area of +/–30°. This will be capable of detecting a hovering helicopter at 2.5 n miles (4.6 km), it will have a range against surface targets of 4.5 n miles (8 km) although target recognition will be possible at 1.4 n miles (2.7 km) and identification at 0.75 n miles (1.35 km) or 21 and 12 seconds before impact respectively. Behind it are the missile electronics compartment with an inertial measuring unit supported by a laser altimeter and a Global Positioning System (GPS) – Navstar receiver, the warhead, the propulsion section with fuel tank (which contains a controllable oxydiser to adjust cruise speed) and Microturbo SA MBR 240 (also TRI 10) integrated propulsion system (which would form an intregal part of the fuselage structure) based upon a 10 daN turbojet engine with efflux vents at the base of the wings.

Behind the turbojet are the solid-propellant rocket booster, a battery, the actuation systems and the bobbin with glass fibre thread. The warhead, being developed by EADS/LFK's Schrobenhausen factory, will weigh 25 kg and will be a combined shaped charge/blast fragmentation unit with an 'intelligent' fuze which identifies the target material upon impact by monitoring *g*-forces using a shock-wave sensor. The warhead has a 'special explosive' and three layers of pre-formed fragments which give it a lethal radius of 500 m, and it includes a small rocket booster which separates it from the missile body 6 m from the target, which is then struck at M1.4. The warhead is lethal to 500 m and can penetrate up to 4 m of concrete.

The K-130 class corvettes will have four twin vertical launchers for Polyphem but for export purposes there will be a deck-mounted quad launcher (first shown at IDEX in March 1999), in which the missiles would be in a shaped container to reduce the radar cross section, the cells being raised to a 45° angle to launch the missile. The operator station will feature two high-quality vertical and horizontal 20 in displays with radio links, a command system interface and Global Positioning System Navstar reference. The system will include a system control computer with embedded simulator and the Kongsberg control console will be similar to that used in MSI 90 (qv) and MSI 2005 (qv) with a MMI consisting of two display screens, buttons, a joystick and tracker ball.

The surface-ship version is launched with the booster at a typical angle of 60° to the vertical and 60° off-angle to an altitude of up to 400 m at a speed of 180 m/s, at which point the turbojet takes over. The missile can operate between 20 and 600 m and is initially guided through a mixture of inertial navigation and GPS which allows the missile to follow a predetermined trajectory into a designated target area through two way points until the search phase when the operator uses the IR sensor. When a target is detected the tracker is locked on and the missile can make a manoeuvring approach to the target with various angle and altitude options. The operator will be able to guide it to within 10 to 15 cm of the desired impact point.

Triton will be a rocket-powered weapon with totally redesigned aerodynamic surfaces, with the wings being of folding conformal shape of Russian design. The sustainer would be behind the warhead in line with the wing's leading edge. The internal configuration is otherwise similar to the ship-launched version. The missile will be in a launcher-container each containing four missiles installed in pairs, the container fitting into a 21 in (533 mm) torpedo tube. The bobbin arrangement is modified with one each on the missile and the launcher and a third at an unidentified intermediate location to reduce strain on the line. The missiles can be fired at 20 second intervals and all four can be in the air simultaneously.

The missile can be launched in conditions up to Sea State 5. Tests have shown that the Triton engagement sequence will last a total of 70 seconds and that, while a gas bubble is created when the booster is ignited to carry the missile to the surface, this takes four minutes to reach the surface, by which time the ASW helicopter or maritime patrol aircraft will have been shot down.

Improvements to Triton are already being considered, including a longer version with a range of up to 32 n miles (60 km) with two per launcher container. Consideration is also being given to replacing the existing electro-optical sensor with one using a zoom lens to improve long-range target recognition. This sensor might also be retrofitted into Polyphem.

Operational status

The naval versions of Polyphem/Triton are under development, although it is claimed three Triton systems have been sold to Israel (1) and five Polyphem systems are likely to be ordered for the K-130 class. The German Government has restricted sales of Triton to German-designed submarines.

Note: (1) This sale is reported but not confirmed.

Specifications

	Polyphem	Triton
Length	3.0 m	2.0 m
Weight	145 kg	120 kg
Diameter	228.6 mm	228.6 mm
Speed	M0.65	M0.58
Range	33 n miles (61 km)	8 n miles (15 km)
Guidance	Global Positioning and Inertial Navigation Systems with fibre optic terminal guidance	

Note: * The manufacturer claims that Polyphem's range can be extended to 54 n miles (100 km)

Contractor

EADS/LFK - Lenkflugköppersysteme GmbH.

UPDATED

Kongsberg NSM (Nytt Sjomals Missil or New Sea target Missile)

Type

Ship-launched missile - surface-to-surface.

Development

In the mid-1990s, Norwegian naval staff requirement SMP 6026 designated the successor for its Penguin (qv) missile as the Nytt Sjomals Missil (NSM or New Sea target Missile). It defined the requirement as retaining the Penguin's electro-optic seeker solution but with greater range (officially intermediate range but this was defined as a minimum of 100 km or 54 n miles) and incorporating stealth technology.

The NSM will be the Royal Norwegian Navy's next-generation anti-ship weapon system for its in-service 'Hauk' class and new 'Skjold' class fast attack craft and new 'Nansen' class frigates. The planned in-service date for the first operational missiles is 2005.

A development contract was placed with Kongsberg in December 1996 and a co-operative agreement was signed with Aerospatiale (now MBDA Missile Systems - France) in June 1997, making the company responsible for the integrated propulsion system. An agreement was signed with EADS/TDW for the development of a new warhead for NSM, and with EADS/LFK for a joint programme if the missile were to be selected for use by the German Navy.

In 2001 an extended range version was proposed, with a maximum range increased to 250 km (135 n miles).

Description

The NSM is a long-range, cost-effective and flexible anti-ship missile system with multiplatform capability. It can be launched from naval vessels, helicopters (NH 90), aircraft (F-16 or P-3) and land mobile units. The design features advanced technology to minimise the missile's signature, while its sea-skimming ability and random manoeuvres in the terminal phase improve its survivability.

The high-resolution, passive imaging, infra-red seeker provides a high degree of discrimination and target selection, rendering efficient operation in confined as well as open waters. The missile can also be used against land targets. The advanced mission planning system provides an efficient tool for selecting optimum trajectories. The missile's multisensor navigation system ensures accurate navigation and target detection. With its very low signature, extreme sea-skimming ability and capability to fly over land, the missile will be very difficult to detect.

Salvo capability and specific manoeuvres in the terminal phase, ensures high penetration in the target area.

Specifications

Length: <4.1 m
Wing span: 1.4 m
Wing span, folded: 0.7 m
Weight (with booster): 410 kg
Propulsion: Microturbo TRI-40 turbojet
Terminal guidance: passive, imaging infra-red seeker
Cruise speed: high subsonic (M0.85)

Operational status

In development, with a planned operational debut in 2005.

Contractor

Kongsberg Defence & Aerospace AS.

NEW ENTRY

Kongsberg Penguin Mk2 Mod7N

Type

Ship-launched missile - surface-to-surface.

Development

Summary: fire-and-forget, passive, IR seeker, programmable trajectory, evasive manoeuvring.

Development of the Penguin, ship-launched, anti-ship missile began in the early 1960s after a request by the Royal Norwegian Navy to the Norwegian Defence Research Establishment (NDRE) for a feasibility study into a cost effective, easy-to-operate anti-ship missile system. The study was completed in 1966 and the development task was handed over to Kongsberg Våpenfabrikk A/S (now Kongsberg Defence & Aerospace AS). The first generation Penguin missile (Mk1) became operational in 1970 and was designed in particular for operations in littoral waters against hard targets of the Soviet North Fleet. Development of the second-generation Penguin (Mk2) started in 1978. It entered service in 1980 offering increased range, optional dog-leg trajectory and an enhanced seeker system.

The development of the air-launched Penguin Mk3 for the Royal Norwegian Air Force started in 1980. This version featured a new-generation seeker with enhanced target recognition capability and decoy resistance. Penguin Mk3 features a single-stage motor which provides a range in excess of 55 km. The new missile - also designated AGM-119A - entered service in 1989.

Talks with the US Navy about a helicopter-launched version of the Penguin missile started in 1983. As the development programme of the third generation, all digital Mk3 progressed and successful firings were made, it became the basis for the new helicopter-launched version, eventually designated Mk2 Mod7 (AGM-119B). The main difference being foldable wings and a two-stage motor.

The latest version of the ship-launched Penguin is a further development of Mk2 Mod7, designated Penguin Mk2 Mod7N, for coast defence roles.

Description

The Penguin Mk2 Mod7N surface-to-surface system consists of the missile, a launching system and a control cabinet which interfaces either with a dedicated or the existing fire-control system.

The missile is of conventional aerodynamic design with a rolling airframe, cruciform wings and canard configuration.

Internally, the missile is divided into guidance, warhead and motor sections. The guidance section features the infra-red (IR) seeker, a gas-operated actuation mechanism, a three gyro gimballed inertial platform with accelerometers, an autopilot and thermal batteries. The warhead is a semi-armour-piercing unit fitted with a delayed action impact fuze.

The motor section is a two-stage (boost/sustain) solid propellant rocket motor. The Penguin Mk2 Mod7N missile is an all-digital, fire-and-forget missile designed to autonomously locate, identify and home on targets. The missile does not require

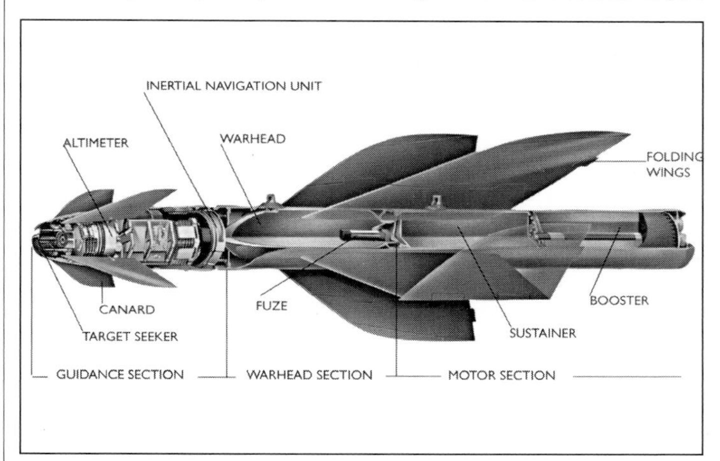

Penguin Mk2 Mod7 missile with foldable wings 0055094

Penguin missile fired from a US Navy MK3 'Swift' class 65 ft FPB 0055092

any in-flight guidance. All data needed to achieve a successful result, such as target position, or bearing on target, desired flight profile and seeker mode, are transmitted to the missile prior to launch. Using the inertial navigation system for mid-course guidance, the missile is placed correctly in a pre-determined position, relative to computed target position, with the necessary data to activate the high-resolution IR-seeker.

The Mk2 Mod7N missile is installed on deck in a weather-proof canister launcher. A launcher may consist of a single canister, a pair or two pairs stacked. The weapon control system is typically able to handle eight missiles.

The weapon control system can either be an existing general fire-control system where Penguin functions and controls will be integrated, or a Penguin dedicated system adapted to the existing ship sensors.

The Penguin Mk2 Mod7N surface-to-surface system requires data from the target sensors, navigation sensors and the ship attitude sensors. No specific radar other than the ship' normal navigation radar is needed.

When the target is acquired the fire-control computer calculates the bearing to the predicted point of impact. The bearing data is then fed through the missile control system into the missile's inertial navigation system. The operator can select from a set of seeker settings, trajectory options and launch modes. The missile follows the pre-programmed course at the selected altitude either direct to the target or it may make an indirect approach via a freely selectable waypoint, with changes in course and altitude in flight. Once the seeker acquires the target it switches to tracking mode for the terminal phase and guides the missile in to hit the target at a point close to the waterline. The operator may also programme the missile to perform randomly generated, high-G, evasive manoeuvres as it approaches the target.

Operational status
The ship-launched Penguin is in operation with the Royal Norwegian Navy, the Turkish Navy, the Royal Swedish Navy and the Hellenic Navy. The air-launched Penguin Mk3 (AGM-119A) is in service with the Royal Norwegian Air Force on the F-16 Fighting Falcon combat aircraft. Penguin Mk2 Mod7 (AGM-119B) has been purchased by the US Navy for use on the SH-60B Seahawk helicopter. It has also been acquired by the Hellenic Navy for the S-70B Aegean Hawk and by the Royal Australian Navy for the SH-2G Super Seasprite naval helicopters. The company states that it has produced 76 ship-launched Penguin systems, while unofficial estimates suggest that total production of the ship-launched missile is between 850 and 900. Penguin remains in production and the ship-launched system is in service with four navies.

Specifications
Penguin Mk2 Mod7N
Length: 3.02 m
Diameter: 0.28 m
Span:
 Deployed: 1.42 m
 Folded: 0.76 m
Launch weight: 395 kg
Total weight launcher/missile: <675 kg
Speed: High subsonic
Range: <3 to >18 n miles
Warhead weight: 120 kg (43 kg HE)
Guidance:
 Mid-course: Inertial programmable flight path
 Terminal: Passive IR

Contractor
Kongsberg Defence & Aerospace AS.

UPDATED

MBDA Ulixes

Type
Ship-launched missile — surface-to-surface.

Development
Ulixes is a development programme for a successor to the Teseo Mk 2 (OTOMAT Mk 2 for export) anti-ship missile. The missile is being proposed for the Italian Navy's requirement as well as for other navies.

The new Italian Navy requirement extends the operational application of the classical anti-ship weapon beyond the littoral scenario and it is aimed towards the development of an intermediate range weapon. The new missile, vertically launched, is intended to be capable of operating against ships as well as other surface targets with simplified pre-mission planning and short reaction time. Penetration and target selection capability are the key operational drivers.

Description
The missile will have a dual-mode radar (Ku band) and imaging infra-red seeker with data fusion. It will have a range in excess of 250 km, will incorporate a number of stealth features and will be equipped with GPS and datalink.

Operational status
Under study/development.

Specifications
Diameter: 48 cm
Max speed: M0.9
Range: > 250 km

Contractor
MBDA Missile Systems.

VERIFIED

Taiwan Hsiung Feng II

Type
Ship-launched missile - surface-to-surface.

Development
Summary: the Hsiung Feng II contains an imaging infra-red as well as an active radar seeker.

In the mid-1970s, Taiwan obtained Israeli Gabriel short-range anti-ship missiles. However, Taiwan sought self-sufficiency and developed its own Hsiung Feng I missile which resembled Gabriel. Much of the work was undertaken by the Chung Shan Institute of Science and Technology Institute. After the operational test and evaluation, the missile went into production as Hsiung Feng IA in 1981. It was used as the main armament of the Second World War former American destroyers which were upgraded under the Wu Chin I and III programmes.

Meanwhile, Taiwan had been seeking to import medium-range anti-ship missiles but the manufacturers of Exocet, Harpoon and Otomat all refused to provide products. Around 1983/4, Taiwan decided to produce its own medium-range missile based upon experience gained in developing the Hsiung Feng I. This was a scaled-up Hsiung Feng with turbojet engine replacing the rocket motor. Hsiung Feng II entered service in May 1993 with the commissioning of the frigate *Cheng Kung*.

Description
Hsiung Feng II has a wider cylinder than Hsiung Feng I, with a rounded nose and clipped delta-platform aerodynamic surfaces arranged in an 'X' configuration. There are two broad ribs along the side and at the top is a protruding structure with an infra-red seeker and a long cable duct which is along most of the upper body. A tandem booster with slim, rectangular fins is also fitted. Internally the missile has an active radar seeker with a planar-array which is complemented by the imaging infra-red seeker. Behind is the 225 kg self-forging fragment warhead.

The launcher operation is likely to follow that of most conventional medium-range anti-ship missiles, with the target acquired by long-range surface search radars. These radars contain over-the-horizon targeting facilities through a Ta Chen datalink. The launcher and target locations are inserted into the inertial guidance system, the missile is launched and adopts a low-level cruising altitude. The target is acquired by both the radar and the infra-red sensors and the returns are compared with an onboard library. If significant deviations in one sensor's returns occur, then the guidance system assumes it is being jammed and gives priority to the other as the missile enters the sea-skimming terminal phase.

Operational status
A total of 50 Hsiung Feng II systems with 800 missiles have been ordered or delivered. They are in service aboard some of the Taiwanese Navy Gearing destoyers and on their *La Fayette, Cheng Kung* and *Tien Tan* class frigates.

Specifications
Length: 4.845 m
Diameter: 40 cm
Wing span: 1.2 m
Weight: 775 kg
Speed: M0.85
Range: >100 km
Guidance: inertial with active radar and imaging infra-red homing

Contractor
Chung Shan Institute of Science and Technology.

UPDATED

Launch of Hsiung Feng II missile (Harrison Chen)

SHIP-LAUNCHED MISSILES

SURFACE-TO-AIR MISSILES

CATIC PL-9N

Type
Ship-launched missile - surface-to-air.

Development
Summary: IR seeker; possibly electro-optic director.

The PL-9N missile system was revealed in 1991 and may be associated only with the Type 88C fire-control system. It is clearly a navalised version of the PL family of air-to-air missiles based upon the Soviet equivalent of the Western Sidewinder (AIM-9). The PL-9 family has been developed to meet a variety of roles including naval surface-to-air, ground-launched surface-to-air and air-to-air; development is reported to have begun in 1988.

Description
The PL-9N is a short-range surface-to-air missile system which features IR guidance and laser fusing in the missile and probably some form of electro-optic director. The missile is likely to follow the design of the earlier PL missiles. In the nose will be the all-aspect infra-red seeker with the gas-powered actuation system behind it. The 10 kg HE blast/fragmentation warhead is reported to have an active laser proximity fuze behind it. The solid propellant rocket motor occupies the remainder of the airframe.

The missile has its own fire-control console, with the target acquired by the ship's radar through which designation is transferred to an electro-optic director. It seems likely that the fire-control console displays the target image on its display system. Whether or not the missile launcher traverses towards the target automatically or under the operator's control is not known.

The operator probably assigns a missile or missiles to the target and initiates the appropriate missile seekers. When these indicate they have acquired the target the missile or missiles are launched.

Operational status
Not known.

Specifications
Length: 2.90 m
Diameter: 16 cm
Wing span: 81 cm
Launch weight: 120 kg
Speed: n/k
Range: possibly 5 km
Altitude: n/k
Guidance: infra-red

Contractor
China National Aero-Technology Import and Export Corporation (CATIC).

VERIFIED

The PL-9N surface-to-air missile on a four-round launcher

Fakel SA-N-5 'Grail' (Strela 2) — SA-N-8 'Gremlin' (Strela 3) — SA-N-10 'Gimlet' (Igla)

Type
Ship-launched missile - surface-to-air.

Development
Summary: the missile has an infra-red seeker.

Development of the Strela 2 (Arrow 2) portable surface-to-air missile began in 1959. The missile entered service with the army in 1966 and is best known by its NATO codename of 'Grail' (SA-7). During the late 1970s it entered service with the navy, receiving the NATO codename SA-N-5 'Grail' to provide an air defence capability for smaller warships and auxiliaries. In 1972 an improved version with greater range, improved warhead and filtered seeker, entered service as Strela 2M and has probably been issued to the navy.

Strela 2 suffered tactical shortcomings, especially in relation to its seeker and, during the late 1960s and early 1970s, the design was developed to incorporate technological improvements. The new missile, Strela 3 (NATO codename SA-14 'Gremlin'), entered service around 1975 and is believed to have supplemented or replaced Strela 2 systems in some vessels. The naval version has been allocated the designation SA-N-8 and many ships carry a lightweight multiround launcher. A substantial redesign of Strela 2 led to the Igla which appeared in the early 1980s and was eventually given the NATO designation SA-16 'Gimlet'. A naval version of this system was produced with new launcher and given the NATO designation SA-N-10 'Gimlet' and it is possible that a launcher was experimentally installed in a 'Kilo' class submarine.

Description
Missiles are housed in glass fibre launcher-containers and probably share the Strela 2 ('Grail') configuration with four small fins of cruciform configuration and two pop-out trailing fins behind the exhaust. In Strela 2 the 9M43M missile uses an uncooled lead sulphide infra-red seeker, operating in the 1.7 to 2.8 μm range and with a spinning reticle, centre-null tracking system with a 1.9° field of view. In Strela 2M (SA-7B 'Grail' Mod 1) the seeker has a filter for improved performance.

Once the target has been acquired, the operator activates the thermal battery and when the seeker has acquired the target, an audio signal is received. The operator then partially pulls the trigger and activates the missile gyros which are armed up in 4 to 6 seconds. The operator uses stadia reference marks in his optical sight to compute the lead angle to the target, then fully engages the trigger to launch the missile. The booster burns for 0.05 seconds to eject the missile from the launch tube at a speed of 28 m/s and to spin it, at a distance of 6 m the sustainer then ignites and the trailing fins pop out. The missile homes on the 'hottest' part of the aircraft and probably has an impact fuze on its 1.10 kg armour-piercing RDX warhead. In Strela 2M, the warhead has improved fragmentation uniformity.

Compared to the 9M32M missile in Strela 2, the 9M39 missile used in the Strela 3 system has an uprated rocket motor, a more powerful warhead (2 kg HE fragmentation with impact and graze fuzing) and a cryogenically cooled infra-red seeker with proportional guidance to deal with approaching, receding and manoeuvring targets. It is believed the missile incorporates IR signal processing to defeat common countermeasures such as flares and modulated 'hot brick' type decoys.

The naval version is mounted in a manually operated framework with four launcher containers fixed side by side and supported on a central pedestal. The mounting, which probably weighs about 1,000 kg, is unstabilised. The most common fitting consists of two launchers per ship but single, triple or quadruple fittings have also been encountered.

The 9M313 missile used in Igla is fatter than its predecessor but retains the warhead of the 9M39. Guidance is by proportional navigation using the cooled seeker unit with maximum target bearing angle for a launch being ±40°. The SA-N-10 launching system is reported to be a four-round system which may be loaded, possibly semi-automatically, from below decks.

SA-N-8 'Gremlin' launcher

Operational status
It is believed that some 170 Russian-built vessels have Strela systems.

Specifications
Strela 2 ('Grail')
Length: 1.45 m
Diameter: 7 cm
Weight: 9.97 kg
Speed: M1.7
Range: 5.50 km (2.6 n miles)
Altitude: 18-4,500 m

Strela 3 ('Gremlin')
Length: 1.40 m
Diameter: 7.50 cm
Weight: 9.90 kg
Speed: M1.7
Range: 600 m-6 km (3.25 n miles)
Altitude: 10-5,500 m

Igla ('Gimlet')
Length: 1.55 m
Diameter: 8.0 cm
Weight: 10.80 kg
Speed: M1.67
Range: 600 m-5 km (2.7 n miles)
Altitude: 10-3,500 m

Contractor
Fakel Experimental Design Bureau.

IAI, Rafael Barak-1

Type
Ship-launched missile - surface-to-air.

Development
Summary: the fire-control radar is supplemented by a thermal imager.

The Barak ('Lightning') missile system, designed as a point defence system to protect warships of fast attack craft size and above, was first revealed in June 1981, following an agreement between Israel Aircraft Industries Ltd (IAI) and the Rafael Armament Authority. It was originally conceived as a conventional lightweight weapon such as Seawolf with two eight-round launchers, each of which had a fire-control radar. Ballistic test firing of the missile in this configuration was carried out from a 'Saar' type fast attack craft during the early 1980s.

The conventional configuration was found to have problems and a new configuration, based upon a vertical launch missile and a mast-mounted fire-control radar, made its appearance as Barak-1.

Missile development continued throughout the 1980s and in August 1991 the first successful trials were conducted in the 'Saar 4.5' class fast attack craft *Nirit*. Development of the system was completed in 1993.

In March 1996, the Israel Navy announced that Barak-1, launched from a 'Saar' class fast attack craft, had successfully intercepted a sea-skimming Gabriel missile simulating a hostile threat to the ship.

Description
The Barak-1 is designed to be a relatively low-cost point defence missile system. It is designed to protect ships against both manned aircraft and anti-ship missiles and consequently has a quick reaction time, typically 3 seconds including 0.6 seconds to turnover. It consists of the missile, the launcher and the fire-control system.

Externally the Rafael-made missile is a long, slim cylinder with a sharply pointed nose. Along the top is a narrow, rib-like aerial which is part of the Command to Line Of Sight (CLOS) guidance system.

Internally the nose contains the seeker and guidance system. Behind it is the 22 kg HE fragmentation warhead, with a proximity fuze behind it. This fuze, by IAI, is a dual-sensor unit featuring electronic and infra-red sensors, the latter also acting as an altimeter.

The fire-control system is based upon the Elta EL/M-2221GM I, J and K-band (X-Ka band) monopulse coherent tracking and illumination radar which is supplemented, on the right-hand side, by a Rafael thermal imager. It features a dish antenna with front feed on four 'legs', the antenna having an elevation of –25 to +85°. Search acquisition and tracking may be conducted in either I/J (8 to 20 GHz) or K (20 to 40 GHz) bands and it can track the target or targets while controlling two missiles. The system may also be used for controlling guns, possibly with the assistance of a separate ballistic computer.

Upon acquisition of the target/targets by the ship's search radar the fire-control radar designates the targets. The system automatically calculates the level of threat from each target, allocates the missile or missiles and automatically launches them. In the anti-ship missile role the Barak-1 leaves the launcher and is turned over towards the target by the thrust vector control system which is automatically discarded upon completion of the task. The missile is acquired and controlled by the fire-control radar which then guides it towards the target. The missile is capable of engaging targets 2 m above the sea and can manoeuvre at 25 *g*.

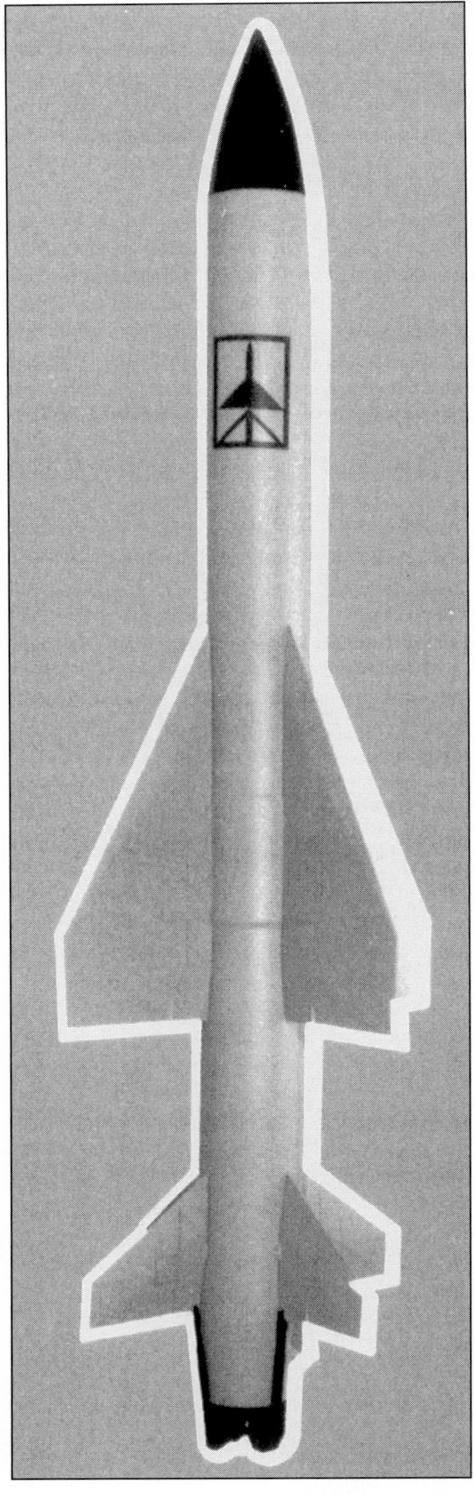

The Barak-1 missile

Operational status
Development of the system was completed by 1993 and the systems were sold to Chile, Israel and Singapore. In July 1997, there were reports that South Korea was on the verge of ordering Barak but the country's severe economic problems may have caused the plan to be postponed and then cancelled. It appears some 17 Barak-1 systems and 1,500 missiles have been sold. A Super Barak with longer range has been proposed.

Specifications
Length: 2.17 m
Diameter: 17 cm
Wing span: 68.40 cm
Weight: 97.90 kg
Speed: M2
Range: .5-12 km (6.5 n miles). Range against missiles probably about 5 km
Guidance: CLOS

Contractors
Israel Aircraft Industries, MBT Weapon Systems Division.
Rafael Armament Development Authority.

UPDATED

Kentron Umkhonto

Type
Ship launched missiles - surface-to-air

Description
This anti-aircraft missile, with its associated vessel-installed subsystems, is supplied as a missile group which allows easy integration into the weapon suite of modern naval vessels.

The 'Umkhonto' uses an Infra-Red (IR) homing seeker which originated from the U-Darter air-to-air missile, in service with the South African Air Force (SAAF). The remaining missile subsystems originated from the SAHV-3 Command to Line of Sight missile, which has been successfully integrated with the South African Crotale 4000 equivalent firing unit. All subsystems were substantially upgraded to incorporate latest technologies and provide improved performance.

The missile operates in a Lock-On-After-Launch (LOAL) mode, with inertial mid-course guidance. It receives target coordinate updates from the launch vessel during the inertial phase, enabling it to make corrections to its course as required to counter target manoeuvres.

The missile group comprises of the following:

(a) the Missile Round, consisting of the missile inside its sealed launch canister
(b) the SAM System controller, which provides for the control over the missile group
(c) the Missile Control Panel, which provides the operator interface in the control room
(d) the Missile Sequencer Subsystem, which controls the missile during the launch phase
(e) the Telecommand Transmitter Unit, which provides target inertial information to a number of missiles in flight
(f) the Gas Supply Subsystem, which supplies compressed air for cryogenic cooling of the seeker prior to launch.

The missile group requires the input of three-dimensional target data from the combat suite sensors, plus the input of a firing command, which may be manually or automatically generated.

The missile is launched from a concentric launch canister, which contains the plume and provides sliding surfaces for the sabots. Vertical launching, together with the autonomous flight control of the missile, provides for 360° cover and the capability to launch a number of missiles (up to eight or more) at different targets simultaneously, giving the system a very high saturation level.

The missile, which weighs 125 kg at launch, is 3.3 m long and has a body diameter of 180 mm and a wingspan of 400 mm. It uses a high-performance, low-smoke composite propellant grain, housed in a wound carbon-fibre casing. It carries a 22 kg blast/fragmentation warhead, and an active proximity fuze, giving a high kill probability against all types of targets including sea-skimming missiles.

Apart from the IR seeker, the guidance system uses a digital autopilot, and inertial pack containing the sensors needed for mid-course guidance phase, and a set of electromechanical servos. The latter control the tail-mounted, aerodynamic control fins, as well as the thrust vectoring vanes in the motor nozzle during the launch phase. In-flight power is supplied by a set of thermal batteries.

The cylindrical launch canister is sealed and pressurised for maximum protection of the missile. During launch, the exhaust gas is deflected in the canister and ejected vertically alongside the missile as it leaves the canister. The empty canister is removed for reloading and is replaced by a full missile round.

Serviceability of the missile group is maximised by a built-in test system controlled by the Missile Control Unit. This ensures that all parts of the group, including the missiles, are tested on a regular basis with a minimum of operator intervention. The same system is used for fault localisation during maintenance and quick maintenance turn-around is obtained by replacement of Line Replaceable Units in the on-board equipment. The missile canister is not opened while on board.

Future upgrades of the system will take the form of replacement of the IR seeker with an active radar seeker, and at the same time, incorporation of a larger rocket motor. The resulting missile will have the advantages of full all-weather operation, coupled with longer interception range (20 km). In all other respects, the new missile will be a direct replacement of the IR missile, using the same auxiliary systems and a similar, but slightly larger launch canister. In particular, the on-board infrastructure will be virtually identical so that the lower cost, shorter range IR missile may still be used.

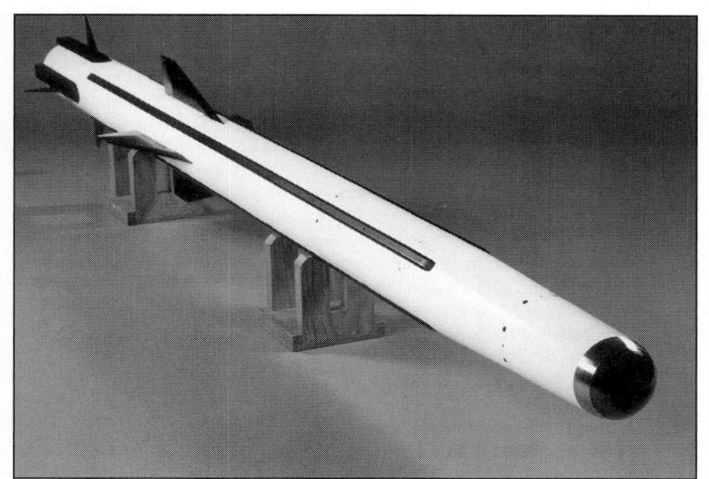

The Umkhonto surface-to-air missile 0088176

A future upgrade of the IR seeker to full Imaging Infra-Red (IIR) is also foreseen.

Operational status
Development. On order for South Africa's MEKO A200 class patrol corvettes and Finland's Squadron 2000 surface combattant programme.

Specifications
Length: 3.32 m
Diameter: 180 mm
Wingspan: 500 mm
Weight: 125 kg
Warhead: 22 kg blast fragmentation
Maximum altitude: 24,000 ft
Horizontal range: 9,000 m (aircraft) or 12,000 m (helicopters)
Time of flight to 7 km: <16 s
Minimum launch interval: 2 s

Contractor
Kentron Division of Denel (Pty) Ltd.

VERIFIED

Lockheed Martin RIM-72 Sea Chaparral (Chapfire)

Type
Ship-launched missile - surface-to-air.

Development
Summary: Infra-Red (IR) seeker.

A requirement for a lightweight air defence system which could meet 'Styx' type anti-ship missiles, led to RIM-72A Sea Chaparral being trialled and deployed in the 'Gearing' FRAM I class destroyers in 1972. All these mountings were removed by October 1973 and subsequently the systems were sold to Taiwan.

In 1992 there was a proposal to add a Hellfire missile launcher capability to Chaparral systems as Chapfire. This launcher would have two Hellfire rails replacing two Chaparral rails, a Raytheon FLIR target acquisition unit and a laser designator/range-finder for Hellfire. It was intended as a low-cost shipborne air/surface defence system for cargo vessels in coastal waters and is known as the Rapid Deployment Integrated Defense System. In August 1992, Hellfire test rounds were fired from a Chaparral launcher against ground and towed aerial targets. Plans also exist to modify the mount both for remote operation and the addition of a 20 mm Vulcan gun.

Description
Externally the RIM-72 Chaparral is a long, slim cylinder with ogive nose, swept-back delta-shaped fins in a cruciform configuration and large trapezoid-wings. One pair of wings is provided with rollerons to reduce roll rate. In the nose is the AN/DAW-1 guidance section with all-aspect IR seeker.

Early warning of the target is provided by the ship's radar and the CIC directs the operator as to which direction he should turn the mounting. He acquires the target optically, tracks it and selects and activates a missile for the engagement. When the missile's IR detector has acquired the target an audio tone sounds and the

A Sea Chaparral launcher in close up

operator launches the missile. Proportional navigation guidance commands are generated from seeker tracking rates and used to control the missile's flight path.

In August 1987, a Rosette Scan Seeker (RSS) was type classified for land-based Chaparral. This is software controlled and can be programmed to meet new threats. There is also a 50 per cent increase in the target acquisition range. It is not known if this has been incorporated into Taiwanese missiles. In December 1990, a dual-mode IR/radio frequency seeker underwent proof-of-principle testing.

Operational status

Some 20 launchers were produced and all were sold to Taiwan. The Sea Chaparral is no longer in production, perhaps 250 naval missiles were produced and the systems may continue to be updated.

Specifications

Length: 2.91 m
Diameter: 12.70 cm
Wing span: 63 cm
Weight: 86.20 kg
Speed: M1.2
Range: 9 km (4.75 n miles)
Guidance: Optical aiming, IR homing

Contractor

Lockheed Martin Corporation.

VERIFIED

MBDA Mistral

Type

Ship-launched missile - surface-to-air.

Development

The MBDA (formerly Matra) Mistral portable surface-to-air missile system began development in December 1980 and by 1983 the first test firings had begun. These were completed in March 1988 and in 1989 the French Army received the first systems, which were designated Mistral.

Since the start of development, consideration was given to a naval version and the development of two launching systems (SADRAL and SIMBAD) began. In October 1986 trials of the SADRAL system were successfully conducted and the naval applications of Mistral began to be marketed. The system was ordered by the French Navy; but an order from Finland in 1989 was the first export success for the naval version in a Finnish mounting. Subsequently, orders for SIMBAD and SADRAL mountings were received from the United Arab Emirates and, in 1990 from Norway. Qatar, Singapore, Brazil, Cyprus have also ordered naval versions. In 1996 Thailand placed an order for SADRAL/Mistral and Indonesia for SIMBAD/ Mistral. Since then another South American country has ordered SIMBAD. A Mk 2 version of Mistral has been developed.

Description

The Mistral missile is a slim cylinder with a sharp, pyramid-shaped nose. At the front of the missile are two pairs of canard control surfaces while at the rear are four folded tail surfaces and a small booster.

In the missile nose is a SAGEM infra-red seeker under infra-red transparent magnesium fluoride panels. The seeker is cooled at the same time as the gyro is run up, within 2 seconds.

Behind the seeker, gyro and actuation system is a Manurhin 2.95 kg high-explosive warhead with impact and laser proximity fuze. The warhead consists of 1 kg of explosive and nearly 2 kg of tungsten balls for improved lethality.The whole system has a reaction time of as little as 5 seconds.

Three mounting options are available: SIMBAD *(Système Integré de Mistral Bitube d'Auto-Défense)*; SADRAL *(Système d'Auto-Défense Raprochée Anti-aérienne Léger)*; and SIGMA (Stabilised Integrated Gun Missile Array)

SIMBAD is a manually operated system which has two missile launcher-containers, a missile selector and a prelaunch electronics box. Replaceable battery/coolant units are used to provide up to 45 seconds of prelaunch power to the missile and to cool the seeker detector cells for lock on. SIMBAD has an operating radius of 1.50 m, may be installed on any 20 mm gun mounting and can be deployed from –5 to +55° in elevation.

When the target is designated the operator tracks it, designates a missile and activates the battery/coolant unit. This cools the seeker and warms up the missile. If the target is within range the seeker locks on and the operator receives an audio signal allowing the missile to be launched.

SADRAL is a remotely operated system on a two-axis stabilised mount and is designed to provide self-protection for larger warships such as frigates and aircraft carriers. It is used as a point defence system and has to be integrated with a radar or an electro-optic search and track system.

SADRAL has six launcher-containers and a television camera in addition to a SAGEM Murène thermal imaging camera for night or bad weather operations. The launcher angular velocity is 1.5 rad/s in both axes with acceleration of 2 rad/s^2 in azimuth and 1.5 rad/s^2 in elevation.

SIGMA is a combination of three Mistral missiles with a small calibre (30/25 mm) on the MSI-Defence Systems stabilised mount. The installation of the system does not require any deck penetration.

Mistral launched from a SADRAL

The missile is ejected from the launcher-container by the booster and once clear the control and tail surfaces are deployed. The sustainer ignites some 15 m away from the launcher-container and burns for 2.5 seconds as the missile homes in on the target. The maximum total flight time is 14 seconds.

Operational status

Production of Mistral 1 has been completed with15,000 rounds ordered for 23 countries of which at least 1,500 are for naval use by 12 countries. Production of Mistral 2 has begun. Some 120 Mistral naval systems have been ordered .

Specifications

Weight: 18.40 kg
Seeker: passive infra-red, fire-and-forget guidance
Cool down: 2 s
Max speed: M2.5
Warhead: 3 kg

Contractor

MBDA (France).

VERIFIED

MBDA Seawolf (GWS 25/26)

Type

Ship-launched missile - surface-to-air.

Development

Full-scale development of a quick-reaction, all-weather, point-defence, surface-to-air missile system to replace Seacat, began by BAe Dynamics (now MBDA) in July 1968. In 1975, sea trials of the system (now designated Seawolf) began and were completed by the end of 1977 by which time the system had begun production. In May 1979 it entered service in the 'Type 22' frigate HMS *Broadsword* as GWS 25 Mod 0, with GWS 25 Mod 3 and Mod 4 entering service in the late 1980s. The system also successfully demonstrated on exercises its ability to destroy anti-ship missiles, with the first success in November 1983 against an Exocet. However, it failed to find an export customer, partly because both the launcher and tracker were regarded as too heavy by potential customers.

In order to achieve faster reaction times and all round cover, Vertical Launched Seawolf (Seawolf VL) had been developed as GWS 26 Mod 1 and was completed in mid-1990. In October 1993, Matra BAe Dynamics announced a follow-on contract of £100 million for 450 Seawolf VL missiles. These will equip 'Duke' (Type 23) class frigates.

Seawolf VL received its first export order from Malaysia in December 1993.

BAE Systems, Avionics, Radar and Countermeasures Systems has been selected to undertake a project definition study for the mid-life update of the Seawolf point-defence missile system, fitted on board the Royal Navy's 'Type 22' and 'Type 23' frigates. Subcontractors include MBDA and BAE Systems (formerly GEC-Marconi Electro-Optics, Sensors Division).

Description

The Seawolf system consists of an air search radar, one or two trackers, a missile and a command subsystem. In GWS 25 the air search radar is the BAE Systems Type 967 D-band system whose waveguide antenna is mounted back-to-back with that of the Type 968 surface search antenna. In GWS 26 the air search radar is the BAE Systems Type 996 E/F-band radar with stabilised planar-array antenna.

The tracker in the GWS 25 Mod 0 is the BAE Systems Type 910 which is normally dedicated to a launcher and can control two missiles simultaneously. This last feature and the ability to track both target and missile, is achieved through time-division multiplex techniques. The Type 910 is an I-band monopulse radar which has a range of about 20 km against missile-size targets. The Cassegrain antenna is used for differential tracking while two attached dish antennas are used for transmitting or receiving guidance commands to the missile over a J-band command link. The tracking loop incorporates electronic angle tracking so that antenna movement is unnecessary when correcting missile line of sight guidance for small angles. In addition, the system includes an electro-optic differential tracker using a BAE Systems V334 low-light television camera. From April 1987 this was augmented by a BAE Systems THIM (Thermal Imager Modification), this being a repackaged V3800.

Vertical Launched Seawolf provides the principal air-defence capability for the Royal Navy Type 23 'Duke' class frigates 0002011

Targets are detected by the air search radar and are assigned through the ship's command system to the Seawolf command subsystem. There the threat is evaluated and the target is assigned to a tracker-launcher combination. Both slew towards the target, the tracker searches in elevation until it acquires and locks on to the target to establish a radar line of sight. When the target is within range, one missile is automatically launched and, if a salvo engagement is required, a second will follow it moments later. After launch the missile is acquired by the wide-angle antenna and gathered into the target line of sight. Its position relative to the centre of the tracking beam is monitored by the radar. The shipborne guidance computer generates the necessary steering commands to keep the missile in the centre of the beam until it intercepts the target. The missile's rearward facing antennas help to provide a good ECCM capability. The television camera is used for low-level tracking with flares showing the missile's location.

Operational status

Some 33 systems and over 2,000 Seawolf missiles have been produced or ordered. Seawolf GWS is in service on the four former Royal Navy 'Type 22' frigates in service with Brazil (Batch 1) and two Batch 2 ships sold to Romania. VL Seawolf is in service on Royal Navy Type 23 'Duke' class frigates and is being installed on Malaysia's two 'Lekiu' frigates.

Specifications

Length: 2 m (Seawolf); 3 m (Seawolf VL)
Diameter: 18 cm
Wing span: 70 cm
Weight: 82 kg (Seawolf); 140 kg (Seawolf VL)
Speed: >M2
Range: 6 km (Seawolf); 7 km (Seawolf VL)
Guidance: CLOS (Seawolf); ACLOS (Seawolf VL)

Contractor

MBDA (UK).

UPDATED

Raytheon FIM-43 Redeye/FIM-92 Stinger

Type

Ship-launched missile - surface-to-air.

Development

Summary: FIM-43 Redeye - infra-red seeker; FIM-92 Stinger - infra-red seeker.

In 1959, Convair Pomona (now Raytheon Missile Systems) began development of a shoulder-launched IR-guided surface-to-air missile and in October 1965 production of the FIM-43A Redeye began.

In 1966, the US Army and General Dynamics Pomona Division began to study new design concepts for a portable surface-to-air missile system development of the FIM- 43A Redeye, the Advanced Seeker Development Programme (ASDP).

In 1972, ASDP led to a second-generation missile system, XFIM-92A Stinger, with improved sensor and performance. In 1979, production of the system began and the first deliveries were made to the US Army in 1980. Some 16,000 were produced until mid-1987 when production ceased. The basic Stinger became famous in the Afghanistan conflict with 270 confirmed kills against Soviet fixed-wing aircraft and helicopters.

The US Navy purchased a number of Stingers from 1987 and a naval launching ring was designed and deployed.

Description

The FIM-43 Redeye missile is a long slim cylinder with a rounded nose and two rectangular nose vanes and four tailfins in cruciform configuration canted at 15°. These pop out when the missile is launched. Internally it has an IR detector in the nose, actuators, guidance mechanism and a 2 kg HE fragmentation warhead with impact fuze. Behind is the Atlantic Research M115 two-stage, solid propellant ejector sustainer rocket motor.

When the operator acquires the target the battery coolant unit is activated, this sends Argon gas to cool the missile IR detector. The sighting system, with ×7.5 magnification and 25° field of view, is then used to track the target until the missile seeker has acquired lock on. Audio and visual indicators are generated in the launcher unit and the missile is pyrotechnically initiated. The rocket motor's first stage carries it out of the launcher-container tube and the aerodynamic surfaces deploy. After 1.6 seconds, by which time the missile is 6 to 7 m ahead of the operator, the sustainer ignites and the warhead is armed while the seeker constantly measures the difference between the gyro line of sight and the IR source. Throughout the engagement the seeker is continuously repositioned to remain aimed at the target.

The basic Stinger (FIM-92A) has a 4.1 to 4.4 μm waveband infra-red, conical scan reticle seeker with signal processing electronics. These process the IR data to determine the relative angle then predict an intercept point by using a proportional navigation guidance technique. In Stinger-POST (FIM-92B) the reticle seeker unit is replaced by an image scan optical processing system with IR and UV (Ultra-Violet) detectors and two microprocessors integrated into the signal processor.

The Stinger-RMP (FIM-92C) is similar but has a facility allowing the microprocessors to be reprogrammed to meet new threats. It also employs a rosette pattern image scanning technique for more efficient target discrimination.

To engage the target the operator depresses the trigger to the first stage causing the impulse generator to energise the BCU. This cools the seeker and generates prelaunch electrical power to activate the missile and ignite the ejector motor. When the missile seeker has acquired the target, an audio signal is sent to the operator who depresses the seeker uncage bar and inserts superelevation and lead data. The operator then depresses the trigger fully, causing the BCU to activate the missile battery and give the missile full energy autonomy. The umbilical connector to the grip-stock is retracted and the BCU sends a pulse which ignites the ejector motor.

The initial thrust rolls the missile airframe and starts the fuzer timer. The missile then breaks both frangible covers. As the missile leaves the tube control surfaces, the tailfins deploy and the ejector motor, which has burned out within the tube, is jettisoned. When the missile has reached a predetermined safe distance from the operator the sustainer ignites and the missile seeks the target.

A US Navy Stinger team

The total time elapsed between the initiation of the firing sequence and motor ignition is 1.7 seconds.

For naval use a firing ring gives the operator a place to rest the launcher when it is not in use.

The fire-and-forget versatility of Stinger allows for platform application of multiple missiles by use of Air-to-Air Stinger (ATAS) launchers and Standard Vehicle Mount Launchers (SVML). These applications increase available firepower per gunner and take advantage of platform features such as night sights, automatic target tracking and external target cueing. Examples include the Avenger, Light Armored Vehicle-Air Defense (LAV- AD) and Dual Mount Stinger (DMS). Applications to naval platforms, such as small boats and frigates, are also being developed.

Operational status

By 1999, 90,000 Redeye and Stinger missiles had been produced, including 44,000 Stinger RMP which remain in production together with Stinger Block I, while Stinger Block II is entering production. Naval and marine customers for Stinger are Denmark (FIM-924), Germany (FIM-92A/C/D, but designated Fliegerfaust-2 FIF-2), the Netherlands (FIM-92A/C/D), Taiwan (FIM-92C/D), and the United States (FIM-92A/B/C/D).

The European customers use FIM-92A/B while the US uses all three versions. The Redeye is known to be used by the Greek Navy in 'Gearing' class destroyers and 'Kortenaer' class frigates.

Specifications

	FIM-43	FIM-92
Length	1.28 m	1.52 m
Diameter	7.0 cm	7.0 cm
Weight	8.20 kg	10.10 kg
Speed	M1.6	M2.2
Range	3.3 km	8 km
Altitude	3 km	3.5-3.8 km
Guidance	IR passive homing IR/UV passive	IR passive homing (FIM-92A)

Contractor

Raytheon Missile Systems.

UPDATED

Raytheon Standard Missile 2 and 3

Type

Ship-launched missile - surface-to-air.

Development

The Standard Missile (SM) family has evolved from several missile programmes of the late 1940s, which produced systems designed to meet US Navy (USN) area air defence requirements. Development of what became the SM began in December 1964. Shortly after the SM-1 entered service, development began of a longer ranged version, SM-2, for use with the Aegis radar system. Production of the SM-2 began in 1977 and full details of the programme may be found in *Jane's Naval Weapons Systems* and *Jane's Strategic Weapons Systems*.

The SM-2MR (Medium Range) Block III/IIIA version (with improved capability against high-performance aircraft and cruise missiles) entered USN service in 1990. It was followed up by a Block IIIB version (RIM-66M-5), which has an Infra-Red (IR) seeker developed US Navy's Missile Homing Improvement Program (MHIP) and was also applied to the RIM-7R Seasparrow (qv). This variant went into production in 1995, being converted from Block II and III weapons, and entered service in 1996.

The RIM-156A SM-2ER (Extended Range) Block IV system was developed to provide a dedicated vertical-launch extended range version for Aegis-equipped ships, that is, the 'Ticonderoga' Baseline 2 onwards and 'Arleigh Burke' classes. These missiles had a dual-mode RF and IR seeker. Although Block IV reached Low Rate Initial Production (LRIP) in May 1995, volume production was abandoned in favour of Block IVA.

SM-2 Block IVA not only met conventional missions but was seen as providing the 'lower tier' (endo-atmospheric) levels of the US Navy's Theatre Ballistic Missile Defense (TMBD) role. A Block IV missile converted to Block IVA configuration with an Imaging Infra-Red (IIR) seeker was successfully tested in July 1999, with the first fully controlled flight test of a Block IVA occurring in the second quarter of 2000. LRIP missiles were scheduled to reach the fleet by December 2001.

However, the TBMD programme was cancelled on 14 December 2001, following integration problems with shipboard systems and rising costs. What was the 'upper tier' (exo-atmospheric) weapon for the 'Ticonderoga' class cruisers is now known as the Aegis ballistic missile element of the USN's sea-based ballistic missile defence programme and will now be the SM-3 (provisionally designated RIM-300A).

Description

The Block IIIB version incorporates the multimode guidance upgrade with an infra-red seeker that has been developed under the USN's MHIP. The infra-red seeker is mounted to the side of the forward fuselage beneath an ejectable cover.

The Block IV is designed to provide hemispheric defence in a severe electronic warfare environment. The missile features improved cross-range, downrange and very high-altitude performance even against reduced signature, high-speed manoeuvring targets. The Block IVA missile had the capability of engaging ballistic missiles, being designed to sustain the higher gravitational forces involved in a theatre ballistic missile engagement. The IIR seeker, located beneath an ejectable cover, provided target discrimination and precise aim point selection. The IIR seeker features a staring focal plane array operating in the 3 to 5 μm bands, and a cooled sapphire dome. After the seeker fairing is ejected, the missile rotates to bring the seeker to bear and comes under IR guidance until the target comes within view of the forward-looking proximity fuze.

The SM-3 is similar to SM-2ER Block IVA, having GPS/inertial navigation and a hit-to-kill EX-142 kinetic warhead, equipped with a long wavelength IR seeker. The single-colour IR seeker has a 256 × 256 element focal plane array in the long-range band with a claimed tracking accuracy measured in microradians. When the target is acquired, the kinetic warhead is released. During 2002, the SM-3 development missiles scored three successful intercepts, on 25 January, 13 June and 21 November.

The Standard family may be operated from all the US Navy's 3-T Guided Missile Launch Systems (GMLS) as well as the Mk 41 vertical launch missile system. However, it should be noted that not every launcher version is able to operate every Standard Missile version.

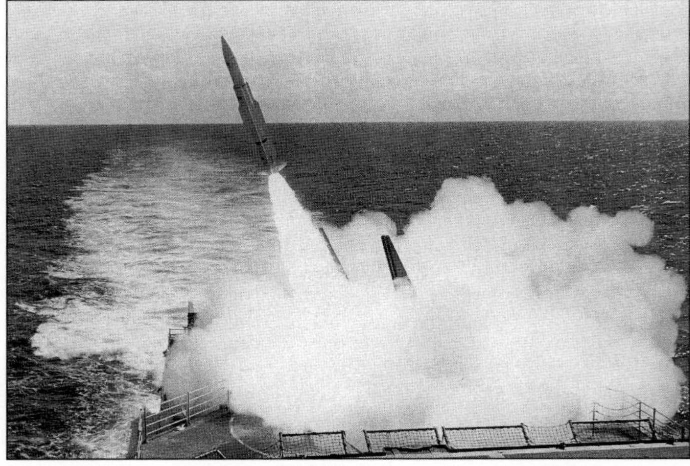

An SM-2MR is launched from a Mk 26 launcher on the cruiser USS Ticonderoga (US Navy)

Specifications

	SM-2/IVA	SM-3
Length (incl booster)	6.55 m	6.55 m
Diameter	35 cm	34.8 cm
Weight	1,497 kg	1,501 kg
Speed	M = 3.0	M = 3.0
Range	130 n miles	N/A
	240 km	N/A
Altitude	33,000 m	N/A

Operational status
SM-2MR Block IIIB is in service; SM-3 is under development.

Contractor
Raytheon Missile Systems.

UPDATED

Raytheon RIM-7R Seasparrow

Type
Ship-launched missile - surface-to-air.

Development
The RIM-7R was developed as part of the Seasparrow family of missiles, originating from 1968. For more detailed information on the system and its non EO-guided versions, see *Jane's Naval Weapons Systems* and *Jane's Strategic Weapons Systems*.

Description
The Seasparrow missile, is based on the AIM-7 Sparrow air-to-air missile and has been enhanced for launching from ships. All versions up to the RIM-7P have a semi-active radar seeker with a digital computer. The RIM-7R has a dual-mode semi-active radar and infra-red seeker. This seeker was developed from 1988 under the US Navy's Missile Homing Improvement Program (MHIP) for use on both the Seasparrow and Standard SM-2 (qv) Block IIIB missiles. The RIM-7R completed development in 1995 and passed an operational evaluation.

An integral part of most ships' Seasparrow missile system is the Mk 16 Low-Light Level TV (LLLTV) camera made by Ball Aerospace (qv). The camera augments the tracking radar and provides day and night imaging for search, surveillance, target identification and acquisition, fire control and kill assessment. The latest version Mk 16 Mod 2 has day and night CCD sensors and dual field of view.

Operational status
While the RIM-7 Seasparrow missile family remains in service with 12,000 reportedly being produced, together with some 210 systems, ordered or delivered to 16 navies, the RIM-7R did not enter service, as funds were diverted to other programmes.

A Seasparrow missile being fired 0002008

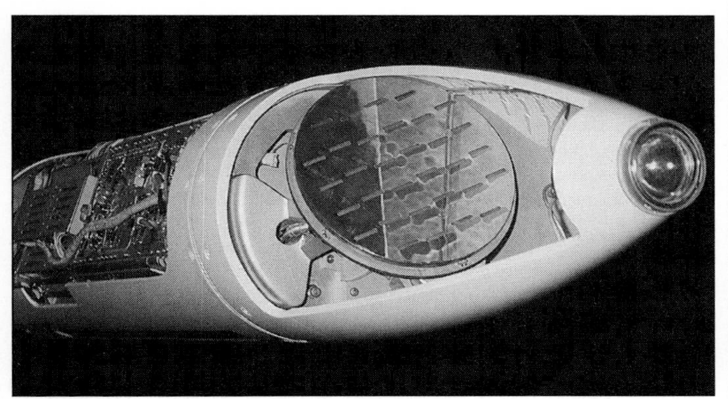

This view shows a dual-mode semi-active radar and infra-red seeker, proposed by Raytheon and exhibited in 1994, and subsequently developed for the RIM-7R (Duncan Lennox) ***NEW***

Specifications
RIM-7R
Length: 3.66 m
Diameter: 20.30 cm
Wing span: 1.02 m
Weight: 231.50 kg
Speed: M2.5
Range: 14.5 km (8 n miles)
Guidance: semi-active radar and infra-red

Contractor
Raytheon Missile Systems.

UPDATED

Raytheon RIM-162 Evolved Seasparrow Missile (ESSM)

Type
Ship-launched missile - surface-to-air.

Development
In June 1995, Hughes Missile Systems Company (now Raytheon Company) was awarded a US$167 million contract to design and develop an Evolved Seasparrow Missile (ESSM). Companies teamed with Raytheon include AWA Defence Industries Australia (now BAE Systems Australia), AlliedSignal (Canada), Terma (Denmark), RAM-System GmbH (Germany), ECON, ELFON, HAI and Intracom (Greece), Thales Naval Nederland (Netherlands), NAMMO Raufoss (Norway), Indra (Spain), Roketsan and Kalekalip (Turkey), Alliant Techsystems and United Defense (USA).

It is anticipated that production will exceed 4,000 missiles with the potential market being for 140 US and 115 foreign warships. Ships scheduled to receive ESSM include the US Navy's 'Arleigh Burke' Flight IIA destroyers, German F124 frigates, the Dutch LCF frigate, Spanish F-100 frigates, the Australian 'Anzac' class and the Turkish 'Barbaros' class frigates.

The first test-firing of RIM-162 took place on 23 July 2002 from the USS Shoup (DDG 86), an 'Arleigh Burke' class destroyer, with other successful launches conducted from the same vessel on 13 and 16 March 2003. On 21 January 2003, the first non-US firing took place from the 'Anzac' class frigate, HMAS Warramunga, of the Royal Australian Navy.

Description
The initial ESSM programme will consist of a kinematic upgrade to the RIM-7P Block II to produce a tail control version with a tail control rocket motor with 25.4 cm booster, which will double the speed as well as improving both manoeuvrability and range. The improved electronics will include a fast digital autopilot with inertial measurement unit.

An RIM-162 ESSM launched from the Mk 41 vertical launch system aboard the USS Shoup. The missile successfully intercepted a BQM-74 drone (Raytheon) ***NEW***/0522940

ESSM will include mid-course guidance and advanced modes of operation including launch-on-search, delayed illumination and home-all-the-way. Potentially this will enable each tracker/illuminator to control up to three missiles rather than the existing one missile. The missile is to have a quick-start capability allowing it to avoid the four-minute warm-up of the present generation and reduced smoke from vertical launches to avoid obscuring shipborne electro-optical sensors.

The second phase of the ESSM programme will involve an evolutionary development of the RIM-7R (qv) with a multimode seeker.

Operational status

Under development.

Specifications

RIM-162
Length: 3.7 m
Diameter: 25.4 cm
Weight: 282 kg
Range: 45 km

Contractor

Raytheon Missile Systems.

NEW ENTRY

Raytheon/RAMSYS RIM-116 RAM

Type

Ship-launched missile - surface-to-air.

Development

Summary: Two versions of the RAM exist: (1) RAM Block 0 has a dual-mode radar and infra-red seeker and (2) RAM Block 1 has an improved imaging infra-red seeker. An incremental upgrade to the RAM Block 1 is currently under development by the US and German navies adding capability to engage surface craft, helicopters and aircraft.

The US and West German navies launched RAM as a joint programme in 1977, the US Navy requiring a surface-to-air missile system to supplement Phalanx and Seasparrow. The original contractors were General Dynamics (now Raytheon Company) for the USA and RAMSYS GmbH, a consortium which comprised DASA (now EADS Deutschland) BGT and Diehl for Germany (now Diehl, EADS Deutschland and BGT). Production in both countries began in 1989 and the system entered service in 1993, aboard the USS *Peleliu* amphibious assault ship.

The Raytheon Sea RAM trials installation on board HMS York is a potential solution to the FIAC threat (R Scott/Jane's) ***NEW***/0118030

A RIM-116A RAM missile emerges from a Mk 49 launcher on the USS David (R Ray)

RAM performed well in the initial SSDS (Ship Self-Defense Systems) trials for the US Navy and is likely to be selected for all SSDS platforms. It will be fitted on all aircraft carriers, all 'Arleigh Burke' (DD 51) class and 12 'Spruance' (DD 963) class destroyers, all 'Tarawa', 'Wasp', and 'San Antonio' (LPD 17) amphibious warfare ships, plus the Littoral Combat Ship.

The US and German navies completed qualification testing on RAM Block 1 in August 1999 and achieved authorisation for production in January 2000. RAM Block 1 testing was described by OSD as the most stressing and realistic testing a US Navy missile system has ever been subjected to.

During 2002, new software allowed interception of manned aircraft and small vessels. Raytheon has proposed merging RAM and the Mk 15 Phalanx CIWS (qv) into a dual-layer self-defence system called RAIDS (RAM And Phalanx Integrated Defense System).

Another development involving fitting an 11-cell RAM launcher in place of the 20 mm Vulcan cannon of the Phalanx IB CIWS (the US Navy's Phalanx Surface Mode Upgrade), known as Sea RAM, was initiated in 1999 and tested in 2000. In February 2001, a Sea RAM system was installed on board a UK Royal Navy (RN) Type 42 destroyer, HMS York, for a 10-month trial (but no live firing) against a capability gap in the RN's ability to defend itself against fast attack craft and Fast Inshore Attack Craft (FIAC) in the littoral.

The most recent development has seen a RAM equipped with a new digital autopilot successfully perform controlled manoeuvres from the US Navy's China Lake facility on 12 June 2003.

Description

RAM is a lightweight, quick-reaction anti-ship missile system for close-in defence. The system consists of the missile, a launcher and a control panel in the ship, combat direction/combat information centre.

The RIM-116A missile is based upon the Sidewinder air-to-air missile but with significant modifications. It is a slim cylinder with two movable delta-wings and four folding, truncated delta fins. In the nose are two RF antennas and a Raytheon reticle scan infra-red seeker for terminal guidance, the same as that used by the Stinger portable surface-to-air missile. Behind this is a dual-band passive radio frequency seeker for mid- course guidance. Most of the rear of the missile is based upon Sidewinder, using the same 9.09 kg blast/fragmentation WDU-17B warhead with Mk 13 Mod 2 safety and arming mechanism and the Raytheon Mk 20 Mod 0 fuze.

Targets are designated by shipboard electronic or electro-optical sensors and this data is used to assign a launcher, to turn it in the target's direction and to elevate it to the correct angle for efficient interception. Radar and electro-optical sensors provide details of target location, distance and speed, while the ESM system inputs data on the target's radar frequency as well as correlating radar data on location. The missile motor is initiated and as it drives the missile along the firing cell a rifling band ensures that it spins for effective guidance as it emerges from the launcher.

As the missile leaves the launcher, the fins and tail surfaces emerge while the autopilot and control system maintain its line of sight course towards the target. The RF seeker is activated and once this acquires the target it controls the guidance system which can make the appropriate course alterations. The IR seeker has also been activated and when a sufficient signal-to-noise ratio is achieved this takes over guidance control for the terminal phase. The missile is capable of manoeuvres up to 20 *g* in any direction.

The Block I missiles (RIM-116B) feature the Infra-Red Mode Upgrade (IRMU) in which an 80-channel linear array with improved seeker processor electronics replaces the narrow field of view 'hot spot' graticule seeker to a wide field of view scanning array configuration giving the missile an IR fall-back option if RF acquisition is lost. Other advantages are improved performance in heavy rain or fog, a lower altitude approach, reduced radar cross-section and IR signature as well as increased manoeuvrability.

Operational status

Some 600 launchers have been produced and some 230 systems have been acquired, or ordered. RIM-116 has been ordered by the navies of Germany, Greece, South Korea and the United States.

Specifications
Length: 2.82 m
Diameter: 12.70 cm
Wing span: 44.5 cm
Weight: 73.60 kg
Speed: supersonic
Range: n/k
Guidance: infra-red

Contractors
RAMSYS GmbH.
Raytheon Missile Systems.

UPDATED

Thales Crotale

Type
Ship-launched missile - surface-to-air.

Development
Summary: infra-red tracker; optional thermal imager for Modular Crotale Navale.

A Crotale prototype was first launched in 1965 and in late 1968 trials of the complete system began. In 1971 the first systems were delivered to South Africa and the following year the system was ordered by the French Air Force.

In 1972, the French Navy issued a requirement for a point defence system and Thomson-CSF now Thales Air Defence's Crotale Navale system was selected and a development-production contract was awarded in December 1974. The first prototype system was installed in the trials ship *Ile d'Oleron* in May 1977 with the second prototype being installed in the new destroyer *Georges Leygues*. The first operational installation was in the 'Tourville' class destroyer *Duguay-Trouin* in 1979.

The system has been fitted into the 'Georges Leygues' (F 70 ASW) class and retrofitted into the 'Clemenceau' class aircraft carriers and 'Tourville' class destroyers. It has also been exported to Saudi Arabia.

The Crotale Navale system is intended for larger warships and from the late 1970s the manufacturers began to develop a system more suitable for ships as small as 200 tonnes. The lightweight system appeared in 1982 as the Modular Crotale Navale System and was subsequently modified to engage sea-skimming missiles. The modifications involved the addition of an infra-red tracker and the replacement of the infra-red fuze with an RF proximity fuze for very low-altitude engagements. The system was purchased for the United Arab Emirates and entered service early in 1990 and it is believed to have been selected for retrofitting into some Chinese destroyers.

Crotale NG (Nouvelle Génération) uses the latest VT1 hypervelocity missile (the naval version which is designated CN2) and in 1990 the system was selected by the French Navy for the new 'La Fayette' class frigates. In October 1992, it was revealed that Oman had become the first foreign customer for Crotale CN2 which entered service in December 1994 with the commissioning of the French frigate *La Fayette.*

Description
The Crotale system consists of the missile, a combined director and missile-launching mounting, a fire-control room with supporting electronics and a supervising console in the combat information centre.

The missile can manoeuvre at up to 25 *g* in each axis and can reach M2.3 in 2.5 seconds. The missiles are stored in alloy launcher-containers, eight of which are mounted on the lower coaxial mounting of the 6.50 tonne DCN launcher-director. The sensors are on the upper mounting. The director's sensors are a two-axis stabilised Castor radar antenna, a missile guidance antenna, an infra-red localiser and a TV for low-level tracking. The localiser is designated *Ecartometrie Differentielle Infra-Rouge* (EDIR). The J-band monopulse radar has a peak power of 80 kW to give it a range in excess of 16 km (8.5 n miles) and has an accuracy of 0.3 mrad in tracking and 0.10 m in missile-to-target differential deviation measurement. The RIC has a deviation measurement accuracy of 1 mrad and is designed to gather the Crotale missile for the command link.

The radar tracker is supplemented by an infra-red tracker developed by SAGEM and designated SEID (*Système d'Ecartometrie Infra-rouge Differentielle*). SEID, which operates in the 8 to 12 μm waveband, has a minimum range of some 500 m and a maximum range in excess of 20 km. It is designed to track both the Crotale and its target with an accuracy of 0.1 mrad to improve anti-missile performance.

Target designation is carried out by ship sensors such as the search radar, fire-control radar, electro-optical surveillance system or by optical sight. When the target has been designated through the ship's fire-control system, the director automatically slews in the direction of the threat and the onboard fire-control radar acquires and tracks it from a maximum distance of 20 km. The system automatically selects the missile or missiles for the engagement and the launch sequence is begun by the operator. Upon launch the missile is gathered within the broad field of the infra-red localiser which tracks the missile with the aid of the tail flare and locks it within the radar beam. The missile flies slightly off the radar's line of sight to prevent obscuring the sensor. Using the radar data, the fire-control computer continually calculates an interception course and relays guidance instructions over the radio link. The system reaction time from target acquisition to launch is a minimum of 4.5 seconds and an average of 6 seconds.

Missile guidance is performed by an improved line of sight system which optimises the manoeuvrability of the missile. In the case of a very low-altitude target

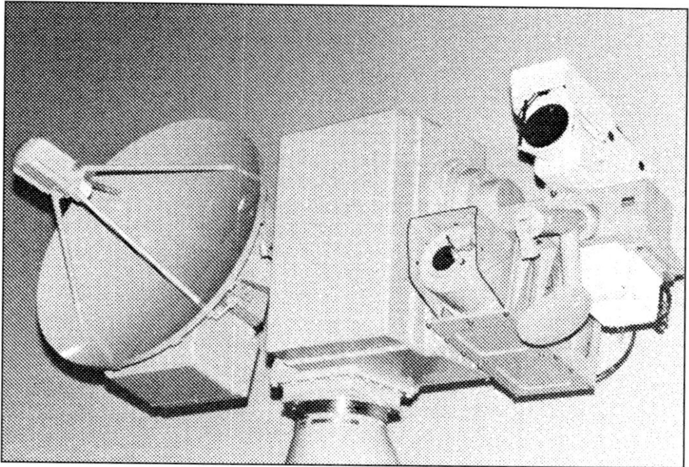
Crotale Navale System

(sea-skimmer), tracking of the target and of the missile is performed by the SEID tracker. The missile is fired and guided as in the radar mode with the terminal phase involving a 1° dive.

The maximum range against a helicopter-type target is 13 km but against high-speed manoeuvring targets, such as fixed-wing aircraft or missiles, it is between 6.50 and 10.50 km. The system is capable of intercepting four separate targets simultaneously.

Modular Crotale Navale uses the same missile as the original system but the launcher and the director are separated while the electronics may be dispersed about the ship. The launcher has four or eight containers together with the infra-red localiser.

As an optional replacement to SEID, Thales is offering its Castor thermal imager which has optional fields of view of 3 × 9°, 2.7 × 8.1° and 4 × 9°.

Crotale CN2 is based upon the VT1 which can manoeuvre at up to 35 *g*. The fire-control system, which can control two missile launchers and a 100 mm gun, is based upon a lightweight radar director. This is based upon a 12-18 GHz J-band radar with four-horn antenna on one side and a daylight TV camera with SAGEM IR tracker on the other side. The launcher itself has six launcher tubes on either side of the mounting and what is described as 'a TV escartometer' which helps automatically to bring the missile into line of sight with the target.

Operational status
Crotale is in production. It is in service on board French 'George Leygues' and 'Tourville' class destroyers and 'La Fayette' class frigates (CN2); Omani 'Qahir' class corvettes (Crotale NG); Saudi Arabian 'La Fayette' class (CN2) and 'Madina' class frigates; and United Arab Emirates' 'Lurssen' class corvettes (Modified Crotale).

Specifications

Missiles	**R400**	**VT-1**
Length	2.90 m	2.34 m
Diameter	16 cm	16.50 cm
Wing span	54 cm	n/k
Weight	84 kg	73 kg
Speed	M2.4	M3.5
Range	700 m -13 km	11 km
Altitude	4 km	

Director
Azimuth ambit: 360°
Elevation ambit: –25 to +85°
Rate of turn
Azimuth: 2 rad/s
Elevation: 3 rad/s
Missile mounting
Rate of turn
Azimuth: 1.5 rad/s
Elevation: 1 rad/s
Electrical power requirements: 440 V 60 Hz, 115 V 60 Hz, 115 V 400 Hz

Contractor
Thales Air Defence SA (France).

UPDATED

Thales Starburst

Type
Ship-launched missile - surface-to-air.

Development
Summary: Starburst has laser guidance and the NML launcher can be fitted with a thermal imager.

In 1986, Shorts (now Thales) began development of a new version of their Javelin missile. It incorporated a laser beam guidance system developed in parallel for the company's Starstreak missile (qv).

Description

For naval applications, Thales provides the laser-beam riding Starburst missile with a lightweight launcher, the Naval Multiple Launcher (NML). The NML was designed to provide close in/inner layer protection of warships, merchant ships and offshore installations against low-level attack by aircraft, helicopters or anti-ship missiles. Shorts has also demonstrated to a NATO customer the system's capability against small surface vessels. Ship fitting is achieved with no through deck penetration and no external power supplies are required. The all-up weight of the system is 450 kg with a height of 211 cm. The system has 360° all-round rotation and can be elevated from –10 to +35°. The NML can be in transit with the three missiles fitted giving firing reaction times of less than 4 seconds.

NML can be fitted with the Starburst thermal imager and standard IFF equipment. In addition it can be integrated with surveillance and command and control systems. See also entry for Starburst in the Land Air Defence Missiles section.

Operational status

Starburst is in service with the Royal Navy and the Malaysian Navy.

Specifications

Missile
Length: 1.39 m
Diameter: 7.60 cm
Wing span: 27.50 cm
Weight: 15.20 kg
Speed: M1
Range: 300-4,500 m
Launcher
Weight: 330 kg
Height: 1.98 m
Elevation: –15 to +45°

Contractor

Thales Air Defence Ltd (UK).

UPDATED

SHIP CLOSE-IN WEAPON SYSTEMS

SURFACE-TO-AIR MISSILES/GUNS

KBP Kashtan (SA-N-11 'Grisson'/CADS-N-1)

Type
Surface-to-air missile/gun.

Development
Summary: The electro-optic sensors associated with the system include an infra-red localiser to initiate the missile guidance and a TV system which includes an automatic target tracking capability.

Production of a new short-range air defence system, the 2S6M Tunguska, began in 1987. This consisted of two radar-directed 30 mm guns supplemented by a new short-range air defence missile system which was given the NATO codename SA-19 'Grisson'. The naval version of the system, which was given the NATO designation CADS-N-1 (Close Air Defence System-Naval) was first observed when the Project 1144 Atomic-Powered Missile Cruiser *Kirov* (now *Admiral Ushakov*) was commissioned in October 1988. Sea trials of the system, which has the Russian designation Kashtan, were conducted in a Black Sea Fleet 'Tarantul II' class missile corvette (Missile Cutter in Russian terminology). The system has not been retrofitted into the remainder of the 'Kirov' class but was selected for the Project 1143 'Tblisi' (now 'Kuznetsov') class aircraft carriers, Project 1155.1 Large Anti-Submarine Ships ('Udaloy II' class destroyers) and Project 1154 Large Anti-Submarine Ships ('Neustrashimy' class frigates). Installations appear to have favoured anti-submarine vessels rather than surface-warfare group forces.

Description
The Kashtan system combines two six-barrel automatic guns for close-in defence with the SA-N- 11 'Grisson' missile system for theatre air defence. The missile is claimed to be able to engage 3 to 4 cruise missiles for one combat module and has a range of up to 8 km. It is based on a modular design with one command module and up to six combat modules, depending on ship type. The combat module is on a deck mounting on which it can turn 360° in azimuth. On either side of the module is a GSh-6-30K 30 mm gun. These guns are gas operated with pneumatic charging and liquid cooling. They are mounted on arms that swivel in elevation, the whole system rotating for direction.

Eight SA-N-11 missiles are contained in two groups of four launcher containers. The combination of gun, missile and radar system on one mounting is believed to be unique. The missile is command guided, with a flare in the tail for automatic missile tracking. The integrated guidance system has milimetre waveband radar and a TV/optical system with automatic target tracking. Combined signal processing of the two leads to automatic selection of the optimum mode.

The system is used in conjunction with the Hot Flash/Hot Spot radar/electro-optic director, which it is believed includes a thermal imaging sight, TV system, laser rangefinder and a laser designator. It is believed that the EO system can be used for passive surveillance, tracking and target designation, as well as augmenting the radars in heavy ECM conditions.

Targets are probably designated through the ship's prime air search radar. The target is acquired through the search radar and at ranges of 1.5 to 8 km it will be engaged with missiles using radar or TV tracking, with the missiles probably launched in pairs for increased kill probability. Guns will engage targets at closer ranges down to 500 m.

Operational status
Some 30 of these systems were produced for the Russian Navy but production has ceased. Kashtan continues to be marketed for export and at least six have been sold to India.

The Kashtan shipborne air defence gun missile system

Specifications
SA-N-11
Length: 2.56 m
Diameter: 17 cm
Wing span: n/k
Weight: 57 kg
Speed: M2.6
Range: 8 km (4.50 n miles)
Guidance: Radar to command to line of sight
Tracking: Millimetre-wave radar or TV image storage

Contractor
KBP Instrument Design Bureau.

UPDATED

Polish Navy Wrobel II (ZU-23-2MR)

Type
Ship-launched missile - surface-to-air.

Development
Summary: the missile has an infra-red seeker.

Wrobel II is a naval version of the Polish Army's combined twin 23 mm gun and SA-7B 'Grail Mod 1' missile system, ZUR-23-2S Jod. Prototypes received sea trials on minesweepers in the late 1980s and were first commissioned in 1990. Wrobel II is now being offered for export.

Description
Wrobel II is an improved 23 mm gun mounting with SA-7B 'Grail Mod 1' (Russian and Polish designation Strela 2M) short-range surface-to-air missiles. It is mounted on a barbette which contains the turret training machinery.

The ZU-23-2MR 'Wrobel II'

The mounting weighs 2.50 tonnes and consists of the gunner's cabin with elevation and training gears, two 23 mm 2A14 guns, two launcher-containers for missiles, a night sight and an optional GP-02MR tachometric sight.

The missiles are housed in 4.71 kg launcher-containers and the missiles are slim cylinders with four small fins and two pop-out trailing fins behind the exhaust. The uncooled lead sulphide IR seeker operates in the 1.7 to 2.8 μm range. The seeker uses a spinning recticle, centre-null tracking system with a 1.5° field of view.

Wrobel II is not linked to the ship's fire-control system and the operator is believed to receive verbal instruction for engagement, using either the missile alone or both weapons in combination. The missiles can engage targets, at altitudes of 50 to 2,300 m at elevation angles of 20 to 60°. For attacks on oncoming targets they can be used at distances up to 2.80 km and for retreating targets at distances up to 4.20 km.

The guns, which have a practical firing rate of 400 rds/min, can engage targets at angles up to 90°. They can engage air targets below 2.30 km and surface targets up to 2 km.

Operational status

Over 25 Wrobel II mountings have been ordered or produced. The system is in service on Polish Navy 'Mamry' class minesweepers/hunters, 'Lublin' tank landing ship/minelayers and on two training ships.

Specifications

Weapons: 23 mm gun; Grail (Strela 2M) missile
Traverse: 360°
Elevation
Gun: −10 to +90°
Missile: +20 to +60°
Mounting weight: 2.50 t
Rate of fire: 400 rds/min
Range
Gun: 2 km (1 n mile)
Missile: 4.20 km (2.25 n miles)
Altitude: 2.3 km

Contractor

Polish Navy.

VERIFIED

SHIP CLOSE-IN WEAPON SYSTEMS

GUNS

Bofors 40 mm Mk 3

Type
Close in weapon system.

Development
By the early 1980s, Bofors began giving consideration to improving their 40 mm multipurpose gun, to integrate the fire-control system on the mounting and to improve both the accuracy and the range of the rounds. Development began in 1982 and the system was revealed in September 1984 with a naval mounting. The first sea firing trials were conducted from the 'Landsort' class minehunter *Vinga* in May 1989 and further operational trials were conducted in 'Landsort' class minehunters, in the minelayer *Carlskrona* and 'Göteborg' class corvettes.

In 1990 the Swedish Navy ordered the system to equip the fast attack craft *Smyge* and trials of this system began in the third quarter of 1991. Since 1994, Bofors has offered the Mk3 as a two stage upgrade.

Description
The fire-control system is based upon the Bofors MicroBOF microprocessor used in the SAK 57 Mk 2 with MicroBOF 8086 as the central processing unit and MicroBOF 8087 as the ballistic processor. The numbers used and their functions depend upon customer requirements but they provide ballistic calculations and interface with ship systems. The sensor fit also depends upon customer requirements but can include target acquisition and tracking radars, laser rangefinder, ×7, ×12 optical and ×4.9 electro-optic sights, television tracker, IR tracker and a gyro-reference system. The gun can be integrated with a search radar or electro-optical/optical designator in the auto-combat mode. In this mode, selection of the ammunition, proximity fuze setting, length of burst and initiation of the firing sequence is fully automatic even in manned systems. Manual override or manual control is available with the operator selecting the sensors. Alignment, test and simulation modes also exist.

Operational status
In production and in service aboard Swedish Navy 'Goteborg' class corvettes (some), 'Smyge' experimental patrol craft and a 'Landsort' class minehunter. It is also being fitted to Brazilian Navy 'Niteroi' class frigates.

Specifications
Calibre: 40 mm
Length of barrel: 70 calibres
Elevation: –20 to +80°
Training speed: 90°/s
Elevation speed: 55°/s
Mounting weight
Without sensors: 3,500 kg
With sensors: 4,000 kg
Firing dispersion: 0.7 mrad
Power requirements: 3 × 440 V, 60 Hz or 3 × 380 V, 50 Hz
Power consumption
Without sensors: 8 kVA (average), 26 kVA (peak)
With sensors: 12 kVA (average), 30 kVA (peak)
Crew: 1
Range: 3 km (1.61 n miles) anti-surface to surface; 6 km (AA)
Rate of fire: 330 rds/min
Ammunition: 2.80 kg (3P)

Contractor
Bofors AB.

UPDATED

A Sea Trinity mounting undergoing firing tests. The mounting has none of the sensors which will be incorporated in operational tests

FABA Meroka

Type
Close in weapon system.

Development
The Spanish Navy began to examine the options for a CIWS during the early 1970s and by 1975, they began to consider a system based upon a 20 mm multibarrel gun being developed for the Spanish Army. By the end of 1975 the former CETME received a contract for the mount and system integration and a prototype, called Meroka (acronym of more-barrel-gun), appeared in 1977. In October 1978 the Spanish Navy placed an order through EN Bazán (now Bazán SA) (the main military shipyard in Spain and the regular prime contractor for the Spanish Navy) to the former Lockheed Electronics Co (LEC) for the development and production of 20 Meroka Fire-Control Systems (FCS) sets.

The first production system (Model 0) was delivered in 1983 and was extensively tested and debugged by Fabrica de Artilleria of EN Bazùn as Model 1 on board the *Cadarso* fast patrol craft. From 1983 to 1986 a total of 20 FCS sets was delivered by LEC to Bazán.

In parallel FABA started the production of the Meroka naval gun mount departing from the basic development done by CETME and the integration, modification and setting up of the FCS delivered by LEC. This first operational configuration was called Naval Meroka Weapon System (NMWS) Model 2 and the first system was delivered in 1985 and commissioned on board the first 'Santa Maria' class frigate (FFG) in 1986.

Since then FABA has developed several FCS improvements to the LEC design, such as automatic designation and acquisition control, integration of an ENOSA/Opgal FLIR TV camera and a proprietary videotracker, designated as Model 2A. From 1986 to 1990 a total of 19 of these CIWS was delivered to the Spanish Navy for the Spanish aircraft carrier, four FFG frigates, five DEG frigates and one land sited training facility. In 1991 LEC was closed and the Spanish Navy acquired the manufacturing rights from Lockheed. FABA then started a major upgrade in 1992 called Model 2B. This version entered service in the last two commissioned FFG frigates *Navarra* and *Canarias*, respectively in 1994 and 1995. This is the version currently in production and is being introduced at several levels for upgrading former systems depending on funding availability.

In 1997 a further enhanced version, called Model 3, was ordered where the system has been fully integrated in the ship's combat system with a new generation multipurpose console. Unit integration has been simplified making the system easier to maintain.

Description
The Meroka system currently in service is unique in its configuration and operation. The weapon operations, such as loading, feeding and unlocking are carried out pneumatically by using compressed air from reservoir bottles on the mounting, the bottles being recharged from a redundant high-pressure air service on board. All the weapon operation cycles are directly controlled by a dedicated digital data processor, including Built-In-Test (BIT) diagnostics. The weapon itself consists of 12 barrels arranged in two superimposed rows of six and with a common breechblock.

The gun is in a hydraulically driven turret with an ammunition feeding mechanism and an internal cylindrical drum of ammunition with 720 rounds ready for use plus 720 extra rounds in three on-mount fast reloading boxes. The mounting includes the hydraulic, pneumatic and main electrical power drive, as well as some of the sensor electronics for the new FABA SPG-M2B I-ENOSA/Opgal 8 to 12 μm FLIR, in two independently gyrostabilised platforms boresighted to the gun axis.

The Meroka Model 2B close in weapon system 0002022

This non-deck penetrating mounting weighs 5.1 tonnes loaded, being 4.2 m long, 3.4 m high and with a swept radius of 2.9 m.

The Meroka does not need any dedicated or autonomous surveillance and designation subsystem and can accept any 1-D to 3-D target data from any source in the combat data system, so long as it includes bearing. A serial digital datalink to provide weapon system status and condition tell-back to the combat data system is included. The system automatically adjusts the scan pattern to achieve the required designation, accuracy and required scan, therefore a minimum target data accuracy of ±500 m and 1° is recommended for rapid acquisition.

A realistic surveillance performance for sea-skimming missile detection range performance is around the visual horizon, that is 18 km (10 n miles) with suitable positioning of the surveillance antennas on the ship's mast, frequency band selection and signal processing techniques. The target can be detected by any of the ship's air search radars, an infra-red search and track system (IRST) if available, the electronic support measures or a manually operated Target Designation Sight (TDS).

Detection, classification, identification, Threat Evaluation and Weapon Assignment (TEWA) functions are continuously performed through the command and weapon control system (C^2). Meroka engagement range for locking targets is around 9 km (5 n miles) with an average time of 2 seconds for the turret to slew to face the target. Acquisition by the SPG-M2B radar then typically takes 0.5 seconds and the system initiates tracking at a range of some 7 km (4 n miles). After the initial automatic radar acquisition, an automatic IR video tracking window is generated around the target image and the system performs a data fusion between both sensors to get the maximum servo-control accuracy, also in adverse low-elevation (radar multipath) conditions.

The IR image in the TV monitor of the system console is used to confirm target acquisition as well as confirming its identity (for engagement safety) while ballistic and prediction data are processed by the main FCS computer. Within 2 seconds, the ballistic super-elevation and lead angles are generated and the weapon is ready to fire at the target once the ammunition effective range, from 1,800 m to 200 m, is reached.

The full-scale production version Model 2A, within the former LEC Meroka tracking radar (a navalised angle monopulse major modification of the former US Army AN/VPS-2) is still in service in some ships but it is no longer available for production and will be completely superseded by the Model 2B upgrading programme currently in progress. The Model 2B shares the camera and its corresponding cryogenics and automatic videotracker. However it features a new and more powerful radar of 7,000 m range with a new frequency agile band, a new main FCS data multiprocessor with digital servo-control and system built-in-test diagnostics with synthetic target simulation. There are other improvements to the turret design focusing on the mechanics, hydraulics and pneumatics.

In the latest Model 3, currently being developed, a new multipurpose console fully integrated in the combat system replaces the present control panels and the system maintenance has been simplified.

Operational status

Since 1986, 21 systems have been in service in the Spanish Navy, upgraded to Model 2B. Model 3 is being developed.

Specifications

Calibre: 20 mm
Length of barrel: 120 calibre
Muzzle velocity: 1,250 m/s
Traverse: ±180°
Elevation: −20 to +85°
Traverse speed: 115°/s
Crew: 1 console supervisor
Effective range: 1,800-2,000 m
Rate of fire: 1,440 (9,000 in a salvo) rds/min
Ammunition: subcalibre Meroka-APDS or standard 20 mm for training, exercises and so on)

Contractor

IZAR Systems Division (FABA).

UPDATED

Mauser MLG 27 mm light naval gun during firing trials 0120739

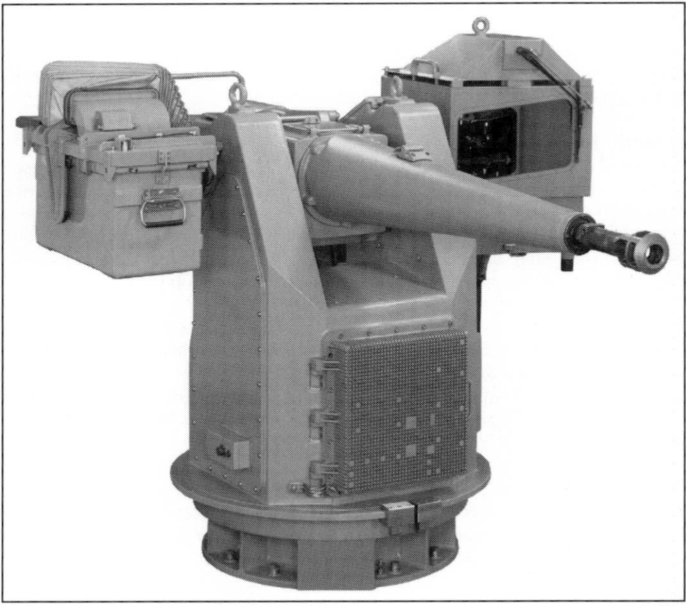

Mauser light naval gun system MLG 27 mm 0120738

Mauser naval gun system MLG 27 mm 0055135

Mauser naval gun system MLG 27 mm

Type

Naval gun based weapon system.

Development

The gun mount for this system is based on the Coulevrine developed by Navasl Guard SA from 1990. Prototypes were shown at the Euronaval exhibition in 1996 and a mount, armed with a 20 mm Rheinmetall Rh 202 gun, was ordered for the UAE Coast Guard 'Protector' class patrol boat, with a further four on order.

After Mauser-Werke Oberndorff Waffensysteme GmbH took an interest in 1995, the mount was developed as the MLG 27 (Marine Leicht Geschütz) armed with the company's 27 mm BK 27 revolver cannon, with the MLG 30 variant having a 30 mm cannon. In August 1998, the company was contracted to develop the MLG 27 for the German Navy and qualification trials on the auxiliary, FGS *Schwedeneck*, were completed in December 2000. A formal contract for 83 MLG 27 systems was announced in October 2002, with first deliveries due from late 2003.

Description

The system is equipped with an EO-sensor unit including a thermal imaging sight, a CCD-day light camera and an eye-safe laser range-finder based on the STN ATLAS Multi-Sensor Platform MSP 500.

The weapon system includes the Mauser BK 27 mm with a rate of fire of 1,700 rpm and is planned to replace the 20 and 40 mm gun based systems in the German Navy. For high efficiency a special FAPDS ammunition will be fired. It is intended to use the new weapon system on a wide range of ships and boats.

System operation requires target data from an external data source. After receiving these data the gunner starts the target engagement by lock on. The electro-optical sensor unit is stabilised in two axes and allows optical monitoring. By using a multifunctional video tracker and a suitable fire-control system the weapon system is able to engage sea, air and land targets effectively.

Operational status

In production for the German Navy, to replace 20 mm and 40 mm gun mounts on ships of the 'Brandenberg', 'Bremen', 'Sachsen', 'Ensdorf', 'Frankenthal', 'Berlin' and 'Elbe' classes, plus the new K 130 class of corvettes.

Specifications

Calibre: 27 mm
Length of barrel: 1,700 mm
Muzzle Velocity: 1,100 m/s
Number of cannons:1
Traverse: 340°
Elevation: –15 to +60°
Traverse speed: 60°/s
Elevation speed: 60°/s
Range: 2 n miles (4 km) against surface targets
Mounting weight: 850 kg
Crew: none on mounting
Rate of fire: 1,700 rds/min
Ammunition: TP, FAPDS
Ammunition capacity: 90 rounds

Contractor

Mauser-Werke Oberndorf Waffensysteme GmbH.

UPDATED

Mauser-Werke, EADS Drakon EO

Type

Close-in-weapon system (CIWS).

Development

In 1984, Mauser began studying the requirements for a CIWS and selected the 27 mm aircraft cannon as the basis for the new system. In 1988, the company began construction of a firing prototype with four guns which was completed in 1989. The quad 27 mm CIWS was later designated AMS 27-4.

In 1992, Mauser and Signaal (now Thales Naval Nederland) signed an agreement for development, delivery and testing of the AMS 27-4 which had been renamed MIDAS by October of that year. In 1994, Signaal withdrew from the agreement because by then the German Navy had selected RAM rather than MIDAS to meet its CIWS requirements. A redesign of the mounting was taking place early in 1995 and the new mounting was renamed Drakon.

The latest version, Drakon EO is being co-developed with EADS and integrates the Najir 2000 electro-optic fire-control system.

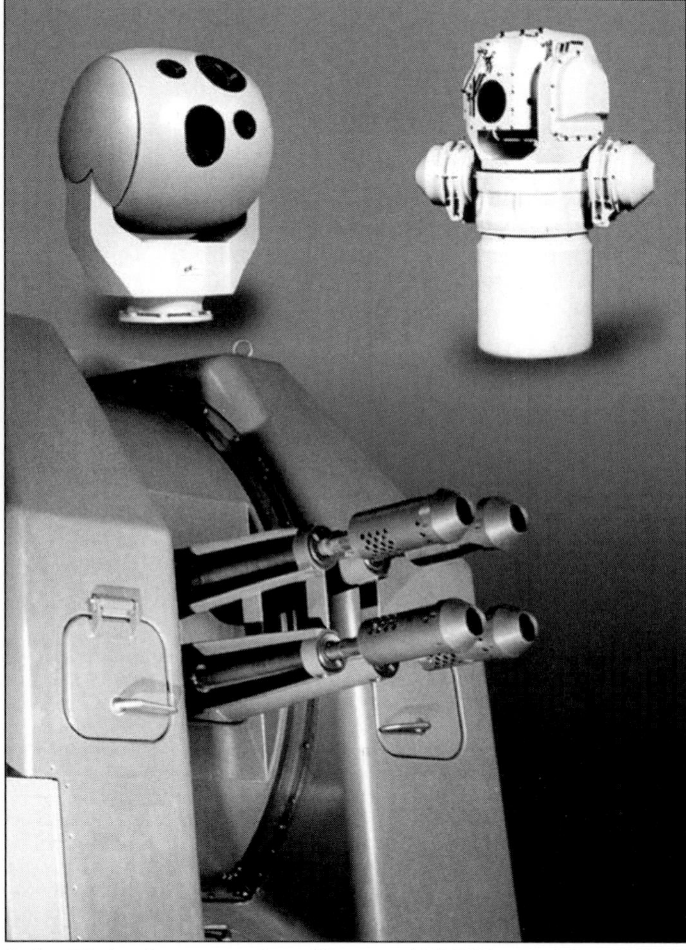

Drakon EO close in weapon system with sensors, Najir 2000 on the left and VAMPIR ML 11 on the right 0002021

Description

The Drakon EO CIWS is a compact lightweight system installation requiring no deck penetration. It can be fitted to a wide variety of vessels from frigates to mine countermeasures vessels. It can be used either as an autonomous subsystem or be integrated into the ship's combat system. As a stand-alone system it is provided with Najir 2000 fire-control system and VAMPIR ML 11 IRST. Najir 2000 has a dual waveband thermal imager, CCD TV camera and eye-safe laser range-finder.

Drakon employs four automatic gas-operated 27 mm revolver cannons, operating independently each with a rate of fire of 1,800 rds/min. The gun is based on the BK 27 mm aircraft cannon of the Tornado. Magazine capability is sufficient to permit 12 target engagements before reloading.

Target data are obtained from VAMPIR and are transmitted to Najir 2000 which controls the gun mount. The whole weapon system is controlled by a multifunction standard Calisto console.

Drakon uses 27 mm APFSDS (Armour-Piercing Fin-Stabilised Discarding Sabot) ammunition causing warhead detonation at distances up to 2,000 m.

For longer range air defence, Drakon EO provides for the optional integration of a surface-to-air missile system such as Stinger or Mistral.

Operational status

Ready for production.

Specifications

Calibre: 27 mm
Length of barrel: 1,700 mm
Muzzle velocity: 1,150 m/s
Number of cannons: 4
Traverse: 360°
Elevation: –15 to +70°
Traverse speed: 175°/s
Elevation speed: 175°/s
Mounting weight: 4.6 t
Crew: none on mounting
Rate of fire: 7,200 rds/min
Ammunition capacity: 1440 rds
Ammunition: HE, AP, SAPHE, TP, APFSDS
VAMPIR ML 11
Spectral band: 3-5 or 8-12 μm
Detection range (sea-skimmer): 10 km typical
Number of threats processed simultaneously: >50
Reaction time between first detection and TD: 2 s
TD accuracy: 3 mrad
Najir 2000
TV camera: CCD
IR3 camera: 8-12 μm
IR2 camera: 3-5 μm
Laser range-finder: eye-safe 1.54 μm
Total elevation travel: –35 to +85°
Max velocity in bearing: 120°/s

Contractors

Mauser-Werke Obendorf Waffensystem.
EADS Systems and Defense Electronics.

UPDATED

Octec LEAP ballistic predictor

Type

Predictor for ballistic calculations for video gun sights and electro-optical fire-control systems.

Description

LEAP is a family of single printed circuit-card ballistic predictors designed for the full range of gunfire control applications.

The units are available in 6U, 3U and miniature formats which makes them ideal for installation in a range of systems and equipments from full fire-control systems to video-based weapon sights.

Prediction software is available for all calibres of weapon and all operational applications including infantry weapons, land based air defence and naval fire control.

Operational status

Operational in a naval gunfire control application.

Contractor

Octec Ltd.

VERIFIED

Raytheon Phalanx CIWS

Type
Close in weapon system.

Development
In 1970 a contract was awarded to General Dynamics, Pomona (now Raytheon Company) for a closed-loop fire-control system. Prototypes were developed and delivered in January 1973. In November 1974, a remotely operated system was installed in the hulk of the destroyer *Cunningham* and used in lethality testing against a wide variety of missiles including Walleye laser-guided bombs. In July 1977 the system, Mk 15 Close In Weapon System (CIWS) and named Phalanx, was approved for service use with a full production contract awarded to General Dynamics in September 1979.

The first systems entered service in April 1980 in the aircraft carrier USS *America* (CV 66) and its use was subsequently extended to all surface warships of frigate size and above, all major amphibious warfare vessels and many important auxiliaries. The first export customer was Saudi Arabia and the system entered service in September 1981 with the commissioning of the corvette *Badr*. In February 1987 GE Aerospace, Ordnance Systems (later Lockheed Martin Armament Systems Division) was awarded a second source contract for Phalanx.

In 1994, work began to integrate improved sensors and controls to produce the Block 1B. The Block 1B configuration is capable of engaging surface craft and slow-speed aircraft while maintaining its primary role of anti-ship missile defence.

Raytheon and Thales Naval Nederland have agreed to collaborate on a development programme to produce a high-energy (200 kW) laser to replace the 20 mm gun. Raytheon is also working on a low energy soft-kill countermeasures system.

Description
The Mk 15 CIWS is a 'closed-loop' (that is, tracking both the target and the stream of rounds) weapon system, designed to provide the innermost layer of defence against anti-ship missiles.

Phalanx uses a Ku-band pulse Doppler sensor with a tracking antenna at the front and a search antenna at the top. The shared transmitter, receiver and servo-mechanisms are all in the same housing. The system is capable of search, detection, threat evaluation, acquisition, track and fire modes which can be conducted automatically through a digital computer with manual override.

In search mode the radar uses unambiguous Doppler/ambiguous range and switches PRF to achieve range resolution. It uses unambiguous Doppler/ambiguous range and a derived search range to acquire the target as well as bearing and speed to search the range, angle, speed envelope provided by the search radar. Angle tracking is accomplished via monopulse techniques with PRF switching to track across blind ranges.

Target data is correlated according to range, range rate and angular position relative to the mounting. When the target is evaluated as a threat the tracking antenna, which has a rate integrating gyro, is adjusted to the target's elevation to minimise delay between target handover and track acquisition. An error signal between the gun's current position and the target's location provides gun control.

The Raytheon Systems Company Mk 15 Phalanx 0089895

Phalanx is normally set up to scan a particular sector and will automatically engage fast air targets unless the 'hold fire' button is pressed. Targets can be detected at 5.6 km (3.00 n miles) and acquired at 4.30 km (2.30 n miles). When the gun fires, the radar tracks the centroid of outgoing projectiles, predicts their point of closest approach to the target and corrects the aim of the following projectiles. This technique uses variable PRF with selected spectral frequency line tracking to measure the stream of projectiles' angular error. Firing usually begins at 1.85 km (1.00 n miles) with a maximum probable kill at 460 m. System reaction time is reported to be 3 seconds.

The Block 1B Surface Mode Upgrade gives Phalanx a day/night engagement capability for small surface craft or low-speed aircraft by the integration of a Thales Optronics High-Definition Thermal Imager (HDTI 5-2F) and SRI Electro-Optics automatic video tracker as well as manual acquisition controls. HDTI is an 8 to 12 μm system with fields of view of 5 × 3° and 2 × 1.3°.

Although designed for autonomous missile engagement, Phalanx targets may be designated by other ship sensors through the command and weapon control system. It may also be used for surface-to-surface engagements through another fire-control system or its own FLIR. In the SSDS (United States Navy, Ship Self-Defence Program) the Phalanx radar can be used to cue the RIM-116A Rolling Airframe Missile (qv).

Operational status
Some 850 Phalanx systems have been delivered or ordered by the navies of 19 countries. The Block 1B Surface Mode Upgrade completed US Navy testing and deployed on USS *Underwood* in September 1999. The first production unit was delivered in May 2000.

Specifications
Gun
Calibre: 20 mm
Barrel length: 53 calibres
Muzzle velocity: 1,030 m/s
Rate of fire: 3,000 rds/min (Block 0, Block 1 Baseline 0), 4,500 rds/min (Block 1 Baseline 1)
Range: 1.47 km (0.75 n miles) (effective, horizontal)
Mounting
Weight
Block 1: 5.42 t
Block 0: 6.18 t
Traverse: 310°
Elevation: –25 to +85°
Training speed: 126°/s
Elevation speed: 92°/s
Power requirement: 440 V, 60 Hz, 3 phase. 18 kW search, 70 kW transient

Contractor
Raytheon Missile Systems.

UPDATED

Thales Goalkeeper

Type
Close-in weapon system.

Development
In 1976 Thales Naval Nederland began to develop the Close-In Weapon System (CIWS) which would meet the requirements of the RNLN. A prototype was completed in 1981 and firing trials began the following Spring. In May 1983 the system, named Goalkeeper, was ordered by the RNLN. The following year it was ordered by the Royal Navy. The first ship with the system was the 'Kortenaer' frigate *Gallenburgh* (now the Greek *Adrias)* in September 1984.

During 1990, the US Navy evaluated the Goalkeeper system. Tests were conducted with the Harpoon missile. Also during 1990 the existence of a containerised system was revealed, together with plans for a version which required little or no deck penetration.

Description
Goalkeeper is an autonomous CIWS in which the entire engagement sequence from search to destruction is carried out automatically. It consists of a General Electric Sea Vulcan 30 gun, an I-band search radar, I/K-band tracking radar, TV camera, separate transmitter and receiver cabinets, waveguide drier, mount control electronics cabinet, system interface cabinet, weapon control console and a monitor and keyboard.

The radar sensors may be supplemented by electro-optics on the mount and, for passive alerting and cueing, the Signaal infra-red search and track system IRSCAN can be integrated.

The Sea Vulcan 30 is based upon the electrically powered GAU-8/A 30 mm seven-barrel gun. This weapon, although conventionally fed, operates on the Gatling principle. The warhead kill capability of Goalkeeper against anti-ship missiles is achieved by use of MPDS rounds, produced by NWM de Kruithoorn, Netherlands. This MPDS round has a high-density tungsten alloy penetrator. For 'soft' targets, FMPDS, E1 and TP types of ammunition can be used.

The I-band search radar uses a 2,050 × 280 mm linear array antenna with integrated sidelobe suppression. The antenna rotates at 60 rpm and has a

The Goalkeeper weapon system showing the gun, radar and electro-optic tracker

horizontal beamwidth of 1.7° and a vertical beamwidth of 60°. The radar is powered by a water-cooled, synthesiser-driven, travelling wave tube transmitter with high output power for greater frequency and pulse repetition frequency flexibility in the face of ECM threats.

To ensure detection of small targets in dense clutter, the radar includes digital pulse compression and FFT techniques. Digital moving target detection is used to provide accurate ranging. Dual-receiver channels are used to speed up Doppler processing, plot extraction and track build-up. This two-dimensional track data is used for threat evaluation including target priority, followed immediately by target assignment for the tracking radar and weapon. Continuous search, while engaging a target, allows fast reaction against subsequent targets.

The I/K-band monopulse tracking radar has a 1 m diameter Cassegrain antenna. The pencil-like K-band beam is used to provide accurate and continuous tracking data on sea-skimming targets. By automatic, sustained comparison of the signal-to-noise ratios of the two beams' returns, tracking can be maintained in the most demanding environment.

This system has been tested against missile targets flying as low as 3 m. Digital fire-control processing, including curved path prediction, is used to predict the point of impact and automatic calibration and closed-loop correction of the point of impact are used to compensate for bias errors, including inaccurate ballistic data. There is an automatic kill assessment subsystem for use in the case of multiple attacks to optimise the system's effectiveness.

Upon detecting the target, the threat evaluation and target designation module in the search radar automatically determines the threat priority and begins the engagement sequence. The target is designated to the tracking radar and acquired by both the I- and K-band system. When the target is within range the gun opens fire.

By applying dedicated acquisition, tracking and prediction techniques in combination with destructive FMPDS ammunition supported by radar and E-O sensors. Goalkeeper is believed to provide excellent performance against surface targets.

Since the introduction of the system some modifications have been developed, such as integration with an IRST system, multi-Goalkeeper co-ordination function and automatic ammunition loading device.

Operational status

Thales has sold over 55 Goalkeeper systems to five navies, Netherlands, Qatar, South Korea, UAE and the UK.

The Goalkeeper weapon system in a mast-mounted configuration 0129176

Specifications

Calibre: 30 mm
Muzzle velocity: 1,020 rn/s (TP, HEI) 1,150 rn/s (MPDS)
Traverse: unlimited
Elevation: –25 to +85°
Mounting weight: 6,800 kg
Crew: 0
Range: 200 - 3,000 m (1.61 n miles)
Rate of fire: 4,200 rds/min
Ammunition: 369 g (TP, HEI, MPDS, EMPDS)

Power requirements
Gun mount: 440 V 60 Hz, 3-phase (10 - 90 kVA)
Below decks: 440 V 60 Hz, 3-phase (36 kVA); 115 V 400 Hz, 3-phase (5.5 kVA); 115 V 60 Hz, 3-phase (2.5 kVA); 115 V 60 Hz, single-phase (0.1 kVA)

Contractor

Thales Naval Nederland.

UPDATED

SHIP COUNTERMEASURE SYSTEMS

LASER DAZZLE SYSTEMS

Raytheon Sea Lite beam director

Type

High energy laser system (naval).

Development

Sea Lite, a pointing and tracking device for a high-energy laser system, was originally built by Hughes Aircraft Company (now the Raytheon Company) for the US Navy. The Sea Lite was designed to evaluate the effectiveness of a high-energy laser to defend Navy ships against high-velocity, highly manoeuvrable anti-ship missiles.

The Sea Lite system has been under test at White Sands Missile Range in New Mexico. It is part of the Nautilus programme to evaluate the effectiveness of lasers as tactical air defence against short-range rockets. In February 1996, the US Army successfully engaged and destroyed a short-range rocket in flight using a high-energy laser system. The Nautilus programme is a joint US and Israeli project managed by the US Army Space and Strategic Defense Command and the Israel Ministry of Defense Directorate of Defense Research and Development.

The high-energy laser is the MIRACL, Mid-Infra-Red Advanced Chemical Laser, a megawatt class continuous wave deuterium fluoride laser developed by TRW in the 1970s. Land-based tests in 1989 demonstrated the ability of a laser beam to destroy a supersonic missile, an M2.2 Vandal, in flight. Tests against subsonic targets were completed in 1987. Recent advances in reducing the size of the laser and in the fabrication of a large high-energy window, have made the laser system viable for ship defence. The material for the window is a custom designed fluoride-based glass which can withstand the thermal stresses generated by the intense heat from the transmission of the laser beam.

In 1990 an upgraded version of Sea Lite was constructed as a testbed for the development of ground-to-space capability for anti-satellite applications. This needed Sea Lite to track and engage targets at much greater ranges. Previously, the Sea Lite target tracking had used a small separate infra-red sensor attached to the side of the main telescope. The system was modified to provide full aperture tracking through the large output telescope. This resulted in significantly improved system performance critical for long-range operation.

A planned development programme includes the installation of adaptive optics to compensate for atmospheric distortion of the laser beam. The adaptive optics will include a wavefront sensor and a deformable mirror and control system. These alterations will compensate for distortion of the laser beam as it propagates through the atmosphere. The wavefront sensor will measure the current atmospheric conditions between the system and target. It will then use a deformable mirror intentionally to distort the laser beam. This distortion is to anticipate and compensate for the distortion that will occur as the beam travels to the target. Thus a more tightly focused and precisely pointed beam will be obtained at the target.

In July 1996, Sea Lite successfully illuminated and actively tracked a boosting Black Brant ballistic missile during a USAF Phillips Laboratory experiment at White Sands Missile Range, New Mexico.

The Raytheon Sea Lite beam director

Description

Sea Lite is a pointing and tracking device for a high-energy laser weapon. It includes the MIRACL chemical laser and a beam director. The main components of the director are a system of mirrors, sensors, alignment and stabilisation equipment and a computerised controller for automatic operation. The large-aperture telescope, which is gimbal mounted, has a 1.8 m diameter glass primary mirror and eight smaller water-cooled metal mirrors. The target tracking subsystem uses infra-red and visible sensors. It is designed to accept the MIRACL beam and focus it at a specific point on a moving target. The system can be mounted in place of a shipborne 5 in gun.

For the July 1996 test, Sea Lite incorporated a low-power pulsed laser and was initially directed by tracking radar. Once the missile plume appeared in Sea Lite's infra-red sensor, a passive track loop was closed around the leading edge of the missile plume. The illuminated laser was directed at the missile's nose cone and the Sea Lite track mode was changed from passive plume track to active missile body track. During this phase, Sea Lite demonstrated the capability to direct a high-energy laser beam at a specific spot on the nose cone for the few seconds that would have been necessary to disable the missile.

Operational status

In service as a test bed at the US Army Space and Strategic Defense Command's High Energy Laser Test Facility, White Sands Missile Range, New Mexico.

Contractor

Raytheon Company.

UPDATED

US Navy NRL, Lockheed Martin MATES self-protection system

Type

Laser dazzle system (naval).

Description

MATES is the acronym for the Multiband Anti-ship cruise missile defense Tactical Electronic warfare System. This is an early development programme designed to provide a system to defend the ship against infra-red-guided surface-skimming cruise missile threats. It is an adjunct to the US Navy's AIEWS programme and will run concurrently with that programme.

Requirements for the system would be to handle multiple targets, have optical augmentation, demonstrate effective threat sorting and discriminate between IR and non-IR threats.

Operational status

Lockheed Martin Naval Electronics & Surveillance Systems - Akron was awarded a US$5 million contract from the US Navy to develop an advanced technology demonstrator, designed around a laser-based countermeasure that interrogates and jams the infra-red seeker. The MATES programme has been reallocated to the Office of Naval Research (ONR) SHIELDS programme.

Contractors

US Navy, Office of Naval Research (ONR).
Lockheed Martin Naval Electronics & Surveillance Systems - Akron

UPDATED

SHIP COUNTERMEASURE SYSTEMS

LASER WARNING SYSTEMS

EADS COLDS NG laser ESM system

Type
Laser warner.

Development
The Common Opto-electronic Laser Detection System (COLDS) has been developed since the early 1980s by the now European Aeronautic Defence and Space company's EADS/LFK business unit (then DaimlerChrysler Aerospace) to provide a reliable EMI/EMC resistant multispectral laser warning system for all naval platforms. It has been trialled in Canada, Finland, Germany, the United Kingdom and the United States.

In 1991, a new version of COLDS was evaluated by the United States Army Tank Automotive Command. Tracor Aerospace (now BAE Systems), DaimlerChrysler Aerospace AG's (now EADS Deutschland's) United States licensee, has been awarded a contract by LTV now Lockheed Martin Vought Systems to supply an upgraded version of COLDS, which is to be tested by Tank Automotive Command as part of the Vehicle Integrated Defense System.

The COLDS systems have been delivered since 1993 to Canada, Finland and Germany. It is installed on different ship types for these customers and the system is integrated with the ship combat management systems for releasing automatic countermeasures to improve the ship self-defence capability against laser threats.

The COLDS NG (New Generation) is designed for application on board ships. The system detects laser light pulses from laser range finders and target designators with a high efficiency in all bearings across Band 1 and Band 2. For the laser detection, COLDS NG does not have to be directly hit by the laser beam.

COLDS NG can provide an entire ship self-defence capability, including detection of laser emissions from coastal firing stations which cannot be detected by radar; providing timely alerting to ship systems from seaborne or airborne laser sources; providing laser threat information; identification of laser threats via a user-definable threat library; and 'just-in-time' initiation of countermeasures after threat alert.

Two versions of the system are available:- COLDS NG 70, for ships of up to 70 m in length; and COLDS NG 140 for ships of up to 140 m in length.

Description
The chief features of COLDS are: fibre-optic coupled heads; multiple wavelength coverage; pulsewidth discrimination; PRF measurement and pulse code interval; optical BITE.

The improved version of COLDS will have an expanded frequency coverage to allow it to detect carbon dioxide laser rangefinders (operating at 10.6 μm) such as those fitted to the latest MBTs like the Challenger 2 and M1A1. Earlier versions had a lower limit of 1.9 μm which is suitable for ruby and Md:YAG laser range-finders and designations.

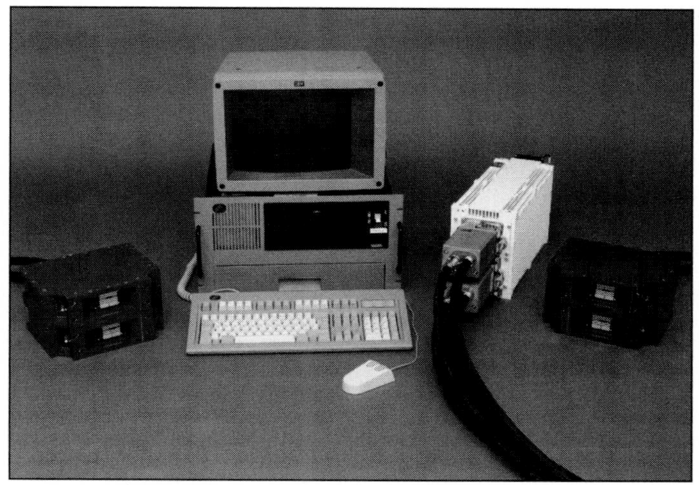

COLDS system Band I and Band II

Operational status
Serial production types already delivered and in service with the navies of Canada, Finland and Germany.

Specifications
Waveband: 0.4-1.7 μm; 2.0-6.0 μm optional; 5.0-12.0 μm optional
Angular cover: 360° in azimuth; ±45° in elevation
Angular resolution: 3 or 1.6° in azimuth; elevation optional
Dynamics range: 77 dB (signal voltage)
Sensorhead dimensions: 270 × 200 × 220 mm
Sensor unit weight: <10 kg (each)
Electronic unit dimensions: 400 × 160 × 220 mm
System weights:
COLDS NG 70: 45 kg
COLDS NG 140: 75 kg

Contractor
EADS Systems and Defence Electronics.

UPDATED

SHIP WEAPON CONTROL SYSTEMS

FIRE CONTROL

AMS Medusa Mk 3 optronic fire-control system

Type
Naval fire-control system.

Description
Medusa Mk 3 is a simple and easy to use optronic system. It can be used as a FCS for remote control of medium/small calibre guns, as an optical aid to more complex fire-control systems, or as a surveillance system for maritime patrol and in general for sea control operations. The system consists of an unmanned, self-stabilised director which mounts a TV camera. An optional IR camera can be mounted as well.

The director is controlled from a control desk provided with BIW monitor, keyboard and joystick. The monitor displays the TV image and alpha-numeric symbology relevant to the aimed target.

Medusa Mk 3 performs the following tasks:
- Optronic surveillance and search
- Target detection, recognition and identification
- Target tracking by means of the automatic video tracker
- Control and firing of one gun
- Video recording/play back.

Target range (for firing) can be obtained either by an optional eye-safe LRF, or connecting the system to an active sensor suite (for example, search radar, ARPA display and Directive Target Indication).

Operational status
The Medusa Mk 3 is a fully developed item, currently in service onboard several patrol boats of the Italian Customs Guard.

The Medusa director

Medusa Mk 3
0101467

Specifications
Director
Self-stabilised
Training limit: unlimited (slip ring)
Training max speed: 2 rad/sec
Elevation limits: –40 to +85°
Elevation max speed: 2 rad/sec

TV camera
Target: CCD 1/2 in
Sensitivity: 10 to 200,000 lux
Lenses: zoom, automatic iris

3-5 μm IR camera
Detector: InSb FPA
Two optical FoVs
Closed-cycle cooling system

8-12 μm IR camera
Detector: IR CCD 288 × 4
Two optical FoVs
Closed-cycle cooling system

Laser range-finder
Type: eye-safe 1.54 μm
(RAMAN shift-cell)
Range: 300 to 20,000 m
Range accuracy: better than 5 m
Repetition rate: Type 1 single shot
1 Hz continuous
Type 2 1 Hz continuous
3 Hz 1 minute
8 Hz burst

Installation data
Director (TV camera included)
Dimension (h): 740 mm (above deck) × 153 mm (below deck)
Diameter: 1,200 mm
Weight: 165 kg

Computer unit
Dimension (h w d): 245 × 400 × 400 mm
Weight: 25 kg

Contractor
AMS (formerly Alenia Marconi Systems), Naval Systems Division.

UPDATED

AMS NA 10

Type
Naval fire-control system.

Development
The NA 10 is a development of the NA 9 with improved electronics while retaining the original fire-control radar. In Italian Navy service the new system was designated Argo 10 whereas NA 10 was used for the commercial version. The Italian Navy received its first systems in 1971 and the Argo 10 entered service when the 'Audace' class destroyer *Ardito* was commissioned in December 1972 (it was later replaced by an NA 30 system). The system was subsequently installed or retrofitted into the cruiser *Vittorio Veneto* (replacing NA 9), 'Lupo' class frigates, 'Sparviero' class FACs, 'Stromboli' class replenishment tankers and, the latest, 'Cassiopeia' class OPVs from 1988. The system was exported to Libya, Peru, Taiwan and Venezuela.

The first version of the NA 10, the Mod 0, was adapted for the Seasparrow surface-to-air missile system and became the Albatros Mk 1 Mod 1. However, this did not prove satisfactory and a digital version of the family, still retaining the RTN 10, was developed in the early 1970s. It was part of the Albatros Mk 2 fire-control system for the Aspide air defence missile and entered service when the Peruvian frigate *Meliton Caravajal* was commissioned in October 1978.

Description
The NA 10 system is designed for surface engagement, air engagements or shore bombardments and can control three guns of two calibres simultaneously. It uses an RTN 10X fire-control radar which was supplemented in earlier versions by two optical periscope sights. Electro-optical equipment which could be mounted on

The Orion 10 radar antenna and TV camera used in NA 10

the director included IR camera, TV camera and laser range-finder while a CW illuminator could also be installed. The system is operated through the Main Control Unit (MCU). A TV display may be added when electro-optical units are used. The MCU is used for target acquisition, tracking and engagement, either manually or automatically. Two periscope sights may also be used to provide supplementary target data input to the radar and permit a second target to be engaged.

There are four versions of the NA 10:

Mod 0 – For larger warships with periscope sights.

Mod 1 – For FACs with lighter director (no laser range-finder, CW illuminator or IR sensor) and without periscopic sights. The electronics are all collocated with the MCU.

Mod 2 – An improved version of the Mod 0.

Mod 3 – An improved Mod 1 designed for FACs with anti-ship missiles and one or two guns. The electronics are separated from the MCU.

Operational status

Some 60 NA 10 systems have been produced and production is now complete. NA 10 is in service with the navies of Italy ('Cassiopeia' class offshore patrol vessels, 'Sparviero' class hydrofoil-missile, 'San Georgio' class LPD, 'Stromboli' class replenishment tankers), Taiwan ('Lung Chang' class fast attack craft-missile) and Venezuela (Modified 'Lupo' class frigates, 'Constitucion' class fast attack craft and 'Almirante' class coastguard patrol craft).

Contractor

AMS (formerly Alenia Marconi Systems), Naval Systems Division.

UPDATED

AMS NA-18L

Type

Naval fire-control system.

Description

The NA-18L is an optronic director specifically suitable as main gun fire-control system onboard medium and small size vessels, as well as secondary fire-control system for larger vessels. Up to three sensors can be fitted on the NA-18L Director: TV, laser and IR. The sensors carry out the acquisition and tracking functions under the control of the FCS processor and of the operator at the console. In addition, the NA-18L may be connected to the Ship Search or Navigation Radar for simultaneous presentation of the radar video, for surveillance purposes with resultant self-designation capability.

The system is capable of controlling up to three guns of two different calibres against the target under tracking. It can be provided with a dedicated MAGICS 2-MFC console or can be controlled by any console of the Combat Management System.

NA-18L main features are:

- An unmanned self-stabilised pedestal mounting an optronic sensor suite (TV, IR and laser sensors) for target automatic tracking
- A state-of-the-art computer complex
- A mono-monitor or bi-monitor multifunction display console, including raster scan, high resolution, colour monitor(s).

NA-18L main operational functions are:

- Electro-optic autonomous search, manually by joystick or by preprogrammed search patterns, with self-designation capability
- Radar surveillance presentation (search/navigation radar)
- Target recognition and identification
- Designation management
- Automatic tracking (and prediction models) of air/surface/shore targets
- Stabilised gun orders computation for up to three guns of two different calibres simultaneously
- Video recording and play back.

Operational status

The NA-18L is a fully developed system, currently in service.

Specifications

Azimuth training limit: unlimited (slip ring)
Azimuth max slewing speed: 2 rad/sec
Elevation limits: –40 to +85°
Elevation max slewing speed: 2 rad/sec

TV camera
Target: 1/2 in CCD
Sensitivity: 10 to 200,000 lux
Lenses: fixed focus or zoom, automatic iris

IR camera
Type 1
Spectral band: 8 to 12 μm
Detector: IR CCD 288 × 4
FOV: 3.3 × 2.4° (narrow)
8 × 6° (wide)

Type 2
Spectral band: 3 to 5 μm
Detector: InSb FPA 384 × 256
FOV: Continuous zoom from 1.3 to 2.7°

Laser range-finder
Type: eye-safe 1.54 μm (RAMAN shift cell)
Range: 300 to 20,000 m
Range accuracy: better than 5 m
Instrumental range: 300 to 20,000 m
Repetition rate: Type 1, 18 Hz
Type 2, 20 Hz

Installation data
Director (TV, IR, laser included)
Dimension (h): 865 mm (above deck)
Diameter: 406 mm (below deck)
Weight: 200 kg

Contractor

AMS (formerly Alenia Marconi Systems), Naval Systems Division.

UPDATED

AMS NA-25

Type

Naval fire-control system.

Description

The NA-25 is a radar and optronic fire-control system capable of controlling medium calibre guns in the anti-aircraft and anti-surface roles as well as small calibre guns in the CIWS role. Up to three guns of different calibres can be controlled at the same time. The fire-control radar associated to the system is the ORION RTN-25X, a J-band fully coherent equipment which is characterised by anti-nodding, ECCM and anti-clutter features together with high-tracking accuracy. A set of two optronic sensors (TV camera/IR camera) can be mounted on the radar director to enable firing assessment and to provide alternative lines of sight on the same target.

In its standard configuration, the NA-25 is provided with its own operator console, the MAGICS 2-MFC dual-monitor display system. Alternatively, the NA-25 can be provided without its standard console in order to enable the integration with one of command and control system consoles; this is possible, provided that the console to be used is equipped with typical FCS devices (particularly range hand-wheel and joystick) and that the TV/IR presentation is supported as well.

The NA-25 features are:

- A versatile display system composed of one or more interchangeable consoles, according to the NA-25 configuration, which incorporates two high resolution colour monitors to present the raw radar (search or tracking radar) and TV/IR videos, the tactical situation and supplementary information
- A powerful state-of-the-art computer complex
- The ORION RTN-25X tracking radar, is a monopulse, fully coherent tracker radiating a coded waveform. Matched compression filters and anti-clutter filters, together with additional ECCM features allow high-tracking accuracy in a dense ECM environment
- TV, IR/laser (option) sensors: to provide an alternative line of sight on the same target and scene monitoring and kill assessment.

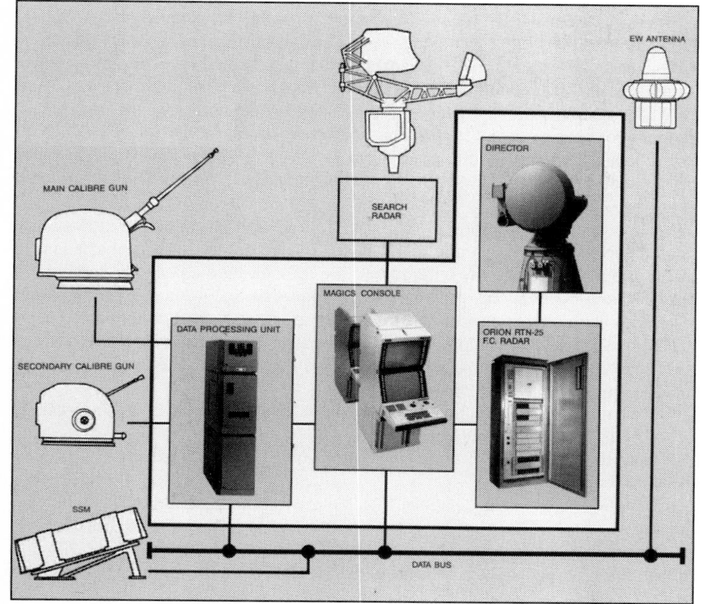

A diagram of a typical NA-25 system

Configurations
- FCS, provided with one MAGICS 2-MFC console and implementing weapon control functions only
- FCS (integrated with IPN-S C2 system), controlled by one IPN-S console and implementing weapon control functions only
- C2IFCS, provided with one MAGICS 2-MFC console and implementing either weapon control functions and command and control functions.

Operational features
FCS functions:
- Autonomous radar and TV/IR search
- Search Radar presentation
- Automatic air/missile/shore and surface targets tracking
- Control of three guns with different calibres
- Line of sight, line of fire stabilisation
- Simultaneous surface firing on TVVS data.

C2 functions:
- System tracks handling
- Target identification
- Track history
- Weapon control
- EW management
- Data link management
- Navigation
- Maps/Patterns management
- Data recording/playback
- Training.

Operational status
Several NA-25 fire-control systems and RTN-25X tracking radars are in production for the Italian and other Navies.

Specifications
Radar
Frequency: J-band (formerly X-band)
Antenna type: Cassegrain
Transmitter: fully coherent with TWT, coded waveform, frequency agility
Receiver: superheterodyne, double conversion, MTI

Tracking method: monopulse
IR camera
Type 1: 8 to 12 μm band, IR CC 288 × 4
Type 2: 3 to 5 μm band, InSb FPA

Laser range-finder
Type: eye-safe 1.54μm (RAMAN shift cell)
Accuracy: better than 5 m

TV camera
Type: CCD
Lens: fixed focus or zoom
Sensitivity: 10 to 200,000 lux

Served pedestal
Training limits
Azimuth: unlimited (slip ring)
Elevation: –22 to +84°

Training
Azimuth max speed: 2.8 rad/sec
Elevation max speed: 2 rad/sec

Contractor
AMS (formerly Alenia Marconi Systems), Naval Systems Division.

UPDATED

AMS NA-30

Type
Naval fire-control system.

Description
The NA-30 is an advanced modular weapon control system designed to control surface-to-air missile system and guns in a sophisticated threat environment. Automatic in its operation, the NA-30 system is able to control, in co-ordinated tire reaction mode, a surface-to-air missile system and medium/short range guns (up to three gun outputs).
Main features
- A versatile display system (not provided when NA-30 is integrated with the IPN/S C2 system), which incorporates two high-resolution colour monitors to display raw radar (its own tracking radar or the search radar), TV/IR videos and supplementary information
- A powerful state-of-the-art computer suite
- The ORION RTN-30X tracking radar, a monopulse, fully coherent tracker radiating a coded waveform. The extensive range search and tracking accuracy in a dense ECM environment is ensured by its high average power, and the receiver applied technologies, such as: Surface Acoustic Wave (SAW), Charge Coupled Device (CCD) and microprocessor techniques which provide range tracking, angular errors extraction together with multipath filtering and clutter cancellation
- An optronic sensor suite, TV, IR and laser (option), can be mounted to provide alternative lines of sight on the same target and scene monitoring and kill assessment.

Operational features
- Radar and optronic autonomous search with automatic/manual self-designation
- Surveillance on ship's search radar video, with self-designation
- Automatic engagement of evaluated priority target up to firing action
- Automatic co-ordination of weapons (SAM and guns) for a combined reaction
- Simultaneous control of up to three guns with different calibres and a Surface-to-Air Missile (SAM) system
- Automatic air/missile/surface/shore targets tracking with specialised target matched models.

Operational status
Several NA-30 family of fire-control systems and RTN-30X tracking radars are in service or in production for the Italian and other foreign Navies.

Specifications
Radar
Frequency: I-band (X-band)
Antenna type: cassegrain
Transmitter: fully coherent with TWT coded waveform, frequency agility
Receiver: superheterodyne, double conversion, MTI
Tracking method: monopulse

IR camera
Type 1: 8 to 12 μm band, IR CCD 288 × 4
Type 2: 3 to 5 μm band, InSb FPA

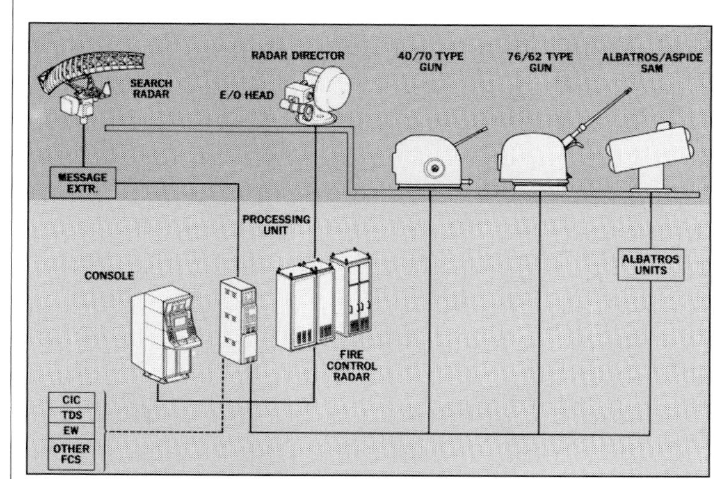

A diagram of a typical NA-30 fire-control system

Laser range-finder
Type: eye-safe 1.54 μm (RAMAN shift cell)
Accuracy: better than 5 m

TV camera
Type: CCD
Lens: fixed focus or zoom
Sensitivity: 10 to 200,000 lux

Servoed pedestal
Training limits:
Azimuth: unlimited (slip ring)
Elevation limits: –22 to +84°
Azimuth max slewing speed: 2.8 rad/sec
Elevation max slewing speed: 2 rad/sec

Contractor
AMS (formerly Alenia Marconi Systems), Naval Systems Division.

UPDATED

BAE Systems Sea Archer 1 (GSA 7)

Type
Naval fire-control system.

Development
The Sea Archer 1 system was developed as a private venture by Sperry Gyroscope (now BAE Systems) from the Sapphire system in the mid-1970s. It was designed to provide fast attack craft and larger warships with an electro-optical fire-control system. The first customer was Oman and the system entered service in March 1977 with the commissioning of the Brooke Marine Type Large Patrol craft. In the UK Royal Navy it was designated GSA7 when it was fitted in the 'Peacock' Class Hong Kong patrol craft.

Description
Sea Archer 1 was designed as the primary gun control for fast attack craft/fast patrol boats but can be used as a supplementary weapon control on larger vessels.

The Sea Archer 1A is an enhanced version which provides complete integration of a TV subsystem with below-decks control by joystick giving an improved surveillance capability and the option of manned or unmanned director on deck. It also has provision for controlling two different types of gun-mounting. Infra-red sensors may also be integrated in the Sea Archer 1A system.

Sea Archer 1A gunfire control system

The system consists of two basic elements: the MSI-Defence Systems Optical Fire Director (OFD) with a combination of electro-optic tracking sensors and a laser range-finder; and a gun control console containing a predictor and operator control panel. In the simplest configuration, an aimer uses binoculars fitted to the stabilised OFD to acquire and track the target. To assist in target acquisition the system can accept information derived from the vessel's surveillance radar to slew the OFD on to the target bearing. Electro-optic sensors, IR or television, are fitted in other configurations to give below-decks operation and an auto-tracking unit is available if required.

Information regarding the target range, training and elevation is automatically transmitted to the weapon control console. The computer, utilising ballistic and meteorological data already entered into the system, then calculates gun aiming order. To reduce reaction time to a minimum, gun slew occurs as soon as the target is acquired and offset adjustments are then made when a valid prediction solution is computed. Typically this process takes 2 seconds. The gun is fired by means of the console foot push when a valid prediction is achieved and the display indicates that the target is within effective gun range.

The computation techniques and electronics interfaces have been successfully proved with 30 mm, 40 mm and 76 guns.

The system is capable of three modes of gunfire control as follows:
(a) Air defence
In this mode the system is designed to achieve minimum reaction time against fast moving air targets. Firing can be achieved from the director or the control console.
(b) Surface
Splash spotting corrections, observed by the director aimer, are applied via the control console.
(c) Naval gunfire support
In this mode the gun is used to bombard shore targets. The target co-ordinates are inserted manually at the control console and gun orders are calculated and corrected for ship movement by dead reckoning. Spotting corrections, observed by a shore- or ship-based observer, are applied at the control console.

Three versions of the Sea Archer 1A system now exist: Mod 0, which offers dual-mode alternate ballistics; Mod 1, which includes a thermal imager and auto-track capabilities Mod 2, which controls different calibre guns simultaneously and has Mod 1 features.

Operational status
Some 30 systems have been delivered for use by the navies of Brunei, Oman, Thailand and the UK. UK ships have now been sold to the navies of Ireland and the Philippines. The system is no longer in production.

Specifications
Electro-optical fire director
Arc: unlimited
Elevation: –20 to +70°
Laser: Nd:YAG
Beamwidth: 2 mrad
Repetition rate: 20 Hz
Power requirements: 440 V, 50/60 Hz, 3 phase 3-5 kVA; 24 V DC

Contractor
BAE Systems.

UPDATED

BAE Systems Sea Archer 30 (GSA 8)

Type
Naval gun fire-control and surveillance system.

Development
The Sea Archer 30 system was developed in response to a UK MOD requirement of the early 1980s for an electro-optical fire-control system to control the 4.5 in gun in 'Type 22' and 'Type 23' frigates. Contracts were placed in 1985 by the MOD for use by the Royal Navy, which has designated the system GSA8. The first system was installed in HMS *Cornwall* in 1988 and installation has continued through the 'Type 23' class.

Description
The Sea Archer 30 system is designed to detect and track air and surface targets and engage them with a medium-calibre gun. It is designed to operate in four modes: surveillance, air defence, surface engagement and shore bombardment. The system consists of an electro-optic director and a director power unit and director electronic unit, a gun control unit and a console.

The sensor package comprises a thermal imager, a television camera and a laser range-finder. The thermal imager uses the SPRITE detector, TICM II (Thermal Imaging Common Modules) and a closed-cycle cooling engine. It operates in the 8 to 12 μm band and has a 3 and 17° field of view. The television camera is a daylight monochrome camera with fields of view of 3 to 18°. It is a CCD camera with a 450,000 element detector and Peltier cooler. The laser range-finder operates at 1.54 μm wavelength and gives range to 20 km.

Video tracking uses one of three techniques: centroid tracking for well-defined bounded targets, correlation for unbounded targets in difficult clutter conditions and the use of an edge tracker where there is a strong tracking edge. Auto-track processing is enhanced by enclosing the target in a tracking window which bounds

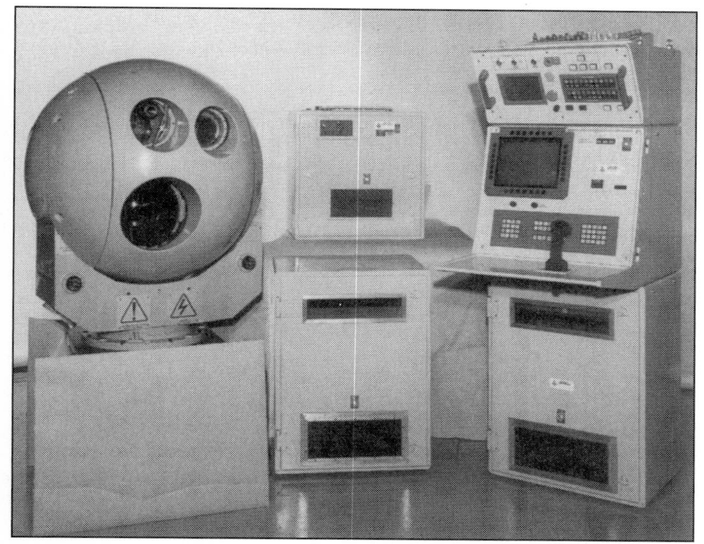

The Sea Archer 30 system components (l to r); the director, director power unit (top), predictor, gun control desk unit (top) and director electronics unit

the target area. This may be adapted in width and height and automatically adjusts to encompass the target details.

Operational status

Sea Archer 30 is fitted in four UK Royal Navy Batch 3 'Type 22' frigates, 16 'Type 23' frigates and three Shore Systems. The UK Royal Navy received 23 GSA 8 systems, the last of which was delivered in 1997.

Specifications

Director
Height: 1,100 mm
Swept radius: 465 mm
Weight: 270 kg (incl sensors)
Pointing accuracy: better than 3 m at 10 km in sea state 5
Elevation motion: 110° (–35 to +75°)
Training motion: 610° (±305°)
Elevation velocity: 90°/s
Training velocity: 90°/s
Elevation acceleration: 150°/s^2 sustained until max velocity is reached
Training acceleration: 150°/s^2 sustained until max velocity is reached

Thermal imager
Wavelength: 8-12 μm
Output: CCIR 625 line 50 Hz
Field of view: 17 to 3° (diagonal)
Magnification: ×3.6 to ×20
Detector: SPRITE
Cooling: integral Stirling engine

TV camera
Type: Charge Coupled Device (CCD)
Output: CCIR 625 Line 50 Hz
Field of view: 17 to 3° (diagonal)
Magnification: ×3.6 to ×20
Lens max aperture: f1.4

Laser range-finder
Wavelength: 1.54 μm
Beam divergence: 1.50 mrad max
Pulse repetition frequency: 12.50 Hz
Peak power: 10 MW
Range resolution: 5 m

Contractor

BAE Systems.

UPDATED

Ball Aerospace Mk 16 LLLTV

Type

Low-Light Level TV (naval).

Development

During the early 1970s, Ball Aerospace designed and developed the original Isocon tube-based Mk 16 Mod 0 Low-Light Level TV (LLLTV) camera for the Raytheon/RIM-7 Seasparrow surface-to-air missile system (qv). The camera augments the tracking radar and provides day/night search, surveillance, target identification and acquisition, fire control and kill assessment. Deliveries began in 1976.

In 1992 Ball was awarded a development contract for a new CCD camera to replace the initial design. The first production model of this new Mk 16 Mod 2 camera was delivered in February 1996.

Description

The Mk 16 Mod 2 contains two sensors, a CCD (Charge-Couple Device), day sensor and a night sensor which consists of a 25 mm third-generation image intensifier coupled to a 16 mm CCD. The camera uses a lens iris for light control and automatic gain control circuitry to maintain a constant video output signal. It is coupled to a dual field of view large aperture lens.

The Mk 16 is mounted above decks and is designed to withstand severe electromagnetic and marine conditions. Removal of the rear cover permits access to 13 replaceable subassemblies for ease of repair.

To date over 120 Mk 16 Mod 2 cameras have been produced, to replace the Mk 16 Mod 0, for Denmark, Italy, Norway and the USA.

Operational status

In production and in service. Over 160 Mk 16 LLLTVs have been produced for the navies of Denmark, Italy, Norway and the USA as part of the Seasparrow missile system. A total of 15 systems supports law enforcement activities in the US Coast Guard's 'Bear' class (WMEC-901) medium-endurance cutters.

To date over 120 Mk 16 Mod 2 cameras have been produced, to replace the Mk 16 Mod 0, for Denmark, Italy, Norway and the USA.

Specifications

Dimensions (envelope): 15 (diameter) × 44 in
Weight: 150 lb (incl 67 lb Seasparrow lens)

Optics
Diagonal field of view: 11° wide, 2.6° narrow
Seasparrow lens: Image format 40 mm, WFOV 11°, NFOV 2.6°
ICCD (night) sensor: 25 mm GEN III coupled to 16 mm CCD
CCD (day) sensor: 16 mm CCD

Resolution/Sensitivity (TVL/RH faceplate illumination)
CCD (day) sensor
at ≥0.1 ft/c: 700 TVL
at ≥0.01 ft/c: 600 TVL
at ≥0.001 ft/c: 200 TVL
ICCD (night) sensor
at >10^{-5} ft/c: 600 TVL
at >10^{-6} ft/c: 400 TVL
at >10^{-7} ft/c: 200 TVL

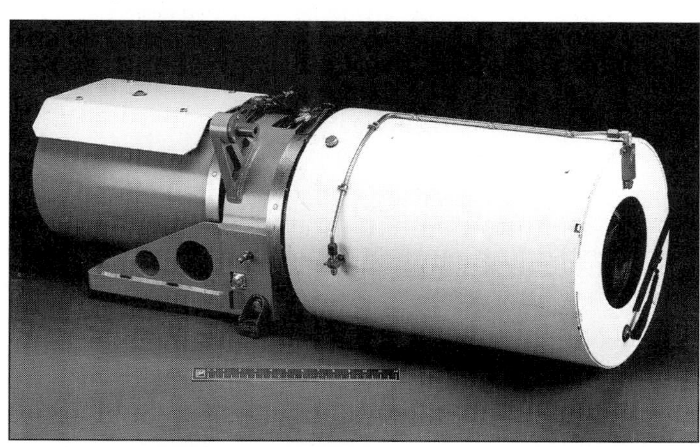

The Mk 16 Seasparrow Low-Light Level TV system 0002018

Spectral response
CCD sensor: 550-1,050 nm with peak approx 750 nm
ICCD sensor: 600-900 nm with peak approx 850 nm

Video
Line rate: 875 per EIA RS 343
Bandwidth: ≥15 MHz
Aspect ratio: 4:3
Output: composite, differential, 95 Ω, drives up to 750 ft of RG 22B/U cable
Geometric distortion: <3%

Contractor
Ball Aerospace & Technologies Corporation.

VERIFIED

Brashear LSEOS Mk IV On-Gun electro-optic director

Type
Naval fire-control system.

Development
Brashear LP was awarded an US$8.1 million contract in July 2002 to build an additional 35 Lightweight Shipboard Electro Optics Systems (LSEOS) Mk IV On-Gun Electro-Optics directors for the US Navy Phalanx Block IB Close-In-Weapon-System (CIWS). The LSEOS Mk IV manufactured by Brashear LP is the tracking pedestal for the Thermal Imager on the Phalanx Block IB Upgrade. The Block IB Weapon System provides the capability for CIWS engagement of surface or low, slow airborne targets; and enhances anti-ship warfare fighting capability for high-speed manoeuvrable targets.

The Phalanx Weapon System is on every US Navy combatant and used by over 20 allied nations. The Raytheon Company of Tucson, Arizona made the award. The Phalanx CIWS is a rapid-fire, computer-controlled radar and 20 mm gun system that automatically acquires, tracks and destroys enemy threats that have penetrated all other ship defense systems.

This is the 4th award Brashear LP has received for the Phalanx Block IB Upgrade program. There is a high probability that over the next five years Brashear LP will build as many as 300 systems for the upgrade.

Description
The LSEOS Mk IV combines rugged, high-precision electro-optical instrumentation with the flexibility to serve on a variety of vessels and weapon platforms. It offers good close-in fire-control performance in a small, reliable, lightweight, easy to maintain system. It features high-quality gyrostabilised precision and the system uses rugged, field proven designs. The open architecture permits mounting of nearly any sensor type. The director was designed and proven to withstand continuous gun shock in a full naval environment. It's high-performance isolation and decoupling system allows tracking with heavy gun shock. The lightweight pedestal minimises impact on gun servo performance.

The LSEOS Mk IV is available as part of a fully integrated fire-control system or as a component for OEM applications. The upper and lower mounting points accommodate a variety of electro-optic sensors: FLIRs, TVs and lasers.

Operational status
Available.
Part of the Block IB Surface Mode Upgrade for the Raytheon Mk 15 Phalanx CIWS
- Batch 1:10 units August 1995
- Batch 2: 10 units September 1999
- Batch 3: 22 units January 2002
- Batch 4: 35 units July 2002
- Designed into Oerlikon Contraves' 35-1,000 Millenium gun
- Rolling Airframe Missile (RAM) compatible

Specifications
Mount type: side-mounted Stewart platform
Elevation: 15° (expandable)
Azimuth: 15° (expandable)
EOD LOS slew rate (gyro stabilised)
- **Elevation:** >1.5 rad/sec
- **Azimuth:** >2.0 rad/sec

EOD LOS slew acceleration (gyro stabilised)
- **Elevation:** >6.0 rad/sec^2
- **Azimuth:** >8.0 rad/sec^2

Servo bandwidth: >30 Hz
Accuracy: ± 0.2 mrad
Orthogonality: <± 1.0 mrad
Load weight at performance: 44 kg/97 lb
G-Shock design point: 30 G, half sine, 11 ms
MIL qualifications:
- MIL-STD-46 1 (EMI/EMC)
- MIL-STD-8 I OE (environmental)
- MIL-STD-2 I OC (climatic extremes)

Electrical interface
Power requirements: ±5 V, +28 V, ±29.5 V. ±15 V, +55 V DC
Position feedback: multispeed resolver
Communications: RS-422
Weight (mount, no sensors): 64 kg (143 lb)

Contractor
Brashear LP.

UPDATED

Brashear systems LSEOS Mk IIA and Mk IIB

Type
Naval fire-control system.

Development
Brashear LP first began developing stabilised naval pedestals for NSWC Dahlgran's Laser Fire Director Program with a system onboard the USS *Briscoe* DD-977. This was followed by 32 AN/SDV-I Optical Surveillance System (OSS) pedestals for the United States Coast Guard (USCG). The systems consist of a stabilised pedestal for Low Light Level Television cameras and Infra-Red (IR)/ White Light searchlights. The USCG Program lead to the development of the Lightweight Shipboard Electro Optical System (LSEOS).

Over 50 ships and 4 navies are benefiting from the technology and adaptability of the Brashear LSEOS Mk IIA and Mk IIB. Brashear is the preferred shipboard electro-optics provider and Brashear electro-optics are also used in conjunction with radar/sonar systems. The example of the latter configuration is the US Navy Phalanx Block IB CIWS Upgrade where over 40 ships to date have been upgraded with Brashear electro-optics.

Description
LSEOS is used for fire control as well as surveillance, covert tracking, and damage assessment. Two platforms of the LSEOS product line can be used on vessels as small as patrol boats to the larger frigates.
- The LSEOS MK-JIA is designed to provide a stand-alone electro-optical fire control system for smaller warships
- The LSEOS MK-IIB can integrated into a combat suite for larger warships

The basic system consists of
- Electro-Optical Director (EOD)
- Operator Control console (OCC)
- Weapons Control Module (WCM)
- Systems Control Module (SCM)
- Bridge Surveillance Display Unit (BDSU)

The SCM provides the interface between the operator positions (0CC and BDSU) and the EOD. The SCM also provides the ship's system data (Speed Log and Anemometer) to the Fire Control Computer for use in the calculation of ballistic solutions to be fed to the weapon via the WCM. Ballistic solutions are available for weapons systems ranging from 20 mm to 76 mm.

The EOD is designed to operate in conditions up to Sea State *5* and can support 4 sensors. These sensors include an Eyesafe Laser Rangefinder (ESLRF), a TV camera with zoom and an infrared or thermal imager sensor. ESLRF are typically 10 Hz systems and FUR/Thermal imager are in either the *3-5* or 81-2 um range. TV's are typically 10:1 zoom either color or black/white. The LSEOS can receive target information from 2-D or 3-D radar or can provide Azimuth and Elevation data for integrated combat systems. Targets can be manually tracked or auto-tracked.

An LSEOS/Sea Hawk family director 0052059

Operational status
Available. Existing customers as follows:

Country	Type of system	Weapon
Saudi Arabia	MK-IJA	FCS for 30 mm
Republic of China	MK-JIA	FCS for 40 mm
Malaysia	MK-IIB	76 & 33 mm
Egypt	MK-JIB	76 mm

Specifications
Angular coverage
Azimuth: Continuous (slip rings)
Elevation (relative to ship's deck): –30 to +85°
EOD LOS slew rate (gyro stabilised) AZ and EL: 75°
EOD LOS slew acceleration (gyro stabilised) AZ and EL: 150°/s^2
Operation average: 1.7 kVA
Peak operation: 3.4 kVA

Contractor
Brashear LP.

UPDATED

DCN CTA/CTD/CTM

Type
Naval fire-control system.

Development
From the early 1960s DCN began to develop electronic computer-based fire-control systems using Thomson-CSF (now Thales) hardware and CPM software. These systems appear to have been designed initially to complement the SENIT tactical data handling systems.

The first of the new systems to enter service was the analogue CTA in 1967, followed in 1973 by the digital CTD. CTA and CTD usually complement tactical data handling systems which are essentially French copies of the US NTDS system, for example SENIT. As development of the new generation of SENIT systems continued, DCN (Direction des Constructions Navales) conducted a series of development programmes into infra-red trackers (*Pointeur Infra-Rouge pour l'Artillerie Navale* or PIRANA) to augment radar fire-control systems for guns. Three versions were developed and the last, PIRANA III, was selected for service in the new CTM (*Conduit de Tir Modulaire*).

The system was installed in the last three ships of the 'Georges Leygues' class destroyers with the *Primauguet* commissioning in November 1986. It was also installed in the 'Cassard' class destroyers, while between September 1989 and March 1991 it was retrofitted to the 'Suffren' class destroyers.

Description
CTA is designed to control guns of various calibres between 40 and 127 mm in operations against air, surface and shore targets. The targets are usually designated through the ship's SENIT system, although in frigates they are designated by the ship's search radar through the combat information centre.

It uses an analogue computer with CS-1 software language and four-axis optical directors. The optical trackers are supplemented by a TV camera and these are increasingly being augmented by IR trackers. The CTA is supported by DRBC 32A or DRBC 32C I-band monopulse radars and DRBC 31 I/J-band radar. The DRBC 32 family has a peak power of 80 kW and pulse-width of 4 µs. They can track a 0.10 m^2 target at 15 km (8 n miles). All are on stabilised mounts but both Cassegrain and offset antennas are used.

The CTD makes use of one of the two Univac 1230 computers which support the SENIT 3 system to conduct ballistic calculations. Its sensors are a single DRBC 32B or DRBC 32D radars and TTAb electro-optic director in the Type F 65 *'Aconit'* class while in other ships it uses the lightweight two-axis TTAc director.

The CTMS is supported by a variety of sensors including the lightweight TTAd director, the DRBC 33A (the French Navy version of the Thomson-CSF (Thales) Castor IIC) I-band pulse-Doppler radar with a peak power of 30 kW and a TV camera. The DRBC 33A is a fully coherent monopulse radar with a peak power of more than 30 kW. The PIRANA system (DIBC-1A in the French Navy) consists of an IR camera attached to the DRBC 33 radar director, which gives simultaneous tracking in dual wavebands (3 to 5 and 8 to 12 µm). The CTM systems in the 'La Fayette' class frigates have a laser range-finder.

In many of the systems using the TTAd and TRA remotely operated directors, the Murène infra-red camera and laser range-finder may be used.

Operational status
Around 25 systems have been produced and they continue to be manufactured.

Contractor
DCN International.

UPDATED

DCN OP3A/SARA

Type
Naval fire-control system.

Development
In the late 1980s the French Navy began work on a self-defence system for its major warships under a programme designated *Amélioration de l'Autodéfense Anti-missile* (AAA). The system was to be based upon optical and electro-optical sensors and short-range weapons and the requirement was defined during the early 1990s.

The system was redesignated *Opération de l'Amélioration de l'Autodéfense Anti-missile* (OP3A) and hardware orders were placed from November 1993 beginning with the SAGEM VIGY 105. The system has been selected for 'Georges Leygues' class destroyers, for the 'La Fayette' class frigates, Landing Ships Dock, and underway replenishment tankers. The first system was installed on the destroyer *Jean de Viennes*

The OP3A will be marketed by DCN International as the *Système d'Autodéfense Rapprochée Anti-ùerien* (SARA).

Description
OP3A consists of a fire-control station, electro-optical sensors and 'hard' and 'soft' kill elements. Some ships will receive only some elements of the system.

The fire-control station will be on the bridge where there will be a skylight to provide the commanding officer with a panoramic view from a raised seat. The seat will have controls which will give threat priorities to the self-defence officer who will sit behind him and be in charge of the *Système de Traitement de l'Information de la Defénse Vue* (STIDAV). This is based upon two Calisto consoles and a real-time processor with software based upon that developed for SENIT 8.

Some ships will have the SAGEM VAMPIR MB infra-red surveillance system but for target designation and fire-control purposes most will have two VIGY 105 electro-optic systems. In some vessels and classes these will be replaced by the SAGEM TDS 90. This is the British MSI-Defence TDS with a SAGEM SAS 90 sight.

The training limits are a 340° arc and the elevation limit is –20 to +65°. Target designation accuracy is 3 mrad. The sensors are a reflex sight or a SAGEM ALIS 8 to 12 µm thermal imager.

The 'hard kill' element consists of the MBDA Mistral point defence air defence missile and a 30 mm gun. The Mistral is in the SADRAL remotely operated launcher version and ships will carry two, although some vessels will retain their SIMBAD launchers. The gun will be produced by ECN Ruelle and is a licence-built version of the Otobreda Mk 30 Compact. The 'soft kill' element will consist of a new Dassault Electronique (now Thales) ECM system and an improved CS Défense Dagaie decoy system in the AMBL-1C version.

Operational status
In production with 19 French Navy ships scheduled to receive the OP3A system to a greater or lesser degree. In May 1996 it entered service on the destroyer *Jean de Viennes*.

Contractor
DCN International.

UPDATED

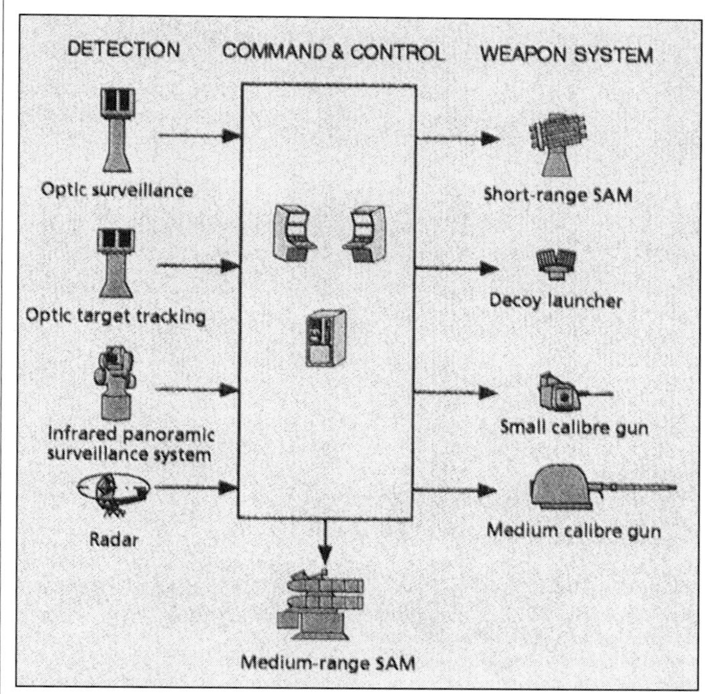

A diagram of the SARA system

EADS Lynx-IR

Type
Fire-control (naval).

Development
The original Lynx system is an optical sighting system which entered service in 1982. Lynx-IR is an electro-optic director intended as an upgrade to Matra's Panda and Lynx systems.

Description
The Lynx-IR is a stand-alone system which has a sighting system with height-adjustable electro-optic unit which includes dual field of view infra-red camera, TV camera and a monitor. The infra-red imager is made by SAGEM and operates in the 8 to 12 μm waveband with four element CMT detector, serial parallel scanning and split Stirling cooling. A 3 to 5 μm thermal imager is optional. Ranges claimed for this system are 14 km (detection) and 2.1 km (recognition). The sighting system is mounted on a rotating assembly and fixed pedestal.

Lynx-IR is capable of day/night air and surface surveillance, target designation to a weapon system and direct control of small- or medium-calibre weapons, with computation of parallax and gravity drop corrections. It is operated by one standing operator who carries out surveillance by means of a monitor within the electro-optic unit.

Operational status
The 8-12 μm thermal imager version is in production with first delivery in 1996.

Specifications
Traverse: 360°
Elevation: –15 to +40°
Elevation laying speed: 1 rad/s
Thermal imager: 8-12 μm (3-5 μm optional)
Fields of view
CCD camera: 4-16°
Infra-red imager: 4°
Weight: 205 kg

Contractor
EADS Systems and Defense Electronics.

UPDATED

Lynx-IR from Matra

EADS Najir

Type
Naval fire-control system.

Development
During the 1980s Matra developed Najir, a more sophisticated version of the Panda/Naja family with both day and night capability, which could also be operated remotely. The French Navy designated Najir DMAc and it entered service with the commissioning of the destroyer *Cassard* in July 1988.

In the early 1990s Matra began to develop a remotely operated version of Najir as Najir Mk 2, the manned version becoming Najir Mk 1. In October 1994 it was announced that Pakistan would be the first customer to upgrade the 'Amazon' (Type 21) class frigates. A new electro-optical version, the Najir 2000 is being developed.

Matra Najir 2000 0002017

Description
Najir Mk 1 has similar functions to its predecessors, Panda/Naja, but the manned element supplements the laser range-finder (originally the CILAS TMY 113 and later an Ericsson Nd:YAG unit) with a thermal imager (originally a CT 15 but later a SAGEM Murène operating in the 8 to 11 μm waveband). The latest production Najir features Radamec TV cameras. The manned element has, in addition to three sensors, a servo cabinet, a ballistic unit based upon two Motorola 68000 microprocessors and a power supply box. Optional extras include a remote-control console and 625-line television/infra-red monitors.

Najir Mk 2 is designed to control fire from two guns of two calibres and consists of a line of sight stabilised mounting with electro-optic sensors, a processing unit, a power supply unit and a control console. The two-axis stabilised pedestal is identical to that used in the Sadral decoy launching system and has a swept radius of 80 cm. The sensors are the SAGEM Murène thermal imager, a Radamec HK 202 TV camera and an eye-safe CILAS THS 304 laser range-finder.

Najir 2000 has a new lighter weight sensor head which can accommodate up to four electro-optic sensors and with an elevation angle pointing capability of +85°. Sensors can include two infra-red cameras, TV camera and eye-safe laser range-finder. The processing unit includes servo and system management processors and dual-channel videotracker.

Najir 2000 is designed to be fully integrated in a combat system and can be operated by a single operator seated at a multifunction console. The system can carry out the following missions: sectorial search (manual or automatic); autotrack; aid to identification; 'three-dimensional' TD delivery; simultaneous control of two guns. It can also engage a surface target from a bearing/range T D information delivered by an external system (search radar).

Operational status
Some 21 Najir Mk 1 units have been produced. Production of Najir Mk 1 and Mk 2 continues with 11 ship fits of the latter ordered. Najir 2000 has entered production.

Specifications
Najir Mk 1
Height: 1.35 m
Width: 1.50 m
Weight: 560 kg
Elevation: –20 to +70°
Pointing velocities
Azimuth: 90°/s
Elevation: 60°/s
Acceleration: 2 rad/s^2
Najir 2000
Total elevation travel: –35 to +85°
Max velocity in bearing: 120°/s
Max acceleration in bearing: 200°/s
Sweeping radius: 0.53 m
Weight
Sensor head: 200 kg
Processing unit: 295 kg
Power supply unit: 80 kg

Contractor
EADS Systems and Defense Electronics.

UPDATED

Electro-Optic Systems Electro-Optic Fire-Control System (EFCS)

Type
Naval fire-control system.

Description
The Electro-Optic Fire Control System (EFCS) is a full-solution fire-control system suitable for a wide variety of weapon stations. The EFCS is a fully integrated day/night system which incorporates an eye-safe laser rangefinder, a ballistic computer, CCD cameras, an intensified CCD camera and a flat panel display. The products provide an almost instantaneous new aiming point that takes into account range, target movement and environmental conditions. It is claimed that the effective range of weapons such as 20-35 mm cannons, automatic grenade launchers and 0.50 calibre machine guns can be increased, with a significant improvement in accuracy.

The EFCS is adaptable to most direct fire weapon stations and a single unit can be programmed to handle multiple weapons and a wide variety of munitions. The common module architecture meets modern logistics support requirements and allows a low-cost upgrade path through module replacement.

Operational status
EFCS now has thermal imaging and has been type-classified by the US Navy and is being used for the US Army Common Remotely Operated Weapon Station (CROWS).

Specifications
Dimensions: Sensor/Processor Unit: 316 × 235 × 167 mm
Weight: 12 kg
Laser: eye-safe Class 1 (AS 2211:1991)
Range: 5,000 m wavelength: 1.54 µm
Pulse repetition rate: 12 ppm
Accuracy: ±5 m
Magnification: ×27
Night sight: GEN II+ or GEN III ICCD to customer specification
Power supply: 24-28 V DC

Contractor
Electro Optic Systems Pty Ltd, Australia.

UPDATED

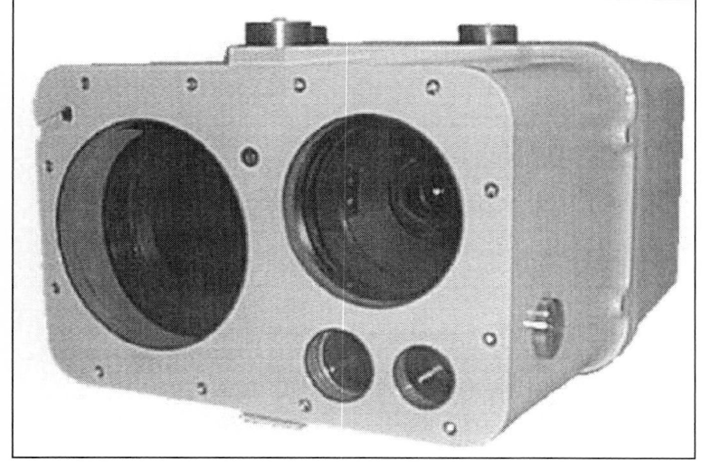

Electro-Optic Fire-Control System (EFCS) 0019392

Elop Multisensor Stabilised Integrated System (MSIS)

Type
Naval fire-control system.

Description
The MSIS is designed for installation aboard naval vessels to provide target detection, recognition, acquisition identification and tracking. It will also provide fire control of guns. It consists of a FLIR, TV camera and laser range-finder/designator and provides multisensor capabilities for day/night operations under all weather conditions. The sensor will detect, recognise and track the full range of surface targets from large tankers down to rubber boats, as well as all types of helicopters, fixed-wing aircraft and sea-skimming missiles.

The MSIS features a four-axis gimballed platform and offers wide-angle pointing coverage, combined with a high level of stabilisation. The equipment possesses high flexibility for interfacing with shipboard acquisition and fire-control systems, being of low weight and small size. A typical patrol boat configuration weighs only 87 kg and consists of:

(a) a stabilised ball shape turret weighing 60 kg and incorporating a thermal imager, TV camera and laser range-finder

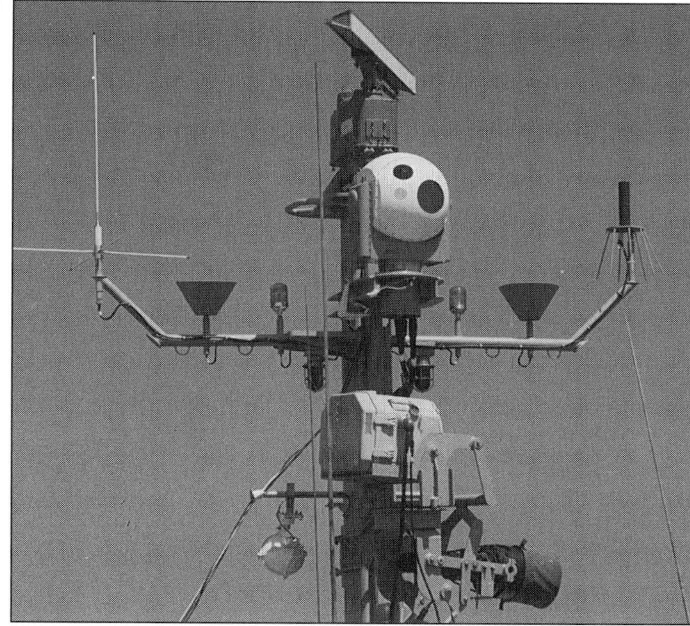

Multisensor Stabilised Integrated System (MSIS) at the masthead of a Super Dvora fast attack craft

The MSIS turret mounted on the mast of a Super Dvora fast attack craft
NEW/0547540

(b) electronic control and power supply units
(c) two operators' control stations
(d) VCR for automatic recording
(e) power converter.

Operational status
In service on the Israel Navy 'Hetz' (Saar 4.5) missile fast attack craft, 'Eilat' (Saar 5) corvettes and 'Super Dvora' fast attack craft (gun). The system is also operational on Israeli Navy Eurocopter SA 366G Panther shipborne helicopters. Some 180 MSIS systems have been ordered or delivered both for the Israel Navy and for unidentified foreign customers.

Specifications
Weight:
turret 60 kg
total system 87 kg
Turret dimensions: 780 mm (h) × 600 mm (d)
Line of sight stabilisation: <20 µrad

Angular coverage:
Elevation -35 to +85°
Azimuth: n × 360°
Slewing rate: 60°/s
Reporting accuracy: 0.8 mrad in Sea State 3
Tracking accuracy: better than 150 mrad

Thermal imager
Detector: second-generation 8-12 µm MCT, 480 × 4 elements or third-generation 3-5 µm FPA
Spectral band: 8-12 µm or 3-5 µm
Cooling: integral Stirling engine

Laser range-finder
Wavelength: 1.064 μm or 1.54 μm eye-safe

TV camera
Type: B/W CCD

Contractor
Elop Electro-Optics Industries Ltd.

UPDATED

FABA DORNA

Type
Naval fire-control system.

Development
In 1991 the Spanish Defence Ministry, through the *Subdirección General de Tecnologia e Investigacion de Defensa,* signed a research and development contract with FABA (IZAR Systems Division) of Pta2.49 billion (US$21.9 million) for the design, development and manufacture of an operational prototype whose requirements had been defined by the *Estado Mayor de la Armada.* During this phase INISEL was a subcontractor of FABA. In 1996 shipborne trials of DORNA aboard a Spanish Navy 'Santa Maria' class frigate were successfully passed.

In 1998 FABA carried out the process of adapting the prototype to the state-of-the-art and to the industrial manufacturing procedures. Again new trials were passed, this time on a 'Baleares' class frigate. After those trials the Defence Ministry signed a contract with FABA for the manufacturing of DORNA systems for the F100 frigate programme.

Description
DORNA is a modular fire-control system designed to control 40 mm, 76 mm and 5 in calibre guns, among others, in warships larger than 300 tonnes in the air defence, anti-surface vessel and gunfire support roles. It may also be used as part of a command and weapon control system in larger warships. It is intended to be a flexible system that may be used with a wide variety of sensors to acquire, track and provide ballistic calculations to intercept targets in all weather and electronic warfare conditions.

It is a modular, distributed architecture design based upon VME bus, Motorola CPUs and 802.3 Ethernet LAN and uses Ada and C software languages.

The system consists of four major subsystems, connected by means of a redundant LAN. The tracking subsystem consists of a sensor group on a stabilised platform; sensors are an X-band (Ka-band in the prototype) tracking radar, a high definition television camera, an infra-red camera and an Nd:YAG laser range-finder. The system is capable of automatic tracking in both radar and electro-optic modes. The weapon control subsystem performs the ballistic calculations and interfaces with the gun. The ship's data acquisition subsystem provides the system with environment information from platform sensors as well as performs the interface with the vessel's combat system (on ships with a combat system). The tactical management subsystem is a multifunction console with two high-resolutions displays.

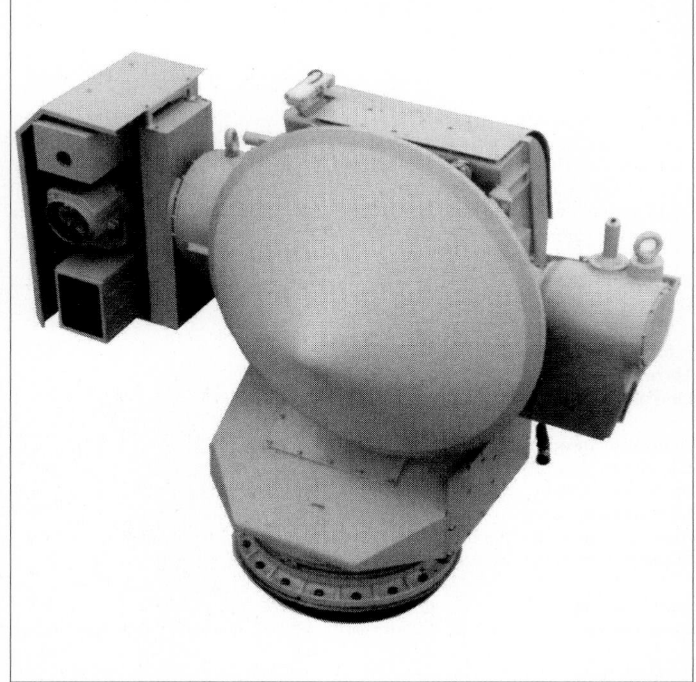

The DORNA fire-control director 0041547

Operational status
In addition to the prototype, four units are in production for the F100 frigate programme. The first DORNA system is already installed on board the frigate *F101 Don Alvaro de Bazan* of the Spanish Navy.

Contractor
IZAR Systems Division, FABA.

VERIFIED

Kollmorgen Mk 46 Mod 1

Type
Naval fire-control system.

Description
The Mk 46 Mod 1 is an electro-optical surveillance and gun control system which has four modes: search/surveillance, navigation, weapon slaved and director slaved. It feeds azimuth, elevation and range data into the Mk 34 gun fire-control system for the engagement of surface and air targets. The director is 71 cm high and 76 cm wide with a nominal azimuth coverage of 360° and an elevation coverage of –25 to +80°. The sensor package consists of a CCD television camera, a two field of view 3 to 5 μm focal plane array thermal imager and an eye-safe laser range-finder.

The director is controlled from a standard Q70 console, a handgrip controller and trackball line of sight controller.

Operational status
Kollmorgen has supplied systems to date for the US Navy's 'Arleigh Burke' class guided-missile destroyers, the USS *Yorktown* and other platforms.

Contractor
Kollmorgen Electro-Optical.

UPDATED

Ship fire-control system 0041399

Kongsberg MSI-80S

Type
Naval fire-control system.

Development
Development of the MSI-80S (Multi-Sensor Integration) began in the early 1970s to meet a Royal Norwegian Navy requirement for a low-cost fire-control system for the 'Hauk' class fast attack craft. The system was developed by Kongsberg Våpenfabrikk. The company changed its name to Norsk Forsvarsteknologi (NFT) but has now changed again to Kongsberg Gruppen. The system was unusual in relying largely upon passive target detection. This proved feasible because the associated Penguin ASV missiles use an IR seeker.

A production contract was awarded in May 1975 and the system entered service in the KNM *Hauk* in August 1977. Some of the technology has been incorporated in the NAVKIS tactical data handling system.

Description
MSI-80S is a mainframe system used for detecting and engaging multiple targets by means of low-cost sensors. The sensors used are the Decca (now Thales) TM 1226 navigation radar and an electro-optic director. The system can control six ASV missiles, four torpedoes and two light guns (20 mm/40 mm) for air defence.

The electro-optic sensor associated with MSI-80S in the 'Hauk' class FAC

The MSI-80S is based upon the KS500 fire-control computer which is housed in the display system which is similar to that used in NAVKIS. The director consists of a stabilised platform manufactured by NFT, a low-light television camera, a laser range-finder, an infra-red scanner, optical sights and electronic warfare sensors. All sensors are collimated to the TV camera which is a V0084 with 2.66° field of view and an Ebiscon image intensifier. The camera, produced by Marconi-Elliott Avionics Systems Ltd (now BAE Systems), automatically adjusts to all light levels from bright sunlight to starlight. The range-finder is an Saab UAL 10102 Nd:YAG laser with a power output of 5 MW.

The MSI-80S system is now being upgraded to accommodate the controls of an advanced, wire-guided torpedo.

Operational status

The system is no longer in production but some 15 were manufactured and used only by the Royal Norwegian Navy.

Contractor

Kongsberg Gruppen AS.

UPDATED

MSI-Defence Systems Director Aiming Sight (DAS)

Type

Naval fire-control system.

Description

The Director Aiming Sight (DAS) is a power-driven, line of sight stabilised director. It comprises a deck-mounted pedestal with rotating head and a platform which carries a seated operator. A binocular is mounted at eye level on an elevating arm. Electro-optical sensors may be added (thermal imager, laser range-finder, TV) together with auto-track and on-mount prediction facilities up to a balanced load of 150 kg.

Two direct-drive servo systems are employed. One trains the rotating head on to the target bearing, while the other positions the elevation arm and sensors to acquire the target in elevation. The director is 1.82 m high, 0.75 m across and 1.02 m long.

The sight is capable of being driven in the following modes:

(a) Local. The director is controlled manually by a seated operator who tracks the target through the binocular. Director movement is controlled by a thumb operated joystick. Outputs from the system can be fed to the ship's fire-control system, or to gun or missile mounts.

(b) Remote. With the addition of a suitable on-mount sensor package, control can be exercised from a remote console weighing 182 kg.

(c) Remote Follow. The director can be slaved to external fire-control radar. The operator can act as an observer, verbally recording fall of shot and other information.

(d) Remote Target Indication. The director may be automatically rotated onto a bearing by a surveillance radar, while the target is acquired by operator joystick movement.

Operational status

In production, in service.

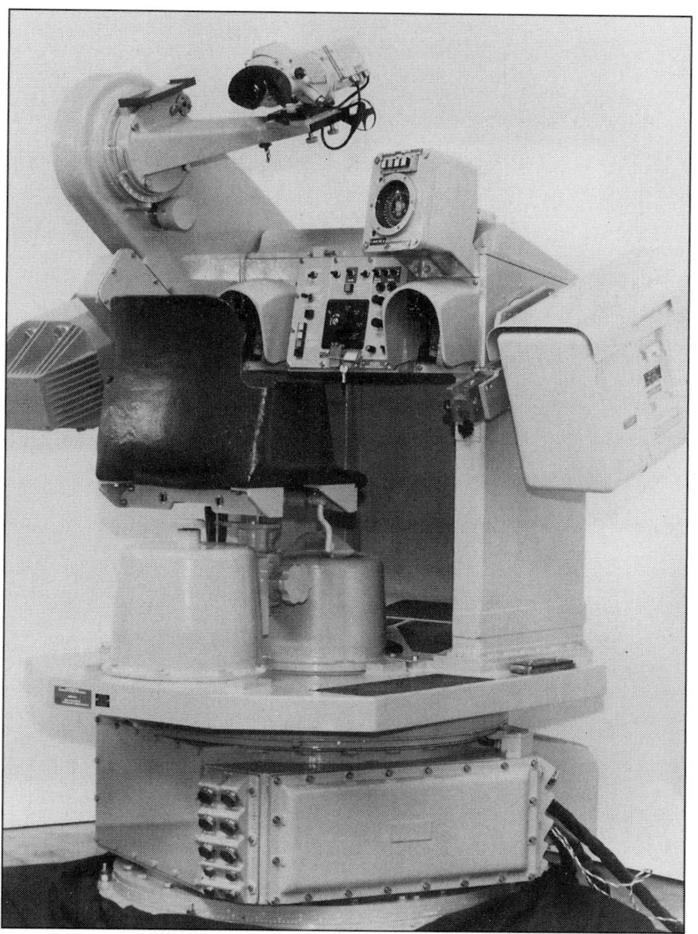
The DAS – Director Aiming Sight from MSI-Defence Systems

Specifications

Overall height: 1.82 m
Binocular access height: 1.73 m
Weight
- **Director:** 650 kg (without sensors)
- **Sensors:** 150 kg (max)
- **Below deck:** 182 kg

Azimuth: 360°
Elevation: –20 to +70°
Acceleration
- **Azimuth:** 100°/s^2
- **Elevation:** 100°/s^2

Speed
- **Azimuth:** 70°/s
- **Elevation:** 100°/s

Power requirements: 440 V, 3 phase 60 Hz, 750 VA average 2 kVA; 115 V, 1 phase 400 Hz 250 VA; 115 V, 1 phase 60 Hz; maintained supply 90 W; 24 V, DC at 10 A

Contractor

MSI-Defence Systems Ltd.

VERIFIED

MSI-Defence Systems Target Designation Sight (TDS)

Type

Naval fire-control system.

Development

The Target Designation Sight (TDS) is manufactured and marketed exclusively by MSI-Defence Systems Ltd. Currently in service with nine navies, including the Royal Navy, the TDS is used for the visual acquisition and tracking of both surface and air targets, often as a Quick Pointing Device (QPD).

Description

The unit consists of a sight head, to which most service types of binoculars or direct view sensors can be fitted, mounted on a pedestal. The head is telescopically supported and counterbalanced to allow easy tracking between low and high angles of sight and easy height adjustment to suit individual operators. An illuminated bearing scale at the top of the pedestal, enables the operator to align the sight with a given target bearing. No deck penetration is required for fitment.

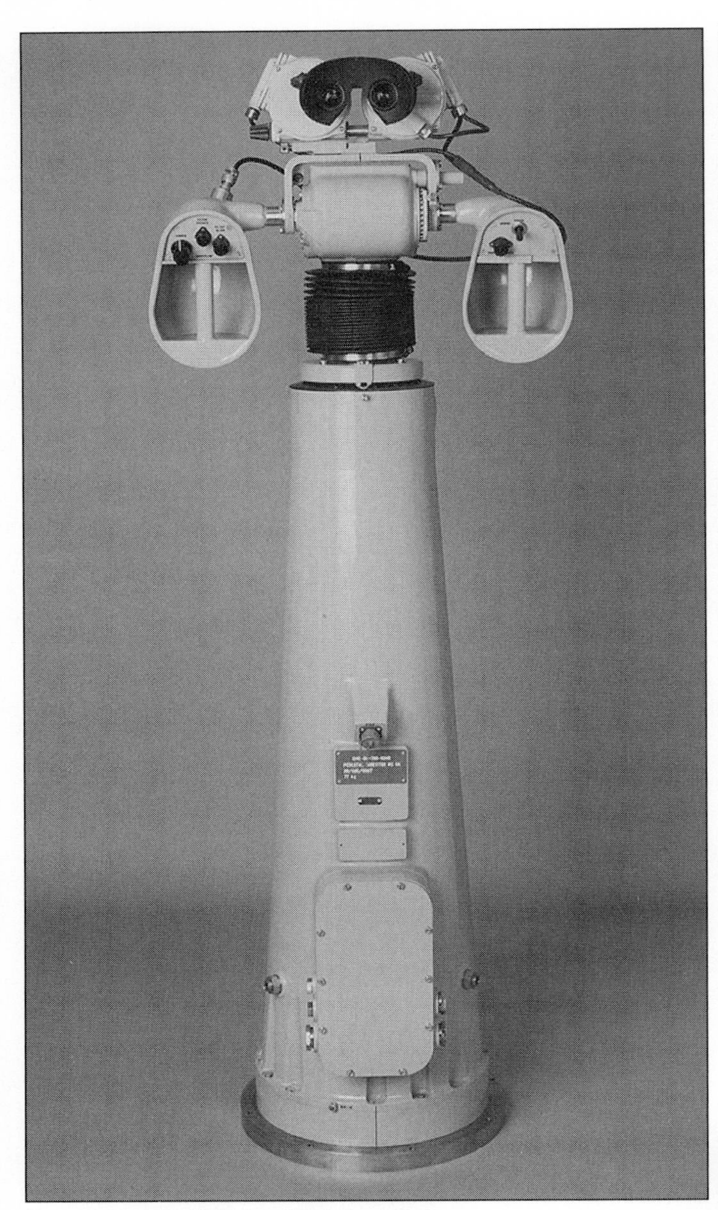

Target Designation Sight from MSI-Defence Systems

Using the protected handgrips, the operator rotates and elevates the sight head, bringing the sight to bear on the target. On acquiring the target, he depresses the designated alarm switch on the right handgrip. The target training and elevation information is then transmitted automatically to the weapon control centre by synchro elements. The TDS can be used to provide target acquisition data for either radar or electro-optical trackers. It can be used as a coarse tracking device in reversionary mode during periods of radar silence.

The TDS is manufactured to individual customer requirements. Any variation necessary to integrate the sight with existing equipment can be incorporated. If required, intercommunication and heated suit facilities can be provided for the TDS operator. Additional control switches may also be fitted.

Operational status

In service and in production. It is a standard fit on the Royal Navy's 'Type 23' frigates and is in service with 12 navies.

Specifications

Height
Extended: 2,030 mm
Closed: 1,500 mm
Swept radius: 300 mm
Weight: 92 kg
Training limits: ±170°
Elevation limits: –15 to +75°
Sector Value
Training: 360°
Elevation: 180 or 360°
Transmission accuracy: ±30 min

Contractor

MSI-Defence Systems Ltd.

UPDATED

M-Tek Triton electro-optic day/night tracker

Type

Fire-control (naval).

Development

The Triton electro-optical tracking and fire-control system is being developed for the South African Navy as an option for its 'Minister' class strike craft upgrade programme. It is also being considered for army anti-aircraft systems.

Description

Triton's sensors comprise an 8 to 12 μm thermal imager (the Eloptro TS20), a dual field of view (2 and 8°) TV camera (developed by M-Tek), an eye-safe laser range-finder (Eloptro LR40), a TV auto-tracker and a fibre optic sight line gyro unit. This configuration is stated to give an acquisition range for a fighter aircraft of better than 15 km, range-finding to 10 km and 5 m accuracy. Triton can also accommodate a range of other sensors. The most basic version has only a TV camera and a single shot laser range-finder with monochrome console display.

The direct-coupled, brushless, rare-earth, permanent magnet motors allow angular accelerations up to 4 rad/s^2 with rates up to 2 rad/s. Commercial-Off-The-Shelf (COTS) items used include VME-based industrial cards for digital control and video imaging.

Below deck, the operator console has a colour monitor, three raw video monitors and a video recorder. The colour monitor adds a full range of overlay information to the image, and cross-hairs for target selection. Two multifunction hand controllers allow manual tracking control and are used for all inputs. The system can provide target position and velocity information for most gun or missile fire-control systems.

Operational status

One system has been produced for trials. The system will equip the MEKO A 200 frigates for the South African Navy.

Specifications

Elevation: –20 to +85°
Azimuth: 360°
Accuracy (sightline positioner): 0.3 mrad
Angular acceleration: up to 4 rad/s^2
Fields of view: 2° and 8° (TV camera); 2.7° and 9.6° (thermal)
Laser range-finder: 10 Hz in 1.54 μm band
Seeker head
weight: 225 kg
height: 90 cm

Contractor

M-Tek.

UPDATED

The Triton electro-optic tracker from M-Tek

Norinco Type 88C

Type
Fire-control system (naval).

Development
The Type 88C (ZKJ-3) was revealed at the Paris Air Show in 1991, although the designation might suggest that it had been in development for three years. It appears to be an export-only system whose development is complete but no more data is available. The first ship known to be fitted with the system was a Chinese 'Houjian/Huang' class (Type 037/2) fast attack craft laid down in 1989 and commissioned in January 1991.

Description
The Type 88C includes a digital computer which interfaces with a variety of electro-optic, gun, missile and EW systems through a local area network of undefined standard. The system can accept data from shore-based radars, although it is unclear whether or not this is by datalink or voicelink. The system can control up to three weapons.

The prime electronic sensor is the Type 88C search radar which features an elliptical antenna with front horn feed on a stabilised mounting. The manufacturer states that the Type 88C radar has a track-while-scan capability and features frequency agility and MTI modes with two rotation rate options. The radar acquires the target and designates it for the remotely operated electro-optical sensor. This is on a two-axis stabilised platform with separate search and tracking TV cameras, infra-red sensor and a laser range-finder. These sensors are in two packs, one on each side of the mounting. The below-deck mounting features a console with two screens and joystick/button MMI. As a back-up there is the manned JM833 optical director which can control two or three guns.

The Type 88C can control up to three weapons from the following:

(a) PL-9N surface-to-air missile
(b) Type 76A twin 37 mm gun
(c) Type 69 twin 30 mm gun
(d) a new combined gun-missile system
(e) a 122 mm multiple rocket launcher system
(f) chaff dispensers.

Operational status
Development of the Type 88C is believed to be complete and the system is being marketed. The ZKJ-3 has been exported to Thailand.

Contractor
China North Industries Corporation (Norinco).

VERIFIED

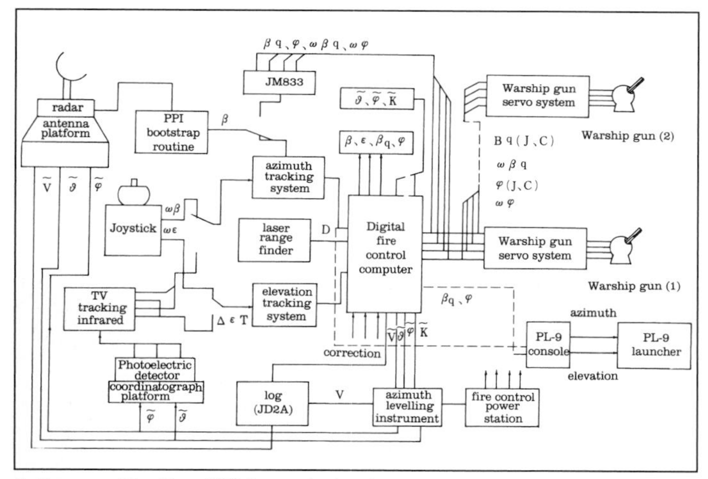

A diagram of the Type 88C fire-control system

Oerlikon Contraves Seaguard/TMK-EO

Type
Fire-control system (naval).

Development
Oerlikon Contraves developed the modular Seaguard CIWS and FCS product range in the early 1980s. Within this product range is a three-axis Tracker Module (TM) able to be fitted with a range of radars (Ku-, X- and Ka-band), together with electro-optical sensors (TV, FLIR and laser range-finder), in a combination appropriate for customer requirement and usage.

Description
A particular example of a TM is that of the TMK-EO where a Ku-band radar is combined with a BAE Systems V3901 FLIR operating in the 8 to 13μm waveband and a SAAB Nd:YAG laser range-finder operating in the 1.06 μm wavelength. A TV

TMK-EO MultiSensor (Radar/EO) fire-control system

alignment/observation camera and a video tracking unit complete the electro-optical element of TMK-EO.

The normal operating firing mode for CIWS engagements with the TMK-EO utilises advanced adaptive multisensor target data processing techniques, whereby the target data from radar and EO sensors are fused to provide the optimum target data. In addition, the EO sensors can be used as a fully three-dimensional target data source for reasons of ECM resistance or redundancy.

The COSYS range of Combat Systems from STN ATLAS-Elektronik also includes TMs from Oerlikon Contraves.

Operational status
In service. Variants of Seaguard tracker modules with EO sensors are fitted in the 'Yavuz' and 'Barbaros' classes of MEKO 200 frigates of the Turkish Navy and are entering service in the Indian Navy with the BEL Shikari Air Defence System.

Contractor
Oerlikon Contraves AG, Zurich.

VERIFIED

Radamec 460 Series naval Target Designation Sight (TDS)

Type
Fire-control system (naval).

Description
The RADAMEC 460 Series Target Designation Sight (TDS) is designed for visual identification and acquisition of airborne and surface targets from naval vessels. Variants exist for single mountings on port, starboard, fore or aft configurations, or as dedicated pairs or sets. The TDS can be supplied and fitted with a number of different sighting devices according to operational requirements.

The design of the sight head features a yoke configuration incorporating a standard platform with a quick release dovetail slide clamp. This enables a selection of sighting devices to be used, such as binoculars, intensified monocular night sight or optical ring sight. A direct view thermal imager can also be fitted, the cabling being accommodated within the pedestal. The advantage of the dovetail clamp for sight head mounting is that it allows for rapid change from one type of sight to another and storage of sensitive or valuable equipment when not in use.

The TDS can also be supplied with a small environmentally sealed video monitor. This additional feature enables the operator to slave control a remote electro-optical sight while directly viewing its image. This monitor is mounted separately and in addition to the TDS's standard sighting device supplied.

The TDS is also fitted with a coarse rifle sight for fast target acquisition.

The TDS is available in fixed or adjustable height variants. The height adjustable variant can be set over a range of 1,500 to 1,850 mm by operation of a clamp release on the side of the pedestal.

Control of the TDS is by handgrips which allow smooth operation in both training and elevation. The yoke has a counterbalance spring in the elevation axis to balance the sight head payload. A friction device is incorporated to provide positive resistance for easier operation against ship motion. A 'designator manned' toggle switch and a 'target acquired' push button are incorporated in the left hand grip.

Additional controls can be located in the right handgrip to provide emergency control of a gun.

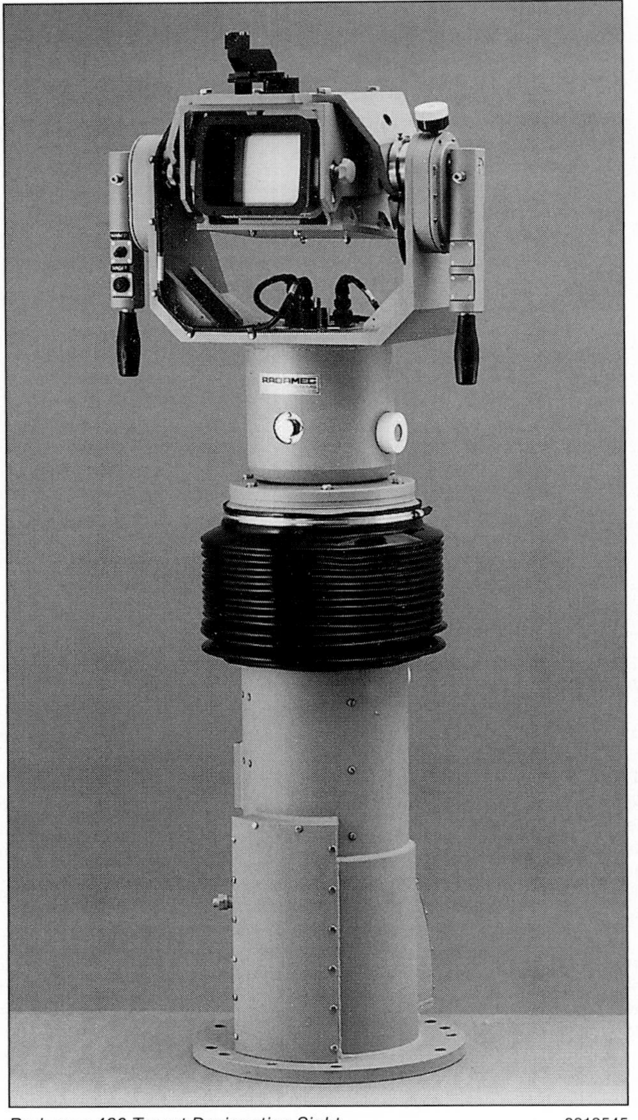

Radamec 460 Target Designation Sight 0019545

Coarse synchros are fitted to both axes for position reporting to a command or weapon control system. The standard output interface is synchro although a serial interface can be provided. Both axes also have external calibrated angular scales and are fitted with push-pull locks to locate them in their zero positions when not in use.

The TDS can be supplied with an electrical connector interface for head-set communication.

Operational status
In service with the Finnish and Turkish navies. Selected by the Royal Malaysian Navy.

Specifications
Target designator
Type: HK999-642
Axis configuration: Elevation over azimuth
Azimuth travel: ±170°
Elevation travel: +85 to –30°
Position tellback: Synchro format
Coarse ratio: 1:1
Angle reporting errors
Training and elevation: 15 arc min
Height adjustment: 10 arc min
Signal/reference voltage: 90/115 V
Reference frequency: 400 Hz ± 10%
Synchro type: 11M1M3
Sight manned: Isolated switch closure
Target acquired: Isolated switch closure
End stops: Mechanical (rubber bumpers)
Transit locking pin: Nominal mid-travel position in azimuth 0° elevation
Height adjustment range: 1,500 mm minimum
To sight head: 1,850 mm maximum
Weight: 145 kg typical

Contractor
Radamec Defence Systems Ltd.

VERIFIED

Radamec Series 2400

Type
Fire-control system (naval).

Development
System 2400 is an electro-optical fire-control system evolved from Radamec's Series 2000 surveillance and weapon control system developed during the mid-1980s. This series now also includes System 2500 and 2600 (qv).

Description
System 2400 is an electro-optical fire-control system designed for precision target tracking, ballistic prediction and weapon control. For small- and medium-calibre naval guns in anti-air, surface- and shore-bombardment engagements, the system can also be used as a remote observation and tracking sensor for navigation and surveillance.

The system offers the following features - use as a primary sensor and control system for control of any quantity of gun mountings of either common or different calibres and shell ballistics; passive detection, positive identification, tracking and engagement of surface and air targets at ranges out to 15 km and beyond; digital prediction of gun aim-off data; fully computer-controlled engagement of targets; on-screen splash marker for line and range spotting; fully stabilised line of sight coupled with digital video auto-tracking; fully automatic search and acquisition; electronic alignment of sensors and weapons.

The system comprises three major subsystems:

Electro-Optical Tracking System - this consists of an electro-optical director located above decks and a below-decks cubicle with the electronics, processors and power supplies for the tracking system. The director comprises a dual-axis gyrostabilised director head carrying TV, thermal imager and high-power eye-safe laser rangefinder.

Weapon Interface Cubicle - this contains the predictor computers, weapon interface electronics and system interface processors.

Operator's Control Console - this includes two TV monitors displaying video images from the sensors and a joystick for manual control of the director and tracking gate.

System 2400 is configured for all system functions to be controlled by one person. It includes the following sensors:

Daylight TV - solid-state camera with a ×1 to ×10 continuous zoom lens.

Thermal imager - dual switched field of view lens operating in the 8 to 12 μm spectral band. A 3 to 5 μm imager is available for high-temperature/humidity environments. Capable of detecting temperature differences of less than 1°. Sensitivity can be adjusted to enhance contrast between objects when their thermal signature is close to that of the general background.

Eye-safe laser range-finder which is completely autonomous with the transmitter, receiver and processor contained in the same housing.

Operational status
In production and in service. Series 2400 systems have been sold to Finland, India (coastguard), South Korea and Malaysia.

Contractor
Radamec Defence Systems Ltd.

VERIFIED

Radamec System 2400 fire-control system 0006462

Radamec System 1400/1500

Type
Surveillance, tracking and fire-control system (naval).

Development
The RADAMEC 1400/1500 is described as a cost-effective family of electro-optical systems designed for EEZ patrol vessels, support vessels and other similar applications.

Description
The RADAMEC 1400/1500 is a family of systems designed around a common set of baseline modules that allow system capability from basic surveillance up to full ballistics control of one or two guns to be procured in a single specific configuration or built up by incremental acquisition.

The basic building blocks comprise an Electro-Optical Director (EOD) capable of mounting a suite of up to three sensors and a below-decks System Electronics Cabinet (SEC) containing a number of discrete electronic circuits, firmware and software.

Starting at the most basic, system configurations are: -

- electro-optical surveillance system, comprising an EOD with thermal imager and TV camera;
- addition of autotracking capability by fitting circuit card to SEC;
- addition of ranging capability by fitting laser rangefinder to EOD;
- addition of basic fire control capability with gun aim-off by lead angle calculation;
- upgrade to full fire control capability by fitting predictor with ballistics and meteorological compensation.

The system is capable of carrying out a wide range of operator-defined surveillance search patterns, will accept target indications from an external source and (with the addition of the intelligent digital tracker) automatic detection, acquisition and tracking.

In common with all Radamec systems, 1500 can be controlled from any combat system multifunction console or from Radamec's own dedicated console. Radamec can supply its own Windows based MMI software for insertion into an MFC.

Operational status
In service with Royal Australian Navy on 'Huon' class minehunters and two of the Irish Naval Service's ex-Royal Navy (RN) 'Peacock' class patrol vessels (P51/2 and P41/2). Selected for the Philippines Navy 'Jacinto' (ex-RN 'Peacock') class patrol vessels and an undisclosed Asian navy.

Specifications
Director
Azimuth coverage: continuous 360°
Azimuth slew speed: >1.25 rad/sec
Elevation movement: +85° –35°
Elevation slew speed: >1.25 rad/sec
Position reporting accuracy: <0.2 mrad

Thermal Imager
Type: MCT FPA
Detector elements: 384 × 288
Spectral band: 3-5 µm (8-12 µm available)
Fields of view (2 × switched FOV)
Narrow: 1.2° × 2.0°
Wide: 5.4° × 7.2°

TV Camera
Type: 3-CCD colour
Fields of view (continuous zoom)
Narrow: 1.3° × 1.7°
Wide: 12.0° × 17.0°
NETD: 0.05 K

Radamec System 1500 0006463

Laser Range-finder
Type: KTP-OPO shifted ND Yag
Operating wavelength: 1.54 µm
Pulse repetition frequency: 5 Hz, 1 Hz, single shot
Maximum range: 20,000 m
Minimum range: 280 m
Transmitter energy: 1 mJoule

Contractor
Radamec Defence Systems Ltd.

UPDATED

Radamec System 2500

Type
Fire-control system (naval).

Development
Introduced in 1997, the 2500 is Radamec's top-of-the-range EO fire-control system. Evolving from the successful 2400 (qv), the system has a new direct-drive director with multifaceted stealth shape and more advanced processors.

Description
The RADAMEC 2500 is a high performance electro-optical tracking and fire control system for use with small- and medium-calibre guns in air defence, surface-to-surface and shore bombardment roles. The system can track five targets simultaneously and conduct an engagement on one; it is claimed to be capable of detecting high performance combat aircraft at ranges between >10 km (100% humidity, 30°C tropical) and >30 km (25% humidity, 15°C temperate).

The system director platform (Radamec's Type 499) employs direct drive technology in a yoke configuration, with lightweight composite materials used to reduce weight. The electro-optical director incorporates stealth designs with reduced radar cross-section and thermal signature.

The system is automatic in operation but an operator override enables manual control to be implemented quickly. There are four tracking modes: centroid (for air targets), correlation (for surface engagements), edge-tracking (against sea-skimming missiles) and combined centroid/correlation (where the tracker automatically selects the best image according to conditions). In centroid or edge-tracking, there are three sub-modes that can be selected according to target/background parameters.

In common with all Radamec systems, 2500 can be controlled from any combat system multifunction console or from Radamec's own dedicated console. Radamec can supply its own Windows-based MMI software for insertion into an MFC. The fire-control unit is the same as that supplied for the 1500 system (qv).

Operational status
In service with Royal Australian Navy 'Adelaide' class (FFG-7) frigates and Royal Brunei Navy on 'Brunei' class OPVs and 'Waspada' class FACs. Selected for the UK Royal Navy Type 45 'Daring' class (AAW) destroyers and Romanian Navy ex-RN Type 22 frigates.

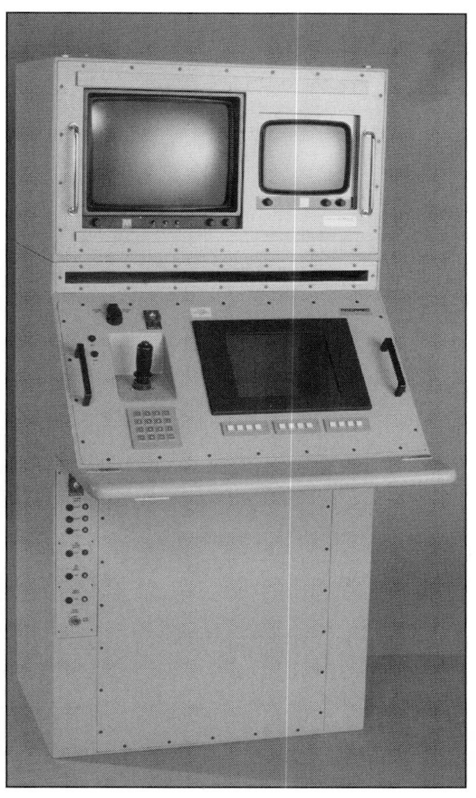
Radamec operator's control console
0019394

Specifications

Director
Azimuth coverage: continuous 360°
Azimuth slew speed: >3 rad/sec
Elevation movement: continuous, customer defined software limited
Elevation slew speed: >2 rad/sec
Position reporting accuracy: <80 µrad

Thermal Imager
Type: MCT FPA
Detector elements: 640 × 486
Spectral band: 3-5 µm (8-12 µm available)
Fields of view (4 × switched FOV)
Narrow: 1.1° × 1.45°
Wide: 8.6° × 11.3°
NETD: 0.015K

TV Camera
Type: 3-CCD colour
Fields of view (continuous zoom)
Narrow: 1.3° × 1.7°
Wide: 12.0° × 17.0°
Scene Illumination: 20 Lux

Laser Range-finder
Type: ND Yag diode pumped
Operating wavelength: 1.54 µm
Pulse repetition frequency: 10 Hz, 1 Hz, single shot
Maximum range: 20,000 m
Minimum range: 300 m
Transmitter energy: <7.8 mJoule

Contractor

Radamec Defence Systems Ltd.

UPDATED

Reutech Systems RTS-6400

Type

Electro-optical/radar tracker (naval).

Description

RTS-6400 naval electro-optic radar tracker is being developed as part of a South African National Defence Force (SANDF) technology retention project. It is intended to control anti-aircraft and anti-surface weapons in medium-range and close-in engagements.

The missile detection radar and electro-optical tracker are both being developed by Reutech. Sensors will include: an X-band air-cooled monopulse Travelling Wave Tube (TWT) radar with a digital, programmable, Doppler signal processor and composite dual offset Gregorian reflector antenna; triple field of view thermal imager with manual and automatic exposure control; dual field of view TV camera, eye-safe laser range-finder and a video auto-tracker. The system is capable of automatic, manual and memory tracking employing sensor fusion technology. It has a fibre-distributed data interface, accepting 1-D, 2-D, 3-D and 4-D designation data. The stabilised mounting is designed to operate in conditions of Sea State 5 and 25 mm/h rainfall.

RTS-6400 is designed to detect low-flying aircraft at 24 km and sea-skimming missiles at 16+ km. It will acquire and track targets automatically and carry out autonomous surveillance with a helical scan search facility for aircraft.

Operational status

Under development.

Specifications

Radar frequency: X-band (NATO I/J-band)
Radar detection range: fighter: >38 km
missile: >18 km
Tracking accuracy
Range: better than 5 m RMS
Angles: better than 0.6 mrad RMS (1.5 mrad incl multipath with radar)
Acquisition time: better than 3 s (1.5 s typical)
Instrumented range
Air and surface mode: 60 km
Positioner
Velocity: >5 rad/s
Acceleration: >6 rad/s^2
TV camera
Fields of view
Narrow: 2.5 × 1.9°
Exposure range: 2 lx (WFOV)/8 lx (NFOV) to 100,000 lx
Thermal imager
Spectral band: 3-5 µm
Fields of view
Narrow: 2.2 × 1.5°
Wide: 9 × 7°
Laser range-finder
Pulse repetition frequency: >20 Hz
System data rate: 50 Hz
Platform motion (amplitude of roll/pitch)
Operating: 27/7.5°
Survival: 40/9°

Contractor

Reutech Radar Systems (Pty) Ltd.

UPDATED

Saab 9LV 100

Type

Naval fire-control system.

Development

During the 1970s Philips Elektronikindustrier AB (PEAB), now Saab Systems and Electronics (via CelsiusTech and SaabTech), developed an electro-optic air defence system, 9 AA 100/Kalle and in 1977 it revealed a naval version as 9LV 100. It entered service in August 1981 with the commissioning of the Bahrain fast attack craft *Al Riffa*. Development of the system has continued to the present day through the 9LV 200 and 9LV Mk 3.

Description

The system controls one or two 30 to 76 mm guns against air and surface targets as well as surface-to-surface missiles. It can track up to two air or surface targets and can receive designation from search and navigation radars or designators as well as having the capability of being slaved to other types of directors. It is small enough to be used on vessels down to 50 tonnes as well as larger vessels, that is 300 to 400 tonnes with limited armament. It can also act as a supplement to more comprehensive fire-control systems such as the 9LV 200 Mk 3E.

The main elements of the 9LV 100 are an electro-optic director weighing 140 kg with CCD TV camera, eye-safe laser and optimal infra-red camera (8-12 µm or 3-5 µm), a control and display console with a 41 cm (16 in) screen, gun and director controls, two bridge pointers for designation and the fire-control computer.

Operational status

Approximately a dozen systems have been produced for the navies of Bahrain, Oman and Sweden.

Specifications

Director
Azimuth: 360°
Elevation: –25 to +80°
Angular speed: 85°/s
Angular acceleration: 400°/s^2
TV camera
Type: CCD
Field of view: 2.3-23°
Scanning: 625 lines, 25 frames/s
IR camera
Wavelength: 8-12 µm or 3-5 µm
Field of view: 52 × 35 mrad; 157 × 105 mrad
Cooling: Stirling closed-cycle
Detector: CMT
Laser range-finder
Type: Nd:YAG
Wavelength: 1.54 µm
Peak power: 1.5 MW
Pulse repetition frequency: 10 Hz

Contractor

Saab Systems and Electronics.

UPDATED

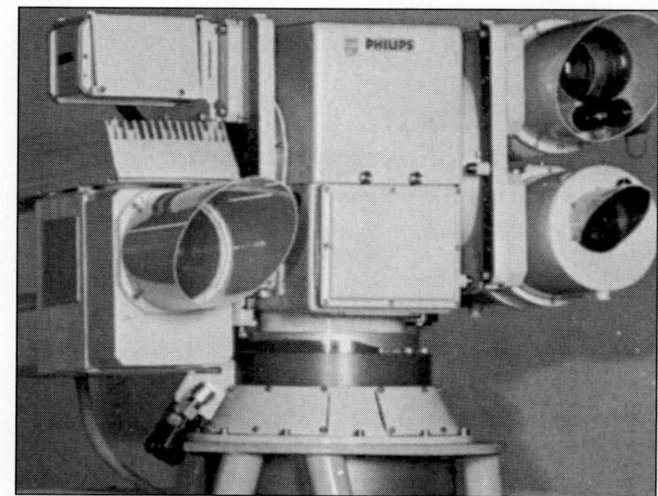

9LV 100 electro-optic director with IR camera, TV camera and laser

Saab 9LV 200 Mk 3

Type
Naval fire-control system.

Development
The 9LV system was originally devised by Philips Electronikindustrier (later CelsiusTech, SaabTech Systems and, now, Saab Systems and Electronics) as a systematic solution to combat requirements for naval warfare. The system matches high-performance requirements, for blue-water operation as well as for littoral warfare, with minimum life-cycle cost. The manufacturers state that 9LV has world class performance in weapon control for missiles, torpedoes or guns.

Development of the fully distributed architecture used in the Mk 3 generation started in 1985, with the first systems commissioned in 1989. This third generation, 9LV 200 Mk 3, is now in operation with seven navies in different configurations ranging from patrol boats to frigates, such as the Swedish Corvette *Göteborg*, the Danish Standard Flex 300 Multirole ships as well as the Australian and New Zealand 'ANZAC' class frigates with deliveries to Finland, Oman and Pakistan the 9LV 200 Mk 3 is considered the market leader in this new generation of naval combat management systems. The success and versatility of the system architecture has also been proven in the Swedish A19 'Gotland' class submarine, and in the retrofit of the Pakistan Type 21 'Tariq' class frigates.

Saab Systems and Electronics is continuously developing and enhancing the 9LV system. Based on the distributed Mk 3 concept, an evolved version is currently under development. The evolved version, 9LV 200 Mk 3E, reuses the open architecture and SW components, in the form of more than 2 million lines of Ada software. As part of continuing development, the latest technological improvements are added to the system. A high-performance fibre optic Local Area Network (LAN) is used for the distribution of data and radar video. The LAN follows the 100 MHz FDDI dual attachment standard with built-in redundancy.

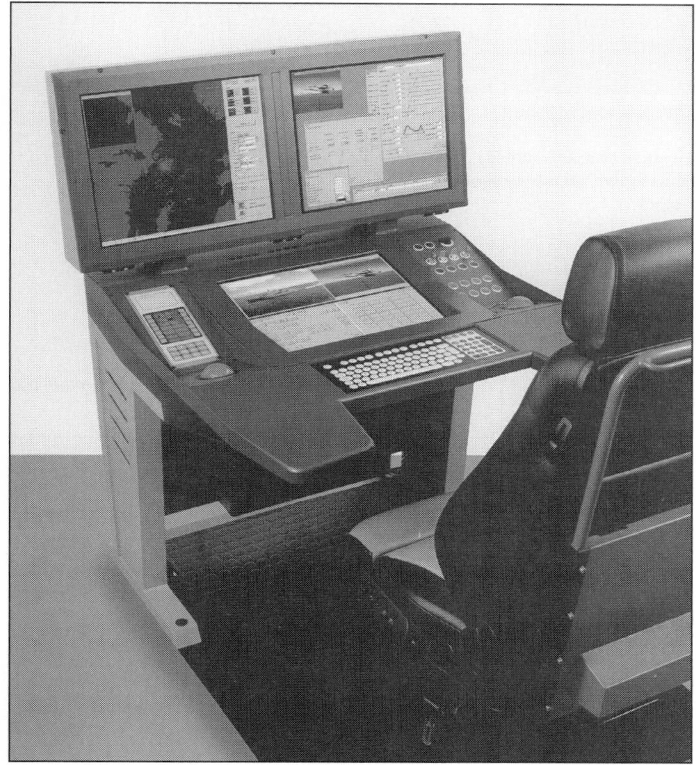

Saab Bofors multifunction console for use with 9LV 200 Mk 3E 0019389

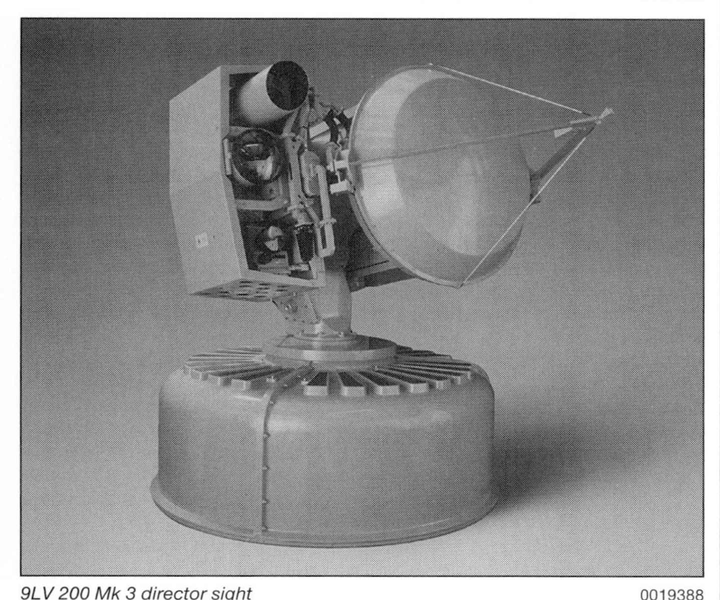

9LV 200 Mk 3 director sight 0019388

Main stream COTS technology is used for a new state-of-the-art generation of multifunction consoles. The full colour flat screens provide excellent viewing angle, brightness and flicker-free presentation of live video (TV/IR and Radar) overlaid by synthetic data such as tracks and sea charts. The console software uses the Windows NT operating system, which adds the standard capabilities of a windows based graphical user interface. Commonality with the standard office environment is exploited to reduce training efforts and increase the operator efficiency.

Current development is also focused on the CEROS 200 radar director, which soon will be capable of controlling the Enhanced SeaSparrow Missile (ESSM).

A new unique method is introduced to enhance tracking of low flying targets. This patented CHASE technology eliminates the nodding problem and tracks a target as accurate as in free space.

Another current development is stealth-adaptation, being developed for the new Swedish stealth corvette *Visby*.

Description
The 9LV 200 Mk 3 systems are based on a distributed architecture and all components are designed for reuse. The extensive library of applications covers AAW, ASuW, ASW, EW, MW and MCM, efficient integration of sensors/weapons and a wide range of command functions. Several tactical data links have been integrated, for example Link 11. This large re-use library enables Saab to tailor each system to the ship configuration without the programme risk normally associated with new software.

An essential feature of the AAW/ASMD capability is the ability to optimise the employment of all available engagement resources. This optimising function is able to operate in a fully automatic mode and takes into account all of the ship's weapons including missiles, guns and soft kill systems. It provides an integrated engagement plan that utilises the best possible combination of available weapons throughout the entire engagement window, rather than a traditional 'layered defence' with successive engagements by each weapon in turn.

For fire-control purpose a CEROS 200 director is available with servo- controlled pedestal, tracking and integrated ballistic calculations. The director can control both guns and surface-to-air missiles, for example the NATO SeaSparrow, and is claimed to provide outstanding accuracy and reaction time.

The CEROS 200 weighs between 700 kg and 800 kg, depending on version. The prime sensor is a Ku/J-band tracking radar with stabilised Cassegrain mono-pulse antenna and travelling wave tube transmitter. It uses high-performance FFT processing and operates in the 15.5 to 17.5 GHz frequency range. It permits the combination of pulse-Doppler and/or MTI operation, with batch-to-batch frequency agility. Pulse-to-pulse frequency agility is retained for use when clutter rejection is not required.

The radar can be supplemented by a wide range of electro-optical sensors including a CCIR-compatible TV-camera, an infra-red camera and a laser range-finder.

Operational status
Over 170 9LV systems have been contracted to more than 15 navies worldwide.

More than 55 Mk 3 systems have been ordered or delivered for the navies of Australia, Denmark, Finland, New Zealand, Oman, Pakistan, Singapore and Sweden.

Specifications
Tracking radar
Frequency range: 15.5-17.5 GHz
Antenna
Type: Cassegrain, multimode feed, 1 m Ø
Gain: 41 dB
Beam width: 1.4°

Transmitter
Type: Grid-pulsed helix TWT
Peak power: 1.5 kW, 4% duty cycle ('Extended Range' version is also available)
No of frequencies: 128

Receiver
Type: 3-channel amplitude monopulse
Compressed pulse width: 0.2 μs
Signal processing: 32-point FFT/4-pulse FIR
Noise figure: 10 dB ('Extended Range' version is also available)

TV camera
Type: CCIR 625/25, 2:1 interlace
Field of view: 2.3-23°
Scene luminance: 10-300,000 cd/m²

IR
Several types available in both the 8-12 and 3-5 μm wavelength bands.
Laser range-finder
Type: Nd:YAG
Wavelength: 1.06 or 1.54 μm
Peak power: 4 MW (1.06 μm) or 1.5 MW (1.54 μm)

Contractor
Saab Systems and Electronics.

UPDATED

Saab Bofors Dynamics EOS-400/450

Type
Fire-control system (naval).

Description
The Saab Bofors EOS-400 is an optronic fire-control system for ship defence. The system has been in service with Finland and Brazil since 1985.

The Saab Bofors EOS-450 is the latest generation with state-of-the-art electronics and high-performance sensors for long detection range and special features for combating fast manoeuvring targets, for example sea-skimming missiles. Automatic tracking of air and surface targets, combined with laser-ranging and ballistic prediction, provides accurate control of one or several guns of varying calibres and ammunition types.

Above deck, the system consists of a gyrostabilised platform with three sensors, a high-resolution CCD TV camera, a multi-element IR camera and an eye-safe laser range-finder. Below deck the operator console and the electronic unit are installed as one cabinet. The operator console is equipped with a 20 in high-resolution colour display.

The new TV/IR tracker is based on digital correlation and utilises advanced image processing. During acquisition on target indication data or in surveillance mode, the tracker provides automatic target indication and acquisition. In case the target is momentarily concealed, the tracker performs automatic prediction and relock.

Sensor data fusion, that is simultaneous TV and IR tracking, in combination with background suppression using Moving Target Indication (MTI) assures best possible performance against all kinds of targets even in cluttered background conditions. The tracker is capable of tracking up to four targets simultaneously which ensures that target switching can be performed very quickly.

Its small size and light weight makes it easy to install on any kind of vessel. Modularity makes it simple to configure either as an autonomous fire-control system or as a remote target sensor controlled by any combat management system, for example via a local area network or datalink.

A target simulator, generating realistic targets, can be implemented for operator training and system test.

Operational status
EOS-400 is in service with Brazil and Finland. EOS-450 is in service with Mexico.

Specifications
EOS-450
Director
Stabilisation: two-axis rate-gyro
Angular range:
(Traverse) Training unlimited
Elevation –30 to +85°
Angular speed: 175°/s
TV camera
Type: second-generation multi-element 288 × 24
Spectral bandwidth: 8-12 μm (optional 3-5 μm)
Field of view:
1.5 × 2°
5.4 × 7.2°
Laser range-finder
Type: Raman shifted Nd-YAG
Wavelength: 1.54 μm (eye-safe)
TV/IR tracker
Type: digital correlation
Target acquisition: automatic
No of targets: 4 simultaneously

The mulltisensor head of the Saab Bofors EOS-450 0019547

The multisensor head of the Saab Bofors EOS-400 fire-control system

Physical characteristics
Director with basic sensors
Height: 1.4 m
Width: 0.7 m
Depth: 0.7 m
Weight: 200 kg

Contractor
Saab Bofors Dynamics AB.

VERIFIED

Saab CEROS 200

Type
Fire control (naval).

Development
The CEROS 200 Air Defence fire-control director forms an integral part of the 9LV 200 Mk 3E Combat Management System. It can also be offered as a stand-alone fire-control system or target tracking subsystem. The radar and the electro-optic sensors can be used in any combination to provide multiple target tracking capability.

Description
One CEROS 200 director can control all weapons to create a three-layer defence: long range, with surface-to-air missiles of both semi-active and command to line of sight types; medium range, for fire control of guns of any calibre; short range, for use with close in weapons.

CEROS 200 supports dual-target tracking within the radar beam or electro-optic sensor field of view. This allows simultaneous transfer of secondary target co-ordinates; missile alerts; hit evaluation; kill assessment; surveillance within either sector and horizon search. Target tracking data can be accepted from a number of sources – C2/C3 systems, 2-D/3-D surveillance radars, ESM and bridge pointers.

The director adapts its scan-to-acquire programme according to the source of designation data. Transmitter waveform and signal processing automatically adapt accordingly. Acquisition and initiation of tracking are automatic but there is the capacity for intervention and/or support from an operator. The director has direct-drive motors with integrated hydrostatic bearings for improved accuracy, speed and acceleration.

The radar is coherent using a TWT (Travelling Wave Tube) as a power amplifier. Depending on the clutter situation and the electronic countermeasures threat, the radar selects its frequency-agility pattern between: 32-pulse bursts and Pulse-Doppler signal processing, four-pulse bursts and MTI (Moving Target Indicator) processing, or pulse-to-pulse agility. The Pulse Repetition Frequency (PRF) and pulse-width are selected depending on the range to the target. The transmitted pulses are frequency coded and pulse compression in the receiver ensures high-range resolution in all modes.

CHASE (CelsiusTech High-Accuracy Sea skimmer Estimator) is Saab's patented algorithm for processing complex radar target return signals from a very low-flying target, that is a sea-skimming missile, and its image below the sea surface.

The electro-optic sensors of the CEROS 200 radar and electro-optic sight

Detection relies on the very wide bandwidth of the radar receiver and signal processor. The radar tracking accuracy of the CHASE method is stated to be better than 0.2 mrad in calm sea and 0.4 mrad in rough sea.

The director pedestal carries the following electro-optic sensors - TV camera, thermal imager and laser range-finder. The TV and infra-red cameras operate together with an advanced video tracker enabling edge, centroid or correlation mode of tracking. Together with the laser this gives the capability of pure electro-optic tracking. The sensors can be selected according to the user's preference. The IR camera is available in both 3 to 5 μm and 8 to 12 μm wavebands. The laser can be eye-safe if required.

CEROS 200 is available in a baseline version and a version which has the addition of an X-band radar channel for continuous wave illumination of a target for guidance of the semi-active Seasparrow surface-to-air missile and a longer range radar is available for use with the Evolved Seasparrow missile. A Stealth version is optimised for low-radar cross-section over a multi-octave frequency band, enabling the director to be mounted above deck on board extreme Stealth vessels.

Operational status

In production and in service on HMAS *Anzac* frigate (Australia), HMNZS *Te Kaha* frigate (New Zealand), HMS *Göteborg* (Sweden), HMS *Flyvefisken* large patrol craft and HMS *Thetis* frigate (Denmark), RNO *Al Bushra* offshore patrol vessel (Oman) and OHJV *Rauma* fast attack craft (Finland).

Specifications

Tracking radar
Frequency range: Ku-band, 15.5-17.5 GHz
Pulse compression: frequency coded
Antenna
Type: Cassegrain with multimode feed
Gain: 41 dB
Beam width: 1.5°
Transmitter
Type: Grid-pulsed helix travelling wave tube
Output power: 1.5 kW peak, 4% duty cycle
Number of frequencies: more than 100
Transmit patterns:
Pulse-Doppler: 32-pulse batches
Moving Target Indicator: 4-pulse batches
Frequency agility: pulse-to-pulse frequency agility
Receiver
Type: 3-channel amplitude monopulse
Compressed pulse-width: 0.2 μs
Noise figure: 10 dB
TV camera
Type: CCIR 625/25, 2:1 interlace
Diagonal field of view: 2.3-23°
Scene luminance: 10-300,000 cd/m^2
Thermal imager: several types can be offered in both the 3-5 μm and 8-12 μm wavebands
Laser range-finder
Type: Nd:YAG
Wavelength: 1.06 or 1.54 μm
Peak power: 4 MW (1.06) or 1.5 MW (1.54)
Director pedestal
Type: 2-axis, elevation over azimuth
Angular speed: 2 rad/s
Angular acceleration: 9-10 rad/s^2
Dimensions and weight
Height above deck: 2 m approx
Diameter: 1.6 m approx
Weight: 700-800 kg depending on version

Contractor

Saab Systems and Electronics.

UPDATED

SAGEM EOMS

Type

Electro-optical multifunction system.

Development

Development of the EOMS started in 1997 to integrate into one single equipment the functions performed by two of SAGEM's already operational systems, the VIGY 105 electro-optical director (EOD) and the VAMPIR MB, dual-spectrum infra-red search and track.

Description

The EOMS is a one-unit Electro-Optical Multifunction System designed to cost-effectively improve the self-defence capability of any surface combatant. With the same gyrostabilised head, the EOMS performs:

- single-spectrum IRST function: in this mode, the EOMS performs passive panoramic search, automatic detection of any air or surface targets, track-while-scan of targets and accurate target designation to the combat system
- electro-optical director function: in this mode the EOMS performs 3-D target tracking upon target designation, trajectory filtering and ballistic computation and control of small- and medium-calibre guns.

The panoramic surveillance head comprises a fixed sensor housing and an accurate line of sight pointing device. The IR, TV and eye-safe laser range-finder are fixed in the housing, while a multispectral line of sight is pointed and stabilised with a mirror.

The system uses a high-efficiency IR sensor, the SAGEM IRIS thermal imager, operating in the 3-5 or 8-12 μm wavebands. The IRIS thermal imager is built around an IRCCD 288 × 4 detector, which can be used either in imagery or scanning mode.

The IRST function integrates the latest achievement in the field of IR data processing algorithms, taken from the VAMPIR MB, which SAGEM state is the sole second-generation qualified IRST in service.

Operational status

The EOMS is now deployed by the Belgian Navy 'Wielingen' class frigates, where two EOMS per ship have been provided. It is also in production for the UAE Tarif programme and for the Finnish Navy Squadron 2000 programme.

Specifications

Stabilised sensor head
Training range: n × 360°
Elevation range: –20 to +65°
Weight: 190 kg

Dimensions
Diameter: 500 mm
Height: 1,635 mm

Sensors
IR camera: IRIS IRCCD, 8-12 μm or 3-5 μm, 288 × 4
Black and white TV camera
Eye-safe laser range-finder: 1.54 μm, 20 Hz

Contractor

SAGEM SA, Aerospace & Defence Division.

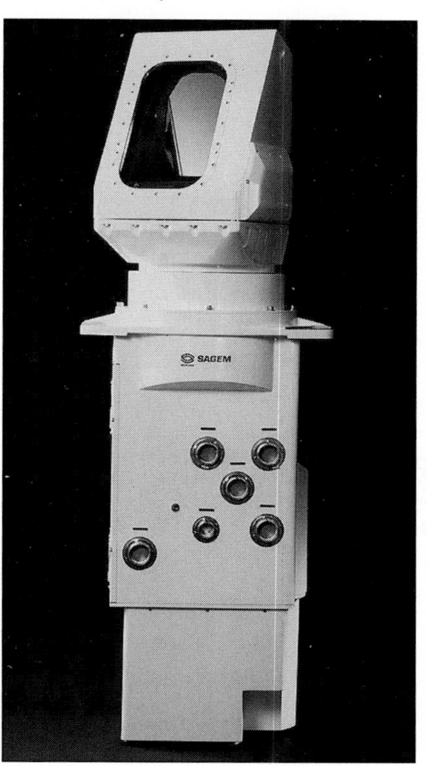

SAGEM EOMS, electro-optical multifunction system for surface ships

SAGEM Pirana infra-red tracker

Type
Fire-control system (naval).

Description
The Pirana infra-red tracker has been developed for use on combined electro-optical naval fire-control equipment in which it is employed with TV and a laser range-finder for the direction of anti-aircraft weapons, or for use on its own as an ancillary to autonomous weapons. The equipment provides a means of passive search, acquisition and target tracking, especially useful in low-altitude, low-elevation engagements. It has two IR channels for simultaneous tracking at two different wavelengths.

The main modules of the infra-red unit are: the receiver optical assembly, analysis section, IR detector, cooling, and tracking electronics module. Power supply unit and control panel are located below deck.

Operational status
Experience with the Pirana I and II resulted in the Pirana III version which is operational with the French Navy in the frigates *Cassard*, *Suffren* and *Duquesne* and fitted to the *Jean Bart*. This version is a bispectral equipment with two channels working simultaneously, together with appropriate signal processing.

Specifications
Dimensions: 75 × 41 × 41 cm
Weight: 50 kg (tracking telescope)
Optical aperture: 170 mm
Field of view: ± 8 mrad
Wavelength: 4-5 and 8-12 μm bands
Range: 10-20 km, depending on type of target
Cooling: Joule-Thomson cycle to about 80 K
Power supply: 115 V, 400 Hz, 1 kVA

Contractor
SAGEM SA, Aerospace & Defence Division.

Pirana infra-red tracker mounted on the frigate Cassard

SAGEM TDS 90

Type
Target designation sight (naval).

Description
TDS 90 is a very fast target designation sight for surface ships. It has been specially designed to fulfil the following missions: target acquisition for day or night operation; designation and tracking of airborne and surface threats; transmission of target parameters to the command and control system; secondary fire-control system.

TDS consists of an optical reflex sight, mounted on a very ergonomic pedestal. The head is telescopically supported and counterbalanced to allow easy tracking and easy height adjustment; the reflex sight provides reticle and elevation/azimuth information to the operator. No deck penetration is required for fitment. The TDS offers RS-422 and synchro interfacing capabilities, with target information delivered either in absolute or relative co-ordinates.

Operational status
In production and operational on board the French Navy 'La Fayette' class frigates and the Norwegian Navy 'Hauk' class fast patrol boats.

Specifications
Height: 1,720 to 2,230 mm
Overall diameter: 610 mm
Weight:105 kg
Training limits: ±170°

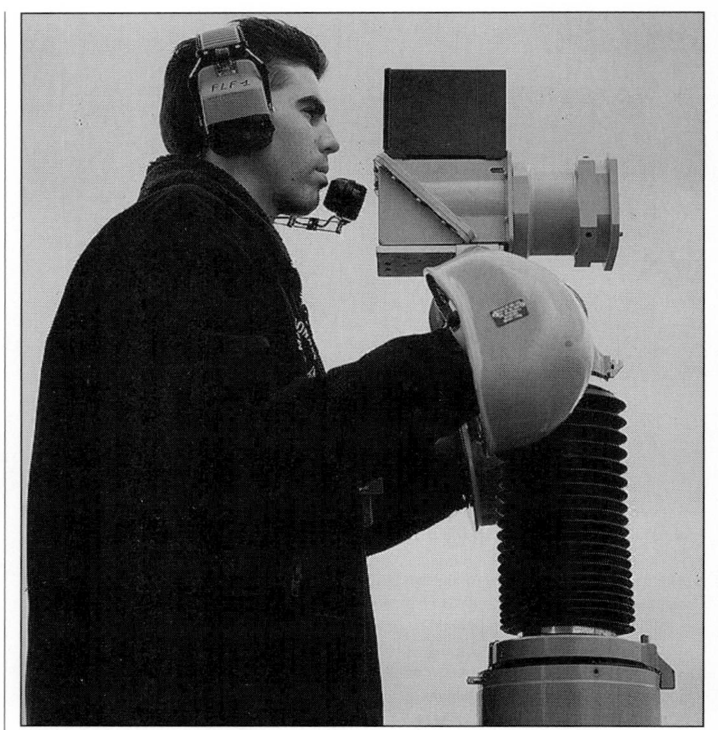

SAGEM TDS 90 target designation system

Elevation limit: −15 to +75°
Target designation accuracy: 3 mrad
Target acquisition and transmission time: < 3 s
Night designation: thermal imager or night goggles

Contractor
SAGEM SA, Aerospace & Defence Division.

SAGEM VIGY 20

Type
Surveillance and fire-control system (naval).

Development
The VIGY 20 is SAGEM's latest development in the field of electro-optical day and night maritime surveillance, identification and fire-control systems. The VIGY 20 programme began in 1997 and the first production order was received in 1998.

Description
Modular and configurable with a variety of sensors, VIGY 20 is a compact and lightweight system, combining multimission capability and high performance. The system can house a large variety of sensors, depending on the operational and climatic conditions. Key features of the VIGY 20 are:
- use of high-efficiency sensors
- easy installation: compact and lightweight system
- modularity and easy maintenance through rapid replacement of subassemblies without adjustment

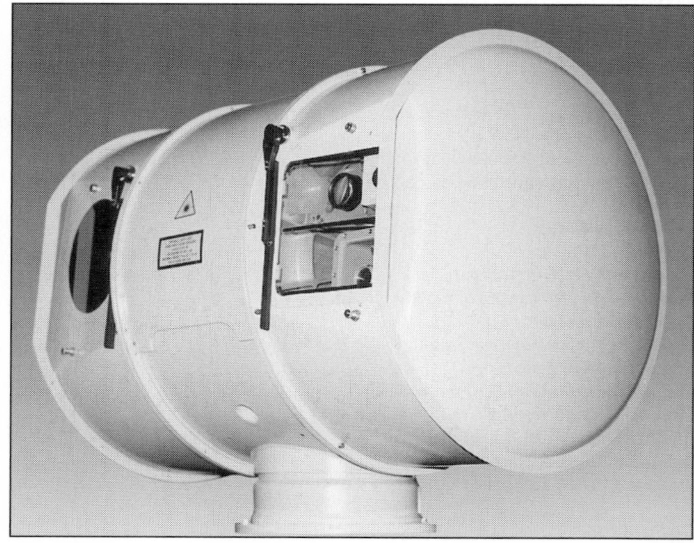

The SAGEM VIGY 20 lightweight surveillance and fire-control system 0129308

- direct drive motors providing very fast reaction time and accurate tracking
- high-precision fire control.

Operational status

The VIGY 20 is now in production. Several units have been delivered to the Norwegian Navy 'Hauk' class fast patrol boats and 'Norkapp' class coastguard vessels. Two units have also been delivered to a Middle East Navy. The VIGY 20 will also equip the new Norwegian 'Nansen' frigates.

Specifications

Stabilised sensor head
Training range: unlimited (n × 360°)
Elevation range: –30 to +85°
Azimuth velocity: 2.5 rad/s
Elevation velocity: 2.5 rad/s
Mass: 150 kg

Sensors
IR camera: IRIS IRCCD 8-12 μm, 288 × 4 or MATIS 3-5 μm, 384 × 256, 2 × 2 microscanning
Black and white TV camera
Colour TV camera
Eye-safe laser range-finder: 1.54 μm, 3-8 Hz

Contractor
SAGEM SA, Aerospace & Defence Division.

UPDATED

STN ATLAS Elektronik Optronic Multi-Sensor Platform (MSP 500)

Type

Fire-control and reconnaissance system (naval).

Development

The Optronic Multi-Sensor Platform began as a private venture development by STN ATLAS Elektronik and Zeiss Optronik in 1995. A preproduction model underwent sea trials on board the Federal German Navy's F122 frigate *Emden* in 1995. Three preproduction systems have been handed over to the German Navy and are being trialled on board F122 frigate *Koln* and minesweeper *Weiden* and one system is installed at a naval test centre.

The system was selected for retrofit to the German Navy surface fleet beginning in 1998.

Description

The Optronic Multi-Sensor Platform (MSP 500) is a lightweight, two-axis-stabilised, electro-optical system, which can be used in land-based applications (coastal surveillance), airborne (helicopters) and shipborne as a part of naval command and fire-control systems. The stabilisation system is digital with coarse/fine control. Sensors include daylight camera, thermal imager and laser range-finder. Applications include observation, detection and identification during day and night; manual or automatic search and acquisition; ranging of surface or aerial objects; manual or automatic tracking of targets; stand-alone fire control and integration into command and control systems.

Operational status

In production. The system will be installed on all German Navy surface vessels, completing in 2003.

Specifications

Stabilised platform
Elevation range: –40 to +85°
Azimuth range: n × 360°

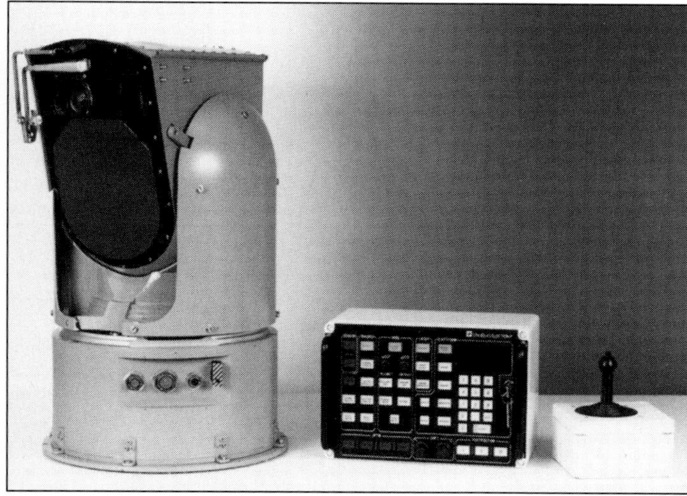

The Optronic Multi-Sensor Platform 0002009

Daylight camera – zoom objective
Fields of view: 2.4 × 3.3° to 24 × 33°
CCD sensor: 752 × 576 pixel
Thermal imager
Detector: 2nd generation CMT PV, 96 × 4 IRCCD array
Spectral band: 7.5-10.5 μm (optional 3-5 μm 3rd generation)
Fields of view
Wide: 5.2 × 7
Narrow: 1.5 × 2° (zoom and alternative fields of view on request)
Resolution: 576 × 756 active pixel (full CCIR format)
Laser range-finder
Type: Nd:YAG-Raman
Wavelength: 1,543 nm
Features: adjustable distance window, first/last echo switchover, echo storage for up to 10 echoes
Pulse length: 4 ns
Pulse rate: 1 Hz/6 Hz
Pulse energy: 10 ± 3 mJ
Pulse power: 2 MW
Tracker
Correlation, centroid and complementary modes of operation
Grey-level based centroid tracking
Sub-pixel tracking error calculation
Lost track approximation of the trace
Video interfaces: CCIR/STANAG 3350B
Data interface: MIL-Bus 1553B, RS 422
Power: 115 V/60 Hz
Power requirements: typical < 800 VA
Weight: < 150 kg

Contractors

STN ATLAS Elektronik GmbH.
Zeiss Optronik GmbH.

VERIFIED

Thales LIOD Mk 2

Type

Naval fire-control system.

Development

LIOD (Lightweight Optronic Director) Mk 1 was developed in the late 1970s and entered production during the early 1980s. In the late 1980s a major update, LIOD Mk 2, was developed and is currently in full scale production.

Description

LIOD MK 2 is a fully, comprehensively qualified system, both in terms of environmental testing and functional and performance testing in accordance with the MIL Standards for shock, vibration, temperature, humidity and EMC.

LIOD MK2 Sensor system for automatic, electro-optic tracking of air and surface targets 0019391

Like LIOD Mk 1, LIOD Mk 2 is available in two configurations – as a sensor subsystem and as a complete fire-control system. In the latter configuration the sensor equipment is extended with an operator console, including an interface section to connect the subsystem to one or more weapons, ship's reference and compass/log facilities and a C^2 system for remote target designation.

As a sensor subsystem, LIOD Mk 2 is primarily designed to conduct automatic or rate-aided tracking of an air, surface or shore target providing track data for the simultaneous control of different types of gun. The system can track targets moving at up to 800 m/s and can track larger ships at 20 km and fighter-size aircraft at 10 km.

LIOD Mk 2 comprises an electro-optic director which carries sensors consisting of TV and IR cameras and a Nd:YAG or eye-safe laser range-finder. The daylight CCD TV camera has automatic filters to adjust for light conditions. The IR camera applied as standard is the Thales (formerly Signaal) USFA0 UP 1043 but alternative compatible IR sensors can be offered.

The LIOD Mk 2 sensor subsystem is capable of performing the following functions:
(a) optical surveillance for air and surface targets
(b) target acquisition based on 2-D or 3-D handover data
(c) target tracking using gated TV/IR video either automatic or rate-aided for one air or surface target
(d) engagement monitoring
(e) system status monitoring.

The standard interfaces are compatible with Thales' SEWACO FD architecture. Other types of interfaces are available if required.

Operational status
A total of 65 LIOD systems (Mk 1 and Mk 2) systems have been ordered or produced for the navies of Indonesia, Ireland, South Korea, Malaysia, Nigeria, Turkey and the United States. LIOD Mk 2, the current production model is in service and in production.

Specifications
LIOD Mk 2
Director
Weight: 225 kg
Traverse: n × 360° (unlimited)
Elevation: –30 to +115°
Speed: 130°/s
Acceleration: >300°/s²
TV camera
Field of view: 2-20° (diagonal)
IR camera
Field of view: 1 × 2° and (optional) 7.5 × 15°
Spectral band: 8-12 μm
Laser range-finder
Beam divergence: 1.5 mrad
Pulse length: 20 ns
Max pulse repetition frequency: >10 Hz

Contractor
Thales Naval Nederland.

UPDATED

Thales LIROD Mk 2

Type
Naval fire-control system.

Development
The LIROD (Lightweight Radar Optronic Director) is a parallel development of the LIOD electro-optical fire-control system. LIROD is biased to radar tracking rather than electro-optic tracking. An improved version was marketed from 1993 as LIROD Mk 2. The first export sales of this version were to Indonesia for the 'PB 57' Nav V patrol boats.

Description
LIROD, in all versions, is a lightweight fire-control system based upon a K-band tracking radar and a TV camera. It is designed to control simultaneously different dual-purpose guns while tracking one air or one surface target. It may be used either as a stand-alone unit or to complement the Thales family of fire-control systems. LIROD Mk 2 is claimed to be extremely efficient against low-altitude sea-skimming missiles.

LIROD Mk 2 features a new radar with TWT transmitter and elliptical parabolic antenna with monopulse cluster. The antenna is 1 m high and 0.40 m wide and is on a 475 kg director with a swept distance of 1.60 m wide and a height of 1.96 m. The radar features horizontal polarisation and operates on a frequency of 35 GHz. LIROD Mk 2 additional features: kill assessment support, sector search, auto sensor selection (TV/Radar), pre-action calibration, miss distance indication (measuring shells of different types of guns), high level of ECM resistance, using modern FFT processing for data extraction in combination with special track filters to enhance the performance in the case of low-flying targets. The only electro-optical sensor is a TV camera providing video format to CCIR standards (625 lines, 25 picture/s, 2:1 interface), which is applied for observation and for angular

LIROD Mk 2, the latest version of Thales' Lightweight Radar/Optronic Director (LIROD) 0019390

tracking parallel to the radar angular tracking, for fire control in combination with the radar range tracking to build a three-dimensional solution. Growth potential: IR camera and laser. The manufacturers claim the Mk 2 can track a fighter-size target at 20 km with radar and 10 km electro-optically.

Operational status
Some 40 LIROD Mk 1 systems have been delivered or ordered for the navies of Argentina, Canada, Indonesia and Thailand. LIROD Mk 2 is the present production configuration.

Specifications
Director Mk 2
Traverse: 360°/s
Elevation: -30 to +85°
Traverse speed: 120°/s
Elevation speed: 80°/s
Radar: K-band
TV camera
Type: Daylight variable FDV
Field of view: 2-20°
Mk 2: 115 V 60 Hz (3 kVA); 440 V 60 Hz (1.0 kVA)
Weight
Mk 2: 625 kg (subsystem)

Contractor
Thales Naval Nederland.

UPDATED

Thales Mirador

Type
Naval fire-control system.

Development
Mirador is a lightweight electro-optical surveillance, tracking and fire-control system from Thales, Nederland (formerly Signaal). It is available in both maritime and land-based versions. The naval version is designed for use on a range of platforms from small patrol craft to large carriers. It can be integrated with Thales' SEWACO or other combat system or can be used as a stand-alone fire-control system.

Description
The Mirador stand-alone system consists of a stealthy director, a processing cabinet controlling the weapons and a control console.

The director is a carbon-fibre shell structure of minimal weight, including a state-of-the-art IR camera, (standard 8-12 μm, optional 3-5 μm) a TV tracking camera, a colour TV camera and a low light TV camera for night use, for observation and an eye-safe laser range-finder.

The system is capable of performing the following functions:
(a) optical surveillance
(b) automatic acquisition after remote or local target designation
(c) automatic target tracking (surface, air and shore)

Mirador, the latest electro-optical fire-control system from Thales 0002015

(d) fire-control computations
(e) gun control for different types of guns.

The operator console features a touch sensitive screen with two video displays and a very user-friendly human-machine interface. This allows easy situational awareness and rapid reaction during critical operations.

Operational status

In production for the Royal Netherlands Navy and Hellenic Navy. Delivered to Venezuela (six land based version systems in 2000) and the Bangladesh Navy.

Specifications

Target tracking range: 20 km (large size surface targets), 15 km (fighter aircraft)
Max target tracking speed: 1,000 m/s
Director
Azimuth coverage: 360°
Azimuth speed: >5 rad/s
Elevation movement: –30 to +120°
Elevation speed: >4 rad/s
Colour TV camera
Type: colour, with zoom (×12)
Spectral band: visible light
Field of view: 4.4 to 3.3°
Video output: CCIR
Tracking TV camera
Type: monochrome, fixed focus
Spectral band: visible light
Field of view: 2 × 1.5°
Video output: CCIR
Thermal imager
Type: Stirling cooled
Spectral band: 8-12 μm
Fields of view
Narrow: 3 × 2.25°
Wide: 9 × 6.75°
Detector elements: 288 × 4 CMT
NEDT: 0.1 K
Laser range-finder
Type: Nd:YAG Raman shift
Wavelength: 1.54 μm
Instrumented range limit: 20 km
Repetition rate: 3 Hz average, 8 Hz burst
Video track unit: gated contrast centroid tracking
Dimensions:
Optronic director (w × h × d): 496 × 1047 × 752 mm
Control console: 745 × 1068 × 445 mm

Contractor

Thales Naval Nederland.

UPDATED

Thales Sting EO

Type

Naval fire-control system.

Description

STING EO is a multipurpose, lightweight, naval fire-control radar and electro-optical tracking system which can provide track data for control of a broad range of naval guns against air and surface threats. Derived from Thales' (formerly Signaal) STIR family, STING EO utilises advanced signal processing techniques.

Sting EO, the electro-optical version of Thales' naval fire-control system 0002016

The electric servo drives with brushless motors provide fast and accurate director control and the system is suitable for installation in naval vessels of any size. The system provides automatic target acquisition through its I-band radar or K band radar on remote one-, two-, or three-dimensional designation whereafter I/K-band tracking is performed. Target data can be simultaneously derived from radar and optronic. Automatic best data source selection (I/K) ensures optimal track accuracies supported by TV/IR camera tracking and laser range-finding. Under control of a command and control system, STING EO offers supportive functions such as sector search (with automatic target detection), missile launch detection, projectile position measuring during gunfire, kill assessment support and automatic protection against radiation hazard. Facilities for calibration, tilt and alignment corrections are also provided.

STING EO is equipped with dual radar receivers (I- and K-band). The integrated K-band radar ensures tracking of sea-skimming missiles while the addition of electro-optic sensors offers the possibility of passive detection under radar silence conditions. The electro-optic sensors consist of a combination of TV camera, IR camera and a laser range-finder along with a TV/IR auto-tracking unit. STING EO is also available in a basic version without IR camera and/or laser range-finder.

STING EO has been designed to operate under extreme environmental and severe ECM conditions. It applies burst to burst frequency agility, PRF stagger along with the dual band and electro-optic capabilities. STING's patented foundation support protects the director against ship-induced shocks and vibrations.

Operational status

In production. STING is in service on Qatari 'Vita' class patrol craft and the Sultan of Oman's Navy 'Qahir' class corvettes and is being installed on Qatari 'Vita' class patrol craft and Turkish Navy 'Kilig' class FPBs.

Specifications

Director
Train movement: unlimited
Elevation: –30 to +120°
Max speed
Training: 3.0 rad/s
Elevation: 2.5 rad/s
Dimensions (diameter × height): 2,200 × 2,412 mm
Weight (with TV, IR camera and laser range-finder): 850 kg

TV camera
Field of view: 2-20° (diagonal)
IR camera
Field of view: 1 × 2°, 75 × 15°
Laser
Beam divergence: 1.5 mrad
Max PRF: >10 hZ
Power consumption:

Voltage	Frequency	Phase	Power
115 V	60 Hz	3 ph	4.7 kVA
115 V	60 Hz	1 ph	0.3 kVA
440 V	60 Hz	3 ph	2 kVA

Contractor

Thales Naval Nederland.

UPDATED

US Navy/Raytheon SSDS

Type
Fire-control system (naval).

Development
The US Navy's Ship Self-Defence System (SSDS) programme arose from its decision to improve the self-defence posture of its surface fleet following the attack upon the frigate USS *Stark* during Desert Storm. Severe damage was caused by an Exocet anti-ship missile and the US Navy began to consider means of integrating ship defence resources to prevent a repetition of such an incident.

An SSDS Mod 1 demonstrator, based upon the RIM-116A RAM missile, the Mk 15 Phalanx CIWS, the AN/SLQ-32 EW system, and electro-optical sensors, was installed in the USS *Whidbey Island* (LSD 41) by June 1993. Success of these trials led to acceleration of SSDS with the first installation on board USS *Ashland* in 1996.

Description
The demonstrator SSDS Mod 1 installed in the USS *Whidbey Island* used the Raytheon AN/SPS-49 long-range 2-D air search radar, a prototype Spar AN/SAR-8 infra-red search and track device and AN/SLQ-32 EW system as the sensors. The SLQ-32 also controlled a Loral Hycor Mk 36 SRBOC chaff/IR launcher as the 'soft kill' element of the system while the 'hard kill' elements were a RIM-116 RAM point-defence surface-to-air missile (qv) and two Mk 15 Phalanx CIWS (qv).

SSDS Mod 2 will retain most of the hardware of Mod 1 but will also include the Seasparrow surface-to-air missile system (qv).

SSDS Mod 3 is still to be defined but is likely to involve Evolved Seasparrow, a new multifunction radar and EW system, as well as being compatible with the co-operative engagement concept. The last is a system which would permit ship and airborne elements of a task group to track anti-ship missiles and to engage them co-operatively.

The Radiant Mist programme is for a shipborne acquisition, tracking and identification system with multiple electro-optic sensors as a complement to radar. An operational derivative of the demonstrator may form part of SSDS. A technology demonstrator provided by Rockwell Tactical Systems Division was trialled during 1993 using a medium-wave IR camera based on second-generation technology with staring focal planar-array for initial acquisition of targets and then handed over to a laser radar (ladar) for range-finding and identification. The system could acquire targets at 5 n miles (10 km).

Operational status
Production of SSDS is underway and the total US Navy requirement is likely to be about 100 systems. The first SSDS Mod 1 installation was on USS *Ashland* (LSD-48) in FY96, and all eight of the 'Whidbey Island' class and four 'Harpers Ferry' class LSDs now have the SSDS Mk 1 system installed. The Mk 2 Mod0 system has been installed on the aircraft carrier USS *Nimitz* (CVN 68). The Mk 2 Mod 1 system has been delivered to the aircraft carrier USS *Ronald Reagan* (CVN 76), and it is intended that all LPD and LHD classes should receive this version.

Contractors
US Navy.

Raytheon Integrated Defense Systems.

UPDATED

SHIP WEAPON CONTROL SYSTEMS

INFRA-RED SEARCH AND TRACK

Elop Advanced IRST systems

Type

Infra-Red (IR) search and track.

Development

Since 1999, Elop has been expanding its activities in IR Search and Track systems for naval, airborne and ground applications as the result of a number of contracts for development and supply of advanced IRST systems.

Description

The IRST systems are based on utilisation of Elop's extensive experience in state-of-the-art IR sensors and stabilisation systems together with intelligent image processing algorithms to provide IRST systems with high probability of detection and very low False Alarm Rates (FAR). Depending on the application, the IRST systems utilise 3-5 and/or 8-12 μm wavelengths with both linear and focal plane array detectors. A common image processing module is being used for the systems in order to optimise the utilisation of the algorithmic knowledge base.

Operational status

A number of systems are under development for various applications and will enter operational service over the next few years.

Contractor

Elop Electro-Optics Industries Ltd.

VERIFIED

Lockheed Martin Shipboard Surveillance Infra-Red Search and Track system (SIRST)

Type

Infra-Red (IR) search and track (naval).

Development

Lockheed Martin Electronics & Missiles was awarded a US$14.9 million contract in August 1996 for the engineering and manufacturing development of a Shipboard IR Search and Track (SIRST) system for the US Navy, increasing to US$45 million on the award of the second phase EMD contract. The system is intended to improve detection of low-flying aircraft and sea- skimming missiles.

Description

The SIRST mast-mounted sensor head will employ a Raytheon 3 to 5 μm infra-red scanner. The system may eventually be extended to dual waveband and use FPA detectors. The IRST head will rotate at 60 rpm to provide continuous 360° surveillance and track refresh. Field of view is expected to be about 2 to 2.5°. Below-deck equipment will comprise an NDI signal processor, system control and display, which will be integrated with the vessel's combat data system.

Operational status

Under development.

Specifications

Spectral band: 3-5 μm
Scanning speed: ≥360°/s

Contractor

Lockheed Martin Missiles and Fire Control.

UPDATED

SAGEM VAMPIR (DIBV-1A)/VAMPIR MB (DIBV-2A)

Type

Infra-red (IR) search and track (naval).

Development

In the mid-1970s the Direction Technique Des Constructions Navales (DTCN), now DGA/SPN, began examining alternatives to radar for detecting anti-ship missiles. A number of electro-optical programmes were pursued with the objective of providing reliable, panoramic, low-level detection, even in periods of radar silence.

This led to VAMPIR (Veille Air-Mer Panoramique Infra Rouge) which achieved production status and entered service as DIBV-1A when the destroyer *Cassard* was commissioned in July 1988.

The early VAMPIR systems were purely for surveillance but from 1982 DTCN began developing an IR Search and Track (IRST) system for weapon control. Now a second-generation of IRST VAMPIR MB has been developed and is in service as DIBV-2A.

VAMPIR MB forms part of OP3A, the French Navy Ship Self-Defence System (SSDS) which has been installed on the 'Georges Leygues' class destroyers

VAMPIR MB (centre) with a pair of VIGY 105 EOD on the mast of the ASW frigate Jean de Vienne
0019548

For details of the latest updates to *Jane's Electro-Optic Systems* online and to discover the additional information available exclusively to online subscribers please visit jeos.janes.com

Jean de Vienne, Montcalm and *Dupleix.* It is also installed on the aircraft carrier *Charles de Gaulle.*

Description

VAMPIR MB is an IRST which provides panoramic surveillance. It is used to detect, identify and designate air and surface targets while also providing a secondary navigation capability. It consists of a rotating sensor head 550 mm in diameter and 1.50 m high, upon a fixed-sensor module (which includes a signal processing unit, servo-control and power supply units) as well as a display system. It can track up to 50 targets simultaneously and is reportedly capable of detecting missiles at ranges up to 27 km (14.5 n miles) and fighter-type aircraft at 25 km (16 n miles).

The sensor head has a two-axis gyrostabilised plane mirror which reflects radiation on to dual spectrum infra-red sensor modules. This consists of a second-generation, IR Charge Coupled Device (IRCCD) focal planar-array (288 × 4) which operates in the 3 to 5 and 8 to 12 μm bands. The sensor provides a 360° bearing coverage in azimuth and an instantaneous 5° field of view in elevation. Elevation coverage of the system is –20 to +45°.

Data is presented in strip format upon the display screen with three strips per spectral band representing the scene and a zoom capability used for close-ups of potential threats. Alphanumeric data is used for target data as well as presenting data on the ship's bearing.

VAMPIR MB is usually operated in the remote mode with surveillance and target designation data automatically transferred to weapon systems through the ship combat system. A fallback stand-alone mode is available for direct transmission of data to the weapon systems.

Operational status

Production of VAMPIR is complete. Some seven VAMPIR MB have been ordered by France. VAMPIR MB is now operational on board 'Georges Leygues' class ASW frigates. It is also fitted on the French Navy *Charles de Gaulle* aircraft carrier and replaces the first generation on board the 'Cassard' AAW frigates. The VAMPIR MB has also been selected to equip the Franco-Italian 'Horizon' class frigates.

Specifications

VAMPIR MB
Detector type: second-generation IRCCD focal planar-array (288 × 4)
Spectral bandwidth: 3-5 and 8-12 μm
Instantaneous field of view: 5° (elevation)
Bearing coverage: 360°
Rotational speed: 1.4 rps
Elevation coverage: –20 to +45°
Delivered track accuracy: max 0.5 mrad
False alarm rate: <1/h (typical)
Number of simultaneous tracked targets: 50
Cooling: closed-cycle Stirling
Weight of the turret:180 kg
Power:115 V, 60 and 400 Hz
Typical detection ranges
Subsonic missiles: 9-16 km
Supersonic missiles: 14-27 km
Fighter aircraft: 10-25 km
Helicopter: 7-10 km

Contractor

SAGEM SA, Aerospace & Defence Division.

Thales IRSCAN

Type

Infra-red search and track (naval).

Development

IRSCAN is a fast reaction infra-red search and track system. It is designed to detect both air and surface targets. The company has developed a version which will be retrofitted into the Royal Netherlands Navy's (RNLN) Goalkeeper systems (qv).

IRSCAN has been extensively tested in a variety of different climates including South Africa, Caribbean, the Middle East, the Arctic and Southeast Asia.

Description

IRSCAN is a modular, lightweight, omnidirectional passive IRST system which may be used autonomously, with fire-control systems such as LIOD, MIRADOR and LIROD, or with weapon systems such as Goalkeeper. In the RNLN it will be used to detect and track targets in support of the Goalkeeper system.

It consists of a stabilised sensor head, an air drier, an electronics cabinet and a control console, all linked by fibre optic cable. The system has a typical detection range of 12 km against missiles and 15 km against aircraft. A typical reaction time is 2.5 seconds.

The sensor head has a 1,024 element staggered line array with CMT elements covering the 8 to 12 μm range. It has a scanning speed of 78 rpm and an elevation coverage of 14.6°.

The system uses a variety of special techniques for filtering and signal correlation. This is designed to ensure highly reliable and accurate target detection, together with tracking performance coupled with an extremely low false alarm rate of less than one per hour.

Thales' IRSCAN infra-red search and track system 0002014

Upon detection, targets are tracked. When certain criteria are satisfied, a fully automatic alert is generated for the weapon system or the ship's command system. In addition to the automatic function, a three axis stabilised, high-quality picture is presented.

As an option IRSCAN can be provided with a man/machine interface for manual target selection of air and surface targets.

Operational status

Two IRSCAN systems were ordered for two Royal Netherlands Navy auxiliary oilers and delivered by February 1995. IRSCAN is also fitted to one RNLN 'Karel Doorman' class frigate for trials and four new 'Vita' class fast attack craft of the Qatari Navy. One system is leased to the Swedish Navy.

Specifications

Detection capability
For incoming aircraft and supersonic missiles: typically 20 km
For subsonic missiles: typically 12 km
Target designation accuracy: < 1 mrad
Internal track capacity: > 500 tracks real time
Alert generation: automatic alert generation on 32 tracks, classified threatening
False alert rate: < 1 false alert/h
Elevation coverage: 14.6°
Scanning speed: 78 rpm
Spectral band: 8-12 μm
Detector: 1,024 elements staggered line array (CMT plus CMOS readout technology)
Detector cooling: closed-cycle Stirling

Dimensions and weights
Stabilised sensor head
Width: 480 mm
Height: 1,035 mm
Depth: 366 mm
Weight: 100 kg
Power requirements: 115 V; 60 Hz; 3 phase 115 V; 60 Hz; 1 phase anti-condensation heating
Max power consumption: 4.55 kW

Contractor

Thales Naval Nederland

UPDATED

Thales SIRIUS

Type

Infra-red search and track (naval).

Development

The Canadian/Dutch development programme of the long-range Infra-red Search and Track system (IRST) SIRIUS made a restart in April 2000. DRS Technologies became Thales Naval Nederland's new Canadian subcontractor responsible for the Processing Cabinet. For the processing hardware high-performance COTS (Commercial-Off-The-Shelf), boards were selected to provide future growth potential.

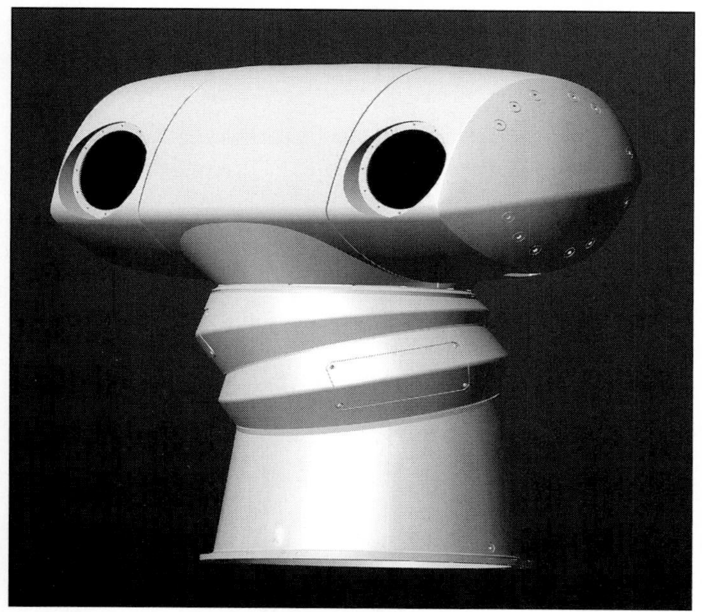

The SIRIUS dual-waveband infra-red search and track system showing the stealthy sensor cover developed (Thales Naval Nederland) 0002012

System integration takes place at Thales Naval Nederland, followed by cold and warm water trials to prove the system's performance. After the trials the first system will be delivered to the RNLN.

At the end of 2002 series contracts for the Dutch LCF frigates and the Canadian CPF frigates are expected.

Description

SIRIUS is a hi-spectral (3 to 5 and 8 to 10 um) Long Range IR Surveillance and Tracking system (LR-IRST) intended to reinforce the horizon search capabilities of surface ships against sea-skimming anti-ship missiles. SIRIUS is designed to deliver optimal performance under any circumstances, both in area defence and self-defence scenarios. It can be integrated with any combat system, in close co-operation with any sensor system, ranging from simple track radars or autonomous close-in weapon systems up to active phased array radars.

The operational output of SIRIUS consists of prioritised tracks, alerts and video. Upon detection, targets tracks are automatically established and reported to the combat system.

SIRIUS video provides detailed hi-spectral IR images during day and night, revealing objects hidden for both radar and the human eye. With this video SIRIUS significantly supports operations in liuoral areas.

The sensor head is fully stabilised over 3 axes by means of the Thales wedge platform and rotates at 60 rpm.

It provides dual-band coverage, adaptive signal processing, bird clutter suppression and modem tracking and classification algorithms.

The use of large lens apertures (150 mm) and Time Delay Integration (TDI) on the detectors achieve extremely high sensitivity. The MWIR detector has 300 elements vertically and 10 horizontally (TDI). The LWIR detector has 300 elements vertically and 8 horizontally (TDI). The detectors use CMOS readout technology. The detectors are cooled to 80 K by means of the Thales Delft Optronics closed-cycle Stirling coolers. The detector output is in 16 bits analogues to digital converted and sent over a fibre optic link to the Processing Cabinet.

The detection range for supersonic sea-skimming missiles is in most cases horizon limited.

Operational status

Under development. Sea trials of a production-standard unit mounted on a Canadian 'Halifax' class frigate began in January 2003, with production deliveries expected in 2004. Planned for 'De Zeven Provincien' class destroyers of the RNLN.

Specifications

Detection ranges (typical)
Supersonic sea-skimming missiles: 35 km
Subsonic sea-skimming missiles: 21 km
Fighter aircraft: 30 km
Elevation coverage: 3.8°
Scanning speed: 60 rpm
Spectral bands: 3-5 μm, 8-12 μm
Detectors: 300 ×10, 300 × 8 elements
Cooling: closed-cycle Stirling
Weight stabilised sensor head: 280 kg
Weight below-deck equipment: 930 kg
Power consumption: 8 kVA

Contractor

Thales Naval Nederland.

Subcontractor

DRS Technologies Canada.

UPDATED

Thales Optronics ARISE

Type

Infra-red search and track (naval).

Development

In 1987, Thorn EMI Electronics, Electro-Optics Division (now Thales Optronics) was asked by the Admiralty Research Establishment (ARE) (now the Defence Science and Technology Laboratory (DSTL)) to develop an IR surveillance sensor to be known as ARISE. The experimental system was delivered in October 1990. After initial trials with an 8 to 12 μm detector, the system was fitted with a second-generation dual-waveband (3 to 5 and 8 to 12 μm). IRCCD detector was delivered to DSTL for further trials. The IRCCD detector was developed by BAE Systems Infra-Red Ltd. DSTL is said to be evaluating the algorithms for the signal processing. ARISE is now being upgraded with a new detector to form part of a multisensor fusion testbed for operation with a multifunction radar.

Description

ARISE is an experimental Infra-Red Search and Track system (IRST) that uses an IR scanner to detect sea-skimming missiles. It is based on the 8 to 12 μm IR technology developed for Thales Optronics' land-based air defence system, ADAD with second-generation IRST technology improvements. It has a second 3 to 5 μm channel which can operate simultaneously to determine the optimum configuration.

Operational status

Experimental programme.

Specifications

Spectral bandwidth: 3-5, 8-12 μm
Detector type: long linear

Contractor

Thales Naval Nederland.

UPDATED

The ARISE sensor head

SHIP WEAPON CONTROL SYSTEMS

SURVEILLANCE

BAE Systems Australia LRTS multirole thermal imaging sensor

Type

Surveillance system (naval).

Description

The BAE Systems Australia LRTS thermal imaging sensor operates in the 3 to 5 μm spectral band for long-range, high-resolution performance in warm, high- humidity environments. Its small, lightweight configuration is suitable for land, sea and airborne operations and is applicable to the full range of military and commercial platforms for new or retrofit installations.

The system incorporates 3 to 5 μm starting array detector technology which simultaneously images the entire infra-red scene with over 311,000 individual detectors, without an opto-mechanical scanning system. The system offers a selection between two or three fields of view providing navigation, surveillance and targeting/identification capabilities.

Automated functions and flexible digital interfaces permit integration with a range of tactical platforms and systems including fixed- and rotary-wing aircraft, military vehicles and maritime forces.

Operational status

In production. In service with the Royal Australian Navy including the FFG Frigate E-O tracking system and under consideration by several military forces around the world.

Specifications

Fields of view
Narrow: 2.9 × 2.2°
Intermediate: 11.3 × 8.6°
Wide: 37.7 × 28.5° (optional)
Spectral band: 3-5 μm
Detector: platinum silicide (PtSi), 486 × 640 pixels
Video format (frame rate): RS-170 (30 Hz) or CCIR (25 Hz) available
Imager module weight: 17 kg

Contractor

BAE Systems Australia

VERIFIED

Ball Aerospace All-Light-Level Marine Television Camera (ALLMTV)

Type

Surveillance system (naval).

Description

The All-Light-Level Marine Television Camera (ALLMTV) provides imaging at sea in a wide range of light conditions. Drawing on more than 30 years of naval camera experience, Ball designed this camera for performance and high reliability in shipboard environments.

The heart of the new camera is a charge-coupled device (CCD) sensor developed specifically for military applications. This solid-state device provides higher resolution, better haze and fog penetration and a × 75 improvement in reliability over the tube-based sensors, commonly used in older cameras.

The CCD is coupled with a GEN III intensifier, which multiplies ambient light up to 30,000 times. A combination of intensifier gating and lens iris light control enables operation from bright sunlight to starlight. A proven, solid-state sensor and freedom from mechanical coolers and scanners, enable round-the-clock imaging for search, surveillance, target identification, fire control, and navigation.

A sealed housing and a sun shield protect the ALLMTV camera and its zoom lens. The housing is pressurised with either nitrogen or dry air to keep out the salt atmosphere.

The entire inner camera and lens assembly can be removed and replaced using common hand tools. Once removed from its housing, the camera's open design allows ship technicians easy access to all the modules. Every module is shipboard replaceable, reducing downtime and ownership costs.

Ball also offers a variety of other full MIL-Spec and ruggedised cameras for military, paramilitary and specialised civil mission requirements.

Operational status

In service with the US Navy as the Mk20 ALLTV.

Specifications

Weight (approx): 21.4 kg (47 lb)
Dimensions (approx): 203 mm (8 in), diam × 483 mm (19 in), length
Sensor type: 18 mm GEN III image-intensified, high-resolution CCD, 1032 pixels per line
Video formats: 525 line, RS-170A, 30 Hz
625 line, CCIR, 25 Hz
875 line, RS-343A, 30 Hz
Horizontal resolution: ≥500 TVL/PH
Signal to Noise
525-line rate: 40 db
625-line rate: 40 dB
875-line rate: 35 dB
Spectral band: 550 to 950 nm
CCD sensor format: 12.4 (H) × 9.3 (V) mm
Lens type: 10:1 zoom minimum
Field of view range: 25.0° wide to 2.5° narrow (diagonal)
F number: 3.8
Power consumption: 28 V DC, ~ 20 W average
excluding optional: 30 W peak
Heaters
Operating temperature: –20 to +60°C
With optional heaters: –55 to +60°
Storage temperature: –46 to 75°C
Vibration: per MIL-STD 167
Shock: 60 g
Humidity: 100%

Contractor

Ball Aerospace & Technologies Corp.

VERIFIED

DRS NMMS

Type

Surveillance system (naval).

Development

On 14 April 1988, the frigate USS *Samuel B Roberts* (FFG 58) was mined while operating in the Persian Gulf and in the aftermath of that incident the NMMS was developed. The Chief of Naval Operations ordered the US Navy to investigate means by which non-MCMVs could avoid floating mines and, among various optical and electro-optical solutions proposed by manufacturers, was the Boeing (formerly McDonnell Douglas) Mast-Mounted Sight (MMS) target location and marking system. The Boeing electro-optical systems product operation has since been acquired by DRS. The MMS had been under development since October 1984 to meet a US Army helicopter target location and designation system requirement.

The NMMS (Navy Mast-Mounted Sight) appeared late in 1988 and was a navalised MMS. In March 1989, the prototype system was installed at the

The DRS NMMS

manufacturer's expense in the USS *Cromelin* (FFG 37) and was operated while the ship was in the Gulf. Another two systems were simultaneously used in drug interdiction operations in 'Oliver Hazard Perry' (FFG 7) class frigates.

These trials were a success and in December 1989 a contract for three systems was awarded and was later followed by more. By the spring of 1994, the NMMS had been used in US Navy cruisers, destroyers and frigates.

Description

NMMS is a fully integrated stabilised electro-optical surveillance system and can be used for target detection and recognition, floating mine detection, assisting with shipboard helicopter landings and for search and rescue missions. NMMS consists of three subsystems: the turret, the control/display panel and the external electronics.

The turret consists of a post and a ball-shaped shroud, both made of graphite-epoxy composite. Patented Boeing technology – known as the 'soft mount' – isolates the sensors and boresight from ship vibration and pitch, roll and yaw motion. The resulting stabilisation increases target detection and recognition ranges in day or night operation.

The sensor package comprises a TV and a thermal imaging sensor. NMMS also has a video tracker which provides manual or automatic tracking of targets. There is also a digital scan converter which enhances the TIS image and provides ×2 and ×4 electronic zoom capability.

External to the turret is the control/display system consisting of the master controller which is linked to the keyboard, display screen and handgrip. Additionally there is the system power supply and processor.

Operational status

NMMS is a derivative of MMS which is currently in production with over 400 systems delivered. The US Navy has purchased 15 NMMS systems and rotates them on an 'as required' basis. To date, NMMS has acquired over 150,000 operational hours on all classes of surface warfare ships. In addition, the Royal Australian Navy operates two NMMS systems in HMAS *Adelaide* and *Sydney.*

Specifications

System weight: 113.4 kg
Turret
Weight: 72.57 kg
Diameter: 64.77 cm
Stabilisation: 2 axis, <20 μrad jitter
Azimuth: ±190°
Elevation: ±30°
TV camera
Type: silicon vidicon
Frequency range 0.65-0.9 μm
Output: 875 line video
Field of view: 2° narrow, 8° wide
Thermal imaging sensor
Type: 120 element MCT scanning
Spectral band: 8-12 μm
Field of view: 3° narrow, 10° wide

Contractor

DRS Technologies, Electro-Optical Systems Group.

UPDATED

DRS Thermal Imaging Surveillance System (TISS)

Type

Surveillance system (naval).

Development

TISS, a derivative of the Mast-Mounted Sight, is a multisensor, electro-optical system with daylight and infra-red sighting and eye-safe laser ranging capabilities. TISS is designed for naval surface and air surveillance including such targets as mines, swimmers, periscopes, boats, ships and aircraft. In October 1995, Boeing (formerly McDonnell Douglas and now DRS) was selected to meet the US Navy's requirement for a thermal imaging sensor system. Under the terms of the initial contract, a single EX 8 Mod 0 TISS prototype was delivered for evaluation, integration and test in mid-1996. Engineering development models have already been produced with a variety of sensor payloads. Boeing began delivery of the first of 24 TISS units to the US Navy in February 1998. Thirteen TISS units were scheduled for delivery to Korea from June to October 1998. TISS is approved for sale to all NATO countries plus Australia, Egypt, Israel, Japan, South Korea and New Zealand, with export approval for other countries pending.

Description

TISS is a fully integrated stabilised electro-optical surveillance system and can be used for target detection and recognition, floating mine detection, assisting with shipboard helicopter landings and for search-and-rescue missions. TISS consists of three subsystems: the turret, the control/display panel and the support electronics. Patented DRS stabilisation technology isolates the sensors from platform vibration and motion, providing maximum day and night operational range for the sensors.

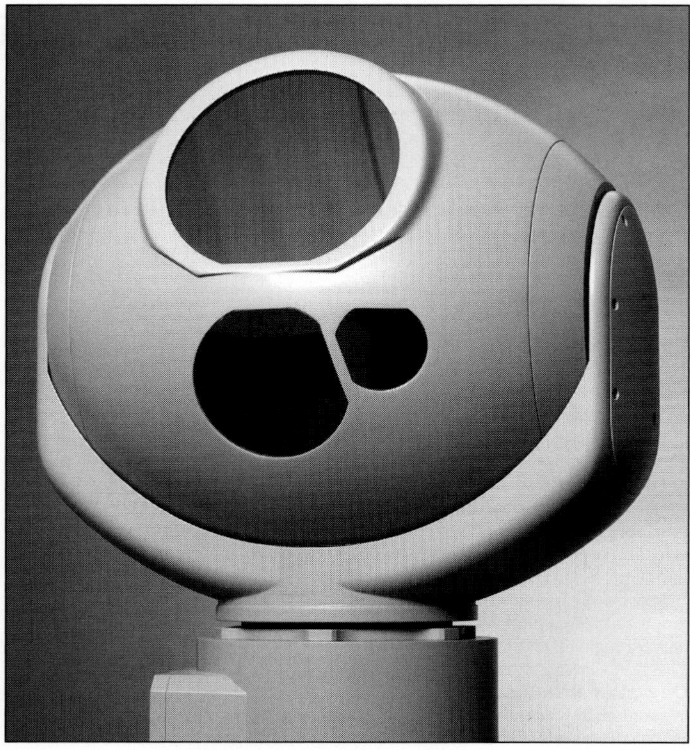

The DRS Thermal Imaging Surveillance System (TISS) 0002007

The sensor package consists of a Videospection dual field of view CCD TV sensor, an Amber 3 to 5 μm indium antimonide (InSb) thermal imaging sensor, a Raytheon Class 1 eye-safe laser range-finder and boresight. The associated electronics allow precise manual or automatic target tracking.

External to the turret is the control and display panel subsystem with switches, controls, keyboard, handgrip, indicators and display. The system support electronics consist of a pedestal electronics unit, a system electronics unit, an automatic video tracker and a power supply/power conditioner.

TISS integration into a US Navy fire-control and combat system is in progress.

Operational status

In production, with more than 24 systems delivered to the US Navy. Thirteen units were delivered to Korea from June to October 1998.

Specifications

System weight: 363 kg max

Turret
Weight: 70 kg max
Diameter: 44.45 cm
Width: 52.07 cm
Stabilisation: 2 axis, <15 μrad jitter
Azimuth: ±270°
Elevation: +75, –35°

TV sensor
Type: low-light silicon CCD
Spectral range: 0.65-0.95 μm
Output: 875 line video
Fields of view (h × v)
Narrow: 1.6 × 1.2°
Wide: 6.0 × 4.5°

Thermal imaging sensor
Type: 480 × 512 staring InSb
Spectral band: 3-5 μm
Target temperature range: –60 to +100°C
Field of view
Narrow: 1 × 1°
Wide: 2.9 × 2.9°

Laser range-finder
Type: Class 1A eye-safe Nd:YAG
Wavelength: 1.54 μm
Range: 20,000 m
Durability: 10^6 shot life

Contractor

DRS Technologies, Electro-Optical Systems Group.

UPDATED

DRS TISS II next-generation thermal imaging sensor system

Type

Surveillance system (naval and air).

Description

The Next Generation TISS is a stabilised, multi-sensor Surveillance and Targeting System (S&TS). It features a Boeing-patented stabilisation technology that results in very high stability, even when exposed to dynamic environments. This enables the use of high-performance sensors with long-range capabilities onboard a variety of platforms. These sensors are boresighted with an internal tool that allows in-mission precision alignment. A high-performance tracker rounds out the system. The result is world-class detection, identification, tracking and engagement in daylight, at night, and in adverse weather conditions. This is all contained in a package that is small and light enough to mount on the mast of a ship or the arch of a patrol boat.

The system traces its heritage through the AN/SAY-1 TISS to the Mast Mounted Sight, which is the above rotor S&T system onboard the US Army OH-58D Kiowa Warrior. A US Navy shipboard version of this system, the Naval Mast Mounted Sight (qv) has been in use in the Fifth Fleet for more than 10 years and was heavily utilised during Operation Desert Shield/Storm. More than 25 AN/SAY-1 TISSs have been delivered to the US Navy. Other production programs in the TISS family include US Navy VISUAL, US Army Mast Mounted Sight Upgrade and Republic of Korea (ROK) Light Helicopter System.

Sensors

Since TISS was integrated into the US Surface Fleet in 1995, industry has made significant advancements in Infrared (IR) sensor technology. The new Forward Looking Infrared (FLIR) sensors are more reliable and offer better resolution and sensitivity. DRS manufactures a sensor called the Boeing Common FLIR, named so for two reasons. First, the sensor is offered across several product lines, helping to reduce cost and risk. Second, the Integrated Dewar Assembly (IDA) is 'drop in' replaceable. The same FLIR housing and optics support 320 × 240, 640 × 480 or 1,024 × 1,024 pixel format staring Focal Plane Arrays (FPAs). This enables cost-effective, requirements-based sensor configuration and upgrade. Laser technology has also advanced. New design diode pumped lasers enable eye-safe ranging at high repetition rates with significantly higher reliability than flashlamp designs. The new generation of designation capable lasers offers higher performance and tighter packaging. To take advantage of improved sensors, higher performance optics are also available. A new FLIR sensor is baselined with a greater than 700 mm Equivalent Focal Length (EFL) lens and the narrow TV optic is over 625 mm EFL. Options to this new baseline extend the EFL to over 1,400 mm and the aperture to over 200 mm. These sensor subsystems have already been integrated into the TISS family of products and are available in the next-generation TISS.

Electronics

Computing and packaging advances have moved through the electronics industry since the original TISS was designed. New production systems in the TISS family have incorporated these changes and they are now built into the next-generation TISS. The result is a more compact support electronics configuration that offers higher performance, higher reliability and lower cost.

Better Integration

The next-generation TISS is capable of combat system and/or gun fire-control system integration. It is compliant with the US Navy's EO Interface to gun fire-control system specification (WS-32791 C). It has completed integration to the Mk 86 GFCS and is undergoing integration to the Mk 92 GFCS. Next-generation TISS also offers a Boeing-developed gun control system. This system is featured onboard the Phantom, a Boeing S&TS demonstration, test and evaluation platform. This system is capable of directing minor calibre gun systems or launcher subsystems.

The next-generation TISS offers the improvements in packaging and system integration efforts that were undertaken for various programmes in the TISS family. This means that the next-generation TISS offers flexibility in its sensor suite and tracker subsystem. The FLIR sensor can be midwave or longwave, staring or scanning. A variety of FPAS are available for the 'common' FLIR, enabling export compliant solutions that can he upgraded as ITAR restrictions evolve. Different optics configurations are available for the FLIR and TV sensors, fixed or zoom. The laser sensor is available in two eye-safe range-finding configurations (flashlamp or diode pumped) and one designating configuration. The tracker configuration is offered in two standard high-performance formats and can be further customer defined. Other customer-defined sensor and system requirements can also be accommodated.

Operational status

Available.

Specifications

Turret
Weight: 130 lb (59.1 kg)
Diameter: 18 in (45.7 cm)
Line Of Sight (LOS) stability: <15 μrad
Field of regard (azimuth): ± 270°
Field of regard (elevation): +85, –30°
Slew rate (azimuth and elevation): 60°/s
Slew acceleration (azimuth and elevation): 60°/s²
Stabilisation design: 2 axis/4 gyro

FLIR

Parameter	Midwave option 1	Midwave option 2	Longwave option 3
Wavelength	3.8-4.8 μm	3.8-4.8 μm	65 - 10.25 μm
FPA Type	Staring	Staring	Scanning
FPA detector material	HgCdTe	HgCdTe	HgCdTe
FPA size	640 × 480	320 × 240	240 × 4
Detector	27 μm	30 μm	25 × 28 μm
Focal length	450 mm/ 700 mm/1.4	450 mm/ 700 mm/1.4 m	350 mm
NETD	<0.025°	<0.025°	<0.1°
FOV's (narrow, wide)	2.2 × 1.65°/ 6.6 × 4.95°	1.2 × 0.92°/ 3.67 × 2.75°	18 × 24°/3.9 × 4.2°/ 1 × 1.5°
Operating temp	95 ±5°	95 ±5°	95 ±5°
Cool down time	< 7 min	< 7 min	< 10 min
Reliability	>4,000 hr MTBF	>4,000 hr MTBF	>2,500 hr MTBF

Laser

Parameter	ESLRF option 1	ESLRF option 2	LR/D option 3
Wavelength	1.54 μm	1.57 μm	1 .064 μm
Pulse rate	1 Hz	>10 Hz	8-20 Hz
Power	7.8 mJ/pulse	FOUO CLASSIFIED	FOUO CLASSIFIED
Beam divergence	<800 μrad	FOUO CLASSIFIED	FOUO CLASSIFIED
Range performance	20,000 m	FOUO CLASSIFIED	FOUO CLASSIFIED
Range accuracy	± 5.0 m	FOUO CLASSIFIED	FOUO CLASSIFIED
Range resolution	± 5.0 m	FOUO CLASSIFIED	FOUO CLASSIFIED
Source	Nd:YAG Flashlamp	Nd:YAG diode pump	Nd:YAG Flashlamp
Detector	InGaAs	InGaAs	Si
Reliability	>1 million shots	>100 million shots	>5 million shots

TVs

Parameter	Narrow TV (NTV)	Wide TV (WTV)
Field Of View (FOV) (diag)	2.2°	8.0°
Focal length	317 mm/650 mm/ 1.4 m	63 mm
Aperture	77 mm	21 mm
F stop	3.0	3.0
Resolution	>800 TVL/PH	>800 TVL/PH
CCD	1,134 × 484 pixel	1,134 × 484 pixel
Spectral band	600 - 950 nm	600 - 950 nm
Low light level	0.5 lux	0.5 lux
Reliability	>7,000 hrs MTBF	>7,000 hrs MTBF

Boresight tool
Mechanical alignment of sensors required: No
Pre-mission boresighting: Yes
During mission boresighting: Yes
Boresight accuracy: <1 pixel
Boresighting cycle time: ~1 min
Compatible with midwave and longwave FLIR: Yes
Compatible with 1.54/1.57/1.064 mm laser: Yes
Compatible with visible and near IR TV: Yes
Reliability: >35,000 hrs MTBF

Tracker

Parameter	Tracker option 1	Tracker option 2
Auto detect	No	Yes
Auto acquire	No	Yes
Auto track	Yes	Yes
Multiple tracks	2	No
Offset tracking	Yes	No
Image processing capability	1 GFLOP	4 GFLOP
Reliability	>7,000 hrs MTBF	>7,000 hrs MTBF

Contractor

DRS Technologies, Electro-Optical Systems Group.

UPDATED

Indigo Systems Motorised Video Surveillance Platform (MVSP)

Type

Long-range tracking and wide-area surveillance.

Development

Indigo Systems has integrated its InSb mid-wave IR (3-5μ) cameras with the Motorized Video Surveillance Platform (MSVP) to provide a tracking or wide-area surveillance system for long range target detection. The system is designed for fixed site or mobile installations requiring 24-hour continuous operation in harsh and marine/coastal environments.

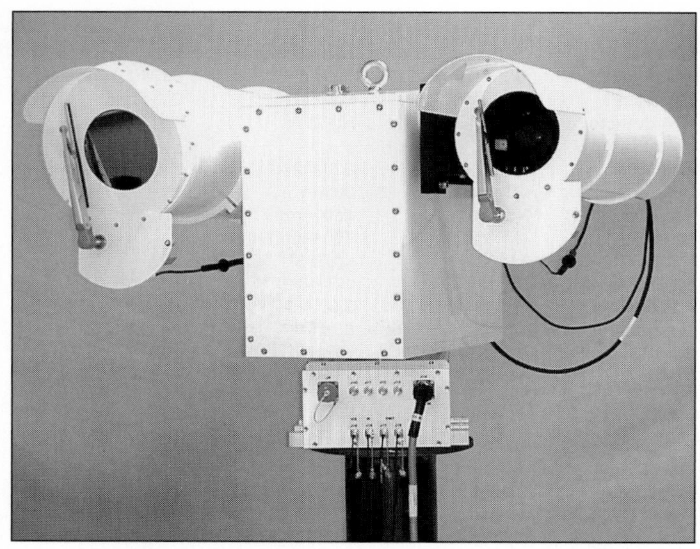

The Indigo Systems Motorised Video Surveillance Platform (MVSP) ***NEW***/0595417

Description

Large-format, 640 × 512 and mid-format 320 × 256 InSb detectors are offered with choice of dual- or triple-FOV lenses for either camera. The dual-FOV offers 75 and 250 mm focal lengths and provides human target detection at greater than 5 km and the triple-FOV lens includes 60, 180 and 500 mm focal lengths. At 500 mm the large-format camera enables human target detection at greater than 14 km.

Up to four E-O sensors can be installed and accurately boresighted on one MVSP. Typically, the sensor suite includes the Indigo Systems TH-10 IR camera system (qv), companion long-range zoom Visible CCD camera with matching and slaved FOV, and an eye-safe laser range-finder with 20 km range. Each E-O sensor is housed in an O-ring sealed and nitrogen-purged enclosure.

The MVSP system has a modular and flexible configuration architecture permitting the user to easily upgrade or change the sensor suite as needed to meet specific mission requirements.

Indigo Systems has also developed the Integrated Command and Control System (ICCS) for MVSP surveillance systems. The ICCS provides all-system integration and full operator control with an easy-to-use Graphical User Interface in a Windows operating environment. Included is a joystick with quick access-enabled function buttons. Besides full control of all EO sensors, the ICCS provides advanced positioning functions. These include auto video tracking with four algorithms including correlation, radar or external sensor interface 'go to' queuing; and geo-referencing to GPS, Lat/Lon (WGS84 Datum) and UTM co-ordinates. For long distance and multiple site networking, each local ICCS control server enables a Web-based centralised command centre over the user's IP network.

Operational status

Deployed and available.

Specifications

Position: 360° continuous azimuth rotation via sliprings
+ 60° elevation
Velocity: <0.1 to 100°/sec in azimuth
<0.1 to 60°/sec in elevation
Acceleration: 100°/sec2 in azimuth
60°/sec2 in elevation
Environment: Outdoor/Marine to MIL-STD-810 (IP-67)
- covering humidity, salt fog, solar, blowing sand/dust and rain
Temperature: –30°C to +60°C
Weight: 129 kg
Dimensions: 94 × 71 × 117 cm
Power: 120/230 V AC, 50/60 Hz, <600 W consumption
Video output: NTSC or CCIR
IR Detector: 3-5μ InSb with integral Stirling cooler
IR Detector size: 30 × 30μ for 320 × 256 format; 25 × 25μ for 640 × 512 format
NEDT: <25 mK

Visual camera

Detector: 768 × 493 interline transfer CCD (CCIR optional)
Optics: 12 to 660 mm with ×2 extender (24 to 1,320 mm)
Enhancements: Contrast stretcher and five-position filter wheel

Laser range-finder

Laser type: Erbium-glass, 1.54μ, 8 millijoules typical
Range: 20 km
Resolution: ±5 m standard (2 m optional)

Contractor

Indigo Systems Corporation (USA).

NEW ENTRY

L-3 Communications WESCAM 14PS-MAR

Type

Surveillance system (naval).

Description

The WESCAM™ 14PS-MAR is a sealed, marine rugged, high-performance, multisensor surveillance system for shipborne mission applications. It is available as a baseline tri-sensor configuration (sensors 1, 2 and 3) or as a quad or penta-sensor version (sensors 4 and 5).

Features

- Marine-rugged, high-performance, multisensor surveillance system for shipborne mission applications
- Excellent long-range multisensor performance for night, day or poor weather conditions
- Patented long-range spotter
- Proven fielded performance with the US Navy
- Optional thermal control unit extends performance range, including full solar loading.

The 14PS-MAR's high-performance payloads include:

- **Sensor 1** - high-sensitivity InSb (Indium Antimonide) 3 to 5 mm FLIR with patented 6 field of view optics
- **Sensor 2** - low-light colour CCD TV camera with ×20 zoom lens
- **Sensor 3** - colour daylight TV camera with 955 mm long-range spotter lens
- **Sensor 4** - eye-safe laser range-finder
- **Senosr 5** - laser illuminator.

The high-sensitivity FLIR, low-light day TV, and laser illuminator provide high-performance capability for night-time operations, and result in more lives saved and less maritime property damaged or lost. This system maximises efficient wide-area patrolling while maintaining the capability to covertly recognise and identify from a long stand-off range.

The 14PS-MAR is designed to provide reliable imaging for pilotage, day or night.

Contractor

L-3 Communications WESCAM Inc.

UPDATED

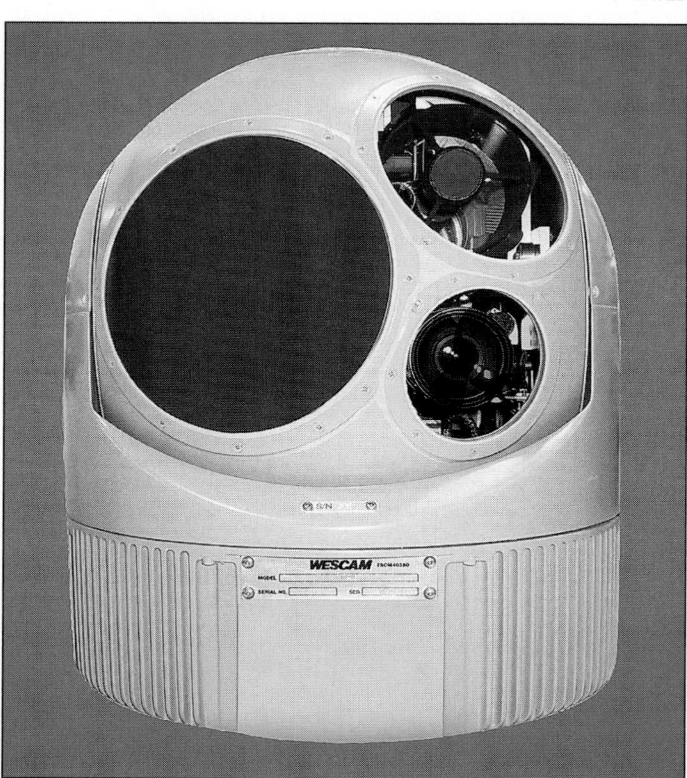

The WESCAM 14PS-MAR shipborne sensor 0143681

Radamec System 1000N

Type

Surveillance system (naval).

Development

The Radamec lightweight System 1000N surveillance system was developed from their Series 2000 systems and was first shown in 1991. It was trialled on board UK Royal Navy 'Type 42' destroyer, HMS *Liverpool* in 1995 and was fitted to HMS *Invincible* for operation in the Mediterranean and Persian Gulf.

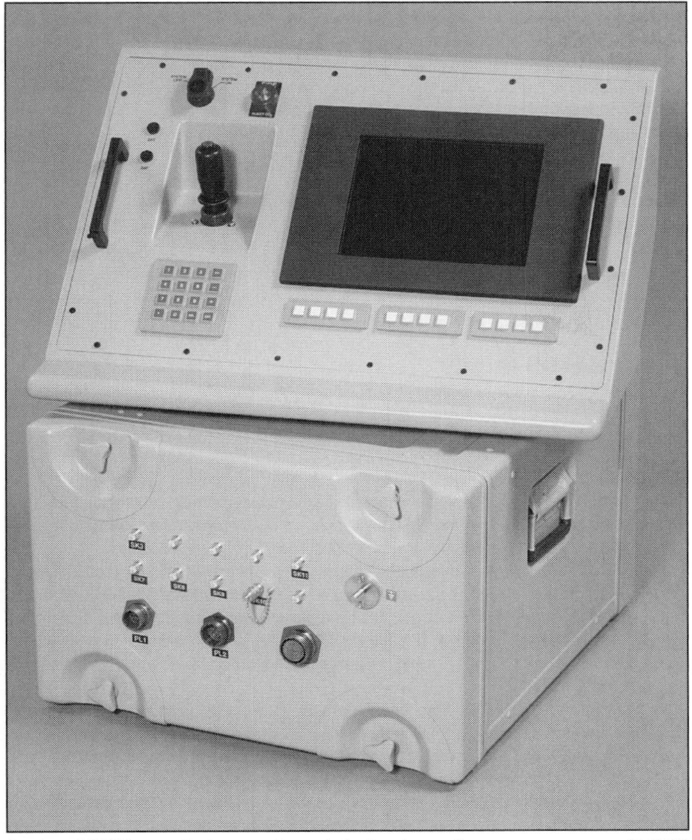

System 1000N operator's console and electronics unit 0019546

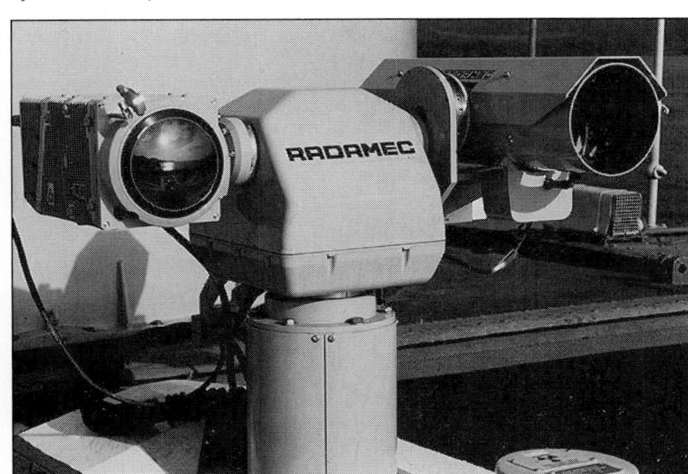

The Radamec System 1000N naval electro-optical surveillance system

Description

System 1000N is a lightweight electro-optical surveillance system designed for maritime and naval applications. The system offers the following features: covert detection and positive identification of targets at ranges in excess of 10 km; 24 hour operation in all but the most adverse weather conditions; stabilised line of sight.

The system consists of two major assemblies:

Electro-Optical Director: this is located above decks on a purpose-built pedestal and comprises a dual-axis gyrostabilised director head with TV camera and thermal imager.

Operator's Control Console: this is located below decks and includes the electronics, power supplies, controls and displays for the system. The controls include a joystick to steer the director and a touch-sensitive display panel to operate the sensors and other system functions. There are two TV monitors to display the output of both sensors. The director can also be programmed to search within sectors defined by left and right bearing limits. The console can be supplied as a single integrated unit or in modular form for installations where space is restricted.

Options include video auto-tracking; eye-safe single shot laser range-finder; compass and GPS interface; video cassette recorder; electronic alignment of sensors; secondary video monitors and remote control from a command system.

Operational status

In production and in service with the UK and a Middle East customer.

Contractor

Radamec Defence Systems Ltd.

VERIFIED

Radamec EOS

Type

Surveillance (maritime).

Development

This new marine EOS (Electro-Optical Surveillance) system is designed to commercial maritime standards for police, coastguard and naval vessels employed in EEZ patrol, search-and-rescue and other similar applications.

Description

The RADAMEC EOS is a commercial standard E-O system comprising a lightweight electro optical director (<65 kg) and compact below decks electronics unit/console. The system utilises COTS modules packaged specifically for the maritime environment and is designed for ease of maintenance without recourse to third line workshop facilities.

The system is capable of carrying out operator-defined surveillance search patterns and will accept target indications from an external source.

Additional options include: -

- Video autotracker;
- Low power laser rangefinder;
- Digital output to remote controlled small calibre gun turret with manually injected offset correction.

In common with all Radamec systems, EOS can be controlled from a multifunction console (MFC) or from Radamec's own dedicated console. Radamec can supply its own Windows-based MMI software for insertion into an MFC.

Operational status

In production.

Specifications

Director
Azimuth coverage: continuous 360°
Azimuth slew speed: >1 rad/sec
Elevation movement: –30° to +89°
Elevation slew speed: >1 rad/sec
Sightline stabilisation: 100 μrad rms

Thermal Imager
Type: InSb FPA
Detector elements: 320 × 240
Spectral band: 3-5 μm (8-12 μm available)
Fields of view (2 × switched FOV)
Narrow: 1.65° × 2.2°
Wide: 8.2° × 11.0°
NETD: 0.2 K

TV Camera
Type: CCD colour
Fields of view (continuous zoom)
Narrow: 2.0° × 2.7°
Wide: 36.0° × 48.0°
Digital Zoom: ×2 and ×4
Scene Illumination: 3 lux

Laser Range-finder (option)
Type: Erbium glass
Operating wavelength: 1.54 μm
Pulse repetition frequency: 0.167 Hz, intermittent
Maximum range: >5,000 m
Minimum range: 50 m
Transmitter energy: 8 mJ

Contractor

Radamec Defence Systems Ltd.

NEW ENTRY

SAGEM VIGY 10

Type

Surveillance system (naval).

Development

The VIGY 10 is a second-generation day/night naval surveillance system. It allows continuous operation of the ship in all conditions of weather and light, to cover a wide variety of missions: day/night surveillance, navigation and piloting and assistance in search and rescue operations.

Description

Compact and modular, the VIGY 10 can fit a large range of sea-going and inland water craft including fast patrol boats, coastguard vessels and even smaller boats.

The VIGY 10 comprises two units:

- the stabilised sensor head, containing one or two sensors in their own water and salt spray proof housing

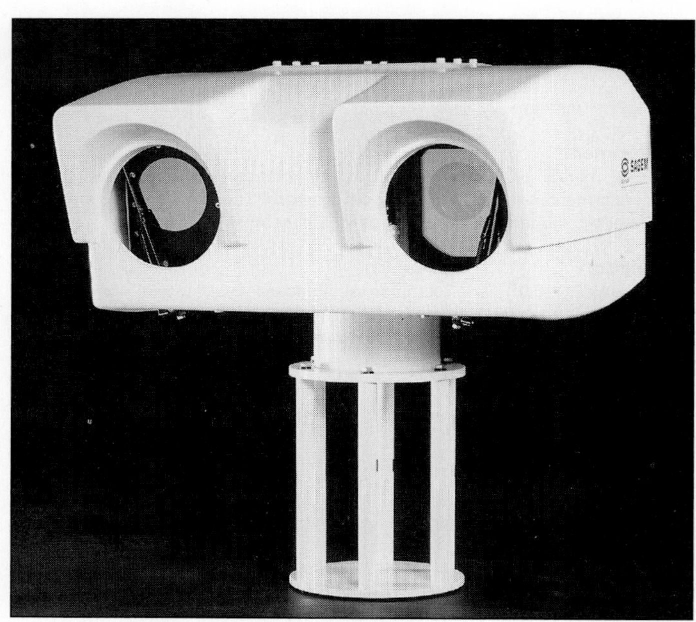

VIGY 10 naval Surveillance System Sensor head 0055129

- the display and control system, comprising:
- a monitor, capable of displaying the images according to three configurations, left or right sensor alone and simultaneous display of both
- an electronic unit
- a keyboard unit.

As an option, VIGY 10 can interface with a radar or be integrated in a general level surveillance system.

Specifications

Stabilised sensor head
Max elevation travel: ±30°
Max traverse travel: ±170°
Max elevation speed at full load: 30°/s
Max traverse speed at full load: 30°/s
Sensors
Thermal imager unit: 8-12 μm class, 3-5 μm class
Black and white or colour unit (zoom, autofocus)
Image intensifier unit
Laser range-finder

Operational status

The VIGY 10 is now in production. The first unit has been delivered to the French customs. Two other units have been purchased by a Far East country.

Contractor

SAGEM SA, Aerospace & Defence Division.

VERIFIED

Thales Optronics CYCLOPS stabilised thermal imaging surveillance system

Type

Surveillance system (naval).

Description

The CYCLOPS is a high-performance surveillance system designed for harsh marine environments. It utilises a second-generation high-resolution thermal imager to allow operation by day and night and with enhanced vision in haze and mist conditions when compared with the naked eye and TV sensors. The stabilised mounting platform is of robust construction and is suitable for weatherdeck location on a wide range of vessel types.

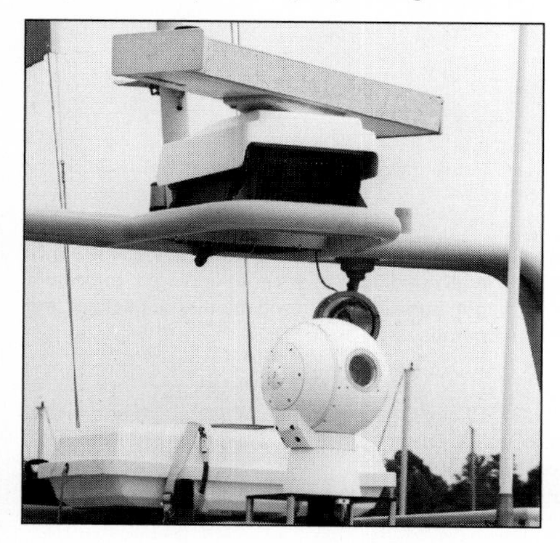

The CYCLOPS sensor mounting 0055131

The system is intended for vessels with a wide range of duties, including area surveillance, search and rescue, pilotage and navigation, collision avoidance, sea traffic monitoring and fishery protection.

Contractor

Thales Optronics.

VERIFIED

Vistar IM405 family

Type

Surveillance system (naval).

Development

Vistar Night Vision Ltd is an independent manufacturer which produces maritime sensors and surveillance systems, usually with image intensifiers, thermal imagers or both under the name Vistar® trade name.

In 1987, the company teamed with the then Marconi Radar and Control Systems which provided the TIVS thermal imager for a system for sea trials in 1988. Marconi marketed the Vistar IM405 electro-optic surveillance system that was purchased by the US Navy for the 'Cyclone' (PC1) class coastal patrol craft, the first of which entered service in October 1992. Fourteen systems were supplied for this programme.

Subsequently the system was marketed by Thorn EMI with the LITE thermal imager, seven systems under the name Marine Electro-Optical Surveillance System (MEOSS). Seven of these systems were purchased by the British Customs and Excise Service and installed in patrol boats during 1993 and 1994.

In 1995 Pilkington Optronics acquired Thorn EMI's Electro-Optic Division and began marketing as 'Hydra' with both the LITE imager and the Pilkington High Definition Thermal Imager (HDTI). The first Hydra was installed in the USS *Shamal* (PC 13) which was commissioned in 1996. Two more HDTI based systems were destined for an undisclosed customer.

The French SAGEM Aerospace & Defence Division also uses units of the Vistar system with one of their thermal imagers. In 1994, the company unveiled a Vistar derivative, VIGY 10, with redesigned display and control system and using the SAGEM ALIS thermal imager. Four ALIS thermal imagers fitted inside Vistar housings capable of being retrofitted to the Vistar IM405 systems on the 'Cyclone' class craft have been supplied as alternative sensors.

Vistar still markets the IM405 with a choice of sensors. Two systems using HDTI have been supplied direct to the US Navy and five for commercial use to other customers.

In early 2002 Vistar revealed it was developing a new product, Vistar 350.

US Navy 52 m 'Cyclone' Class patrol boat is fitted with an IM405 system 0055130

Description

All these products are designed as electro-optical surveillance systems for patrol boats down to 10 m in length and 15 tonnes displacement. They may be integrated (like Hydra) with a medium calibre (25/40 mm) gun to act as a fire-control system.

Most consist of a lightweight (circa 50 kg loaded) electro-optical director and a control console with monochrome display and MMI consisting of push-buttons, switches and a joystick. The director is usually gyrostabilised in two axes and can carry two or more sensors, of which one is normally a Vistar SM8190 image intensified camera. The naval image intensifiers use the XX1332, a 50 mm second-generation tube with micro-channel plate to improve resolution and sensitivity, with output image amplified up to 250,000 times. Other options are daylight TV cameras, laser range-finders and so on. The director can carry a 30 kg payload.

In Vistar IM405 the tracking head (which has a glass/carbon composite protective cover) is the Vistar SM8268 with optional two-axis stabilisation. The other sensor was a BAE Systems (formerly GEC-Marconi) Thermal Imager based upon the TIVS and operating in the 8 to 12 μ waveband range with ×2 and a ×5 magnification. This imager is no longer in production. It featured parallel Cadmium Mercury Telluride (CMT) detectors with cooling options of close-cycle Sterling engine or mini-compressor and was a 625 line CCIR system with 525 line EIA option. The control and display console weighs some 9 kg and has a 22 cm (9 in) display which may be made to show two images simultaneously on a split screen.

The MMI consists of joystick and function switches. A repeater screen for the bridge may be included. The Vistar IM405 is still available, but with different imagers.

MEOSS was similar but with a Thorn EMI (now Thales) LITE thermal imager based upon TICM 1 direct-view modules adapted for remote view and with a redesigned cover. The detectors are similar to those of the BAE Systems (formerly GEC-Marconi) thermal imager with cooling by mini-compressor or Sterling engine, while a slightly more sophisticated control and display system is used to interface with the craft's radar and GPS systems, with optional repeater screen and remote-control box for the helmsman. Hydra is essentially MEOSS renamed, but with the option of a Pilkington HDTI, although alternative sensors may also be used.

VIGY 10 (qv) is an evolutionary development of Vistar IM405 with both the director cover and control/display elements redesigned. The director has options of monochrome and colour TV cameras, image intensifier, laser range-finder and thermal imager (3 to 5 and 8 to 12 μm). No sensors other than SAGEM thermal imagers appear to be specified for the system. The control and display console, with joystick, push-button and switch MMI is 40 cm wide, 53 cm deep and 53 cm high.

Operational status

Over 20 Vistar IM405 systems have been ordered and delivered for marine use with the programme continuing. Some 10 MEOSS systems have been delivered and two Hydra systems.

Country	Class	Type	System	Weapons
Norway	Nordkapp	PB	MEOSS	57 mm
UK	Brooke Marine 33 m	PB	MEOSS	
	Brooke Marine 26 m	PB	MEOSS	
	Vosper 32 m	PB	MEOSS	
	Rosythe 26 m	PB	MEOSS	
USA	Cyclone (PC1-PC13)	PB	Vistar IM405	25 mm

Specifications

Vistar IM405 sensor head
Weight: 50 kg
Height: 42 cm
Width: 63 cm
Length: 60 cm
Slew rate: 30°/s
Azimuth coverage: 340°
Elevation coverage: -30 to +30°
Fields of view:
16 × 12° (image intensifier)
9.6 × 5.6°, 4.0 × 2.3° (thermal imager)
Power supply: 28 V, 5A DC

MEOSS/Hydra sensor head
Weight: 51.5 kg
Height: 43.8 cm
Width: 25.8 cm
Depth: 40.7 cm
Azimuth coverage: 340°
Elevation coverage: -30 to +30°
Power supply: 19/32 V, 6A DC

VIGY 10 sensor head
Height: 44 cm
Width: 63 cm
Depth: 59 cm
Azimuth coverage: 340°
Elevation coverage: -30 to +30°

Contractors

Thales Optronics (MEOSS/Hydra).
SAGEM SA (VIGY 10).
Vistar Night Vision Ltd (subcontractor for systems above).

UPDATED

SHIP WEAPON CONTROL SYSTEMS

THERMAL IMAGERS

BAE Systems optronic mast sensor

Type
Submarine mast sensor.

Description
The increasing need to supply accurate information day and night in all weather conditions requires high-performance electro-optic sensors and processing. The BAE Systems electro-optical systems for periscopes and non-hull penetrating masts incorporates fully stabilised, high-performance daylight, low light, colour and thermal imaging sensors. Processing includes automatic target detection, image enhancement and other processing features. The data from the multiple sensors may then be correlated to provide optimum information. Fibre optics and signal multiplexers are used to minimise the system complexity.

The TICM mini-scanner and electronic modules are utilised, operating over the 8 to 13 μm wavelength. Staring focal plane array camera technology can be employed to provide a miniature IR sensor in the 3 to 5 μm range. In the visible spectrum, use is made of high-resolution CCD cameras for colour, monochrome or low light.

Operational status
In production.

Contractor
BAE Systems, Avionics, Basildon.

UPDATED

BAE Systems optronic mast sensor

BAE Systems V3900 thermal imaging sensor

Type
Thermal imager (naval).

Description
The V3900 series thermal imaging sensor is a compact, two-unit equipment comprising a sensor head and small processing electronics. It is based on the UK Class II Thermal Imaging Common Module (TICM II). The V3901 is integrated into the GSA8 electro-optical tracker used to control the Vickers 4.5 in gun mounting in the Royal Navy's 'Type 22' and 'Type 23' frigates.

The standard 8 to 13 μm spectral bandwidth sensor head incorporates a telescope with two fields of view and an advanced mini-scanner with an integral split Stirling-cycle cooling engine.

Operational status
Units have been supplied to, or are on order for, both the UK MoD and other customers, including Hong Kong and Malaysia. The V3901 version is in service with navies worldwide.

V3900 thermal imaging sensor and PEU

Specifications
Spectral bandwidth: 8-13 μm
Independent field of view: 0.06 mrad
Telescope: ×19 or ×5 (switched)
Field of view
Narrow: 3.16 × 2.1°
Wide: 12 × 8°
Cooling: integral split Stirling
Weight: 26 kg (sensor head), 8 kg (processing electronics unit)

Contractor
BAE Systems, Avionics, Basildon.

UPDATED

BAE Systems V4500 Series thermal imaging sensor

Type
Thermal imager (naval).

Description
The 3 to 5 μm or 8 to 10 μm spectral bandwidth V4500 Series naval thermal imaging sensor is a lightweight single unit using focal plane staring array technology. The unit incorporates a telescope with two fields of view and a Stirling-cycle cooling engine.

Operational status
In service.

The BAE Systems, V4500 thermal imaging sensor for naval applications

Specifications

Dimensions: 504 × 169 × 195 mm
Weight: 15 kg max
Spectral band: 3-5 μm MWIR or 8-10 μm QWIP
NETD: <0.02°C
Detector: Staring FPA
Field of view (telescope): selectable by lens
Cooling: closed-cycle Stirling
Video formats: 625/50 Hz CCIR, 525/60 Hz RS-170 or RGB
Control interface: RS-422 serial link
Environmental: DEF STAN 07-55 MIL-STD equivalent
Weight: 13 kg

Contractor

BAE Systems, Avionics, Sensors and Communications Systems., Basildon.

VERIFIED

Galileo Avionica NTG-500SG thermal imaging system

Type

Thermal imager.

Description

The Galileo NTG-500 is a high-performance thermal imaging system using an eight-element SPRITE detector and closed-cycle cooler. Output is to a standard TV monitor. The NTG-500 covers a wide range of operative requirements, both in naval and ground applications, for surveillance, observation and aiming. It can be remotely operated and can be integrated in automatic fire-control systems.

The NTG-500 consists of a thermal camera and remote-control panel if required. The two main units are cable connected. The thermal camera is self-contained and outputs to a standard CCIR TV 625/50 monitor.

Operational status

In development.

Specifications

Spectral band: 7.5-10.5 μm
Fields of view
  Narrow: 2.5 × 1.9° or 3.2 × 2.4°
  Wide: 8° × 6°
Reticle (optional): electronic
Detector: 288 × 4 IRCCD
IR lines: 576
Display standard: CCIR 625/50 TV monitor
Cooling system: linear Stirling
Power supply: 24 V DC
Power consumption: 150 W
Weight: 18.5 kg
Dimensions: 340 × 266 × 500 mm

Contractor

Galileo Avionica.

UPDATED

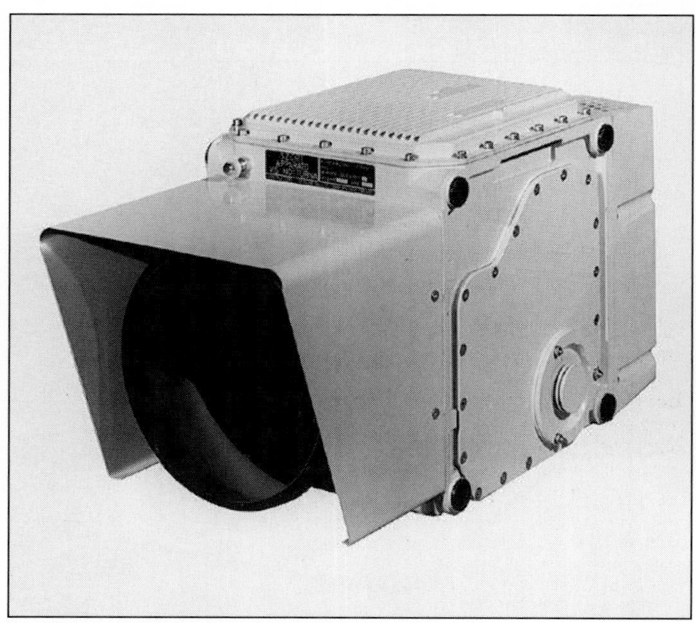

The NTG-500 thermal imaging system from Galileo

Northrop Grumman Advanced Maritime Infra-Red Imaging System (AMIRIS)

Type

Thermal imager.

Description

AMIRIS represents the new generation of marine infra-red imaging systems. Using a cryogenically cooled detector operating in the 3 to 5 μm wavelength band of the infra-red spectrum, the AMIRIS provides high performance at sea or dockside. Where image intensified systems fail due to emitted or reflected light and near infra-red energy, AMIRIS relies on thermal energy in the scene and provides a view that would otherwise have been obscured.

Effective in day or night-time applications, AMIRIS allows the mariner to identify radar targets and detect objects that would normally be obscured, for instance by mist or smoke, or invisible to the naked eye. The imager provides an aid to navigation, even in difficult backlit harbour settings or when underway.

Applications are as follows:

- aids to navigation
- enhanced man overboard recovery
- in port security
- underway security
- radar target confirmation/identification
- non-radar reflecting target detection/identification
- high-speed craft aid to navigation
- low radar cross section target detection
- iceberg detection
- drug intercept/interdiction
- land-based perimeter security.

Operational status

On order for the Kuwait Coast Guard as part of an integrated bridge system for 10 'Al Shaheed' class patrol boats under construction by Ocea of France.

Specifications

Weight:
camera: 18.2 kg
gimbal: 18.2 kg
Angular range:
azimuth 360°
elevation ± 30°
Slew rate:
azimuth up to 60°/s
elevation up to 30°/s
Resolution: > 300 TV lines (0.4 mrad IFOV)
Power requirement: 20 V DC at 7 amps continuous, 10 A peak

Contractor

Sperry Marine (division of Northrop Grumman).

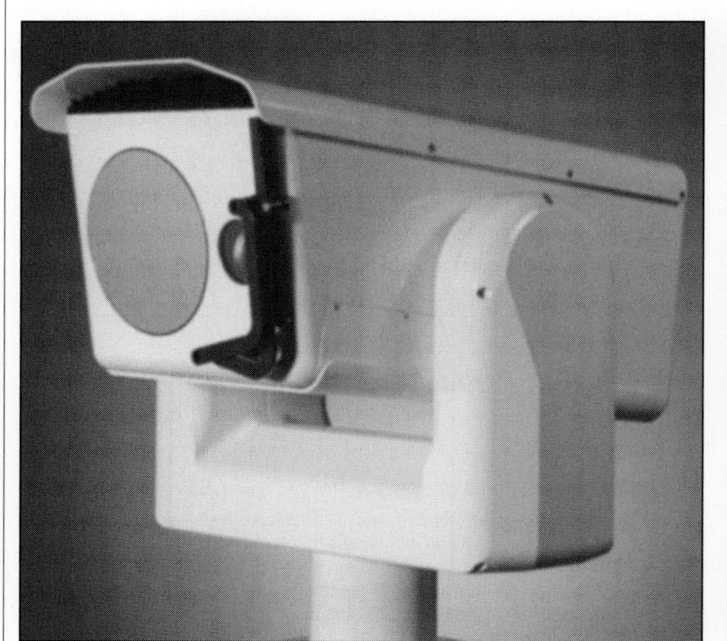

The AMIRIS thermal imager 0116610

Opgal Submarine Nite Eye

Type
Thermal imager (submarine).

Description
The Submarine Nite-Eye is a thermal imaging system for coastal surveillance and naval target detection at night and in adverse weather conditions. It is a passive observation system, providing real-time visual images. The design is miniaturised and modular and is a version of Opgal's M2TIS Modular Thermal Imaging System. Nite-Eye can be fitted to a range of submarine periscopes.

The system has a standard video output with intermediate real image. A separate afocal attachment is available for field of view expansion.

Operational status
Nite Eye is being incorporated into the periscope of the Israel Navy's three 'Dolphin' class submarines.

Specifications
Spectral bandwidth: 8-12 μm
Optical aperture: 102 mm
Focal length (NFOV): 120 mm
Fields of view
Narrow: 2.75 × 1.9°
Wide: 10.5 × 6.5°
Cooling: cryogenic closed-cycle
Sensor dimensions: 16 × 12 × 40 cm
Sensor weight: 7 kg

Contractor
Opgal Optronic Industries Ltd.

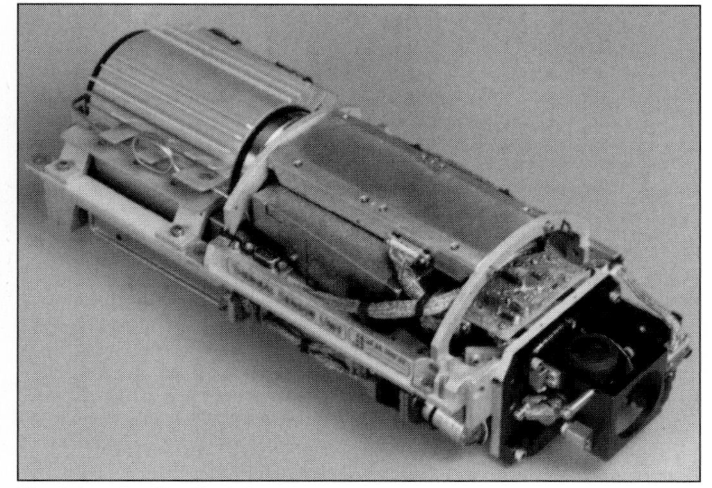

The Submarine Nite Eye infra-red sensor 0002005

Raytheon Radiance HS configurable IR camera

Type
Thermal imager.

Description
Radiance HS is a modular FLIR sensor designed for integration into gimbal payloads and research applications. It uses a snapshot-mode indium antimonide (InSb) focal planar-array, miniaturised electronics and a linear Stirling cooler. Custom cold filters and cold shields are available.

Radiance HS uses Raytheon's AE173 focal planar-array featuring selectable frame rates and simultaneous pixel integration which the manufacturer refers to as snapshot integration. This allows stop-action analysis of events as short as 2 μs and microscanning to produce high-resolution imagery. Operating in the 3 to 5 μm spectral band, Radiance HS is designed for long stand-off detection in high-humidity environments such as naval surveillance and range tracking.

Interfaces include NTSC or PAL video outputs, parallel differential and serialised (hot link) high-speed 12-bit digital data out, and external trigger signals for synchronising the camera frame rate. The camera operates at up to 120 frames/s in full-frame mode and allows microscanning to boost system resolution. Exact frame rate limits are a function of selected integration times. Radiance HS is designed to work with a range of optics from f/2.0 to f/4.0

Operational status
Available.

Specifications
Amber 256 × 256 (AE173)
Detector technology: 256 × 256 InSb
Input power: 18-32 V DC, 50 W

The Radiance HS infra-red camera fitted with a 100 mm lens 0002000

Video interfaces: NTSC or PAL, S-Video
Dynamic range: 65 dB FPA ave/45 dB per pixel
Remote control: RS-422 or RS-232C
Spectral band: 3.0-5.0 μm
Cooling method: Stirling closed-cycle
NETD: <0.025 K at 23°C
Data interface: parallel and serial digital
Resolution: 12-bits
Video synchronisation: (Gen Lock) external sync input
Frame rate: 120 Hz (256 × 256 frame mode), 1400 Hz (64 × 64 frame mode)
Dimensions (h × w × l): 12.95 × 14.48 × 17.15 mm
Weight: ≤4.1 kg
Operational temperature range: 0 to +50°C (case temperature)
Storage temperature range: –54 to +65°C (camera only)

Lens configurations

Focal length	25 mm	50 mm	100 mm	75/250
FOV degrees	17.5°	8.8°	4.4°	5.9/1.8°
FOV mrad	1.20	0.60	0.30	0.40/0.12
Dimensions (l × d)	7.24 × 5.72 cm	4.42 × 5.72 cm	9.45 × 9.65 cm	25.4 × 13.34 cm
Weight	0.28 kg	0.22 kg	0.61 kg	2.95 kg

Contractor
Raytheon Company.

UPDATED

SAGEM Murene thermal imager

Type
Thermal imager (naval).

Description
Murene is a thermal imaging camera designed for naval applications and is based on the design of the French SMT common modules. Murene is an 8 to 12 μm single-block camera in a waterproof casing with sunshade and window with wiper and washer. It has a CCIR output of 625 lines and its own power supply, preheating system and control board. Options include temperature-sensitive autofocus, automatic control (gain and accentuation of contour lines) and an electronics reticle.

Operational status
In production for the French Navy. The Murene thermal camera is also fitted on the Sadral turret and the Najir sight.

Specifications
Spectral bandwidth: 8-12 μm
Cooling: closed-circuit
Cooling down time: approx 10 min
Fields of view (v × h)
Narrow: 1.9 × 2.9°
Wide: 5.7 × 8.7°
Electronic zoom: 0.96 × 1.44
Integrated built-in test module scanning technique: semi-parallel
Frame format: 2/3
Number of TV lines: 506

Frame rate: 25 Hz
Typical NETP: 0.1°
Power supply: 115 V/60 Hz
Power consumption (approx): 115 V (while operating)
Dimensions (l × w × h): 955 × 210 × 310 mm
Weight: approx 53 kg

Contractor
SAGEM SA, Aerospace & Defence Division.

Thales Albatross 3 to 5 μm thermal imaging tracking and observation camera

Type
Thermal imager.

Description
The Albatross tracking and observation camera is based on new FPA technology which benefits from the better transmission characteristics of the 3 to 5 μm spectral window, especially in tropical climate zones. The large two-dimensional staring array combines high sensitivity with the advantages of the absence of a scanning mechanism: high MTBF, low weight, small dimensions and low power consumption. A highly uniform image is obtained by a special Non-Uniformity Correction (NUC), using a Scene Based Thermal Referencing technique.

The typical 20 mK sensor sensitivity together with a 2° FOV offers a range of more than 20 km both in North Atlantic as well as tropical areas, meeting the operational requirements in passive day- and night-tracking applications of many users.

The Albatross camera meets stringent military requirements including NBC and ECM, and is suitable for both military and civilian applications including long-range tracking and identification, search and rescue operations and security surveillance of vulnerable areas.

Operational status
Full production.

Specifications
Narrow field of view (h × v): 2.0 × 1.5°
Wide field of view (h × v): 7.2 × 5.4°
NETD: 20 mK typical
Noise equivalent irradiance: 4 pW/m^2 typical
MTF: 50% at 3.5 1p /mrad
Line of sight stability: 0.1 mrad
Environmental conditions: naval environment
Weight: 6.2 kg
Power consumption: 25 W
Dimensions (l × h × w): 410 × 150 × 115 mm
Video output: CCIR
Electrical interface: RS 422

Contractor
Thales Optronics.

UPDATED

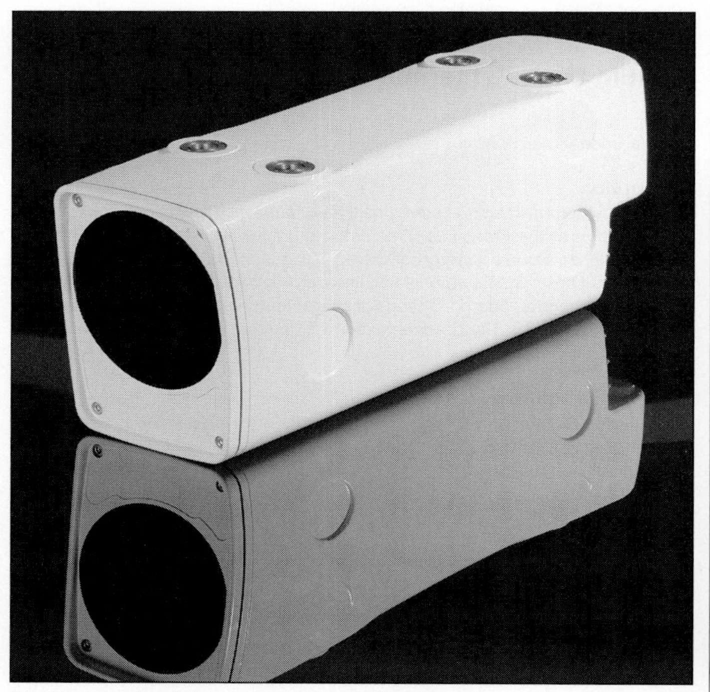

The Albatross Gen 3 thermal imager is used for observation and tracking as part of the Mirador weapons control system (Thales Optronics) ***NEW***/0554154

Thales Optronics HDTI 5-2F

Type
Thermal imager (naval).

Description
The HDTI 5-2F is a version of Thales' High Definition Thermal Imagers, and is an 8 to 12 μm FLIR with fields of view of 5 × 3° and 2 × 1.3°. It has been designed specifically for upgrading the US Navy Phalanx CIWS. The Block 1B Surface Mode Upgrade will give Phalanx a day/night engagement capability for small surface craft or low-speed aircraft. The imager incorporates a fibre optic video link in order to provide high-clarity video over long distances to various locations on the ship.

Operational status
In production for the US Navy.

Contractor
Thales Optronics.

VERIFIED

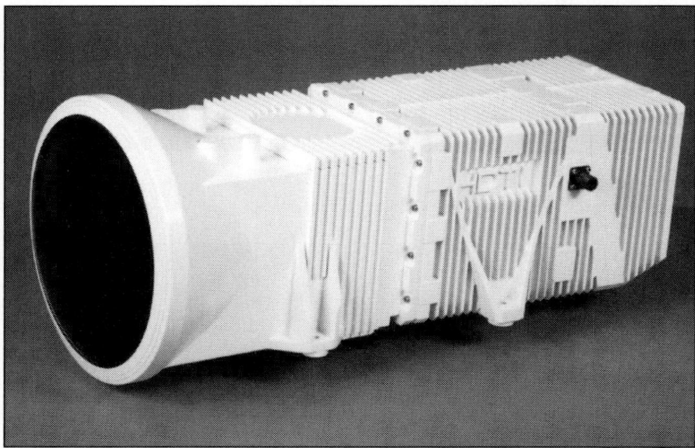

The HDTI 5-2F long-range thermal imaging sensor developed for the Phalanx close-in weapon system 0002003

Thales UP 1043/01

Type
Thermal imager.

Description
The UP 1043 family of thermal cameras has been designed for passive day and night tracking applications and operates in the 8 to 12 μm band. The UP 1043/01 camera provides a focusing range of 100 m to infinity and 35 × 18 mrad field of view. The narrow field of view enables tracking to an accuracy of 0.1 mrad.

Powered from an 18 to 32 V DC supply, the camera can operate in temperatures ranging from –35 to +63. Various functions, including an aiming mark, can be adjusted by means of an RS-422 control interface. The camera's output is a CCIR video signal which can be displayed and recorded on standard video equipment. For stand-alone observation purposes the necessary infrastructure can be provided.

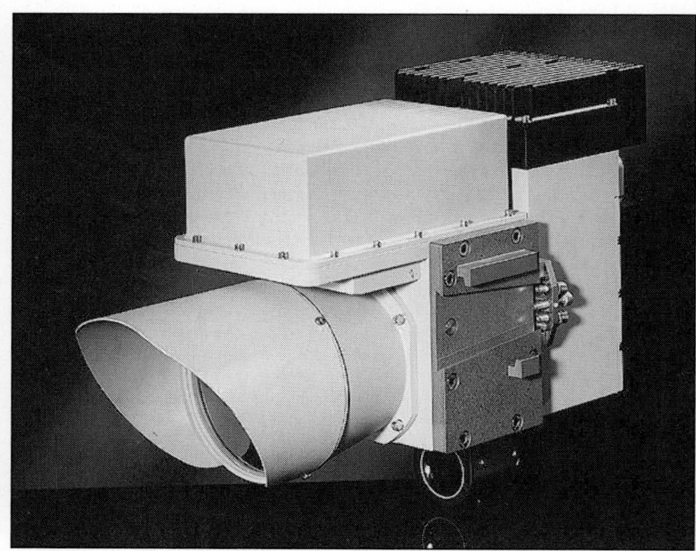

The UP1043/02 thermal imaging camera from Thales

Operational status

In production and in service in Asia, Europe and Middle East. UP1043/01 is incorporated in the baseline version of Signaal's LIOD Mk 2 electro-optical director.

Specifications

Spectral bandwidth: 8-12 μm
Field of view: 35 × 18 (mrad)
Instantaneous field of view: 0.14 × 0.14
Focusing range: 100 m to ∞
Instantaneous FOV: 0.14 × 0.14 mrad
NETD: <0.2K at 22 ±7°C ambient
NEI: <5 nW/m^2 at 22°C ambient
Video output: CCIR
Control: RS-422
Supply voltage: 18-32 V DC
Input power: 50 W max
Operating temperature range: −35 to +63°C
Storage temperature range: −40 to +70°C
Weight: 18 kg
MTBF (at least): 1,500 h
Dimensions (l × w × h): 596 × 199 × 393 mm

Contractor

Thales Optronics.

UPDATED

Thales UP1043/02

Type

Thermal imager.

Description

The UP 1043 family of thermal cameras has been designed for passive day and night tracking applications and operates in the 8 to 12 μm window. The UP 1043/02 camera features a dual field of view; the narrow FoV enables tracking to an accuracy of 0.1 mrad while both fields of view have a focusing range of 100 m to infinity.

Powered from an 18 to 32 V DC supply, the camera can operate in temperatures ranging from −35 to +63°C. Various functions, including an aiming mark, can be adjusted by means of an RS-422 control interface. The camera's output is a CCIR video signal which can be displayed and recorded on standard video equipment.

Operational status

In production and in service in Asia, Europe and Middle East.

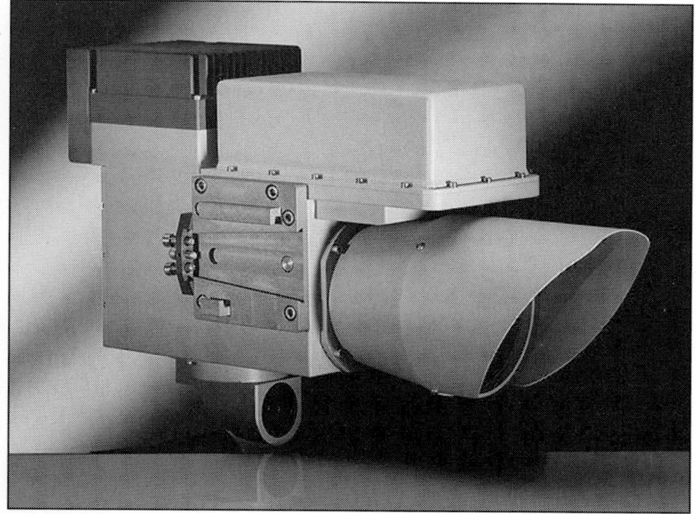

The UP1043/01 thermal imaging camera from Thales

Specifications

Spectral bandwidth: 8-12 μm
Fields of view
narrow (h × v): 35 ×18 mrad
wide (h × v): 260 × 130 mrad
Focusing range
NFOV: 100 m to ∞
WFOV: 10 m to ∞
Instantaneous FOV
NFOV: 0.14 × 0.14 mrad
WFOV: 1.0 × 1.0 mrad
Noise Equivalent Temperature Difference (NETD): <0.2K at 22°C ±7°C ambient
Noise Equivalent Irradiance (NEI): <5 nW/m^2 at 22°C ambient
Video output: CCIR
Control: RS-422
Supply voltage: 18-32 V DC
Input power: 50 W nominal
Temperature ranges
operating: −35 to +63°C
storage: −40 to +70°C
Dimensions: 596 × 393 × 199 mm
Weight: 20 kg

Contractor

Thales Optronics.

UPDATED

Thales UP1043/07

Type

Thermal imager.

Description

The UP 1043/07 thermal camera has been designed as a direct replacement of the UA 9053 camera which is no longer in production. The UP 1043 series is already in use in modern digital tracking systems and the UP 1043/07 version is the specific variant for installation in existing systems which use a UA 9053 camera.

The UP 1043 includes a single, low-speed scanner for both horizontal and vertical scanning, dual-piston Stirling split cooler and digital electronic signal processing. This has resulted in a higher reliability and a longer operational life when compared to the UA 9053. Other improvements are low noise, low vibration, low-power consumption and a better line of sight stability.

Operational status

In production and in service.

Specifications

Spectral bandwidth: 8-12 μm
Field of view: 35 × 18 mrad (h × v)
Focusing range: 100 m to ∞
Instantaneous FOV: 0.14 × 0.14 mrad
Noise Equivalent Temperature Difference (NETD): <0.2K at 22 ±7°C
Noise Equivalent Irradiance (NEI): <5 nW/m^2 at 22°C
Video output: CCIR
Control: analogue
Supply voltage: 18-32 V DC
Input power: 50 W nominal
Temperature ranges
operating: −35 to +63°C
storage: −40 to +70°C
Weight: 19 kg
MTBF (calculated in accordance with MIL-HBK-217): >1,500 h

Contractor

Thales Optronics.

UPDATED

LAND SYSTEMS

Electro-optic countermeasures
Electronic countermeasures
Laser warners

Air defence missiles
Vehicles
Vehicle sights
Static and towed
Static and towed sights
Portable
Portable sights

Air defence guns
Vehicles
Vehicle sights
Static and towed
Static and towed sights

Anti-armour missiles and munitions
Vehicles
Vehicle sights
Portable
Portable sights

Armoured fighting vehicles
Vehicle turrets
Fire control
Gunner's sights
Commander's sights
Driver's sights

Infantry weapon sights
Illuminating
Passive - crew-served weapons
Passive - personal weapons

Observation and surveillance
Air defence sensors
Forward observations
Laser range-finders
Image intensifier binoculars
Image intensifier cameras
Image intensifier goggles
Image intensifier monoculars
Area surveillance
Thermal imagers

LAND SYSTEMS – SECTION SUMMARY

This section includes electro-optic systems reported as used by land forces or developed for land applications. Systems are grouped in the following subsections according to their type:

Electro-optic countermeasures

Electronic countermeasures
Land-based tactical infra-red jammers, laser deception or laser damage systems.

Laser warners
Land-based laser warning systems.

Air defence missiles

Vehicles
Self-propelled air defence missile vehicles and turrets with infra-red or laser-guided missiles, electro-optic fire-control systems or optional electro-optic fire-control systems for weapons control, commander or driver.

Vehicle sights
Electro-optic sights used on self-propelled air defence missile vehicles.

Static and towed
Land-based static or towed air defence missile systems with infra-red or laser-guided missiles, electro-optic fire-control systems or optional electro-optic fire-control systems for weapons control.

Static and towed sights
Electro-optic fire-control systems or optional electro-optic fire-control systems for land-based static or towed air-defence missile systems

Portable
Land-based portable air defence missiles with infra-red or laser guidance or electro-optic fire-control elements on the launcher.

Portable sights
Sighting and guidance control systems for land-based portable air defence missiles (see also land-based air defence missiles – portable, where most sighting and guidance systems are described along with the missile system itself).

Air defence guns

Vehicles
Self-propelled air defence guns and turrets with electro-optic fire-control systems or optional electro-optic fire-control systems for weapons control, commander or driver.

Vehicle sights
Electro-optic sights used on self-propelled air defence guns.

Static and towed
Land-based static or towed air defence guns with electro-optic fire-control systems or optional electro-optic fire-control systems for weapons control.

Static and towed sights
Sighting systems for land-based static or towed air defence guns with electro-optic fire- control systems or optional electro-optic fire-control systems for weapons control.

Anti-armour missiles and munitions

Vehicles
Anti-amour missiles, launcher/turret and guidance systems which are mounted on both wheeled and tracked chassis for mobile capability.

Vehicle sights
Electro-optic sights used on self-propelled anti-armour missile vehicles.

Portable
Land-based portable anti-armour missile or munitions systems with infra-red or laser guidance or electro-optic fire-control elements on the launcher.

Portable sights
Sighting and guidance control systems for land-based portable anti-armour missiles (see also land-based anti-armour missiles – portable, where most sighting and guidance systems are described along with the missile system itself).

Armoured fighting vehicles

Vehicle turrets
Turrets used on armoured fighting vehicles which include an electro-optic fire-control system or optional electro-optic fire-control systems for weapons control.

Fire control
Integrated armoured fighting vehicle fire-control systems that include ballistic computers, as well as electro-optic components and generally offer a full fire-control solution for stationary or moving vehicles and targets (see also gunners' sights for closely related systems).

Gunner's sights
Armoured vehicle gunner's sights which include electro-optic components (see also fire control for closely related systems).

Commander's sights
Armoured vehicle commander's sights which include electro-optic components.

Driver's sights
Armoured vehicle driver's sights which include electro-optic components (see also driver's goggles for closely related systems).

Infantry weapon sights

Illuminating
Infantry weapon sights (excluding missile fire control) that include infra-red illuminators, or laser markers that are capable of continuous operation – sights that are passive except for a laser range-finder are included in the section on passive sights.

Passive – crew-served weapons
Infantry weapon sights (excluding missile fire control) that are fully passive or contain laser range-finders for use on crew-served weapons. Generally, these are larger, long-range sights that are too large or heavy for personal weapons.

Passive – personal weapons
Infantry weapon sights (excluding missile fire control) that are fully passive or contain laser range-finders for use on personal weapons.

Observation and surveillance

Air defence sensors
Scanning infra-red sensors for the detection of air threats for alerting rather than direct weapon control purposes.

Forward observation
Combined electro-optic imagers and range-finders specifically designed for forward observation for artillery fire control.

Laser range-finders
Laser range-finders and or designators used or designed to be used in land-based weapon control systems of all kinds.

Image intensifier binoculars
Image intensifiers available with binocular eyepieces – many systems are available with a variety of eyepieces and camera attachments.

Image intensifier cameras
Image intensifiers configured as cameras – many systems are available with a variety of eyepieces and camera attachments.

Image intensifier goggles
Image intensifiers configured as head-mounted goggles (see also pilot's goggles in the Air Systems section for closely related systems), although those systems are reported as specifically qualified for air use.

Image intensifier monoculars
Image intensifiers configured as hand-held monoculars (alternatively called pocketscopes) – many systems are available with a variety of eyepieces and camera attachments.

Area surveillance
Systems configured for surveillance and observation but not specifically associated with any weapons.

Thermal imagers
Thermal imaging cameras and modules used in land-based systems (see also related systems in the Naval and Airborne Thermal Imaging sections).

ELECTRO-OPTIC COUNTERMEASURES

ELECTRONIC COUNTERMEASURES

BAE Systems AN/VLQ-8A infra-red countermeasures set

Type

Infra-red countermeasures system for land vehicles.

Description

The AN/VLQ-8A infra-red countermeasures device provides protection to armoured vehicles against a wide variety of Anti-Tank Guided Missiles (ATGM), employing infra-red guidance. The system consists of a small, lightweight, robust transmitter with an integral electronic control unit which contains the operator signal and built-in test circuitry.

Operational status

The AN/VLQ-8A is operationally qualified on the US Army M1 Abrams and M2/M3 Bradley fighting vehicles. More than 1,000 systems have been delivered to the US Army.

Specifications

Weight: 11.4 kg
Length: 406 mm
Width: 267 mm
Height: 254 mm

Contractor

BAE Systems North America, Information and Electronic Warfare Systems.

AN/VLQ-8A missile countermeasures system

BAE Systems AN/VLQ-6 Missile Countermeasures Device (MCD)

Type

Land electronic countermeasures.

Development

The AN/VLQ-6 Missile Countermeasures Device (MCD) is an anti-tank guided missile countermeasure system developed by Lockheed Martin Information

Bradley fighting vehicle with BAE Systems AN/VLQ-6 MCD device mounted on roof of vehicle

Systems, now BAE Systems North America, for installation on a wide range of vehicles, tracked and wheeled, to improve their battlefield survivability.

Description

MCD is a multithreat jammer which protects the vehicle against a wide range of ground- or air-launched anti-tank guided missile threats.

It is normally mounted on the roof of the vehicle and operates directly from the vehicle's 28 V DC power, either in an open loop stand-alone mode or integrated with other vehicle warning and/or self-protection equipment. In the standard version the display and control box would normally be positioned near the vehicle commander.

As the MCD system has an azimuth field of view of 60°, one or more jammers would normally be fitted to cover the required protection zones. As an option, the roof-mounted system can be provided with external armour plate and an optional gimbal system to allow slewing in response to threat warning sensor inputs. Another option is a pop-up optical assembly for enhanced survivability.

Operational status

Production. During the 1990-91 Gulf War conflict the US Army purchased about 1,400 of these new missile jammers.

Specifications

Field of view
Horizontal: 60°
Vertical: 15°
Weight: 12.7 kg (without armour)
Jammer size
Height: 305 mm
Width: 457 mm
Depth: 508 mm
Power: 600 W

Contractor

BAE Systems, North America Information and Electronic Warfare Systems.

BAE Systems Type 405 laser decoy system

Type

Land electronic countermeasures.

Development

The Type 405 is a range of laser decoy systems designed to seduce laser-guided munitions away from their targets.

Description

The complete decoy system consists of a laser warner with fibre optic-coupled sensor heads, a signal processor, a pointing system and a solid-state laser transmitter.

In operation, the Type 405 detects and analyses the laser beam which is illuminating the target. It will then replicate the offensive laser beam characteristics and illuminate a decoy target to seduce the laser-guided munition on to the decoy. Having successfully carried out this function, the decoy system reverts to standby, ready to counter the next attack.

The warner is easy to retrofit and will interface with existing vehicle optics.

Operational status
Fully developed.

Contractor
BAE Systems, Avionics, Radar and Countermeasures Systems, Edinburgh.

VERIFIED

EADS EIREL infra-red jamming system

Type
Land electronic countermeasures.

Development
The EIREL infra-red jamming system was developed as a private venture by Matra Défense Equipements & Systèmes but was adopted by the French Army for installation on its Giat Industries AMX-10RC (6 × 6) armoured vehicles which subsequently took part in Operation Desert Storm, the liberation of Kuwait, early in 1991.

According to Matra, the main characteristics of the EIREL infra-red jammer are that it provides permanent protection against a wide range of threats, it can be quickly installed on most armoured fighting vehicles and it is very reliable.

Description
The scanner-type jamming device is normally mounted on the roof or side of the tank to cover the vulnerable frontal arc with the control box being mounted inside the turret. In the case of the AMX-10RC application it was mounted on the left side of the turret roof.

The EIREL infra-red jammer has two operational modes for different threats and can also be used in conjunction with an alarm detector. It is normally powered from an onboard 28 V DC power supply although different versions exist according to the power supply available on the vehicle. It is claimed to be very reliable with an integrated back-up mode providing permanent protection.

In most anti-tank guided weapons, the missile is slaved to the gunner's line of sight and for this purpose the missile is fitted with a flare in the rear so that its position with respect to the target can be sensed from the launcher. As soon as the missile moves away from the target the deviation is detected and correction instructions are sent to the missile through the control wire.

When the target is fitted with the EIREL infra-red jamming system, this will substitute for the missile flare. The launcher then no longer measures the missile-to-target deviation but that of the jammer-to-target. The missile is no longer guided and quickly moves away from its course and drops without reaching its target.

EADS EIREL infra-red countermeasure system with control box below

There are two methods of operation. When the vehicle is stationary the jammer emits in a fixed direction, typically over the frontal arc and in line with the main armament. This method is used when it is known where the threat is coming from. The incoming missiles can be jammed as soon as possible.

When the vehicle is moving, the jammer emits while carrying out an optimised horizontal scan so as to considerably increase the protected area. This method is used in cases of uncertain threat.

Specifications
Height
Upper part: 450 mm
Lower part: 175 mm
Depth: 195 mm
Weight: 20 kg
Power supply: 28 V DC
Consumption: 15-30 A according to type

Status
Production as required. In service with the French Army on Giat Industries AMX-10RC armoured vehicles.

Contractor
EADS Systems & Defence Electronics.

Giat Industries Decoy S infra-red jammer

Type
Land electronic countermeasures.

Development
This is similar in concept to the EIREL infra-red jammer and was also fitted to a number of French armoured vehicles, including AMX-30 B2 MBTs, deployed to the Middle East in 1990 to take part in Operation Desert Shield/Desert Storm.

Description
Although different in appearance it performs a similar function to the EIREL infra-red jammer, although exact details of the system have yet to be released.

Operational status
Production complete. In service with the French Army.

Contractor
Giat Industries.

VERIFIED

Giat Industries AMX-30 MBT fitted with Decoy S infra-red jammer and Ruggieri Spider close in vehicle defence system

Giat Industries Kit Basic de Contre-Mesures (KBCM)

Type
Land electronic countermeasures.

Development
Giat Industries, with some funding from the French procurement agency (DGA), has developed a Kit Basique de Contre Mesures (KBCM), (or basic countermeasures kit) Defensive Aids System (DAS). The system design takes into account the considerable development in anti-tank threats such as top attack, sophisticated guidance systems and improved terminal effectiveness of warheads.

The Giat AMX-10 RC 105 mm turret fitted with the KBCM defensive aids system
0089885

The KBCM can be installed on high-value armoured fighting vehicles such as Main Battle Tanks (MBTs) and reconnaissance vehicles to improve their battlefield survivability against various direct threats. These include Anti-Tank Guided Weapons (ATGWs), laser-guided munitions and MBT fire-control systems.

The system has already been installed on the turret of a Giat Industries AMX-10RC 6 × 6 105 mm armed reconnaissance vehicle and is currently being evaluated by the French Army.

A typical DAS comprises a man/machine interface and central processing unit coupled to the existing vehicle's combat system and databus.

To this are added infra-red jammers, a missile launch detector, three laser-warning detectors covering a full 360° and the well-established Galix system, which can launch a wide variety of smoke grenades which create a screen in the visual and infra-red range (thermal imager, laser, homing missile).

The man/machine interface would display all threats and the appropriate countermeasures that could be taken. The system can operate in manual, semi- or fully-automatic modes. In the latter case, the system would engage and defeat those targets which it determines are the greatest threat to the platform.

While Giat Industries is overall prime contractor and system integrator for the KBCM, other contractors are involved, including SAGEM for the laser warning sensors and EADS for the missile warning system.

Further growth potential, according to Giat Industries, would include a network of several vehicles fitted with DAS linked to a Fast Information, Navigation, DEcision and Reporting System (FINDERS®) type command-and-control system to provide well co-ordinated protection to a complete armoured unit.

Giat Industries states that some key parts of its DAS are in production and service. Galix, for example, is used by, or being built for, a number of countries. These include France, Saudi Arabia, Sweden and the United Arab Emirates. FINDERS is standard equipment on all Giat Industries Leclerc MBTs built for the home and export markets. Galix can launch various grenades such as smoke and screening as selected by the operator.

Operational status
In development.

Contractor
Giat Industries.

VERIFIED

Iraqi ATGW decoy system

Type
Land electronic countermeasures.

Description
During the 1990-91 Gulf War conflict, numerous Iraqi T-72 MBTs of the Republican Guard Divisions were observed to be fitted with a roof-mounted electro-optical jammer, or 'Dazzler' as it was also referred to by the Allies. This was first observed on a T-72 MBT during a defence equipment exhibition held in Baghdad in 1989.

This device was mounted on a small platform on the left side of the turret roof next to the gunner's hatch cover and pointed to the front of the vehicle. The exact origins of the device are not clear, although it could be Chinese. Reports indicate that it was effective in decoying some types of wire-guided anti-tank guided weapon.

Some Iraqi T-72 MBTs were also fitted with a roof-mounted laser warning system which may have been of Polish origin, as Poland supplied a number of the Iraqi T-72 MBTs from its production facilities.

Close-up of an Iraqi T-72 MBT showing roof-mounted ATGW decoy system with optical cover closed

Operational status
Probably still in limited service with the Iraqi Army, pre-operation 'Iraqi Freedom' in 2003.

Contractor
Iraq.

Israel Military Industries ARPAM defensive aids system technology demonstrator

Type
Electronic countermeasures (land).

Development
ARPAM (Active Armour Protection Against anti-tank guided Missiles) Defensive Aids System Technology Demonstrator is designed for fitting on Main Battle Tanks (MBTs), but in trials it has been fitted on an M113 series Armoured Personnel Carrier (APC).

Description
ARPAM consists of two elements - detection and early warning and response to the threat. The detection and early warning systems that are fitted are the LWS-2 laser warning system from Amcoram (qv) and Israel Military Industries Piano passive electro-optical system (qv). The LWS-2 (standard equipment on the Israeli Merkava Mk 3 MBT) has three systems installed on the turret to give 360° coverage. It gives the crew an audio-visual analysis of optical source threats from laser range-finders/designators and infra-red searchlights.

Piano is mounted above the turret and constantly scans through 360°. It detects incoming missiles and provides directional information. Scanning rate can be adjusted for different platforms and it can also detect air-to-air and surface-to-air missiles.

The response unit is the POMALS (Pedestal Operated Multi-Ammunition Launching System) with multitube launchers for smoke grenades and the Violin Mk 1 jammer also from Israel Military Industries. POMALS consists of two pedestals normally mounted on either side of the vehicle with six or more launch tubes for smoke grenades, chaff and flares, decoys or other ammunition. A salvo of three grenades is claimed to create a smokescreen 50 m wide, 8 m high at a distance of 80 m from the MBT in 2 seconds. This is said to last for 30 to 90 seconds depending on windspeed and other environmental conditions. The system can be operated automatically or manually.

The ARPAM technology demonstrator mounted on an M113 series armoured personnel carrier

The Violin Mk 1 jammer is installed on the turret roof below Piano and would cover 180° at the front of the turret. A second system could be added to give 360° coverage. It is armour-protected (including optics).

Information from the LWS-2 and Piano sensors is sent to a central processing unit which selects the best response to the threat. For first-generation missiles this might be smoke grenades, or electro-optical jamming for second-generation missiles.

Operational status
Under development.

Contractor
Israel Military Industries.

Lockheed Martin Outrider combat protection system

Type
Land direct energy laser system.

Description
The Outrider combat protection system consists of an electro-optical system integrated into an HMMWV armament carrier for use in scout, reconnaissance and surveillance mission roles as well as in support of light armour units such as TOW-HMMWVs (hunter/killer operations). Its Electro-Optical CounterMeasure (EOCM) and target acquisition capabilities protect friendly forces by suppressing fire-control devices.

The system consists of an integrated passive sensor suite (FLIR and low-light level TV) and an active sensor that detects and engages targets at extended ranges. Using an integrated Global Positioning System (GPS), electronic compass and tactical map display, Outrider can hand-off digitised target data via SINCGARS radio to adjacent units, command and control, intelligence and standoff attack support elements.

The basic system also includes a low-power laser but can be enhanced by the addition of a laser range-finder/designator. It also features stabilised line of sight and 1553B Bus interface.

Operational status
Not stated.

Contractor
Lockheed Martin Electronics & Missiles.

The Lockheed Martin Outrider observation and countermeasures system

Lockheed Martin Stingray laser detection system

Type
Land direct energy laser system.

Development
Lockheed Martin has completed work on an advanced technology demonstration contract awarded in 1992. Two field-ready Stingray systems have been built and delivered and are being tested by the army to develop tactics and doctrine for the system. Between August and November 1994, at the US Army White Sands Missile Range, both Stingray systems were successfully trialled.

Description
Stingray is a directed-energy system designed to protect frontline forces by detecting, locating precisely and defeating enemy optical and electro-optical fire-control systems using a low-energy laser. Consisting of a sensor assembly, laser transmitter, control console and supporting electronics, Stingray can be employed from a stationary position or on the move against moving or stationary targets in exposed or hidden positions. The system can be operated manually, or in automatic and semi-automatic modes. It has a built-in simulation system which provides safe, realistic training.

Stingray showing the gunner's sight on the right with additional viewers on the upper left

Stingray provides target co-ordinates for direct fire, indirect fire and/or countermeasures. The semi-automatic mode provides man-in-the-loop target verification. The manual mode provides an additional sensor for enhanced target identification. An eye-safe training laser is used to replicate faithfully all modes of operation.

Operational status
Development for the US Army Communications Electronics Command is complete. Future production and status are under consideration, but activity at Lockheed Martin has ceased.

Contractor
Lockheed Martin Electronics & Missiles.

Northrop Grumman HELWEPS High-Energy Laser Weapon System

Type
High-energy laser weapon.

Development
HELWEPS (High-Energy Laser Weapon System) is a design for a deployable laser weapon, using experience gained from TRW's MIRACL laser developed in the late 1970's. HELWEPS formed the basis for a feasibility study in 1993 undertaken by the US Navy's Space and Naval Warfare Systems Command (SPAWAR) which suggested that a laser-based point defence system could be both effective and affordable. It concluded that a shipborne laser could be fitted on an AEGIS cruiser in the space taken by the Mk 45 Mod 2 gun and magazine.

Description
HELWEPS is a chemical laser using ethylene, hydrogen and fluorinated nitrogen. The full system would include an integral electro-optic tracker and generator and would weigh 25 tons.

Operational status
Under development.

Contractor
Northrop Grumman Space Technology.

UPDATED

Northrop Grumman MIRACL

Type
High-energy laser weapon.

Development
MIRACL, Mid-Infra-Red Advanced Chemical Laser, is a megawatt class continuous wave deuterium fluoride laser developed by TRW in the 1970s.

Tests have been carried out against various targets at White Sands Missile Range in New Mexico. An airborne target drone, a Northrop BQM-74, flying at 500 kt at an altitude of 1,500 ft, was successfully targeted and downed in September 1987. In November 1987, a Ryan BQM-34S Firebee target drone was targeted and downed. Land-based tests in 1989 demonstrated the ability of the laser to destroy a Vandal supersonic missile in flight.

MIRACL is the laser used for the Nautilus programme (qv), a joint US/Israeli project managed by the US Army Space and Strategic Defense Command. In a test conducted at White Sands Missile Range in February 1996 as part of the Nautilus programme, MIRACL with the Raytheon Sea Lite beam Director destroyed a short-range 122 mm artillery rocket in flight. This is said to be the first time this has been done by a laser and also the first time a warhead has been exploded in flight by any weapon system. Only a small fraction of MIRACL's laser power is said to have been used for the test.

Recent advances in reducing the size of the laser and in the fabrication of a large, high-energy window have made the laser system viable for ship defence.

The material for the window is a custom-designed fluoride-based glass which can withstand the thermal stresses generated by the intense heat from the transmission of the laser beam.

Description
MIRACL is a deuterium fluoride laser which can emit at several discrete wavelengths in the band 3.6 to 4.2 μm. It has been reported that MIRACL operates at 3.8 μm and has achieved an output of 2.2 MW.

Operational status
Under development, it is installed at the White Sands Missile Range and is being used for a variety of high-energy laser lethality demonstrations.

Contractor
Northrop Grumman Space Technology.

UPDATED

Northrop Grumman Tactical High-Energy Laser (THEL) programme

Type
Air defence experimental laser programme.

Description
In July 1996, TRW was awarded a US$89 million contract for the Tactical High Energy Laser (THEL) Advanced Concept Technology Demonstrator (ACTD) to follow on the Nautilus concept-evaluation programme. THEL is being sponsored by the US Army Space and Strategic Defense Command (SSDC) and the Israeli Defense Ministry. The contract is to provide a demonstrator of a transportable deuterium fluoride laser. The Israel Defence Force plans to deploy up to 13 operational fire units of a weapon resulting from the THEL ACTD. The US Army is considering applying THEL to other types of targets.

THEL consists of three major subsystems - laser, pointer-tracker and C³I (command, control, communications and intelligence). The deuterium fluoride chemical laser uses optics which incorporate uncooled single-crystal silicon mirrors. The pointer-tracker includes a Contraves beam director, a Ball beam-alignment and stabilisation system, an off-axis tracker from IAI MBT and a shared-aperture tracker that uses an Amber infra-red camera and a solid-state laser. The THEL shelter-based installation will consist of eight containers, one for pointer-tracker, one for C³I and the rest for the laser and power supply.

Operational status
Under development. The Nautilus programme is continuing testing to provide data for THEL. The ACTD programme reached a milestone when, in June 2000 a test firing resulted in the destruction of a Katyusha type missile in flight.

Contractor
Northrop Grumman Space Technology.

UPDATED

Russian Shtora-1 armoured fighting vehicle defence system

Type
Land electronic countermeasures.

Development
The Shtora-1 armoured fighting vehicle defence system has been developed by the Zenit Research and Production Corporation to increase the battlefield survivability of vehicles from attack from Anti-Tank Guided Weapons (ATGW) with a semi-automatic command to line of sight guidance system as well as missiles and artillery projectiles that use laser illumination.

The system can be installed on vehicles as they are built or back- fitted to older vehicles. The first known application of the system is the T-90 MBT which entered service with the Russian Army in 1993. It is believed that Russian T-80 tanks and Ukrainian T-84 now carry Shtora-1.

Close-up of laser warning system on T-80U series MBT with protective covers open (Christopher F Foss) ***2001***

Part of the Shtora-1 electro-optical countermeasures system installed on a T-80 command tank

Description
The complete Shtora-1 system comprises four key components:

(a) electro-optical interference station comprising jammer, modulator and control panel with a total weight of 80 kg. Typically an MBT would have two of these with one jammer being mounted either side of the main armament pointing forward
(b) bank of grenade dischargers mounted either side of the turret firing forward, capable of firing grenades dispensing an aerosol screen, weight 115 kg
(c) laser warning system consisting of four laser warning receivers with precision and coarse heads weighing 20 kg each
(d) control system with a total weight of 15 kg comprising control panel, microprocessor and manual screen laying panel, this processes the information from the sensors and activates the aerosol screen laying system.

The laser warning system detects the threat laser system and automatically orients the turret in the direction of the threat. It then triggers the grenade launchers which create an offboard aerosol screen said to be effective over a frequency band of 0.4 to 14 μm covering both laser range-finders/designators and infra-red homing weapons.

The jammer, which is designated the TShU1-7 (made by Elers-Electron), introduces a spurious signal into the guidance circuitry of the incoming ATGW through the use of coded pulsed infra-red jamming signals continuously generated. It is claimed to be effective against Western ATGWs such as the TOW, HOT, MILAN and Dragon as well as Eastern ATGWs such as the AT-3 'Sagger'.

The TShU1-7 has a specified life of 1,000 hours, an MTBF of 250 hours and a radiation source life of 50 hours. It operates from a 27 V DC power supply with the infra-red jamming source consuming 1 kW of power.

In addition to being used to jam incoming ATGWs, the manufacturer claims that the TShU1-7 system has a target illuminating capability, including for night vision devices.

Shtora-1 has three methods of operation:

(a) fully automatic
(b) semi-automatic, target designation
(c) manual and emergency mode.

According to the manufacturer, the installation of the Shtora-1 system on an MBT reduces the target hit probability by the following factors:

(a) TOW and Dragon ATGWs, Maverick, Hellfire and Copperhead laser seeker systems by four to five times
(b) MILAN and HOT ATGWs by about three times
(c) artillery and tank projectiles fired from systems with laser range-finders by about three times.

Operational status
In production. Installed on T-80U, T-84 and T-90 MBTs and offered for installation on other armoured vehicles. Russia has exported the turbine-powered T-80U to

Cyprus and South Korea, while the Ukraine has exported the diesel-powered T-80UD to Pakistan. Neither of these two customers are understood to have the Shtora-1 system installed. Both manufacturers of the T-80 series are offering it as an option. As far as it is known, all production of this system has so far been undertaken only in Russia.

Specifications

Dimensions: 280 × 350 × 350 mm
Weight: not more than 30 kg
Power supply
  Dimensions: 280 × 350 × 120 mm
  Weight: not more than 15 kg
Control panel
  Dimensions: 100 × 70 × 50 mm
  Weight: not more than 0.30 kg
Max continuous operation: 6 h
Power consumption: 2.2 kW
Search field of view
  Horizontal: 360°
  Elevation: –5 to +25°
Direction-finding accuracy: 1.7-1.9°
Aerosol screen
Launchers: 12
Warm-up time: less than 3 s
Effective duration: about 20 s
Spectral range: 0.4-14 μm
Electro-optical station radiation sector
  Vertical: 4°
  Horizontal two-module design: 40°
  Horizontal one-module design: 20°
Spectral range: 0.7-2.5 μm
Radiation sector: 2 ±0.5°

Contractor

Electronintorg Ltd.

ELECTRO-OPTIC COUNTERMEASURES

LASER WARNERS

Amcoram LWS-2 laser warning system

Type
Land laser warner.

Development
The Merkava Mk 3 MBT, first shown in 1989, is fitted with the LWS-2 advanced threat warning system which has been developed by Amcoram, a member of the Aryt Group. LWS-2 also forms part of the ARPAM defensive aids technology demonstrator from Israel Military Industries (qv).

Description
The system provides an alert whenever infra-red or laser radiation is aimed at the vehicle from any direction. The system detects the type of radiation including infra-red searchlight, laser range-finder and a laser designator, with additional options being available according to the customer's specific operational requirements. The RS-232 output is to the main tank computer.

The complete LWS-2 laser warning system comprises three radiation sensors which are positioned around the vehicle, typically on the turret, to give a full 360° coverage; a data processing unit; a command, control and display unit which also includes an operation and test switches indicator; and interconnecting wiring harnesses.

The sensing element is tailored to the type of vehicle for which it is intended while the display unit provides the commander with the following information:
(a) threat identification
(b) visual directional display of threat source, clock dial style or digital display
(c) audio alert to the commander and/or crew.

The display unit includes an indication of the type of radiation detected, multiple radiation sources detection, source direction and system failure alert.

The display unit also includes the following operation and command controls: main power switch, audio mute switch, day/night illumination intensity switch, test push-button and additional/prior threat.

Operational status
In production. In service with the Israel Defence Force on the Merkava Mk 3 MBT.

Contractor
Amcoram Ltd.

Sensing element of the Amcoram LWS-2 laser warning system installed on an MBT

ATCOP LTS 1 laser threat sensor

Type
Laser warner (land).

Description
The LTS-1 laser threat sensor can be used on main battle tanks, bridges or key installations. It can distinguish between laser range-finders and laser designators and can be coupled with acoustic alarms (or intercom set for armoured fighting vehicles), smoke generators and other countermeasures.

Operational status
Available. In service.

Specifications
Waveband: 0.8-1.06 μm
Field of view
Azimuth: 360°
Elevation: −15° to +90°
Resolution (azimuth): 15°
Operating voltage: 12 or 24 V DC nominal
Power consumption: 8 W nominal
Dimensions
Head (diameter × height): 165 × 35 mm
Control box: 80 × 130 × 55 mm

Contractor
Al Technique Corporation of Pakistan (Pvt) Ltd.

VERIFIED

The ATCOP LTS 1 laser threat sensor 0004741

Avitronics Laser Warning System for Combat Vehicles (LWS-CV)

Type
Laser warner (land).

Description
The Avitronics Laser Warning System for Combat Vehicles (LWS-CV) is designed to provide crews with vitally important situational awareness of laser emissions associated with anti-armour threats.

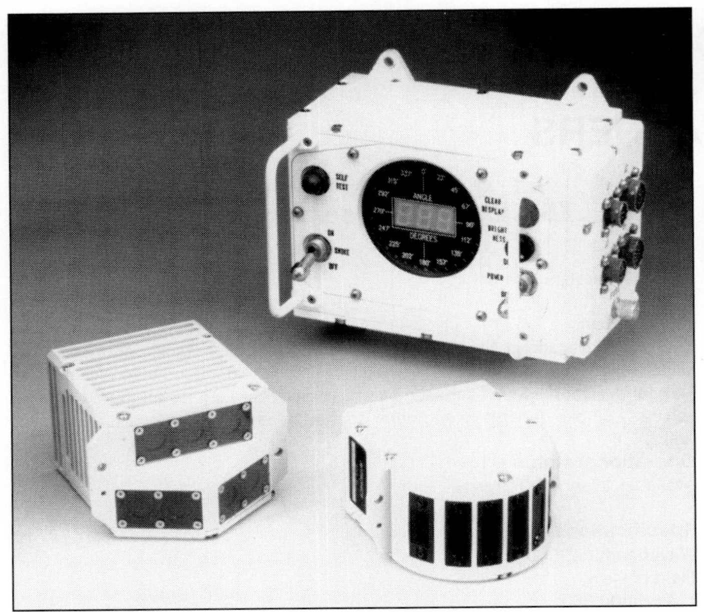

LWS-CV laser warning controller with integrated display and LWS-300 sensor (left), LWS-200 sensor (right) 0041556

The LWS-CV forms the basic building block of the Avitronics Land Electronic Defence System (LEDS) and comprises a Laser Warning Controller (LWC) and a number of laser warning sensors. Each sensor consists of a laser detector array and dedicated hardware and software for laser detection. LWS-CV provides 3,600 azimuth coverage of a platform by using four LWS300 sensors. Full hemispherical coverage and anti-reflection capability can be provided by adding a LWS-500 sensor.

By using synthetic intelligence and threat library mapping, the LWC processes raw threat energy data detected by the sensors. Apart from a stand-alone mode of operation, the LWC can also interface with on-board host battle management or fire-control system displays. The level and detail of threat information displayed is a function of the display option selected by the customer. Standard interface options are available for interfacing with a host system. The sensors communicate with the LWC via a high-speed serial link. The LWC also provides interfaces for audio warning, on-board laser blanking, countermeasure dispensing and automatic directing of commander's sights, directed dispensers and gun control systems.

Features

- Detects all known lasers associated with anti-armour threats (up to 8 threats simultaneously)
- Provides threat direction
- Provides spectral band information
- Low false-alarm rate, <1 in 16 hours (under operational conditions)
- Single pulse probability of intercept >99%
- Easy installation and integration
- Stand-alone or integrated mode of operation
- Cost efficiency by scalability
- Full-range threat management option (including identification and library linked prioritisation)
- Large detection envelope counter peripheral laser techniques
- Unique anti-reflection capability that is extremely efficient in typical high-clutter land scenarios
- User definable threat library tools
- First line flash-RAM reprogram mable via diagnostics port
- Enhanced sensitivity for the detection of lasers used for missile guidance (Beamriders).

Operational status

Available.

Specifications

LWS-300 sensor
Wavelength coverage: 0.5 - 1.8 pm
Threat coverage: Doubled NdYAG, Ruby, GaAs, NdYAG, Raman-shifted NdYAG and Erbium Glass lasers
AOA accuracy: AZ. 15° RMS (against all threat types)
Spatial coverage: AZ. 360° (110° per sensor), EL. 60°
Probability of intercept: >99% for a single pulse
Dimensions: 115 × 90 × 76 mm
Mass: 1.2 kg per sensor
Power requirements: power supplied by LWC

LWS-500 sensor
The LWS-500 sensor provides combined anti-reflection capabilities and high-angle threat warning.

Spatial coverage
Anti-reflection: AZ 360°
EL: –20° to +50°

Spatial coverage
Vertical: 100° Conical (excluding Beamriders)
Dimensions: 100 × 100 × 120 mm
Mass: 0.80 kg

Laser warning controller
Dimensions: 188 × 89 × 131 mm
Volume: 2.2 l
Weight: 2.5 kg
Interface: RS-232, 422, 485, MILBUS 1553 and ARINC
Power requirements: +28 V DC (715 mA/25 W)

Contractor

Avitronics.

BAE Systems LWR-98GV (2) Laser Warning Receiver

Type

Land laser warner.

Description

The LWR-98GV (2) is a low-cost laser warning receiver system developed by BAE Systems in Austin, Texas. The system is designed to operate as a stand-alone warning receiver or as an integrated receiver and countermeasures dispensing system. The system is comprised of four detector units, an operators control panel and interconnecting cables. Each detector provides a 90° field of view (FOV) and a +/- 45° FOV in elevation. The four detectors are mounted in a configuration that provides a 360° FOV for the system. The control panel has a dual function on/off knob with controls for the intensity of the 3.0 × 1.5 in display. The display provides a scrollable menu of settings with an arrow indication of laser threat by quadrant. Threats are classified, declared and displayed in two categories, laser range-finder and laser designator. The operator can adjust system sensitivity, audio tone – on/off, dispense mode – auto/manual and countermreasure dispense-all round or to a quadrant via the control panel. The system detects laser energy in the 0.4 to 1.8 μm band with low false alarm rate.

The system has been integrated Light Vehicle Obscuration Smoke System (LVOSS) developed by the US Army Program Manager Smoke/Obscurants and was demonstrated in January 1999.

Status

Believed available.

Specifications

Dimensions (l × h × d)
Control panel: 190 × 89 × 127 mm
Detectors: 76 × 28 × 23 mm
Weight (system): < 5.5 kg
Spectral band: 0.4 to 1.8 μm

Contractor

BAE Systems, North America.

VERIFIED

BAE Systems Type 453 laser warning receiver

Type

Laser warner.

Description

The Type 453 laser warning receiver is designed for use in armoured fighting vehicles. It provides a countermeasure to laser-guided weapon systems and provides an audible alarm and visual display showing the direction from which the threat has originated. Output from the receiver can be combined with a defensive aids system.

The system uses a number of dispersed sensors to detect incident laser radiation. This counters the problem of detecting a very narrow beam which at any instant would be illuminating only a small part of the vehicle. The sensor head configuration can be tailored to provide spherical or hemispherical cover and presents the laser threat bearing to the crew as sectorial information. The number of sensors required is tailored to the particular platform.

The sensor heads are completely passive. Laser radiation is routed to the central processing unit by fibre optic cables. This feature eliminates risk of false alarms being generated by radio frequency interference.

Using advanced scatter rejection techniques, the system can discriminate between direct and indirect hits and hence give a reliable direction of arrival of the threat. The system can analyse the incident laser source and compare it with an onboard threat library to allow identification.

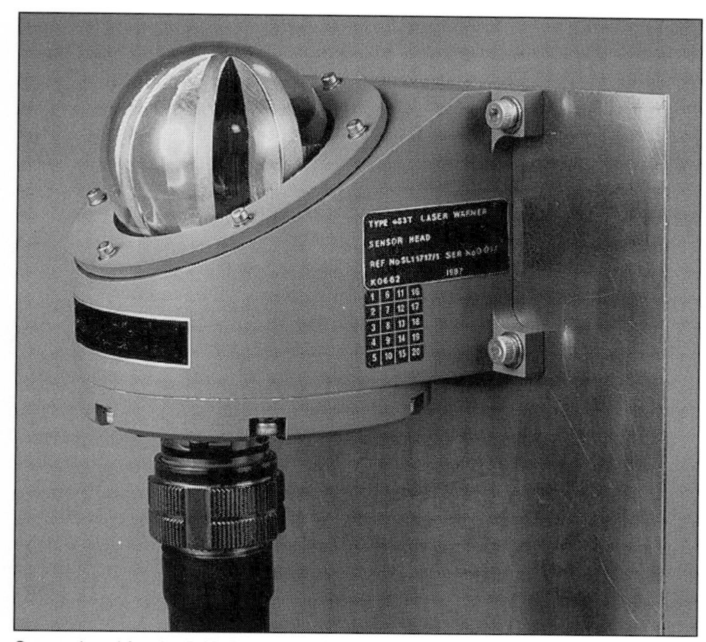

Sensor head for the BAE Systems Type 453 laser warner

Operational status
In advanced development for armoured fighting vehicles. Successfully trialled in the UK and US.

Specifications
Sensor heads
Spectral band: 0.3-1.1 μm
Option: 0.3-1.8 μm
Azimuth: 360°
Elevation: 180°
Sector resolution: 45°
Angular coverage: complete coverage depending on platform requirement
Threat detection: simultaneous multiple sources
PRF coverage: detects from single shot to continuous wave sources
Pulse-width detection range: down to 5 ns
Wavelength discrimination: to 50 nm
Dynamic range: up to 63 dB
Display
Alarm: audible tone, visual display of direction
Audible alarm duration: 2 s min or while being illuminated (laser target designators)
Visual display: 10 s min or while being illuminated
Alarm mute: push-buttons to cancel audible alarm
Interface: compatible with all standard databusses and defensive aids systems

Contractor
BAE Systems, Avionics, Sensors and Communications Systems, Edinburgh.

VERIFIED

BAE Systems Type 480 laser warner

Type
Land laser warner.

Description
The Type 480 laser warning system is one of a range of laser warning systems designed for use in various platforms, including armoured fighting vehicles. The Type 480 provides a countermeasure to laser-assisted weapon systems and provides a visual display and audible alert of the threat. The system analyses the laser radiation to determine the direction of arrival, the wavelength, the PRF and whether or not the source is aimed at the vehicle. It consists of the sensor heads, a central processor and a threat warning display, with appropriate interfaces to the platform systems. Using advanced scatter rejection techniques, the system can discriminate between direct and indirect hits and give a reliable direction of arrival of the threat. It can also analyse the incident laser source and, by comparison with an onboard threat library, identify the threat.

Operational status
Fully developed and under evaluation. The Type 480 has been offered for the British Army's next-generation reconnaissance vehicle.

Specifications
Angular coverage: complete coverage depending on platform requirements
Threat detection: simultaneous multiple sources
Spectral band: 0.35-11.5 μm
PRF coverage: detects from single shot to continuous wave sources
Pulse-width detection range: down to 5 ns
Wavelength discrimination: to 50 nm
Interface: compatible with all standard databusses and defensive aids suites
Dynamic range: up to 63 dB

Contractor
BAE Systems, Avionics, Sensors and Communications Systems, Edinburgh.

VERIFIED

Fotona LIRD-1

Type
Land laser warner.

Development
The Fotona (previously Iskra Electro-Optika) Laser Irradiation Detector and warner (LIRD) has been designed for use in both vehicle and shipborne applications and entered production in 1986.

Description
The purpose of the LIRD is to reduce vehicle vulnerability to the numerous laser-associated weapon threats by providing the crew with a warning that the vehicle is being irradiated by a pulsed laser from a laser range-finder or a laser illuminator/designator. The crew can then take appropriate self-protective action (manoeuvre and/or deployment of a smoke screen) or counterfire.

The basic system (LIRD-1) consists of two separate units: a detector head unit and indicator unit with appropriate cables and mounting parts. The detector head unit consists of two detection system modules: direct detection module and indirect detection module. The direct detection module senses the laser beams which directly hit the vehicle and the detector head unit. The indirect detection module senses the target off-laser beams which indirectly irradiate the instrument. Signals from the detector head unit are routed via a signal cable to the processing circuitry in the indicator unit. Its processing circuitry generates an audible alarm and supplies the display with signals for visual indications. The display on the indicator unit has 16 LEDs which indicate the direction of incoming irradiation and four LEDs for laser threat characterisation.

LIRD-1A is a laser irradiation detector which enables automatic discharging of smoke grenades from a tank. It is composed of the basic LIRD-1 set with a slightly modified indicator unit and smoke discharging system. The automatic smoke discharge does not take place immediately but only after a few seconds. This time delay allows the tank commander the option to cancel automatic smoke discharging if required.

Operational status
In production. It is reportedly being offered as part of an upgrade package for the T-55M tank.

Fotona LIRD-1A laser warning receiver mounted on an armoured vehicle

Specifications

Detector head unit

Direct detection module

Spectral band: 660-1,100 nm (LIRD-1/1A); 700-1600 nm (LIRD- 3/3A)
Number of receivers: 8

Receiver field of view
Azimuth: 45, 60, 160°
Elevation: 80 (–20, +60°)

Total field of view
Azimuth: 360°
Elevation: 80°

Indirect detection module
Spectral range: 1,060 ±50 nm
Number of receivers: 1

Field of view
Azimuth: 360°
Elevation: 6 (–10±3°)
Dimensions: 150 (diameter) × 200 mm
Mass: 3.2 kg

Indicator unit
Angular resolution in azimuth
Forward-looking receivers: 15°
Sideways-looking receivers: 15/30°
Backward-looking receivers: 105°
No of resolution sectors in azimuth: 16
No of resolution sectors in elevation: 1
Threat characterisation: Laser Range-finder (LR), Laser Illuminator (LI), Indirect Irradiation (IND), High-Power Irradiation (HPI)

Duration of warning signal
Audio: approx 2 s or while being irradiated
Visual: approx 8 s or while being irradiated
Operating voltage: 24 V (15-30 V)
Mass: 1.6 kg
Dimensions: 62 × 120 × 190 mm

LIRD-1A
Time delay between laser irradiation and automatic smoke discharge: 0.5-5.0 s adjustable to customers' requests

Contractor
Fotona dd.

VERIFIED

Israel Military Industries Piano passive electro-optical warning system

Type
Land electro-optic warner.

Israel Military Industries passive electro-optical warning system

Description
Piano is a missile detection system that spots incoming threats under day and night conditions so increasing crew survivability. It provides early warning against anti-tank missiles, ground-to-air missiles and air-to-air missiles.

Piano's passive system detects electro-optical emissions from incoming missiles, instantly displays a visual warning and sounds an alert. Built-in system intelligence provides extra reliability with a very low false alarm rate.

A single Piano provides a full 360° detection capability with the adjustable scanning rate supplying the correct level of protection for every platform. Piano emits no electromagnetic or electro-optical signals that can be identified or tracked by enemy forces.

Operational status
Under development, a prototype has been produced. Piano forms part of the ARPAM technology demonstrator being developed by Israel Military Industries (qv).

Contractor
Israel Military Industries.

Moked Third Eye laser warning system

Type
Land laser warner.

Development/Description
The Third Eye laser warning system was developed by Moked Engineering (1969) to meet the requirements of the Israel Defence Force (IDF) and it has been installed on their MBTs for some years.

It has been designed for the instantaneous detection of laser range-finders, designators and infra-red searchlights. It indicates the direction and type of threat on a display screen provided for the tank commander. An audio warning is also provided through the vehicle intercom net.

The sensor is mounted on a 45 cm high mast that is installed on the tank turret and comprises four sensor groups, each of which consists of infra-red and laser sensors. It has been designed to provide a credible warning at all combat ranges and be insensitive to combat by-products such as explosions, flash or smoke.

It can differentiate between the various types of laser and in addition has an arrow composed of Light Emitting Diodes (LEDs) that blink indicating the direction of the detected threat. This signal will continue to blink even after the laser is turned off therefore enabling the defender to operate countermeasures against target designators, range-finders and infra-red sources.

According to the manufacturer, the Third Eye system has been in operational use with the IDF and has proven its performance and reliability under field conditions.

Moked Third Eye laser warning system with display box (lower left) and (inset, top left) M60 series MBT which is fitted with this system

In addition, a complementary sophisticated training system, especially designed to train Third Eye users without exposing them to damaging laser beams, is available. The training system's operation is based on data transfered through the communications system of the AFV on which it is mounted.

The Moked Third Eye laser warning system is also being offered for the retrofit system market.

Operational status
In production and in service. The system is in operational service with the IDF Armour Corps.

Specifications
Infra-red detection range: for all effective combat ranges
Blocking time: 0.3 s
Wavelength peak: 0.9 μm
Laser detection range: for all effective combat ranges
Types: for all state-of-the-art laser range-finders and laser target designators
Display: multidirectional arrows
Weight
Sensor: 2.5 kg
Control and display unit: 1 kg
Cables: 2 kg
Power: 24 V ± 4 V 250 mA

Contractor
Moked Engineering (1969) Limited.

UPDATED

PCO SSC-1 OBRA self-covering and laser warning system

Type
Land laser warner.

Development
The PCO SSC-1 OBRA system is designed for T-series tanks (especially the T-72 MBT) and may easily be adapted to other applications such as armoured vehicles, ships and command centres.

Description
The OBRA system detection waveband covers all currently used lasers from Nd:YAG to CO^2 (older ruby lasers can also be detected). Main functions of the system are as follows:
(a) optical and acoustic warning of impulse range-finder or laser illuminator irradiation
(b) indication of direction and type of laser source
(c) selection and firing of smoke grenade countermeasures
The system has three basic modes of operation:
(a) automatic - system indicates the direction and type of laser source and fires smoke grenades
(b) semi-automatic - system indicates the direction and type of laser source and selects countermeasure type - commander manually fires the smoke grenades
(c) manual - system indicates the direction of laser source - commander decides countermeasure type and when to fire.
The system comprises the following main units: four detection heads, integrated signalling- controlling display unit, distribution box. A smoke grenades launcher unit is available as an option.

Operational status
In production. In service in Poland and elsewhere. Offered for export.

OBRA SSC-1 self-covering and laser warning system 0055133

Specifications
Wave length range: 0.6 μm to 11.0 μm
Detecting angle in azimuth: 0 to 360°
Detected direction sectors: 16
Front zone: 7 sectors ×12°
Rear zone: 8 sectors ×30°
1 sector ×36°
Angular range of detection in elevation: –6 to +20°
Power supply: 27 V + 3 V/–9 V

Contractor
PCO SA, Poland.

PCO SSP-1 OBRA-3 Universal Self Covering laser warning system for vehicles

Type
Land laser warner.

Development/Description
The OBRA-3 system is designed for various kinds of military vehicles (such as BMP-1 or other armoured fighting vehicles), ships, command centres and so on.

OBRA-3 system detection waveband range covers all currently used lasers from Nd:YAG to CO2. Older ruby lasers can also be detected. The system is equipped with four to eight identical detection heads dependent on vehicle size. Other features include:
- optical and acoustic signalling of detected radiation
- identification of:
- laser source type (impulse range-finder or laser illuminator)
- laser source direction
- update of the direction of the laser source
- elimination of reflected radiation
- selection of launchers and firing the smoke countermeasures
- indication of time elapsed from the moment of illumination
- operation in three basic modes; automatic, semi-automatic, manual

Operational status
Under development, a prototype has been produced. For future use with Polish Army vehicles.

Specifications
Laser spectral detection band: 0.6 mm to 11 mm
Angular range of detection:
Horizontal: 0 to 360°
Vertical: –6 to +30°
Ambient operating temperature: –30 to +50°C

Manufacturer
PCO S.A., Warszawa, Poland.

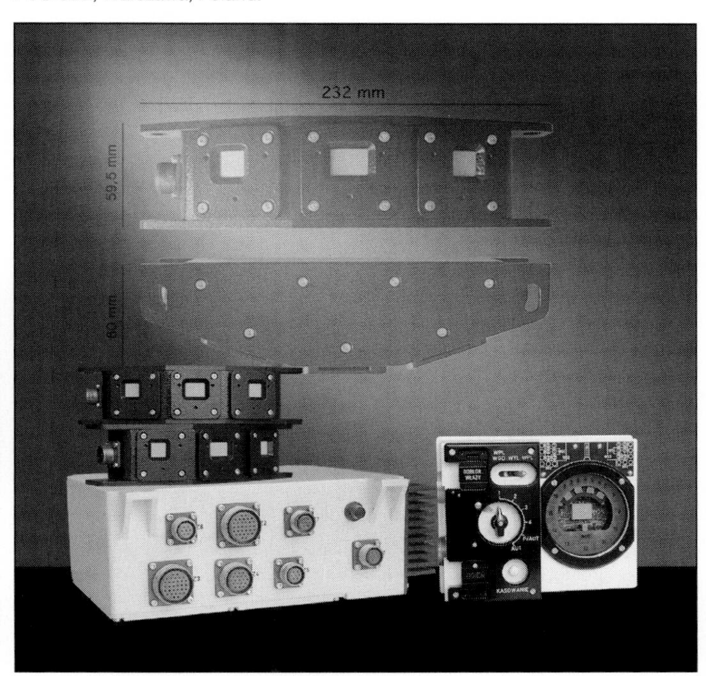

The PCO SSP-1 OBRA-3 laser warning system 0089915

Raytheon AN/VVR-1 Laser Warning Receiver System (LWRS)

Type

The AN/VVR-1 Laser Warning Receiver System (LWRS) detects, recognises and reports laser threats with precise ±1° Angle-of Arrival (AoA), providing the valuable seconds necessary to manoeuvres, slew the turret to combat incoming threats, deploy smoke or initiate other countermeasures.

The AN/VVR-1 identifies whether your vehicle has been targeted by a laser range-finder, designator, or beamrider. Consisting of four sensor heads and an interface unit, the LWRS provides 360° coverage about the vehicles and reports the threats through existing or stand-alone audio/visual displays.

As a Horizontal Technology Integration (HTI) programme, the AN/VVR-1 is built upon proven, transferable technology. Raytheon Optical Systems, Inc has been producing the AN/AVR-2A(V) laser detecting set since 1990 for a variety of helicopter platforms. The core technologies from this field-tested system are integrated into a ruggedised package designed to meet the rigorous requirements of ground combat vehicle operations. The result is the AN/VVR-1 Laser Warning Receiver System.

Specifications

Laser spectral coverage: 0.5-1.6 μm
Detects: designators, beam-riders, range-finders
FOV: 360° azimuth, ±55° for beam-riders
Response time: <100 ms

Contractor

Raytheon Optical Systems, Inc.

The AN/VVR-1 laser warning receiver system 0041607

Raytheon Model 218S Laser Warning Receiver System (LWRS)

Type

Land laser warner.

Description

The Model 218S is intended to provide warning of laser illumination for armoured fighting vehicles. It can determine whether the laser is a range-finder, beam-rider or designator.

Operational status

In production for the US Light Armoured Vehicle.

Contractor

Raytheon Optical Systems, Inc.

Romanian SAILR laser and radar illumination warning system

Type

Land laser warner.

Description

SALLR is a laser and radar illumination warning system designed to provide individual protection for armoured vehicles (MBTs, ICVs, APCs) versus various antitank weapon systems that use active radiolocation and/or laser illumination.

The system determines the source type and direction, gives the crew early warning and information about an enemy threat and allows the crew to take appropriate self-protective measures: manoeuvre and/or smoke screen generation. The system components are: a laser detector head unit, radar detector head unit, processing and counteraction unit and the interconnection cables.

The processing and counteraction unit generates the warning signal of the enemy threat and displays the source type and direction by special symbols.

The system enables automatic, semi-automatic or manual discharge of smoke grenades. The system is designed in a modular way which makes it easy to install and is protected against external factors such as trees and accidental collisions.

Operational status

Not stated.

Specifications

Laser detector head unit
Spectral band: 800 to 1540 nm
Total field of view
azimuth: 0 to 360°
elevation: –5 to +90°
Resolution: 15°
Radar detector head unit
Spectral range: 33+37 GHz
Total field of view
azimuth: 0 to 360°
elevation: +20 to +60°
Resolution: 30°
Processing and counteraction unit
Operating voltage: 24 V
LCD display unit: 128 × 128 pixels
Threat types: laser: single; pulsed radar
Decision time: 50 ms
Temperature
40 + 50° operating
50 to + 60°C storage
Relative humidity: 95% ± 3% at + 50°C
Size (diam × h):
detection unit – 400 × 430 mm
processing and counteraction unit – 260 × 160 × 100 mm
Weight:
detection unit - 65 kg
processing and counteraction unit - 6 kg

Contractors

Manufacturer – Military Equipment and Technology Research Agency.

Exporter - Romtehnica, Romania.

Romanian Warning System on Laser Illumination (WSLI)

Type

Land laser warner.

Development

The Warning System on Laser Illumination (WSLI) has been designed for installation on tracked and wheeled armoured vehicles to enable vehicle crews to take action to avoid being hit.

Description

The four main components of the WSLI are the power supply and signalling unit mounted inside the vehicle, detector units mounted outside the vehicle (normally on the turret) and the interconnecting cables. In addition there is a portable test device.

The system is not influenced by solar radiation, high-powered searchlights, electrical dischargers, artillery blasts, fires or radio equipment operating nearby.

Operational status

Believed to be in production and in service with the Romanian Army.

Warning System on Laser Illumination (WSLI) with signalling unit in centre and three circular detection units

Specifications

Dimensions

Detector: 130 mm diameter × 120 mm high

Signalling unit: 225 × 200 mm

Detected radiation spectrum

Horizontal plane: 12 directions

Vertical plane: 3 directions

Signalling duration: 11 s

Signalling mode: optical for direction and warning duration, acoustic for every detected pulse

Contractor

Romanian State Factories.

Thales Cerberus laser warning and defensive aid system

Type

Laser detection and Defensive Aid System (DAS).

Description

Developed by Helio (now Thales AFV Systems) the Cerberus laser warning and DAS system for armoured vehicles provides detection and warning of laser attack and the means for manual or automatic discharge of grenades to provide a smoke screen. The system comprises three main assemblies: a set of detector arrays, a control/display box and grenade dischargers. Each detector array has seven sensors, six of which cover an arc of 180° in the horizontal plain and one that provides overhead detection cover. Up to 12 separate detector arrays may be used on the vehicle to increase the detection capability, although a typical installation will have four to six detector arrays.

The detectors will respond to laser energy sources including:

- Frequency-doubled Nd: YAG
- Ruby
- Gallium Arsenide

Thales Cerberus laser warning and integrated grenade discharge systems-firing
0019552

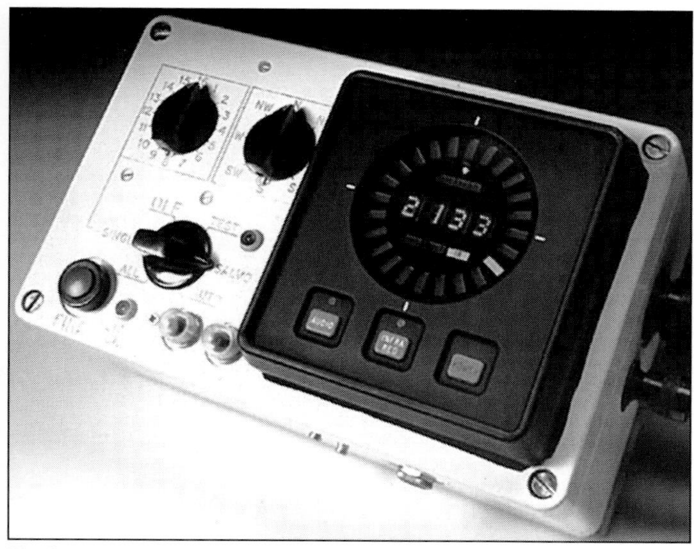

The Cerberus control box
0088172

- Nd: Glass
- Nd: YAG
- Er: Glass (eye-safe)
- Raman-shifted Nd: YAG (eye-safe).

They can differentiate between single-pulse lasers (range-finders) and multipulse lasers (target markers/designators).

The grenade discharger control box has two operational functions. These are to display information about the laser energy detected and to control the discharge of grenades.

The displays on the Cerberus control box include:

(a) a four digit display showing the bearing of the detected energy relative to the vehicle axis or the axis of the gun if the system is mounted on the turret

(b) a ring of 24 indicators that show the direction of any detected energy to within ±7.5°

(c) indicators for single-pulse lasers (range-finders), multipulse lasers (target markers) and overhead laser attack.

In addition, an audible warning is provided which can be integrated into the vehicle intercom system. The control box incorporates a self-test routine that is activated each time the system is switched on. Grenade discharge is initiated either automatically or manually. The controls comprise two discharge pattern selector switches, a mode selector and a firing/test button. The company's FVG66 and FVG76 grenade dischargers are of modular design and can be installed singly or in groups to suit each vehicle application and/or customer's specific requirements. In addition to this, Cerberus is able to operate with any other standard grenade discharger system.

Operational status

In service, in production.

Contractor

Thales AFV Systems (formerly Helio Ltd).

UPDATED

Thales Optronics LWD 2 series

Type

Laser warner (land).

Description

The LWD 2 series Laser Warning Detector is a development of Thales Optronics' (formerly Avimo's) LWD 21 and is designed for vehicle mounting. The system consists of a control box, a junction box and a number of semi-circular detector arrays. All assemblies are hermetically sealed. Most applications would require between four and eight detector arrays. The number of detector arrays depends on application and can vary from a single array to a maximum of 12.

The detector arrays each have an azimuth field of view of more than 180° with a resolution of 15° and can be deployed on the front, sides and rear of a vehicle either parallel or normal to the vehicle axis.

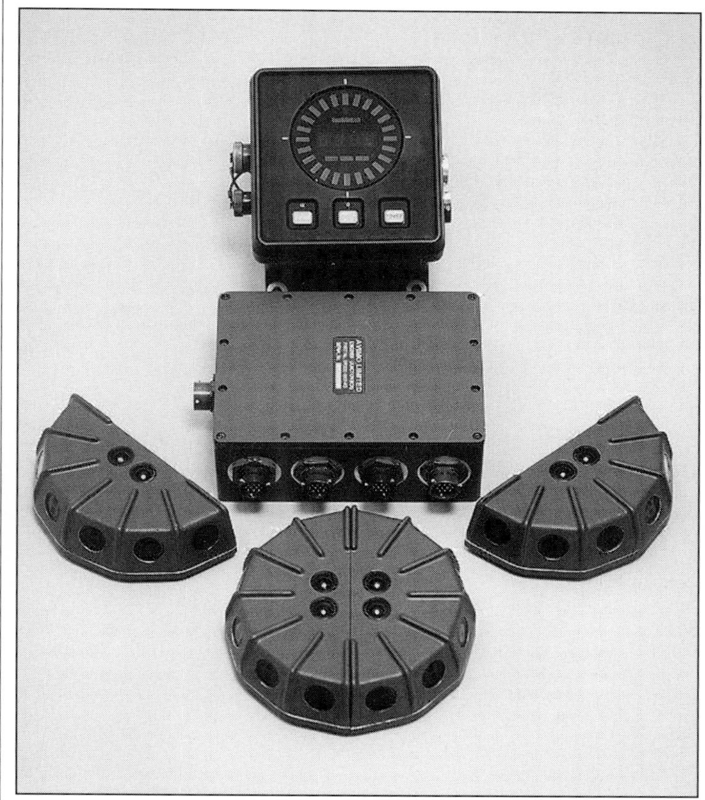

The elements of the Thales LWD 2 Laser Warning Detector

The junction box performs the functions of interconnecting the detector arrays to the control box and collating the data from the arrays before transmitting it to the control box. Switches within the junction box allow it to be set to recognise any particular configuration of arrays. The junction box can be sited inside the vehicle, or outside with a suitable armour cover.

The control box contains the processing electronics, the operator controls and the display panels. The box provides visual and audible warning of a detected threat and indicates whether a single-pulse rangefinder, or multiple pulse rangefinder/designator has been detected. Directional information is indicated by a ring of 24 LEDs and complemented by a four-digit display which gives the bearing in degrees or mils (preset).

An overhead indicator warns of threats from above when detector arrays with overhead facility are in use. Two operator switches are provided: power, infra-red select and audio alarm select. An external socket provides RS-422 communication both into and out of the control box to allow the system to interface with countermeasure devices or to be externally controlled. A second socket provides an alarm output in the form of a relay contact and an audio tone for introduction into the vehicle's internal communications system.

Operational status

The LWD system is in production and has been delivered to several customers for installation on a number of different platforms.

Specifications

Detection (each array)
Azimuth field of view: >180°
Vertical field of view, standard: –12 to +47°
Vertical field of view, overhead option: –12 to +90°
Azimuth resolution: ±7.5°
Spectral band: 0.4-1.6 μm
Detector response time: <0.2 s, typically <0.1 s
Environmental operating temperature range: –40 to +55°C
Power supply voltage: 18-32 V DC
Max power consumption: depends on array quantity, max 20 W
Nato Stock No: NSN S865-99-083-3364

Contractor

Thales Optronics (formerly Avimo Ltd).

VERIFIED

AIR DEFENCE MISSILES

VEHICLES

Aselsan ATILGAN pedestal-mounted low-level air defence missile system

Type

Air defence missile system (vehicle).

Development

Under a contract awarded by the Turkish Undersecretariat of Defence Industries (SSM), Aselsan designed, developed and integrated the ZIPKIN and ATILGAN low-level air defence missile systems on two different vehicle types.

One of the prototypes, ATILGAN, is based on the M113 A2 tracked APC chassis and manufactured to meet the requirements of the Turkish Armed Forces.

ATILGAN has been subjected to the following functional qualification and performance tests:

- Missile Booster Firing Tests
- Aircraft and Helicopter Tracking Tests with predefined scenarios
- Machine Gun Firing Tests
- Missile Firing Tests against Stationary Targets
- Missile Firing Tests against Drone Targets
- Field Operational Tests.

During the live firing tests held in August-September 1998, ATILGAN performed on-the-move and stationary firings against drones and fixed targets. The result was 100 per cent success. ATILGAN scored direct hits in all firings. The acceptance committee comprising the TAF and the SSM representatives, approved and accepted ATILGAN. This finalised the development phase of the programme.

The PMADS programme represents the largest system design and integration ever accomplished in Turkey.

Description

ATILGAN is a fully automated firing unit for very short-range air defence missiles, providing autonomous as well as co-ordinated operation with $C^3$1 systems and other air defence assets. Its modular turret incorporating LRU subsystems, can be integrated on various types of carrier vehicles.

The main mission of ATILGAN is the low-level air defence of stationary and moving forward troops, convoys and tactical bases in the battlefield. The vehicle carries a crew of three; a driver, gunner and commander, all under armour protection. A gyrostabilised one-man turret launcher assembly is fitted on the top of the decking, that carries eight ready to fire missiles and provides on-the-move surveillance, acquisition and fire capabilities. An additional eight reload rounds can be carried within the vehicle in their launcher containers.

Although originally based on the Stinger missile, the flexible system design allows many SHORAD missiles to be integrated on the Pedestal-Mounted Air Defence System (PMADS) systems. The system also carries a 12.7 mm (0.5 in calibre) M3 heavy machine gun mounted on an elastic cradle for self defence and missile dead zone coverage. The machine gun has full remote functions commanded on the system control unit and can be fired either in bursts or in continuous mode, selectable from the system control unit.

The sensor suite incorporates a second-generation, two field of view focal plane array thermal imager and a daylight TV camera with zooming capability for passive day/night surveillance, acquisition and tracking. Also a multipulse laser rangefinder is integrated for target ranging.

The ATILGAN pedestal-mounted low-level air defence system 0089899

PMADS's fire-control computer provides automated system functions such as:

- remote control of all subsystems
- turret control and stabilisation
- automatic slewing of turret to the target co-ordinates pre-assigned by C^3I system
- automatic target tracking
- automatic target type (FW/RW) recognition
- 'Target in Range' warning if the target is within the missile engagement envelope
- automatic super elevation and lead angles insertion before firing,

The fire-control computer also has a flexible hardware and software architecture for evolving mission requirements.

All system functions are commanded from a unique system control unit, which can also be easily dismounted from the vehicle and carried for remote operation up to 50 m away from the platform.

Operational status

A serial production contract for ATILGAN was signed in November 2001 with delivery of 70 systems between 2003-07.

Contractor

Aselsan Inc, Microwave and System Technologies Division, Ankara.

Aselsan ZIPKIN pedestal mounted low-level air defence missile system

Type

Air defence missile (vehicle).

Development

Under a contract awarded by the Turkish Undersecretariat of Defence Industries (SSM), Aselsan designed, developed and integrated the ZIPKIN and ZIPKIN low-level air defence missile systems on two different vehicle types.

One of the prototypes, ZIPKIN, is based on the Land Rover 130 (4 × 4) light vehicle chassis and manufactured to meet the requirements of the Turkish Armed Forces.

ZIPKIN has been subjected to the following functional qualification and performance tests:

- Missile Booster Firing Tests
- Aircraft and Helicopter Tracking Tests with predefined scenarios
- Machine Gun Firing Tests
- Missile Firing Tests against Stationary Targets
- Missile Firing Tests against Drone Targets
- Field Operational Tests.

During the live firing tests held in August-September 1998, ZIPKIN performed firings against drones and fixed targets. The result was 100 per cent success. ZIPKIN scored direct hits in all firings. The acceptance committee comprising the TAF and the SSM representatives approved and accepted ZIPKIN. This finalised the development phase of the programme.

The PMADS Programme represents the largest weapon system design and integration ever accomplished in Turkey.

Description

ZIPKIN is a fully automated firing unit for very short-range air defence missiles, providing autonomous as well as co-ordinated operation with $C^3$1 systems and other air defence assets. Its modular turret incorporating LRU subsystems, can be integrated on various types of carrier vehicles.

The main mission of ZIPKIN is the low-level air defence of fixed assets, like radar, air bases and harbours. The vehicle carries a crew of two; a driver and a gunner. A pedestal-mounted launcher assembly is fitted to the rear decking which carries four ready-to-fire missiles. The gunner is seated in the vehicle cab, however, he can also control the system at a remote location 50 m away from the vehicle.

Although originally based on the Stinger missile, the flexible system design allows many SHORAD missiles to be integrated on the PMADS systems. The system carries a 12.7 mm (0.5 in calibre) M3 heavy machine gun mounted on an elastic cradle for self defence and missile dead zone coverage. The machine gun can be remotely commanded on the system control unit and can be fired either in bursts or in continuous mode, selectable from the system control unit.

The sensor suite incorporates a second-generation, two field of view focal plane array thermal imager and a daylight TV camera with zooming capability for passive day/night surveillance, acquisition and tracking. Also a multipulse laser rangefinder is integrated for target ranging.

PMADS's fire-control computer provides automated system functions such as:

- remote control of all subsystems
- turret control
- automatic slewing of turret to the target co-ordinates pre-assigned by C3I system
- automatic target tracking

ZIPKIN pedestal-mounted low-level air defence system 0089882

- automatic target type (FW/RW) recognition
- 'Target in Range' warning if the target is within the missile engagement envelope
- automatic super elevation and lead angles insertion before firing.

The fire-control computer also has a flexible hardware and software architecture for evolving mission requirements.

All system functions are commanded from a unique system control unit, which can also be easily dismounted from the vehicle and carried for remote operation up to 50 m away from the platform.

Operational status

A serial production contract for ZIPKIN was signed in November 2001 with delivery of 78 systems between 2003-07.

Contractor

Aselsan Inc, Microwave and System Technologies Division, Ankara.

BGT HF/KV hypersonic surface-to-air missile

Type

Air defence missiles - vehicles.

Development

In mid-1998, the German company of Bodenseewerk Geratetechnik (BGT) showed a full-scale mockup of its new HFK/KV (Hochgeschwindigkeits-Flugkorper - Kill Vehicle) hypersonic missile concept for future short-range air defence systems.

Since 1990, BGT has been working under contracts from the German MoD to design and develop concepts of hypervelocity missiles. At the end of 1993 and in early 1998, three successful flight tests of M6 demonstrator missiles were carried out. An additional two flights carried out at the end of 1995 and in 1997 demonstrated the concept of lateral thrust-controlled missiles. Although currently a technology demonstration programme, the HFK/KV is seen as a potential replacement for the current missiles used in the Euromissile Roland self-propelled surface-to-air missile system used by the German Army (on tracked Marder 1 chassis) and Air Force (on MAN 8 × 8 and 6 × 6 wheeled chassis). System studies of missile concepts have been underway by BGT since mid-1996. During a test flight late in 1998, a rocket motor exploded 0.2 seconds into the first test flight. The all-up round stays within the weight and dimension restraints established by the Roland weapon but is designed to reach speeds of approximately M6 after 1 to 2 seconds following ignition and post booster burnout and separation and ranges of more than 12 km with a terminal speed of M3.

Description

The 2.8 m long, 60 kg launch weight two-stage weapon is accelerated to its maximum speed in little over a second by a 44 kg weight high-energy solid propellant rocket booster. This separates after burnout, allowing the 90 mm narrower diameter, less than 1 m long kill vehicle to continue its trajectory towards the target under inertial and/or seeker guidance. This 16 kg kill vehicle is unpowered but light and highly manoeuvrable. It is equipped with an imaging infra-red seeker, inertial reference unit and aerodynamic control actuation system. The HFK/KV is a hit-to-kill missile and carries a 5 kg enhanced warhead. The nose-mounted infra-red seeker is of the imaging type whose multi-element detector uses a scan mechanism to create an image. The infra-red dome is covered by a protective shroud during the boost phase to protect it from the worst effects of

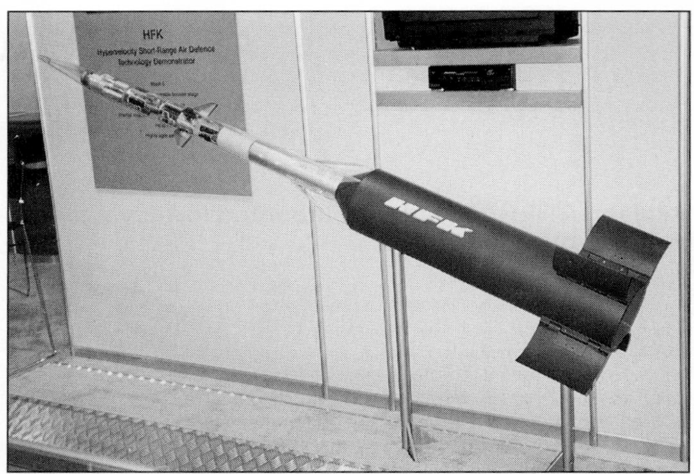

Full-scale model of the BGT HF/KV hypersonic surface-to-air missile (Christopher F Foss) 0069504

aerodynamic heating and would only be uncovered for the final 2 to 3 seconds of flight.

Operational status

Technology demonstrator. No plans for volume production.

Contractor

Bodenseewerk Geratetechnik (BGT).

CPMIEC Hongying-5C low-altitude surface-to-air missile system

Type

Air defence missile (vehicle).

Development

Summary: launcher - infra-red tracker, laser range-finder and TV.

The CPMIEC Hongying-5C (or HN-5C) is the designation used to describe a vehicle-mounted Hongying-5 system. Evaluation trials were completed in June 1986 and production is believed to have taken place for the Chinese armed forces.

The main roles of the HN-5C are stated to be: protection of mechanised units; air defence for forward units; and air defence for high-value targets such as airfields.

Description

The first HN-5C version shown was an HRB-230 (4 × 4) cross-country vehicle with a forward fully enclosed control cab which contains the fire-control equipment. To the rear of the cab is a pedestal upon which a bank of four HN-5 or HN-5A missiles is mounted either side in the ready to fire position.

Between the two banks of missiles is the fire-control electro-optics which consist of an infra-red tracker, a range-finder and a TV camera. Detecting and tracking the target and launching the missiles can be accomplished either manually or automatically.

The weapon assembly can also be mounted on other chassis as it only weighs in the order of 2,000 kg. The WZ 525 version had turret elevation limits of –2 to +80°, a traverse of 360° and total turret system weight of 2,100 kg. The crew was two with

The Hongying-5C (HN-5C) is essentially an HRB-230 truck chassis. On the rear is mounted a pedestal with four Hongying-5A (HN-5A) SAMs either side

the maximum effective engagement range stated as being 4,200 m between altitude limits of 50 and 2,300 m. Typically, eight reserve missiles can be carried within the launch platform.

Operational status

Production as required. In service with the Chinese People's Liberation Army.

Specifications

HN-5C turret assembly
Crew: 2
Combat weight: 2,000-2,100 kg
Typical elevation limits: –2 to +80°
Traverse: 360°
Missile type: single-stage low-altitude
Missile weight: incl launcher 16.0 kg
Max missile operational altitude: 2.3 km
Min missile operational altitude: 50 m
Max missile slant range: 4.2 km
Min missile slant range: 800 m
Max target speed: 150 m/s head-on 260 m/s tail-on
Armament: 2 banks of 4 ready to fire HN-5 family missiles with up to 8 reloads in launch platform hull

Contractor

China National Precision Machinery Import & Export Corporation.

DRS Avenger pedestal-mounted Stinger self-propelled air defence system

Type

Air defence missile (vehicle).

Development

Summary: target acquisition uses FLIR. Stinger missile has infra-red/ultra-violet homing seeker.

In the early 1980s, the Defense Systems Division of the Boeing Aerospace Company developed the Avenger air defence system as a private venture. Total time from concept through to delivery to the US Army for trials was only 10 months. The product line has since been acquired by DRS Technologies

The Avenger consists of a 4 × 4 High-Mobility Multipurpose Wheeled Vehicle (HMMWV) with a turret mounted in the rear with eight missiles in the ready to launch position. The turret can also be deployed as a fixed stand-alone unit.

Target acquisition is either by direct vision using the optical sight or through the use of a Forward-Looking Infra-Red (FLIR) system. Mounted either side of the turret are four Stinger SAMs which are identical to those used in the portable version.

During 1984, the Avenger system was evaluated by the US Army Air Defense Board and, during this evaluation, 171 of the 178 fixed- and rotary-wing aircraft targets were successfully engaged by the system during day and night operations.

In 1986 the US Army issued a Request For Proposals (RFP) for a Pedestal-Mounted Stinger (PMS), or Line of Sight-Rear (LOS-R) as one of the five key parts of the Forward Area Air Defense System and the Boeing Aerospace Avenger was selected. The extensive trial series included firing, target acquisition and tracking and environmental tests.

In August 1987, the Defense Systems Division of Boeing was awarded a contract by the United States Army Missile Command to commence production of the PMS air defence system. The initial contract was for US$16.2 million for the first option buy of 20 systems. With options the contract value reached US$232 million on 325 fire units over a five year period together with associated logistic support. The second option, covering 39 systems, was exercised in 1988 (with deliveries from July 1989 through to June 1990), the third for 70 firing units in 1989, the fourth for 72 firing units in March 1990 and the fifth for 72 firing units in May 1991. A US$436 million multiyear production contract for 600 Army systems and 79 Marine Corps systems over five years was approved with the 1991 budget. By June 1993, 1004 Avenger systems had been ordered by the US Army. First production PMS systems were delivered in November 1988, the system becoming operational with the US Army in 1989 and was the first shoot-on-the-move air defence weapon to enter production for the US Army.

The US Army's current acquisition plan is for 1,004 systems with the US Marine Corps, requiring up to 275 units. The US National Guard is also formulating a requirement. As the Stinger has been sold to a number of foreign customers, Boeing believes that foreign military sales could eventually bring the total production figure to over 1,800 units. By mid-1993 four NATO countries had the Avenger in their five-year procurement plans. In addition to firing the original Stinger missile model, the Avenger is also able to launch follow-on models including the Stinger POST.

Main subcontractors to Boeing Aerospace are: Lockheed Martin Armament Systems - computer and remote-control unit; Lockheed Martin Ordnance Systems - electric turret drive as used in the M2 Bradley Infantry Fighting Vehicle; CAI - CA-562 optical sight; DBA - auto-tracker; FN Herstal SA - 12.7 mm M3P machine gun; Raytheon Systems Company's - forward-looking infra-red system; Texstar - canopy; Raytheon - CO_2 laser range-finder.

Raytheon Systems Company supplies the AN/VLR-1 FLIR system to enable Stinger to acquire targets at night and in bad weather.

The system can be installed on other types of chassis, tracked and wheeled and is also fully air portable.

For the US Army PMS system, Stinger missiles are standard, but its design is such that it can accommodate other sensors and other weapon systems including Hellfires, HYDRA 70 70 mm unguided rockets and other infra-red seeking or RBS 70 laser-guided missiles.

Description

The design of Avenger is modular so that it can accept advances in technology such as the replacement of Stinger by a laser beam-rider missile, new sensors, advanced fire-control system, enhanced position locator, reporting system, user system and hand-held computer, HVRs or a larger calibre weapon.

The gunner is seated in the electrically powered turret which can be traversed by 360°. If required, the complete fire unit can be removed from the HMMWV and used as a stand-alone system. The Stinger pods, which are able to accommodate any Stinger model without modification, can be elevated from –10 to +70°. The gunner looks through a sight glass on which is seen the projection of a driven graticule display. The graticule indicates the aiming point of the missile seeker, confirming to the gunner that the missile seeker is locked on to the target being tracked and planned to engage.

Sensor package mount includes a CAI optical sight, Raytheon AN/VLR-1 thermal imager, DBA Automatic Video Tracker (AVT) and a Raytheon CO_2 eye-safe laser range-finder, thus enabling the system to acquire and track targets under a wide range of operational conditions.

The thermal imager, with an electrically operated optics cover is mounted on the left launch arm beneath the missile pod. This is a dual field of view system operating in the 8 to 12 μm waveband with SPRITE detector. The gunner's foot pedal is used to select the field required. The gunner tracks the target either by direct vision using the optical sight or through the use of the thermal imager system.

The AVT provides an automatic tracking mode. The imager video target-to-bore sight error signals determine the azimuth and elevation repositioning required by the turret drive system in order to maintain turret positioning on the target.

The laser range-finder is mounted on the left-hand launch arm behind the imager. Target range is displayed on a hand-held display in the turret. Target range is processed by the Avenger control electronics for use in the automated fire-permit and fire-control algorithms. The Avenger's FCS processes data from the LRF and displays an advisory fire-permit symbol in the sight and thermal imager display. The fire-permit function incorporates Stinger's engagement boundaries to ensure that targets are only engaged when within missile range. The electric turret drive is gyrostabilised so as to maintain automatically the missile pod aiming direction regardless of the vehicle's movement.

The gunner has a hand controller on which the missile and gun controls are located. In addition, the tracking control can be transferred to the automatic tracking systems, one of which uses signals from the uncaged missile seeker and the other, data from the imager's video auto-tracker, to track the target until the gunner is ready to fire. This allows the gunner to concentrate on target identification. The firing sequence is fully automated and the gunner has only to pull the fire trigger to initiate the launch sequence and immediately select and prepare the next missile for firing.

For self-protection and for coverage of the Stinger dead zone, an M3P 12.7 mm MG with 200 rounds of ready-use ammunition is attached to the right-hand launch beam as supplementary armament. Mounted either side of the turret is a pod of four Stinger low-altitude surface-to-air missiles.

In addition to the eight missiles in the ready to launch position, an additional eight Stingers are carried in reserve. Reloading takes less than 4 minutes.

Operational status

In production. In service with the US Army and US Marine Corps.

Specifications

FIM-92C Stinger Missile
Length: 1.52 m
Diameter: 70 mm
Wing span: 91 mm
Launch weight: 10.1 kg
Propulsion: solid fuel ejector and boost/sustainer rocket motors
Guidance: passive infra-red/ultra-violet homing seeker
Warhead: 3 kg HE fragmentation

The DRS Avenger Stinger air defence system 0004738

Max speed: M2.2
Max effective range: > 4.50 km
Min effective range: 200 m
Max effective altitude: 3.80 km
Min effective altitude: effectively ground level
Reload time: 4 min
Platform
Crew: 2
Configuration: 4 × 4
Combat weight: 3,900 kg
Weight of turret module: 1,134 kg
System
Length: 4.953 m
Width: 2.184 m
Height: 2.59 m
Turret module
Length: 2.13 m
Width: 2.159 m
Height: 1.778 m
Ground clearance: 406 mm
Track: 1.81 m
Wheelbase: 3.30 m
Angle of approach/departure: 69/45°
Max speed: 105 km/h
Range: 563 km
Fuel capacity: 94 litres
Max gradient: 60%
Side slope: 40%
Vertical obstacle: 560 mm
Fording: 760 mm
Engine: V-8 6.2 litre, air-cooled diesel
Transmission: automatic, 3 forward and 1 reverse gears
Transfer box: 2 speed
Suspension
Front and rear: independent, double A-arm, coil spring
Steering: power-assisted
Turning circle: 14.63 m
Brakes
Front: hydraulic
Rear: disc
Tyres: 36 × 12.5-16.5
Electrical system: 24 V DC
Armament: 2 × 4 Stinger SAMs; 1 × 12.7 mm MG
Turret traverse: 360°
Weapon elevation: –10 to +70°

Contractor
DRS Technologies, Electro-Optical Systems Group.

General Dynamics Light Armored Vehicle Air Defense (LAV-AD)

Type
Air defence missile (vehicle).

Development
Summary: sensor suite has thermal image, daylight TV and laser range-finder; the missile has an infra-red seeker.

In December 1987, the US Marine Corps awarded two companies, FMC Corporation and Lockheed Martin contracts to develop a gun/missile/rocket hybrid system. Each company built two prototypes based on the Light Armored Vehicle (8 × 8) chassis which was provided by the Marine Corps and these commenced trials in August 1990.

The main role of the LAV-AD is to engage fixed-wing aircraft and helicopters, with a secondary role to engage ground targets using its cannon. Typically the Stingers would be used to engage targets out to 6,000 m with the cannon engaging targets out to 2,000/2,500 m.

Although the prototype systems had a single pod of four Stinger SAMs and a pod of seven HYDRA 70 rockets, production systems will have two pods each of four Stingers with the HYDRA 70 rocket pod being optional.

Following trials with the four prototype vehicles, in June 1992, the US Marine Corps selected the Lockheed Martin LAV-AD to meet its future requirements. Lockheed Martin Armament Systems became General Dynamics Armament Systems in 1996.

The initial US Marine Corps requirement is for 21 Light Armored Vehicle – Air Defense and LAV-AD entered production in 1996.

In addition to General Dynamics, other members of the LAV-AD team are the Diesel Division of General Motors of Canada for vehicle integration, Raytheon for the FLIR/TV sight and Lockheed Martin Defense Systems for the all-electric turret drives.

Description
The sensor suite fitted to the LAV-AD includes a thermal imager, made by Raytheon, daylight TV, eye-safe laser range-finder and automatic tracking and fire-control system.

General Dynamics Armament Systems Light Armoured Vehicle-Air Defense (LAV-AD) 0004739

The General Dynamics LAV-AD is based on a modified LAV (8 × 8) chassis. The turret installed on the LAV-AD is based on the Blazer two-person power-operated turret and is armed with the GAU-12/U 25 mm Gatling Gun and eight Stinger surface-to-air missiles.

In addition to the eight missiles in the ready to launch position, each version carries a further eight missiles in reserve which are loaded manually. Each version also has a 7.62 mm machine gun for local protection and two banks of four electrically operated smoke grenade dischargers.

Turret controls are all electric with manual controls provided for emergency use. A stabilisation system is fitted as standard which allows the system to engage targets while on the move.

The turret includes laser-safe ballistic turret windows and an armoured structure which contains numerous oblique angles for improved ballistic protection.

Operational status
In production for the US Marine Corps.

Specifications
Weight
Combat (complete system): 13,410 kg
Turret: 2,676 kg
Crew: driver, gunner and commander
Armament: 1 × 25 mm GAU-12/U Gatling Gun with cyclic rate of fire of 1,800 rds/min, 8 × Stinger SAMs (ready use)
Ammunition: 990 × 25 mm rounds (385 ready to fire); 16 × Stinger SAM (8 ready to fire)
Firing rate: 1,800 shots per minute
Turret traverse: 360°
Weapon elevation/depression: +65/–8°
Turret acceleration
Azimuth and elevation: 2 rad/s^2
Turret velocity
Azimuth and elevation: 1 rad/s
Sight: FLIR, TV and eye-safe laser range-finder
Digital fire control: full solution, fire on the move
Sensors: temperature, pressure, wind and vehicle tilt

Contractors
General Dynamics Armament Systems.

General Dynamics, Thales Blazer air defence turret

Type
Air defence missile/gun (vehicle turret).

Development
Summary: the missile has an infra-red seeker; the fire-control system has thermal imaging/TV, laser range-finder.

As a private venture, General Dynamics Armament Systems (formerly Lockheed Martin Armament Systems) and Thales, Air Defence Systems have teamed to develop the Blazer air defence turret.

The first complete prototype was shown for the first time in June 1994 and, in late 1994 firing trials with both the 25 mm cannon and Matra Mistral surface-to-air missiles were carried out in France.

The all-electric turret can be installed on the M2 Bradley chassis, the MOWAG Piranha (8 × 8), the Textron Marine and Land Systems Cadillac Gage LAV-300 (6 × 6) and LAV-150S (4 × 4) and similar wheeled and tracked vehicles.

The Blazer Turret is armed with eight Stinger missiles and a rapid-fire GAU-12/U 25 mm Gatling gun 0004747

Description

Blazer consists of a 25 mm GAU-12/U Gatling Gun and four Raytheon Missile Systems Stinger or MBDA Mistral fire-and-forget surface-to-air missiles on a two-person turret.

The basic Blazer turret features a digital fire-control system, eye-safe laser range-finder, thermal imaging/TV stabilised sight and a crew of two consisting of the commander and gunner. The system is fitted with the Thales TRS 2630 digital radar, the Mistral or Stinger missiles or, as a growth option, command to line of sight missiles.

The two-person turret houses both gunner and commander, each capable of full system operation including acquisition, tracking weapon selection and firing.

Main armament of the Blazer turret comprises the Lockheed Martin 25 mm GAU-12/U Gatling gun which fires the Bushmaster family of ammunition at a cyclic rate of fire of 1,800 rds/min.

In addition, four or eight infra-red seeker missiles are mounted above the gun cradle and integrated into the Blazer's fire-control system. Command guided missiles such as Starstreak can also be fitted as a result of the system's pointing accuracy.

A thermal imaging/TV sight is included for viewing and auto-tracking. The system has demonstrated day/night capability and the ability to track and fire while the vehicle is moving at up to 50 km/h over uneven terrain. In 1985 testing in the US, both gun and missile kills were recorded under such conditions with a prototype system.

The Thales RS-2630 2-D radar has a range of 17 km, IFF, automatic track-while-scan and data exchange for netting capability. The 25 mm cannon can engage targets up to 2,500 m and the missiles at 6,000 m. Depending on the chassis, additional missiles can be stored internally. Electronically operated smoke dischargers are located in banks of four on either side of the turret.

Operational status

Preproduction prototype.

Specifications

Crew: 2 (commander/gunner)
Armament: 25 mm 5-barrel GAU-12/U, 4 or 8 fire-and-forget SAMs (Stinger, Mistral)
Traverse: 360°
Elevation/depression: +65/–8°
Sight: thermal imaging/TV and eye-safe laser range-finder
Fire control: digital, full fire on the move solution
Sensors: temperature, pressure, wind and vehicle tilt, Thales TRS 2630 acquisition radar

Contractors

General Dynamics Armament Systems.
Thales.

Hägglunds Lvrbv 701 RBS 70 low-altitude surface-to-air missile system

Type

Air defence missile (vehicle).

Development

Summary: the missile uses laser beam-riding guidance.

In February 1983, the Swedish Defence Matériel Administration awarded a contract to Hägglunds Vehicle AB for the conversion of a number of obsolete Ikv-102 and Ikv-103 self-propelled infantry cannon into Lvrbv 701 RBS 70 surface-to-air missile carriers. First vehicles were delivered to the Swedish Army in 1984 and production continued till 1986.

The modifications have been extensive and include the replacement of the engine and transmission, extending the crew compartment, improving protection and fitting new communications equipment and observation devices.

Lvrbv 701 anti-aircraft vehicle with Saab Bofors RBS 70 surface-to-air missile system in operating position (FMV)

Description

The gun has been removed from the glacis plate and the position plated over. The driver sits at the front of the hull on the left side and has a single-piece hatch cover and fixed periscopes for observation to the front and sides. The commander sits to the right of the driver and has observation periscopes and a single-piece hatch cover which opens to the right rear.

In the centre of the hull roof is a large circular opening with a two-piece hatch cover that opens to either side. The Saab Bofors RBS 70 surface-to-air missile system is carried inside the hull and raised up when required for action. The IFF system is mounted on the stand above the missile launcher tube. Reserve missiles are stowed in the hull rear above the engine and transmission compartment with access via the front edge of the magazine, which folds upwards.

Operational status

Production complete. In service with the Swedish Army.

Specifications

RBS 70 missile
Length: 1.32 m
Diameter: 106 mm
Wing span: 320 mm
Launch weight
Mk 2: 16.50 kg
Propulsion: solid propellant rocket motor booster and solid propellant sustainer rocket motor
Guidance: laser beam-rider
Warhead: HE fragmentation with impact and laser proximity fuzes
Max speed: close to M2
Max effective range
High-speed targets Mk 2: 6 km
Low-speed targets Mk 2: 7 km
Min effective range: about 200 m
Max effective altitude
Mk 2: 4 km
Min effective altitude: ground level

Contractor

Hägglunds Vehicle AB (now owned by Alvis).

KBP, Ulyanovsk Pantzyr-S1 self-propelled air defence system

Type

Air defence missile (vehicle).

Development

Summary: the launcher has a thermal imager and an infra-red localiser.

Development has been completed of a self-propelled air defence system known as the Pantzyr-S1 that is intended for defence of high-value strategic targets such as airfields and communications nodes against a variety of air targets including precision-guided munitions.

A typical battery would comprise five fire units and one command and control vehicle on the same chassis.

Description

The system is mounted on the rear of a Ural-53234 (8 × 8) 20,000 kg forward control cross-country truck chassis. A large turret is mounted on the top of a shelter-type compartment which has two banks of six ready to fire 57E6 surface-to-air missile container-launchers on either side. Also mounted with the missiles are two model 2A72 30 mm cannon. Once the missiles have been fired the system has to be reloaded from an external source.

Typically the missiles would be used first to engage targets at long range and then the cannon to engage any leakers at shorter ranges. Mounted on top of the turret is the surveillance radar. Below this and between the two banks of SAMs is the infra-red sensor package and dual-waveband tracking radar. The thermal imaging sight, the 1TPP1, is made by NPO GIPO, the State Institute of Applied Optics in Kazan and there is also an infra-red localiser.

Target engagement can be fully automatic with a capability to engage two targets simultaneously.

The missile has a maximum altitude capability stated to be 6,000 m, a minimum altitude capability of 3 to 5 m and a range estimated to be 12,000 m or more. A contact and radar proximity fuzing system is fitted to the 16 kg HE-fragmentation warhead.

A navalised version has also been developed.

Operational status
Ready for production and offered for export.

Specifications
Pantzyr S1 system
Combat weight: 20,000 kg
Crew: 3
Armament:
2 × 30 mm 2A72 automatic cannon
12 missile container launch canisters
Ammunition: 750 × 30 mm rounds, 12 missiles
Max gun range: 4,000 m
Max gun altitude: 3,000 m
Guidance system: multiband surveillance and target tracking radars, electro-optical tracking
Reaction time: 4-6 s
Chassis: Ural-53234 (8 × 8) truck
Missile
Length: 3.2 m
Diameter: 170 mm
Weight: (missile in container launcher) 90 kg
Warhead: 16 kg HE-fragmentation
Max velocity: 1,100 m/s
Max altitude: 6,000 m
Min altitude: 5 m
Max range: 12,000 m
Min range: 1,000 m

Contractors
Ulyanovsk Mechanical Plant.
KBP Instrument Design Bureau.

Kentron SAHV-3

Type
Air defence missile static or towed.

Description
The SAHV-3 missile was developed as part of the Crotale upgrade programme for the South African National Defence Force. The missile uses Automatic Command to Line of Sight (ACLOS) guidance. It is electrically and mechanically compatible with the Crotale firing unit and uses a command receiver transponder unit that is designed to Crotale STRESA standards. It can however, be adapted to be used with different tracking radars, depending on the client's specific needs. The missile may be vertically launched, utilising thrust vector control for initial tip-over, or it can be launched from a directed launcher. The inertial navigation capability of the missile enables it to fly into the gathering beam to simplify gathering.

The missile is supplied as a missile group which includes the missile sequencer subsystem. The SAHV-3 missile provides an effective counter to high-speed aircraft attacking from very low altitudes to over 24,000 ft. It is also capable of countering attacks from armed helicopters fitted with short- to medium-range guided missiles or from air-to-surface or ship-to-ship missiles at very low altitudes.

Missile control is by a digital autopilot and a set of linear electromechanical servos driving independent, tail-mounted control fins. Propulsion is by a high-energy, low-smoke composite rocket motor, enabling the missile to reach 8 km in 10 seconds and 12 km in 17 seconds.

A comprehensive, built-in test function, controlled by autopilot and accessible without opening the sealed launch/transportation canister, provides for low-maintenance operation.

Operational status
Advanced development.

Specifications
Length: 3.13 m
Diameter: 180 mm
Wing span: 400 mm
Mass: 115 kg
Warhead: 22 kg blast fragmentation
Maximum altitude: 24,000 ft
Horizontal range: 12 km
Time of flight to 8 km: 10 s
Manoeuvrability: 40 g

Contractor
Kentron, Division of Denel (Pty) Ltd.

LFK European vehicle-mounted Low-Level Air Defence System (L-LADS)

Type
Air defence missile (vehicle).

Development
Summary: missile - dual-mode infra-red/ultra-violet seeker; pedestal - thermal imager or low-light level TV, laser range-finder.

The LFK Lenkflugkörpersysteme GmbH, an EADS company, European private venture L-LADS, has been designed as a low-weight modular air defence system to fill the gap between portable missile weapons and complex heavy wheeled or tracked air defence missile systems.

Computerised operation of the missile launcher, sensors and an interface to a battlefield command, control, communications and intelligence system allows night and all-weather quick reaction capability and function in a multitarget environment. Typical system employment would be for point defence or escort (using its shoot-whilst-moving capability).

The system is air-transportable with two units in a CH-53G helicopter, three units in a C-130 Hercules transport or four units in a Transall C-160 transport. A continuous evolution of the system concept and hardware configuration is being performed.

In 1993, the prototype equipped with GPS and thermal imager fired four missiles and was used twice for German MoD Stinger acquisition trials. In 1994, the prototype was modified to accept cueing data by a radio link from an air defence radar. In 1996, the prototype took part in very low temperature Stinger firings.

Description
The vehicle mounting the system is the Mercedes-Benz GD 250 (4 × 4) 750 kg light all-terrain vehicle. The vehicle is fitted with a specially made stabilised pedestal on its rear decking which has direction controls for azimuth and elevation. Other light vehicles such as the Peugeot P4 and Land Rover can also have the pedestal mount fitted on their cargo platform.

The pedestal can be fitted with a variety of sensors including either a thermal imager for target acquisition (at night or in adverse weather) or a Low-Light Level TV (L-LLTV) (for limited night capability) and a laser range-finder mounted beside the primary target acquisition system to determine the distance to target-in-range decision.

On either side of the mount are two vertically fixed standard Air-To-Air Stinger (ATAS) launcher modules, weighing 43.6 kg each and with a total of four ready to launch Stinger missiles.

The individual launcher module incorporates the launcher structure, a mechanical interface, an electronics package and a cooling system. It is operated by the fire-control system of the carrier via interface electronics and can be used with all the Stinger missile versions.

The field of fire is –10 to +70° in elevation and 360° in traverse. Storage for four additional Stinger container-launcher tubes is available beneath the pedestal mount. The Stingers also retain their shoulder launch capability using the grip-stock assembly.

The vehicle also carries an antenna and communications system for preassignment tasking via the C^3I network, a Northing Gyro to obtain a north reference and a Global Positioning System (GPS) unit to provide accurate position co-ordinates for defence networking the LLADS.

Variants
The standard Stinger ATAS module has also been adapted by LFK for use with the 40 mm Bofors L/70 air defence gun. A boxed set of two ATAS modules and target acquisition system is mounted on the top of the gun mount by the ammunition feed area.

Operational status
Advanced development.

Specifications
FIM-92C Stinger
Length: 1.52 m
Diameter: 70 mm
Wing span: 91 mm
Launch weight: 10.1 kg
Propulsion: solid fuel ejector and boost/sustainer rocket motors
Guidance: passive infra-red/ultra-violet homing seeker
Warhead: 1 kg HE fragmentation
Max speed: M2.2
Max effective range: >4.5 km
Min effective range: 200 m
Max effective altitude: 3.80 km
Min effective altitude: effectively ground level

The European Low-Level Air Defence System (L-LADS) mounted on a Mercedes-Benz GD 250 vehicle ***1997***

Mercedes-Benz GD 250 launch platform
Dimensions (h × w × l): 2.65 × 2.01 × 4.65 m
Weight: 2,850 kg
Missiles: 4 or 8 ready to fire Stinger
Thermal imager: 8-12 μm wavelength with 3 × 4° and 9 × 12° fields of view
Traverse: 360°
Slew rate traverse: 70°/s
Acceleration rate elevation: 110°/s
Acceleration rate traverse: 130°/s
Elevation: –10 to +70°
Slew rate elevation: 70°/s
Reaction time of system: < 5 s
Acceleration rate elevation: 110°/s
Reload time (two persons with 4 missiles): less than 2 min

Contractor
LFK Lenkflugkorpersysteme GmbH, Friedrichshafen.

Lockheed Martin Chaparral Chassis Service Life Extension Programme (CCSLEP)

Type
Air defence missile (vehicle).

Development
Summary: the Hellfire missile has a semi-active laser homing seeker. The Chaparral missile has a passive infra-red homing seeker; fire control to include thermal imaging sensor.

The CCSLEP, which has also been referred to as the Advanced Chaparral and Roadrunner, is the result of the joint efforts of the US Army Missile Command (MICOM), the US Army Tank Automotive Command and a Joint Contractor Team led by Lockheed Martin as the systems integrator. Apart from the development of a pedestal-type trailer-mounted version there is also the tracked Universal Carrier XM1108 CCSLEP and the wheeled M1047A Light Assault Vehicle (LAV) CCSLEP.

All the CCSLEP versions made their debut in 1993.

Description
Each version is offered with the MultiPurpose Launcher System (MPLS) to fire Chaparral, Chaparral RSS or Hellfire Modular missiles, as well as the 70 mm HYDRA 70 unguided rocket system. Additional weapon applications identified for future integration on the CCSLEP include the AMRAAM, Stinger, tail-guided Chaparral, Sparrow and even TOW ATGW missiles once they have evolved into a fire-and-forget design. The MPLS has also been demonstrated previously on the Bradley IFV chassis.

The term CCSLEP refers to the joint usage of existing air defence assets from proven US Army inventories in a new weapon usage role. The MPLS is based on the standard M54 Chaparral launcher and allows for the universal mounting of various pedestals equipped with launch rails specific to the required missile system(s).

There are four Chaparral and four Hellfire missiles in the ready to fire position on the MPLS plus, with the mobile versions, an onboard capability to stow another eight Chaparrals and four Hellfires, to bring the total onboard capacity to 20 rounds.

The individual CCSLEP mobile fire units are also able to detect, identify and fire on close-in targets by using their own onboard integrated sensor systems.

Sensor packages under investigation for use on the CCSLEP include the McDonnell Douglas (now Boeing) Nighthawk, which essentially provides the capabilities of the Bell OH-58D helicopter Mast-Mounted Sight (MMS) in a smaller and lighter package and the Thales Optronics infra-red search and track Air Defence Alerting Device (ADAD).

Operationally the Chaparral RSS missiles would be used at the longer engagement ranges out to 9 km plus with the Hellfire missiles utilised for close-in deep clutter engagements of targets out to around the 6 km mark.

CCSLEP chassis fitted with launcher armed with Chaparral and Hellfire missiles and HYDRA 70 rockets. The Thales Optronics ADAD is mounted on top of launcher

As the complementary missile on the launcher, the Hellfire can also be used against armoured and non-armoured ground targets. The HYDRA 70 rocket system is offered on the CCSLEP for use in the suppressive role against advancing troops and vehicles.

Operational status
Prototype stage.

Specifications
Rockwell Hellfire Modular missile family
Type: multipurpose with semi-active laser homing seeker
Max range: 8 km (typical aerial target range 6 km)
Min range: 1.50 km
Speed
AGM-114A/B/C: M1.4
AGM-114F: M1 plus
Anti-ship: M1 plus
M1047A version
Crew: 2 (driver, vehicle commander/gunner)
Combat weight: 13,364 kg
Unloaded weight: 12,159 kg
Length: 6.39 m
Width: 2.50 m
Height
Max: 2.69 m
Armament
Main: 1 multipurpose launcher system with typically 4 Chaparral missiles, 2 Hellfire missiles and 1 pod of 7 × 70 mm HYDRA 70 unguided rockets
(secondary) 1 × 7.62 mm M240 machine gun
(smoke grenade dischargers) 2 × M257 smoke grenade launchers
Ammunition: up to max 20 missile rounds carried (12 Chaparrals, 8 Hellfires)

Contractor
Lockheed Martin, Electronics and Missiles.

Lockheed Martin M48 Chaparral/M48A1 Chaparral/ M48A2 and M48A3 Improved Chaparral low-altitude self-propelled surface-to-air missile system

Type
Air defence missile (vehicle).

Development
Summary: the missile has an infra-red seeker on MIM-72C/D/E and a rosette scan seeker on MIM- 72G. The launcher has FLIR.

The Chaparral low-altitude surface-to-air missile system was initiated with the modification of the US Navy Sidewinder 1C (AIM-9D) air-to-air proportional navigation guidance infra-red homing missile for ground-to-air launch. Study and evaluation of the Chaparral began in 1964 at the Naval Weapons Center, China Lake, California. The following year a development contract was awarded by the US Army Missile Command (MICOM) to Lockheed Martin (formerly Loral Aeronutronic). First production missile systems were delivered to the US Army in 1969. Lockheed Martin produced the M54 launch and control station, improved missile guidance sections and test equipment as well as being responsible for overall system integration.

The Chaparral is one of the US Army air defence assets being replaced in the active force component by the FAAD systems. In 1984, the US Army Air Defense Artillery School presented a plan to transfer Divisional Chaparral systems to the Army National Guard.

A version known as Sea Chaparral (qv) has also been developed by Lockheed Martin using a modified M54 launcher and is in use with the Taiwanese Navy aboard its larger warships.

A US Army Chapparal launcher vehicle 0055134

Chaparral self-propelled air defence system used by Taiwan showing thermal imaging system between missiles on their launcher arms

By 1 September 1990, 714 Chaparral systems had been built, of which 596 had been purchased by the US Army.

In 1983, the Pentagon announced a letter of offer to Egypt for the sale of 25 M48A2 Improved Chaparral self-propelled air defence systems.

In 1986, Portugal ordered five Chaparral systems, 28 MIM-72F missiles, two AN/MPQ-54 Forward Area Alerting Radars (FAARs). These have been delivered.

In 1986 Taiwan ordered 41 Chaparral fire units with final deliveries made during 1989. Currently it has 45 Chaparral land systems and 22 Sea Chaparral systems.

Description

A standard configuration Chaparral fire unit consists of two main elements, a tracked carrier and the M54 missile launch station. The carrier is designated the M730 and is based on the M548 tracked cargo carrier which in turn uses components of the M113A1 armoured personnel carrier.

The original missile, designated the MIM-72A (based on the AIM-9 Sidewinder), has a launch weight of 86.9 kg, is 2.91 m long and has a diameter of 127 mm, a wing span of 715 mm and is fitted with an 11.2 kg high-explosive warhead. Between 1970 and 1974 an improved all-aspect missile called the MIM-72C was developed, which entered service in July 1978. It weighs 85.7 kg, the other dimensions remain the same and includes an M817 radar proximity fuze developed by Harry Diamond Laboratories. A 12.6 kg M250 HE blast/fragmentation warhead developed by Picatinny Arsenal and an AN/DAW-1B all-aspect infra-red seeker with infra-red counter-countermeasures developed by Lockheed. Effective launch range is increased to over 9 km. Later versions of this missile with smokeless motor are designated the MIM-72E. Non-CCM versions (AN/DAW-1 seekers) are designated MIM-72F and MIM- 72H. All versions are powered by a single-stage solid propellant rocket motor. The battlefield signature of the system has been reduced by the adoption of the M121 smokeless motor for the later missiles.

To provide a night and bad weather capability and to improve daylight performance in smoke and haze, the US Army has retrofitted launchers with a thermal imager device, with auto-track features. The imager contains a 180 element cadmium mercury telluride 8 to 12 μm wavelength detector, has an 18 × 20° wide field of view and a 2 × 2.7° narrow field of view. It can operate in either a wide or narrow field of view to optimise the infra-red detection capability of the receiver and improve the thermal image for the gunner. The optics on the receiver magnify the image and the infra-red target video obtained is presented on the video display located in the mount. At the moment of missile launch a small protective cover will briefly close over the FLIR optics to protect the sensitive optical elements. The system also contains a lightweight Mk XII IFF subsystem for target friend/foe identification.

With the RISE power train and FLIR the designation changes to M48A3. With just the FLIR it is M48A2 Improved Chaparral.

In a typical daylight target engagement, early warning is provided either by the AN/MPQ-49 Forward Area Alerting Radar or by a visual sighting. Once the gunner detects the target he moves the turret to acquire and maintain the aircraft in the centre of his sight or FLIR field of view. The turret can be traversed by 360° and the launch rails have an elevation of +90° and a depression of –9°. As the gunner tracks the target, an audio tone in his headset notifies him when the target is within infra-red sensing range.

The gunner then launches a fire-and-forget missile which operates under its own internal power. Proportional navigation guidance commands are generated from seeker tracking rates and used to control the missile flight path. The rate of fire of the basic system is four missiles per minute with a full reload time of 5 minutes. The single-shot kill probability was assessed at 0.5 against targets with velocities between 0 and 550 kt with the basic missile but this has been substantially improved with later missile versions. The Chaparral surface-to-air missile has maximum range limits from 500 m to beyond 12 km and effective altitude limits of less than 15 m to greater than 3 km. The missile is armed after 180 to 340 m of flight.

A new guidance section called the Rosette Scan Seeker (RSS) and designated AN/DAW-2 was developed for the Chaparral missile under a contract placed by MICOM with Lockheed Martin. Development started in 1982 and the seeker was type classified in August 1987. Missiles fitted with the RSS guidance section are known by the designation MIM-72G. Aeronutronic was awarded a production contract for 441 RSS guidance sections plus depot test equipment for delivery in 1990 to 1992. Within the contract, options exist for 422 additional RSS units. Hughes Aircraft Company, Tucson (now Raytheon) was awarded a contract as a second source producer of 721 RSS guidance sections for delivery between 1990 and 1992.

The MIM-72G has demonstrated launch lock on and intercept of helicopter-type targets out to beyond 9 km range, launch and lock on of fixed-wing targets at 16 km range with final intercept occurring at 9 km and validation computer simulations against some tactical ballistic missile types of launch lock on at approximately 22 km with a projected intercept point at 9 km range.

Target acquisitions at 50 per cent longer range than the current guidance have been demonstrated. Flight tests, including contact hits on a helicopter target beyond 8 km launch range and on fixed-wing targets beyond 12 km, have been successfully conducted.

Supporting the Chaparral Air Defence System for the Egyptian Army is the TRACKSTAR radar system, a self-contained 360° D-band 60 km range AN/MPQ-49 derivative integrated radar/command and control (C^2) system. It automatically broadcasts cueing, fire distribution and IFF data via its VHF radio or hard-wire datalinks to Integrated Weapon Display (IWD) operator control and processor units that are mounted in the Chaparral fire units.

Operational status

In service with 10 countries. Production complete, (over 700 systems built with some 21,700 missiles produced) but can be resumed if further orders are placed. Lockheed Martin is offering upgrades to several overseas customers to further improve operational capabilities and enable launchers to fire other missiles such as Hellfire II. The US Army Chaparral Air Defense Systems have been declared Excess Defense Articles and are available for Foreign Military Sale. Over 4,000 missiles and 115 launchers have been exported.

Specifications

MIM-72G missile
Length: 2.91 m
Diameter: 127 mm
Wing span: 630 mm
Launch weight: 86.20 kg
Propulsion: solid propellant rocket motor
Guidance: proportional navigation with passive IR homing
Warhead: 12.60 kg HE blast/fragmentation with proximity fuze
Max speed: M1 plus
Max effective range
Helicopter target: 8 km
Aircraft target: 9 km
Min effective range: 500 m
Max effective altitude: 3 km plus
Min effective altitude: about 15.0 m

Contractor

Lockheed Martin, Electronics and Missiles.

MBDA Roland low-altitude surface-to-air missile systems

Type

Air defence missile (vehicle).

Development

Summary: the Roland 2 launcher has an infra-red localiser and the Roland 3 upgrade has a sight with a thermal imager, laser range-finder and an infra-red localiser.

In 1964, Aerospatiale of France and Messerschmitt-Bölkow-Blohm of Germany began design work on a low-altitude surface-to-air missile system which eventually became known as Roland. Aerospatiale had overall responsibility for the clear-

Roland 2 of the French Army deployed in firing position (Michael Jerchel)

weather version, called Roland 1 and MBB (now LFK GmbH, part of EADS) and overall responsibility for the Roland 2 all-weather version. At a later stage, the joint company Euromissile (now part of MBDA) was established to market this and other missiles produced by the two companies. Currently the all-weather version (formerly called Roland 2) is offered together with the latest variant known as Roland 3.

The French Army had a requirement for 181 firing units based on the AMX-30 MBT chassis. Of these, 181 have been funded to date, with all 98 Roland 2 and 83 Roland 1 vehicles now delivered. The first Roland fire units were delivered to the French Army in December 1977.

In June 1981, the German Army officially took delivery of the first of 140 Roland SAM systems. These are installed on Marder tracked vehicles.

In 1984, the Spanish Defence Ministry selected Roland for its mobile battlefield low-level air defence system with six Spanish companies participating including Inisel, Santa Barbara and Ceselsa.

In December 1983, the Roland shelter variant was selected to protect the NATO, US and German air-bases in Germany. All 95 units are manned by German Air Force personnel. A further 20 Roland fire units were procured by the German Navy to protect three of its airbases.

France and Germany are carrying out the ROLAND VMV upgrade programme which will maintain the systems in service until 2015 and beyond. The programme involves replacement of the current optical sight by the GLAIVE infra-red sight assembly in order to provide a third operational mode for Roland - visual, radar and infra-red. Furthermore, the system will be upgraded to the M3S configuration by replacement of the search and track radars.

Description

The Roland weapon system has been designed to provide protection both for static and partially mobile facilities in the rear combat zone, as well as for mobile units in the forward combat zone, with the capability to detect and identify targets while on the move.

The weapon system includes an all-round search radar with a range of 16 km, height cover of more than 3 km and integrated IFF. The search radar is the pulse Doppler Siemens MPDR 16 D-band surveillance radar. Once the target has been detected it is interrogated by either a Siemens MSR-400/5 (German vehicles) or an LMT NRAI-6A (French vehicles) IFF system. An optical sight/goniometer and a target tracking/guidance radar are employed for target acquisition and tracking. These devices track the target and fix the missile's position, or calculate its deviation from the line of sight (radar/sight target). At any time during the missile's flight, it is possible to switch between the sight/goniometer and the target tracking/guidance radar. The tracking radar is a two-channel, monopulse Doppler microwave Thomson-CSF (now Thales) Domino 30 system; one channel tracks the target and the second locks in on a microwave source on the missile.

A total of 10 missiles can be carried on the Roland weapon system. Four missiles are carried in each of the drum magazines built into the vehicle and two on the launchers, ready to fire. The missiles fired by the launchers can be replaced in a few seconds by automatic reloading from the magazines. If necessary, all 10 missiles can be fired one after the other in just a few minutes.

The Roland missile has a range of up to 8,000 m at heights in excess of 3,000 m. The missile is kept on the line of sight by a command transmitter which receives the appropriate commands from a computer and transmits them to the missile. The Roland missile has both proximity and impact fuses to detonate the warhead.

Carrier vehicles used are the tracked vehicles Marder APC (Germany), AMX30 (France) and the wheeled MAN 10 t mil GIW (Germany). For the Crisis Reaction Forces, Roland is to be manufactured for France and Germany as an airloadable shelter version for transporting by Transall C-160.

Roland 3 missile has an increased speed (570 m/s) and range (8 km). The system can also fire the VT-1 hypervelocity missile used in the MBDA Crotale NG. A new VT1-R missile is being developed which will have a range of >11 km at a maximum speed of M3.5. Other improvements being developed include the Glaive sight, a digitised operating and control system with multifunction displays and MIL-Bus system.

The GLAIVE sight is of modular design and is for use in heavy ECM environments in place of the radar sensors. It has an IRCCD thermal imager, a Raman type eye-safe laser range-finder, CCD TV channel and automatic target and missile tracker.

Operational status

In production. In service with 10 countries. In August 1988, Euromissile stated that the Roland order book stood at 644 firing posts (231 AMX-30, 148 Marder 1, 234 Shelter and 31 US Army) and 25,500 missiles.

In 1995, the numbers of Roland 1/2 systems in service fell to 100 for France, while Germany maintains 144 for the army and 115 air force/navy/NATO systems.

Between 1996 and 1999, MBDA (was Euromissile) delivered air transportable systems to Germany and France, 10 to the KRK of the German Air Force and 20 to the FAR of the French Army.

The French and German Roland upgrade began in 2001.

Specifications

Roland 2/Roland 3 Missiles
Length: 2.40 m
Diameter: 160 mm
Wing span: 500 mm
Launch weight
Roland 2: 66.5 kg
Roland 3: 75.0 kg
Propulsion: solid fuel booster and sustainer rocket motors
Guidance: command control
Warhead
Roland 2: 6.5 kg HE hollow charge fragmentation with contact and proximity fuzing
Roland 3: 9.2 kg HE hollow charge fragmentation with contact and proximity fuzing
Roland VTAR: 13 kg HE hollow charge fragmentation with contact and proximity fuzing
Max speed
Roland 2: 500 m/s
Roland 3: 570 m/s
Roland VTAR: 1,250 m/s
Max effective range
Roland 2: 6.3 km
Roland 3: 8 km
Roland VTAR: 11 km
Min effective range
Roland 2: 500 m
Roland 3: 500 m
Roland VTAR: 700 m
Max effective altitude
Roland 2: 5.5 km
Roland 3: 6 km
Roland VTAR: 6 km
Min effective altitude
Roland 2: 10.0 m
Roland 3: 10.0 m
Roland VTAR: 10.0 m
Reload time: 6 s from magazines

Contractor

MBDA.

Norinco PL-9 low-altitude surface-to-air missile system

Type

Air defence missile (vehicle).

Development

Summary: the missile has an infra-red (IR) seeker; the launcher employs electro-optic tracking and laser rangefinder.

This low-altitude SAM system was first displayed in model form in 1989. It comprises a four-rail launcher assembly cupola for PL-9 missiles and associated target acquisition radar and electro-optical instruments mounted on the rear decking of a Norinco WZ 551D (6 × 6) APC.

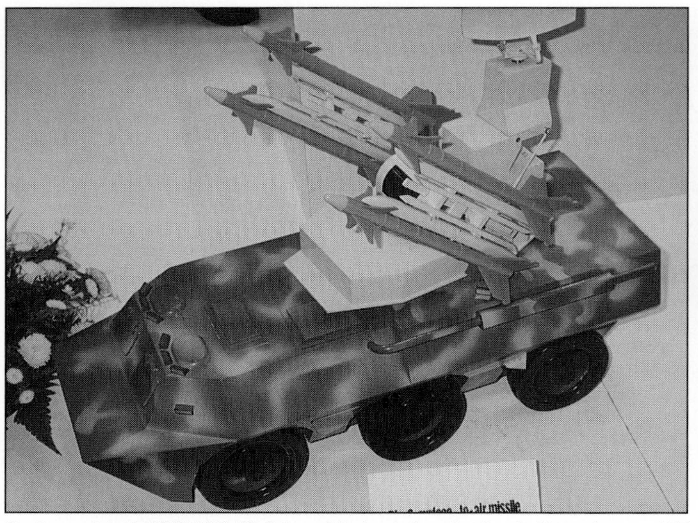

Scale model of NORINCO PL-9 low-altitude surface-to-air missile system on WZ 551D (6 × 6) armoured personnel carrier (Christopher F Foss)

The launcher can also be refitted to fire other types of surface-to-air or modified air-to-air missiles of similar configuration to the PL-9, according to the customer's requirements.

Description

Each WZ 551D carries two vehicle drivers, three system operators and four ready to fire PL-9 missiles. The target acquisition radar has a detection range of around 18 km and an altitude capability up to 6 km. The passive electro-optical target sighting/tracking system with TV monitor has an operational range of 15 km and a laser range-finder unit with a 10 km range capability.

The missile used is the dual-purpose PL-9 air-to-air and ground-to-air passive IR terminal homing weapon which appears to be roughly similar in physical appearance to the American AIM-9L Sidewinder air-to-air missile. It is capable of off-axis launch and uses an all-aspects cryogenic liquid nitrogen gas-cooled seeker head unit which utilises proportional navigation guidance techniques.

Maximum effective range is 5.5 km at targets up to the same altitude limit. The single-shot kill probability for a single missile launch at an approaching target is 0.8.

The PL-9 surface-to-air missile is also used in the Brigade (Regiment) Level 390 Integrated Gun Missile Air Defence System.

Operational status

In production and in service.

Specifications

PL-9 Missile
Length: 2.994 m
Diameter: 160 mm
Wing span: 808 mm
Launch weight: 120 kg
Propulsion: solid propellant rocket motor
Guidance: all-aspects passive IR seeker with proportional navigation
Warhead: 10.0 kg HE fragmentation
Max speed: M2.0
Max effective range: 5.5 km (max range 8.5 km)
Min effective range: n/a
Max effective altitude: 5.5 km
Min effective altitude: n/a
Reload time: n/a

Contractor

China North Industries Corporation (NORINCO).

Nudelman 9K31 Strela-1 (SA-9 'Gaskin') low-altitude surface-to-air missile system

Type

Air defence missile (vehicle).

Development

Summary: the missile has an infra-red (IR) seeker; the launcher has an IR night sight on some vehicles in the battery.

The first SA-9 launchers were produced in 1966 with the system attaining operational status in 1968. The SA-9 'Gaskin' (US/NATO designations) low-altitude clear-weather surface-to-air missile system's first recorded combat use was in May 1981.

The SA-9 'Gaskin' was issued to the anti-aircraft batteries of RFAS motorised and tank regiments on the basis of four systems per battery to give a total of 16 per division. Replacement by the SA-13 'Gopher' system is almost complete.

SA-9 'Gaskin' SAM system based on BRDM-2 (4 × 4) amphibious chassis with four missiles ready to launch

Description

The system consists of a 9P31 BRDM-2 Transporter-Erector-Launcher with the normal turret removed and replaced by one with four ready to launch SA-9 container-launcher boxes. These are normally lowered to the horizontal when travelling to reduce the overall height of the vehicle. The original version of the Strela-1 missile was known as the 9M31 (US designation SA-9a, NATO codename 'Gaskin' Mod 0) and used an uncooled first-generation lead sulphide (PbS) IR seeker operating in the 1 to 3 μm waveband region. This was supplemented by the 9M31M variant (Russian Strela-1M system, US designation SA-9b, NATO codename 'Gaskin' Mod 1) which has an improved PbS seeker operating in the 1 to 5 μm waveband region to provide greater target sensitivity and lock on ability. The 32 kg M1.8 missile is 1.80 m long, 0.12 m in diameter and has a wing span of 0.36 m. The minimum range of the 9M31 is 800 m and the maximum range 4.2 km within altitude limits of 30 to 3,500 m. The minimum range of the 9M31M is 560 m and the maximum range 8 km (increasing to a possible 11 km when used in a tail-chase engagement) within altitude limits of 10 to 6,100 m. When engaging a head-on target the system has a considerably reduced range. The SA-9 is fitted with a 2.60 kg HE fragmentation warhead and proximity fuze. The warhead has a lethal radius of 5 m and damage radius of 7.60 m.

One SA-9 vehicle in each battery has been fitted with 'Flat Box-A' passive radar detection antennas, one either side of the hull above the front wheel housings, one under the left launch canisters pointing forward and one mounted on a small frame above the rear engine deck plate pointing rearwards to give 360° coverage. The vehicle with no 'Flat Box-A' system has an IR system for the first two to use at night.

Operational status

Production complete. In service with a number of countries.

Specifications

SA-9 missile
Length: 1.803 m
Max diameter: 120 mm
Max wing span: 360 mm
Launch weight: 32.0 kg
Propulsion: single-stage solid propellant rocket motor
Guidance
9M31: 1-3 μm waveband uncooled PbS passive IR homing seeker
9M31M: 1-5 μm waveband cooled PbS passive IR homing seeker
Warhead: 2.60 kg HE fragmentation with contact and proximity fuzing
Max speed: M1.8
Max effective range
9M31: 4.2 km
9M31M: 8 km (increasing to 11 km in tail-chase against slow manoeuvring target)
Min effective range
9M31: 800 m
9M31M: 560 m
Max effective altitude
9M31: 3.50 km
9M31M: 6.10 km
Min effective altitude
9M31: 30 m
9M31M: 10 m
Max target speed: 300 m/s

Contractor

Nudelman OKB-16 Design Bureau.

Nudelman 9K35 Strela 10 (SA-13 'Gopher') low-altitude surface-to-air missile system

Type
Air defence missile (vehicle).

Development
Summary: the missile has an infra-red (IR) seeker; Strela 10M3 has dual-mode optical and IR seeker.

The fully amphibious NBC-equipped SA-13 'Gopher' mobile SAM system with a range-only radar entered operational service in 1975. In the Russian Army it has now virtually replaced the far less capable SA-9 'Gaskin' system on a one-for-one basis to improve the mobility of the anti-aircraft batteries in the Motorised Rifle and Tank divisions.

There are two versions of the SA-13 Transporter-Erector-Launcher And Radar (TELAR) variant of the MT-LBu vehicle in service, designated TELAR-1 and TELAR-2 by the US Army. Appraisal of both does not show any significant structural differences but it is known that the TELAR-1 carries four 'Flat Box-B' passive radar detection antenna units, one on either corner of the vehicle's rear deck, one facing aft and one between the driver's vision ports at the front, whereas the TELAR-2 has none. The TELAR-1 is apparently used by the SA-13 battery commander.

Description
The complete system is known as the ZRK-BD 9K35 Strela 10 in Russian service. The 39.5 kg SA-13 missile (index number 9M37) is 2.20 m long, 120 mm in diameter with a 400 mm wing span and has a maximum speed of M2. It carries a 4 kg HE warhead and is fitted with either an improved passive PbS all-aspects IR seeker unit, which operates in two individual frequency bands in the 1 to 5 μm waveband to give high discrimination against IR countermeasures such as flares and decoy pods, or a cryogenically cooled passive indium antimonide all-aspects IR seeker unit. Normally the TELAR carries four ready to fire SA-13 missile container-launchers and four reloads in the cargo compartment but it has also been seen on numerous occasions with either SA-9 'Gaskin' container-launcher boxes in their place or a mixture of the two. This enables the battlefield features of both missiles to be utilised to the full by allowing the cheaper SA-9 (Strela 1) to be used against the 'easier' targets and the more expensive and sophisticated SA-13 (Strela 10) against the 'difficult' targets.

The missile mix also allows a choice of IR seeker types on the missiles for use against extremely low-altitude targets as well as in adverse weather conditions.

Comparison of the missile characteristics are given in the accompanying table.

Missile	Strela 10M (9M31M)	Strela 10M2 (9M37M)
Weight		
Missile in container-launcher box	53.4 kg	70.2 kg
Missile	32 kg	39.5 kg
Dimensions		
Container-launcher box	1,900 × 290 × 290 mm	2,330 × 290 × 290 mm
Guidance	Optical aiming	Optical aiming
IR seeker types	Uncooled lead sulphide (PbS) Near-IR homing type with no countermeasures capability	Uncooled lead sulphide (PbS) Near-IR homing type with counter-countermeasures capability against IR decoys or cooled indium antimonide (InSb) mid-IR homing type with counter-countermeasures capability against IR decoys
Control method	4 movable canards and 4 rotor-controlled surfaces for roll stabilisation	
Warhead type	HE controlled fragmentation (2.6 kg HE and 460 fragments)	4 kg HE fragmentation rod (2.7 kg HE and 100 rods)
Lethal radius	5 m	5 m
Fuzing	Impact and active xenon lamp proximity	
Max missile speed	M1.8	M2
Max target speed	300 m/s	420 m/s

SA-13 'Gopher' SAM system with launcher arms elevated (Michael Jerchel)

SA-13 is 800 m and the maximum effective range 5,000 m, with altitude engagement limits of 25 to 3,500 m. The 450 to 10,000 m range circular parabolic 9S86 'Snap Shot' radar antenna is located between the two pairs of missile canisters and is a simple range-only set to prevent wastage of missiles outside the effective range of the system. An IFF system is also fitted which determines target identification at ranges up to 12 km and altitudes from 100 to 5,000 m and ranges of up to 10 km at an altitude of 25 m.

Apart from the Strela 10M and Strela 10M2 the Russian Army has deployed another version of the SA-13 known by the designation Strela 10M3. This is designed for use in the mobile battle and to defend troops on the march from low-level attacks by aircraft and helicopters, precision-guided munitions and other flying vehicles such as reconnaissance RPVs.

The major change is the adoption of a dual-mode guidance system for the missile seeker - optical 'photocontrast' and dual-band passive IR. The missile accommodating this system is the 9M333. Target acquisition range using the optical 'photocontrast' channel is between 2,000 and 8,000 m while, for the IR channel, it is between 2,300 and 5,300 m. Altitude engagement limits are from 10 m up to 3,500 m at ranges of 200 to 5,000 m. Average missile speed is 550 m/s. Kill probability is 0.6.

An upgrade package with an updated 9S86 Snap Shot radar from the Russian Retia company is being offered for export and has already been delivered to the Czech Army.

Operational status
Production as required. In service with a number of countries.

Specifications
Length: 2.20 m
Max diameter: 120 mm
Max wing span: 400 mm
Launch weight
9M37/9M37M: 39.5 kg
9M333: 42 kg
Propulsion: single-stage solid propellant rocket motor
Guidance
9M37/9M37M: passive IR seeker
9M333: dual-mode passive 'photocontrast'/IR seeker
Warhead: 4 kg HE fragmentation with contact and proximity fuzes
Max speed: M2
Max effective range: 5,000 m
Min effective range
9M37/9M37M: 800 m
9M333: 200 m
Max effective altitude: 3,500 m
Min effective altitude
9M37/9M37M: 25 m
9M333: 10 m
Max target speed: 420 m/s

Contractor
Nudelman OKB-16 Design Bureau.

Oerlikon Contraves ADATS missile system

Type
Air defence missile (vehicle).

Development
Summary: the missile has laser beam-riding guidance; the launcher has electro-optics module with FLIR, TV and laser range-finder.

In 1979, Oerlikon Contraves, with prime subcontractor Lockheed Martin, commenced development of a low-level missile system to defeat both air and ground threats. In June 1986, after an exhaustive evaluation, ADATS was selected by the Canadian Forces to fulfil its Canadian Forces Low-Level Air Defence System (CF LLADS) requirements. Oerlikon Aerospace Inc, now Oerlikon Contraves was formed in 1986 to carry out the CF LLADS contract and to take technical and, subsequently, commercial responsibility for selling the ADATS missile system worldwide.

In 1987 ADATS was one of the four weapon systems that competed in the US Army's Forward Area Air Defense — Line of Sight — Forward-Heavy (FAAD-LOS-FH) competition. In November 1987, ADATS was selected as the winner. The programme was halted in 1992, because of US DoD budget cuts, after only eight ADATS had been delivered.

In 1993, the Royal Thai Air Force purchased the shelter-based configuration of ADATS, without radar, to be integrated into its existing air defence system for

Canadian Forces' ADATS firing in Suffield, Alberta, Canada 0004754

defence of airbases. The ADATS fire unit receives information on targets from a Skyguard fire-control station.

For the CF LLADS programme ADATS is installed on a modified M113A2 APC platform. For the FAADS-LOS-FH programme it was installed on the M3 Bradley.

ADATS fire units have conducted successful live firings against both air and ground targets, in adverse weather and battlefield conditions. Oerlikon Aerospace states that in over 200 live missile firings, ADATS has achieved a success rate of over 80 per cent against the full spectrum of low-level threats.

A shipborne design, in either stand-alone or integrated (with the platform's sensors) configuration, is currently being proposed for defence against sea-skimming missiles. The latter configuration is currently being marketed as 'Sea Sprint'.

At the Roving Sands exercise in Fort Bliss, Texas in June 1996, Oerlikon Contraves demonstrated the integration of ADATS systems with Patriot batteries and theatre level air defence assets using the built-in C^3 and networking capabilities operating through an encrypted Link 11 interface. This allowed ADATS units to remain silent yet obtain theatre-level information from networked sensors.

An ADATS Mk 2 is currently under development and will include the following product improvements:

(a) E-O module will incorporate a new FLIR Vidicon tube and a new CCD-TV module
(b) new radar electronics cabinet for enhanced maintainability
(c) new multipurpose consoles which can alternately perform radar and E-O functions, with improved man/machine interface and automation which will permit single-user operation
(d) improved automatic target tracker
(e) new C^3I system using high-throughput digital radios
(f) new RF fuze and guidance electronics for the ADATS missile
(g) new turret electronics assembly
(h) new hydraulic power supply
(i) new air conditioning unit.

Description

ADATS is a multipurpose, all-weather low-level missile system designed to defeat both air, ground and surface threats. The air threats, including attack helicopters flying nap of the earth at ranges of up to 10 km, can be engaged at very low altitudes. It can also engage landing craft, patrol boats and similar threats in an oil rig or island defence scenario.

As well as the M113 and Bradley armoured vehicles, the ADATS missile system has also been adapted to fit the Mowag Piranha 10 × 10 armoured vehicle.

The ADATS missile system was conceived and designed to optimise performance in the low to very low-level zone where threats try to slip under radar surveillance and radar-guided systems. Actual engagement therefore takes place using the electro-optical system. The electro-optical module, built by Lockheed Martin, includes an infra-red sensor that permits an operator to detect and track targets at night and in adverse weather; a TV sensor for daytime use; a laser guidance beam assembly to guide the missiles; and a laser range-finder. The FLIR and TV sensors, both online and operating simultaneously, are available to the E-O operator at all times. The E-O system also includes a boresight module that automatically aligns these precision optics. The ADATS E-O system advances technology derived from Lockheed Martin's Target Acquisition Designation Sight/Pilot Night Vision Sensor (TADS/PNVS) flown on US Army Apache helicopters. The E-O system on ADATS is not affected by ground clutter and radar multipath effects, thus it can engage all threats to zero altitude.

The X-band 25 km range, pulse-Doppler surveillance radar is used to provide surveillance and cueing information. It is fitted with search-on-the-move, track-while-scan and automatic threat prioritisation for up to 10 target capabilities. Targets are automatically interrogated by the onboard IFF system.

The ADATS missile uses laser beam-riding guidance and is guided to the target by a carbon dioxide laser which is detected by two receivers located in the guidance fins at the rear of the weapon. This guidance system, combined with the smokeless solid fuel rocket propellant, further enhances the system's survivability on the battlefield. The missile is dual-fuzed with both an impact and proximity fuzing mechanism, which allows the 12 kg shaped warhead to be effective against both aircraft and armoured vehicles.

In operation the ADATS is operated by a crew of two, a radar operator (who is also the system commander) and an E-O operator (who is also the gunner). The E-O operator can also operate both console stations during periods of low activity.

The ADATS missile system mounted on an M113 armoured vehicle 0019549

ADATS has a built-in C^3 (Command, Control and Communications) system which allows any ADATS unit in a network to obtain and exchange surveillance data with other units. One ADATS unit with its radar on can provide fire-control data to other units which can remain totally passive, thus minimising the risk of being detected or neutralised by anti-radiation missiles.

Operational status

Production as required. In service with the Canadian Forces — a total of 36 ADATS was delivered — and Thailand (shelter-mounted version with Skyguard fire-control unit as target information source). A total of eight ADATS was delivered to the US Army before the US DoD cut funding in 1992.

Specifications

ADATS missile system
System performance
Missile intercept range: 10 km
Missile ceiling: 7,000 m
Missile velocity: M3+
Azimuth: 360°
Elevation: continuous, tracking and interception continuous from –4.5 to +85°
Missile
Length: 2.05 m
Diameter: 152 mm
Launch weight: 51 kg
Propulsion: smokeless solid propellant rocket motor
Guidance: laser beam-rider using digitally coded CW CO_2 laser
Warhead: 12 kg HE fragmentation/shaped charge with range gated laser proximity fuze with variable fuze delay device, a nose-mounted crush fuze
Armour penetration: greater than 900 mm
Radar type: X-band pulse-Doppler: 25 km range at 38 rpm scan rate, 17 km range at 57 rpm scan rate
Electro-optics module
Thermal imager spectral band: 8-12 μm
Field of view
Wide: 7.7°
Narrow: 3.0°
TV
Waveband: 0.7-0.9 μm
Field of view
Wide: 3.8°
Narrow: 0.9°
Laser range-finder
Type: Nd:YAG
Wavelength: 1.06 μm

Contractor

Oerlikon Contraves Inc.

VERIFIED

Saab Bofors RBS 70/M113 low-altitude surface-to-air missile system

Type

Air defence missile (vehicle).

Development

Summary: missile - laser beam-riding guidance.

In March 1988, the then Bofors announced that it had test fired its latest vehicle-mounted application of the RBS 70 missile system, the RBS 70/M113 combination.

Saab Bofors RBS 70/M113 SAM system ready to engage target

Designed to meet a Pakistan Army requirement for a mobile SAM system to protect mechanised units in the field, the conversion is ready for production in Pakistan which also manufactures some parts of the RBS 70 missile under licence.

Description
The M113 variant chosen by Pakistan for the conversion is the M113A2. The operating crew consists of four: the fire (and vehicle) commander, missile operator, loader/radio operator and vehicle driver. The fire commander is seated to the rear in the centre front of the troop compartment. Above is a cupola which can traverse through 360° and mounts a 12.7 mm calibre heavy machine gun.

The RBS 70 missile platform is hinged to the compartment roof on its right side and held upright in the travelling position by two torsion springs. Behind this and on the vehicle's right wall, overhanging the track, is the missile store for six RBS 70 standard container-launcher tubes.

Specifications
RBS 70 Mk 2 missile
Length: 1.32 m
Diameter: 106 mm
Wing span: 320 mm
Launch weight
Mk 2: 16.50 kg
Propulsion: solid propellant rocket motor booster and solid propellant sustainer rocket motor
Guidance: laser beam-rider
Warhead: HE with combination of pre-fragmentation with tungsten spheres and shaped charge plus armour-piercing with impact and laser proximity fuzes
Max speed: >M2
Max effective range: >8 km
Min effective range: about 200 m
Max effective altitude: >5 km
Min effective altitude: ground level
Platform
Crew: 4
Combat weight: 11,600 kg
Length: 4.863 m
Width: 2.686 m
Height
Transport mode: 2.04 m
Deployed mode: 3.44 m
Max speed
Cross-country: 23 km/h
Road: 67 km/h
Armament: 1 × RBS 70 missile launcher
Ammunition: 6 × RBS 70 missiles
Armour: 12-44 mm

Contractor
Saab Bofors Dynamics AB.

STN ATLAS Elektronik Atlas Short-Range Air Defence (ASRAD) system

Type
Air defence missile vehicle turret.

Development
Summary: the turret system is fitted with a variety of missiles with infra-red (IR) seekers and a sensor package with IR camera, TV camera and eye-safe laser rangefinder.

STN ATLAS Elektronik has developed the Atlas Short-Range Air Defence (ASRAD) weapon platform as a private venture. Based on modular construction, it can be mounted on a wide variety of tracked and wheeled platforms.

Following work with Bodenseewerke Geratetechnik (BGT), the system is designed to be compatible with a variety of missile types such as the Raytheon Stinger (Basic, Post and RMP), Kolomna KBM Igla-1, MBDA Mistral and Saab Bofors RBS 70. The ASRAD on a Wiesel 2 chassis has been selected by the German Army for its LeFlaSys air-portable mobile air defence system.

Description
The ASRAD air defence missile system carries SHORAD (Short Range Air Defence) missiles with IR seeker heads, but can also be retrofitted for laser beam-riding missiles. It includes: pedestal with traverse and elevation drive; sensor package with 8 to 12 μm (or 3 to 5 μm option) thermal camera, TV and eye-safe laser range-finder with accompanying stabilised Line of Sight (LOS); system electronics including missile interface electronics; multipurpose launcher for Stinger, Mistral and Igla (SA-16), with capacity for RBS 70 Mk 2 (ASRAD -R) or similar laser beam-riding missiles in preparation.

In the normal operational mode, the ASRAD weapon platform receives its target cueing data from a platoon command post equipped with a search radar (with integral IFF) and/or an IR Search and Track (IRST) system such as the Thales Optronics ADAD. IRST's other fire units can also pass targeting data. The ASRAD operator has his/her own air picture display as an overlay to his/her IR or TV monitor picture.

The operator presses the 'Target Allocate' button on the right joystick, then the pedestal automatically moves on to the target's azimuth bearing. If the cueing data is sent by a 3-D radar or an IRST, the pedestal moves in both traverse and elevation and the target is immediately visible in the IR or TV monitor picture.

If the cueing data is sent by a 2-D radar, the pedestal moves only in traverse to the bearing of the target and the operator has to perform a manual elevation search until the target on the monitor is discovered.

When the operator recognises the target on the monitor, the operator moves the track gate by means of the joystick over the target and initiates the auto-track mode. Since the target is tracked automatically and the operator presses the 'engagement' button then the ASRAD system commences to measure automatically and calculate all the necessary fire-control solution data on the chosen target while activating one of the onboard missiles. The target data are displayed on the monitor.

Missile lock on, as well as superelevation and lead angle setting are automatically performed and displayed on the monitor. The lock on is also acoustically announced to the operator.

When the target enters the missile engagement envelope, this is also presented on the monitor and the gunner can then press the 'fire' button.

Operational status
Preproduction vehicles completed. In production for the German Army (Wiesel 2 tracked vehicle). Deliveries 2000 to 2003. The ASRAD-R preproduction system on an M113 vehicle has been tested successfully including four firings in 1999.

Specifications
ASRAD System
Weight: approx 320 kg
Pedestal:
(traverse) 360°
(elevation) −10 to +70°
Sensor LOS (relative to the platform)
(traverse) ≥±15°
(elevation) ≤+4 to ≥16°
Stabilisation accuracy (of the LOS Sensor):
better than 0.05 mrad
LOS accuracy:
(pedestal) ≤0.2°
(pedestal aiming velocity) ≥56°/s
Sensors: 8-12 μm waveband IR camera, TV camera and eye-safe laser range-finder
Control box: fitted with monitor, keyboard and joystick; remote control up to 100 m
Interfaces: for data transmission; GPS, inertial and north-finding navigation system; radio telephone; and vehicle intercommunication
Weapons:
(main) 2-4 ready to fire Stinger, Mistral, Igla-1, Starburst or RBS 70 SAMs
(secondary) light 7.62 mm machine gun
Power supply: 18-32 V DC vehicle supply

Contractor
STN ATLAS Elektronik GmbH.

Tarnow Sopel self-propelled gun/missile/air defence system

Type
Air defence missile (vehicle).

Development
Summary: SOPEL has infra-red (IR) homing missiles and may have a sighting system with thermal imager, low-light level TV and laser range-finder.

SOPEL self-propelled air defence system from above with both ammunition feeding hatches and rear door open (Tomasz Szulc)

In early 1994, two key members of the Polish defence industry, the Mechanical Plant Tarnow and the Steel Works Stalowa Wola, completed the prototype of a new self-propelled gun/missile air defence system called SOPEL.

Poland had planned to introduce the new Russian 2S6 Tunguska self-propelled air defence system, but the collapse of the Warsaw Pact and problems with deliveries from Russia caused reopening of the contest to replace the existing ZSU-23-4 SPAAGs of the Polish Army armed with four 23 mm cannon.

SOPEL is the first contender to meet this requirement while the other options remain at the conceptual stage.

Description

The prototype of the SOPEL, is in fact the OPAL chassis fitted with the new electrically driven turret. The armament, installed in the central part of the turret, consists of a twin barrel 23 mm cannon which is an improved version of the well-known ZU-23 anti-aircraft gun which has been produced for more than 20 years in Tarnow for the home and export markets.

The second part of the armament of the SOPEL are two mounts for tube-launched fire-and-forget surface-to-air missiles which are mechanically linked with the gun elevation device. For first trials the old Strela-2M (SA-7b Grail) were used but the new Polish GROM missile is the long-term solution. Both types of missile are of the IR homing fire-and-forget type.

A sighting system has yet to be selected. The prototype is fitted with a simple tachometric sight. It is expected that a sophisticated stabilised system with thermal camera, low-light level TV and laser range-finder will be developed. The turret is large enough to accommodate all the improvements and the current sight is obviously an interim solution.

Two weapons have been considered, first the twin barrel water-cooled 2A38 cannon which is already installed in the Russian 2S6 and Slovakian STROP self-propelled air defence system and second an unspecified 345 mm gun system.

Tests of modified variants of SOPEL complete with new missiles, new sighting system and better mobility are expected in the future.

Operational status

Prototype. Not yet in production or service.

Specifications

Crew: 3
Weight: 16,000 kg
Length: 7.61 m
Width: 3.15 m
Height: 3.02 m
Max speed
  Road: 60 km/h
Range: 500 km
Engine: 268 hp diesel
Armament
  Guns: 2 × 23 mm
  Muzzle velocity: 970 m/s
  Range horizontal: 2.5 km
  Range vertical: 1.5 km
  Rate of fire: 1,800 rds/min
  Magazine capacity: 2 × 250
  Missiles: 2 × 9K32M
  Range: 2,900-4,200 m
  Speed: 500 m/s
  Guidance: passive IR
Gun control equipment
Turret power control: powered/manual
Turret traverse: 360°
Gun elevation/depression: +80/–8°
Max rate power traverse: 80°/s
Max rate power elevation: 50°/s

Contractor

Tarnow Mechanical Works.

Thales Air Defence ASPIC automated mobile firing unit for Very Short-Range Air Defence (VSHORAD)

Type

Air defence missile vehicle turret.

Development

The ASPIC system is a fully automated fire unit for very short-range surface-to-air missiles. It is designed to defend Vital Point (VP) sites or troops on the march and operates with automated target tracking and engagement facilities to shorten reaction times. If required the automated functions can be switched to manual control at any time.

Description

The two axis ASPIC servo-controlled turret assembly can be mounted on a variety of chassis with up to eight ready to fire missiles (depending upon missile and vehicle type). The missiles used include the Mistral, Stinger, Starburst, Starstreak, RBS 70 or any other suitable VSHORAD type. The vehicle is also normally fitted with a land navigation and north-seeking system.

The Fire Unit assembly is fitted with a fire-control system based on a TV tracking package. This includes a TV camera and a TV angle deviation measurement device, a digital computer and a vertical gyro. This allows for fast and automatic target acquisition phase, accurate target tracking, optimisation of the lead angle and validation of the designated missile infra-red or other guidance-type seeker lock on to the assigned target. An IR camera provides for night-time and reduced visibility engagement capability.

ASPIC can be integrated into automatic C^3I networks and the systems can be alerted by IR Search and Track (IRST) equipment such as the Air Defence Alerting Device (ADAD) from Thales Optronics or radars such as the Thales early warning command and control system CLARA.

If the ASPIC is operated in the stand-alone mode, the target designation is provided by an infra-red search and track device on an optical target designator called ARES. The unit uses the operator-driver as the target 'pointer'. In order to do this, he wears the ARES as a helmet-mounted optical target designator system. This allows for fast deployment and almost instantaneous reaction time from the operator-gunner at the deployment site.

The ASPIC system can be integrated onto a 1.5 tonne class of vehicle and can be transported by road, rail, sea and air to be deployed in a theatre of operations.

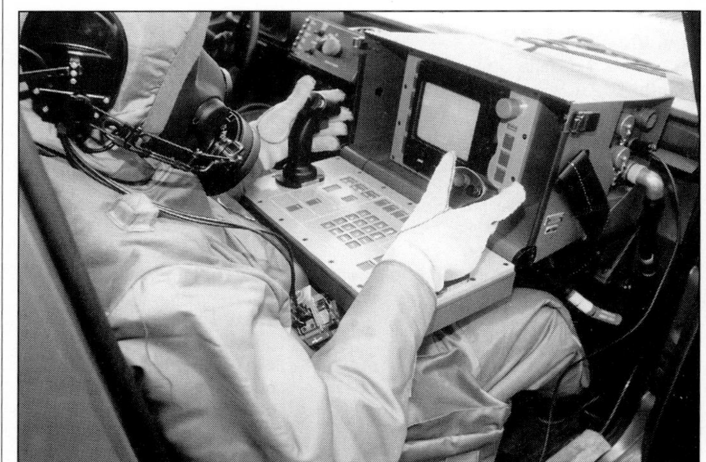

The control console of the Thales ASPIC system ***NEW***/0080360

Thales Starburst missile being fired from the ASPIC VSHORAD system 0004746

Operational status

In production. In service with France (30 systems ordered in mid-1994 for French Air Force use with Mistral missiles) and several other armed forces including those of Chile and Gabon. The French Air Force systems are equipped with thermal imaging cameras supplied by BAE Systems.

Contractor

Thales Air Defence.

UPDATED

Thales Air Defence Crotale low-altitude surface-to-air missile system

Type

Air defence missile (vehicle).

Development

In 1964 South Africa placed a development contract with the French company, Thomson-CSF (now Thales Air Defence), for a mobile all-weather, low-altitude, surface-to-air missile system. The Electronic Systems Division of Thomson-CSF was prime contractor for the complete system including the radar and electronics. Matra was responsible for the missile.

The South African government paid 85 per cent of the development costs of the system, which it calls the Cactus and the remaining 15 per cent was paid by France. Seven platoons were delivered to South Africa by 1973.

The French Air Force ordered the Crotale (Rattlesnake) system for airfield defence and by late 1978, 20 batteries had been delivered.

As produced, Crotale is normally mounted on a P4R (4 × 4) vehicle and can also be shelter-mounted for use in static defence. The first Crotale, produced in 1969, was called the 1000 series. This was followed by the 2000 series in 1973, the 3000 series (originally designed for the French Air Force) in 1975, 4000 series in 1983 and the evolved version based on the 3000 series by the French Air Force in 1989. The evolved system includes a better ECM performance and the passive tracking (by FLIR) of targets in both day and night conditions.

Description

Marketing of the Crotale is now being concentrated on the Crotale NG (New Generation) and on the improvement of delivered Crotale systems to cope with the modern battlefield threat. These improvements include fully automatic electro-optic target and missile tracking mode and also increased ECM resistance and enhanced operational characteristics (using new planar antenna, associated data processing and ECCM devices and hypervelocity VT-1 missile).

The basic system uses the R440 missile and has been designed to combat targets flying at a speed of M1.2 at an altitude of 50 to 3,000 m and an equivalent radar area of 1 m^2 fluctuating. The acquisition unit carries out target surveillance, identification and designation. Mounted on the top of the vehicle is a Thomson-CSF (now Thales) E-band Mirador IV pulse Doppler radar. Thirty targets can be processed per antenna revolution with up to 12 targets simultaneously tracked by the system.

The firing unit has a J-band monopulse 17 km range, single target tracking radar mounted concentrically with the launcher turret, which carries four ready to launch missiles, two each side. The system also has an I-band 10° antenna beamwidth command transmitter, IR gathering system with a ±5° wide field of view (and in French Air Force systems a further narrow field of view mode for passive operations), an integrated TV tracking mode as a low-elevation back-up, an optical designation tripod-mounted binocular device.

The radar can track one target and guide one or two missiles simultaneously. The missiles, fired 2.5 seconds apart, are acquired immediately after launch by the 1.1° tracking beam of the radar with the help of IR detection and radar transponders during the gathering phase. There is also a TV tracking mode possibility. Guidance signals are transmitted to the missiles by a remote-control system.

Improved Thales Crotale acquisition unit

Operational status

Over 7,000 missiles produced. In service with Abu Dhabi, Bahrain, Chile, Egypt, France, Libya, Pakistan and Saudi Arabia.

Specifications

R440 Missile
Length: 2.89 m
Diameter: 150 mm
Wing span: 540 mm
Launch weight: 84 kg
Propulsion: solid propellant rocket motor
Guidance: command control
Warhead: 15.0 kg HE fragmentation with contact and proximity fuzing
Max speed: 930 m/s
Max effective range: see text
Min effective range: see text
Max effective altitude: 5-5.5 km (depending upon target velocity)
Min effective altitude: 15.0 m
Reload time: 2 min (full 4-round load)

Contractor

Thales Air Defence.

UPDATED

Thales Air Defence Crotale NG low-altitude surface-to-air missile system

Type

Air defence missile (vehicle).

Development

The Crotale NG (New Generation) low-altitude SAM system began development in 1988. Crotale NG differs from the original Crotale system in having all the acquisition, tracking, firing and computer units mounted in a single vehicle. The modular all-weather system is divided into six main subsystems which can be carried on a variety of platforms ranging from a simple shelter to armoured vehicles.

The VT-1 missile was developed for Thales and Crotale NG system by the Missiles Division of the American company LTV (now Lockheed Martin Tactical Defense Systems) and successfully completed a series of test firings in March 1989.

The first prototype shelter system was used in 1988 for tracking trials, while the first complete system trials using the shelter unit took place in 1990 with the VT-1 missile. System integration was undertaken in France with the initial batch of 1,000 VT-1 missiles and 42 reusable launch pod containers being built in the USA. Production of these began in mid-1989 with the series being shipped in 1992.

In September 1991, Thomson-CSF (now Thales) signed an agreement with the Euromissile consortium (Aerospatiale and LFK, part of EADS) to start producing the VT-1 missile from around 1997-98 onwards. Apart from its use with Crotale NG, the weapon is also being integrated into the Euromissile Roland 3 programme.

Elements of the Crotale NG system such as the VT-1 missile, electro-optical module and acquisition unit can also be introduced into the Crotale 4000 system, either as new-build items or retrofitted into existing Crotale systems.

In 1988, the Crotale NG was selected by the Finnish Army as its low-level air defence weapon, 20 units being delivered.

By mid-1991, the French Air Force had selected the shelter form of Crotale NG to protect its airbases.

Crotale New Generation for Finland mounted on SISU XA- 180 (6 × 6) APC

Crotale NG also forms the basis of the Daewoo Pegasus air defence system being developed in South Korea.

Description

The Crotale NG electrically driven turret weighs around 4,800 kg and includes a surveillance radar with associated IFF subsystem, a cupola housing a tracking radar, electro-optical equipment including a day and night thermal imager, a daylight only TV camera, video tracking, an IR localiser and eight ready to fire missiles in two packs of four container-launcher tubes.

The electro-optical systems comprise: a Castor double field of view (8.1 × 5.4° wide and 2.7 × 1.8° narrow) thermal imaging camera (with a maximum acquisition range of up to 19 km which reduces to around 10 km in optical visibility conditions); a Mascot day-use single 2.4 × 1.8° field of view CCD TV camera (with a range of up to 15 km); a video tracker for automatic tracking of the target and missile and a large field of view IR localiser to track the missile in its initial few seconds of flight.

The standard option surveillance radar is the Thales TRS 2630 E- band frequency-agile pulse compression Doppler model with a 40 rpm planar antenna, improved ECCM features (including strobe-on-jam, low sidelobes, wideband frequency agility and constant false alarm rate) and search-on-the-move capability. Detection range against high-performance aircraft is 20 km and around 8 km on a hovering helicopter. Altitude coverage is from ground level to around 5 km. An automatic track-while-scan capability provides track details on up to eight targets while simultaneously evaluating the threat.

The tracking radar fitted is of the frequency-agile monopulse Doppler type with improved ECCM features (including low sidelobes, wideband frequency agility, multimode burst-to-burst or pulse-to-pulse frequency agility, constant false alarm rate and jammer tracking), operating in the J-band with a range of up to 30 km on target types from hovering helicopters to M2 plus aircraft.

The VT-1 missile has a maximum range of 11 km, a minimum range of around 500 m and an altitude engagement limit of very low to 6 km.

Maximum missile speed is M3.6, which is achieved by using an improved version of the Sidewinder air-to-air missile solid propellant rocket motor developed by Morton Thiokol specifically for the VT-1 project. The weapon has a flight time of 10 seconds to 8 km range and is capable of manoeuvring under load factors of up to 35 *g* at this distance.

Missile guidance is by an IR deviation measurement system or narrow radar beam using the multisensor guidance principle which Thomson-CSF (now Thales) used in its naval Crotale system. This principle involves using all the sensors to send their data to the onboard computer which then processes it, after filtering out such interferences as clutter, decoys or jamming in a few milliseconds, to determine the guidance control commands to be uplinked to the missile.

In a normal engagement mode, both the radar and the electro-optical systems operate together and constantly check each other. The tracking systems define the observation window that displays the target, the missile to be tracked and the false targets already detected.

Within this defined window are the measured windows associated with each target to be tracked, the physical measurements made by the different sensors are correlated in each measuring window. All the collected data packages are sent to a 2-D digital filter and the best estimate of target and missile positions is extracted. Any control orders to the missiles are passed through the narrow beam, frequency-agile remote-control guidance radio uplink channel of the radar system.

A colour console displays the alphanumeric data on the targets, the TV and thermal images and the video images from the tracking radar, the surveillance radar scope and information on the available missile. All the operator has to do is follow a computer- generated menu displayed on the console and select the desired functions by pressing buttons.

The target engagement cycle from detection to interception is entirely automatic with the gunner only pressing buttons twice to ensure the safety of friendly aircraft. Reaction time is very short at 5 seconds or less with the total engagement time for target detection and final interception at 8 km range being estimated at approximately 15 seconds. Re-engagement time is 1 or 2 seconds depending upon whether the target is isolated or in a group. It is theoretically possible for a single firing unit to engage two separate groups of four aircraft each and destroy all of them at a distance of between 500 and 11,000 m. Reloading the two missile packs takes around 10 minutes.

Operational status

In production and in service with Finland and France.

Specifications

VT-1 Missile
Length: 2.29 m
Diameter: 165 mm
Wing span: n/a
Launch weight: 75 kg
Propulsion: solid propellant rocket motor
Guidance: command control
Warhead: 14 kg HE fragmentation with contact and proximity fuzing
Max speed: M3.6
Max effective range: 11,000 m
Min effective range: 500 m
Max effective altitude: 6,000 m
Reload time: 10 min

Contractor

Thales Air Defence.

UPDATED

Thales Air Defence Starburst low-level close air defence self-propelled missile system

Type

Air defence missile (vehicle).

Development

Thales (formerly Shorts) self-propelled Starburst close air defence missile system is a private venture version of the man-portable Starburst SAM. The missile incorporates the laser optical command guidance technology of the high-velocity follow-on, Starstreak, significantly to increase the missile's effectiveness.

Description

Thales provides two self-propelled systems: ASPIC for lightweight vehicle missions and the armoured SP system (for Alvis Stormer, M113 types of vehicles) for more forward roles. Both systems offer a high degree of automation and eight missile launchers, the capability of integrating Automatic Target Tracking (ATT). The systems also contain high-power dual-magnification daylight and thermal imaging sensors.

The ASPIC System can be integrated onto a 1.5 tonne class of vehicle and can be transported by road, rail, sea and air to be deployed in a theatre of operations.

The armoured SP system is designed to operate in forward contact zones in missions where crew and equipment protection is a high priority. The system comprises four main modules – panoramic sight, eight missile launchers, control console and servo and power control unit. It can be fitted on wheeled and tracked vehicles of typically 10 tonne class and above.

The launcher may be slewed in elevation and azimuth either automatically by the fire-control system or manually by the aimer. Reloading of the complete basic load can be carried out in three minutes from the 12 rounds carried internally.

Both ASPIC and SP can be integrated into automatic C^{3}I networks and the systems can be alerted by Infra-Red Search and Track (IRST) equipment such as the Air Defence Alerting Device (ADAD) from Thales Optronics or radars such as the Thales Air Defence Systems early warning command and control system CLARA which can co-ordinate up to 12 self-propelled Starburst systems.

The panoramic sight system comprises the optical sight, the laser guidance transmitter unit and servo systems, all contained in an armoured housing. The optical sight provides the gunner with low- and high-magnification fields of view for target surveillance, acquisition and tracking. The thermal sight is collimated with the optical sight and gives the aimer a low- and high-magnification direct image of the target. The laser transmitter unit, collimated to the sight axis, provides missile guidance data. As these laser emissions are only transmitted post-launch this reduces the probability of detection by the target.

The Starburst laser beam-riding missile consists of a two-stage motor, prefragmented blast warhead and dual-mode impact/proximity fuze. Twist and steer commands are sent to the forward-mounted steering control surfaces whilst ballistic stability is provided by the rear fins, which also house the two interconnected laser transceiver guidance units. The latter act as the relays between the laser beam transmitted from the aiming unit and the missile's forward-positioned electronics and control section.

Thales Starburst missile being launched from an Alvis Stormer armoured personnel vehicle 0004749

Each of the transceiver units incorporates a laser receiver, a signal processor and a transmitter in a small cylindrical pod. The reasons for two electrically interconnected pods being fitted are system redundancy and the prevention of any possible screening effects acting upon the guidance signals.

The transmitter is mounted in the nose of the pod and relays the command uplink data to the missile's forward-mounted electronics. The optical data signals are detected by small pop-up detectors connected to the electronics which, apart from software changes, is essentially unchanged from that used on the Javelin.

In combat, when the system receives a target indication from the ADAD or CLARA type system, the sighting system is slewed automatically on to the target bearing and the aimer alerted to the presence of a target. Alternatively, the target may either be acquired by the aimer using the sight in an automatically controlled sweep mode or by the vehicle commander carrying out his own visual search outside the vehicle.

On acquisition of the target the aimer selects 'Systems On' and tracks the target, using the joystick to place an aiming graticule in his sight on to the target. This tracking enables lead angles to be generated. When the target is within engagement range, the aimer presses the trigger and the first-stage motor of the selected missile is ignited, propelling the weapon clear of its container-launcher. After a short period of coasting the second-stage motor ignites, boosting the missile to supersonic speed in less than 1 second.

Throughout the engagement the aimer's sole task is to maintain the target in the centre of the aiming graticule and the missile is then automatically guided to hit the target. On completion of the engagement, the aimer selects another target and the fire-control system automatically allocates the next available missile to be fired.

On reaching the target the missile's warhead is detonated either by impact or the proximity fuzing circuit. If, after launch, it is realised that the target is in fact a friendly aircraft, then the aimer has the facility to command the missile to self-destruct.

Operational status
Available on request.

Specifications
Missile
Type: 2 stage, low altitude
Length: 1.394 m
Diameter: 197 mm
Weight
Missile in canister: 15.2 kg
Propulsion: 2-stage solid propellant
Guidance: beam-riding laser
Warhead type: 2.74 kg HE fragmentation with contact and proximity fuzing
Max speed: M1.5 approx
Max effective range: well in excess of 4 km
Launcher: 8-round turret system on vehicle chassis

Contractor
Thales Air Defence.

UPDATED

Thales Air Defence Starstreak low-altitude self-propelled high-velocity missile system

Type
Air defence missile (vehicle).

Development
Designed to fulfil the British Army's General Staff Requirement (GSR) 3979 to replace a regiment of Tracked Rapier for the battlefield role of engaging late-unmasking close support aircraft and ATGW-equipped hovering helicopters, Starstreak is a new High-Velocity Missile (HVM).

In late 1986, Shorts Brothers (now Thales Air Defence) was awarded a £225 million fixed-price contract to cover the development, initial production and supply of the Starstreak HVM weapon to the UK MoD.

The two Starstreak High-Velocity Missile Regiments are the 12th Air Defence Regiment, Royal Artillery with three batteries and the 47th Air Defence Regiment, Royal Artillery also with three batteries.

The first production Self-Propelled (SP) Starstreak systems on the Vickers Alvis Stormer chassis were completed in 1993. The first contract was for a total of 135 systems for the British Army and the systems were accepted into service in 1995.

The current production SP HVM is fitted with a roof-mounted passive Thales Optronics Air Defence Alerting Device (ADAD). In 2001, Thales Air Defence was awarded a £70 million contract to provide the British Army's Starstreak/Stormer with a Thermal Sighting System (TSS). The contract covers design, development and integration of the TSS into the system. The TSS will provide a night capability equivalent to that of daytime, while enhancing the detection and recognition of targets in poor visibility.

TSS makes use of STAIRS-C (Sensor Technology of Affordable Infra-Red Systems) technology developed by Thales Optronics (Glasgow and Staines) and the then Defence Evaluation and Research Agency (now QinetiQ) facility at Malvern. This is the first production application of STAIRS-C technology.

Production Thales Air Defence Starstreak HVM system on a Stormer APC showing Thales Optronics ADAD on top of launcher

Description
The Stormer in the Starstreak launcher configuration carries a crew of three, driver, gunner and commander, with eight ready to fire rounds in two armour-protected servo-controlled containers on the vehicle roof. Collocated with these is the ADAD. Forward of the launcher is the gunner's surveillance, firing and target tracking turret which is fitted with an Avimo servo-controlled target acquisition and tracking sight.

The ADAD provides for detection of targets, their prioritisation, operator alerting and automatic pointing of the Avimo (now Thales AFV Systems) weapon sight at the priority target in azimuth and elevation.

It is based on existing infra-red system technology and offers a 24 hour operational capability for Starstreak which is totally independent of optical visibility, being able to 'see' through battlefield smoke, haze and mist.

The ADAD system comprises three lightweight modules:
(a) Scanner Infra-red Assembly (SIA)
(b) Electronic Pack Processor Unit (EPPU)
(c) Electronic Pack Remote Display Unit (EPRDU).

A guidance beam transmitter is also housed in the sight unit and this is collimated to the target sightline. A total of 12 reload rounds is carried within the hull and these can be used to reload the missile containers or to provide a shoulder-launch or lightweight multiple launcher (LML) capability off-vehicle. For these roles an aiming unit is carried within the vehicle.

The integration of Starstreak with the Thales ASPIC automatic fire unit provides a lightweight self-propelled HVM system. This can be fitted to a wide range of vehicles for easy transportation and deployment into a theatre of operations.

Operational status
In production. In service with the British Army.

Specifications
Starstreak Missile
Length: 1.397 m
Diameter: 127 mm
Wing span: n/a
Launch weight: n/a
Propulsion: 2-stage solid propellant booster/sustainer rocket motor
Guidance: laser beam-riding
Warhead: triple kinetic energy/HE submunitions
Max speed: M3-M4
Max effective range: 7 km
Min effective range: 300 m
Max effective altitude: n/a
Min effective altitude: n/a

Contractor
Thales Air Defence.

UPDATED

Toshiba Type 93 Kin-SAM self-propelled low-altitude surface-to-air missile system

Type
Air defence missile (vehicle).

Development
Summary: the missile has infra-red (IR) tracking and target acquisition.

In FY90 the Japanese Defence Agency (JDA) contracted Toshiba to develop a mobile version of its Type 91 Kin-SAM man-portable SAM system. The result is the Type 93 system, which is now in low-rate production for the Japanese Ground Self-Defence Force.

Description
The Type 93 is based on a light 4 × 4 vehicle chassis with an eight-round (Type 91 missiles in two pods of four) pedestal launcher assembly mounted on the rear

decking. Between the pods is located the target acquisition and tracking unit with both IR and optical sights.

Operational status
In production (small numbers only per year). In service with the Japanese Ground Self-Defence Force.

Contractor
Toshiba (Tokyo Shibaura Electric) Company Limited.

United Defense Limited Partnership (UDLP) Bradley Linebacker, M6

Type
Air defence missile (vehicle).

Description
Summary: the Stinger is an air defense missile with an infra-red/ultra-violet dual-mode passive homing seeker.

M6 Linebacker is a modified M2A2-ODS Bradley Fighting Vehicle and replaces the dual TOW anti-tank/armour missile system on the Bradley turret with the Stinger missile system allowing firing of Stinger missiles from under armour and while on the move using the Bradley turret stabilisation system. A further significant feature is the Linebacker's ability to slew-to-cue to targets handed off from the air defence network and forward area air defence command and control system.

The M6 Bradley Linebackers will replace Bradleys equipped with the Stinger Man-Portable Air Defence System (MANPADS) in active air defense units as planned for the number of M6s procured. The Bradley MANPADS required the stinger team to dismount to engage any low-level serial targets.

Following a competition, in 1995 Boeing Defense and Space Group was awarded a development contract by the US Army Missile Command to develop and build Stinger missile kits for retrofit on Bradley M2A20DS vehicles (M2A2 Bradleys with improvements resulting from lessons learned in Operation Desert Storm). These vehicles designated Bradley Linebackers used the Hughes (now Raytheon) four-round Stinger Standard Vehicle Mounted Launcher (SVML) in addition to retaining the MANPADS mission capability with six stowed Stinger missiles and gripstocks in the rear of the vehicle for dismounted operations.

Eight prototype Bradley Linebackers were successfully tested during 1996 and 1997 in the Army's Advanced Warfighting Experiment (AWE). Subsequently United Defense Limited Partnership, in July 1997, was awarded a production integration contract for 85 M6 Bradley Linebacker Systems from the Army's Tank Automotive Command. The Stinger Missile Systems are being supplied to United Defense through a contract awarded to Boeing Defense and Space Group by the US Army's Missile Command. The first 85 vehicles were produced in 1998 with distribution to Air Defense units starting that same year. An additional 14 Linebackers were contracted with United Defense in 1998 for delivery in 1999. All 99 Linebackers have been delivered to the US Army.

Operational status
The First Unit Equipped (FUE) took place in mid-1998 with subsequent fielding to Air Defense batteries through 1999 and 2000. M6 Bradley Linebackers will replace the Bradley MANPADS vehicles on a one for one basis, as the M6 Linebackers are fielded. Some MANPADS Bradleys will remain operational.

Specifications
Bradley Linebacker, M6 under armour configuration, with 2,727.3 kg armour tile kit fitted
Crew: 5
Combat weight: 29,940 kg
Air-transportable weight: 19,960 kg
Length: 6.55 m
Width: 3.61 m

The United Defense Bradley Linebacker, M6 0098883

Height: 2.97 m
Max speed
Road: 61 km/h
Max road range: 400 km
Armament
Main: 1 × 25 mm M242 cannon, 1 × 4-round Stinger launcher with 6 stowed Stingers

Contractor
United Defense Limited Partnership (UDLP), Ground Systems Division.

Yugoimport SAVA low-altitude surface-to-air missile system

Type
Air defence missile (vehicle).

Development
Summary: the missile has an infra-red seeker; the launcher has an electro-optic sight.

The fully amphibious SAVA mobile SAM system is an indigenously developed variant of the former Soviet SA-13 'Gopher' system. It is intended to provide protection for highly mobile armoured units against attack by low-flying aircraft, helicopters and Remotely Piloted Vehicles (RPVs).

Description
The four-round launch assembly is mounted on a modified BVP M80A APC chassis that acts as the Transporter-Erector-Launcher (TEL) vehicle and is fitted with target acquisition and identification subsystems, electro-optic sight and launcher control systems.

The missile used appears to be similar to an early model 9M31 (SA-13) and is equipped with an improved all-aspects lead sulphide (PbS) passive IR homing seeker, operating at two individual frequency bands of the 1 to 5 µm near-IR spectrum region. A 4 kg HE fragmentation rod warhead is fitted, together with an impact fuze and an active xenon lamp proximity fuzing system. Altitude engagement limits are said to be between 25 and 3,500 m, maximum engagement range 5 km and maximum target detection range 10 km. Reaction time from initial target detection to missile launch is less than 6 seconds. The single shot kill probability is said to be better than 0.6.

Operational status
It is not certain whether this system has entered production.

Specifications
SAVA missile
Missile type: low-altitude air defence
Guidance: passive 1-5 µm near-IR uncooled all-aspects PbS homing seeker with proportional navigation
Weight
Missile in container-launcher box: 70.20 kg
Missile: 39.50 kg
Length: 2.20 m
Diameter: 0.12 m
Wing span: 0.40 m
Warhead
Type: HE fragmentation rod
Weight: 4 kg (2.7 kg HE)
Propulsion: Solid propellant rocket motor
Speed: M2
Range limits: 800-5,000 m
Altitude limits: 25-3,000 m
Max target detection range: 10 km

Contractor
Yugoimport SDPR, Yugoslavia.

SAVA low-altitude surface-to-air missile system launching a missile

AIR DEFENCE MISSILES

VEHICLE SIGHTS

Raytheon air defence FLIR/DAY TV sight system

Type
Air defence missile (vehicle) sight.

Development
Raytheon Systems Company has developed the air defence FLIR/Day TV for General Dynamics Armament Systems (formerly Lockheed Martin) for incorporation into that company's turret for the US Marine Corps Light Armored Vehicle – Air Defense (LAV–AD) programme and for the Thales Blazer vehicle.

Description
The totally integrated sight system has a two-axis stabilised digital line of sight director for fire-on-the-move. The sensor suite includes a second-generation thermal imager, day TV and eye-safe CO_2 laser range-finder for passive target acquisition. The system has the facility for high-resolution video output for detailed remote viewing and digital fire-control interface.

The assembly is suitable for use against both manned and unmanned fixed- and rotary-wing threats, with a secondary role against ground targets.

Operational status
In production. The first systems were delivered to US Marine Corps late in 1997.

Specifications
Fields of view
FLIR
Narrow: 2.6 × 3.5°
Wide: 7.8 × 10.5°
TV
Narrow: 1.7 × 2.2°
Wide: 5.0 × 6.7°
Field of regard
Azimuth: ±22.5°
Elevation: +60 –10°
LOS stabilisation (AZ/EL): <100 μrad
Weight: 295 lbs

Contractor
Raytheon Systems Company.

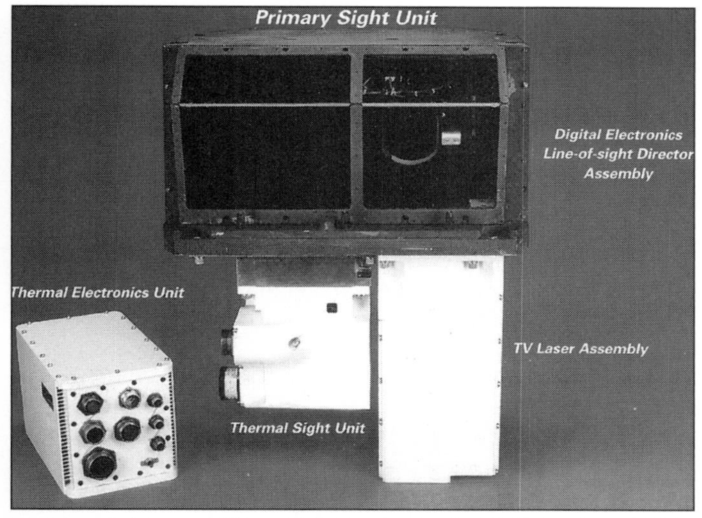

Raytheon air defence FLIR/Day TV sight system for the US Marine Corps LAV-AD 0004730

Reutech ETS 2400 tracking system

Type
Electro-optic/radar tracking system.

Description
The ETS 2400 tracking system is an electro-optic/radar tracking system for anti-air, missile and surface applications. System features include sensor fusion, automatic, manual and memory tracking and autonomous search capability. The electro-optic sensors include a dual field of view 8 to 12 μm thermal imager, dual field of view TV camera and high-repetition rate eye-safe laser range-finder. A 3 to 5 μm thermal imager is optional.

The ETS 2400 electro-optic/radar tracking system from Reutech Systems 0004737

The Ka-band radar has composite Cassegrain reflector antenna, air-cooled magnetron transmitter and three-channel monopulse self-calibrating and equalising receiver. A TWT (Travelling Wave Tube) transmitter is optional.

The signal processor uses MTI (Moving Target Indicator) and non-coherent processing and non-parametric thresholding techniques.

A number of ECCM (Electronic Counter-CounterMeasures) features have been incorporated including: frequency selection; random PRF (Pulse Repetition Frequency) stagger; low antenna sidelobes; multispectral operation; large instantaneous dynamic range; track-on jam; and adaptive tracking filters.

Operational status
Available.

Specifications
Radar frequency: Ka-band
Radar detection range (fighter): >14 km
Tracking accuracy
range: better than 5 m RMS
angles: better than 0.5 mrad RMS
Acquisition time: better than 3 s (1.5 s typical)
Instrumented range: 20 km
Positioner
Velocity: >4 rad/s
Acceleration: >5 rad/s
TV camera
Fields of view
Narrow: 1.8 × 1.4°
Wide: 7.3 × 5.6°
Exposure range: 2 lx (WFOV)/8 lx (NFOV) to 100,000 lx
Thermal imager
Fields of view
Narrow: 2.2 × 1.5°
Wide: 7.3 × 5.6°
Spectral band: 8-12 μm
Laser range-finder
Pulse repetition frequency: <20 Hz
System data rate: 50 Hz

Contractor
Reutech Systems.

SAGEM Glaive

Type
Air defence missile (vehicle) sight.

Description
The Glaive integrated sight is being developed by SAGEM for Roland 3 and to upgrade ROLAND 2 low-altitude surface-to-air missile system. Its modular design enables the Glaive sight to be adapted to day/night sighting systems that require

the following functions – stabilisation, tracking, sectorial surveillance, target designation/acquisition, target flight path analysis.

The Glaive sight performs the following main functions: aiming (speed and absolute position) and two-axis stabilisation of all optical channels; automatic passive surveillance by stabilised infra-red sector scanning; range-finding and computation of targets' radial speed; automatic internal boresighting; binocular display of video signals via very high-resolution micromonitor. There is a built-in test facility and interface is performed through MIL-STD-1553 bus.

The Glaive sight integrates the following optical channels: a dual field of view IRCCD thermal channel with digital video output; high-power, high-frequency eye-safe laser range-finding channel; CCD TV channel based on high-resolution CCD matrices with two simultaneous fields of view; missile tracking channel based on very high-resolution CCD matrices with two simultaneous fields of view in the near infra-red/visible spectral band.

Operational status

In production, under contract for the French Ministry of Defence. Glaive will equip the Roland 3 Upgrade.

Specifications

Range: up to 20 km for aircraft and up to 10 km for helicopters
Elevation aiming: −10, +80°
Azimuth aiming: −10, +10°
Stabilisation accuracy: <40 µrad, peak
Repeatability of the line of sight of measure: <1 pixel
Size of surveillance sector: >20°
Sector analysis time: <1 s
Aiming acceleration: >100 rad/s^2
Aiming speed: >4 rad/s
IR channel
 Spectral band: 8-12 µm
 Detector: 288 × 4 IRCCD
 Video output: 12-bit digital
 IR pupil diameter: 120 mm
Laser range-finding channel
 Wavelength: 1.54 µm
 Type: eye-safe
 Frequency: 12.5 Hz
 Laser pupil diameter: 150 mm
TV channel
 Resolution: 768 pixels
 Video output: 8-bit digital
Missile tracking channel
 Wavelength: 1 µm
 Resolution: 768 pixels
 Video output: 8-bit digital

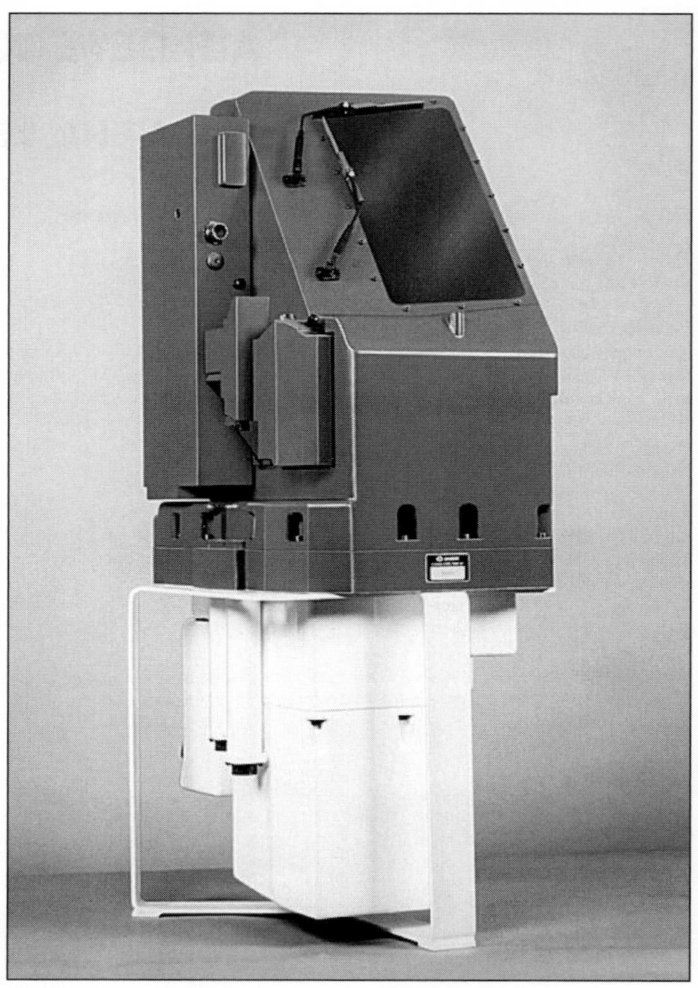

The Glaive day/night sighting system for the Roland air defence system

Contractor

SAGEM SA, Aerospace & Defence Division.

AIR DEFENCE MISSILES

STATIC AND TOWED

CPMIEC FM-80 low-altitude surface-to-air missile system

Type
Air defence missile (static or towed).

Development
Summary: the system has an infra-red localiser for the missile launch and a TV tracker which can be used with or instead of the radar tracker.

The CPMIEC FM-80 all-weather low- and very low-altitude surface-to-air missile system was developed over the period 1978-88 and is designed as a shelter- or vehicle-mounted air defence system for close-in use around airfields, army field units and other Vital Point (VP) communications hub installations.

The development of the FM-80 appears to have been aided considerably by a technology transfer package from France involving the MBDA (formerly Thomson-CSF) Crotale SAM system. The resultant FM-80 system is very similar in physical and technical characteristics to the shelter version of Crotale.

Description
FM-80 can operate in one of three missile guidance operating modes. The exact mode is chosen by the operator to fit the operational scenario. The modes are:

(a) IR - radar, where the target is tracked by the radar, the missile is launched and gathered by the IR localiser and the radar is used to track and measure the missile's angular deviation so as to generate the command radio guidance system signals to guide the weapon to the intercept point

(b) TV - IR - radar, where the target is tracked by the passive TV system, the missile is launched and gathered by the IR localiser and its range and angular deviation determined by the radar. The command radio guidance system then generates the control orders to guide it to the target intercept point

(c) Manual Operation, used where the tracking radar's target channel is subjected to jamming so the operator tracks the target using the handgrip of the TV tracker. The range and angular deviation of the target is measured by the radar and the missile is guided by the command radio guidance system.

The search unit includes an E/F-band 60 rds/min pulse Doppler search radar. The firing unit function receives the target designation data (from either the search unit or the optical aiming device), enabling the tracking radar to search in elevation. When the radar or passive TV tracking system acquires the target and goes to automatic tracking, the firing unit computes the firing solution. Once the engagement parameters are met, the designated missile is fired. The IR localiser determines the weapon's angular deviation and it is gathered through the radio remote-control commands.

Once the missile has entered into the main radar beam, the IR localiser hands over the guidance to the radar. This then measures the relative angular deviation of the missile from the target and flight control commands are generated to guide the missile to the intercept point with the target.

The firing unit includes a monopulse J-band tracking radar with an operational range up to 18 km, rotational limits in azimuth of 360° and in elevation of –5 to +70° and relative accuracy of 0.1 mrad. The missile control transmitter operates in the I/J-bands using a 10° antenna beamwidth and an operational range of 12 km plus; a TV tracking system with an operational range, in fair weather, better than 15 km, a 3° field of view and a tracking accuracy of 0.3 mrad; an Infra-Red (IR) localiser with a 10 × 10° wide field of view and 4 × 5° narrow field of view.

The minimum operational range is 500 m and maximum operational range is 8.6 km against a 400 m/s target, 10 km against a 300 m/s target and approximately 12 km against helicopters. Operational altitude limits are from 15 to 5,500 m.

Missile being launched from top right canister with top of tracking radar between two launchers

Operational status
Production. In service with People's Republic of China armed forces.

Specifications
Length: 3 m
Diameter: est 150 mm
Wing span: est 550 mm
Launch weight: 84.50 kg
Propulsion: solid propellant rocket motor
Guidance: command
Warhead: HE fragmentation with proximity fuze
Max speed: 750 m/s
Max effective range
400 m/s target: 8.6 km
300 m/s target: 10 km
Helicopter: 12 km
Min effective range: 500 m
Max effective altitude: 5.5 km
Min effective altitude: 15 m

Contractor
China National Precision Machinery Import & Export Corporation.

Former Soviet, Iraqi modified SA-2 'Guideline' surface-to-air missile system

Type
Air defence missile (static or towed).

Development
Summary: the missiles have infra-red terminal guidance.

Description
Iraq announced in mid-1989 that it had fitted a Soviet-supplied SA-2 'Guideline' surface-to-air missile system with an infra-red terminal guidance system. According to Iraq this increases the missile's ability against highly manoeuvrable targets at high altitude, especially those using electronic countermeasures.

The nose-mounted seeker senses the thermal radiations of the target in the final stages of missile guidance, enabling the missile to transfer from the normal radio-command guidance system to infra-red homing guidance. If, however, the thermal sensing is lost at this stage then guidance can quickly revert to radio command.

Operational status
The status of this system is uncertain after operation 'Iraqi Freedom'.

Contractor
Former Soviet Factories.

Iraqi modified SA-2 'Guideline' SAM fitted with infra-red terminal seeker
(Christopher F Foss)

HKV, Raytheon Norwegian Adapted HAWK (NOAH) low- to medium-altitude surface-to-air missile system

Type

Air defence missile (static or towed).

Development

Summary: the system has a tracking adjunct system which includes the HEOS electro-optic sensor.

A contract for the development of what became known as the NOAH surface-to-air missile system was awarded to HKV, a joint venture company formed by Raytheon (formerly Hughes Aircraft Company) and Norsk Forsvarsteknologi (now Kongsberg Gruppen), in January 1984. A total of six NOAH battery sets was ordered in February 1984 for airfield defence. All six batteries were operational by mid-1989.

Description

As well as the AN/MPQ-46 High-Power Illuminator radar, M192 launcher and MIM-23B I-HAWK missile, the NOAH ground equipment also includes:

Acquisition Radar and Control System (ARCS)

The ARCS unit provides the target surveillance, target identification, threat evaluation and weapon assignment information for the NOAH Fire Unit.

The 3-D trailer-mounted I/J-band range gated Doppler frequency AN/MPQ- 64A is a computer-controlled low-altitude radar system designed to detect targets infiltrating the coverage provided by long- and medium-range air defence radar sensors. It uses independent 2 × 1.8° search, verify (by a 5.5° electronic backscan separation from the search beam) and track pencil beams with a phase beam scanning antenna to provide the target range, bearing and elevation cueing data.

The radar antenna rotates mechanically in azimuth to provide a full 360° coverage with a 20° selectable elevation coverage between limits of –10 to +55°. It can track up to 60 plus targets at ranges up to 75 km and is fitted with an integral Mk XII IFF subsystem. Acquisition range on a low radar target cross-section fighter-sized target is said to be 40 km.

The scan data update rate is once every 2 seconds (that is, an antenna rotation rate of 30 rpm). Target Doppler signal processing is used to reject clutter caused by natural land, sea and rain returns as well as chaff countermeasures.

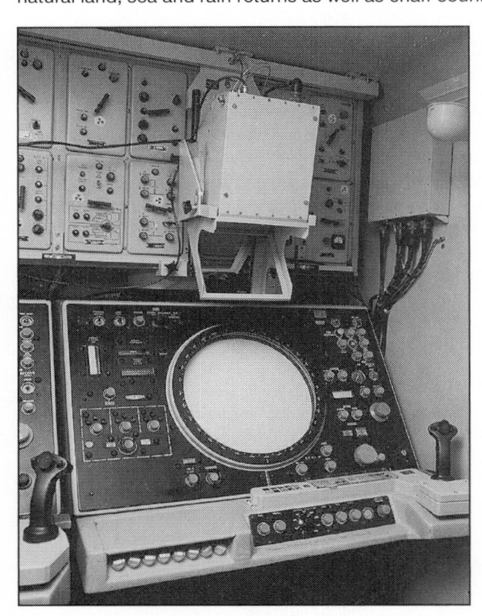

Close-up of the STN Atlas Elektronik HEOS (HAWK Electro-Optical Sensor) equipment with display, control panel and joystick

Norwegian HAWK SAM being launched (left) with Raytheon AN/MPQ-64A radar on right which has an instrumented range of up to 75 km

Elevation coverage has been successfully demonstrated on fixed- and rotary-wing targets operating effectively at velocities of 0 to 724 km/h at altitudes from ground level up to 1.8 km.

The radar can also perform both automatic jammer tracking and sector search in elevation and azimuth and has the ability to conduct jammer burn-through against countermeasures systems.

Norwegian Tracking Adjunct System (NTAS)

This is the early version of the STN Atlas Elektronik HEOS (Hawk Electro-Optical Sensor) system and is an integrated passive day/night infra-red sensor unit mounted above and between the two AN/MPQ-46 High-Power Illuminator radar dishes. Fitted at the rear of the radar dishes are the NTAS power supply and system electronics boxes.

The unit is designed for use either when the illuminator radar is silent or in a heavy ECM environment when it is being jammed to provide accurate raid-size assessment, hostile-act verification, passive tracking and kill assessment.

The acquisition and tracking range is in excess of 100 km and all the system controls, including the joystick, main and auxiliary control panels and visual display, are located in the FDC at the FCO console position.

Operational status

Production complete. The six batteries that were in service with the Royal Norwegian Air Force have now been replaced by NASAMS.

Contractor

HKV.

IAI Arrow 2

Type

Anti-tactical ballistic missile.

Development

The Arrow (Chetz) Anti-Tactical Ballistic Missile (ATBM) system is a joint US/Israeli programme. The initial demonstrator contract was signed in July 1988, with Israel Aircraft Industries (IAI) as prime contractor with US companies as support, to produce four prototypes of the theatre defence missile. A programme of flight tests was carried out between 1990 and 1992.

A follow-on programme, Arrow Continuation Experiments (ACES), Arrow 2, is being jointly funded by Israel and the USA to develop a smaller and less expensive missile system. The first Arrow 2 launch occurred on 30 July 1995. The third test, in August 1996, was successful with the target missile, an Arrow 1 modified to match

The Arrow 2 anti-missile missile is a joint US/ Israeli development
0004744

a Scud missile, destroyed. During the fourth test in March 1997, the Arrow 2 seeker successfully acquired and locked onto its target, a modified Arrow 1. The Elta EL/M-2080 Green Pine phased-array radar was used to monitor target and missile.

Description

It is believed that Arrow is a two-stage solid propellant missile, with an overall length of 7.5 m, a body diameter of 1.2 m and a launch weight around 2,000 kg. It is estimated that the second stage has a length of 2.5 m. Arrow 2 has a dual mode active radar and imaging infra-red seeker developed by Lockheed Martin. The imaging infra-red seeker uses the Amber AE 173 256 × 256 element indium antimonide (InSb) focal plane array. The range capability has been described as around 90 km.

Elta Electronics Industries, an IAI subsidiary, has developed a combined solid-state active phased-array EL/M-2080 L-band early warning and fire-control radar, known as Green Pine.

Rafael was awarded the contract to develop the Arrow's warhead.

Operational status

In limited production. Initial deployment has taken place with final development and integration taking place simultaneously. A total of three batteries are to be procured. The AWS is to be extended beyond the Arrow-2. In March 1997 the Israeli government announced that if it can get additional funding then AWS will be extended to cover a follow-on programme informally known as 'Arrow Beyond 2000.' This began in 1998 and will extend through 2006. Total AWS costs budgeted until 2005 are stated to be US$1.6 billion, although this figure is expected to be exceeded. In 1997, Turkey began preliminary enquiries about the possibility of obtaining an AWS network for its defence.

Specifications

Arrow 2
Length: 6.3 m
Body diameter: 800 mm
Launch weight: 1,300 kg
Warhead: HE fragmentation
Guidance: Inertial, command and IIR/active radar
Propulsion: 2-stage solid propellant
Range: 90 km

Contractor

IAI Ltd.

Kentron, Oerlikon Contraves SAHV-IR/35 mm Skyguard air defence system

Type

Air defence missile (static or towed).

Development

The South African company Kentron has combined with the international company Oerlikon Contraves to offer an integrated missile/anti-aircraft gun air defence system that extends the coverage of a Skyguard AA gun system.

Description

The system is based on a battery comprising a Skyguard fire-control radar unit, two twin- barrelled Oerlikon Contraves 35 mm GDF-005 anti-aircraft guns, and one (with the option of a second) eight-round SAHV-IR containerised missile launcher based on a modified GDF-005 gun platform to ease logistical support.

The missile is able to engage both high-speed aircraft targets at altitudes from below 31 m to well over 6,144 m. It is also capable of countering attacks from anti-tank helicopters equipped with short-range ATGWs.

The missile launcher is aimed at the predicted intercept point beyond the attacking aircraft's weapon release range by the Skyguard fire-control radar so that the SAHV-IR can be launched. Once this happens the radar is free to acquire a second target.

The SAHV-IR missile flies under inertial control of its digital autopilot while its seeker scans a volume of airspace around the calculated target position. When the two-colour, narrowband seeker detects the target it locks on automatically. The scan pattern for this lock-on-after-launch seeker mode makes allowances for any

Artist's impression of Kentron SAHV-IR/Oerlikon Contraves 35 mm Skyguard air defence system deployed in firing position 0044361

Eight-round SAVH-IR missile launcher deployed in the firing position. The carriage is the same as that of the GDF-005 towed anti-aircraft gun system

target manoeuvres and the seeker type ensures rejection of IR countermeasures. For very close-range engagements the seeker's lock on before launch mode is automatically activated.

The weapon uses an active proximity fuze and a 22 kg prefragmented warhead. Missile flight control is achieved by a set of linear electromechanical servos driving independent tail-mounted control fins. Propulsion is by a high-energy, low-smoke solid composite propellant rocket motor.

Operational status

Advanced development.

Specifications

System
Range: 9 km
Altitude: >4.5 km
Fire-control radar: Skyguard/Skyshield
Anti-aircraft guns: 2 twin-barrelled 35 mm GDF-005
Missile launcher: 1 (or optional 2) 8-round modified GDF-005 gun platform
SAHV-IR Missile
Length: 3.32 m
Diameter: 180 mm
Wing span: 500 mm
Weight
At launch: 123 kg
In container-launcher: 200 kg
Propulsion: smokeless composite solid propellant booster sustainer rocket motor
Guidance
Mid-course phase: inertial
Terminal: 2-colour passive infra-red with proportional navigation
Warhead: 22 kg HE-prefragmented with active laser proximity fuzing
Max speed: M2.5
Max manoeuvrability: 40 *g*
Max effective range: 9 km
Min effective range: 1 km
Max effective altitude: 8 km
Min effective altitude
Fixed-wing aircraft: 100 m
Helicopters: 5 m
Time of flight
To 7 km: <16 s

Contractors

Oerlikon Contraves AG.
Kentron Division of Denel (Pty) Ltd.

VERIFIED

Lockheed Martin, Raytheon THAAD Theatre High-Altitude Area Defense missile system

Type

Theatre ballistic missile.

Development

Summary: the THAAD interceptor missile has a mid-wave infra-red seeker.

The THAAD system is envisaged as a transportable battery of weapons capable of hit-to-kill collisions with incoming tactical and theatre ballistic missiles at heights of as much as 20 to 150 times greater than those defended by Patriot and ranges of up to 300 to 400 km. This would then allow the current air defence systems to preserve their primary mission of anti-aircraft defence.

In September 1992, the US Army awarded the contractor team, led by Lockheed Martin Missiles & Space, a US$689 million contract to develop the THAAD system. Raytheon Equipment Division is the prime contractor for the ground-based radar. The main subcontractors are TRW Inc (for software), Texas Instruments

Incorporated (for the solid-state modules), Digital Equipment Corporation (for signal and data processors), EBCO (for the National Missile Defence variant radar turrets), Datatape (for tape recorders) and Raytheon (for the National Missile Defence variant travelling-wave tubes).

In October 1993, the US Army reconfigured the THAAD missile to ensure maximum hit-to-kill capability against incoming ballistic missile threats. The missile diameter has been increased and more propellant added to give a heavier missile that still flies at the required speeds.

Current plans indicate that 1,422 missiles, 99 launchers and 18 GBR radars will be procured. Total costs are estimated at US$9 billion for the THAAD and US$5.4 billion for the Raytheon GBR radar.

The initial part of the 48 month Demonstration/Validation (Dem/Val) phase involves the building of two US Army palletised loading system-based truck-mounted launchers, two Tactical Operations Centre (TOC) shelters and 20 missiles.

The missiles are being used at the White Sands Missile Range for flight tests. Tests started in April 1995, with the seventh test held in March 1997. The last four tests resulted in failure to intercept the target and this may result in delays to the programme.

Problems with further firings in 1998 have resulted in a cost sharing agreement between the Department of Defense and Lockheed Martin. Successful firings were achieved in 1999.

Description

The THAAD missile is a single-stage, solid-fuelled, weapon capable of both exo- and endo-atmospheric intercepts. The missile is 5.8 m long and weighs less than a Patriot round. While on its launcher, primary power will be provided by lead acid batteries automatically recharged by a tactically quiet generator. The THAAD missile will employ thrust vector technology for manoeuvring and a liquid Rocketdyne Divert-and-Attitude-Control System (DACS) for terminal manoeuvring of the forecone, which separates from the missile body before impact. A target object map and predicted intercept point will be provided to the missile before launch, though it will be able to receive in-flight updates.

The interstage at the forward end of the THAAD booster contains a separation motor that ejects the KKV (Kinetic Kill Vehicle), which is integrated into a biconic structure.

During fly-out, the seeker window is protected by a two-piece clamshell shroud. Just before seeker acquisition, metal bladders in the nosecone inflate to eject the shroud. The mid-wave (3 to 5 μm) infra-red seeker is based on a platinum silicide staring focal plane array, but Lockheed Martin is planning to replace this with an indium antimonide array, Raytheon's AE 173 256 × 256 element which is used in the seeker for IAI's Arrow 2 ballistic missile interceptor. The seeker is mounted on a platform that is stabilised in two axes and employs an all-reflective optical system. The seeker looks through a rectangular uncooled sapphire window mounted on the forecone. Guidance is hit-to-kill.

A THAAD missile launch
0004743

A naval version of the THAAD has been considered for the Navy Upper Tier System (NUTS) ATBM requirement.

Each operational THAAD battery will comprise nine launchers (each carrying 10 missiles), two TOCs and a GBR.

The extremely fast THAAD missiles are expected to engage targets out to 200 km plus and intercept missiles as high as 150,000 m in altitude. The latter is especially needed in order safely to engage weapons with nuclear, biological or chemical warheads.

In a scenario known as 'shoot-look-shoot', THAAD weapons engaging two possible targets should have a single shot kill probability of 0.9, the terminal phase of the attack employing onboard infra-red sensors to close in on the target missile.

The THAAD system must also be able to cue other weapon systems and interface with other air defence data information networks to allow battle management and command, control and communications tasking of highly complicated attack scenarios in a distributed manner.

The THAAD system is intended to provide the upper tier of a layered defence of high-value targets, such as airfields with complementary weapon systems including the Raytheon Patriot PAC-3.

Operational status

Under development. Current US Army requirement is for 80 THAAD launchers and 1,272 missiles in an estimated US$8.6 billion programme. Lockheed Martin delivered the User Operational Evaluation System (UOES) demonstrator including four launchers, two systems and two THAAD radars in 1997. Successful test firings were made in 1999. The first THAAD unit is scheduled to enter service in 2004, but this date may have to be extended.

Contractors

Lockheed Martin Missiles and Space (prime).
Raytheon Systems (prime for TMD ground radar).

MBDA Jernas low-altitude surface-to-air missile system

Type

Air defence missile (static or towed).

Development

Summary: the launcher has an electro-optical surveillance and tracking system.

The private venture development of Jernas low-level air defence system started in 1992. It is a lightweight air defence system based on Rapier FSC but optimised for deployment in the Middle East and other hot regions. Jernas can be configured to customer requirements ranging from a passive day/night availability obtained from the launcher operating in isolation through to the full all-weather, dual-engagement capability of a complete system.

Description

Jernas comprises four main elements: eight-round launcher with its own electro-optical surveillance and tracking system; surveillance radar; radar tracker and an optional Tactical Operations Cabin (TOC) which can be mounted on the rear of a suitable vehicle. The launcher uses the eight-round, BAE Systems Blindfire monopulse tracking radar and the BAE Systems 3-D Dagger surveillance radar of Rapier FSC.

Jernas can fire the Rapier Mk 2A impact fuzed and Rapier Mk 2B crush/proximity fuzed missile variants. The complete Jernas system can engage two different targets at once, one with the electro-optical tracker and the other with the radar tracker.

The Dagger surveillance radar allows over 75 targets to be detected and tracked simultaneously. It also automatically prioritises the primary threat tracks. The Dagger can also drive a number of launcher units to increase operational flexibility.

A typical engagement for the complete Jernas fire unit occurs as follows. The fire unit monitors all potentially hostile targets in its assigned airspace using its Dagger surveillance radar. The highest priority target is automatically acquired and tracked

The Jernas air defence system, derived from the Rapier FSC, is optimised for use under extreme climatic conditions
0004745

by both the radar and electro-optical trackers. To engage this target the operator simply presses the fire button and the missile is launched. It is then automatically commanded to intercept using the radar tracker.

Simultaneously, the system will automatically acquire and track the second priority target using the electro-optical tracker. Thus, once the first missile is in flight, this second target is engaged. This operation is once again fully automatic, requiring the operator simply to press the fire button for the second time.

Operational status
Ready for production. Rapier FSC is in service.

Contractor
MBDA (UK).

MBDA Rapier Darkfire low-level surface-to-air missile system

Type
Air defence missile (static or towed).

Development
Summary: Rapier Darkfire has electro-optical tracking.

The Rapier low-level surface-to-air missile system was developed from the early 1960s onwards and the first production units were delivered in July 1970. The system achieved initial operational capability with the British Army and Royal Air Force in 1973.

By 1995, the total signed orders for towed Rapier and Tracked Rapier exceeded 700 fire units and 25,000 missiles, of which more than 12,000 missiles have been fired during development, training and combat.

In 1985, troop trials began of the Field Standard B2 (FS B2) electro-optical system, known as Rapier Darkfire. This involved the introduction of an infra-red tracker which replaced the optical unit, a new six-round launcher incorporating an improved planar-array Racal/Northern Telecon (formerly STC) pulse Doppler surveillance radar, an automatic code changer for IFF and the Console Tactical Control system for the fire unit commander. First deliveries were made in 1988.

A total of 48 FS B2 launchers was delivered to the British Army, the last being in 1992 under the £312 million contract placed in the mid-1980s. A three-month contract was also let by the MoD in 1992 to examine the feasibility of carrying the FS B2 system on the Hägglunds Bv 206 all-terrain vehicle. A number of options were considered including towing the launcher and radar tracker on skis. In this configuration the system would be used by the Royal Marines in any operations in the northern hemisphere's cold regions.

Description
The Rapier Darkfire launcher has six missiles in the ready to fire position, compared to four in the Rapier system, and has a new 3-D Racal/Northern Telecon surveillance radar which provides increased acquisition accuracy and range, helicopter detection mode and considerably improved ECCM (including a counter-anti-radiation missile mode). MBDA developed the Mk 2 Rapier missile which is now the only Rapier missile version in production and is fully compatible with all versions of the Rapier missile system including Optical Rapier, Blindfire Rapier, Tracked Rapier, Rapier Darkfire and Rapier FSC. Production of the Rapier Mk 1 missile has now been completed by MBDA.

The Rapier optical tracker has been replaced by an Electro-Optical Tracker (EOT) with infra-red and TV, providing a day/night and poor visibility capability. The complete system can be plugged into the standard radar tracker using the Blindfire Interface Unit (BIU) which is located on the Radar Tracker in place of the Azimuth Control Unit, whose functions it also incorporates, when all-weather capability is required.

The new Console Tactical Control (CTC) unit provides a colour TV display of surveillance radar data including aircraft tracks with their identification and electronic interference. Also on the display are range intervals and a number of functional markers. The operator, therefore, has a complete picture of the air scenario from which to conduct threat assessment routines and fight the air battle.

Rapier Darkfire deployed in field with six-round launcher (centre) and electro-optical tracker (left)

The display's secondary function is to display test and diagnostic data stored in the processor. The Console Target Tracking (CTT) unit provides the operator's display with a target's thermal image, which is produced by the EOT.

Operational status
In service with the British Army.

Specifications
Rapier launcher
Weight: 1,227 kg
Length: 4.064 m
Width: 1.765 m
Height: 2.134 m
Mk 1 Missile
Type: single stage, low altitude
Length: 2.24 m
Diameter: 133 mm
Wing span: 381 mm
Launch weight: 42.6 kg
Propulsion: 2-stage solid propellant
Guidance: semi-automatic optical command line of sight, thermal command line of sight or automatic command line of sight using Blindfire radar
Warhead: 1.4 kg HE, semi-armour-piercing with crush fuze
Max speed: 650 m/s
Max range: 7 km
Min range: 500 m (optical/Blindfire Rapier)
Max altitude: approx 3 km
Min altitude: <15 m
Launcher: Mobile trainable 4-round trailer-mounted

Contractor
MBDA (UK).

MBDA Rapier Field Standard-C (FSC) low-level surface-to-air missile system

Type
Air defence missile (static or towed).

Development
Summary: Rapier FSC has dual electro-optical and radar tracking.

In 1986, British Aerospace Stevenage (now MBDA) was awarded a contract for the design, development and initial production of the Rapier FSC (previously known as the Rapier 2000) air defence weapon for delivery to the British Army and Royal Air Force Regiment in the mid-1990s and beyond.

Main improvements of Rapier FSC over the existing system can be summarised as higher rate of fire, greater operational flexibility, new and more effective missiles.

Rapier FSC has been designed to counter low- and ultra-low-level air threats including fast ground attack aircraft, pop-up helicopters, RPVs and cruise missiles, under all weather conditions and in an ECM and NBC environment.

The contract, as announced in late 1986, covers an initial production order sufficient to replace two of the British Army of the Rhine's (BAOR) towed Rapier air defence batteries and three squadrons of the Royal Air Force Regiment Rapiers also deployed in Germany. It was originally expected that over 100 Rapier fire units would be purchased by the UK but this has been cut back to a total of 57. The system became operational in 1996.

Description
Rapier FSC can engage targets through a full 360° out to a range of 8 km and up to an altitude of at least 3 km.

The fire unit, or launcher trailer, has eight missiles in the ready to launch position with automatic infra-red tracking which, with manual acquisition and computerised tactical control facilities, can provide an engagement capability by day and night.

By day, an optical acquisition facility can be used to acquire and designate targets for engagement. The IR tracker, mounted between the two banks of four missiles, has a passive scanning mode which can be employed to search for, acquire and track targets by day or night. This provides the operator with a remote viewing system that allows work from a protected position. A planar-array transmitter mounted on the turntable sends secure guidance commands to the missile in flight.

The BAE Systems (formerly Siemens Plessey) surveillance radar trailer can be added to provide a fully automatic engagement capability and a BAE Systems Blindfire radar tracker trailer to provide all-weather operations.

The BAE Systems (formerly Siemens Plessey) 3-D surveillance radar, also known as the Dagger, has a multibeam planar-array antenna providing extremely low sidelobes and good multitarget discrimination. It has a compact high-power transmitter employing Travelling Wave Tube (TWT) technology, wideband receiver unit and high-speed digital processing. It rejects clutter, is resistant to ECM and protects against anti-radiation missiles. It has integral IFF equipment.

Dagger tracks large numbers of targets simultaneously while carrying out automatic identification. It can detect hovering helicopters and very small targets such as RPVs. Dagger can detect multiple targets with their range, bearing and elevation to aid acquisition by the trackers. The most important tracks are displayed on an associated tactical control unit.

The Rapier FSC low-level surface-to-air missile system firing a Rapier Mk 2 missile

The Blindfire radar gives Rapier FSC an all-weather capability by day and night and incorporates its own missile command link with frequency management techniques being used to evade hostile ECM. This gives Rapier FSC a dual-fire capability. The Blindfire radar has narrow beams and low sidelobes to provide higher accuracy target and missile tracking with multitarget discrimination and ultra-low-level tracking capability.

A typical Rapier FSC target engagement would take place as follows: The surveillance radar would first detect, interrogate and form tracks on hostile aircraft. Information describing target tracks is passed to a threat assessment algorithm which allocates the target to be engaged according to predefined threat priorities. Once the target has been allocated, the operator has the choice of selecting either a radar tracker or an electro-optical tracker engagement.

In the case of the former, the radar tracker is directed to acquire the highest priority target and after acquisition the target is automatically tracked. The operator then activates the fire button and the missile is launched. Once launched the radar gathering sensor guides the missile into the tracker beam where it is tracked differentially until target impact. The missile guidance commands are transmitted from the radar tracker command transmitter. The missile is guided towards the target using Automatic Command to Line Of Sight (ACLOS) with the target being destroyed either by a direct hit or proximity fuzed fragmentation warhead.

With the first missile in flight, the fire unit is directed to acquire the next highest priority threat and on acquisition a second missile is launched and guided towards this target using guidance commands from the command transmitter on the fire unit.

The ability of Rapier FSC to engage two targets was demonstrated early in 1991. The first missile was launched and guided by the radar tracker against a Jindivik target and seconds later, while the first target engagement was still in progress, a second missile was launched under control of Rapier FSC's electro-optical tracker against a second target that was simulating a ground attack aircraft.

Operational status

In service with Royal Air Force and British Army. Further orders for the system were placed by the UK MOD to the value of £200 million.

Specifications

Launcher
Weight: 2,400 kg
Length: 4.1 m
Width: 2.2 m
Height: 2.6 m
Elevation limits: –10 to +60°
Traverse: 360°
Number of missiles: 8 ready to fire
Max rate of fire: 7 rds/s
Reload time: 2 min
Target tracker: passive IR
Max tracker range: >15,000 m
Target tracking: automatic and manual
Launcher operation: autonomous
Built-in test (BIT): fully automatic with continuous system monitoring
System interface: MIL-STD 1553B
Interface type: fibre optic

Mk 2 Missile
Length: 2.43 m
Diameter: 140 mm
Max span: 410 mm
Launch weight: 42.6 kg
Propulsion: 2 solid propellant stages
Warhead: HE-fragmentation with active multimode laser proximity and contact fuzes
Max speed: M2
Manoeuvrability: >30 *g* throughout dynamic range
Range: 7 km
Altitude: 0-3,000 m

Contractor

MBDA (UK).

Raytheon MIM-23B I-HAWK low- to medium-altitude surface-to-air missile system

Type

Air defence missile (static or towed).

Development

Summary: the infra-red sensor (HEOS) is used as an adjunct by Netherlands and Germany. The Northrop Grumman Tracking Adjunct System is a video system being upgraded to include an infra-red sensor.

The HAWK (Homing All the Way Killer) semi-active radar-seeking medium-range SAM system commenced development in 1952 with the development contract to Raytheon for the missile. Northrop Grumman was to provide the launcher and loader, radars and fire control. By 1973, this had been upgraded to the MIM-23B Improved Hawk (I-Hawk), which has been successively modernised under various improvement programmes.

Netherlands and Germany have introduced into service the STN Atlas Elektronik HAWK Electro-Optical Sensor (HEOS). A total of 93 systems has been ordered (83 for Germany and 10 for Netherlands). The HEOS is positioned with the radar in between its two antennas. It consists of a thermal imaging sight with the thermal sensor operating in the 8 to 12 μm infra-red range. It has a day and night capability and two fields of view. Tracking range is believed to be over 100 km. The HEOS can be fitted on to any I-HAWK system that can accept the Northrop Grumman Tracking Adjunct System.

During the early 1970s to early 1980s, the I-HAWK system was also sold to a number of countries in the Middle East and Asia. To maintain their HAWK system's viability, the Israelis have upgraded it to the PIP Phase II standard with the addition in the mid-1970s of a Super Eye electro-optical TV system for detection of aircraft at 30 to 40 km and identification at 17 to 25 km.

The most recent change to the missile modified the warhead and fuzing circuit to improve performance against Tactical Ballistic Missiles (TBMs). This missile, which is believed to be designated MIM-23K, has been approved for production by the US Army. By early 1995, 300 plus HAWK rounds had been upgraded to this standard. These will probably be transferred to the US Marine Corps for the TBM role as the US Army does not intend to field an ATBM capability for its own HAWKs.

By 1995, Northrop Grumman had built over 500 tracking adjunct systems with production continuing.

Description

Ground equipment components of the HAWK systems include:

AN/MPQ-50 Pulse Acquisition Radar (PAR)

PAR is the primary source of high- to medium-altitude aircraft detection for the battery. The C-band frequency allows the radar to perform in an all-weather environment. The radar incorporates a digital MTI to provide sensitive target detection in high clutter areas and a staggered pulse repetition rate to minimise the effects of blind speeds.

CW Acquisition Radar (CWAR)

Aircraft detection at the lowest altitudes, in the presence of heavy clutter, is the primary feature the CWAR brings to HAWK. The CWAR and PAR are synchronised in azimuth for ease of target data correlation. Other features include FM ranging, Built-In Test Equipment (BITE) and band frequencies. FM is applied on alternate scans of the CWAR to obtain target range information. During the CW scan, range rate minus range is obtained.

High-Power Illuminator (HPI)

The HPI automatically acquires and tracks designated targets in azimuth, elevation and range rate. It serves as the interface unit supplying azimuth and elevation launch angles, computed by the ADP, to up to three launchers. The HPI J-band energy reflected off the target is also received by the HAWK missile for guidance. A missile reference signal is transmitted directly to the missile by the HPI. Target track is continued throughout missile flight and after intercept, HPI Doppler data are used for kill evaluation. The HPI receives target designations from the BCC and automatically searches a given sector for rapid target lock on.

Close-up of the Northrop Grumman's Tracking Adjunct System (TAS), mounted on top of the HAWK's HPI radar system

AN/MPQ-51 Range-Only Radar (ROR)

This is a K-band pulse radar that provides quick response range measurement when the other radars are denied range data by enemy countermeasures. During a tactical engagement, the radar is designated to obtain ranging information which is used in the computation of the fire command. The ROR reduces its vulnerability to jamming by transmitting only when designated. The ROR is not retained in the Phase III system.

Northrop Grumman Tracking Adjunct System (TAS)

The Northrop Grumman Tracking Adjunct System (TAS) is used in the HAWK-PIP Phase II upgrade for the HPI radar. TAS was derived from the US Air Force's TISEO (Target Identification System, Electro-Optical) device and provides a passive tracking capability with remote real- time video presentation. The day-only TAS is designed to complement the illuminator and can be used either with or independently of the radar line of sight. Manual or automatic acquisition and tracking modes, rate memory and preferential illumination are the key features of the system.

It comprises a two field of view closed-circuit TV camera system which is mounted on a gyrostabilised platform and enhanced by a ×10 magnification telescope.

It is currently a day-only system that is being upgraded to:

(a) improve its daytime performance (in terms of increased range and haze penetration capabilities)
(b) add an automatic target search capability
(c) add an infra-red focal plane array for day/night usage.

The fully functional day/night system is then designated Improved TAS (ITAS). Final development of the ITAS ended in 1991 with the field demonstration and trials phase in early 1992.

Production of TAS devices for the US Marine Corps began in 1980 and exports have been made to seven overseas I-HAWK users. By early 1999 over 500 had been produced.

Operational status

I-HAWK production is complete.

Specifications

Missile
Type: single stage, low- to medium-altitude
Length: 5.08 m
Diameter: 370 mm
Wing span: 1.19 m
Launch weight: 627.30 kg
Propulsion: dual-thrust solid fuel booster sustainer rocket motor
Guidance: semi-active radar homing with proportional navigation
Warhead: 75 kg HE blast/fragmentation with proximity and contact fuzing
Max speed: M2.7
Max effective range
High-altitude target: 40 km
Low-altitude target: 20 km
Min effective range
High-altitude target: 1.5 km
Low-altitude target: 2.5 km
Max effective altitude: 17.7 km
Min effective altitude: 60 m
Launcher: mobile, triple-round trainable, trailer-mounted

Contractor

Raytheon Systems Company.

Saab Bofors RBS 90 low-altitude surface-to-air missile system

Type

Air defence missile (static or towed).

Development

Summary: the missile has laser beam guidance. The system has a thermal imaging sight.

Bofors was awarded a development contract in 1983 by the Swedish Defence Matériel Administration (FMV) for the RBS 90 day/night missile system. Contracts were also placed with Ericsson Microwave Systems AB for complementary search and tracking radars and a thermal imaging/video camera system for the weapon and to Hägglunds Vehicle AB for the conversion of its Bv 206 articulated tracked vehicles as the fire units. First production systems were delivered to the Swedish Army late in 1991 with the first units operational in 1993. The RBS 90 system is due to supplement, rather than replace the RBS 70, and will be deployed at the divisional level in battalion units of one battery RBS 90 and two batteries RBS 70. The RBS 90 battery comprises a radar platoon with an Ericsson PS-90 (Giraffe 75) G-band radar and two platoons, each with three fire units.

Description

The basis of the fire unit is two Bv 206 tracked vehicles.

The Fire Control Vehicle accepts data transmission from the PS-90 central surveillance radar, carries the system electronics, the operator's weapon and radar simulator and a roof-mounted Ericsson PS-91 HARD (Helicopter and Aircraft Radar Detection) H/I-band low probability of intercept 3-D search and acquisition radar. The radar has an effective range of 8 to 10 km against a hovering helicopter and 16 to 20 km against fixed-wing aircraft. The radar has an integrated IFF antenna mounted back-to-back with the microwave unit.

This radar provides aircraft bearing, elevation and range for the commander to select targets for engagement from up to eight being tracked. The information provided is sufficiently accurate for the target to be acquired by the launcher's own electro- optic tracking sensors. These are a single field of view, 8 to 12 μm thermal imager, made by Saab Bofors, and a TV camera. In poor visibility a target can be tracked by radar until it becomes visible through these electro-optics.

The fire unit is autonomous, but can also be co-ordinated by a central surveillance radar. The search and acquisition radar is used to give more accurate target data because of its 3-D function which shortens the system reaction.

The missile operator is seated in front of a video TV screen which includes outputs from either the thermal imager or daylight TV camera tracking devices on the remotely controlled, twin tube launcher. All the missile operator has to do is keep the sight cross-hairs aiming mark on the designated target handed over to him by the commander.

Saab Bofors RBS 90 low-altitude surface-to-air missile system

Operational status
Production. In service with the Swedish Army.

Specifications
RBS 90 system
Missile
Type: RBS 70 Mk 2
Length
Launch tube: 1.75 m
Diameter
Launch tube with end caps: 0.32 m
Weight
Launch tube plus missile: 26.50 kg
Fuzing: laser proximity and impact fuzes
Warhead: combined effect tungsten pellet fragmentation and shaped charge
Max effective range: 7 km
Min effective range: 300 m
Max effective altitude: 4 km
Min effective altitude: ground level
Missile guidance: laser beam
Launcher
Weight
Sight unit: 80 kg
Tripod stand: 90 kg
Two-off ready to fire RBS 90 missiles in container-launcher tubes: 26.50 kg each
Power converter: 35 kg
Dimensions
Sight unit: 645 × 603 × 446 mm
Tripod stand, deployed: 1.271-1.436 m high × 2.179 m wide
Power converter: 460 × 350 × 400 mm
Sight sensors: 3 × 4° field of view TV camera, 4 × 6° field of view thermal imaging system operating in the 8-12 μm waveband region
Guidance beam transmitter: low-energy laser
Power converter voltage: 3-phase 230 V/50 Hz
Engagement control station
Crew: 2
Search and acquisition radar: X-band 3-D pulse Doppler
Elevation coverage: 0-35°
Target velocity coverage: 5-500 m/s
Antenna revolution rate: 40 rpm
IFF: available (back-to-back to antenna arrangement)
Power output: 8 W (average), 65 W (peak)
Max radar range
Aircraft: 20 km
Hovering helicopters: 9 km
Min radar range: 1 km

Contractor
Saab Bofors Dynamics.

Taiwan Tien Kung I low- to medium-altitude surface-to-air missile system

Type
Air defence missile (static or towed).

Development
Summary: the missile's radar seeker is being supplemented with an infra-red seeker.

In 1981, the Chung Shan Institute of Science and Technology started development of the Tien Kung I (Sky Bow I) missile system based on the Patriot. Firing trials were completed in 1985-86.

In September 1993, the first fully operational Sky Bow I system was deployed to replace a HAWK unit in northern Taiwan. The Taiwanese Defence Ministry placed an order in mid-1994 with Raytheon for three fire units and some 200 missiles (to PAC-2 standard). The first unit was due to be delivered in 1996.

Description
The Tien Kung I's physical appearance and basic operational parameters remain essentially similar to those of Patriot. For the target acquisition, tracking and mid-course missile guidance requirements, the army has deployed the CSIST/GE ADAR-1. This is a semi-trailer- mounted 500 km range Chang Bei 'Long White' multifunction, phased-array radar with associated fire-control computer system and the CS/MPG-25 continuous wave dish antenna illuminator radar that are tied into the main phased-array radar.

Each Tien Kung I battery (fire unit) is said to have one Chang Bei and two CS/MPG-25 illuminator radars which each have three or four four-round missile launchers attached.

The Chang Bei radar was developed by the Chung Shan Institute with technological assistance from General Electric Company's RCA Electronic Systems department.

The CS/MPG-25 is solely a Chung Shan Institute development and was derived from the I-HAWK AN/MPQ-46 HPI radar but is estimated to be something like 60 per cent more powerful in output. Improved EW, ECM and IFF capabilities were also introduced.

Tien Kung I SAM being launched (DTM)

Before the Chung Shan Institute developments, operational Tien Kung I batteries were fitted with interface electronics to operate with the fire-control radars of I-HAWK batteries. Post deployment of the radars, the Tien He (interface) system was introduced for both systems to co-ordinate their actions during an engagement.

There is no Track-Via-Missile (TVM) homing capability as this technology was not included as part of the package released to Taiwan. Despite this, the basic radar seeker-equipped version is being supplemented in service by a variant fitted with a passive all-aspects liquid nitrogen-cooled infra-red indium actinide seeker. This provides the battery with the option of firing more than one missile type during a single or multiple target engagement. This variant was tested successfully in April 1985 against a HAWK missile target.

Operational status
In production. In service with the Taiwanese Army.

Specifications
Type: single stage, low- to medium-altitude
Length: 5.30 m
Diameter: 410 mm
Weight: approx 900 kg
Warhead: est 90 kg HE fragmentation with proximity and contact fuzing
Guidance: inertial with command updates and semi-active radar, active radar or passive IR homing seeker options
Propulsion: single-stage solid propellant sustainer/booster rocket motor
Max speed: est M3.5
Range: 6 km plus
Launcher: Mobile trainable 4-round semi-trailer

Contractor
ChungShan Institute of Science and Technology.

Toshiba Type 81 Tan-SAM low-altitude surface-to-air missile system

Type
Air defence missile (static or towed).

Development
Summary: the missile has an infra-red terminal guidance; a thermal imager is fitted in the fire-control system for Type 81 Tan-SAM-Kai.

The Tan-SAM missile system was developed to satisfy a requirement by the Japanese Ground Self-Defence Force (JGSDF) and in 1979 the operational tests of the complete system were undertaken. The JGSDF standardised the Tan-SAM as the Type 81 short-range surface-to-air missile system in late 1980 and began placing yearly production contracts with its manufacturer Toshiba Electric.

The Japanese Maritime Self-Defence Force also generated a requirement for the Type 81 Tan-SAM and bought two fire units in FY89, FY90 and FY92 budgets for the protection of naval bases.

Total procurement of the three services has been 57 fire units for the Ground Self-Defence Force, 30 fire units for the Air Self-Defence Force and six for the Maritime Self Defence Force.

Description
The missile is single stage fire-and-forget with four centrebody wings and four movable tail- mounted fins. The guidance system uses an autopilot for the first part of the flight and then switches to an infra-red seeker for terminal homing.

Before launch, the scanning angle (in degrees) of the seeker head is preprogrammed by the fire-control computer. These data are calculated from the continuously updated information on the target position. The FCS also controls the launcher movements so that a round cannot be fired directly at the sun by accident.

Once the missile is in flight and has reached the point at which the seeker is activated, the infra-red seeker scans the preprogrammed area of the sky to find the target. The guidance unit then locks on and the missile continues to follow the shortest course to intercept. At the target, either the HE fragmentation warhead's contact or radar proximity fuze is activated to detonate the explosive. The lethal radius of the warhead is 5 to 15 m depending on the target type. A self-destruct circuit has not been fitted.

The in-flight seeker lock on feature allows a Tan-SAM fire unit to launch either two missiles simultaneously or successively while the first is still homing on to the target. Thus, theoretically, up to four targets can be engaged by a single fire unit, however, in practice this is doubtful as a single missile cannot be guaranteed a 100 per cent hit probability. The actual hit probability of a Tan-SAM missile is officially stated to be 75 per cent, even in cloud.

Type 81 Tan-SAM low-altitude surface-to-air missile launcher with missiles on upper arm only and stabilisers lowered (Kensuke)

The FCS module is mounted on the rear of a truck and consists of a 30 kW generator unit and the system control cabin. On top of the cabin roof is a 1 m wide, 1.2 m high 3-D phased-array pulse-Doppler radar antenna which is mechanically steered in azimuth and elevation. The radar search range is around 30,000 m and an integral IFF interrogation facility is fitted. The antenna rotates at 10 rpm and sweeps 360° in azimuth and 15° in elevation during a full rotation. In a sector search it automatically sweeps 110° in azimuth and 20° in elevation. Up to six targets can be tracked at any one time.

In 1983, Toshiba began development of the improved Type 81 Tan-SAM-Kai missile. Production of the missile started in 1996. The major improvements include an active radar seeker head, a mid-course update link in the missile which allows the fire-control system to send flight corrections in order that manoeuvring aircraft can be engaged successfully, and the fitting of a thermal imaging optical guidance capability to the fire-control system in place of the current tracker, so as to improve effectiveness in countermeasures conditions.

Operational status

Type 81 Tan-SAM: production complete. In service with the Japanese Ground Self-Defence Force (57 fire units), Japanese Air Self-Defence Force (30 fire units) and Japanese Maritime Self-Defence Force (six fire units). Not offered for export.
Type 81 Tan-SAM-kai: in low-rate production.

Specifications

Type: single stage, low altitude
Length: 2.70 m
Diameter: 160 mm
Wing span: 600 mm
Launch weight: 100 kg
Propulsion: solid fuel sustainer rocket motor
Guidance: preprogrammed autopilot with passive IR terminal homing
Warhead: HE fragmentation with contact and radar proximity fuzes
Max speed: M2.4
Max effective range: 7 km
Min effective range: approx 500 m
Max effective altitude: approx 3 km
Min effective altitude: 15 m
Launcher: trainable 4-round static or vehicle-mounted module

Contractor

Toshiba (Tokyo Shibaura Electric) Company Limited.

AIR DEFENCE MISSILES

STATIC AND TOWED SIGHTS

STN Atlas Elektronik HAWK Electro-optical Sensor (HEOS)

Type

Air defence missile sight.

Description

The HAWK Electro-Optical Sensor (HEOS) is an enhancement for the standard HAWK anti-aircraft missile system for all Hawk systems with an integrated Tracking Adjunct System (TAS) interface. The system is used for passive target detection and tracking. It is used with the illumination radar of the Hawk weapon system and reduces the active operation time of the radar by allowing acquisition and tracking of targets before radar activation and missile firing. It also increases effectiveness against multiple-spaced targets.

HEOS consists of a thermal imaging sight based on US common modules and operating in the 8 to 12 μm waveband, a two-axis stabilised mirror and remote-control functions, with associated tracker and video processing electronics, power and display facilities. It is the same sensor that is used in German Jaguar tanks.

The thermal imaging sight is mounted on the illumination radar and moves with it but can aim independently within the limits of the aiming sectors of its mirror. The sensor signals are fed to a digital scan converter with CCIR output and frame integration and zooming functions. They are also fed in parallel to the auto-tracker which uses either a correlation or centroid track, whichever gives the best result. Recursive reference is used to protect the track against brief disturbances. Track errors, control signals, mode information and operating information are evaluated in the system electronics which then controls the line of sight.

The HEOS Hawk Electro-Optical Sensor is mounted between the antennas of the illumination radar

Operational features of the HEOS system include:

(a) in ECM environments HEOS can be used to distinguish between enemy jammers and the true target; radar energy stays on the target if the radar is slaved to the HEOS
(b) real-time positive visual identification and kill evaluation
(c) protection against anti-radiation missiles by passive acquisition and tracking of target before radar activation and missile firing
(d) increased effectiveness against critically spaced and multiple targets since HEOS allows preferential radar illumination on one selected target.
(e) false target recognition whereby HEOS provides spectral recognition which allows the operator to distinguish between true aerial targets and false targets caused by spurious energy returns.

Operational status

In service. Production completed. A total of 83 systems has been supplied to the German Air Force and 10 systems to the Royal Netherlands Air Force.

Specifications

Spectral band: 8-12 μm
Narrow field of view: 1.5° elevation × 2.4° azimuth
Wide field of view: 4.5° elevation × 7.2° azimuth
Detector: CMT, 120 elements
Cooling: Stirling
Frame rate: 50 Hz
Interlace: 2:1 vertical
Display: high-resolution 22.86 cm (9 in) raster scan, CCIR format
Stabilisation: 2-axis
Stabilisation quality: <0.05 mrad
Aiming sectors
  Elevation: ±10°
  Azimuth: ±15°
Auto-tracker: centroid and correlation with recursive reference
Digital scan converter
  Data rate: 20 MHz
  Format: 8 bit
  Scan rate: 50 Hz
  Output: CCIR interface
Power supply: 115 V/3 phase/400 Hz
Power consumption: 1,000 V A approx
Weight (sensor): 70 kg

Contractor

STN Atlas Elektronik GmbH.

AIR DEFENCE MISSILES

PORTABLE

Arsenal Central Design Bureau UA-424 optical seeker

Type

Man-portable surface-to-air missile seeker.

Development

Arsenal Central Design Bureau of the Ukraine has been actively engaged in the research, development, testing and production of optical seekers for man-portable air defence missile systems for 30 years.

The Igla-1 optical seeker
0041454

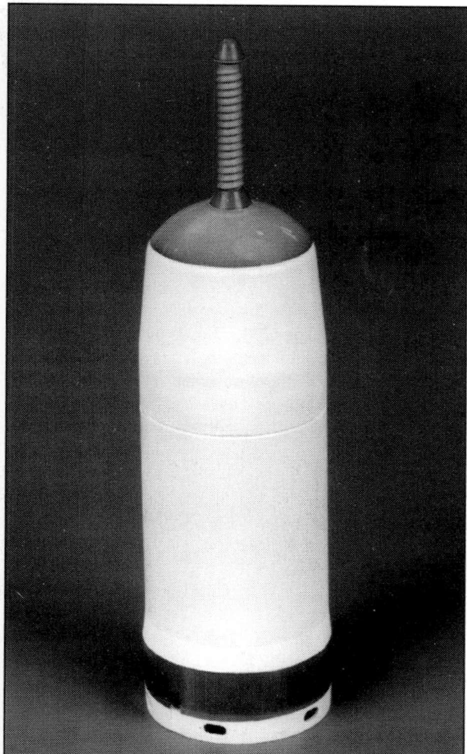

The 'Strela-3' optical seeker
0041452

The seekers for 'Strela-3' and 'Igla-1' systems have been developed and produced in series. One of the latest developments is the UA-424 seeker that could be used for modernisation of the 'Igla' and 'Strela' types portable air defence systems.

Description

The UA-424 seeker operates in the IR mid-wave band and is effective for targets at all aspects. The tracking system uses an astatic gyroscope with a rotor-lens. Lock on and launch are automatic once the operator has acquired the target and selected fire. Rejection of man-made and natural interference are features of the device. There is some built-in recognition of target proximity to direct the missile to hit the most vulnerable zone. A probe on the nose improves aerodynamic performance.

Specifications

Operating band: mid-wave
Range: up to 5 km in the forward hemisphere
Altitude: up to 3 km
Autotracking angular rate: >12°/s
Field of regard: ±40°
Aspect capability: all aspects
Length: 180 mm (without aerodynamic probe)
Diameter: 70 mm
Weight: 0.9 kg

Contractor

Arsenal Central Design Bureau, Ukraine.

CPMIEC Hongying-5 Series portable anti-aircraft missile system

Type

Man-portable surface-to-air missile.

Development

Summary: the missile has an infra-red seeker.

The Hongying-5 (or HN-5) is a product-improved version of the SA-7 'Grail' portable surface-to-air missile system. Development of the Hongying-5 (HN-5) was started in 1975 and finished in April 1985. In January 1979, at the same time as the Hongying- 5 development was underway, the Hongying-5A (HN-5A) missile system was under development. Preliminary work had been started in 1975 and the final design certification was given in November 1986.

The HN-5A is capable of making tail-on engagements against jet aircraft or head-on engagements against propeller-driven aircraft and helicopters under visual aiming conditions. The weapon has seen combat use in a number of regional conflicts. The main improvements of the HN-5A over the original HN-5 include a greater detection range of the infra-red homing seeker (made possible by providing cooling to the detector), reduced susceptibility to background sources, such as bright clouds, by the incorporation of a background noise rejection device into the seeker, and a larger high explosive. The latest production version is the HN-5B.

Transfer of HN-5 component technology has been made to Pakistan for use in the production of that country's Anza Mk I portable SAM system (qv).

The HN-5B missile weapon system 0004734

The HN-5A very low-altitude anti-aircraft missile system 0004733

Description

The system includes a launch tube which serves as an aiming device and launcher, a grip-stock firing unit (designated SK-5A) mounted under the forward part of the launcher which provides launch information and ensures correct firing of the missile and, lastly, a thermal battery mounted on the forward part of the grip-stock to provide power.

The 13 kg missile itself is composed of four sections: the infra-red seeker section which is fitted with both cooling and background noise rejection devices; the control actuator which contains a gas generator; the warhead and fuze and the rocket motor with rear fins attached.

Operational status

Production of the original HN-5 is complete and it is in service with the Chinese armed forces and North Korea. The HN-5A/B is in production and in service with the Afghanistan Mujahideen, Chinese armed forces, Iran, Iraq, North Korea (licence-built), Myanmar, Pakistan and Thailand.

Specifications

HN-5B
Type: single stage, low altitude
Combat weight: 16 kg
Operational altitude: 2,500 m
Operational slant range: 4.40 km
Max velocity of target
In head-on engagements: 150 m/s (550 km/h)
In tail-on engagements: 260 m/s (950 km/h)
Missile diameter: 72 mm
Missile length: 1.44 m
Guidance: passive infra-red homing

Contractor

China National Precision Machinery Import & Export Corporation.

CPMIEC QW-1 Vanguard low-altitude surface-to-air missile system

Type

Man-portable surface-to-air missile.

Development

Summary: the missile has a passive infra-red homing seeker.

CPMIEC has developed the all-aspect QW-1 Vanguard portable system for use against high-speed jets, propeller-driven aircraft and helicopters at low and very low altitudes. Compared to its HN-5 predecessor the QW-1 has major improvements in terms of speed of response, target aspect capabilities, seeker sensitivity and propulsion system.

Description

The QW-1 Vanguard missile has a passive infra-red seeker in the nose section, along with the circuitry that receives the target signal from the seeker and creates the proportional navigation guidance commands that are sent to the missile's control section. The missile's control section contains the two missile control vanes and all the elements needed to fly the missile to the target. The warhead section of

QW-1 Vanguard portable surface-to-air-missile 0004732

the missile contains the HE chemical energy fragmentation warhead and fuzing system. A jettisonable booster motor and dual-thrust solid propellant sustainer rocket motor make up the propulsion section.

When a target is engaged the firing trigger is depressed part way to activate the electrical battery/coolant bottle so that it provides power and gas to the missile's systems. Seeker lock on is signalled to the operator by an audible alarm. The lead angle is set and the firing trigger is depressed all the way. Between 0.3 to 0.8 seconds later the booster motor sends the missile out of the firing tube to a safe point where it is jettisoned and the dual-thrust sustainer motor then powers the missile in flight.

The missile is guided by proportional navigation to the impact point with the target whereupon a target adaptive homing circuit cuts in just before impact to ensure maximum damage is caused.

Operational status

In production. In service with the Chinese armed forces. Offered for export.

Specifications

Type: 2-stage low altitude
Length
Launcher tube: 1.532 m
Weight (system): 16.5 kg
Propulsion: solid fuel booster and dual thrust; solid fuel sustainer rocket motors
Guidance: infra-red passive homing
Warhead: HE-fragmentation
Slant range: 5-500 m
Operational atitude: 500-5,000 m
Reaction time: 3 s

Contractor

China National Precision Machinery Import & Export Corporation.

CPMIEC QW-2 low-altitude surface-to-air missile system

Type

Man-portable surface-to-air missile.

Description

The CNPMIEC has developed a further system in its series of man-portable passive infra-red homing fire-and-forget low-altitude surface-to-air missiles, the Qianwei ('Advanced Guard-2') (QW-2). QW-2 is a new-generation successor to the QW-1. In many respects the all-aspect proportional guidance QW-2 is similar to the Russian Kolomna KBM Igla-1 (SA-16 'Gimlet') system. It has the same diameter as the Igla-1 (0.072 m) but it is slightly shorter and lighter. The QW-2 missile also has a different seeker design which is stated to be highly resistant to a variety of countermeasures. The seeker elements are used to detect hot metal surfaces and engine plume exhaust using FM tracking logic, with the addition that the latter can recognise and distinguish chemical flare countermeasures and filter them from the guidance picture.

Operational status

In production. In service with the People's Liberation Army. Offered for export.

Specification

Length: missile – 1,590 mm
Weight
(system) 18 kg
(missile) 11.32 kg
Diameter: missile – 72 mm
Propulsion: 2 solid fuel rocket stages
Guidance: infra-red passive homing
Control: proportional guidance

CNPMIEC QW-2 low-altitude surface-to-air missile system 0024889

Warhead weight: 1.42 kg
Slant range: 500-6,000 m
Operational altitude: 10-3,500 m
Reaction time: <5 s

Contractor
China National Precision Machinery Import & Export Corporation.

IICS Anza Mk I low-altitude surface-to-air missile system

Type
Man-portable surface-to-air missile.

Description
The Anza (Lance) Mk I passive, infra-red, homing, portable, surface-to-air missile system was developed by the Dr A Q Khan Research Laboratories at Kahuta, Pakistan for the Pakistan armed forces. In general appearance and performance it resembles the Chinese HN-5 system and both are seen as derivatives of the Strela-2.

Anza Mk 1 low-altitude SAM system deployed in the firing position

Operational status
Production. In service with the Pakistan Army.

Specifications
Type: 2-stage, low altitude
Length: 1.44 m
Diameter: 72 mm
Weight: 9.8 kg
Propulsion: solid fuel booster and solid fuel sustainer rocket motor
Guidance: uncooled lead sulphide infra-red homing seeker
Max effective slant range: 4,200 m
Min effective slant range: 1,200 m
Warhead: HE fragmentation
Average missile cruise speed: 500 m/s
Max missile manoeuvring: 6 *g*

Contractor
Institute of Industrial Control Systems (Pakistan).

UPDATED

IICS Anza Mk II low-altitude surface-to-air missile system

Type
Man-portable surface-to-air missile.

Description
The Anza (Lance) Mk II passive infra-red homing portable surface-to-air missile system was developed by the Dr A Q Khan Research Laboratories at Kahuta, Pakistan for the Pakistan armed forces. In general appearance and performance it resembles the Chinese QW-1 Vanguard system but with differences in detail.

The effective engagement limits are estimated to be 500 to 5,500 m. The passive infra-red seeker has been considerably improved over the Anza Mk I version so as to provide an all-aspects engagement capability.

Operational status
Entered production in 1994. In service with the Pakistan Army.

Specifications
Type: 2-stage, low altitude
Length: 1.477 m
Weight: 10.68 kg (missile at launch with booster motor)
Propulsion: solid fuel booster and solid fuel sustainer rocket motor
Guidance: cooled indium antimonide infra-red homing seeker
IR detector: InSb cooled
Warhead: HE fragmentation (containing 0.55 kg HE) with contact and graze fuzing
Average missile cruise speed: 600 m/s
Max missile manoeuvring: 16 *g*

Contractor
Institute of Industrial Control Systems (Pakistan).

UPDATED

Anza Mk II low-altitude SAM system deployed in the firing position

Kolomna Igla (SA-18 'Grouse') low-altitude surface-to-air missile system

Type
Man-portable surface-to-air missile.

Development
Summary: the missile has a dual-mode infra-red seeker.

The Igla (NATO designation SA-18 'Grouse'), was developed by Kolomna MKB and entered service in 1983. The Igla is designed to engage low-flying manoeuvrable and non-manoeuvrable targets and stationary hovering helicopters. In head-on engagements the system can be used against targets with speeds of up to 360 to 400 m/s while against receding targets the maximum target speed is reduced to 320 m/s.

The Igla is an updated Igla-1 system with a dual-channel IR seeker and target identification unit to defeat sophisticated IR decoys. Commonality remains with the Igla-1 in terms of the missile body, rocket, motor, onboard and ground power supply units, as well as the equipment used in servicing the missile systems and personnel training.

A twin-round version, called Dzhigit is being offered for export.

Description
The basic 9K38 Igla missile system uses the 9K39 missile. The missile nose section has a two channel cooled infra-red seeker head operating in the 3.5 to 5 μm wavelength region, made by Lomo plc of St Petersburg. An aerodynamic spike is fitted to improve the speed of the missile.

The Igla launch sequence is started upon visual detection of a target by the gunner through the sight system. The built-in IFF radar interrogator then identifies it. If hostile, the gunner depresses the launch button and the seeker locks on. Within 0.8 seconds the launch signal is generated and the missile's onboard power supply unit, a miniature turbogenerator, is spun up by gas flow from a small powder cartridge.

Approximately 0.6 seconds later, when the onboard power supply is at its optimum operating value, the missile booster ignites to eject the missile from its launcher-container. The main rocket motor ignites after the weapon has travelled some 5 to 6 m from the launcher. The fuzing system is not fully activated until two conditions have been met: the main rocket motor is ignited and the missile must travel a distance of between 80 to 250 m from the launcher.

Guidance to the target is by proportional navigation using the dual-channel passive IR seeker to generate the missile guidance commands. The seeker has a target identification logic unit which is sufficiently sensitive to detect different decoys being launched from supersonic targets with emission levels higher than that of the target itself and having launch period intervals as short as 0.3 seconds.

At the very last moment before hitting the target the programmed identification logic unit orders a shift of the missile aiming point from the engine exhaust area towards the central section of the fuselage. The missile impacts the target and a delay action fuzing circuit is initiated which detonates both the high explosive warhead and the remaining solid sustainer rocket motor propellant after penetration of the airframe has been achieved.

Ring Sights Defence Ltd (UK) provides a version of its LC-40-100 sight (qv) that can be installed on a 9K38 Igla portable system.

Operational status
In production. In service with Brazil (56 grip-stocks and 112 missiles ordered for the Brazilian Army in mid-1994), Russian Federation and Associated States and Yugoslavia (Serbia/Montenegro). Also used by some other states in Asia, Europe and South America.

SA-18 'Grouse' (Igla) man-portable surface-to-air missile deployed in the firing position

Specifications
Type: 2-stage, low altitude
Length
Launch tube: 1.708 m
Missile: 1.70 m
Diameter
Missile: 72 mm
Weight
Total launch assembly in firing position: 18.4 kg
Missile (at launch): 11 kg
Warhead: 1 kg HE chemical energy fragmentation (with additional unused solid propellant rocket fuel) with contact and delay action fuzing circuits
Propulsion: solid fuel booster and 2-grain solid fuel sustainer rocket motor
Guidance: 2-channel 3.5-5 μm wavelength passive infra-red homing
Max firing range
Approaching target: 4.50 km
Receding target: 5.20 km
Min firing range
Approaching target: 500 m
Receding target: 800 m
Max target engagement speed
Approaching target: 360-400 m/s
Receding target: 320 m/s
Max effective target altitude
Approaching target: 3 km
Receding target: 3.50 km
Min effective target altitude: 10 m
System deployment time: 10 s
Missile preparation time: 5 s
Launcher: portable single-round disposable with grip-stock and battery/bottle unit

Contractor
Kolomna MKB.

VERIFIED

Kolomna Igla-1 (SA-16 'Gimlet') low-altitude surface-to-air missile system

Type
Man-portable surface-to-air missile.

Development
Summary: the missile has an infra-red seeker.

Known by the Russian designation Igla-1 (Russian for needle), the American designated SA-16, NATO designated 'Gimlet' was developed by Kolomna MKB and entered service in 1981. It is designed to engage low-flying manoeuvrable and non-manoeuvrable targets and stationary hovering helicopters. In head-on engagements the system can be used against targets with speeds of up to 360 m/s while against receding targets the maximum target speed is reduced to 320 m/s.

There are at least two other members of the Igla-1 family, the 9K310-1 system (which may well be called the Igla-M1) and the Igla-1E (which is an export version) built under licence in Bulgaria and North Korea.

The Igla-1 has seen combat use in Angola and with Iraq during the Gulf War. In the last case it proved to be the most potent of the Iraqi man-portable SAM systems, downing a number of Coalition fixed-wing aircraft.

Description
The 9K310-1 version of the Igla-1 missile has a single-channel cooled passive infra-red seeker operating in the 3.5 to 5 μm waveband. The seeker is currently produced by Lomo plc of St Petersburg. Just before impact the seeker logic system shifts the aiming point from the engine exhaust region towards the central fuselage area at the junction with the wings. An aerodynamic spike is fitted to improve the speed and range of the missile.

The warhead is a 1 kg HE chemical energy fragmentation warhead (plus up to 0.6 to 1.3 kg of remaining solid propellant fuel that is detonated by the warhead) and a detonating system. The latter comprises a remote arming device, contact and grazing fuzing circuits and a self-destruct mechanism which operates after 14 to 17.5 seconds of flight if the missile has not hit the target.

A launch booster provides the missile with its initial velocity to drive it out of the launch tube. Several metres into the flight and a safe distance from the gunner, the solid propellant rocket motor sustainer ignites and accelerates the missile up to its maximum flight speed. Further propellant is used to sustain the velocity.

Operational status
Production. In service with Angola, Bulgaria, Bosnia Herzegovina Croatia (Hrvatsko Viceje Obrane (HVO)), Croatia, Cuba, Czech Republic, Finland (as 86 Igla), Hungary, Iraq, North Korea, Nicaragua, Peru, Russian Federation and Associated States, Saudi Arabia, Slovakia, Syria, Federal Republic of Yugoslavia (Serbia and Montenegro) and UAE.

SA-16 'Gimlet' (Igla-1) surface-to-air missile system deployed in the field

Specifications
Type: 2-stage, low altitude
Length
Missile (fins folded): 1.673 m
Launch tube: 1.70 m
Diameter
Missile: 72 mm
Weight
Total launch assembly in march condition: 18.7 kg
Total launch assembly in firing position: 16.65 kg
Launch tube (without missile): 3 kg
Grip-stock: 1.7 kg
Battery/coolant unit: 1.3 kg
Missile (at launch): 10.8 kg
Warhead: 1 kg HE chemical energy fragmentation with contact and grazing fuzing circuits
Propulsion: solid fuel booster and 2-grain solid fuel sustainer rocket motor
Guidance: single-channel 3.5-5 μm wavelength passive infra-red homing
Seeker head bearing angle limits: ±40°
Average missile speed: 570 m/s
Max firing range slope: 5 km
Min firing range slope: 500 m
Max firing range
Approaching target (jets): 2 km
Approaching target (helicopters and piston-engined aircraft): 2.50 km
Receding target (jets): 2.50 km
Receding target (helicopter and piston-engined aircraft): 3 km
Max target engagement speed
Approaching target: 360 m/s
Receding target: 320 m/s
Max effective target altitude
Approaching target (jets): 2 km
Approaching target (helicopters and piston-engined aircraft): 3 km
Receding target (jets): 2.50 km
Receding target (helicopters and piston-engined aircraft): 3.50 km
Min effective target altitude: 10 m
Missile self-destruct time: 14-17.5 s
System deployment time: 13 s
Missile preparation time: 5 s
Battery/bottle unit lifetime after activation: usually 30 s
Launcher: portable single-round disposable with grip-stock and battery/bottle unit

Contractor
Kolomna MKB.

VERIFIED

Kolomna Strela-2/Strela-2M (SA-7 'Grail') low-altitude surface-to-air missile system

Type
Man-portable surface-to-air missile.

Development
Summary: the missile has an infra-red seeker.

Development of the Strela-2 was completed in 1965 and the SA-7a started service in 1966. Because of its fairly primitive 1.7 to 2.8 μm wavelength lead sulphide seeker head with a 1.9° field of view and 6°/s tracking rate, it was only effectively able to engage a target when it was fired from directly behind at the very hot exhaust area. This tail-chase situation resulted in it only being able to engage aircraft flying at less than 925 km/h with the preferred target being one flying at 462 km/h or less.

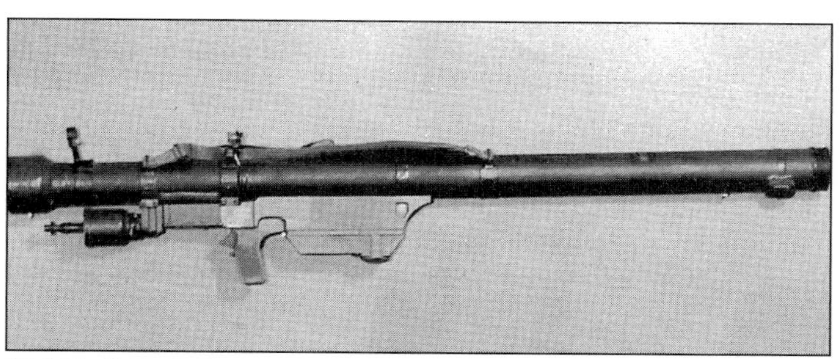

SA-7 'Grail' (Strela-2M) man-portable SAMsystem

This early type of uncooled seeker was also easily saturated by false targets as it did not have any filter system to screen out spurious heat sources. Thus the missile seeker could not be pointed within 20° of the sun (as it would home on to this rather than the target) or fired at an elevation of less than 5°.

The acquisition range varied, depending upon aircraft type and background, from 600 to 2,100 m. It could also be saturated by solar reflection from clouds and go wildly off course. These faults also made the Strela-2 very vulnerable to countermeasures such as infra-red decoys and flares and made it unable to engage low-flying targets.

In order to rectify these faults, in 1971, the then Soviet Army fielded the Strela-2M, (US designation SA-7b, NATO designation 'Grail' Mod 1). This included a more sophisticated seeker with a filter to exclude spurious and countermeasure heat sources and an improved warhead to give greater uniformity in the fragmentation pattern. Although still a tail-chase weapon the SA-7b can be fired from up to 30° either side of the target's tail and still have a good chance of hitting it.

In the mid-seventies an improved version of the Strela-2M was produced. This, known by the US designation SA-7c and NATO designation 'Grail' Mod 2, has a new grip-stock with a vertical handhold and a small paddle assembly just behind the thermal battery. The paddle arrangement is a more sophisticated RF detector to replace the previous helmet-mounted system.

Description
To operate the system the gunner visually identifies and acquires the target. He then loads a missile in its disposable glass fibre container on to the grip-stock and, pointing the launcher at the target, pulls the trigger back to its first stop to start the short-life battery and energise the seeker head's tracker unit. This contains a folded reflective optical system that is sensitive to heat and also acts as a space-stabilised gyroscope to aid missile stability in flight. Once the seeker is energised and uncovered, a red light on the launcher's optical sight is lit. When the seeker detects the target, a green light is activated on the sight and an audible warning is sounded by a small alarm under the rear of the grip-stock near the gunner's ear. Depressing the trigger, the first-stage solid propellant booster motor fully expels the missile at a speed of 28 m/s. The booster burns out in 0.05 seconds, before the tail of the missile leaves the tube, to protect the gunner from being burned.

Because the seeker points at the hot target, guidance proceeds by minimising the difference between the direction that the head is pointing and the weapon's trajectory. There are two variable incidence control fins for this purpose.

Egypt has reverse engineered the Strela-2M under the designation Sakr Eye and China has produced an equivalent system under the HN-5 designation. The basic HN-5 (equivalent to the Strela-2) was followed by the HN-5A which is equivalent to the Strela-2M with some further improvements, including seeker cooling to enhance sensitivity.

The former Yugoslavia has produced its own variant, the Strela-2M/A, and Bulgaria, the Czech Republic and Slovakia, Poland and Romania build a Strela-2M copy under licence.

KBM and LOMO are offering a new, dual-band seeker, the 9E46M, that can be retrofitted in place of the old 9E46. The new seeker is based on the 9E4120 seeker used in the Igla (SA-18 Grouse) and is more resistant to infra-red countermeasures.

Operational status
Production complete. In service with a number of countries.

Specifications
Type: 2-stage, low altitude
Length
Missile (fins folded): 1.44 m
Launcher: 1.50 m
Diameter
Missile: 72 mm
Launcher: 100 mm
Launch weights
Strela-2: 9.2 kg
Strela-2M: 9.85 kg
Launcher weights
Strela-2: 4.17 kg
Strela-2M: 4.95 kg
Propulsion: solid fuel booster and solid fuel sustainer rocket motor
Guidance: infra-red passive homing
Warhead: 1.15 kg HE smooth fragmentation with contact and graze fuzing
Max speed
Strela-2: 385 m/s
Strela-2M: 580 m/s

Max effective range
Strela-2: 3.60 km
Strela-2M: 4.20 km
Min effective range
Strela-2: 800 m
Strela-2M: 800 m
Max effective altitude
Strela-2: 2 km
Strela-2M: 2.30 km
Min effective altitude
Strela-2: 50 m
Strela-2M: 50 m (can be down to 15 m but may be seduced by horizon or ground-radiated heat)
Launcher: portable single-round disposable with grip-stock
Reload time: 6 s

Contractor
Kolomna MKB.

VERIFIED

Kolomna Strela-3 (SA-14 'Gremlin') low-altitude surface-to-air missile system

Type
Man-portable surface-to-air missile.

Development
Summary: the missile has an infra-red seeker.

The Strela-3 (Russian for arrow) was developed by the Kolomna MKB and given the US designation SA-14 and NATO designation 'Gremlin' when it entered operational service in 1978. It replaced earlier SA-7 series weapons on a one-to-one basis and is designed to engage low-flying manoeuvrable and non-manoeuvrable targets and stationary hovering helicopters.

In head-on engagements the system can be used against targets with speeds of up to 310 m/s, while against receding targets the maximum target speed is reduced to 260 m/s. A competent gunner should be ready to engage a second target within 35 seconds of starting the engagement against his initial target.

Description
The Strela-3 missile system uses the 9M36-1 missile which has a single-channel cooled passive infra-red seeker operating in the 3.5 to 5 µm waveband (designated in Russian TGS) and autopilot that receives the target signal from the TGS and creates the guidance commands that are sent to the missile's control section.

The missile also has a 1 kg HE chemical energy fragmentation warhead and detonating system. The latter comprises a remote arming device, contact and grazing fuzing circuits and a self-destruct mechanism which operates after 14 to 17 seconds of flight if the missile has not hit the target.

The launch booster provides the missile with its initial velocity to drive it out of the launch tube. Approximately 0.3 seconds into the flight and some 5.5 m from the gunner with the rear fins deployed, the sustainer ignites and accelerates the missile to its maximum flight speed. This is followed by a second propellant to sustain the flight.

Operational status
Production complete. In service with Angola, Bulgaria (licence-built - qv), Cuba, Czech Republic, Finland, Hungary, India, Iraq, Jordan, Nicaragua, Peru, Poland, Russian Federation and Associated States, Slovakia, Syria, UAE (Abu Dhabi Royal Guard) and the Federal Republic of Yugoslavia (Serbia/Montenegro).

Specifications
Type: 2-stage, low-altiude
Length
Missile (fins folded): 1.42 m
Launch tube: 1.50 m
Diameter
Missile: 72 mm
Weight
Total launch assembly in firing position: 16 kg
Missile (at launch): 10.3 kg
Warhead: 1 kg HE chemical energy fragmentation with contact and grazing fuzing circuits
Propulsion: solid fuel booster and 2-grain solid fuel sustainer rocket motor
Guidance: single-channel 3.5-5 µm wavelength passive infra-red homing
Seeker head bearing angle limits: ±40°
Average missile speed: 470 m/s
Max firing range
Approaching target (jets): 2 km
Approaching target (helicopters and piston-engined aircraft): 4.50 km
Receding target (jets): 4 km
Receding target (helicopters and piston-engined aircraft): 4.50 km
Min firing range (depending upon target type and velocity)
Approaching target: 500-600 m
Receding target: 600-1,100 m

SA-14 'Gremlin' (Strela-3) man-portable surface-to-air missile system captured in Angola by UNITA forces late in 1987

Max target engagement speed
Approaching target: 310 m/s
Receding target: 260 m/s
Max effective target altitude
Approaching target (jets): 1.5 km
Approaching target (helicopters and piston-engined aircraft): 3 km
Receding target (jets): 1.8 km
Receding target (helicopters and piston-engined aircraft): 3 km
Min effective target altitude: 15-30 m (can be lower but missile may be seduced by horizon and ground radiative heat effects)
Missile self-destruct time: 14-17 s
System deployment time: 10-12 s
Missile preparation time: 5 s

Contractor
Kolomna MKB.

VERIFIED

MBDA Mistral MANPADS low-altitude surface-to-air missile system

Type
Man-portable surface-to-air missile.

Development
Summary: the missile has an infra-red (IR) seeker and the launcher has a thermal imaging night sight, either MALIS or MITS2.

Mistral MANPADS originates from a 1980 French military evaluation trial for a third-generation very short-range missile system from which Matra was selected to develop and produce the system. First production systems were delivered to the French Army and French Air Force during late 1989 and after OPVAL trials, Mistral was fielded in the three French services.

Current French Army deployment is to group the Mistral launchers in three corps-level air defence support regiments. These will each deploy a number of batteries that will comprise four to six sections, each of six launchers and a Thomson-CSF Samantha alert system. All sections will normally adopt a triangular field configuration with a pair of launchers at each apex of the triangle, approximately 2.5 km from the other two pairs. The French Air Force uses its MANPADS launchers for defence of its airfields.

Matra BAe Dynamics has developed a C² system for MISTRAL co-ordination using the SHORAR radar from Oerlikon Contraves. This system has been acquired in large numbers by Hungary.

Twenty four countries have acquired MISTRAL based air defence systems either for vehicle mounted land air defence or shipborne air defence.

Description
The Mistral system comprises the missile in its container-launcher tube, the vertical tripod stand, a prelaunch electronics box, a daytime-only sighting system and the battery/coolant unit. A thermal sight MALIS, by SAGEM or MITS 2 by Pilkington Optronics for night-time firing and an IFF interrogation system may be added. A weapon terminal designated AIDA can link MANPADS to a C² system via cable.

The missile has a booster motor to eject it from the launch tube and a sustainer motor to accelerate it to its maximum speed of M2.5. Flight control is by movable canard control surfaces on the missile forebody. The 3 kg HE fragmentation warhead uses tungsten balls to increase penetration of the target. It has both contact and proximity fuzes. The Matra Défense proximity fuze is an active laser type.

The cooled passive IR seeker is derived from technology used on the MBDA R550 Magic 2 air-to-air missile programme and has a multi-element sensor with digital processing that allows head-on, non-afterburning jet combat aircraft to be acquired at ranges of 6 km or so and light combat helicopters with reduced IR signatures at ranges of more than 4 km.

The Mistral MANPADS 0004750

An infra-red transparent magnesium fluoride pyramidal-shaped seeker cover was used so as appreciably to reduce the drag factor normally found at the upper end of the speed range with more conventional cover shapes. This increases the Mistral's manoeuvring capabilities considerably during the terminal phase of the flight.

It takes less than 60 seconds to assemble the Mistral system in the ready to fire state at a firing site. Once an alert is designated in azimuth the gunner acquires it in elevation and begins tracking it. All the aiming data are displayed luminously and continuously which allows the gunner to follow the prelaunch sequence. The safety lever is then released and the seeker activation lever engaged. This causes the battery/coolant unit to energise and release the detector coolant. After 2 seconds the seeker is ready to lock on to the target. When the gunner depresses the firing trigger, the booster motor ignites and the missile accelerates to a muzzle velocity of 40 m/s. Before the entire missile emerges, the motor burns out in order to protect the gunner from burns. Once free of the launch tube and at 15 m from the launcher, the booster motor is jettisoned and the 2.5 second burn composite-fuelled sustainer motor fires to accelerate the missile to its maximum speed. The weapon is then guided to the target's exhaust plume by the onboard infra-red homing system. Maximum flight time is 14 seconds. As soon as a round is fired the expended launch tube is discarded and a new one can be fitted in approximately 10 seconds. Total engagement time from firing sequence initiation to weapon launch is less than 5 seconds, without early warning of a target, and around 3 seconds if a warning is provided. This allows a single firing post to undertake multiple engagements of targets if required.

Operational status

In production (over 15,000 rounds ordered since series production started in 1989) and in service. Mistral has been ordered by 24 countries. Most recent customers are Brunei, Indonesia, New Zealand, Thailand and an undisclosed South American country.

Specifications

Type: 2-stage, low altitude
Length
Missile: 1.86 m
Container-launcher tube: 2 m
Diameter
Missile: 92.5 mm
Wing span: 200 mm
Weight
Missile (launch): 19 kg
Container-launcher (with missile): 24 kg
Propulsion: solid fuel ejector rocket motor with solid fuel sustainer rocket motor
Guidance: infra-red passive homing
Warhead: 3 kg HE fragmentation with contact and active laser proximity fuzes
Max speed: M2.5
Max effective range: 5-6,000 m depending upon target type
Min effective range: 300 m
Max altitude: 4.5 km
Min altitude: 5 m
Launchers: portable or vehicle-mounted single-round disposable, vehicle-mounted twin- round disposable

Contractor

MBDA (France).

Raytheon FIM-92 Stinger low-altitude surface-to-air missile system

Type

Man-portable surface-to-air missile.

Development

Summary: the FIM-92A missile has an infra-red seeker. The FIM-92B/C has a dual-mode infra-red/ultra-violet seeker. The Block II upgrade will probably have an imaging infra-red seeker. The launch units may have thermal imaging or image intensifying sights.

General Dynamics Pomona Division (now Raytheon Systems) began Stinger development in 1972 as a successor to the Redeye missile system. The second-generation portable XFIM-92A Stinger design had a more sensitive seeker head and a better kinetic performance when compared to its predecessor, with the addition of a forward aspect engagement capability to its flight envelope and an integral IFF system.

In 1979, the first production systems were delivered and the first military units achieved initial operational capability status in February 1981 with the basic FIM-92A Stinger version.

In mid-1977, General Dynamics was awarded a full-scale engineering development contract for the next generation of Stinger. This involved the fitting of a microprocessor-controlled Passive Optical Seeker Technique (POST) homing head which used an infra-red and ultra-violet rosette-pattern scanning seeker. Limited procurement of this FIM-92B Stinger-POST version began in 1983 alongside the earlier variant with the production of both ending in 1987. Operational deployment of Stinger-POST systems to the US Army began in July 1987. A total of 15,669 Basic Stinger and just under 600 Stinger-POST missiles was made. The last Stinger-POST rounds were produced by August 1987.

In 1984, General Dynamics began development of the Stinger- Reprogrammable MicroProcessor (RMP) system. Production of this FIM-92C model began in November 1987. A final total of 44,000 Stinger-RMP missiles is expected for the US Army. Additional rounds are being procured for Foreign Military Sales and the other US armed forces.

In FY92, an upgrade contract was placed to improve the Stinger RMP performance. Known as the Block I upgrade, modifications include a new Honeywell ring laser gyro roll sensor, lithium battery and computer memory and software upgrades. The programme is underway and will involve upgrading all the remaining FIM-92A and FIM-92B missiles in the inventory to this standard of FIM-92C, involving some 13,000 rounds; it was to be finished by FY99.

Raytheon Systems Company is working under a demonstration/validation contract for a second upgrade programme, Block II. This involves the development and installation of a 128 × 128 infra-red staring focal plane array seeker. A prototype of the seeker has been built and it successfully tracked a simulated cruise missile at range in excess of 3 km during trials in 1996. The seeker will be incorporated into the standard Stinger airframe.

Description

The Stinger missile is a two-stage solid propellant rocket motor type and in its FIM-92A version is fitted with a second-generation cooled passive infra-red conical scan reticle seeker head with discrete electronic components to provide signal processing. They process the infra-red energy received from the target in the 4.1 to 4.4 μm waveband to determine its relative angle and then, by using a proportional navigation guidance technique, continually predict an intercept point.

In the FIM-92B version the reticle seeker unit is replaced by one which uses an optical processing system. This has two detector materials, one sensitive to infra-red and the other to ultra-violet, together with two microprocessors which are integrated into micro-electronic circuitry for the signal processing phase. The latest Stinger-RMP takes this one stage further by introducing a microprocessor reprogramming facility into the circuitry to allow for new threat characteristics and guidance tailoring.

In all cases the seeker output is sent as steering data to the guidance assembly which converts it into guidance signal format for the control electronics. This module then commands the two movable (of four) forward control surfaces to manoeuvre the weapon on to the required intercept course. The control concept used is known as the single- channel rolling airframe type and, as such,

Stinger man-portable SAM system
(Directorate of Matériel Royal Netherlands Army, Communication Office)

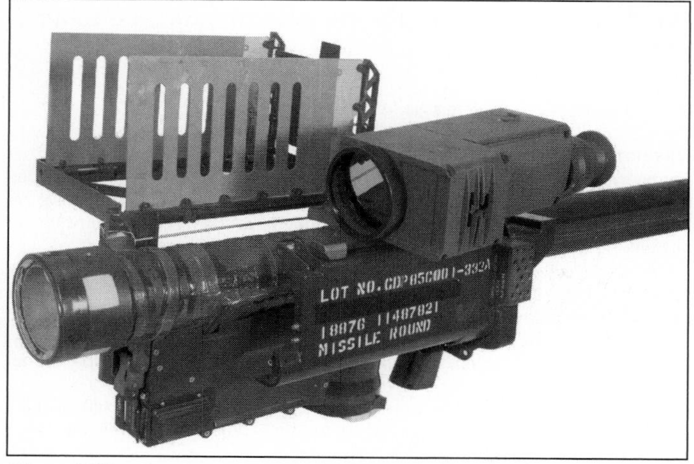

Stinger SAM system fitted with the AN/PAS-18 Raytheon Stinger night sight

considerably reduces both the missile weight and manufacturing costs. As the weapon nears its target, the seeker head activates its Target Adaptive Guidance (TAG) circuit within 1 second of impact to modify its trajectory away from the exhaust plume towards the critical area of the target itself. The fuzing system allows for both contact activation as well as missile self-destruction after 20 seconds of flight time following the launch.

A typical tactical engagement follows this sequence of events. Once alerted to a target the gunner shoulders the system, inserts the battery/cooling unit into its grip-stock and unfolds the IFF antenna. The front protective covers of the launcher tube are removed to reveal the infra-red or dual-mode transparent frangible disc with the seeker immediately behind. Target aquisition is through the launcher sight with optional interrogation of the target using the AN/PPX-1 IFF system. This can be done without having to activate the weapon. The azimuth coverage of the 10 km range IFF system is essentially the same as that of the optical sight enabling the gunner to associate responses with the particular aircraft he has in view. An audio signal 0.7 seconds after the IFF challenge switch is depressed provides the gunner with the cue as to whether the target is friendly or an unknown for possible engagement.

Engagement is initiated by depressing the impulse generator switch. The battery/cooling unit releases argon gas coolant to the detector as well as electrical power for gyros, electronics, and starting the missiles ejector and thermal battery. Unit life is at least 45 seconds.

When sufficient energy is received by the detector an audio signal is produced. Total time required for tracking and missile activation is about 6 seconds. Depressing the seeker 'uncage-bar' and, using the open sight, the superelevation and lead data required for guidance are calculated. Depressing the firing trigger activates the missile battery which can operate for around 19 seconds. After a brief time delay, an umbilical connector to the grip-stock is retracted and a pulse is sent to ignite the ejector motor. Total time to motor ignition from depression of the firing trigger is only 1.7 seconds. Upon ignition the initial thrust imparts roll to the missile and starts the fuze timer system. The missile and its exhaust then break through the frangible discs at either end of the launcher tube.

Before the missile completely clears the end of the tube, the ejector motor burns out in order to protect the gunner from the rocket blast, and two movable control surfaces spring out. Once it clears the tube, the two fixed and the four fixed-and-folded tailfins open out and the ejector motor is jettisoned. The missile then coasts to a predetermined safe distance from the gunner where the fuze timer ignites the combined boost/sustain rocket motor. When the correct acceleration rate is reached (after 1 second of flight) the M934E6 proximity fuzing circuit for the warhead is armed and the self-destruct timer started. The weapon flies a proportional navigation path to the interception point set during acquisition. Even if the target is using 8 *g* manoeuvres, the missile is still capable of engaging it.

Once the gunner has depressed the trigger and the missile has left the launch tube then another round can be used. This takes less than 10 seconds.

Magnavox Electronic Systems, Electro-Optics Division (now part of Raytheon Systems Company) has developed the Stinger Night Sight (AN/PAS-18) which has been manufactured in small quantities against special orders. The sight has been adopted by the US Marine Corps as its definitive Stinger Night Sight. Three units were delivered for trials in August 1991. A US$10.15 million fixed-price contract was placed in November 1993 for the sight with completion of production by January 1996. In April 1994, a US$4.5 million contract was placed by the Danish MoD for AN/PAS-18 sights, spares and depot equipment to be used by the Danish Army and Navy.

The 3 to 5 μm thermal imaging night sight has modifications to the scanner and objective lens to accommodate the Stinger launch envelope. An RS-170 TV format output is provided for remote viewing and/or videotaping for training and post engagement analysis. A quick-release mounting bracket allows the sight to be mated to the next round in under 10 seconds.

ITT Defense F4960 Stinger Night Sight

The ITT F4960 Stinger Night Sight was a third-generation image intensifier system based on the AN/PVS-4 weapon sight technology. It incorporated a 60 mm, f1.2 objective lens (which provides a ×2.26 magnification and a 23.5° circular field of view) with a 25 mm Gen III+I F4844 image intensifier tube. The combination of the lens and image intensifier tube allowed acquisition of targets at ranges of up to 7 km and identification at 4.5 to 5 km even under starlight conditions. A total of 150 examples of the sight has been supplied to the US Marine Corps as the interim Stinger Night Sight system. The system is no longer being marketed by ITT Night Vision.

Operational status

In service in over 20 countries. Basic Stinger production is complete with 15,669 rounds built; Stinger-POST production is complete with just under 600 rounds built; Stinger-RMP (over 40,000 rounds built) production had been halted but, following an order from Taiwan for 1,300 rounds and 334 launchers to equip 74 Boeing Avenger vehicles, has been restarted.

Production of Stinger Block I (FIM-92D) began in 1995 with up to 10,000 earlier model missiles being retrofitted to this standard. Parallel production of 12,000 Stinger-RMPs at the European Stinger Production Group with the participation of Germany, Greece, Netherlands and Turkey, is continuing.

Specifications

Type: 2-stage, low altitude
Length
  Missile: 1.52 m
Diameter
  Missile: 70 mm
Wing span: 91 mm
Weight
  Missile: 10.1 kg (at launch)
  Launcher: 13.3 kg (plus missile)
  Launcher: 15.7 kg (complete)
  Battery coolant unit: 0.4 kg
  Beltpack IFF system: 2.6 kg (incl connecting lead)
  Grip-stock: 2 kg
Propulsion: solid fuel ejector and boost/sustainer rocket motors
Guidance: FIM-92A passive IR homing; FIM-92B/C passive IR/UV homing
Warhead: 3 kg HE fragmentation with contact fuze
Max speed: M2.2
Max range: 8 km
Max effective range
  FIM-92A: >4 km
  FIM-92B/C: >4.50 km
Min effective range: 200 m
Max altitude
  FIM-92A: 3.50 km
  FIM-92B/C: 3.80 km
Min altitude: effectively ground level
Launcher: portable single-round disposable with reusable grip-stock

Contractor

Raytheon Systems Company.

Raytheon/Terma Industries Dual-Mount Stinger

Type

Air defence system.

Development

Summary: Stinger missile has infra-red seeker (FIM-92A) or dual-mode infra-red/ultra-violet seeker (FIM-92B/C).

Dual-Mount Stinger (DMS) is a man-portable weapon platform accommodating two ready to fire Stinger missiles and incorporating a visual cueing device. It has been developed by Raytheon Systems Company and Terma Industries Grenaa (formerly Per Udsen Company) of Denmark for the Danish Low-Level Air Defence System (DALLADS).

The Danish Army plans to procure over 100 DMS systems and the Danish Navy is evaluating the system for protection of its patrol boats. DMS has undergone full operational test and evaluation, full-scale production beginning in November 1995.

Dual-Mount Stinger selected for the Danish Low-Level Air Defence System 0004731

In the DALLADS configuration, it is designed for mounting on the ground, but can be modified for installation on a light vehicle.

Description

The Dual-Mount Stinger (DMS) system is a lightweight weapon system designed for use as a rapidly deployable stationary ground mount. It can also be readily mounted on a light utility truck or deck of a small patrol boat. DMS is capable of launching up to two unmodified Stinger missiles with rapid reload.

The DMS system consists of a tripod base or stand, a pedestal mount with adjustable gunner's seat, and an elevation assembly. The elevation assembly has provisions for mounting a thermal day/night sight, a zoom telescopic visual sight with target-cueing indicator, a control panel and two Stinger launch stations. Each launch station will utilise an affixed interface assembly for missile activation and launch functions, and to communicate with the system control panel.

Missile power and coolant gases are provided by an external gas bottle and power supply. Power for operation of the Danish version of DMW Weapon Interface Unit (WIU) is provided through an external Control and Warning System (CWS) interface with back-up power provided by internal batteries.

Operational status

In production and fielded as part of the Danish Low-Level Air Defence System.

Specifications

Height: 1.32 m
Operating circle: 2.4 m
Weight without missiles, WIU,WT,VCD and sight: 82 kg
Max weight of portable unit: 35 kg
Max movement in elevation: –20 to +65° fixed stop at –20, 0, +20, +40 and +60°

Contractors

Raytheon Systems Company.
Terma Industries Grenaa AS.

Saab Bofors RBS 70 series low-altitude surface-to-air missile system

Type

Man-portable surface-to-air missile.

Development

Summary: the missile is laser beam-riding and on the field sight a COND (Clip-On Night Device) thermal imaging night sight (qv) can be mounted to give 24 hours capability.

A development contract for the RBS 70 low-altitude surface-to-air short-range missile system was placed with Bofors in mid-1969. The first delivery of RBS 70 systems for trial purposes was made in late 1973 with user trials conducted between 1974 and 1975. In June 1975, the first production orders were placed for the RBS 70 missiles, sights, stands and PI-69 IFF sets. The first production day-only RBS 70 sets were delivered to Swedish Army training units in 1976 with the first operational units being formed the following year. The first production radar sets were delivered in 1979.

In the late seventies, Bofors completed development of the RBS 70 system by introducing a bigger observing angle for the missile laser receiver that enlarges the engagement envelope by between 30 and 50 per cent, depending upon the tactical situation. The Swedish Army started an upgrading and modernisation programme in 1998.

Description

The basic RBS 70 missile is a two-stage solid propellant rocket motor powered type. The rear of the missile body houses a receiver unit which senses deviation from the laser centre, which coincides with the line of sight, and a small computer which converts these deviation signals into guidance pulses that command the

RBS 70 surface-to-air missile fitted with a Clip-On Night Device – nightfiring 0053767

RBS 70 Clip-On Night Device (COND) 0010864

missile automatically to follow the centre of the laser beam. Maximum engagement altitude is more than 5,000 m while the minimum altitude is effectively ground level.

The Missile Mk 1 is essentially the same as the basic missile but uses a laser guidance sensor unit which increases the rearward field of view from 40 to 57°. This considerably enlarges the available engagement envelope. This missile is no longer in production.

The Missile Mk 2 has a larger sustainer and warhead. The new sustainer increases the missile velocity profile and the maximum range and altitude (up to 7,000 m as opposed to 6,000 m of the other round types against slow moving targets and from 3,000 to 4,000 m altitude). The basic fire unit comprises two major parts: a stand and the sight. The RBS 70 sight with its uncooled laser diodes only needs Lithium batteries as a power supply.

The latest target data receiver has a built-in threat evaluation providing priority of the assigned targets and also presenting first and last time to fire.

On firing, the laser guidance unit is activated and a booster motor on the missile is ignited to propel it out of the tube. For operator safety the motor cuts out before the end of the missile leaves the tube. The booster motor is jettisoned at a point several metres from the muzzle and the round's four centre-body fins and four rear cruciform control surfaces unfold. The sustainer motor ignites and the guidance receiver on the missile starts to sense the modulated laser guidance beam. This missile is now stabilised and flies in a cross-wing configuration. The onboard computer translates the received signals into commands for moving the electrically operated control surfaces. Once at maximum velocity the sustainer cuts out and the missile continues on course in coasting mode. To ensure a hit, the missile operator has only to keep the target in the middle of the cross-hairs of his gyrostabilised sight by using a thumb joystick.

A complete RBS 70 system comprises a surveillance radar for target detection and includes combat control funtions.

If no search radar is available an observation post has to be established to provide early warning. The missile operator then has to search for the target. When the system has been slewed on to the rough bearing of the target, the weapon's safety-catch is released which activates the electronics for missile launch some 4 seconds later and commences fine aiming with a gyro-stabilised reticle sight. The IFF equipment, if fitted, is automatically activated at the same moment and this transmits interrogation signals. If a friendly response is received, the firing circuit is overridden and visual signal lamps in the arming sight indicate that this has happened. The missile operator then discontinues the action and resets the safety device. The ×7 magnification sight used for the fine aiming has a 9° aperture. The sight reticle is designed to enable manual firing using scale markings which allow the operator to estimate when to fire against various targets.

The RBS 70 Clip-On Night Device (COND), made by FLIR Systems (formerly SaabTech), has been developed for attachment to the front of the day sight unit. This weighs 24 kg (including an integral battery power pack) and utilises an 8 to 12 μm waveband infra-red scanner unit with a 23 element detector for target imaging. Thales Optronics supplies the scanner and electronic control unit, similar to the Class I common modules supplied for the RBS 56 BILL ATGW. The thermal picture generated from the unity magnification equipment of the sight is injected directly into the front end of the day sight allowing the laser beam to pass through. No alignment is necessary.

Operational status

In production (the missile Mk 2 only – in 2002 replaced by the new Bolide missile with enhanced coverage and small target capability). By 1995, total production of the RBS 70 system amounted to well over 1,300 launchers and 14,000 missiles. In service with 13 countries, including Australia, Thailand and Venezuela. In some countries the RBS 70 is in use with all services, navy, army and air force.

Specifications

Bolide missile
Type: 2-stage, low altitude
Length
Missile: 1.32 m
Container-launcher (with end caps): 1.745 m
Diameter
Missile: 106 mm
Container-launcher: 152 mm
Wing span: 320 mm
Weight
At launch: 16.50 kg

Container-launcher with missile: 26.50 kg
Sight with pads and carrying harness: 35 kg
Propulsion: solid fuel booster and solid fuel sustainer rocket motor
Guidance: modulated laser beam-riding
Warhead: HE with combination of prefragmentation with tungsten spheres and shaped charge, with impact and active laser proximity fuzing
Max speed: >M2
Max range
Head-on: >8 km
Crossing: 4 km plus
Min range: about 200 m
Max altitude: >5,000 m
Min altitude: ground level
Reaction time: <5 s
Deployment time: 30 s
Reloading time: approx 5 s
Launcher: portable or mobile single-round trainable stand

Contractor
Saab Bofors Dynamics.

Sakr Factory Sakr Eye low-altitude surface-to-air missile system

Type
Man-portable surface-to-air missile.

Development
Summary: the missile has an infra-red seeker and the launcher has an optical sight with an optional night sight.

The Egyptian Army used the Strela-2/Strela-2M SA-7 'Grail' portable low-altitude surface-to-air missile for many years but when, in the late 1970s, they could no longer obtain spare parts or replacement missiles, a reverse engineered and improved version of the Strela-2M (SA-7b) was produced and subsequently called the Sakr Eye. It is manufactured by Sakr Factory for Developed Industries at Almaza.

It was first shown in public in late 1984. Full production started in 1986 following extensive Egyptian Army trials with initial operational capability being achieved in 1987-88. Egyptian sources indicated that it cost US$180 million to develop Sakr Eye with all funding coming from the Egyptian Defence Ministry. Sakr Eye was deployed with the Egyptian units assigned to the Coalition Forces in the Gulf War of 1991.

Description
The missile is a fire-and-forget type and consists of an infra-red homing seeker, guidance and control, warhead and propulsion sections. The grip-stock combines the firing mechanism and the logic circuits and when attached to the launch tube carries out the firing sequence in either the manual or automatic modes.

In the absence of an integrated IFF system a typical target engagement sequence is as follows: the gunner acquires the target and aligns the weapon using the open sight and then selects either manual or automatic firing mode; audio and visual cues are given to the gunner to indicate when the target is within the weapon's engagement envelope. The trigger is then squeezed according to the selected mode and the missile is ejected from the launch tube by a small rocket motor. Then, after a short delay, the booster accelerates the missile until the sustainer motor is ignited to propel it to the target. If a target is not encountered within 16 seconds Sakr Eye destroys itself. Homing is by means of a passive infra-red seeker which was developed by Teledyne and is more sensitive than the original. The HE warhead is detonated by a contact fuze that also provides for graze initiation of the warhead.

The Sakr Eye ammunition container has two missiles. It takes 10 seconds to be prepared for action and can engage aircraft travelling at a maximum speed of 280 m/s in a pursuit or flying at a maximum speed of 150 m/s head-on. Successful trials have also been carried out with Sakr Eye fitted with the US CA-563 optical sight. This incorporates increased magnification (×3 with a 22° field of view) with advanced sighting techniques. A night vision module can also be incorporated into the CA-563 optical sight. Optional equipment includes a Thales PS-340 IFF unit, which is attached to the right side of the launcher with the interrogator electronics unit hanging on the operator's belt, and a night vision sight.

Operational status
In production. In service with the Egyptian Army.

Specifications
Type: 2-stage, low altitude
Length
Missile: 1.40 m
Diameter
Missile: 72 mm
Weight
Missile: 9.9 kg
Launcher: 5.1 kg
Complete weapon: 15 kg (without CA-563 sight); 18 kg (with CA-563 sight)
Propulsion: solid fuel booster and solid fuel sustainer rocket
Warhead: HE smooth fragmentation with contact and graze fuzing

Sakr Eye SAM system fitted with night sight and IFF system

Max effective range: 4.4 km
Max effective altitude: 2.4 km
Min effective altitude: 50 m (150 m for a helicopter)
Launcher: portable single-round disposable with grip-stock
Reload time: 6 s

Contractor
Sakr Factory for Developed Industries.

Thales Air Defence Javelin low-level surface-to-air missile system

Type
Man-portable surface-to-air missile.

Development
The Thales Air Defence (formerly Shorts) Javelin close-range air defence weapon was developed under contract to the UK MoD from 1979 as a follow-on to the Shorts Blowpipe system and first production Javelins were completed in 1984.

The Javelin has been designed to counter a wide range of low-level air defence targets and it employs semi-automatic command to line of sight (SACLOS) guidance, rather than infra-red detection, to engage its target. Its range enables it to engage and destroy high-speed attacking aircraft before they are able to release their weapon load. The Javelin can also be used against helicopters and has a secondary surface-to-surface capability. By January 1985, the total Javelin order book was over £160 million. By mid-1993, the Javelin had been replaced in the British Army by Starburst in front-line and reserve unit operational use. The Javelin stocks remaining have been reserved for training use only.

In February 1996, it was announced that Javelin had been bought by Peru, thought to be between 200 and 500 launchers and missiles.

Description
The Javelin SAM system consists of two main components: the missile, sealed within its launching canister and the aiming unit.

Man-portable Thales Javelin SAM system deployed with Thales Optronics Air Defence Alert Device (ADAD) in the foreground

The missile is 1.4 m long with the fuzes in the nose and the warhead in the centre. The guidance and control functions are in the forward part of the body and the rocket motors are in the rear.

The aiming unit contains a stabilised sighting system which provides manual target tracking and automatic missile guidance through a solid-state TV camera.

Digital data from the camera are fed to a microprocessor and the resultant guidance demands are transmitted to the missile by radio. BAE Systems was awarded its first production contract, worth over £5 million, for the advanced television guidance system used in the Javelin in May 1984. The automatic gather and guidance system comprises a miniature solid-state CCD (Charge-Coupled Device), TV camera and zoom lens, sophisticated signal processing electronics and a two-axis subminiature gyro assembly. The camera unit and associated data extraction equipment is produced by the company's Sensors and Communications Systems unit, Basildon and the gyro assembly by the Gyro Division, Rochester. The complete electro-optical and gyro subsystem is contained within the operator's lightweight aiming unit which makes extensive use of high-density electronic packaging involving multilayer hybrid microcircuits.

The aimer acquires a target using the monocular sight and switches on the system, selecting the frequency of the guidance transmitter and the mode of the fuze (proximity or impact). This activates the tracking electronics and projects an illuminated stabilised aiming mark (a red circular reticle) into the field of view. The target is tracked briefly with the aiming mark to establish a lead angle, the safety catch is released and the trigger pressed. Range is indicated stadiametrically in the aimer's eyepiece which has a magnification of ×6 compared with ×5 of a Blowpipe.

The camera detects the missile flares, computes optimum guidance commands using digital techniques and transmits these guidance demands to the missile. The TV guidance datum line is collimated with the aiming mark which is maintained on target by the gunner using the thumb joystick.

Operational status

Production complete. In service with eight countries (a total of 12 services) including Botswana, Canada, Dubai, Jordan (may have ordered system in 1988), Peru, South Korea, Oman and the United Kingdom (training only). Over 16,000 Javelin SAMs have been completed.

Specifications

Type: 2-stage, low altitude
Length: 1.39 m
Diameter: 76 mm
Wing span: 275 mm
Weight
Aiming unit: 8.9 kg
Missile: 12.7 kg
Missile in canister: 15.4 kg
Missile in field-handling container: 19 kg
Missile in field-shipping container: 43 kg
Propulsion: 2-stage solid propellant
Guidance: Semi-Automatic Command to Line Of Sight (SACLOS)
FOV
Monocular: 180 mils
Magnification: ×6
TV FOV
Wide: 230 × 180 mils
Narrow: 36 × 36 mils
Warhead type: 2.74 kg HE fragmentation with contact and proximity fuzing
HE content of warhead: 0.6 kg
Max speed: about M1.5
Max effective range
Against helicopters: about 5.5 km
Against jet aircraft: about 4.5 km
Min effective range: about 300 m
Max effective altitude: 3 km
Min effective altitude: 10 m
Power supply: 27.5-35.5 V DC supplied by canister thermal battery and 3× 12 V rechargeable batteries in aiming unit. A disposable Lithium battery pack is available.
Launcher: Portable single round with grip-stock or on a Lightweight Multiple Launcher (LML) with three missiles, also available as a naval version or with a vehicle mount.

Contractor

Thales Air Defence.

UPDATED

Thales Air Defence Starburst low-level surface-to-air missile system

Type

Man-portable surface-to-air missile.

Development

In 1964, Thales Air Defence Systems (formerly Shorts Missile Systems) began a private venture development of a portable surface-to-air missile system for defence against aircraft in the forward battle area. The missile, later called Blowpipe, was

Starburst SL (shoulder-launch) firing 0004751

successfully fired in 1965 and three years later attracted UK MoD funding. Development of the system was completed in 1972 and during that year the system was ordered for the British Army. Blowpipe was exported to 11 countries.

In 1978, Short Brothers began development of a follow-on system, originally named Blowpipe Mark 2, with improved Semi-Automatic Command to Line of Sight (SACLOS) guidance system, a new warhead and more powerful motor. The new missile, later renamed Javelin, was compatible with the Blowpipe launcher system and entered service in 1984. Javelin has been exported to six countries.

From 1986, the manufacturer began development of a new version of Javelin known as S15 or Advanced Javelin. This programme was carried out as an operational emergency for UK MoD, who had recognised the requirement for a missile with greater immunity to countermeasures. Shorts incorporated a laser beam-riding guidance system which was developed in parallel for the company's Starstreak. The first S15 firing trials were conducted in 1986 and development of the system, named Starburst was completed in 1989 during which year the British Army received its first deliveries. The system was deployed with the British Army during Operation Desert Shield/Desert Storm and in Bosnia.

Description

Thales provide the Starburst missile along with a wide range of firing units which vary from the shoulder-launched system to a range of three lightweight launchers, Lightweight Multiple Launcher (LML), Vehicle Multiple Launcher (VML) and the Naval Multiple Launcher (NML). In addition, Thales Air Defence Systems provide two self-propelled systems, ASPIC for lightweight vehicle missions and the armoured SP system for more forward roles.

The laser beam-riding missile is sealed in a lightweight canister which also acts as a launch tube. It is delivered to customers in a high strength composite aluminium container and a field handling container. The missile in its package requires no maintenance and Thales claim a proven 10-year shelf life for Starburst. The missile in its ready-to-fire container is 1.39 m long and weighs 15.2 kg. The missile consists of a two- stage motor, prefragmented blast warhead and dual-mode impact/proximity fuze. The rear fins house the two interconnected laser transceiver guidance units. The latter act as the relays between the laser beam transmitted from the aiming unit and the missile's forward-positioned electronics and control section.

Each of the transceiver units incorporates a laser receiver, a signal processor and a transmitter in a small cylindrical pod. The reasons for two electrically interconnected pods being fitted are system redundancy and the prevention of any possible screening effects acting upon the guidance signals. The transmitter is mounted in the nose of the pod and relays the command uplink data to the missile's forward-mounted electronics. The optical data signals are detected by small pop-up detectors connected to the electronics unit.

The Starburst Shoulder-Launched (SL) system provides the air defender with a firing unit which is suitable for rapid reaction troops, assault and paratrooper type missions. It consists of a Starburst missile in its sealed canister and a Starburst sighting system. The sighting system weighs 8 kg and clips onto the missile without any preparation. This unit comprises an electro-optically stabilised sighting device, a ×6 monocular and the laser guidance transmitter assembly. The unit can be easily fitted with a Simrad image intensifier for night operations.

Starburst Lightweight Multiple Launcher (LML) firing 0004756

The Lightweight Multiple Launcher (LML) was designed by Thales in the early 80s to give the UK a lightweight, three missile, rapid firing capability to deal with multiple target scenarios. The system entered service in the UK in 1985 as the first unit of its kind. The LML is made of two assemblies. The tripod base is made of composite materials and lightweight alloys and provides an adjustable support tripod which can be erected on the ground or, by repositioning its legs, installed in a trench position. The traverse and elevating structure gives 360° all round operation from –8 to +45° in elevation. The standard SL sighting system is fitted to this assembly and from a fully packaged state, the complete system is ready to engage targets within 2 minutes.

The packaged LML weighs approximately 34 kg. For night-time operation, the LML can be fitted with an 8 to 12 μm Starlite thermal imager. This is based on Thales Optronics LITE family of imager and has a field of view of 8 × 6°.

In addition the LML can be fitted with an IFF (Identification Friend-or-Foe), for example the Thales SB14 which can meet all current IFF (Mark XII, Mode 4) requirements for target engagements.

All Starburst systems have been developed by Shorts in parallel with their Starstreak programme for UK MoD and both systems are operated side by side on UK air defence operations. There is a high degree of equipment commonality between the two systems, including missile guidance equipment, firing unit assemblies as well as supporting training and test equipment. This will enable Starburst customers to upgrade to Starstreak when UK MoD releases the system for overseas sales.

Operational status

In production. Over 10,000 missiles have been built. Starburst is in service in UK, under the name Javelin S15, with the regular army, the Airmobile Brigade, the Territorial Army and the Royal Navy. The system is in service overseas with Canada, the Kuwaiti Air Force and all three services in Malaysia. It has also been selected by the Emiri Guard in Qatar.

Specifications

Type: 2-stage, low altitude
Length: 1.394 m
Diameter: 197 mm
Weight
  Aiming unit: 8.5 kg
  Missile in canister: 15.2 kg
Propulsion: 2-stage solid propellant
Guidance: beam-riding laser
Warhead type: 2.74 kg HE fragmentation with contact and proximity fuzing
Max speed: approximately M 1.5
Max effective range: well in excess of 4 km
Launcher: portable single-round with grip-stock, 3-round Lightweight Multiple Launcher

Contractor

Thales Air Defence.

UPDATED

Thales Air Defence Starstreak low-level air defence weapon system

Type

Air defence missile (portable).

Development

In December 1986, the UK Ministry of Defence (MoD) awarded Thales Air Defence Systems (formerly Shorts) a £225 million contract for the development, initial production and supply of the Starstreak high-velocity missile in three versions (a self-propelled version, a three-round lightweight launcher and a single-round portable launcher system) to replace the tracked Rapier SAM system then in service, especially in the forward battlefield area. The system is designed to provide the user with a rapid and lethal response to low-level and late unmasking targets such as armoured attack helicopters.

The Armoured Starstreak model will be the first to be fielded followed by the lightweight launcher and finally the single-round system. Major subcontractors to Thales on the Starstreak programme are Royal Ordnance for the rocket motor, Alvis for the Stormer full-tracked APC and Avimo for the gunner's sighting system.

By the time it was awarded the contract by the MoD, Thales had already carried out over 100 test firings of the missile since 1982 as part of a technology demonstrator programme.

The UK MoD placed a £28 million order for Starstreak missiles in June 1995 and followed this with a further £37 million contract for additional missiles in October 1995.

Thales has also entered into a collaboration programme with the Lockheed Martin Corporation to market Starstreak for US customers, and is also associated with Lockheed Martin and McDonnell Douglas Helicopter Systems in funded trials by the US Army of Starstreak as an air-to-air weapon on the Apache helicopter (qv).

In October 1995, the UK MoD accepted into service the vehicle-mounted Starstreak Self-Propelled High-Velocity Missile system (SP HVM). The other versions are expected to be accepted for service with the British Army in the future. A total of 135 SP HVMs were ordered with final deliveries in 1997.

On 20 May 2003, Thales Air Defence announced that Starstreak system had been ordered by South Africa for the first phase of the Ground Based Air Defence (GBAD) programme - Local Warning Segment (LWS). Starstreak, using the LML, will provide a fully integrated man-portable and air-droppable air defence capability for South African troops deployed on peace-keeping operations. The missiles will be supported by Page radar systems, also supplied by Thales. The contract was awarded by Kentron (part of South Africa's Denel Group, the GBAD prime contractor).

Description

All three versions of the Starstreak being procured by the UK MoD will use the same basic missile which consists of a two-stage solid propellant rocket motor assembly with a payload separation system mounted on the front end of the second-stage motor. This supports three winged darts which each have guidance and control circuitry, a high-density penetrating explosive warhead and delay action fuzing.

The aiming unit contains an optical head with an optics stabilisation system, aiming mark injector unit and aimer's monocular sight. In combat, the aimer acquires the target in the monocular sight. A space-stabilised aiming mark is injected into the centre of the field of view of the aimer who then tracks the chosen target by moving the launcher assembly so as to maintain the target in coincidence with the aiming mark. This permits lead angles in both azimuth and elevation to be generated and ensures that the missile is brought on to the target at the end of its boost phase.

After this pre-launch tracking phase is completed, the aimer presses the firing trigger causing the first-stage rocket motor to ignite. The Starstreak booster accelerates the missile to a high exit velocity while its canted exhaust nozzles impart sufficient roll on the weapon to create a centrifugal force that unfolds a set of flight stabilising fins. The first-stage motor then separates from the main missile body.

At a safe distance from the gunner, the main second-stage rocket motor cuts in to accelerate the missile to an end-of-boost velocity which is in the region of M3 to M4. As the motor burns out, the attenuation in thrust triggers the automatic payload separation of the three darts which, upon clearing the missile body, are independently guided in a fixed formation by their individual onboard guidance systems using the launcher's laser guidance beam.

The darts ride the laser beam projected by the aiming unit which incorporates two laser diodes, one of which is scanned horizontally and the other vertically to produce the required 2-D information field. Each dart then uses its onboard guidance package to control a set of steerable fins so as to hold its flight formation within this information field. Separation of the darts also initiates arming of the warheads.

Man-portable version of Thales Starstreak HVM showing sighting system

All the operator has to do after the launch is to continue to track the target and maintain the sight aiming mark on it. Maximum effective range is around 7 km which is the maximum distance at which the darts can retain sufficient manoeuvrability and energy to catch and penetrate a modern 9 *g* manoeuvring target.

As Starstreak does not rely on a heat source for guidance, it can engage targets from all angles including head-on. In addition, its laser guidance system significantly reduces its vulnerability to countermeasures. A Single Shot Kill Probability (SSKP) of 0.96 has been mentioned in connection with the system.

The basic Starstreak HVM is a clear weather system only, but target information can come from a number of other sensors such as the Thales Optronics Air Defence Alerting Device (ADAD) which was ordered by the British Army in 1987. This is operational with the Starburst missile system and is also mounted on the Stormer Starstreak vehicle.

The Starstreak firing platforms can be equally alerted by the Thales Air Defence Systems CLARA early warning command and control system which can co-ordinate up to 12 weapons systems. Thales have also developed a range of additional equipment for the Starstreak platforms, including thermal imagers, IFF and a full range of front line and base training and test equipment.

Operational status

In production. The missile has been accepted into service with the British Army in vehicle-mounted configuration, Self-Propelled Starstreak. Ordered by South Africa.

Specifications

Type: 2-stage high-velocity low-altitude missile
Length
Missile: 1.397 m
Missile in canister: 1.44 m
Diameter
Missile: 127 mm
Missile in canister: 274 mm
Propulsion: 2-stage booster-sustainer solid propellant rocket motor
Guidance: beam-riding laser
Warhead: triple kinetic/HE submunitions
Max speed: between M3 and M4
Max effective range: 7 km
Min effective range: 300 m
Launcher: portable shoulder-launched single, trainable stand-mounted triple (Lightweight Multiple Launcher LML) or Stormer APC-mounted octuple (Self-Propelled SP) with all using disposable containers

Contractor

Thales Air Defence.

UPDATED

Toshiba Type 91 Kin-SAM low-altitude surface-to-air missile system

Type

Man-portable surface-to-air missile.

Description

The missile is believed to be of standard configuration using separate booster and sustainer rocket motors in the conventional way. The homing system, however, is of a dual-mode imaging type and uses both infra-red and visual wavebands. All the operator has to do is to lock the head on to the target whereupon a high-resolution Charge Coupled Device (CCD) 'memorises' its appearance and causes the weapon to follow an all-aspects attack flight profile which is extremely resistant to any defensive countermeasures that may be employed.

Maximum engagement range will be between 3 and 5 km.

Operational status

Low-rate production. In service since 1991 with the Japanese Ground Self-Defence Force. Not offered for export.

Contractor

Toshiba (Tokyo Shibaura Electric) Company Limited.

Yugoimport Strela-2M/A low-altitude surface-to-air missile system

Type

Man-portable surface-to-air missile.

Development

Summary: missile has infra-red seeker.

Yugoslav-built Strela-2M/A surface-to-air missile deployed in firing position

The Strela-2M/A (Yugoslav military designation S-2M/A) is a locally built derivative of the former Soviet Strela-2M portable surface-to-air missile system.

Description

The S-2M/A varies from the former Soviet model in that the electronic systems in the single-channel passive infra-red seeker have been miniaturised, allowing the warhead section to be enlarged and increased in weight by 20 per cent. All other tactical and technical details are identical to the Strela-2M.

Operational status

Production complete. In service with Bosnia, Croatia, Czech Republic, Serbia/Montenegro and Slovenia.

Specifications

Type: 2-stage, low altitude
Length
Missile (fins folded): 1.44 m
Launch tube: 1.50 m
Diameter
Missile: 72 mm
Weight
Total launcher assembly at firing position: 15 kg
Total launcher assembly in travelling position: 16 kg
Launch tube: 3 kg
Grip-stock plus battery: 1.95 kg
Battery: 0.66 kg
Missile at launch: 9.85 kg
Propulsion: solid fuel booster and solid fuel sustainer rocket motor
Guidance: single-channel passive infra-red homing
Warhead: 1.32 kg HE smooth fragmentation with contact and grazing fuzing
Average missile speed: 500 m/s
Max firing range
Receding target: 4.20 km
Approaching target: 2.80 km
Max target engagement speed
Receding target: 260 m/s
Approaching target: 150 m/s
Max effective target altitude: 2.30 km
Min effective target altitude: 50 m (can be down to 15 m but missile may be seduced by horizon and ground radiative heat effects)
Launcher: portable single-round disposable with grip-stock

Contractor

Yugoimport SDPR.

AIR DEFENCE MISSILES

PORTABLE SIGHTS

FLIR Systems BORC Compact Thermal Imager

Type
Thermal imager.

Description
BORC is a new clip-on thermal imager for Saab Bofors Dynamics RBS 70 Air-defence missile system. Replacing the previous COND sight, BORC is a third-generation thermal imager based on Quantum Well Infrared Photodetector (QWIP) technology. The detector is a 320 × 240 element focal plane array and is sensitive in the long-wave 8 to 9 μm band. It is cooled to its operating temperature by an integrated Stirling rotary microcooler. The image generated by the built-in display is injected into the RBS 70 day sight through its gyrostabilised beam steering mirror. The operator guides the missile to the target using the thermal image in the same way as the system is normally operated with the day/visual image.

Operational status
In development.

Specifications
Weight: 12 kg

Detector
Type: QWIP
Spectral band: 7.5-9.3 μm
Elements: 320 × 240
Cooling: Integrated Stirling rotary microcooler
Field of View: 12 × 9°
Frame rate: 50 Hz
Focus: Fixed focus to infinity with temperature compensation.
Power supply: Lithium batteries

Environmental specifications
Operation: −30 to +60°C
Storage: −40 to +65°C

Contractor
FLIR Systems.

NEW ENTRY

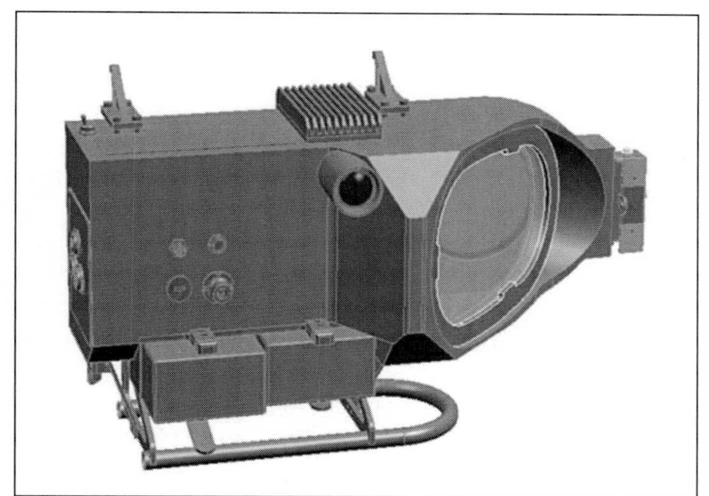

FLIR Systems BORC Compact Portable Sight ***NEW***/0547498

Pod precision targeting upgrade

Type
Air defence missile sight – (portable) upgrade.

Description
The pod precision targeting upgrade consists of an electronics control unit, a ×14 zoom lens and a Helmet Mounted Display (HMD). The system is intended to provide added capability to weapon systems such as the Bradley Fighting Vehicle and the Avenger Vehicle. A Smart Reticle™ displayed on the HMD shows symbology to aid aiming. The processing gives lead angle and superevaluation capability. The system can be adapted to a variety of missile systems.

Manufacturer
Pod - a divison of Recon/Optical Inc.

Raytheon AN/PAS – 18 Stinger Night Sight

Type
Air defence missile sight (portable).

Description
The AN/PAS-18, previously known as the Wide-Angle Stinger Pointer or WASP, is a thermal imaging sight that mounts on the Stinger Missile round to provide a 24 hour mission capability. The SNS can be fitted to the shoulder-launched Stinger MANPADS (man-portable air defence system) and to the Dual-Mount Stinger developed by Raytheon and Per Udsen for the Danish Low Level Air Defence System.

It operates passively in the same portion of the infra-red spectrum as the Stinger seeker and allows the gunner to perform target acquisition and weapon firing during total darkness and under reduced visibility conditions. A true infra-red day/night sight, the Stinger Night Sight does not rely upon ambient lighting and can easily detect and track targets in a cluttered daytime environment or on the darkest of nights. Power is supplied from a standard Mil-Spec battery that is the only consumable required for system operation. This system also features an RS-170 TV format output for remote viewing and/or videotaping for training and post-engagement analysis. Aligned at the factory, the system readily transitions from round to round without loss of boresight. The quick-release mounting bracket allows the Stinger Night Sight to be mated to the next round in under 10 seconds.

Operational status
Production of the AN/PAS-18 (Stinger Night Sight) started in mid-1991 and over 1,000 systems have been delivered. The sight is in service with the US Marine Corps, the Danish Army and Navy and the Royal Netherlands Army.

In May 1997, Raytheon was awarded a contract for Stinger Night Sights for the German Very Short Range Air Defence (VSHORAD) system. The contract is for an initial 60 systems for the German Air Force with options for additional systems for the army and navy. The SNS will be used on both the German Stinger MANPADS and SA-16 missile VSHORAD systems. Adaptation of the SA-16 to accommodate the SNS will be done by Euroatlas GmbH.

Specifications
Field of view (h × v): 20 ×12 °
Spectral band: 3.0-5.0 μm
Mission time: 12 h at 50% duty cycle
Range: well in excess of Stinger flight envelope
Size: 13.5 × 5.0 × 6.2 in
Weight: 5.50 lbs (max)
Battery type: lithium, BA 5847/U (Deployed), Ni/Cd (Training)
Reticle: electronic, programmable, matches open sight assembly

Contractor
Raytheon Systems Company.

AN/PAS-18 Stinger Night Sight

RBS 70 Clip-On Night Device (COND)

Type
Man-portable surface-to-air missile.

Description
The Clip-on Night Device (COND) provides the RBS 70 with a night capability. COND has no electrical interface with the sight and is fitted without the need for special alignment. It operates in the 8 to 12 μm waveband. COND uses LITE thermal imaging modules from Thales Optronics.

Operational status
In service. The Norwegian Army has retrofitted all its RBS 70 portable air defence systems with COND.

Specifications
Spectral band: 8-12 μm
Detector: 23 element
FOV: 12 × 8°
Magnification: ×1
Magnification with day sight: ×7
Detection range: the same as day system
Quick-Lock mounting: no need for harmonisation

Contractor
FLIR Systems (formerly SaabTech).

UPDATED

The COND for the RBS 70 air defence missile 0004752

Ring Sight LC-40-100-9K38 for MANPADS

Type
Air defence missile sight (portable).

Description
Ring Sights produces a variant of its LC-40-100 sight suitable for use with the IGLA 9K38 MANPADS anti-aircraft missile system. The sight clamps on the IGLA barrel at the position of the daytime only open sight without mutual interference. Zeroing adjustments allow the LC-40- 100 to be aligned with the open sights. The system has a ring and bars graticule lit by LEDs which is adjustable in brightness to allow use with NVGs.

Ring LC-40-100 sight for MANPADS 0010869

Operational status
Sold as a night sight to an undisclosed South American country.

Specifications
Length: 130 mm
Width: 170 mm
Height: 130 mm
Weight: 500 gm
Power: 2 × AA batteries

Contractor
Ring Sights Holding Company Limited.

Thales Optronics LITE night sights for low-level air defence weapons

Type
Air defence missile sight (portable).

Description
Thales Optronics produces a range of LITE night sights for low-level air defence systems that are part of the LITE family of thermal imagers which is based on the LITE hand-held thermal imager. LITE imagers are themselves based on UK TICM I thermal imaging modules and have been in service since 1993.

MITS 2 is the missile night sight made by Thales Optronics for the MBDA Mistral portable air defence system. It is available in either Joule- Thomson or engine-cooled version. MITS 2 allows the operator of the Mistral to engage targets by using signals injected into the viewer of the thermal sight.

The StarLITE night sight fixes to the Starburst portable air defence missile system from Thales Air Defence. It has been fitted with an optical relay so that it can be fitted to Starburst's existing day/night tracker unit. It has been adapted so that target images can be displayed with missile aiming marks in the operator's normal eyepiece.

Thales Optronics have also provided LITE modules for FLIR Systems' Clip-On Night Device (COND) for the RBS 70 missile (and also for the RBS56 night sight). The Japanese Air Self-Defense Force has also ordered a LITE night sight for its FIM-92 Stinger MANPADS. LITE can also be fitted on other portable air defence weapon systems.

Operational status
In production and in service.

MITS 2 night sight on MBDA Mistral 0004735

Left: clip-on night sight for Bofors RBS 70, right: StarLITE night sight on Thales Starburst missile system 0004845

Specifications

Spectral band: 8-12 μm
Detector: CMT
Field of view: selected to match missile system
Cooling: Joule-Thomson high pressure air using 0.33 litre bottle or integral Stirling cycle microcooler
Display: direct view or indirect view with CCIR or RS-170 video output
Dimensions (typical): 350 × 135 × 185 mm
Weight (typical): 3.5 kg
Power requirement: 8 W (Stirling), 4 W (Joule-Thomson)

Contractor

Thales Optronics Ltd.

UPDATED

Thales Optronics StarLITE night sight

Type

Thermal imager (land).

Description

StarLITE is primarily intended to be used with the Thales range of air defence missiles, such as Starburst. The imager takes advantage of recent advances in modern material technology, miniaturised electronics and the development of compact, reliable and low power microcoolers. The system operates in the 8 to 12 μm waveband and is said to provide detection of head-on fighter aircraft at 9 km and helicopters at 7 km. The clip-on Starlight operates independently of the Starburst laser guidance unit. Automatic gain and level control are provided.

Operational status

In service with Kuwait.

Specifications

Dimensions: 250 × 200 × 150 mm
Weight: 6 kg
Field of view: 8 × 6°
Frequency: 8 to 12 μm
Time to operation: 3 min switch on, 30 s standby

Contractor

Thales Optronics.

UPDATED

AIR DEFENCE GUNS

VEHICLES

BAE Systems Marksman twin 35 mm anti-aircraft turret

Type
Air defence gun (vehicle turret).

Development
The Marksman is an armoured turret with twin 35 mm Oerlikon KDA cannon which can be fitted to any main battle tank hull or large armoured personnel carrier hull without modification. The system uses the Marconi 400 series combined surveillance and tracking radar and two SAGEM Type VS 580-VISAA roof-mounted optical sights together with a laser range-finder. Vickers Defence Systems manufactures the armoured turret, Oerlikon Contraves the 35 mm cannon and SFIM supplies the gyrostabilised sights and inertial reference system. In 1991, the first Marksman turrets were delivered to Finland and installed on T-55 tank chassis for protection of the armoured brigade.

Operational status
In service with Finland on T-55 MBT chassis. Tested on eight MBT chassis types.

Contractor
BAE Systems, Operations, Land and Naval Systems (Armaments), Leicester.

Bofors Combat Vehicle 90 air defence system

Type
Air defence gun (vehicle).

Description
The Combat Vehicle 90 (CV 90) family of full-tracked armoured vehicles has been developed for the Swedish Army. The driver has three integral periscopes, the centre periscope can be replaced by a passive periscope for night driving. The gunner has a modular sight with common outlook for day sight, night sight and laser range-finder. The CelsiusTech UTAAS fire control system which includes thermal imager and laser range-finder is fitted to the Swedish Army 40 mm version.

Hägglunds Vehicle is responsible for the vehicle chassis and Bofors is responsible for the complete turret with weapons, sensors, fire-control computer. Sensors include the Thomson-CSF Gerfaut TRS 2620 search radar.

Operational status
Development complete. Entering production for the Swedish Army and first production vehicles were to be delivered early in 1997.

Contractor
Bofors Defence AB.

Bofors Defence TRIDON 40 mm L/70 self-propelled air defence gun

Type
Air defence gun (vehicle).

Description
TRIDON consists of the Bofors 40 mm L/70 anti-aircraft gun mounted on the VME 825B (6 × 6) all-terrain vehicle. There are three versions under development. TRIDON 1 would be the basic clear-weather system while TRIDON 2 would have an optical sight, laser range-finder, infra-red camera and muzzle velocity radar. TRIDON 3 has a pedestal sight that includes an optical sight, laser range-finder, TV camera and infra-red camera and also a muzzle velocity radar. TRIDON 4 can have an optical sight on top of the control cabin or on the gun with the laser range-finder, TV camera and infra-red camera either on the control cabin roof or on the gun mount. An all-weather tracking radar can be mounted on the gun mount as well as the muzzle velocity radar. It is being developed under contract to the Swedish Defence Matériel Administration (FMV). It is being developed for the Swedish Army and also for the navy for use in a coastal defence role. A prototype, using the Bofors 40 mm L/70 anti-aircraft gun, has been built and tested.

Operational status
Under development.

Contractor
Bofors Defence (part of United Defense LP).

Daewoo Flying Tiger (Biho) twin 30 mm self-propelled anti-aircraft gun system

Type
Air defence gun system (vehicle).

Description
The Flying Tiger twin 30 mm self-propelled anti-aircraft gun system has been developed for the South Korean Army. Raytheon won a US$5 million contract to provide their Electro-Optical Tracking System (EOTS). EOTS consists of a TV system with a maximum range of 7 km and an infra-red thermal camera with a similar range and eye-safe laser range-finder.

Prime contractor is Daewoo Heavy Industries, with Raytheon as subcontractor for the electro-optics package.

Operational status
In production.

Contractor
Daewoo Heavy Industries Ltd.

Former Czechoslovak and Soviet state factories ZSU-23-4 Quad 23 mm self-propelled anti-aircraft gun

Type
Air defence gun (vehicle).

Development
The ZSU-23-4 comprises four AZP-23M 23 mm cannon mounted on a light armoured vehicle. The driver can have an infra-red TVN-2 periscope for night driving. The commander can have a TKN-1T infra-red periscope for night use. This has a range of 200 to 250 m. The fire-control system has J-band radar, sighting device, computer, line of sight and line of elevation stabilisation units.

It became operational in 1966 and production continued until 1983. In addition to being produced in the former USSR, it was also made under licence in the former Czechoslovakia. Total production was between 6,000 and 7,000 units.

Operational status
Production complete. In service.

Contractors
Former Soviet factories.
Czech and Slovak state factories.

Japanese Type 87 twin 35 mm self-propelled anti-aircraft gun system

Type
Air defence gun (vehicle).

Development
The Type 87 consists of the modified chassis of the Type 74 main battle tank fitted with a new turret armed with twin 35 mm Oerlikon Contraves KDA cannon. Mounted externally on the left side of the tracking radar is a flat box which is believed to contain a laser range-finder, optical tracker and possibly an LLLTV. Standard equipment includes a turret- mounted laser warning device. Full-scale engineering of the total gun system began in 1982.

Operational status
In service with the Japanese Ground Self-Defence Force.

Contractor
Japan.

NORINCO twin 37 mm self-propelled anti-aircraft gun system

Type
Air defence gun (vehicle).

Description
The Norinco twin 37 mm self-propelled anti-aircraft gun system was first disclosed in 1988. It consists essentially of a modified Type 69 MBT fitted with a new two-person turret armed with the Type 76 twin 37 mm gun system originally developed for naval applications. As well as surveillance radar, the fire-control system has an electro-optic tracker which includes a laser range-finder. The electro-optic tracker has a magnification of ×6 and a 10° field of view with traverse and elevation speeds being 60°/s. The laser range-finder has a maximum range of 8,000 m.

Operational status
Development complete. Ready for production.

Contractor
China North Industries Corporation (Norinco).

NORINCO Type 80 twin 57 mm self-propelled anti- aircraft gun system

Type
Air defence gun (vehicle). .

Description
The Type 80 twin 57 mm self-propelled anti-aircraft gun system consists of a modified Type 69-II MBT chassis fitted with a Chinese version of the turret installed on the former Soviet ZSU-57-2 twin 57 mm self-propelled anti-aircraft tank. The driver has an infra-red periscope for night driving.

Operational status
Production as required. In service with the People's Liberation Army (PLA).

Contractor
China North Industries Corporation (NORINCO).

Oto Melara 76 mm self-propelled OTOMATIC air defence tank

Type
Air defence gun (vehicle).

Description
The OTOMATIC 76 mm AA tank is a tank with modifications for the installation of the auxiliary power unit. The field tested fire-control equipment includes an optical sight system utilising a low-light level TV camera, a TV tracker and a Nd:YAG laser range-finder for use in both air and ground defence. Other sensors include a search radar, IFF system and a tracking radar. A panoramic sight for the commander is also installed.

In the latest design configuration, use is foreseen of the Attila panoramic sight incorporating a second generation thermal channel and a laser range-finder. The original sight with an LLLTV will be discontinued. Automatic target tracking will be performed via the thermal video.

Operational status
The system successfully completed technical and operational evaluation by the Italian Army in July 1992.

Specifications
Laser range-finder: Nd:YAG transmitter, 1.06 μm
Repetition rate: 10 Hz
Beam divergence: 1.2 to 3.5 mrad
Daylight TV camera: silicon target VIDICON, video format to CCIR

Contractor
Oto Melara S.p.A.

Oto Melara Quad 25 mm self-propelled anti-aircraft gun system (SIDAM 25)

Type
Air defence gun (vehicle).

Description
The Quad 25 mm, SIDAM 25, consists of four 25 mm KBA cannon mounted on an M113 series APC upgraded to the improved M113A2 configuration. The turret is fitted with a day clear-weather and low-light level TV camera sighting system capable of tracking targets automatically. The fire-control system includes an Officine Galileo MADIS sight, an Alenia laser range-finder, FIAR TV and displays. For all-weather and night operations a video- compatible thermal imager unit can be connected to the electro-optic sight together with a passive IR night and day sight.

Other major subcontractors include Oerlikon-Italiana for the four 25 mm KBA cannon. The Quad 25 mm SIDAM 25, is a joint development between the Italian Army and Oto Melara with first production systems completed in 1989. The Italian Army requirement was for a total of 280 SIDAM 25 systems. Production of the system has now been completed but marketing still continues for the export market.

Operational status
Production complete. In service with the Italian Army.

Contractor
Oto Melara S.p.A.

Yugoimport Model 30/2 twin 30 mm self-propelled anti-aircraft gun system

Type
Air defence gun (vehicle).

Description
The Model 30/2 twin 30 mm self-propelled anti-aircraft gun system was developed in the former Yugoslavia and first shown in 1991. It is the chassis of the locally built BVP M80A mechanised infantry combat vehicle fitted with a two-person power-operated turret armed with twin 30 mm M86 single-feed automatic cannon. The gunner has an Oerlikon Contraves GunKing sight which has a computer, periscope sight, laser range-finder and TV screen. The driver has three periscopes, the forward one of which can be replaced by a passive periscope for night driving.

Operational status
Prototype. Volume production has not started and, as a number of the key components were of foreign origin, it is probable that production will not start in the near future.

Contractor
Yugoimport SDPR.

AIR DEFENCE GUNS

VEHICLE SIGHTS

Galileo Avionica Madis sighting and drive system

Type
Air defence gun (vehicle) sight.

Development/Description
The Madis sighting and drive system was developed for the Otobreda SIDAM 25 self-propelled anti-aircraft gun system. Composed of three modular units, the system can also be installed in other anti-aircraft vehicles either in total or in part.

The modules of the Madis are:

(a) Daylight optical head which contains the visual telescope. This is self-stabilised by a gyro directly coupled to the line of sight to allow for sighting operations on the move. The unit can also operate without the gyro as a servoed optical sight.
Provision is made for compensation of image rotation due to the azimuth angle.
The head can also accommodate a laser range-finder and a TV camera with integration of the visual, laser and TV channels into a single output path to ensure the required parallelism.

(b) Night optical head with an integrated low light level TV camera. The line of sight can be servo-stabilised to enable sight operations to be conducted on the move.

(c) Power servo system for the turret and weapons. This consists of an axial pump and hydraulic motor with an asynchronous motor as the prime mover. The unit also contains the first stage on the reduction gearing and, on its upper part, the fire control computer and operations panel.

Operational status
Production as required. The Italian Army has taken delivery of 280 systems and production is now complete. In service with the Italian Army (SIDAM 25 self-propelled anti-aircraft gun).

Specifications
Daylight optical head
Elevation range: −10 to +85°
Azimuth range: 45°
Magnification
Telescope: ×5
Field of view
Telescope: 12°

Night optical head
Elevation range: −10 to +75°
Azimuth range: 45°
Field of view: 4.76°

Contractor
Galileo Avionica.

UPDATED

The Otobreda SIDAM 25 SPAAG is fitted with the Madis sighting and drive system

Kentron AA-EOT electro-optical tracker

Type
Air defence gun (vehicle) sight.

Description
The AA-EOT is a two-axis stabilised, electro-optical tracker for mobile point defence systems. Its first application is the Kentron ZA-35 twin 35 mm self-propelled anti-aircraft gun system which is still at the prototype stage. It provides a high-quality video image as well as laser range-finding for manual and automatic target tracking in day and night conditions.

The AA-EOT is designed for, and can be fitted with, second-level stabilisation (through the use of gysocopes on the gun-mount). Its modular design allows for upgrading with new sensor technology or for sensors selected by the customer.

The modes of operation are:

(a) 'independent' mode – whereby the operator performs the surveillance operation independently of the base movement

(b) 'dependent' mode – whereby the sightline is slaved to that of an external source, for example, acquisition radar, tracking radar or another optical tracker

(c) 'auto-track' mode – whereby the sightline is locked on to the target by means of an auto-tracker. This is also the 'engage' mode, when weapons are fired under control of the sight.

There are options for the electrical interfaces with a fire-control computer: RS-422, RS-485 and MIL-STD-1553B, while both the control panel with its sight control buttons and displays and the hand controller can be configured according to the customer's requirements.

Operational status
Ready for production. Offered for export.

Specifications
Weight (optical assembly)
Without sensors: 150 kg
With sensors, max: 220 kg
Dimensions
Optical assembly: 520 mm cylindrical diameter × 850 mm long × 675 mm high
Elevation/depression: 90° range of look angles (configured to customer requirements, for example −30 to +60°)
Azimuth: 360°
Sensor choice: CCD TV camera with zoom or fixed optics; LLLTV with fixed optics; thermal imager with switchable fields of view; laser range-finders (single pulse or rapid pulsing); missile goniometers; customer's own sensors
Auto-tracking: by means of video tracker; centroid tracker is standard; options include correlation and/or edge trackers

Kentron AA-EOT electro-optical tracker (back) with key subsystems in front

Stabilisation: better than 0.2 mrad peak-to-peak (one level of stabilisation), better than 0.075 mrad peak-to-peak (two levels of stabilisation)
Line of sight position sensing accuracy: 0.1 mrad RMS error
Gyro: high reliability, low drift dynamically tuned
Power supply: 18-32 V DC or 115 V 400 Hz AC

Contractor
Kentron Division of Denel (Pty) Ltd.

Raytheon EOTS Electro-Optical Tracking System

Type
Air defence gun (vehicle) sight.

Description
Raytheon's Electro-Optical Tracking System EOTS is a sensor suite combined with a tracking system and has been developed for air defence applications on combat vehicles. The EOTS panoramic sight consists of a thermal imager, a near infra-red TV sensor, and a high-repetition rate Raman shift eye-safe Laser Range-Finder (LRF). The sensor suite also features a dual-mode automatic video tracker (DMT), a precision digital two axis stabilised pointing mirror (70° elevation) and a digital scan converter (DSC) for TV-compatible output video. The system is also offered for naval applications.

Operational status
Ready for production. Raytheon has been awarded a US$5 million contract from the Republic of Korea for EOTS to be mounted on the Daewoo Heavy Industries anti-aircraft gun system. Initial production systems were to be delivered in 1998.

Specifications
Elevation range: –10 to +75°
Azimuth range: 360°
Line of sight stabilisation: =50 μrad
Video out: NTSC (RS-170)
Symbology, reticles and data display: programmable
Thermal sensor
Spectral band: 8-12 μm
Detector type: CMT - PV 240 × 4 elements
Magnification
Narrow: ×10
Wide: ×3.3
Fields of view
Narrow: 2.5 × 3.3°
Wide: 7.5 × 10°
NETD: <0.25°C
Laser range-finder
Laser type: Raman shifted Nd:YAG
Wavelength: 1.54 μm
Beam divergence: 500 mrad
Repetition rate: SS, 15 Hz
Range: 200-20,000 m

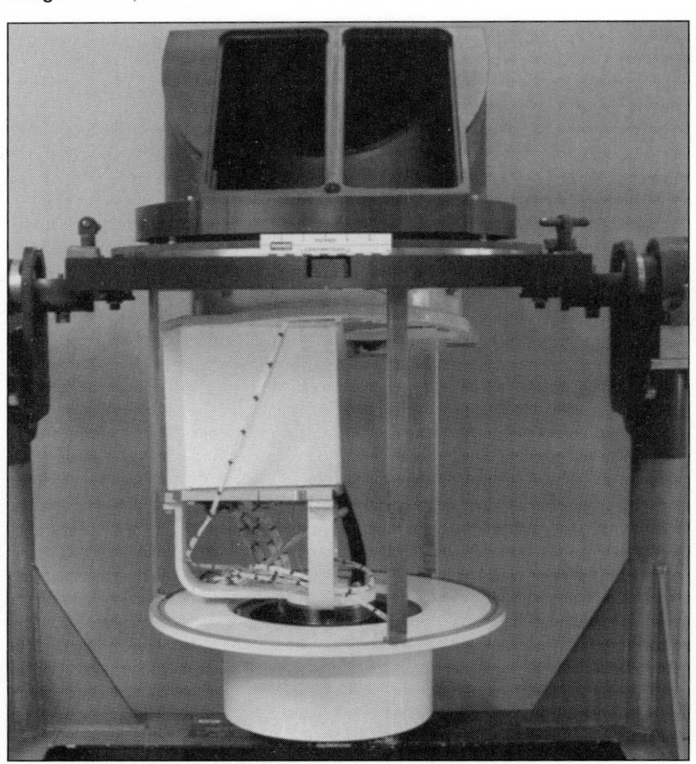
Sight sensor unit for the Raytheon Electro-Optical Tracking System (EOTS)

Power: 35 mJ
False alarm rate: <1/100
TV sensor
Spectral band: 0.4-1.1 μm
Fields of view
Narrow: 2.5 × 3.3°
Wide: 7.5 × 10°
IRIS: auto/manual
Type: CCD
Sensitivity: 0.5 lux
Tracker
Dual mode: simultaneous centroid/correlation
Switching: auto/manual
Acquisition: auto/manual
Coast: auto/forced

Contractor
Raytheon Systems Company.

Rudi CÄƒjavec computerised fire-control system for light anti-aircraft guns

Type
Towed anti-aircraft gun.

Development
Rudi CÄƒjavec Defence Electronics has developed, to the prototype stage, a computerised fire-control system that can be fitted to light anti-aircraft guns of 20, 23, 30, 35 and 40 mm calibre. For trials purposes it has already been installed on a former Czechoslovak twin 30 mm M53/59 self-propelled anti-aircraft gun system.

Description
Main components of the system include a central computer which receives and issues information from a number of subsystems including the gunner's optical sight which also includes a laser range-finder, computer control panel, gunner's controls, target data receiver and gun drive and position sensors. There is a second channel which can contain a thermal imager or image intensifying sight. If required the target data receiver can also receive information from an optical target indicator and a search acquisition radar.

Operational status
Development complete. Ready for production.

Specifications
Optical sight
Wide field of view: 20Â°
Narrow field of view: 7Â°
Night channel: passive 2nd-generation or thermal channel
Laser protection filter
Independent line of sight
Two-axis deviator
Laser range-finder: Nd:YAG, max range
Computer: 16-bit, 2 ballistics
Computer inputs: target distance, meteorological data and parallax data
Computer outputs: lead angle, angular tracking and ballistic corrections

Contractor
Rudi CÄƒjavec Defence Electronics.

SaabTech UTAAS fire-control system for anti-aircraft guns

Type
Air defence gun (vehicle) sight.

Development
The UTAAS sight has a modular structure and is intended for use in combat vehicles and anti-aircraft gun systems. The first self-propelled application for this system is the 40 mm air defence version of the Combat Vehicle 90 family of vehicles developed to meet the requirements of the Swedish Army.

The sight can also be retrofitted to existing towed anti-aircraft guns such as the Bofors 40 mm L/60 and L/70 and the Chinese twin 37 mm systems. The UTAAS system was one of the three systems installed on Chinese-supplied twin 37 mm light anti-aircraft guns and trialled with the Pakistan Army in mid-1989.

Description
UTAAS is based on the independent line of sight principle. The gunner can retain the target in the centre of the reticle during the entire aiming operation and does not need to re-aim between laser range measurements and firing. The gun movements are controlled by the fire-control computer while the gunner controls the line of sight from the control handle. Higher degrees of stabilisation can be achieved if the gun is equipped with gyroscopes.

The sight consists of the following parts:
(a) Top module
The top module is the only part of the sight outside the protection of the turret. The top module contains an electronically controlled head mirror in a servo-gimbal with two degrees of freedom. The multispectral front window is equipped with a defroster and a washer/wiper.
(b) Adaptor module
The adaptor module contains a direct outlook observation channel and a beam splitter that deflects the thermal radiation to the imager, while transmitting the laser and visual wavelengths. The observation channel is equipped with laser filters for eye protection.
(c) Optics module
A folded daylight telescope and a laser beam expander are combined in the optics module. The day channel and the laser use the same main objective. The laser and visual radiation are separated with a beam splitter to provide an interface for the laser range-finder on the opposite side of the ocular. A ballistic reticle and a target range display are superimposed on the image seen through the ocular. The ocular is prepared for mounting of a video camera with a beam splitter. The image seen by the gunner is made available on a CCIR standard video signal for training purposes.
(d) Laser range-finder
The laser range-finder is a hermetically sealed unit attached to the optics module, available with different performance specifications for tank or anti-aircraft purposes.

SaabTech Systems UTAAS sight

All UTAAS versions can be fitted with a wide selection of thermal imagers. The thermal radiation from the target is transmitted through the UTAAS sight top module and adaptor module to the thermal imager. Depending on the thermal imager system selected, the image can be presented to the gunner in the sight or in one or several independent monitors. The system can be interfaced with a search radar and an optical target designator.

Operational status

Series deliveries in progress for the Swedish Army CV 9040 family of vehicles, contract signed for the Norwegian CV 9030 vehicle in 1993. So far no firm orders have been placed for the UTAAS sight for towed anti-aircraft guns.

Specifications

Day channel
Magnification: × 8
Field of view: 8°
Transmission: 40%
Eye protection filter (1.064 μm): 45 dB
Exit pupil: ≥8 mm
Eye relief distance: ≥25 mm
Observation outlook
Fields of view
Azimuth: 35°
Elevation: 15°
Eye protection filter (1.064 μm): 45 dB
Line of sight deflection range
Elevation: –20 to +50°
Azimuth: ±18°
Laser range-finder (ground targets)
Wavelength: 1.064 μm or 1.54 μm
Extinction ratio: >32 dB
Pulse repetition frequency: 1 Hz for 3 s, 12 pulses/min
Laser range-finder (air targets)
Wavelength: 1.064 μm or 1.54 μm
Extinction ratio: >42 dB
Pulse repetition frequency: 4 Hz for 10 s, 1.5 Hz continuously
Thermal imager (typical)
Fields of view
Narrow: 4°
Wide: 15°
Temperature range: –40 to +70°C
Typical detection range (target 2.4 m wide, 3.6 m high for example T-72 tank)
Detection (50%): 4 km
Detection (90%): 2.49 km
Identification (50%): 2.11 km

Contractor

SaabTech Systems AB.

AIR DEFENCE GUNS

STATIC AND TOWED

Allied Ordnance of Singapore 40 mm L70 Field Air Defence Mount (FADM)

Type

Towed anti-aircraft gun.

Description

The 40 mm L70 Field Air Defence Mount (FADM) is an improved version of the standard Bofors 40 mm L70 towed anti-aircraft gun. The on-mount electro-optical fire-control system contains a FLIR and laser range-finder. The first fully operational L70 FADM prototype was completed in 1995 and was been successfully tested.

Operational status

In production.

Specifications

Electro-optic fire-control system
Laser range-finder
Wavelength: 1.54 μm (Nd:YAG)
Pulse repetition rate: continuous, 3 Hz
Accuracy: 7 m
Range: 6 km
Thermal imager
Spectral band: 8-12 μm (CMT detectors), upgradeable to 3 - 5 μm
Cooler: integral Stirling cycle
Cool down time: 4 min
Narrow field of view: 1.5 × 2°
Wide field of view: 4.5 × 6° (pupil 76 mm)
Detection: 10 km on aircraft
Tracking: 7 km on aircraft

Contractor

Allied Ordnance of Singapore (Pte) Ltd.

LIW eGLaS 35 35 mm anti-aircraft gun

Type

Towed anti-aircraft gun.

Description

The eGLaS 35 is a towed 35 mm anti-aircraft gun system being developed by the LIW Division of Denel. It has one GA 35 cannon. The gunner's sight is a roof-mounted stabilised periscope fitted with a laser range-finder. There is an optional stabilised optical sight with thermal imager and an auto-tracker.

Operational status

Development complete. Not yet in production or service.

Contractor

LIW, a Division of Denel PTY (Ltd).

NORINCO P793 37 mm anti-aircraft gun

Type

Towed anti-aircraft gun.

Development

Summary: electro-optical sighting system.

Description

The Type P793 is a twin-barrelled anti-aircraft gun. It has an electro-optical sight system known as the JM 831 which has a ×5 magnification and a 12° field of view. It is stated to be a self-powered electro-optical aiming sight which takes 3 to 4 seconds to be ready for firing once a target has been detected and is effective with target velocities of 60 to 350 m/s.

Operational status

In production and service with the Chinese Armed Forces and with other countries.

Contractor

China North Industries Corporation (NORINCO).

Oerlikon Contraves Skyshield 35 Ahead Air Defence System (ADS)

Type

Air defence gun system.

Description

Oerlikon Contraves' Skyshield 35 Ahead Air Defence System (ADS) has been developed by the company as a variant for its twin 35 mm GDF series gun systems. A typical Skyshield 35 ADS would consist of two 35 mm Revolver Gun Mounts (RGM) each armed with the Oerlikon Contraves 35/1000 rapid fire revolver cannon and a new modular Fire-Control Unit (FCU). The FCU consists of two parts, the unmanned Sensor Unit (SU) and the detached Command Post (CP). The sensor unit has X-band pulse-Doppler search and tracking radars plus an electro-optics module with thermal imager, TV and laser range-finder.

The Skyshield 35 Revolver Gun Mount will fire all standard Oerlikon Contraves 35 mm ammunition plus the new Ahead (Advanced Hit Efficiency And Destruction) ammunition, development of which has been successfully completed and demonstrated.

Development of the Skyshield 35 Ahead ADS is complete. System trials have taken place.

Operational status

Development complete, ready for production.

Contractor

Oerlikon Contraves AG, Zurich.

UPDATED

The Fire-Control Unit of the Skyshield 35 Air Defence System 0004753

Saab Bofors 40 mm L/70 and L/60 modernisation package

Type

Air defence gun upgrade.

Description

The Bofors upgrade packages for its 40 mm L60 and L70 light anti-aircraft guns include an integrated fire-control system which converts the gun to an autonomous firing unit. This fire-control module consists of the SaabTech UTAAS sight with fire-control computer. The basic UTAAS sight (qv) includes an optics module and laser range-finder but it can also be fitted with a variety of thermal imagers. Further upgrading options are available including a muzzle velocity radar and optical target designator.

Operational status

Development complete; ready for production.

Contractor

Saab Bofors Dynamics.

AIR DEFENCE GUNS

STATIC AND TOWED SIGHTS

Oerlikon Contraves Gun King sight

Type
Towed anti-aircraft gun.

Development
The Gun King computerised multidivergence laser sighting system was developed as a private venture by Oerlikon Contraves for installation on a wide range of small to medium calibre air defence guns as well as self-propelled air defence systems.

The Gun King can be installed on new weapons such as the Oerlikon Contraves twin 35 mm GDF series weapons, or retrofitted to older weapons such as the Soviet ZU-23 and Chinese twin 37 mm systems. The Gun King was one of three sighting systems evaluated by Pakistan on its Chinese-supplied twin 37 mm anti-aircraft guns in 1989. Over 450 Gun King sights have been ordered.

Description
The Gun King sight has five key components: a periscope with a laser range-finder as a tracking unit; a collimator for autonomous target acquisition; an operator's control unit; a sight electronics unit with a computer and the servo system and drives.

Once the target has been found by means of an external search radar or the collimator, the operator tracks it through the periscope. The integrated laser range-finder measures the target distance continuously to provide three-dimensional information to the computer system. There is a common optical path in the periscope for the laser beam and the operator's line of sight. Spectral beam-splitting is effected by a system of lenses and prisms, specially coated to ensure operator safety. For engaging ground targets, the laser beam is narrowed to eliminate terrain clutter and for training purposes a TV camera may be mounted on the periscope. All information relating to meteorological parameters and muzzle velocity is entered into the computer via a keyboard and an alphanumeric display to compensate for its influence on the intercept calculation point. This enables the operator to concentrate on target tracking once the weapon is in position.

The gun and sight are operated by a control yoke which provides full hands-on control since all actuators are integrated in the yoke grips. The Gun King allows computer-assisted tracking of the target and the operator only has to control the gun if the target manoeuvres.

The high-speed computer not only calculates all the fire-control data and ballistics, but also the gun drive electronics. At the moment of optimum hit probability, the operator receives an acoustic alarm to commence firing.

Close-up of Oerlikon Contraves Gun King sight from rear installed on Oerlikon Contraves GDF-005 twin 35 mm anti-aircraft gun

For control, modern DC power electronics and DC servo drives are used. The DC power supply is fitted with buffer batteries to smooth out transient peak loads. The battery capacity ensures full operational readiness of the complete system and up to five combat cycles can be accomplished before the power supply engine has to be started to recharge the batteries.

Monitoring of the system function is effected online, even during the combat phase, and a quick test enables functional checking of the system during start up. When a fault does occur, the faulty Line-Replacement Unit (LRU) is localised with the memory resident functional and diagnostic unit without external aids. The faulty LRU is then substituted in the field and returned to the rear for repair.

Operational status
In production. In service with a number of countries including Canada, Cyprus and Saudi Arabia.

Contractor
Oerlikon Contraves AG, Zurich.

VERIFIED

Oerlikon Contraves Gunstar Fire-Control Unit – FCU

Type
Gun fire-control system.

Description
The Gunstar electro-optical fire-control unit is designed for low level air defence applications with small to medium calibre anti-aircraft guns. A Gunstar unit may be operated independently or integrated with a higher order command and control system. The system includes a FLIR sensor, digital optical sight and a laser range-finder. Operational capability includes:

(a) Target acquisition by the FCU, including target designation, target assignment and gate-lock-on
(b) Fire control by the FCU, including angular target tracking, three-dimensional target measuring and filtering and lead angle determination
(c) Target acquisition by the gun operator
(d) Gun aiming, including firing.

The major components of the complete system are the tracker unit (illustrated), data processing unit, tracker control unit, video processing unit, operator's control unit, digital optical sight, tracker mount, laser range-finder and FLIR sensor. The system is mounted on a two wheeled trailer. The operator is provided with a touch panel display for interaction with the system, a video monitor for display of the target, a joystick and intercom station. Video and digital signals between the data processing elements and the operator control unit are transmitted via an optical link.

Operational status
Development complete, in production.

Specifications
Tracker mount
Azimuth range: continuous rotation capability
Elevation: –20 to +200°
Readout accuracy (bearing and elevation): 0.05 mrad

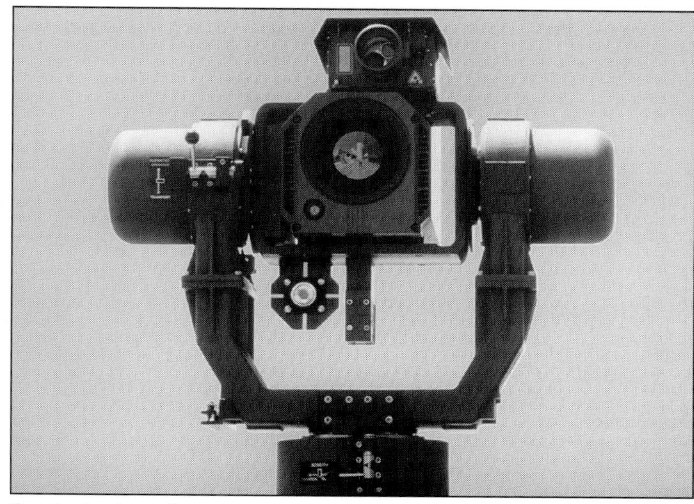

Gunstar tracker unit 0010862

Max velocity both axes: >1.5 rad/s
Max acceleration both axes: >1.5 rad/s^2
Laser range-finder
Type: Eyesafe class 3a
Max range: 20 km
Range resolution: 5 m
FLIR sensor
FOV (wide/narrow): 20/5°
Spectral band: 8-12 μm

Contractor
Oerlikon Contraves AG, Zurich.

VERIFIED

PCO DL-1 high repetition laser range-finder

Type
Air defence gun – laser range-finder.

Description
The DL-1 laser range-finder was developed to meet the requirements of the Polish Army for an anti-aircraft fire-control range-finder to use against fast moving airborne targets. The DL-1 has a high repetition rate and can also be used as a part of anti-missile systems.

When linked to a fire-control computer it is possible to predict the trajectory of air targets and use the information for automatic target tracking and firing. Where there are multiple targets in the measurement gate, the range-finder transmits ranges of the nearest and furthest target to the computer and the number of targets.

Operational status
Available.

Specifications
Range of measurements: 300 to 19,900 m
Accuracy of distance measurement: ±5 m
Wavelength: 1,06 μm
Maximum ranging rate: 20 pps
Data output/input: RS-422A, 9,600 Baud
Supply voltage: 20 to 30 V DC

Contractor
PCO SA, Poland.

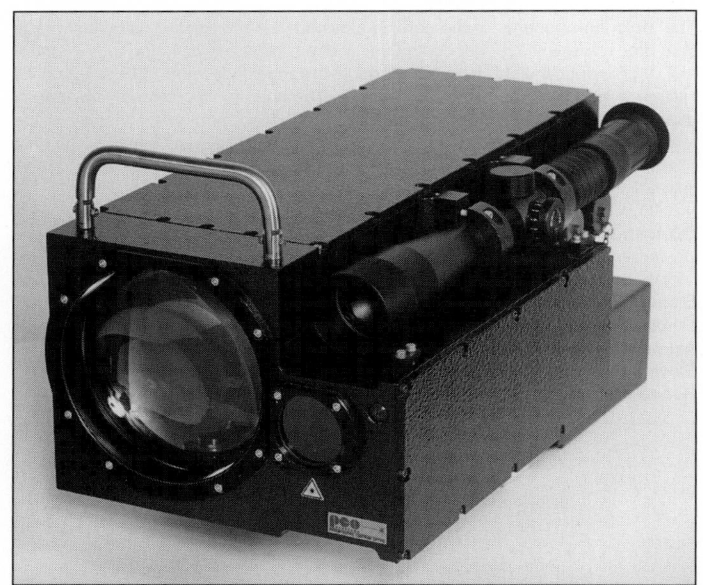

PCO DLI high repetition laser range-finder 0010861

Radamec System 2000FAA

Type
Air defence gun (static and towed) sight.

Description
The Radamec System 2000 FAA is a compact, lightweight electro-optical fire-control system designed for precision target tracking, ballistic prediction and control of land-based anti- aircraft engagements. The System 2000 FAA is capable of controlling any small/medium calibre gun mounting in batteries of up to four.

It uses standard modules from Radamec's Series 2000 family of electro- optical systems, there being two main elements: the electro-optical director and its control electronics. The system is provided with a TV camera for daylight use, a thermal imager for night-time or poor visibility, and a laser range-finder for accurate measurement of target range. A small strike aircraft is said to be detectable at ranges in excess of 15 km, night or day.

In its standard configuration, the electro-optical director is fitted to a purpose-built trailer and the support electronics and operator's control console are housed within a container mounted on a Land Rover 110 long-wheel based vehicle.

Other configurations available include the electro-optical director, support electronics and operator's control console mounted on a single vehicle or container, or supplied as separate units for location in the customer's own vehicle, container or fixed installation.

System 2000 FAA is configured for all system functions to be controlled by one operator from a purpose-built control console. Under normal conditions, the system is automatic in operation. The operator can override any of the automatic functions and revert to manual control at any time.

In autoscan mode, the system is capable of carrying out a wide range of operator-defined surveillance search patterns. In target indication mode, it will accept target indication information from external sources such as radars and target designators. In both these modes, the system is capable of automatic detection and acquisition. The system has a number of basic operational tracking modes configured for all target/background combinations.

The system provides fire control for anti-air engagements and can also be used to control guns in surface engagements. The system will inform the operator of the optimum time to fire the weapons to minimise wastage of ammunition.

Operational status
In production.

Contractor
Radamec Defence Systems Ltd.

Ring Sights LC-40-100 anti-aircraft sight

Type
Towed anti-aircraft gun sight.

Development
The LC-40-100 was originally developed as a replacement for cartwheel sights on anti-aircraft guns and cannon. The RAF is adopting the LC-40-100 as a sight for guns firing sideways from helicopters. For this application Ring Sights has developed the basic sight so that it can be used by day and night, with night vision goggles in the latter case.

This development underlines the need for ground troops to be able to engage air targets at night as well as by day so the basic LC-40-100 is also available with LED graticule lighting so that the gunner can acquire and engage air targets using night vision goggles.

Ring Sights Defence LC-40-100 anti-aircraft gun sight from gunner's side

In February 1996, Ring Sights was awarded a contract for 82 LC-40-100-9K38 sights for the Soviet Igla 9K38 surface-to-air missile by OIP Sensor Systems (now Thales) of Belgium. These are to be used in conjunction with Thales LORIS image intensifying night vision goggles. These were delivered by May 1996.

Description

The basic sight is ordinarily fitted with a Betalight for low-light use: if night vision goggles are to be used the LC-40-100 NVG should be specified.

Both types can have interfaces for particular weapons, (for example, the .50 in calibre Browning) or to use as a putting on/emergency sight for more sophisticated sighting systems. The standard types have integral zeroing. The optic can be supplied bare for integration into other systems.

The graticule pattern can be supplied to the customer's needs to suit the application, cartwheel, ellipse(s), lead and elevation for engaging ground targets in emergency and boresight mark. The maximum horizontal lead is ±150 mils: vertically ±136 mils is possible.

The optic is solid glass and the graticule image is generated internally. There is considerable eye freedom, as a result of the large aperture and image brightness allowing the gunner to wear night vision goggles or NBC kit. Eye-position is not constrained as it is with traditional cartwheel sights where the eye, backsight, foresight and target have to be lined up. The LC-40-100 can also be built in to other sights.

The Ring Sight LC-40-100-9K38 was developed specifically for the 9K38 Igla (SA-18) shoulder-launched anti-aircraft missile system. It uses the same reticle design as the open sight on the Igla and has been designed to be used in conjunction with Thales Optronic MUNOS night vision goggles, although it can also be used by day with the naked eye. It is designed so that it can be moved quickly from a fired launcher to an unfired one without realignment.

Operational status

In production. In service with undisclosed countries.

Specifications

	Optic only	With zeroing	Without zeroing
Aperture	40 mm	40 mm	40 mm
Length	110 mm	190 mm	150 mm
Width	42 mm	65 mm*	65 mm*
Height	58 mm	120 mm	80 mm
Weight	0.5 kg	1.7 kg	1.1 kg

*110 mm for night vision goggles version

Contractor

Ring Sights Defence Ltd.

UPDATED

Saab LVS fire-control system

Type

Anti-aircraft gun fire-control system.

Development

The Saab Avionics LVS modular fire-control system is designed for upgrading motor-controlled small- and medium-calibre anti-aircraft guns such as twin 23, 30 and 57 mm and the 40 mm L70. The system has been successfully trialled with the Swedish Army on an L70 gun and entered service with the Swedish Army in 1994 and is also in service with the Royal Thai Army. More than 250 LVS systems have been ordered.

Description

The main features of the LVS system can be summarised as follows: it measures the three-dimensional target trajectory independently of gun movements; computes aim of angles digitally; controls gun direction automatically and has a short reaction time.

Only limited modification of the gun is required to fit the LVS system and a small amount of training required to use it.

The LVS system comprises two main components, the Target Acquisition and tracking Unit (TAU) and the Computer and Control Unit (CCU). Other parts include the Gun Interface Unit (GIU), Azimuth Absolute Transducer (AAT), Control and Display Unit (CDU) and the Gyro and Pendulum Unit (GPU).

The operator acquires and tracks the target using the TAU which contains a ×7 telescopic sight combined with a laser range-finder. Angular movements of the line of sight are measured by means of gyros. The three-dimensional target position and speed are continuously computed.

Using this information, the system calculates the aim-off angle and controls the gunlaying system. Corrections for wind, temperature, air pressure, types of ammunition and muzzle velocity are also calculated and added to the aim-off angle.

During the initial tracking phase, the gun is automatically slaved to the TAU sightline and in the final phase, when the target is normally within range, the gun is automatically steered to computed aim-off angles relative to the actual line of sight.

The LVS system is designed to be effective against different types of air targets such as aircraft and helicopters. Moreover there is a special mode for stationary targets.

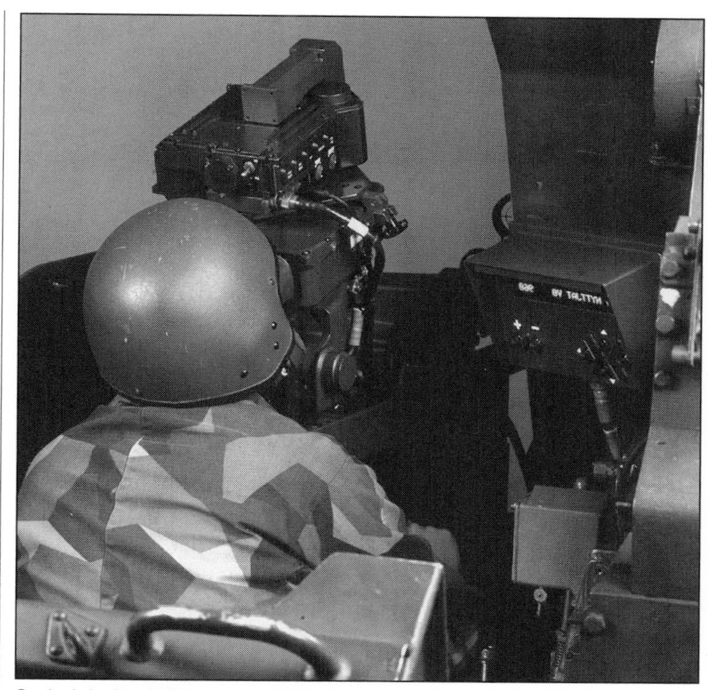

Saab Avionics LVS anti-aircraft fire-control system on Bofors 40 mm L/70 anti-aircraft gun and fitted with thermal imager for night target engagement

While the standard LVS system is a day-only fire-control system, for night engagement of targets the sight can be fitted with a thermal imager. A Training Simulator Unit (TSU) can be mounted on the TAU and used by the gunner to train in a realistic way in different target scenarios. The LVS system can be directed by an external surveillance radar.

Operational status

In production and in service.

Contractor

Saab Avionics AB.

SAGEM SAS 90 M

Type

Air defence optronic fire-control system.

Development

The SAS 90 M is an electro-optic fire-control system which has been developed by SAGEM and the French Ministry of Defence. It has already been fitted to 20, 23 and 40 mm guns.

Description

The modular design of the SAS 90 anti-aircraft fire-control system enables it to meet a wide range of requirements. It can be configured from a simple reflex sight version to a complete day/night automatic fire control for servoed guns.

The primary acquisition sensor for the SAS 90 is a CCD TV or infra-red camera associated with a video-tracking system.

SAS 90 mounted on Giat twin 20 mm anti-aircraft gun

The set of subassemblies which make up the SAS 90 0004765

The system also enables:

- range measurement either by stadiametry or by the addition of an optional medium-rate laser range-finder
- continuous lead angle computation
- automatic gun control by coupling to the weapon servoes
- automatic relay to an external target designation system.

The SAS 90 family comprises the following basic modules:

- a reflex sight with projected aiming reticles or a TV monitor for operator's interface
- an electro-optic sight including the CCD TV or thermal imager, an optional laser range-finder and a 2 axis aiming mirror
- an electronic unit performing video-tracking trajectory and lead angle computation, data input/output and gun control
- a control panel.

Operational status

In production.

Specifications

Sight
Reflex sight
Field of view: 16 × 20°

CCD camera (optional)
2 automatic fields of view: 4°, 2°
Double deviation mirror: ±20°
Resolution: 768 pts/line
Sensitivity: <1 lx
Output: CCIR
IR camera
Spectral band: 8-10 μm or 3-5 μm
Detection range: 10 km
Fields of view: 2 automatic, 6°, 3° or 12°, 2.5°
Cooling: Stirling microcooler
Double deviation mirrors: ±20°

Laser range-finder (optional)
Wavelength: 1.54 μm
Medium rate: 3 Hz
Accuracy: 7 m
Range: >7 km
Electrical
Power supply: 18 to 32 V DC
Power consumption: 80 W

Contractor

SAGEM SA, Aerospace and Defence Division.

UPDATED

SAGEM SAS-90 V

Type

Air defence fire-control system.

Development

The SAS-90 V has been developed by SAGEM for modernisation of gun batteries.

Description

The complete configuration of the system includes:

- One central system which enables early detection, acquisition and automatic tracking of aerial targets by day and by night and which provides each battery gun with precise data on the tracked target
- On each gun, a modernisation kit whose final purpose is to allow precise aiming of the gun

The SAS-90 V (central system) 0085464

- One set of cables linking the central system to the different guns
- One central power supply unit.

The central system can drive up to six remote guns. It can be set up on the ground or integrated into a vehicle and comprises:

- an optronic director two-axis platform equipped with FLIR, TV camera and medium-rate laser range-finder
- a remote control unit including TV monitor, joystick and control panel.

The gun modernisation kit comprises:

- azimuth and elevation gun angular position encoders
- an electrical slip ring
- a computation unit with operator's interface
- an interface with the firing circuit
- a local power supply, if required.

Operational status

In production.

Specifications

Central system 2-axis aiming platform
traverse: n x 360°
elevation: −23°/+90°
IR camera: 3-5 or 8-12 μm
field of view: 2 automatic
cooling: Stirling microcooler
detection range: 10 km
TV camera
field of view: 3° × 2.25°
detection range: 10 km
Laser range-finder
eye-safe: 1.54 μm
Repetition rate: 3-6 Hz
Range: > 7 km
Power supply: 28 V DC
Power consumption: 100 W
Gun modernisation kit:
Azimuth and elevation electronic units: ELD screen, 15-key panel, weight: < 5 kg

Contractor

SAGEM SA, Aerospace and Defence Division.

UPDATED

STN Atlas Elektronik, Zeiss Optronik AOZ 2000 autonomous electro-optic sight unit for direct firing

Type
Gun sight.

Description
The AOZ 2000 Autonomous Electro-optic Sight is designed for the aiming, ranging, fire-control computing and manual alignment of gunnery systems for direct firing. The sight unit is mounted on the gun and allows engagement of targets from a stationary or moving position by day or night. Targets can be designated in relationship to an index position if the sight is used in combination with a reconnaissance system.

The sight sensors are a daylight TV camera, a second-generation thermal imager and an eye-safe laser range-finder. There is also an external air data sensor for measuring pressure, temperature and crosswind, as well as a cant sensor.

Operational status
Available.

Specifications
Thermal imager
Type: 2nd-generation IRCCD
Spectral band: 7.5-10.5 μm
Fields of view (approx)
narrow: 3.3 × 4.4°
wide: 11.5 × 4.4°
TV camera
Output: CCIR
Laser range-finder
Type: Nd:YAG with Raman cell
Wavelength: 1.54 μm
Class: 3A (IEC 825-1)
Power supply: 18 – 32 V DC onboard or external lithium battery pack
Weight: <15 kg (basic version); <20 kg (extended version)

Contractors
STN Atlas Elektronik GmbH.
Zeiss Optronik GmbH.

UPDATED

ANTI-ARMOUR MISSILES AND MUNITIONS

VEHICLES

Boeing, Lockheed Martin ground-launched Hellfire II

Type
Anti-armour missile (vehicle launched).

Development
Summary: the missile has a semi-active laser seeker.

Hellfire was originally developed as the main weapon for the Boeing (formerly McDonnell Douglas) AH-64 Apache attack helicopter, but the system has also been adapted for ground use. In the early 1990s, test launches occurred from HMMWV trucks, CUCV trucks, modified M113 APCs and a modified Sea Chaparral launcher.

In 1995, Lockheed Martin and Boeing Defense and Space Group gave a firing demonstration of a mobile ground-launched Hellfire II system based on an M998 series HMMWV vehicle with four ready to fire missiles, five reload missiles and secondary weapons.

In 1995, Lockheed Martin and Rockwell formed Hellfire Systems to carry out joint production and marketing of the AGM-114F and AGM-114K laser-guided variants of Hellfire II.

Description
The missile has a range of 500 to 6,000 m using onboard laser designation, or 500 to 9,000 m using remote designator. The ground-launched system has a fully stabilised Electro-Optical Targeting and Surveillance system (EOTS) which allows fire on the move.

The Hellfire II missile uses semi-active laser guidance but with an improved seeker which has been hardened against electro-optical countermeasures. The missile contains a digital autopilot to increase launch speeds from 300 kt to M1.1 and a pulsed rocket motor for increased range. The system features a tandem warhead that can defeat advanced enemy reactive armour and improved target discrimination and tracking.

Operational status
The Hellfire II missile is in production and in service with Israel, Sweden, the US Army and the US Marine Corps. On order for Egypt, Greece, South Korea, Netherlands, Norway, Saudi Arabia, Taiwan, the United Kingdom and the United Arab Emirates.

Specifications
Guidance: semi-active lasing homing
Weight: 101 lb (AGM 114 k)
Diameter: 178 mm
Length: 1.626 m

Contractors
Boeing Company, Defense and Space Group, Electronic Systems Division.
Lockheed Martin Electronics & Missiles.

VERIFIED

The Ground-Launched Hellfire II has a stabilised E-O sight which enables tanks to be engaged 0004767

Euromissile HOT ATM (Anti-Tank Modular) system

Type
Anti-armour missile (vehicle launched).

Description
The HOT ATM (Anti-Tank Modular) system, developed by Euromissile (Aerospatiale and LFK EADS), combines the HOT anti-tank missile system with a multisensor platform and an information management centre. It can be installed on any type of carrier, but Euromissile has already equipped the German Wiesel light vehicle with HOT ATM. Emphasis has been placed on flexibility of observation and firing modules, modularity and simplicity of man/machine interfaces in the design.

The electro-optic guidance system includes a Thales Optronics Castor thermal camera and Mascot day TV camera as well as a CILAS laser range-finder and 1 μm infra-red localiser. There are two viewing channels to allow two targets to be engaged at the same time. The system allows the co-ordination of observation and fire control, the transmission of reconnaissance data between ATM units and to and from command posts.

Operational status
Available.

Contractor
Euromissile.

UPDATED

Euromissile HOT Mephisto ATGW system

Type
Anti-armour missile (vehicle launched).

Development
Summary: the launcher has an infra-red tracker and thermal imaging night sight.

The HOT Mephisto system has been designed by Euromissile to meet the requirements of the French Army which uses it mounted on the VAB (4 × 4) APC, but it can be fitted to many other types of APC such as the Panhard VCR (4 × 4). It has also been installed for trials to a MOWAG Piranha (8 × 8) APC.

Over 80,600 HOT missiles have been ordered together with 762 vehicle and 716 helicopter launcher systems by 17 countries.

Description
The Mephisto system consists of a launcher and four ready to launch missiles. The HOT missile is wire guided and tube launched with Semi-Automatic Command to Line Of Sight (SACLOS) guidance. The optical assembly comprises a periscope, the rotating head of which is fitted with a gyrostabilised mirror which enables the gunner to observe and fire regardless of the movement of the vehicle. When the launcher is retracted the periscope can be used for forward observation.

A thermal imaging sight called Mephira, made by Thales Optronics, is fitted on the periscopic sight allowing day and night observation and target acquistion and tracking to the full range of the HOT missile. Mephira is an 8 to 13 μm thermal

MBDA HOT Mephisto system on French Army VAB (4 × 4) with launcher raised

imaging sight with cadmium mercury telluride detectors and series- parallel scanning. In 1992, an infra-red tracker unit operating in the 1 μm waveband was added to the Mephisto system. Together with the thermal imager, the localisation unit will operate in a bispectral mode.

Operational status
Production as required. In service with the French Army.

Specifications
Crew: 1 (gunner)
Armament: 4 HOT ATGW
Sighting: APX M590 gyrostabilised with ×12 magnification (5° field of view) and ×3 (18° field of view)
Thermal imager: 8-13 μm CMT with series-parallel scanning
Infra-red tracker: 1 μm
Traverse: 360° electric at 30°/s max
Elevation: –10 to +10° electric at 5°/s max
Length
Module: 1.45 m
With magazines: 2.70 m
Width
Module: 1.60 m
Weight
Module with 4 missiles: 1,150 kg
Module plus 8 HOT ATGWs: 1,900 kg

Contractor
Euromissile.

UPDATED

Euromissile HOT UTM 800 ATGW turret

Type
Anti-armour missile (vehicle launched).

Development
Summary: the launcher has a thermal imaging night sight and an infra-red tracker.

This turret has been designed as a private venture by Euromissile for AFVs in the 5,000 to 12,000 kg class, such as the M113, Giat Industries VAB (4 × 4 and 6 × 6) and Panhard VCR (6 × 6) APCs. For trials purposes it has also been installed on a MOWAG Piranha (8 × 8) vehicle.

Description
The one-man UTM 800 turret has four HOT ATGWs ready to launch and, in the case of the Panhard VCR (6 × 6), 10 missiles carried in reserve.

A Castor infra-red thermal imager sight fitted to the launcher allows day and night observation and firing out to the 4,000 m range of the HOT missile. Castor is a dual field of view thermal imager made by Thales Optronics and operates in the 8 to 12 μm waveband. It has cadmium mercury telluride detectors with serial-parallel scanning.

In 1992, an improved turret system was tested successfully on the Austrian Pandur armoured vehicle. This uses a bispectral tracking system (operating at 1 and 10 μm wavelengths) to give improved resistance to IR jammers.

Operational status
Production as required (over 150 have been produced). In service with Cyprus, Iraq and Qatar. The status of the Iraqi systems is uncertain.

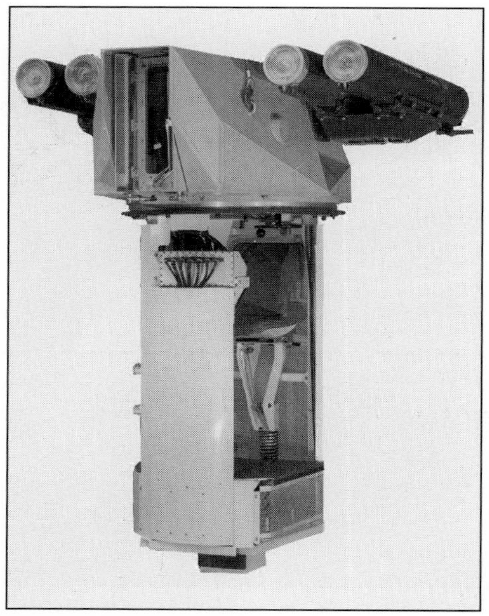

Euromissile HOT UTM 800 turret with four HOT ATGWs ready to launch

Specifications
Crew: 1 (gunner)
Armament: 4 HOT ATGWs
Sighting: APX M509 with ×12 magnification (5° field of view) and ×3 (18° field of view), night/day thermal imaging Castor sight
Control
Traverse: 360° electric at 30°/s max
Elevation: –10 to +22° electric at 5°/s max
Weight
With 4 missiles: 900 kg
Power supply: 27 V DC, rotary base junction box

Contractor
Euromissile.

UPDATED

IAI Nimrod

Type
Anti-armour missile (vehicle launched).

Development
Summary: the missile has semi-active laser terminal guidance.

IAI MBT Weapon Systems Division is producing the Nimrod long-range ground-launched semi-active laser-guided ATGW in response to an export requirement. The Israel Defence Force has also expressed interest. Nimrod can also be used as an anti-ship weapon. In June 1995, MBT announced that total sales of Nimrod had reached US$100 million.

Description
The missile is transonic and is powered all the way to the target. The gunner preselects a flight trajectory mode which may be a direct, high-cruising or low-cruising trajectory, the cruising altitude being constant and between 300 and 1,500 m. Mid-course guidance is provided by an integral inertial reference, and terminal guidance by a semi-active laser homing seeker for the last 15 to 30 seconds of missile flight. The target can be illuminated either by a ground- or airborne-based laser designator.

The gimballed and stabilised seeker head acquires, tracks and homes in on its target using localised proportional navigation. It is said to have a look angle of more than 30°. In the terminal flight phase the weapon adopts a dive angle of approximately 45° to impact the armoured target on its vulnerable upper surfaces.

The missile is stored in a sealed canister which also acts as the launcher. It has five main sections: seeker, guidance and control, warhead, solid propellant rocket motor and servo. It is roll stabilised in flight. Time into action at a launch site is less than 3 minutes without the site having to be surveyed for alignment or levelling,

Sequence of photographs showing Nimrod engaging a target tank

having a direct line of sight to the target. The weapon can be fired in single-round, ripple or salvo modes.

Operational status

Production complete. In service with an unidentified export customer. Recent information has indicated that Nimrod may be in service with the Israel Defence Force with target acquisition being carried out by special forces.

Specifications

Type: semi-active laser homing with inertial mid-course
Length: 2.60 m
Diameter: 0.17 m
Wingspan: 0.40 m
Launch weight: 98 kg
Warhead: 15 kg HE hollow charge
Range: 800-26,000 m
Armour penetration: 800 mm
Speed: 300 m/s

Contractor

IAI Ltd, MBT Weapon Systems Division.

KBP 9K116 Bastion system (AT-10 'Stabber')

Type

Anti-armour missile (vehicle launched).

Development

Summary: the missile guidance is laser beam-riding; the launcher has a day/night sight.

The 9K116 Bastion (AT-10 'Stabber') missile system entered service around 1985 and is a single-piece 100 mm calibre laser beam-guided projectile for use with the 100 mm tank gun of the T-55 medium tank. T-55 tanks built to the AT-10 system configuration have been fitted with the Volna (Soviet), Merida (Polish) or Kladivo (Czech and Slovakian) fire-control system.

T-55 variants known to be fitted with the system include: T-55MV, T- 55AMV, T-55AM2PB (all Soviet models), T-55AM2B (produced in the former Czechoslovakia) and T- 55AM2P (produced in Poland).

Description

The principle components of the 9K116 beam-riding missile system for the equipped tanks are:

(a) the 3UBK10-1 guided round (comprising the 18.4 kg 9M117 missile mounted within a cartridge case)
(b) a vehicle fire-control system which comprises a 1K13 day/night telescopic sight (with integral 1K13BZ laser emitter) to replace the T-55 series standard TPN-1M-22-1 telescopic infra-red sight. The sight is vertically stabilised with its line of sight slaved to the stabilised 100 mm gun for the target acquisition phase.

For an engagement, the sighting system is activated to switch the main armament stabilisation system off and reduce the maximum turret traverse speed so as to avoid any distortion effects.

The 9M117 is fired electrically and after the firing, the commander's laying device is turned off and the laser is activated to emit the modulated infra-red laser guidance beam. The missile flies automatically within a guidance zone formed by this projected beam. All the gunner has to do during the remainder of the engagement is to keep the sight on the target until the missile has impacted.

During the whole of the semi-automatic guidance phase the sightline is independently stabilised from the gun in both azimuth and elevation. A modulator encodes the laser beam so that each reference point of the guidance zone is given a precisely timed sequence of frequencies. The correlation between the duration of the different frequencies is determined by the position of each point within the guidance zone.

100 mm AT-10 'Stabber' ATGW with rear fins and forward control surfaces unfolded as in flight (Christopher F Foss)

The 6 m diameter of the guidance zone is kept constant throughout the flight by progressively zooming the beam. A receiver in the rear section of the missile transmits information on the flight position within the guidance zone to the onboard guidance system which then transmits flight control commands to the four forward-mounted flip-out control surface fins.

When the tank is static the system can be used to engage moving and stationary targets, static positions or slow-moving aerial targets. Effective engagement limits are stated to be from 100 to 4,000 m, with the weapon having a flight time of 12 seconds to the latter range. In the case of a miss, a self-destruct device automatically destroys the shaped charge warhead within 26 to 41 seconds of firing.

Standard combat load for the T-55 is six AT-10s.

Operational status

In production. In service with the Russian Federation, other ex-Soviet republics and possibly some other countries.

Specifications

Calibre: 100 mm
Guidance: laser beam-riding
Total round weight
3UBK10: 26 kg
3UBK12: 23 kg
Missile weight: 18.4 kg
Max armour penetration: 650 mm
Range limits: 100-4,000 m

Contractor

KBP Instrument Design Bureau.

KBP 9K120 Refleks system (AT-11 'Sniper')

Type

Anti-armour missile (vehicle launched).

Development

Summary: the missile has laser beam-riding guidance.

The 9K120 Refleks (AT-11 'Sniper') is a two-part 125 mm calibre laser-guided projectile version of the 9K120 missile system for use with the 125 mm main gun of the T-80U and T-80UD MBT variants and some versions of the T-72 and more recent T-90 MBT.

Description

The missile is the same as used in the Svir version but the heavier overall round weight indicates a larger propellant charge. The laser guidance emitter is contained within the gunner's stabilised primary sight assembly. The 4.2 kg shaped charge warhead can penetrate up to 700 mm of armour.

The major difference between the Refleks and Svir variants is that the Refleks can be selected to fly a second, more sophisticated flight profile for certain engagements such as a stationary firing against a target over low-lying obstacles.

In this case, immediately after launch, the Refleks round climbs to a height of 4 to 5 m above the ground and some 2 m directly above the centre of the laser guidance beam. It remains there for the main part of the flight. At approximately 500 m from the target (a distance derived from data automatically fed into the fire-control system by the laser range-finder) a modulation of the laser guidance beam automatically commands the missile back down into the centre of the beam for the final attack.

Operational status

In production. In service with the Russian Federation and other former Soviet states. Offered for export.

Specifications

Calibre: 125 mm
Guidance: laser beam-riding
Weight
Total round: 28 kg
Missile: 17.2 kg
Max armour penetration: 700 mm
Range limits: 100-5,000 m

Contractor

KBP Instrument Design Bureau.

KBP 9K120 Svir system (AT-11 'Sniper')

Type

Anti-armour missile (vehicle launched).

Development

Summary: the missile has laser beam-riding guidance.

The 9K120 Svir (AT-11 'Sniper') is a two-part 125 mm calibre laser-guided projectile version of the 9K120 missile system for use with the 125 mm main gun of

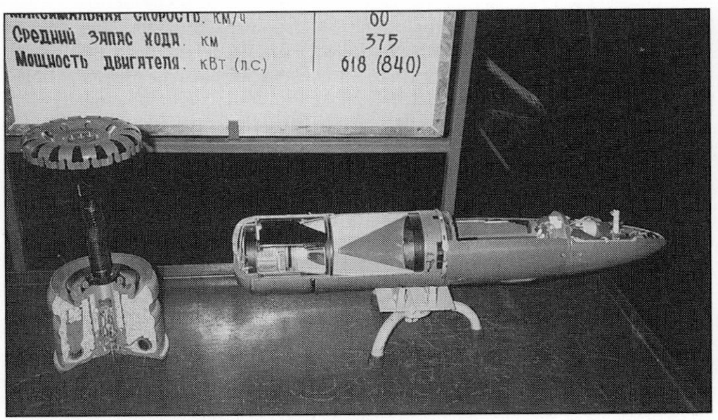

Svir AT-11 system with cutaway booster section (left) and cutaway missile (right)

the T-80U and T-80UD MBT variants and some versions of the T-72 and more recent T-90 MBT. A standard combat load of six Svir rounds is carried. The 9K120 system entered service in 1984.

Description

The missile is the same as used in the Refleks version but the overall round weight is lower, indicating a smaller reduced propellant charge. The system uses the coded laser guidance beam 1K13 sight assembly which is fitted in place of the gunner's TPN-1-49 secondary night sight. The 4.2 kg shaped charge warhead can penetrate up to 700 mm of armour.

The Svir operates on the same principles as the Bastion (AT-10 'Stabber') family with the missile effectively riding in the centre of the laser funnel that is created by the 1K13 sight's laser emitter. The frequency of the beam is modulated in different sectors around the funnel so that if the missile deviates, the missile's onboard guidance system can detect the abnormal movement and correct the flight trajectory so that it moves back into the centre of the guidance beam. The guidance system uses a timer so that the laser funnel is periodically altered in diameter. This means that the missile effectively sees a relatively constant laser tunnel diameter as it moves along its flight trajectory.

Operational status

In production. In service with the Russian Federation and other former Soviet states. Offered for export.

Specifications

Calibre: 125 mm
Guidance: laser beam-riding
Weight:
Total round: 24.3 kg
Missile: 17.2 kg
Max armour penetration: 700 mm
Range limits: 100-5,000 m

Contractor

KBP Instrument Design Bureau.

Kentron ZT-3 Swift family

Type

Anti-armour missile (vehicle launched).

Development

Summary: the missile is laser guided; the launcher for the upgraded ZT-35 has a thermal imaging night sight.

The ZT-3 entered service in 1987 and is a laser beam command-guided Semi-Automatic Command to Line Of Sight (SACLOS) weapon which is tube launched from a helicopter, ground or vehicle-mounted launcher assembly.

Description

The vehicle-launched ZT-3 is launched from a Ratel Mk III ICV vehicle fitted with a turret containing a 7.62 mm self-defence MG, missile fire-control system and surveillance/target acquisition-tracking optics. A three-round box-shaped launcher is mounted above the turret, but the turret can also be mounted on other suitable vehicles.

Once launched the missile is tracked by use of a pulsed infra-red source in the base of the missile and a laser is used to send encoded pulsed guidance commands. The latter's beam is sensed via a rear-mounted sensor which then uses a goniometer to measure the deviation from the beam centre. Its autopilot system uses these data to correct its flight profile so that its actual flight path remains aligned to the line of sight of the gunner's aiming system.

ZT-35 upgraded Swift

Kentron has developed an upgraded version of the ZT-3 Swift ATGW. The improvements include a new HEAT warhead with standoff nose probe, a digital autopilot that can receive a data download before launch to give greatly improved missile gathering, and the addition of a thermal imaging passive night sight channel to the launcher to allow night target acquisition.

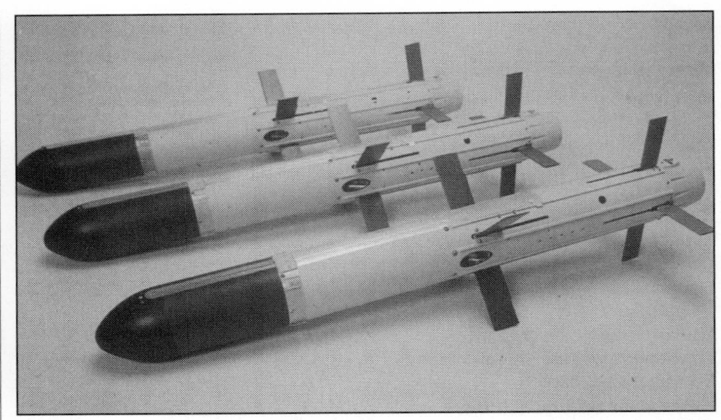

Three ZT3 Swift anti-tank missiles, with wings and fins deployed in their flight positions

The accuracy at a range as low as 150 m is said to be better than 90 per cent, while at ranges of between 100 and 5,000 m it is said to be better than 95 per cent. A version with a precursor charge and laser proximity fuze is being tested.

Operational status

Production of the ZT-3 Swift has been completed. In service only with South African National Defence Force.

Specifications

Type: laser-guided SACLOS
Diameter: 0.127 m
Warhead: hollow charge with contact fuze
Range: > 5 km
Armour penetration: 650+ mm
Speed: 330 m/s

Contractor

Kentron Division of Denel (Pty) Ltd.

Kolomna Khrizantema

Type

Anti-armour missile (vehicle mounted).

Description

Khrizantema is a dual anti-armour/anti-bunker weapon being developed by the KBP Instrument Design-Making Bureau. The weapon can be mounted on a BMP-3 tracked infantry vehicle and fires 9M123 supersonic missiles that have radar and laser command-guidance receivers. Radar operation is automatic with semi-automatic laser. The missile can carry either tandem charge or fuel-air explosive warheads. Range of the missile is 6 km.

Operational status

Prototypes. At the stage of industrialisation.

Contractor

Kolomna MKB.

The Khrizantema dual anti-armour/anti-bunker weapon system mounted on a BMP-3 infantry fighting vehicle 0004761

LFK, Aerospatiale, Italmissile Polyphem

Type
Fibre-optic guided missile.

Development
The programme was started by France and Germany in 1992 and joined by Italy in 1994 to prove the feasibility of a new weapon system with long-range guided missiles. In 1994, the governments of the three countries agreed common military requirements for a future fibre optic guided weapon system with a range of 60 km for army application. The industrial programme team consists of LFK Lenkflugkorpersysteme GmbH (a subsidiary of EADS European Aeronautic Defence and Space Company), Aerospatiale Matra Missile and Italmissile. The name of the programme is TRIFOM (trinational fibre optic missile). In 1997 the first experimental phase of TRIFOM ended with a successful 16.4 km live firing at the Landes Test Centre (CEL) in France. The second experimental phase of TRIFOM will end with a 30 km live firing in 2001 at the Meppen test range in Germany.

EADS 'LFK Lenkflugkorpersysteme' is developing the Polyphem missile also for naval and helicopter applications. Therefore, surface vessels and submarines can be armed with Polyphem, while the submarine version is called TRITON. Adaptations for different helicopters have been studied.

Description
Polyphem is a multimission weapon system that can be installed on various platforms - ships of all sizes, submarines, medium-size helicopters and ground vehicles. An optical fibre is unwound behind the guided missile in flight which enables interference-resistant data transmission of images from a high-resolution infra-red camera in the missile's front section to the command centre. The imaging infra-red seeker uses a platinum silicide focal plane array in the 3 to 5.5 µm waveband which has a detection range of up to 8 km, 3 km in adverse visibility. The missile electronics unit also incorporates an Inertial Navigation System (INS) and Global Positioning System (GPS) as well as a laser altimeter.

The missile is powered by a turbojet engine in the mid-section with a shaped-charge/fragmentation warhead. The bobbin in the tail carries the fibre optic wire which links the missile to the firing station. The firing station is designed to fulfil mission planning before engagement. It stores a digitised map and displays it in real time during missile flight.

Operational status
Under development. The German naval applications (surface vessels and submarines) will be in series production in 2005.

Specifications
Range: up to 60 km
Missile speed: 200 m/s
Flight altitude: 20-400 m variable
Weight: 140 kg

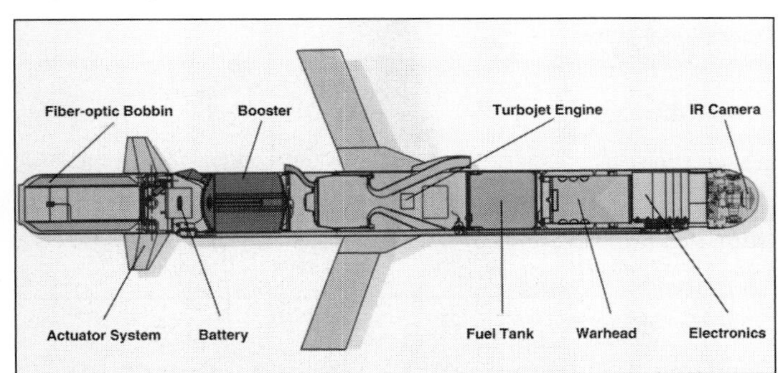

Diagram of the Polyphem missile 0077434

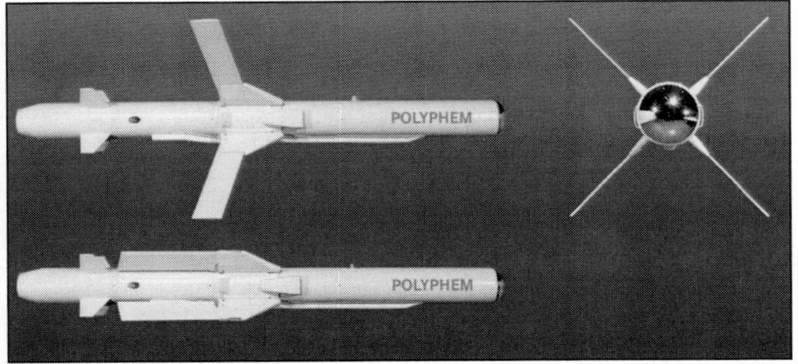

Polyphem fibre-optic guided missile 0077436

Length: 3.3 m
Warhead: over 20 kg shaped-charge/fragmentation
Fibre-optic wire: 16 channel 128 kbit/s uplink and 240 Mbit/s downlink (1 video, 32 data channels)
Imaging infra-red seeker
Detector: PtSi focal plane array
Spectral band: 3-5.5 µm

Contractors
MBDA.
EADS - LFK Lenkflugkorpersysteme GmbH.
Italmissile.

Lockheed Martin LOSAT (Line-Of-Sight Anti-Tank) weapon system

Type
Kinetic energy missile.

Development
The Line-Of-Sight Anti-Tank (LOSAT) weapon system is being developed by Lockheed Martin Missiles and Fire Control under contract to the US Army to provide a deployable long-range anti-tank system.

LOSAT is a hit-to-kill weapon that uses the kinetic energy of direct impact to defeat advanced armour, helicopters, bunkers and other targets. It consists of the Kinetic Energy Missile (KEM) and its fire-control system integrated into an expanded-capacity High-Mobility Multipurpose Wheeled Vehicle (HMMWV). LOSAT was previously proposed for the AGS Armoured Gun System programme which was cancelled in March 1996. The US Army is evaluating LOSAT on the HMMWV as a replacement for AGS to improve the anti-armour capability of early entry forces, and as the anti-tank weapon system for the future Brigade Combat Team.

The missile is currently in the Advanced Combat technology demonstration phase. Principal subcontractors include Raytheon for the electro-optical subsystem of the fire-control system, Hercules for the solid rocket motor and BAE Systems Atlanta for the target ranging system.

Description
Each extended capacity HMMWV can carry four KEM missiles. The KEM missile travels at the hypervelocity speed of more than a mile a second and destroys its targets upon impact without the use of conventional ordnance or explosive warhead, blasting a penetrator rod through even multiplate armour. The penetrator rod is located in the forward area of the missile.

The fire-control system is based on the Improved Target Acquisition System (ITAS) by Raytheon, and uses a second-generation thermal imager.

The entire attack sequence is claimed to take only a few seconds at target ranges of over two miles. The thermal imaging sensor, which is mounted on the launch platform, tracks a target. Immediately after it is initialised and launched, the KEM missile receives guidance updates from the targeting system on board the launch vehicle until the target is hit.

The LOSAT weapon system is a hit-to-kill weapon which uses the kinetic energy of direct impact to defeat armour and other targets 0004766

LOSAT HMMWV is air-transportable by C-130J (2 LOSAT HMMWV), C-17 (7), C-5 (10), C-141 (3) aircraft and CH-47D (1, external load), V-22 (1, external load) helicopters.

Operational status

Under development, technology demonstrator phase.

Specifications

Missile launch weight: 80 kg
Missile length: 2.84 m
Missile diameter: 0.162 m
Missile range: > 4,000 m (max)

Contractor

Lockheed Martin Missiles and Fire Control.

Raytheon EFOG-M Enhanced Fiber Optic Guided Missile

Type

Fibre optic guided missile.

Development

In May 1995, Raytheon Systems Company was awarded a six year contract to build an Advanced Technology Demonstrator of the Enhanced Fibre Optic Guided Missile (EFOG-M) System for the US Army MIssile COMmand (MICOM). EFOG-M is the main weapon in the Rapid Force Projection Initiative (RFPI) Advanced Concept Technology Demonstration (ACTD) programme which is to provide US forces with the weapons and sensors needed to increase effectiveness against an entrenched force. Its primary role on the battlefield is anti-tank but with a secondary anti-helicopter capability.

As part of the EFOG-M ATD, Raytheon will produce one Mobile Simulator, two Stationary Simulators, 300 EFOG-M missiles, 12 EFOG-M Fire Units (FU) and three Platoon Leader Vehicles (PLV). In 1999, an entire company of EFOG-M - three platoons consisting of 256 missiles, 12 FU and three PLV's will be provided to the Army's XVIII Airborne Corps for a two year Extended User Evaluation.

Subcontractors to Raytheon are BAE Systems (for the platinum silicide-based imaging infra-red seeker camera); Southern Research Technologies (for the auto-tracker, seeker gimbal and integration); SCI Technologies Inc (the fibre optic datalink and cable payout dispenser, gunner's console); and Systems and Electronics Inc (HMMWV modification).

Description

The EFOG-M fire unit consists of a 'Heavy Humvee' (HMMWV 4 × 4) vehicle, a launch pack with eight missiles, a gunner's station, ground-based computer, fibre optic electronics and command, control and communications equipment.

Guidance is by a platinum silicide 3 to 5 μm Imaging Infra-Red (IIR) seeker and a Global Positioning System/Inertial Navigation Unit (GPS/INU). Kearfott Guidance & Navigation has been selected to supply their MILNAV navigation system. A two-way datalink operates over a fibre optic line that spools out as the missile flies to its target. The missile receives steering signals back through the link from the gunner's station. The gunner performs target selection and acquisition on a video screen and locks the automatic tracker onto the target image displayed on the console. The tracking commands are sent to the ground station computer which then sends steering commands up the fibre optic datalink to steer the missile. The missiles can be launched in front of the target rather than from the side as is more usual when attacking armour. This is because the missile has a fully developed top attack capability. It follows a non-ballisitic flight path, both to allow launch from a concealed position and to confuse detection of the launch site.

The missile will weigh about 48 kg and will have a range of up to 15,000 m. It uses a solid propellant rocket booster and sustainer and an explosively formed penetrator warhead. Reloading of a complete pack of eight missiles is expected to take less than 8 minutes.

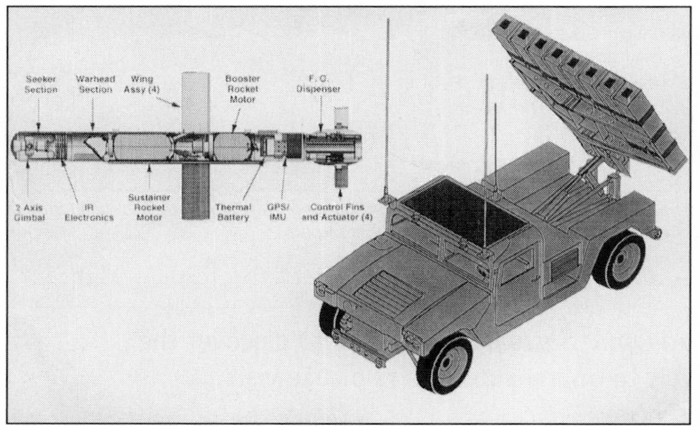

Diagram showing cross-section of the EFOG-M missile and launch vehicle

Operational status

Under development. Classed as an Advanced Technology Demonstrator (ATD) programme. Successful test flights were made in 1998 and 1999.

Specifications

Missile
Seeker: imaging infra-red
Type: 640 × 640 PtSi focal plane array
Spectral band: 3-5 μm
Warhead: shaped charge
Sustainer and booster: solid propellant
Navigation: INU/GPS
Fibre diameter: 240 μm
Length: 1.94 m
Diameter: 0.166 m
Wingspan: 1.14 m
Weight
Missile: 51.3 kg
Missile in canister: 78.5 kg
Max range: 15 km
Min range: 1 km
Speed (flyout): 100 m/s

Contractor

Raytheon Systems Company.

Saab Bofors AAAW 2000 Advanced Anti-Armour Weapon

Type

Anti-armour weapon (vehicle launched).

Development

The Advanced Anti-Armour Weapon 2000 (AAAW 2000) is being developed by Saab Bofors Dynamics for a Swedish Army requirement for a smart off-route weapon. The system uses the warhead technology from the AT-12 shoulder-launched anti-armour weapon, which has been cancelled. A prime contractor has yet to be selected. Trials were due in 1998, with an in-service date around 2002. The Royal Norwegian Army has a similar requirement.

Description

AAAW 2000 weighs 20 kg and is 1,100 mm long. It uses a passive multisensor which detects sound, analyses the sound acoustically, decides whether it denotes a tracked vehicle. If so, the rocket is triggered by an infra-red sensor.

Operational status

Under development.

Specifications

Weight: 20 kg
Length: 1,100 mm

Contractor

Saab Bofors Dynamics.

Systems & Electronics TOW Under-Armour (TUA) turret

Type

Anti-armour missile (vehicle launched).

Development

Summary: the TUA launcher has an optical and thermal imaging sight and the artillery targeting variant also has a laser used as a range-finder, locator and designator.

Systems & Electronics Inc developed the TOW Under-Armour (TUA) turret to allow the crew to launch and guide the TOW missile system from defilade while also protected by armour. Only 1 m^2 of TUA is exposed to hostile fire and this area is completely separate from the crew location. The TUA TOW-tube elevating launcher fires any of the TOW missile variants.

The original TUA turret was designed for the US Army M901/M901A1 Improved TOW Vehicle (ITV) utilising the M113A1 APC chassis. Over 3,200 ITVs were produced for the US Army and National Guard.

The TUA has also been fitted to the OtoMelara VCC-1 and LAV (8 × 8) chassis for Saudi Arabia, to the Light Armored Vehicle (LAV) chassis for the US Marine Corps and to the YPR-765 PRAT tracked vehicle for the Royal Netherlands Army. Altogether more than 4,000 TUA turrets were built.

Description

In addition to the two missile launching tubes, the elevating launcher includes the gunner's optics (TOW day sight and thermal night sight) a ×4 magnification wide field of view and a ×12 magnification target acquisition sight. TOW missile

The M981 FISTV version of the TOW Under-Armour turret 0004768

guidance is accomplished with the standard Missile Guidance Set (MGS) mounted inside the crew compartment. The elevated launcher interfaces with the vehicle using a standard M27 cupola, which makes it compatible with most US and allied armoured combat vehicles.

The TUA has fully powered 360° traverse movement with its 'hammer head' having an elevation of +34° and depression of –30°. It requires only 20 seconds for the launcher to be elevated and the target to be engaged. The time from first TOW missile impact to second round triggering, with up to 12.5° target separation, is 4.25 seconds.

The protected gunner identifies and tracks the target with the sight's narrow field of view via the image transfer assembly. He can select either the day sight or night sight by remote control. The wide field of view allows the gunner to scan the terrain and locate targets. The guidance and sights of the TUA are identical to the standard TOW ground launcher system and these components can be removed from the vehicle and fitted to a ground TOW launcher tripod carried inside the vehicle.

Systems & Electronics also developed the US Army's M981 FIre Support Team Vehicle (FISTV). This is configured to duplicate the appearance of the M901 ITV but is designed to locate and designate targets. The same elevating platform is used and the TOW optics are replaced by the AN/TVQ-2 Ground Laser Locator/ Designator (GLLD). The FISTV carries a north-seeking gyroscope, a land navigation system and extensive communications equipment.

The system provides target identification, acquisition and designation to artillery units, enabling them to achieve 'first-round fire for effect' hits with both conventional and laser-guided smart munitions. Over 1,300 FISTVs were produced for the US Army and the Egyptian government has procured 25 Artillery Target Locating Vehicles (ATLVs), the export variant of the FISTV.

Operational status

In service. ITV/TUA is in service with Egypt (52), Greece (24), Jordan (50), Kuwait (58), Netherlands (304), Pakistan (24), Saudi Arabia (224) and Thailand (18).

Specifications

ITV/TUA
Crew: 1 (gunner)
Armament
Main: twin TOW ATGW launcher
Secondary: 1 × 7.62 mm MG
Smoke grenade dischargers: optional
Ammunition
TOW: 2 + 10
7.62 mm: 2,000
Control
Traverse: 360°
Traversing rate: 35°/s
Elevation/depression: +34°/–30°
Optics
Acquisition sight: ×2.8 magnification, 25° field of view
TOW day sight: ×13 magnification, 5.5° field of view
TOW AN/TAS-4 infra-red: ×4 magnification, 3.4 × 6.8° wide
Night vision sight: Field of view ×12 magnification, 1.1 × 2.2° narrow field of view
Squad leader's periscope (M901 vehicle): 360° traverse, –20 to +10° elevation/depression, ×4 magnification, 12.5° field of view
Launcher height above vehicle top (M901)
Stowed: 1.08 m
Raised: 1.52 m
Power supply: 24 V DC

Contractor

Systems & Electronics Inc.

UPDATED

US Department of the Army Long Fibre Optic Guided (Long FOG) missile system

Type

Anti-armour missile (vehicle launched).

Development

In 1995, the US Army announced the start of another technology demonstrator study programme to run parallel with EFOG-M, called LongFOG. This was for an extended-range fibre optic guided missile fitted with a low-signature variable-thrust turbojet engine to extend the effective engagment range to 40,000 m. The guidance system incorporates a GPS/INU package for the mid-flight phase and IIR for the terminal phase.

Other technologies to be fielded include a composite airframe (for low IR and Radar Cross-Section (RCS) signatures) and variable geometry control surfaces. The programme was due to run from FY95-99 with similar modelling simulations and flight testing to the EFOG-M project.

Status

Technology demonstration.

Contractor

US Department of the Army Missile Command (MICOM).

ANTI-ARMOUR MISSILES AND MUNITIONS

VEHICLE SIGHTS

Raytheon Improved Bradley Acquisition Subsystem

Type
Fire-control system.

Description
The Improved Bradley Acquisition Subsystem (IBAS) is in low rate initial production by Raytheon Systems Company for the Bradley A3 fighting vehicle, under a development contract awarded in February 1994. It will upgrade the current capabilites of the Bradley ISU (Integrated Sight Unit).

IBAS upgraded Target Acquisition Subsystem and Missile Control Subsystem provides enhancements which include the integration of the second-generation HTI (Horizontal Technology Integration) FLIR (qv) with daylight television and direct view optics, automatic dual target tracking, a laser range-finder and a two-axis stabilised head mirror. This improves the ability of the heavy anti-tank weapon system to defeat heavy armoured vehicles and to destroy fortifications and other targets such as helicopters. The HTI FLIR uses SADA II (Standard Advanced Dewar Assembly) technology and is being developed for the US Army.

Other features include dual missile trackers, automatic boresighting capability, electronic zoom, frame integration and biocular display. IBAS is compatible with all TOW missiles and will also provide enhanced accuracy of the Bradley's 25 mm gun while on the move.

Operational status
In production. In service with the US Army.

Specifications
HTI FLIR fields of view
Wide: 7.5 × 13.3°
Narrow: 2 × 3.6°
Direct view optics fields of view
Wide: 9° circular
Narrow: 3° circular
Eye-safe laser range-finder accuracy: ±5 m
MTBF: 330 h
Databus: MIL-STD 1553

Contractor
Raytheon Systems Company.

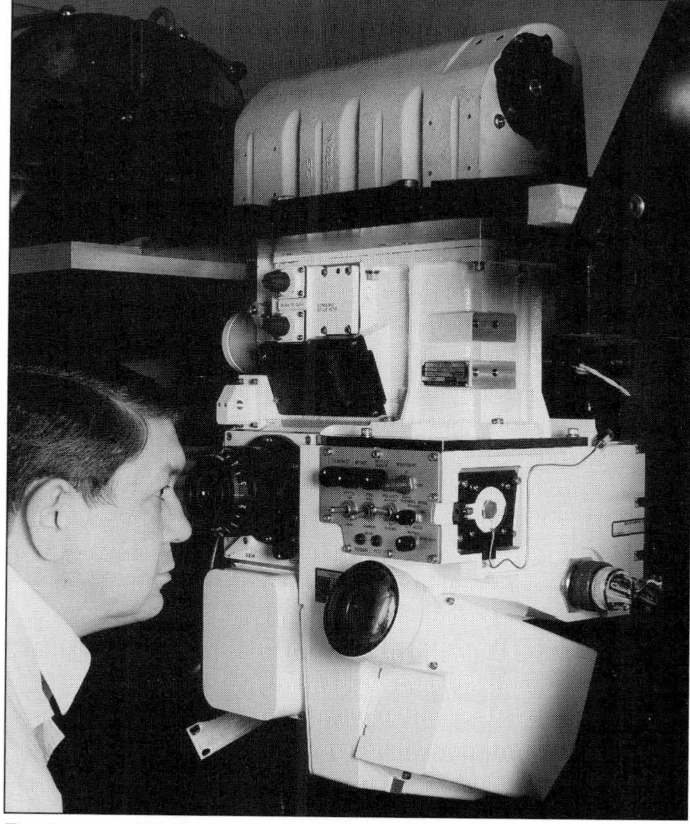

The Raytheon GITS integrated TOW sight, which includes a Raytheon Thermal Sight and TOW missile guidance system

Raytheon GITS integrated TOW sight

Type
Anti-armour missile – vehicle sight.

Development
The Raytheon GITS integrated TOW sight provides an anti-tank capability for light armoured vehicles. GITS features a Raytheon thermal sight and missile guidance system integrated into the Delco Systems Operations light armoured vehicle turret. The Delco turret provides TOW anti-tank capability as well as fire control for the 25 mm Bushmaster cannon and 7.62 mm machine gun.

The Delco turret has been successfully demonstrated on Desert Warrior, Piranha, LAV-25 and M113 combat vehicles and is in production for export versions of the Warrior. With two single TOW missile launchers symmetrically mounted to each side of the turret, gunners can engage heavy armour targets at ranges up to 3,700 m (more than 2 miles).

The LAV-25 turret with TOW and GITS has also been fitted on the prototype of the Warrior Reconnaissance Vehicle.

Description
The modular GITS is designed as an upgrade for small turrets to provide an under-armour TOW capability. It contains an improved Raytheon HIRE infra-red sensor (qv) with a 60-element Cadmium Mercury Telluride (CMT) detector (similar to the AN/TAS-4, -5 and -6). The sensor's fields of view are large for increased gunner acquisition, tracking and firing accuracy. Symbology is software-controlled with stadiametric or ballistic reticles available.

The module subassemblies include:

(a) a coaxial short wavelength tracker derived from the M2/M3 Bradley system
(b) a digital TOW guidance unit derived from the M2/M3 Bradley unit
(c) single field of view visual optics for target acquisition and tracking
(d) a two field of view HIRE thermal sensor for target acquisition and tracking in degraded visibility and at night
(e) full TOW 2A tracking and guidance software

The system is compatible with the Raytheon ELITE eye-safe laser range-finder (qv).

Operational status
In service on Alvis Desert Warrior in Kuwait.

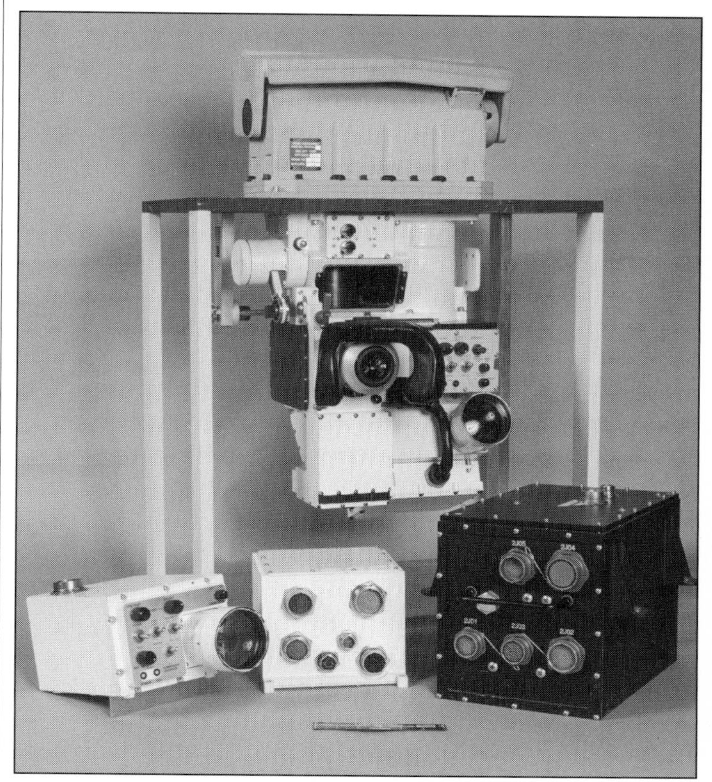

Raytheon GITS integrated TOW sight. Components are, from left to right, HIRE commander's display, HIRE electronics unit and TOW guidance set

Specifications
Elevation range: –20 to +60°, mechanically coupled
Dual window
Thermal: germanium
Visual: quartz
Boresight range
Visual and TOW: ±6 mrad
Thermal sensor: ±10 mrad azimuth and elevation
Unity channel
Field of view
×1 magnification: 10 × 5°
TOW and visual module
Magnification: ×9.8
Field of view: 6.3° (circular)
Short wavelength tracker
Field of view
Narrow: 0.5°
Wide: 4°
Thermal sensor, long wavelength tracker
Magnification: ×12 and ×4
Field of view
Narrow: 1.7 × 4.1°
Wide: 5.1 × 12.3°
Spectral band: 8-12 μm
Detector: 60-element CdHgTe, parallel scanned 240-line common module based
Reticles: NATO, stadiametric (programmable)
Symbology: range, ammunition, faults and so on (programmable)
Displays: primary for gunner, remote for commander
Power: >110 W

Contractor
Raytheon Systems Company.

Raytheon ISU day/night gunner's Integrated Sight Unit

Type
Gunner's sight.

Development
The Raytheon day/night thermal Integrated Sight Unit (ISU) was developed as part of the TOW Weapon Subsystem for use on the gunner's station of the M2/M3 Bradley IFV.

In October 1990, Raytheon Systems Company was awarded a US$77.8 million contract for the first year of a multiyear procurement contract for 1,398 TOW 2 Weapon Subsystems for the US Army. The Subsystem includes the ISU. Total value of the contract is estimated at around US$140 million with an option on another 710 subsystems that is worth approximately US$52 million. Production contracted for FY96 was 80 systems, with delivery in February 1997.

The Integrated Sight Unit for the Bradley Fighting Vehicle's TOW Weapon Subsystem

In mid-1993, Raytheon was awarded a US$1 million contract to deliver 12 eye-safe laser range-finder units for the first phase of a programme to upgrade the Bradley's ISU. Raytheon has begun deliveries of 198 Bradley Eye-safe Laser Range-Finders (BELRF) as part of the US Army's Operation Desert Storm upgrade, and the US Army plans to provide funds for a further 367 bringing the total value of the contract so far to US$11 million. The contract contains further year-by-year options for total quantities of 1,120 to 2,240 units. The Bradley's TOW weapon subsystem is not currently equipped with a laser range-finder. The BELRF will allow gunners to engage targets at the maximum range of the TOW missile and give accurate and automatic elevation commands for the 25 mm Bushmaster gun.

Description
The Integrated Sight Unit has magnifications of ×4 and ×12 with an optical relay for the vehicle commander and includes an eye-safe laser range-finder which provides range information with an accuracy of ±5 m.

The sight enables the gunner accurately to aim and fire the turret's M242 25 mm Bushmaster cannon, 7.62 mm M240C machine gun and TOW ATGWs.

Operational status
Production (over 6,000 produced). In service with Saudi Arabia and US Army (on the Bradley M2 IFV/M3 CFV).

Contractor
Raytheon Systems Company.

ANTI-ARMOUR MISSILES AND MUNITIONS

PORTABLE

Bharat Dynamics/DRDL Nag

Type
Anti-armour missile - portable.

Development
The Nag (Serpent) anti-tank missile system has been developed by the Defence Research and Development Laboratory DRDL at Hyderabad and Bharat Dynamics. Development is understood to be complete. User trials began in 1999 and production is expected in the near future.

Description
The Nag will be a third-generation, top attack, fire-and-forget missile system with a range of 4,000 m. It is intended for ground, armoured vehicle and helicopter applications. It is claimed that it will be the first missile of its type to have a complete glass fibre composite structure. Initial guidance is provided from the launcher's target acquisition system, although for terminal guidance the Nag will have either a millimetre-wave active seeker or an imaging infra-red seeker, both of which are still under development. The infra-red seeker incorporates a focal plane array sensor with a combination of charge-coupled devices and cadmium mercury telluride arrays. Both types of seeker are stated to be resistant to countermeasures.

A tandem charge warhead will be fitted. The missile is propelled by booster and sustainer motors, both of which are ignited simultaneously, using low-smoke composite-modified double-base propellant. The sustainer motor exhausts through four canted, side-mounted venturi.

Operational status
Development complete. The Indian Army expects to order Nag into production in the near future.

Contractors
Bharat Dynamics.
Indian Defence Research and Development Laboratory.

UPDATED

Dynamit Nobel Panzerfaust 3

Type
Anti-armour weapon — portable.

Description
Summary: The reusable firing device with the computerised sight of the Panzerfaust 3 weapon system extends the range of the different ammunition available for use up to 600m against stationary and moving targets.

The Panzerfaust 3 is based on the recoilless Davis Gun Principle. The warhead is placed outside the launch tube allowing interchange of the warhead in terms of weight, shape and calibre to meet different tactical requirements without modifying the launch and propulsion system.

All the different Panzerfaust 3 HEAT models with mono or tandem shaped charge are equipped with a multi-purpose warhead. By extending a spike at the front of the warhead, the weapon is ready to be fired against armour targets. The spike is retracted when using the Panzerfaust 3 against light armoured vehicles, field fortifications or reinforced masonry.

A variety of models for different tactical missions are currently available.

Panzerfaust 3-T600 in firing position 0004758

The Panzerfaust 3 weapon system with Simrad IS2000 computerised sight 0004760

Operational status
The Panzerfaust 3 is in production and in service with the German Army and five other armies since 1990. More than 300,000 units of Panzerfaust 3 have been produced and more than 50,000 live firings have successfully been carried out by army personnel. The computerised sight is available on the market.

Specifications
Calibre: launch tube 60 mm
warhead 110 mm
Length: dependent on type of ammunition
Weight: dependent on type of ammunition
firing device with computerised sight 3.6 kg
Range - moving/stationary targets: 600 m (with computerised sight)

Contractor
Dynamit Nobel GmbH, Explosivstoff- und Systemtechnik.

UPDATED

GD-OTS Dragon medium anti-armour missile system

Type
Anti-armour missile - portable.

Development
Summary: the launcher has an infra-red sight.

Initial production of the Dragon Weapon System was carried out by the Boeing Company (formerly the McDonnell Douglas Corporation). In December 1993, CMS

Dragon improved Night Tracker

Inc purchased all rights and assets associated with the Dragon system from McDonnell Douglas. Subsequently CMS was acquired by General Dynamics - Ordnance and Tactical Systems (GD-OTS). The system has been in service with the US Army and Marine Corps since 1973.

Description

The Dragon medium anti-armour missile system is a portable system capable of defeating main battle tanks, even those equipped with explosive reactive armour. All versions of the missile have Command to Line Of Sight (CLOS) wire guidance and both a day tracker and an infra-red tracker.

Dragon II includes an improved warhead which provides an 85 per cent increase in armour penetration performance. Existing inventories of the Dragon I missile can be upgraded with the Dragon II warhead.

SuperDragon, the export version, involves additional improvements, including a range increase to 2,000 m by introducing aerodynamic refinements to the missile body to reduce drag and weight, and some reorientation of the rocket thrusters. SuperDragon's average velocity is more than 174 m/s. The warhead has been altered to a tandem configuration with extendable probe and integral precursor charge. This provides the capability to defeat reactive armour. All existing Dragon missiles can be upgraded to the SuperDragon configuration.

Dragon IIT was introduced during 1995 and is a short-range missile with the SuperDragon tandem warhead developed in response to a requirement from Turkey. Because of the weight of the round and its extended standoff probe carrying the initial charge, the range is limited to 750 m.

Operational status

Dragon I is no longer in production. Dragon II is in production and in service with the US Army and Marine Corps and a number of other countries. SuperDragon is in production.

Specifications

Dragon I
Range: 65-1,000 m
Time of flight: 11.2 s-1,000 m
Length: 1.15 m
Weight
With day tracker: 14 kg
With infra-red tracker: 20.7 kg
Dragon II
Range: 65-1,000 m
Time of flight: 11.5 s-1,000 m
Length: 1.15 m
Weight
With day tracker: 15.4 kg
With infra-red tracker: 22.1 kg
SuperDragon
Range: 65-2,000 m
Time of flight: <11 s-2,000 m
Length: 1.15 m
Weight
With day tracker: 17.9 kg
With infra-red tracker: 24.6 kg

Contractor

General Dynamics - Ordnance and Tactical Systems.

Instalaza ALCOTAN-100 weapon system

Type

Anti-armour missile - portable.

Description

The ALCOTAN-100 weapon system is composed of an electronic fire-control unit, named VOSEL, and a family of ammunitions to defeat targets at medium range.

The fire-control unit provides the ALCOTAN-100 with a high single-shot hit probability. Its main components are a vision unit with a passive day/night capability; telemetry unit with a range of up to 2,000 m; transverse velocity unit to provide data for evaluating angular displacement of the target; ammunition interface through which ammunition type and temperature data are received and from where signals to activate the fuze and fire are sent; computing unit which manages information received and calculates the eventual aiming point; presentation and vision unit with ×3.5 magnification providing the user with the aiming point; and a control unit with easy-to-use operative handgrips.

To operate the ALCOTAN-100, the fire-control unit on the ammunition container-launcher is turned on and the target is tracked using the vision unit. The computing unit evaluates the firing data received and calculates the future aiming point. In less than 0.3 seconds, this information is sent to the presentation unit and displayed to the user, who aims the weapon and fires.

The ammunition unit is composed of a projectile and its Davis gun-launch motor, both located in a disposable container-launcher. There are three types of warhead: anti-tank, anti-tank/anti-personnel and anti-bunker, all of them with fuzing systems optimised for each case.

Operational status

Development complete. In production. The weapon system will enter service with the Spanish Army in 2002.

Specifications

Weapon system
Calibre: 100 mm
Length (firing configuration): 1.15 m
Weight (with VOSEL fire control unit): 15 kg
Launch velocity: 235 m/s
Effective range: 600 m anti-tank

Firing control unit
Weight: 5 kg
Magnification: ×3.5
Night vision: image intensifier
Min ambient illumination: 0.1 mlx
Range-finder: Eye-safe, class 1

Contractor

Instalaza S.A.

UPDATED

The Instalaza ALCOTAN-100 weapon system ready to fire 0101832

The VOSEL fire-control unit for the ALCOTAN-100 0101837

Israel Military Industries MAPATS anti-armour missile system

Type
Anti-armour missile - portable.

Development
Summary: the missile has laser beam-riding guidance; the launcher has a night sight.

The MAPATS anti-armour missile system exhibits elements of the Soviet 9K11 Malyutka (AT-3 Sagger) and US TOW systems, so much so that during its early stages it was codenamed 'Togger'.

Description
The MAPATS system consists of a laser beam-riding missile and a launcher. After launch, the missile follows a laser beam which is directed at the target and maintained there by the operator. An optical sight, with the laser slaved to its axis, is used for target tracking. In flight the missile detects the presence of the beam by using a rear-mounted sensor, detects any deviation from the beam axis and generates correction manoeuvres so as to stay on the line of sight. The system is immune to jamming and entirely autonomous.

The components of the MAPATS system are launch tube and missile, tripod, traverse and guidance unit, and the night vision system. On firing, the missile leaves the launch tube, the ejector motor falls away, and the laser beam modulator starts to transmit signals to the sensor in the missile. The laser beam cross-section is kept constant by a zoom system, creating a constant corridor which coincides with the line of sight. Beam intensity on the target is initially very low, increasing as the missile approaches.

Operational status
In production and in service with the Israel Defence Force and other countries.

Specifications
Calibre: 148 mm
Length: 1.45 m, missile
Weight
Missile: 18.5 kg
Tripod and battery: 14.5 kg
Traversing unit: 25 kg
Guidance unit: 21 kg
Launch tube: 5.5 kg
Night vision system: 6 kg
Missile in container: 29.5 kg
Launch velocity: 70 m/s
Flight velocity: 315 m/s
Time of flight: 19.5 s to 4,000 m
Max range: 5 km
Firing angles:
Traverse: 360°
Elevation: –20 to +30°
Warhead: shaped charge, 3.6 kg
Armour penetration: >800 mm

Contractor
Israel Military Industries.

MAPATS missile on tripod mount

Kawasaki Type 87 Chu-MAT anti-tank missile system

Type
Anti-armour missile - portable.

Development
Summary: the missile uses semi-active laser guidance.

Development of the Type 87 Chu-MAT was begun by the Japanese Defense Agency's Technical Research and Development Institute in 1976. Development was handed over to Kawasaki Heavy Industries in 1980, with a small batch of prototype missiles commencing trials during 1982. A complete prototype system was delivered in May 1985 and a second prototype in February 1986. Subsequent procurement was delayed until 1989. Production has continued at a slow rate since then (about 24 launchers each year).

In 1990, development of a new fibre optic heavy anti-tank missile known as the XATM-4 commenced. An XATM-5 light man-portable anti-tank missile is also under development.

Description
The Type 87 Chu-MAT medium-range anti-tank missile system employs a semi-active laser guidance system. The missile, powered by a single-stage solid propellant motor, is held in a container/launcher tube which is usually mounted on a low tripod served by a two-person crew from the prone position. A further crew member operates a laser target designator which may be positioned up to 200 m from the launcher, although it is possible to place the launcher on the laser designator mounting. The latter procedure can encounter problems caused by the ionised gas exhaust from the missile interfering with the laser beam. The system may be carried on a Type 73 Jeep.

The laser target designator employs a Nd:YAG laser to provide target illumination for the missile seeker head. The seeker head is manufactured by the NEC Corporation of Tokyo.

Operational status
In production and in service with the Japanese Ground Self-Defence Force. Not offered for export.

Contractor
Kawasaki Heavy Industries Ltd.

VERIFIED

KBP 9K111 Fagot anti-tank guided missile system

Type
Anti-armour missile - portable.

Development
Summary: the missile has infra-red mid-course guidance; the 9M135M3 launcher has a thermal imaging sight.

The 9K111 Fagot anti-tank guided missile system (NATO codename 'Spigot') was first shown in 1980. The Fagot SACLOS system is similar to MILAN and it rapidly replaced the older manual command to line of sight systems.

Description
Fagot was first used on a 9P135 ground launcher, after which it appeared on the BMP-1 and BMD- 1 infantry vehicles. The 9P135 launcher, when tripod-mounted, is capable of being moved in fine control in azimuth and elevation by gears.

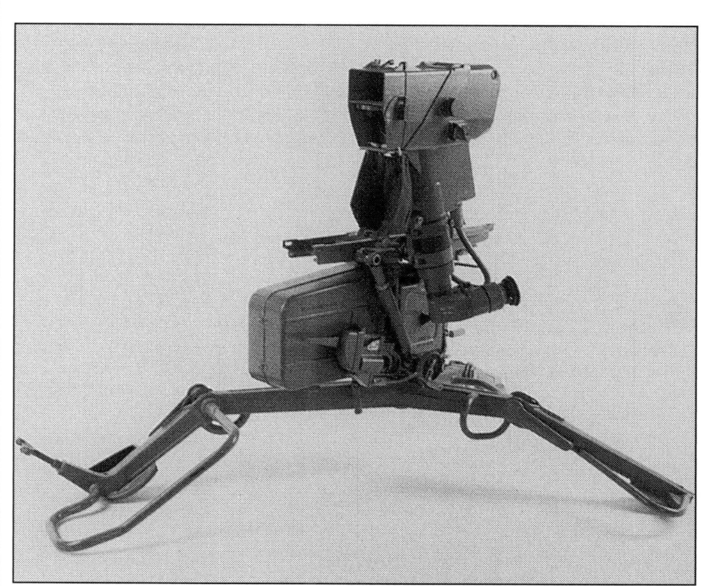

Launcher for 9K111 Fagot anti-tank guided missile system

There are three current types of missile, the 9M111-2, the 9M111M and the 9M113 Faktorlya.

The control system uses three optical channels; one is visible light and is used for aiming, the other two are infra-red and track the flare at the rear of the missile during flight, deriving the deviation from the line of sight and generating corrections. The two infra-red channels have different fields of view, one being for ranges to 1,000 m, the other for ranges above 1,000 m. The change from one channel to the other is performed automatically. This ensures that the FoV is no wider than necessary for control, reducing the possibility of infra-red jamming.

The State Institute of Applied Optics (NPO GIPO) in Kazan has developed the Mulat (1PN86-1) thermal sight for the Fagot missile system. It operates in the 8 to 13 μm waveband and weighs 9 kg with air bottle and battery. A modernised version uses the 1 PN65 thermal sight.

Operational status
Production complete in Russia. In service with Afghanistan, Algeria, Angola, Bulgaria, Croatia, Cuba, Czech Republic, Ethiopia, Finland, Hungary, India, Iraq, Kuwait, Mozambique, Poland, Polisario Front, Russian Federation and Associated States (CIS), Slovakia, Slovenia, Syria and the Federal Republic of Yugoslavia.

Specifications
Operation: SACLOS, wire-guided
Length
Missile: 1.10 m
In canister: 875 mm
Weight
Launcher: 22.5 kg
Complete round 9M111-2: 13 kg
Complete round 9M111M: 13.4 kg
Missile 9M111M: 7.3 kg
Warhead: 2.5 kg
Warhead diameter: 120 mm
Range: 9M111-2, 75-2,000 m; 9M111M, 75-2,500 m, 9M113, 70-2,500 m
Velocity: 240 m/s
Armour penetration: (armour at 90°) 9M111-2, 400 mm; 9M111M, 450 mm, 9M113, 550m

Contractor
KBP Instrument Design Bureau.

KBP 9K113 Konkurs anti-tank guided missile system

Type
Anti-armour missile - portable.

Development
Summary: the launcher can be fitted with a thermal imager.

The 9K113 Konkurs (Contest) anti-tank guided missile system entered service in 1977 and was given the NATO codename of 'Spandel'. It has been licence-produced in various countries. Although production of the 9K113 Konkurs is still continuing, it is intended that the replacement for the system will be the Kornet laser-guided missile system.

Description
In its ground-launched form, the 9M113 Konkurs missile is launched from the same 9P135M-1 launcher as the 9M111 Fagot missiles. It may also be launched from various vehicle mountings for which there is a special system using a 9P148 launcher. The operational range of the missile is from 75 to 4,000 m using Semi-Automatic Command to Line Of Sight (SACLOS) wire guidance.

The State Institute of Applied Optics (NPO GIPO) in Kazan has developed the Mulat (1PN86-1) thermal sight for the Konkurs missile system. It operates in the 8 to 13 μm waveband and weighs 9 kg with air bottle and battery. The Mulat sight extends the recognition range to 2,000 m and detection range to 3,600 m.

On target, the 9M113 Konkurs missile's 135 mm diameter warhead can penetrate 750 to 800 mm of armour set at 90°. Against armour set at 60° the figure is 400 mm.

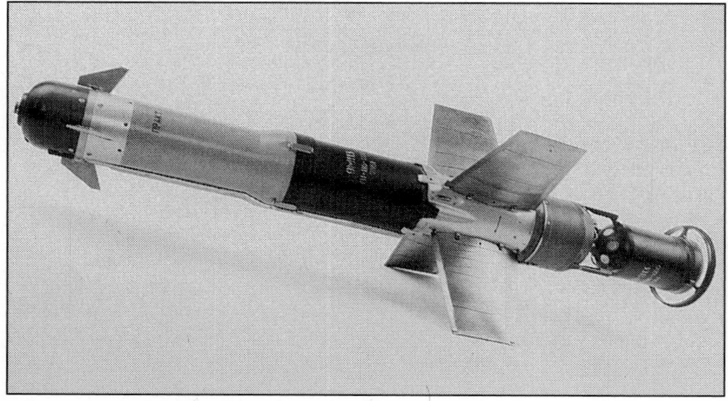

9M113 Konkurs anti-tank guided missile

Operational status
In production. In service with Afghanistan, Algeria, Bulgaria, Czech Republic, Hungary, India, Iraq, Poland, Russian Federation and Associated States (CIS), Slovakia and Syria. Offered for export sales.

Specifications
Operation: SACLOS, wire-guided
Weight:
Launcher: 22.5 kg
Complete round: 25.9 kg
Missile: 14.58 kg
Warhead: 2.7 kg
Warhead diameter: 135 mm
Length, missile: with gas generator, 1.15 m
Range: 75-4,000 m
Velocity: 270 m/s
Armour penetration: 750-800 mm armour at 90°; 400 mm armour at 60°

Contractor
KBP Instrument Design Bureau.

KBP 9K115 Metis/Metis-M anti-tank guided missile system

Type
Anti-armour missile - portable.

Development
Summary: the missile has infra-red tracking; the launcher can have a thermal imaging sight.

The 9K115 Metis is a short-range (40 to 1,000 m) system which first appeared in the early 1980s. The latest version, the 9M131 Metis-M missile has an improved range and a choice of larger diameter warheads.

Description
In principle, Metis is similar to Fagot and Konkurs but smaller and with revised Semi- Automatic Command to Line Of Sight (SACLOS) electronics and a simpler design. It is normally fired from a 9P151 tripod ground mounting but there are also reports of a 9P152 wheeled carriage. The guidance system is the 9C817, powered by a thermal battery attached to the front of the launch tube before launch.

The 9M115 Metis missile is roll stabilised in flight and the tracking flare is mounted eccentrically so that it traces a circle as the missile rolls. The infra-red tracking system derives deviation from the centre of this circle and steering sense from the movement of the flare. This is necessary because there are only two control surfaces and thus their position must be accurately known for the correct instructions to be determined and transmitted.

The Metis-M has a larger diameter warhead (130 mm as compared to 94 mm on the original Metis missile) and may have either a tandem warhead or a 4.95 kg fuel-air explosive warhead for attacking bunkers and similar targets.

The 9M151 launcher can be elevated ±5° and traversed 20° either side of a fixed arc, although full traverse through 360° is possible. A ×6 optical sight is used for target tracking. A 1PN86-VI Metis-2 thermal imaging sight developed by the State Institute of Applied Optics (NPO GIPO), Kazan can be mounted on the launcher.

Maximum range of the Metis-M is 1,500 m although the missile guidance is effective only to 1,000 m; time of flight to 1,500 m is 9 seconds. On target the 9M115 warhead can defeat up to 460 mm of armour.

9P151 launcher for 9K115 Metis anti-tank guided missile system with thermal imager fitted (T J Gander)

Operational status
In production and in service with the Russian Federation and former Soviet states. Offered for export sales.

Specifications
Operation: SACLOS, wire-guided
Warhead diameter: Metis, 94 mm; Metis-M, 130 mm
Length, complete round: Metis, 740 mm; Metis-M, 910 mm
Weight, launcher: 10 kg
Weight, complete round: Metis, 6 kg, Metis-M, 13.8 kg
Range: Metis, 40-1,000 m; Metis-M, 80-1,500 m
Velocity: Metis, 223 m/s; Metis-M, 200 m/s
Armour penetration: Metis, 460 mm; Metis-M, up to 900 mm

Contractor
KBP Instrument Design Bureau.

KBP Kitolov/Kitolov 2 mortar projectile

Type
Anti-armour weapon.

Description
Kitolov-2 (120 mm) and Kitolov-2M (122 mm) Cannon-Launched Guided Projectiles (CLGP) have been developed by KBP Instrument-making Design Bureau from their Krasnopol 152 mm CLGP, with a much smaller semi-active laser seeker and new autopilot unit. The Kitolov-2/2M CLGP is intended for use against light armoured vehicles and field fortifications although given its terminal dive approach with top attack it may also be used to engage main battle tanks.

Kitolov is inertially guided for the mid-phase but terminal guidance is by semi-active laser seeker developed by LOMO in St Petersburg. Targets are illuminated using a laser designator/range-finder; the designation range for tank-type targets is 7,000 m. The seeker in the nose of the projectile locks onto the illuminated target and the projectile guidance electronics make the necessary flight corrections to guide the projectile towards the target. Shortly before impact the projectile makes a top attack on the target's upper armour at an angle between +35 and +45°. Four nose and four tailfins are used for guidance.

The warhead is High-Explosive Fragmentation (HE-FRAG). Range of fire for the Kitolov-2 is 9 km and 12 km for the Kitolov-2M. Kitolov-2 is used in the NONA SP system and Kitolov-2M in the D30 gun and the 2S1 SP system.

Operational status
Available.

Specifications
Calibre: 120 mm (Kitolov-2) 122 mm (Kitolov-2M)
Length: 1,220 mm
Weight
  Projectile: 26.5 kg (Kitolov-2), 28 kg (Kitolov-2M)
  Warhead: 11 kg (Kitolov-2), 12 kg (Kitolov-2M)
  Explosive: 5.3 kg (Kitolov-2), 5.5 kg (Kitolov- 2M)
Range: 9 km (Kitolov-2), 12 km (Kitolov-2M)
Warhead type: HE/Frag
Laser designator range (tank type targets): 7 km
Guidance: free flight (initial phase), inertial (mid-phase), semi-active laser homing (terminal phase)

Contractor
KBP Instrument Design Bureau.

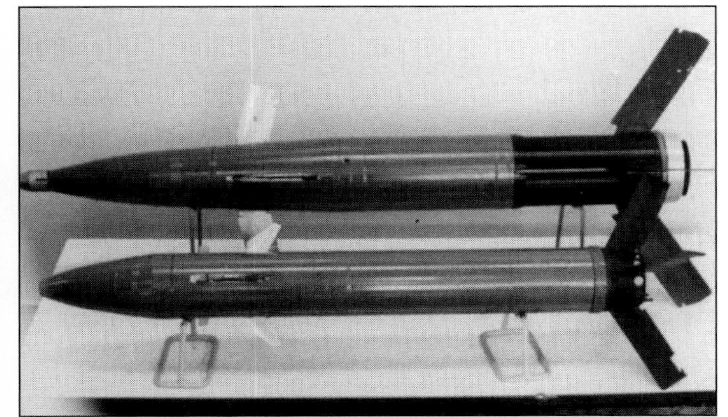

The Kitolov semi-active laser-guided mortar in the foreground, with the Krasnopol behind 0004763

KBP Kornet anti-tank guided missile system

Type
Anti-armour missile - portable.

Development
Summary: the missile has laser beam-riding guidance; the launcher has a thermal imaging sight.

First shown in the West in October 1994, the Kornet third-generation anti-tank guided missile system is intended to be the replacement for the 9K113 Konkurs system.

Description
The Kornet long-range anti-tank guided missile system uses a laser beam-riding missile with Semi-Automatic Command to Line Of Sight (SACLOS) guidance. The system also includes a launch tripod which incorporates the optical sight and guidance control equipment, and a thermal imaging sight. The thermal sight, designated 1PN79-1 is made by the State Institute of Applied Optics (NPO GIPO), Kazan. The operator tracks the target with either the optical or thermal imaging sight until launch and the missile rides the laser beam until impact, automatically staying on the line of sight. Missile velocity has not been released but is understood to be of the order of 240 m/s. Maximum range is stated to be 5.5 km in daylight and 3.5 km at night.

Kornet can be fitted with either a tandem warhead or an HE/Incendiary (thermobaric) warhead.

Operational status
In production and offered for export.

Specifications
Operation: SACLOS, laser beam-riding
Warhead diameter: 152 mm
Length: 1.20 m complete round
Weight: launcher 25 kg; complete round, 29 kg, infra-red sight 11 kg
Max range: 5.5 km (day), 3.5 km (night)
Velocity: *c* 250 m/s (not confirmed)
Rate of fire: 2-3 rds/min
Armour penetration: up to 1.20 m

Contractor
KBP Instrument Design Bureau.

KBP Krasnopol-M guided projectile

Type
Artillery projectile.

Description
The Krasnopol 152 mm and Krasnopol-M 155 mm Cannon-Launched Guided Projectiles (CLGP) have been developed by the KBP Instrument-making Design Bureau, Tula, Russia and use semi-active laser terminal guidance. Krasnopol is intended for use against armoured fighting vehicles and fortifications. It can be stowed in standard ammunition racks on SP artillery systems since it comes in two parts which are carried inside protective transport containers and are joined together just before firing. Krasnopol-M has been designed to fit standard storage racks without the need to separate the projectile into two sections.

Initial guidance is free flight. As the projectile approaches the target, the forward observer illuminates the target using a laser target designator/range-finder; the designation range for tank-type targets is 7 km. Mid-course guidance is performed by inertial navigation. In the terminal phase the semi-active laser homing seeker in the nose of the projectile locks onto the illuminated target and the projectile guidance electronics make the necessary flight corrections to guide the projectile towards the target. The laser seeker has been developed by LOMO of St Petersburg. Krasnopol-M uses the smaller laser seeker and guidance unit developed for the 120 mm Kitolov-2. Shortly before impact the projectile makes a top attack on the target's upper armour, at an angle between +35 and +45°. The nose-mounted seeker has a footprint of about 1,000 m. Nose fins are used for guidance with a tailfin assembly providing flight stabilisation.

The warhead is High-Explosive Fragmentation (HE-FRAG). Range of the 152 mm Krasnopol CLGP and the Krasnopol-M is 20 km. Firing systems include M109, G6 and 2S19 SP howitzers.

Operational status
In production. In service with the Russian Federation and other former Soviet nations. Krasnopol-M is ready for production.

Specifications
Krasnopol
Calibre: 152 to 155 mm
Weight
  Projectile: 45 kg
  Warhead: 19 kg
Length: 1.3 m
Range: 20 km

The Krasnopol CLGP

Laser designator range: (tank type targets) 7 km
Rate of fire: 4-5 rds/min
First shot kill probability: 0.9
Max target speed: 36 km/h
Max height above sea level: 3,000 m
Warhead type: HE/Frag
Krasnopol-M
Calibre: 155 mm
Weight
Projectile: 45 kg
Warhead: 19 kg
Explosive: 6 kg
Length: 960 mm
Range: 20 km
Warhead type: HE/Frag
Max velocity: 700 m/s

Contractor
KBP Instrument Design Bureau.

Lockheed Martin, Diehl PGMM Precision Guided Mortar Munition

Type
Mortar munition.

Development
The US Army's Armament Research and Development and Engineering Center (ARDEC) has chosen a team of Lockheed Martin Electronics and Missiles and Diehl of Germany to build an advanced technology demonstrator as Phase II of its 120 mm precision-guided mortar programme, part of the Rapid Force Projection Initiative. Seven protoype 120 mm rounds are to be built. The PGMM is based on Diehl's Bussard projectile.

Description
The PGMM has an effective range of up to 15 km and is intended for targets requiring a precision attack, such as bunkers and other high value targets. Guidance will be by a man-in-the-loop semi-active laser seeker. The imaging infra-red seeker developed by Lockheed Martin incorporates a cooled 256 × 256 Cadmium Mercury Telluride focal plane array. It has a 500 m square footprint which may be increased to 1 km square.

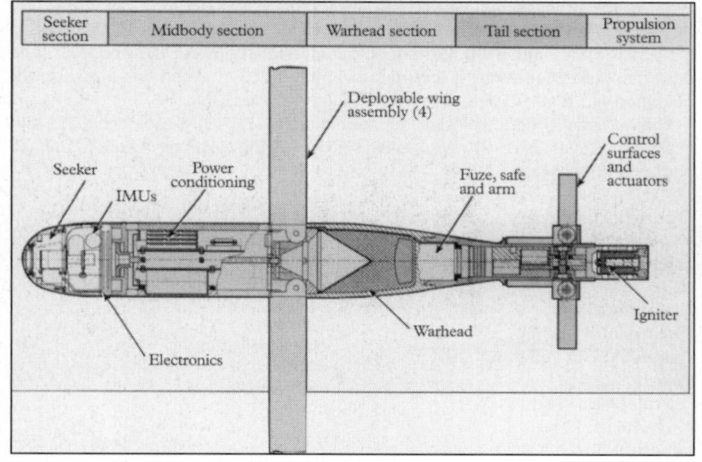

Diagram of the PGMM showing the dual-mode seeker in the nose 0004846

After launch the mortar deploys wings and four stabilising tail fins. At apogee, four wings unfold giving a controlled glide path to the target.

PGMM has a blast fragmentation warhead. PGMM can be either ground or vehicle launched.

Operational status
Under development.

Specifications
Weight: 17.2 kg
Length: 965 mm
Calibre: 120 mm
Guidance: dual-mode imaging infra-red and semi-active laser
Range: 500 - 1,500 m

Contractors
Diehl GmbH.
Lockheed Martin Electronics and Missiles.

MBDA Eryx short-range anti-tank missile

Type
Anti-armour missile – portable.

Development
Summary: the missile has an infra-red beacon and the launcher has the Mirabel infra-red sight.

Eryx was designed for use by forward infantry, providing them with a weapon effective against new types of armour at ranges of up to 600 m. Mass production began in 1991. In 1989, it was announced that a Memorandum of Understanding had been signed by Canada and France with the intention of producing Eryx as a co-operative venture.

Eryx on daylight firing (MBDA/2002) ***NEW***/0547553

Eryx anti-tank missile being launched using MIRABEL thermal sight

The Eryx anti-tank missile can be used within confined spaces (MBDA/2002) ***NEW***/0547552

To date, the French Army has ordered 400 Eryx firing posts and 4,700 missiles. In February 1996, Canada ordered a further 1,600 missiles bringing the total to 425 firing posts and 6,100 missiles. Orders for Norway stand at 424 firing posts and 7,200 missiles. The Brazilian Army has ordered Eryx, as has the Malaysian Army.

Description
The Eryx missile is wire guided with a 135 mm shaped warhead capable of penetrating 900 mm of armour. The missile in the launch tube clips onto a shoulder-mounted firing post which contains the ignition, detection and timing systems, together with a remote control. The missile can be emplaced and made ready to fire in less than 5 seconds. During flight (4.2 seconds to 600 m) the user is required only to keep his sight on the target, guidance being automatic. The missile carries an infra-red beacon which is detected by the sight unit, corrections derived, and steering commands sent to the missile by means of a wire link which is unspooled as it flies. The application of a new concept, the direct thrust flight control which is efficient even at low speed, allows the launching to be achieved using a small propulsion unit. The missile can thus be used in confined spaces. After launch the rocket motor accelerates to the flight speed of 300 m/s.

A Thales Optronics MIRABEL thermal imager can be added to the Eryx launcher for firing at night or under poor visibility conditions.

Operational status
In production and in service with Brazilian, Canadian, Chadian, French, Malaysian and Norwegian armies. Over 20,000 missiles produced and on order.

Specifications
Diameter: 160 mm
Length: 925 mm
Weight
Missile: 13 kg
Firing post: 4.5 kg
Range: 50-600 m
Time of flight: 4.2 s to 600 m
Max velocity: 300 m/s
Armour penetration: 900 mm

Contractor
MBDA.

UPDATED

MBDA MILAN 1/2/3

Type
Anti-armour missile - portable.

Development
The Missile d'Infanterie Leger Anti-char (MILAN) was introduced by Euromissile (now part of MBDA), an international consortium of the French EADS/Aerospatiale Matra Missiles, Missile Division, and the German EADS/Lenkflugkorpersysteme GmbH. MILAN has been in production since 1973. Well over 332,000 missiles and 10,000 firing posts have been sold to 43 countries.

MILAN was designed to be used by infantry and is a SACLOS wire-controlled missile. Flares on the missile emit an infra-red signature enabling the system computer to measure the error between its position and the line of sight. Missile velocity is twice that of most early portable missiles, allowing the MILAN missile to reach 1,500 m in 10 seconds and 2,000 m in 12.5 seconds. In the Milan 3 version a flashlamp emits infra-red pulses instead of the signature of the flares.

Description
The complete MILAN weapon system is made up of two units: a round of ammunition, consisting of a missile, factory loaded into a sealed launcher/container tube; and a combined launching and guidance unit, comprising a launcher combined with a periscopic optical sight and an infra-red tracking and guidance system. The whole is mounted on a tripod. The round of ammunition comprises an assembled missile, factory loaded, with wings folded, into a sealed tube which serves the dual purpose of storage/transport container and launching tube. The container/launcher tube is fitted with mechanical and electrical quick-connection fittings and a self-activating battery is mounted on the outside, providing electrical power for the firing installation. MILAN has a night-firing capability through the addition of the MIRA thermal imaging device adopted by the French, German, UK and other armies, or the later MILIS infra-red thermal device. MIRA consists of a case weighing 9 kg, which can be mounted on the standard firing post. Target detection is possible at a range of over 4,000 m and firing at 2,000 m. Euromissile offer a new thermal imager known as MILIS. Having better performance and the same interface as MIRA, MILIS weighs less than 7 kg and is fitted with an integrated Stirling microcooler.

MILAN 1

The MILAN 1 missile is an assembly of the following main components: an ogival head containing a shaped charge and fuze; a two-stage solid propellant motor discharging through an exhaust tube to a central nozzle located at the rear of the missile; and a rear part containing the jet spoiler control system and guidance components. The guidance components include: a gas-driven, turbine-operated gyro; an infra-red flare or flashlamp; a spool carrying the two guidance wires in one cable; a decoder unit; and a self-activating battery for internal power supply. The missile is launched from its tube by a booster charge gas generator which is contained in the tube and burns for 45 ms. Initial velocity is 75 m/s. The recoil effect is compensated but part of it is used to eject the tube to the rear of the gunner to a distance of 2 to 3 m. The two-stage propulsion motor burns for 12.5 seconds and increases the velocity of the missile, at first rapidly, then more slowly to 210 m/s. The operator must keep his sight cross-hairs on the target throughout the engagement. Guidance is achieved by means of a single jet spoiler operating in the sustainer motor exhaust jet. The jet spoiler operates on guidance command signals generated automatically by the launcher/sight unit (by measurement of the angular departure of the missile from the reference directions of the infra-red tracker in the sight unit) and transmitted to the missile via the guidance wires which unwind from the missile. The guidance commands are decoded by a transistorised decoder unit within the missile. The self-activating battery which provides internal power is designed for long-term storage and use in worldwide temperature conditions. For safety the missile is locked inside its tube and the solid propellant gas generator cannot be ignited until the missile is unlocked by the gunner. In addition the sustainer motor ignites when the missile is released from the tube and the wings have unfolded. The fuze cannot arm until the sustainer motor is ignited and an electrical safety device functions when the missile has flown approximately 20 m. MILAN 1 production is complete.

MILAN 2 with MIRA night sight

MILAN 2
Introduced in 1984, MILAN 2 has a warhead assembly intended to improve performance against the frontal arc of main battle tanks. The improvement was calculated to be some 65 per cent and penetrations into steel armour of 1.06 m have been recorded. The MILAN 2 missile does not affect the interface with the firing post, the overall missile weight remains the same and the tubular container is unchanged. Existing firing posts can therefore fire both MILAN 1 and MILAN 2 missiles.

MILAN 2T/3
Introduced in 1993, MILAN 2T/3 is an improved missile using a tandem warhead and an extended nose probe. The probe contains the 30 mm diameter precursor shaped charge which is designed to detonate reactive armour protection and thus clear the line of attack for the 117 mm diameter main shaped charge. No modification is required to existing launchers.

MILAN 3
MILAN combat efficiency has been enhanced by a newly developed tracking system in combination with the T2 warhead (MILAN 2T/MILAN 3 missile) providing the weapon system with the following features:
- Warhead with improved penetration capability, even against reactive armours.
- Tracking system fully resistant to sources of visible radiation (jamming by pyrotechnic radiation sources, fire, lights and so on) and compensation for a short-time severance of the IR link (due to smoke or natural obstacles such as trees and houses).

Included in MILAN 3, these improvements allow for an extension of the previous MILAN mission profile.

The improved MILAN 3 system consists primarily of an improved "new generation" localiser and a MILAN 3 missile featuring a new warhead and flash lamp. After the missile is launched, the light emitted by the flash lamp is picked up by the new localiser's CCD sensor. The various image data (flash pulses and the entire background and jamming environment) are read out from the CCD and passed on to an alignment computer for further processing. The computer determines the missile's metric displacement from the line of sight, taking into account various criteria relating to optimal jamming resistance.
MILAN 3 jamming resistance is based on the following techniques:
- Differentiation of pictures
- Windowing (dynamic FOV narrowing)
- Thresholding
- Algorithms of comparison (object evaluation)
- Prediction Algorithms

The techniques described give MILAN 3 a considerably enhanced jamming resistance. Despite the new improvements, the system remains compatible with earlier versions with the following features:
- The MILAN 3 firing post is capable of firing previous types of ammunition.
- 2nd and 3rd level maintenance test equipment modified to MILAN 3 standards may be used on both MILAN and MILAN 3 firing posts.
- The MIRA night sight as well as any other MILAN peripheral equipment may be used on MILAN 3 without any modification.

Operational status
Available. In service with 43 countries.

Specifications

Type	M1	M2	M2T/3	M3
Munition in Tactical Pack				
Weight in carrying mode:	12.23 kg	12.23 kg	12.62 kg	12.62 kg
Weight in firing mode:	11.52 kg	11.52 kg	11.91 kg	11.91 kg
Length in carrying mode:	1.26 m	1.26 m	1.26 m	1.26 m
Length in firing mode:	1.2 m	1.2 m	1.2 m	1.2 m
Diameter:	133 mm	133 mm	133 mm	133 mm
Missile				
Weight:	6.73 kg	6.73 kg	7.12 kg	7.12 kg
Length:	769 mm	918 mm	918 mm	918 mm
Diameter, wings folded:	125 mm	125 mm	125 mm	125 mm
Wing span:	267 mm	267 mm	267 mm	267 mm
Warhead				
Weight:	2.67 kg	2.7 kg	3.12 kg	3.12 kg
Diameter:	103 mm	115 mm	117 mm	117 mm
Explosive filling:	1.36 kg	1.79 kg	1.83 kg	1.83 kg
Cone diameter:	101 mm	112.9 mm	112.9 mm	112.9 mm

Velocity: at launch, 75 m/s; at 2,000 m, 210 m/s
Time of flight (to max range): 12.5 s
Range (effective): 25-2,000 m
Chance of a hit: 0-250 m, average 75%; 250-2,000 m, greater than 98% (manufacturer's figures)

Contractors
MBDA.

Mectron MSS-1.2 anti-tank missile system

Type
Anti-armour missile - portable.

Development
Summary: the missile has laser beam-riding guidance; the launcher has a thermal imaging sight.

In the late 1980s, MAF a long-range anti-tank weapon system of advanced SACLOS type, was developed by Otobreda, in co-operation with Orbita of Brazil, in order to meet a requirement of the Brazilian Army. A number of systems were delivered with further production to be undertaken by Orbita. Development and production has since been taken over by another Brazilian company, Mectron, who have also taken over the MAA-1 missile programme for the Brazilian Air Force.

Mectron has been awarded a contract for the development of the MSS-1.2 laser-guided anti-tank missile which is based on the earlier MAF system. Initial deliveries of 40 missiles commenced during 1999, 10 of which were to be fitted with high-explosive warheads, the rest with telemetry. By 2001 the MAF/Mectron MSS 1.2 system was undergoing trials with the Brazilian Army. The MAF missile electronics will also be digitised. The total Brazilian Army requirement could be 4,000 missiles and 600 firing posts.

Description
The MAF missile system consists of a missile in its container launcher and a firing post. Guidance is by laser beam-riding, the missile being subsonic. The firing post consists of a tripod, projector unit and sights.

The guidance beam is obtained from a modulated laser operating in the near infra-red spectrum. The beam is generated in the launcher unit and the receiver is in the rear of the missile body. Missile guidance is by a servo loop which corrects the missile trajectory as a function of the deviations from the line of sight. The sighting system has both an optical sight and a thermal imaging camera.

The missile configuration is of the canard type (roll-free) with a conical nose and tail. Folding stabilisers are mounted at the rear. The warhead is a shaped charge. The maximum range is in excess of 3,000 m. At night this reduces to 2,000 m.

Operational status
Advanced development. MSS-1.2 deliveries for the Brazilian Army contract were completed in 1999.

Specifications
MAF missile
Diameter: 130 mm
Length: 1.38 m
Launcher container: 1.412 m
Missile weight at launch: 14.5 kg
Firing post weight:
Max range: 3 km fine weather, 2000 m all weather
Min range: 70 m
Max speed: 290 m/s
Time of flight: 16 s-3 km

Contractor
Mectron Engenharia Industria e Comercio.

Norinco Red Arrow 8 guided weapon system

Type
Anti-armour missile - portable.

Development
Summary: the launcher has an infra-red tracker for flares in the missile tail.

Red Arrow 8 is a second-generation guided missile system intended for use by infantry against tanks and other armoured targets with a range of 100 to 3,000 m.

Description
It is a crew-portable weapon, fired from a ground tripod mount; it can also be configured for mounting in or on a variety of wheeled and tracked vehicles.

The system uses a tube-launched, optically tracked, wire-guided missile controlled by a Semi-Automatic Command to Line Of Sight (SACLOS) system based on an infra-red flare in the tail of the missile. The position of the flare is detected by the sight unit and corrections are automatically generated and signalled down the wire so as to fly the missile into the axis of the line of sight.

In general appearance the system is similar to MILAN, having a sight unit on to which the missile transport and launch tube is attached before firing. The tripod resembles that used with the TOW system.

Operational status
In production and in service with the People's Liberation Army.

Specifications
Operation: SACLOS, wire-guided
Warhead diameter: 120 mm
Missile length: 875 mm
Missile weight: 11.2 kg at launch

Red Arrow 8 launch unit showing sights

Wingspan: 320 mm
Launch tube diameter: 255 mm
Launch tube length: 1.566 m
Launch tube weight, with missile: 24.5 kg carry; 22.5 kg firing
Sight unit weight: 24 kg
Tripod weight: 23 kg
Effective range: 100-3,000 m
Armour penetration: >800-900 mm
Velocity: 70 m/s initial; 200-200 m/s max

Contractor
China North Industries Corporation (Norinco).

Raytheon BGM-71 TOW anti-tank weapon system

Type
Anti-armour missile - portable.

Development
Summary: the launcher has an infra-red tracker and the TOW 2 launcher has an infra-red sight.

TOW is a crew-portable, vehicle-mounted or helicopter-launched, heavy anti-tank weapon system. Design work started in 1962, the first firings were carried out in 1963 and the weapon entered service with the US Army in 1970. Since then it has been proved in action and has been sold to more than 40 countries.

TOW, produced by the Raytheon Systems Company, is recognised as the most successful and widely used anti-tank missile system in the world. According to US Army statistics, the missile has a cumulative reliability record of over 93 per cent in over 12,000 test and training firings conducted since 1970.

In order to defeat tanks fitted with explosive reactive armour blocks, the US Army started an upgrade programme for TOW 2 in December 1984 and the BGM-71E TOW-2A variant entered service in 1987.

In 1991, the BGM-71F TOW 2B ATGW entered service.

A development contract for the US Army's successor for the TOW missile system, FOTT (Follow On To TOW) was expected to be placed in 1998. TOW 2A and 2B will form the bulk of the US Army inventory of heavy anti-tank missiles to 2007.

Description
The TOW missile is a tube-launched, optically tracked, wire command link guided missile. The weapon system includes tripod, traversing unit, launch tube, optical sight and missile guidance set.

The optical sight is used to track the target and to detect the infra-red signal from the missile in flight. The sight contains a ×13 telescope, boresighted to an infra-red tracker, with a field of view of 4°.

The latest version of the TOW missile is the BGM-71F. This consists of the launch motor, the flight motor, four control surfaces, the infra-red source, two wire dispensers, a battery, a digital electronics unit, a gyro, safety and arming devices and top attack, dual EFP warheads.

The two-wire command link is dispensed from two spools at the back of the missile. These wires carry steering commands from the launcher to the missile, and they are applied, together with signals from the gyro, to the four control surfaces. The infra-red source(s) provides a beacon which is detected by the infra-red tracker in the sight to determine the missile's position. Multiple sources preclude countermeasures and allow tracking through fog and other obscurants. The infra-red lamp current is modulated to allow discrimination against a background of other strong infra-red emitters, such as the sun.

TOW 2 missile launcher

The TOW 2 launcher is a modified version of the original launcher, with the addition of the infra-red AN/TAS-4 sight to track targets and function as a totally independent fire-control sensor. All types of TOW missile - basic TOW, ITOW, TOW 2, TOW 2A, TOW 2B or any future TOW missile contemplated by the US Army - can be fired from this modified launcher.

When the target is in range the trigger is pressed. This activates the missile batteries and gas is released to spin the gyro up to speed. Approximately 1.5 seconds later the launch motor is fired. This burns out completely within the tube but the missile acquires sufficient momentum to coast until the flight motor ignites. The missile wings and control surfaces are extended from the missile body. The infra-red sources on the missile start to operate and the two command-link wires are dispensed from the internal spools. The first stage of arming the warhead occurs. The flight motor is activated at the end of the 12 m coasting period and the warhead is fully armed after about 60 m.

The infra-red sensor tracks the signal from the modulated lamp in the missile and detects any deviations from the line of sight path to the target. It provides continuous information over the wire link and to the missile guidance set which produces signals which are delivered, with those from the gyro, to the control surface to correct the flight path and keep the missile on the line of sight.

TOW 2B (BGM-71F) is designed for top attack. It features a dual-mode sensor and a new armament section with two warheads substantially different from those used in other TOW versions which form two explosively formed projectiles with substantial pyrophoric effects after penetration. The missile is programmed to fly over the target and the dual-mode laser ranger/magnetic sensor system triggers the two warheads to shoot downwards at the proper instant.

Operational status
BGM-71E and BGM-71F are in production. More than 600,000 TOW missiles have been produced for the US Army, Marine Corps and 43 countries.

Specifications
Designation: BGM-71A (basic TOW); BGM-71C (ITOW); BGM-71D (TOW 2); BGM-71E (TOW 2A); BGM-71F (TOW 2B)
Guidance principle: automatic missile tracking and command to line of sight guidance from optical target tracker (SACLOS)
Guidance method: wire command link controlling aerodynamic surfaces
Propulsion: 2-stage, solid propellant motor
Warhead: HEAT
Missile velocity: 200 m/s
Max range: 3,750 m
Min range: 65 m
Crew: 4

	Length	Width	Height	Weight
Launcher, tubular, guided missile, M151E2	2.21 m	1.143 m	1.118 m	78.5 kg
Launcher, with AN/TAS- 4				87.5 kg
Launcher, with TOW 2 Mods				93 kg
Tripod, retracted	1,064 mm	645 mm	569 mm	9.5 kg
Basic TOW, BGM- 71A	1,174 mm	221 mm	221 mm	22.5 kg
ITOW, BGM- 71C	1,174 mm	221 mm	221 mm	25.7 kg
TOW 2, BGM- 71D	1,174 mm	221 mm	221 mm	28.1 kg
TOW 2A, BGM- 71E	1,174 mm	221 mm	221 mm	28.1 kg
TOW 2B, BGM- 71F	1,168 mm	221 mm	221 mm	28.1 kg

Contractor
Raytheon Systems Company.

Raytheon, ENOSA LightWeight Launcher (LWL) for TOW missiles

Type
Anti-armour missile - portable.

Description
The LWL is a lightweight launcher for the TOW anti-tank missile which has been developed and produced in co-operation with Raytheon Systems Company, ENOSA and the Spanish Ministry of Defence. A joint venture company, Gyconsa has been formed between Raytheon and Indra Group, of which ENOSA is part. As well as missile guidance and control, LWL is also capable of surveillance, detection and engagement of targets. It incorporates a second-generation 8 to 12 μm thermal imager with one eyepiece for visible and thermal images. The thermal imager is based on a 240 × 4 cadmium mercury telluride focal plane array. Missile tracking is in the near and long-wave infra-red wavebands. Boresighting is automatic and integrated. Automatic tracking and a laser range-finder are optional.

Operational status
In production. In service with the Spanish Army. Evaluated by the Austrian Army.

Specifications
Recognition range: >6 km in any battlefield conditions
Fields of view
Day sight: 6.0° (×10)
Thermal camera
Narrow: 2.2 × 1.37° (×15.3)
Wide: 6.6 × 4.12° (×5.1)
Entrance pupil diameter
Day sight: 80 mm
Thermal camera: 110 mm
Infra-red detector: CMT, 240 × 4 element, 8-12 μm
Cooling: split cycle Stirling 0.4 W
Total weight: 73.3 kg (incl tripod and traversing unit)

Contractors
Empresa Nacional de Optica SA (ENOSA).
Raytheon Systems Company.

The LWL Lightweight TOW Launcher, produced by ENOSA of Spain and Raytheon of the USA

Raytheon, Indra Group MACAM

Type
Anti-armour missile - portable.

Development
Summary: the missile has an imaging infra-red seeker and the launcher has an infra-red sight.

MACAM 3 (third-generation) is being developed with Gyconsa as prime contractor under contract to the Spanish Ministry of Defence. Gyconsa is a joint venture between Raytheon Systems Company of the USA and the Indra Group of Spain.

Feasibility studies commenced during 1992, based on a previous demonstration and validation test of the fibre optic missile concept conducted by Hughes (now Raytheon) in 1988. Definition studies extended until mid-1994, with the design and development phase commencing at the beginning of 1995. This phase extended until the end of 1999.

The MACAM anti-tank weapon system in operation

Description
MACAM is a portable shoulder-launched anti-armour system with an integral day/night sight. The missile has an imaging infra-red seeker and a fibre optic datalink for fire-and-forget guidance.

The eyepiece provides a visible light image from the day sight, infra-red imagery from the night sight, or missile seeker imagery, all as selected by the operator. The day/night sight provides two selectable fields of view. The night sight and the missile seeker operate in the medium-wave infra-red band (3 to 5 μm).

The electronics receive the missile seeker video information through the fibre optic datalink and operator commands from the aiming and control unit. Functions performed by the launcher electronics include system executive and system modes control, target tracking and the generation of missile guidance commands.

The MACAM missile is issued as a sealed round containing the missile, plus a replaceable cryogenic gas bottle to cool down and operate the missile's infra-red detector before launch. If the missile is activated but not fired the external gas bottle can be changed in the field. The launch container is compatible with the TOW launch tube.

The missile itself contains a seeker module, the missile electronics, an armament module, a flight motor, a control module, a launch motor and a fibre optic dispenser.

The seeker module is a sealed unit containing the gimbal-mounted infra-red sensor based on a staring focal planar-array detector. Cooling is provided by an external gas bottle before launch and from an internal bottle during flight. The detector assembly is mounted on a two-axis rate-stabilised platform.

Video processing, target tracking and missile guidance are performed by the launcher electronics via the fibre optic datalink so only onboard tasks, such as the internal mode control and timing, seeker video electrical-to-optical conversion and autopilot loop closure are carried out by the missile electronics.

The warhead is a tandem configuration with a small precursor charge to initiate reactive armour prior to the detonation of the main charge. The charges are canted downwards so that they do not need to defeat the nose-mounted seeker module prior to encountering the main target.

The fibre optic dispenser pays out the optical fibre during flight and will support missile ranges up to 5,000 m, although the maximum effective combat range is of the order of 2,500 m.

There are two launch modes: Lock On Before Launch (LOBL), and Lock On After Launch (LOAL). LOBL is the primary mode.

For LOBL the operator uses the day/night sight to carry out surveillance and target search. Using the selectable fields of view the operator can detect and then recognise a target. The operator activates the missile once a target has been selected and the imagery in the sight automatically switches to missile seeker video in approximately 10 seconds. Using a two-dimensional thumbswitch the operator positions a tracker gate over the desired target and commands track.

At this time the operator can select the trajectory required. The elevated trajectory is the normal choice against armoured vehicles as it provides a top attack terminal trajectory. Also available is a direct trajectory for use against hovering helicopters or bunkers.

Once lock on is achieved, the operator can fire the missile which will be automatically guided towards the target on a fire-and-forget basis. The sight will continue to display the live video to the operator so the operator can intervene to either improve the aimpoint, select an alternative target not previously seen, re-establish control if a lock is broken for various reasons, or abort an attack entirely.

During the elevated trajectory the missile climbs to a mid-course elevation of about 2,000 m and levels off. The missile follows a proportional navigation law in the horizontal plane and a fixed altitude in the vertical plane. As the missile approaches the target, the look-down angle continues to increase. When this angle reaches a defined limit the terminal trajectory commences with proportional navigation utilised in both planes. The system automatically stores the last video images to allow the operator to conduct a real-time battle damage assessment.

The secondary LOAL mode is used when the operator knows where a target is but for various reasons cannot, or chooses not to, initiate automatic tracking before launch. Typical LOAL situations thus include the operator being able to see the target in the launcher sight but not in the seeker video, the operator may not wish to risk exposure, or the operator knows where the target is but cannot establish a line of sight because of obstacles. In these cases the operator selects the LOAL mode and fires the missile in an elevated trajectory. Once at the mid-course altitude the seeker head is depressed to provide an optimum search footprint on the ground. The operator watches the live seeker video ready for when a target or target area is recognised. The track gate is then positioned over the desired point and track is commanded. From this point onward the operation proceeds as with the LOBL mode.

Operational status
Advanced development. Planned in-service date for the Spanish Army and Marines was 2002 but the present status remains uncertain. The requirement for the Spanish government has been stated to be 10,000 to 14,000 missiles, although a proportion of this total would be intended for export sales.

Specifications
Provisional calibre: missile, <147 mm; launch tube, 174 mm
Length: missile, 1.05 m; launch tube, 1.2 m
Weight: system, <25 kg; missile, 13.8 kg; launcher, <9 kg
Max range: 5,000 m
Effective combat range: 150-2,500 m

Contractors
Gyconsa - a joint venture of:
Indra Group.
Raytheon Systems Company.

Raytheon, Lockheed Martin Javelin portable anti-tank missile system

Type
Anti-armour missile - portable.

Development
Summary: the missile has an imaging infra-red seeker; the launcher has a thermal imaging sight.

The Javelin portable anti-tank missile system (formerly the Advanced Anti-tank Weapon System - Medium (AAWS-M) is being produced for the US Army Missile Command/US Marine Corps Systems Command. Development and production are being performed by the Raytheon (formerly Texas Instruments)/Lockheed Martin Javelin joint venture. The joint venture comprises Lockheed Martin Electronics and Missiles, Orlando, Florida and Raytheon Systems Company, Texas.

In February 1989, the US Army selected the Raytheon/Lockheed Martin (formerly Martin Marietta) Javelin joint venture to develop the AAWS-M portable medium-range anti-tank system. A three year full-scale development contract was signed in June 1989. A restructured engineering and manufacturing development programme was completed in early 1994 with an operational and technical test programme of 148 successful engagements out of 165.

In 1994, the army approved a joint army/joint venture cost-reduction programme, projected to lower programme costs considerably during its 11 years of production.

Lockheed Martin will produce the missile rounds and Raytheon the Command Launch Units (CLUs) for qualification under the engineering and manufacturing development contract.

The first army units were equipped with the system in June 1996. The unit received 36 missiles and nine CLUs. Over 30,000 missiles and 3,000 CLUs will be produced for the US Army and Marine Corps over 10 years and production for international customers is projected at an additional 40,000 units. A three year full-rate production contract valued at US$745 million was awarded in 1997. Total programme value is estimated at US$4 to 5 billion.

In August 2000, the US Army awarded the Raytheon-Lockheed Martin Joint Venture a US$1.236 billion multi-year contract for Javelin. The contract covered a four-year period and included 11,805 missiles, 2,968 CLUs, 1,990 classroom and field tactical trainers, and other associated equipment. A contract modification in March 2003 added another 378 CLUs plus associated equipment. The US Department of Defense has approved the Javelin system for foreign military sales in 16 countries. In January 2002, Lithuania became the first export customer, with Jordan following later that month. Taiwan ordered Javelin in July 2002 and Ireland in November 2002. The UK announced its selection of Javelin for the British Army and Royal Marines in January 2003, while Australia is expected to confirm a Javelin order in the near future.

Description
The Javelin is a shoulder-launched, portable anti-tank missile system designed to replace the Dragon. Javelin has a range over twice that of the Dragon and is effective against current and projected armour threats in degraded battlefield environments.

The Javelin fire and forget missile in the hands of a US Army gunner (Lockheed Martin) 0053717

Javelin weapon system being fired 0004724

The system includes the missile with a fire-and-forget, imaging infra-red seeker which incorporates a 64 × 64 cadmium mercury telluride long-wave infra-red focal plane array, produced by Raytheon. The command launch unit consists of a day sight for use in clear conditions and an imaging-infra-red sight for use at night or in reduced visibility.

The gunner uses the CLU day or dual-field of view thermal imaging sight display for surveillance and target acquisition, locks on to the target using the missile seeker video displayed through the CLU, then fires the missile. The missile's infra-red seeker guides it to the target, leaving the gunner free to seek cover, move to a new position, assess battlefield damage or reload and fire again.

The minimum-smoke tandem launch/flight motor design allows safe fire from enclosed areas and provides a low launch signature. The dual shaped charge warhead provides required lethality against modern armour.

Operational status
Full-rate production. On order for or in-service with Ireland, Jordan, Lithuania, Taiwan, the UK and the US Army and US Marine Corps.

Specifications
Total weight: 22.3 kg
Missile
Seeker: imaging-infra-red
Spectral band: 8-12 μm
Detector: 64 × 64 CMT staring focal plane array
Guidance: lock on before launch, automatic self-guiding
Weight: 11.8 kg
Length: 1,081 mm
Diameter: 127 mm
Range: 65 m min, 2,500 m max
Warhead: tandem-shaped charge
Propulsion: 2-stage solid propellant
Command Launch Unit
Magnification: ×4 (day sight), ×4 and ×9 (thermal sight)
Weight: 6.4 kg

Contractors
Lockheed Martin Electronics & Missiles.
Raytheon Systems Company.

UPDATED

Saab Bofors Dynamics RBS56 BILL 1

Type
Anti-armour missile – portable.

Development
The RBS 56 BILL medium-range anti-tank system was developed by Bofors under a contract from FMV (the Swedish Defence Materiel Administration) in July 1979, with the first production contract being awarded in 1985.

The RBS56 BILL 1 is a portable, medium-range, tube-launched, wire-guided ATGW, which is capable of being fired from either ground or vehicle mounts. It integrates fully over-fly, top attack technology with an elevated flight, a 30° canted warhead and an interactive dual-purpose sensor system.

The Swedish FMV has modified its contract for BILL and remaining deliveries will now be the BILL 2 missile.

Description
BILL is a man-portable, top-attack, medium-range, anti-tank missile system designed to overcome the technological advances made in special armour protection and to have the ability to combat any known or projected armour. The system consists of a day sight, an optional clip-on thermal imaging sight made by Saab, a tripod and the missile in its launch tube. The BILL system uses an advanced SACLOS technique with the wire-guided missile using coded signals (from a laser beam at the rear of the missile) between the missile and sight unit for guidance, making the system immune to jamming. The system was designed for the missile to fly 0.75 m above the line of sight. The BILL guidance system brings the missile under control immediately after launch, giving an extremely high hit probability, both at short ranges and against fast-moving targets.

The missile warhead is canted 30° to the horizontal, incorporating a warhead ignition system to detect the target and initiate the shaped charge at the right moment to give maximum effect. There is also an impact fuze for igniting the warhead in the event of a direct hit.

Located at the rear of the launch tube is a gas generator to propel the missile at a velocity of 72 m/s before the sustainer motor accelerates the missile to 250 m/s. The sustainer motor burns for approximately 2 seconds, or 400 m downrange.

Bofors RBS56 BILL 1 ATGW system in its standard tripod-mounted infantry version and fitted with the BILL Night Sight (BNS) 0023441

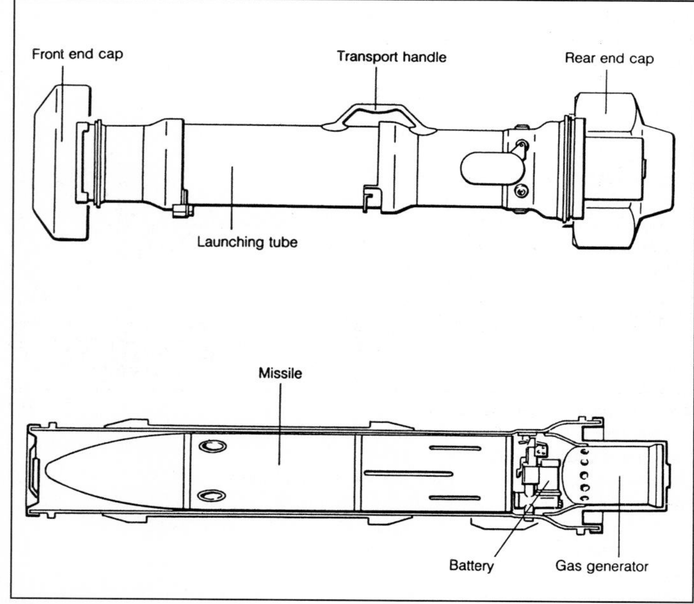

RBS56 BILL 1 missile container (top) and missile in launching tube (bottom)

Saab Bofors RBS56 BILL 1 ATGW system infantry version on its tripod with BILL Night Sight (BNS) alongside

The missile then continues in unpropelled flight, but throughout that time it is gyrostabilised in roll, keeping the warhead pointing downwards. Low parts of the target vertical section can be engaged by aiming low and utilising the impact fuze. In some engagement situations, where it would be an advantage to operate without the sensor system, the gunner can switch it off before missile launch. The elevated flight path adds an important element of ground clearance to that of the line of sight, thus avoiding many terrain obstacles which might otherwise be limiting factors.

Operational status
No longer in production. Superceded by BILL 2. In service with Austria, Brazil and Sweden.

Specifications
Type: wire-guided SACLOS, Over-fly Top Attack (OTA)
Length: 900 mm
Diameter: 150 mm
Wingspan: 410 mm
Weight
Missile at launch: 10.9 kg
Day sight travelling: 6.0 kg
Thermal imager: 9.2 kg
Warhead: approx 100 mm diameter canted HEAT
Range
Stationary target: 150-2,200 m
Moving targets: 150-2,200 m slow crossing; 300-2,200 m 10 m/s crossing speed; 600-2,200 m 20 m/s crossing speed
Speed
Launch: 72 m/s
Max: 250 m/s
Terminal: 135 m/s
Flight time
300 m: 2.1 s
1,000 m: 5.2 s
1,500 m: 7.9 s
2,200 m: 13 s

Contractor
Saab Bofors Dynamics AB.

Saab Bofors Dynamics RBS56 BILL 2

Type
Anti-armour missile – portable.

Development/Description
The Bofors BILL 2 is a further development of the RBS56 BILL 1 concept. Externally the BILL 2 is identical to the BILL 1 and uses the same launch tube, day and night sight and tripod.

Internally however, there have been many changes, including more compact electronics and the fitting of two canted HE shaped charge warheads – the forward one is 80 mm diameter and the aft one 102 mm diameter. It requires no lock on or cooling down time before launch, giving it very short reaction times.

Saab Bofors BILL 2 fitted with data-gathering equipment overflying a Leopard 2 MBT during trials in 1995

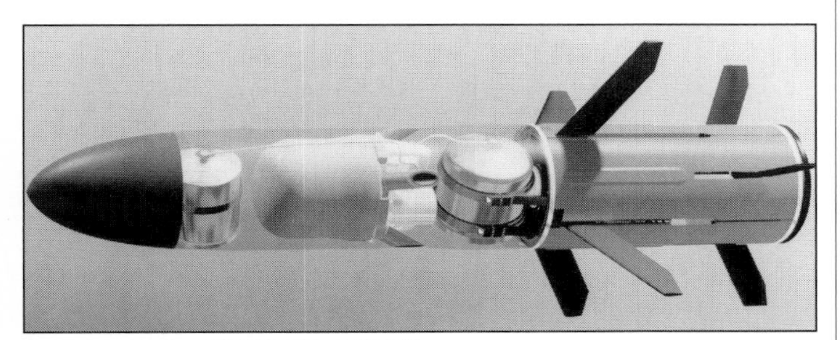

Cutaway of Bofors RBS56 BILL 2 ATGW 0023440

Once launched the missile's refined Over-fly Top Attack (OTA) technology gives it an average flight altitude of 1.05 m above the target. The advanced sensor system incorporates an optical sensor; magnetic sensor and algorithms determine the exact position for the two shaped charges to be ignited.

The vertical jets produced, with suitable time delay and compensation for dynamic effects, strike at almost the same spot to penetrate the target. The charge is designed to neutralise any Explosive Reactive Armour (ERA) modules present before the main charge jet hits.

Other enhancements include variable missile altitude from above, to on the Line Of Sight (LOS), selection of ignition conditions, impact or dual-purpose sensor system.

The BILL 2 gunner has a number of different firing mode selections available suitable for the various targets to be engaged.

In Mode 1, (Basic Mode), he aims at the turret roof and the missiles refined OTA technology gives it an average altitude of 1.05 m above the target. The two interactive, dynamically compensated, vertically striking shaped charges attack the MBT from above striking down through the turret roof avoiding the heavily protected frontal arc. The fuse system consists of both a dual-sensor proximiity fuse and an impact fuse.

Using Mode 2, (Non-armoured target) the gunner has a direct attack mode which can be used to engage other types of battlefield targets. In this mode the missile flies on the line-of-sight. The sensor system is disconnected and an impact fuze function is used. The operator aims at the spot on the target he wants to hit.

Using Mode 3 (soft target mode) the gunner aims at the target and the missile flies 1.4 m above the line-of-sight. The magnetic sensor is disconnected. The optical sensor together with special algorithms ensure that the warhead is ignited at the right moment above the target.

Mode 3 can be reprogrammed to counter future threat scenarios and in accordance with customers operational requirements.

The first complete flight trials of the BILL 2 successfully took place in January 1996. Late in 1996, the FMV (Swedish Defence Materiel Administration) decided that remaining deliveries of the Bofors RBS 56 BILL Anti-Tank Guided Weapon (ATGW) to the Swedish Army will be in the latest improved BILL 2 configuration.

Operational status

First weapons were available in 1999. The Swedish FMV decided to change remaining orders of BILL 1 to BILL 2.

Specifications

Type: wire-guided SACLOS, Over-fly Top Attack (OTA)
Weights
Missile at launch: <10.9 kg
Missile in launch tube travelling: 20 kg
Missile in launch tube ready to fire: 18 kg
Day sight: 6 kg
Thermal imager: 8.5 kg
Tripod: 11.6 kg
Warhead: 2 canted HE shaped charges
Range
Moving target <5 m/s: 150-2,200 m
Moving target <10 m/s: 300-2,200 m
Moving target ≤20 m/s: 600-2,200 m

Contractor

Saab Bofors Dynamics AB.

Saab Bofors Dynamics Strix smart mortar projectile

Type

Anti-armour weapon.

Development

The Strix 120 mm guided mortar projectile was jointly developed and manufactured by Bofors AB and Saab Dynamics. In 1991, the Swedish Defence Matériel Administration (FMV) signed a contract for the procurement of Strix for the Swedish Army. Deliveries began in 1994.

Strix was the first 'smart' terminally guided mortar to go into production. It can be fired from a standard 120 mm mortar and uses an imaging infra-red terminal seeker and a hollow charge warhead. It is a top-attack anti-tank weapon for infantry which is completely autonomous in operation.

Description

Strix comprises projectile, sustainer, launch unit and hand-held programming unit. The projectile has a nose-mounted imaging infra-red seeker. It creates a digital image of a target area which is continuously updated during the terminal phase of the flight. The image is fed to a processor unit where target acquisition and tracking is performed using advanced signal processing algorithms.

The electronic elements of the processor unit consist of large scale integrated circuits mounted on ceramic multilayer boards. The unit contains analogue and digital circuits as well as high-performance general purpose microprocessors and memories. Analogue signals are amplified, filtered and converted to digital form before collection in an image memory. The target signature is extracted from the background by image processing techniques and the correct target is selected and tracked. With the help of the continuously updated target position in the seeker field of view, the correct moment and direction of trajectory correction pulses are calculated, resulting in initiation signals for the thruster rockets. Via power amplifiers in the power unit and via the safety unit, these pulses are routed to the thruster rocket igniters. The gyro unit measures the roll angle which is used to select the correct thruster rocket, bearing in mind that the projectile is spinning.

Strix is armed with 12 thruster rockets, each of which can be fired individually to steer the mortar. The shaped charge warhead can penetrate explosive reactive armour. The fuze is of the Point-Initiating Base-Detonating (PIBD) type.

The basic projectile can be used over ranges from 1,000 to 5,000 m. If ranges of more than 5,000 m are required a rocket motor sustainer unit is employed.

Before launch the Strix projectile is programmed using a hand-held unit which enters time of flight until target seeker activation, projectile velocity and angle of descent at the start of the guidance phase, one of two activation levels, and one of three control rocket temperatures. The programming unit is connected to the projectile by a cable. At a preset height above the target, the seeker is activated and a target is selected for tracking. Once a target has been selected the error vector between the centre of the target and the projected impact point is continuously monitored. As soon as it exceeds a preset value, one or several thruster rockets are fired in a direction to bring the value of the error close to zero.

By continuous calculation of the predicted impact point relative to the predicted target position at impact, it is possible to use proportional navigation which avoids any influence of target movement, wind effects and so on. The tracking technique makes it possible to make several course corrections in rapid succession. If necessary all 12 thruster rockets can be used with full control during the last few seconds of flight.

Operational status

In production and in service with the Swedish Army and export delivery to the Swiss Army.

Specifications

Calibre: 120 mm
Length: 842 mm (projectile), 295 mm (launch unit), 197 mm (sustainer)
Weight: 18.2 kg (projectile), 3.6 kg (sustainer)
Muzzle velocity: 180-320 m/s
Range: 1,000 to 5,000 m, 5,000 to 7,500 m (with sustainer)

Contractors

Saab Bofors Dynamics.

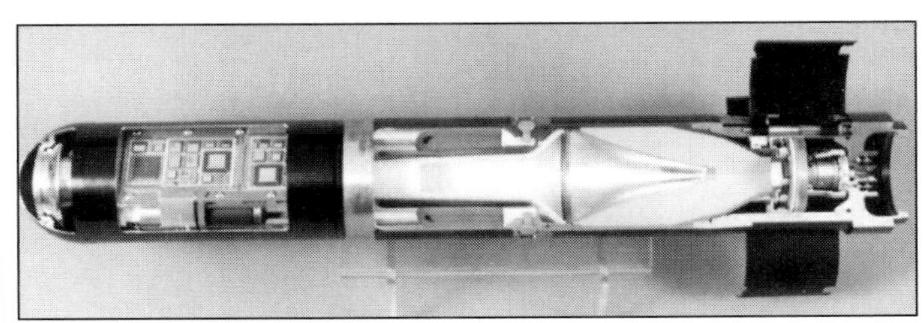

The Strix smart mortar projectile 0004762

ANTI-ARMOUR MISSILES AND MUNITIONS

PORTABLE SIGHTS

Azimuth TOW-SLIK (System Level Improvement Kit)

Type
Anti-armour missile launcher (portable).

Description
The TOW-SLIK (System Level Improvement Kit) is designed to be integrated into a TOW unit to improve the ability of the commander to control and co-ordinate the fire of his launchers. It can be integrated in the field without any extra equipment.

TOW-SLIK incorporates the following elements: laser range-finder; electronic compass; either hand-held MP3 computer with a Global Positioning System (GPS) for navigation or MC tactical computer; and software modules for mission preparation, target acquisition and engagement.

TOW-SLIK generates target grid co-ordinates so that the commander can make target allocations to the sub-units and this data serves as the basis for co-ordination of the engagement and also enables each TOW launcher to translate these co-ordinates into aiming commands. The azimuth and range of newly detected targets can be translated by each launcher into geographic or grid co-ordinates for transmission to the command or to co-operating land and air units.

Operational status
Available.

Contractor
Azimuth Ltd.

SLIK mounted on a launcher showing all major modules
0101559

Elop A-TIM night sight for Fagot anti-tank missile launcher

Type
Anti-armour missile sight (portable).

Description
The A-TIM compact and lightweight thermal imaging sensor has been developed as an add-on night sight upgrade to the Soviet 9K 111 Fagot (AT-4 'Spigot') anti-tank missile launcher.

A-TIM is a separate battery powered unit with dual field of view telescope which clips on to the day sight without requiring any changes to the launcher. The sensor is automatically boresighted to the day channel.

Operational status
Ready for production.

Specifications
Spectral band: 8-12 μm
Detector: CMT, more than 100 elements
Cooler: integral closed-cycle
Aperture: 100 mm
Fields of view
Narrow: 2.4 × 1.5°
Wide: 8.4 × 5.3°
Magnification: ×10
Weight: 10 kg incl battery
Dimensions (l × h × w): 275 × 336 × 195 mm
Power supply: 40 W 28 V DC

Contractor
Elop Electro-Optics Industries Ltd.

UPDATED

FLIR Systems BIRC portable Missile IR sight

Type
Anti-armour missile sight-portable.

Description
BIRC is a third-generation thermal imager based on Quantum Well Infrared Photodetector (QWIP) technology. The detector is a 320 × 240 element focal plane array and is sensitive in the long-wave 8-9 μm band. It is cooled by an integrated Stirling rotary microcooler. BIRC has two fields of view to enable easy surveillance and target acquisition in the wide field of view and accurate long-range target identification in the narrow field of view. The image from the built-in display is input into the day sight and BIRC is quickly fitted in the field without any alignment.

Operational status
In production.

The FLIR Systems BIRC weapon sight 0130953

Specifications
Weight: <8.5 kg

Detector
Type: 320 × 240 QWIP
Spectral band: 7.5-9.3 μm
Cooling: Integrated Stirling rotary microcooler

Fields of View
WFOV: 4.6 × 3.5°
NFOV: 2.3 × 1.7°
Frame Rate: 50 Hz

Optics
FOV change: <0.5 s
Focus: Manual from 50 m to infinity; automatic parfocus 500 m to infinity

Interfaces
Remote control: RS-422
Video: Standard CCIR Monochrome Pal

Power Supply
Voltage: 10 V-16 V DC; 24 V DC MIL-STD-1275B via power adapter
Power Consumption: <15 W (stand-by) <25 W (operating)

Environmental Specifications
Operation: –30 to +60°C
Storage: –45 to +65°C

Features:
- Automatic/Manual gain and offset
- BITE

Contractor
FLIR Systems.

VERIFIED

Instalaza VN38-C night vision equipment for C90 weapon system

Type
Anti-armour weapon sight.

Description
The VN38-C night vision device has been designed by Instalaza, S.A. to provide the family of C90 systems with day and night operational capability. The C90, in service with the Spanish Army and many other armies, is a light anti-tank weapon which fires a projectile with a shaped charge warhead. Series M3 of the C90 family incoporates the adaptor for the VN38-C as standard but, it can be fitted to previous versions of the C90, using a template supplied by Instalaza.

The VN38-C is placed in front of the optical viewfinder of the C90 system in an adaptor device. The device has been designed so that it does not at all modify the boresighting, stadiametry and lateral prediction of the C90's optical viewfinder. The aiming process is performed using the optical viewfinder of the C90 system.

The VN38-C uses GEN II+ or GEN III image intensifiers.

In order to provide small units, or the individual soldier, with a better night vision capability, the VN38-C can also be used separately as a night observation and surveillance device. In this case the collimator and eyepiece are interchanged in their respective housings. In this configuration, the eyepiece provides a ×3.3 magnification and the collimator is used as a handle.

VN38-C fitted on the C90 weapon system 0004797

VN38-C in observation and surveillance configuration 0004798

Operational status
In production. In service with the Spanish Army and several other armed forces.

Specifications
Image intensifier: GEN II+/GEN III 18 mm tubes
Sensitivity: up to 0.1 mlx, >500 μA/lm
Resolution: 36 lp/mm
Field of view: 13°
Optics: adapted to the optics of the C90 optical viewfinder
Coupling: self-locking adaptor; coaxial coupling with the optical axis of the C90 viewfinder; removed by pushing the retaining element
Collimator and eyepiece fitting: 60° twist-lock
Surveillance magnification: ×3.3
Eyepiece dioptre adjustment: +3 to −4 dioptres
Power: 2 × AAA batteries
Weight: 750 g (incl batteries, collimator, adaptor and eyepiece)

Contractor
Instalaza S.A.

VERIFIED

Photonics day/night laser sighting system for missile launchers

Type
Anti-armour missile sight.

Description
The day night laser sighting system is especially designed for operation with man-portable anti-armour multirole direct-fire weapons systems such as the Carl Gustaf missile launcher. It is designed to be mounted on the standard weapon mount.

The system uses a semiconductor eye-safe laser range-finder which is provided with an adjustable bracket for mounting a monocular day sight or a monocular night sight. The two types of sights are interchangeable without losing the sight alignment. An optical display on the back of the range-finder projects the measured range figures to the observer's eye. A remote trigger for initiating the range measurement can be fixed on the handle of the weapon.

Photonics day/night laser sighting system for Carl-Gustaf missile system 0053710

Detailed view of the Photonic day night laser sighting system 0089884

Operational status
In service.

Specifications
Laser range-finder module
Range: 50 m to 2,500 m
Accuracy: +1 to −3 m
Wavelength: 910 nm
Eyesafe: Class 1, ANSI Z136.1 (1993)
Beam divergence: 0.3 × 1.5 mrad
Receiver: Silicon APD
Range logic: multi-echo memory (3 echos)
Range gate: 50, 100, 200, 300, 400, 500, 700, 1,000 m
Repetition rate: 15 measurements per min
Aiming telescope: ×3
Eye relief : 80 mm
Display: 4 digits
Display resolution: 5 m
Repetition rate: 15 measurements per min
Power supply: 4 × 1.5 V alkaline, size AA or 4 × 1.5 V lithium, size AA

Mount: boresighting range
Elevation:+1 to −10 mrad, 0.33 mil/click
Lateral:+1 to −10 mrad, 0.33 mil/click
Alignment weaver rail/weapon mount: 0.1 mm
Boresighting periscope – periscopic distance: 100 mm
Operating temperature: −35 to +55°C
Storage temperature: −40 to +70°C
Shock: 30g/ 11 ms
Immersion:1 m water / 2 h
Humidity: 95% RH/40°C/120 h
Weight:1.25 kg

Night vision sight
Dimensions: 150 × 120 × 95 mm
Weight (with batteries): 1.3 kg
Field of view: 10.5°
Magnification
NS-ZF3.3-80: ×3.3
NS-ZF4-80: ×4

Daysight
Dependent on choice of weapon manufacturer.

Contractor
Photonic Optics.

VERIFIED

Raytheon TOW Improved Target Acquisition System (ITAS)

Type
Anti-armour missile sight.

Development
The TOW Improved Target Acquisition System (ITAS) is a modification to the existing TOW 2 weapon launcher for the HMMWV and ground-mounted TOW. It was developed to address user operational issues identified during Operation Desert Storm. ITAS allows US light forces which use the M220A2 TOW ground launcher either in stand-alone mode or mounted on HMMWVs to engage enemy armour at greater ranges by day and night and in adverse weather conditions. A development contract was awarded in 1993 to Raytheon by the US Army Missile Command (MICOM). Following delivery of a prototype, preproduction testing began in January 1995 and a Low-Rate Initial Production (LRIP) contract (25 kits) was awarded in September 1996 allowing the US Army's 82nd Airborne Division to field ITAS in September 1998. Two more LRIP contracts followed (74 and 102 kits) with the first multi-year buy in FY00 for 122 kits. The US Army requirement is for 1,841 units.

The Improved Target Acquisition System ITAS on a ground-launched TOW missile system

Description
The ITAS includes a second-generation thermal imaging sight, a laser range-finder and an improved fire-control system. The dual field of view thermal imager is based on a Standard Advanced Dewar Assembly (SADA II) focal plane array. The dual-spectrum missile tracker uses the Bradley TOW ISU Xenon Beacon Tracker as the primary tracker with the addition of a gunner-aided thermal tracker. Other improvements include automatic boresighting and advanced signal processing. ITAS was developed using Ada software and contains missile and Video Thermal Tracking (VTT) capability, with an open architecure to allow for growth.

Operational status
In production and in-service with the US Army.

Specifications
Thermal imager fields of view
Wide: 6 × 8°
Narrow: 2 × 2.7°
Direct view optics fields of view
Wide: 8° circular
Narrow: 2.7° circular
Eye-safe laser range-finder
Accuracy: ±5 m
MBTF: 700 h

Contractor
Raytheon Systems Company.

UPDATED

Saab Bofors Dynamics BILL Night Sight (BNS)

Type
Anti-armour missile sight.

Development
The BILL Night Sight, has been designed as an add-on thermal imaging night sight, primarily for use with the BILL portable medium-range anti-tank missile system, made by Saab Bofors Dynamics.

BILL Night Sight can be adapted to various types of day sights by including mechanical Quick-Lock mounts in the equipment.

Description
The BILL Night Sight is mounted on top of the day sight with the thermal image mirrored into the front lens of the day sight. It can be mounted rapidly while maintaining sightline accuracy, without the need for harmonisation.

The front lens is coated with a diamond-hard anti-reflection carbon coating which has increased resistance against mechanical abrasion.

The thermal image is displayed in red (hot) and black (cold), but can be reversed by inverting the polarity of the image to improve target recognition and false targets rejection.

The BILL Night Sight now exists in two different versions:

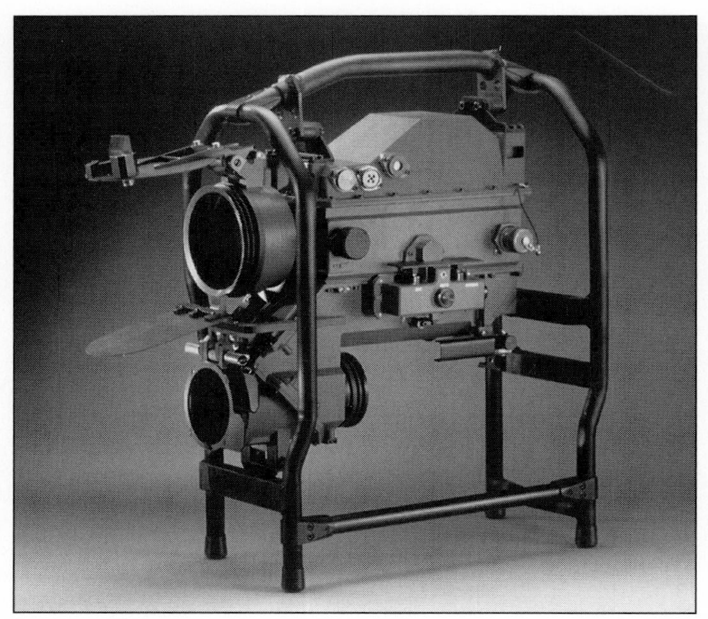

The BILL TI Sight 0010866

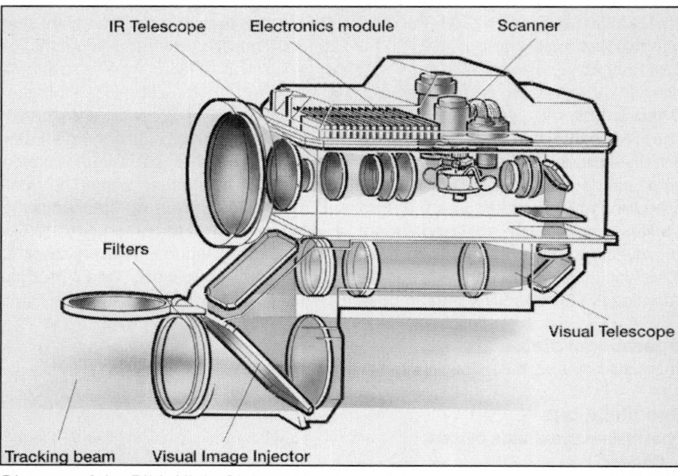

Diagram of the BILL Night Sight 0004727

BILL NS (BNS)

BNS has two fields of view - wide for surveillance and narrow for more detailed observation and aiming and is powered by a battery pack and a high-pressure air bottle for cooling. BNS is equipped with low battery and gas pressure indicators.

BILL TI (BTI)

TI stands for Thermal Imager which is a better name since the TI will be used also during daytime. This version uses a closed-cycle Stirling cooling engine, has a single field of view and a power consumption of 10 W. It can be powered either by a standard Ni/Cd or lithium battery or via a built-in vehicle power converter.

Other versions

Other versions are formed by using only the top part of the sight. It can be used as a separate direct-view sight by adding a reticle and an eyepiece, or a CCD-camera can be used for video conversion.

Operational status

Both sights are in production, the BILL NS for the Austrian Army, the Brazilian Marine Corps and the BILL TI for the Swedish Army. Both contracts were signed during 1997.

Specifications

Spectral band: 8-12 μm
Field of view
BNS wide: 12.5 × 5.5°
BNS narrow: 6 × 3°
BTI: 6.2 × 2.8°
Magnification: ×1
Magnification with daysight: ×7
Weight
BNS (incl battery and bottle): 9.2 kg
BTI (excl battery, incl eye cup): 8.75 kg
Power consumption
BNS: 4 W
BTI: 10 W

Contractor

Saab Bofors Dynamics.

Saab Bofors Dynamics BILL Infra-Red Camera (BIRC)

Type

Anti-armour missile sight infra-red camera.

Description

The BILL Infra-Red Camera (BIRC) is designed to be used together with either the BILL Day Sight or the BILL Simulator Sight and is mounted on top of it (clip-on unit). The BIRC can, besides its normal use during night-time, also be used in daylight. This is especially useful in bad weather conditions and for camouflaged targets.

With a daylight blocking filter on the BIRC it is possible for the gunner to select either visual or thermal image, and with a blue filter the gunner can select a mixed visual and thermal image, for example the target will appear in 'reds'.

The detector of the camera is a focal plane array of 320 × 240 Quantum Well Infra-Red Photodetector (QWIP) detector elements designed to operate in the Long-Wave Infra-Red (LWIR) spectral band at around 8 to 12 μm. This is generally considered to be the best waveband for a relatively dry, dusty and smoky battlefield environment and the temperature range of the target. QWIP technology offers good uniformity across the array and excellent picture quality. Fixed pattern noise, that produces an overlaid pattern on the image, is very low especially when compared with uncooled detectors.

The third-generation clip-on thermal imager offers a significant improvement compared with other scanning systems when it comes to resolution and image quality, providing detection at up to more than 10 km and recognition at up to 2 km. The identification range is of vital importance because it gives the gunner the possibility to identify the target that he intends to engage at maximum weapon range.

BIRC is powered by either an attached Li-Ion battery or by an external power source. A built-in Sterling cooler cools the IR-detector elements.

The BIRC is based on four major parts:

- IR-optics
- IR-detector
- Display
- Visual optics

Operational status

In service.

Contractor

BIRC fitted to a missile sight 0122626

SAGEM, Euromissile MILIS

Type

Anti-tank missile (portable) sight.

Development

SAGEM has developed MILIS, a third-generation thermal sight for the MILAN anti-tank weapon system, in co-operation with Euromissile (now part of MBDA).

Description

MILIS is a member of the SAGEM family of MATIS third-generation thermal imagers, based on the latest generation of cooled focal plane arrays, operating in the 3 to 5 μm waveband. It is specially designed in co-operation with Euromissile for the MILAN anti-tank weapon system, and can be mounted on the MILAN sight in less than 10 seconds, without any bore-sighting and adjustment. High thermal

The SAGEM MILIS thermal imager on the Euromissile MILAN anti-tank weapon system (Euromissile) 0129315

sensitivity and resolution give the system a recognition range superior to the MILAN missile maximum range. It has a dual field of view and can be powered directly from its own battery or from an external source. A standard TV output allows the connection to a monitor for observation on vehicle or for instruction.

Operational status
In production for several countries.

Specifications
Weight: <7.6 kg
Spectral band: 3-5 μm
Cooling: integrated Stirling cycle microcooler
Cool-down time: 6 min (typical)
Display: integrated LCD
Video output: CCIR
Operating time: > 5h
Weight: <7 kg

Contractor
SAGEM, Aerospace and Defence Division, France.

UPDATED

Thales Optronics Milan-LITE thermal imaging sight

Type
Anti-armour missile sight (portable).

Description
The Milan-LITE is a clip-on thermal sight for the Milan portable anti-tank missile system manufactured by Euromissile. It mounts on top of the weapon's tracker/day sight. It can be mounted without any need for adjustment during operation.

The Milan-LITE sight for the Milan anti-tank weapon system 0004726

In standard mode the Milan-LITE sight has automatic image control, with manual control if required during surveillance and target recognition. It uses an electrically driven microcooler and is powered by its own battery pack.

Operational status
In service with the Egyptian army.

Specifications
Spectral band: 8-12 μm
Detector: CMT
Field of view: 6.2 × 3.1°
Cooling: closed-cycle Stirling microcooler
Weight (incl battery): 8 kg
Power consumption: 8 W

Contractor
Thales Optronics.

UPDATED

Thales Optronics MIRA and MEPHIRA thermal imaging night sights

Type
Anti-armour missile sight (portable).

Development
MIRA (MILAN Infra-Red Attachment) is an add-on sight unit which can be rapidly fitted to the existing MILAN missile firing post sight to allow night operation.

A variant model, known as MEPHIRA, has been developed by Thales and is being introduced into service with French forces. This version is optimised for use with HOT missiles and is installed on VAB-HOT launchers.

Description
MIRA is automatically harmonised with the optical path of the standard sight and reflects the infra-red picture into the optical sight so that the normal day sight eyepiece is still used for viewing. Using series-parallel scanning and a cadmium mercury telluride detector, the sight picture is produced by an array of light emitting diodes.

The sensor may be cooled either by an HP compressed gas bottle (Joule-Thomson) or by a built-in cooling unit (Stirling cycle).

Operational status
In production and in service with the French, German and UK armies and other forces.

Specifications
Spectral band: 8-13 μm
Detector type: CMT
Field of view: 6 × 3°
Angular resolution: 0.17 mrad
NETD: 0.1°C
Sight axis harmonising: automatic
Scanning: series-parallel
Image frequency: ≤22 Hz
Detection range: 6 km typical on vehicle
Reconnaissance range: 3 km typical on vehicle
Identification range: 2 km typical
Temperature range: –40 to +52°C
Start-up time: <20 s (JT); <4 min (Stirling) from cold, <30 s from standby condition
Dimensions: 530 × 290 × 160 mm
Weight: 8 kg JT version with bottle and power; 9 kg Stirling version

Contractor
Thales Optronics.

VERIFIED

The MIRA sight fitted to the MILAN missile system

Thales Optronics Mirabel

Type
Thermal imager (land).

Description
The Mirabel thermal imager is integrated with MBDA's Eryx short-range anti-armour missile weapon system (qv). This thermal imager has been developed by Thales Optronics, Canada (formerly AlliedSignal, Canada) in co-operation with Thales Optronics of France.

Weighing approximately 3 kg, the system's modular design allows for a wide variety of applications and it is suitable for retrofit on existing weapon sights, primarily for portable shoulder-launched, short-range, anti-tank weaponry applications. It has been developed as a 'snap-on' device for existing direct view optic sights without any need for boresight alignment.

Operational status
In production. First production batch was delivered in 1996. Eryx is in service with the Canadian, French, Malaysian and Norwegian armed forces.

Specifications
Weight: <3.4 kg
Dimensions (l × w × h): 265 × 185 × 205 mm
Magnification: ×1
Field of view: 8 × 6°
Range: identification within the 600 m range of the Eryx missile
Power supply: 8-16.5 V battery pack; 5 h autonomy
Focus: athermalised (manual control available)
Spectral band: 8-12 μm
Display: LED

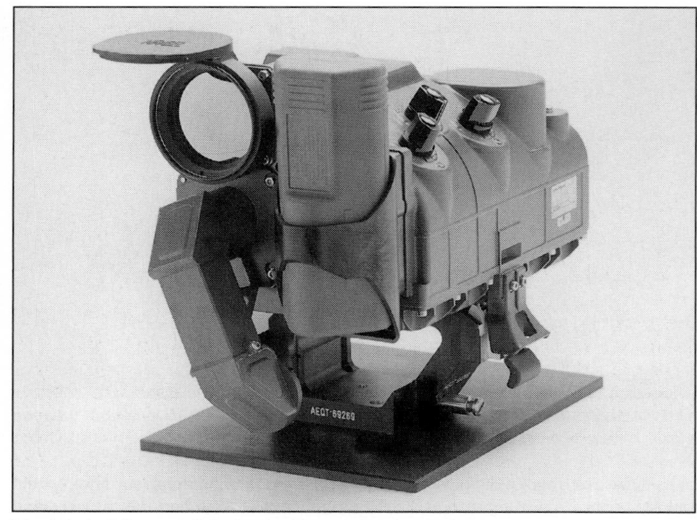

The Mirabel thermal sight for the Eryx anti-tank missile system 0053709

Contractors
Thales Optronics, Canada.
Thales Optronics, France.

VERIFIED

ARMOURED FIGHTING VEHICLES

VEHICLE TURRETS

Alvis Warrior 30 mm turret

Type
Armoured fighting vehicle turret.

Description
Summary: the gunner and commander have image intensifying night sights.

This turret is fitted to Alvis Warrior vehicles. Both gunner and commander have Thales Optronics Raven sights giving magnifications of ×1 for observation and ×10 for target engagement. The commander has seven ×1 magnification forward sloping window periscopes and a binocular ×10 magnification (No 57 Mk 2) with lever introduced ×1 magnification, limited traverse and 30 mm graticule. The Thales Optronics (was Avimo) NVC53C combined/day night assembly can be fitted as an option. The gunner has two ×1 magnification forward sloping window periscopes and a No 52 Mk 1 binocular day/passive night sight with fixed eyepiece systems and ×1 and ×10 magnifications. An optional sight is the Avimo NVL53 combined day/night assembly with an integrated laser range-finder.

The objective systems elevate and depress with the gun.

Operational status
Production complete. Over 520 of these turrets have been manufactured although there has been no recent production. In service with Belgium (on Scimitar), Honduras (on Scimitar) and the UK (on Scimitar).

Contractor
Alvis Vehicles Ltd.

AV Technology LLC, Multi Gun Turret System (MGTS), 12.7 mm/40 mm UGWS II turret

Type
Armoured fighting vehicle turret.

Description
Summary: the gunner can have an image intensifying sight or a thermal imaging sight and laser range-finder.

The one-man turret is suitable for any armoured vehicle with a 1 m or greater turret ring mount. For target acquisition and sighting purposes the turret is equipped with seven unity vision blocks, each with blackout capability, as well as a modified M36E1 gunner's passive day/night sight. With the all-electric, stabilised, fire-control system, a Day Night Range Sight (DNRS) is installed in the one-man MGTS (see entry for AV Technology two-man MGTS). Space is allocated in the turret to allow for the fitting of a variety of radio and intervehicle communications systems. It is also equipped with solid-state electronics while supplemental NBC protection is offered as option. A laser range-finder can be fitted.

AV Technology one-man MGTS in 7.62 mm/12.7 mm configuration mounted on the Dragoon Armoured Security Vehicle (4 × 4) 0059448

For target acquisition and sighting purposes the turret is equipped with seven unity vision blocks, each with blackout capability, as well as a modified M36E1 gunner's passive day/night sight. With the all-electric, stabilised, fire-control system, a Day Night Range Sight (DNRS) is installed in the one-man MGTS (see entry for AV Technology two-man MGTS).

Space is allocated in the turret to allow for the fitting of a variety of radio and intervehicle communications systems. It is also equipped with solid-state electronics while supplemental NBC protection is offered as option.

Variants
The company has also produced a prototype variant armed with a 30 mm cannon and designated AV-30. It can be fitted to any vehicle which can accommodate provisions for a 0.86 or greater turret ring mount opening. The turret is equipped with a modified M36E1 passive day/night sight and may be fitted with a thermal sight with laser range-finder. The gunner is provided all around vision through seven direct view unity vision blocks equipped with blackout capability.

The turret fire control also incorporates an all electric, fully stabilised, digitally controlled traverse and elevation system. The system comprises GDLS M2 Bradley AFV components including the Digital Electronic Control Assembly (DECA) which provides a built-in-test capability.

Operational status
Production – over 1,375 produced. In service with Turkey and the US Marine Corps.

Contractor
AV Technology, LLC.

UPDATED

AV Technology LLC, Multi Gun Turret System (MGTS), single/twin/combination machine guns

Type
Armoured fighting vehicle turret.

Description
Summary: the gunner has an image intensifying night sight or thermal sight.

The one-man MGTS can be fitted to any vehicle with provisions for a 1 m or greater turret ring mount. For target acquisition and sighting purposes the turret is equipped with seven unity vision blocks, each with blackout capability, as well as a modified M36E1 gunner's passive day/night sight.

There is a choice of fire-control systems for the one-man MGTS. With the first system the turret is electrically traversed with manual fine lay. The second more sophisticated system provides all electric, fully stabilised, digitally controlled traverse and elevation. With the latter fire-control system, a Kollsman thermal imaging Day/Night Range Sight (DNRS) is installed.

Operational status
In production. In service with the Canadian Forces (Grizzly), US Air Force, UN peacekeeping forces and Turkish armed forces.

Contractor
AV Technology, LLC.

AV Technology LLC, Multi Gun Turret System (MGTS), 25 mm/30 mm/35 mm

Type
Armoured fighting vehicle turret.

Description
Summary: the gunner has a thermal sight with commander's link and laser range-finder. The commander may have an independent thermal sight.

The two-man MGTS is a drop-in replacement turret system for the LAV-25 and similar light armoured vehicles. It can be fitted to any vehicle with provisions for a turret ring mount of 1.38 m or greater.

Target acquisition and general observation is achieved through nine M27 periscopes, three at the gunner's position and six at the commander's position. Target acquisition and range is determined through the Kollsman Day/Night Range

Sight (DNRS) (qv). The DNRS is equipped with a thermal imager, visual channel, unity vision window and laser range-finder. Both the gunner's and commander's positions are equipped with their own thermal displays. An optional capability on the two-man MGTS is an independent thermal sight which allows the commander to search out secondary threats while the gunner is occupied.

The Kollsman Primary Direct Fire-Control System (PDFCS) (qv) can also be integrated. This is similar to the DNRS but with the addition of a ballistic computer and meteorological and cant angle sensors.

The turret has an optional integrated TOW ATGW system, controlled through the same Digital Electronic Control Assembly (DECA) unit that drives the main weapon. The side-mounted TOW units are sighted and tracked through the DNRS using the same controls as the main weapon.

Operational status

Production as required. In service with several countries.

Specifications

Optics: 1 × DNRS thermal sight with 60-element FLIR, eye-safe laser range-finder and ballistic reticle, 9 × M27 periscopes with blackout screens and optional laser filters.

Contractor

AV Technology, LLC.

AV Technology's two-man MGTS turret fitted to Steyr-Daimler-Puch Pandur (6 × 6) armoured personnel carrier as supplied to the Kuwait National Guard 0006920

AV Technology LLC two-man 90 mm turret

Type

Armoured fighting vehicle turret.

Description

Summary: the gunner has an image intensifying sight. A laser range-finder and thermal sight are optional.

Dragoon Light Forces Vehicle fitted with AV Technology two-man 90 mm turret

The AV Technology two-man 90 mm turret has been mounted on a number of armoured vehicle types including the LFV-90 mm (two-man turret) variant of the (4 × 4) Dragoon Light Forces Vehicle, the US Marine Corps LAV, and Steyr Pandur (6 × 6).

For observation purposes the gunner has a modified M36E1 passive day/night sight capable of mounting a Northrop Grumman Laser Systems SIRE (Sight Integrated Laser Range-finder) system. The commander has seven M27 periscopes for all-round vision. Other types of gunner's sight assemblies, including thermal imaging, are available.

Operational status

Production as required. In service with a number of countries.

Specifications

Optics

Commander: 7 × M27 periscopes

Gunner: 1 × modified M36E1 sight, with optional SIRE sight integrated laser range-finder system

Contractor

AV Technology, LLC.

Cockerill Mechanical Industries C25 25 mm turret

Type

Armoured fighting vehicle turret.

Description

The C25 is a joint development between CMI, which is responsible for the turret and systems integration, and Oerlikon Contraves, which is responsible for the 25 mm KBB cannon. For trials purposes, the turret has been installed on the SIBMAS (6 × 6), M113A1-B, Puma and Warrior armoured vehicles. The C25 turret has been evaluated by several european countries, including Switzerland. The fully stabilised turret is equipped with the Officine Galileo (now Galileo Avionica) Janus fire-control system and a thermal sight. For the same projects and based on the design of the C25 turret configuration, CMI has designed and manufactured a 30 mm turret with the same overall design features. The main armament is the Mauser 30 mm Model F cannon; as an alternative it can be fitted with the 30 mm Boeing Company Bushmaster II Chain Gun. The turret successfully underwent customer evaluation in 1993 and has been named the C30. The turret weapon system has been designed to achieve high-hit probabilities while firing on the move against moving ground targets, as well as in the anti-aircraft mode. Although this turret has been tested on a number of chassis, by early 2000 it had yet to enter volume production.

Operational status

Available.

Contractor

CMI, Cockerill Mechanical Industries SA.

Cockerill two-man C25 25 mm turret installed on an M113 series APC

Cockerill Mechanical Industries C30 30 mm turret

Type
Armoured fighting vehicle turret.

Description
Summary: the gun sight has a thermal imager.

The C30 is based on the design of the C25 turret, with all its main features included. The major difference in design is that the Oerlikon Contraves KBB cannon on the C25 is replaced by a Mauser 30 mm Model F cannon on the C30. The C30 has two sights and 12 periscopes with the current configuration using the Officine Galileo Janus (now Galileo Avionica) fire-control system which in this case is reported to include a thermal imager.

Operational status
Available.

Contractor
CMI, Cockerill Mechanical Industries SA.

Cockerill C30 turret with Mauser 30 mm Model F cannon and protected against attack from 14.5 mm projectiles on front and sides

Cockerill Mechanical Industries CSE 90 mm turret

Type
Armoured fighting vehicle turret.

Description
Summary: the gun sight may have an image intensifying or thermal sight. It may also have a laser range-finder.

The CSE 90 mm turret has been designed as a private venture by CMI for export. The turret has been installed on all the 162 SIBMAS Armoured Fire Support Vehicles delivered to the Malaysian Army. For trials, the CSE 90 mm turret has also been installed on the Vickers Defence Systems Valkyr (4 × 4), Alvis Vehicles Warrior, Steyr- Daimler-Puch Pandur, an M113A2 APC and an M113A1-B APC.

Cockerill CSE 90 mm turret 0077825

The commander has seven observation periscopes and the gunner four. Both have a roof-mounted sight with magnifications of ×8 and ×1 with a projected graticule for aiming the gun. Optional equipment includes commander's cupola with vision periscopes for improved all-round observation, and various types of sights for the commander and gunner including day, image intensifier, thermal sight, laser range-finder and ballistic computer. Various thermal sights have been evaluated on the turret.

Operational status
Production as required. In service with Malaysia on 162 SIBMAS 6 × 6 vehicles.

Contractor
CMI, Cockerill Mechanical Industries SA.

Cockerill Mechanical Industries LCTS 90 mm turret

Type
Armoured fighting vehicle turret.

Description
Summary: the gunner has a thermal sight.

This two-person, power-operated turret recently developed by Cockerill Mechanical Industries has been fitted to a number of vehicles, including the Pandur (6 × 6) for trials purposes. The turret is equipped with a stabilised day/thermal imaging sight and ballistic computer with a video display for the gunner. The commander has a stabilised panoramic sight (allowing various uses such as armament firing, 360° observation, automatic realignment to line of fire and target designation to gunner).

Operational status
In production. In service with Qatar (on 8 × 8 Piranha II chassis from Alvis Vehicles Ltd of the UK), Kuwait (on 6 × 6 Pandur chassis from AV Technology International of the USA) and Saudi Arabia (installed on the General Motors Defense Piranha 8 × 8 LAV chassis under a US DoD FMS programme, for the Saudi Arabian National Guard).

Contractor
CMI, Cockerill Mechanical Industries SA.

Eloptro MSS Modular Sighting System

Type
Modular upgrade for light turret systems.

Description
Eloptro's Modular Sighting System was designed to meet the requirement for a low-cost optronics upgrade package for light turret systems. The system provides night aiming and range measurement capability and is used in conjunction with the existing mechanically slaved gunner sight.

The Modular Sighting System comprises the NM-10 Night Module and LM-10 Laser Range-finding Module. The modules are designed to be externally mounted to any light turret system and are provided with a custom-designed mounting interface which enables accurate integration with the trunnion axis.

The NM-10 Night Module sensor incorporates an image intensifier tube (Gen II+ or Gen III) with CCD camera, enclosed in a protective armour hood. During night operations the CCD image is transmitted to the gunner and commander display monitors. The night control unit enables the gunner to control and boresight the night module. An armour flap with adjustable iris allows the gunner to boresight the unit during daylight conditions.

The LM-10 Laser Module incorporates Eloptro's proven LH-40 Eye-safe Laser Range-finder enclosed in a protective armour hood with integrated boresighting mechanism. The laser control unit enables the gunner to operate the laser module and obtain target range data. Range data is also transmitted to the commander via the laser display unit.

Operational status
In production.

Specifications
Image intensifier tube: 18 mm Gen II+ or Gen III
Field of view: 4.3° H × 3.2° V
System resolution (GEN II IIT): Moonlight: 2,75 lp/mrad; Starlight: 0,87 lp/mrad
Boresight adjustment range: –10 to +10 mrad, elevation and traverse
Boresight adjustment accuracy: 0.2 mrad, elevation and traverse
Power supply: +18 to +32 V DC
Power consumption: <60 W, excluding monitors
Mass: <45 kg with mount
Dimensions: 290 × 280 × 300 mm (l × w × h)
Operating temperature: +55 to –10°C
Shock/vibration: Information available on request

LM-10 Laser Module specifications
Measuring range: 80 to 10,000 m
Range resolution: ± 5m
Standard measuring rate: 10 measurements/min
Target range discrimination: 50 m
Laser Type: erbium glass, Class 1 eye-safe
Boresight adjustment range:–10m to +10 mrad, elevation and traverse
Boresight adjustment accuracy: 0.2 mrad, elevation and traverse
Power supply: +18 to +32 V DC
Mass: <40 kg
Dimensions: 240 × 280 × 220 mm
Operating Temperature: +50 to –20°C
Shock/vibration: Information available on request

Contractor
Eloptro (Pty) Ltd.

GDLS-MO 105 mm Low-Profile Turret (LPT)

Type
Armoured fighting vehicle turret.

Description
Summary: the gunner has a thermal sight with a video link to the commander.

The General Dynamics Land Systems - Muskegon Operations (GDLS-MO) Low-Profile Turret (LPT) has been operated on a Centurion Mk 5, GM Defense LAV III, ASCOD IFV and an installation kit has been designed for the M60.

The gunner has a GDLS Compact Modular Sight (CMS) with thermal imager and laser range-finder, with the image provided to the commander through a video link.

Operational status
Prototype.

Specifications
Optics
Primary: GDLS CMS lightweight modular (×8 magnification) day, 2 FOV, FLIR, trunnion encoder, laser range-finder, video link to commander
Panoramic: Optional, mounted on gun pod
Auxiliary, commander and gunner: 7 mechanically driven auxiliary sights with two eyepieces

Contractor
General Dynamics Land Systems – Muskegon Operations.

VERIFIED

General Dynamics Low Profile Turret (LPT) during firing evaluations on ASCOD chassis 0099501

Giat Industries 105 TGG 105 mm turret

Type
Armoured fighting vehicle turret.

Description
Summary: the gun sight has a laser range-finder, while image intensifying or thermal sights are optional.

The 105 TGG turret was developed originally for use on the AMX-10RC (6 × 6) wheeled reconnaissance vehicle. It has also been trialled on the Krauss-Maffei Puma ACV and can be adapted for other AFV types.

The commander has six periscopes and a panoramic SAGEM M389 telescope with magnifications of ×2 and ×8 which has full contrarotation. The gunner has two periscopes and an M504 telescope which is the main part of the COTAC fire-control system. The M504 has a magnification of ×10 and is combined with an optical compensator (with control electronics for automatic input of fire corrections) and an M550 laser range-finder unit. Night vision equipment for observation and identification such as a low-light level camera with a range of up to 4 km or an infra-red sight with a range of up to 1 km can be fitted as an option.

For Operation Desert Shield/Desert Storm, the AMX-10RC vehicles of the French Army were upgraded to include thermal imaging sights.

Operational status
Production complete, but can resume if orders are placed. In service with France, Morocco and Qatar.

Contractor
Giat Industries.

VERIFIED

Giat Industries 105 TML 105 mm turret

Type
Armoured fighting vehicle turret.

Description
The TML 105 turret is a lightweight three-man turret armed with a 105 mm NATO standard gun, which provides light wheeled or tracked vehicles (13-tonne class or over) with the firepower of a Main BattleTank. It draws on experience gained in combat with the Giat Industries' 105 mm turrets and benefits from the know-how developed for the Leclerc MBT.

Its modular design makes it suited to meet various operational requirements.

The NBC collective protection is provided by the sealed construction of the turret. The turret is fitted with the SAVAN 15 stabilised sight derived from the Leclerc MBT and associated with a digital computer, as well as with an IR CCD thermal imager. The tank commander's station is fitted with a Thales VIGY 15 stabilised panoramic day sight.

Optional equipment includes a laser warning and countermeasures system and the FINDERS® on~board Battlefield Management System.

Specifications
Combat weight: 5.2 to 5.6 t depending on versions
Dimensions
Overall: 7.8 × 2.6 × 2.2 m
Basket: 2.14 × 0.95 m
Ballistic protection
14.5 mm AP multihits and shell splinters
Medium-calibre rounds and shaped charges (infantry rockets), with add-on armour.
Firepower
Main armament: 105 mm/S 1-calibre high-pressure NATO gun.
Secondary armament: 7.62 mm coaxial machine gun,
7.62 mm roof- and pintle-mounted machine gun (optional)
GALIX close-defence system
Control: power-stabilised systems through a digital computer
Gun elevation/depression: –60 to ±180°
Turret traverse: n × 360°
Communications
Capability of integrating 2 HF and/or VHF radio sets

Operational status
Under development. Successfully trialled by several armies.

Contractor
Giat Industries.

VERIFIED

The Giat TML 105 turret 0089886

Giat Industries 25 mm DRAGAR turret

Type
Armoured fighting vehicle turret.

Description
Designed as a modular one-man turret with in-chassis commander's station, the DRAGAR turret meets the growing requirement for infantry fighting vehicles (IFV) combining extreme mobility with superior payload, high level of protection and firepower.

The DRAGAR turret is light and compact to preserve cross-country, carrying and amphibious capabilities. The modular ballistic protection, low profile and lack of highly inflammable hydraulic system (all-electric drive) ensure enhanced battlefield survivability.

In all weather conditions, the DRAGAR turret provides the dismounted combat units with effective fire support. The 25 M 811 cannon delivers 50% more firepower than any 20 mm weapon. The DRAGAR turret enables accurate on-the-move firing and provides for self-defence against low-level air threats. It is fully interoperable with NATO standard 25 × 137 mm ammunition (STANAG 4173).

The highly versatile DRAGAR turret has been fielded on the AMX-10, Piranha and AIFV armoured vehicles. It has completed successfull trials on the MI 13 and its derivatives, RN-94, VAB and Vextra. Basic features such as ballistic protection, sighting system or close defence are tailored according to user's requirements. Focusing on Life Cycle Cost, wide use is made of mass-produced commercial components.

The DRAGAR turret is currently in full-scale production. Some 540 units have been sold to date. Due to its modular design, the DRAGAR turret continuously benefits from technological advances. Expansions include defensive aid suites, battlefield IFF, in-chassis aiming video control, firing at the commander's station and FINDERS® on-board battlefield management system. The DRAGAR turret is being offered to equip the future French Army's wheeled IFV.

Operational status
In serial production.

Specifications
Crew: 1 gunner.
Combat weight: from 1,500 up to 2,000 kg depending on selected ballistic protection (for example 1,800 kg with 14.5 mm AP ballistic kit)
Basket diameter: 980 mm.
Basket height: 1,250 mm
Traverse: 360°, with 'no fire/no go' zones management and primary firing sectors control.
Elevation/depression: +45 to –8°
On-the-move observation and firing:
Day/Night sight: image intensified or thermal 360° panoramic ring of 4 day periscopes and 2 rear vision blocks
Laser range-finder: (optional)
Multiband screening smoke grenade launchers: Galix or similar
Options: FINDERS® on-board battlefield management system (expansion).
In-chassis aiming video control and firing at the commander's station (expansion)
Frequency hopping radio set (optional)
Ease of maintenance: very fast gun disassembling, built-in tests

Contractor
Giat Industries.

VERIFIED

Giat 25 mm DRAGAR turret 0089887

Giat Industries Toucan II turret

Type
Armoured fighting vehicle turret.

Description
Summary: the gunner has an image intensifying night sight.

The two-person Toucan II turret was designed by Giat Industries for the AMX-10P IFV.

The commander has an M371 sight with magnifications of ×1 and ×6, a direct sight for anti-aircraft use and an external sight for direct fire. The gunner has an OB 40 day/night periscope with a day magnification of ×6 (10° field of view) and a night magnification of ×5 (7° field of view). The gunner's OB 40 sight can be replaced by an M406 day sight with magnifications of ×2 and ×6 or an OB 37 image intensification sight with a magnification of ×6. The commander and gunner have seven periscopes for all-round observation.

Operational status
Production complete. In service with France, Greece, Qatar, Saudi Arabia and UAE.

Contractor
Giat Industries.

VERIFIED

Giat Industries TS 90 90 mm turret

Type
Armoured fighting vehicle turret.

Description
Summary: the gunner has an optional image intensifying night sight and laser range-finder.

The TS 90 modular/90 mm gun turret is suitable for use on any AFV weighing over 7,500 kg.

The commander has seven (three M556 and four M554) periscopes and the gunner five (three M556 and two M554). The gunner also has an M563 telescopic sight with a magnification of ×5.9 for laying the main armament.

Optional equipment includes: replacement of the gunner's telescope with a day/night telescope (day magnification ×5 with a 10° field of view and night magnification ×7 with a 7° field of view); a laser range-finder; replacement of commander's forward periscope with a periscopic gunsight and a magnification of ×5 to enable the commander to aim and fire both main and coaxial weapons; and a fire-control system.

A SAGEM (was SFIM Industries) SOPTAC 11 fire-control system can be fitted which includes day sight, night sight, laser range-finder and computer.

Operational status
Production complete. In service with a number of countries including France (army and gendarmerie), Indonesia, Ivory Coast, Oman, Saudi Arabia and Singapore.

Contractor
Giat Industries.

VERIFIED

GM Defense – Delco LAV-25 turret

Type
Armoured fighting vehicle turret.

Description
Summary: the turret is equipped with an image intensifier and the optical sight can be replaced with a variety of thermal imaging sights and laser range-finders.

Delco Systems Operations is the prime contractor for the two-person turret installed on the LAV-25 (8 × 8) vehicle built for the US Marine Corps. While the US Marine Corps LAV-25 was the first production application for this turret, it has since been sold to a number of other countries, both on the LAV-25 and on the Alvis Desert Warrior.

The commander has seven M27 unity periscopes for all-round observation and an M36E1 sight for aiming. The gunner has an M27 unity periscope and an M36E1 sight. The M36E1 sight has a day channel with a magnification of ×7 and a 10° field of view, passive night image intensifier channel with a magnification of ×8 and an 8° field of view and a unity channel with 60° horizontal × 10° vertical field of view for observation.

The M36E1 can be replaced by various other sight models such as: the Delco Improved M36 sight (DIM36); the Delco LAV-25 thermal sight; the GM Raytheon (formerly Hughes Electronics) (GMHE) Integrated TOW Sight (GITS) (qv); the DIM36 with laser range-finder; and the Delco LAV-25 thermal sight (qv) with laser range-finder.

Turret options include a Raytheon (HIRE) thermal sight (qv) which contains accurate stadia and ballistic reticles and has both wide and narrow fields of view. It

The Delco LAV-25 turret

is compatible with the existing M36E1 sight and uses a biocular video viewer or remote CRT display which is fully operable from the commander's position.

The TOW missile system can be added. The TOW lightweight sight and tracker is the GITS, which is obtained by expanding the Delco thermal sight with add-on modules. It is shared with the 25 mm cannon.

A new turret, the LAV-30 has also been developed, incorporating many sub-systems common to the LAV-25.

Operational status

In production and in service with Australia, Canada, Kuwait, Saudi Arabia and USA. A single turret has been procured by Finland for trials, with a possible production order to follow.

Specifications

Daylight channel

Magnification: ×8
Field of view: 10°

Thermal channel

Narrow field of view: 4 × 2°
Wide field of view: 12 × 5°

Unity channel

Magnification: ×1
Horizontal field of view: 60°
Vertical field of view: 10°

Contractor

GM Defense – Delco Systems.

GM Defense – Delco, Royal Ordnance 120 mm Armoured Mortar System (AMS)

Type

Armoured fighting vehicle turret.

Description

Summary: the gun sight is fitted with a thermal imager and laser range-finder.

The 120 mm Armoured Mortar System can be fitted to a wide variety of light armoured vehicle chassis. The full-solution fire-control system has: an integrated day/thermal imaging sighting system; an integrated laser range-finder which provides automatic range inputs to the fire-control computer; a differential Global Positioning System (GPS) aided Turret Attitude Sensor System (TASS).

Operational status

Ready for production, on order for the Saudi Arabian National Guard.

Contractors

BAE Systems, Royal Ordnance Division.
GM Defense – Delco Systems.

Hägglunds two-man 30 mm gun turret

Type

Armoured fighting vehicle turret.

Description

Summary: the gunner has a thermal sight and laser range-finder.

The two-man 30 mm gun turret was originally designed for use on the Combat Vehicle 9030 and can also be installed on a wide range of other armoured vehicles. The gunner on the right of the turret has a Saab Tech Universal Tank and Anti-Aircraft System (UTAAS) sight with thermal imager, laser range-finder and a ballistic computer. The commander has an SFIM Industries TJ71H periscopic sight and remote thermal display. Around the turret hatches are 11 periscopes (seven for the commander and four for the gunner).

Operational status

In production. 104 produced between 1997-2000. In servvice with the Norwegian Army (CV9030N), Finland (CV9030) and Switzerland (CV9030CH).

Contractor

Hägglunds Vehicle AB (now part of Alvis plc).

UPDATED

Hispano-Suiza Lynx 90 90 mm turret

Type

Armoured fighting vehicle turret.

Description

Summary: laser range-finder; there is an optional image intensifying night sight.

The Lynx 90 turret is suitable for a variety of armoured fighting vehicles such as the Panhard AML (4 × 4) and the Panhard ERC (6 × 6) armoured cars. A laser range-finder such as the Cilas TCV 107 is mounted over the main armament just in front of the mantlet. A SAGEM SOPTAC 11 fire-control system with laser range-finder, computer and passive night vision equipment for the commander and gunner can also be fitted.

In 1983, the Lynx 75/90 turret was announced. Optional equipment includes a day/night sight and a laser range-finder.

Operational status

Production as required. In service with the French Army (on Panhard chassis) and countries in Africa, the Middle East and South America.

Specifications

Optics

Commander: 8 × L794 periscopes in cupola and optional passive night periscope as replacement for one of the day periscopes
Gunner: 4 × L794 periscopes plus day sight. The latter can be replaced by SFIM Industries day/night telescope model TJN 2-90B

Contractor

Hispano-Suiza.

UPDATED

KUKA 20 mm two-man Marder 1A3 turret

Type

Armoured fighting vehicle turret.

Description

Summary: a thermal imaging sight is standard in the Marder 1A3.

This two-person turret was developed and manufactured by KUKA (now part of Rheinmetall Landsysteme) for the Marder 1 IFV. It has been introduced into the German armed forces with more than 2,100 units in service. A dual-control system allows both the commander and the gunner to operate the turret and fire the weapons, thereby allowing the commander to override the gunner's system. Under-armour observation and sighting are accomplished by eight periscopes for the commander and three periscopes for the gunner, as well as two combined sighting and observation periscopes type PERI-Z 11 A1. In the latest turret configuration (Marder 1A3) there is one type PERI-Z 11 and ancillary sight device from the thermal imager for the commander with combined sighting and observation periscope and one PERI Z 11 A1 and a Zeiss thermal imager for the gunner.

Operational status

In service with the German Army (2100 Marder 1 A3 IFV).

Contractor

Rheinmetall Landsysteme GmbH (formerly KUKA).

KUKA two-man turret E4

Type
Armoured fighting vehicle turret.

Description
Summary: the gunner's station has a two-axis stabilised 360° periscope with day sight, night sight and laser range-finder. The commander's station has the PERI Z17mod.

The turret E4 has been primarily designed to engage ground targets but can engage combat helicopters as well.

A PERI Z 17mod at the commander's station and an STN ATLAS 'FAUST' 360° periscope at the gunner's station are fitted for sighting and observation. The PERI Z17mod has a ×8 magnification and a 1:1 periscope. The 360° periscope has ×2 or ×8 magnification and is equipped with a thermal imager and laser range-finder. Three 1:1 periscopes at the commander's station and four 1:1 periscopes at the gunner's station provide additional observation aids. Ballistic calculations are carried out by the fire control computer. Required data are displayed in the gunner's sight and on two CRTs. Turret and weapon are slaved to the sight. The commander can also fire the weapon using the fire-control system.

Operational status
Ready for production on receipt of orders. Has been tested on the Swiss MOWAG Trojan infantry combat vehicle.

Contractor
Rheinmetall Landsysteme GmbH (formerly KUKA).

The KUKA E4 two-man turret 0053708

Kurganmashzavod BMP-3 100 mm turret

Type
Armoured fighting vehicle turret.

Description
Summary: the fire-control system includes an image intensifying night sight and laser range-finder. A thermal imaging sight is optional.

The BMP-3 turret is designed for combat vehicles, naval and riverine warfare vessels and other platforms. It is in use on the BMP-3 ICV, which is in service with Kuwait, the Russian Federation and Associated States and UAE (Abu Dhabi).

The automated fire-control system consists of day and night sights, laser range-finder, the armament stabilisation system and a ballistic computer with data input sensors. This allows for both the commander and gunner to engage ground and low flying aerial targets while the platform is either stationary or on the move.

An upgraded BMP-3 turret 0077969

The fire-control system may be enhanced by fitting the Namut (qv) thermal imaging sight to the gunner's position. The Namut sight is a joint development by SAGEM, Peleng and Kurganmashzavod. The vehicles for the UAE are fitted with Namut.

Operational status
Production as required. In service with Azerbaijan, Cyprus, Korea (South), Kuwait, Russia and the United Arab Emirates.

Contractor
Kurganmashzavod JSC.

Kurganmashzavod Modified BMP-2 30 mm turret

Type
Armoured fighting vehicle turret.

Description
Summary: the gunner may have a thermal sight.

The Modified BMP-2 turret is for combat vehicles, naval and riverine warfare vessels and other platforms. The fire-control system allows for target designation and weapon laying by the turret commander. The commander has a monocular daytime periscopic sight for ground and high-angle fire. The gunner has a combined day/active-passive night binocular sight. The gunner's sight can be retrofitted with a Sanoet-1 (qv) thermal imaging sight, developed by SAGEM, Peleng and Kurganmashzavod.

Operational status
Production as required. In service with Afghanistan, Algeria, Angola, Armenia, Azerbaijan, Belarus, Czech Republic, Finland, Georgia, India (local production), Indonesia, Iran, Iraq, Jordan, Kazakhstan, Kuwait, Russia, Sierra Leone, Slovakia, Sri Lanka, Sudan, Syria, Tajikistan, Togo, Turkmenistan, Ukraine, Uzbekistan and Yemen. It is also fitted to the Egyptian-built Fahd-30 (4 × 4) armoured personnel carrier (of which at least 30 have been built).

Contractor
Kurganmashzavod JSC.

The BMP-2 infantry fighting vehicle can be upgraded with the thermal imaging Sanoet-1 gunner's sight 0004773

Kvaerner Eureka Armoured Launching Turret (ALT) for TOW missile systems

Type
Armoured fighting vehicle turret.

Development
The turret was designed and developed by Kvaerner Eureka and the Royal Norwegian Army Material Command during the 1980s to launch the American Raytheon TOW anti-tank missile. It has been selected by the Royal Norwegian Army (on an M113), the Swiss Army on the Panzerjäger 90 MOWAG Piranha (6 × 6) chassis, the Canadian Forces (on M119) and the Turkish Army (on the Turkish Armoured Combat Vehicle). It has also been trialled on the Finnish Patria Vehicles Oy XA-186 AFV.

With this turret system the gunner is under full armour protection and new missiles can be loaded from the rear without the loader being exposed to small arms fire and shell splinters.

At Eurosatory 1998 Kvaerner Eureka and Delco Defense Systems Operations of the USA announced a co-operation agreement regarding the joint marketing and sale of the ALT turret. Delco will market, co-produce and support Kvaerner's ALT in selected international markets.

Description

The turret is of all-welded armoured steel construction with a single-piece hatch cover, and a single periscope mounted in the front of the turret for observation. On either side of the turret is the elevating armoured TOW launcher.

The Armoured Launching Turret (ALT) has in its forward part, the electro-optic sight and guidance systems of the normal infantry version of the TOW system without any modifications. It is possible to reassemble these with the tripod, traverse and elevating mechanism and the launcher tube of the original infantry TOW system and deploy the system away from the vehicle in a few minutes. The new hand controller used in the turret is very similar in operation to the original TOW controller. The turret is capable of firing Basic TOW, Improved TOW and TOW-2 missiles.

Time between the first missile impacting the target and the second missile being armed and ready for launching is 2 seconds. Reload time is about 40 seconds for both missiles with the turret in the loading position. Fire-control interlocks prevent the arming of more than one missile at a time. The vehicle can be driven at full speed with both launchers armed, coming to a halt for firing.

Kvaerner Eureka Armoured Launching Turret on Swiss Army Panzerjäger 90, MOWAG Piranha (6 × 6) armoured vehicle, with two TOW missiles ready to launch
0022559

M113 with Kvaerner Eureka Armoured launching Turret with four TOW missiles in ready to launch position

FMC-Nurol Armoured TOW Vehicle (ATV) based on the Turkish Infantry Fighting Vehicle chassis

Variant

Improved Armoured Launching Turret for TOW missile systems.

By mid-1994, Kvaerner Eureka completed development and testing of its improved TOW launching turret designs. These have been fitted with enhanced protection and firepower features. The first type incorporates add-on armour according to customer requirements and four ready to fire TOW missiles. The four-round TOW launcher underwent trials in Austria during 1995. The missile used was the TOW-2B.

The second type incorporates the same add-on armour capability and includes the addition of a 12.7 mm heavy machine gun and two seven-round pods of 2.75 in rockets for attack and self-defence. Both the gun and rocket pods have a maximum elevation capability of 30°. Two TOWs are fitted in the ready to launch position.

These upgrades can be supplied as either new build or as add-on modification kits to existing turrets. The four-round TOW turret has also been modified to accept the lightweight launcher sight for TOW developed by Gyconsa of Spain if specified by the customer. The modified turret was the type trialled in Austria.

Operational status

Production as required (527 built to date). In service with Canadian Forces (72 licence-built), Royal Norwegian (97), Swiss (310) and Turkish (48) armies. The Canadian systems were originally installed on M113 series full-tracked armoured personnel carriers but they are now being installed on a wheeled chassis.

Specifications

Basic version
Crew: 1 (gunner)
Armament: 2 × TOW launchers
Ammunition: 2 TOW ATGWs ready to launch
Control:
(traverse) n × 360° at 9°/s, slew; n × 360° at 2°/s, track
(elevation) –15 to +15° at 3°/s
Length: 1.5 m
Width: 1.06 m
Height: 1.1 m

Contractor

Kvaerner Eureka A/S, Defence Products Division.

Rheinmetall Landsysteme one-man turret E8

Type

Armoured fighting vehicle turret.

Development/Description

The one-man turret E8 is designed as a low-profile turret which can carry either a Mauser 30 mm MK 30F or, alternatively, The Boeing Company 30 mm Bushmaster II as main armament. The integration of other 30 mm or even 25 mm machine cannons is possible as well. The turret has 200 rounds of ready-use ammunition consisting of 100 AP rounds and 100 HE rounds in magazines located on the left and right hand side of the main weapon. The gunner can select either single-shot or burst modes of fire.

A 7.62 mm machine gun is mounted coaxially with the main armament. Cocking and firing mechanisms are remotely operated from under armour protection. Both weapons can be cocked and fired manually in an emergency mode. The armament of the turret is completed by eight smoke grenade launchers.

The E8 one-man turret consists of a turret shell welded of high hardness armour steel plates and a turret basket. Using add-on armour and internal spall-liner elements the turret can be optimised to the customer required ballistic protection level. The turret is equipped with electric drives for turret traverse and weapon elevation and is further characterised by the integrated Thales Optronics Sabre day/night or day/day sight and three more periscopes all together providing the gunner's all round vision. The state-of-the-art FLARM fire-control system from STN ATLAS Elektronik is fitted as the main sighting and observation device together with stabilised electric drives for traverse and elevation from the Swiss Company Curtiss-Wright Antriebstechnik. The FLARM consists of a two-axis stabilised optoelectronic sensor unit carrying a thermal imager TIM, a day-sight CCD camera and a laser range-finder. Images of TIM or CCD camera are provided on a flat display. Turret and weapon are slaved to the independent line of sight. Sensor unit and weapon stabilisation system are controlled by the fire-control electronics, so that superelevation and lead angle resulting from the ballistic parameters are

KUKA 30 mm one-man turret E8, prototype No 1
0053707

automatically realised. Further turret options include a laser warning system with a directional indicator and integrated command and control system.

Specifications

Crew: 1 (gunner)
Armament:
(main) 1 × 30 mm Mauser MK 30 automatic cannon (see text)
(secondary) 1 × 7.62 mm machine gun and smoke grenade launchers
Optics: Thales SABRE day/night or Thales SABRE day/day sight 3 × periscopes (option) STN ATLAS Elektronik FLARM fire-control system (see text)
Control: traverse 360°, elevation –10 to +45°, stabilised electric drives slaved to the optoelectronic independent line of sight
Weight: (approx) 2,300 kg, depending on fire-control equipment and ballistic protection level
Power supply: 24 V DC
Bearing diameter: 1,380 mm

Operational status

Prototype. Fitted to ELBO Kentaurus armoured infantry fighting vehicle which is undergoing company trials as a private venture project.

Contractor

Rheinmetall Landsysteme GmbH.

Steyr-Daimler-Puch SP30 weapon station

Type

Armoured fighting vehicle turret.

Development

The SP30 two-man weapon station was designed by Steyr-Daimler-Puch for installation on the ULAN armoured infantry fighting vehicle ordered by the Austrian Army, with the first vehicles completed in May 2001.

Description

The gunner's fire-control system is a modular thermal and visible target acquisition system with ground-to-ground and anti-aircraft capabilities within the framework of air defence of all troops. The system consists of:

(a) gunner's optical sight with thermal channel electrically controlled head mirror with mechanical support (Automatic Target Tracking system) and gunner's display (TV monitor)
(b) integrated laser range-finder
(c) digital Fire-Control Computer (FCC) with cant sensor providing improved firing performance by automatic determination of superelevation and azimuth lead as a function of ammunition type, range, turret azimuth rate, cant angle, altitude, temperature and crosswind. The following types of ammunition can be processed: cal. 30 mm: TP/HE, APDS, FAPDS, MPDS, APFSDS and MP; cal. 7.62 mm: Ball/AP
(d) Automatic Target Tracking (ATT) provides tracking of ground and air targets
(e) Computer Control Panels (CCP) for commanders and gunner's station
(f) commander's station remote display (TV monitor giving identical image to gunner's thermal display) and commanders optical sight
(g) a turret position indicator gives the driver, commander and gunner the turret position relative to the hull. The commander has a further five and the gunner one episcopes around their hatch covers for general observation of the surrounding terrain. Turret traverse and weapon elevation are carried out by the commander using electric power controls. The gunner has a manual back-up system.

Operational status

In production. On order for Austrian Army (on ULAN AIFV). First production ULAN were completed in May 2001.

Specifications

Optics
Gunner: laser integrated periscopic sight
Field of view: ×8 magnification, 8° FOV
Laser range-finder: unity prism, ×1 magnification
Laser type: Erbium glass (eye-safe)
Wavelength: 1.54 μm

Thermal imaging channel:
Fields of view: narrow ×7 magnification, 1.3 × 2.1° FOV; wide ×2.4 magnification, 4.6 × 7.3° FOV
Detector: 240 pixel MCT – PV
Spectral band: 8 – 12 μm
NETD: 0.25°C

Commander periscope sight:
Fields of view: ×6 magnification, 8.5° FOV and ×1 magnification, 20 × 40° FOV

Contractor

Steyr-Daimler-Puch Spezialfahrzeug AG.

Textron 105 mm Low Recoil Force Turret

Type

Armoured fighting vehicle turret.

Development

The 105 mm Low Recoil Force Turret is installed on the Textron Marine and Land Systems Stingray light tank and has also been installed on an M41 light tank chassis and the LAV-600 armoured car for trials. Thus far, Thailand has been the only customer for the Stingray light tank with 106 vehicles in service.

The gunner has the latest proven stabilised Raytheon Systems Company, Sensors & Electronic Systems, HIRE day/thermal night sight with integrated laser range-finder which is currently in volume production for a number of other applications. The gunner also has an auxiliary day telescope as a back up. The commander is provided with a monitor which enables him to have the same thermal picture of the target as the gunner. The commander's sight is electrically linked to that of the gunner and the commander has the ability to override the gunner.

Operational status

Production as required. In service with Thailand (on Stingray light tanks).

Contractor

Textron Marine & Land Systems.

Textron 25 mm turret

Type

Armoured fighting vehicle turret.

Description

Summary: the gunner has an image intensifying sight and the commander may also have one.

This turret has been developed as a private venture by Textron Marine and Land Systems (formerly Cadillac Gage) and was installed on the LAV-300 (6 × 6) and LAV-150 (4 × 4) vehicles entered in the LAV competition.

The gunner has an M36E1 day/night sight and the commander has eight periscopes for all-round observation. As an option, the commander could have six periscopes and an M36E1 day/night sight.

Operational status

Production as required. In service with several undisclosed countries.

Contractor

Textron Marine & Land Systems.

Textron LAV-105 mm weapon system

Type

Armoured fighting vehicle turret.

Development

Summary: the gunner has a thermal sight and laser range-finder. The commander also has a thermal sight.

The LAV-105 was developed for the US Marine Corps Light Armoured Vehicle (LAV). The Marine Corps has no plans for production of this system and it is now being offered for installation on a variety of other vehicles.

The commander and gunner each have a single-piece hatch cover that opens to the rear and four periscopes to view to the rear and sides. The gunner has four periscopes and a Raytheon Day/Night Range Sight (DNRS) stabilised thermal day/night sight with integral laser range-finder. The commander has four periscopes and the Raytheon DNRS remote monitor unit.

The Raytheon DNRS comprises five key subsystems, Raytheon Infra-Red (HIRE) 240-line common thermal image with dual field of view, Raytheon laser range-finder, Raytheon Line of Sight Stabilisation Platform (LSSP), commander's remote display for viewing the day and night sight images and with commander override for all the gunner's controls and the Line Of Sight Electronics Unit (LOS-EU). The DNRS has a unity window and a ×10 magnification narrow field of view for the day sight.

Operational status

Development complete. Ready for production. No funded US Marine Corps requirement. Still being marketed by Textron Marine and Land Systems.

Contractor

Textron Marine & Land Systems.

Thales AFV cupolas

Type
Armoured fighting vehicle cupolas.

Description
Summary: Thales (formerly Helio) cupolas can all have day/day or day/night image intensifying sighting systems.

The FVC114 and Cupolas No 16 and 27 (HVM or TRV) are designed to give a protected position for surveillance and target acquisition and, by using one of the weapon fit options, the ability to engage both ground and airborne targets.

A variety of small calibre weapons, such as the 7.62 mm GPMG (General Purpose Machine Gun) or Browning .50 in (12.7 mm) calibre machine gun, can be mounted on the FVC114, while the Cupolas No 16 and 27 are designed for mounting the 7.62 mm only. The weapon on FVC114 and Cupola No 16 can be loaded, fired and serviced under armour.

The sighting options include Thales Optronics' Day/Day, Day/Night second- or third-generation image intensified or thermal imaging sighting systems with optional eye-safe laser range-finder (DNGS series) or Thales Optronics Sabre sights. The No 44 Mk 4 periscope of Cupola No 27 can be replaced by an image intensifying viewing device.

Operational status
The FVC114 is in production and Cupolas No 16 and 27 are in production and in service with a number of countries.

Contractor
Thales AFV Systems (formerly Helio Ltd).

UPDATED

Thales FVT turrets

Type
Armoured fighting vehicle turrets.

Description
Summary: Thales AFV Systems (formerly Helio Ltd) one-person turrets can all have day/day or day/night image intensifying sighting systems.

Thales FVT800, 840, 900, 925 and 940 turrets form a family of one-person turrets designed specifically for Ground, Explosive Ordnance Disposal (EOD) and Air Defence roles on wheeled and tracked AFVs.

The FVT800 turret has a .50 in (12.7 mm) machine gun and 7.62 mm coaxial General Purpose Machine Gun GPMG (optional) and is available in several variant.

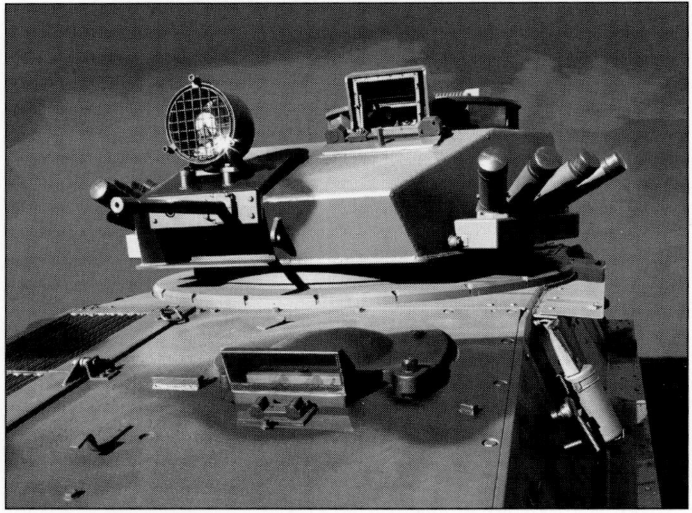

A Thales FVT 800 turret 0088173

The FVT840 turret has a 40 mm Automatic Grenade Launcher (AGL) with co-axial 7.62 GPMG and is available in several variants.

The FVT900 turret has a 20 mm cannon and 7.62 mm coaxial GPMG (optional) and is available in several variants.

The FVT925 turret has a 25 mm cannon and 7.62 mm coaxial GPMG (optional) and is available in several variants.

The FVT940 turret has a 40 mm Automatic Grenade Launcher (AGL) with a .50 in M2-HB QCB machine gun mounted on a cradle assembly and is available in several variants. Helio turrets have optional manual or power traverse and elevation, operating at speeds from 0.5 to 800 m/s.

A wide selection of sighting systems is available, including Thales Optronics (formerly Avimo) day/day, day/night second- or third-generation image intensifiers or thermal sighting systems with optional eye-safe laser range-finder (DNGS series) and Thales Optronics Sabre sights. The multipattern reticules are designed for ammunition and targeting as appropriate. Unit Vision Periscopes give the crew member almost 360° of vision without rotating the turret. These can be fitted with a blackout blind and/or replaceable laser filter.

Operational status
The FVT840,940 is in production, the FVT800, 900 and 925 are in production and in service with several countries.

Contractor
Thales AFV Systems (formerly Helio-Ltd).

United Defense 25 mm two-man turret

Type
Armoured fighting vehicle turret.

Description
Summary: the gunner and commander have thermal sights.

The United Defense LP two man-turret was developed and produced for the Bradley fighting vehicle. The M 242, 25 mm 'Chain Gun' with dual feed is the primary armament. An M 240 (7.62 mm) machine gun is coaxially mounted and provides close in and anti-personnel protection. A dual TOW anti-tank guided missile launcher is mounted on the left side. The TOW launcher is retracted into a stowed position when not in use. The sighting and controls for both the guns and the missiles are integrated to optimise the soldier machine interface. The thermal imaging day/night sight provides 4 × 4 and ×12 power, selectable viewing for both commander and gunner. An all-electric turret drive/ stabilisation system provides precision target acquisition and fire on the move capabilities.

The turret has undergone numerous upgrades since initial production of the A0 in 1981. The A1 turret incorporated the TOW II missile system with improved lethality and hit probability. The A2 turret, introduced in 1986, was uparmoured with steel applique and provisions for armour tiles. The ODS (Operation Desert Storm) up grades to the A2 turret include the Bradley Eye-safe Laser Range-Finder (BELRIF) and the precision Lightweight GPS Receiver (PLGR).

The A3 turret entered low-rate production in 1998 and leverages state-of-the-art technology (digital electronics, 1553 databus and second-generation FLIR) to provide significant improvements in situational awareness and fire control. FAC B2 software coupled with EPLARS and a precision Inertial Navigation System (coupled to GPS) assures situational awareness. The fire-control computer and the Improved Bradley Acquisition System (IBAS) and the Commanders Independent Viewer (CIV) provides full solution hunter killer-type fire control. Both the IBAS (which serves as the gunner's sight and incorporates an eye-safe LRF) and the CIV provide second-generation FLIR and visual channels and line of sight stabilisation.

Rebuild and upgrade of turrets to the A2-ODS and A3 configuration continues. Production of new turrets for Bradley is now complete but can be resumed if further orders are received.

Operational status
Production as required. A total of 6,785 M2/M3 units produced. In service with Saudi Arabia and the United States, both on Bradley chassis. This turret has been installed for trials purposes on the United Defense Infantry Fighting Vehicle Light.

Contractor
United Defense Limited Partnership (UDLP), Ground Systems Division.

ARMOURED FIGHTING VEHICLES

FIRE CONTROL

ATCOP IFCS 69 Integrated Fire-Control System

Type
Tank fire-control system.

Description
The Integrated Fire-Control System IFCS 69 integrates various sensors with a ballistic computer to compute gun correction for typical MBT and AFV ammunition. It is specifically designed for T-55 and T-59 tanks and offers precision firing and gun control from static-to- static and static-to-moving targets.

The system includes a ballistic computer, azimuth tracking and tilt/cant sensors, temperature and cross-wind sensors and gun controller. It can be coupled to the ATCOP GNS 1 gunner's night sight and ATCOP TR 2 laser range-finder. Typical time to acquire, range, track and fire to a moving target is 6 seconds. Operating temperature is –30 to +35°C. Interface is available to customer's existing laser range-finders and gunner's night sights. Ruggedised to MIL-STD 810D.

Operational status
Available.

Contractor
Al Technique Corporation of Pakistan (Pvt) Ltd.

VERIFIED

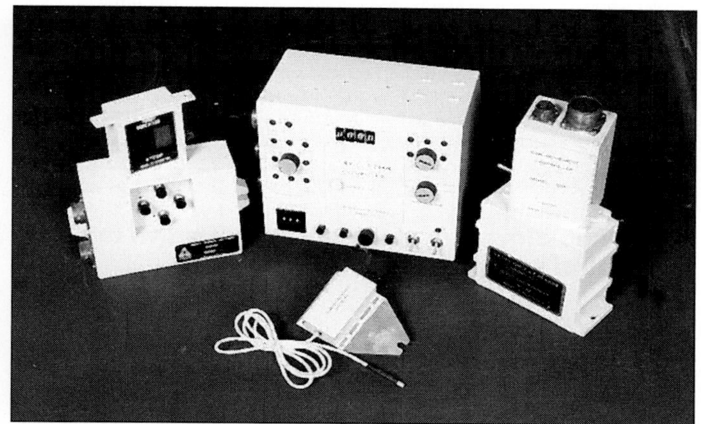

IFCS 69 Integrated Fire-Control System 0004769

Bharat tank fire-control system Mk34-1B

Type
Land fire-control system.

Development
The tank fire-control system Mk 1B (AL4421) developed by the Bharat Electronics Tank Electronics Support Centre, Madras is a computerised follow-on to the Mk 1A for the Vijayanta MBT. It incorporates a Thales Optronics tank laser sight assembly.

Bharat Electronics is involved in developing the tank fire-control system for the Arjun MBT project which is believed to be a further development of the Mk 1B (AL4421) and includes a day/thermal night sight with integrated laser range-finder, computer and various sensors.

Description
The aim of the Mk 1B (AL4421) is to reduce the target engagement time cycle and increase the first-round hit probability of both a moving tank engaging a static target and a static tank engaging a moving target at any speed. Night firing is possible with a night sight adaptor mounted on the existing infra-red assembly. If required the FCS can be adapted to fit any tank model.

The 105 mm L7 main gun can be fired using one of the following options:

(1) The complete FCS and its subsystems in either:
  (a) 'autolay' mode – with the gun driven directly by the Motorola MC6800 ballistic computer
  (b) 'stab' mode – in which the gunner uses the existing GCE weapon power controllers to bring the aiming mark on to the target. The engagement time is about 8 seconds against static targets and 10 seconds against moving targets
  (c) 'manual' mode – whereby the gunner operates conventional weapon control back-up traverse and elevation handwheels to align the aiming mark with the target. Engagement time against static targets is better than 10 seconds in this mode. Changeover between the modes is instant, with the computer-generated aiming mark also displaying the target range as measured by the laser range-finder.

Tank Laser Sight used with the Bharat Electronics Tank Fire-Control System Mk 1B

(2) The 1.064 μm Nd:YAG laser range-finder on its own by means of a unique look-up table in the sight graticule.
(3) By reversion to use of the original coaxially mounted 12.7 mm calibre ranging machine gun facility. This has a maximum range of 1.8 km and fires three-round bursts of tracer ammunition.

A Muzzle Reference System (MRS) is integrated with the sight optics so that the gunner can align, within 10 seconds, the sight and gun axes without moving from his seat.

The fire-control and gun-control systems are fully integrated, with both the system power supplies and the GCE weapon controller interfaces housed in the computer unit.

The computer is normally programmed for 105 mm APDS and HESH ammunition types but only requires simple software changes for new ammunition types. Crosswind, line wind, ammunition charge temperature and barrel wear parameters have to be entered manually.

Movement limits of the sight mount, armoured hood and sector gear assembly are –7° in depression to +18° in elevation.

Operational status
In production. In service with the Indian Army (Vijayanta MBTs).

Specifications
Dimensions
Tank laser sight: 333 × 289 × 528 mm
Aiming mark electronic unit: 311 × 175 × 170 mm
Ballistic computer and power supply unit: 310 × 420 × 345 mm
Tilt sensor assembly: 235 × 170 × 55 mm
Traverse encoder assembly: 260 × 190 × 230 mm
Elevation encoder assembly: 110 × 160 × 100 mm
Weight
Tank laser sight: 21.3 kg
Aiming mark electronic unit: 6.7 kg
Ballistic computer and power supply unit: 22 kg
Tilt sensor assembly: 5.5 kg
Traverse encoder assembly: 9.6 kg
Elevation encoder assembly: 0.8 kg
Sight mount: 15 kg
Armoured hood: 27 kg
Linkage fire control: 8.3 kg
MRS light source assembly: 10 kg
MRS mirror assembly: 0.45 kg
Gunner's tank laser sight
Magnification: ×9.4

Field of view: 8.5°
Laser range-finder type: Nd:YAG
Wavelength: 1.064 μm
Operating range: 300-9,990 m
Range accuracy: ±10 m (90% of shots)
Acquisition sight
Magnification: ×1
Computer
Max range processing: 9.99 km
Range processing accuracy: 10 m
Ammunition types: HESH and APDS (others with software changes)
Trunnion tilt: 250 mils max
Trunnion tilt accuracy: 0.25 mil
Elevation accuracy: 0.2 mil
Traverse displacement accuracy: 0.02 mil
Crosswind: 25 m/s
Line wind: 25 m/s
Control system
Ballistic computation accuracy: better than 0.1 mil
Gun positioning accuracy: 0.2 mil

Contractor
Bharat Electronics.

VERIFIED

CEIEC GM-09 tank fire-control system

Type
Land fire-control system.

Description
The GM-09 tank fire-control system was developed in the early 1980s and is in service with the People's Liberation Army on some of its T-series MBTs. It is also being offered for export as an upgrade for T-54/T-55 tanks.

The GM-09 comprises a central control unit which incorporates an 8080 microprocessor. Connected to this are a number of sensors and control systems: azimuth rate sensor for the turret; an elevation rate sensor for the main armament with an in-built ammunition selection indicator unit; a gun trunnion tilt sensor; a laser range-finder; and a modified gunner's telescopic sighting system which has an optical system attached to display the computed lead to the gunner by means of an aiming point on a mini-CRT.

During combat, the gunner stops the tank temporarily when a target is observed, sights it and then tracks it for 3 seconds while the laser range-finder is triggered. The computer derives the parameters of the target's movements from the incoming data of target range and angular increments and computes the predicted azimuth and firing angle by utilising any control panel and sensor inputs. This firing solution is simultaneously transmitted to the automatic gunlaying device on the weapon control handle and the gunner aims and fires at the target.

The entire process from observing the target to firing the weapon takes not more than 10 seconds with computation taking about 1 second.

Operational status
Production as required. In service with the People's Liberation Army and other countries.

Specifications

Unit	Weight	Dimensions
Central control unit	19 kg	500 × 400 × 150 mm
Power supply unit	11 kg	280 × 210 × 240 mm
Azimuth rate sensor	1.8 kg	200 × 120 × 150 mm
Elevation rate sensor	1.3 kg	130 × 130 × 80 mm
Trunnion tilt sensor	2 kg	140 × 135 × 90 mm
Modified control handle	8 kg	180 × 200 × 200 mm
Laser range-finder power supply	2.25 kg	176 × 130 × 90 mm
Laser range-finder	7 kg	344 × 118 × 141 mm
Gunner's telescope spot injection system	–	200 × 60 × 40 mm
Laser range-finder range	–	300-6,000 m
Accuracy		±10 m
Tracking speed		
Azimuth	not less than 40 mils/s	
Elevation	not less than 10 mils/s	

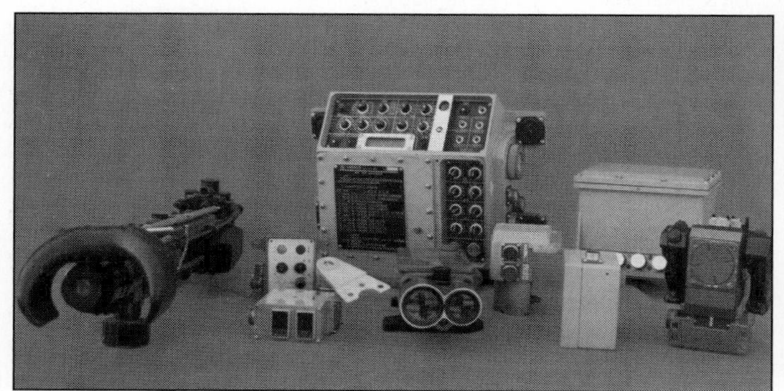

Main components of the CEIEC GM-09 tank fire-control system

Contractor
China National Electronics Import & Export Corporation (CEIEC).

VERIFIED

Chung Shan Institute M48H advanced fire-control system

Type
Land fire-control system.

Description
The Taiwanese Chung Shan Institute of Science and Technology has developed an advanced fire- control system for the M48H hybrid MBT. This uses new M60A3 tank hulls purchased from General Dynamics Land Systems Division before the production line closed down, modified M48A5 type turrets and a locally produced United Services Ordnance Institute of Taiwan version of the 105 mm M68 rifled tank gun.

The fire-control system incorporates a Nd:YAG laser range-finder, a ballistic computer and Raytheon's AN/VSG-2 Tank Thermal Sight.

Operational status
In production. In service on M48H MBTs of the Republic of China Army.

Contractor
Chung Shan Institute of Science and Technology.

UPDATED

Czech and Slovakian state factories Kladivo tank fire-control system

Type
Land fire-control system.

Description
The Kladivo fire-control system (FCS) contains a laser range-finder and a laser warning device. The first conversion of T-54/T-55 tanks with the Kladivo FCS were observed in service with Czechoslovakia in 1984 and since then Bulgaria, Hungary, Poland and the Russian Federation and Associated States have updated many of their T-55s to the same standard with similar systems.

The Kladivo FCS comprises a laser range-finder mounted externally above the mantlet of the 100 mm gun, ballistic fire-control computer, wind velocity sensor, mast-mounted meteorological sensor (that also incorporates a laser warning device), operating control switch, data input devices, commander's periscopic sight, ammunition selection switch and power supply.

The laser may be triggered up to 10 times a minute but its performance is degraded in rain, snow and fog. The armoured meteorological mast, mounted to the rear of the turret roof, contains ambient air and pressure sensors as well as the laser warning device.

The computer is capable of estimating target speed and direction for the establishment of the fire-control solution lead angle.

Czech T-55AM2 MBT showing laser range-finder over 100 mm gun

The commander's periscope is installed in the commander's cupola and allows independent observation of the target by the commander and the automatic hand-off of the target to the gunner.

The loader operates the ammunition selection switch and has a choice of five different conventional ammunition types, with the actual selection being made by the tank commander on the basis of the target type. The ammunition types are HE, HEAT, incendiary, APFSDS and machine gun. The power supply unit maintains a constant voltage for the system.

The installation of the Klavido FCS has enabled the upgraded T-54/T-55 vehicles to engage moving targets out to an effective range of 1.6 km with a high first- round hit probability.

Operational status

Production as required. In service with the armies of the Czech Republic, Hungary and Slovakia. Being offered as an upgrade for T-54/T-55 tanks.

Contractor

Czech and Slovak state factories.

VERIFIED

Elop Knight family of Advanced Tank Fire-Control Systems (ATFCS)

Type

Land fire-control system.

Development

The Knight Mk I modular ATFCS has been developed as a retrofit/upgrade package for the M48, M60 and Centurion MBTs. Its modular nature also allows easy adaptation to other turret types with the minimum of interfacing changes. A Knight Mk II with more sophisticated systems is available for new build MBTs.

The Knight Mk III is a new fire-control system designed for use on the Merkava Mk III MBT and combines improvements in Line Of Sight (LOS) stabilisation, an Automatic Target Tracking (ATT) system and precision slaving of the tank's weapon systems to the LOS. The system also incorporates TV, thermal imaging channel and computers into a central control station.

Description

Knight is fully integrated with the turret weapon control/drive system and actively controls the turret dynamics. It has stabilised LOS, a short target acquisition cycle, a day and night operation capability and a high hit probability for stationary and on-the-move engagements.

The major modes of operation are:

(a) 'stabilised' where the gun axis is slaved to the LOS
(b) 'slave' where only the gunner's LOS is slaved to the gun axis.

The main components of the complete Knight system are:

(a) ballistic computer
(b) Dual-axis stabilised head mirror
(c) Eyesafe Laser Range-Finder (LRF)
(d) Thermal imaging camera
(e) High Definition TV (HDTV) channel
(f) Display units (CRT and flat panel)
(g) Central computer
(h) Meteorological mast
(i) Dynamic cant angle sensor
(j) Gun resolver control panels
(k) sensors

The Knight fire-control system family 0134534

Operational status

Production as required. In service with Israel Defence Force on M60 and Merkava series MBTs.

Contractor

Elop Electro-Optics Industries Ltd.

UPDATED

Elop Lansadot family of armoured vehicle fire-control systems

Type

Land fire-control system.

Description

The Lansadot compact modular fire control family (Mk I, Mk II, Mk III and Mk IV) has been developed by Elop and Elbit for updating various types of AFVs. No vehicle modifications are required for M41 light or M47 medium tanks, and only minor modifications for the AMX-13 light tank, AMX-30 MBT, Sherman medium tank (all models), Saladin armoured car, T-54/T-55 MBTs, Light Armoured Vehicles (LAVs) and Armoured Personnel Carriers (APCs).

Depending upon model the Lansadot can be capable of day and night observation, target identification, laser range-finding and accurate firing against stationary and moving targets.

The main components are:

(a) gunner's peritelescope sight with integrated laser range-finder, passive image intensifier night vision elbow, and unity prism for daylight observation purposes
(b) ballistic computer system with computer, computer control unit and cant angle sensor. The computer provides a single reticle for day/night lasing and aiming and automatically compensates for ballistic parameters.

Operational status

Production as required. In service with unspecified countries.

Specifications

(Lansadot Mk I)
Gunner's peritelescopic sight
Elevation: +35° (optional +60° version for anti-aircraft application)
Depression: –18°
Day channel field of view
×8 magnification: 8°
×1 magnification: 40 × 14°
Night channel field of view
×7 magnification: 7.5°
Laser range: 300 to 9,900 m
Laser range accuracy: ±5 m
Computer compensation ranges (day and night)
Line of sight
Horizontal deflection: ±30 mils
Superelevation: 0-70 mils

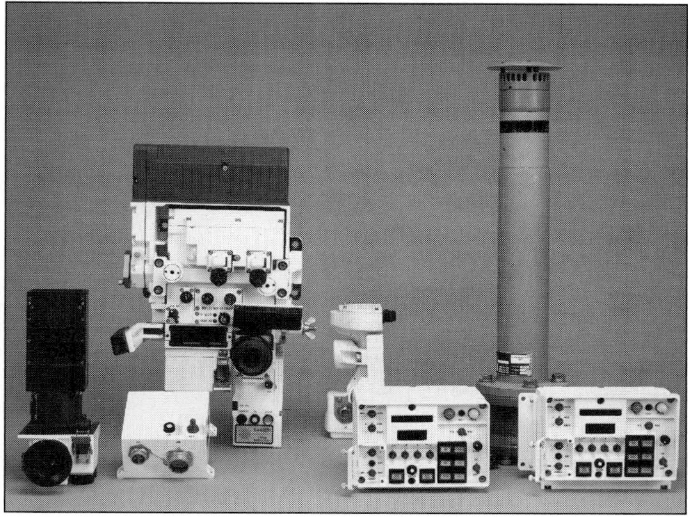

Lansadot Mk I fire-control system, gunner's peritelescope in night configuration

Cant angle: ±20°
Max moving target velocity: 72 km/h
Crosswind: ±19 km/h
Ambient air temperature: –19 to +69°C
Altitude: –400 to +4,500 m

Contractors
Elop Electro-Optics Industries Ltd.

UPDATED

Elop Thermal Imaging Stand Alone System (TISAS)

Type
Land fire-control system.

Description
TISAS is intended as a fire-control system upgrade package for the T-series of MBTs and is designed to be fitted with minimum modification to the tank.

Features
- High-performance thermal sight
- Electrical and mechanical head mirror configurations
- Mechanical back-up capability
- Full ballistic solution for the thermal and day channels
- Commander monitoring and operation, using display.

Upgrade Capabilities:
- Add-on ballistic sensors
- Additional new ammos in both thermal and day channels.

Advantages:
- Modular and compact, fitted to the T-72 turret drop-in, no machining required
- Retains current operation procedures
- Currently in production
- Excellent reliability and maintainability.

Applications:
- Upgrades of T-55, T-72, T-80 and T-90 tank families.

Specifications
LOS elevation: –15 to +25° (stabilised and slaved to TPD-K1 day sight)
Thermal channel: second-generation high performance
Ballistic tables: 90 mils in elevation +1 to 15 mils in azimuth
LOS and ballistic: 6 ammos, environmental corrections
Ballistic solution: accuracy 0.1 mil

Contractor
Elop Electro-Optic Systems Ltd.

NEW ENTRY

The TISAS family of fire-control systems from Elop **NEW**/0547550

ENOSA Improved Mk 7 Laser Tank Fire-Control System (LTFCS) for M48A5E MBT

Type
Land fire-control system.

Development
ENOSA has built under licence from Hughes Aircraft Company (now Raytheon) an Improved LTFCS for the locally upgraded M48A5E MBT. ENOSA also produces the LTFCS for the Spanish Army AMX- 30E MBT.

Description
The system has full-solution day/night and shoot-on-the-move capabilities with automatic lead, using the following components:
(a) an M48A5/M60A3 digital solid-state ballistic computer to determine the gun aiming deflection and superelevation angles. The following inputs are made automatically from sensors:
(i) target range
(ii) cant angle
(iii) target angular velocity
(iv) crosswind
these parameters can be input manually:
(i) range
(ii) ambient air temperature
(iii) altitude
(iv) crosswind
(v) gun jump correction
(vi) zeroing correction
(vii) gun barrel wear.
Up to four ammunition type parameters are also stored.
(a) M32/M35 periscope assemblies
(b) automatic turret cant angle, turret tacheometer (to measure target velocity for lead angle) and relative crosswind velocity sensors
(c) modified M1 Abrams 1.064 μm Nd:YAG digital laser range-finder unit with ×8 magnification and 7° field of view day telescope, ±10 m range and 30 m dual-target resolution capabilities
(d) gunner's control unit with auxiliary manual input, boresighting, ammunition zeroing and self-test facilities
(e) passive night sight with either:
(i) monocular image intensifier elbow with ×7 magnification and 7.3° field of view. Light gain is 1,000 minimum with a focus range of 50 m to infinity. Resolution is 0.23 mil at 0.1 lux
(ii) monocular or biocular display 60-element Cadmium Mercury Telluride (CMT) linear array split, Stirling cooled, thermal imager unit, operating in the 7.5 to 11.8 μm wavelength spectral region. The wide field of view capability is 6.1° elevation and 12° azimuth. The narrow field is 1.7° elevation and 4° azimuth. Focus range is 75 m to infinity with NATO and laser cross graticules superimposed on the display
(f) gunner's and commander's ammunition selector units. Four basic ammunition selections can be made together with a stationary or moving selector option.

Operational status
Production as required. In service with the Spanish Army (on 164 M48A5E MBTs).

Contractor
Empresa Nacional de Optica SA (ENOSA).

VERIFIED

Main components of ENOSA Mk 7 LTFCS including (bottom left) passive night vision elbow and (bottom right) laser range-finder unit

ENOSA MK-10 fire-control system for armoured vehicles

Type
Fire-control system (land).

Development
The MK-10 is an integrated fire-control system with thermal imager and laser range-finder for Infantry Combat Vehicles (ICVs) and Light Armoured Vehicles (LAVs). ENOSA, in collaboration with a vehicle manufacturer, has developed, fielded and tested the MK-10 Fire-Control System for one- or two-person turrets on armoured fighting vehicles and light armoured vehicles with 25 to 105 mm guns. This FCS package is largely based on non-developmental in-service hardware items and is composed of units, subassemblies and components that are installed on in-service tanks such as upgraded M48A5 and AMX-30 and many other related tank fire-control systems. The MK-10 is stated to provide ICVs and LAVs with fire-control performances comparable to modern main battle tanks, such as automated operation, full-solution fire control and 24 hour engagement capability.

Description
MK-10 is a compact, disturbed-reticle type, fire-control system. The core elements of the system are a gunner's sight, integrating day, thermal and laser channels, slaved to the gun and a ballistic computer, that generates a full ballistic solution for the displacement of the aiming reticle, based on various automatic or manually entered parameters.

Shoot-on-the-move is possible if a gun/turret stabilisation system is separately provided to the vehicle. Better stabilisation accuracy is achievable in lighter vehicles than with the heavy gun of main battle tanks.

Vehicle installation is easily carried out on small turrets as the gunner's sight adapts to a standard M-36 periscope aperture.

System components include:

(a) The head mirror unit includes a one-axis mirror mechanically slaved to the gun in elevation. If the computer is off or failing, it will behave as a stand-alone aiming periscope. When the fire-control system is on, an electrical mechanism inside the head will offset the line of sight by the superelevation angle provided by the computer.

The MK-10 fire-control system

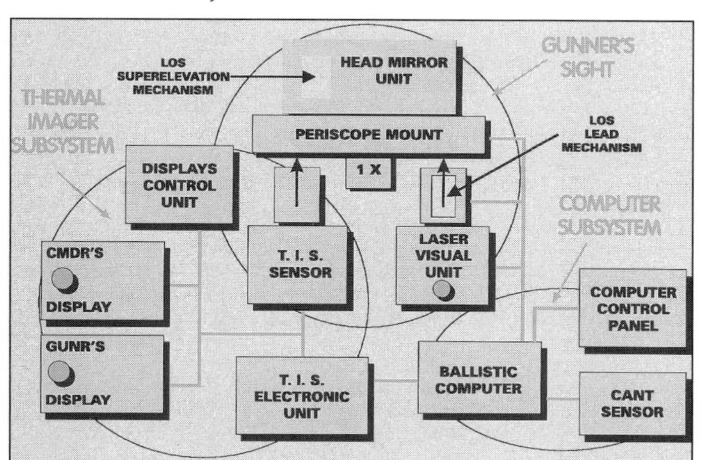

System block diagram of the MK-10 fire-control system from ENOSA

(b) The thermal imager is an update of the original US common modules family of thermal sights that provides improved performance (vertical resolution, sensitivity and image quality) and increased modularity and flexibility for sight reconfiguration. The gunner and the commander have separate display units. Lead ballistic angle provided by the computer is injected in this channel through the displacement of a software generated aiming reticle.

(c) The integrated day sight/laser range-finder, with a single aperture for day vision and laser emission and reception, and laser electronics integrated in the same LRU largely accounts for the compactness of the system. Also included is an optical mechanism that moves in azimuth at the same time as the scene and the laser/aiming reticle to receive the lead correction from the computer.

(d) The ballistic computer, controlled from a small keyboard.

Operational status
In production for the Spanish Pizarro ICV.

Contractor
Empresa Nacional de Optica SA (ENOSA).

VERIFIED

ENOSA MT-01 fire-control system for T-72 MBT

Type
Fire-control system (land).

Development
The MT-01 is an integrated tank fire-control system with thermal imager and laser range-finder for the upgrade of T-72 MBTs. Since 1995, ENOSA has collaborated with a T-72 manufacturer to develop the MT-01 fire-control system which upgrades the fire power of the T-72 in service or the new T-72 M3 vehicles. The upgrade package is based on in-service non-developmental hardware items and is composed of units, subassemblies and components that are installed on in-service tanks such as upgraded M-48A5s and AMX-30s. The MT-01 integrates to the turret with minimal modifications to layout and human factors: care has been taken to avoid armour cutting and to keep the original location of operator's eyepieces, handles and control units. The MT- 01 fire-control system provides T-72 tanks with fire-control performances comparable to latest-generation NATO tanks, such as automated operation and a fire on the move director-type system, full-solution fire control and 24 hour engagement capability.

Description
MT-01 is a fully stabilised, director-type fire-control system. The core elements of the system are a gunner's primary sight, primary stabilised in two axes, and integrating day, thermal and laser channels, to which the gun/turret movement is slaved and a ballistic computer/electronics subsystem, that receives information from the sensors and from the control panels to manage system operation and modes and to generate ballistic correction for automatic gun aiming.

The existing gun/turret drive and stabilisation subsystem in the vehicle is adapted and interfaced to the MT-01 so that weapon driving, ballistic computing

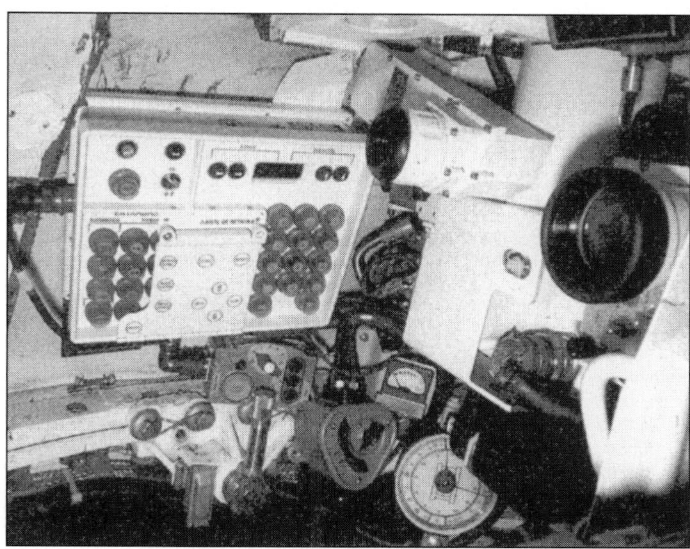

The gunner's position of the MT-01 fire-control system from ENOSA

and sighting functions are integrated into the automatic control system, where the only essential task of the operator is to keep the aiming reticle on target.

Vehicle installation requires removal of the original TPDK-1 fire- control sight and TPN-1 active infra-red sight. Part of this volume is occupied by the new gunner's main sight and associated controls.

System components include:

(a) The stabilised head mirror assembly allows the operator, even with the tank moving, to determine a director line of sight for firing, to observe and detect targets, and to use the laser range-finder. Precision resolution sensors are used to slave the gun/turret to mirror orientation and to produce a reference for coincidence firing.

(b) The thermal imager is an update of the original USA common modules family of thermal sights and provides improved performance (vertical resolution, sensitivity and image quality) and increases modularity and flexibility for sight reconfiguration. The gunner and the commander each have control/display units.

(c) The laser range-finder, integrated with the day sight, can be operated from either the gunner's or commander's position.

(d) The ballistic computer, controlled from a keyboard panel, and various sensors provide the system with an expanded envelope, ballistic control for gun aiming, from stationary or moving tank, against stationary or moving targets.

Operational status

Development complete with major components in service. Ready for production.

Specifications

Head mirror unit
LOS travel: –10 to +20°
Stabilisation accuracy: <0.2 mrad RMS
Static position accuracy: 0.075 mrad
Day sight
Magnification: ×8
Field of view: 7°
Laser range-finder
Material: Nd:YAG
Wavelength: 1.06 μm
Range: 200-9,990 m
Resolution: 10 m
Dual target discrimination: 30 m
Thermal imager
Spectral band: 8-12 μm
Fields of view
Wide: 5.2 × 12.3°
Narrow: 1.7 × 4.1°
Focus: 50 m to infinity
Detector: CMT, PC, 60 element, linear
Scanning: 4; interlace, 240 IR raster lines
Cooling: split Stirling closed-cycle

Contractor

Empresa Nacional de Optica SA (ENOSA).

VERIFIED

Fotona EFCS-3 family of tank fire-control systems

Type

Land fire-control system.

Development

The EFCS-3 family of fire-control systems was originally developed by Fotona for use with the T-series MBTs. The family has since been expanded to include other MBT designs.

Description

The EFCS-3 is a full-performance system with day/night operation facilities, shoot-on-the-move capability and independent Line Of Sight (LOS).

The family members are:

(a) EFCS-3-72 – this improves stabilisation of both the main armament and the gunner's field of view. It also introduces movement of the vehicle as an additional ballistic calculation parameter and fully integrates with the existing automatic gun loading system. The original fire-control equipment remains as a reserve. The distance information from the original gunner's TPD-K1 sight can be input into the system as an alternative measuring source

(b) EFCS-3-72/B – apart from the EFCS-3-72 features, this introduces the additional facilities of a new gyrostabilisation concept, a dynamic trunnion tilt sensor for severe movement conditions and tilt independent horizontal target tracking

(c) EFCS-3-55 – this uses the existing gyros to enhance stabilisation of both the main armament and the gunner's field of view. An improved gun triggering mechanism is also fitted. Software is available for both 100 and 105 mm main guns without any hardware modifications. With few modifications the system can be adapted to the T-62 and Type 59 MBTs

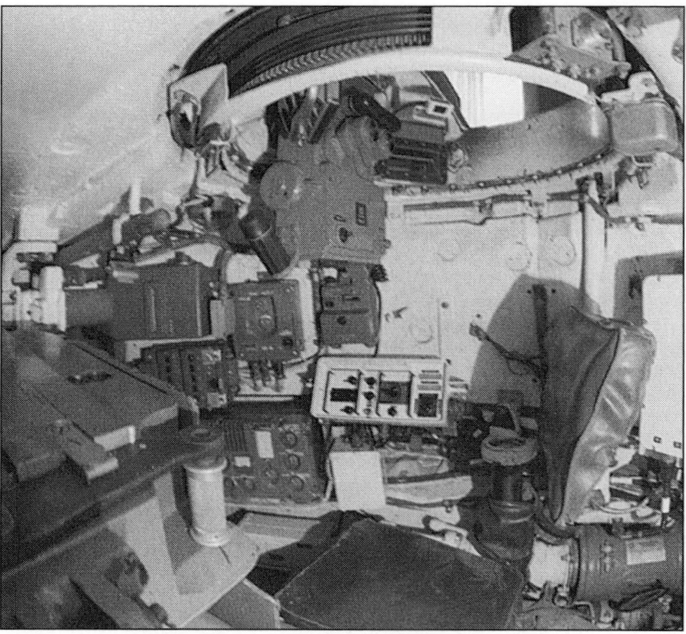

The EFCS-3 fire-control system fitted on M-84 tanks 0021470

(d) EFCS-3-CH – except for the interfacing electronics and a few mechanical adaptations the T-series FCS can be installed in the Chieftain MBT. Almost all the existing FCS equipment is retained as a reserve or integrated into the EFCS-3 system. Improvements are made in all the existing FCS modes of operation together with the addition of a 'silent' watch mode whereby the battlefield can be observed and the laser range-finder and armament used while the main engine and stabilisation is turned off. In its full system configuration the EFCS-3-CH offers target engagement possibilities for the commander even while the vehicle is moving

(e) EFCS-3-M60 – this system is for the M60A1 MBT and comes together with a stabilisation kit to allow precise gun positioning while the tank is moving. The existing FCS equipment, including the mechanical ballistic computer, remains in use as a reserve. The output of the original optical range-finder is fed to the EFCS-3 computer as an alternative range source. The commander has the same firing possibilities as the gunner in all target engagement situations.

The main components of the EFCS-3 are:

(a) Top protection cover – which shields the servo- stabilised mirror and sight optics.

(b) Stabilised gunner's day/image intensifying night laser sight. It has a gyrostabilised view as the vehicle moves and maintains the aiming LOS to the target at all times.

(c) Rate gyro box (for T-72) – this is a two-axis angular velocity sensor

(d) Roll sensor - two-axis gun angular velocity sensing

(e) Meteo ballistic sensor – automatic sensing of cross wind, air pressure and temperature

(f) Gun elevation sensor – this senses the precise gun elevation angle

(g) Electronics box – this contains the circuitry for the mirror control and stabilisation

(h) Tilt sensor (for T-72) – this senses the trunnion tilt

(i) Gunner's control block modification kit (for T-55) – this provides target tracking, laser triggering, range selection and wiping, flushing and heating controls

(j) Fire-control computer – this has full ballistic performance capabilities, controls all the system components and acts as the two-axis gun controller. BITE and self-test functions are standard

(k) Commander's panel – this allows ammunition selection and displays the ballistic input and output data. It is also used to prepare the meteo- ballistic sensor

(l) Commander's remote display – this displays computer range data and other control data

(m) Laser irradiation detector and warner – this provides immediate warning against single-pulse laser range-finders, repetitive-pulsed laser illuminators/designators, indirect laser light and IR searchlights. It provides distinct direction determinations and is automatically coupled with smoke grenade dischargers. It can also be interfaced with a radar warning receiver.

Operational status

Production as required. In service with Slovenia and other undisclosed countries. Offered for export.

Specifications

Stabilised gunner's day/night laser sight
Stabilised top mirror
Elevation limits: –10 to +20°
Azimuth deflection: –45 to +45 mils
Elevation deflection: –22.5 to +67.5 mils
Day channel
Magnification: ×10
Field of view: 6°
Dioptre range: –4 to +4

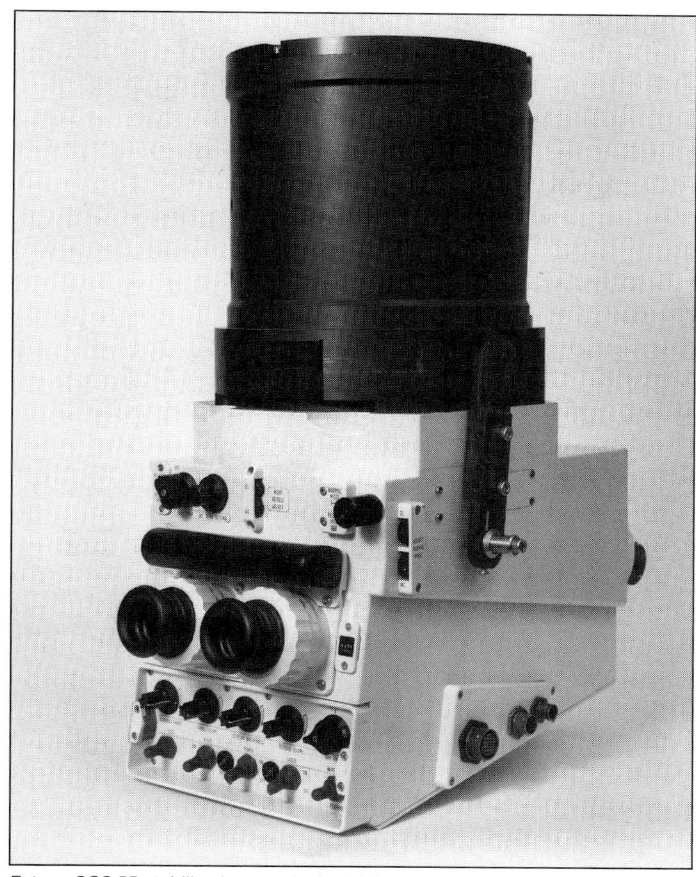

Fotona SGS-55 stabilised gunner's day/night laser sight for the EFCS-3 tank fire-control system family

Night channel
Image intensifier: Gen II
Magnification: ×7.5
Field of view: 7.5°
Focus adjustable: 50 m to infinity
Eyepiece setting: –4 to +4 dioptres
Resolution: (USAF test target, illumination 1 mlx): 0.35 mil
Laser range-finder
Type: Nd:YAG
Wavelength: 1.06 μm
Range discrimination: 40 m
Accuracy: ±7.5 m
Range display: 200-9,995 m 2 targets
Min range setting: 200-3,000 m
Max range: 6,000-10,000 m
Fire-control computer
Input range: manual or laser
Ammunition: up to 7 types including coaxial machine gun
Max azimuth angular velocity: ±80 mils/s
Input meteo-ballistic data
Air temperature: –35 to +65°
Powder temperature: –35 to +65°
Air pressure: 600-1,133 mbar
Crosswind velocity: –40 to +40 m/s
Deviation of nominal muzzle velocity: –50 to +50 m/s
Trunnion tilt: –15 to +15°
Min firing gate
Azimuth: ±0.3 mil
Elevation: ±0.2 mil
Computing accuracy: 0.1 mil
Max computed output angles
Azimuth: ±45 mils
Elevation: –22.5 to +67.5 mils

Contractor
Fotona d d.

VERIFIED

Galileo Avionica JANUS fire-control system

Type
Land fire-control system.

Development
The JANUS dual-purpose fire-control system is derived from the Officine Galileo (now Galileo Avionica) P56 anti-aircraft model (which is used in a range of towed light anti-aircraft gun systems) and is designed for use on light armoured vehicles with turrets mounting 20 to 30 mm automatic cannon in both autonomous and netted anti-aircraft operational modes, as well as for use against ground targets.

Galileo JANUS fire-control system sight assembly and control panel

Description
The system comprises the following components:
(a) directly coupled with the gun; periscopic optical aiming sight assembly with fixed eyepiece, graticule illumination and dimming, control panel and dual-axis servoed scanning mirror; the sight assembly can support a low frequency LRF with a single output path.
(b) 16-bit ballistic microcomputer with five main software subprograms for kinematic, extrapolation, ballistic, lead angle and crossing range estimation tasks.
(c) commander's control panel, which is used to control the computer and manually input the target information on type, speed and range
(d) passive second- or third-generation night vision image intensifier unit which is inserted into the eyepiece of the aiming sight when required for night time engagements
(e) gun elevation and traverse transducers which are used to sense the angular position of the gun and turret for the fire-control computer's calculations

The system has two firing modes: ground target engagement, where the target has to be visually observed in the sight in order to produce an angular tracking speed and the fire-control parameters entered manually into the computer; anti-aircraft engagement, where the lead angles are calculated from the target speed only and manually introduced into the system. The target crossing range is automatically calculated by the fire-control computer by processing the target speed and the relevant aiming angular velocity.

Operational status
Ready for production. Fitted on the Cockerill C25 25 mm and C30 30 mm turrets.

Specifications
Weight
Optical sight assembly: 33 kg
Computer unit: 12.5 kg
Commander's control panel: 1.5 kg
Sight assembly
Traverse: ±25°
Elevation: –9 to +80°
Magnification: ×5
Field of view: 12°
Image intensifier
Focal length: 137 mm
Discovery range
Gen II tube: 2.7 km
Gen III: 4 km
Turret
Servo speed: 45-100°/s
Acceleration: 60-150°/s^2
Power supply: 24 V DC vehicle system

Contractor
Galileo Avionica.

UPDATED

Galileo Avionica TURMS laser tank fire-control system

Type
Land fire-control system.

Development
The Tank Universal Reconfigurable Modular System (TURMS) is a third-generation day/night laser fire-control system for use with MBTs and armoured fighting vehicles.

The latest TURMS configurations are designed to meet the operational requirements of the IVECO/Otobreda consortium C1 MBT and the Czech Republic T-72 MBT upgrading. Further TURMS configurations are available for upgrade packages for other tank types made in the West and the several models of MBT manufactured in the countries of the former Warsaw pact (T-72, T-80, T-84, T-90, M-84 and so on). TURMS has been ordered by the Czech Republic to refit its T-72 MBT fleet.

For the VCC-80 infantry combat vehicle, Galileo has provided a TURMS package incorporating a thermal night viewing capability plus an additional unity direct view facility in the gunner's sight. The maximum elevation of both this, and the commander's head mirror assemblies has also been increased from +20 to +60° in order to allow air defence engagements using the vehicle's 25 mm Oerlikon Contraves KBA cannon.

For the B1 Centauro 8 × 8 wheeled tank destroyer, light intensifiers are used in the TURMS-type commander's sights which are then integrated with a SEPA-designed computerised fire-control system.

Further fits are envisaged for the TURMS equipment family. These include the use of the gunner's sight variant with a lateral sight head arrangement as part of an upgrade package (known as the TURMS RXL) for the Italian Army's Leopard 1A1 MBTs, and the adoption of a day/night version of the commander's panoramic sight with a 60° elevation capability by Otobreda for use on its private venture Otomatic 76 mm self-propelled air defence gun, with a fire-control system based on the Italian Armed Forces standard MARA microprocessor made by Alenia.

The system comprises the following components:

(1) a gunner's monocular periscopic laser sight system of:
  (a) a common primarily stabilised head mirror
  (b) a ×10 magnification day visual channel with a 5° field of view
  (c) a second-generation thermal imaging unit operating in the 8 to 14 μm InfraRed (IR) waveband driving a green/black CRT display with a narrow and wide field of view capability
  (d) an Nd:YAG laser transceiver module with a 10,000 m maximum range
(2) a commander's sight for which is supplied the day/night thermal IR ATTILA Periscope Modular System. The ATTILA task is to observe, detect and identify and it has autonomous firing capability being composed of:
  (a) a multispectral primary stabilised panoramic head
  (b) a two field of view day visual channel (×4 and ×12 magnification)
  (c) elevation range up to 60° or higher for AA or anti-helicopter use
  (d) visual unit (eyepiece set)
  (e) second-generation thermal IR vision unit with two fields of view
  (f) laser range-finding unit (optional).
(3) an Otobreda or Marconi ballistic digital computer which performs all the fire computations, controls the turret servo-mechanisms and manages the operation of all the ballistic units (optical sight, laser range-finder) and the sensors. It also permits the reconfiguration of system operation from the automated level to back-up manual modes when partial failures occur. The ballistic computation range is from 300 to 4,000 m
(4) meteorological, tank attitude and powder temperature sensors
(5) gunner's control panel
(6) commander's control panel
(7) muzzle reference set
(8) interconnecting cable set.

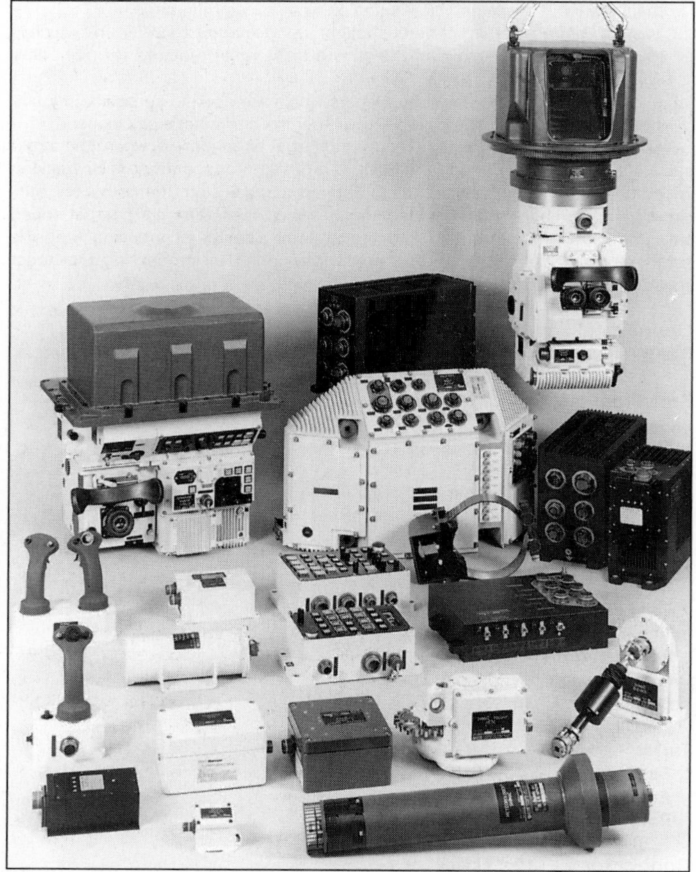
TURMS laser tank fire-control system components 0004770

The TURMS system can be used even on the move against stationary and moving targets in both day and night conditions to provide a very high first round hit probability.

Operational status
In production. In service with the Italian Army (versions are used on the C1 MBT (250 required), B1 tank destroyer (400 required) and the VCC-80 infantry combat vehicle) and on order for the Czech Republic's T-72M4 upgrade (now expected to cover 140 vehicles, down from an original requirement of 350).

Contractor
Galileo Avionica.

UPDATED

Kearfott Navigation and Target Acquisition System (NTAS)

Type
Land fire-control system.

Description
Kearfott Guidance & Navigation Corporation has developed a Navigation and Target Acquisition System (NTAS) which supports observation crews for artillery and other remote weapons fire support.

The key elements of the system are the MILNAV® Vehicle Reference Unit (VRU), a Commander's Display Unit (CDU), a Velocity Motion Sensor (VMS), a Laser Range-Finder (LRF) and a GPS receiver. The LRF or the GPS can be either customer furnished or supplied by Kearfott.

Target position is calculated in three dimensions and is a result of the computation of the LRF sighting information and the MILNAV (vehicle position, heading, pitch angle to target).

The MILNAV VRU performs all navigation, attitude, pointing and north finding functions with heading accuracy to 1 mil RMS and attitude accuracy to better than 0.5 mil RMS. The VMS provides forward velocity aiding for optimal system performance. The GPS input channel is provided to achieve enhanced performance and to provide for initialisation and alignment on the move. In addition to the normal alignment and alignment on the move, the system has the capability of stored heading alignment and extended alignments, as well as outputs, using a variety of spheroid zones and extended zone coordinate systems.

The CDU is a rugged, colour, liquid crystal, daylight readable man/machine interface device. This CDU displays the menus required after initialisation and is the readout device for vehicle navigation, attitude and driver information, as well as target selecting location data. Information typically displayed includes:

- Azimuth to target
- Elevation to target
- Range to target
- Target easting
- Target northing
- Target altitude
- Vehicle heading
- Distance to go
- Steer to angle
- Present position, northing, easting and altitude
- Absolute and relative position
- Direction relative to north.

The LRF provides 'range-to-target' measurement information to the CDU for processing this data.

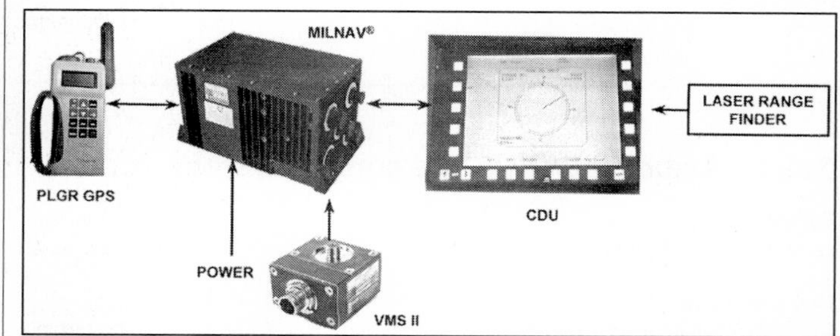

The Kearfott NTAS system 0100717

The NTAS system can be integrated into a variety of vehicles and/or systems for observation and target sighting. All system elements are mounted within the cupola (except the VMS). The VRU and the LRF are mechanically coupled together so the VRU can measure LRF pitch and azimuth as targets are identified.

For applications where varying turret orientation is required during navigation, the system will accept azimuth and elevation digital resolver outputs to facilitate VMS aided navigation.

For applications where the LRF is not mounted in the cupola, the resolver input for azimuth and elevation can be used to provide viewing angle for the LRF input

Specifications

Horizontal target accuracy with P(y) code GPS: 15 m CEP
Vertical target accuracy with P(y) code GPS: 15 m PE
Laser accuracy: ±5 m at < 10 km
Pointing accuracy: 1.0 mils RMS
Roll and pitch: 0.5 mils RMS
Navigation accuracy with P(y) code GPS
  Horizontal position: 10 m CEP
  Vertical position: 10 m PE

Navigation accuracy without GPS (VMS input)
  Horizontal position: 10 m CEP at ≤4 m, 0.25 m CEP at >4 km
  vertical position: 6.7 m PE at ≤10 km, 0.067% DT at >10 km

Contractor

Kearfott Guidance and Navigation Corporation.

Kearfott T-55 tank thermal sight/fire-control system

Type

Land fire-control system.

Description

Kearfott has developed a passive target acquisition sight system for modernising T-55 MBTs that contains a full-solution, thermal fire-control system. The system is relatively small in size and can be installed without cutting tank armour.

The system contains the following subassemblies and modules:

(a) gunner's Infra-Red (IR) thermal sight – featuring dual field of view optics and an advanced SPRITE detector which provides recognition at ranges exceeding 3.5 km. Also included is the head mirror, IR receiver, cryogenic cooler and T- 55 turret mounting adaptor
(b) signal electronics assembly containing the video processing, reticle and alphanumeric generator. Built-In test electronics are also included
(c) display electronics assembly containing the system's power conditioning, power supplies and resolver servo electronics
(d) Nd:YAG laser range-finder mounted and integrated into the head mirror assembly
(e) digital ballistic computer with internal cant angle sensor, laser range-finder and trunnion resolver superelevation offset interface electronics
(f) gunner's and commander's control and display units
(g) gun trunnion resolver, which is electrically servo linked to the head mirror axis resolver and interfaced to the digital superelevation interface electronics
(h) meteorological mast sensor containing crosswind, temperature and barometric pressure sensors
(i) target tracking switch, which is used to enter the velocity of a moving target and used by the ballistic computer in calculating the gun's horizontal deflection.

In operation the computer automatically recalculates the superelevation and horizontal deflection each time the laser range-finder is fired or when any input data are updated. With each recalculation the output from the ballistic computer is used automatically to offset the reticle in elevation. When the gunner repositions the main armament to align the reticle and sets the horizontal deflection, the gun can be fired.

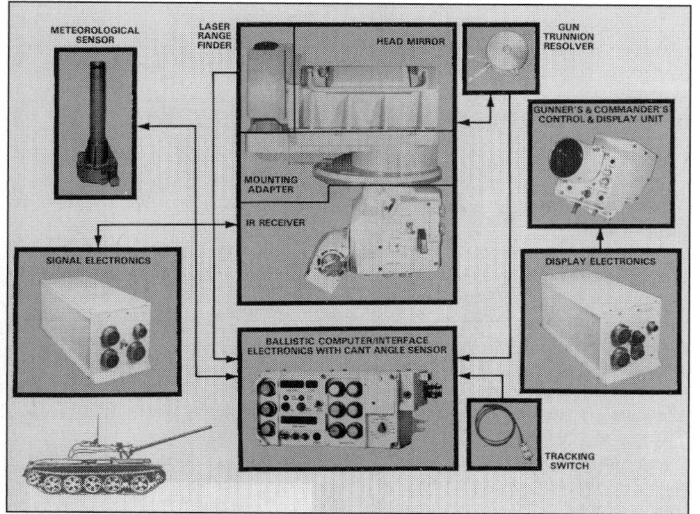

Main components of the Kearfott T-55 tank thermal sight/fire-control system

Operational status

Production as required. Offered for export.

Contractor

Kearfott Guidance and Navigation Corporation.

Kollsman PDFCS Primary Direct Fire-Control System

Type

Land fire-control system.

Description

The PCFDS, providing a full fire control solution, further enhances the DNRS system by adding a ballistic computer and meteorological and cant angle sensors. The PDFCS provides improved probability of hit while on the move. Either one- or two-axis Line-Of-Sight (LOS) stabilisation is available. In addition to the fully integrated ballistic computer, the PDFCS stabilised sight provides gyro-stabilised day and thermal imagery. The PDFCS offers automatic fire-control correction for moving targets from a moving platform. Meteorological sensors automatically provide correction for crosswind, temperature, and atmospheric pressure conditions. When stationary, cant angle sensors sense the turret roll angle and automatically provide correction to the main weapon before firing.

Operational status

In production. Forty seven systems have been delivered to Kuwait and Qatar, installed in the CMI LCTS 90 mm turret equipped Pandur 6 × 6 vehicles. Saudi Arabia has ordered 130 PDFCS systems to be integrated on the DDGM LAV 8 × 8 vehicle.

Contractor

Kollsman Inc.

UPDATED

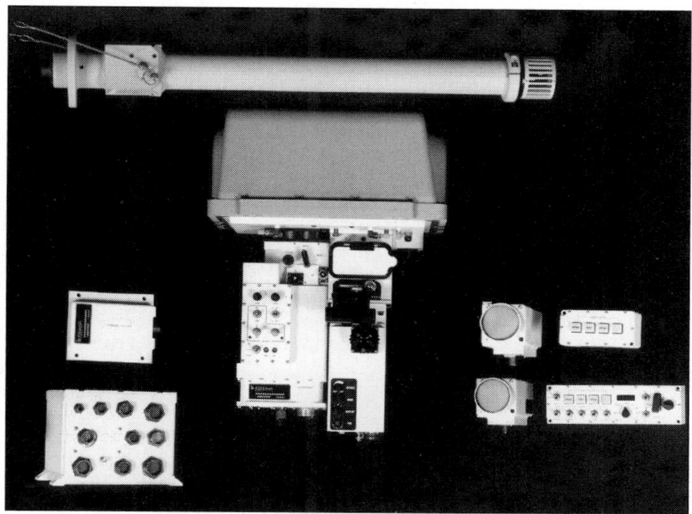

Kollsman Primary Direct fire-control system with line of sight stabilisation 0097105

LIW (DENEL) LMT-105 medium turret

Type

Armoured fighting vehicle turret.

Description

The LMT-105 uses a LIW proprietary GT-7 low-recoil force gun which fires the NATO standard 105 mm ballistic system. The gunner is equipped with a primary stabilised periscopic sight with an integrated laser range-finder. There is a day channel and a thermal imaging night channel. The commander has a primary stabilised periscopic sight with optical night vision. 360° viewing is available. He can also view the gunner's thermal imagery.

Operational status

Development complete. Ready for production.

Contractor

LIW (DENEL).

UPDATED

LIW Tiger fire-control system for T-72 MBTs

Type
Land fire-control system.

Development
LIW together with Kentron and Eloptro, all part of the Denel Group, have developed and tested a computerised fire-control system for the T-72 MBT called the Tiger New-Generation Fire- Control System (NGFCS).

The Tiger computerised fire-control system has been designed to improve the fighting capability of the T-72 MBT under day and night conditions. It is a drop- in design and requires no modifications to the T-72 turret.

The main improvements to a T-72 MBT when fitted with the Tiger fire-control system include improved first-round hit probability at 2 km against both static and moving targets, and shorter reaction time between target detection and firing.

Description
The Tiger fire-control system includes Eloptro's GS72T/N gunner's sight and CS65N commander's sight, each with image intensifier and laser range-finder; fire-control computer; sensor subsystem; and man/machine interface.

The gunner's GS72T/N sight is stabilised in both azimuth and elevation to better than 200 µrad peak-to-peak with the T-72 MBT moving over rough terrain. It has an elevation range from −13 to +28° and a traverse range of 5° and is steered by a two–axis hand-controller which forms part of the Tiger man/machine interface subsystem. A mechanical back-up linkage is provided to slave the sight to the main armament in the event of a failure.

The primary stabilised sight incorporates a direct-view optics elbow which has a single field of view ×8 magnification, a dual field of view thermal imaging elbow and an integrated Nd:YAG laser range-finder which can measure out to 10 km to an accuracy of ±5 m. The thermal image is displayed to the gunner and the commander. The thermal imager can be replaced by an optional image intensification elbow with a single ×6.4 field of view.

Information such as range to target, ammunition type, ready to fire and critical system status is displayed on the eyepiece of the day telescope. A custom-designed ballistic graticule is incorporated in the day telescope and is used for ranging and implementing ballistic offsets in the back-up mode.

The Tiger commander's panoramic sight is designated CS65N and is primary stabilised in both azimuth and elevation to better than 200 µrad peak-to-peak with the T-72 MBT moving over typical terrain.

The sight head includes direct-view optics channel and image intensifier module for night operation.

The direct-view optics channel has a two field of view day telescope (×3 and ×10 magnification). The image intensifier module uses a third-generation image intensifier and has a single field of view (×3). The image intensifier image is displayed on the commander's display unit.

The Tiger digital fire-control computer receives data from the sight and sensors, performs ballistic calculations and aims the 125 mm smoothbore gun onto the target.

The Tiger sensor subsystem measures turret position, trunnion position, dynamic tilt and cant angle, vehicle speed and direction, air temperature, barometric pressure and propellant charge temperature, crosswind and downwind.

The GS72T/N gunner's sight is mounted in the existing T-72 night sight position in the turret with its electronics units being mounted in the space occupied by the existing TPDK-1 gunner's sight. The installation of the Tiger CS65N commander's

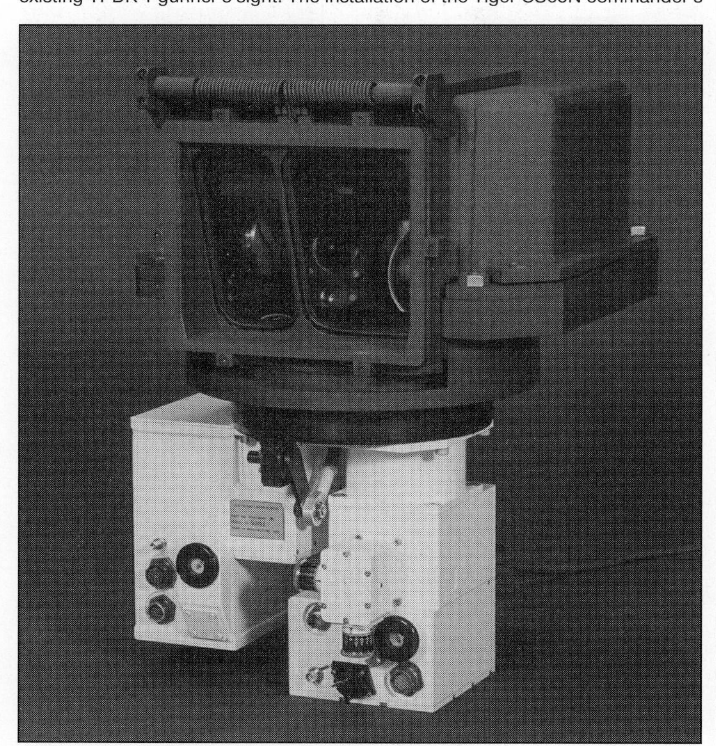
Close-up of the GS72N gunner's sight from the front

The Tiger fire-control system for T-72 tanks 0021471

sight requires that the existing commander's cupola be removed and replaced by a brand new cupola, developed by LIW, into which the new sight is mounted.

The commander's new cupola has a pop-up and rotate hatch to avoid interference with the new sight. The original T-72 only had four periscopes but the new cupola has six which gives a significant increase in all-round visibility for the tank commander.

LIW is looking at further growth potential for the Tiger fire-control system which would include an auto-tracking capability, laser warning devices, land navigation system, muzzle reference system and all-electric gun drives and stabilisation system.

Operational status
Ready for production. Trials of the system have been carried out in Malaysia, Poland, South Africa and Syria.

Specifications
GS72T/N gunner's sight
Stabilisation: better than 200 µrad
Elevation: −13 to +28°
Traverse: 5°
Direct-view optics elbow
Magnification: ×8
Nd:YAG laser range-finder
Range: 10 km
Accuracy: ±5 m
Thermal elbow
Detector: 120 element
Spectral band: 8 to 12 µm
Cooling: closed-cycle split Stirling cryogenic cooler
Commander's CS65N panoramic sight
Stabilisation: better than 200 µrad
Elevation: −13 to +28°
Traverse: 360°
Direct-view optics channel
Dual field of view day telescope
Magnification: ×3 and ×10
Image intensifier module
Field of view: ×3

Contractor
LIW, a Division of Denel Pty (Ltd).

VERIFIED

Mitsubishi Type 90 MBT fire-control system

Type
Land fire-control system.

Description
Mitsubishi Electric has developed a fire-control system for the Type 90 MBT which is capable of engaging static and moving targets. Both the vehicle's gunner and commander have full fire- control facilities with the latter having an override authority. The fire-control system includes a gunner's periscopic sight which is stabilised in azimuth, commander's periscopic sight with 180° traverse with dual-axis stabilisation and a digital fire-control computer.

The gunner's sight has a daylight channel, thermal sensor and an Nd:YAG laser range-finder. The sight is produced by Nikon Corporation with thermal imager made by Fujitsu.

The fire-control system includes an advanced auto-tracking capability that is based on the output of the thermal imager. The automatic tracker is effective against ground targets and can be used when the Type 90 MBT is stationary or on the move.

The commander's sight is a dual-axis stabilised day sight which is manufactured by Fuji Photo Optical Company. He has a monitor with the gunner's thermal display.

The digital fire-control computer compensates for range, wind, temperature, gun tube bend (information from the muzzle reference system) and cant angle (trunnion tilt).

Operational status
Production. In service with Japan on the Type 90 MBT.

Contractor
Mitsubishi Electric Company (MELCO).

VERIFIED

Norinco ISFCS-212 Image-Stabilised tank Fire-Control System

Type
Land fire-control system.

Description
The ISFCS-212 fire-control system is believed to be used on a number of the People's Liberation Army T-series MBTs and is also being offered as a retrofit package for other AFVs. The system consists of:

(a) a ballistic computer with control display/panel and step motor driver
(b) two-axis gun stabiliser with gyro set and actuating motors package
(c) sensor package with crosswind velocity, turret angular rate and tilt measuring devices
(d) laser range-finder
(e) gunner's sight with stabilised field of view and interchangeable second-generation image intensifying night elbow and unity vision periscope
(f) commander's sight interface to allow combat direction by overriding the gunner's system.

The ISFCS-212 has two modes of operation: image-stabilised and graticule automatic setting. In the first the gunner uses his controls to guide his line of sight systems. These generate the gun pointing directions which are fed into the gun stabiliser system. The loop is then closed by the gun positions being fed back to the sight so that the weapon continues to follow the target.

The distance is then measured by the laser range-finder and the ballistic computer calculates the firing solution using the range information from the laser, the target's relative angular rate signals, the trunnion tilt sensor signal, crosswind velocity data, the selected ammunition type and the manual settings. The solution position information is sent to the stabiliser system to set the gun's elevation and lead angle in azimuth. Once the gun reaches this position the control unit generates a fire permitting signal for the gun firing circuit and the gunner can press the trigger.

In its second operating mode the image-stabilised gyro unit is locked, so that the field of view is no longer stabilised. The turret angular sensor generates an azimuthal angular rate signal for the target. Then, following ranging with the laser, the computer calculates the firing solution which is sent via the step motor driver to the gunner's sight. Here it automatically sets the ring graticule and all the gunner has to do is align this onto the target and fire the main armament.

Operational status
Production as required. In service with the People's Liberation Army and other undisclosed countries.

Ballistic computer of ISFCS-212 fire-control system

Specifications
System operating range: 200-3,990 m
Processing range accuracy: ±10 m
Target tracking rate:
Azimuth: −20 to +20 mils/s
Elevation: −10 to +10 mils/s
Target tracking rate processing accuracy:
Azimuth: ±0.4 mil/s
Elevation: ±0.4 mil/s

Contractor
China North Industries Corporation (Norinco).

VERIFIED

PCO Drawa-T Thermal Imaging Fire-Control System TIFCS (for T-72)

Type
Land fire-control system.

Development
Drawa-T TIFCS has been designed especially for the T-72 tank family after experience of the earlier Merida FCS for the T-55. As a result the standard T-72 day sight TPD-K1 was updated by PCO and became a vital part of the Drawa FCS. Installation of the TIFCS Drawa-T is possible without modification of other onboard equipment and without the need to cut into the armour. In this way the cost of modernisation is kept to a minimum. The manufacturers claim that T-72 tanks with Drawa TIFCS installed have a high probability of first round hit on stationary or moving targets. This was confirmed during government trials in Poland and acceptance trials by end-users of this system in Polish T-72 and PT-91 tanks.

Description
Drawa-T consists of the following main units:

(a) Day Sight with laser range-finder – TPD-K1 sight modernised by PCO, with stabilised LOS. Principal modernisation features:
 –continuous measurement of LOS position angle
 –automatic application of lateral correction
 –increased range finder measurement range
 –modification of the sight to interact with ballistic computer
(b) Thermal sight – provides observation, recognition, aiming and distance measuring to the target by day and night. The thermal sight's periscope head including stabilised mirror is coupled with the gun either electrically or mechanically. The commander has a thermal sight monitor
(c) BE/M Electronic Unit with Ballistic computer – uses fire control algorithm to compute solution and coupling of Lines of Sight (LOS) of both sights with the gun

PCO Drawa-T fire-control system 0023003

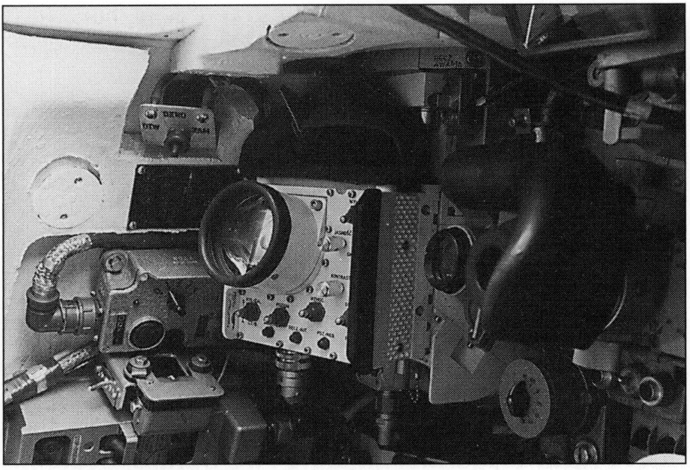

PCO Drawa-T thermal imaging fire-control system installed in the turret of a main battle tank

(d) Commander's control panel and monitor provide MMI with the computer and laser range-finder. The panel has it's own Built-In Test (BIT)
(e) Drawa-T includes sensors to provide the ballistic computer with data which influence shell trajectory such as elevation angle of the gun in relation to the tank turret body, tilt and pitch of gun, angle of turret rotation, cross-wind, tank groundspeed, angular velocity of target, ambient temperature, temperature of propellant charge.

The overall system includes BIT which checks correct operation of each sensor and the system.

Drawa-T operates in three basic modes:
(a) Automatic Mode – basic mode used for firing while moving or stationary with engine running and stabilisation system of periscope mirror and gun operating – sight mirrors coupled to gun electronically
(b) Manual – used in special conditions (that is covert/silent), when the stabilisation system of the periscope mirror and gun is not operating. The sight mirrors are mechanically coupled with the gun
(c) Emergency Mode – in case of damage to the FCS or tank, which makes automatic calculations of corrections and/or transmission to the sights and/or gun stabilisation system impossible. In such cases special aiming marks are introduced into the sight's eyepiece. Option for a lower cost solution: Passive Night Sight PCN-A with Gen II Super or third-generation image intensifier replaces thermal imager.

Operational status
In production and service with the Polish Army. Offered for export.

Specifications
Thermal sight
Spectral band: 8-12 μm
Fields of View:
Narrow: 3° × 2°
Wide: 10.5° × 7°
Detector type: CMT
Number of detection elements: 120
Optional passive night sight
Magnification: ×5.4
Field of view: 5.6°
Day sight with range-finder
Magnification: ×8
Field of view: 9°
Laser range-finder type: Nd: YAG
Max range: 10 km
Computer output data
Sight angle: 0 to +100 mils
Azimuth lead angle: ±40.0 mils
Main input ballistic data
Gun position angle: ±25°
Gun trunnion tilt: ±15°
Gun elevation angle: 500 mils
Turret traverse angle: 0 to 6,000 mils
Air temperature: –50 to +65°C
Target velocity: ±05 to 150 km/h
Number of ammunition types: 6

Contractor
PCO S.A. Poland.

PCO Merida tank Fire-Control System (FCS)

Type
Land fire-control system.

Development
With the assistance of Polish research centres, PCO developed and began production of the first Merida Fire-Control System FCS for the T-55 tank family in the mid-1980s. The design allows installation to the FCS Merida without modification of existing tank equipment. It is possible to install the system at field workshop level. The main improvements in combat capability for the T-55 tank FCS have better first hit probability on stationary or moving targets, better rate of fire and the benefit of a passive night sight.

Description
FCS Merida consists of the following main units:
(a) Integrated day/Image intensifying night sight with laser range-finder. The sight has LOS stabilisation with electronic coupling of sight mirror and gun. Ballistic corrections and output range-finder data are displayed on commander and gunner display. The existing TSz telescopic sight has been left in the tank as the reserve sight
(b) Electronic unit with ballistic computer. Sensors data updates are every 0.5 seconds
(c) Commander's control desk - plays the role of co-ordinator for the whole system as it controls the performance of the computer and also allows the possibility of immediate manual introduction of data, controlling of input and output data as well as testing the performance of the whole system before effecting operational tasks
(d) Sensors for gun elevation angle, turret traverse angle, meteo data, gyroscope, charge temperature and tank velocity
(e) Gun laying panel: gun trigger, select turret stabilisation, initiate range-finding, select tank turret traverse velocity.

The system includes Built-In Test (BIT) of sensors and the whole system. A summary description of FCS Merida operation is as follows:
(a) After placing the aiming mark on target, range and angular velocity are computed
(b) Sensors data are used to compute aiming angles
(c) Computed aiming angles are used to correct the mirror pointing angle in elevation and to illuminate and LED in a linear array indicating azimuth correction
(d) Firing is effected after re-aligning the LED generated mark at the target
(e) On firing, the target angles are stored to enable the periscope mirror to be quickly pointed to observe effect of fire.

All Polish army-55 tanks due for modernisation have been equipped with Merida FCS.

Operational status
Production as required. Offered for export.

Specifications
Integrated day/night sight with range-finder
Magnification
Day: ×3.5 and ×7
Night: ×7

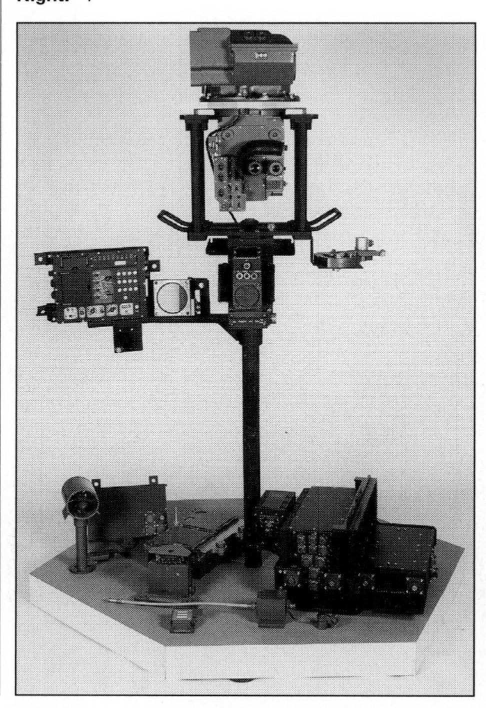

PCO Merida tank fire-control system
0023005

Field of view
Day: 7°
Night: 7°
Image intensifier type: Super GEN II or GEN III
Laser range-finder type: Nd: YAG
Max range: 10 km
Computer output data
Sight angle: 0-100 mils
Azimuthal lead angle: ±25.0 mils
Main input ballistic data
Gun position angle: ±25°
Gun trunnion tilt: ±15°
Gun elevation angle: −100-350 mils
Turret traverse angle: 0-6,000 mils
Air temperature: −35 to +65°C
Charge temperature: −35 to +65°C
Target velocity: ±05 to 150 km/h
Number of ammunition types: 6

Contractor
PCO S.A, Poland.

Raytheon full-solution tank fire-control system

Type
Land fire-control system.

Description
The full-solution tank fire-control system has a thermal sight and laser range-finder and provides updated technology for later M-series and similar type MBTs to enhance their target acquisition, hit probability and shoot-on-the-move capabilities.

The system consists of the following sub-units:
(a) Raytheon Systems Company AN/VSG-2 Tank Thermal Sight (TTS) with integrated Nd:YAG laser range-finder. The latter has a range display in the gunner's and commander's thermal displays, automatic ballistic solution entry, continuous 1 pps duty cycle, first/last pulse logic and a single reticle for laser boresighting and gun solution
(b) commander's extension for TTS
(c) computer electronic unit with the computer having the capability to determine full ballistic solutions for all standard NATO and US round types from data input from the laser range-finder and other sensors. It can also be reprogrammed easily in the field for new round types
(d) computer control panel
(e) electronic interface unit
(f) gunner's ammunition type selection panel
(g) gunner's control panel
(h) output unit
(i) commander's ammunition type selection panel
(j) power supply
(k) sensor package including rate/tach, cant and wind measuring units.

Operational status
Production as required.

Contractor
Raytheon Systems Company.

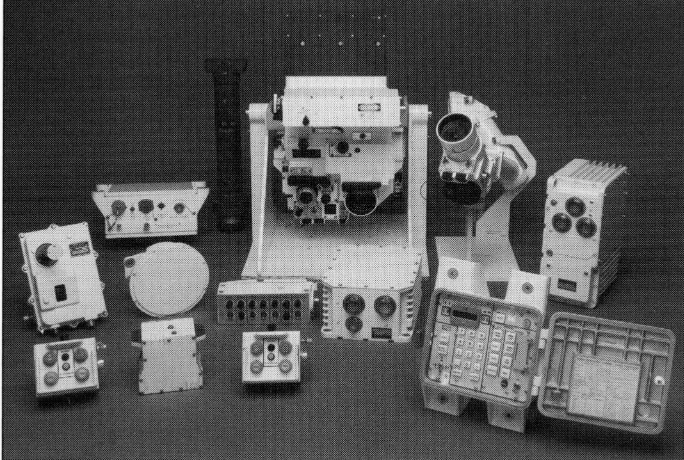

Raytheon full-solution tank fire-control system showing main components

Reutech High-Integration Technology Tank Fire-Control System (HITT-FCS)

Type
Land fire-control system.

Description
The Reutech (formerly ESD) HITT-FCS has been designed to be a cost-effective solution to the requirement for advanced fire-control systems in tanks and other armoured fighting vehicles.

The main components are:
(a) Gunner's sight – this is a periscope assembly incorporating sight elbows for both day and night channels. The day channel includes an Nd:YAG laser range-finder with a maximum range of 9.99 km and accuracy of ±5 m. An adjustable minimum range gate is provided along with multiple target indication and controls. Magnification of the day channel is ×8 with an 8° field of view. The night channel uses second-generation image intensifier technology that provides a performance of 1.18 Lp/mrad at 10^{-1} lx for target contrast of 48 per cent. Magnification of the night channel is ×7.1 with a 7.2° field of view. The night channel may be upgraded to a thermal infra-red version, accompanied by minor periscope modifications.
(b) Operator panel – this incorporates the ballistic processor and sight offset drive electronics. A numeric keypad and 2 × 24-line character LCD are used for calibration and testing requirements. The operational controls (including input facilities for six ammunition types) are located separately with LED displays. Drive electronics for the implementation of graticule offset in the thermal imager display are included.
(c) Meteorological and cant sensors.
(d) Mechanical linkages and back-up range drum – the mechanical linkage couples the sight periscope mirror and main armament together for synchronisation purposes while a back-up range drum is provided especially for indirect fire missions.

Operational status
Development complete. Ready for production.

Specifications
Crosswind/headwind range: ±25 m/s
Accuracy: 1 m/s
Air temperature range: −40 to +60°C
Accuracy: ±3°C
Air pressure range: 600-1,300 mbar
Accuracy: 10 mbar
Cant range: ±30°
Accuracy: 5 mils at 3°
Sensor type: meteorological (dual-axis wind)

Contractor
Reutech Systems.

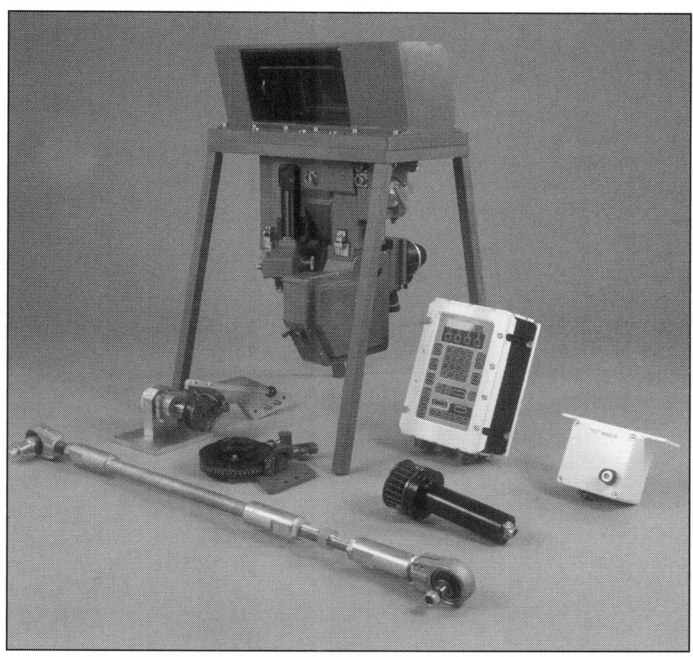

Main components of the Reutech HITT fire-control system

SaabTech Universal Tank and Anti-Aircraft System – UTAAS

Type
Land fire-control system.

Development
The UTAAS fire-control system has been developed by Saab (formerly CelsiusTech) for use with both new build and retrofitted AFVs and air defence guns. Its first application was the Combat Vehicle 90 series of armoured vehicles developed to meet the requirements of the Swedish Army. It has also been chosen for the CV 9030 of the Norwegian Army.

Description
UTAAS is a sight and fire-control system with modular structure available in a variety of configurations which depend upon the customer's requirements for the sophistication level.

UTAAS with an advanced tank fire-control computer

This is an advanced fire-control system to combat moving ground targets and helicopters. The UTAAS, with laser range-finder and fire-control computer, is supplemented by a cant angle sensor and interface to the gunlaying system to provide an independent line of sight.

The system is specifically designed to shorten the reaction time and achieve a high hit probability against moving targets when firing from a moving vehicle. The aiming operations of the gunner are made easier by the fact that the system generates an independent, secondary stabilised line of sight. The gunner needs only to track the target in the reticle centre and fire the laser. The fire-control system then calculates the firing solution and guides the barrel to the calculated super-elevation and lead angle while the line of sight remains on the target.

The superelevation and lead angle are calculated as a function of the:

(a) target range
(b) target speed
(c) ammunition type
(d) muzzle velocity deviation and charge temperature
(e) gun jump
(f) air temperature
(g) air pressure
(h) wind velocity
(i) trunnion tilt.

Operational status
In production. In service with the Swedish Army on the CV 9040 (the ICV version as well as the FO and AD versions). It has also been ordered for the CV9030N of the Norwegian Army.

Specifications
Elevation limits: –20 to +50°
Traverse: ±18°
Field of view (direct outlook)
Azimuth: 35°
Elevation: 15°

Close-up of the UTAAS from the operator's side

Day channel
Magnification: ×8
Field of view: 8°
Transmission: >40%
Thermal channel (typical)
Field of view
Narrow: 2 × 3°
Wide: 6 × 9°
Laser range-finder
Wavelength: 1.064 μm
Extinction ratio, ground target: >32 dB
Extinction ratio, air target: >42 dB
Pulse repetition frequency, ground target: 1 Hz
Pulse repetition frequency, air target: 4 Hz for 10 s, 2 Hz continuously

Contractor
SaabTech Systems.

SABCA SAIPH modular fully integrated MBT fire-control systems

Type
Land fire-control system.

Description
The SABCA SAIPH modules are designed to act as building blocks in the production of a series of modern fire-control systems used in the upgrade of various MBT types. Most of the modules were developed for the SABCA upgrade of the Belgian Army's Leopard 1 MBT.

The SAIPH system comprises the following subsystem modules:

(a) dual-axis head mirror – with built-in motors and resolvers. The mirror substrate is made with a metal matrix composite that confers low weight, stiffness and long-term flatness stability properties. It is slaved to the gun position in elevation and stabilised by the error signals of the gun/turret stabilisation electronics. The mirror is driven in elevation and azimuth by the ballistic computer. SAIPH is also compatible with a director type, dual-axis stabilised head and different electronics
(b) 1.064 μm or eye-safe 1.54 μm wavelength laser range-finder, depending upon customer requirement
(c) high-resolution Thermal Imaging System (TIS) – this is based on the advanced Common Modules (TICM class II) with SPRITE detector. Diffraction-limited optics with large aperture and long focal length give the TIS system an identification range of more than 2 km.
(d) automatic Muzzle Reference System (MRS) - this is based on an electro-optical transceiver in the gunner's primary sight. Gun bending or droop, turret distortion or any alignment error are measured by the head mirror using a zero-method and then corrected. When activated the MRS automatically boresights the tank in less than 3 seconds. The system can be operated during day, night or NBC conditions
(e) near Infra-Red (IR) Charge Coupled Device (CCD) camera – this gives clear and detailed images during the day with better contrast and atmospheric penetration than visible light
(f) colour CCD camera (option) – available for short-range observation
(g) direct day vision optics (option) – day vision through classical optics and eyepiece assembly is available as an option
(h) digital ballistic computer
(i) miniature high-resolution displays – a remote display unit is provided to the commander
(j) menu-driven control units – with this and the fitting of additional control handles the commander can override the gunner and engage targets himself
(k) non-standard sensors with digital interface – these include crosswind velocity, cant, ambient air pressure, ambient air temperature, power temperature and tracking speed
(l) Built-In Test Equipment (BITE)

The available SAIPH system configurations are:

(a) SAIPH – Leo 1 Mk 1
This is the upgrade of the original Leopard 1 fire-control system with SAIPH modules, of which the TIS and MRS are the most important. The existing ballistic computer, analogue non-standard sensors, gunner's control unit and commander's control unit are retained
The gunner's primary sight assembly integrates a dual-axis mirror drive, direct vision day optics with dual Field Of View (FOV), a near-IR CCD day camera with dual FOV, a high-resolution TIS, a miniature high-resolution display and an automatic MRS transceiver
The commander has a remote display unit and a remote-control unit. Near IR and TIS images are provided for both the gunner and the commander
(b) SAIPH – Leo 1 Mk 2
A new configuration with digital ballistic computer, non-standard sensors and digital control units. The gunner's primary sight has a direct day vision channel (the same as the Leo 1 Mk 1)
(c) SAIPH – Leo 1 Mk 3
Same as the SAIPH Leo 1 Mk 2 but with no direct day vision; the gunner's primary sight is fully electro-optical. Any direct day vision capability is provided through the coaxial TZF assembly

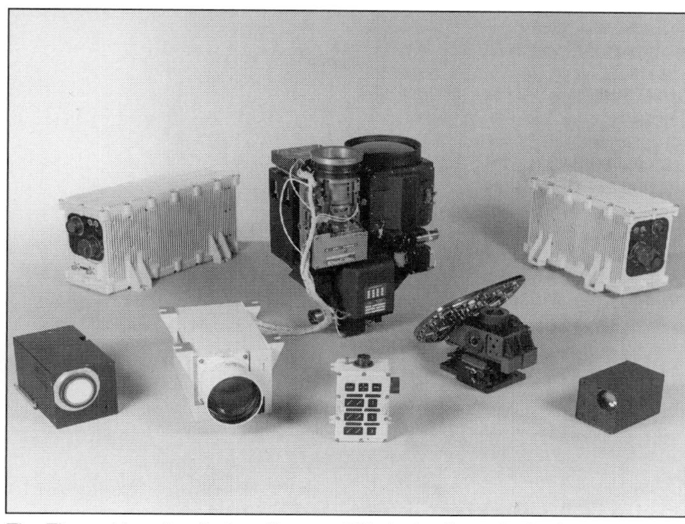

The Thermal Imaging System (imager, TIS electronics unit, display unit, remote display unit and a smart remote unit) (centre and left) and the line of sight control units (main electronics unit, dual-axis stabilised mirror and Muzzle Reference System)

(d) SAIPH – M60
Same configuration as for the Leo 1 Mk 2 and Mk 3
(e) SAIPH – T-72
Same as above. The gunner's primary sight is repackaged to fit the T-72, but no major turret modifications are required
(f) SAIPH – T-55
Same as above, but the TIS is replaced by an image intensifier device. The MRS is optional.

In each configuration the gunner's primary sight acts as the integration unit for all the optical channels. If required special customer requirements based on the SAIPH modules can be produced.

Operational status
In production. SAIPH configurations offered for export.

Specifications
Operational limits
Ballistic range limits: 500-4,000 m
Ammunition types: 4 plus machine gun
Target angular rate
Azimuth: 0-50 mils/s
Elevation: 0-20 mils/s
Elevation lead angle: –100 to +100 mils
Line of sight elevation: –10 to +20°
Air temperature: –40 to +50°C
Powder temperature: –40 to +50°C
Windspeed: 0 to 72 km/h
Turret cant: –15 to +15°
Gun wear: 0 to 100%
Jump and throw off: independent for each ammunition type
MRS correction (autoboresight range): –4 to +4 mils
MRS reproducibility:
–40 to +50°C: 0.2 mil accuracy
20°C temperature range: 0.05 mil accuracy
Thermal Imaging System (TIS)
Spectral band: 8-12 μm
Cooler and detector: 8-element SPRITE (CMT), closed-cycle silent compressor
Field of view (h × v)
Narrow: 3.5 × 2.3°
Wide: 7.9 × 5.15°
Focus range: 50 m to infinity
Video output: CCIR 625 lines
Horizontal sampling: 790 pixels
Static operational ranges (conservative):
Detection
Wide FOV: >5 km
Narrow FOV: >10 km
Recognition
Narrow FOV: 3.5 km
Identification
Narrow FOV: 2 km
System power supply: 19-30 V DC
System MTBF: >1,500 h, >6,500 h without TIS

Contractor
SABCA (Societé Anonyme Belge de Constructions Aeronautiques).

SABCA Vega and Vega Plus fire-control systems for the T- 72M1 MBT

Type
Land fire-control system.

Description
SABCA of Belgium has developed the VEGA and VEGA Plus add-on fire-control systems for the T- 72M1 MBT variant.

The vehicle installation does not require the tank to return to a workshop and can be performed in the field in two days. All the work can be manually performed by a team of three.

The systems are designed to fit within the volume left by removing the TPN-1 gunner's sight with the existing head of the TPN-1 being used and modified to accommodate the 8 to 12 μm waveband Thermal Imaging System (TIS) and digital ballistic fire-control computer (which is integrated into the TIS Electronics Unit (TISEU)).

The system is mechanically coupled to the main 125 mm calibre gun and uses the existing electrical couplings to the gunner's main TPD-K1 sight laser range-finder and ammunition selector.

A Smart Coupling Mechanism (SCM) has been incorporated into the TPD-K1 mechanical link to keep the line of sight aimed at the target while the main gun undergoes its loading sequence. As the sequence starts, the SCM releases the mechanical link between the gun and the sight's head mirror. The mirror is locked and remains aimed at the target while the loading is carried out. After the loading is completed the gun returns to its initial position and the mirror's mechanical link is re-engaged.

The VEGA system consists of the following Line Replaceable Units (LRUs):
(a) Gunner's Thermal Sight (GTS) which consists of single-axis head mirror assembly with SCM; thermal imager and telescope with two fields of view; high-resolution, miniature TV Display Unit; TIS Control Unit; TIS Electronics Unit (TISEU); ballistic computer; interface with laser range-finder, sensors and ammunition selector; interface with automatic gun loading system; and Built-In Test Equipment (BITE). The VEGA Plus system incorporates all the VEGA subassemblies given above with all of the following options: remote display and control unit for the commander; tracking speed sensor (for moving targets) and vertical sensor; atmospheric conditions sensor (wind velocity, ambient pressure and temperature).

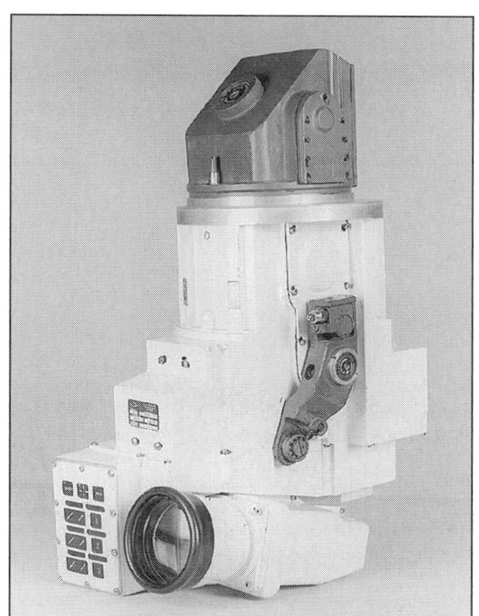

The gunner's thermal sight for the Vega and Vega Plus, with integral display and menu-driven control panel

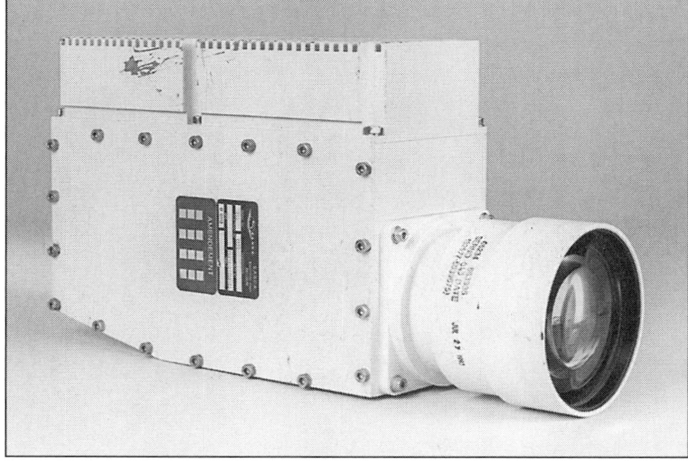

The optional commander's remote display unit for the Vega and Vega Plus

Operational status
Ready for production.

Specifications
Ammunition types: computer has ballistic tables available to compute lead angles for 3 ammunition types
Ammunition range limits for computer
HESH: 100-4,000 m
HEAT: 100-4,000 m
APFSDS: 6 km
Elevation lead angles: –5 to +55 mils
Sight elevation limits: –8 to +14°
Jump and throw-off: 3 independent jump and throw-off corrections, adjustment ranges –3 to +3 mils
Power supply: 19-30 V DC vehicle supply
Target azimuth angular rate: –30 to +30 mils/s
Ambient air temperature range: –30 to +60°C
Ambient air pressure range: 690 to 1,240 mbar
Ammunition powder temperature range: –30 to +60°C
Crosswind velocity range: 0-108 km/h
Headwind velocity range : 0-108 km/h
Turret cant: –15 to +15°

Thermal imaging system
Field of view
Narrow ×5.5 magnification: 3.3 × 5.0°
Wide ×1.8 magnification: 10 × 15°
Performance (static 2.3 × 2.3 m NATO target)

	narrow FOV	**wide FOV**
Detection	>7 km	>3 km
Recognition	>2.7 km	n/a
Identification	>1.6 km	n/a

Focus range: 50 m to infinity
Spectral band: 8-12 μm
Detector and cooler: integrated cooler detector assembly, 8-element SPRITE (CMT) with closed-cycle Stirling cooler
Video output: CCIR 625 lines
Number of IR lines: 512
Horizontal sampling: 790 pixels

Contractors
SABCA (Societé Anonyme Belge de Constructions Aeronautiques).
ZTS Dubnica.

SAGEM Sanoet-2

Type
Fire-control system (land).

Development
SAGEM, in conjunction with UKBM-Nizhny Tagil, the Russian designer of the T-72 MBT family, and Peleng-Belomo, the Russian designer of stabilised sights, has developed the Sanoet-2 Improved Fire-Control System (IFCS) for the T-72 MBT family.

Description
The original TPN-1 gunner's sight of the T-72 is replaced by a new two-axis stabilised sight with an integrated SAGEM MATIS thermal imager without any modification to the turret.

The Sanoet-2 system is stated to improve the accuracy of the weapon stabilisation, in elevation by 20 per cent, in azimuth by 100 per cent and be usable in three operating modes: stabilised day/night observation for the commander; firing-on-the-move by the commander and firing-on-the-move by the gunner.

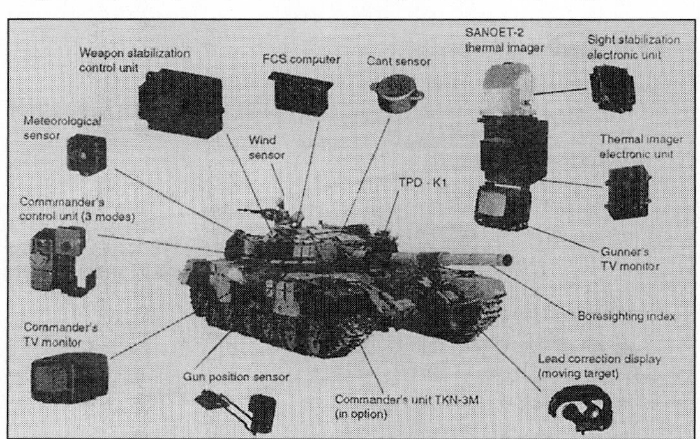

Components of the Sanoet-2 fire-control system

Operational status
Available for retrofit of T-72.

Specifications
Sight
Mirror angular displacement
Elevation: +20 to –8°
Azimuth: +15 to –15°
Thermal imager: SAGEM MATIS
Spectral band: 3-5 μm
Fields of view: 12 × 8°, 2.5 × 1.7° and electronic zoom 1.25 × 0.85°

Contractor
SAGEM SA, Aerospace and Defence Division.

UPDATED

STN ATLAS Elektronik EMES-18 tank fire-control system

Type
Land fire-control system.

Description
The EMES-18 fire-control system is designed for use with the Leopard 1A5 105 mm main and coaxial armaments for the engagement of stationary and moving targets under day or night conditions with the firing platform itself either moving or stationary.

The main subsystems of the EMES-18 are:

(a) stabilised Gunner's Primary Sight (GPS) with the following sub-units:
   (i) Nd:YAG 1.06 μm wavelength laser range-finder transmitter with an operating range of 200 to 9,900 m
   (ii) integrated thermal imaging sight. Target markers are superimposed on the thermal image created which is interjected into the daylight channel optical path
   (iii) stabilised, head mirror (azimuth and elevation) which is used for the daylight visual channel (with a ×12 magnification and 5° field of view), laser transmission/reception and thermal imaging systems. This ensures the synchronisation of the optical axes and allows both the gunner and commander to aim at the target using either the daylight or the thermal imaging system and perform the required laser ranging. The identification and tracking of the target in daylight is achieved by means of a high-quality day telescope

(b) commander's monocular GPS eyepiece assembly which, together with the gunner's eyepiece system, allows either operator to control the Line Of Sight (LOS) via the use of hand controls. A reticle, range value and system status information are superimposed on the telescope image

(c) digital fire-control computer which contains ballistic information for up to seven ammunition types and a computation range up to 4 km

(d) vertical sensor unit to eliminate automatically cant angle error from the ballistic computations.

In operation the commander identifies a target, slews the turret round to its azimuth and hands it over to the gunner to engage. The gunner aims through the sight, performs the laser range-finding task and starts the tracking procedure. The fire-control computer then takes all the manual and automatic input fire-control parameters such as the cant angle correction value, selected ammunition type, powder temperature, ambient atmospheric conditions, target speed and vehicle attitude and continuously calculates the superelevation and lead angles for the armament. The relevant command signals are transmitted to the gun-control system, which relays to the weapon while not disturbing the LOS and the offset aiming mark is generated for the sight system.

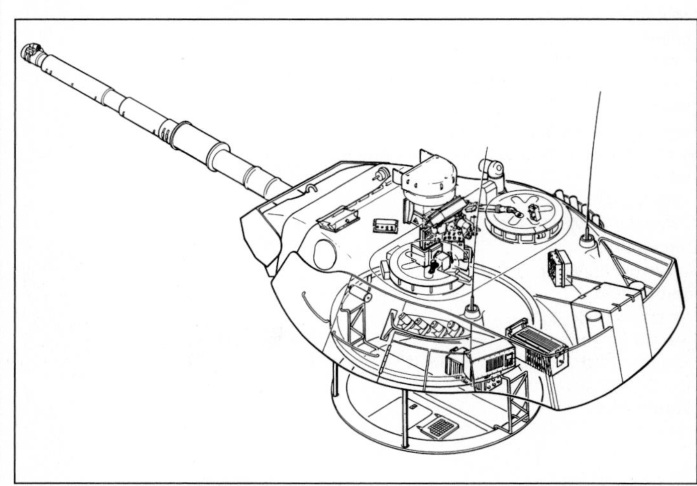
STN ATLAS Elektronik EMES-18 tank fire-control system in Leopard 1 MBT showing main components 0008543

Operational status
Production as required. In service with Denmark (120 modernised Leopard 1A5, plus 110 Leopard 1A3 from Germany); Germany (Leopard 1 being phased out); Greece (75 Leopard 1A5 from Germany); Norway (78 modernised Leopard 1A5 and 92 Leopard 1 from German stocks being upgraded with EMES 18 fire-control systems). Canada bought 123 German Leopard 1A5 to replace the turrets of 114 Canadian tanks to include the EMES-18 fire-control system.

Contractor
STN ATLAS Elektronik GmbH.

STN ATLAS Elektronik MOLF modular tank laser fire-control system

Type
Land fire-control system.

Description
The MOLF main and coaxial armament day/night fire-control system is based on the technology used in the EMES-18 system for the Leopard 1 MBT family but is designed as a modular retrofit kit for modernising a much wider variety of tanks such as Indian Arjun and Vijayanta MBTs, the T-series including the T-62, the M41, M48 and M60A1 and the AMX-30.

The various sets of modules available are:

(a) Single-axis stabilised system set for rectangular turret openings (as used on the M41 light tank)
(b) Two-axis stabilised system set (as used on the M41, M48, M60A1 and AMX-30)
(c) Single- or two-axis stabilised system set for round turret openings (as used on the T-series).

Depending on system configuration each set will normally comprise the following components:

(a) monocular gunner's primary sight with:
  (i) stabilised mirror head
  (ii) mechanical interface
  (iii) periscope/sight package with commander's eyepiece assembly, integrated Nd:YAG 1.06 μm wavelength laser range-finder and ×12 magnification day sight, 8 to 12 μm waveband thermal imaging sight with ×12 and ×4 magnification channels, thermal imaging power supply unit and optional image intensifier sight. The laser range-finder has a working range of 200 to 9,900 m with an accuracy of ±10 m.
(b) fire-control electronics and sensor package with computer control panel, digital ballistic computer that can handle up to eight separate ammunition types, cant sensor, air data sensor (for crosswind, temperature and pressure), gun elevation sensor and turret rate sensor
(c) functional system interface electronics unit.

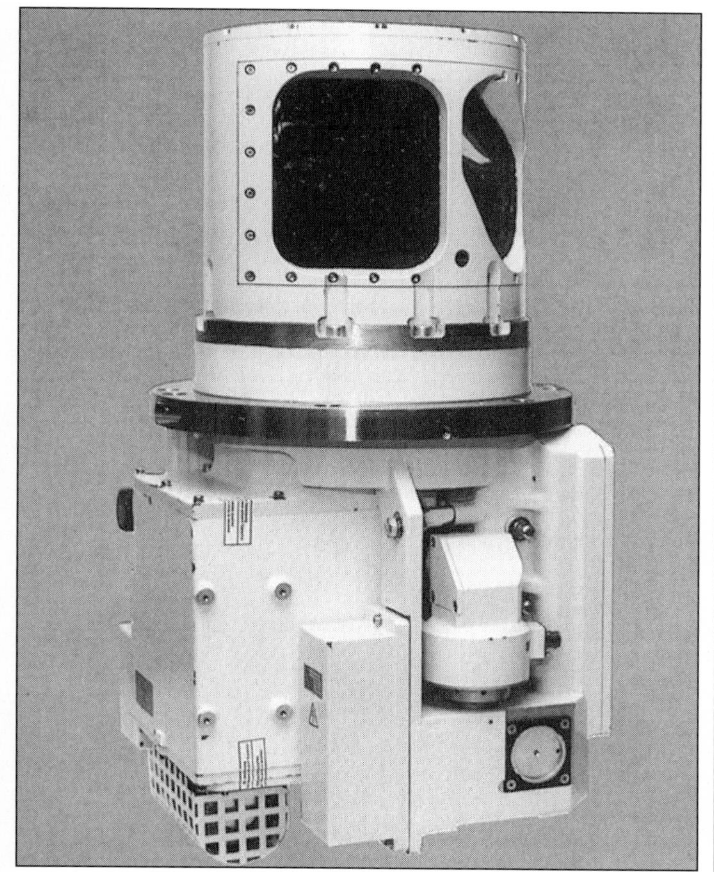

The version of MOLF fire-control system for T-62 tanks

The modular design of the system allows all kinds of aiming and line of sight stabilisation to be achieved, namely:

(a) mechanical linkage mirror-gun
(b) electrically slaved mirror drive in elevation
(c) electrically slaved mirror drive to stabilised main gun in elevation and independently stabilised mirror in azimuth
(d) primary stabilised LOS in both axes in combination with a slaved gun stabilisation system.

This makes it possible for the retrofitted tank, while it is either stationary or moving, to fire on a stationary or moving target with a high first-round hit probability.

The sighting system design allows for full utilisation of the fire- control system by day or night and in haze, fog or smoke. All the gunner has to do is acquire the target with his sight, find its range with the laser range-finder and then, while the computer continuously calculates the corrected superelevation and lead angles from the manually and automatically input fire-control parameters, keep his LOS on it until the firing circuit is enabled and commence firing with the selected armament.

Operational status
Production as required. Deliveries against total orders of 400 MOLF systems for installation in upgraded M48A5 MBTs started in 1992. The orders were placed by Greece in 1991 and 1993. The MOLF system has been picked for the CFE M60 tank modernisation as the NATO Standard LTFCS/SS.

Specifications
Laser range-finder
Type: Nd:YAG
Wavelength: 1.06 μm
Range: 200-9,900 m
Accuracy: ±10 m
Thermal imaging sight
Spectral band: 8-12 μm
Magnification: ×12, ×4

Contractor
STN ATLAS Elektronik GmbH.

Thales Optronics LRS 5 fire-control system

Type
Land fire control.

Description
In its basic configuration the LRS 5 is a monobloc gunner's sight including day sight, image intensifying night sight, laser range-finder, ballistic computer and mechanical link to the gun. Laser firing and moving target tracking switches are integrated into the handles of the turret. There is visual identification of all system parameters. The LRS 5 can be fitted to most types of armoured vehicles and tanks with modifications to the following: mounting plate; armoured cover; mechanical parallelogram between the gun and the output mirror.

Options include a thermal sight, Commander Display Box (CDB 5), Slaved Commander Sight (SCS 5) and eye-safe laser range-finder.

Operational status
Production as required. In service with six countries.

Specifications
Optical

Periscope	**Day**	**Night**
Magnification	×8	×4
Field of view	7°	7°
Entrance pupil	49 mm	100 mm
Dioptre adjustment	–5 to +5	None - Fixed (biocular)
Resolution	0.04 mils	0.3 mils at 1 mlux 0.6 mils at 0.1 mlux

LRS-5 fire-control system integrated in an AMX- 13 0004772

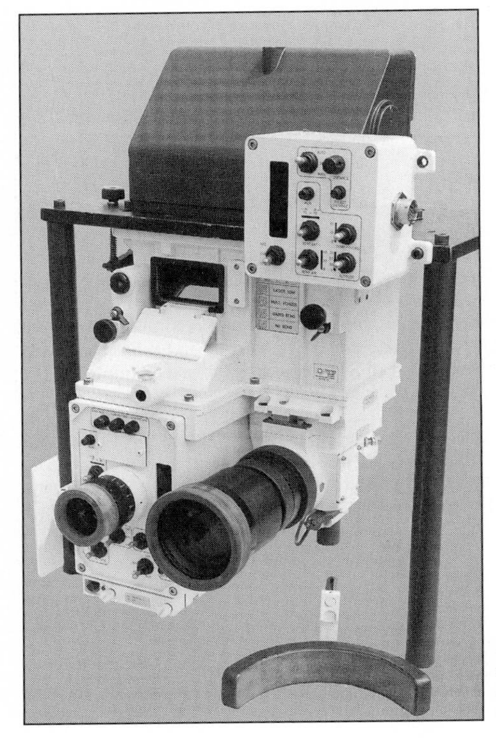

LRS 5 fire-control system from OIP Sensor Systems
0004775

Optical Periscope	**Day**	**Night**
Exit Pupil	6 mm	6 mm
Eye relief	35 mm	35 mm
Laser safety	> 50 dB	Protected by tube

Episcope field of view

Azimuth	> 16°
Elevation	> 8°

Laser
Wavelength: 1064 μm
Output energy: 4 mJ
Pulse repetition rate: 1 shot every 5 s
Receiver field: 0.7 mil
Range resolution: 5 m
Range accuracy: ± 10 m
Max/min range: 9,995/200 m

Electronics
Programmed ammunition: 4 to 6 types
Elevation ballistic angle: up to 50 mils
Azimuth lead angles: –28 to + 28 mils
System accuracy: 0.15 mil (1 σ value)

Total system
Weight: < 30 kg
Dimensions: 200 × 550 × 400 mm
Power supply: 18 to 30 V DC
Power consumption: 1.5 A
Operating temperature: –30 to + 55° C

Contractor
Thales Optronics (formerly OIP Sensor Systems).

UPDATED

Thales Optronics Mithras thermal imaging fire-control system

Type
Land vehicle fire-control system.

Description
Mithras is a modular range of thermal imaging fire-control systems suitable for a wide range of new-build or retrofit armoured fighting vehicles from light reconnaissance vehicles to main battle tanks. At the heart of Mithras is a second-generation thermal imager, which can be mantlet or trunnion mounted within an armoured barbette appropriate to the vehicle turret configuration.

Each Mithras system is tailored to integrate with, and take full advantage of, existing vehicle subsystems and sensors. The optimum configuration is then achieved through consideration of the operational requirement, the concept of use, the support infrastructure and affordability.

AFV fitted with the Mithras system 0053714

Three basic levels of Mithras configuration are available:
Mithras Level 1 – an entry-level surveillance system. In its baseline, lowest-cost configuration, Mithras Level 1 provides a rugged, vehicle mounted, thermal imaging surveillance system, bringing the advantage of long waveband IR imaging to vehicles perhaps equipped only with image intensified night sights.
Mithras Level 2 – an adjunct thermal fire-control system. With the addition of a three-board processing unit linked to the elevation and azimuth offset outputs of an existing fire control system, Mithras Level 2 can provide a slave-mode thermal-fire control engagement capability.
Mithras Level 3 – full fire-control and navigation system. The system can be expanded to include options such as a laser range-finder, an automatic muzzle reference system and a global positioning and navigation system. Full ballistic computation can then be undertaken, enabling independent fire-control operation, with display of own and target positional information within the thermal scene.

Operational status
Production as required.

Specifications
TI Sensor: second-generation 8-12 μm waveband long linear array
Laser rangefinder: Nd:YAG or eye-safe Erbium glass/Raman shift

Contractor
Thales Optronics.

Thales Optronics MK-72 thermal observation system for T-72 modernisations

Type
Fire-control system.

Description
The MK-72 long-range thermal observation system has been developed by Thales (formerly OIP Sensor Systems) for upgrading the fire-control system of T-72 main battle tanks. The modular system can be integrated with the existing laser/day sight with no requirement to cut armour shielding and only minor modifications to the internal turret arrangement.

The system consists of a thermal camera with a display unit for the gunner, a stabilised periscopic mirror head, integrated ballistic computer, systems electronics module and commander's display unit. Alignments of the camera assembly are electronically controlled.

Operational status
Available.

Specifications
Magnification: ×4/×12
Field of view
  Wide: 5.1 × 12° ±5°
  Narrow: 1.7 × 4° ±5°
Time change FOV: 1 s
MRTD: ≤0.03K at 0.5 lp/mrad
Focus: 50 m to infinity
Range
  Identification: 1,500 m
  Recognition: 3,300 m
  Detection: 7,600 m
Spectral band: 7.5-12 μm

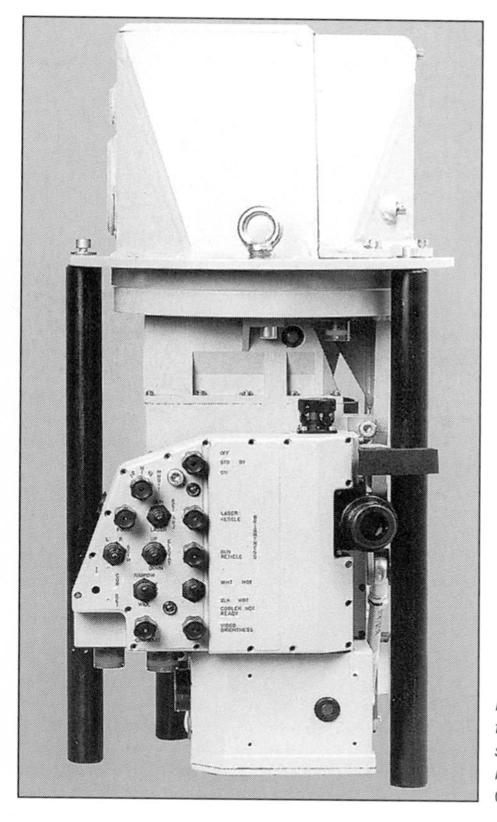

MK-72 long-range thermal observation system for T- 72 tank modernisations
0004771

Dimensions (h×w×d)
Thermal camera incl stabilised mirror head: 600 × 230 × 265 mm
Commander's display: 130 × 11 × 250 mm
Electronics module: 295 × 250 × 280 mm
Weight (total): 65 kg
Mirror head line of sight angles: –8 to +22°

Contractor
OIP Sensor Systems.

UPDATED

Yugoimport Rudi Cajavec SUV-T55A tank fire-control system

Type
Land fire-control system.

Description
The tank Fire-Control System (FCS) type SUV-T55A is designed for use on T-55 and T-62 MBTs, as well as for retrofitting to modified T-55 (with 105 mm main guns) and Chinese Type 59 MBTs. The system claims a high first-round hit probability at long ranges by both day and night against both static and moving targets while the platform vehicle is stationary, and against both stationary and moving targets in the daytime while it is moving. Typical engagement conditions for these scenarios are:

Platform	Target
stationary	stationary (size 2.3 × 2.3 m)
stationary	moving (size 4.6 × 2.3 m, speed 40 km/h)
moving (speed 25 km/h)	stationary (size 4.6 × 2.3 m)
moving (speed 25 km/h)	moving (size 4.6 × 2.3 m, speed 40 km/h)

The modular SUV-T55A system comprises the following subsystems:
(a) day/night gunner's stabilised sight with independent line of sight target observation, integrated Nd:YAG 1.064 μm laser range-finder and second-generation passive night vision image intensifier tube assembly (optional thermal imager)
(b) manual input computer panel (which displays target range and other system data)
(c) combined meteorological sensor assembly (which measures ambient air temperature, ambient air pressure, crosswind velocity and headwind velocity)
(d) fire-control computer which calculates the firing solution from manually input and automatic input data sources
(e) loader's display panel (which displays ammunition types and selection)
(f) gunner's control handle
(g) amplifier box unit

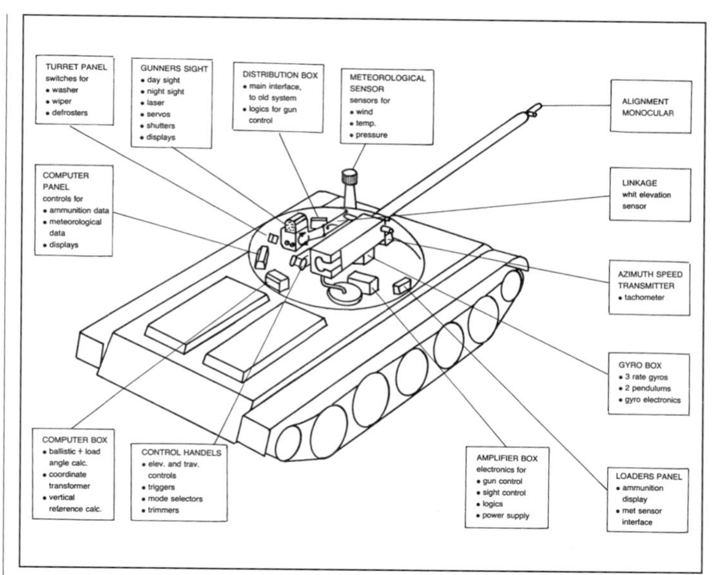

Schematic of T-55 MBT showing position of main components of SUV-T55A tank fire-control system

(h) amplifier input/output logic and signals distribution box
(i) gun linkage mechanism
(j) gyro box
(k) azimuth speed transmitter
(l) elevation sensor.

A typical moving target engagement cycle would be:
(a) the tank commander detects the target and presses the target acquisition button; the turret swings around to align the gunner's line of sight with the commander's cupola position (typical time taken for the acquisition phase is 2 seconds)
(b) the gunner takes over the engagement by tracking the target and operates the laser range-finder to establish range (typical time taken for tracking phase is 6 seconds); this is then displayed in the optics and fed to the computer; a firing solution is computed from all the data inputs and the calculated superelevation and lead angles continuously fed to the gunlaying system; the sight deflection system simultaneously counter-rotates in order to maintain the line of sight to the target; the gunner then fires the main armament (typical time taken for firing phase is 6 seconds)
(c) further firing of rounds against the target can thereafter take place very rapidly as the lead angles are still being calculated.

The engagement procedure is the same whether the firing platform is moving or stationary.

Variants
SUV-M84 intended for the Yugoslav M-84 MBT. All the main tactical usage and technical data of the SUV-T55A remain valid for the SUV-M84.
SUV-T72 intended for upgrading of the T-72 MBT and derivatives. The system is based on the SUV-T55A and SUV-M84 models with the only difference being that the integrated day/night sight has been split into two sub-units: a gunner's day sight with built-in laser range-finder and a gunner's night sight with either a passive night image intensifier or thermal imager.
SUV-60 intended for upgrading of the US M60 MBT.
SUV-CH intended for upgrading of the UK Chieftain MBT.

Operational status
Not known post Kosovo. In service with unspecified countries.

Specifications
Day/night gunner's sight
Magnification
Day: ×7 and ×3
Night: ×8.5
Field of view
Day: 9 and 20°
Night: 5.5°
Laser range-finder type: Nd:YAG
Wavelength: 1.064 μm
Max range: 10 km
Accuracy: ±5 m
Image intensifier type: Gen II tube
Thermal imager option
Spectral band: 7.5-11.8 μm
Field of view
Wide: 5.1 × 12°
Narrow: 1.7 × 4°
Detector type: CMT (Cadmium Mercury Telluride)
Number of elements: 60
Cooling system: Stirling closed-cycle
Display: CRT
Computer
Number of ammunition types: 6

Temperature
Air: 15°C (range –45 to +50°C)
Powder: 15°C (range –45 to +50°C)
Air pressure: 750 mm Hg (range –300 to +100 mm Hg)
Velocity
Crosswind: ±40 m/s
Headwind: ±40 m/s
Tilt: ±15°
Range limits
Laser: 200-6,000 m
Manual: 200-9,400 m
Lead angle: ±45 mils
Superelevation: –10 to +70 mils
Gun control system
Target tracking velocity
Max azimuth speed: ±80 mils
Max elevation speed: ±40 mils
Gun/turret velocity
Max azimuth speed: ±300 mils
Max elevation speed: ±80 mils

Contractor

Yugoimport SDPR.

ARMOURED FIGHTING VEHICLES

GUNNER'S SIGHTS

Aselsan Day and Night Thermal Sight System (DNTSS)

Type
Gunner's sight.

Development
This system is also manufactured in the US by Raytheon Systems Company, the licensor. Aselsan were co-producing the Day and Night Thermal Sight System (DNTSS) with Raytheon under contract to the Turkish Army with 650 systems ordered. The baseline DNTSS is also under development for integration into gyrostabilised spherical gimbals for airborne platforms.

Description
Aselsan's DNTSS is a modular sight combining thermal sensor and day sensor and visible display which provide observation and aiming. The gunner views the scene through a dual field of view direct view optics telescope, dual field of view daylight television and dual field of view thermal sensor. The thermal imager is based on a second-generation focal plane array detector with linear drive cryogenic cooler. The commander also has the capability to view the TV or thermal image independently through a separate video display unit.

The DNTSS integrates into the Turkish Land Forces Armoured Infantry Fighting Vehicles (AIFV). The thermal sight uses second-generation focal plane array (FPA) technology and can be integrated into a fire-control system without any modification to the base equipment.

Operational status
In production. First deliveries were made to the Turkish Army in March 1994.

Specifications
Weight: 163.64 kg (360 lb)
Detector: 240 × 4 CMT focal plane array
Spectral band: 7-12 μm

Head mirror
Elevation: –20 to +60°

Direct view optics
Field of view: 6° (narrow), 18° (wide), circular
Magnification: ×8 (narrow), ×2.7 (wide)

Thermal imager
Fields of view
Narrow: 1.77 × 4°
Wide: 5.31 × 12°
Magnification: ×9.3 (narrow), ×3.1 (wide)
Power consumption: 215 W max at 24 V DC
System MTBF: >1,000 h

Contractor
Aselsan Inc Microelectronics, Guidance and Electro-Optics Division.

VERIFIED

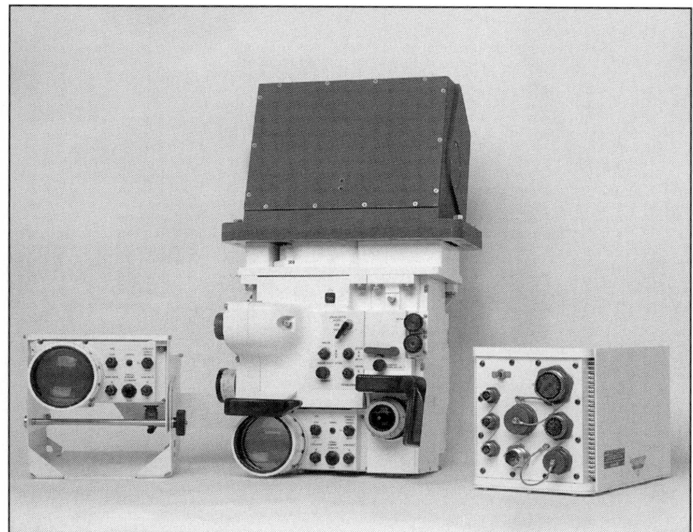

The Aselsan Day and Night Thermal Sight System (DNTSS) 0053712

ATCOP GNS 1

Type
Gunner's sight.

Description
GNS-1 is an image intensifying gunner's night sight for T-59 tanks using second-generation tubes. It enables night firing either as a stand-alone system or interfaced with the ATCOP Integrated Fire-Control System IFCS 69. Although designed for T-tanks, it can also be retrofitted to other vehicles.

Operational status
In production. In service with the Pakistan Army.

Specifications
Magnification: ×7
Field of view: 6°
Focus adjustment range: 100 m to infinity
Dioptre adjustment: ±5 dioptres
Transmission accuracy: 0.6 mil
Elevation range setting: 0 to 25 mil
Azimuth range setting: 20 mil on left and right each
Accuracy of elevation/azimuth setting: 0.2 mil
Input voltage/current: 24 V/100 mA
Dimensions: 285 × 248 × 406 mm
Weight: 24 kg (approx)
Gun linkage: mechanical standard with computer interface optional
Environmental specs: MIL-STD 810D

Contractor
Al Technique Corporation of Pakistan (Pvt) Ltd.

VERIFIED

GNS 1 gunner's sight
0004783

ATCOP thermal gunner's sights for T-series tanks

Type
Gunner's sight.

Description
These sights provide both the gunner and commander with the capability to conduct surveillance, target acquisition and engagement mission in all weather, both day and night. These sights are being retrofitted to T-series tanks under an upgrade programme.

Both electronic and mechanical interfaces to these sights for fire control system and gun of T-series tanks are developed by ATCOP.

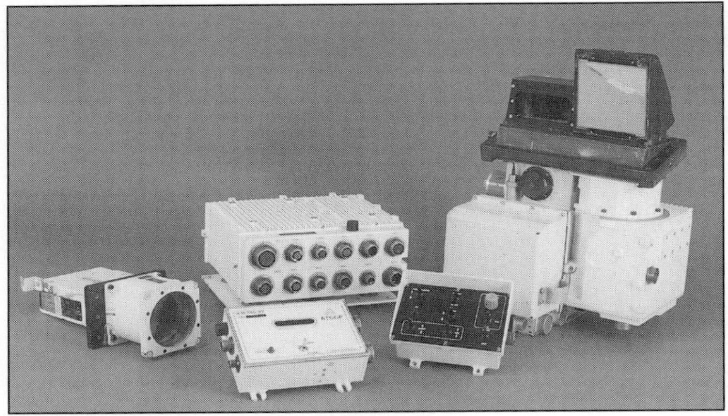

The ATCOP thermal gunner's sight system for T-series tanks 0053711

Specifications

	Uncooled	Cooled
Detector:	256 × 128	
	Uncooled staring	8 elements SPRITE
Spectral Band:	8 to 14 μ	7.5 to 11.5 μ
Optics:		
narrow FOV	4 × 2°	4 × 2.69°
wide FOV	12 × 6°	4 × 8.96°
Environment:		
operational	–30 to +55°C	–40 to +60°C
storage	–40 to +70°C	–45 to +90°C

Operational status
Test completed.

Contractor
AL Technique Corporation of Pakistan (Pvt) Ltd.

VERIFIED

BAE Systems SS100/SS110 night sights

Type
Gunner's sight.

Development
The SS100 (British Army designation SPAV L2A1) and SS110 (British Army designation SPAV L3A1) first-generation binocular vision, dual field of view, image intensifier sights were designed respectively for use with the Fox CVR(W) and Scorpion CVR(T) armoured vehicles. The equipment is turret mounted in front of the gunner and alongside the main armament.

Description
A dual-role capability is achieved by the use of two objectives, mounted one within the other. The outer objective gives a magnification of ×5.8, while the inner, for surveillance, has a magnification of ×1.6.

When the higher magnification is in use, a shutter isolates the low-power objective and when the low-power objective is selected, an iris diaphragm isolates the high-power lens.

BAE Systems SS100 sight installed in CVR(W) Fox with sight mounted to right of 30 mm RARDEN cannon

The image intensifier tube is of the first-generation 25 mm cascade type and is protected from the effects of muzzle flash by a shutter which is electrically operated by the gun firing circuit. Automatic brightness control for the tube is also fitted.

An illuminated ballistic ring graticule with brightness control is injected automatically into the optical system when the high magnification is used, and this is used for laying the Royal Ordnance 30 mm L21 RARDEN or 76 mm L23A1 main armament.

Operational status
SS100 production as required. In service with the British Army and other countries.

Specifications
(SS100 and SS110)
Weight
Total: 59 kg
Dimensions
Sight unit: 1,120 × 355 × 355 mm
Electronic control box: 108 × 165 × 295 mm
Field of view
×5.8 magnification: 8°
×1.6 magnification: 28°
Power supply: 28 V DC vehicle system

Contractor
BAE Systems, Avionics, Tactical Systems Division, Basildon.

BAE Systems SS122 series armoured vehicle day/night sights

Type
Gunner's sight.

Description
The SS122 and its associated series of derivatives (SS123 - SS126) are designed to be fitted to the gunner's and commander's turret positions of most current armoured fighting vehicles.

The SS122 consists of an optical system which provides the operator with a ×9.3 magnification fixed-focus day system that has detection, recognition and identification ranges in clear atmosphere of 30, 15 and 7 km respectively, a ×9 magnification variable focus night system with detection, recognition and identification ranges respectively of 1,800, 900 and 450 m at 10^{-2} lx, and a unity magnification fixed-focus general daylight observation facility.

Switching from the 25 mm three-stage cascade-type image intensifier passive night vision system to day vision is instantaneous, allowing the sight to provide a 24-hour target engagement capability. The eyepiece has integral laser absorption filters to protect the operator's eyes.

Elevation of the line of sight is performed by a tilting head mirror. In the case of the gunner's versions this is linked mechanically to the main armament while with the commander's variants it is manually operated and fully independent of the gunner's sight.

If required, a Type 520 Nd:YAG or LV353 Nd-doped laser range-finder module can be integrated into the sight assembly as an optional extra. When fitted to a sight in the vehicle, the range information can also be transmitted to either the other operator's sight or a remote display unit.

Operational status
Production, with over 1,200 produced to date. In service with at least 11 countries.

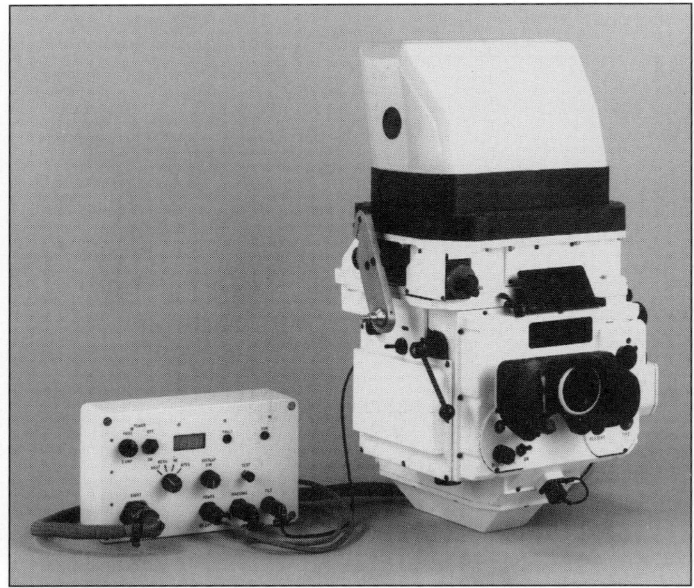

BAE Systems SS122 series armoured vehicle day/night sight

Specifications
Weight
Sight: 44 kg
Control box: 4 kg
Laser range-finder module: 3 kg
Dimensions: 360 × 350 × 625 mm
Power supply: 28 V DC vehicle system
Unity day sight
Magnification: ×1
Field of view
Horizontal: 25°
Vertical: 10°
Day vision channel
Magnification: ×9.3
Field of view: 6.4°
Passive night vision channel
Magnification: ×9
Field of view: 6.5°
Focus range: 30 m to infinity
Dioptre range: fixed (−1.75)

Contractor
BAE Systems, Avionics, Tactical Systems Division, Basildon.

BAE Systems SS180 armoured vehicle day/night sight

Type
Land fire-control system.

Description
The SS180 is designed as a commander's and/or gunner's elevatable sight for use on light armoured fighting vehicles and armoured personnel carriers. The commander's variant can be fitted with a circular slew ring with a ±30° azimuth arc rotation and indexing plunger to lock the sight in line with the main armament.

To allow for an uninterrupted target engagement capability, the SS180 can be switched instantly by hand lever from day to night vision operation. Both modes feature injected illuminated ballistic graticules with boresight and brightness adjustment for targeting purposes.

An electromechanical shutter protects the second-generation image intensifier tube during daylight operations while absorption filters in the eyepiece and unity magnification vision channels provide laser protection for the observer.

The SS180 can be supplied fitted with a third-generation image intensifier tube.

Operational status
Production. In service with unspecified countries.

Specifications
Elevation range: −10 to +55°
Unity day reflecting telescope
Magnification: ×1
Field of view
Horizontal: 30°
Vertical: 14°
High magnification refracting day vision channel
Magnification: ×5
Field of view: 8°

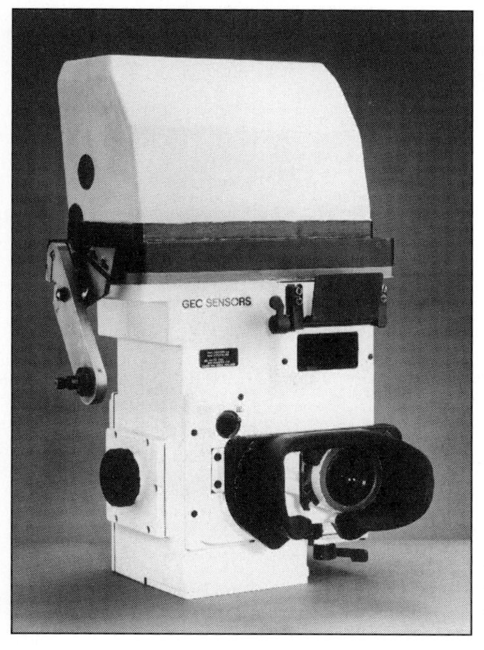

BAE Systems SS180 armoured vehicle day/night vision sight from operator's side

Night vision channel
Magnification: ×5
Field of view: 8°
Operational ranges in starlight
Detection: 1.35 km
Recognition: 750 m
Identification: 520 m
Power supply: 28 V DC vehicle system

Contractor
BAE Systems, Avionics, Tactical Systems Division, Basildon.

BAE Systems Uncooled Gunner's Thermal Sights

Type
Gunner's sight.

Description
BAE Systems Uncooled Gunners Thermal Sights utilise advances in uncooled Focal Plane Array (FPA) technology to produce thermal sighting systems stated to be at a price level previously only associated with image intensification systems.

The systems are of modular design, incorporating various sight configurations to fit most vehicles. They comprise under-armour, indirect view sensors fitted with dual field of view optics. This feature provides both gunner and commander with the means to carry out detection, identification and engagement of targets in the 24-hour battlefield. The sight head units are mechanically linked to the main armament to provide a means of accurately aiming the weapon. Electronically linked or stabilised options can be provided.

The systems have an external interface for connection to the vehicles existing fire-control system, either directly through a serial link or via an interface unit.

The systems also incorporate a stand-alone reversionary (back-up) mode which enables the weapon to be aimed using internally generated ballistic reticles, should the vehicle's fire-control system fail.

Operational status
Available.

Specifications
Typical performance NATO std target
Detection: >5.0 km
Recognition: >2.0 km
Identification: >1.0 km
Spectral waveband: 8 - 14 μm
Typical fields of view:
narrow 2°
vertical × 4° horizontal
wide 6° vertical × 12° horizontal
Detector type: uncooled micro-bolometer
Detector configuration: 256 × 128 with 2:1 microscan
NETD: <120 mK
Time to operation: <15 seconds @ 20°C
Video format: CCIR or RS 170
Power supply: 18 - 32V (MIL STD 1275)
Temperature (operating): −30 to +55°C

Contractor
BAE Systems Avionics, Tactical Systems Division.

EADS PZB 200/IRS 100 LLLTV aiming and observation system with IR scanner

Type
Gunner's sight.

Development
The PZB 200/IRS 100 LLLTV aiming and observation system with infra-red scanner combines the advantages of LLLTV and IR technology so that the thermal information detected by the latter is combined with the video signal supplied by the LLLTV, giving a congruently superimposed blinking infra-red signal on the monitor. The integrated IR signal allows even low contrast and camouflaged targets that can hardly be detected by the LLLTV to be observed.

The LLLTV system has been series produced and installed on Leopard 1, M48, OF-40, Centurion and Leopard 2 (as an interim fit) MBTs. It has also been field tested on AMX-30, TAM, Ikv-91 and Kurassier tanks. Existing PZB 200 systems can easily be retrofitted with the IRS system.

The PZB200 is being offered as a potential upgrade to surplus Royal Netherlands Army Leopard 1-V MBTs.

Description
The complete system comprises the following subassemblies:
(a) Pick-up unit – which consists of two components, the LLLTV camera and IR scanner, mounted coaxially with the main gun on special mounts attached to

PZB 200/IRS 100 LLLTV aiming and observation system

the gun shield. The PZB 200/IRS 100 system can be easily boresighted to the gun during the day or night.

The PZB 200 LLLTV camera picks up the night scene, intensifies it and supplies the scene as a TV picture via cable to the monitor(s) of the gunner and/or commander. For high aiming accuracy, a fixed optical reticle in the objective is projected with the scene image on to the first photocathode of the system pick-up tube combination and used as a reference for the subsequent manually or fire-control computer-controlled electronically generated reticle. The camera is fitted with an automatic light control. When the IRS 100 IR scanner is switched on, targets are detected by their thermal radiation differences in the 3 to 5 μm waveband region. The IR radiation picked up by the IR objective is depicted via an oscillating mirror on to a thermoelectrically cooled PbSe detector. The useful signals are then digitised, temporarily stored and converted into standard TV format. They are reconverted into analogue signals and superimposed on the LLLTV signal so that all the image scene information is congruently displayed on the monitor

(b) Monitor – which displays the LLLTV picture, thermal information and reticle
(c) PZB 200 control unit – which comprises the voltage supply, signal distribution, built-in test circuits and various control switches
(d) IRS 100 control unit – which comprises the built-in test circuit and various control switches
(e) Super elevation device – which generates the electronic reticle that is correspondingly adjustable to the ballistic data of the main armament. This reticle is moved by reference to the fixed optical reticle, and by manual input of target range (500 to 2,000 m in steps of 100 m) and type of ammunition (standard APDS, HEAT and HEP/HESH rounds with option for any other type) into the device together with automatic information from the cant angle sensor. It is set accordingly in elevation and azimuth with automatic drift and parallax compensated. It also has various controls and switches for selection of ammunition type, target range and so on
(f) Cant angle sensor – which automatically transmits the ballistic correction value for the cant angle (up to +10°) of the vehicle to the super elevation device, following alignment of its bubble level which is coupled to a potentiometer and which has its level axis orientated in parallel to the trunnion axis
(g) Interconnecting cable set
(h) Lens protection tube.

Operational status

Production as required. In service.

Specifications

Monitor screen size: 90 × 120 mm
Power supply: 24 ± 6 V DC
PZB 200 LLLTV
Focus range: 100 m to infinity
Optimum light range: 10^{-4} to 10 lx
Field of view: 64 × 48 mils
Waveband range: 0.4 to 0.8 μm
IRS 100 IR scanner
Focus range: fixed
Field of view: 48 × 16 mils
Detector type: PbSe
Spectral band: 3-5 μm

Contractor

European Aeronautic Defence and Space Company.

Elop Day/Night Thermal Sighting System (DNTSS)

Type

Gunner's sight.

Description

The DNTSS is designed as a low-cost modular thermal sight assembly that is suitable for upgrading tanks and armoured fighting vehicles such as T-series MBTs, AMX-13, M41, M48, M60 and LAVs.

The system combines an integral laser range-finder module and a thermal imaging elbow with full ballistic solution.

The system provides a commander's remote display, automatic line of sight deflection according to ballistic data correction, add-on ballistic sensors and video recording and data transmission.

Operational status

Production as required. In service with undisclosed countries.

Specifications

LOS elevation: –15 to +60°
Direct view optics
Field of view
  Day channel: 8° (× 8)
  Unity channel: 25 × 14°
Magnification: ×8
Field of view: 8°
Day unity sight
Field of view
  Azimuth: 25°
  Elevation: 14°
TV day channel
Magnification: ×4
Field of view: 2.3 × 3°
Thermal channel
Spectral range: 3-5 μm
Fields of view
  Wide: 9.6 × 4.8°
  Narrow: 2.8 × 1.4°
Laser range-finder
Type: Er Glass
Wavelength: 1.54 μm
Operating range: 200-9,995 m
Accuracy: ±5 m
Discrimination: 50 m
Range gate: Min 300-5,600 m
Multiple echo logic: optional min range gate or first/last

Contractor

Elop Electro-Optics Industries Ltd.

UPDATED

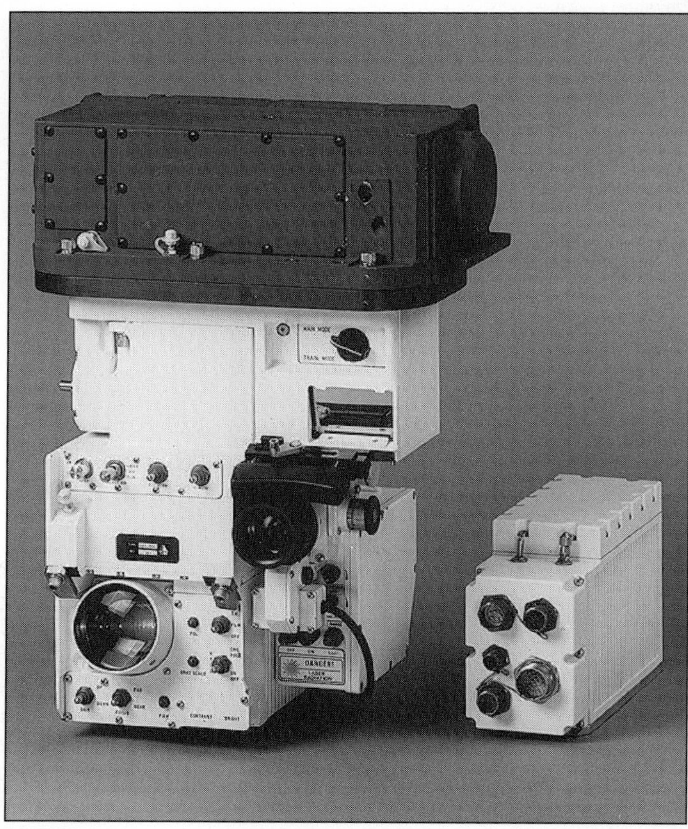

DNTSS Day/Night Thermal Sighting System from Elop

Elop MSZ-2 gunner's day/night periscope

Type
Gunner's sight.

Development
The Elop MSZ-2 periscope is a modular integrated electro-optical fire-control system for use by gunners on tanks and other AFVs such as the M24, M41, M48, M50 Sherman, M51 Sherman, Centurion, Saladin and Scorpion. It is used to upgrade the ranging and aiming capabilities of the vehicles by incorporating a laser range-finder and passive night vision capabilities into a relatively small size.

Description
Two basic versions are available, with a number of intermediate variants. The simplest model consists of the Elop Mini-Laser Range-finder with a ×1 magnification prism and an image intensifying night elbow. The elbow contains a standard NATO graticule (or any other required graticule) and a laser aiming mark for night range-finding. The graticules are projected and collimated on the image intensifier tube to give the accuracy needed for aiming and range-finding at night.

The most sophisticated version is a fully computerised model with the same basic systems but also a ballistic computer, sensors and a rotating wedge module for the automatic insertion of ballistic compensation data, providing a single higher accuracy graticule and full fire-control system solutions in a faster reaction time.

The day/night periscope for a vehicle commander is identical to the gunner's model but is designed to utilise a ×8 magnification day elbow in place of the mini-laser.

Operational status
Production as required. In service with the Israel Defence Force and other unspecified countries.

Specifications
Unit
Prism device
Magnification: ×1
Field of view
Horizontal: 40°
Vertical: 14°

Elop MSZ-2 gunner's day/night periscope
0059660

Passive night vision elbow
Magnification: ×7.2
Field of view: 7.5°
Image intensifier tube: 18 mm, GEN II or III

Power supply
Regular: 24 V DC vehicle
Emergency: two 1.5 V batteries

Daylight system
Magnification: ×8
Field of view: 8°

Eye-safe laser range-finder
Wavelength: 1.54 μm
Operating range: 200-9,900 m
Range accuracy: ±5 m
Power supply: 24 V DC vehicle

Contractor
Elop Electro-Optics Industries Ltd.

VERIFIED

Elop Tades – advanced second-generation thermal imaging elbow sight

Type
Gunner's sight.

Description
Tades is a gunner's sight intended for use as a MBT and AFV long-range fire-control system. It can also be applied as a mobile air defence sight. The system employs an advanced 480 × 4 second-generation thermal imaging detector.

Main features are as follows:
- State-of-the-art 480 × 4 second-generation TDI Detector Technology
- Wide fields of view combined with very high-sampling resolution
- Three fields of view plus electronic zoom
- High-reliability linear Closed Cycle Cooler
- Powerful signal processing based on proprietary ASIC
- Real-time algorithms for image enhancement
- CCIR or RS170 standard video output or 8/12-bit digital video output
- Flexible optical, mechanical and electronic interfacing, for ease of upgrade to various Armoured Fighting vehicles (AFV)
- Software generated on-screen graphics
- Designed to meet the environmental requirements of MBTs.

Elop's Tades is a Gen II thermal imaging elbow sight (Elop) ***NEW***/0547551

For details of the latest updates to *Jane's Electro-Optic Systems* online and to discover the additional information available exclusively to online subscribers please visit jeos.janes.com

Specifications
Dimensions sensor unit: 180 × 190 × 450 mm
Weight sensor unit: 10 kg
Electronics unit: 6 kg
Spectral band: 8 - 12 μm

Fields of View
NFOV: 3.0 × 9°
MFOV: 10 × 6.25°
WFOV: 230 × 14.35°
Electronic zoom: ×2
Detector: MCT-PV, TDI, 480 × 4 elements
Cooler: High Reliability, Linear Closed Cycle, Split Configuration
Video output: CCIR or RS-170 Composite video or 8/12-bit Digital Video
Serial communication interface: RS-422, Fast Ethernet (optional)
Power: 85 W @ 24 V as per MIL-STD-1275

Contractor
Elop Electro-Optic Industries Ltd.

NEW ENTRY

Elop Thermal Elbow Sight (TES)

Type
Gunner's sight.

Description
The Thermal Elbow Sight (TES) is designed as a modular thermal sight for installation in main battle tanks (T-72 and others).

The TES modules include thermal elbow, electronics unit, gunner's display unit, commander's display unit (optional) and an external head mirror assembly coupled mechanically or electrically to the vehicle's fire-control system.

Operational status
Production as required. In service with undisclosed countries.

Specifications
Spectral band: 8-12 μm
Detector: CMT
Cooler: split closed-cycle Stirling
Aperture: 110 mm

Field of view
Narrow: 3 × 2°
Wide: 10.5 × 7°
Video output: CCIR, 625 lines/RS-170, 525 lines

Weight
Thermal elbow: 10 kg
Display unit: 3.5 kg
Electronic unit: 7 kg
Power supply: 28 V DC

Contractor
Elop Electro-Optics Industries Ltd.

VERIFIED

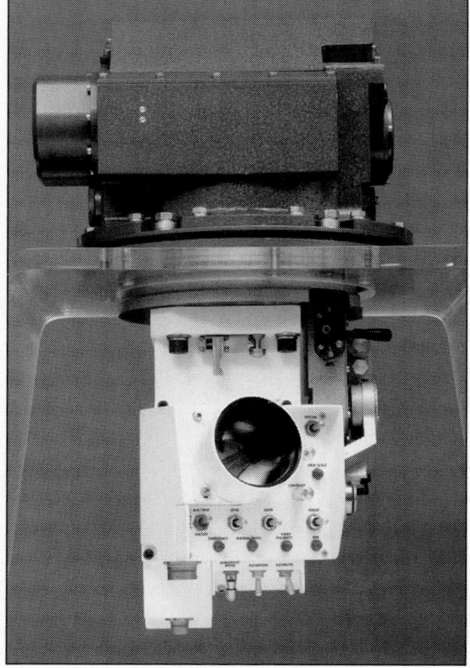

The Elop thermal elbow sight
0059665

Elop Thermal Gunner's Elbow (TGE)

Type
Gunner's sight.

Description
The Thermal Gunner's Elbow (TGE) may be used to upgrade MBT or LAV sights and can act as a standard module for AFV and tank fire-control systems or act as a stand-alone thermal imager for LAV turrets.

The TGE subsystems comprise:
(a) thermal imaging elbow
(b) gunner's display unit
(c) optional commander's display unit.

Operational status
Production ceased. In service with undisclosed customers.

Specifications
Spectral band: 8-12 μm
Aperture: 110 mm
Fields of view
Narrow: 2.8 × 1.4°
Wide: 9.6 × 4.8°
Cooler: Integrated closed-cycle
Video output: CCIR, 625 lines
Weights
Elbow: 12.7 kg
Display: 4 kg
Dimensions
Elbow: 230 × 180 × 295 mm
Display: 140 × 180 × 235 mm

Contractor
Elop Electro-Optics Industries Ltd.

UPDATED

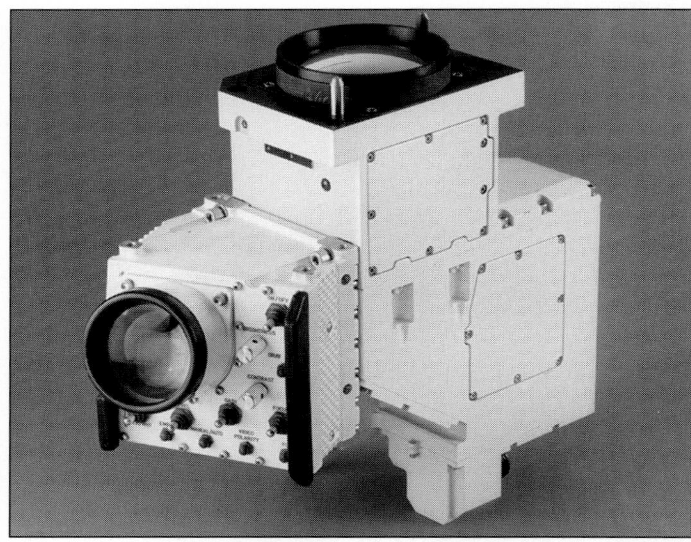

The Elop Thermal Gunner's Elbow (TGE) 0059666

Eloptro LE-30 Laser Elbow

Type
Gunner's sight.

Description
The LE-30 Laser Elbow is an integrated day sight and laser-ranging system designed to be fitted to existing tank sights. Together with the NE-30 Night Elbow, it is used in Eloptro's GS-30 gunner's sight.

The LE-30 allows the gunner to accurately lay the main weapon either manually or with the aid of an integrated fire-control system. A projected reticle channel is provided, displaying both fixed and adjustable reticle in the eyepiece. The fixed reticule indicates the laser aiming mark and ballistic scales, whilst the adjustable reticle displays the NATO cross.

Operational status
In production, as required. In service with undisclosed customers.

Specifications
Maximum range: 9,995 m
Range resolution: 5 m
Standard measuring rate: 10 measurements/min
Fast measuring rate: 30 measurements/mm

The LE-30 Laser Elbow 0110164

Multiple targets: displays first and last targets
Target discrimination: 50 m
Range gate: 300 - 9,990 m
Laser type: Nd-Yag
Magnification: ×8
Field of view: 8°
Focus: fixed at infinity
Eye relief: 23 mm
Eyepiece dioptre adjustment: –4 to +2
Laser safety filter: against 1,064 nm
External power: 20 - 30 V DC @ 3A
Mass (incl battery): <7 kg

Contractor
Eloptro, Division of Denel (Pty) Ltd.

UPDATED

ENOSA PP - 03 aiming periscope

Type
Gunner's sight.

Description
The ENOSA PP - 03 aiming periscope has been designed and developed to equip turrets of armoured vehicles fitted with machine guns or small calibre automatic guns. The system consists of two modules, a tilting top body and a lower body with a biocular sight for observation and a monocular aiming telescope. For night operation the lower body is replaced by a passive elbow (Model CP-25) incorporating an image intensifier.

An important feature of this periscope is that the two optical systems are available at the same time. The wide field biocular is used for observation while the magnifying monocular is used for aiming. In the aiming system a stadiametric reticle with lines or strokes enables aiming and lead angle prediction for moving targets.

Operational status
Production as required. In service with the Spanish Army and other countries.

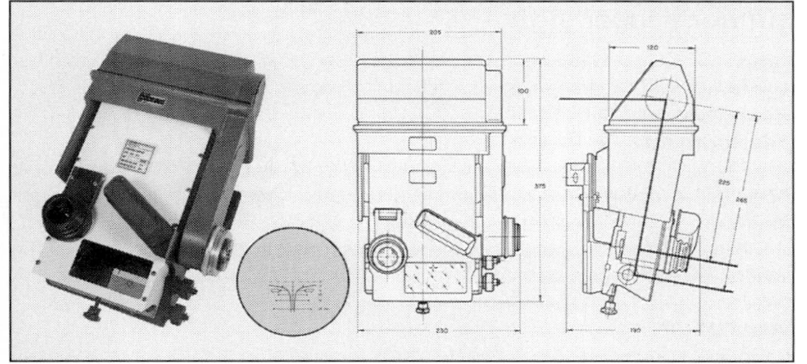

ENOSA PP - 03 aiming periscope 0023001

Specifications
Upper body elevation range: –20 to +70°
Observation biocular sight
Magnification: × 1
Field of view: 23° (horizontal) × 11° (vertical)
Aiming telescope
Magnification: ×3
Field of view: 12°
Exit pupil diameter: 3.8 mm
Eye relief: 20 mm
Power supply: 24 V (vehicle supply)
Total system weight: 13 kg

Contractor
Empresa Nacional de Optica SA (ENOSA).

VERIFIED

Fotona Thermal Imaging Gunner's Sight (TIGS)

Type
Gunner's sight.

Description
The Thermal Imaging Gunner's Sight is designed as a replacement for the existing TPN-1 gunner's night sight in T-55, T-62 and T-72 tanks without turret modifications. The mechanical linkage connects the gun and mirror of the TIGS optical deviator. The system can function as a stand-alone instrument or as part of a fire-control system.

As a stand-alone instrument the firing procedure is as follows:
(a) acquisition of the range to target either by laser range-finder or by aiming reticle
(b) the built-in computer calculates the ballistic angles, taking into account range to target, ammunition type, gun jump correction and boresight offset
(c) the gunner leads the gun to the firing position by aligning the aiming reticle and target
(d) the gun is fired.

In the case where TIGS functions as part of the fire-control system the aiming procedure is completely subordinated to the fire-control system. The man/machine interface is through programmable alphanumeric and graphical characters received from the fire-control system computer via the serial interface. The shape of the aiming reticle changes according to the phase of the firing procedure.

Operational status
Available. TIGS has been fitted to a prototype T-72 Moderna tank upgrade package being offered by ZTS Dubnica of Slovakia.

Specifications
Accuracy
Elevation, gun-LOS: 1 sigma = 0.4 mrad
Azimuth, gun-LOS: 1 sigma = 0.3 mrad
Line of sight alignment range: ±3 mrad
Laser marker alignment range: ±3 mrad
LOS elevation: –5 to +19°
Elevation lead angle: up to 25 mrad

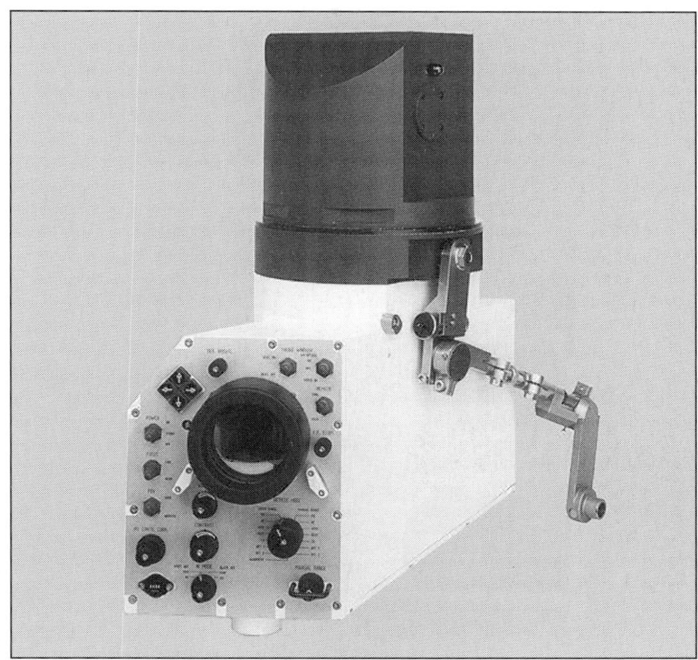

The Thermal Imaging Gunner's Sight from Fotona

Ammunition types: ballistic tables for 4 ammunition types are used to compute lead angles
Jump corrections: 2 mrad
Power supply: 19-30 V DC

Thermal imaging module
Fields of view
Wide: 9.2° (horizontal) × 5.4° (vertical)
Narrow: 3.4° (horizontal) × 2° (vertical)
Spectral band: 8-13 µm
Detector type: CMT photoconductive
Optics: Ge, F/N 1.7
Focusing range, NFOV: 20 m to infinity
Focusing range, WFOV: 5 m to infinity

Temperature resolution MRTD, NFOV
At 0.5 cy/mrad: <0.06 K
At 2 cy/mrad: <0.6 K
Detection range for MBT: >6,000 m
NFOV switch time: <0.5 s
Display: biocular, green
Temperature window/contrast: 2.5 to 40°C
Temperature offset/brightness: ±45°C

Contractor
Fotona dd Slovenia.

Galileo Avionica OG-P101 periscope sight

Type
Gunner's sight.

Description
The Galileo Avionica (formerly Officine Galileo) OG-P101 observation/aiming sight is designed to replace, with suitable modifications, the US-designed and built M32, M34 and M36 periscopes on armoured fighting vehicles.

For night vision and firing it uses a second-generation light intensifier tube assembly.

Operational status
Production as required. In service with the Italian Army.

Specifications
Dimensions: 470 × 205 × 295 mm
Elevation range: –10 to +70°
Fields of view
×1 magnification: 26 × 10°
×8 magnification: 9°

Contractor
Galileo Avionica.

UPDATED

Galileo OG-P101 periscope sight

Galileo Avionica P170/P204 day/night gunner's periscope head

Type
Gunner's sight.

Development
The P170/P204 was developed for use as the gunner's periscope in the IVECO Type 6616 armoured car primary fire-control system. It can also be adapted to replace M32, M34 and M36 periscopes installed in other types of armoured vehicles.

Description
It comprises a main body and various elbow systems. The body contains the head prism assembly, a unity power optical system and a mechanical lever linkage arrangement to the weapon.

Operational status
Production as required. In service with the Italian Army.

Specifications
Weight: 8.8 kg
Dimensions: 175 × 130 × 340 mm
Elevation/depression: –10 to +70°
Magnification: ×1
Field of view
Azimuth: 26°
Elevation: 10°
Elevation arm
P170: right
P204: left

Contractor
Galileo Avionica, Avionic Systems and Equipment Division, (formerly Alenia Difesa).

UPDATED

Alenia P204 periscope head with P170/P204 day sight elbow

Galileo Avionica P170L/P204L day/night gunner's laser sight

Type
Gunner's sight.

Development
The P170L/P204L are the laser elbows derived from the P170/P204 sight to meet the technical requirements of observing, aiming and laser ranging in day and night conditions.

The only difference between the two periscopes is that the position of the weapon linkage lever is on the right in the P170L and on the left in the P204L.

Description
In physical appearance the complete sights are similar to the P204 with the main components being:

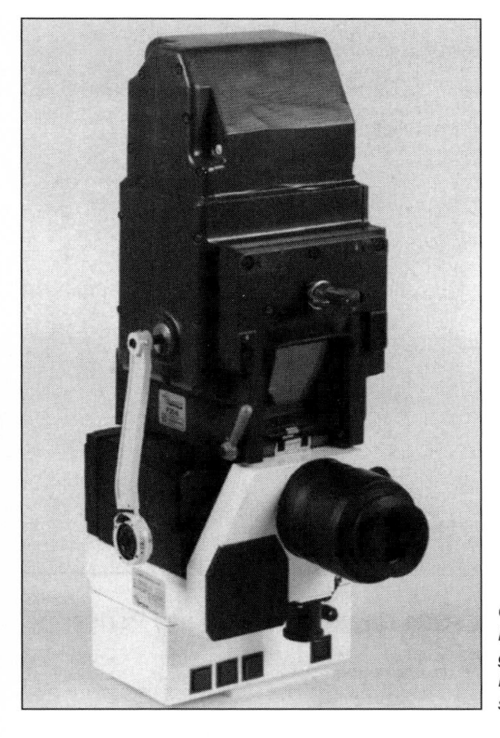

Galileo Avionica P170/ P204 day/night gunner's periscope with P170L/P204L laser sight body

(a) the P170/P204 periscope head assembly with a prism, ×1 magnification optical window and weapon linkage unit
(b) the daylight elbow which includes the objective, right angle prism, reticle assembly, eyepiece laser assembly, laser optics and display module. The actual daylight elbow itself is fully interchangeable with the passive night vision elbow.

The integrated MTL-8 Nd:YAG 1.064 μm wavelength laser transceiver module uses a common aperture arrangement for the transmitted pulse, received echo and the visual image channels.

Operational status

Production as required. In service with undisclosed countries.

Specifications

P204L TXRX Laser Unit (MTL-8)
Weight: ≤2 kg
Dimensions: 96 × 48 × 86 mm
Type: Nd:YAG
Wavelength: 1.064 μm
Laser electronic unit (optional)
Weight: 1.6 kg
Dimensions: 210 × 150 × 95 mm
Range: 300-9,990 m
Accuracy: 10 m
Discrimination: 30 m
Range gate: selectable in range 300-3,500 m
P204L laser sight (monocular)
Weight: 8.2 kg
Dimensions: 210 × 300 × 265 mm
Magnification: ×8
Field of view: 8.5°
Focus range: ≥250 m
Dioptre range: –3 to +3

Contractor

Galileo Avionica, Avionic Systems and Equipment Division (formerly Alenia Difesa).

UPDATED

Galileo Avionica P265 passive night vision elbow

Type

Gunner's sight.

Development

The P265 night elbow aiming periscope has been designed to be directly interchangeable with the daylight elbow of the P170/P204 periscopes and can also be used to replace the infra-red elbow used in the M32 and M36 MBT periscopes.

Description

The main components comprise:
(a) elbow assembly with the objective, high-reflectance mirror, a 25 mm second-generation image intensifier tube and eyepiece unit

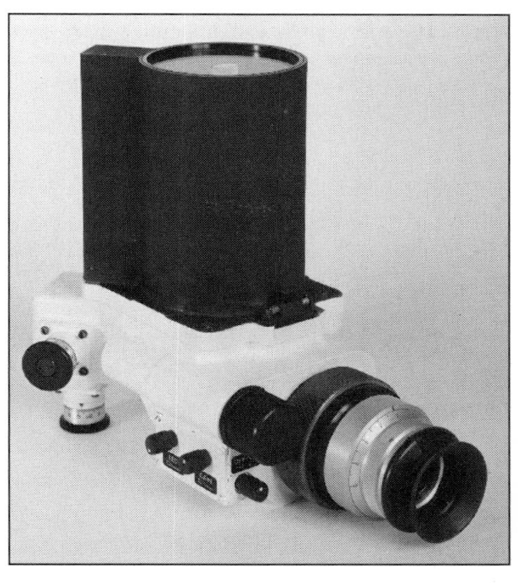

Galileo Avionica P265 passive night vision elbow

(b) reticle projection device assembly which includes an objective, mirror, pentaprism, reticle unit and a high emitter diode.

Range of the image intensifier in starlight conditions is 1.3 km with the projected reticle intensity being adjustable to suit the conditions. If the main power supply is interrupted for any reason then a 3 V DC lithium battery automatically cuts in.

Operational status

Production as required. In service with undisclosed countries.

Specifications

Weight: 8 kg
Dimensions: 345 × 175 × 278 mm
Magnification: ×7
Field of view: 8.3°
Resolution (USAF target 85% contrast, 90% reflection)
1 lx: 0.3 mil
10 lx: 0.4 mil
Focal range: 20 m to infinity
Dioptre range: –5 to +5
Power supply: 24-28 V DC
Emergency battery: 3 V DC lithium

Contractor

Galileo Avionica, Avionic Systems and Equipment Division, (formerly Alenia Difesa).

UPDATED

GM Defense – Delco Systems thermal sight for the LAV-25

Type

Gunner's sight.

Description

GM Defense – Delco has developed a compact thermal sight for its LAV-25 turret. Designed to fit directly into the existing LAV M36E1 gunner's and/or commander's

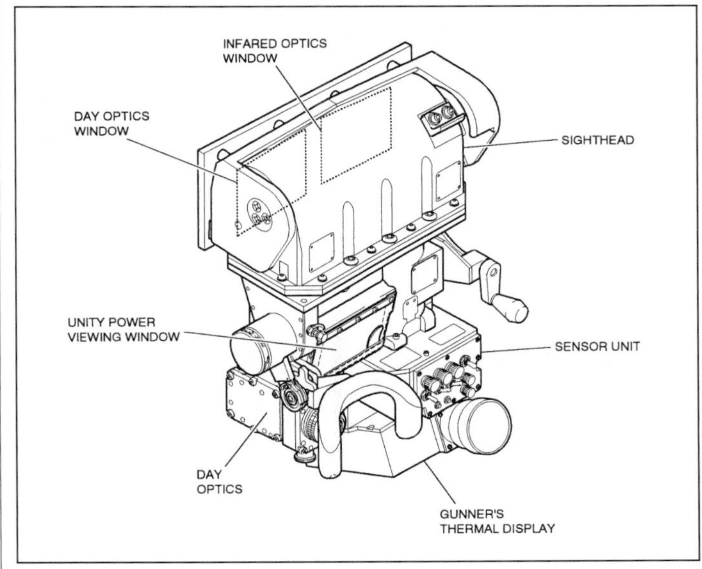

Diagram of the thermal sight for the LAV-25

weapons aiming sight mount, it is a serial scan imager, the output of which is connected to a television display producing an image in a monocular eyepiece.

The complete sight assembly also has a unity channel periscope, a ×8 magnification day channel and a mechanical gun linkage device.

Operational status

In production. The LAV-25 turret is in service with Australia, Canada, Kuwait, Saudi Arabia and USA.

Contractor

GM Defense - Delco Systems.

Hensoldt PERI-Z16 tank modular sighting system periscope

Type

Gunner's sight.

Description

The PERI-Z16 tank periscope is a modularised sighting system suitable for a wide variety of combat vehicles up to and including MBTs. It can be configured as a basic gunner's sight up to a sophisticated and integrated sensor of a fire-control system.

The following subassemblies are used to make up the appropriate PERI-Z16 periscope configuration:

(a) Head assembly with:
- (i) elevating head mirror
- (ii) elevation traverse range −355 to +1,244 mrad
- (iii) appropriate mechanical interfaces for different modules
- (iv) mechanical interface to the weapon drive
- (v) sealing window

(b) PERI-Z11 daylight aiming elbow with:
- (i) ×2 and ×6 switchable magnification
- (ii) monocular eyepiece
- (iii) customer-designed reticle for ground and air targets
- (iv) offset angle regulation by the mechanical link between the weapon and elevation mirror
- (v) adjustable reticle illumination
- (vi) protective filters against laser and bright sunlight

(c) Direct-view prism block with:
- (i) unitary magnification
- (ii) 711 mrad horizontal and 195 mrad vertical field of view
- (iii) mechanical interface identical with the NAE 200 image intensifier Night Aiming Elbow

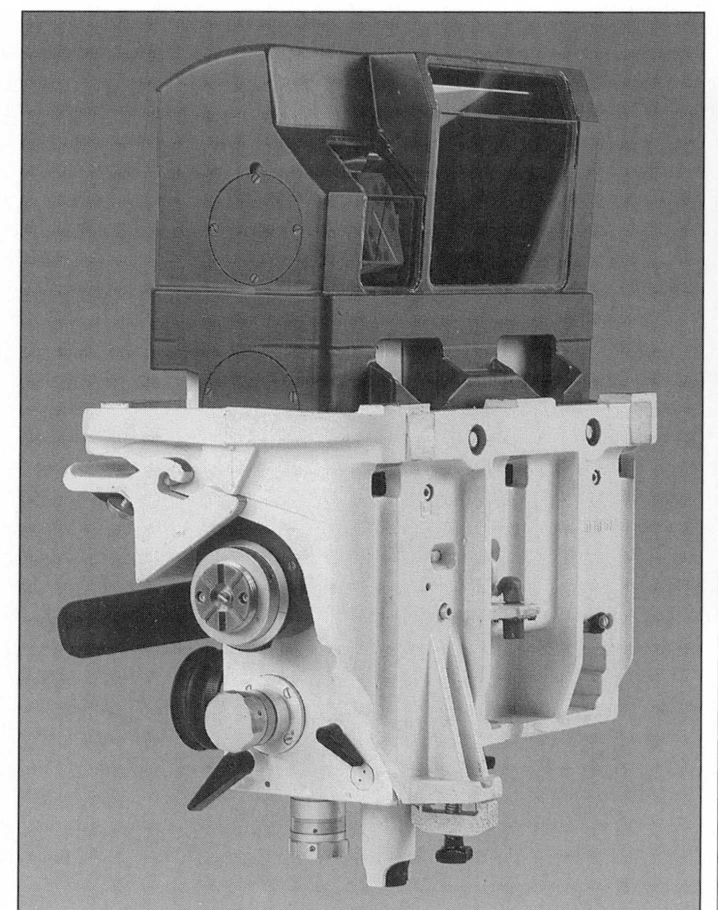

PERI-Z16 tank modular sighting system periscope

(d) NAE 200 image intensifier Night Aiming Elbow with:
- (i) ×8 magnification
- (ii) second- or second +-generation image intensifier tube
- (iii) objective focusing
- (iv) mechanical interface identical with direct-view prism block

Operational status

Production as required – over 3,000 basic PERI-Z16 periscopes in service.

Contractor

Hensoldt Systemtechnik GmbH.

UPDATED

Kollsman Day/Night Range Sight (DNRS)

Type

Gunner's sight.

Description

Kollsman's DNRS adds an Eyesafe Laser Integrated Periscope (ELIP) to the basic DNS system (qv) to improve first-round hit capability. The ELIP attaches to the Day Sight Interface Unit, and provides the gunner with a narrow field of view, day periscope. A ballistic reticle bore-sighted to the turret's weapons is displayed in the day sight and also provided to the gunner's and commander's remote displays. This provides for day, night and obscured battlefield lasing and firing of the weapons.

Status

The DNRS is in production; 43 have been delivered to Kuwait on the AV Tech two-man MGTS turret equipping Pandur 6 × 6 vehicles; 200 are on order for the HITFIST turret of the Italian DARDO IFV programme. Final deliveries due 2004.

Specifications

The Specification for the ELIP is:
Transmitter: Er:Glass
Wavelength: 1.54 μm
Pulse energy: >8mJ
Repetition rate: 1/6 pps, l pps burst
Beam divergence: <1 mrad
Range: 200 to 9,995 m
Accuracy: ± 5 m
Optical Magnification: ×8 or ×12
Weight: 6 kg
Power consumption: <40 W

Contractor

Kollsman Inc.

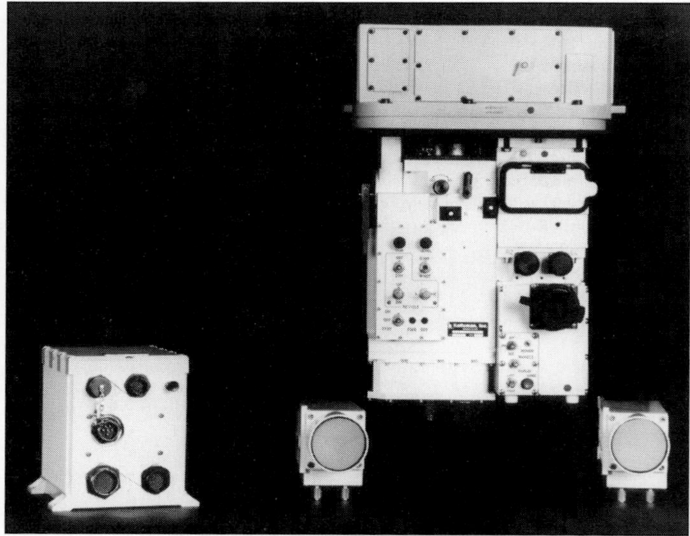

Kollsman's Day Night Range Sight (DNRS) 0097107

Kollsman Day/Night Sighting (DNS) systems

Type

Gunner's sight.

Description

The Kollsman Day /Nigh Sighting (DNS) system was developed as a compact, low cost, highly modular approach to satisfy the fire control needs of both armoured and light armoured vehicles. The DNS is the most basic configuration of this high

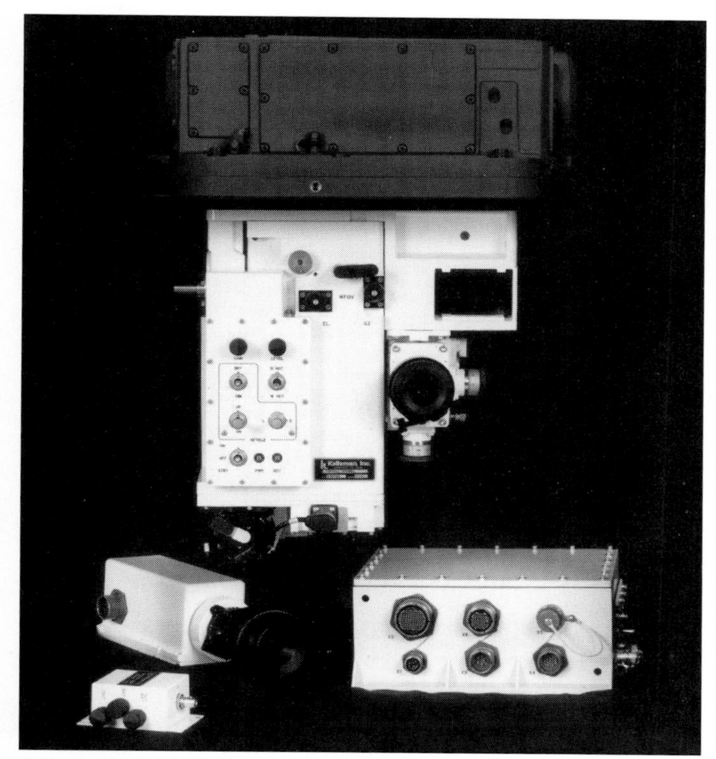

Kollsman Day/Night Sight system from operator's side 0097106

performance system, The system's enhanced thermal camera allows the gunner and commander to acquire targets day or night at distances far exceeding thc range of the main weapon. The system is designed to accommodate main weapons from 20 to 120 mm. An 8 to 12 micron thermal camera provides the armoured vehicle gunner the capability to detect, recognize, and engage threats at night and under obscured battlefield conditions. A remote display of the thermal scene is provided to the commander. The DNS consists of:

a. Periscope head with mirror mechanically linked to the main weapon.
b. Periscope mount including unity (×l) window.
c. 8× Day Periscope.
d. Thermal Channel utilising US common module FLIR.
e. Gunner's and Commander's Thermal Display.
f. Thermal Channel Power Conditioner.

Operational status
The Kollsman DNS has been demonstrated in the FNSS 1-man Sharpshooter Turret on M113A3 tracked vehicles.

Specifications
Elevation range (Mechanical head): –20 to +60°
Day periscope magnification: × 8
Thermal spectral range: 8 - 12 μm
Thermal field of view: WFOV 3.4 × 6.8° (2.8 magnification)
NFOV 1.1 × 2.2° (8.4 magnification)
Detector: 60 element US common module CMT
Video format: RS-170 or CCIR

Contractor
Kollsman Inc.

VERIFIED

Norinco Type 79-II tank gunner's sight

Type
Gunner's sight.

Description
The Type 79-II (or WC 532) tank gunner's sight incorporates a second-generation image intensifier tube and is designed for fitting to T-series MBTs. The external fitting comprises a casing for a periscope attachment. Internally there are the control and power regulation boxes.

Operational status
In production. In service with the People's Liberation Army (PLA).

Specifications
Weight: 22 kg (total)
Length: 490 mm
Width: 248 mm

Norinco Type 79-II tank gunner's sight

Height: 406 mm
Magnification: ×7
Field of view: 6°
Dioptre range: –5 to +5
Focus: 100 m to infinity
Power supply: 26 ±4 V vehicle supply

Contractor
China North Industries Corporation (Norinco).

VERIFIED

PCO 1PN-22MZ modernised passive day-night sight for BMP-1 AFV

Type
Gunner's sight.

Description
The 1PN-22MZ modernised passive day-night sight is a modernised version of the active Russian-type sight 1PN-22M2 for BMP-1 armoured fighting vehicles.

Application of SuperGEN or GEN II+ image intensifier tubes ensures fully passive operation of the device in the night channel and allows the gunner to provide detection, aiming and firing from the BMP-1 weapon under night conditions at longer distances than the active sight.

Operational status
Available.

Specifications
Magnification: day path ×6, night path ×6,7
Field of view: day path 15° , night path 6° 30′

Manufacturer
PCO S.A, Poland.

The PCO 1PN-22MZ modernised passive day-night sight 0089917

PCO CDN-1 day-night periscope sight

Type
Gunner's sight.

Development/Description
The CDN-1 day/night periscope sight is a viewing assembly designed for upgrading the BRDM-2A/B wheeled armoured vehicle. It provides the vehicle gunner with vision of the terrain and any adjacent vehicle(s) in both day and night conditions. It can also be used for observation, target recognition and aiming of the vehicle's main and coaxial machine gun armament in the ground and anti-aircraft roles. It is a direct drop-in replacement for the original Russian PPN-61B periscope system and is fitted with an image intensifier tube. The latter can either be a GEN II+ or SuperGEN assembly.

Status
In production. In service with the Polish Army. Offered for export.

Specifications
Magnification: ×3.7
Field of view: day and night channels 10.5°
Resolution
  day channel: 17 arc secs
  night channel: 72 arc secs
Image intensifier tube: GEN II+ or SuperGEN
Dioptre range: –5 to +5

Manufacturer
PCO S.A, Poland.

PCO CDN-1 day-night periscope sight
0089905

PCO TPN-1P passive night sight for T-55 and T-72 tanks

Type
Gunner's sight.

Development/Description
The TPN-1P passive night sight is an aiming device for T-55 and T-72 tanks that makes possible both the aiming and firing of the main and coaxial machine gun armaments at night. It can be fitted with either GEN II+ or SuperGEN image intensifier tubes.

The tubes are equipped with an iris diaphragm that allows protective covering of the image intensifier photocathode in bright or sudden intense light situations.

A fluid based aiming line reticule in both the vertical and horizontal planes can be introduced into the assembly by the use of a special spanner.

Status
In production. In service with the Polish Army. Offered for export.

Specifications
Magnification: ×5.6
Field of view: 7°
Resolution: 60 arcsecs
Dioptre range: ±4

The PCO passive night sight for T-55 and T-72 tanks
0089904

Vertical angular range: –7 to + 18°
Power supply: 22-29 V DC
Power consumption: 50 mA

Contractor
PCO S.A, Poland.

PZO 1PN-22-M1M day/night gun-sight

Type
Gunner's sight.

Description
This PZO is designed as an upgrade for existing night vision gunner' sights in AFVs. The upgrade includes a change in the optical path and replacement of first-generation intensifiers by second-generation microchannel plate tubes.

Operational status
Available.

The PZO 1PN-22-M1M day/night gun-sight 0089894

Specifications

Field of view: day channel 14, night channel 5
Magnification: day channel ×5.7, night channel ×5.4
Resolution: day channel 11 arc secs, night channel 70 arc secs
Dioptre adjustment: –3 to +5
Range of vision of 2.3 × 2.3 target, with 50% contrast between target and background:
Illumination 50 mlx: 2,500 m
Illumination 5 mlx: 1,500 m

Contractor

Polskie Zaklady Optyczne - PZO Warszawa.

Raytheon Advanced LAV sight

Type

Gunner's sight.

Description

Based on the Delco DIM-36 periscope mirror assembly head, Raytheon's Advanced Light Armoured Vehicle (LAV) sight incorporates a Raytheon HIRE thermal sensor, a Raytheom Eye-safe Laser Integrated Telescope Equipment (ELITE) range-finder and a head mirror assembly. It accommodates a first- or second-generation thermal sensor and can be integrated into existing LAV-25 turrets and is fitted to the LAV-120 Armoured Mortar System.

HIRE 2 is a high-performance thermal imaging system with a 240 × 4 FPA for use in light armoured vehicles with biocular displays for gunner and commander and dual field of view telescope.

ELITE is a Class 1 eye-safe laser range-finder with laser components common to the Bradley ODS laser and an integrated telescope. The telescope incorporates a high-magnification sighting optic with a separate unity window which permits direct viewing at low magnification. As well as the integrated direct-view optics, it has etched reticle pattern with backlighting and range display data.

The head assembly includes integral head mirror assembly, armoured housing enclosure, ballistic cover for windows and mechanical linkage. The mid-body assembly integrates the sight components and includes a unity window.

Operational status

Production as required. In service.

Specifications

HIRE
Detector: first-generation Common Module based thermal sensor; HIRE 2 detector 240 × 4 FPA based sensor
Field of view
Wide: 12.3 × 5.1°
Narrow: 4.1 × 1.7°

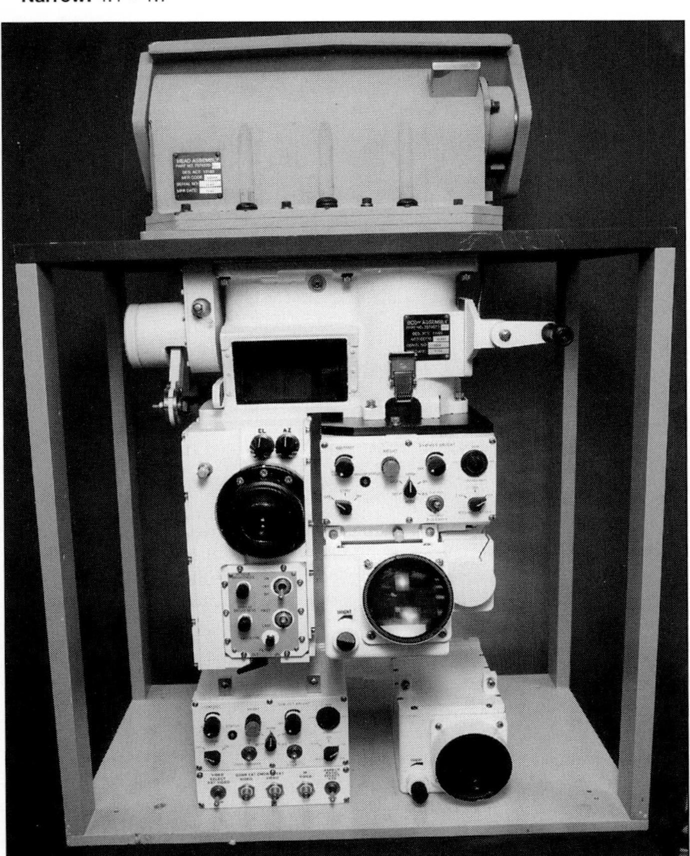

Raytheon Advanced LAV sight 0023042

ELITE
Pulse repetition rate: 1 Hz
Magnification: ×8 to ×10

Contractor

Raytheon Systems Company. ***VERIFIED***

Raytheon AN/VSG-2 Tank Thermal Sight (TTS)

Type

Gunner's sight.

Development

The AN/VSG-2 biocular thermal imaging infra-red Tank Thermal Sight (TTS) was developed as part of the US Army upgrade programme for the M60 MBT to replace the passive image intensification M35E1 and M36E1 sights from mid-1979 onwards.

Upgraded tanks fitted with the sight are known as the M60A3 TTS. In addition to the US Army's M60A3 fleet, AN/VSG-2 is also used by the Republic of China (Taiwan) on its indigenous M48 Hybrid MBT design, Saudi Arabia (M60A3 TTS) and Turkey (on its M48A5T2 upgrade MBT).

Description

The 120-element Cadmium Mercury Telluride (CMT) detector unit not only improves the night fighting capability of the vehicle but also allows the gunner and the commander to see the external scene on their own individual 40 mm image intensifier tube displays.

Incorporated into the US Army version is the AN/VVG-2 Ruby laser range-finder unit, while the Taiwanese variant uses an Nd:YAG type system.

Operational status

In service with Saudi Arabia (M60A1 to M60A3 upgrade), Taiwan (M48H MBT), Turkey (M48A5T2 MBT) and other countries.

Specifications

Weight
Head assembly: 28.86 kg
Gunner's display: 15.23 kg
Commander's display: 16.36 kg
Power converter unit: 11.14 kg
Field of view
×1 magnification: 7.8 × 15°
×8 magnification: 2.6 × 5°
Spectral band: 7.6-11.75 μm
Power supply: 18-30 V DC vehicle system

Contractor

Raytheon Systems Company. ***VERIFIED***

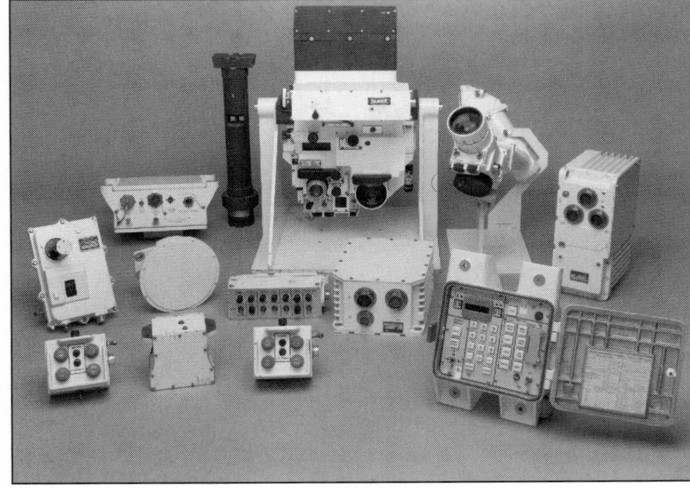

Raytheon AN/VSG-2 Tank Thermal Sight (TTS) and complete fire-control system

Raytheon Armored Gun System (AGS) gunner's primary sight

Type

Gunner's sight.

Development

Under a US$6.2 million production subcontract with M8 Armored Gun System prime contractor United Defense, LP, Raytheon Systems Company was to provide the AGS Gunner's Primary Sight (GPS). This is based on a sight developed for the

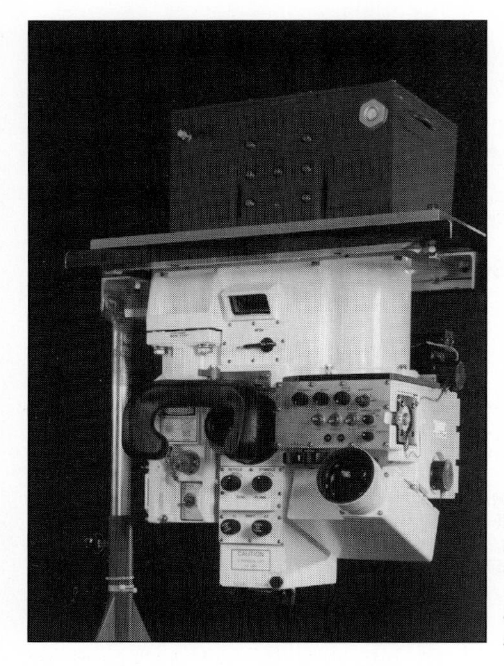

Raytheon Systems Company Gunner's Primary Sight (GPS) for the US Army Armored Gun System (AGS) armed with a 105 mm gun

US Marine Corps LAV-105. The AGS programme has been terminated by the US Army and with it the possibility of foreign military sales.

Description
The modular AGS GPS will allow a day or night shoot-on-the-move capability, using a unity window and a dual field of view ×3.3 wide field magnification and ×10 narrow field magnification hard optic TV camera day sight unit and Raytheon HIRE infra-red equipment night sight. The AGS GPS has modularity features that allow for applications to other armoured vehicles, such as the M60 MBT. For main battle tanks such as the M60 a full-solution fire-control system can also be offered.

The AGS GPS comprises the following subsystems:
(a) Raytheon HIRE – with a 240-line first-generation common module based thermal imager, programmable reticles and symbology and dual field of view capability
(b) Line Of Sight stabilisation platform – based on the system as used in the South Korean K-1 MBT and fitted with an armour housing for ballistic protection and two-axis gyrostabilised head mirror
(c) Lower Sight Assembly (LSA) – with the TV camera day sight and unitary window
(d) Laser range-finder – the same unit as produced for the M1 Abrams MBT
(e) Commander's Remote Display (CRD) – which remotely displays the day and night sight images and can be used to override all the gunner's controls
(f) Gunner's Sight Electronics Unit (GSEU).

The sight is compatible with the Raytheon M1-based eye-safe laser range-finder.

Operational status
The AGS programme has been cancelled but the system is still being marketed for export.

Specifications
HIRE
Fields of view
Narrow: 4.1 × 1.7°
Wide: 12.3 × 5.1°
Laser range-finder
Wavelength: 1.064 μm
Type: Nd:YAG

Contractor
Raytheon Systems Company.

VERIFIED

Raytheon Combat Vehicle Thermal Targeting System (CVTTS)

Type
Gunner's sight.

Development
The CVTTS is a family of systems designed to be adaptable to a wide variety of light armoured vehicles, MBTs and gun systems. It is M36 sight mount compatible.

Three sights, configured as remote surveillance sensors, have been delivered to CECOM to facilitate development of advanced SCOUT products and techniques. The sight is in production for multiple customers (US and International), in infantry fighting vehicle, main battle tank and air defence applications. A commander's panoramic sight is under development.

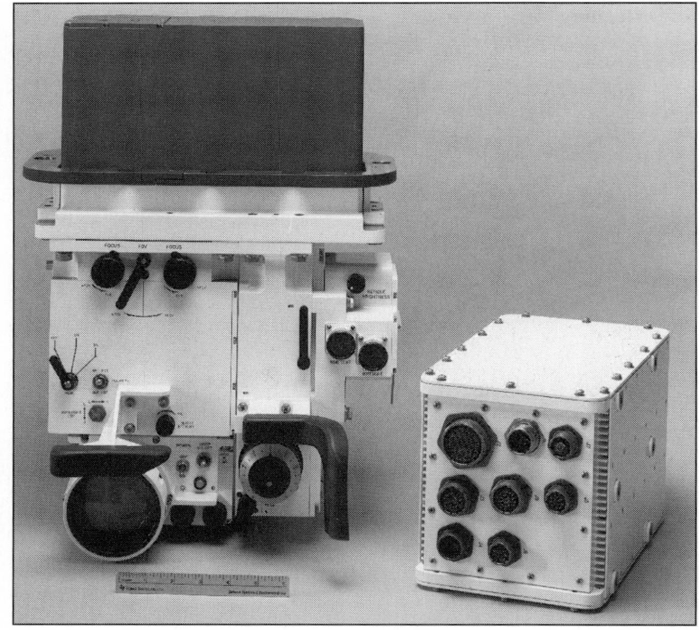

Raytheon Combat Vehicle Thermal Targeting System (CVTTS)

Description
The CVTTS is a modular system comprising the following subassemblies:
(a) Thermal sensor unit incorporating second-generation focal plane array. The dual field of view thermal imager uses standard RS-170 video format with reimaging afocal optics, servo-controlled scan and thermal references. The system is DC restored with automatic gain/level and has linear drive cryogenic cooling
(b) Visible sensor unit with dual field of view telescope with projected reticle and optical integrated TV camera. Gunner and commander can display thermal or visible scenes independently
(c) Line of sight director which can be mechanically or electronically linked, following the movement of the weapon. the director can have an optional dual-axis stabilised head-mirror

Range information can be provided by a simple stadiametric range-finder in the reticle or by an optional eye-safe laser range-finder integrated with the visible sight unit. Other options include a digital fire-control computer, control panel and sensors.

Operational status
Production. The Focal Plane Array (FPA) CVTTS is in production for multiple customers (US and international).

Specifications
Fields of view
Narrow: 4 × 1.77°
Wide: 12 × 5.3°
Magnification: ×9.3 (NFOV), ×3.1 (WFOV)

Contractor
Raytheon Systems Company.

VERIFIED

Raytheon Gunner's Primary Tank Thermal Sight (GPTTS)

Type
Gunner's sight.

Development
The Gunner's Primary Tank Thermal Sight (GPTTS) was developed to provide the tank gunner with a fully integrated visible and thermal imaging target aiming gyrostabilised sight. Direct vehicle fit applications include the K-1 MBT and the M1 Abrams. With modification it can also be fitted to the Leopard 1 and AMX-30 MBTs.

Description
The GPTTS comprises the following subsystems:
(a) optional ×8 magnification, 6° field of view and unity magnification, 17° field of view visible channel assembly
(b) dual field of view (2.58 × 5° narrow and 7.7 × 15° wide) biocular eyepiece thermal imaging channel assembly
(c) two-axis line of sight stabilised head assembly
(d) Raytheon carbon dioxide (CO_2) laser range-finder module with first/last pulse logic and a multiple target return indicator
(e) optional Charge-Coupled Device (CCD) standard TV format camera with through-the-sight video recording capability and remote viewing devices.

Raytheon Gunner's Primary Tank Thermal Sight (GPTTS)

Operational status
Production. In service with South Korea. Production blocks 2 and 3 are complete for 588 K-1 MBTs. ILS and spares production ongoing.

Contractor
Raytheon Systems Company.

VERIFIED

Raytheon gyrostabilised Gunner's Primary Sight Subsystem (GPSS)

Type
Gunner's sight.

Development
The Raytheon two-axis stabilised Gunner's Primary Sight Subsystem (GPSS) was developed for the South Korean Main Battle Tank (Type 88, K-1) in order to provide it with a shoot-on-the-move capability in both day and night engagement situations. Raytheon M1 production hardware is used as the Thermal Imaging System (TIS) and for the Laser Range-Finder (LRF).

Description
The initial 210 production systems incorporated the Raytheon-designed Leopard 2 MBT, two-axis head mirror and a vertical sensor used to furnish roll and pitch dynamics to the computer subsystem. Subsequent orders (for 276 systems) incorporated a fourth-generation improved two- axis head mirror and updated electronics with the vertical sensor.

This system is compatible with the Raytheon M1-based eye-safe laser range-finder.

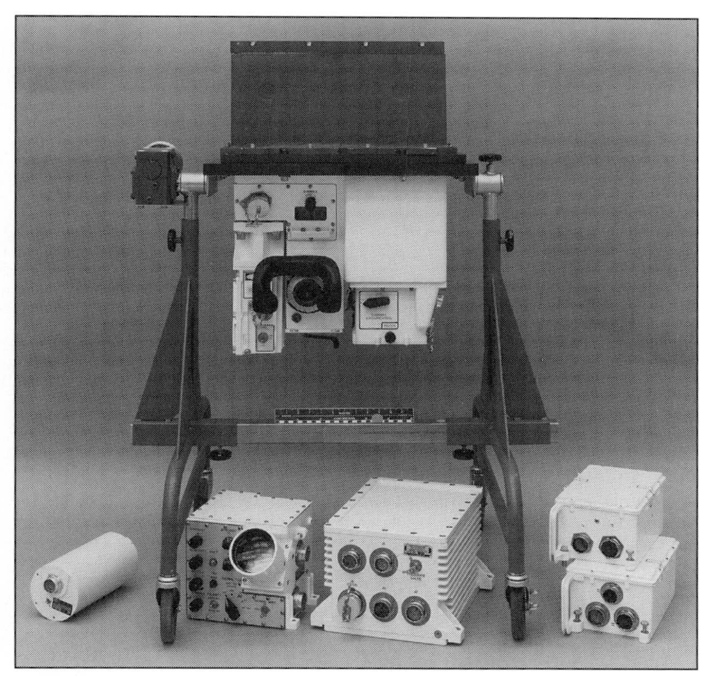
The Raytheon two-axis stabilised Gunner's Primary Sight Subsystem for the South Korean K-1 MBT

Operational status
In service with the South Korean Army (Type 88 MBT). In total, 500+ tanks are in service with this system; spares are also being produced.

Contractor
Raytheon Systems Company.

VERIFIED

Raytheon Korean Advanced Gunner's Sight (KAGS)

Type
Gunner's sight.

Description
The KAGS has been developed by Raytheon Systems Company to update the existing GPTTS (Gunner's Primary Tank Thermal Sight) for the South Korean K-1 MBT.

KAGS is equipped with second-generation FLIR technology in a 240 × 4 element focal plane array. This detector is stated to increase the detection and recognition ranges by more than 25 per cent over common module-based imagers. The thermal imager has both automatic and manual mode gain and level controls and automatic built-in test detection and correction for 'dead and noisy' detectors. The system also includes an eye-safe CO_2 laser range-finder and dual field of view day TV. Scene stabilisation is provided by the two-axis stabilised head-mirror used in the GPTTS.

Both thermal imager and TV provide output in standard RS-170 format, with display on the monitor.

Operational status
Completed development.

Specifications
Second-generation FLIR
Detector: 240 × 4 elements
Fields of view
Wide: 5.73 × 13°
Narrow: 1.85 × 4.20°
Day TV
Fields of view
Wide: 5.1 × 6.6°
Narrow: 1.7 × 2.2°
Laser range-finder
Type: eye-safe CO_2
Accuracy: ±10 m
Range: 3,000 m (in 90°F/80% rh conditions)
Two-axis stabilised head-mirror
Field of regard
Elevation: +20 to −10°
Azimuth: ±3°
Azimuth coverage for muzzle reference sensor alignment: −9°
Position accuracy: <0.25 mil
Stabilisation accuracy: <0.08 mil elevation and <0.02 mil (RMS) azimuth axes (with M1A1 or K-1 specification base-motion disturbance inputs)

Contractor
Raytheon Systems Company.

VERIFIED

Raytheon M1A2 Gunner's Primary Sight Line Of Sight (GPS-LOS) subsystem

Type
Gunner's sight.

Development
Under production subcontracts with General Dynamics, Raytheon will deliver 1,250 GPS-LOS subsystems for the US Army M1A2 Abrams and foreign applications. Deliveries began in late 1993. Under previous contracts Raytheon delivered 24 prototype and 17 pilot production systems for the development programme.

Description
The GPS-LOS subsystem is a two-axis stabilised head mirror that significantly improves the first-round hit probability. M1 and M1A1 versions of the Abrams tank are equipped with a single-axis stabilised head mirror. The two-axis system (Dual-Axis Head Assembly – DAHA) on the M1A2 enables faster target acquisition, improved gun pointing and, because the azimuth axis is inertially stabilised, the gunner can detect, recognise and acquire additional targets at greater ranges. A built-in test facility is controlled by microprocessor to provide an intelligent test and isolation capability.

Operational status
Production. In service with the US Army, Kuwait and Saudi Arabia. More than 1,400 DAHA systems have been delivered.

Raytheon M1A2 Gunner's Primary Sight Line Of Sight (GPS-LOS) subsystem

Specifications
LOS excursion range
Elevation: –16 to +22°
Azimuth: ±5°
Stabilisation accuracy: ≤100 µrad
Plumb and synchronisation: ≤70 µrad
Boresight retention: ≤100 µrad

Contractor
Raytheon Systems Company.

VERIFIED

ROMZ T01-K01 night aiming and surveillance system

Type
Gunner's sight.

Description
T01-K01 is a tank night aiming and surveillance system with passive and active modes. It consists of the TPN4 sight, a communication unit (BK), an interface unit (UWP) for inputting windage and elevation parameters and parallelogram and a projector (L4A). It can be used for both the main armament and the machine gun if fitted. There are heating elements in the eyepiece and protective glass in the head lens.

Specifications
Dimensions
Height: 614 mm
Width: 226 mm
Length: 312 mm
Weight of night sight TPN4: 32 kg
Weight of the system: 76 kg

Optical magnification: ×6.8
Field of view: 5.4°
Dioptre adjustment: ± 4
Resolution: 42 secs of arc
Line of sight limits: –70 to +20°
Effective identification range (side projection of tank) at an atmospheric transparency of 0.85
Active: 1.5 km at an illumination < 0.003 lx
Passive: 1.2 km illumination of > 0.003 lx
Operating voltage: 27 V DC
Power consumption (sight): 75 W
Power consumption (illuminator): 460 W

Contractor
Rostov Optical-Mechanical Plant Joint Stock Company (ROMZ).

UPDATED

Saab Bofors Dynamics Type FV gunner's day/night sight

Type
Gunner's sight.

Description
The Type FV gunner's day/night sight is a component of the Saab Bofors fire-control system for armoured fighting vehicles. It is fitted with day and low-light level night vision channels and has a miniature laser range-finder integrated into the system.

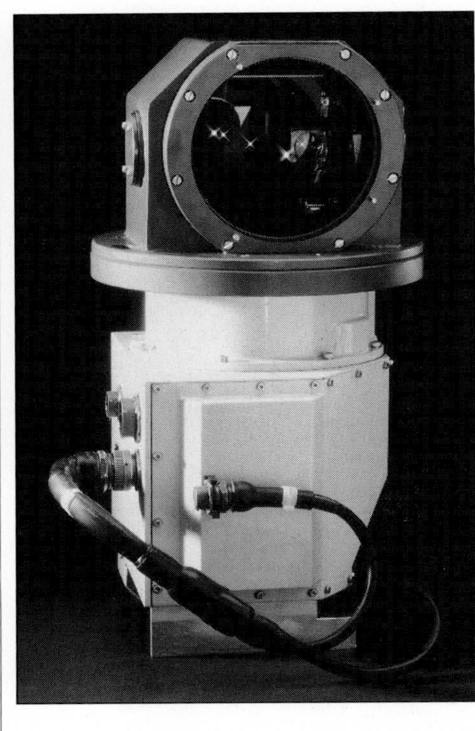

Saab Bofors Type FV day/night T-series gunner's sight

All the optical channels are aligned with their directions in azimuth and elevation controlled by a servo-driven gyrostabilised top mirror to ensure a stabilised line of sight.

Tracking is performed independently of the movements of the turret and gun, so that all the gunner has to do is keep the line of sight on the target. He can then measure the range by use of the laser, even when the gun has already been realigned, by the fire-control system on to the target's predicted future position as the sight's line of sight contrarotates with the lead angle and superelevation.

Operational status
Production as required. In service with several undisclosed countries (on T-series tanks).

Specifications
Field of view
×7 magnification day channel: 9°
×8.5 magnification night channel: 5.3°
Dioptre range: –5 to +5
Laser range-finder
Wavelength: 1.064 µm
Operating range: Typically 6 km

Contractor
Saab Bofors Dynamics.

UPDATED

SAGEM Stabilised Aiming, Vertical sensing And Navigation (SAVAN) gunner's multichannel stabilised sights

Type
Gunner's sight.

Development
The SAVAN (Stabilised Aiming, Vertical sensing And Navigation) family of gunner's primary sights has been developed to equip armoured vehicles, providing high-accuracy fire control in all combat conditions (day and night, all weather operations, vehicle stationary or on the move).

The SAVAN sights are compact and modular, which makes them suitable for all fire-control applications: MBT's or infantry combat vehicles, vehicles in development or in upgrade phase, trunnion or roof-mounted applications.

The SAVAN sights are in service on the Leclerc MBT for the French army, the United Arab Emirates (UAE) Armed Forces and on the Challenger 2 MBT for the British Army and Sultan of Oman's Army. They have also been selected for use on the new TML 105 'fire on the move' turret and by several MBT designers for use on T-72 or equivalent MBTs.

Description
The current family of SAGEM SAVAN sights comprises three main members:
- SAVAN 20, a day/night Gunner Primary Sight with a very large stabilised mirror
- SAVAN 15, a day/night Gunner Primary Sight with flat stabilisation mirror assembly particularly suitable for roof-mounted applications

SAVAN 15 day/night gunner primary sight

- SAVAN 10, a day Gunner Primary Sight with flat stabilisation mirror assembly

All systems can be provided with:

- Customer specified electro-optic sensors
- one or two magnification direct day channel
- ×2 or ×3 magnification, 3 to 5 μm or 8 to 12 μm latest-generation thermal imager
- one or two magnification high resolution TV day channel
- eye-safe or Nd:YAG laser range-finder
- TV monitor
- Symbol generator
- automatic muzzle reference system
- Customer-specified databus interfaces (for example 1553, Digibus, RS-422)
- Customer-specified inertial functions
- two-axis stabilisation of the line of sight
- vertical reference stationary or on the move
- gun and turret stabilisation reference
- fully inertial or GPS assisted north seeking and land navigation
- A fire-control computer which automatically takes into account all the necessary parameters and makes use of a second-generation thermal imager
- range and speed of target
- ammunition characteristics
- meteorological data
- motion, tilt and roll of carrier
- Automatic tracking aid optimised for ground and air targets
- A full set of equipment required in modern fire-control systems (man/machine interface, meteorological sensor, gun elevation sensor and turret azimuth sensor)

Several options are available: technical and tactical training systems; commander day-only VIGY 15 or day/night stabilised panoramic sight - MVS 580; battle management system.

Operational status
Production. In service with France, Oman, UAE and UK.

Contractor
SAGEM SA, Aerospace and Defence Division.

UPDATED

SAGEM TJN2-71 day/night optical sight

Type
Gunner's sight/commander's sight.

Description
The TJN2-71 periscopic sight is intended to be fitted both to cupolas armed with small calibre weapons and to larger scale turret commander stations. A number of versions are available with different installation heights and elevation interfaces.

The TJN2-71 sight is also selected to fit the gunner station of the SOPTAC 39 FCS. The TJN2-71 has the same interface as TJ-71, and comprises an integrated prism head with three paths, a ×1 magnification day periscope, a ×6.8 magnification day sight and a ×4 magnification night sight.

The SAGEM TJN2-71 day/night periscopic sight for turrets and cupolas

The night sight is fitted with a second-generation, 25 mm image intensifier tube with a range of 500 to 1,000 m according to atmospheric conditions. A third-generation image intensifier is available as an option. The day and night sight paths have a common eyepiece. The sight can be used for day observation with a wide field of view and there is the possible application for defensive anti-aircraft fire, target designation and day and night firing.

Operational status
Production as required. In service on the Giat DRAGAR turret.

Specifications
Elevation range: −15 to +55°
Alignment range:
Elevation: ±10 mils
Azimuth: ±10 mils
Periscopic day sight
Magnification: ×1
Field of view
Horizontal: 43°
Vertical: 20°
Daylight sight channel
Magnification: ×6.8
Field of view: 7°
Night sight channel
Magnification: ×4
Field of view: 10°
Range: 500 to 1,000 m (according to atmospheric conditions)

Contractor
SAGEM SA, Aerospace and Defence Division.

UPDATED

SAGEM, SKBM, Peleng Namut thermal sight

Type
Gunner's sight.

Development
The Namut thermal sight is a joint development between SAGEM of France (for the sight's thermal imager), SKBM-Kurgan of the Russian Federation and Associated States (for the BMP-3 IFV), Peleng-Belemo of Belarus (for the Obzor stabilised sight section) and the UAE, which is the customer for a number of BMP-3 vehicles to which the sight is being fitted.

BMP-3 of the United Arab Emirates with the Namut thermal sight on turret rear

Description

The Namut sight has the following subsystems:

- Sight section – this is a two-axis stabilised assembly with a mirror angular displacement fully compatible with the 100 mm/30 mm and 100 mm ATGW armament requirements in both elevation and azimuth
- Thermal imager – based on the SAGEM Athos 8 to 12 μm thermal imager or 3 to 5 μm MATIS.

Operational status

Production. In service with the UAE (on the BMP-3).

Specifications

Athos thermal imager
Spectral band: 8-12 μm
Magnification: ×3, ×10
Fields of view
Wide: 9 × 6°
Middle: 3 × 2°
Narrow: 1.5 × 1°

MATIS thermal imager
Spectral band: 3-5 μm
Magnification: ×3, ×10
Fields of view
Wide: 12 × 8°
Narrow: 2.5 × 1.7°
Electronic zoom: 1.25 × 0.85°

Contractor

SAGEM SA, Aerospace and Defence Division.

UPDATED

SAGEM, SKBM, Peleng Sanoet-1 infra-red sight

Type

Gunner's sight.

Description

The SAGEM group has developed the Sanoet-1 sight in collaboration with the infantry fighting vehicle design company SKBM-Kurgan of the Russian Federation and Associated States (RFAS), and stabilised sight design company Peleng-Belemo of Belarus. Sanoet-1 is a day/night thermal gunner's sight equipped with a SAGEM MATIS thermal imager. There is an optional TV output for the commander. The Sanoet-1 can replace the existing gunner's sight of a BMP-2 in less than 2 hours without modification to the turret.

Sanoet-1 thermal sight for the retrofit of BMP-2 infantry fighting vehicle

Operational status

Operational on RFAS BMP-2 tanks.

Specifications

Sight
Mirror angular displacement: +30 to –7.5°
Micromonitor: standard CCIR 625 lines
Optical day channel
Magnification: ×6
Field of view: 10°
Thermal imager: SAGEM MATIS
Spectral band: 3-5 μm
Fields of view: 12 × 8°, 2.5 × 1.7° and electronic zoom 1.25 × 0.85°

Contractors

Kurganmashzavod JSC.
Peleng-Belemo.
SAGEM SA, Aerospace and Defence Division.

UPDATED

SAGEM SAVAN 7 compact day and night gunnery sight for fighting vehicles

Description

SAVAN 7 has been designed by SAGEM for applications such as IFVs, LAVs or APCs, either as original equipment or for mid-life upgrade.

The system is an integrated gunnery sight which gives the user the ability to perform general surveillance and target recognition, identification and engagement at extended ranges by day and night. In its basic version, aiming of the line of sight is realised by direct coupling of the head mirror to the gun through a mechanical linkage.

In addition to a dual field of view day optical channel and a latest generation thermal imager, SAVAN 7 incorporates a two magnification CCD TV channel and an eye-safe laser range-finder. A separate monitor enables display of the TV or TI image for the gunner.

SAVAN 7 provides a full fire-control capacity. Accurate firing and fast reaction time is achieved trough automatic management of the aiming marks.

Remote control of the sight can be achieved by a second operator (commander's over ride mode) through an additional TV and TI control and display unit.

The following options are possible:

- Choice of several types of thermal imager (cooled 8 to 12 μm and 3 to 5 μm TI or uncooled TI)
- Super elevation kit for automatic Line of Sight deviation of the computed ballistic offset
- Stabilisation of the video channels (TV and TI).

Status

In production. Ordered for export upgrade programme.

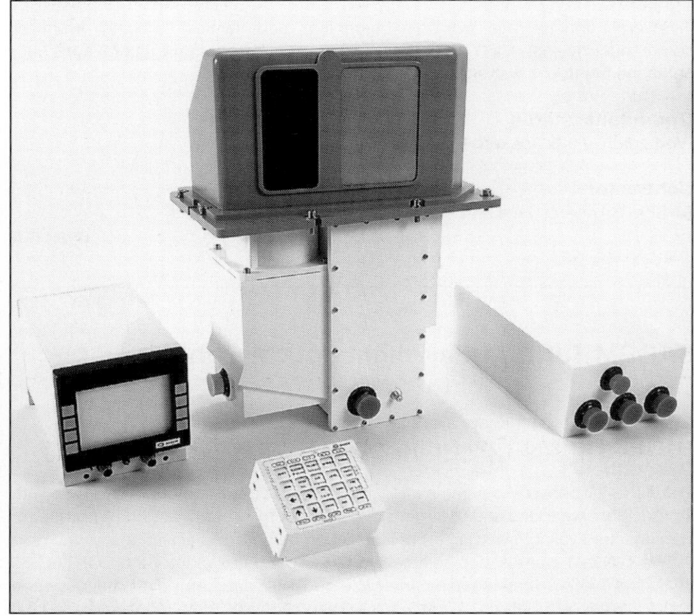

Sight with electronics unit and man machine interface 0129311

SAVAN 7 rear view of sight 0129312

Specifications

Optical channels
Day Channel
Direct day optics with two magnifications:
Unity channel (X1)
High magnification channel (X6) with monocular eyepiece (FOV > 7°)

Night Channel (basic version)
Third-generation thermal imager:
Spectral band: 3 to 5 μm
Fields of view: (12 × 8° and 2.5 × 1.6°)
Recognition range : 3,500 m

CCD high-resolution TV channel
Automatic contrast adjustment
Fields of view: 9 × 6.75° and 3° × 2.25°

Laser range-finder
Wavelength: Eye-safe, 1,54 μm
Range: 80 m to 10 km
Range accuracy: ±5 m
Repetition rate:12 pulse per minute (1 Hz burst)

LOS aiming
Elevation/Depression range: –10 to +50°
Graticule:
Engraved graticule (stadia and ballistic patterns) on high magnification day channel.
Electronically inlaid graticule on TV/TI channels with automatic deviation according to target range and ammunition.

Control and display
Gunner's and Commander's control box
Display by standard or relaxed view TV monitor for both operators
Interface: Interface available to turret weapon system and battlefield management system

Manufacturer

SAGEM SA, Aerospace and Defence Division.

UPDATED

Thales Optics Raven combined day/night sight

Type

Gunner's sight.

Development

The Raven combined day/night sight was developed by Thales Optics to provide commanders and gunners of MBTs and light armoured vehicles with a 24-hour day/night capability. It can be used for surveillance, target acquisition, firing main armament and observation of fall of shot. Raven was originally designed to fit the British Army's Warrior vehicle but can be fitted to most armoured fighting vehicles.

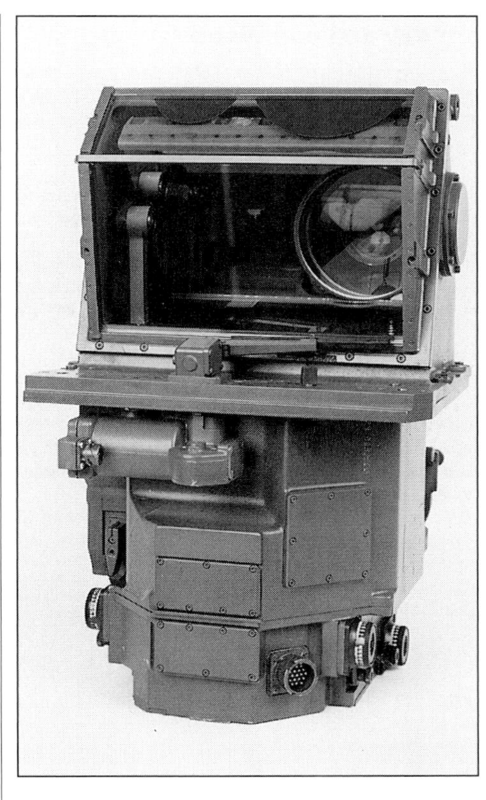

Raven combined day/night vehicle sight
0004787

Description

The Raven day/night sight can be installed from under or above armour and interface directly with the gun via a linkage on the right side of the sight.

The sight has graticule adjusters that enable the sighting system to be boresighted to the main armament and separate adjustments are provided for boresighting the ×1 and magnified day and night channels.

The Raven sight comprises three viewing channels, a fixed focus magnified day channel, a dual magnification second- or third-generation image intensified night channel and a unity power day periscope.

The magnified day and image intensified night channels are viewed through common eyepieces with selection of day or night surveillance being effected by a single changeover lever. A focus control and light control iris are also provided in the night channel. The unity power periscope is viewed through a window above the eyepieces.

In order to provide the facility for main armament engagement the sight embodies three separate graticules, one in each channel. All are separately adjustable in elevation and azimuth to achieve accurate boresighting. A variant with a CRT Fire-Control Interface is available.

Operational status

In production. In service with the British Army (on Warrior MICV – two sights, one for the gunner and one for the commander (latter with additional traverse)).

Specifications

Dimensions (h × w × d): 560 × 350 × 450 mm
Weight: 50 kg
Max elevation of line of sight: +47°
Max depression of line of sight: –12°

Day sight
Magnification: ×1 and ×8
Resolution: 25 cycles/mil (full daylight to twilight)
Recognition range: 3 km
Detection range: 5 km
Focus range: fixed at infinity

Night sight
Magnification: ×2 and ×6
Resolution: 2.7 cycles/mil (starlight)
Recognition range: 1 km
Detection range: 2.5 km
Focus range: 20 m to infinity
Power supply: 28 V DC (vehicle supply)

Contractor

Thales Optics Ltd.

Thales Optics Sabre day and combined day/night sights

Type
AFV gunner's sight.

Description
The Sabre sighting system has been designed and developed by Thales Optics to fulfil the market requirement for an inexpensive, easy to operate, highly compact vehicle sight able to meet exacting performance requirements.

Sabre is suited for use by the commander or gunner of an AFV for surveillance or target acquisition. The sight can be easily retrofitted to existing vehicles from above or from below armour, and represents a significant upgrade in overall vehicle capability.

Sabre is available in two separate build options:

(1) A periscope day sight with unity magnified (×1) day channel and magnified (×6) day channel

(2) A periscope day/night sight with switchable ×1.5 and ×6 magnified day channel and ×4 magnified night channel. This sight can be fitted either with an image intensifier (II) tube (second or third generation) or with a thermal imager (uncooled 3-5 μm, cooled 8-12 μm).

Both sight options view the scene via an elevating mirror, the line-of-sight of which is harmonised to the armament boresight by a 2:1 gearing mechanism.

Operational status
Production. In service with unspecified countries.

Specifications
Dimensions:
Below armour (h × w × d): 192 × 241 × 145 mm (high profile); 136 × 241 × 145 mm (low profile)
Above armour: (h × w × d): 258 × 211 × 84 mm
Fields of view/magnification: ×6, 6°; ×1, 30° H × 15° V
Day/night option FoVs: ×4, 8.44°; ×1.5, 24°
Day recognition range (×6): 4,000 m
Night recognition range (×4): 600 m (GEN II), 900 m (GEN III)
Elevation range: –15 to +50°
Boresight adjustment range: ± 15 mils
Power supply: +28 V DC vehicle supply

Contractor
Thales Optics Ltd.

UPDATED

Sabre day/night sighting system
0004786

Thales Optronics Battle Group Thermal Imaging (BGTI) system

Type
Land fire-control system.

Description
The Thales Optronics Battle Group Thermal Imaging (BGTI) system was selected by the UK MoD to meet the requirement to upgrade the Warrior and Scimitar CVR (T) vehicles in August 2001. The BGTI Group 1 system incorporates a STAG FC Gunners Sight, a Commanders Crew Station, an integrated Navigation/Far Target Location subsystem and a Drivers waypoint display. For the Warrior Repair and Recovery vehicles, the BGTI Group 2 system utilises low-cost uncooled TI and a common BGTI CRT display. The BGTI system offers the user a significant enhancement to conventional Image Intensified-based sights and improves the situational awareness of the crew. It allows vehicles to detect, recognise and engage targets at longer ranges and under almost all weather conditions than the current II systems.

In addition, BGTI will enable the vehicle crew to rapidly establish its own position and the co-ordinates of other targets on the battlefield as an inertial navigation system is fitted as standard, with the driving route map controlled by the commander and displayed to the driver.

The BGTI system comprises the following subsystems:

- Surveillance Targeting Acquisition and Gunnery – Fire Control (STAG FC) gunners sight with day channels, eye-safe LRF, integrated day camera, Gen II Catherine FC TI and stabilised sight head.
- Commanders Crew Station with 10.4 in (26.6 cm) SVGA Flat Panel Display (allowing access to the Bowman Communications and Battle Management System), magnified direct vision optics, high-performance CCD colour camera and CRT display.
- Gyro-based Navigation subsystem incorporating a navigation processor and turret sensors with navigation and target data displayed on the Commanders FPD.
- Drivers 6 in display showing waypoint data.

Applications:

- Suitable for all vehicles up to 40 mm weapon calibre.
- Upgrade of all crew stations.
- STAG FC Gunners Sight can installed in most LAVs.
- Weapon aiming using ballistic fire control.
- Long-range surveillance and target location.

Operational status
In production for British Army. Trials of first production systems took place in Summer 2003. The BGTI Group 1 element covers about 350 Warrior AFVs (infantry combat and observation post variants) and 150 Scimitar vehicles with Group 2 covering approximately 100 Warrior repair and recovery variants. Installation is due to commence early in 2004, with kits being supplied by Thales Optronics for vehicle conversions in UK, Canada and Germany. The system will be installed at the same time as the Bowman communications system, with the object of converting one brigade at a time.

Specifications
Gunner's station
Thermal imager
Detector: 2nd generation 288 × 4 array
Field of view: dual FoV (9 × 6° and 3 × 2)

Day channels
Fields of view: ×8 Magnification, 7.5 × 5.5°; ×1, 30 × 25°
Laser range-finder
Type: erbium glass, 1.54 μm
Maximum range: > 8,000 m
Minimum range: 300 m
Range accuracy: ± 5 m
Stabilised sight head
Stabilisation: Single axis stabilisation in elevation allowing stable viewing on the move

Commander's station
Day channels
Fields of view: ×8 magnification, 6°; ×1, 30 × 12°
Colour camera: 3 chip, colour camera (4 > 34°)
Flat panel display: 26.6 cm (10.4 in), SVGA (800 × 600)
Interfaces: CCIR I PAL video, RS-232/RS-422
Gunner's/commander's CRT display: CCIR 625 line 50 Hz interlaced, biocular viewing
Navigation: Far target location accuracy < 300 m at target range of 6 km
Power: Total system power 132 W

Contractor
Thales Optronics Ltd.

UPDATED

Thales Optronics Challenger 2 gunner's primary sight

Type
Gunner's sight.

Description
The gunner's primary sight is a modular system comprising the sight body containing the visual sighting channel, the head unit with a two-axis stabilised aiming mirror, a 4 Hz Nd:YAG laser range-finder and a display monitor used to view the thermal image through the monocular eyepiece. The GPS is accompanied by a

Gunner's primary sight from rear with the stabilised head electronics unit

Line-Replaceable Unit (LRU) containing the electronics for the stabilised head. The sight head has a low profile with a small front window area. The modular design of the sight allows replacement of the laser range-finder and monitor and sight electronics in the field. The GPS and thermal imaging systems are interfaced to the fire- control computer via a MIL-STD-1553B databus.

Operational status
In service on the Challenger 2 MBT for the British Army (386) and the Sultan of Oman's Army (38).

Specifications
Visual magnification: ×3° wide, ×10° narrow
Day scene attenuation
Weight: 51 kg
Dimensions: 566 × 381 × 394 mm
Stabilisation: 2-axis
Elevation: –25 to +45°
Azimuth: ±7°
Stability: 100 μrad (peak)
Laser range-finder
Type: Nd:YAG
Wavelength: 1.064 μm
Operating range: 200 - 10,000 m
Accuracy: ±5 m
Discrimination: 30 m
Gunner's display monitor
Display: 2 in CRT injected through visual optics
Stabilised head electronics unit
Interface: MIL-STD-1553B
Dimensions: 560 × 145 × 230 mm
Weight: 20 kg

Contractor
Thales Optronics Ltd.

UPDATED

Thales Optronics LRS 7 fire-control system

Type
Day/night periscopic sight.

Description
The Thales (formerly OIP Sensor Systems) LRS 7 is a monobloc day/night periscopic gunner's sight for use on smaller AFV turrets. The system is capable of day and night operation detecting, recognising, identifying and aiming at targets as well as having an anti-aircraft capability.

The main functions incorporated into the system are a ×7 magnification day sight, a ×7 magnification night sight with second-generation image intensifier, an output mirror mechanically slaved to the main weapon and an episcope. The graticule configuration used is the same for both day and night operations.

Optional equipment includes the fitting of a fully integrated ballistic computer and/or a 1.064 μm wavelength Nd:YAG laser range-finder.

Operational status
In production. In service with unspecified countries. This system was developed specifically for the export market.

Specifications
Optical

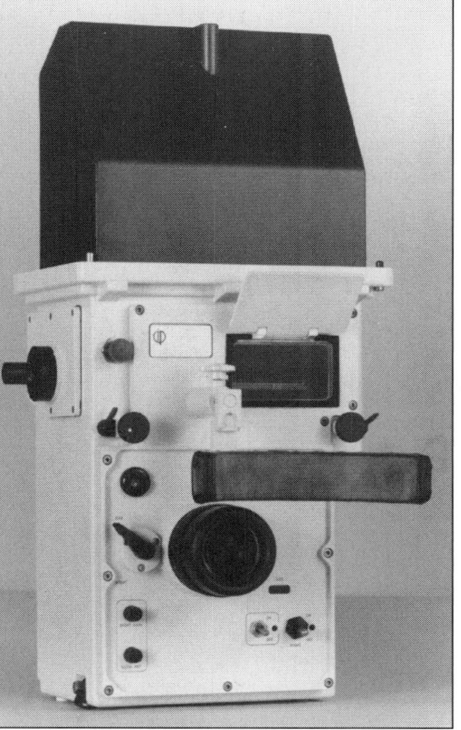

OIP Sensor Systems LRS 7 fire-control system

Periscope	**Day**	**Night**
Magnification	×7	×7
Field of view	8°	8°
Entrance pupil	46 mm (approx)	90 mm
Dioptre adjustment	–5 to +5	–5 to +5
Resolution	0.04 mils	0.3 mils at 1 mlux 0.6 mils at 0.1 mlux
Exit pupil	6 mm	6 mm
Eye relief	35 mm	35 mm
Laser safety at 1064 μm	>50 dB	Protected by tube
Episcope field of view	**Instantaneous**	**Total**
Azimuth	>16°	>40°
Elevation	>8°	>32°

Total System:
Weight: 20 kg maximum
Alignment day/night channels: = 0.1 mils
Boresight range: ± 5 mils (total 10 mils)
Line of sight angle range output mirror linkage: –10 to +60°
Gun angle display accuracy: ≤2°
Elevation angle accuracy: ≤ 1.0 mils
Elevation angle plumb travel: = 1.2 mils
Power supply: nominal 24 V DC at 0.5 A

Contractor
Thales Optronics.

UPDATED

Thales Optronics M-DNGS day night gunnery sight

Type
Gunner's or commander's sight.

Description
M-DNGS is a family of gunner's/commander's sights designed to fit a wide range of turrets and vehicles where space is at a premium including Scorpion 90, BMP series, MRAV/GTK, AML60/90, M113-150 series, CV90 series, VAB, BTR.

Available in a number of options, each offering increasing levels of performance and functionality at a modest extra cost. Head mirror drive from four different points means that DNGS is adaptable for most vehicles, and suitable for installation in both gunner's and commander's stations. Options include: wash/wipe facility, eye-safe laser range-finder, fire-control system, tactical navigation and GPS interface.

Operational status
In production.

Specifications
DNGS image intensified
Intensifier type: Gen 2, 2+ or 3
Night channel (second-generation)
Magnification: ×4
Field of view: 8.4°
TDNGS (8-12 μm)

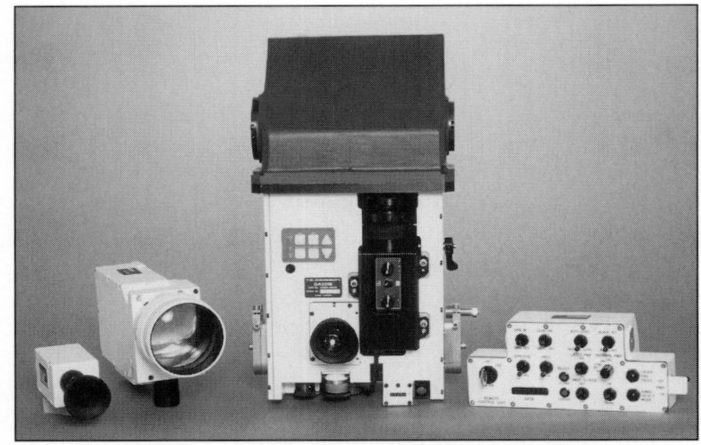

The DNGS thermal day/night gunners sight with integrated eye-safe laser range-finder and control unit 0053713

Detector type: thermal, uncooled microbolometer
Night channel
Field of view (h × v):10.75 × 8.1°

Fire-control system
Ballistic, meteorological and range/tracking rate correction for up to 12 ammunition types against static or moving targets
GPS and TACNAV compatible
TDNGS (3-5 μm)
Detector type: thermal cooled - PtSi, InSb or QWIP

Night channel
Field of view (h × v): 8.6 × 8.6° (Pt Si)
11 × 8.25° or 3.67 × 2.65° (InSb)
11 × 8.25° or 3.67 × 2.65° QWIP)

Day channel all variants
Fields of view: 2 switchable 28 × 7°
Magnification: ×1.5 and ×6
Laser protection: standard against Nd:YAG. Other wavelengths optional

Common to day and night channels
Graticule: stadia and ballistic patterns specific to armament. Variable brightness
Elevation/depression range: +50 to –10° (–15° optional)
Environmental: meets Def Stan 07-55
Electrical supply: 28 V DC Def Stan 61-5 Part 6
Physical dimensions
Below armour W × H × D: 211 × 261 × 84 mm
Above armour W × H × D: 240 × 190 × 140 mm

Contractor
Thales Optronics Ltd (Taunton).

UPDATED

Thales Optronics Surveillance Targeting Acquisition and Gunnery (STAG) sights

Type
Gunner's sight.

Description
The Surveillance Targeting Acquisition and Gunnery (STAG) family of high-performance sights is based on second-generation FPA thermal imaging and uncooled detector technology. The modular design allows the easy interchange of the key modules such as the thermal imager, sight head and laser elbow. Integrated fire control, navigation and battlefield management system interfaces are available in order to provide a total integrated turret system solution. An integrated eye-safe laser range-finder is a standard fit on the High-Performance STAG FC (Fire Control) and is an option on the other STAG variants.

STAG FC
This new sight incorporates second-generation (288 × 4) thermal imaging and provides a day/night capability including an integrated eye-safe laser range-finder and a low-cost stabilisation option. This sight provides the high performance expected from a second-generation thermal imager within a compact envelope. STAG FC is incorporated into the Battle Group Thermal Imaging (BGTI) system.

MINI STAG (U)
This variant of the Thales Optronics Sabre sight incorporates one of the highest performing uncooled detectors currently available. It provides a high-reliability, day/night capability for weapons up to 25 mm calibre. The compact sight can be fitted into the smallest of turrets and the modular design will allow future uncooled technology upgrades. This sight is also available with 3-5 μm cooled detector modules.

STAG (T)
This variant of the STAG family is suited for installation on to vehicles where turret space is at a minimum. This thermal imaging sight is in telescopic configuration and can be mounted externally on the vehicle. This sight incorporates 8-12 μm uncooled thermal imaging modules and the roof/mantlet-mounted system allows a larger aperture to optimise the range performance of the sight.

Operational status
In production for the British Army's BGTI programme for Warrior and Scimitar AFVs.

Specifications
STAG FC
Dimensions: 560 × 321 × 300 mm
Weight: 39 kg
Visual FOV: 7.5 × 5.5°
Visual magnification: ×8
Detector type: 288 × 4 FPA
Thermal imager FOV: 9 × 6.7° and 3 × 2.3°
Electronic magnification: 1.5 × 1°
Nav/BMS interface: option
Fire-control interface: option
Integrated LRF: ebium glass eye-safe LRF
Elevation/depression: +45 to –14°
Serial interface: RS-422
Power (24 V at 22°): 75 W

MINI STAG (U)
Dimensions: 450 × 241 × 145 mm
Weight: 10.5 kg
Visual FOV: 8.4° and 24 × 16°
Visual magnification: ×4 to ×1.5°
Detector type: micro-bolometer (320 × 240)
Thermal imager FOV: 16 × 12°
Nav/BMS interface: option
Fire-control interface: option
Integrated LRF: eye-safe LRF option
Elevation/depression: +40 to –15°
Serial interface: RS-422
Power (24 V at 22°C): 10 W

Contractor
Thales Optronics.

UPDATED

Thales Optronics Thermal Observation and Gunnery Sight (TOGS I and 2)

Type
Gunner's sight.

Development
The Thermal Observation and Gunnery Sight (TOGS) is a high-performance thermal fire-control system developed for the UK MoD for the Challenger 1 and Challenger 2 main battle tanks.

Description
The system uses the UK TICM II imager (Thermal Imaging Sensor Head or TISH), with a dual field of view telescope. This is mounted in a servo trunnion unit enclosed within an armoured barbette. To enable rapid cooldown, the imager is cooled by high-pressure pure air provided by the Coolant Supply Unit (CSU), which incorporates reservoir air bottles for silent watch. Computer-generated aiming

TOGS 2 system showing TISH, CSU, electronics and displays

marks are overlaid on the thermal picture together with symbology containing system status and BITE messages. Thermal imager adjustment controls are provided for both the commander and gunner together with high-resolution biocular viewers.

TOGS 2 is an improved and simplified version of TOGS which is being fitted to the new Challenger 2 main battle tank currently entering service with the British Army. The TISH is mounted on the mantlet above the 120 mm gun. The thermal image can be displayed either on monitors built into the gunner's and commander's primary sights or on relaxed view displays.

Operational status

TOGS 1 is in service with the Jordanian Army on Challenger 1, now known as the Al Hussein. TOGS 2 is in service on Challenger 2 for the British Army (386) and the Sultan of Oman's Army (38).

Specifications

Thermal Imaging Sensor Head (TISH)
Dimensions: 510 × 284 × 310 mm
Weight: 19.5 kg
Spectral band: 8-13 μm
Field of view
Wide: 16 × 10°
Narrow: 4.75 × 3°

Contractor

Thales Optronics Ltd.

UPDATED

Thales Optronics TIS-LSL thermal sight for Leopard 1

Type

Gunner's sight.

Development

The TIS-LSL (Thermal Infra-red Sight - Laser Sight Leopard) was developed as part of the modernisation programme for the Belgian Leopard 1 main battle tanks. It now includes a thermal imaging channel. The system is also being marketed as part of the integrated SABCA fire-control system.

Description

As well as the dual field of view thermal imaging sight, TIS-LSL includes a dual magnification day sight and a laser reception channel which runs through the day objective, optimising the alignment between both channels. A muzzle reference system is integrated in the system along with a CCD camera which records the visual image and commander's display.

TIS-LSL can be adapted as a stand-alone gunner's sight in which case the TEMPO laser range-finder from CILAS is added. TEMPO is available in eye-safe form if required.

Operational status

In service on 132 Belgian Army Leopard 1A5 (BE) upgraded tanks.

Specifications

Daysight
Magnification: ×7, ×14
Fields of view: 5° (×7), 3° (×14)
Focus: at 1,000 m
Thermal channel
Magnification: ×12 (NFOV), ×5 (WFOV)

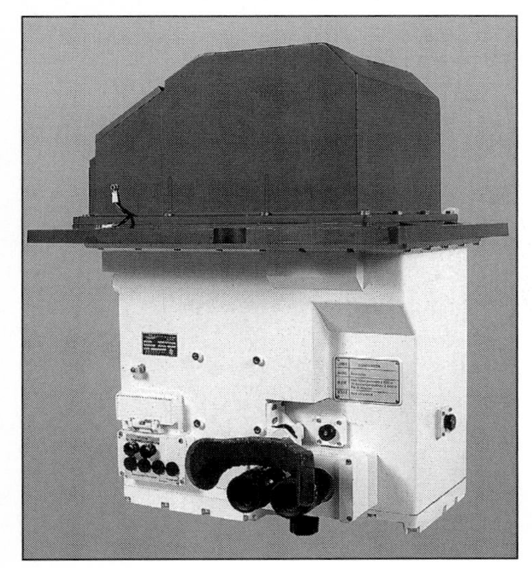

TIS-LSL thermal infra-red sight for the Leopard 1 MBT
0004784

Fields of view (azimuth × elevation)
Narrow: 3.5 × 2.3°
Wide: 7.7 × 5.1°
Focus: 50 m to infinity
Range
Identification: 2,000 m
Recognition: 3,000 m
Detection: 5,000 m
Laser
Distance discrimination: 30 m
Angular discrimination: 0.5 mil

Contractor

Thales Optronics (formerly OIP Sensor Systems).

UPDATED

Thales Optronique Castor thermal imaging system for armoured vehicles

Type

Gunner's sight.

Description

The Castor thermal imaging system, for use on armoured fighting vehicles, is intended for observation and firing under all conditions or for complete day/night fire-control systems. It can be installed on:

(a) tanks such as the M48, T-series, AMX-30 B2 in either an armoured housing attached to the gun shield and coupled to the fire-control system, with TV screen displays inside the turret for the commander and gunner (as on the AMX-30 B2 as the French Army DIVT 16 system), or inside the mantlet itself
(b) reconnaissance vehicles such as the AMX-10RC with the Castor unit within the turret
(c) artillery observation vehicles such as the AMX-10 SAO and AMX-10 VOA. In the latter case the Castor unit, known by the French Army designation of DIVT 17, is mounted outside the turret in an armoured housing and is slaved to the laser range-finder's elevation system. A graduated graticule is provided with the system coupled to the laser range-finder and onboard computer to calculate the target's co-ordinates. The image is displayed on a TV monitor inside the turret
(d) on turrets such as the MBDA UTM 800 as mounted on VAB and MOWAG vehicles or the Giat Lancelot mounted on AMX-10P and MOWAG Piranha vehicles for launching HOT ATGWs
(e) on other surface-to-surface or surface-to-air firing units for weapons such as the Thales Air Defence Crotale or MBDA Mistral
(f) the T-series MBT family.

Castor is based on a modular concept. If required, an additional control unit module can be fitted with the vehicle commander's unit acting as the master controller.

The system is available with either a monofocal or bifocal lens system giving a variety of fields of view. An electronic magnification capability (×2) is also available. Target detection/recognition range of Castor is up to 4 km.

A modular system for image processing, image detection and tracking is associated with the camera.

Operational status

Production. In service with the French Army (on AMX-30 B2 MBTs and AMX-10 VOA artillery observation vehicles) and other unspecified countries. The associated modular system for image processing entered serial production in 1993.

Specifications

Spectral band: 8-12 μm
Detector type: Cadmium Mercury Telluride (CMT) photovoltaic assembly, serial-parallel scanned

Giat Industries AMX-30 B2 MBT fitted with DIVT 16 thermal imaging system

Target detection/recognition range: Up to 4 km
Fields of view

lens variant	wide field	narrow field
1	6 × 9°	2.7 × 4°
2	6 × 9°	2 × 3°
3	5.4 × 8.1°	1.8 × 2.7°

Weight:
Optical unit: 20 kg
Electronic unit: 9 kg
Control unit: 2 kg
Dimensions
Optical unit: 452 × 181 × 302 mm
Electronic unit: 270 × 180 × 152 mm
Control unit: 105 × 155 × 135 mm

Contractor
Thales Optronique, France.

UPDATED

Zeiss Optronik PERI-ZL gunner's gyrostabilised observation and sighting periscope with integrated laser range-finder

Type
Gunner's sight.

Description
The PERI-ZL is designed to allow an MBT gunner to fire at a stationary or moving target while the vehicle itself is either stationary or moving.

The main subassemblies are:
(a) PERI-ZL sight assembly
(b) commander's combined sighting telescope attachment
(c) electronics unit which performs voltage generation, control electronics and operation monitoring tasks
(d) standardised Zeiss laser modules.

The PERI-ZL uses main subassemblies from the PERI-R17 stabilised periscope on the Leopard 2 MBT. Any deviations of the line of sight from the nominal elevation and azimuth directions caused by platform movements are compensated for by the sight's primary gyrostabilisation system.

The sight can be integrated into the MBT's fire-control system where the ballistic computer determines the target's lead and superelevation angles on the basis of the range determined by the sight's laser range-finder. Dynamic lead takes into account the relative movements of the vehicle and the target.

The gunner uses a single handle to control the line of sight in elevation and azimuth. Displays are present for the range measured, multi-echo, firing and laser readiness. The gunner has an observation channel with switch-selectable magnifications of ×3 and ×10, a brightness control and a minimum range control.

The vehicle commander can use a light tube telescope attachment on the right side of the sight which has a monocular eyepiece combining the sighting telescope and display images seen by the gunner in the left and right eyepieces respectively.

Operational status
No longer in production. In service with the Swiss Army.

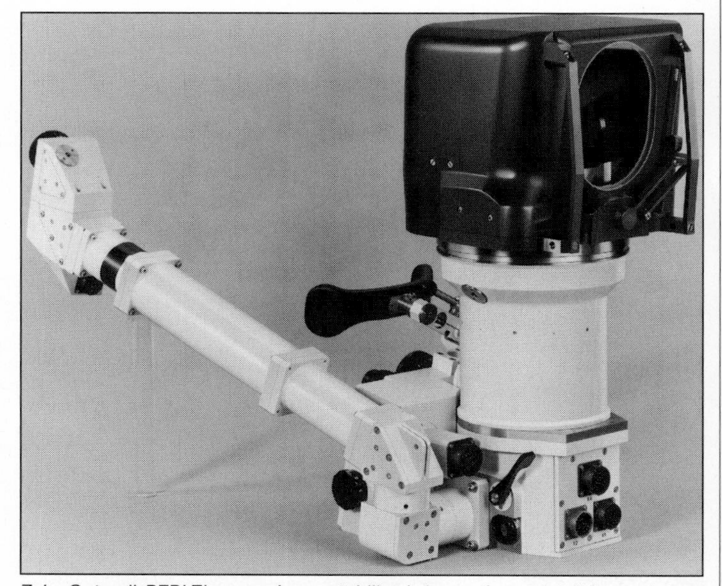

Zeiss Optronik PERI-ZL gunner's gyrostabilised observation and sighting periscope with integrated laser range-finder

Specifications
Elevation range: –12 to +20°
Traverse: ±5°
Field of view
×3 magnification: 20°
×10 magnification: 5.6°
Dioptre range: –4 to +4
Displays: Target range; multiple echo (yellow LED); laser readiness (green LED); firing readiness (red LED)
Laser range-finder
Type: Nd:YAG
Wavelength: 1.064 µm
Operating range: 400-10,000 m
Accuracy: ±10 m
Min range gating: 400-4,500 m

Contractor
Zeiss Optronik GmbH.

VERIFIED

Zeiss Optronik WBG-X thermal sight for armoured vehicles

Type
Fighting vehicle sight.

Description
The WBG-X thermal sight uses standardised US Army common modules developed by Raytheon, operating in the 8 to 14 µm waveband, together with appropriate afocals to achieve the required combat performance. The target aiming reticle is projected into the visual channel and, for adaptation to eye

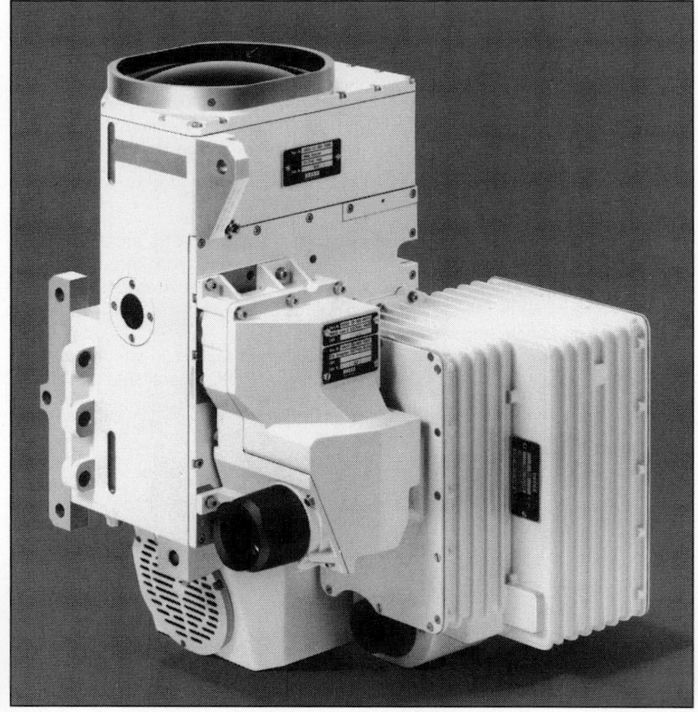

WBG-X thermal sight configured for the Leopard 2 MBT

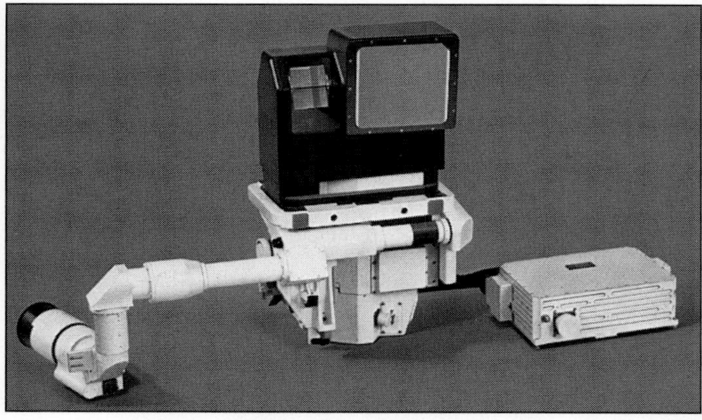

WBG-X configured for the Marder armoured personnel carrier 0004788

sensitivity, the red image provided by the system LED array is converted into a green image by an image intensifier tube.

The German Army has adopted the WBG-X for use on various vehicles:

(a) Leopard 2 MBT – the thermal sight is a component of the gunner's EMES 15 primary sight assembly. The gunner views the thermal image through the biocular eyepiece of the EMES 15. The eyepiece need not be changed when switching from the day to the thermal channel. A light pipe attachment transmits the thermal image to the PERI-R17 panoramic telescope of the commander. Thus, he can view either the thermal image or the panoramic visual image through the same eyepiece

(b) Leopard 1 MBT – the thermal sight with mirror head assembly is retrofitted to the loader station. The elevation mirror of the head assembly is slaved to the stabilised weapon system. The ballistic correction data are fed into the servo chain. The image is projected into the primary sight of the gunner and into the biocular eyepiece of the commander by means of light pipes.

(c) Marder 1 armoured personnel carrier – the thermal sight is installed at the gunner's station in place of the PERI-Z11. The gunner observes through an eyepiece and the commander through a biocular. The PERI-Z11 remains unchanged

(d) Luchs (8 × 8) reconnaissance vehicle – the thermal sight is installed in the same way as in the Marder 1. In either case, integration of the thermal sight in the vehicle does not reduce the space available for the crew.

Operational status

Production. In service with the armies of Denmark, Germany, Netherlands, Norway, Sweden and Switzerland.

Specifications

Dimensions

Basic instrument: 400 × 130 × 200 mm

Power supply: 405 × 205 × 205 mm

Weight

Basic instrument: 18 kg

Power supply: 13 kg

Power supply: 24 ± 6 V DC

IR channel

Fields of view

Wide: 15 × 7.5°

Narrow: 5 × 2.5°

Spectral band: 8-14 μm

Detector type: CMT, 120 elements

Cooling system: Stirling closed-cycle

Scan frequency: 42 Hz

Line count: 240

Interlace: 2:1

Day channel

Magnification: ×2 and ×6

Contractor

Zeiss Optronik GmbH.

VERIFIED

ARMOURED FIGHTING VEHICLES

COMMANDER'S SIGHTS

BAE Systems SS141, SS142 commander's night vision periscopes

Type

Commander's sight.

Description

The SS141 commander's biocular fixed-head passive night vision periscope can be fitted into 130 × 70 mm apertures on most turreted light armoured vehicles. It uses a second- generation 20/30 image intensifier tube with automatic brightness control. It has a typical target detection range of 1.32 km, recognition range of 330 m and identification range of 170 m against a 2.3 m^2 target in clear starlight (10^{-2} lx).

The use of an integrated 0, 2, 4 and 6 neutral density filter selector allows daylight use in an emergency.

The SS142 commander's night vision sight is virtually identical to the SS141 but is fitted with an elevating head assembly for use against low-flying airborne targets such as helicopters and requires a 170 × 90 mm fitting aperture.

Operational status

In service with unspecified countries.

Specifications

Weight
SS141: 7 kg
SS142: 8.2 kg
Elevation range
SS141: Fixed
SS142: –10 to +20°
Fields of view
SS141/142 ×2.7 magnification: 13.3°
Power supply: 28 V DC vehicle system

Contractor

BAE Systems, Avionics, Tactical Products Division.

UPDATED

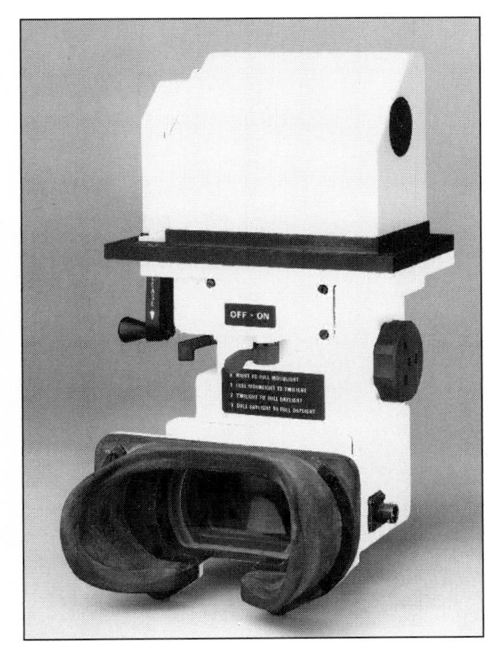

BAE Systems SS142 commander's night vision periscope from commander's side

Bofors Defence Systems Lemur

Type

Drivers sight.

Description

The Lemur roof-mounted electro-optical stabilised sight has been developed as a private venture project for installation on new-build vehicles as well as older vehicles as an upgrade. The sight can be used by the vehicle commander or gunner and has already been installed on prototypes of the Hägglunds Vehicle

The Bofors Defence panoramic sight (prototype) 0089190

CV120 tank and the MOWAG (10 × 10) ACV fitted with the Bofors Defence CV9040B 40 mm turret assembly.

In both of these applications, the Lemur is installed at the commander's station to give the vehicle a hunter-killer capability. In a typical engagement the target would be acquired initially by the vehicle commander and then handed over to the gunner to conduct the actual engagement while the commander searches for another target.

The sight is mounted on a pedestal. The height can be adapted to allow an unobstructed view over the commander's hatch or any other obstruction on the turret roof. The design is modular with the choice of sensor depending upon the users exact operational requirements. For example, if used as a primary sight a laser range-finder is then included.

All versions use the same stabilised platform with the armoured cover and sight windows adapted to meet the size of sensor and elevation requirement. The sight is armoured to withstand machine gun fire and shrapnel from artillery shells. The Lemur can be traversed through 360° with the basic versions having elevation from –20 to +35°. If required the window can be enlarged to give an elevation up to +55°.

The second-generation thermal imager operates in the 7.5 to 10.5 µm IR waveband region with the standard detector having 288 × 4 elements. There are two fields of view: wide (9 × 6.75°) and narrow (3 × 2.25°). The TV camera has a selectable field of view of 16 × 12° with the 1.064 µm wavelength laser range-finder having range limits of 200 to 9,995 m. The latter may be extended if required. The digital serial communications interface uses a CAN-bus or RS-422 link with the video output having 625 lines.

Specifications

Physical characteristics
Weight: 160 kg
Dimensions: 450 × 450 × 450 mm
Traverse range: 360°
Elevation range: –20 to +35° (up to +55° possible with enlarged window)
Thermal imager
Detector: 284 × 4 elements
Wavelength: 7.5-10.5 µm
Wide field-of-view: 9 × 6.75°
Narrow field-of-view: 3 × 2.25°
TV Camera
Field-of-view: 16 × 12° (selectable)
Laser range-finder
Max range: 9,995 m (can be extended)
Min range: 200 m
Resolution: 5 m
Beam width: 1 mrad
Wavelength: 1.064 µm
Pulse repetition frequency: 1 Hz continuously
4 Hz intermittent operation

Electrical interface
Communication: CAN-bus or RS-422
Video outputs: CCIR624-3BG (625 lines 50 Hz)
Power supply: 18-32 V

Operational status
Prototype systems undergoing trials with Swedish Army installed on CV 9040 infantry fighting vehicle.

Contractor
Bofors Defence AB (a United Defense company).

Electro Optic Systems CCS – Crew Commander's Sight

Type
Commander's sight.

Description
The Crew Commander's Sight (CCS) is a full-solution fire-control system suitable for a wide variety of weapon stations. CCS is a fully integrated day/night system which incorporates an eye-safe laser range-finder, a ballistic computer, CCD cameras, an intensified CCD camera and a flat panel display. The system can be used with a helmet mounted display instead of the flat panel display. The product is able to rapidly designate a new aiming point that takes into account range, target movement and environmental conditions. It is claimed that the effective range of weapons such as 20 to 35 mm cannons, automatic grenade launchers and 0.50 calibre machine guns can be increased, with a significant improvement in accuracy.

CCS is adaptable to most direct fire weapon stations and a single unit can be programmed to handle multiple weapons and a wide variety of munitions. The common module architecture meets modern logistics support requirements and allows a low-cost upgrade path through module replacement.

Operational status
CCS is a variant of EFCS which has been type-classified by the US Navy and is to be evaluated by the US Army for the Common Remotely Operated Weapon System (CROWS).

Specifications
Dimensions:
Sensor/processor unit: 316 × 235 × 167 mm
Weight: 9 kg
Laser: Eye-safe Class 1 (AS 2211:1991)
Range: 5,000 m
Wavelength: 1.54 μm
Pulse repetition rate: 12 ppm
Accuracy: ±5 m
Magnification: ×10
Night sight: GEN II + or GEN III ICCD to customer specification
Power supply: 24-28 V DC

Contractor
Electro Optic Systems Pty Ltd, Australia.

Crew Commander's Sight (CCS) 0023043

Elop CPS (Commander's Panoramic Sights)

Type
Tank commander's sights.

Description
Elop has developed and produces a range of Commander's Panoramic Sights (CPS), optimised for use on Main Battle Tanks (MBTs) and AFVs. The CPS product line includes a variety of day and/or night channels; day optical direct view and/or high-resolution TV channels, stabilised or non-stabilised to Line-of-Sight (LOS).

The CPS provides the MBT/AFV commander with high quality viewing capability for target acquisition and 'hunter-killer' operation. It is designed as a roof-mounted, stand-alone sight, interfaced to the main weapon fire-control system, and is compatible with most tanks and AFVs.

The CPS can also function as a back-up to the gunner's sight (providing redundancy). A laser range-finder/designator is optional.

Status
In production and service with the Israel Defence Forces.

Contractor
Elop Electo-Optics Industries Ltd.

NEW ENTRY

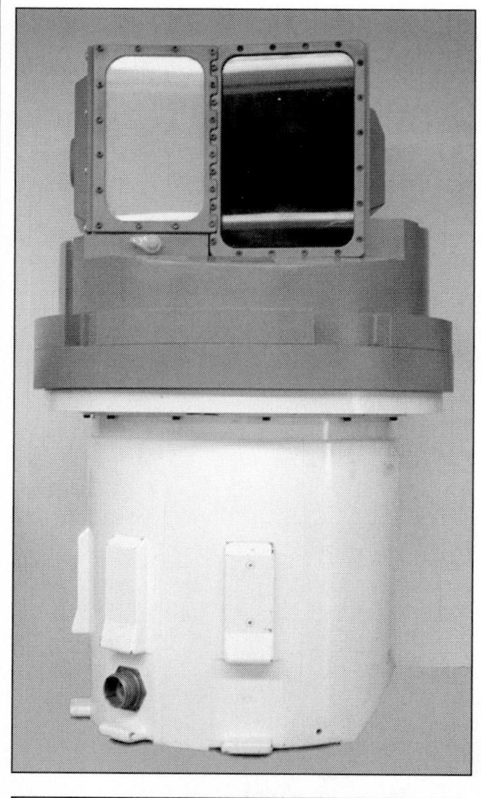

An alternative configuration of Elop CPS
NEW/0547549

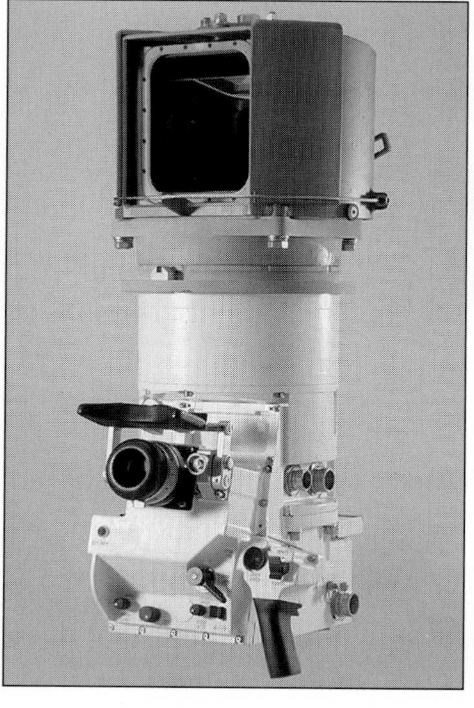

This configuration of an Elop CPS features a panoramic day/night sight
NEW/0547548

Elop Stabilised Panoramic Day/Night Sight (SPDNS)

Type
Commander's sight.

Description
The SPDNS is designed as a stand-alone, roof-mounted panoramic stabilised day/night commander's sight that is compatible with most modern tank and AFV designs.

The SPDNS contains an electrical stabilised head-mirror, counter- rotation optics and interfaces for hunter-killer operational modes while interfaced to an advanced fire-control system. It has been ergonomically designed for ease of operation, minimum volume and low silhouette. There is a built-in flash shutter, selectable sun filter and laser protection filter.

Future growth options will include full integration with advanced fire-control systems, an integrated laser range-finder/designator and a thermal imaging module.

Status
Final development.

Specifications
Elevation range: −14 to +35°
Traverse range: 360°
Stabilisation: 100 μrad
Day channel
Magnification ×14 (NFOV): 5.5°
Magnification ×4 (WFOV): 14°
Night channel
TV day channel
Magnification ×14 (NFOV): 1 × 1.5°
Magnification ×4 (WFOV): 3 × 4°

Contractor
Elop Electro-Optics Industries Ltd (Israel).

VERIFIED

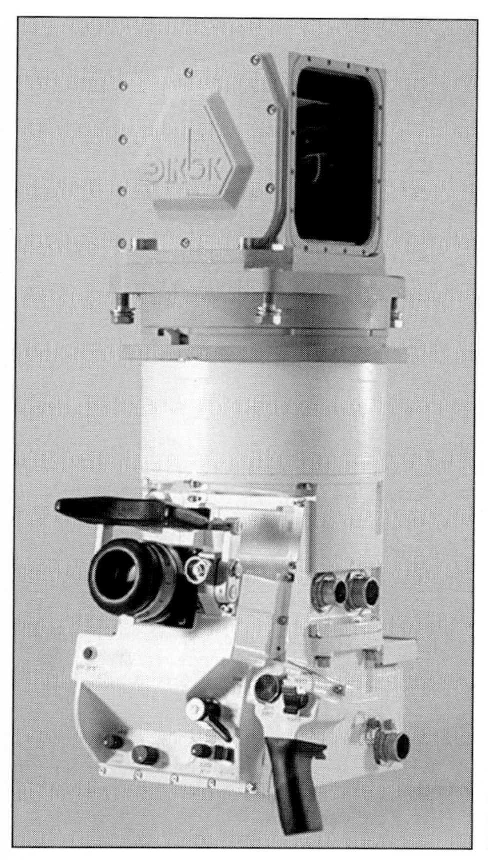

The Elop Stabilised Panoramic Day/Night Sight (SPDNS)
0059664

Eloptro CS-30 commander's sight

Type
Commander's sight.

Description
The CS-30 is a low-cost electronically slaved panoramic day sight designed for use in Main Battle Tanks (MBTs) and Armoured Fighting Vehicles (AFVs). The system increases the commander's ability to observe, detect and identify targets and enables automatic target designation to the gunner.

The CS-30 comprises a panoramic sight head, mounted on the turret, with integrated electronics and control panel. A hand controller is used to steer the line of sight in elevation and traverse. The CS-30 provides a direct view optical channel with three magnifications. The system's modern digital architecture offers complete integration with the fire-control computer, facilitating mode selection, exchange of line of sight position data and built-in test data.

The sight mount complies to a NATO standard which facilitates retrofitting to existing turrets. Since there is no mechanical link, the sight can easily be customised for optimum line of sight and eyepiece positioning.

Operational status
In production, in service.

Specifications
Weight: 25 kg
Magnification: ×3, ×6 and ×9
Elevation range: −18 to +22°
Traverse range: 360° (without limit)
Positioning accuracy: <0.3 mrad
Boresighting capability: <0.5 mrad
Field of view: 14, 7 and 4,5°
Eyepiece dioptre setting: −4 to +2
Eye relief: 22 mm
Power supply: 18-32 V DC
Secondary stabilised in azimuth
Selectable brightness filter
Laser protection filter

Contractor
Eloptro (Pty).

Eloptro CS-35 stabilised commander's sight

Type
Stabilised commander's sight.

Description
The CS-35 stabilised commander's sight is an upgraded version of Eloptro's production CS-30 commander sight. The sight is a two-axis, primary stabilised, panoramic day sight enabling surveillance, target tracking, target acquisition and automatic target designation to the gunner, while the armoured vehicle is in motion.

The CS-35 comprises a panoramic sight head, mounted on the turret, with integrated control electronics and control panel. The CS-35 is steerable in elevation and traverse by means of a hand controller. The CS-35 provides a Direct Vision Optical (DVO) channel with three magnifications.

The CS-35 configuration is offered as a low-cost stabilisation upgrade path for the CS-30 commander sight.

Operational status
Production as required.

Specifications
Weight: 25 kg
Elevation range: −18 to +35°
Traverse range: 360° (without limit)
Stab accuracy: 500 μrad peak to peak
LOS drift: <20°/h
Positioning accuracy: <0.3 mrad
Boresighting capability: <0.5 mrad
Magnification: ×3, ×6 and ×9
Field of view: 14, 7 and 4,5°
Eyepiece dioptre setting: −4 to +2
Eye relief: 22 mm
Power supply: 18-32 V DC
Secondary stabilised in azimuth
Selectable brightness filter
Laser protection filter

Contractor
Eloptro (Pty).

Fotona Comtos-55 commander's takeover set for T-55 tank

Type
Tank commander's sight.

Description
The Comtos-55 commander's takeover set for the T-55 tank is equipped with an EFCS3-55 fire-control system which enables the commander to override all the gunner's controls and to perform observation, target tracking, ranging, target selection and gun/machine gun firing instead of the gunner.

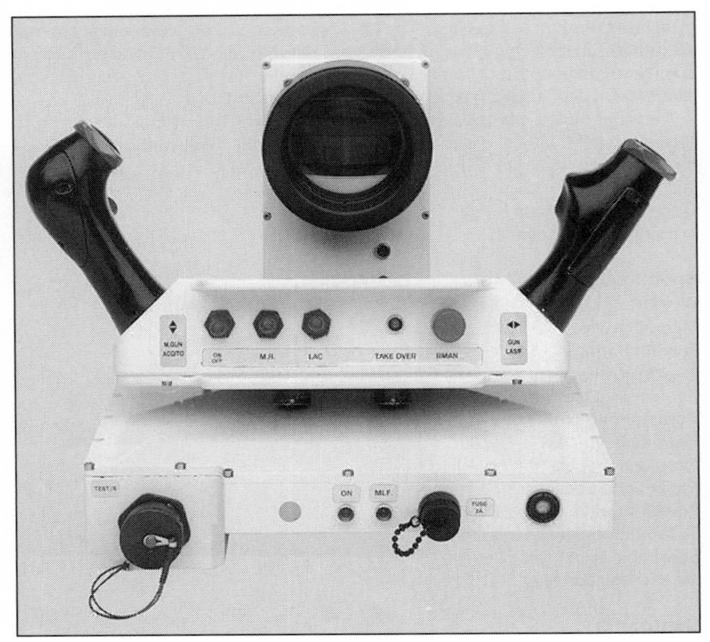

The Comtos-55 commander's takeover set for the T-55 tank

An interconnection box with a takeover push-button connects the EFCS3 system and the T-55 installation with the commander's block. The following functions are available via this block; azimuth and elevation control for turret and gun movement, laser range-finder control (triggering and target selection), LAC Lead Angle Calculation on/off, manual range control and gun/machine gun trigger. When the takeover button is depressed only the original acquisition function is available on the commander's block.

A CCD camera, resolution 752 (H) × 582 (V), is mounted on the eyepiece of the stabilised gunner's sight.

Contractor

Fotona d d (Slovenia).

Meopta TKN-3 P tank viewing unit

Type

Commander's sight.

Description

The TKN-3 P viewing unit is a binocular periscopic telescope for tank or armoured personnel carrier commanders. At night an image intensifier (Gen II plus) is introduced into the optical path to the user. By day the system is purely optical. A range-finding scale is available in the day mode. The optics are heated when required to prevent condensation.

Operational status

Available.

Contractor

Meopta Prerov, a.s.

VERIFIED

Meopta TKN-3 P commander's tank viewing unit
0023044

PCO LISWARTA commander's night vision periscope

Type

Commander's sight.

Description

The PCO commander's night vision periscope is a binocular day/night viewing system for use in a wide variety of armoured vehicles for example: T-55, T-55A, T-72, T-72M tanks, infantry fighting vehicles BWP-1, BWP-2. Also fitted to the GOŽDZIK howitzer and WZT-2, WZT-3 technical support vehicles. The periscope provides the commander with direct vision of the terrain around the vehicle by day or night. LISWARTA replaced a Russian manufactured periscope but retains the same electrical switches. The periscope comprises two independent viewing channels, each equipped with an objective lens system and eyepiece. There is one 18 mm image intensifier tube in the night channel. The tube has automatic gain control and a stabilised power supply.

Operational status

In production. In service in Poland and other countries.

Specifications

Dimensions: 350 × 305 × 200 mm
Magnification: ×4.9 (for day and night channel)
Field of view
Day: 10°
Night: 8°
Image intensifier tube: GEN II+ or Super GEN II
Power supply: 22-29 V DC

Contractor

PCO S.A, Poland.

VERIFIED

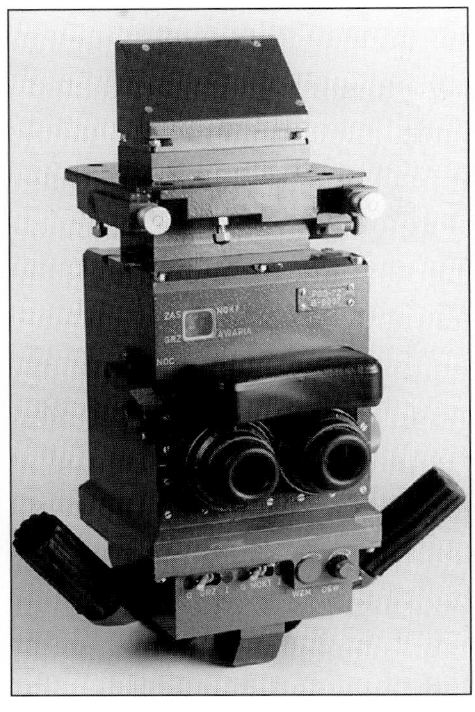

LISWARTA commander's day/night periscope
0023045

PCO TKN-1Z passive commander's periscope

Type

Commander's sight.

Description

The TKN-1Z passive commander's periscope is a small monocular night vision device adapted for the T-55 tank. It enables observation of the road, surrounding terrain and movement of neighbouring tanks during starlight conditions. TKN-1Z is modernised version of active TKN-1 periscope. This device does not need any infra-red or other artificial light source therefore it provides maximum security against any detection.

TKN-1Z periscope uses either a GEN II+ or SuperGEN image intensifier tube.

Operational status

In production. In service with Polish Army.

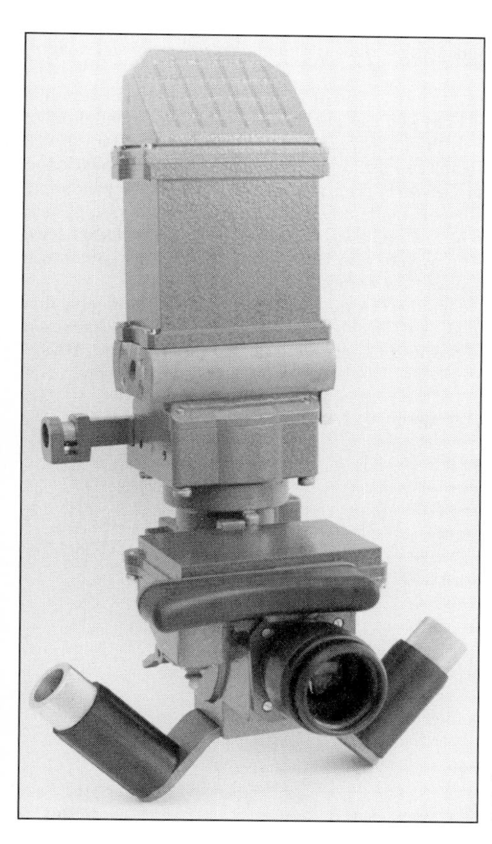

The PCO TKN-1Z passive commander's periscope
0089902

Specifications
Magnification: ×3
Field of view: 10°
Ambient operating temperature: –50 to +65°C;
Power supply: 18 to 30 V DC

Contractor
PCO SA, Poland.

VERIFIED

PCO TKN-3Z tank commander's periscope

Type
Day/night commander's sight.

Development/Description
The TKN-3Z tank commander's periscope is a viewing assembly designed for use by the commander of a T-72 series MBT. It provides the vehicle commander with vision of the terrain and surrounding vehicle(s) in both day and night conditions. It is a modernised direct drop-in replacement for the original Russian TKN-3 periscope system and is fitted with an image intensifier tube. The later can either be a GEN II+ or SuperGEN assembly and has a built-in flash and bright-light protection facility.

Status
In production. In service with the Polish Army. Offered for export.

The PCO TKN-3Z tank commander's periscope
0089903

Specifications
Magnification: day channel ×4.9, night channel ×4.2
Field of view: day channel 9° 30′ min, night channel 11° 30' min
Image intensifier tube: GEN II+ or SuperGEN
Power supply: 22-29 V DC

Manufacturer
PCO S.A, Poland.

VERIFIED

Raytheon CITV Commander's Independent Thermal Viewer

Type
Commander's sight.

Description
The Commander's Independent Thermal Viewer (CITV) is a second-generation thermal imaging system that is a key feature of the M1A2 Abrams main battle tank improvement programme. It was developed by Raytheon Systems Company for General Dynamics Land Systems and the US Army.

CITV provides the tank commander with a high-quality viewing capability for searching for targets while the gunner engages previously identified targets. This 'hunter-killer' concept is stated to increase the effective rate of fire for the Abrams by over 30 per cent. A 360° independent rotation permits the commander to search the scene without turret movement, thus increasing survivability and providing longer silent watch operations. CITV provides automatic non-verbal electronic target cueing of the gunner's sight system and can also function as a back-up main gun sight for redundancy.

CITV incorporates the Horizontal Technology Integration (HTI) FLIR, developed by Raytheon. HTI FLIR (qv) is a US Army programme to develop a common thermal sensor for ground-based battlefield platforms. It is an 8 to 12 μm scanning system based on SADA II (Standard Advanced Dewar Assembly) technology which uses a 480 × 4 focal plane array.

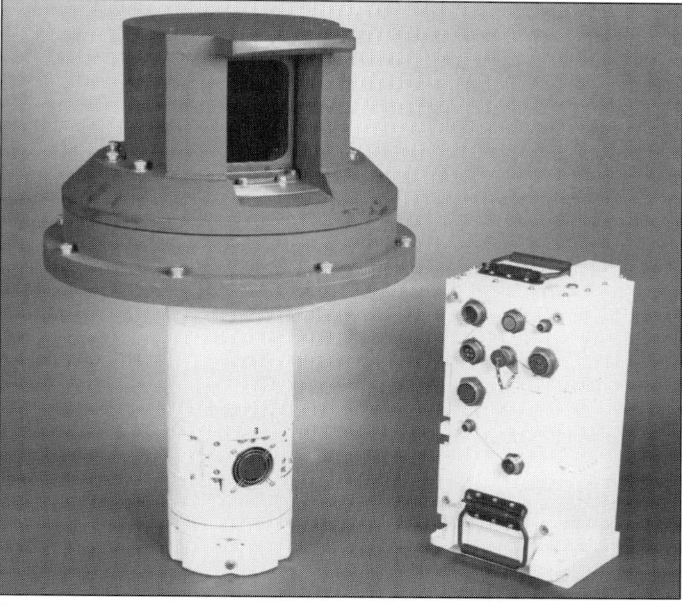

Raytheon Commander's Independent Thermal Viewer (CITV) for the M1A1/M1A2 MBT

CITV fitted on the Abrams main battle tank 0004781

The CITV comprises:
(a) a gyrostabilised head sensor assembly
(b) commander's combined hand controller grip/parameter setting panel
(c) electronics box
(d) commander's remote CRT display unit.

Operational status
Raytheon was awarded a multiyear contract for initial production of CITV in April 1997 by the US Army Communications & Electronics Command (CECOM). In-service on US Army M1A2 MBT's.

Specifications
Weight: 181.8 kg
Elevation range: –12 to +20°
Azimuth range: 360°
Fields of view
Narrow: 2.6 × 3.4°
Wide: 7.7 × 10.4°
Min resolvable temperature (NFOV)
at 2.7 cy/mrad: 0.19°C typical
at 5.0 cy/mrad: 1.42°C typical
Stabilisation jitter (RMS, 1 σ)
Azimuth: 100 μrad
Elevation: 100 μrad
MTBF: 7,720 h

Contractor
Raytheon Systems Company.

VERIFIED

Raytheon Commander's Independent Viewer (CIV)

Type
Commander's sight.

Description
The Commander's Independent Viewer subsystem is currently in production by Raytheon Systems Company for use in the Bradley A3 fighting vehicle. It will upgrade the current capabilities of the Bradley Fighting Vehicle by providing 360° panoramic viewing of the battlefield enhancing situational awareness and weapon effectiveness.

Commander's Independent Viewer (CIV) fitted on the Bradley fighting vehicle
0004777

The baseline CIV includes the integration of the second-generation Horizontal Technology Integration (HTI) FLIR with daylight television and preplanned product improvements for interface to the Improved Bradley Acquisition System (IBAS) autotracker, into a two-axis stabilised platform. IBAS (qv) is also being developed by Raytheon. The HTI FLIR is a US Army programme to develop a common thermal sensor for ground-based battlefield platforms. It is an 8 to 12 μm scanning system based on SADA II (Standard Advanced Dewar Assembly) technology which uses a 480 × 4 focal plane array. The system also features electronic zoom, automatic boresighting and optional eye-safe laser range-finder.

These improvements increase the commander's ability to locate, identify and defeat threats, both stationary and on the move. The CIV provides battlefield oversight in silent watch and provides target handoff to the gunner via the IBAS.

Operational status
In production. In-service on US Army M2A3/M3A3 Bradley fighting vehicles.

Specifications
HTI FLIR
Spectral band: 8-12 μm
Detector: 480 × 4 SADA II CMT
Fields of view
Wide: 13.3 × 7.5°
Narrow: 3.6 × 2°
Daylight TV
Fields of view
Wide: 10 × 7.5°
Narrow: 2.7 × 2°

Contractor
Raytheon Systems Company.

UPDATED

Raytheon CVTTS-S Combat Vehicle Thermal Targeting System for Surveillance

Type
Surveillance system land.

Development
The Combat Vehicle Thermal Targeting System – Surveillance (CVTTS-S) is a surveillance variant of the Combat Vehicle Thermal Targeting System currently in production.

Description
The CVTTS-S uses a second-generation focal plane array thermal imaging sight and an optional visible sight that can be mounted on a remotely operated pan and tilt assembly. The Scout vehicle variant of the CVTTS-S has a small hand-held control panel. A remote-control station is available which includes a high-resolution monitor, an optional video recorder and a range of controls for the sights and the pan and tilt assembly. The standard output of the system is RS-170 with other formats available.

Optional equipment for the CVTTS-S includes an additional monitor, tripod, laser range-finder, vehicle mounting hardware, manual pan and tilt, voltage converters, extended distance remotability and application-specific on-screen menus and/or characters.

Operational status
Available.

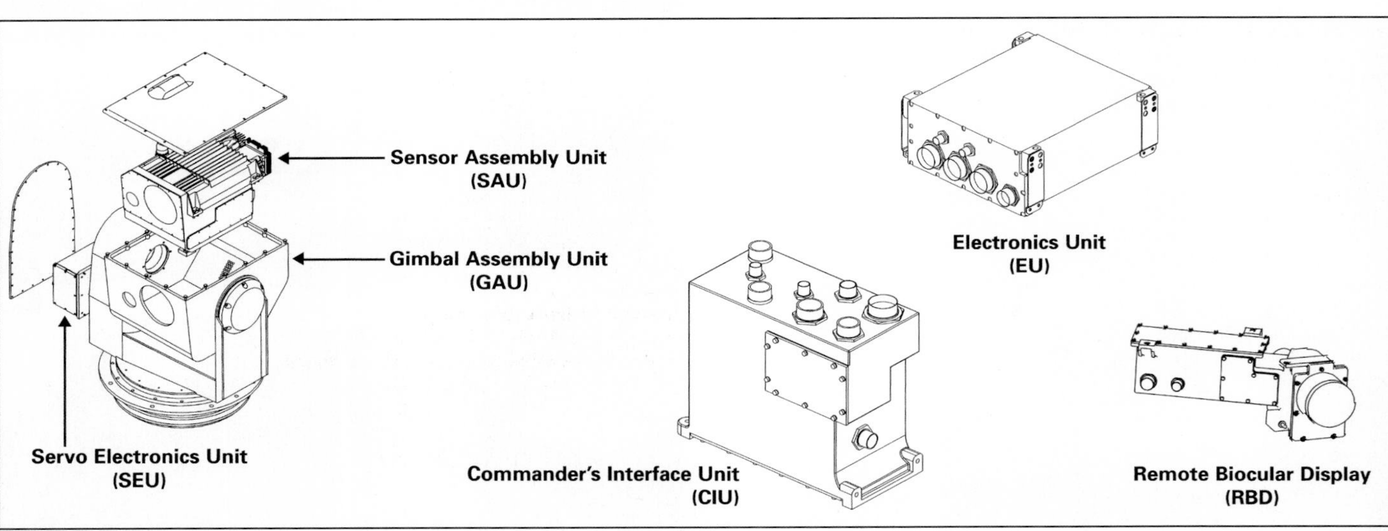

Diagram showing the components of the CIV for the Bradley A3
0004778

The Thermal Sight Unit of Raytheon's CVTTS- S

Specifications

Detection range (WFOV): >5,000 m
Recognition range (NFOV): >4,000 m
Field of view, horizontal and vertical
Narrow: 2.5 × 1.9°
Wide: 7.5 × 5.6°
Aspect ratio: 4 × 3
Spectral band: 7-12 μm
Input power: 18-33 V DC
Dimensions: 9.1 × 11.4 × 11.9 in
Weight: <30 lb

Contractor

Raytheon Systems Company.

VERIFIED

Raytheon/Samsung Thales CPS 1 Commander's Panoramic Sight

Type

Commander's sight.

Development

The CPS 1 Commander's Panoramic Sight is in development to support the Korean Army K1A1 (Up Gun) programme. The sight is being developed co-operatively by Samsung Thales and Raytheon Systems Company. The sight is a fully stabilised 360° panoramic sight that provides independent target acquisition and cueing to the gunner while the commander continues to search for targets. The sight is readily adaptable to the M1A1, M1A2, Challenger MBT, Leclerc and infantry fighting vehicles.

Description

The CPS 1 features direct view optics and a second-generation 240 × 4 focal plane array thermal imager. This solution minimises the use of valuable turret and hull space, providing the commander with multiple target acquisition, fire control and maximum night fighting capability.

Standard features:
(a) Direct view optics/2 fields of view
Narrow 2.1 × 2.8°, Wide 8 × 10.7°
(b) Magnification ×10, ×3
(c) Second-generation MCT thermal imaging technology (8-12 μm}
(d) Stabilised head-mirror
(e) 360° azimuth
(f) +35 to –20° elevation
(g) Dual electronic zoom
(h) Automatic gain and level balance controls
(i) (Multilingual symbology)
(j) BIT/BITE
(k) RS-170 video

Options
(a) Thermal only
(b) Day TV/ICCD
(c) Eye-safe laser range-finder
(d) Digital data transmission
(e) Dual target tracker

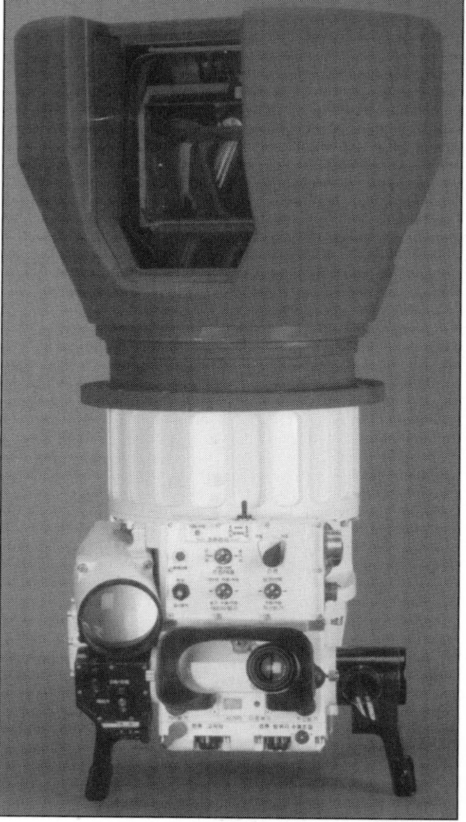

The CPS 1 Commander's Panoramic Sight
0055258

Operational status

In production for the RoK Army K1A1 programme.

Contractors

Raytheon Systems Company.
Samsung Thales.

VERIFIED

ROMZ PNK-4C

Type

Commander's sight.

Description

The PNK-4C is a combined gyrostabilised (elevation) system for day and night aiming and surveillance. It is intended as a commander's sight for use in rotating turrets of T-72, T-80, T-82 tanks.

Specifications

Optical channel
Field of view: 7.2° (V) and 27.7° (H)
Magnification: ×1 or 7.6
Image intensifying channel
Field of view: –7.7°
Magnification: ×5.1
Dioptre adjustment for day and night channels: ±4
Ambit with gyrostabiliser (elevation): –10 to +20°
Tracking rate with gyrostabiliser: 0.05 to 3°/s
Gyrostabilisation angular accuracy:
normal conditions: 1 min of arc
at –50°C after 2 mins operation: 6.5 mins of arc
after 30 mins operation: 1 min of arc
Weight
Main device: 43 kg
Power unit: 4.2 kg
Indicator of barrel position: 10.8 kg
Run-up time: 2 min
Identification range (side projection of tank):
At night atmospheric transparency of 0.85 for 1 km in passive mode and illumination of >0.003 lx: > 700 m
At night in active mode (IR projector with luminosity of 10 Mcd) and at ambient illumination of <0.003 lx: > than 800 m

Contractor

Romz, Rostov Optical-Mechanical Plant (JSC).

VERIFIED

SAGEM HL-70 commander's gyrostabilised panoramic sight

Type
Commander's sight.

Description
The HL-70 gyrostabilised panoramic sight is in production for fitting to the Giat Industries Leclerc MBT programme for the French Army. It has a day optics system (eight periscopes) with switchable magnifications of ×2.5 and ×10 across a 5° and 20° field of view, and an image intensifier CCD night optics module. In the commander's eyepiece are projected symbols which indicate the status of the tank (for example, type of ammunition, firing range, firing authorisation). Optional equipment includes a laser range-finder, TV camera to broadcast the image to the gunner and a micromonitor display.

The sight is shrouded in separately controlled armour and uses a digital bus for electrical interfacing with the Leclerc weapon system.

Operational status
In production for the French Army (Leclerc MBT). More than 400 systems have been delivered.

Specifications
Elevation range: –20 to +40°
Azimuth range: 360°
Fields of view
×2.5 magnification: 20°
×10 magnification: 5°
Night vision channel: second-generation image intensifier CCD

Contractor
SAGEM SA, Aerospace & Defence Division.

UPDATED

Close-up of Giat Industries Leclerc MBT turret showing commander's HL-70 gyrostabilised roof-mounted panoramic sight

The export version of the Leclerc main battle tank for the UAE is fitted with the HL 80 commander's sight

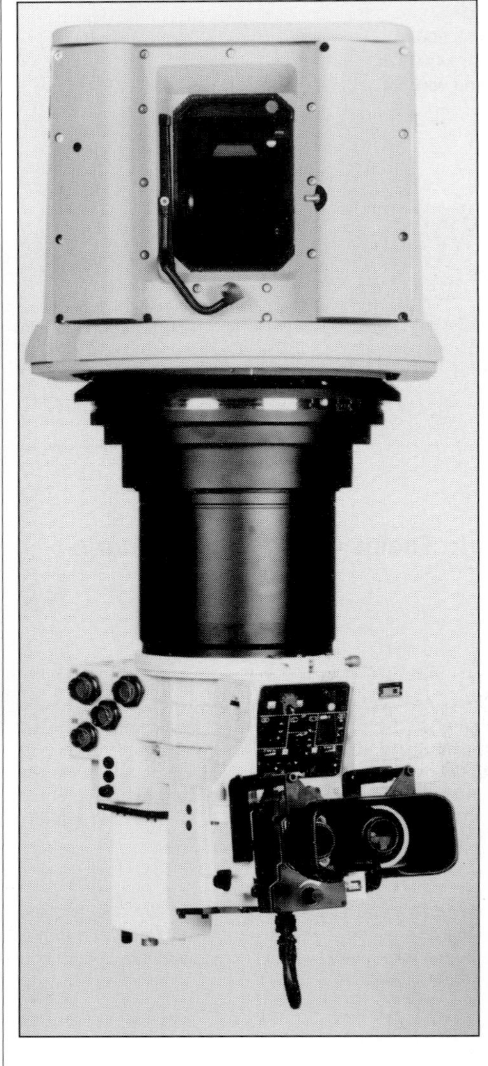

The HL 80 sight for Export Leclerc MBT
0055261

SAGEM HL-80/HL-120 commander's gyrostabilised panoramic sight

Type
Commander's sight.

Description
The HL-80/HL-120 gyrostabilised panoramic sight is in production for fitting to the Leclerc export MBT programme. It has a day optics system with switchable magnifications of ×2.5 and ×10, a thermal imager and a built-in Avimo (now part of Thales Optronics) HL-58 laser range-finder. In the commander's eyepiece are projected symbols which indicate the status of the tank, for example, type of ammunition, firing range and firing authorisation. Optional equipment includes a TV camera to broadcast the image to the gunner and a micromonitor display.

The sight uses a digital bus for electrical interfacing with the Leclerc weapon system.

The HL-120 is a similar sight fitted with a second-generation SAGEM Iris thermal imager.

Operational status
In production and in service with the UAE for the Leclerc export MBT. The last batch has been ordered for the French Leclerc MBT. Around 400 have been ordered with more than 300 delivered.

Specifications
Elevation range: –20 to +40°
Azimuth range: 360°
Fields of view
×2.5 magnification: 20°
×10 magnification: 5°
Night vision channel: 8-12 μm thermal imager
HL-80: ALIS thermal imager
HL-120: IRIS thermal imager

Contractor
SAGEM SA, Aerospace & Defence Division.

UPDATED

SAGEM MVS 580 commander's multichannel sight

Type
Commander's sight.

Description
The tank commander's Multichannel Versatile Sight (MVS) 580 is the latest derivative of the VS 580 family of sights. It utilises the main components of the VS 580 Commander's Panoramic Sight (CPS) with an 8 to 12 μm waveband thermal imager in the stabilised head.

The stabilisation and servo assemblies are similar to the VS 580 CPS. It can be fitted with an optional laser range-finder, built-in test equipment, autotracker, a micromonitor for the thermal channel and data display and a head-down display monitor. Two options are available for the thermal imager: either low cost or high performance.

Operational status
Ready for production. Selected for two major programmes. Included on Slovakian T-72M2 upgrade proposal.

Specifications
Dimensions
Turret roof hole diameter: 280 mm
Depth, under roof: 660 mm
Height, outside: 440 mm
Elevation range: –35 to +35°
Azimuth range: 360° (without limit)
Scanning rate: 0.1 mil/s to 1,000 mils/s
Residual stabilisation: 0.03 mil (APG course)

M2 Moderna T-72 tank fitted with the SAGEM MVS 580 commander's panoramic sight 0055260

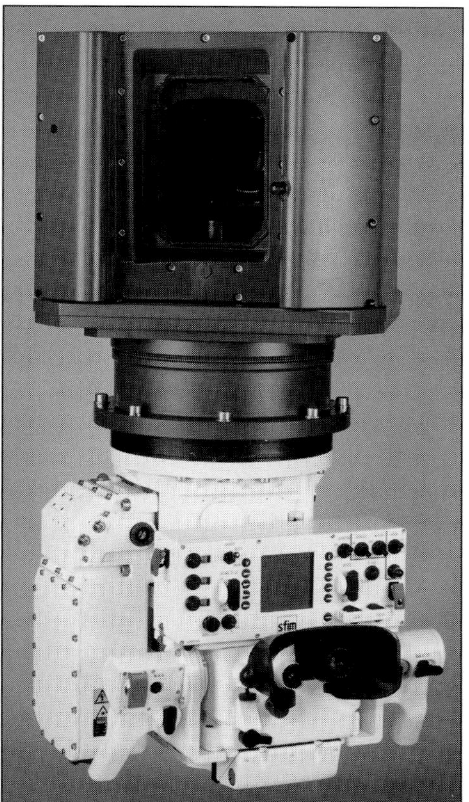

The SAGEM MVS 580 gyrostabilised commander's sight 0055259

Angular output accuracy: 0.1 mil
Day channel: similar to VS 580 CPS with 16.5° and 5° fields of view
Night channel:
Spectral band: 8-12 μm
Recognition range (STANAG)
Low-cost thermal imager: 2,000 m
High-performance thermal imager: 4,000 m
Optional devices: laser range-finder, micromonitor for thermal imager display, auto-tracker

Contractor
SAGEM SA, Aerospace and Defence Division.

UPDATED

SAGEM VS 580 family of gyrostabilised sights

Type
Commander's sight.

Description
The VS 580 family of gyrostabilised commander's sights is intended for the detection, recognition, identification and acquisition of targets from a moving combat vehicle by means of accurate line of sight stabilisation. VS 580 sights are compatible with a large number of fire-control systems. A variant has been developed for anti-aircraft combat vehicles.

All the VS 580 sight variants comprise four subsystem assemblies:

- a sight upper assembly mounted on the turret roof which contains the gyrostabilised panoramic sight head and its related electronics
- an intermediate assembly which houses the laser range-finder
- a lower telescopic assembly within the turret confines containing the optical viewing system and associated hand controls
- an electronics unit.

Operational status
More than 2,000 are in service worldwide. Users include South Korea (K-1 MBT), United Arab Emirates (OF-40 MBT), Slovakia (T-72 Moderna) and the UK and Oman (Challenger 2). The PSPC variant of the VS 580, developed with Alenia Difesa (Officine Galileo), is dedicated to the Italian Army's B1 Centauro (8 × 8) tank destroyer.

Specifications
Weight: 100 kg
Elevation range: ±35° (–10 to +60° anti-aircraft variant)
Azimuth range: 360° (without limit)
Day channels
Fields of view
×3.2 magnification: 16.5°
×10.5 magnification: 5°
Laser range-finder
Wavelength: 1.064 or 1.54 μm

Contractor
SAGEM SA, Aerospace and Defence Division.

UPDATED

SAGEM VS 580 sight-mounted on a Challenger 2 MBT (Oman) 0023046

Zeiss Optronik ATTICA third-generation thermal imaging cameras

Type
Commander's/gunner's sight.

Development/Description
The ATTICA product family are third-generation thermal imagers using a focal plane array detector with standard CCIR, RS-170 and digital video output.

Developed by Zeiss Optronik GmbH for various applications, the different performances of the camera system allows tailoring for individual user roles such as surveillance, reconnaissance, identification and targeting.

The system comprises a set of basic modules that can be combined with dedicated assemblies such as lens, mount, control unit, monitor and so on, to form the required infra-red imaging assemblage. The modules that comprise the ATTICA system are:

(a) IR objective - an athermalised JR objective with 2 field of views, which features a minimum number of optical components for the 3 to 5 μm or the 8 to 12 μm waveband
(b) detector/cooler assembly - which consists of various detector arrays based on Cadmium Mercury Telluride (CMT), Platinum Silicide (PtSi), Indium Antimonide (InSB) or Gallium Arsenide (GaAs) that are sensitive in the 3 to 5 μm and in the 8 to 12 μm waveband. An integrated cooler dewar ensures that the assembly is optimised for minimum power consumption
(c) driving and readout electronics - a highly integrated electronic module that generates all clock impulses and DC voltages required for the operation of the IR detectors and read-out circuit. In addition it synchronises the multiplexer and the processes taking place in the electronics with the synchronising signals produced by the system electronics
(d) videoelectronics - a high-performance video processing board with a data through put of more than 200 MB/s to do the image processing and the control the camera by means of RS-422 or CAN-Bus interface.

Operational status
Prototypes available.

Status
Development complete, ready for production.

Contractor
Zeiss Optronik GmbH.

UPDATED

Specifications: Zeiss Optronik ATTICA third-generation thermal imaging cameras:

	ATTICA-P256D	ATTICA-C384	ATTICA-C384μ	ATTICA-G640
Detector type	PtSi	CMT or InSb	QWIP or CMT	GaAs
Number of elements	256 × 256	384 × 288 384 × 256 320 × 240	384 × 288 320 × 240	640 × 480
Detector frame rate	25 Hz	100 Hz	100 Hz	25/50 Hz
Spectral waveband	3 – 5 μm	3 – 5 μm	8 – 12 μm	8 – 12 μm
Cooling method	rotary integral cooler	closed-cycle split linear Stirling cooler; rotary integral cooler	closed-cycle split linear Stirling cooler; rotary integral cooler	closed-cycle split linear Stirling cooler
Field of view	7 × 7°	3.6 × 4.8°/11.2 × 14.8°; 1.5 × 2°.4.6 × 6.2° others on request	3.6 × 4.8°/11.2 × 14.8° others on request	3.6 × 4.8°/11.2 × 14.8° others on request
Power input	5 V and 12 V	18 to 32V	18 to 32V	18 to 32V
Power dissipation	<10 W nominal	<35 W nominal	<35 W nominal	<80 W nominal
Dimensions	220 × 100 × 100	250 × 160 × 220	250 × 160 × 220	250 × 160 × 220
Mass	≈1.7 kg	≈5 kg	≈5 kg	≈7.5 kg
Video interface	8 bit digital	CCIR, RS-170, VGA: 8 bit digital	CCIR, RS-170, VGA: 8 bit digital	CCIR, RS-170, VGA: 8 bit digital
Mass	≈1.7 kg	≈5 kg	≈5 kg	≈7.5 kg
Data interface	RS-422	RS-422, CAN-BUS	RS-422, CAN-BUS	RS-422, CAN-BUS

Zeiss Optronik HDIR second-generation high-definition thermal imaging camera

Type
Commander's/gunner's sight.

Development/Description
The HDIR camera system is a very high-resolution second-generation thermal imager with video output according to HDTV standard. Developed by Zeiss Optronik for various identification and targeting applications where the highest level performance is required. The system comprises a set of basic modules that can be combined with dedicated assemblies such as afocal, mount, control unit, monitor and so on, to form the required infra-red imaging assemblage. The modules that comprise the HDIR system are:

(a) a scanning mirror – for both scanning directions and distortion-free scanning at high-scan efficiency
(b) IR imaging – an athermalised IR imager which features a minimum number of optical components. The image includes an intermediate image plane and provides virtual diffraction-limited performance on axis (image centre)
(c) detector/cooler assembly – this comprises 576 × 7 detector Cadmium Mercury Telluride (CMT) elements that are sensitive in the 7.5 to 10.5 μm waveband region. An integrated cooler dewar ensures a unit which is optimised for minimum power consumption
(d) driving and readout electronics – a highly integrated electronic module that generates all clock impulses and DC voltages required for the operation of the IR detector and readout circuitry. In addition they synchronise the IR multiplexer and the processes taking place in the electronics with the synchronising signals produced by the external system electronics. The module converts analogue signals to digital signals and feeds them to the digital scan converter
(e) digital scan converter – which performs image homogeneity, reformatting and video output generation
(f) system electronics – that provide master system control and ensures communication with a control unit
(g) afocal and TRA – the afocal defines the field of view of the systems. The thermal reference is used for scene related adaptation of the operating point and compensation of the variance of the detector characteristics.

Operational status
Prototype.

Specifications
Detector type: CMT
No of elements: 576 × 7
Waveband: 7.5-10.5 μm
Cooling method: split Stirling closed-cycle with integrated dewar
No of lines: 1,152
Pixels per line: 1,920
Scan frequency: 25 Hz
Fields of view (examples): 2.1 × 3.7°; 2.1 × 14.6° and 2.6 × 4.7°
Display: HDTV monitor
Power supply: 18-32 V DC

Contractor
Zeiss Optronik GmbH.

VERIFIED

ARMOURED FIGHTING VEHICLES

DRIVER'S SIGHTS

ATCOP DNS 3

Type
Driver's sight.

Description
The DNS 3 is a passive image intensifying night sight using second-generation image intensifying tubes. Primarily designed for T-series tanks, it can also be configured for other vehicles. It is effective to a range of 175 m. Features include a 'relaxed view' type eyepiece, unity magnification and bright source protection.

Operational status
Available.

Specifications
Magnification: ×1
Field of view: 27°, increased by rotation to 90°
Operating temperature: –30 to +52°C
Operating voltage: 24 V DC or 2 × 1.5 V AA batteries
Dimensions: 290 (H) × 152 mm (W)
Environmental specs: MIL-STD-810D

Contractor
Al Technique Corporation of Pakistan (Pvt) Ltd.

UPDATED

DNS 3 driver's night sight

BAE Systems Driver's Night Vision System (DNVS)

Type
Driver's sight.

Description
The Driver's Night Vision System (DNVS) from BAE Systems is a passive, multispectral, wide-angle indirect view observation system designed to enhance the operability of an Armoured Fighting Vehicle (AFV) by providing a 24-hour vision capability. The system consists of an armoured sensor unit, which is fitted to the hull or turret, a display and control unit, which are mounted internally and an optional low light reversing camera mounted at the rear of the vehicle.

The Driver's Sensor Unit (DSU) is a dual-channel device and houses an uncooled Thermal Imager (TI) operating in the 8-12 μm portion of the IR range and an optional auxiliary sensor. The auxiliary sensor can be either a colour day (CCDTV) camera or a low-light camera. The DSU is designed for both the UK and US uncooled imagers produced by BAE Systems.

The Driver's Night Vision System from BAE Systems ***NEW***/0547635

A wash/wipe system is provided to clear both channels of the sensor, and switchable heaters are incorporated in the windows to provide de-icing under very cold conditions.

The system is controlled using a membrane switch panel which is back illuminated for night-time operation and a small rugged Flat Panel Display is used to display the imagery. The control panel can be mounted separately or around the bezel of the display.

Among the key benefits ascribed to the DNVS by the company are its 24-hour all weather operating capability; its modular construction fits most AFVs; the near Instantaneous operation; and its low through-life costs.

A variant of the Driver's Sensor Unit is also being supplied as part of the Indirect Vision System for the British Army's new Titan AVLB and Trojan AEV being provided by Alvis Vickers under the Engineering Tank System (ETS) programme.

Operational status
In production for the ETS programme.

Specifications
Thermal Imager
Spectral band: 8 to 14 μm
Field of view: 50° horizontal
Detector type: uncooled micro-bolometer
Resolution: Not less than 320 × 240
System NETD: Typically <50 mK
Video format: CCIR or RS 170
Power consumption: 8W
Temperature (operating): –32 to +55°
Recognition of a Man: 100 m

Daylight Camera
Format: 1/3 in Colour CCD
Resolution: 752 × 582
Sensitivity: 0.8 to 100,000 Lux
Field of View: 64° horizontal

Low Light Camera
Resolution: 752 × 582
Sensitivity: 0.001 to 100,000 Lux
Field of View: 60° horizontal

Options available
Integrated Auxiliary Sensors
- CCD Colour Daylight Camera
- Low-Light Camera

Reversing Camera
Steerable platform to provide >110° horizontal FOV

Contractor
BAE Systems Avionics Ltd, Land Vehicle Systems.

NEW ENTRY

BAE Systems SS130 passive night driving periscope

Type
Driver's sight.

Description
The SS130 night driving periscope has been developed to fit into many armoured vehicles in service with the UK and other armed forces and can also be adapted to view through the windscreen of conventional vehicles such as Land Rovers.

It comprises an objective channel tube with a microchannel image intensifier tube and a biocular fixed focus eyepiece. The latter enables the driver to use both eyes for driving. A thermostatically controlled heater is also fitted.

The provision of an iris diaphragm allows the emergency use of the sight during daylight hours. Power is provided through the vehicle supply although a trickle-charged battery is fitted within the sight for back-up purposes.

Operational status
Production. In service with the British Army and other countries.

Specifications
Dimensions: 235 × 175 × 340 mm
Weight: 7 kg
Magnification: ×1
Field of view
Horizontal: 48.7°
Vertical: 40°
Power supply: 24 V DC vehicle system
Performance
Detection: 1.1 km
Recognition: 275 m
Identification: 130 m

Contractor
BAE Systems, Avionics, Sensors and Communications, Basildon.

VERIFIED

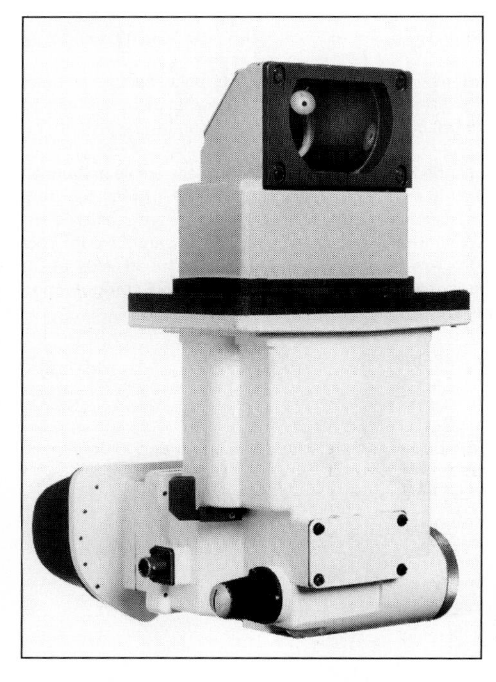

The BAE Systems SS130 night driving periscope
0004791

Econ Industries Polyphimos driver's periscope

Type
Driver's sight.

Description
The Polyphimos driver's periscope is a night vision viewer for the drivers of closed-hatch combat vehicles. It is a biocular display viewer with an integrated image intensifier. Current models include the Polyphimos II based on the AN/VVS-2 design and Polyphimos III based on the AN/VVS 503 design. With the use of the appropriate vehicle-mounting adaptor either model can be fitted to a variety of main battle tanks and armoured vehicles.

Operational status
Available.

Specifications
Model: Polyphimos II
System resolution: 0.81 lp/mrad

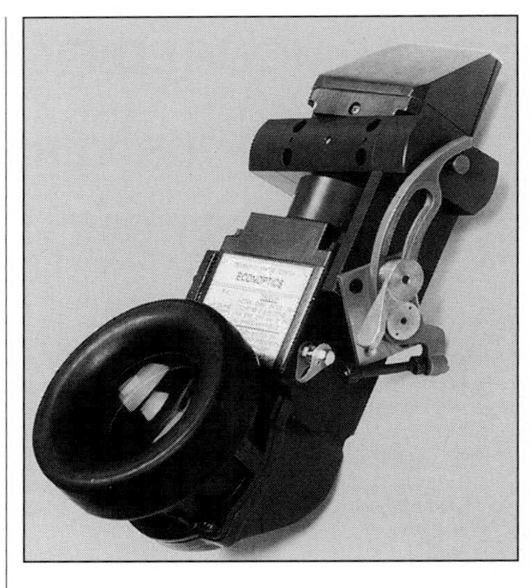

The Polyphimos II driver's night periscope from Econ Industries

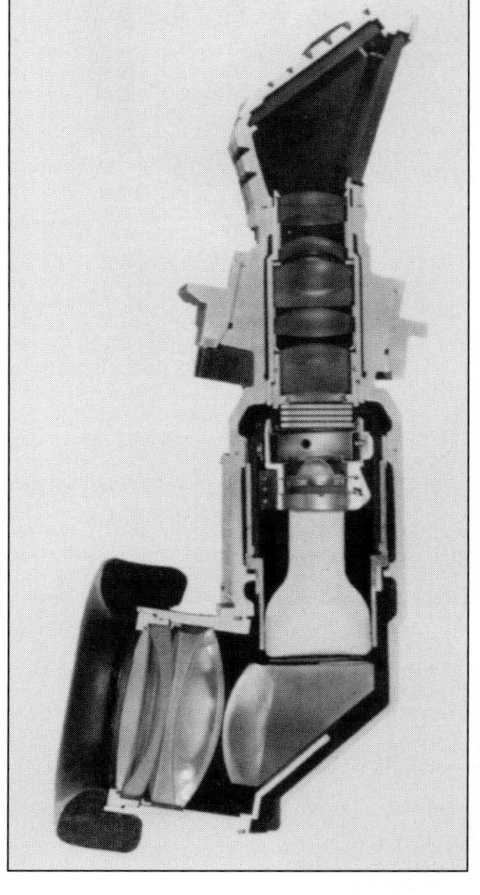

Cutaway view of Polyphimos II
0055262

Image intensifier: 25 mm GEN II with automatic gain control and bright source protection
Field of view: 44 × 35°
Depth of focus: 4 m to infinity
Magnification: ×1
Linear distortion: <13% from an on-axis pupil at 50 mm eye relief
Eyepiece focus: 2.0 dioptres on-axis from 50 mm eye relief

Contractor
Econ Industries S A.

VERIFIED

ENOSA PCN-150 night driving periscope series

Type
Driver's sight.

Description
The PCN-150 series of night driving periscopes is supplied according to the type of head (prism, swivel or tilting model) required by the type of armoured vehicle to be fitted.

PCN-151 driver's periscope from the rear

The current family members are:

(a) PCN-151 with internal azimuth swivel mounting for use on the BMR-600 IFV
(b) PCN-152 which can be used to replace the M24 periscope on the M48 series MBT
(c) PCN-153 which is used to replace the M19 periscope on the Spanish Army M113 TOA APC
(d) PCN-154 which can be used to replace the OB 31A periscope on the AMX-30E MBT.

All are fitted with a second-generation 25 mm image intensifier tube with automatic and manual gain controls. The driver views through a biocular type screen which is fitted with an electro-optical protection system to prevent the illumination level becoming too high.

Operational status
Production as required. In service with the Spanish Army.

Specifications
Dimensions
PCN-151: 155 × 140 × 410 mm
PCN-152: 155 × 200 × 410 mm
PCN-153: 155 × 200 × 396 mm
PCN-154: 155 × 193 × 410 mm
Magnification: ×1
Field of view
Horizontal: 44°
Vertical: 35°
Resolution: 1.25 mrad
Depth of focus: 4 m to infinity
Image intensifier tube: 25 mm GEN II
Power supply: 24 V DC
Emergency power supply: 1 × 2.7 V BA1567/U battery

Contractor
Empresa Nacional de Optica SA (ENOSA).

VERIFIED

ENOSA PCN-160 night driving periscope

Type
Driver's sight.

Description
The PCN-160 is a night driving biocular periscope for light wheeled or tracked armoured vehicles with a second-generation image intensifier. The main components of the periscope are a head prism, an objective, an image intensifier, a biocular eyepiece, a motorised shutter for tube protection and a converter to power the tube from either the vehicle supply or batteries. The image intensifier tube is protected against bursts of high-intensity luminosity. The device can be powered from the vehicle supply or by battery.

Operational status
Production as required. In service with the Spanish Army.

Specifications
Magnification: ×0.85 ±0.1
Horizontal field of view: ≤48°

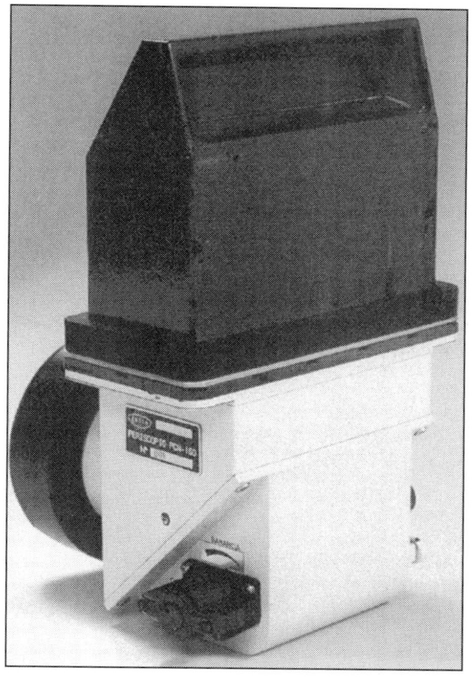

ENOSA PCN-160 night driving periscope

Verticle field of view: ≤44°
Dioptre setting: –2 ±0.3 dioptres
Focus range: 2.5 m to infinity
Resolution: ≤0.7 lp/mrad (0.1 lx)
Resolution: ≤0.5 lp/mrad (0.001 lx)
Image intensifier: GEN II
Eyepiece: biocular
Power supply, vehicle service: 24 V DC
Power supply, battery: 3 V DC

Contractor
Empresa Nacional de Optica SA (ENOSA).

UPDATED

Fotona COmbined day/night DRIver's periScope (CODRIS/CODRIS-E)

Type
Driver's sight.

Description
CO(mbined) DRI(ver's) (peri)S(cope) provides the driver of an armoured vehicle with the capability of driving during day and night without a need to change his viewing device under different light conditions. CODRIS combines a wide angle

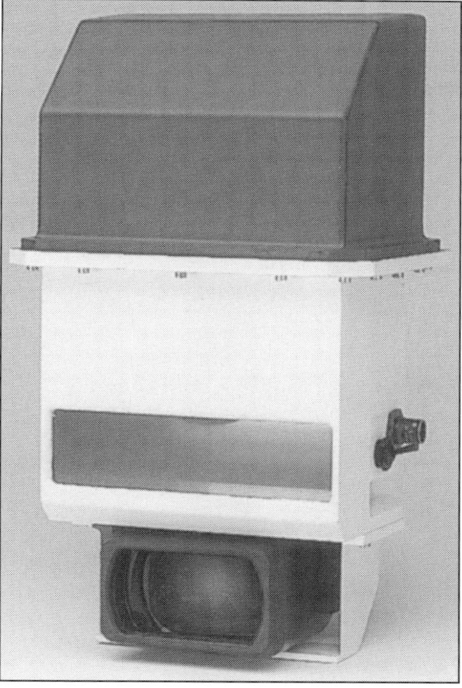

CODRIS showing driver's viewer
0055265

day channel prism with an image intensifying night channel equipped with a GEN II image intensifier tube. Both channels are mutually independent and non-excluding.

The night channel can be rotated manually in azimuth by ±20° providing a horizontal field of view, wide enough for safe night driving. The night image is displayed via a large aperture biocular ensuring simultaneous vision for both eyes.

The day channel is protected against incoming laser irradiation by a special filter.

Operational advantages inherent to the combined day/night design are:

(a) No breaking of the driver's compartment seal, no exposure to NBC contamination from the outside environment during exchange periscopes
(b) No damage to the equipment due to the exchange of day and night periscopes under battlefield conditions
(c) No need for additional storage space for the unused periscope
(d) If the night channel is temporarily blinded due to high light levels (explosions and so on) driving is still possible using the day channel.

Specifications

		CODRIS	CODRIS-E
Day channel			
Field of view	horizontal:	80°	120°
	vertical:	25°	
Magnification		×1	
Protection against incoming laser pulses	A filter is built-in to protect the driver's eyes against incoming laser irradiation. The attenuation of the filter is equal to or greater than 6 × 10^{-4} at the wavelength of 1.06 μm		
Night channel			
Field of view	total horizontal:	80°	
	instantaneous horizontal:	40°	
	rotation in azimuth:	±20°	
	vertical:	30°	
Magnification		×0.9	
Objective focal length		29 mm	
Relative aperture		1.3	
Focus depth		10 m to infinity	
Resolution (contrast = 0.9)		2.0 mrad/lp (at 10^{-2})	
		1.6 mrad/lp (at 10^{-1})	
Biocular focal length		50 mm	
Image diameter		30 mm	
Image intensifier tube		20/30, GEN II	
Protection		Automatic shutdown in case of overexposure	
Power supply		18 to 32 V DC	
Operating temperature		–30 to +52°C	

Contractor
Fotona dd, Slovenia.

VERIFIED

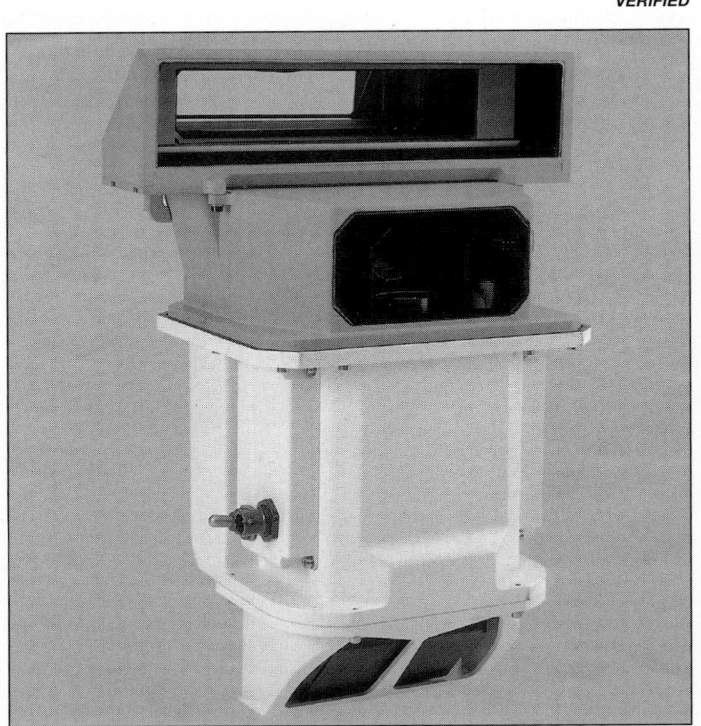

A Fotona CODRIS-E variant showing sight window 0058940

Fraser-Volpe M19A1 driver's night periscope

Type
Driver's sight.

Description
The M19A1 periscope sight is a sealed unit which enables the driver to see at night. Using an image intensifier in the sight combined with an infra-red illuminator, the terrain in front of the vehicle is clearly visible. Construction is in two parts, the head assembly and the body assembly. The head projects above the armour and contains optics to direct the incoming scene to the intensifier and biocular eyepiece. The device has been adapted to a number of combat vehicles in the US Army inventory. These include the M113 personnel carrier, the M114 command and reconnaissance vehicle, and M47 and M48 MBTs.

Operational status
In service.

Specifications
Dimensions (including head assembly) (h × w × d): 368 × 212 × 76 mm
Weight (including head assembly): 6.35 kg
Magnification: ×1
Objective focus (nominal): infinity
Resolution: <4 mins of arc
Dioptre adjustment: –0.75 to –1

Contractor
Fraser-Volpe Corporation.

VERIFIED

Galileo Avionica P192 night driving periscope

Type
Driver's sight.

Description
The P192 is designed to be fitted on various types of armoured vehicle including the Fiat 6616, VCC 1, Marder 1, M41, M47, M48, M60, Leopard 1 and M113.

It consists of a head prism, S 25 microchannel image intensifier tube and a biocular eyepiece assembly. The image intensifier is fitted with a gain control that automatically cuts out when the photocathode illumination level becomes too high.

A filter allows observations to be made within the luminance range of 10^{-3} to 10^{4} lx.

The biocular arrangement enables the driver to view the scene with both eyes through the single large-diameter eyepiece and allows the vehicle to be driven at speeds of up to 40 km/h in starlight conditions without any artificial illumination.

Operational status
Production as required. In service with unspecified countries.

Specifications
Weight: 7 kg
Magnification: ×0.9
Field of view
Horizontal: 48°
Vertical: 40°

Contractor
Galileo Avionica (formerly Alenia).

UPDATED

P192 night driving periscopes showing different head arrangements

Hellenic Aerospace AN/VVS-2(V) driver's night vision viewer

Type
Driver's sight.

Description
The AN/VVS-2(V) night vision driver's viewer is a through-the-hatch periscope for tanks and other armoured vehicles, including M-151, M-109, M-113, M-60, M48A5, Leonidas and BMP-1. The viewer uses a third-generation image intensifier tube. This is a licence-built version of the Northrop Grumman AN/VVS-2 (V) (qv).

Operational status
Available.

Specifications
Weight: 7 kg
Height: 45 cm (15 cm above hull, 30 cm below hull)
Field of view (w×h): 800 × 680 mrad
Azimuth: ±45°
Elevation: fixed
Focus: fixed
Depth of field: 5 m to infinity

Contractor
Hellenic Aerospace Industry SA.

The Hellenic Aerospace AN/VVS-2(V) night vision driver's viewer 0004793

Institute of Optronics Pakistan DNVP-1 driver's night vision periscope

Type
Driver's Night Vision Periscope.

Description
The driver's night vision periscope is a biocular display viewer used for closed hatch, as well as open hatch driving of combat vehicles. It consists of an entrance prism assembly, objective lens assembly, an eyepiece assembly and a main housing assembly which includes an image intensifier/magnifier assembly and a rechargeable battery assembly.

The viewer can be used by drivers of tanks, armoured personnel carriers and other combat vehicles. The apparatus picks up reflected light from targets of interest under moonlight, starlight and sky glow conditions, intensifies this light and produces an image viewed through the eyepiece.

Operational status
In production.

Specifications
Dimensions (l x w x h): 150 × 170 × 362 mm
Weight: 7 kg
Magnification: ×1
Slew angle: ±35° (azimuth)
Field of view: 45°
Operating Temperature: –54 to + 52°C
Image intensifier tube: 25 mm
Objective lens focus: fixed at 15 m

Driver's night vision periscope DNVP-1A 0089871

Eyepiece dioptre: fixed at –2.0
Resolution: 1.25 mrad
Power: AA Batteries

Contractor
Institute of Optronics Pakistan.

VERIFIED

Kintex TVN-2B night driving device for armoured vehicles

Type
Driver's sight.

Description
The TVN-2B active night driving device is designed for land viewing during night driving of armoured vehicles. The complete system set comprises the TVN-2B device, infra-red illuminating light and separate parts, and a toolkit.

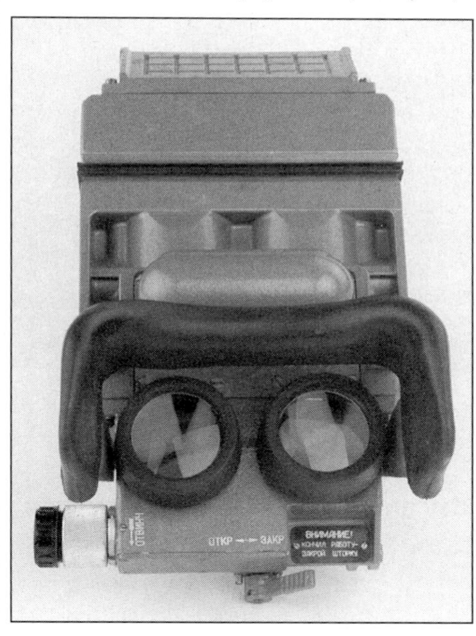

TVN-2B night driving device for armoured vehicles 0004790

Operational status
Production as required. In service with the Bulgarian Army and other undisclosed countries.

Specifications
Type: active
Weight: 5.2 kg
Magnification: ×1
Field of view: 30°
Wavelength coverage: near IR
Bright vision range: 60 m
Periscope effect: 212 mm

Contractor
Kintex.

Meopta NV-3P night vision device

Type
Driver's sight.

Description
The NV-3P passive night vision periscope is designed for drivers of armoured vehicles to drive at light levels as low as 1×10^{-3} lx or full darkness with additional floodlighting. The stereoscopic optical system of the periscope with two second-generation image intensified channels enables easy observation of the terrain without causing eye fatigue. The outer optical surfaces of the device and the eyepieces are protected from condensation and ice accretion by electronic heaters.

Operational status
Available.

Specifications
Dimensions: 300 × 177 × 146 mm
Weight: 6 kg
Magnification: ×1 to ×1.2
Field of view: 33° V × 35° H
Effective viewing range at 3×10^{-3} to 5×10^{-3} lx (starlight): 180 to 400 m
Effective viewing range at 3×10^{-2} to 5×10^{-3} lx (moonlight): 400 to 500 m
Dioptre adjustment: ±3 dioptres
Nominal focus of objectives: 30 m
Power input: 3 W (90 W with heating on)

Contractor
Meopta Prerov, a.s.

UPDATED

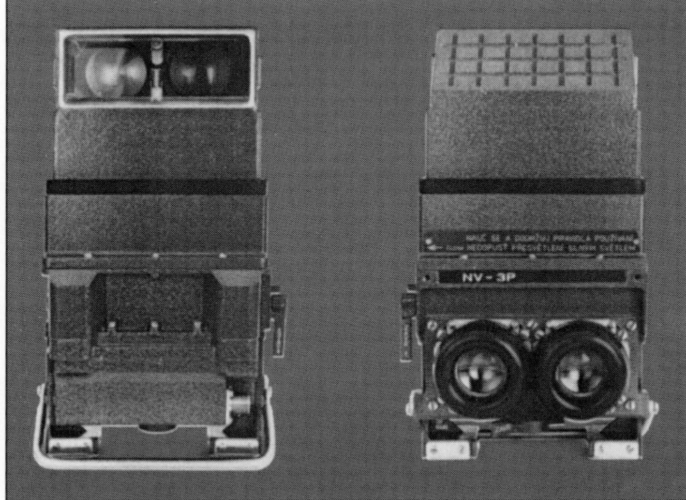

Meopta NV-3P night vision device for drivers 0023048

MES VG/DIL 186 day and night driver scope system

Type
Driver's sight.

Description
The VG/DIL 186 driver scope system has been designed to satisfy the technical and operational requirements of driving an armoured fighting vehicle in both day and night conditions.

The VG/DIL 186 driver scope system 0055264

The available models are the VG/DIL 186-C1 for the IVECO/Otobreda C1 Ariete MBT and the VG/DIL 186-B1 for the IVECO/Otobreda Centauro B1 (8 × 8) tank destroyer.

The driver scopes consist of the following modules:
(a) binocular VO/IL 186 passive night vision goggles which use two, second-generation, 18 mm, microchannel wafer-type image intensifier tubes that have automatic brightness control and bright source protection
(b) type MES 82/1 or M17/1 daylight iposcope
(c) goggles and iposcope interface.

This arrangement allows the vehicle to be driven in the following ways:
(a) at night with the hatch closed and the goggles and interface unit installed on the iposcope
(b) at night with the hatch open and the driver wearing the goggles only
(c) during the day with the hatch closed and the driver using the iposcope only.

Operational status
Production. In service with the Italian Army on C1 Ariete MBT and Centauro B1 tank destroyer.

Specifications

Type	VG/DIL 186-B1	VG/DIL 186-C1	VO/IL 186
Magnification	×1	×1	×1
Weight	5.00 kg	4.50 kg	0.90 kg
Field of view			
Horizontal	38°	38°	38°
Vertical	30°	35°	30°
Dioptre range	–6 to +2	–6 to +2	–
Focus range	25 cm to infinity	25 cm to infinity	25 cm to infinity
Resolution (USAF target 90% contrast)			
10^{-3} lx	3.2 mils	3 mils	3 mils
10^{-2} lx	2.2 mils	2 mils	2 mils
10^{-1} lx	1.7 mils	1.5 mils	1.5 mils
Power supply	2.2-3 V DC	2.2-3 V DC	2.7 V battery

Contractor
Meccanica per l'Elettronica e Servomeccanismi SpA (MES).

VERIFIED

Norinco Type TDPN-2 driver's night viewer

Type
Driver's sight.

Description
The Norinco Type TDPN-2 driver's night viewer is a second-generation biocular single-tube, image intensifier system of the periscopic type for use in tanks or other AFVs.

Norinco tank driver's night viewer Type TDPN-2

The biocular display allows the driver to view the image intensified screen at a distance of 50 mm from the large diameter eyepiece with both eyes while remaining seated in the normal driving position.

In physical appearance, components and operation the TDPN-2 is almost identical to the Israeli Elop company's Compact Driver's Night Viewer and may well be a licensed copy.

Operational status

Production as required. In service with the People's Liberation Army and other undisclosed countries.

Specifications

Weight: 7 kg
Length: 360 mm
Width: 162 mm
Depth: 186 mm
Field of view: 35°
Traverse adjustment: 30° left and right
Focus setting: fixed
Magnification: ×1

Contractor

China North Industries Corporation (Norinco).

VERIFIED

Northrop Grumman AN/VVS-2 and AN/VVS-1924 passive driver's night vision viewer

Type

Driver's sight.

Description

AN/VVS-2 was initially developed for use on the M60 MBT family but is now used in the M1 Abrams under the designation AN/VVS-2(V)2A, the M60A3 (AN/VVS-2(V)1A version) and the M2/M3 Bradley vehicles as the AN/VVS-2(V)3.

It uses a second-generation image intensifier tube system which provides protection from bright lights, as well as shell bursts and flares.

For other US AFVs such as the M41/M47 and the AAV P7, the AN/VVS-1924 has been developed as a private venture. This is an exact replacement for the M19 and M24 active infra-red viewers. No vehicle modifications are required and the installation can be performed in the field if necessary. Any vehicle that uses the AN/VVS-1924 mount is also field-retrofitted with a modified tank hatch insert that will accept the internal azimuth rotation mount. AN/VVS-1924 provides a rotatable 45° field of view and incorporates a similar second-generation image intensifier tube.

AN/VVS-2 consists of an objective lens assembly, a biocular eyepiece assembly, an entrance window housing assembly, a main housing assembly (which includes an electronic power control and an electric power adaptor), the low-light level image intensifier tube and a mount assembly. It provides the same night vision capabilities to the driver as the M32E1 and M35E1 gunner's night vision periscopes and the M36E1 commander's night vision periscope systems.

Operational status

In production. AN/VVS-2 series is in service with the US Army on the M1A2 (M1-M1A1), M113, M60A3, Bradley, M109 vehicles. AN/VVS-1924 is in service with Israel and other unspecified countries.

Specifications

Field of view: 45 × 38°
Total coverage: 135° horizontal; 38° vertical
Focus: 4 m to infinity

AN/VVS-1924 driver's passive night periscope

Resolution: 1.2 mils
Magnification: ×1
Distortion: <4%
Power supply: 24 V DC vehicle

Contractor

Northrop Grumman Electro-Optical Systems.

VERIFIED

Northrop Grumman AN/VVS-501 passive night driving viewer

Type

Driver's sight.

Description

AN/VVS-501 was developed for use by the Canadian Forces on the Leopard 1 MBT and the Armoured Vehicle, General Purpose (AVGP) family of wheeled vehicles. On the former it replaces the day vision blocks of the commander's and driver's stations while on the latter it replaces the identical day vision block of the driver's station. When not in use it is kept in a purpose-built stowage case. Other fits include the Leopard 2 MBT.

AN/VVS-501 passive night driving viewer from driver's side

The biocular eyepiece assembly is identical to that used on the US Army's AN/VVS-2 system (qv). The night vision capability is provided by a single second-generation 25 mm microchannel image intensifier tube with automatic gain control and bright source protection.

It can be upgraded with a Gen III tube for a 60 per cent improvement in system resolution, higher sensitivity to low light levels, and better spectral performance in the near infra-red region. The viewer is mounted in the driver's centre day periscope, opening for use during night operations. There are no vehicle modifications required to install the unit.

The AN/VVS-501 is also produced under licence by Thales Optronic Canada – formerly the Electro-Optics division of AlliedSignal Canada.

Operational status
Production. In service with the Canadian Forces (on AVGP vehicles and the Leopard 1 MBT) and other unspecified countries.

Specifications
Field of view: 44°(h), 35°(v)
Field of regard: 132°(h)
Resolution: 0.81 p/mm
Eye relief: 50 mm
Depth of field: 4 m to infinity
Magnification: unity
Power source: vehicle battery DC 28 V or DC 2.7 – 3.0 V Lithium battery

Contractor
Northrop Grumman Electro-Optical Systems.

UPDATED

Optechs Korea Inc AN/VVS-2(V) III, (V)IV, (V)1A, (V)M1924, (V)2-A

Type
Driver's sight.

Description
The night driver's viewer AN/VVS-2(V) is an image intensified night driving periscope for combat vehicles. In December 1987, a development contract was placed by the South Korean Army for use on the K-1 tank. First production tests were completed in September 1989. Production started in September 1990.

Designations for the AN/VVS-2(V) include:
- AN/VVS-2(V) M1924 - M113, M47, M48, M60
- AN/VVS-2(V) 2A - M1, M1A1, M1A2 MBTs
- AN/VVS-2(V) III - Korean self-propelled 155 howitzer
- AN/VVS-2(V) IV - Light armoured vehicle, NBC, Flying Tiger
- AN/VVS-2(V) 1A - M60 MBT, M113A3

AN/VVS-2(V) provides a biocular display for the driver to view with unity magnification through one large-diameter eyepiece. The electronic control assembly disengages power to the image intensifier when the ambient brightness exceeds a specified level, typically 4 Foot Lamberts.

Operational status
In production and in service with the South Korean Army.

Specifications
Field of view: 44 × 35°
Resolution: 1.25 mrad
Depth of field: 4 m to infinity
Magnification: ×1

Contractor
Optechs Korea Inc.

VERIFIED

AN/VVS-2(V) III, IV and 1A 0023049

PCO RADOMKA driver's periscope

Type
Driver's sight.

Description
The PCO driver's passive night vision periscope is a small binocular, wide-angle viewing system for use in a wide variety of armoured vehicles including tanks and AFVs such as the T55 and T72 series of tanks and BWP-1, BWP-2 technical support vehicles.

The periscope provides the driver with a direct and constant view of the road or terrain in front of the vehicle under starlight conditions. It contains two independent viewing channels - each with an objective lens, an eyepiece and an 18 mm image intensifier tube. The system enables stereoscopic viewing. The tubes have a stabilised power supply, automatic bright light protection and automatic gain control.

Operational status
In production and service in Poland and other countries.

Specifications
Periscope dimensions: 293 × 166 × 145 mm
Magnification: ×1
Field of view: 30°
Power supply: 18-30 V DC

Contractor
PCO SA, Poland.

RADOMKA driver's periscope 0023051

PCO ZZT-1 CYKLOP Terrain Display Set

Type
Driver's sight.

Description
The ZZT-1 CYKLOP driver's sight and display is intended for:
- Displaying the terrain in front of and behind the vehicle on the monitor placed at the driver's position, during day and night with dosed hatches, leading to greater safety.
- Improvement in safety during driving in reverse with closed hatches through the ability to view terrain behind the vehicle by day and night.
- Workload reduction for the driver during driving with closed hatches.
- Increasing the safety for crews during training in crossing deep water obstructions through the ability of maintaining a view below the water surface.

Principal components:
- Day - night monochromatic camera – 2 pieces
- Day monochromatic camera

Driver monitor
- Power supply
- Wiper assembly
- Fluid tank
- Connecting cables set

Set of mechanical holders for rescue and air RT3 tubes

Operational status
Prototype.

Specifications

Operational voltage: 18 – 30 V
Power consumption in stable state: <75 W (max <90 W)

	Day – night camera	Portable camera
Field of view	57 × 44° (optional 40 × 30°)	70 × 52.5°
Focusing range	10 m to infinity	10 m to infinity
Usable range of illuminance	5 m lx – 50,000 lx	0.5 lx – 50,000 lx
Dimensions	120 × 130 × 310 mm	115 mm (diam) × 220 mm
Weight	approx 6 kg	approx 3.5 kg

Monitor
Type: 7 in LCD, 854 × 480 pixels
Resolution: 400 lines
TV signal standards: CCIR: RS-170
Dimensions: 80 × 150 × 270 mm
Weight: approx 5 kg

Power supply
Dimensions: 80 × 140 × 170 mm
Weight: approx 1.8 kg
Usable capacity of windscreen tank 2.25 l

Contractor

PCO SA, Poland.

Raytheon AN/VAS-3 Driver's Thermal Viewer (DTV)

Type

Driver's sight.

Description

AN/VAS-3 Driver's Thermal Viewer (DTV) enables armoured fighting vehicle drivers to manoeuvre at normal daytime driving speeds in total darkness, all weather and degraded visibility conditions including dust, smoke and any other battlefield obscurants. It is a non- developmental item that is in production for foreign military sales to Kuwait for installation in that country's M1A2 Abrams tank.

In addition to aiding vehicle mobility, the DTV provides a surveillance capability during silent watch and has a sufficient range for close-up target acquisition. It is designed to be rugged and survivable under extreme environmental conditions.

A fully equipped, highly automated facility is in place at Raytheon for high-rate production. Logistics support equipment is also available.

Operational status

In service. Some 231 systems were produced for foreign military sale to Kuwait for installation on the M1A2 Abrams MBT.

Specifications

Spectral band: 7.5-12 μm
Field of view: 40 azimuth × 20° elevation
Field of regard: 100 azimuth × 40° elevation
Magnification: unity

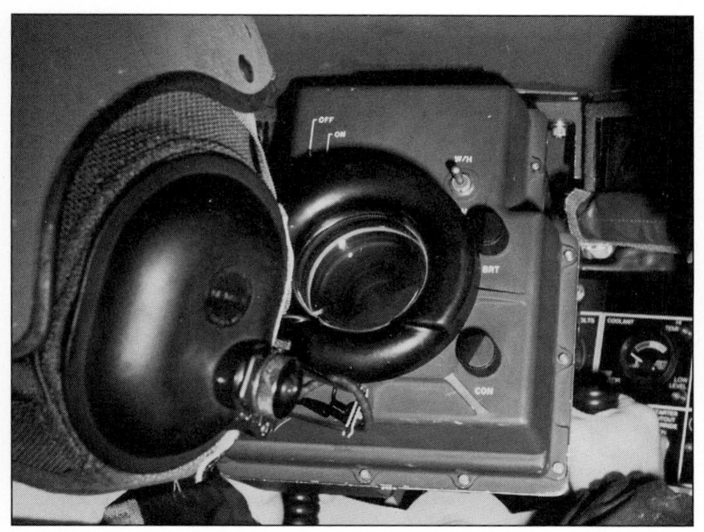

Raytheon driver's thermal viewer shown installed in a Bradley Fighting Vehicle

Detector/Dewar DT-591A: 60-element CMT
Cooling: split Stirling 0.25 W
Frame/field rate: 25/100 Hz
Display: electrostatic CRT
Aspect ratio: 2:1
Display viewer: biocular
Controls
Power: rotary switch off/on standby
Contrast: potentiometer
Polarity: 2-position switch White Hot/Black Hot
Brightness: potentiometer
Indicators (detector ready): green LED
Power, 18 to 32 V DC vehicle: <55 W
Weight: < 12.8 kg
Size: 7.4 l
Vehicle interchangeability: M1, M2/M3, M109, M113, LAV, AGS, with vehicle interface adaptor
Operational readiness: <10 min
Reliability MTBF: >1,080 h
Survivability/vulnerability: designed for ballistic protection, nuclear and directed energy

Contractor

Raytheon Systems Company.

UPDATED

Raytheon AN/VAS-5 NIGHTSIGHT™ Driver's Vision Enhancer (DVE)

Type

Driver's sight.

Description

The DVE is an uncooled 8 to 12 μm thermal imager designed to provide a low-cost driver's night vision option for tracked and wheeled vehicles. It is a designated Horizontal Technology Integration (HTI) system, allowing across-the-fleet insertion of a common DVE in armoured and tactical wheeled vehicles.

The DVE is composed of a sensor module and display control module. The sensor module contains the imaging optics, uncooled detector and vehicle sensor electronics. The display module provides eight bit video and user interface controls and an advanced flat panel display. This standard module set, when used with the appropriate mounting bracket, can be integrated in armoured vehicles such as Bradley, Abrams, LAV and M113. When mated with the pan/tilt for 360° azimuth situational awareness capability, the DVE can be configured to fit tactical vehicles such as HMMWV, HETS, HEMTT and PLS.

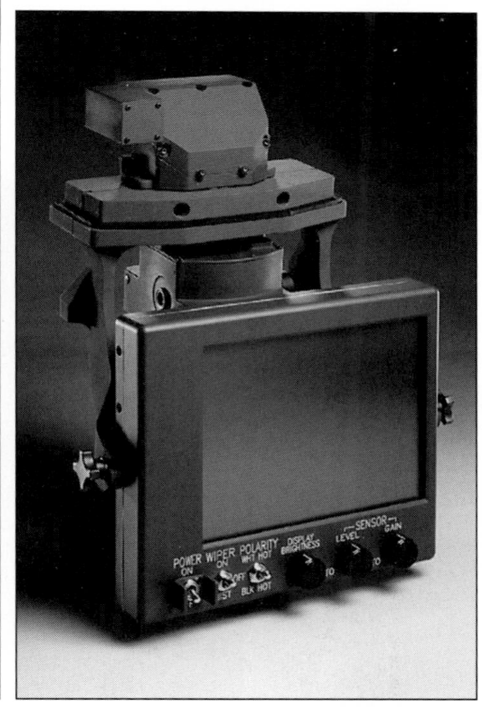

Raytheon AN/VAS-5 NIGHTSIGHT™ driver's vision enhancer
0023050

DVE is part of the NIGHTSIGHT™ family of thermal imaging products for both military and commercial end-users. The product line currently features weapons sights and systems for surveillance and situational awareness (Model 200 and PalmIR hand-held series) NIGHTSIGHT™ products share common uncooled thermal imaging focal plane array and core electronics components.

Operational status
In production. Deliveries of a first order of over 1,189 units began in mid-1996. As a result, a further three year US Army contract was awarded in 1998.

Specifications
Weight
Sensor: 1.41 kg
Display: 3 kg
Spectral band: 8-12 μm
Detector: 320 × 240 uncooled ferroelectric staring array
Field of view (v × h): 30 × 40°
Field of regard: manually slew ±30° azimuth and +15° elevation
Focusing range: 5 m to infinity
Display: RS-170 9.6 in flat panel (640 × 480 active matrix liquid crystal display)
Video output: EIA-RS-170
Digital interface: RS-232/422
Power: 16-32 V DC, 28 V DC nominal

Contractor
Raytheon Systems Company.

UPDATED

Thales Optics passive night vision driving periscope

Type
Driver's sight.

Description
This passive night vision driving periscope is designed to be interchangeable with the existing day periscopes of the following vehicle types without modification: Leopard 1 MBT, Chieftain (ARV, ARRV and MBT), Challenger 2, CRAAV, Challenger 1, Scorpion, Scimitar, Striker, Spartan, Samson, Samaritan, FV430 series, M107, M109, M110, M113 APC series and M60 MBT series. With some modifications it can also fit the AMX-30 MBT and the Panhard AML armoured car.

Assembly is effected by special vehicle mounting brackets and upper casings.

The periscope incorporates a single image intensifier tube and a biocular eyepiece with wide field of view. An iris control is incorporated in the periscope, allowing the operator manually to set the desired image brightness, while an automatic brightness control restricts the effects of bright light sources such as flares, shell bursts and headlights in the field of view.

Operational status
In production. In service with the British Army (the periscope is standard equipment on all British Army tracked vehicles) and several unspecified countries.

Specifications
Weight: 6-10 kg according to the vehicle type
Magnification: ×1
Field of view (h × v): 50 × 40°
Eye relief: 120 mm max, 35 mm min
Power supply: 28 V DC (vehicle supply)

Contractor
Thales Optics Ltd.

VERIFIED

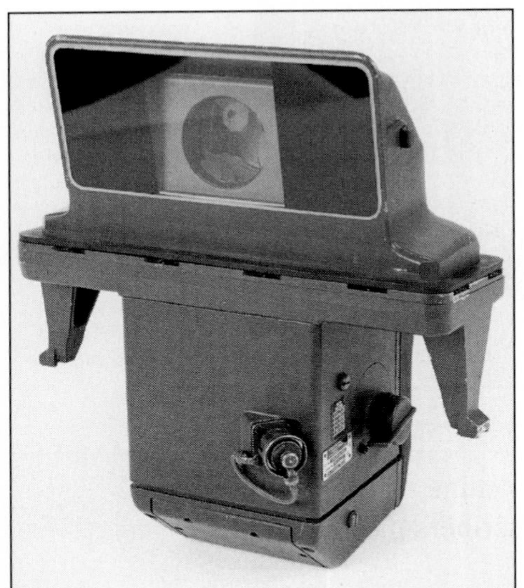

The Thales Optics night driving periscope
0004794

Thales Optronics AN/VVS-501 passive night driving viewer

Type
Driver's sight.

Description
The AN/VVS-501 night vision periscope (built under Northrop Grumman licence) fits a variety of military vehicles and is primarily used for night driving and forward observation. Applications include the Leopard 1 tank, the M113 APC, a variant of the Piranha APC and many other vehicles. The periscope is currently used by the Canadian Forces. The design allows for simple installation by replacing standard vision blocks in either the turret or driver's compartment of most armoured fighting vehicles.

Operational status
Production as requested. In service with Canadian Forces.

Specifications
Weight: 7.2 kg
Magnification: ×1
Field of view: 45 × 38°
Total coverage: 135 × 38°
Power source: battery DC 2.7 V, external 18-32 V
Focus depth: 4 m to infinity

Contractor
Thales Optronics, Canada (formerly the E-O Division of AlliedSignal).

UPDATED

AN/VVS-501 from Thales Optronics

Thales Optronics OB-60 day/night driver's periscope

Type
Driver's sight.

Description
The OB-60 driver's periscope system has been chosen by the French Army to equip its Leclerc MBT. The OB-60 belongs to a family of day/night periscopes which includes large field of view systems (by internal night channel rotation) for the AMX-30 and AMX-10 and compact systems (without internal night channel rotation) for the Leopard 1/Leopard 2 MBT or M113 armoured vehicle family.

This concept of day/night periscopes provides additional operational advantages, particularly in NBC conditions whereby:

(a) no breaking of the driver's compartment seal occurs
(b) there is no need to stop the vehicle to exchange the periscope for a day or night version so that no risk of damage to equipment occurs
(c) there is no need for any storage space for the unused periscope.

The OB-60 combines a conventional day channel prism and a passive night image intensifier channel with the capability to switch instantaneously from one vision mode to the other.

Although the periscope housing remains fixed, the ±30° azimuthal turning ability of the night vision channel provides a 100° horizontal field of view for night vision operations. For the latter, a second-generation image intensifier tube is used whereby the image is visualised into a large aperture biocular to ensure simultaneous vision to both eyes.

Other system capabilities include:

(a) upper window defrosting
(b) lower window defrosting

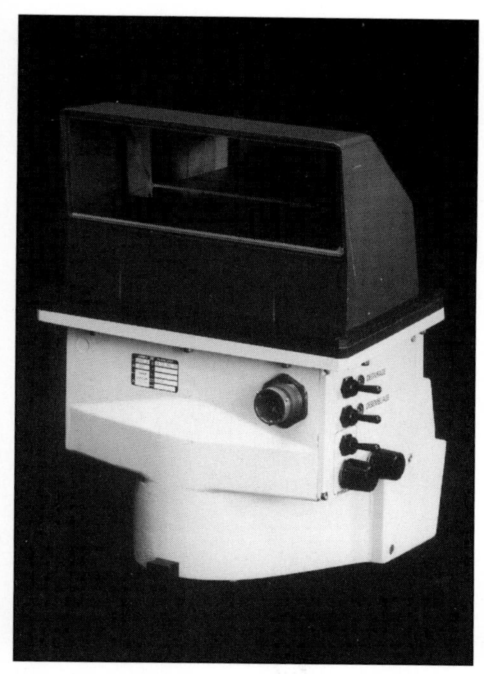

Thales Optronics OB-60 driver's periscope

(c) autolaser filter on day channel
(d) automatic cut-off and shielding of the image intensifier tube in day vision mode
(e) automatic cut-off of the image intensifier tube in event of prolonged exposure to high light levels in night vision mode.

For emergency use the periscope has an image generating unit through which six different message symbols can be displayed in the lower half of the driver's field of view. These symbols are sent by the vehicle commander and relate to emergency driving instructions.

Operational status
In production. In service with the French Army on the Leclerc MBT.

Specifications
Day channel
Field of view
Horizontal: 80°
Vertical: 25°

Night channel
Magnification: ×0.92
Field of view
Full horizontal azimuthal rotation: 100°
Instantaneous horizontal: 45°
Instantaneous vertical: 30° (45° total)
Resolution (contrast = 0.9)
10^{-1} lx: 1.2 mrad/lp
10^{-3} lx: 2 mrad/lp
Focus depth: 10 m to infinity
Biocular focusing: 0.5 dioptre
Gain: >2,000

Contractor
Thales Optronics.

VERIFIED

Thales Optronics Viper Driver's Viewer Aid (DVA)

Type
Driver's thermal sight.

Description
The DVA is a passive infra-red real-time thermal imaging system operating in the 8 to 12 μm spectral band, configured as a biocular display driver's viewer for combat vehicles. The DVA is designed to provide the driver with the capability of continuing normal driving operations during conditions of darkness and in a battlefield environment of degraded visibility, including smoke, haze and dust. The DVA is based on Thales Optronics' proven technology, the hand-held Cobra thermal imager and the AN/VVS-501 night vision periscope systems. The design allows for simple installation by replacing standard vision blocks in either the turret or the driver's compartment of many armoured fighting vehicles. Applications include the LAV family of vehicles, but the system can be adapted for other types of vehicles, by simply changing the adaptor mount.

Operational status
In production. In service in Canada on the LAV.

Driver's Viewer Aid (DVA) from Thales Optronics

Specifications
Weight: <13 kg
Dimensions: w × h × d 351 × 372 × 330 mm
Spectral band: 8-12 μm
Magnification: ×1 and ×3
Field of view: 40 and 20° (×1 magnification)
Field of regard: 100 × 40°
Focus ranges: 5 m to infinity

Ranges
Detection: 760 m (×1); 2,200 m (×3)
Recognition: 260 m (×1); 800 m (×3)
Identification: 130 m (×1); 400 m (×3)
MRTD: 0.54 at 0.3 cy/mrad
Power supply: 45 W, MIL-STD-1275 compatible
Video output: RS-170A
Communication datalink: RS-422
Operating range: –40 to +63°C
Optional equipment: tripod adaptor, battery adaptor

Contractor
Thales Optronics, Canada.

UPDATED

Thales PowerVision Driver's Vision Enhancer (DVE)

Type
Driver's thermal sight.

Description
The PowerVision Driver's Vision Enhancer (DVE) is a lightweight real-time thermal imaging system operating in the 8 to 12 μm spectral band. The system allows normal driving capability during conditions of darkness and degraded visibility, such as smoke, dust and haze. The PowerVision DVE is based on uncooled microbolometer focal plane array technology which offers high reliability with low total life cycle costs.

The DVE features common modules that may be used in both armoured and tactical wheeled vehicles. This modular design allows the system to be installed on most vehicles within minutes. The driver interface is a flat panel display and the position in the vehicle can be customised to offer best man/machine interface. PowerVision DVE is based on Thales Optronics (formerly AlliedSignal) proven technology in driving applications, which include the AN/VVS-501 night vision periscope (qv) and the thermal Viper periscope systems (qv). Applications include the LAV and Bradley families of vehicles, but the system can also be used in other vehicles by simply changing the adaptor mount.

Specifications
Weight: <5 kg
Spectral band: 8-12 μm
Detector: 320 × 240 uncooled microbolometer focal plane array
Magnification: ×1
Field of view: 42 × 31°
Field of regard: 100 × 40° (may be larger for different host vehicles)
Focus: athermalised, focus free 5 m to infinity
MRTD (max value)
at 0.05 cy/mrad: 0.045°C
at 0.10 cy/mrad: 0.075°C
at 0.16 cy/mrad: 0.1°C
at 0.20 cy/mrad: 0.15°C

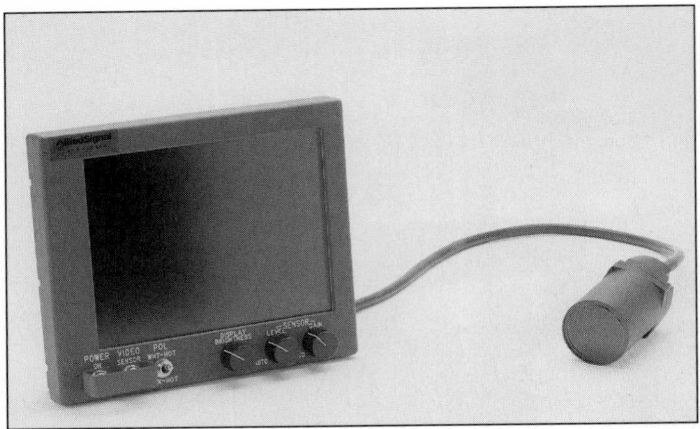

The Drivers' Vision Enhancer (DVE) 0055263

Power supply: 30 W, military vehicle power
Video output: RS-170 or CCIR
Data port: RS-422 for overlay data
Operating range: –46 to +50°C

Contractor
Thales Optronics, Canada.

UPDATED

Thalis NX-129 night vision driver's viewers

Type
Driver's sight.

Description
The NX-129 night vision driver's viewer is a passive image intensifying night driving system for the M113 Armoured Personnel Carriers (APCs). The viewer is supplied with a Gen III, 25 mm, image intensifier tube with integral inverter/expander fibre optic. The NX-129 is part of a family of night driver's viewers developed by Thalis Sensors to be adapted to various types of armoured vehicles.

Operational status
In service with the Hellenic army.

Specifications
Magnification: ×l

Field of view:
44° horizontal
35° vertical
Azimuth range: up to ±45° depending on vehicle application
Depth of field: 4 m to infinity
Resolution: better than 1.0 lp/mrad
Eye relief: 50 mm
Image Intensifier tube
25 mm Gen III with integral inverter/expander fibre optic
Power supply: one (1) 3 V lithium battery or 28 V DC vehicle power

Contractor
Thalis S.A.

VERIFIED

Vectronix NAP5 night driving periscope

Type
Driver's sight.

Description
The NAP5 night driving periscope is available for use on different armoured vehicle types simply by changing the vehicle adaptor and head prism. Known designations include the NAP5-1 for use on the Pz68/88 MBT, NAP5-3 for the M113 APC, NAP5-4 for M109 Howitzer and NAP5-5 for the MOWAG Piranha AFV.

The periscope is used for driving in ambient moonlight, starlight and/or skyglow conditions. This residual night light is transmitted via the prism and an objective to a second-generation image intensifier tube which produces the image of the area being observed for the driver's biocular viewing system which has a fixed dioptre setting of –1.

If a bright light source suddenly appears the image intensification is automatically reduced to avoid the driver being dazzled.

The Vectronix NAP5 night driving periscope 0004792

The periscope is powered from the vehicle's 24 V DC supply with automatic changeover to an internal battery if the main supply is interrupted for any reason.

Operational status
In service with the Swiss Army and several undisclosed countries.

Specifications
Weight
Basic unit: 4.50 kg
Dimensions
Basic unit: 140 × 170 × 310 mm
Magnification: ×1
Traverse range: ±30°
Field of view
Horizontal: 40°
Vertical: 30°
Image intensifier: GEN II 20/30
Resolution: 0.77 lp/mrad
Objective: f/1.3
Depth of focus: 4 m to infinity
Power supply: 24 V DC
Emergency power supply: 1 × 3 V lithium cell

Contractor
Vectronix AG (formerly Leica).

UPDATED

Yugoimport PPV-2 passive night vision periscope

Type
Driver's sight.

Description
The PPV-2 passive night vision periscope was designed for driving and general night observation use on the T-55 MBT, M-84 (licence-built T-72) MBT, BVP M80A MICV, BVP M80AK IFV, M-60P APC and M-60PB anti-tank vehicle.

It uses an 18 mm second-generation light intensifier tube unit.

Operational status
Production as required. In service with the Yugoslav Army.

Specifications
Weight: approx 6 kg
Dimensions: 300 × 168 × 139 mm
Magnification: ×1

Field of view: 34°
Dioptre range: –0.7 to +1
Resolution power
3×10^{-2} lx: 2.2 mrad
1×10^{-3} lx: 2.7 mrad
1×10^{-4} lx: 5.0 mrad
Image intensifier tube: 18 mm GEN II
Power supply: 26 V DC

Contractor
Yugoimport SDPR.

VERIFIED

Zeiss Optronik, Raytheon Driver's Vision Enhancer (DVE)

Type
Driver's sight.

Description
The Driver's Vision Enhancer (DVE) is an uncooled thermal imager with a third-generation staring 320 × 240 focal plane array. Raytheon uncooled technology is used for the Leopard 2 MBT application. It operates in the long wave infra-red spectrum from 7.5-13.5 μm wavelength. The DVE provides drivers of tactical wheeled vehicles and combat vehicles with the capability of continuing operations during darkness, adverse weather or dirty battlefield conditions.

Operational status
One preproduction model has been installed in the German Leopard 2 MBT for trials. Another unit will be trialed in the German Marder APC.

Specifications
Detector: 320 × 240 staring focal plane array
Spectral band: 7.5 - 13.5 μm
NETD: <0.1°
Field of view: 30 × 40°
Field of regard: elevation –15 to +25°, azimuth ±30°
Weight: <7 kg
Power supply: 16 - 32 V DC
Power consumption: 96 W max, 26 W average
Video interface: RS-170 (SMPTE 170M)/CCIR
MTBF: >5000 h (calculated)

Contractors
Raytheon Systems Company.
Zeiss Optronik GmbH.

UPDATED

INFANTRY WEAPON SIGHTS

ILLUMINATING

Aims Optronics laser hit marker

Type
Weapon sight (illuminating).

Description
The Aims Optronics laser hit marker is a laser spot projector which, by means of various adaptors, can be attached to any small arms. When switched on, it delivers a clearly visible red spot indicating the point of aim of the weapons. The firer need take no conscious aim, but with both eyes open merely lays the spot on the target and fires.

The marker is a battery-powered helium laser operating at 632.8 nm to deliver a visible spot. A compact model is available for distances up to 100 m and a newly developed laser beam control offers long-distance aiming for snipers. For military applications a different laser, generating a spot only visible with night sights or night goggles to a range of 300 m can be provided. The marker can be mounted using any conventional telescope mounting rings on any type of small arms including pistols. Depending on the model of marker, the laser beam can be adjusted either mechanically or optically in order to boresight the weapon at any selected range. Special X-Y collimators are available for beam control and long-distance work.

Different power supplies accommodate the user's requirements in action or for training. Complete marksman training systems are available to improve accuracy, follow- through, or to assess the shooter's overall performance.

Operational status
In production.

Contractor
Aims Optronics NV SA.

VERIFIED

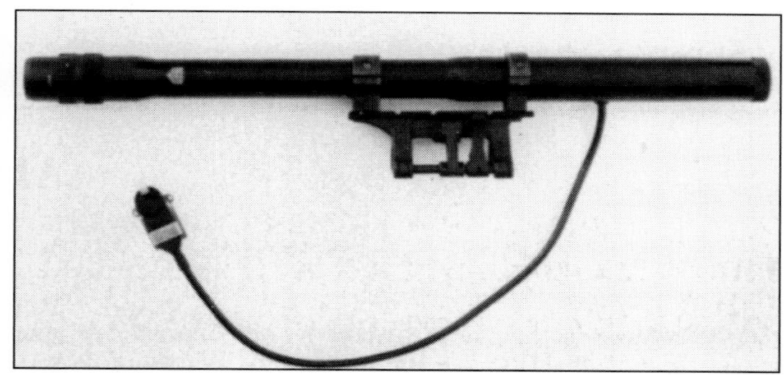

Aims Optronics laser hit marker with mounting bracket

Aselsan IRHN-9396

Type
Weapon sight (illuminating).

Description
Aselsan's IRHN-9396 laser infra-red aiming device provides an aim point for personnel engaged in night operations. The IRHN-9396 does this by projecting an intense, continuous IR laser beam which is invisible to the eye but can be readily seen with night vision goggles. Once it is boresighted to a weapon, the operator simply puts the laser beam on target and fires.

Optical design and optical baffle reduce off-axis detectability to ≥10° and preclude the IRHN-9396 laser beam from being seen by others using night vision devices. The optical baffle is screwed on to the front of the aiming light.

IRHN-9396

Operational status
In production.

Specifications
Dimensions (l × w × h): 101 × 40 × 52 mm
Weight: 185 g (without batteries, remote-control cable and weapon adaptor)

Optical
Beam divergence: 0.3 mrad
Peak wave length: 810-840 nm
Power output: 1.8 mW
Boresight adjusters: azimuth and elevation, click-stop type
Boresight adjustment: 0.28 mrad per click, 60 mrad travel on each axis
Max range: 2.5 km

Electrical
Power requirement: 3 V DC
Power sources: 2 AA size 1.5 V batteries
Construction: hard black anodised aluminium case

Contractor
Aselsan Inc, Microelectronics, Guidance and Electro-optics Division.

VERIFIED

Crimson Trace™ Corporation LG-229 Lasergrips™

Type
Weapon sight-illuminating.

Description
The Lasergrips™ laser sighting system is designed to be used with a variety of hand guns to provide very accurate aiming. The laser projects a beam which produces a bright red-orange dot on the intended point of impact. The device is fitted by first removing the existing grips of the weapon and replacing them with Lasergrips™. The laser is activated via a pressure switch in the left grip position so that it can be operated with either hand. A master on/off switch at the base of the grips allows the activation switch to be isolated.

Fine tuning alignment is easily accomplished in a short time. The user can make precision adjustments to allow for windage and elevation.

The LG-229 is designed for the SIG Sauer P228 and P229 weapons. There are further models designed for a variety of weapons. Variants are produced with either visible or near infra-red lasers.

Operational status
Available.

Specifications
Model LG-229
Output power: <5m W
Wavelength: 633 nm
Dot size: 13 mm approx at 15 m, 7.5 cm at 100 m
Power supply: 2 × DL2032 lithium batteries
Battery life: over 4 h

Contractor
Crimson Trace™ Corporation.

VERIFIED

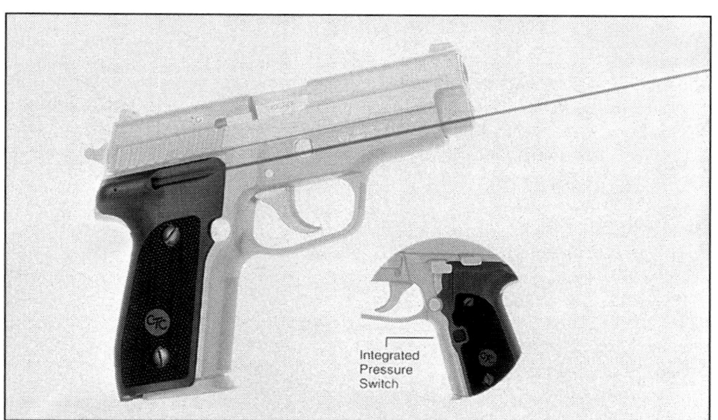

Lasergrips™ on SIG Sauer pistol 0055254

Dedal-200 night vision sight

Type
Weapon sight (illuminating).

Description
The Dedal-200 sight is one of the Dedal range of professional night vision sights. It may be fitted with either GEN I Super+ or GEN II image intensifiers. There are three variants, Mods A, B and C. Mods B and C have automatic brightness control and protection of the intensifier from excessive light levels. There is a built-in IR illuminator for use in very low light levels. Windage and elevation adjustments are provided. The sight comes with a standard weaver mount.

Operational status
Available.

Specifications

Model	mod A	mod B	mod C
Overall dimensions	180 × 82 × 72 mm	187 × 82 × 72 mm	215 × 82 × 72 mm
Weight	0.75 kg	0.78 kg	0.84 kg
Magnification	×1.7 (×3.4)*	×2.6 (×5.2)*	×4.0
Field of view	17°	17°	10°
Objective	64 mm f/1.2	64 mm f/1.2	100 mm f/1.5
Eye relief distance	45 mm		
Dioptre adjustment	+4, −4		
Image Intensifier	GEN Super I+	GEN II+, 18 mm MCP	GEN II+, 18 mm MCP
Photocathode sensitivity	280 μA/lm	350 μA/lm	350 μA/lm
Gain	1,000	25,000	25,000
Resolution	50 lp/mm	40 lp/mm	40 lp/mm
Power supply	2 × 3 V AA batteries		
Operating time	60 h		
Operating temperature	−40 to + 50 °C		

Contractor
Joint Stock Company 'Dedal' Moscow.

VERIFIED

The Dedal-200 night vision sight 0041440

ENOSA ANL-02 laser aiming sight

Type
Weapon sight (illuminating).

Description
The ANL-2 is an auxiliary aiming sight for individual weapons such as rifles or light machine guns. The aiming sight, designed for night-time operations, emits a narrow beam of pulsating light of 820 nm wavelength which is invisible to the naked eye but which is detectable with night vision goggles. A round spot, blinking at 6 Hz appears above the target.

Operational status
Prototype and preproduction complete. Ready for production.

Specifications
Laser diode: GaAs
Wavelength: 820 nm
Beam shape: round
Beam angle: ≤1 mrad
Range: 300 m
Operating mode: pulsating

The ANL-02 laser aiming sight

Homogenisation: micrometric systems in direction and height through displacement of beam
Length: 145 mm
Weight : 350 g
Compatibility: NATO fitting. Optional fittings for different types of weapon
Power supply: 1 9 V battery
Operating time: 25 h
Environmental standards: MIL-STD 810C
Eye safety: ANSI Z-136.1

Contractor
Empresa Nacional de Optica SA (ENOSA).

VERIFIED

Euroatlas RT 5A laser illuminator

Type
Weapon sight (illuminating).

Description
The RT 5A laser illuminator is intended to be used with night vision devices in the case of absolute darkness or in difficult observing conditions. It is also recommended where light barriers, for example bright light sources, are between the observer and his objective or where a dark objective is between bright areas. This laser torch is light and small and can easily be mounted on any vision device by using the adjusting and mounting bracket provided.

The laser diode has a working range up to 1 km at low current consumption. There is an adjusting wheel which permits the illumination field to be changed during operation within the ratio of 1:4.

Operational status
Available.

Specifications
Length: 178 mm
Width: 75 mm
Height: 85 mm without mounting
Weight: 650 g
Wavelength: 840-870 nm
Laser diode: double hetero-structure LB1
Radiant output: 6-10 mW
Pulse duty factor: 5%
Beamwidth: approx 1 mrad
Normal ocular hazard distance: 50 m
Working range: 1 km
Power supply: 6 × AA/LR6 batteries

Contractor
Euroatlas GmbH.

VERIFIED

RT 5A laser illuminator

Fraser-Volpe LAD-LR infra-red laser aiming device

Type
Weapon sight (illuminating).

Description
The LAD-LR is a small lightweight long-range laser aiming device which can be mounted on and boresighted with most infantry weapons, from handguns to machine guns. It projects a small infra-red spot visible only with night vision devices onto a target. Also functioning as an illuminator, the laser beam increases the effective range of night vision devices. There are two illumination levels which can be selected to suit the ambient light conditions. Boresight adjustments in elevation and azimuth are provided.

Operational status
Available.

Specifications
Dimensions (l × h × w): 92 × 57 × 49 mm
Weight: 312 gm
Wavelength: 830 nm
Average output power: 10 mW (high power)
Boresight adjustment: 2.5° on each axis
Adjustment increments: 0.58 mrad
Beam divergence: 0.3 mrad
Laser operating time: >10,000 h

Contractor
Fraser-Volpe Corporation.

VERIFIED

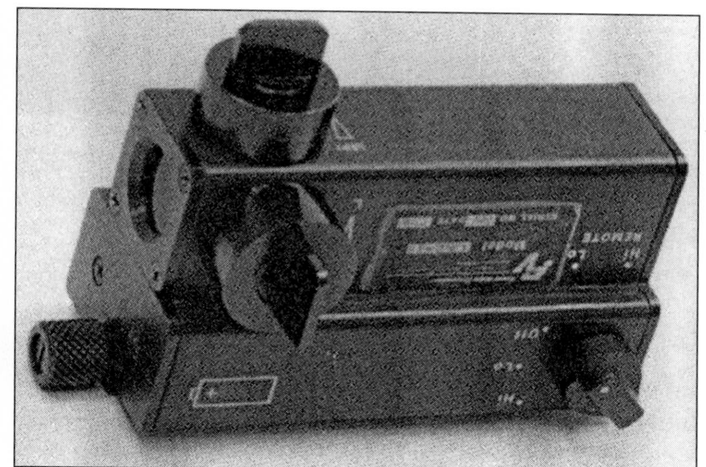

Fraser-Volpe LAD-LR laser aiming device 0023053

Imatronic LS55 laser aiming system

Type
Weapon sight (illuminating).

Description
The LS55 laser aiming system can be fitted to almost all standard issue rifles, shotguns, carbines and sub-machine guns in military, police and special forces use worldwide. It is available in both visible red dot and infra-red versions. The target is acquired using an intense red dot of laser light providing pinpoint accuracy, especially useful in security situations. Faster than conventional optical sighting methods and giving the ability to fire with both eyes open, from the waist or shoulder level, the LS55 improves first-round hit probability.

The LS55 is suitable for any weapon which will take 1 in telescope rings and incorporates elevation and windage to allow zeroing to individual weapons.

Operational status
In production. Ordered by an Asian paramilitary force.

Specifications
Length: 176 mm
Width: 25 mm
Laser output
  Visible: <5 mW (Class IIIa)
  IR: <5 mW (Class IIIb)
Wavelength
  Visible: 670 nm
  IR: 800/850 nm
Beam diameter: 25 mm at 50 m; 50 mm at 100 m; 150 mm at 300 m

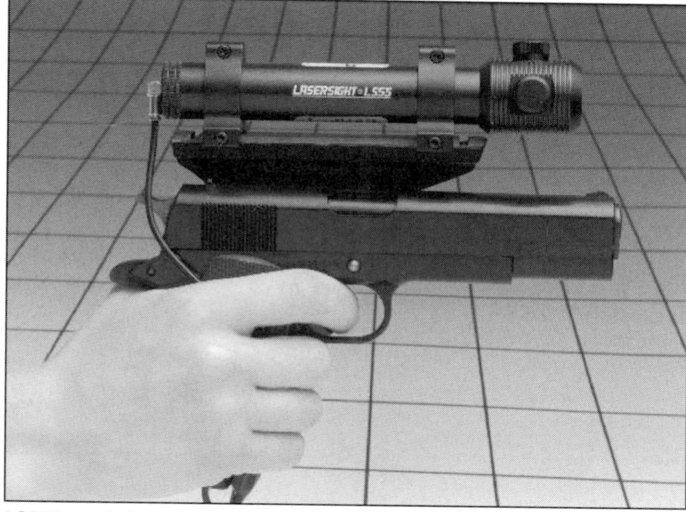

LS55 laser aiming system

Divergence: 0.5 mrad
Range: 300 m
Power supply: 3 × AAA alkaline batteries
Battery life: >8 h continuous use

Contractor
Imatronic Limited.

VERIFIED

Insight Technology AN/PAQ-4C eye-safe infra-red aiming light

Type
Weapon sight (aiming illuminating).

Description
AN/PAQ-4C incorporates lessons learned with aiming lights used in Desert Storm, Afghanistan, other combat operations and extensive field training and evaluation. It forms part of the US Army's Land Warrior programme, to develop baseline systems for the next century's infantry soldier. AN/PAQ-4C is manufactured by Insight Technology Inc.

The unit provides a rapid, accurate aiming point for personnel engaged in night operations. AN/PAQ-4C projects a highly collimated (0.3 mrad) laser beam, invisible to the eye but readily seen with night vision goggles. The 4C is a continuous wave laser. Once boresighted to a weapon the firer simply puts the laser beam on the target and fires. The unit has a minimum Mil-Spec range of 600 out to 1,800 m under ideal conditions. The unit incorporates a preset zero setting (neutral position) at which the beam is in precise alignment with the unit's mounting surface. This, combined with the M16 mounting bracket, enables the unit to be nearly zeroed when first mounted on the weapon. Accurate elevation and windage controls permit fine zero adjustment. This combination enables precise zeroing to be established by firing only a single three round shot group.

Optical design and optical baffles reduce off-axis detectability to less than, or equal to, 6° and preclude the beam being seen by others using night vision devices.

As well as use with M16 series rifles, the AN/PAQ-4C is also compatible with the M4 carbine series, M2, M60, M67 and M72 machine guns and the M249 SAW.

Over 20,400 aiming lights were produced in 1993-1996 and 40,000 in 1996-1999.

Operational status
In production and in service with the US Army and Marines.

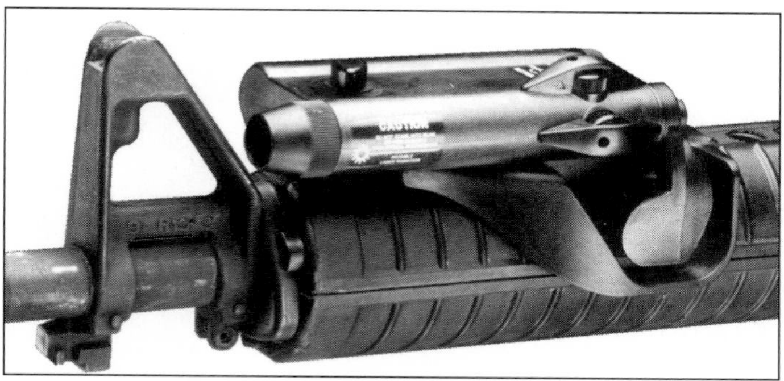

AN/PAQ-4C eye-safe infra-red aiming light mounted on an M16A2 rifle 0097479

Specifications
Dimensions: 136 × 27 × 56 mm
Weight: 125 g
Range: 1,800 m
Laser type: laser diode
Beam divergence: 0.3 mrad
Visual security sensitivity: ±6°
Peak wavelength: 830 nm +20/–10 nm
Input current: 50 mA (max)
Power output: 0.7 mW
Power supply: 2 × AA 1.5 V batteries
Operating temperature range: –54 to + 71°C
Reverse polarity protection: yes

Contractor
Insight Technology Inc.

UPDATED

Insight Technology AN/PEQ-2A ITPIAL Infra-red Target Pointer/Illuminator/Aiming Laser

Type
Weapon sight (illuminating).

Description
The AN/PEQ-2 Infra-red Target Pointer/Illuminator/Aiming Laser (ITPIAL) is a dual-beam multifunction laser which combines the utility and operational effectiveness of aiming lights and illuminators/pointers in a single device.

The ITPIAL features a unique switching capability for both gun-mounted and hand-held operation. A fully integrated mode selector switch allows the user to select one of six modes: aim low power, aim high power, illuminator, dual high power, dual low power and off. Low power is eye-safe and is useful to preclude blooming in close combat situations, while high power enables engagements at extended range and in high light conditions.

The ITPIAL electro-optic system comprises two index-guided laser diodes, diffraction-limited microlens assemblies and full-logic microprocessor controlled electronic drivers. The pointer/illuminator laser beam can be adjusted from a 0.5 mrad beam for long-range pointing to a 10° circular illumination beam. A variety of light-shaping diffusers is available for creating wider illumination angles. The ITPIAL has two independently adjustable beams with two sets of azimuth and elevation adjusters.

The ITPIAL is equipped with an integral push-button switch for use in the hand-held mode which allows the operator to activate the system in either momentary or steady-on mode. When weapon mounted it can be activated using a variety of remote switches.

A rail grabber bracket can be used to attach the ITPIAL to any weapon equipped with a MIL-STD-1913 mounting rail. Suitable brackets are available for the MI6A1/A2 rifles, M4 carbine, M249, M60, M2 machine guns and the AT4 (M136) anti-armour weapon.

ITPIAL is available in three versions – the US Military Standard AN/PEQ-2, the higher power IR/IR Model 5000 and the combination visible/IR Model 7500.

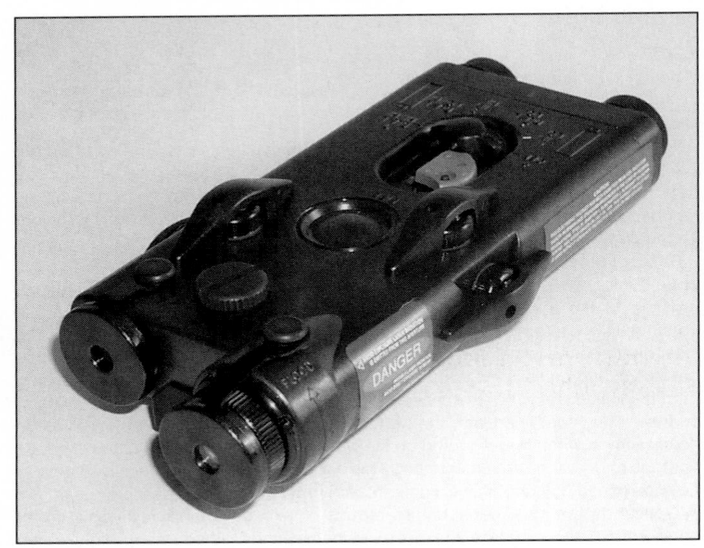

The AN/PEQ-2A infra-red target pointer/illuminator/aiming laser 0097483

AN/PEQ-2 mounted on a weapon 0004837

Operational status
In production. Under contract for the US Navy Special Forces.

Specifications
Weight: 210 g

	AN-PEQ-2	Model 5000	Model 7500
Aiming light			
Divergence (FWHM)	0.3 mrad	0.3 mrad	0.3 mrad
Laser power	50 mW	50 mW	15 mW
Output power	25 mW	25 mW	12 mW
Pointer/illuminator			
Divergence (FWHM)			
Spot	0.3 mrad	0.3 mrad	0.3 mrad
Flood	>10°	>10°	>10°
Laser power	50 mW	100 mW	50 mW
Output power	30 mW	70 mW	30 mW
Peak wavelength			
Aiming light	830 nm	830 nm	635 nm
Pointer/illuminator	830 nm	830 nm	830 nm

Boresight retention: <0.5 mrad
Battery type: 2 × AA alkaline
Reverse polarity protection: yes
Waterproof: to 66 ft

Contractor
Insight Technology Inc.

UPDATED

Insight Technology AN/PEQ-4 Medium- and High-Powered Laser Illuminators (MPLI/HPLI)

Type
Weapon sight (illuminating).

Description
The AN/PEQ-4 is a Class 4 laser pointing and illuminating system capable of providing illumination and pointing capabilities for directing naval gunfire at ranges of 15 km or greater. The AN/PEQ-4 is also being heavily employed in Close Air Support (CAS) and Forward Air Control (FAC) applications. The AN/PEQ-4 can be used as either a hand-held illuminator/pointer or can be weapon-mounted with included brackets and accessory mounts. In weapon-mounted mode, the AN/PEQ-4 can be used to accurately direct gunfire as well as illuminate and designate targets.

The illuminator assembly is a compact, lightweight electro-optical assembly which provides both a highly collimated beam of infra-red energy for target designation and an adjustable focus for target illumination. Output power of HPLI is adjustable from a minimum of 100 mW to a maximum of 1 W. Output power of MPLI is adjustable from a minimum of 100 mW to a maximum of 700 mW. Independent adjusters control azimuth and elevation providing the capability to zero the laser to the weapon. The AN/PEQ-4 (HPLI) receives power from an external 10-30 V DC vehicular power via a 25 ft supply cable. The AN/PEQ-4 (HPLI/MPLI) also incorporates an internal battery capability which utilises 3 DL-123 Lithium or 6 AA commercial batteries.

Operation
The AN/PEQ-4 is designed for operator comfort and convenience and to minimise fatigue for either right- or left-handed operators. It operates using a detachable remote momentary on/off 'deadman' switch for control. It is built with two separate control switches.
Switch 1: Off/Low/Med/High
Switch 2: Off/2 Hz, 5 Hz, 10 Hz, 10 Hz, CW
This allows the operator to adjust the intensity and modulation of the laser. A power indicator light, visible under normal indoor lighting, identifies that the device is in standby mode or is emitting laser radiation or under low battery condition.

Operational status
In service.

Specifications: Insight Technology AN/PEQ-4 Medium- and High-Powered Laser Illuminators (MPLI/HPLI)

	HPLI	AN/PEQ-4	AN/PEQ-4A
Size	23.5 × 7.0 × 5.72 cm	23.5 × 5.72 × 5.72 cm	23.5 × 5.72 × 5.72 cm
Weight (with 6 AA)	459 g	456 g	456 g
Weight (with 3 DL123)	379 g	376 g	376 g
Power output	100 - 1,500 mW	100 - 600 mW	100 - 800 mW
Beam divergence	2 mrad -10°	1 mrad-3°	1 mrad-3°
Wavelength	860 ±10 nm	860 ±10 nm	860 ±10 nm
Pulse rates	CW, 2, 5, 10 Hz	CW, 2, 5, 10 Hz	CW, 2, 5, 10 Hz
Power source	3ea DL123 or 6 AA	3ea DL123 or 6 AA	3ea DL123 or 12 V
External	10-30 V DC	6 AA Batteries	BA 5590
Power selectable	Low, medium, high	Low, medium, high	Low, medium, high
Adjustments	(External) adjustment	(Internal) 1 mrad per click	(Internal) 1 mrad per click
Waterproof	up to 20.1 m	up to 20.1 m	up to 20.1 m
Environmental shock and vibration	MIL-STD-810	MIL-STD-810	MIL-STD-810

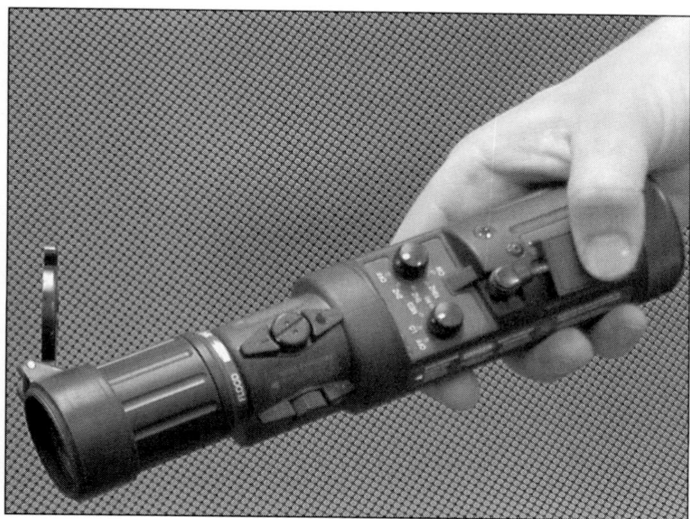

AN/PEQ-4 Medium Power Laser Illuminator (MPLI) – hand-held 0097484

Contractor
Insight Technology Inc.

VERIFIED

Insight Technology AN/PEQ-5 Carbine Visible Laser (CVL)

Type
Weapon sight (illuminating).

Description
The Carbine Visible Laser (CVL) is a US Military Standard visible aiming light which has a unique integral rail grabber bracket built-in to the product that allows it to be mounted to the weapon's rail interface system. It therefore attaches and detaches easily to the M4A1 carbine or weapons equipped with MIL-STD-1913 rails. The CVL allows boresight repeatability within 0.5 mrad. The range of the device is greater than 500 m in low light conditions. It is designed for use with either the naked eye, powered optics or night vision devices. CVL is qualified for submersion in up to 20.1 m of water. It operates on a 1.5 V AA battery and includes a remote cable switch.

Operational status
In production.

CVL mounted on a M4 weapon 0097486

The AN/PEQ-5 Carbine Visible Laser aiming light 0041397

Specifications
Weight: 134 g (with batteries)
Dimensions (l × w × h): 7.2 × 6.45 × 4 cm
Range: 600 m (low-light and night conditions)
Wavelength: 615-655 nm
Beam divergence: 0.5 mrad
Windage and elevation adjustment: 1 cm/click at 25 m
Waterproof: 20.1 m submersible for 130 min
Operating temperature: –60 to +160°F
Power: 1 × 1.5 V AA alkaline battery, 28 h continuous operation at 4.44°C

Contractor
Insight Technology Inc.

UPDATED

International Technologies Lasers AIM-1 laser aiming light

Type
Weapon sight (illuminating).

Description
The AIM-1 family comprises infra-red laser aiming devices for night combat and operations, which may be mounted on a variety of weapons by means of specific adaptors.

The light source of the AIM-1 is a laser diode which emits radiation in the near infra-red spectrum (excluding the AIM-1/V which emit a visible red dot). The AIM-1 emits a dot-shaped beam which marks the target with an infra-red dot, visible with suitable night vision equipment, so allowing rapid and accurate night-time aiming.

The AIM-1 family consists of the following options:

AIM-1/C An infra-red aiming light for infantry personal weapons of various calibres, ranging from pistols to assault rifles.

AIM-1/V A visible red laser aiming light. It is effective for indoor and outdoor operations in twilight and night conditions, urban fighting and anti-terrorist missions. The unit will fit any weapon, from pistols to shoulder arms, and can be operated locally or by remote-control cable.

AIM-1/DLR An infra-red aiming light with longer range for infantry standard heavy weapons such as the M60 or MAG machine gun. Adding an optional adaptor with ballistic compensation, it may be mounted on the Mk 19 grenade launcher, 0.50 machine gun or other crew-served weapons.

AIM-1/MLR A long-range IR aiming light with the same characteristics as AIM-1/DLR but with a different activation switch (on/off/remote) located at the rear side of the unit. Its principal applications are shipboard and airborne door-mounted machine guns.

AIM-1/EXL A long-range aiming light with the same characteristics as the AIM-1/DLR.

AIM-1/SLR An extra long-range aiming light (10 to 12 km) designed to be installed on extremely long-range observation devices such as long-range FLIRs.

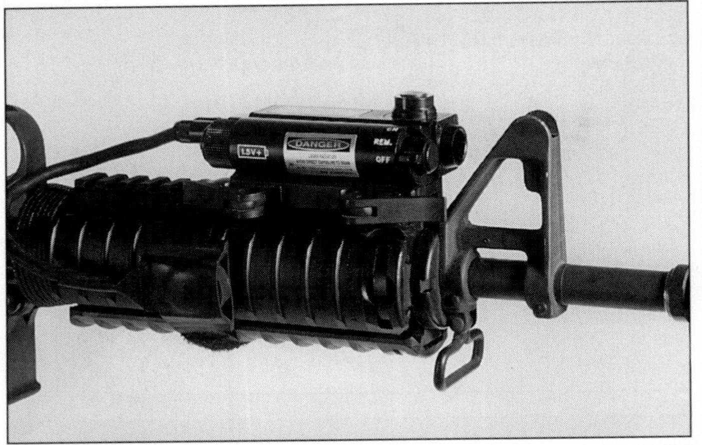

AIM-1/V laser aiming light 0023054

The unit is utilised as a pointing device for long-range targets for assault helicopters and special forces units active deep inside enemy territory.

AIM-1/SLT Miniature super long-range aiming light designed to be integrated in gymbals and payloads. Capable of pointing and designating at distances up to 10 km.

AIM-1/SLX Same as the AIM-1/SLR but with an external power supply option designed to be installed on aerial platforms to increase visibility of target designators or mounted on helicopters. The power supply is obtained directly from the platform power source by cable connected to the rear of the unit.

The new AIM-2000 offers a dual-wavelength laser module transmitting the laser beam through the same optical aperture enabling the user to switch between visible and IR while using only one common set of boresighting knobs – for further details see AIM-2000 entry.

Operational status

In production.

Specifications

Model	AIM-1/DLR	AIM-1/EXL	AIM-1/MLR
Range (typical)	3 km	3 km	3 km
Output power	15 mW	15 mW	15 mW
Beam divergence	0.3 mrad	0.3 mrad	0.3 mrad
Emission wavelength	830 nm (IR)	830 nm (IR)	830 nm (IR)
Power source	3 V (2 × AA alkaline cells)	24/28 V (ext)	3 V (2 × AA alkaline cells)
Boresight + adjustment travel range	Minimum 2.5° end to end in discrete clicks	same as AIM-1/D	same as AIM-1/D
Step (click) size	0.58 mrad	0.58 mrad	0.58 mrad
Operating time	low: 50 h high: 5 h	n/a	low: 50 h high: 5 h
Operational function	same as AIM-1/D	n/a external control	3 position switch off, on, remote
Remote operation	available	available	available

Model	AIM-1/SLR	AIM- 1/SLX	AIM-1V (new)	AIM-1C (new)
Range (typical)	10 km	10 km	300 m	500 m
Output power	50 mW	50 mW	4 – 3 mW	1.5 mW (lower power is optional)
Beam divergence	0.3 mrad	0.3 mrad	0.3 mrad	0.3 mrad
Emission wavelength	830 nm (IR)	830 nm (IR)	630 nm (red)	830 nm (IR)
Power source	3 V (2 × AA alkaline cells)	12/24/28 V (ext)	1.5 V (1 × AA)	1.5 V (1 × AA)
Boresight + adjustment travel range	min 2.5° end to end in discrete clicks	same as AIM-1/DLR	same as AIM-1/DLR	same as AIM-1/DLR
Step (click) size	0.58 mrad	0.58 mrad	0.58 mrad	0.58 mrad
Operating time	low: 50 h high: 3 h	n/a	15 h	30 h
Operational function	3 position switch off, low, high	n/a external control	3 position switch Off, rem, on	3 position switch Off, rem, on
Remote operation	available	available	available	available

Contractor

International Technologies Lasers Limited.

UPDATED

International Technologies Lasers AIM-2000 dual wavelength laser aiming device

Description

The AIM-2000 Laser Aiming Device derived from ITL's combat proven AIM-I family of laser devices offers a dual wavelength laser module that emits either a visible (red) or infra-red laser beam through the same optical aperture that allows boresighting by a single mechanism. The electronically controlled laser module enables immediate selection between the two laser beams according to the operational need:

- Visible-red beam for the unaided eye in low ambient light and shaded environments typical of urban fighting and anti-terror missions
- Infra-red laser beam detectable only to image intensified night vision devices for night fighting activities.

The AIM-2000 fully meets the relevant MIL-specifications, and provides the user with a simple to operate, reliable lightweight device, with good boresight retention and accurate target pointing with a precise and distinct laser dot. In summary, the following features are included:

- Switching from visible (red) to infra-red by a flip of a button
- A single boresight mechanism enables the adjustment of both laser beams simultaneously in one action
- A laser signature reduction mechanism reduces the beam side lobes and scattering to protect the user from being detected
- Powered from a single standard 1.5V AA size battery
- Visible switch position protected by a safety catch to avoid unintentionally operation
- Variable mounting options available for different types of weapons
- Remote switch cable for momentary operation
- Low- or high-power of the infra-red laser can be selected by the operator.

Operational status

Available.

Specifications

Dimensions (w × h × l): 62 × 32 × 102 mm
Weight: 195 gm (excluding battery and remote cable)

	Optical	Visible IR
Output power	1 mW	2 mW (0.4 mW at low position)
Beam divergence	0.5 mrad	0.5 mrad
Wavelength	640 nm	840 nm
Boresight Adjustment	40 mRad in both elevation and deflection	

Increment size: 0.5 mrad (discrete increments – clicks)

Electrical
- 120 mA
- Approx 10 hr of continuous aiming
- Four position selector switch:
  - visible laser beam (via safety catch)
  - power off
  - IR laser beam, low power
  - IR laser beam, high power (via safety catch)

Environmental: Complies with MIL-STD-810
Accessories: Various weapon adapters for different types of weapons.
* All values are typical.

Contractor

International Technologies Lasers Limited.

UPDATED

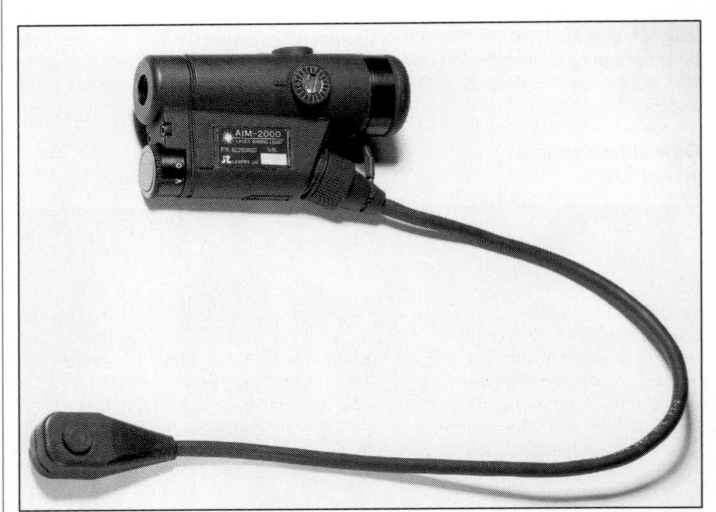

The AIM-2000 dual wavelength laser aiming device 0077946

International Technologies LPL-30/Z long-range laser pointer

Type
Weapon sight (illuminating).

Description
The LPL-30/Z is an infra-red laser long range pointer and illuminator specially designed for field commanders and soldiers on night missions. It enables clear marking to identify and allocate targets at distances up to 4 km (or 10 km using the LPL-30-SL/Z) by marking them with the high power infra-red laser beam.

It can also be used as a high power illuminator by utilising its zoom capability. This variable beam divergence is used in conjunction with image intensified night vision devices when ambient light levels are too low for observation to rely on ambient light illumination only.

LPL-30/Z can also be used for signalling between units at up to 40 km.

The LPL-30/Z can also be supplied with the optional beam flashing of 5 Hz, for ease of detection. Pending the options mentioned, the device can come either with a momentary operating pushbutton or with a momentary 3 position toggle switch (with the flashing configuration).

The more powerful LPL-30-SL/Z is also available.

Specifications
Dimensions: 130 × 40 × 25 mm
Weight: 195 g (excluding batteries)
Output power: 15 mW (LPL 39/Z), 60 mW (LPL-30-SL/Z)
Beam divergence: 0.5 to 350 mrad (20°)
Wavelength: 830 nm
Power supply: 2 × 1.5 V AA alkaline or equivalent
Battery life: 10 h (approx)

Manufacturer
International Technologies (Lasers) Ltd.

UPDATED

LPL-30/Z long-range laser pointer with flash option 0023058

ITT Laser Aiming Module (LAM)

Type
Weapon sight (illuminating).

Description
The ITT Laser Aiming Module is designed to provide precision aiming and can also be used as an illuminator. It contains visible and infra-red lasers for aiming, an infra-red illuminator and a visible white illuminator.

The LAM is simple to operate. The current model of the LAM is designed to be securely attached to the SOCOM Mk23 pistol utilising the mounting grooves on the frame of the weapon. Attachment of the LAM does not affect the functionality of the weapon, and the unit can be detached and reattached to the weapon while maintaining boresight repeatability at better than 0.5 in at 25 m. The LAM can operate in one of four selectable operational modes by rotating the selector switch on the side of the LAM to the desired position. These four modes of operation are:
(a) Visible laser only
(b) Visible laser/flashlight
(c) Infra-red laser only
(d) Infra-red laser/illuminator

Once a mode is selected, the LAM is easily activated in either a momentary or steady-on condition by depressing a switch lever conveniently located in front of the pistol's trigger guard. Internal boresight adjusters enable the LAM to be precisely zeroed to the weapon.

The LAM's Infra-red Laser Aiming Light is optimised for precision aiming and meets Class 1 eye-safety requirements. The Infra-red Laser Illuminator also employs an eye-safe laser that projects a beam of covert illumination to a range of greater than 50 m. The Infra-red aiming and illumination beams are optimised for use with GEN II and GEN III night vision goggles.

For low light and night operations without night vision goggles, the visible aiming light and visible illuminator are provided. Both the infra-red and visible illuminators allow facial recognition at 25 m. When switched to either the infra-red or visible aim-

LAM mounted on a USSOCOM pistol 0055255

and-illuminate mode, the LAM projects a laser aim point in the illumination area for target identification to enable immediate target engagement.

The LAM is designed to withstand the harshest US Special Operations conditions and is tested to meet the most stringent US Military environmental specifications, including immersion to 60 ft of water and 30,000 rounds of weapon fire.

Operational status
Available.

Specifications
Size: 11.43 L × 4.06 W × 5.08 H cm
Weight: 142 g (with batteries)
Range:
IR aiming mode: >200 m
IR illuminator mode: >50 m
Visible aiming mode: >50 m/>700 m night
Focusable visible illuminator: >25 m (facial recognition)
Wavelength:
IR mode: 810-850 nm
Visible mode: 615-645 nm
Target repeatability: ±1.27 cm at 25 m
Attachment/detachment time: <15 s
Windage and elevation adjustment: (Adjustment tool stowed on unit) 30.48 cm min total horizontal and vertical deflection at 25 m
Power
Type: 2 standard 'DL123A' 3 Volt lithium batteries
Battery Life: 300 Illuminators (laser and flashlight simultaneously)
Environmental: Meets full Military specifications
Shock and Vibration: Meets full Military specifications
EMI/EMC: Meets full Military specifications
Water proof: 20 m submersion for 2 h
Durability: 30,000 rounds of weapon fire

Contractor
ITT Industries, Night Vision.

VERIFIED

Night Vision Equipment Company – GCP-2 combat commander's tactical aimer/pointer

Type
Weapon Sight (Illuminating).

Description
The GCP-2 Combat Commander's Tactical aimer/pointer (CCT aimer) combines the best features of NVEC's GCP-1A/B pointer and the US Army's AN/PAQ-4 weapon aimer. It may be used as a hand-held pointer as well as an illuminator for marking and illuminating targets for night strike. A weapon mount and zeroing functions significantly enhance combat utility.

When mounted on an individual or crew served weapon the user has several options. These include aiming the weapon at maximum range, marking a hard target for air strike, and illuminating small targets at closer range to allow identification using night vision equipment. In addition to weapon mounting and hand held use the CCT Aimer may be vehicle mounted, attached to a night vision camera or used with a night sight to extend target acquisition range.

Operational status
Available.

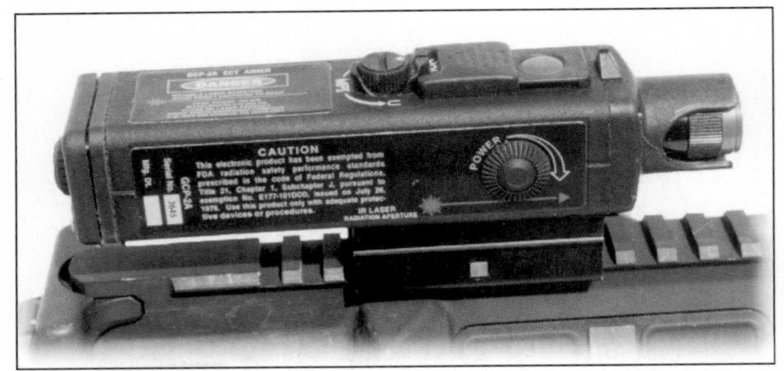

The GCP-2 tactical aimer/pointer 0041417

Specifications

	GCP-2	GCP-2A
Weight:	143 g	143 g
Output power:	50 mW	100 mW
Wavelength:	830 nm	830 nm
Beam modulation:	continuous wave	continuous wave
Range:		
(0.5 mrad (dot) beam)	8,000 m	12,000 m
(2° beam)	1,500 m	3,000 m
(10° beam)	800 m	1,600 m
Power supply:	2 × AA batteries	

Contractor
Night Vision Equipment Company Inc.

VERIFIED

Night Vision Equipment Company – Lasergrips – laser sighting system

Type
Weapon sight (illuminating).

Description
The laser grip is designed to be used with pistol-type hand-held weapons such as those from Beretta, Ruger and SIG Sauer. It projects a collimated laser beam in the visible red or near wave infra-red parts of the spectrum. The infra-red model is designed for use with night vision devices. For boresighting, the unit incorporates a preset zero setting which enables close to zeroed alignment when first fitted. Windage and elevation adjusters allow fine zero adjustment. The equipment is manufactured by the Crimson Trace™ Corporation (qv) and marketed by NVEC.

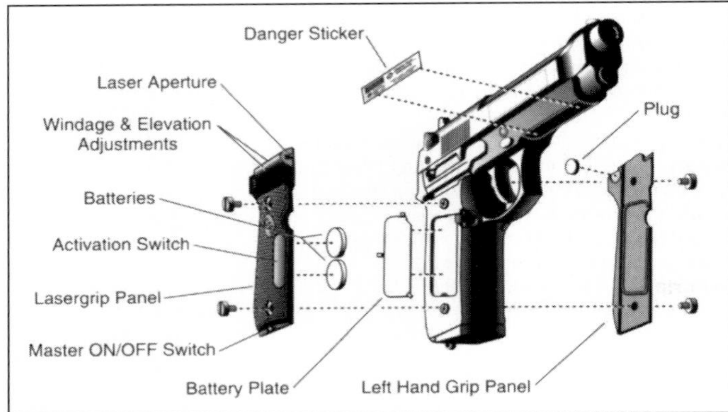

The NVEC Lasergrips laser sighting system 0041414

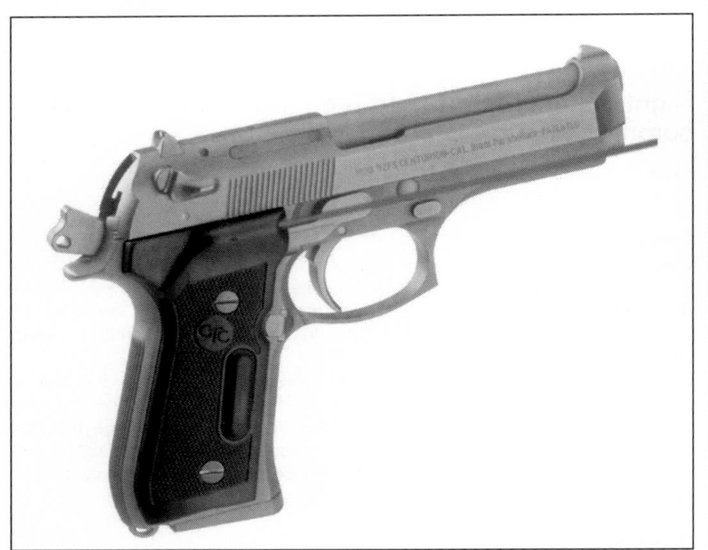

Lasergrips LG-202 fitted to Beretta pistol 0041413

Operational status
Available.

Specifications
Typical system performance

Model	Visible red	Infra-red
Wavelength	633 nm	850 nm
Power output	3 mW	3 mW
Range	300 m	1,000 m (with GEN III NVGs)
Dot size	13 mm approx at 15 m 7.5 cm at 100 m	13 mm approx at 15 m 7.5 cm at 100 m
FDA laser	Class IIIA	Class IIIB

Contractors
Night Vision Equipment Company Inc.
Crimson Trade™ Corporation

VERIFIED

Night Vision Equipment Company Ground Commander's Pointer GCP-1

Type
Weapon sight (illuminating).

Description
The Night Vision Equipment Company (NVEC) Ground Commander's Pointer (GCP-1A/B) is an Infra-Red (IR) pointer and illuminator, invisible to the naked eye but extremely visible through night vision systems sensitive to the 830 nm wavelength. Its adjustable lens permits the focusing of a 0.5 mrad pencil beam at ranges in excess of 8,000 m, as well as intermediate illumination at shorter ranges, ending in a 'fan' 30° wide. There is a selectable eye-safe mode for training purposes. The original GCP-1 pointer is in service with the US special operations community. The similar GCP-2 can be used as a hand-held pointer or may be mounted on a weapon.

Operational status
Available.

Specifications
Dimensions: 150 × 48 × 30 mm
Weight: 128 g
Wavelength: 830 μm
Beam modulation: continuous wave
Power output:
GCP-1A 50 mW
GCP-1B 100 mW

Range	GCP-1A	GCP-1B
(0.5 mR beam)	8,000 m	12,000 m
(2° beam)	1,500 m	3,000 m
(10° beam)	800 m	1,600 m

Contractor
Night Vision Equipment Company Inc.

VERIFIED

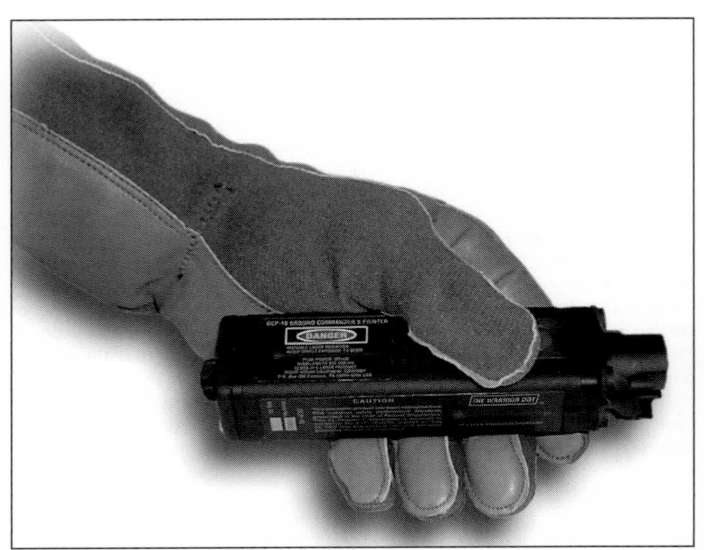

NVEC GCP-1A Ground Commander's Pointer 0041402

Night Vision Equipment Company NVEC Variable Intensity Aiming Light – VITAL

Type
Weapon sight (illuminating).

Description
The VITAL is a family of three infra-red aiming lights of different power levels that may be used on individual or crew-served weapons. Characteristics include selectable beam intensity and projection mode. The beam intensity may be adjusted to best suit the ambient lighting condition, target contrast and range. The projected beam may be selected to operate in steady, fast or slow pulsing modes. The Close Quarter Battle (CQB) illumination modes include a steady aiming light. The selectable two-position CQB illuminator provides covert augmentation lighting for night vision operations in extremely dark environments.

Operational status
Available.

Specifications

	VITAL-1	VITAL-2	VITAL-100
Output power:	0.7 mW	0.7 mW	100 mW
Wavelength:	830 nm	830 nm	830 nm
Beam modulation:	Steady, slow or fast pulse		
Range:	1,000 m	2,000 m	12,000 m
Beam diameter:	0.5 mrad	0.5 mrad	0.5 mrad
Power supply:	2 × AA batteries		
M16 Plant assembly	no		yes

Contractor
Night Vision Equipment Company Inc.

UPDATED

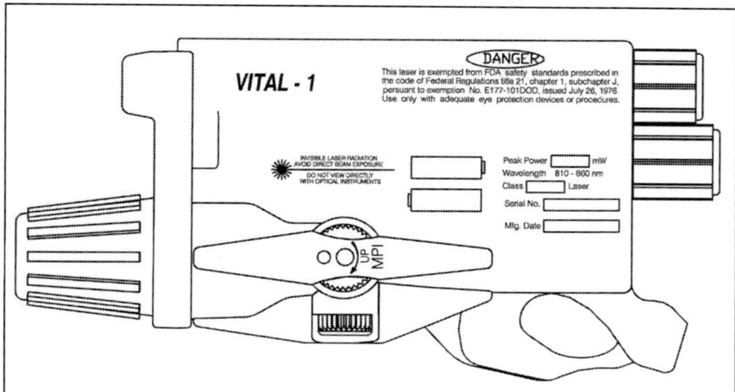

Drawing of VITAL 0041405

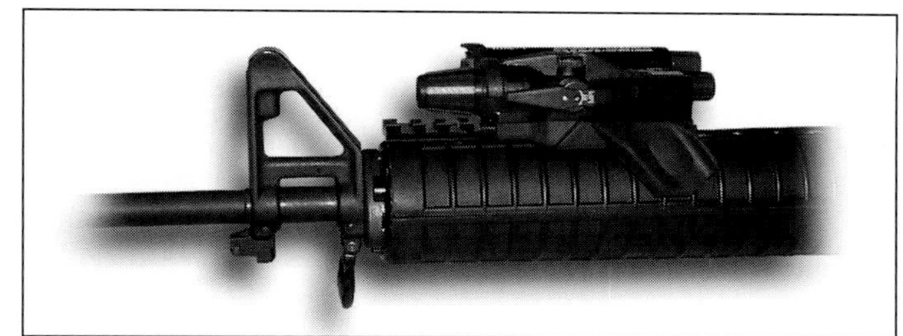
The VITAL variable intensity tactical aiming light 0041404

Night Vision Equipment Company Strike Eagle – day/night weapon aimer Models 509/510/609/610

Type
Weapon sight (illuminating).

Description
The Strike Eagle designed primarily as a weapon sight is a versatile, lightweight unit that adapts for one handed direct viewing, or can be secured to still or video cameras for night photography. The basic unit is an image intensified pocketscope with a ×1 objective lens, eyepiece and 5 mW (Class IIIa) eye-safe laser. Strike Eagle models use commercial specification second- or third-generation image intensifiers.

The sight's visible red dot marks targets out to 1 km. It may be zeroed to a weapon for accurate off-hand aiming in daylight or through the night vision scope at night. The unit is equipped with a weapon mounting foot.

Operational status
Available.

Strike Eagle – day/night weapon aimer 0041407

Specifications

Model	**509**	**610**
Tube generation	II	III
Typical detection range in m for still human target on green grass		
Quarter moon (10^{-2} lux)	236	427
Starlight (10^{-3} lx)	117	310
Overcast (10^{-4} lx)	95	163

All models
Weight: 0.45 kg
Field of view: 25°
Magnification: ×1 (×2, ×3 optional)
Focus range: 25 cm to infinity
Objective focal length: 25 mm (f/1.4, T/1.58)
Dioptre adjustment: +2 to –6
Eye relief: 15 mm (min)
Laser
Output power: 5 mW selectable continuous or pulsed
Wavelength: 635 nm (visible red)
Range: night 1,000 m, day 200 m
Beam: 3.8 cm dot at 23 m
System power supply: 2 × AA batteries

Contractor
Night Vision Equipment Company Inc.

VERIFIED

Northrop Grumman Model 9886A Infra-red Aiming Light (IAL)

Type
Weapon sight (illuminating).

Description
The Model 9886A Infra-red Aiming Light (IAL) is used in conjunction with night vision goggles to provide an aiming point that will allow the user to deliver accurate fire at night. The IAL consists of an aiming light assembly and a carrying bag. Mounting brackets are available for adapting to most weapons currently in use and can be supplied with the aiming light.

A borelight assembly with either a 5.56 or 7.62 mm mandrel can be supplied as an accessory to the IAL. The borelight assembly is a small lightweight infra-red source which is used to provide a fast and accurate means of boresighting the IAL to the bore of any weapon. The IAL incorporates a mounting base that is compatible with standard AN/PVS-4- type adaptor brackets. Additional mounting brackets are available which permit the IAL to be mounted to either the NATO STANAG base or a Weaver-type rail.

Controls and adjustments include a remote on/off switch and boresight adjustments. The boresight adjustment includes azimuth and elevation knobs indexed at 0.5 mm increments.

Operational status
Available.

Specifications
Length: 208 mm
Width: 52 mm
Height: 75 mm
Weight: 345 g with batteries
Optical output power: 3.2 mW
Output peak wavelength: 820 nm
Output beam size: 0.2 mrad (max)
Spectral bandwidth: 12 nm
Range: >400 m
Input DC current: 95 mA
Boresight adjustment: 0.5 mm clicks

Model 9886A infra-red aiming light

Power supply: 4 × AA batteries or 2 × BA5567/U lithium
Operating temperature: –54 to +51°C
Storage temperature: –57 to +65°C

Contractor
Northrop Grumman Electro-Optical Systems.

UPDATED

Octec AIM10 video tracker

Type
Infantry weapon (illuminating) miniature video tracker.

Development
The AIM10 video tracker has been developed by Octec to meet applications where space, weight and power consumption is at a premium. Selected for the US Army OICW and OCSW next-generation infantry weapons programmes the AIM10 provides the sighting systems for these weapons with sophisticated automatic multiple target detection and tracking. The unit which measures 60 × 60 × 10 mm is a fully programmable video tracker and is capable of running all the company's ADEPT family of algorithms.

Operational Status
The AIM10 is a fully developed unit.

Contractor
Octec Ltd.

VERIFIED

Thales Optronics TM series laser target pointers

Type
Weapon sight (illuminating).

Description
This family of small, lightweight, infra-red laser target pointers is specially designed for close-range combat with small arms at night. It enables the user equipped with an image intensifying night vision device to aim at a target with great accuracy under low-light conditions or in complete darkness, without visual recognition by the enemy.

The laser pointers are primarily used on basic infantry weapons such as rifles and sub-machine guns and aligned to the axis of these weapons. When the user activates the pointer for some seconds by means of a remote switch on the weapon, the emitted laser beam projects a bright infra-red spot exactly where the weapon is to be aimed and where the bullet will hit. This enables a precise shot without necessitating aiming through rifle scopes or mechanical sights.

Operational status
In production.

Specifications

	TM-Mini	TM-007	TM-077	TM-Maxi
Dimensions	51 × 49 90 mm	157 × 34 34 mm	115 × 35 18 mm	203 × 30 18 mm
Weight inc batteries	285 g	350 g	230 g	250 g
IR laser output power	0,35 mW	2 mW	5 mW	100 mW
Wavelength	820 nm	820 nm	810 nm	830 nm
Beam diameter	4 mm	5 mm	5 mm	6.7 mm
Beam divergence	0.4 mrad	0.35 mrad	0.5 mrad	0.26 to 86.6 mrad
Batteries	1 × 1.5 V AA	2 × 1.5 V AA	2 × 1.5 V AA	2 × 1.5 V AA
Environmental	MIL-STD-810	MIL-STD-810	MIL-STD-810	MIL-STD-810

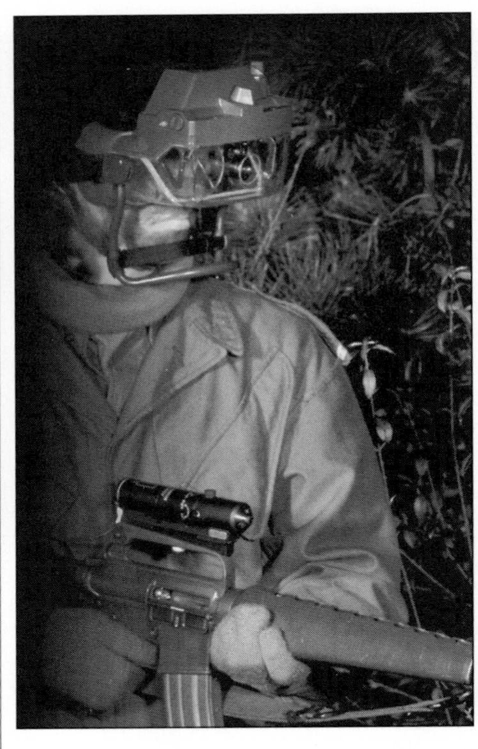

The TM-007 laser target pointer from Thales Optronics

Contractor
Thales Optronics (formerly OIP Sensor Systems).

UPDATED

Transvaro ATOK IR laser target pointer

Type
Weapon sight (illuminating).

Description
The ATOK IR laser target pointer is designed for use with a variety of weapons. Specification details are not stated, but the unit is compatible with night vision goggles. Horizontal and vertical boresighting adjustments are provided. A low-/high-power selection switch and a momentary action switch, are fitted to minimise illumination time on the target.

Operational status
Available.

Contractor
Transvaro AS, Turkey.

VERIFIED

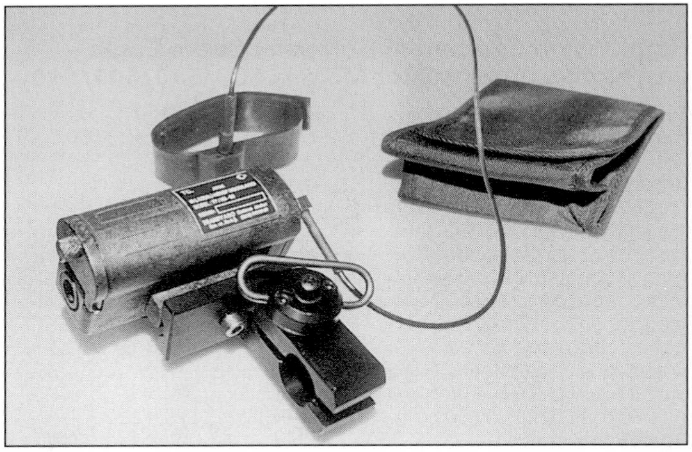

The Transvaro ATOK IR laser target pointer 0055256

INFANTRY WEAPON SIGHTS

PASSIVE - CREW-SERVED WEAPONS

Brashear Small Arms Fire-Control System II (SAFCS II)

Type

Weapon sight (crew-served, fire control).

Description

The Brashear Small Arms Fire Control System II (SAFCS II) is a high quality optical fire-control system that provides both day and night full ballistic solutions. Designed for use with the Mk 19 (40 mm), M2 .50 Cal, long-range sniper rifle, and various other crew-served weapon systems, SAFCS II has demonstrated vastly improved first burst probability of hit. A typical target at 1,000 m can be engaged using the SAFCS II with a first hit probability of 54 per cent compared to only 3 per cent without the SAFCS II.

SAFCS II is being developed to provide precise aiming accuracy to weapons such as the 40 mm Mk 19 automatic grenade launcher, the M2 heavy machine gun, and long-range sniper weapons. A programmable ballistic computer module enables the SAFCS II to be easily adapted to new weapons and ammunition types.

It is a single unit mounted over the host weapon receiver. The unit contains high quality ×5 direct view optics for long-range target surveillance and identification, an eye-safe laser range-finder with an operational range of 4,000 m, an internal scene/laser steering reticle controlled by a small joystick and a computerised system for correction of air temperature, atmospheric pressure, crosswinds and weapon cant. The SAFCS II accommodates a variety of night vision sensors on a MIL-STD-1913 Picatinny rail and provides a video display with annotated systems information and a ballistic solution reticle.

An optional internally-mounted laser designator can provide full-solution fire control during night operations using standard night vision goggles. An integral battery has a life of more than 10 hours and the SAFCS II can be connected to a 14 to 32 V DC vehicle power supply.

Field testing of the SAFCS II by the US Army and Marine Corps with the Mk 19 has demonstrated a SAFCS II hit probability increase from 3 per cent to 54 per cent at 1,000 m.

Features:

Readily adaptable for multiple crew-served weapons and applications:

- Class I eye-safe solid state laser range-finder
- High-resolution video day sight
- Self-contained night sight with un-cooled IR
- 5° diagonal FOV
- <5 kg system weight
- RS-170 output makes multiple display options possible
- Remote operation capability
- Ballistic solutions for Mk 19 GMG, M2 MG, and LRSR
- −32 to +41°C operating temperature, −51 to +71°C storage temperature
- Designed for rugged military use per MIL-STD 810 and MIL-STD 461.

Operational status

Initially, 20 production units were delivered to US Army Picatinny Arsenal for type classification in 2003. Type classification in 2004.

Contractor

Brashear LP.

UPDATED

Brashear SAFCS fire control on the Mk 19 0041474

Euroatlas AN/PVS-4, AN/TVS-5 weapon sights, EURONOD-2 observation sight

Type

Weapon sight (crew-served).

Description

All three of these sights are image intensifying devices using a 25 mm second-generation microchannel plate tube, and are equipped with the same common standardised battery housing and eyepiece/eyeguard assembly which guarantees interchangeability. Only the objective lens assemblies are different and serve the required purpose.

The second-generation technology has built-in automatic gain control protecting the tube against muzzle and detonation flash, which results in a uniform image without tendency to flare. Additional manual brightness control allows the operator to adjust screen luminance to his individual requirements and for contrast enhancement.

AN/PVS-4 and AN/TVS-5 are weapon sights with different magnifications. They use interchangeable projected graticules for a wide variety of weapons and can be easily adapted to the weapon by using appropriate brackets.

EURONOD-2 is an observation sight with high magnification, with a wide range of applications, particularly for artillery fire control.

Operational status

Available.

Specifications

Model	**AN/PVS-4**	**AN/TVS- 5**	**EURONOD-2**
Magnification	×3.7	×5.8	×9.7
Field of view	14°	9°	5.6°
Object lens	95 mm f/1.6(T)	155 mm f/1.6(T)	258 mm f/1.7(T)
Focus range	10 m – ∞	25 m – ∞	75 m – ∞
Length	240 mm	310 mm	480 mm
Width	120 mm	160 mm	240 mm
Height	120 mm	170 mm	240 mm
Weight	1.7 kg	3.6 kg	9 kg
Power supply	2 × AA/LR6	2 × AA/LR6	2 × AA/LR6
Battery life	50 h (20°C)	50 h (20°C)	50 h (20°C)
Operating range	to 600 m	to 1.2 km	to 6 km

Contractor

Euroatlas GmbH.

VERIFIED

AN/PVS-4 rifle sight with G3 rifle mounting

General Canada Lightweight Video Sight (LVS)

Type
Weapon sight (crew-served).

Development
The LVS is a compact video-based full solution day/night fire-control sighting system. It was developed utilising patented CDC video sighting technology, expressly for direct fire crew-served weapon systems.

Description
Incorporating a laser range-finder, day/night vision (Gen II though Gen IV image intensifier) and powerful ballistic computer, LVS provides extremely precise ballistic solutions for a variety of preprogrammed ammunitions. In doing so, it uses inputs from automatic environment sensors as well as direct data entry from the gunner. The LVS utilises a graphical user interface allowing such features as visual marking of the limits of fire, predesignated targets, visual status display, Built-In Testing (BIT) and so on. It has an RS-170 output which can be connected to an integrated display on the weapon or to a remotely viewed display. Target grid positions can be determined using range and onboard azimuth tracking data. The day/night CCD system with ×12 magnification also provides an effective surveillance capability.

The LVS computer also has a number of unique features which allow easy boresighting and quick maintenance. These capabilities are presented in a series of programmes which can be activated by the gunner, or the maintainer.

The LVS is simple for the gunner to use. The target is lased. The computer calculates the required azimuth and elevation offset moving the aiming reticule accordingly. The gunner returns the aiming reticule onto the target and fires. From lasing to firing is less than 2 seconds. Power is provided by either a 28 V vehicle power adaptor, or by a standard NATO battery with a minimum 8 hour duration. The LVS also includes the following features:

- full ballistic solution computation for multiple ammunition types
- air pressure and temperature sensing
- ammunition temperature sensing
- cant and angle of sight sensing
- high resolution display
- easy to use text and graphics overlay
- remote controls and video output – RS-170 interface

Operational status
Developed and available.

Specifications
Magnification: ×12
Image intensification: Gen II through Gen IV
Laser range-finder: eye-safe
Range: 40 – 3,900 m
Range accuracy: ± 2 m

Contractor
General Dynamics Canada.

UPDATED

The General Dynamics Canada Lightweight Video Sight (LVS) 0102864

Hellenic Aerospace Polifimos night vision aiming monocular

Type
Weapon sight (crew-served).

Description
The Polifimos night vision aiming monocular is an integral part of the Hellenic Aerospace Polifimos Artillery Laser Range-finder (qv), to enable aiming at night. The system uses a second-generation 25 mm image intensifier tube and has a large aperture catadioptric objective lens.

Polifimos night vision aiming monocular 0004826

Operational status
Available.

Specifications
Weight: 2 kg
Dimensions (l × diameter): 24 × 12
Magnification: ×4
Field of view: 14°
Focal length: 95 mm
Resolution
at 0.1 lx: 2.7 lp/mrad
at 1.0 lx: 1.7 lp/mrad
Voltage: 3 V (2 × AA 1.5 V alkaline batteries)

Contractor
Hellenic Aerospace Industry SA.

VERIFIED

Institute of Optronics Pakistan AN/TVS-5A crew-served weapon sight

Type
Weapon sight (crew-served).

Description
This licence-produced sight is intended for use on heavier weapons such as recoilless rifles, heavy machine guns and some types of wire-guided anti-tank missile. It can also be tripod mounted for use as a surveillance device.

The sight uses a 25 mm second-generation image intensifier tube and has automatic gain control so that sudden bright lights in the target area will not affect the system. The internally mounted adjustable illuminated graticule can be interchanged to cater for different weapons or ammunition. There is a bright source protection circuit which prevents damage due to inadvertent exposure to daylight and the sight is provided with a cover which allows daytime boresighting and training.

Operational status
In production.

Specifications
Length: 310 mm
Width: 160 mm
Height: 170 mm
Weight: 3 kg

AN/TVS-5A weapon sight on 106 mm recoilless rifle

Magnification: ×6
Field of view: 9°
Object lens: 155 mm T/1.7
Power supply: 2 AA cells
Battery life: 15 h at 20°C

Contractor
Institute of Optronics Pakistan.

UPDATED

Night Vision Equipment Company AN/TVS-5 weapon sight Models 770/780/790

Type
Weapon sight (crew-served).

Description
The Night Vision Equipment Company is one of a number of manufacturers of sights selected by the US Army to manufacture the AN/TVS-5. The specification required vehicle acquisition ranges of 1,000 m in starlight and 1,200 m in moonlight. The Model 780 uses a second-generation image intensifier tube and matches this requirement. It is fitted with a Military Specification image intensifier tube. The 790 is fitted with a third-generation tube which improves the range performance by at least 10 per cent. The 770 and 790 have commercial specification tubes. The sight is suitable for heavy and medium machine guns, recoilless rifles and small cannons.

Reticle adjustments include brightness, elevation and azimuth in 0.25 mil increments. Automatic brightness control is a feature of the intensifiers as well as protection from over bright light sources.

Operational status
In service.

Specifications
Typical detection range in m for still human target on green grass:

Model	770	780	790 (GEN III)
Quarter moon (10^{-2} lx)	1654	1760	1931
Starlight (10^{-3} lx)	1032	1117	1369
Overcast (10^{-4} lx)	398	470	693

(All models)
Dimensions: 310 × 160 × 170 mm
Weight: 3 kg
Field of view: 9°
Magnification: ×6.2
Focus range: 25 m to infinity
Dioptre adjustment: –5 to +4
Eye relief: 34 mm
Power supply: 2 × AA batteries

Contractor
Night Vision Equipment Company Inc.

VERIFIED

NVEC's AN/TVS-5 weapon sight 0023062

Norinco Type JWJ machine gun low-light level sight

Type
Weapon sight (crew-served).

Description
The Type JWJ is a second-generation image intensifying sight intended for use on machine guns and other crew-served weapons and also as an independent surveillance instrument. It is provided with bright light protection circuitry which prevents interference from flashes and bright illumination in the target area. An accessory handle is provided for when the sight is used for hand-held observation.

Operational status
In production.

Specifications
Weight: 2.1 kg
Magnification: ×4.5
Field of view: 10°
Focusing range: 20 m to infinity
Dioptric adjustment: ±2.5 dioptres
Resolution: 0.5 mrad (E - 10^{-1} lx, C - 0.85); 1.2 mrad (E - 10^{-3} lx, C - 0.35)
Night vision range
Person: 500 m
Vehicle: 1 km

Contractor
China North Industries Corporation (Norinco).

VERIFIED

Type JWJ night vision sight

Northrop Grumman AN/TVS-5 second-generation crew-served weapon sight

Type
Weapon sight (crew-served).

Description
The AN/TVS-5 weapon sight was originally designed and manufactured under a US Army contract. It has been continuously improved and can be upgraded from a standard 25 mm second-generation with a 25 mm third-generation drop-in image intensifier.

AN/TVS-5 is suitable for heavy and medium machine guns, recoilless rifles, small cannons and missile launchers. Its reticle is matched to the weapon through a cell housing inserted at the time of manufacture. After mounting, the system is boresighted by adjusting the reticle elevation and azimuth actuators. Automatic gain control circuitry is employed to maintain the viewed scene illumination constancy during periods of changing light level conditions, such as the period from sunset to full darkness.

The tube features automatic brightness control for muzzle-flash protection, illuminated reticle and tube brightness control.

Operational status
In production and in service with the US Army and Marine Corps and 20 other countries.

AN/TVS-5 second-generation crew-served weapon sight mounted on 40 mm Mk 19 grenade launcher

Specifications

Dimensions: 310 × 160 × 170 mm
Weight: 3.67 kg
Field of view: 9.2°
Magnification: ×5.8
Focus range: 25 m to infinity
Dioptre range: +2 to –5
Power supply: 2 × AA batteries
Battery life: 60 h at +20°C

Contractor

Northrop Grumman Electro-Optical Systems.

UPDATED

Northrop Grumman NVS-800 night vision system

Type

Weapon sight (crew-served).

Description

Designed for use on heavy automatic weapons, recoilless guns and other armament of similar size, the NVS-800 sight closely resembles the NVS-700 (small Starlight scope) individual weapon sight but has a greater range capability resulting from the use of a larger objective lens. The eyepiece, 25 mm second-generation image intensifier and associated components are common to both sights.

Operational status

In service with US and other armies.

Specifications

Diameter: 165 mm
Length: 355.6 mm
Weight: 3.856 kg
Field of view: 156 mils
Magnification: ×6 (nominal)
Viewing range
Moonlight: 25-2,000 m
Starlight: 25-1,200 m
Objective lens: 155 mm T/1.7
Power supply: 2 × AA mercury cells
Battery life: 60 h

NVS-800

Contractor

Northrop Grumman Electro-Optical Systems.

UPDATED

Ortek ORT-TS 5 crew-served weapon sight

Type

Weapon sight (crew-served).

Description

The second-generation ORT-TS 5 night vision sight was primarily designed for heavy machine guns, recoilless rifles and other crew-served weapons. It serves such weapons as the M2 0.50 machine gun, the 106 mm M40 recoilless rifle and the B-10 rocket launcher. On weapons where access to the conventional eyepiece is hazardous, an optional right-angled relay is available.

The sight has an internally mounted adjustable illuminated reticle for boresighting. Different reticle patterns are available. Image intensification is performed by a second-generation 25 mm tube. To compensate for differing levels of ambient light, image tube and reticle brightness can be manually adjusted. Power is provided by two standard 1.5 V AA batteries which can maintain operation for a minimum of 50 hours.

In order to facilitate logistics and maintenance, major assemblies and subassemblies are interchangeable with the AN/PVS-4 individual sight and the AN/TVS-5 crew-served weapon sight.

Operational status

In production.

Specifications

Diameter: 165 mm
Length: 356 mm
Weight: 3.8 kg
Field of view: 9° (156 mils)
Magnification: ×5.6
Dynamic range: from 10^{-5} to 10^{-2} FC
Viewing range
Moonlight: 2 km
Starlight: 1.2 km
Focal length: 155 mm
Focus adjustment: 25 m to infinity
Reticle adjustment: ±2.5° (in 0.25 mil increments)
Eyepiece focal length: 26.5 mm
Eyepiece adjustment: –6 to +3 dioptres

Contractor

Ortek Limited.

VERIFIED

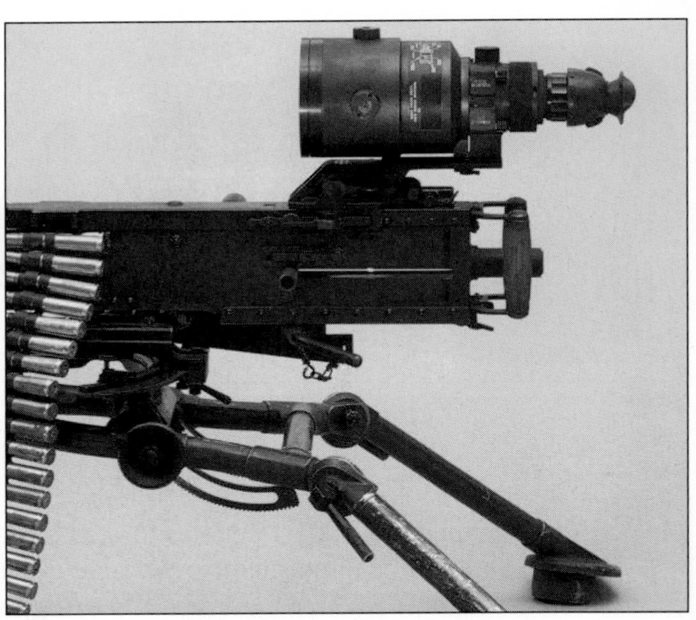

Ortek TS-5 crew-served weapon sight on M2 machine gun

Raytheon Integrated Sight Module ISM

Type
Crew-served weapon sight.

Development
Raytheon has been awarded a US$5.7 million contract by the US Army Communications-Electronics Command (CECOM) Night Vision and Electronic Sensors Directorate (NVESD) to deliver 12 protoypes of the Integrated Sight Module (ISM). This technology demonstration forms part of the US Army's Land Warrior programme to develop the next generation of systems for the soldier on the battlefield.

Description
The Integrated Sight Module will integrate an eye-safe laser range-finder, thermal imager, electronic compass, eye-safe infra-red laser pointer and CCD camera into a lightweight weapon sight.

Operational status
Under development.

Contractor
Raytheon Systems Company.

UPDATED

RH-ALAN NC-2

Type
Weapon sight (crew-served).

Development
The development of the NC series night sights began in 1991. The first model, NC-1, was used by the Croatian Army in military operations in 1992. Model NC-2 has been in serial production since 1993.

Description
The NC-2 is a night sight that can be used with various infantry weapons, as well as artillery pieces for direct aiming. The sight uses image intensifier tubes of second-, second-plus or third-generation with microchannel plate and electronic focusing and has automatic gain control.

Operational status
In production and in service with the Croatian Army since mid-1995.

Specifications
Magnification: ×4
Field of view: 14°
Objective diameter: 100 mm
Objective focal length: 100 mm
Objective f/No: 1
Eyepiece focal length: 25 mm
Eyepiece field of view: 60°
Eye relief: 25 mm

Contractor
RH-ALAN.

VERIFIED

NC-2 night sight mounted on a long-range Croatian army 20 mm sniper rifle

Ring Sights COSIMUN – Collimated Sight for use with MUNOS

Type
Weapon sight (crew-served).

Development
The COSIMUN sight is the result of co-operation between Ring Sights and Thales Optronics (formerly Delft Sensor Systems) to enhance the aiming capability using the MUNOS night vision monocular.

Description
COSIMUN is one of a family of unity power solid glass collimator sight suitable for crew served or personal weapons as well as applications such as MANPADS. The solid glass construction gives robustness and prevents internal misting problems. The device is fixed and boresighted to the weapon and projects an illuminated graticule in the same focal plane as the target. By day the sight may be used stand-alone. At night the sight is used with the MUNOS night vision monocular.

Operational status
In service.

Specifications

Type	COSIMUN MG	COSIMUN 9K38
Application	machine guns	IGLA-SA-18 MANPADS
Dimensions (l × h × w)	170 × 120 × 65 mm	140 × 110 × 175 mm
Aperture	40 × 35 mm	40 × 27 mm
Weight	1.7 kg	1.56 kg

Contractor
Ring Sights Holdings Company Ltd.

UPDATED

The COSIMUN sight from Ring Sights 0010870

Simrad LP101 laser gun sight

Type
Weapon sight (crew-served).

Description
The Simrad LP101 is a complete fire-control system for mounting on direct fire weapons, providing instant target range and aiming information. The built-in ballistic computer is programmed for the type of weapon and ammunition used. Aiming point and time of flight information are displayed in the eyepiece. The time of flight information enables the gunner to estimate the correct lead angle.

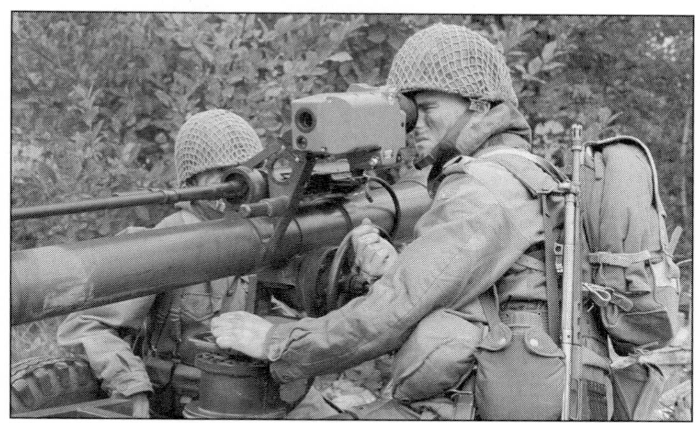

Simrad LP101 laser gun sight on 106 mm RCL gun

The Simrad LP101 is suited for upgrading the 106 mm M40 RCL gun and similar weapons, increasing hit probability and effective range. Night capability can be achieved using Simrad KN200 and KN250 image intensifiers. The add-on principle of these instruments ensures that the eye position of the gunner remains unchanged by day or night and no boresighting is required.

The LP101 can be programmed by the operator for ballistic and environmental data. Parameters displayed as standard are minimum range, boresight marker, four types of ammunition, manual range setting and the dimming of the display. Optional parameters which can be programmed include ammunition stock, compensation for wind, muzzle velocity, air temperature and powder temperature, adjustment of display contrast and switching the time of flight indication on or off.

Operational status
In production.

Specifications
Dimensions: 320 × 195 × 140 mm
Field of view: 200 mrad (11.3°)
Laser type: Nd:YAG (1.064 μm)
Sighting telescope: 4 × 30
Reticle pattern: to customer specification
Range resolution: 5 m
Range
Min: 60 m
Max: 9,995 m
Power supply: 10-15 V
Battery type: Ni/Cd (rechargeable); alkaline optional
Rangings per battery before recharge: up to 400 at 20°C

Contractor
Simrad Optronics ASA.

VERIFIED

Thales Optics Maxi-Kite weapon sight

Type
Weapon sight (crew-served).

Description
Maxi-Kite is a compact, lightweight night vision sight for use as a portable long-range surveillance or crew-served weapon sight. This fully ruggedised system has a ×6 magnification and provides a high-resolution image for observation and long-range weapon aiming. Maxi-Kite is produced by adding an afocal lens assembly to Thales' Kite sight.

Maxi-Kite can be supplied either as a complete sight unit or as an upgrade to the standard Kite individual weapon sight. The sight can be fitted/supplied with a suitable graticule to interface fully with a weapon's ballistic charateristics. The sight is particularly suited for use on sniper rifles, heavy machine guns and shoulder launched or crew-served anti-tank weapons. Interface brackets can also be provided.

Operational status
In production and in service.

The Maxi-Kite night sight 0004823

Thales Maxi-Kite night vision sight fitted on a heavy machine gun

Specifications
Weight: 1.36 kg (excluding batteries)
Dimensions: 360 × 90 × 95 mm
Magnification: ×6
Field of view: 5.5°
Power supply: 2 × 1.5 AA batteries
Eyepiece dioptre: fixed at -1±0.35 dioptre, adjustable at +2 to -6.00 dioptre
Focus range: 25 m to infinity
Eye relief: 30 mm
Range (min): 580 m (recognition of standing man in starlight), 10^{-3} lx using GEN II tube

Contractor
Thales Optics.

VERIFIED

Thales Optronics DFS90

Type
Direct fire sight.

Description
The DFS90 is a direct fire sight for the AS90 self-propelled howitzer. The sight is configured as a mantlet-mounted telescope. It provides the gunlayer with the capability of engaging stationary or moving targets in the direct fire mode via a detachable, monocular eyepiece elbow. Night firing capability is provided by the use of a 18 mm second- or third-generation image intensifier. A ballistic reticle matched to Charge 8 is also incorporated in day and night channels. Other features include a cursor for range setting and adjustment for other charges, trunnion tilt compensator, gun ready lamp and a heated mantlet window.

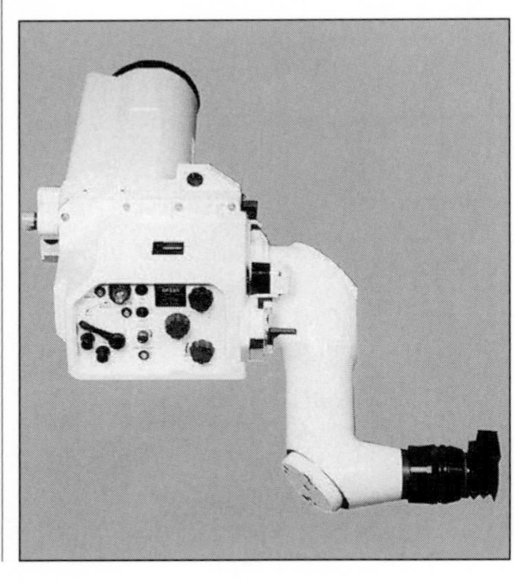

DFS90 direct fire sight for the AS90 self-propelled howitzer

Operational status

In service.

Specifications

Dimensions (l × w × d): 867 × 615 × 275 mm
Day channel
Magnification: ×6.2
Field of view: 117 mrad
OG focus: fixed, 200 m to infinity
Range: detection of AFVs at 2,000 m (with visibility of 8,000 m)
Eye protection: fixed laser protection filter
Night channel
Magnification: ×6.2
Field of view: 117 mrad
Focus: adjustable, 50 m to infinity
Resolution: 0.783 lp/mrad
NATO Stock Number: NSN 1240-99-477-4166

Contractor

Thales Optronics (formerly Avimo Ltd).

VERIFIED

Yugoimport anti-tank gun fire-control system

Type

Weapon sight (crew-served).

Description

The anti-tank gun FCS is designed for use on various types of anti-armour guns including recoilless rifles. It enables the observation of target distance measurement to, target tracking of and firing at both static and moving targets during day or night engagements.

The FCS comprises the following subsystems:

(a) day/night sight with integral 1.064 μm wavelength Nd:YAG laser range-finder unit
(b) built-in solid-state ballistic Digital Computer (DC) which controls the whole system (including the laser firing) and calculates the firing solution lead angle and superelevation figures using laser range data, manually input meteorological and ammunition data (up to four types of information stored). Once the solution is calculated it then generates an electronic aiming mark within the sight field of view
(c) exchangeable Battery Block (BB)
(d) angle speed sensor (for direction and elevation)
(e) interconnecting cable set.

All the gunner has to do during an engagement is acquire the target, fire the laser, lay the weapon to the calculated firing parameters and fire the weapon.

Up to the maximum target distance, which is a function of the weapon type and the ammunition type used (for example, for a 106 mm recoilless rifle it would typically be 1 km), the required superelevation is directly indicated by an aiming mark, offset in the vertical direction so that there is no necessity to display the range and possibly distract the gunner. However, if required, the gunner can press a button and recall the value for display on a four-digit, seven-segment LED display which superimposes itself onto the image viewed through the system eyepiece.

For ranges above the engagement limit it is assumed that only soft targets are being engaged which have a negligible crossing speed so that in most instances it is not necessary to calculate a lead angle. To allow the gunner to fire the weapon with a minimum of delay the range display appears in the gunner's field of view immediately after the LRF is activated. All he has to do then is lay the weapon in elevation by use of the built-in ballistic reticle. If the lead angle is required then it is still possible to observe the target movement during the TOF indicated by the inbuilt computer. The reticle has horizontal lines corresponding to ranges of 200 to 2,000 m and vertical lines for lead angle determination.

Operational status

Production as required. In service with the former Yugoslav armies.

Specifications

Weight: approx 40 kg
Dimensions: 550 × 280 × approx 260 mm
Target speed tracking rates
Traverse: 0-20 mrad/s
Elevation: 0-10 mrad/s
Resolution: 0.001 mrad/s
Power supply: 12 V exchangeable battery block
Day sight
Magnification: ×6.5
Field of view: 7°
Night sight
Magnification: ×7.7
Field of view: 6.5°
Observation distance
At 1 × 10^{-3} lx: >900 m
Laser range-finder
Type: Nd:YAG
Wavelength: 1.064 μm
Automatic laser measuring distance: 180-10,000 m
Automatic laser measuring distance accuracy: ±10 m
Manual laser measuring distance: 100-9,995 m
Manual laser measuring distance accuracy: ±1 m
Resolution: 5 m
Min range gating: 200-3,400 m

Contractor

Yugoimport SDPR.

VERIFIED

INFANTRY WEAPON SIGHTS

PASSIVE – PERSONAL WEAPONS

Aimpoint CompM2/CompML2 weapon sights

Type
Weapon sight (personal).

Description
Aimpoint sights are designed for the 'two eyes open' method of aiming, which greatly enhances situational awareness and target acquisition speed. Thanks to the parallax-free design, the dot follows the movement of the user's eye while remaining fixed on the target, eliminating any need for centring. Also, the sight allows for unlimited eye-relief.

The successful CompM2 and CompML2 family of sights and mounting systems for small calibre firearms such as pistols, sub-machine guns and assault rifles have set a high standard for advanced sighting systems. Present in the CompM2 and CompML2, the new CETechnology is a radical advancement in red dot sighting performance with over 30 times longer battery life than the former XD – sights (CompM and CompML).

CompM2 and CompML2 Reflex Sights are rugged, precision, electronic, optical, red dot sights developed for civilian, military and law enforcement applications. The CompM2 is compatible with first-, second- and third-generation night vision devices, while the CompML2 is optimised for applications, which do not require night vision compatibility.

Operational status
In 2001, Aimpoint AB signed a 5-year contract with US-SOCOM and has already delivered over 100,000 sights to the US Army. CompM2 is the new improved version of CompM, which the US Army has tested and selected for general issue. The French Army has selected the Aimpoint CompML for use with the FAMAS assault rifle. Several Special Forces around the world are also equipped with these sights. CompM and CompML have been replaced by CompM2 and CompML2.

Specifications
Dimensions (l × w): 130 × 55 mm
Weight (md lens covers): 200 g (7.1 oz)
Optical coating
Anti Reflex coating, all surfaces
Multilayer coating for reflection
Band Pass coating for NVD compatibility (CompM2)
Objective diameter: 36 mm
Tube diameter: 30 mm
Material - housing: extruded, high-strength aluminium, anodised
Surface finish: hard anodised, Graphite Grey, matte
Material - lens covers: thermoplastic elastomer, black, non-glare

US Army soldier equipped with an Aimpoint CompM mounted with quick release
0089869

CompM2 with quick-release in combination with night vision device on an M4 carbine
0041432

Optical magnification: ×1 for all models
Eye relief: unlimited, no centring required
Dot size: 4 MOA
Switch, dot brightness: CompM2 has 10 positions:
- 4 NVD, 6 daylight of which 1 extra bright

CompML2 has 10 positions:
- 1 Off, 9 daylight of which 1 extra bright

Battery: one 3 V Lithium battery type 2L76 or DLI/3N
Battery life* (hours): 10.000 h (typically 10 years at CompM2 NVD position)
Adjustment
Range ±2 m at 100 m, in windage and elevation
1 click 10 mm at 80 m = 13 mm at 100 m
Mounting: one wide ring, 30 mm diam, or Aimpoint QR Ring
Max temperature range: –45 to +70°C
Water resistance: submersible to 25 m (80 ft) water depth
* Average values, depending on brightness setting

Manufacturer
Aimpoint AB.

UPDATED

Allen Wasp system P2000SPW/P3000SPW

Type
Weapon sight (personal).

Description
The Wasp system (formerly a product of Marconi Applied Technology) was developed in 1995 and is under evaluation by UK police authorities. The Nite Watch Wasp viewers P2000SPW/P3000SPW are compact night sights for use on a range of semi-automatic and other tactical weapons. Aiming is by the use of a laser target designator, eliminating live zeroing and thus allowing unrestricted use by any number of operators.

Operational status
Under evaluation by UK police authorities.

Specifications
Intensifier type: GEN II or III
Tube type: 18 mm channel plate wafer tube.
Resolution: 1.75 mrad (25 mm lens), 0.575 mrad (75 mm lens)
Field of view: > 40° (typical with 25 mm lens)
Magnification: 25 mm lens ×1, 75 mm lens ×3
Objective lens: (standard C mount) fl.4, 25 mm.
Supply voltage: 2 AAA alkaline batteries
Supply current: 18 mA nominal.
Battery life (typical): > 50 h
Weight (excl carry case and shoulder strap): 380 g

Contractor
P W Allen & Company Limited.

VERIFIED

The Allen Wasp weapon sight

Aselsan M992/M993/M993A

Type
Passive weapon sight (personal).

Description
The M992/M993/M993A night weapon sight is produced by Aselsan under licence from Northrop Grumman of the USA. This individual night weapon sight is compact and lightweight. Because of the offset eyepiece design, a more conventional head position can be used, just as for a day sight. One AA-sized battery is required to power this rugged, water-resistant ×4 sight. A variable-intensity red reticle is optically superimposed on the green intensified image, providing the marksman with a high-contrast aiming capability.

The M992/M993/M993A models differ only in the type of interchangeable image intensifier – either second-plus or third-generation. The M992 and M993/M993A include a splashproof lens which incorporates a large neutral density filter for daylight training and daylight weapon zeroing. A ×6 interchangeable lens is also available for weapons requiring a longer viewing range.

Operational status
In production. These products are also manufactured in the USA by Northrop Grumman, the licensor.

Specifications
Weight: 1.35 kg
Magnification: ×4
Min field of view: 10°
System gain min: 1,200 (GEN II+), 1,500 (GEN III), 2,500 (Advanced GEN III)
Resolution: 3.2 lp/mrad (GEN II+), 3.6 lp/mrad (GEN III), 4.5 lp/mrad (advanced GEN III)
Windage/elevation adjustments
Angular increments nom: 0.25 mrad
F-number: F/1.5
Eye relief: 30 mm
Reticle: duplex cross-hairs or chevron (custom reticle can be quoted on request)
Image intensifier
M992: GEN II+, non-inverting
M993: GEN III, non-inverting
M993A:Adv. GEN III, non-inverting
Power supply: (1 required) 1.5 V AA alkaline or 1.5 V/3.9 V lithium

Contractor
Aselsan Inc, Microelectronics, Guidance and Electro-Optics Division.

UPDATED

The Aselsan M992/M993/M993A

Aselsan M994/M995/M995A

Type
Passive weapon sight (personal).

Description
The M994/M995/M995A night weapon sight is produced by Aselsan under licence from Northrop Grumman (was Litton) of the USA. This individual night weapon sight is compact and lightweight. Because of the offset eyepiece design, a more conventional head position can be used, just as for a day sight. Only one AA- sized battery is required to power this rugged, water-resistant, ×6 sight which provides a high-resolution, intensified image. A variable-intensity red reticle is optically superimposed onto the green intensified image, providing the marksman with a high-contrast aiming capability.

M994/M995/M995A models differ only in the type of interchangeable image intensifier - either second-plus or third-generation. The M994 and M995/M995A include splashproof lens covers which incorporate a large neutral density filter for daylight training and daylight weapon zeroing. A ×4 interchangeable objective lens is also available for weapons requiring a shorter viewing range.

The Aselsan M994/M995/M995A

Operational status
In production. These products are also manufactured in the USA by Northrop Grumman, the licensor.

Specifications
Weight: 1.65 kg
Magnification: ×6
Min field of view: 6.3°
System gain (min): 1.200 (GEN II+), 1.500 (GEN III), 2.500 (Advanced GEN III)
Resolution: 3.2 lp/mrad (GEN II+), 3.6 lp/mrad (GEN III), 4.5 lp/mrad (Advanced GEN III)
Windage/elevation adjustments
Angular increments nom: 0.17 mrad
F-Number: F/1.7
Eye relief: 30 mm
Reticle: duplex cross-hairs or chevron (custom reticules can be quoted on request)
Image intensifier tube
M994: GEN II+, non-inverting
M995: GEN III, non-inverting
M995A: Adv. GEN III, non-inverting
Power requirement (1 required): 1.5 V AA alkaline or 1.5 V/3.9 V lithium

Contractor
Aselsan Inc, Microelectronics, Guidance and Electro-Optics Division.

VERIFIED

Aselsan thermal weapon sight

Type
Weapon sight (personal).

Description
The Aselsan thermal weapon sight is a lightweight, ruggedised, uncooled thermal sight operating in the long wave spectral band. The thermal sight is for use as a primary targeting sight for individual or crew-served weapons or as a hand-held forward observer's scope. It uses standard BB2847/U rechargeable battery or commercially available 9 V battery for hand-held usage. It can also be powered by vehicle power of 7 to 32 V DC.

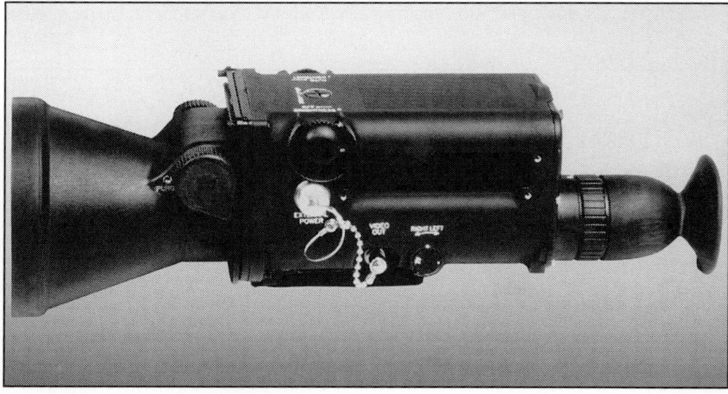

The Aselsan thermal weapon sight 0053702

Operational status
In production.

Specifications
Weight: 1.7 kg incl battery
Field of view: 9°
Magnification: ×3
Detector: 320 × 240 uncooled focal plane array
Spectral band: 7.5 to 13.5 μm
Controls: contrast (automatic or manual), brightness, reticle selection, reticle azimuth and elevation, polarity and objective lens focus
Power supply: lithium ion rechargeable battery

Contractor
Aselsan Inc, Microelectronics, Guidance and Electro-Optics Division, Turkey.

VERIFIED

BE Delft BENWS-9701 night weapon sight

Type
Weapon sight (personal).

Description
The BENWS-9701 night weapon sight is a state-of-the-art night vision system designed to enable accurate night firing of small arms. It uses a second-generation 18 mm non-inverting image intensifier tube. Using suitable mounting brackets it can be adapted to a variety of weapons such as 7.62 mm LMGs, the AK47 and the 5.56 mm INSAS.

Operational status
Available.

Specifications
Dimensions: 255 × 70 × 65 mm
Weight: 850 g
Field of view: 10°
Magnification: ×4
Resolution: 0.4 mrad

Contractor
BE Delft Electronics Ltd, India.

VERIFIED

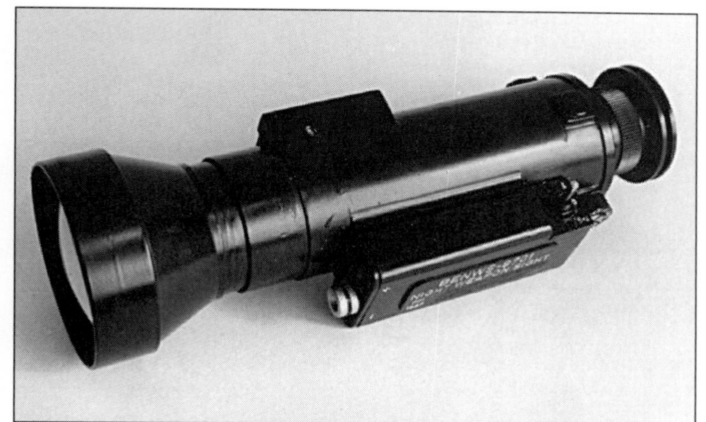

BE Delft BENWS-9701 night weapon sight 0023067

Dedal-300 night vision sight

Type
Weapon sight (personal).

Description
The Dedal-300 sight is designed for security agencies and commercial applications. It will fit rifles equipped with a dovetail mounting seat to European or US standards. A 25 mm second-generation image intensifier provides the night capability. Automatic protection from over bright light sources is provided for the image intensifier. An LED infra-red illuminator enhances performance at close ranges. Windage and elevation adjustments are provided. The sight can be used as a stand-alone monocular and can also be fitted to 35 mm SLR cameras. Like the Dedal-040 it is believed to be built to high-end commercial standards. It is fitted with a standard wearer mount. There are two variants.

The Dedal-300 night vision sight 0041438

Operational status
Available.

Specifications

	Mod. A	Mod. B
Dimensions	225 × 74 × 75 mm	225 × 74 × 75 mm
Weight	0.97 kg	1.1 kg
Magnification	×2.8 (×5.6 with ×2 objective)	×4.4
Field of view	17°	10°
Resolution	≥2.4 Lp/mrad	≥2.8 Lp/mrad
Objective lens		
F-number	f/1.2	f/1.5

Eyepiece lens
Eye relief: 45 mm
Dioptre adjustment: +3 to –4
Image intensifier
Type: GEN II 25 mm MCP
Photocathode sensitivity: 300 μA/lm
Gain: ≥20,000
Resolution: 41-45 lp/mm
System
Operating temperature: –40 to +50°C
Power supply: 2 × 3 V AA batteries
Operating time: 30 h

Contractor
Joint-Stock Company 'Dedal', Moscow.

VERIFIED

DiOP TADS 850 thermal augmented day sight

Type
Weapon sight (personal).

Description
The field-proven Thermal Augmented Day Sight (TADS) is a clip-on weapon night sight, which augments the user's day sight. It provides thermal imaging in the 8-12 μm spectral regions through the user's day sight. Features of the TADS system include a nil retract error and boresight retention under the most severe weapon shock (7.62 mm, 300 Win Mag, Barrett 0.50 in, and so on). The TADS incorporates the latest uncooled sensor technology, eliminating the need for expensive cryogenic coolers. Coupled with an advanced 640 × 480 display technology, images appear 'image intensified'. The unit mounts directly onto the using a tri-mount mounting system, without disturbing the user's day sight and is completely interchangeable with DiOP's image intensified clip-on night sights.

Operational status
In production and in service since July 2002.

Specifications
Weight: 2.58 kg
Spectral band: 8-12μm
Optics: f/1, 85 mm focal length, hard carbon exterior coating
Field Of View: 10.7° H × 8.1° V
Magnification: ×1
MRTD @ 0.8cy/mr: <0.7 K
Focal Plane Array: 320 × 240 microbolometer
Cooling: Thermoelectric
Controls: Polarity, display brightness, non-uniformity correction, power
Focus: Manual
LCD: Internal active matrix display, 640 × 480, green backlight.

The DiOP TADS 850 thermal augmented day sight, mounted on an SR-25 Model K 7.62 mm assault rifle ***NEW***/0594446

Power: two 'D' size Lithium batteries
Video Output: RS-170, NTSC or PAL
Battery Life: >5 hours
Operating Temperature: –25° to 55° C
Storage Temperature: –46° to 71° C
Environmental: sealed to withstand 3 ft immersion in water
Audible Security: inaudible at >5 m
Purge Port: ANVIS type

Contractor
Diversified Optical Products Inc (DiOP).

NEW ENTRY

DRS Land Warrior Thermal Weapon Sight

Type
Weapon sight (personal).

Description
The Land Warrior Thermal Weapon Sight (LW TWS) offers an extremely lightweight sight for use with Helmet Mounted Displays (HMD). Incorporating microbolometer-based uncooled thermal imaging technology, the 320 × 240 pixel LW TWS operates in the 8 to 12 μm spectral range to detect personnel and vehicles during absolute darkness. The system uses external power and incorporates a standard RS-170 video output for use with Helmet Mounted Displays (HMD).

Although the system is compact and lightweight, it offers a wide array of advanced features. The LW TWS houses environmentally sealed pushbutton controls for power, calibration and polarity. As these are the more commonly used commands, they have been positioned in an ergonomic manner to provide operators with convenient access. In addition to the push-button controls, the system has RS-232 serial interface which is utilized for advanced features, such a reticle control for boresight related activity. Due to its compact nature, the system is ideal for Military Operations in Urban Terrain (MOUT), yet rugged enough for long-range operations.

Specifications
Weight (HMD version): 682 g (1.5 lb)
Detector type: Vo^x Silicon Microbolometer
Spectral range: 8 – 12 μm
Video output: RS-170 Standard
Optics (standard): 62 mm F/1 (15° horizontal FOV), 40° HFOV also available
Pixels: 320 × 240
Symbology: RS-232 programmable symbology and reticle
Controls: Push-button controls
NETD: (MMB-3000) 64 – 75 mK (typical)
Power: 2 W

Contractor
DRS Electro-Optical Systems Group.

UPDATED

Econ Industries Andromeda 1

Type
Weapon sight (personal).

Description
The Andromeda 1 is a night vision weapon sight using second- or third-generation image intensifier tubes. It has a range of 600 m. Mounts are available for a range of weapons including G3, M16, FN-FAL, HK11 MG3, FN-MAG, M67 and STRIM. Features include adjustable reticle illumination and an adjustable dioptre eyepiece.

Operational status
Available.

Specifications
Magnification: ×3.3, ×4 optional
Focal length objective: 98 mm
Focal ratio: 1:1.2
Entrance pupil diameter: 80 mm
Focal length ocular: 30 mm
Exit pupil diameter: 9 mm
Eye relief: 35 mm
Focusing range: 10 m to infinity
Field of view: 10.5° (18.5 m at 100 m)
Reticle adjustment in azimuth and elevation: ±12
Eyepiece adjustment: ±3 dioptres
Image intensifier, GEN II or GEN III: 18 mm tube
Power supply: 2.2-3.4 V DC
Battery: 2 1.5 V type AA, alkaline or Ni/Cd
Reticle: brightness adjustable
Dimensions (l×h×w): 150 × 1,210 × 120 mm (no eyeshield)
Weight: 1.3 kg
Operational temperature range: –30 to +55°C
Storage temperature range: –40 to +60°C
Shock in optical axis: 300 *g*
Storage: 2,000 ft altitude
Leak test: 1 m water/30 min

Contractor
Econ Industries SA.

VERIFIED

Andromeda 1 night vision sight from Econ Industries

Econ Industries Andromeda 3

Type
Weapon sight (personal).

Description
The Andromeda 3 from Econ Industries is a combined day/night weapon aiming sight. The sight can be fitted with either second- or third-generation image intensfier tubes. The sight has the F1.2, 98 mm catadioptric objective lens used in all of the Andromeda series of night vision devices. This provides a magnification up to ×4 and a field of view of 10.5°.

The user can select day or night operation by turning the mode selector lever which brings into the optical axis either a third-generation tube for night operation or a fibre optic bundle for day operation. There is a single reticle for both day and night use which is projected via the objective lens and is illuminated by an AlInGaP LED. A filter for laser eye protection is provided.

Operational status
Available.

The Andromeda 3 day/night weapon aiming sight 0004829

Specifications
Magnification: ×3.3, ×4 optional
Entrance pupil: 80 mm
Field of view: 10.5°
Objective lens
Lens type: catadioptric
Effective focal length: 98 mm
Entrance pupil: 80 mm
F-number: 1.2
T-number: 1.7
Eyepiece lens
Effective focal length: 30 mm
Exit pupil: 9 mm
Eye relief distance: >30 mm
Reticle
Type: ballistic or single cross
Adjustment range: ±25 mils in elevation and in azimuth
Brightness: adjustable
Image intensifier: 18 mm GEN III, with automatic gain control and brightness protection
Gain: ×30,000
Power supply: 2 × 1.5 V AA alkaline batteries or 2 × 1.2 V Ni/Cd rechargeable batteries

Contractor
Econ Industries SA.

VERIFIED

Electromagnetica DGE-248 weapon sight

Type
Weapon sight (personal).

Description
The electro-optical DGE-248 sight is designed to replace conventional sights on infantry weapons. It provides a day or night aiming capability. An aiming mark focused at infinity is projected into the user's eye. This allows aiming with both eyes open.

Operational status
Available.

Specifications
Dimensions: 105 × 65 × 50 mm
Weight: 0.218 kg
Magnification: ×1
Field of view: 5°
Power supply: 3.6 V lithium battery

Contractor
Electromagnetica, Romania.

VERIFIED

Elop MALOS Miniature Laser Optical (Day/Night) Sight

Type
Weapon sight (personal).

Description
The MALOS integrates a laser range-finder with a sighting telescope and a fire-control computer with provision for a passive night vision device. It extends the effective range of sniping rifles and various other types of direct fire weapon. Though slightly larger than a conventional sighting telescope it is nevertheless of convenient size and lightweight.

Range measurement is controlled by a single button which can be mounted anywhere on the weapon. Once the range has been measured, the ballistic computer calculates the required elevation for the measured range and displays the required aiming point. Windage and range drums are fitted for boresighting and as manual back-up.

Operational status
In production.

Specifications
Laser range-finder
Range
Min: 50 m
Max: 2 km
Accuracy: ±5 m
Field of view: 0.6 mrad
Output energy: 3 mJ
Repetition rate: 12 ppm
Laser wavelength: 1.06 or 1.54 μm

Optical sight
Magnification: ×6 or ×10
Objective lens diameter: 40 mm
Field of view: 3° 45′ (19.6 m at 300 m)
Exit pupil diameter: 6.7 mm
Eye relief: 75 mm

Physical
Weight, day: 1.2 kg
Weight, day and night: 2 kg
Dimensions: 351 × 85 × 61 mm
Power: 2 × 3 V 2/3 AA batteries (lithium)

Contractor
Elop Electro-Optics Industries Ltd.

VERIFIED

MALOS 0059657

ENOSA VNP-009

Type
Weapon sight (personal).

Description
The VNP-009 was developed as a night sight for personal weapons. The VNP-009 is a second- generation sight using a Type S-25 photocathode, giving variable gain from ×5,000 to ×50,000. The sight can be attached to a variety of weapons. The projected reticle is available with luminous intensity adjustment.

The VNP-009 night sight from ENOSA 0004828

Operational status
Production as required. In service with the Spanish Army and other countries.

Specifications
Magnification: ×3.5
Catadioptric lens assembly: F = 95 mm
Entrance pupil: 80 mm
Aperture: 1:1.2
Resolution: better than 0.45 mrad (contrast 90%), at 0.001 lx
Image intensifier: 18 mm GEN II with S-25 photocathode
Adjustable gain: 5,000-50,000
Focusing: 20 m to infinity
Dioptric correction: +2 to –5 dioptre
Weight: 1.4 kg
Dimensions: 252 × 120 × 106 mm
Power: 4 dry batteries, R6
Operating time: >50 h

Contractor
Empresa Nacional de Optica SA (ENOSA).

UPDATED

Euroatlas EUROVIS-4 weapon sight

Type
Weapon sight (personal).

Description
The EUROVIS-4 is a weapon sight which is adaptable to any weapon by using an appropriate bracket.

It uses an 18 mm image intensifier tube of either second- or third-generation type and features an illuminated and controllable graticule and ×4 magnification which gives a focusing range from 15 m to infinity, enabling man-sized targets to be detected at up to 600 m under normal weather conditions.

The image intensifier tubes use microchannel plates and are provided with automatic gain control against muzzle flash and bright light sources. EUROVIS-4 fulfils all military environmental test requirements.

Operational status
Available.

EUROVIS-4

Specifications
Weight: 1 kg
Magnification: ×4
Field of view: 9°
Objective: 116 mm f/1.8 (T)
Focusing range: 15 m to infinity
Power supply: 2 × AA/LR6
Battery life: >50 h at 20°C
Operating range: to 1.5 km

Contractor
Euroatlas GmbH.

VERIFIED

FLIR Systems SnipIR™

Type
Weapon sight (personal).

Description
The SnipIR thermal imaging rifle sight is intended for armed forces special operations personnel as well as law enforcement SWAT teams. Easily man-portable at just 2 kg, the SnipIR is fully ruggedised to withstand harsh environmental conditions. Mounted on top of and viewed through the weapon's day scope, TIRS combines leading-edge optics and infra-red detector technology. The system's long-range performance and thermal sensitivity provides the ability to detect targets out to 1,500 m.

The cryogenically cooled imager is powered by a single 'D' cell lithium battery and allows the sight to continuously operate for greater than 4 hours. During daylight hours, operators can fuse the visible light image from the day scope with the SnipIR's thermal image.

Operational status
In service.

Specifications
Weight: 2.05 kg (4.5 lb) complete including battery, lens and beamsplitter assembly
Detector: 256 × 256 InSb FPA
Spectral band: 3-5 μm
NETD: 0.020°C at 23°C ambient
Cooling: Closed-cycle Stirling microcooler
Optics: f/4.0
Field of View: 150 mm, 2.9° (h) × 2.75° (v)
On-camera controls: Gain, auto/gain, level, polarity, power on/off, non-uniformity correction, reticle adjust, zero-zero, focus
Video output: RS-170 compatible, BNC connector
LCD display: 256 × 240 pixels
Power requirement: 4.5 W
Battery life: >4 hrs
Battery type: 'D' cell lithium battery
Operating temperature: –23 to +49°C
Storage temperature: –33 to +71°C
Environmental: fully sealed and ruggedised
Audible security: inaudible at greater than 10 m
Weapon mount: compatible with KAC Tri-Mount

Accessories
Carry case: hard shell, lockable with fitted foam insert
DC power adapter: 6 to 28 V DC input
Batteries: 3 each, 3.6 V, lithium

Contractor
FLIR Systems.

UPDATED

FLIR Systems SnipIR 0137873

Institute of Optronics Pakistan AN/PVS-4A individual weapon sight

Type
Weapon sight (personal).

Description
AN/PVS-4A is a licence-produced weapon sight for use on rifles, machine guns and rocket launchers. It uses a 25 mm second-generation image intensifier tube and has an internally adjustable, illuminated graticule projector; the graticule being interchangeable for use with different weapons. The sight has automatic gain control which allows it to cope with large fluctuations in brightness at the target, and a bright source protection circuit prevents damage caused by inadvertent exposure to daylight. A special cover permits daytime boresighting and training.

Operational status
In production.

Specifications
Length: 240 mm
Width: 120 mm
Height: 120 mm
Weight: 1.7 kg
Magnification: ×3.6
Field of view: 14.5°
Objective lens: 95 mm T/1.6
Dioptre adjustment: +2 to –6
Power source: AA cell
Battery life: 35 h at 25°C

Contractor
Institute of Optronics Pakistan.

VERIFIED

Pakistan AN/PVS-4A

International Technologies (Lasers) NVL-11 Mk IV fire-control night sight

Type
Weapon sight (personal).

Description
The NVL-11 Mk IV is a computerised fire-control night sight for rocket launchers and anti-tank weapons such as the RPG-7, Carl Gustaf 84 mm and 106 mm recoilless rifle. The weapon mounted system includes a night scope, laser rangefinder and a computer controlled laser aiming light. It is capable of storing ballistic data for up to five weapon/ammunition combinations and is said to be effective up to 900 m. The system measures the distance to the target and computes the elevation required for the specific weapon and ammunition.

A high-light level detector is installed to avoid damage to the image intensifier from light intensities over 1 lx.

Operational status
Available.

Specifications
Field of view: 13.6°
Scope magnification: ×3
Resolution (tube): 32 lp/mm
Ranging laser wavelength: 840 nm
Output power (min): 100 mW
Beam divergence: vertical line, 3.5 mils high, 0.5 mil wide
Performance
  Range: 20-990 m
  Resolution: 10 m

NVL-11 Mk IV sight-mounted on 84 mm Carl Gustav

Aiming laser wavelength: 830 nm
Peak power output
  High position: 10 mW
  Low position: 1 mW
Beam divergence: 0.3 mil
Ballistics
  Elevation: up to 70 mils
  Accuracy: ±0.5 mil
Dimensions (l × w × h): 190 × 210 × 130 mm
Weight: 2.1 kg excl weapon, adaptor and batteries
Power source: 4 × 1.5 V AA alkaline batteries
Battery life: 3 h ranging (approx 2,000 rangings of 5 s each)
Construction: anodised aluminium

Contractor
International Technologies (Lasers) Limited.

VERIFIED

IRT Luna-Tron Z300 day/night weapon sight

Type
Weapon sight (personal).

Description
The Luna-Tron Z300 day/night weapon sight incorporates a second-plus or third-generation image intensifier tube with a two-mode selector switch that enables user selection between night channel and day channel. The sight can be adapted for several types of personal weapons with STANAG or other adaptors.

Operational status
Fully operational.

Specifications
Day channel
Magnification: ×4.2
Field of view: 6°
Objective lens: 113 mm, f1.6
Resolution: 12 arc s
Dioptre adjustment: +2 to –6 dioptre
Reticle brightness: 10,000 ft lambert (adjustable)
Line width: 0.1 mrad
Boresight adjustment: ±13 mrad
Boresight increments: 0.12 mrad/click
Eye relief: 50 mm
Exit pupil: 6 mm
Night channel
Magnification: ×4.2
Field of view: 9°
Objective lens: 113 mm, f1.6

The Luna-Tron Z300 aiming device for day/night use from IRT

Resolution: 4.1 lp/mrad
Dioptre adjustment: +2 to –6
Image intensifier tube:
NL-300: 18 mm, GEN II+, Gain 15,000
NL-300: 18 mm, GEN III, Gain 25,000
Reticle brightness: 10,000 ft L (adjustable)
Line width: 0.1 mrad
Boresight adjustment: ±13 mrad
Boresight increments: 0.12 mrad/click
Eye relief: 30 mm
Exit pupil: 6 mm
Weight: 1.5 kg
Dimensions: 300 × 100 × 70 mm

Contractor
IRT Infrarot-Technik Eiselt.

UPDATED

ISAP TADS 850 Thermal Weapons Sight

Type
Weapon sight (personal).

Description
The Thermal Augmented Day Sight is an 8 – 12 μm thermal imaging system which clips on to most rifle-mounted day scopes. The sight incorporates an uncooled focal plane array as the detector. The detector converts the received thermal energy to an electrical signal which is processed to drive an internal display. The image on the display is relayed by a visible optical assembly on the day scope. This image is viewed at ×1 magnification, preserving the accuracy of the day scope reticule. A beamsplitter module accomplishes day/night boresighting and night image direction.

Operational status
Available.

Specifications
Dimensions: 82.55 × 269.24 mm
Weight: 1.4 kg
Spectral band: 8-12 μm
Detector: uncooled
Field of view: total,10.75 × 8.1°, instantaneous 0.588 mrad
Detection range: man > 525 m
Battery type: 2 × D size lithium
Environment: militarised and submersible to 66 ft

Contractor
International Security Alliance Partners.

UPDATED

ITT F7000A/F7001A night vision weapon sight

Type
Weapon sight (personal).

Description
ITT Industries Night Vision offers the F7000A and F7001A Night Vision Weapon Sights. These units are rugged, ×4 and ×6 magnification night vision weapon sights which provide excellent observation, target acquisition and aiming capabilities at night. Most importantly, these weapon sights provide superior range performance, accuracy and stability.

The ITT F7001 night vision sight 0132093

The F7000A and F7001A feature an injected ballistic reticle which is fully adjustable for boresighting and has a brightness control for optimum performance under all light conditions. Simple to operate, these weapon sights were designed with consideration of human factors and durability.

User Benefits
- The rugged aluminum housing construction is extremely durable and provides dependable boresight retention during the emission, across al! temperature extremes.
- The external objective lens and eyepiece lens elements are protected with a scratch-resistant coating which permits cleaning without damage to the optical coatings.
- The illuminated reticle is sharply defined with variable intensity for best contrast against a light or dark target.
- The weapon mount interface provides for multiple fore/aft positions to enable all soldiers to achieve a comfortable firing position.
- The weapon sight has a rubber objective lens cap that has a pinhole in the centre, so the cap also may serve as a daylight cover. The lens cap flips up for use and remains secured to the objective lens.
- A rubber eyecup provides light security and cushions the soldiers face against injury from weapon recoil. The eyecup may be removed easily to facilitate cleaning the eyepiece optics. No tools are required.

Kit Supplied

1 ea	F7000A or F7001A weapon sight, with lens cap and eyeguard
1 ea	Soft carrying bag
1 ea	Shipping/storage case
2 ea	'AA' alkaline batteries
1 ea	Lens tissue pack
1 ea	Operator manual
1 ea	Lens anti-fog compound
1 ea	Weapon mount

Specifications

	Model F7000A	Model F7001A
Magnification	×4	×6
Field of view	8.4°	5.6°
Range (clear starlight, 10^{-3} Lux)	1,318 m	1,933 m
Reticle adjustment step size	.2 mil	.13 mil
Eye relief	30 mm	30 mm
Eyepiece focus	+2 to –5 dioptres	+2 to –5 dioptres
Focus range	25 m to infinity	25 m to infinity
Exit pupil	7.5 mm	7.5 mm
Weight	1.2 kg	2.0 kg
Power supply	2 AA batteries	2 AA batteries

Contractor
ITT Industries Night Vision.

UPDATED

ITT F7201A modular day/night weapon sight

Type
Weapon sight (personal).

Description
The F7201 modular day/night weapon sight has interchangeable night vision and daytime eyepieces. The sight includes a high quality scope featuring ×2.5 to ×10 variable power with a large 56 mm focusable objective and detachable glare hood. The reticle, which is forward of the eyepiece and therefore not disturbed by changing eyepiece modules, is of a triplex design with mil dots and range calculations. These are visible during the day or night operation. Weapon zero is set using turret knobs which provide precise adjustment with distinct ¼ minute-of-arc increments that are both audible and tactile.

The ITT F7201 modular day/night weapon sight 0004825

Daytime eyepiece module
The interchangeable daytime eyepiece provides a +2 to −2 dioptre focus adjustment and presents a minimum 5 mm exit pupil at 76 mm of eye relief when set at ×10. Lower magnification settings have a larger exit pupil diameter.
Night vision eyepiece module
The night vision eyepiece module presents a 5 mm exit pupil at 64 mm of eye relief (which compensates for the night module being ½ in longer) and comes with a high- performance third-generation image intensifier. It features an integral covert infra-red illuminator and a gain control knob for manually adjusting the brightness level.

The F7201 can be mounted using standard 30 mm scope rings that accommodate the large 56 mm objective.

Operational status
Available.

Specifications
Scope
Magnification: ×2.5 to ×10
Field of view: 6.8 to 2°
Objective lens
Diameter: 56 mm
EFL: 131 mm
F-number: 2.4 at ×2.5
Focus: adjustable to remove parallax and focus for night vision module
Reticle: triplex with mil dot
Barrel diameter: 30 mm
Length (without hood): 13.25 in (daytime), 13.75 in (night-time)
Weight: 2.0 lb (daytime), 2.9 lb (night-time)
Daytime eyepiece module
Exit pupil: 5 mm at 76 mm eye relief
Dioptre adjustment: +2 to −2 dioptres
Night vison eyepiece module
Scene illumination: 10^{-6} to 1 fc
Spectral response: visible to 0.90 μm (IR)
Image tube
Photosensitivity: 1500 μA/lm
Signal-to-noise: 20:1 approx
Brightness gain: 0-50,000 variable
Resolution min: 45 lp/mm
Reliability min: 10,000 h
System resolution min: 4 cy/mrad at ×2.5, 8 cy/mrad at ×10
Dioptre adjustment: +2 to −3 dioptres
Exit pupil: 5 mm at 64 mm eye relief
Voltage required: 3 V DC
Battery type: 2 × AA 1.5 V alkaline

Contractor
ITT Industries, Night Vision.

UPDATED

Meopta MEO 50S 50K and 50P night sights

Type
Weapon sight (personal).

Description
The MEO 50S is a multipurpose night sight of compact construction designed for personal automatic weapons. The similar MEO 50P can be used with anti-tank weapons like the RPG-7 and the MEO 50K can be used with machine guns. The sight can be fitted with a second- or third-generation image intensifier.

Operational status
Available.

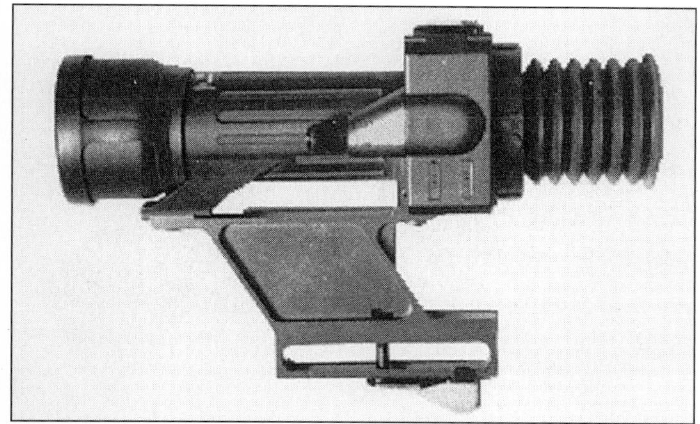

Meopta MEO 50 night sight 0022035

Specifications
Magnification: ×3.2
Field of view: 11°
System gain: 1,200
Target recognition range – man size at 0.001 lx: 220 m
Dimensions (l × w × h):
MEO 50S: 280 × 190 × 115
MEO 50K: 280 × 160 × 160
MEO 50P: 280 × 195 × 115
Weight:
MEO 50S: 1.2 kg
MEO 50K: 1.4 kg
MEO 50P: 1.2 kg

Contractor
Meopta Prerov, a.s.

UPDATED

Meopta ZN ×6 night-time dial sight

Type
Weapon sight (personal).

Description
The ZN ×6 night-time dial sight is a monocular device for observation and aiming personal weapons at reduced light levels and in darkness. The sight is composed of a tube with a mirror objective lens assembly which transmits the light received to a second-generation image intensifer and to the viewer via eyepiece optics. The eyepiece has a reticle with vertical and horizontal corrections and a knob to select the range to target. The reticle can be lit and has adjustable brightness.

Operational status
Available.

Specifications
Magnification: ×6
Field of view: 5°
Entrance pupil diameter: 112 mm
Dioptre adjustment: ±3 dioptres
Resolution: 32 lines/mm
Eye relief: 35 mm
Firing distance adjustment range: 200-1,000 m
Power supply: 2 × 1.2 to 1.5 V batteries
Dimensions: 132 × 290 mm
Weight; 2.5 kg (incl batteries)

Contractor
Meopta Prerov, a.s.

VERIFIED

Meopta ZN ×6 night-time dial sight 0022033

MES Lynx night and daylight weapon sight

Type
Weapon sight (personal).

Description
The Lynx night and daylight weapon sight is designed for use on individual weapons and light anti-tank weapons. At night an image intensifier is introduced into the optical path to allow firing in darkness. This is achieved through a simple lever selection. By day the image intensifer is moved out of the way and the sight reverts to an optical system only.

MES Lynx night and daylight weapon sight 0022034

Operational status
Available.

Specifications
Day and night
Magnification: ×3.3
Field of view: 10.5°
Focus range: 10 m to infinity
Dioptre adjustment: ±4 dioptres
Battery life: >50 h (with rechargeable batteries)
Weight: 1.47 kg

Contractor
Meccanica per Elettronica e Servomeccansimi SpA (MES).

VERIFIED

New Noga Light NL-61 mini night light system

Type
Weapon sight (personal).

Description
Noga Light's device, model NL-61, is a very small and light night weapon sight. Designed according to Israeli special units' requirements, the NL-61 compact design includes remote control operation, different magnifications and dual use.

The sight provides infantry and light support weapons with high-performance observation, target acquisition and aiming capabilities at night.

Operational status
In production.

Specifications
Dimensions (l × w × h): 120 × 80 × 58 mm without eyeguard
Weight: 375 g
Power supply: 1 × 3 V 2/3 A size battery
Battery life: 50 h nominal

	with ×1 magnifier lens	with ×3 magnifier lens	with ×5 magnifier lens
Magnification	×1	×3	×5
Field of view	40°	13°	7.5°
Focus range	5 m to infinity (fixed focus)	5 m to infinity (fixed focus)	30 m to infinity (fixed focus)
Eyepiece dioptre range	+2 to –5	+2 to –5	+2 to –5
Eye relief	25 mm	25 mm	25 mm
Optimal operational range	5-100 m	15-300 m	30-450 m
Boresight adjustment range	±40 mrad	±13 mrad	-
Boresight increment	0.4 mrad/click	0.13 mrad/click	-

Image Intensifier Tube	GEN III	GEN II Super	GEN II
Photocathode sensitivity, at 2,856K	1,800 μA/lm	650 μA/lm	400 μA/lm
Signal to noise ratio (min)	21	17	14.5
Resolution	64 lp/mm	45 lp/mm	36 lp/mm

Specifications *continued*

	with ×1 magnifier lens	with ×3 magnifier lens	with ×5 magnifier lens
Dimensions (without eyeguard) (l × w × h)	120 × 80 × 58 mm	205 × 80 × 79 mm	207 × 83 × 82 mm
Weight	375 kg	630 g (including battery)	790 g (including battery)
Power supply	1 × 3V 2/3 A size battery	1 × 3V 2/3 A size battery	1 × 3V 2/3 A size battery
Battery life	50 h	50 h	50 h

Contractor
New Noga Light Ltd.

UPDATED

New Noga Light NL-74B/NL-76B weapon sights

Type
Weapon sight (personal).

Description
The NL-70 series weapon sights provide high-performance observation, target acquisition and aiming capabilities at night.

The NL-74B/NL-76B weapon sights are lightweight, waterproof, weapon-mounted, self-contained, ×4/×6 magnification, high-performance night vision systems. The sights are used for aiming and firing, or for surveillance when hand-held. The sights use standard AA batteries and can be rapidly mounted on most infantry weapons.

Operational status
In production.

Specifications
Dimensions (l × h × w): NL-74B 255 × 82 × 80 mm (excl daylight cover and mounting adaptor)
NL-76B 425 × 114 × 114 mm (excl eyeguard)
Weight: NL-74B 1 kg excl mounting adaptor, batteries and daylight cover
NL-76B 2.017 kg excl mounting adaptor, batteries and daylight cover
Magnification: ×4 NL-74B; ×6 NL-76B
Field of view: NL-74B 8.3°, NL-76B 5.7°
Focus range: NL-74B 25 m to infinity, NL-76B 50 m to infinity
Eyepiece dioptre range: +2 to –5 dioptres
Eye relief: 30 mm
Exit pupil: 7.5 mm
Image intensifier types: NL-74B 18 mm GEN II, GEN II Super, GEN III, Nl-7B 18 mm GEN II Super
Power supply: 2× 1.5 V AA batteries
Battery life: 40 h nominal at 20°C; 12 h nominal at 0-40°C

Contractor
New Noga Light Ltd.

UPDATED

The Noga Light NL-74B sight 0053704

New Noga Light Nogascope NL-300 and NL-303 day/night weapon sight

Type

Weapon sight (personal).

Description

The Nogascope NL-300 and NL-303 day/night weapon sight combines a daylight telescope and night sight in a single device, with a two-mode selector switch that enables the user to select between night channel and day channel operation. It incorporates a second-plus or third-generation image intensifying tube. With a magnification of ×4.2, the sight is suitable for short- and medium-range weapons. The sight can be adapted for several types of personal weapons with STANAG or other adaptors.

Operational status

In production.

Specifications

Day channel
Magnification: ×4.2
Field of view: 6° day
Objective lens: 113 mm, F1.6
Resolution: 12 arc s
Dioptre adjustment: +2 to –4.5 dioptre
Reticle brightness: 10,000 ft lambert (adjustable)
Line width: 0.1 mrad
Boresight adjustment: ±13 mrad
Boresight increments: 0.12 mrad/click
Eye relief: 50 mm
Exit pupil: 6 mm
Night channel
Magnification: ×4.2
Field of view: 9°
Objective lens: 113 mm, F1.6
Resolution: 4.1 lp/mrad
Dioptre adjustment: +2 to –6 dioptre
Image intensifier tube: NL-300: 18 mm, GEN II+ Gain 15,000; NL-300: 18 mm, GEN III, Gain 25,000
Reticle brightness: 10,000 ft lambert (adjustable)
Line width: 0.1 mrad
Boresight adjustment: ±13 mrad
Boresight increments: 0.12 mrad/click
Eye relief: 30 mm
Exit pupil: 6 mm
Battery: 2 × 1.5 AA alkaline
Battery life: 50 h
High intensity light cut-off: light level above 1 lx
Weight: 1.5 kg
Dimensions: 300 × 100 × 70 mm

Contractor

New Noga Light Ltd.

UPDATED

The Nogascope weapon sight from Noga Light 0004835

Night Vision Equipment Company AN/PVS-4 NVEC Models 700/700HP/750 night weapon sights

Type

Weapon sight (personal).

Description

The Night Vision Equipment Company's version of the US Army AN/PVS-4 image intensified sight comes in three forms, designated models 700, 700HP and 750. The sight is intended for use on weapons such as the M14, M16, M60, the M72A1 rocket launcher and the M203 grenade launcher. All models use 25 mm image intensifiers. The 700 meets the army's original specification which included the requirement for man-sized target engagements at 600 m in moonlight and 400 m in starlight. The 700HP and 750 have higher performance intensifiers (second-generation plus and third-generation) and improve on those ranges by 30 per cent and 60 per cent respectively. The 700HP uses a commercial specification intensifier. Reticle adjustments include brightness control and elevation and azimuth in 0.25 mil increments.

Operational status

Available.

Specifications

Model	**700**	**700HP**	**750**
Tube generation	II	II+	III
Typical detection range in m for still human target on green grass			
Quarter moon (10^{-2} lx)	1,014	1,078	1,283
Starlight (10^{-3} lx)	632	685	839
Overcast (10^{-4} lx)	244	288	424

(All models)
Dimensions: 240 × 120 × 120 mm
Weight: 1.59 kg
Field of view: 14.5°
Magnification: ×4
Focus range: 7 m to infinity
Objective focal length: 95 mm
Dioptre adjustment: –5 to +2
Eye relief: 28 mm
Power supply: 2 × AA batteries

Contractor

Night Vision Equipment Company Inc.

UPDATED

The NVEC AN/PVS-4 night weapon sight 0041411

Northrop Grumman AN/PVS-10 sniper night sight

Type

Weapon sight (personal).

Description

The AN/PVS-10 sniper night sight has been produced under a contract to NV/RSTA (Night Vision/Reconnaissance Surveillance and Target Acquisition) department of the NVESD (Night Vision and Electronic Sensors Directorate) of the US Army. AN/PVS-10 is an integrated day/night sight for the M24 sniper rifle. It has the same 2° field of view, the same reticle pattern, azimuth and elevation operation and rail mounting as the M3A sniper scope.

The sight is switchable between day sight and third-generation image intensifying night sight. Output brightness and reticle illumination are adjustable. The system mounts to the existing rail of the M24 rifle and uses the standard mil-dot reticle.

Operational status
Developed for the US Army Night Vision and Electronic Sensors Directorate.

Specifications
Image intensifier: GEN III 18 mm
Field of view: 2° (1-5° 0.50 cal version)
Magnification: ×8.5 (day and night), ×12.5 0.50 cal version
Resolution
System: 4.0 cy/mrad
Tube: 50 lp/mm
Weight: 2.2 kg
Range
Detection: 800 m
Recognition: 600 m
Hit probability
Day: 0.85 out to 800 m
Night: 0.70 out to 600 m
Reliability (MTBOMF): 1,000 h
Power supply: 2 AA batteries

Contractor
Northrop Grumman (formerly Litton) Electro-Optical Systems.

VERIFIED

Northrop Grumman AN/PVS-12/12A submersible low-profile weapon sight

Type
Weapon sight (personal).

Description
The AN/PVS-12 night vision weapon sight with its ×4 magnification is suitable for a wide range of 5.56 mm individual or crew-served weapons. The standard model meets a 1 m immersion requirement. AN/PVS-12A is an optional model that can be transported at depths of 20 m (66 ft). AN/PVS-12 can use either GEN II or GEN III image intensifier tubes interchangeably.

AN/PVS-12 is constructed from proven high strength aluminium alloy designed to survive rugged military type environments. The external surfaces are a non-reflective matte black finish. It is designed to interface with a wide variety of weapons incorporating STANAG or Weaver type mounts. It can also be mounted to other weapons to meet customer requirements.

Controls and adjustments are simple to operate. The system is activated by an ON/OFF control knob. Reticle brightness is controlled by a separate knob. A focus ring is located on the eyepiece for individual eye adjustment. Range focus is controlled by a knob on the left hand side for distances from 25 m to infinity. Azimuth and elevation adjustments are in increments of 0.2 mrad per click. A soft rubber, light secure, eyeguard protects from weapon recoil. A daylight cover permits boresighting and/or use of the sight during dawn to dusk light level conditions.

Operational status
Available.

Specifications
Length (at zero dioptre): 235 mm (excl eyecup and daylight cover)
Width: 87 mm
Height: 83 mm (excl mounting adapter)
Weight: 1.2 kg (2.6 lb)
Magnification: ×4
Field of View: 8.3°
Focal length (obj): 120 mm
F number: 1.7
Dioptre adjustment: +2 to –5 dioptres
Eye relief: 30 mm
Exit pupil: 7.5 mm II
Image tube: 18 mm GEN II or GEN I
Reticle: (Red LED – 0.07 mrad line width)

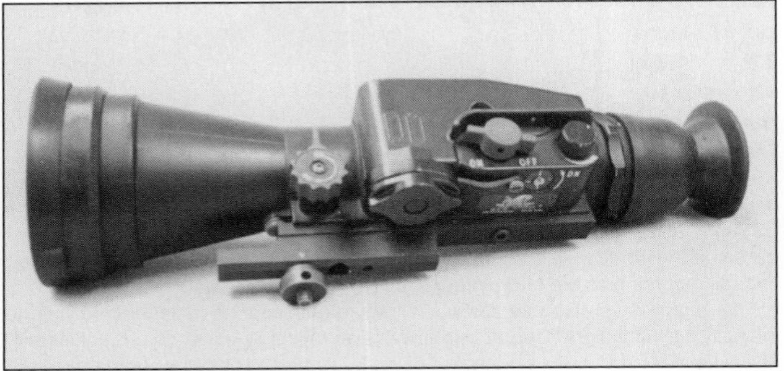

Northrop Grumman AN/PVS-12 0053706

Brightness: Manual Control
Windage and Elevation adjustments:
±18 mrad (±60 MOA) reticle travel in 0.2 mrad (.69 MOA) increments
Power supply: 2× AA Alkaline batteries
Battery life: 40 h. (Nominal) at 20°C
12 h. (Nominal) at 0 to –40°C

Contractor
Northrop Grumman (formerly Litton) Electro-Optical Systems.

VERIFIED

Northrop Grumman AN/PVS-17 mini night vision sight

Type
Weapon sight (personal).

Description
The mini night vision sight has been designed to meet the specific requirements of the US naval Surface Warfare Center for a lightweight, compact and high-performance system to be fielded on the Special Operations Peculiar MODification (SOPMOD) to the M4A1 Carbine. SOPMOD will use Northrop Grumman's (formerly Litton's) latest OMNI V enhanced GEN III image intensifier which will offer performance equal to some ×3 and ×4 fielded systems. Other GEN II and GEN III tubes are available.

AN/PVS-17 weighs less than 1 kg and is submersible to 20 m (66 ft). Mounting on the extended carbine rail has been optimised to allow other systems to be used in conjunction with the sight such as a laser aimer/illuminator.

AN/PVS-17 is operated with a three position OFF/ON/ON knob. The first ON position puts the system in a stand-by mode that can be activated by a momentary push-button located on the top of the unit. The second ON position is for continuous operation. A remote control pad turns on the sight when the ON switch is in the momentary position. Its red LED Dot projected reticle has 5 brightness levels controlled by its own selection knob. This unique momentary action allows presetting of controls before observation, saves battery life and reduces the need for shuttered eyecups.

AN/PVS-17 includes an indicator window and desicant in the sight to verify seal integrity and prevent fogging. When using the sight as a detached observation device a neck strap is provided. An optional adaptor can be used for short duration helmet mounting.

Operational status
Available.

Specifications
Weight: <1 kg
Magnification: ×2.25
Field of view (h × v): 20° × 17°
Focal length: 54 mm
F number: 1.25
T number: 1.35
Focus range: fixed at infinity
Dioptre range: +2 to –2
Eye relief: 24 mm
Image tube: ANVIS Style GEN II or GEN III
Reticle: projected 0.6mrad (2.0 MOA) Red LED Dot
Windage and elevation adjustments – movement per click: 0.15 mrad (0.5 MOA)
Electrical
1 AA alkaline: approx 24 h at 23°C
1 AA lithium: approx 32 h at 23°C; approx 16 h at –20°C
Environmental
Temperature range:
Operating: –40 to +52°C
Storage: –51 to +71°C
Immersion: 20 m (66 ft) for 2 h

Contractor
Northrop Grumman (formerly Litton) Electro-Optical Systems.

VERIFIED

The Northrop Grumman AN/PVS-17 mini night vision sight 0053705

Northrop Grumman AN/PVS-4 second-generation weapon sight

Type
Weapon sight (personal).

Description
Performing the same range of functions as the earlier first-generation AN/PVS-2 Starlight scope but with superior characteristics, the AN/PVS-4 is a light, passive night vision sight originally designed and manufactured under a US Army contract. Using a 25 mm image intensifier tube, the sight is suitable for use with 5.56 mm and 7.62 mm rifles, light and medium machine guns, rocket and grenade launchers. It has been continuously improved and can be fitted with a drop-in third-generation tube that significantly improves performance.

An adjustable internally projected reticle and interchangeable reticle pattern allows the sight to be boresighted to the various weapons without having to move the sight.

Image tube gain and reticle brightness are manually adjustable to compensate for different levels of ambient lighting. Automatic gain control circuitry is employed to maintain the viewed scene illumination constant during periods of changing light level conditions, such as the period from sunset to full darkness. This allows the operator of the sight to use the sight without having to readjust the tube gain control every few minutes during this period.

The tube features muzzle-flash protection which prevents the tube from being damaged by high-intensity short duration flashes of light. The flash protection circuit is designed to recover in time for the observer to see the round hit the target.

Operational status
In production and in service with the US Army and Marine Corps and in other countries.

Specifications
Dimensions: 240 × 120 × 120 mm
Weight: 1.72 kg
Field of view: 14.5°
Magnification: ×3.6
Focus range: 25 m to infinity
Objective focal length: 95 mm
Eyepiece focal length: 26.5 mm
Eye relief: 25 mm
Dioptre range: +2, –5
Power: 2 × AA batteries
Battery life: 60 h at 20°C

Contractor
Northrop Grumman (formerly Litton) Electro-Optical Systems.

UPDATED

AN/PVS-4

Northrop Grumman Aquila mini weapon sight

Type
Weapon sight (personal).

Description
The Aquila mini weapon sight is a ×4 passive individual weapon sight for infantry and special forces use. Aquila III is an upgraded design of the Aquila 1 mini weapon sight and features a new family of objectives and an afocal lens which allows the use of one basic night vision system to be customised for a wide variety of individual and crew-served weapons.

Aquila III is offered with either an 18 mm second- or third-generation image intensifier. The objectives and afocal lenses are the same as the AN/PVS-10 for commonality of spare parts.

The Aquila mini weapon sight

The system can be ordered as either ×4, ×6 or ×4 with screw-on afocal to achieve ×6 magnification. It is constructed of high strength aluminium to withstand rugged environments.

Mounting for a wide range of weapons with common systems such as STANAG or Weaver is available, as are custom mounts. The system operates on two AA batteries.

Operational status
Available.

Specifications
Weight: 1.1 kg
Field of view: 8.3°, 5.7°
Magnification: ×4, ×6
Objective focal length: 125, 178 mm
Eyepiece focus: +2 to –5 dioptres
Graticule: (red LED) 0.2 mil increments
Focus range: 25 m to infinity
Operating temperature: –54 to +52°C

Contractor
Northrop Grumman (formerly Litton) Electro-Optical Systems.

UPDATED

Northrop Grumman M845 Mk II night weapon sight

Type
Weapon sight (personal).

Description
The M845 Mk II is a second-plus generation night vision sight, light in weight and powered by two AA alkaline batteries. It has a red dot variable-intensity aiming reticle for a simple point of aim and quick reaction. The red dot is variable in intensity as well as adjustable for windage and elevation for zeroing purposes. The system is suited for military use at combat ranges out to 300 m.

Operational status
Available. In military and paramilitary use.

Specifications
Dimensions: 260 × 70 mm
Weight: 1.3 kg with batteries
Height above mounting surface: 90 mm
Field of view: 13.5°
Magnification: ×1.55
Resolution: 1.5 lp/mm
Dioptre adjustment: ±2
Eye relief: 50 mm

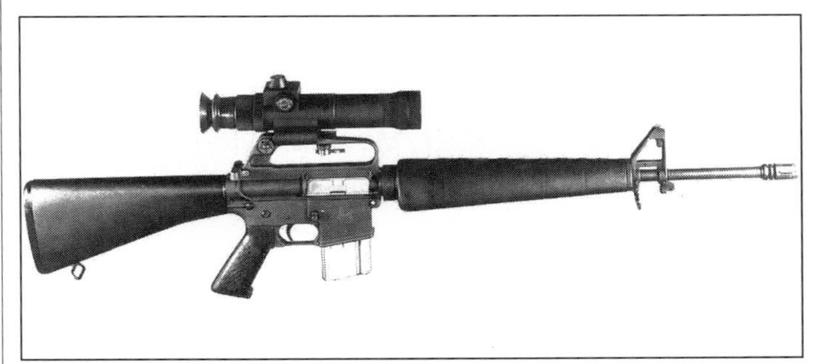

Northrop Grumman M845 Mk II

Gain: × 1,000 (nominal)
Tube: GEN II+ or GEN III
Power supply: 2 × 1.5 V alkaline
Battery life: approx 40 h

Contractor
Northrop Grumman (formerly Litton) Electro-Optical Systems.

VERIFIED

Northrop Grumman M921 submersible night vision sight

Type
Weapon sight (personal).

Description
The Northrop Grumman (formerly Litton) M921 is a submersible second-generation night sight for use on weapons from 5.56 mm rifles to light machine guns or for general surveillance. The submersible design provides watertight protection to a depth of 50 m for missions requiring underwater transport. It offers resistance to saltwater corrosion, since there are no exposed threads and all external surfaces are hard Teflon coated. The M921 can be factory-upgraded with a third-generation intensifier tube.

An open-cross graticule provides a highly effective sight picture and there are click adjustments for azimuth and elevation.

Operational status
In military and paramilitary use.

Specifications
Dimensions: incl eye guard, 191 × 85 mm
Weight: incl battery, 2.1 kg
Height above mounting surface: 109 mm
Field of view: 13°
Magnification: ×3
Resolution: 1.8 lp/mrad
Tube: 18 mm GEN II
Battery: BA1567/U (mercury) or BA5567/U (lithium)
Battery life: approx 12 h

Contractor
Northrop Grumman (formerly Litton) Electro-Optical Systems.

Northrop Grumman M921

Northrop Grumman M937/M938 individual weapon sights

Type
Weapon sight (personal).

Description
The Northrop Grumman (formerly Litton) M937 and M938 are compact, lightweight, battery-powered night vision sights for use on weapons from 5.56 mm assault rifles to light machine guns. The ×4 telescope provides a high-resolution

Northrop Grumman M937

intensified image for aiming at medium- and long-range targets. The two models differ only in the type of image intensifying tube; the M937 uses a second-generation tube, while the M938 uses a third-generation tube for greater sensitivity and increased resolution under conditions of extremely low light.

A variable intensity amber-coloured graticule, which can be superimposed on the green intensified image, provides the marksman with aiming capability. There are precision click adjustments for both azimuth and elevation. Both models include a splashproof lens cover which has a central pinhole for daylight training and boresighting.

Operational status
In military and paramilitary use.

Specifications
Dimensions: 255 × 85 mm
Weight: 1.1 kg with battery
Height above mounting surface: 80 mm
Field of view: 8.5°
Magnification: ×4
Resolution
  GEN II: 2.8 lp/mrad
  GEN III: 3.2 lp/mrad
Battery: 2 × 1.5 V AA alkaline
Battery life: 50-60 h

Contractor
Northrop Grumman (formerly Litton) Electro-Optical Systems.

Northrop Grumman NVS-700 night vision system

Type
Weapon sight (personal).

Description
The NVS-700 second-generation individual weapon sight is extensively used by military and police authorities in many countries. Both it and the NVS-800 crew-served weapon sight use the same 25 mm image intensifier tube and have almost all parts in common except the objective lens and associated fittings. The smaller NVS-700 is suitable for mounting on the 5.56 mm M16 rifle and similar weapons.

Operational status
Widespread military sales to foreign governments.

Specifications
Diameter: 101.6 mm nominal
Length: 292 mm nominal
Weight: 1.814 kg
Field of view: 253 mrad
Magnification: ×3.5 (nominal)
Viewing range
  Moonlight: 25-700 m
  Starlight: 25-450 m

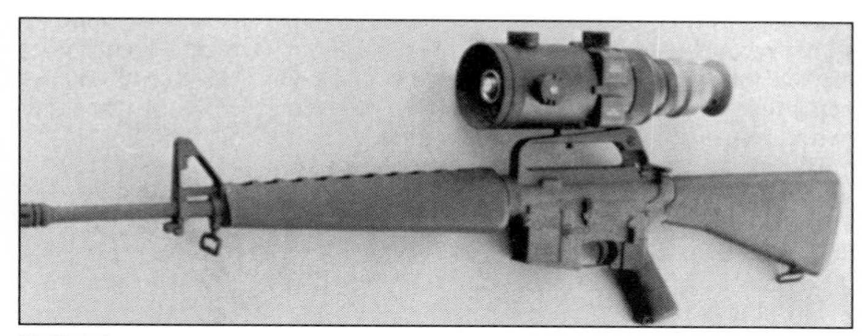

NVS-700

Objective lens: focal length, 95 mm T/1.7; focus adjustment, 25 m to infinity
Eyepiece: focal length, 26.5 mm; focus adjustment, +3 to –6 dioptres
Operating temperature: with arctic kit, –54 to +52°C
Power supply: 2 × AA mercury cells
Battery life: 60 h

Contractor

Northrop Grumman (formerly Litton) Electro-Optical Systems.

Northrop Grumman Ranger™ M992/M993/M994/M995 night weapon sights

Type

Weapon sight (personal).

Description

The Ranger™ series is the latest development from Northrop Grumman (formerly Litton) electronic devices. It is a high-performance, individual night weapon sight, with offset eyepiece design so that a more conventional head position can be adopted, as for a day sight.

The sight also incorporates a ballistically calibrated elevation knob to allow the user to adjust the estimated range using the knob range scale. In addition, the precision knobs provide repeatable clicks for adjustment of both windage and elevation. Only one AA size battery is required to power these water-resistant, ×4 and ×6 sights. A variable intensity red reticle is optically superimposed on the green intensified image, providing the marksman with a high-contrast aiming capability.

The Ranger models M992, M993, M994 and M995 differ only in the type of interchangeable image intensifier and magnification. The M992 and M994 incorporate the Litton-developed second-plus generation image intensifier, while the M993 and M995 use third- generation tubes for greater sensitivity and increased resolution. Magnification is ×4 for the M992 and M993, ×6 for the M994 and M995. A ×3 variant is also now available. All parts except the object lenses are interchangeable. All models include splashproof lens covers which incorporate a large neutral density filter for daylight training and weapon zeroing.

Operational status

Available. Military and paramilitary use.

Specifications

M992/M993
Dimensions: 250 × 103 × 88 mm
Weight: 1.2 kg
Object lens: f/1.5 (T/1.6)

Northrop Grumman Ranger™ M993

Field of view: 10.2°
Magnification: ×4
Resolution: 3.2/3.6
Gain: 1,200/1,500
Eye relief: 30 mm
Dioptre adjustment: +2 to –6
Tube: GEN II+/GEN III
Battery: 1 × AA
Battery life: 24 h alkaline, 48 h lithium
M994/M995
Dimensions: 310 × 123 × 120 mm
Weight: 1.7 kg
Object lens: f/1.7 (T/1.8)
Field of view: 6.3°
Magnification: ×6
Resolution: 4.5/5.0
Gain: 900/1,100
Eye relief: 30 mm
Dioptre adjustment: +2 to –6
Tube: GEN II+/GEN III
Battery: 1 × AA
Battery life: 24 h alkaline, 48 h lithium

Contractor

Northrop Grumman (formerly Litton) Electro-Optical Systems.

Novosibirsk night vision devices and sights

Type

Weapon sight (personal).

Description

The following image intensifying night vision infantry weapon sights are in production in the RFAS, manufactured by the Novosibirsk Instrument Making Plant: 1LH51/52/53; 1LH58; 1LH84; 1LH54; TLB-2; NHB.

Specifications: Novosibirsk night vision devices and sights

Night sights

Model	1LH51	1LH52	1LH53	1LH58	1LH84
Generation	2	2	1	1	2+
Magnification	×3.46	×5.3	×5.9	×3.5	×3.7
Field of view	8.5°	7.6°	5.5°	5°	10°
Identify range					
Tank	700 m	700 m	-	600 m	600 m
Man	400 m	-	-	400 m	400 m
Voltage	6 V	6 V	6 V	6 V	6 V
Battery life	10 h	10 h	10 h	10 h	10 h
Weight	2.1 kg	3.2 kg	15 kg	2 kg	1.3 kg
Dimensions	300 × 210 × 140 mm	333 × 186 × 183 mm	452 × 305 × 301 mm	458 × 186 × 99 mm	295 × 98 × 90 mm
Applications	AK-74, RPK, SVD	RPG-7	anti-tank gun	AK-74, RPK, SVD	AK-74, RPK

Night vision devices

Model	1LH54*	TLB- 2	NHB	1N10	ML	UM8- 2	T3K
Magnification	×5.5/5	×15	×15	×4-20	×4-20	×8	×8/9.9
Field of vision	6°/5°	6°	6°	1.5-8°	1.5-8°	5°	6/7°
Diameter exit pupil	5 mm	7.33 mm	7.33 mm	7-1.4 mm	7- 1.4 mm	2 mm	3/8 mm
Dioptre setting	±5	–3/+12	–3/+12	–5/+10	–5/+10	±10	±5
Dimensions	544 × 255 × 607 mm	565 × 325 × 545 mm	400 × 415 × 580 mm	765 × 42 × 190 mm	765 × 42 × 190 mm	100 × 45 mm	396 × 423 × 438 mm
Weight	18.5 kg	14.8 kg	30 kg	0.4 5 kg	0.5 kg	0.12 kg	14.6 kg

*day/night

RFAS 1LH51

Operational status
In production. In service with the RFAS and former Soviet Union and Warsaw Pact armies.

Contractor
Novosibirsk Instrument Making Plant.

VERIFIED

Ortek ORT-MS4 mini weapon night vision sight

Type
Weapon sight (personal).

Description
The ORT-MS4 is a lightweight second-generation night sight. It is for use on infantry weapons, or as a hand-held night observation device. The ORT-MS4's modular design incorporates an 18 mm second-generation image intensifier.

The instrument is self-contained, battery-powered and is supplied in a storage/carrying case with all necessary accessories. Power is provided by two standard 1.5 V AA batteries which can maintain operation for a minimum of 50 hours.

Operational status
In production.

Specifications
Dimensions: 266 × 70 mm
Weight: 1.16 kg
Field of view: 10°
Magnification: ×3.75
Eyepiece focal length: 27 mm
Focus adjustment: +2 to –4 dioptres
Objective: 100 mm fixed focus
Viewing range
Moonlight: 500 m
Starlight: 350 m

Contractor
Ortek Limited.

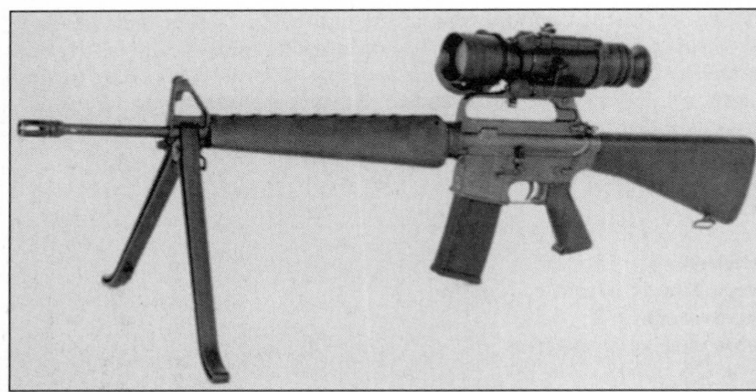

ORT-MS4

PCO PCS-5 mini weapon sight

Type
Weapon sight (personal).

Description
The mini weapon sight is a compact lightweight second-generation night aiming device with a built-in GEN II+ or Super GEN image intensifier for portable infantry

The mini-weapon sight from PCO of Poland
0022038

weapons, particularly Russian types. It is used for battlefield observation, target detection and recognition and aiming of light weapons and anti-tank grenade launchers at night. Mounting holders are available to fit the AKMN-1 (AKMSN-1) automatic rifle, PKMN-1 (PKMSN-1) machine gun, RPG-7N1 (RPG-7 DN1) hand-held grenade launcher and other similar weapons. A 'dovetail' side mounting system is available. Features include dioptre eyepiece adjustments and automatic brightness control. The observation range for a human-sized target, under illumination levels of 3 to 5 mlx (dark night, no moonlight) is given as 300 m. The observation range for a human- sized target, under illumination levels of 30 to 50 mlx (moonlight) is given as 600 m.

Operational status
In production. In service in Poland. Offered for export.

Specifications
Dimensions (l × h × w): 275 × 195 × 75 mm
Weight: 1.5 kg
Magnification: ×2.4 nominal
Field of view: 12.3°
Objective adjustment: 110 mm
Dioptre adjustment: ±6 dioptre
Operating temperature: –40 to +50°C
Power supply: lithium battery 3.4 V (R6/AA)

Contractor
PCO SA, Poland.

UPDATED

PCO PCS-6 passive night vision sight

Type
Weapon sight (personal).

Description
The PCS-6 passive night vision sight is a lightweight, compact sighting telescope with built-in passive GEN II+ or super GEN II image intensifier. It is used for aiming small arms, battlefield observation and detection and recognition of targets in conditions of natural night illumination. The basic version of PCS-6 is designated for the 5.56 mm assault rifle. It can also be adapted to 7.62 mm small arms. The device is equipped with automatic brightness control and an eyepiece with focus adjustment.

Operational status
In production. In service in Poland. Offered for export.

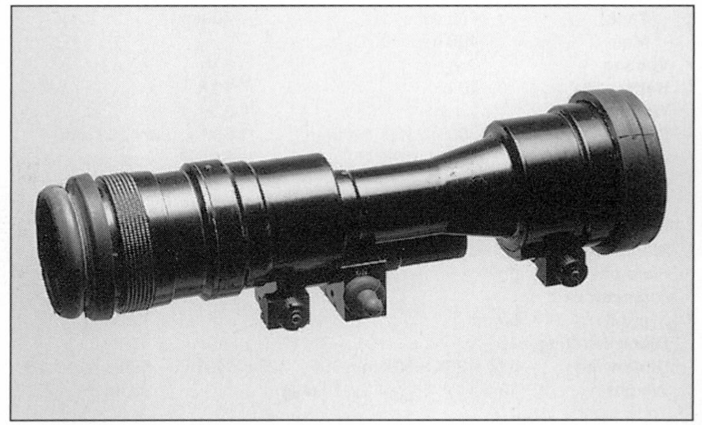

PCO PCS-6 passive night vision sight 0022037

Specifications
Dimensions (l × h × w): 220 × 76 × 78 mm
Weight: 0.8 kg
Magnification: ×2.2
Field of view: 12° 30′
Dioptre adjustment: ±4 dioptres
Power supply: 1 lithium battery 3.6 V (R6/AA)

Contractor
PCO SA, Poland.

Pyser-SGI PNP-XD-4/S, 2+/S, HG/S night vision weapon sights

Type
Weapon sight (personal).

Description
The Pyser-SGI PNP series are image-intensifying monocular night sights incorporating new XD-4, Generation 2 Plus or Hypergen technology in compact bodies with ×3 magnification waterproof lenses and a variety of rail options to suit all weapon types. Can also be utilised as hand-held monoculars and can be fixed to SLRs and camcorders.

Operational status
In production.

Specifications
Dimensions: 100 × 50 × 63 mm
Weight: 395 g without batteries and objective lens
Field of view: 20°
Magnification: ×2
Dioptre adjustment: +6 to –2
F number: f/1.3
Image intensifier: 18 mm PNP-XD4
Illumination: IR LED for zero light use (automatically incapacitated when weapon mounted)
Power supply: 2 × 1 .5V AA alkaline batteries
Battery life: >100 h
Water resistance: waterproofed to IP67 (immersion to 1 m for 30 mins with waterproof lens)

Contractor
Pyser-SGI Limited.

NEW ENTRY

Pyser-SGI small arms laser collimator

Type
Alignment system - laser collimator.

Description
The Pyser-SGI range of small arms collimators allow rapid and accurate zeroing of laser aimers and pointers on all small arms. The benefits are zeroing without live fire and accuracy to 0.25 mil.

Operational status
Proven and available. In service worldwide.

Pyser-SGI small arms laser collimator 0022039

Specifications
Accuracy: 0.25 mil (25 mm in 100 m)

Contractor
Pyser-SGI Limited.

VERIFIED

Raytheon AN/PAS-19

Type
Weapon sight (personal).

Description
The Raytheon AN/PAS-19 is a lightweight, multi-purpose, thermal imaging sight designed for use on individual and crew-served weapons, laser designators and as a general purpose surveillance sight. Operating passively in the infra-red spectrum, the AN/PAS-19 provides the user with the capability of target acquisition and firing in total darkness and under adverse visibility conditions. A ×2/×4 electronic zoom is provided, as is a memory which can store a number of electronically generated graticules. Boresighting the system with an electronic reticle adds to the unit's ease of operation. Push-button controls are incorporated for ease of operation and power is provided by a standard disposable MIL-STD BA 5847/U disposal battery or an external 6 to 24 V DC supply. The system also features a standard RS-170 video format output (CCIR is also available) for remote viewing and/or recording for training and post-engagement analysis.

Operational status
In service. Orders for over 250 systems have been received to date.

Specifications
Field of view: 15.3° horizontal × 7.6° vertical (×1.7)
Electronic zoom: ×2, ×4
Spectral band: 3-5 μm
Cooling: thermoelectric
Instantaneous FOV: 320 horizontal × 160 vertical pixels
Sampled FOV: 640 horizontal × 200 vertical pixels
Image: 51,200 independent pixels
Mission time: 10 h at 50% duty cycle with disposable internal battery
Dimensions: 355 × 114 × 114 mm
Weight (with internal battery): <1.8 kg

Contractor
Raytheon Systems Company.

VERIFIED

Raytheon AN/PAS-19

Raytheon MAG-1200

Type
Weapon sight (personal).

Description
The Raytheon MAG-1200 is a lightweight multipurpose thermal imaging sight designed for use on individual weapons and as a general surveillance sight. It provides the user with the capability of target acquisition and firing in total darkness and under adverse visibility conditions.

A ×2/×4 electronic zoom and electronically generated reticle is provided. Boresighting the system with the electronic reticle adds to the system's ease of

The MAG-1200 thermal imaging weapon sight 0004833

operation. There is an optional 9/15° or 5/15° telescope. Push-button controls are incorporated and power is supplied from a standard throwaway MIL-STD lithium battery that is the only consumable item for system operation.

The system also features an RS-170 video format output (CCIR available) for remote viewing and/or videotaping for training and post-engagement analysis.

The MAG-1200 is the result of several years of evolutionary design of systems qualified for sniper application. With the proper mounting bracket, the system provides accurate fire control for a range of weapons.

Operational status

In service.

Specifications

Field of view (apparent magnification): 9 × 4.5° (×2.9)
Resolution (h × v)
Vertical IFOVs: 320 × 160
Vertical samples: 640 × 200
Spectral band: 3-5 μm
Detector: non-cryogenically cooled lead selenide detector array
Mission time: 10 h at 50% duty cycle with disposable internal battery
Dimensions: 355 × 114 × 114 mm
Weight with internal battery: <1.96 kg
Battery type (6 V DC): lithium, BA5847/U (disposable internal); Ni/Cd (training external)

Contractor

Raytheon Systems Company.

VERIFIED

Raytheon NIGHTSIGHT™ W1000 portable weapons sight

Type

Weapon sight (personal).

Description

Raytheon Systems Company's W1000 portable weapons sight is part of the NIGHTSIGHT™ family of thermal imaging products for military and commercial end-users. The NIGHTSIGHT™ product line features driver's aids (DVE) and systems for surveillance and situational awareness (Model S1000, 200 series and PalmIR 250). NIGHTSIGHT™ products share a common advanced uncooled thermal imaging array and core electronics components to reduce logistics, supportability and acquisition cost.

The NIGHTSIGHT™ Portable Weapons Sight is a ruggedised, long-wave (8 to 12 μm) thermal sight for use as a primary targeting system for weapons or as a hand-held forward observer's scope. The system allows detection of a person up to 690 m with a standard 15° lens and over 1 km with the optional 9° lens. Up to 15 reticle patterns can be stored and recalled by the user depending upon which weapon is in use. The sight uses a standard US Army battery which provides over 14 hours of continuous use or can be powered by commercially available 9 V or AA battery cassettes. In addition, the sight can be powered by vehicle power of 7 to 32 V DC.

The W1000 sight can be mounted on a variety of weapons, including the M16A1/M16A2, M4 carbine, M203, M249 and M60. The unit is equipped for external video monitoring via an RS-170 compatible output.

Operational status

In production.

Specifications

Spectral band: 8-12 μm
Detector: 320 × 240 staring array
Type: uncooled ferroelectric

Raytheon NIGHTSIGHT™ W1000 with electronics modules 0022040

NIGHTSIGHT™ W1000 portable weapons sight from Raytheon

Resolution: 0.8 mrad (15° lens); 0.5 mrad (9° lens)
Field of view
15° lens: 11.3 (elevation) × 15° (azimuth)
9° lens: 6.75 (elevation) × 9° (azimuth)
Focusing range: 5 m to infinity
Recognition range
Man: 600 m (15° lens), 900 m (9° lens)
Vehicle: 1,800 m (15° lens), 2,900 m (9° lens)
Dimensions (l × w × h):
33 × 10 × 11 cm (15° lens)
36 × 11 × 12 cm (9° lens)
Weight: 1.6 kg (standard lens with US Army BA5847/U battery)
Power: 3.5 W nominal
Operating temperature: : –32 to +49°C
Battery: Standard BB-2847 military battery
9 V or AA battery cassettes available
Controls: contrast and brightness (automatic or manual), reticle selection, reticle azimuth and elevation, polarity, dioptre and objective lens focus

Contractor

Raytheon Systems Company.

UPDATED

Raytheon AN/PAS-13 Thermal Weapon Sight (TWS)

Type

Weapon sight (personal).

Development

The US Army has exercised a US$22 million contract option with Raytheon Systems Company for production of the Thermal Weapon Sight, the first infra-red sight for the army designed for use with rifles and other infantry weapons. TWS will replace the AN/PVS-4 and AN/TVS-5 image intensifier night sights currently in use for infantry weapons.

In 1994, Raytheon delivered 27 engineering and manufacturing development units for user testing to the US Army Project Manager, Night Vision/Reconnaissance, Surveillance and Target Acquisition, Fort Belvoir, Viginia. TWS successfully passed testing that included operations in arctic and tropical conditions. TWS has been selected for the US Army's Land Warrior programme. The current P3I programme will add a modular detachable laser rangefinder for disturbed-reticle fire control and a 6/2° high-magnification telescope for

Raytheon Thermal Weapon Sight (TWS) AN/PAS-13 0004834

applications requiring range performance beyond 3 km. New contract options include production of a lightweight version weighing under 3 lbs.

Description
TWS features a second-generation mid-waveband infra-red focal plane array that makes possible long-range target acquisition with a small aperture sensor.

TWS will replace the AN/PVS-4 and AN/TVS-5 image intensifier night sights currently in use for infantry weapons.

There are four TWS configurations which can be selected by using the basic sensor or installing the appropriate telescope on the sealed basic sensor housing:
(a) basic sensor, for applications requiring a wide field of view such as the Stinger anti-aircraft missile
(b) light weapon thermal sight, for weapons such as the M16 and the M4 Carbine
(c) medium weapon thermal sight, for weapons such as the M60 machine gun
(d) heavy weapon thermal sight, for long-range weapons such as the M2 machine gun, Mk 19 automatic grenade launcher and sniper rifles.

TWS incorporates a thumbnail-sized thermoelectric silent cooler; binary optics requiring 40 per cent fewer lens elements; VLSI electronics with lower power CMOS integrated circuits; advanced plastic-composite materials to reduce weight while providing structural strength and dimensional stability over the full range of environmental conditions; LED display. It has both RS-170 and digital video output.

Operational status
In production. Deliveries began in 1996. In 1998, as part of a US Army contract called Thermal Omnibus, Raytheon are to produce 4,020 Thermal Weapon Sight units. Additional options may bring production to more than 19,000 units.

Specifications
Spectral band: 3-5 μm
Detector: 40 × 16 photovoltaic hybrid CMT infra-red focal plane array
Cooling: thermoelectric
Fields of view (azimuth × elevation)
Basic sensor: 30 × 18°
Medium weapon thermal sight
18 × 10.8° (wide)
6 × 3.6° (narrow)
Heavy weapon thermal sight
9 × 5.4° (wide)
3 × 1.8° (narrow)
Weights (with battery)
Basic sensor (no telescope): 3.8 lb
Medium weapon thermal sight: 2.3 kg, 5 lb
Heavy weapon thermal sight: 2.5 kg, 5.5 lb
Range (70% probability of recognition through clear air)
Medium weapon thermal sight: 1.3 km (personnel targets)
Heavy weapon thermal sight: 2.4 km (vehicle targets)

Supported weapons
Medium weapon thermal sight: M16 (A1, A2, A3), M203, M4, M136, M249, M60
Heavy weapon thermal sight: M24, M2, Mk19, Barrett M82A1

Contractor
Raytheon Systems Company.

UPDATED

Ring Sight SLC-9-46-RGGS

Type
Weapon sight (personal).

Development
The LC-9-46-RGGS is a unit power sight specifically developed for the British Army to improve the hit probability of rifle launched grenades. The makers claim a high chance of a first round hit at 150 m.

Description
The sight has a solid glass optic which generates graticule aiming marks entirely within the glass which gives robustness and protection from internal misting. Aiming marks are provided for 50, 100 and 150 m range. Intermediate ranges can be interpreted by the user. By day the graticule is lit with ambient light and in low-light conditions and at night by a tritium source. The displayed graticule is compatible with NVGs. The eye relief is not critical so that the sight may be well clear of the user's eye. The British version is fitted to the SA80 rifle using the SUSAT telescopic sight. Mounts are available for other weapons.

Operational status
In service in quantity with the British Army.

Specifications
Aperture (v × h): 16 × 9 mm
Optic (l × w): 50 × 16 mm
Weight: 110 g
Housing and interface dimensions: weapon dependent

Contractor
Ring Sights Holdings Company Ltd.

VERIFIED

Ring sight LC-9-46-RGGS for rifle launched grenades 0010868

SAGEM OB-50 night firing scope

Type
Weapon sight (personal).

Description
The SAGEM night aiming telescope was designed for observation and aiming and is adaptable to all current types of infantry weapon. It is particularly suitable for rifles, machine guns and light anti-tank rocket launchers.

SAGEM OB-50 night firing scope

It incorporates either a SuperGen or Gen III 18 mm light intensifier tube fitted with double-proximity focus and with built-in automatic gain control. An illuminated micrometer and an eyepiece shade with shutter are also fitted. Various types of graticule are available and designs can be adapted to meet special requirements.

Operational status
In production. In service with the French Army and other special forces.

Specifications
Length: 230 mm
Weight: 900 g with batteries
Magnification: ×3.2 (option × 4)
Resolution: 2.5 lp/mrad
Focus: 30 m to infinity
Field of view: 11°
Tube: 18 mm, SuperGen or GEN III
Eyepiece focus: –4 to +2 dioptres
Power supply: 2 × AA 1.5 V
Operating temperature: –45 to +52°C

Contractor
SAGEM SA, Aerospace & Defence Division.

UPDATED

Seiler VISIONMASTER®

Type
Weapon sight (personal).

Description
The Seiler VISIONMASTER® is a combined day and night weapon sight. The night channel is provided by an add-on fitting which includes an image intensifier. A changeover time of 30 seconds is given. A selection of intensifier standards is available including Generation III tubes from ITT and Generation II+ tubes from Delft. The optics are specifically designed to accommodate wavelengths from the visible to the near infra-red and the manufacturers state that the custom lens coatings used give excellent transmission from 400 to 900 nm. A ranging reticle is included.

Operational status
Available.

Specifications
Weight: Day scope 2.38 kg, night scope 3.1 kg
Length: Day scope 438.15 mm, night scope 463.55 mm
Diameter (main tube): 40 mm
Magnification: ×6 to ×16
Field of view: 3.3
Focus range: 30 m to infinity
Objective lens: 80 mm
Eye relief: 30 to 89 mm at ×6, 49 to 77 mm at ×16

Contractor
Seiler Instrument & Manufacturing Company Inc.

VERIFIED

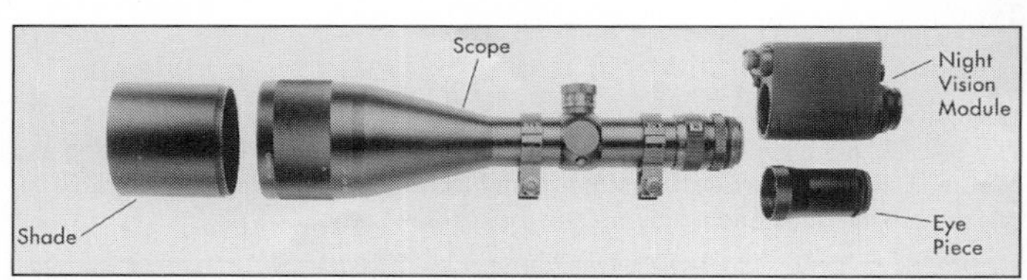

The Seiler VISIONMASTER weapon sight 0080258

Simrad IS2000

Type
Passive weapon sight (personal).

Description
The Simrad IS2000 eye-safe laser gun sight is a fire-control system for direct fire weapons. The instrument provides instant target range and aiming information and it has a built-in computer programmed for ammunition ballistics. Range and aiming information for moving and stationary targets is displayed. Range is shown in the panel display and an aiming point corrected for elevation and lead angle is shown in the head-up display. The IS2000 can easily be reprogrammed by the operator to compensate for changes in battlefield conditions. The IS2000 is particularly suited for upgrading portable direct fire weapons such as the 110 mm Panzerfaust 3 weapon (qv) or the 84 mm Carl Gustav anti-tank gun. The Simrad IS2000 can be used with Simrad GN1 night vision goggles or it can be fitted with a Simrad KN250F image intensifier for 24-hour operation.

Operational status
In production. Offered as part of the Dynamit Nobel Panzerfaust 3 anti-tank weapon.

Specifications
Eye Laser safety: Class 3A (IEC825-1). Option: Class 1 (IEC825-1)
Wavelength: 905 nm
Min/max range: 50/2,000 m
Resolution: 5 m
Objective diameter: 45 mm
Field of view, HUD: 8°
Magnification, HUD: ×1, day sight ×3.5
Battery: Ni/Cd or alkaline
Input voltage: 7 - 15 V
Batteries: 6 AA cells
Measurements before recharging: >1,000 at +20°C
Remote control: yes
Data communication: RS-485
Weight: approx 1.85 kg
Dimensions, casing (l×h×w): 219 × 150 × 112 mm

Contractor
Simrad Optronics ASA.

VERIFIED

Simrad IS2000 mounted on a Panzerfaust 3 anti-tank weapon 0041500

The Simrad IS2000 eye-safe laser gunsight with KN250F image intensifier on 84 mm Carl Gustaf 0041499

Simrad KN200 image intensifier

Type
Weapon sight (personal).

Description
The Simrad KN200 is a battery-powered image intensifier intended as an add-on unit to optical sights on direct fire weapons, laser range-finders and other daylight image devices such as television cameras.

Second- or third-generation tubes can be used and changes require no modification to the instrument. The eyepiece of the existing day sight is also used at night, the graticule pattern remains the same and the position of the operator's eye remains unchanged day and night. It can be mounted and removed within a few seconds and no boresighting adjustments are necessary.

The use of a 100 mm objective aperture yields a good range performance, enabling the operator to detect and engage targets within the practical range of the weapon during night operations. Sudden illumination does not have any effect on the sighting capabilities.

Operational status
In production.

Specifications
Dimensions: 210 × 209 × 120 mm
Weight: 1.56 kg incl battery
Field of view: 10°
Focusing range: fixed or adjustable 25 m to infinity
Resolution: 0.5 mrad/lp (contrast 30%, illuminance 100 lx)
Objective lens: catadioptric, 100 mm f/1.0
Power supply: 2 × 1.5 V AA alkaline or 2 × C cells; lithium optional
Battery life: >80 h at 20°C
Operational temperature range: –40 to +52°C

Contractor
Simrad Optronics ASA.

VERIFIED

The KN200 image intensifier sight 0041501

Simrad KN250 image intensifier

Type
Weapon sight (personal).

Description
The Simrad KN250 image intensifier is mounted as an add-on unit on telescope sights, laser range-finders and other optical daylight imaging devices. As no boresighting adjustment is required, the mounting procedure only takes a few seconds.

The KN250 is available with GEN II or GEN III image intensifier tubes. Change between these tubes requires no modification of the instrument.

The gunner can aim through the eyepiece of the day sight by both day and night. Sudden illumination of the scene does not have any effect on the sighting capabilities.

Operational status
In production.

Simrad KN250 image intensifier

Specifications
Weight: 1 kg approx incl batteries and standard brackets
Magnification: ×1
Field of view: 12°
Resolution: 0.6 mrad/lp (contrast 30%, illuminance 100 lux)
Objective lens: catadioptric, 80 mm f/1.1 (T/1.4)
Focusing: 25 m to infinity
Tube: GEN II or GEN III
Theoretical recognition range (light level 10 mlx, 90% target contrast)

	Man-sized target	**Tank target**
KN250F	420 m	1,770 m
KN252F	450 m	1,920 m
KN253F	500 m	2,160 m

Power supply: 2 × 1.5 V alkaline AA cells
Battery life: >80 h at 20°C
Operational temperature range: –40 to +52°C

Contractor
Simrad Optronics ASA.

VERIFIED

Thales Optics Kite weapon sight

Type
Weapon sight (personal).

Description
Kite is a lightweight, high-performance night sight for infantry weapons. The refractive lens system incorporates an injected graticule to assist with accurate weapon aiming. Graticules and weapon interface brackets to suit the

Kite night sight mounted on a weapon 0004824

characteristics of a wide variety of different calibre weapons can be provided. The sight is configured to accept second- or third-generation tubes and can have either a fixed or adjustable eyepiece. Kite has a rotary ON/OFF switch, graticule brightness adjustment and a large collar focusing control.

Operational status
Kite is in service in over 40 countries and several thousand have also been supplied to the British Army. Over 40,000 sights manufactured to date.

Specifications
Weight: <1.2 kg
Dimensions: 255 × 105 × 80 mm
Field of view: 8.5°
Magnification: ×4
Focus range: 15 m to infinity
Eye relief: 30 mm
Range (min): recognition of standing man at 400 m in starlight (10^{-3} lx using GEN II tube)
Power supply: 2 × 1.5 V AA batteries

Contractor
Thales Optics.

UPDATED

Thales Optics Thermal Weapon Sight (TWS)

Type
Weapon sight (personal).

Description
The thermal weapon sight is of a lightweight, modular design, suitable for both individual and crew-served weapons. Designed on the basis of a common body with optional objectives offering single or dual field of view, dependant upon customers preference. Electronic injected graticules and weapon interface brackets to suit the characteristics of a wide variety of weapons can be provided. The sight is based upon an uncooled detector operating in the 8 to 12 μm wavelength.

Operational status
The thermal weapon sight is not yet in military service.

Specifications
Weight: <1.4 kg (16° FOV objective)
Dimensions: length incl eyeguard 336.5 mm; width 113 mm; height excl weapon bracket 119 mm
Field of view: dependant upon customer's choice of objective lens
Magnification: dependant upon customer's choice of objective lens
Focus range: typically 10 m to infinity
Eye relief: 30 mm
Range, min: detection of a man-sized target >550 m (16° FOV objective lens)
Power supply: commercially available lithium ion batteries plus AA option; auxiliary power supply port

Contractor
Thales Optics.

UPDATED

Thalis NS-467 and NS-685 night vision weapon sights

Type
Weapon sight (personal).

Description
The NS-467 and NS-685 are ×4 and ×6 magnification passive image intensifier sights for use on infantry weapons. The main characteristics of both sights are low weight, small dimensions, excellent boresight retention and performance under all light conditions due to the optimised optics and high-performance US Gen III and European ANVIS image intensifier tubes.

The sights include a Mounting Bracket Assembly that allows direct mounting on Picatinny Rails, precise adjustment in elevation and azimuth and retention of boresighting under harsh military conditions. Brackets are available for a large variety of weapons. The optics of the sights are optimised to take advantage of the characteristics of current high-performance image intensifiers. They are lightweight, offer excellent resistance to adverse conditions and have multilayer anti-reflection coatings. The reticle is ballistic type 200 to 800 m, in 100 m increments, or as per customer request. The sights conform to MIL-STD-810.

The supplied kit includes: one NS-467/NS-685 sight complete with Mounting Bracket Assembly, one soft carrying bag, one shipping/storage case, one users manual, two AA alkaline batteries and one lens cleaning kit.

Thalis NS-467 weapon sight 0134453

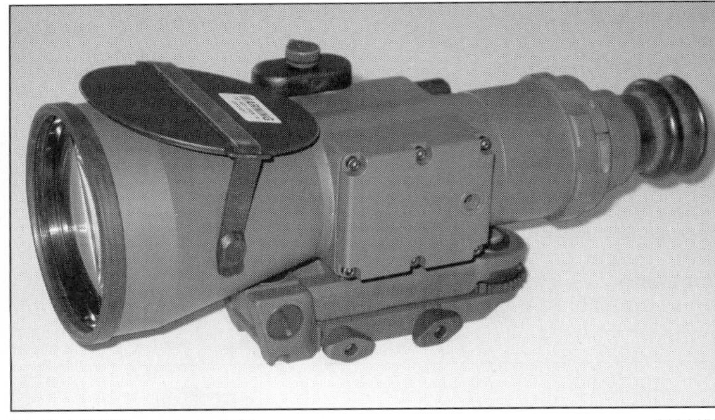
Thalis NS-685 weapon sight 0134455

Operational status
In service.

Specifications

	NS-467	NS-685
OPTICAL DATA		
Magnification	×4	×6
Field of view	10.0°	6.7°
Range focus	15 m to infinity	25 m to infinity
OBJECTIVE LENS		
Type	Petzval, 5 elements	Petzval, 5 elements
EFL	100 mm	150 mm
Free aperture	67.0 mm	83.5 mm
F No/T	1.5/1.6	1.8/1.9
EYEPIECE		
EFL	25 mm	
Dioptre adjustment	+2 to −6 dioptre	
Eye relief	30 mm	
Exit pupil	8 mm	
MOUNTING BRACKET ASSEMBLY		
Interface	MIL-STD-1913	(Picatinny Rail)
Elevation range	50 mils	
Azimuth range	45 mils	
Adjustment	0.25 mils	
Image intensifier tube	18 mm ANVIS Gen III/ Gen II	
Power Supply	2 × AA alkaline batteries (2 × 1.5 V)	
Battery life	More than 70 hours at 20°C	
Dimensions		
Length (incl eyeguard)	239 mm	285 mm
Width	94 mm	102 mm
Height (mounting bracket assembly)	115 mm	113 mm
Weight (incl mounting bracket assy and batteries)	1.3 kg	1.4 kg

Contractor
Thalis S.A.

UPDATED

US Army INOD

Type
Weapon sight (personal).

Description
The Improved Night/day fire-control/Observation Device (INOD) is intended to provide Joint Special Operations Forces with an integral day/night scope for medium and heavy sniper rifles as well as strategic reconnaissance/observation.

There are two versions: medium and large. The devices share a common body and are fitted with a prototype GEN IV image intensifier. A clip on laser range-finder is an option.

Operational status
Development.

Contractor
US Army Night Vision and Electronic Sensors Directorate.

OBSERVATION AND SURVEILLANCE

AIR DEFENCE SENSORS

SAGEM SIRÈNE infra-red search and track system

Type
Air defence sensor.

Description
SIRENE is an infra-red panoramic search and track system developed by SAGEM for the French Ministry of Defence.

The system provides a completely passive air target detection, tracking and indication, and is immune to Electronic Counter Measures. It can be operated as a stand-alone equipment or connected to a surveillance network. It naturally enhances radar surveillance by filling the low altitude radar gap.

SIRENE is suited for use in medium, short and very short ground-to-air weapon systems and units. It can be installed on fixed position (tripods, masts) or gimballed on board vehicles. Its small size and easy and fast deployment make it a suitable designation system for power projection operations.

The basic system includes two main portable operating modules :
- The sensor module comprising the head and associated electronic unit,
- The remote module comprising the processing unit and the operator's display and control unit,

The sensor module can be deployed at up to 200 m from the remote module.
SIRENE is served by a single operator through a MMI displaying the tactical situation and system status on a high-definition color screen. It provides a full 360° IR picture of the scanned area with overlaid symbols on the detected targets. The man-machine dialogue is performed through display, keyboard and track-ball.

Operational status
In pre-production for the French Army.

Specifications
Spectral band: 8-12 μm
Horizontal coverage: 360°
Elevation coverage: 5.4° or 8.6° (adjustable from –10° to +20°)
Refresh rate: 1.4 Hz
Detection ranges
aircraft: 20 km
helicopters: 10 km
cruise missiles: 8 km
Target designation accuracy: ± 1 mrad
Track-table capacity: up to 50 tracks
Electric power supply: 18 – 32 V DC
Interfaces: various interfaces available for compatibility with other systems or surveillance networks

Contractor
SAGEM SA Aerospace & Defence Division.

VERIFIED

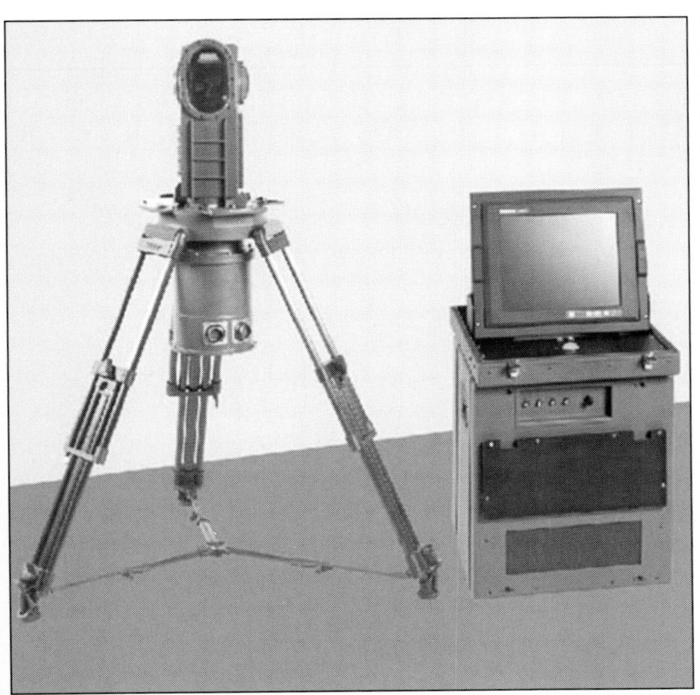

Sirene IR search and track system 0137379

Thales MUNOS Multiple Use Night Weapon Sight

Type
Night vision family of aiming sight.

Description
The night fighting and aiming sight MUNOS family, consists of a range of modern lightweight image intensifier weapon sights such as:
- a monocular ×4 magnifying support Weapon Sight type WS4 for short-range aiming
- a monocular ×6 magnifying support Weapon Sight type WS6 for short and medium aiming
- a monocular ×10 magnifying support Weapon Sight type WS1 0 for long-range aiming.

The MUNOS is a monocular passive night vision weapon sight, which is designed for the XD4 technology generation as well as third-generation image intensifier tubes, and consists of:
- a tube-sub-module, including the eyepiece sub-module
- a body-sub-module, including the optical system and the reticule sub-module. Both sub-modules are integrated in one system.

The MUNOS lightweight, high-performance night vision device can be mounted on individual weapons and is designed and proven to fulfill military requirements.

Different adapters (optional) are available to mount the MUNOS to a variety of individual weapons.

Operational status
Full production.

Specifications

Type	4× weapon sight	6× weapon sight	10× weapon sight
Magnification	4×	6×	10×
Field of view	10°	6.7°	4°
Focal range	15 m to infinity	25 m to infinity	50 m to infinity
Weight (incl batteries)	900 g	1,100 g	2,400 g
Dimensions (l × h × w)	210 × 95 × 78 mm	235 × 95 × 87 mm	360 × 150 × 140 mm

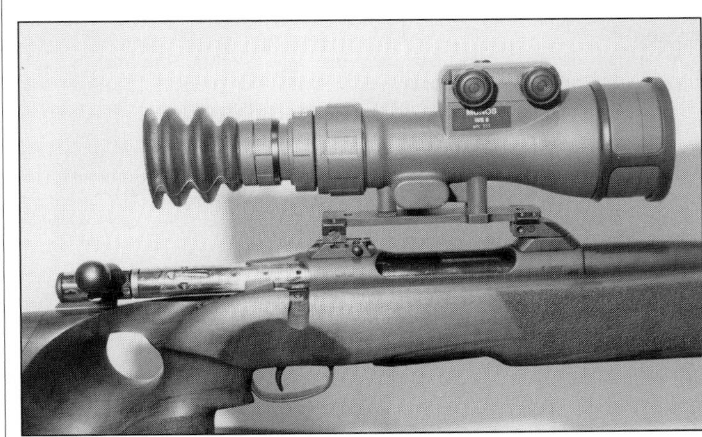

MUNOS WS6 sight on SIGSAUER SSG 2000 weapon 0053701

The MUNOS WS10 ×10 magnification sight 0053703

Resolution at 30 mlux (USAF target 85% contract) HyperGen tube, XD4 or equivalent	0.23 mrad/lp	0.18 mrad/lp
Eye relief	28 mm	
Reticle adjustment range	–15 to +15 mrad	
Reticle adjustment step	0.1 mrad	
Battery type/life	2 standard AA size Alkaline: typical 150 h	
Dioptre adjustment	–6 to +2	

Contractor
Thales Optronics B.V.

NEW ENTRY

Thales Optronics Air Defence Alerting Device (ADAD) air defence sensor

Type
Air defence sensor.

Description
The Air Defence Alerting Device (ADAD) was developed by Thorn EMI Electronics, Electro-Optics Division, now Thales Optronics, as a lightweight, compact, autonomous system which detects rotary- and fixed-wing aircraft through their infra-red emissions and then indicates their bearing and elevation to the operator. The manufacturer claims that it increases the chances of a successful kill by 400 per cent.

ADAD comprises a scanner, a processor and a remote display unit. The scanner provides coverage over a wide field of view in azimuth with an elevation baseline adjustment. It covers 240° in azimuth and –7 to +17° in elevation.

The scanner passes data to the processor, allocates priorities, then transmits alert and cueing signals to the display units. Data and tactical information are entered by a simple alphanumeric keyboard which permits the entry of primary arcs of responsibility and restricted arcs. Each ADAD system can service up to four display units at distances up to 500 m from the processor.

The freestanding system, which is powered by a 300 W/28 V portable generator, can be brought into action within 5 minutes and can be dismantled in half this time. The vehicle mounted system can be brought into action in under 1 min from standby.

A version developed for use with the vehicle-borne Starstreak missile automatically slews the weapon sight onto the target's bearing and elevation. SP HVM Starstreak with ADAD entered service with the British Army in October 1995.

ADAD has been selected by the German Air Mobile Brigade for fitting on their Wiesel 2 light tracked vehicles together with the Stinger surface-to-air missile. The contract has been awarded by STN ATLAS Elektronik as part of Germany's LeFlaSys short-range air defence system. ADAD is fitted on the Ozelot weapon platform which entered service in June 2001.

Operational status
In production for the UK MoD. Entered service with the British Army in 1992 and as part of the Starstreak missile system in 1995. In service with the German Army. Also under evaluation in Poland and Thailand. In 1999, the US AMCOM placed another order for an evaluation ADAD.

Specifications
Spectral band: 8-12 μm
Detector: CMT
Cooling: split-cycle Stirling
Field of regard
Azimuth: 240°
Elevation: –7 to +17°
Dimensions
Scanner (h × d): 985 × 250 mm
Processor (h × w × d): 400 × 400 × 520 mm
Display unit (h × w × d): 142 × 311 × 236 mm
Weight
Scanner: 35 kg
Processor: 26 kg
Display unit: 5 kg
Power: 300 W/28 V

Contractor
Thales Optronics.

VERIFIED

Air Defence Alerting Device (ADAD) 0004821

OBSERVATION AND SURVEILLANCE

FORWARD OBSERVATION

Aselsan FALCONEYE high-performance passive electro-optic target acquisition system

Type

Thermal imager and target acquisition system.

Description

The FALCONEYE is an electro-optical sensor system which comprises an advanced thermal imaging system and a high-performance day TV sensor for sighting. The system has an embedded eye-safe laser range-finder, GPS receiver and digital compass providing the user either co-ordinates of the target or bearing and inclination information to the target. A very accurate Astronomical North Finder capability is optionally available. A laser pointer for illuminating targets for friendly image intensifier weapon sight users is also embedded for optional use.

The FALCONEYE can be easily integrated into any platform or be used with a tripod. All functions are attained through user friendly interfaces.

Operational status

In production.

Specifications

Thermal imaging system
Spectral band: 8-12 μm
Detector: 288 × 4, HgCdTe - FPA
Field of view:
6.0 × 4.5° (wide)
2.0 × 1.5° (narrow)
or
9.0 × 6.75° (wide)
3.0 × 2.25° (narrow)

Day TV sensor
Field of view: >100 (wide); <10 (narrow)
Magnification: ×27

Laser range-finder
Type: Erbium-Glass-1.54 μm (eye-safe)
Range: 20 km
Measurement accuracy: ±5 m

Laser pointer
Wavelength: 830 nm
Power: >10 mW

Digital compass
Azimuth accuracy: 8 mils
Elevation accuracy: 3 mils

Aselsan FALCONEYE thermal imager and target acquisition system
NEW/0522802

GPS receiver
This unit calculates its own position, utilising the GPS satellite network, with a position error of 10 m.
Astronomical North Finder
This optional feature utilises custom software and a device to find a precise geographic north bearing by observing the sky for sun, moon, planets and stars either day or night.

Heading accuracy: 1 mil

Contractor

Aselsan Inc, Microelectronics, Guidance and Electro-Optics Division.

UPDATED

Aselsan MARS-V armoured reconnaissance/ surveillance vehicle

Type

Area surveillance system.

Description

MARS-V is a combination of advanced sensor technologies and computer controlled infrastructure into a single platform. Mounted on 4 × 4 light armoured wheeled vehicle, this system provides multispectral surveillance using optical and radar sensors. MARS-V has elevated sensors for extended coverage and has been designed for continuous monitoring of the battlefield, detecting possible targets and transmitting the collected information to command centres and reaction forces.

MARS-V uses ground surveillance radar and second-generation thermal imager as long-range target detection sensors. Two sensor systems can perform surveillance independently on non-coinciding sectors. When a target is detected by one of the sensors, other sensors can be automatically directed to the target area in order to improve target recognition and classification capability. Target identification can be performed using the Doppler tone generated by the radar, then the thermal imager and Day TV are focused on the appropriate range. Target information collected by the sensors and evaluated/classified in the information processing software, can be transmitted to the command and monitoring centres using various communication equipments integrated in the vehicle. Target informations, including co-ordinate, type and so on, are gathered in formatted reports prepared using the software, can be transmitted securely via the frequency hopping radio, whereas full motion target imagery can be sent through microwave video link. Still images can be included in the formatted reports.

Main system functions, including control of peripheral equipment and electro-optic sensor system, is controlled by the System Control Unit, which is operated by the System Operator.

Operational status

The Modular Armoured Reconnaisance and Surveillance Vehicle (designated as MARS-V) is in use and in production.

Specifications

Sensors: Aselsan ground surveillance radar
2nd-generation thermal imager
Day TV target co-ordinate determination system
Communication equipment: 9600 VHF frequency hopping radio
Video transmission system (MW link)
Field telephone
Command control software: Geographical information system
Database
Formatted message
Target detection range (with radar): 38 km (vehicle convoy)
15 km (personnel)

Operational status

The Modular Armoured Reconnaisance and Surveillance Vehicle (designated as MARS-V) is in use and in production.

Contractor

Aselsan Inc, Microwave and System Technologies Division, Ankara.

UPDATED

Aselsan Rattlesnake laser range-finder/designator

Type
Laser range-finder/designator.

Description
Rattlesnake ground laser designator/range-finder has been developed for the dismounted soldier. It is a coded laser designation system. The system selectively operates at Band 1, Band 2 NATO codes and a no-code mode. System is designed to provide guidance for all types of munition used by NATO such as laser-guided bombs, Hellfire and Copperhead. The system is capable of measuring ranges up to 20 km with high accuracy.

The system can further be enhanced by a Quick mount II and Generation Focal Plane Array Uncooled Thermal Imager.

Operational status
In production.

Specifications
Laser transmitter
Type: Nd:YAG
Wavelength: 1,064 nm
Pulse energy: >80 mJ
Pulse width: 10-20 nsec
Pulse energy stability: <15%
Beam divergence: <0.5mrad

Operating features
Operating modes: Range-finding and designating
Coding: NATO Band 1, 2 (NATO STANAG 3733LAS)
Operating time: 5 min on, 1 min off, 3 cycles max, then minimum 30 min cool down (10 pps at designation)

Range
Designation range: 10 km, depending on weather conditions (NATO STANAG 3850LAS and STANAG 3875 Cat.B)
Range-finding limit: 20 km
Range finding accuracy
Sighting Telescope:
Magnification: ×10
Field of view: >3°
Dioptre adjustment: −4 to +3
Eye-piece angle: 45°
Eye-piece protection: 1,064 nm beam reflector and absorbing filter (1/100,000)
Reticle: Circle at the centre with 1 mrad divisions
Weight: <13 kg (laser designator)
Size: <1.2 l

Electrical
Power consumption: <250 W
Power source: 24 V battery or DC source connection

The Aselsan Rattlesnake LRF/D
NEW/0522803

Environmental
Minimum operating temperature: −30°C
Maximum operating temperature: 50°C
Minimum storage temperature: −30°C
Maximum storage temperature: 55°C

Contractor
ASELSAN Inc, Microelectronics, Guidance and Electro-Optics Division.

NEW ENTRY

Azimuth ATLAS target acquisition systems

Type
Forward observation system.

Description
Azimuth ATLAS systems represent an advance in target acquisition, intelligence gathering and fire-control applications. Available in standard, lightweight and gyro-integrated versions, Atlas systems create a common language between forward observers and firing batteries, aircraft and helicopters.

Atlas Mk2, Atlas LT and Atlas GLT digital integrated target acquisition systems achieve a high accuracy of ±1 mrad at an affordable cost. They enable rapid, precise aiming of maximum fire power on a target.

Systems can be configured with laser range-finders, viewing systems (binoculars, night vision, FLIR and SLS, TV camera). Built-in modems position Atlas systems for wireless digital communications. A GPS option is available. Atlas systems incorporate highly advanced hardware and comprehensive, optimised, battle-proven task software.
Applications:
- target acquisition
- target intelligence gathering and enemy location
- artillery fire adjustment
- mortar fire control
- designation for laser-guided weapons
- air strike co-ordination
- gun laying
- gateway to C41. These elements support any VHF/UHF radio communication.

Operational status
Available.

Contractor
Azimuth Ltd.

The Atlas LT system in forward observation configuration 0101560

BGT, Vectronix, Zeiss Optronik TAS10

Type
Forward observation system.

Description
Bodenseewerk Gerätetechnik GmbH, Vectronix (formerly Leica), and now Zeiss Optronik developed the Manpack Target Acquisition System, TAS10, which carries out the following functions: determination of own position, target acquisition and

The TAS10 in field use

The TAS10 vehicle mounted

target survey and laying of weapon systems. The TAS10 consists of a laser range-finder (eye-safe) with integrated telescope, a fully automatic north-finding gyroscope, an electronic goniometer (for determination of angle of elevation and azimuth) and a display and control panel.

The Raman-shifted eye-safe laser range-finder by Zeiss Optronik works at a wavelength of 1.54 μm. The wire-suspended type north finding gyroscope was developed by BGT and the electronic goniometer by Vectronix. The optional adaption of a GPS receiver and a thermal imaging system to the TAS10 provides autonomy and day/night capability to the system. The equipment incorporates a built-in test which monitors all system functions during system operation.

Operational status

Series production. In service with German Bundeswehr, French and UK armies and several other armies.

Specifications

Weight: 16 kg
Laser rangefinder: up to 20 km
North accuracy: 1.0 mil
Accuracy target survey, X,Y: =10 m, vector of 5 km
Accuracy target survey, Z: =3 m, vector of 5 km
Angle resolution: 1.0 mils
Area of deployment: ±75° latitude
Temperature range: –35 to +65°C
Power supply: 24 V DC, battery or vehicle source

Contractors

Bodenseewerk Gerätetechnik GmbH (BGT).
Vectronix, Defence & Special Projects.
Zeiss Optronik GmbH.

VERIFIED

Diel Eagal 1.5

Type

Forward observation system.

Description

The Eagal 1.5 is a lightweight reconnaissance and target acquisition system designed for use by a forward observer. It consists of a goniometer with servo-assisted, finger-controlled precision drive, a compact telescopic lens to meet customer specifications, a laser range-finder, an image intensifier and a compact tripod. All data are referenced to magnetic north via an electronic fluxgate compass. Eagal can be controlled locally by an onboard operator interface panel or remotely by an IBM-compatible ruggedised computer.

Operational status

Available.

Specifications

Azimuth: 6,400 mrad
Elevation: +533 to –355 mrad
North-finding accuracy: <20 mrad
Slewing speed: stepped in 0.5 mil increments; continuous from 1 mrad/s to 320 mrad/s
Angle readout: incremental shaft encoders with 0.5 mrad resolution; mechanical graduation scales
Levelling: X and Y axis electronic tilt sensor
Levelling accuracy: ± 1 mrad
Weight
  incl payload: 15 kg
  excl payload: 5 kg
Dimensions: 140 × 140 × 200 mm

Contractor

Diel (Pty) Ltd.

VERIFIED

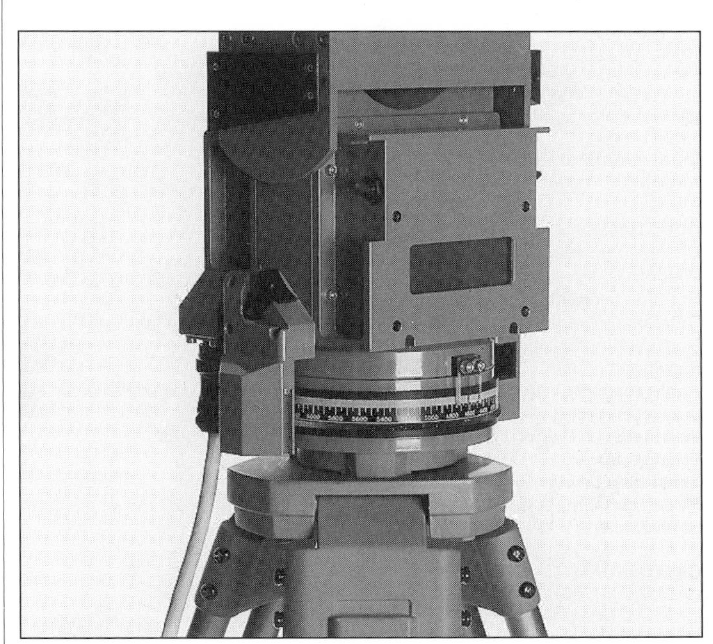

The Eagal 1.5 forward observation system

Diel Eagal 2

Type

Forward observation system.

Description

The Eagal 2 Laser Range-Finder Goniometer (LRFG) is a compact, rugged, lightweight, portable unit used to determine the bearing and elevation of a target. It is designed for use in conjunction with a laser range-finder such as the Eloptro LH30 (qv) to provide full target postional information.

Eagal 2 comprises a positional platform with tripod and laser range-finder mountings. An integrated keypad and liquid crystal display provide information entry and data display. Target elevation and azimuth are determined by locating the target in the graticule of the range-finder. The movement of the LRFG in azimuth and elevation is registered by shaft encoders fitted to the azimuth and elevation axis. Azimuth readings are referenced to either surveyed landmarks or the electronic or tubular compass. Target position information is displayed in mils and metres on the display. Each target acquired is time stamped and stored. The user may allocate the target to a particular storage location. The unit stores up to nine targets and data are preserved in memory even when the unit is powered down.

The Eagal 2 laser range-finder/goniometer 0004816

All stored target information may be retrieved and displayed on the integral display. It may also be accessed from a remote computer via the RS-232 interface.

Operational status
Available.

Specifications
Dimensions: 190 × 115 × 130 mm
Weight: 3.5 kg
Rotation (azimuth): 6,400 mils continuous
Rotation (elevation): 533 mils below horizontal, 355 mils above horizontal
North finding: fluxgate or tubular compass
Slewing: manual
Angle readout: incremental shaft encoders with 0.8 mil resolution, mechanical graduation scales with 10 mil resolution
Interfaces: LCD display, keyboard, RS-485 to LRF, RS-232 to PC
Primary power: 3 PP3 9 V batteries
Secondary power: 10-18 V DC
Power consumption: 70 mA at 12 V DC operational max, 20 mA at 12 V DC standby max

Contractor
Diel (Pty) Ltd.

UPDATED

DRS Nightstar

Type
Forward observation.

Development
Nightstar is an upgraded version of the N/CROS developed by International Technologies Lasers Ltd (ITL) and manufactured under licence by DRS Optronics, Inc. The original N/CROS was used by special forces in operation 'Desert Storm' during the 1990-91 Gulf War.

Description
This hand-held night viewing binocular incorporates an image intensifier tube, laser range-finder and digital compass in a compact lightweight system for infantry units and night operations involving forward observers and reconnaissance patrols. With a range of 20 to 2,000 m, the Nightstar displays range and azimuth data in the user's eyepieces, allowing the identification and relay of fire support data for night time engagement.

Operational status
In service with the US Army and Israel Defence Forces.

Specifications
Dimensions (l × w × h): 250 × 270 × 80 mm
Weight: 1.6 kg
Magnification: ×3
Dioptre setting: +2 to –6
Exit pupil: 7 mm
Eye relief: 15 mm
Objective F: 1.3
Image intensifier
Resolution: 64 lp/mm
Tube type: 18 mm, non inverting, GEN III

Laser
Type: GaAs laser diode
Wavelength: 904 nm
Output power: 2.85 W
Safety: eye-safe (ANSI Z136.1)

Ranging
Sensor type: APD
Range: 20 to 2,000 m
Range accuracy
at 20-200 m: ±2 m
at 200-500 m: ±3 m
at 500-2000 m: ±5 m
Receiver field of view: 3.3 mrad
Measuring time: 1.05 s max

Compass
Sensor type: flux gate
Accuracy: –1.5° RMS
Tilt: up to ±20°

Display
Numeric: 4 digit, 7 segment, 1 for range, 1 for azimuth
Reticle: illuminated hollow cross
General power: 6 AA alkaline batteries
Operation time: 10 h continuous viewing incl 1,000 ranging and 1,000 azimuth measurements

Contractor
DRS Optronics, Inc.

UPDATED

Electromagnetica STLA-M3

Type
Forward observation system.

Description
The STLA-M3 is designed for measurement of target co-ordinates and target speed for artillery aiming. The data can then be transmitted to a command position. It includes a laser range-finder, optical sight and microcomputer. It is portable and can be used on a tripod or mounted on a vehicle. The target co-ordinates can be provided in both polar and rectangular format and simultaneously displayed. Co-ordinates for up to 64 targets may be stored.

Specifications
Dimensions (l × w × h): 280 × 300 × 350 mm
Weight: 13 kg

Sight
Traverse: horizontal ±60°, vertical ±4.5°
Accuracy: +0°, –0.5°
Speed measurement range: 5-99.5 m/s
Laser range-finder
Max range: 20 km
Operating range: 300 m - 20 km
Accuracy: 5 m
Measurement rate: 10/min

Contractor
Electromagnetica, Romania.

Elop Portable Lightweight Laser Designator (PLLD)

Type
Laser designator.

Description
The Portable Lightweight Laser Designator (PLLD - formerly known as PAL) is a day/night laser designation system used as part of a weapon system together with

Lightweight Laser Designator (PLLD) ***NEW***/0547497

laser-guided missiles or projectile bombs. The system can designate stationary or moving targets for all types of munitions; collect target information (range, azimuth and elevation) for target acquisition and mark targets for laser spot tracking sensors.

The components of the system are the designation head with laser range-finder/designator and optical sight, tripod and pedestal. There is also an optional thermal imaging sight, which can be chosen from various imagers including Elop's advanced thermal imaging system, and an optional TV channel. The system is also designed to adapt to mounting on a variety of stabilised platforms for land, naval and airborne application.

Operational status

In production and in service with the Israel Defence Forces and worldwide.

Specifications

Designation range: >10 km
Built-in: Electronic compass
Output Energy: 130 mJ
Pulse repetition frequency; up to 20 pps (continuous)
Beam divergence (90% of energy): 130 μrad
Range: 250-20,000 m
Accuracy: ±5 m
Logic: first/last
Day direct view optics: ×13
Field of view: 5.5°
Designation head weight: 7.9 kg
Power: 24-28 V DC

Contractor

Elop Electro-Optics Industries Ltd.

NEW ENTRY

ENOSA SIRO

Type

Forward observation system.

Description

The Sistema Infra-Rojo de Observacion (SIRO) has been developed as a modular surveillance system for long-range operation which includes a tracking, range-finding and aiming capability.

SIRO infra-red surveillance system 0022042

SIRO 412 sensors consist of a thermal imager, a CCD daylight camera and a laser range-finder. The thermal imager is the ENOSA SVT-041 which uses Raytheon MCTNS modules, while the CCD is the ENOSA CCD 41L which has motorised zoom and focus with automatic iris.The sensors are installed on a tilt and pan head mounted on a tripod.

It is linked to an electronic control unit with ENOSA Nd:YAG laser range-finder and this is linked to a remotely operated display and control system.

Operational status

In production as required. In service with the Spanish Army and Security Forces.

Specifications

SVT-041 thermal camera
Spectral band: 8-12 μm

Fields of view
Narrow: 2.77 × 1.73° (×13)
Wide: 6.92 × 4.32° (×5.2)

Detection range
Vehicles: 10 km
Personnel: 2 km
Focus range: 25 m to infinity
Detector: 60 element
Weight: 15 kg

CCD TV camera
Zoom: ×10
Field of view: 2-22°
Resolution: 380 (horizontal) × 400 (vertical) TV lines
Min illumination conditions: 0.45 lx

Sensor head
Azimuth: ±180°
Elevation: ±25°
Accuracy: >0.1°
Laser range-finder
Type: Nd:YAG
Measuring range: 150 m-20 km
Wavelength: 1.064 μm
Accuracy: ±5 m

Contractor

Empresa Nacional de Optica SA (ENOSA).

VERIFIED

FLIR Systems MilCAM TargetIR™

Type

Binocular thermal imager.

Description

The MilCAM TargetIR™ is designed for military applications such as targeting, special operations, geographical positioning and forward observations. This

The MilCAM Target IR™ binocular unit ***NEW***/0547633

binocular unit incorporates features such as a 320 × 240 InSb FPA with a 30 to 5,000 m eye-safe laser range-finder, a digital orientation module (azimuth and elevation), and an integrated Global Positioning Systems (GPS).

Operational status
In production.

Specifications
Display: Dual-eyepiece, IPD 55-72 mm
Micro LCD screen 800 × 600
Laser range-finder: 1.55 μm wavelength
Digital magnetic compass: 1°azimuth accuracy, 0.5° elevation accuracy
GPS: self and target position solving, C/A, 8 channels, active internal antenna, supports external antenna
Laser Pointer: 15 mW or 60 mW, 0.3 mRad, 830 nm, Class 3b
Dimensions (nominal): 325 × 210 × 115 mm
Weight: <2.7 kg

Thermal imager
Detector: 320 × 240 InSb FPA
WFOV: 11 × 9°
NFOV: 3.7 × 2.9°
FPA cooling: Closed-cycle Stirling microcooler
Accessories: Variable tripods, rechargeable batteries, battery charger, AC to DC converter, communication cable, 2× extender

Contractor
FLIR Systems.

NEW ENTRY

FLIR Systems MilCAM XP™

Type
Thermal imager (hand-held).

Description
The MilCAM XP is a hand-held thermal imager featuring a third-generation InSb FPA detector, designed for diverse military applications. The unit is ultra-lightweight (less than 3 kg) and is extremely compact. The unit is available with a variety of SFOV and DFOV lens options. The DFOV50/250 mm optic makes it possible to detect vehicles at ranges exceeding 9 km.

Operational status
In service.

Specifications
Dimensions (nominal)
360 (l) × 160 (w) × 120 (h) mm
14.2 (l) × 6.1 (w) × 4.7 (h) in
Weight (incl battery): <2.7 kg (5.5 lb)

Thermal Imager
Detector: 256 × 256, InSb FPA
Spectral band: 3-5 μm
Optics: f /4.0
WFOV: 50 mm, 9° (h) × 7° (v) (100 mm with ×2 lens extender)
NFOV: 250 mm, 1.8° (h) × 1.6° (v) (500 mm with ×2 lens extender)
FPA cooling: Closed-cycle Stirling
NETD: 0.025 at 23°C ambient
Video format: RS-170 or CCIR compatible (BNC connector std)

The FLIR Systems MilCAM XP ***NEW***/0137879

Power requirements: 6 V DC/5.5 W
Battery life: more than 2.5 h
Environmental qualifications: fully ruggedised/MIL-STD-810E qualified
Operating temperature range: –20 to + 55°C
Camera controls: on/off/standby, auto and manual gain and level, polarity, low-battery warning, field calibration, reticle select
Mounting: tripod mount, ¼ - 20

Accessories
Storage/Transport case: hard shell, lockable with fitted foam insert
Adapter/Charger: 110 V AC/220 V AC power supply/battery charger
Batteries: NiCd rechargeable, 3 supplied

Contractor
FLIR Systems.

NEW ENTRY

FLIR Systems MIRV™ Miniature Infra-red Viewer

Type
Thermal imager (hand-held).

Description
The MIRV palm-sized miniature infra-red viewer weighs less than 1.6 kg, runs for 5 hours on a single 'D' cell battery and is sealed and ruggedised. It offers four times the thermal sensitivity of uncooled infra-red viewers, while its long focal length optic provides detailed images at long range in all climactic conditions. Detection range for a man-sized target is more than 1,500 m, with recognition range exceeding 500 m. Its target thermal imagery is displayed on a high-resolution miniature liquid crystal display integrated into the MIRV and viewed through the unit's eyepiece. Developed per US Department of Defense specifications, it can be single-handedly operated with minimal fatigue.

The standard MIRV product set includes a wide field of view 75 mm focal length lens for situational awareness and a narrow field of view 150 mm focal length lens for long-range target recognition.

Operational status
In production.

Specifications
Dimensions (l × h × w nominal): 270 × 102 × 102 mm
Weight (incl battery): <1.6 kg including battery and lens
Detector: 256 × 256 IR pixels, InSb FPA
Spectral band: 3-5 μm
FPA cooling: Closed-cycle Stirling
NETD: 0.020°C @ 23°C ambient
Optics: f/4.0, minimum focus distance 1 m

Field of view
WFOV: 75 mm, 5.8 (h) × 5.5° (v)
NFOV: 150 mm, 2.9 (h) × 2.75° (v)
Video format: RS-170 or BNC connector
Power requirements: 4.5 W
Battery life: 'D' cell lithium battery >5 hours
Environmental qualifications: Fully sealed and ruggedised
Operating temperature range: –32 to +49°C (–26 to +120°F)
Storage temperature: –33 to +71°C (–27 to +160°F)
Audible security: Inaudible at greater than 10 m
Purge port: ANVIS type purge port
Camera controls: Gain, auto/gain, level, polarity, power on/off, low-battery warning, non-uniformity correction, reticle adjust, zero-zero, reticle select, focus

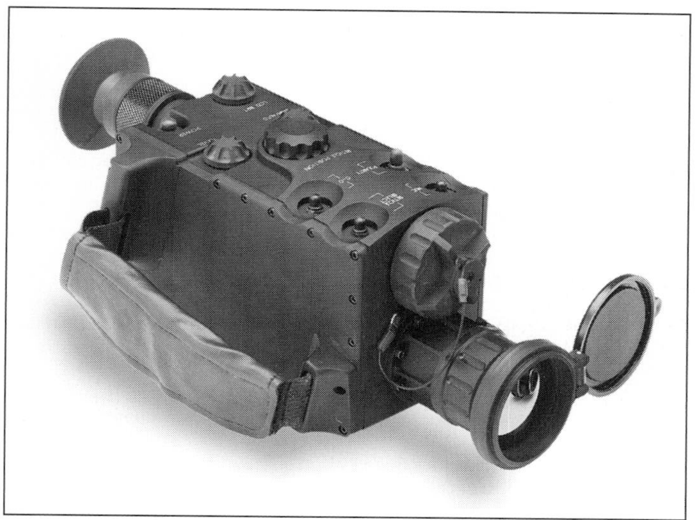

FLIR Systems MIRV ***NEW***/0137878

Mounting: Tripod mount ¼-20 (Picatinny or NATO mount options)
Eye-piece: Miniature LCD dioptre adjustment, protective eyecup for visual security
Colour: Flat black, non-reflecting finish
Storage/Transport case: Hard shell, lockable with fitted foam insert
Lens cap: Attached to camera
DC power adapter: 6-28 V DC input
Lens cleaning kit: Complete cleaning kit in nylon pouch
Batteries: 3 each, 3.6 V, lithium sulfur dioxide
Operator's manual: Covers operation and operator-level maintenance

Contractor

FLIR Systems.

NEW ENTRY

FLIR Systems See Spot III™

Type

Thermal imager (hand-held).

Description

See Spot III is designed for standalone operation or use with a 1.06 μm laser target designator. Developed per US Department of Defense specifications, the See Spot III is fully sealed and ruggedised to withstand the rigors of field applications. Powered by a single 'D' size lithium battery, the See Spot III will operate continuously for greater than 4 hours. With the long-range 250 mm focal length lens, detection range for a man-sized target is in excess of 5,500 m; recognition range for a man-sized target is in excess of 2,500 m. Thermal imagery of the target is displayed on a high-resolution miniature liquid crystal display integrated in the See Spot III and viewed through the unit's eyepiece. With four times the thermal sensitivity of uncooled IR viewers, the See Spot III provides detailed thermal imagery at long range in all climatic conditions.

Operational status

In service.

Specifications

Detector: 256 × 256, InSb FPA
Spectral band: 4.5-4.8 μm and 1.06 μm
FPA cooling: Closed-cycle Stirling
Optics: f/4.0, minimum focus distance 10 m
Field of view: 250 mm efl: 1.8 (h) × 1.2° (v)
Instantaneous field of view: 0.12 mrad
NETD: 0.020°C @ 23°C ambient
On-camera controls: Gain, auto/gain, level, polarity, power on/off, low-battery warning, non-uniformity correction, reticle adjust, zero-zero, reticle select, focus
Video output: RS-170 compatible, BNC connector
Power requirement: 4.5 W

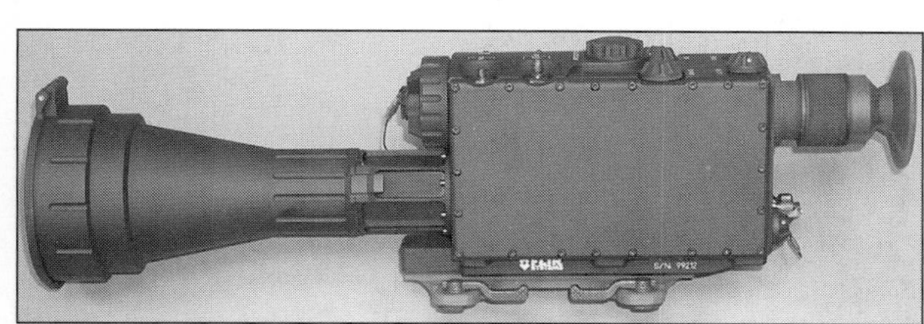

FLIR Systems See Spot III ***NEW***/0137875

Eye-piece: Miniature LCD, dioptre adjustment, protective eyecup for visual security
Battery life: >4 hrs
Battery type: 'D' cell lithium, 3.6 V
Weight: 2.4 kg complete including battery and lens
Operating temperature: –32 to +49°C
Storage temperature: –33 to +71°C
Environmental: Fully sealed and ruggedised
Audible security: Inaudible at greater than 10 m

Accessories

Carry case: Hard shell, lockable with fitted foam insert
Batteries: 3 each, 3.6 V, lithium sulphur dioxide

Contractor

FLIR Systems.

NEW ENTRY

Fotona ARTES-1000

Type

Forward observation system.

Description

ARTES-1000 is a system for battlefield observation and planning, executing and controlling artillery fire. Combined with the telecommunications equipment on the forward observer's post, it is used as an observation and data acquisition instrument, while at the gun position it is used as a device for calculating and displaying the elements of artillery fire.

In the stand-alone mode only the forward observer is equipped with ARTES-1000. He performs all fire-control functions, from battlefield surveillance to ballistic calculations. The results are transmitted to the gun position(s) via radio (voice) communications.

In the battery deployment mode, forward observers equipped with ARTES-1000 survey the battlefield; they issue fire requests and fire orders to the Battery Control Post, where the elements of fire are calculated. The Battery Control Post, defines targets, assigns priorities and selects units for the execution of fire. Here, additional elements of the system, such as PCs, graphic displays, printers and meteo stations can be introduced. At the gun position ARTES-1000 is used as a gun display unit (without laser range-finder). The elements of the system are connected via standard links and/or telephone lines.

As well as battlefield observation and target identification, ARTES-1000 may also be used for fire planning, execution and control and for topographic calculations. ARTES-1000 contains a set of procedures for topographic problem solving. These procedures enable the determination of co-ordinates of unknown points, the determination of own position and orientation (north setting), in case these data cannot be determined by built-in sensors, that is by direct measurements.

ARTES-1000

ARTES-1000 artillery engagement system in the field 0004802

Co-ordinates are displayed in either spherical polar co-ordinate system, cylindrical polar co-ordinate system, or grid co-ordinate system. Co-ordinates can also be calculated relatively to the observer's position (locally), or relatively to any point in the database. Direction can be given similarily, relative to zero direction.

A procedure for tracking moving targets is built in, enabling the prediction of the target's position based on previous measurements.

The system includes the RLD-E eye-safe laser range-finder but can have an optional RLD-3 laser range-finder with a longer range and a thermal imager.

ARTES-1000 contains two RS-232 communication channels for connection to other modules of the system (PC, thermal imager, direct communications) and one RS-232 channel for tests and calibrations.

Operational status
Available.

Specifications
Goniometer module
Azimuth: 0-360°
Elevation: ±27°
  Accuracy: ±1.3°
Elevation leveling: ±10°
  Accuracy: 0.29 mrad

Elevation compass
Accuracy: ±0.5°
Max tilt: ±10°

Laser range-finder
Range: 100-9,998 m (RLD-E); 200-20,000 m (RLD-3)
GPS module
6 channel satellite receiver, built in the handle of ARTES-1000
Acquisition time: 90 s max
Accuracy: 25 m relative spherical error (120 m, if selective availability is enabled)

Communication
Radio or field telephone with proper adaptor
Voice link: encrypted
Modem: 1,200 bps half duplex

Power supply
Goniometer
Built-in batteries: 9 × 1.5 V AA size
External: 12-18 V DC
Consumption: 400 mA, when display illumination, GPS and compass are switched off (otherwise 600 mA)

Range-finder
Battery life: 300 rangings at 25°C

Contractor
Fotona d d.

UPDATED

Galileo Avionica Attila sighting system

Type
Forward observation system.

Description
The Galileo Attila is a family of functional modules, each of which is dedicated to a specific task, that can be integrated according to a customer's operational requirements to produce a modular periscopic system.

For the battlefield surveillance role, Attila can perform automatic co-ordinate acquisition and transmission to a battlefield management system. The system can also be linked to other observation and fire-control systems by serial line or bus interface.

The highest specification Attila configuration is a system that can perform day/night observation, detection, identification and firing at stationary and moving surface and air targets.

The Attila modules include:
(a) multispectral primary stabilised panoramic head to allow operations from a moving vehicle
(b) second-generation thermal infra-red vision unit
(c) image intensifier vision unit
(d) laser range-finder unit
(e) vision unit (eyepiece unit)
(f) remote visual display unit
(g) electronic unit
(h) control unit (also remote)
(i) TV unit
(j) automatic TV tracking unit.

The main characteristics of the system are:
(a) integrated day/night capability with single observation element
(b) wide line of sight elevation range up to +60° (also suitable for anti-helicopter use); high elevation available for specialised AA use
(c) multiple day channel magnification (×4 and ×12)
(d) IR wide and narrow field of view
(e) TV compatible IR image (CCIR standard)
(f) eyesafe laser range-finder (optional Nd:YAG). 1.54 μm.

Operational status
In production. In service with undisclosed country.

Contractor
Galileo Avionica.

UPDATED

The Galileo Attila sighting system 0053699

Galileo Avionica Thermal Unit and Range-finder for Battery Observers (TURBO)

Type
Forward observation system.

Description
Thermal Unit and Range-finder for Battery Observers (TURBO) has been designed to provide the co-ordinates of targets, both day and night. It features data transmission to other units by an RS-422 standard cable and a thermal image remote display. The TURBO consists of a modified VTG 120 thermal imager, a laser range-finder, a goniometric support, tripod, power supply with multiple adaptor, accumulator battery providing 2 hours of power supply, cables, fittings and container. Cooling for the thermal imager is based upon rechargeable bottles of compressed air.

Specifications
Thermal imager
Operating wave band: 8-12 μm
Fields of view: 3.3 × 6.6°; 1.1 × 2.2°
Magnification: ×4; ×12
Laser range-finder
Max range: 20 km type Nd:YAG
Resolution: ±10 m
Pulse Repetition Rate: 2 s
Echo Suppression: 200-12,000 m adjustable
Goniometric support
Traverse: unlimited
Elevation: ±22.5°
Power supply
Dedicated rechargeable battery: 24 V AC
Vehicle: 115 V, 400 Hz AC; 24 V DC
Required power: 80 W

Contractor
Galileo Avionica.

UPDATED

The TURBO observation device viewed from the front

Indigo Systems ThermoCorder™

Type
Handheld thermal imaging and recording system.

Description
The ThermoCorder™ system transforms Indigo Systems' Omega™ (qv) IR camera into a complete, handheld thermal imaging and recording system.

It integrates the Omega™ with a Sony camcorder (DCR-TRV or equivalent) via an intelligent accessory 'hot' shoe interface. This interface allows the camera to be powered via the camcorder battery, and enables the camera's video signal to be input into the camcorder for viewing on the LCD display or viewfinder. The IR imagery can be recorded directly onto the camcorder tape.

All standard features and furnished accessories of the Sony Camcorder are available with ThermoCorder™. The accessory shoe adapter is designed to be compatible with Sony camcorders featuring the Sony Intelligent Shoe.

The Indigo Systems ThermoCorder™ hand-held thermal imaging and recording system **NEW**/0595423

Operational status
Available.

Contractor
Indigo Systems Corporation (USA).

NEW ENTRY

International Technologies (Lasers) N/CROS Mk III night compass range-finder system

Type
Forward observation system.

Description
The N/CROS Mk III is a compact, lightweight hand-held night binocular combining night vision with laser range-finding azimuth and inclination/elevation measurement and optional laser pointing. It incorporates a GEN II+, SUPERGEN, or third-generation image intensifier, an eye-safe laser range-finder and a digital compass.

The N/CROS Mk III has a measurement range of 20 to 2,000 m. Range, azimuth md elevation data are displayed in the eyepieces, allowing the user to identify targets and relay fire support data. The system can store data for up to 10 targets (range and azimuth) and an RS-232 communications port provides an interface to external systems such as GPS (PLGR+96) or other combat computers for immediate co-ordinate calculation. An optional LPL-30 laser pointer is integrated into the system to allow the user to indicate targets to others at distances of up to 4,000 m.

Variants
N/CROS Mk III with the ORYOM day sight - ORYOM is an add-on device which has a × 4 telescope. All the displays of the N/CROS used during the night appear upon selection in the ORYOM field of view and the image intensifier tube automatically shuts off.

N/CROS Mk III with GPS: the N/CROS sends real-time digital data via its RS-232 port to a GPS. The GPS immediately calculates the co-ordinates of the target using the GPS position data and the measured data (range, azimuth and elevation).

A new optional N/CROS Mk IV is already in production.

Operational status
In production and in service with various Defence Forces.

Specifications
Dimensions (l × w × h): 230 × 235 × 90 mm
Weight: 1.7 kg
Field of view: 13°
Magnification: ×3
Focus range: 20 m to infinity
Dioptre adjustment: –6 to +2
Objective lens: 75 mm, f/1.3
Resolution (system): 3.3 lp/mrad, GEN III
Gain: 1,800 lumen/lumen for GEN III
Laser range-finder range: 20-2,000 m
Accuracy: 2 m up to 200 m; 3 m at 200 to 500 m; 5 m at over 500 m

The N/CROS Mk III fitted with the ORYOM device 0077953

Measurement time: 1 s
Laser type: GaAs laser diode
Laser output: 3 W (max)
Wavelength: 905 nm
Beam divergence: 0.3 × 1.5 mrad

Compass accuracy:
±1.5° RMS over 360° @ ± 0 to 20° tilt
± 2.5° RMS over 360° @ ± 20° to 30° tilt

Inclinometer:
accuracy 0.7° over tilt ±0 to 20°
1° over tilt ±20 to 30°
Power source: 6 × 1.5 V AA size alkaline batteries
Battery life: 36 h for typical night mission use
Range, azimuth, elevation display: 4 digit, 7 segment LED type

Contractor
International Technologies (Lasers) Limited.

UPDATED

Northrop Grumman Mk VII hand-held eye-safe laser range-finder

Type
Forward observation system.

Development
The Mark VII is an eyesafe laser range-finder with integrated electronic compass, inclinometer and switchable day/night sight operation. The handheld Mark VII eyesafe laser target locator is designed for use by artillery and aircraft fire support personnel and long-range reconnaissance patrols.

Under a US$2 million, eight month contract from Lockheed Martin, Northrop Grumman – formerly Litton Laser Systems together with Northrop Grumman Integrated Systems and Northrop Grumman Electro-Optical Systems have delivered twelve sets of the Mk VII for the US Air Force Combat Command 'Sure Strike' programme. In the field, a one or two person forward air controller team within visual sight of a target can activate the equipment to determine precisely the target's map co-ordinates. Data are automatically relayed by radio through Lockheed Martin's aircraft system to the pilot's head-up display for acquisition and identification to enable a single pass attack with conventional or guided weapons. The equipment has also been successfully deployed with peacekeeping forces in Bosnia. One hundred and fifty-three Mark VII Laser Target Locators have also been delivered to the US Joint Special Operations Command.

The Northrop Grumman Mk VII laser range-finder 0022045

Description
The unit includes an eye-safe laser range-finder using technology developed for Northrop Grumman Laser System's Mark IV and Mark V laser range-finders. Night vision capability is provided using a Northrop Grumman Electro-Optical Systems third-generation image intensifier. In addition to providing range, azimuth and vertical elevation angle to the target in a display built into the eyepiece, the unit also provides an RS-232 digital output. This allows direct interface with a Global Positioning Satellite (GPS) receiver or digital messaging unit to store and/or communicate absolute position information in latitude, longitude, or map grid co-ordinates.

Operational status
Development completed. Some 192 sets have been delivered.

Specifications
Dimensions: 78 × 187 × 161 mm
Weight: 1.9 kg with battery
Operating temperature: –20 to +50°C
Eye-safe laser transmitter
Type: Nd:YAG with KTP OPO converter
Wavelength: 1.57 μm
Pulse rate: 6 ppm
Max range: 19,995 m
Min range: 20 m
Range increment: 3 m
Accuracy: 3 m
Range computation: Selectable first/last pulse logic with multiple target indicator
Daysight
Magnification: 7.3 × 18
Field of view: 4.5° (80.1 mils)
Reticle: projected, open centre crosshair
Nightsight
Type: Litton image intensifier
Magnification: ×4
Field of view: 8° (140 mils)
Reticle: projected, open centre crosshair
Electronic compass
Type: Magneto-resistive
Accuracy: 9 mil, 1 σ
Electronic inclinometer
Type: Pendulum
Accuracy: 1.3 mil, 1 σ
Data display: projected in eyepiece
Data interface: RS-232 (PLGR, FED NMEA protocol selectable)
Battery type: disposable lithium
External power: RS-232
Night operation per battery: >1,000

Contractor
Northrop Grumman Electronic Systems, Laser Systems Division.

VERIFIED

Northrop Grumman, Advanced Laser Targeting System (ALATS)

Type
Laser designator/target locator.

Description
The ALATS is a lightweight, flexible targeting system which uses Northrop Grumman, Electronic Systems' Ground Laser Target Designator II (GLTD II) as the core of a versatile system. The GLTD II can designate targets effectively at ranges in excess of 5 km; it provides non-eye-safe ranging to approximately 20 km. When a FLIR camera or image-intensified capability is added, the system will operate at night. An eye-safe range-finder (ESL-200) can be added, which, when combined with PLGR provides target location. The combined ESL-200/PLGR can be removed for use a target locator independent of the ALATS. It can also be combined with an angulation/tracking head for target azimuth and elevation information.

Operational status
Components in production.

Specifications
Designator: GLTD II
Laser type: Nd:YAG
Wavelength: 1.064 μm

Pulse energy: >80 mJ
Pulse-to-pulse stability: <15%
Beam divergence: <0.3mrad @ 90% energy
Boresight retention: <0.25 mrads
Sighting optics: ×10

Field of view:
Horizontal: 5.0° nominal
Vertical: 4.4° nominal
Reticle: 0.2mrad Open Cross
Dioptre adjustment: +2 to −6 dioptres
Exit pupil: 5 mm diameter (nominal)
Eye relief: 15 mm (nominal)
Marking: 5 km (typical) See Litton Laser System's GLTD Live Fire Report for long-distance marking
Pulse repetition frequency: Band I/Band II NATO Stanag
Duty cycle: 5-1-5-1-5 (on/off) at 10 pps. Continuous operation under most conditions
Ranging: 200 to 9,995 m (± 5 m nominal)
Range counter logic: selectable first/last
Range discrimination: 35 m
Battery power: 24 V DC Lithium or re-chargeable NiCAD
Weight: <5.7 kg
Size (l × w × h): 29 × 34 × 14 cm

ESL-200 (Eye-safe laser range-finder)
Type: Er glass
Pulse energy: 16 mJ
Dioptre adjustment: +4 to −4
Sighting power: ×7
Field of view: 7.0°
Range: 50 to 9,995 m
Range accuracy: ±5 m
Fire rate: 1 per 6 s
FLIR: customer requested
Tripod and tracking head: customer requested

Contractor
Northrop Grumman, Electronic System, Laser Systems.

UPDATED

The Northrop Grumman Advanced Laser Targeting System (ALATS) 0022046

Northrop Grumman, Ground Laser Target Designator (GLTD II)

Type
Forward observation system.

Description
The GLTD II is a compact, lightweight, portable laser target designator and rangefinder. It is capable of exporting range data via an RS-422 link and importing azimuth and elevation. Developed to enable combat soldiers to direct laser-guided smart weapons, such as Paveway bombs, Hellfire missiles and Copperhead munitions, the GLTD II can be implemented as part of a sophisticated, digitised fire-control system with thermal or image-intensified sights. Versions of the GLTD II are currently in use with various NATO countries and other armed forces around the world.

Operational status
In service.

Specifications
Weight: <5.7 kg
Size(l × w × h): 29 × 34 × 14 cm
Laser
Laser Type: Nd:YAG
Wavelength: 1.064 μm
Pulse Energy: >80 mJ
Pulse-to-Pulse Stability: <15%
Beam Divergence: <0.3 mrad @ 90% energy
Boresight Retention: <0.25 mrad
Modes: Range and Mark (Designate)
Operating Temperature: −322 to +45° C

Mark (Designate)
Marking: 5 km typical
Pulse repetition frequency: NATO Stanag Band I/Band II or programmable
Optional User - Programmed Codes
Minimum Duty Cycle: 5-1-5-1-5 min (On/Off) at 10 pps. Continuous under most conditions
PRF Coding: Push button selectable
Ranging
Range: 200 to 19,995 m, ±5 m
Range Counter Logic: Selectable First/Last
Range Discrimination: 35 m
Display: 5-Digit Red LED in Eyepiece

Sighting Optics
Sighting Optics: ×10 (nominal)
Field of View (h × v): >5 × 4.4°
Reticle: 0.2 mrad open cross
Dioptre Adjustment: +2 to −6
Exit Pupil: 5 mm diameter (nominal)
Eye Relief: 15 mm (nominal)
Tilted Eyepiece: 45°
Operation: manual or remote control
Power supply
Battery Power: 24 V DC power sources Lithium and re-chargeable NiCad
Vehicle Power: 28 V DC (MIL-STD-1 275)

Contractor
Northrop Grumman, Electronic Systems, Laser Systems.

VERIFIED

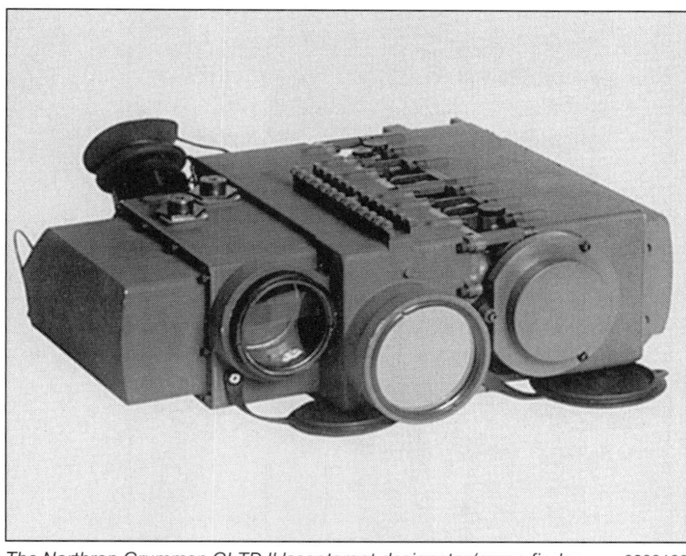

The Northrop Grumman GLTD II laser target designator/range-finder 0088166

PCO PZA-1 Artillery Measuring System

Type
Forward observation system.

Description
The PCO Artillery Measuring System (PZA-1) was designed for use with mortar batteries and other artillery. The system is used to measure the range and bearing to a target necessary for accurate fire. PZA-1 also enables determination of the firing position battery of mortars, howitzers or other artillery for which it is used. The system includes Built-In-Test (BIT)

The system comprises:
- goniometer, for target angle measurement
- magnetic compass
- azimuth attachment device
- laser range-finder
- night vision telescope
- power supply accumulators
- two measurement poles (with illumination)
- base plinth with illumination
- tripods (short and high) and BIT

An option is a programmable computer which will carry out co-ordinate transforms and distance calculations for targets, taking geodesy into account.

The similar AZR-1 electro-optic forward observation system for artillery was developed in collaboration with BGT of Germany. The system includes: eye-safe laser range-finder, north seeking gyroscope, electronic goniometer and low-light level TV camera.

Operational status
Available.

Specifications
System observation parameters
Day channel
Magnification: ×7
Field of view: 1-00
Dioptre adjustment: ±6 dioptres
Night channel
Magnification: ×2.2
Field of view: 12° 20'
Dioptre adjustment: ±4 dioptres
Power supply: 12 V/10 to 28 V from batteries
Goniometer
Angular range
Horizontal: n × 64-00 mils
Vertical: ±4-50 mils
Accuracy – horizontal and vertical: ±0-00.3 mils
Laser range-finder

PCO PZA-1 artillery measuring system 0022047

Measurement range: 30 – 2,000 m
Accuracy: =1 m

Contractor
PCO SA, Poland.

VERIFIED

Raytheon lightweight thermal observation equipment

Type
Forward observation system.

Development
The lightweight thermal observation equipment is a portable third-generation infra-red system for artillery fire direction teams, forward observation and reconnaissance applications. Raytheon has been awarded a US$11.2 million contract from the Danish Army Technical Service to produce 178 of the systems, including logistics back-up, for use by the Danish Army.

Description
The system consists of the portable thermal imager, tripod and angulation head and will be lighter and smaller than existing first-generation systems. The infra-red sensor features a 3 to 5 μm mid-wave indium antimonide (InSb) staring focal plane array, with nearly 100,000 detector elements. It has multiple fields of view, standard video image format, remote control and both analogue and digital video outputs. The camera uses a standard Raytheon linear drive cryogenic cooler in an all-metal integrated focal plane array/dewar/cooler assembly.

Raytheon is also developing medium-wave staring infra-red sensors for thermal imagers for the US Navy attack submarine, the Global Hawk UAV (qv) and AN/AAQ-16 (qv) for the MV-22 Osprey rotary-wing aircraft.

Operational status
Under contract for the Danish Army.

Specifications
Thermal Imager
Spectral band: 3-5 μm
Detector: InSb

Contractor
Raytheon Systems Company.

UPDATED

A Danish soldier operates the Raytheon third-generation infra-red lightweight thermal observation equipment 0004817

Raytheon Long-Range Advanced Scout Surveillance System (LRAS^3)

Type
Forward observation.

Development
The Long-Range Advanced Scout Surveillance System (LRAS^3) Program is sponsored by the Project Manager (PM) Night Vision/Reconnaissance,

Scout 0099734

Surveillance, and Target Acquisition (NV/RSTA) of the Program Executive Office Intelligence, Electronic Warfare & Sensors (PEO IEW&S) of the United States Army. The LRAS^3 system is currently entering the production phase and will be fielded to US Army Scouts. LRAS^3 will provide Scout forces with a long-range reconnaissance and surveillance sensor system that operates outside the range of currently fielded threat direct fire and sensor systems.

Description

The LRAS^3 provides the US Army with real-time acquisition, target detection, recognition, identification and far target location information. This long-range target acquisition capability will improve the survivability of the Scout force and increase the lethality and force effectiveness of combat units. The LRAS^3 sensor can operate in both mounted and dismounted configurations providing a 24-hour, adverse weather target acquisition capability.

The system consists of a second-generation Horizontal Technology Integration (HTI) Forward Looking Infra-red (FLIR) thermal imager, a day video camera, an eye-safe laser range-finder, long-range common aperture reflective optics and a global positioning system interferometer subsystem. The LRAS^3 design also includes a digital port, which will allow it to interface with battlefield command and control.

Operational status

The LRAS^3 is entering the production phase.

Specifications

2nd generation HTI FLIR
Field of view: wide 8.0° × 4.5° narrow 2.6° × 1.5°

Day video camera
Field of view: wide 6.0° × 4.5°, narrow 2.0° × 1.50°

Eyesafe laser range-finder
accuracy: ±5 m

Far target location
Accuracy: 60 m CEP at 10 km

Laser rangefinder: Eyesafe

Accuracy: ±5 m

Contractor

Raytheon Company.

UPDATED

Raytheon Systems HSS Hunter Sensor Suite

Type

Forward observation system.

Development

The Hunter Sensor Suite (HSS) integrates a second-generation thermal imager and Aided Target Recognition (ATR) technology to perform long-range Reconnaissance, Surveillance and Target Acquisition (RSTA). HSS detects, recognises and communicates target positions beyond the range of enemy direct fire weapons, while operating as a hunter vehicle of a hunter/killer team.

In mid-1994, the US Army Communications-Electronics Command (CECOM) awarded Raytheon Systems Company a US$20 million contract for the Hunter Sensor Suite programme. The contract was to run for a period of 48 months and be managed by the US Army Night Vision and Electronic Sensors Directorate (NVESD).

Under the contract terms, Raytheon is to develop a single Advanced Technology Demonstration (ATD) system and integrate this on a 4 × 4 AM General High-Mobility Multipurpose Wheeled Vehicle (HMMWV) as part of the US Army Rapid Force Projection Initiative (RFPI).

Description

The HSS system includes a second-generation thermal imager, eye-safe laser range-finder, a day TV video camera, acoustic cueing system, aided target recognition, Global Positioning System (GPS) and north-seeking module. The system will also be able to receive target information from a number of dismounted sensors.

HSS's long-range is achieved through the use of a large aperture second-generation thermal imager. This uses the US Army's Horizontal Technology Integration (HTI) B-kit FLIR receiver (qv) with higher resolution optics to address US Army long-range ground system requirements. This high-resolution image is available over the tactical radio net, to the field commander via PhotoTelesis image compression and transmission.

System and operational effectiveness is enhanced by aided target recognition. The ATR automatically scans a wide search area and helps the operator detect, track and recognise targets. Automated processing of the onboard GPS, laser rangefinder and north-seeking module allows rapid, precision targeting information to be communicated over the tactical internet.

HSS also includes an acoustic cueing system for enhanced surveillance capability and an auxiliary power unit to maintain a low thermal and acoustic signature during long overwatch missions.

The HSS system operator will be inside the HMMWV with the displays, processors and communications equipment. The sensor package will be mounted on a telescopic mast.

HSS is a modular RSTA system allowing installation in a variety of tactical reconnaissance platforms.

Operational status

Under development for the US Army Night Vision and Electronic Sensors Directorate (NVESD).

Contractor

Raytheon Systems Company.

VERIFIED

SAGEM Day And Night Artillery Observer System (DANAOS)

Type

Forward observation system.

Development

DANAOS is a day/night electro-optical and inertial navigation system for use on field artillery forward observation vehicles. Its small, compact size allows it to be installed on light or heavy, tracked or wheeled armoured vehicles.

DANAOS is able to undertake the following functions:

- battlefield surveillance
- target data acquisition – the system automatically computes the position of the target selected by the operator
- target co-ordinate transmission – after validation by the operator, the position of the target can be automatically transmitted by the radio network to the artillery battalion headquarters.

The system comprises the following subsystems: a day/night NBC compatible one-person LOA 20 Artillery Observation Cupola equipped with a day binocular device; a direct panoramic observation device; an 8 to 12 μm waveband thermal imager, a laser range-finder and SAGEM land navigation system.

A high-level Built-In Test Equipment (BITE) is also integrated into the DANAOS assembly.

Operational status

In production. In service with the French Army, equipping about 80 VAB (4 × 4) vehicles for the forward artillery observation role.

SAGEM Day And Night Artillery Observer System (DANAOS) on vehicle

Specifications

Elevation range: –20 to +40°
Traverse: 360°
Magnification (day observation)
Panoramic direct-view: ×1
Binocular: 2.5 and ×10
Field of view (night observation)
Narrow: 23 mil
Wide: 100 mil
Laser rangefinder wavelength: eye-safe 1.54 μm
Station placing and target co-ordinate determination time: immediate
Target localisation range: 10,000 m
North-seeking accuracy: <1 mil
Navigation accuracy: <0.2% of distance travelled
Target co-ordinates accuracy: 20 m

Contractor

SAGEM SA Aerospace & Defence Division.

UPDATED

STN ATLAS Elektronik BAA observation and reconnaissance equipment

Type

Forward observation.

Description

The STN ATLAS Elektronik BAA is a stand-alone system for observation and reconnaissance which can perform target detection, recognition and acquisition for mobile or fixed ground targets and low-altitude air targets with interface to C³I units.

The system consists of three units:

- the sensor head which contains the OPHELIOS thermal imaging sight, a day sight telescope with CCD camera and an eye-safe laser range-finder. OPHELIOS is a second-generation IRCCD 96 × 4 thermal camera
- the rotating and tilting head which contains the interface and power electronics and the drives for azimuth and elevation. It is connected to the observation and reconnaissance equipment via a two wire bus
- the Control and Display Unit (CDU) which includes monitor, operating panel, built-in joystick and the electronic controller boards. On the joystick panel, functions are integrated including field of view selection, zoom adjustment, sight switchover (CCD camera/thermal imager) and aiming. The joystick itself is used for aiming and laser firing.

The BAA can be mounted on a variety of platforms including vehicles, masts and tripods. Medium-sized targets, for example helicopters, can be detected up to a range of approximately 20 km. The system for the German and Royal Netherlands armies will be mounted on a 1.5 m mast in the Fennek vehicle.

Operational status

BAA has been selected to equip the 450 Leichtes Spähfahrzeug lightweight armoured reconnaissance vehicles for the German and Royal Netherlands armies.

Specifications

Daysight camera
Fields of view: 2-20°
CCD sensor (h × v): 752 × 582 effective pixels
Thermal imager
Spectral band: 7.5-10.5 μm
Detector material: CMT PV
Detector array: 96 × 4 IRCCD

BAA on the Leichtes Spähfahrzeug reconnaissance vehicle, being jointly developed by DAF Special Products and Kraus-Maffei Wegmann for the German and Royal Netherlands armies

Fields of view
Wide: 9.4 × 12.5°
Narrow: 2.7 × 3.5°
Resolution (h × v): 756 × 576 active pixels (full CCIR format)
Laser range-finder
Type: Nd:YAG Raman shift
Wavelength: 1.54 μm (eyesafe)
Ranging accuracy: ±5 m
Aiming sector
Elevation: –30 to +30°
Azimuth: –220 to +220°
Dimensions
Sensor and platform: 505 × 420 × 320 mm
Control and display unit: 350 × 240 × 315 mm
Weight
Sensor and platform: 27 + 19 kg
Control and display unit: 22 kg

Contractor

STN ATLAS Elektronik GmbH.

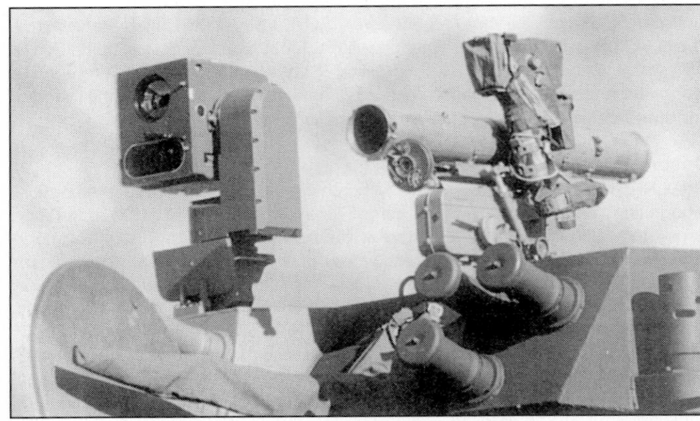

Close-up of STN ATLAS BAA observation and reconnaissance equipment fitted to Polish Army BRDM-2 scout vehicle with AT-4 'Fagot' ATGW launcher to right
0089200

Systems and Electronics STRIKER

Type

Forward observation.

Development

SEI developed the STRIKER fire-support vehicle under the US Army's Warfighting Rapid Acquisition Program (WRAP) to meet a pressing need for a mobile targeting platform for its heavy and light division forward observers. Under the WRAP program, STRIKER progressed from concept to prototype in eight months. The HMMWV-mounted system is currently in full-rate production. The STRIKER-unique mission equipment is also being mounted on the fire-support vehicle variant of the US Army's Interim Armored Vehicle.

Description

SEI's STRIKER is an M1025A2 HM7MWV-mounted fire support/reconnaissance system consisting of a laser designator/range-finder, thermal imager, digital command and control system, blended inertial/GPS navigation and targeting system and self-defense weapon. The system provides mounted, precision far target location and laser target designation for precision guided munitions. The current primary sensors are the AN/TVQ-2 Ground/Vehicular Laser Locator/Designator (G/VLLD) and the AN/TAS-4B night vision sight. The system can also be fitted with the Lightweight Laser Designator Range-finder or other sensors where appropriate. The sensor system can be removed for dismounted operation.

STRIKER's position/navigation subsystem combines inputs from the GIVLLD, inertial navigation unit, GPS receiver, and a vehicle motion sensor to provide self-location, navigation, and far target location data. This data is then automatically formatted within the fire-support tactical data system for immediate digital transmission as a call-for-fire or other message. Position and targeting data can also be incorporated into other battlefield command and control systems as necessary.

Operational status

In full-rate production for the US Army.

Specifications

Far Target Location (NATO std tgt): 50 m CEP @ 5 km/80 m CEP @ 10 km
Laser Designation (NATO std tgt): 5 km (stationary tgt) / 3 km (moving tgt)

Vehicle
Overall length: 485 cm
Overall width: 218 cm

SEI STRIKER 0122464

STRIKER with multisensor package on roof 0053695

Overall height: 259 cm
Power requirements: 19.1 A @ 24 V DC
Silent watch duration: 2 hrs @ 0°C
Weight: 3,608 kg (curb)
4,520 kg (3-person crew and all stowage)
4,676 kg (max GVW)
Max Fording Depth: 76.2 cm

Contractor

Systems & Electronics Inc (SEI), Saint Louis, Missouri, USA.

Thales Optronics LITE artillery observation system

Type

Artillery observation system.

Description

The typical artillery observation system is tripod mounted and is based upon a direct view LITE (qv) thermal imager as the primary sensor. In its artillery configuration the LITE imager uses a ×14/×4.6 dual field of view telescope. LITE can be co-mounted with an eye- safe laser range-finder, electronic angulation head and a global positioning system. The integrated software allows the system to be used in a variety of configurations. Applications for the system extend from long-range surveillance to fire control, monitoring of fall of shot and fire correction.

In addition to a long-range detection capability, the artillery observation system provides information on the range, bearing and elevation to target. All data, including GPS co-ordinates, can be displayed with a user selection of UTM (Universal Transverse Mercator), latitude/longitude or other specific map grids. The system has the ability to co-ordinate all data and output the information to a fire-control centre.

Operational status

In service with the Danish and New Zealand armies.

Specifications

Total system weight: 15 kg
LITE thermal imager
Spectral band: 8-13 μm

The Thales Optronics LITE Artillery Observation System as ordered by the New Zealand Army 0004819

Detector: CMT
Cooling: Integrated Stirling

Contractor

Thales Optronics.

UPDATED

Thales Optronics Mechanised Artillery Observation Vehicles system (MAOV)/NVL8700 series sights

Type

Forward observation.

Description

The NVL8700 series of sights is designed to be used with MAOV to survey targets, to observe fall of shot accuracy and to apply rapid fire corrections directly to an artillery battery computer. The sight is housed in an FVC107 cupola manufactured by Thales AFV Systems (formerly Helio Ltd), which incorporates an azimuth resolver. The sight is able to record range and target bearing/elevation data.

The NVL 8700 series of sights provides options for night vision with either thermal or image intensified elbows. The system incorporates a daylight elbow and a laser range-finder.

Operational status

In production. In service with Finland.

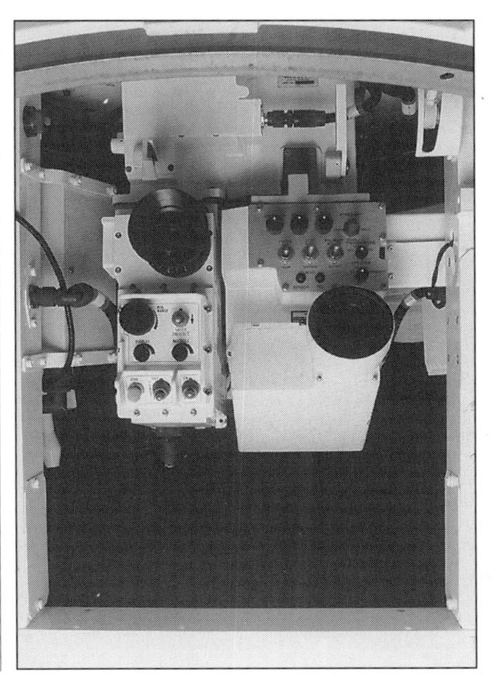

An internal view of the MAOV showing laser range-finder/laser daylight elbow and thermal imaging elbow

Specifications

Head/Body assembly
Magnification (incorporating laser protection): ×1
Unit vision field of view: 25 × 8°
Elevation: –10 to +50°
Elevation increment: 0.4 mils
Backlash error: <0.4 mils
Azimuth drift: <1.1 mils
Laser daylight elbow
Magnification: ×8
Field of view: 8°
Eyepiece focus: +3 to –4 dioptres
Eye relief: 25 mm
Boresight adjustment: ±9 mils
Eye protection: dichroic beam splitter and KG 3 filter
Reticle pattern: graduated cross with central circle
Power supply: 24 V DC ± 6 V
Laser range-finder
Laser type: dye Q switched Nd:YAG
Wavelength: 1.064 μm
Output power: 1 MW typical
Beam divergence: 0.5 mrad ±0.2 mrad
Pulse repetition rate: 12 ppm
Pulse length: 10-17 ns
Range logic, switch controlled: first/last target
Range storage: up to 10 targets
Near/Far target indication: left hand indicator on main display up to 9,995 m
Min/Max range: 250 m/9,990 m, (can range to 19,990 m with ±10 m resolution
Range accuracy: ±5 m
Target discrimination: 30 m
Nato stock number: NSN 1240-99-317-6293

Contractor

Thales Optronics (formerly Avimo Ltd).

VERIFIED

Vectronix SG12 digital goniometer

Type

Forward observation system.

Description

The Vectronix (formerly Leica) SG12 is an electronic angle-measuring instrument. A built-in Digital Magnetic Compass (DMC) provides autonomous orientation to 3 mils PE. Survey software options include resection, fire correction and target determination in grid co-ordinates.

For mortar fire control and artillery forward observers, the SG12 can be equipped with a variety of laser range-finders including Vectronix's (formerly Leica's) Vector. A second payload such as a lightweight thermal imager can also be added. Laser range-finders, GPS receivers and data entry devices can be linked via RS-232, -422 and -485 interfaces.

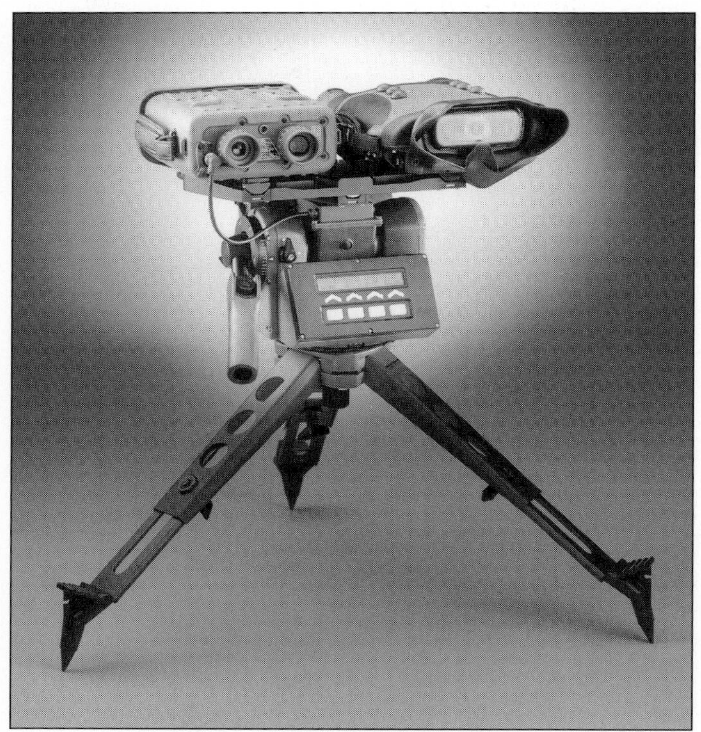

SG12 with dual payload for long-range, day/night observation 0089919

SG12 in use with VECTOR binocular OP system 0022044

Operational status

In production for Belgium, Finland and Switzerland.

Specifications

Dimensions (l × w × h): 240 × 236 × 165 mm
Weight: 3.9 kg (without batteries and payload)
Accuracy
Orientation by digital magnetic compass: 3 mrad
Horizontal direction: 0.7 mrad
Elevation: 2 mrad
Display: LCD dot matrix, 2 × 25 characters
Data interface: RS-232, RS-422, RS-485
Power supply: 10-32 V DC

Contractor

Vectronix AG.

UPDATED

Vision Systems, Thales RAdar Plus Thermal Observation and Recognition (RAPTOR)

Type

Forward observation system.

Description

RAPTOR is a ground surveillance system which combines a ground surveillance radar and a thermal imager into a single sensor. RAPTOR can be used in a wide variety of short- to medium-range ground surveillance applications, both in mobile and fixed configurations for military or industrial surveillance including battlefields, borders and high-value fixed-sites. RAPTOR is a lightweight modular system that can be used in a stand-alone configuration or mounted on a wheeled or tracked vehicle such as a truck, APC or tank.

Both sensors, radar and thermal imager, operate in parallel and are boresighted on a single rotating platform. Both sensors concurrently scan the spatial volume, and the detection is automatically carried out by each sensor. The individual results are combined using data fusion algorithms to generate a single stream of target information.

RAPTOR has two operating modes - scanning mode for target acquisition and staring mode for target identification and tracking. Scanning mode is used for long duration wide area surveillance. RAPTOR acquires and displays targets with no need for operator control. Staring mode provides target parameters including azimuth, range, elevation and velocity as well as visual recognition through the target picture. Details of the target's activity, surrounding and audio Doppler signature are available.

Targets are tracked automatically at the command of the operator enabling the system to monitor and update target status, while the operator is free for other work.

Markers are provided to record important positions (junctions, bridges and so on) in the scanned sector, and instantly return the sensor head to those positions at the push of a button.

RAPTOR can operate with two different range resolutions: normal or expanded. Normal mode is generally used for wide area surveillance, during which targets can be located with an accuracy of 160 m. Expanded mode provides an accuracy in the range of 20 m and is used to pinpoint a target once it has been detected.

System wide parameters can be set via a menu window separate from the normal operating screen. This is used for parameters that only change infrequently and do not need to be constantly displayed to the operator. Battery backed memory is used to store all system information even after the power is turned off, so set up time is minimised.

Any area within the scanned sector can be classified as desensitised, thus inhibiting target detection in this zone. This facility can be used to mask out regions where friendly activity is known to occur or at points where nuisance alarms are common.

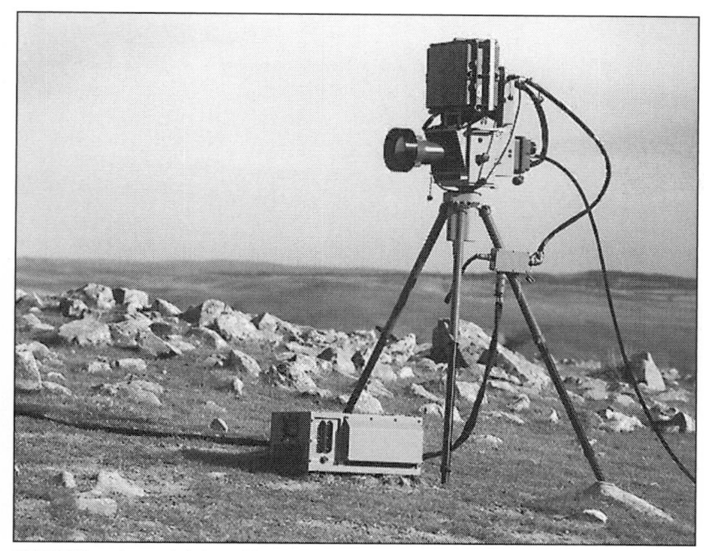

RAPTOR radar and thermal imaging system

Operational status
Available.

Specifications
Power supply: 24 V/168 W
Selection width: 5-180° user selectable
Rotation speed: 10°/s
Display resolution: 512 × 768 pixels
Display format: video standard (suitable for video recording and transmission)
Weight: 80 kg
Detection performance (typical)
Pedestrian
  Detection and recognition: up to 4 km
Light vehicle
  Detection and recognition: up to 6 km
Heavy vehicle
  Detection and recognition: up to 8 km
Radar
Antenna lobe: 4.5° azimuth, 6.5° elevation
Transmitter
Peak power: >200 mW
Frequency: 5 frequencies available
Band: Ku-, J-band
Thermal imager
Resolution: 0.115 mrad
Frame rate: 25 frames/s
Magnification: ×12
Spectral band: 8-12 μm
Cooling: Stirling cycle cooling engine, dry nitrogen gas supply

Contractors
Thales Airborne Systems.
Vision Systems Ltd.

UPDATED

VTÚVM SNĚŽKA reconnaissance and observation vehicle

Type
Forward observation system.

Description
The SNĚŽKA vehicle is designed for artillery reconnaissance and can perform the following functions:
(a) detection, recognition and tracking of moving and stationary targets
(b) observation of artillery shell impacts and explosions
(c) measurement of target co-ordinates and transmission of data to command post.

The vehicle (modified BMP) contains a position manipulator (mast), sensor manipulator (pedestal), radar, wind velocity measurement sensor, day and night TV cameras, thermal imager and laser range-finder. The vehicle was developed for the Czech Army.

Operational status
In service.

Specifications
Manipulator range: elevation +60 to -30°, azimuth ±190°
Day observation TV camera mode detection range: 5 km

The SNĚŽKA vehicle 0053697

The sensor package of SNĚŽKA 0053696

Day sight TV camera mode detection range: 10 km
Night camera detection range: 1.6 km
Thermal imager detection range: WFOV 7 km, NFOV 9 km
Laser range-finder
Maximum range: 20 km
Radio range: voice 20 km, data 14 km
Power supply: 24 V DC

Contractor
Military Institute for Weapon and Ammunition Technology - VTÚVM, Czech Republic.

VERIFIED

OBSERVATION AND SURVEILLANCE

LASER RANGE-FINDERS

ALST ELRF series long-range eye-safe laser range-finder

Type

Eye-safe laser range-finder.

Description

The ELRF series of laser range-finders from ALST Inc are Class 1 eye-safe lasers based on erbium:glass. All models are rugged, sealed, dry nitrogen purged systems with kinematic mounting points for boresight retention under harsh military environments. They are also available in a compact modular form for integration into customer supplied housings. They can be used for a variety of land-based and airborne applications.

ELRF-1 is provided with an InGaAs PIN detector capable of detecting signals to 80 nW and has a range of up to 10 km.

ELRF-2 has a high-sensitivity InGaAs APD (Avalanche PhotoDiode) detector capable of detecting signals down to 20 nW and has a range of up to 20 km.

ELRF-3 is a longer range version, ranging up to 30 km. This performance is achieved using 0.5 mrad beam divergence, a 70 mm receiver aperture and a high-sensitivity APD. A modular version designated ELRF-3M is also available.

ELRF-4 is the longest range version of the ELRF series, ranging up to 50 km, using a 106 mm receiver aperture.

All models are provided with Built-In-Test (BIT) and users choice of RS-232 or RS-422 serial data interfaces. Software for operation via PCs is optional.

ALST offers a high repetition rate option, HR-1 for its erbium:glass laser transmitters. This permits extended periods of pulsed operation at 1 Hz and limited periods of operation at 2 Hz. This option is stated not to add to the complexity, size or weight of the laser range-finder.

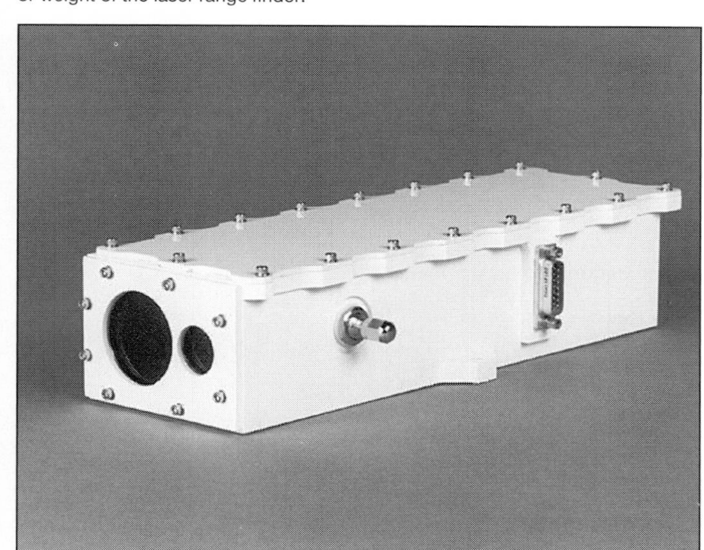

The ALST ELRF-1/2 laser range-finder 0022049

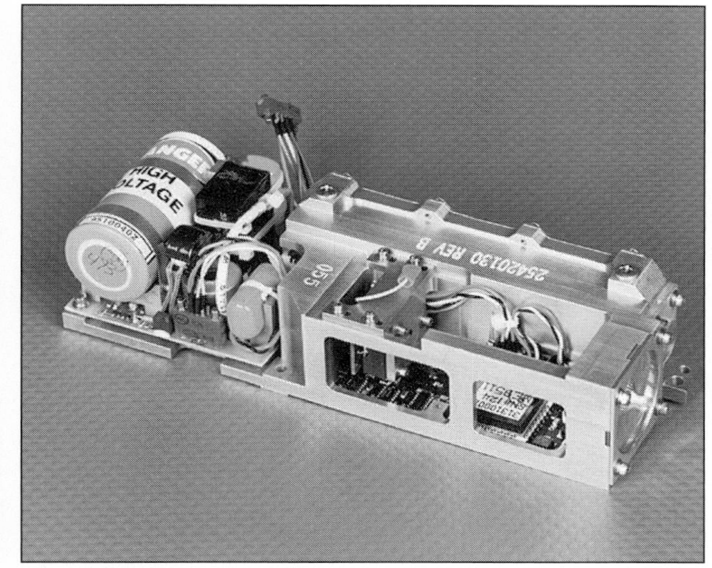

Construction of the ALST ELRF-1/2M laser range-finder 0022050

Operational status

In production.

Specifications

ELRF-1 and ELRF-2

Dimensions: 279 × 117 × 71 mm
Weight: 1.8 kg in housing, 870 g modular
Laser type: Er:glass
Wavelength: 1.54 μm
Energy/pulse: 9 mJ (typical)
Pulse width: 22 ±4 ns
Pulse rate: 12 ppm, 1 Hz with optional HR-1
Beam divergence: 1.0 mrad, 0.5 mrad optional
Receiver aperture: 40 mm
Receiver field of view: 3.3 mrad (ELRF-1), 2.2 mrad (ERLF-2)
Receiver detector: InGaAs PIN (ELRF-1), InGaAs APD (ELRF-2)
Receiver sensitivity: 80 nW (ELRF-1), 20 nW (ELRF-2)
Max range: 10 km (ELRF-1), 20 km (ELRF-2)
Min range: 50 m
Range resolution: 5 m (2 m optional)
Range logic: first/last
Min range gate: zero to max range, 100, 100 m increments
Max range gate: 100 to max range, 100 m increments
Prime power: 12/24/28 V DC

ELRF-3/3M

Dimensions
ELRF-3: 262 × 142 × 104 mm
ELRF-3M: 197 × 108 × 76 mm
Weight
ELRF-3: 3.1 kg
ELRF-3M: <1 kg
Laser type: Er:glass
Wavelength: 1.54 μm
Transmitter energy: 9 mJ (typical)
Pulse rate: 12 ppm, 1 Hz with optional HR-1
Pulse width: 22 ±4 ns
Beam divergence: 0.5 mrad
Output beam diameter: <19 mm
Receiver objective diameter: 70 mm
Detector sensitivity: ≤20 nW
Boresight accuracy (between laser transmitter output axis and module axis): <0.25 mrad
Max range: 29,900 ±5 m
Min range: 50 m
Multitarget resolution: 20 m
Range logic: first target/last target selectable
Min range gate: 0-19,900 m (0-29,800 m optional) in 100 m increments
Max range gate: 100-29,900 m in 100 m increments
HR-1 high repetition rate option
Laser type: pulsed Er:glass
Wavelength: 1.54 μm
Energy/pulse: 8 mJ typical
Pulse width: 22 ±4 ns
Pulse rate: 1.0 Hz for 3 min continuously, followed by 1 min cool down; 2.0 Hz at 50% duty cycle, max continuous. ON time 30 s
Beam divergence: 1.0 mrad standard, 0.5 mrad optional
Prime power: 12/24/28 V DC

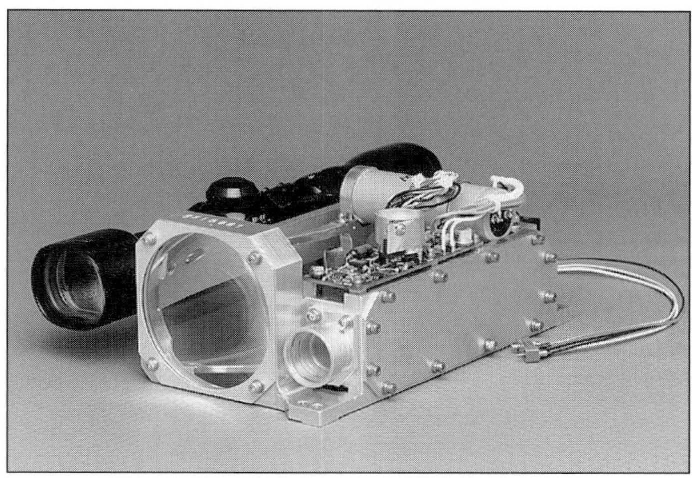

The ALST ELRF-3M laser range-finder 0022051

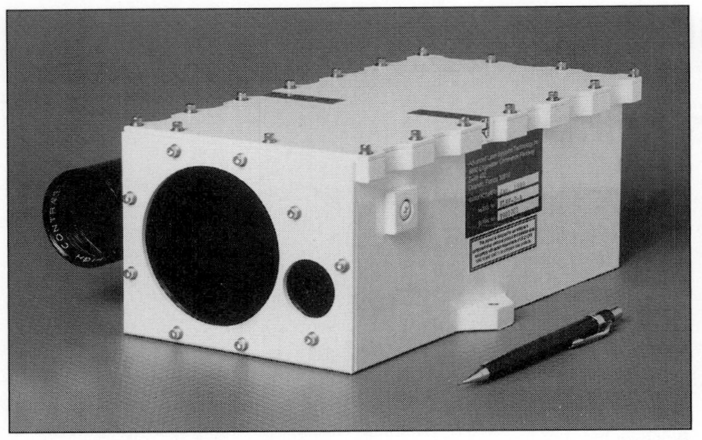

The ELRF-3 laser range-finder in case 0053783

ELRF-4
Dimensions: refer to ICD
Weight: 4.54 kg
Laser type: Er:glass
Wavelength: 1.54 μm
Energy/pulse: 9.0 mJ (typical)
Pulse width: 22 ±4 ns
Pulse rate: 12 ppm, 1 Hz with optional HR-1
Beam divergence: 0.5 mrad
Receiver aperture: 106 mm
Receiver field of view: 1.0 mrad
Receiver detector: InGaAs avalanche photodiode
Receiver sensitivity: 20 nW
Max range: 50 km
Min range: 50 m
Range resolution: 5 m
Range logic: first/last
Min range gate: 0-19,900 m in 100 m increments
Max range gate: 100-49,900 m in 100 m increments
Prime power: 12/24/28 V DC

Contractor
ALST (Advanced Laser Systems Technology) Inc.

ALST Falcon miniature eye-safe laser range-finder

Type
Eye-safe laser range-finder.

Description
ALST Inc's Falcon series of miniature eye-safe laser range-finders is designed for very compact low-cost high-performance applications where a separate sighting or aiming device, such as a thermal imager, video system or night vision system will be used. Applications include reconnaissance, surveillance systems, UAVs, ground vehicles, hand-held sensors, fire-control systems and navigational aids.

The range of system capabilities includes line of sight distance measurement to passive targets at ranges up to 20 km; the ability to detect targets beyond the visibility range under poor visibility conditions; range resolution of ±5 m or optionally ±2 m; and first pulse/last pulse selectable range logic and adjustable range gating, to reduce the possibility of false returns.

Built-In-Test (BIT) and user's choice of RS-232 or RS-422 data interfaces are standard features. Software for operation via PCs is optional.

Falcon modules use PIN detectors (Falcon P-40) providing 80 nW detectable signal as standard, with optional Avalanche PhotoDiode (APD) InGaAs detectors (Falcon A-40) that can detect a signal at 20 nW.

Operational status
Available.

Specifications
Dimensions: 125 × 98 × 66 mm
Weight: 1.1 kg
Laser type Er:glass
Wavelength: 1.54 μm
Energy/pulse: 9.0 mJ typical
Pulse width: 22 ±4 ns
Pulse rate: 12 ppm average
Pulse rate peak: 0.33 Hz
Beam divergence: 0.7 mrad, 0.5 mrad optional
Receiver aperture: 40 mm
Receiver sensitivity: 80 nW (P-40), 20 nW (A-40)
Receiver field of view: 3.3 mrad (P-40), 2.2 mrad (A-40)
Max range: 10 km (P-40), 20 km (A-40)
Min range: 50 m

The Falcon miniature eye-safe laser range-finder from ALST 0022048

Range resolution: 5 m
Min range gate: zero to max range, 100 m increments
Max range gate: 100 to max range, 100 m increments
Receiver detector: InGaAs PIN (P-40), InGaAs APD (A-40)
Power supply: 28 V DC P series, 12 V DC A series, 0.5 A peak
Data interface: RS-232, RS-422, ALST EZ-Link

Contractor
ALST (Advanced Laser Systems Technology) Inc.

ALST LRF-2

Type
Nd:YAG laser range-finder.

Description
ALST Inc's LRF series is Nd:YAG laser range-finders. LRF models use second-generation lasers and feature Built-In Test (BIT), adjustable minimum range gating, compact packaging, 1 Hz operation and a maximum range of 10 km. The LRF models use a silicon avalanche photodiode receiver detector, sensitive to 10 nW.

Operational status
Available.

Specifications
LRF-2
Laser type: Nd:YAG
Wavelength: 1.06 μm
Energy/pulse: 9 mJ typical
Pulse-width: 8 ±1 ns
Pulse rate: 1 Hz
Beam divergence: 1.0 mrad
Receiver aperture: 30 mm
Receiver field of view: 3.0 mrad
Receiver detector: silicon APD
Receiver sensitivity: 10 nW
Max range: 10 km
Min range: 50 m
Range resolution: 5 m
Range logic: first/last
Prime power: 12/24/28 V DC
Dimensions: 279 × 117 × 71 mm
Weight: 1.8 kg housed, 900 g modular

Contractor
ALST (Advanced Laser Systems Technology) Inc.

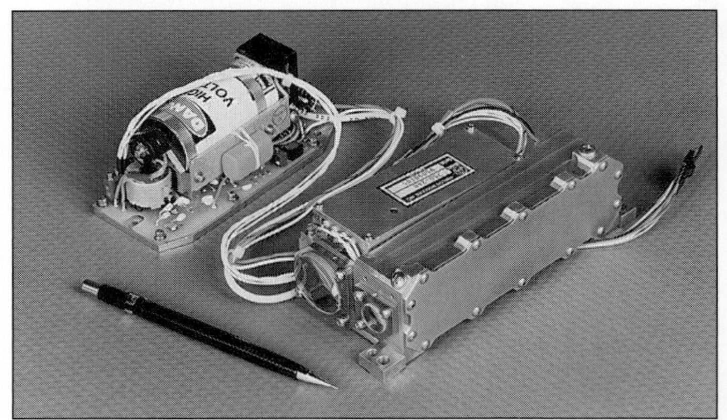

The ALST LRF-2M laser range-finder 0022052

AMS GAQ-4 anti-aircraft system laser range-finder

Type
Land laser range-finder.

Description
The GAQ-4 high repetition rate laser range-finder is intended for use in anti-aircraft fire-control systems and has been adopted as part of the electro-optic fire-control system used in the Italian Army's SIDAM 25 self-propelled anti-aircraft gun. Adaptation to other fire-control systems can be made by performing minor opto-mechanical and electrical modifications to its modular designed components.

The GAQ-4 comprises the following subsystems:
(a) a 1.06 μm wavelength laser transmitter module
(b) a laser receiver module
(c) a laser electronic unit
(d) mounting plates and interconnecting cable set.

Operational status
Production. In service with the Italian Army (SIDAM 25).

Specifications
Weight
Laser transmitter: approx 10 kg
Laser receiver: approx 1.5 kg
Electronic unit: approx 10 kg
Dimensions
Laser transmitter: 420 × 210 × 100 mm
Laser receiver: 120 × 230 × 70 mm
Electronic unit: 330 × 230 × 130 mm
Wavelength: 1.06 μm
Operating range: 300-10,235 m
Accuracy: 6 m RMS
Repetition rate: 20 pps
Power supply: 28 V DC and 115 V, 400 Hz with current drawings of 11 A and 0.6 A respectively

Contractor
AMS (formerly Alenia Marconi Systems).

UPDATED

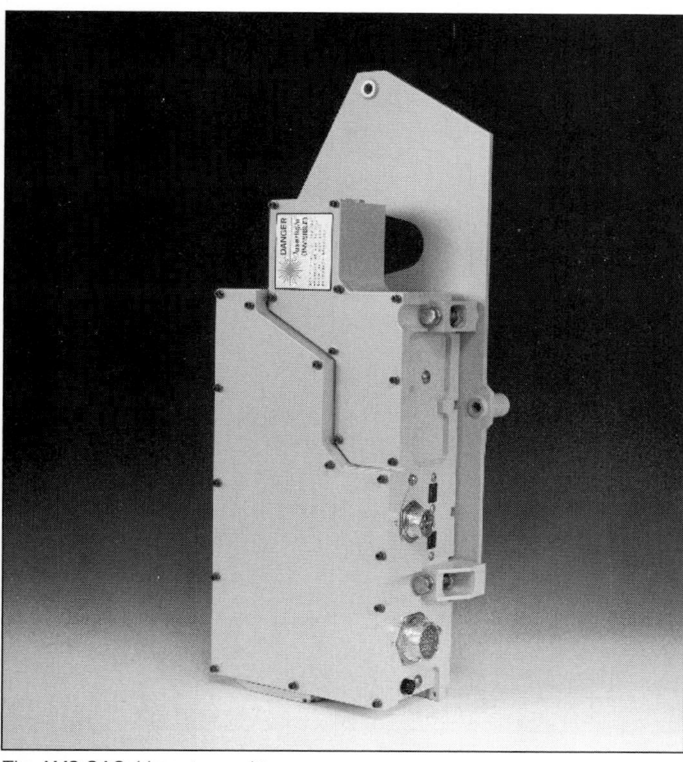

The AMS GAQ-4 laser transmitter

AMS MTL-8 modular laser range-finder

Type
Land laser range-finder.

Description
The MTL-8 modular laser range-finder is available for use in a variety of applications, in particular for use on fighting vehicles. MTL-8 forms part of the AMS P170L/P204L gunner's sights and the Galileo Avionica TURMS fire-control system.

The MTL-8 comprises the following subsystems:

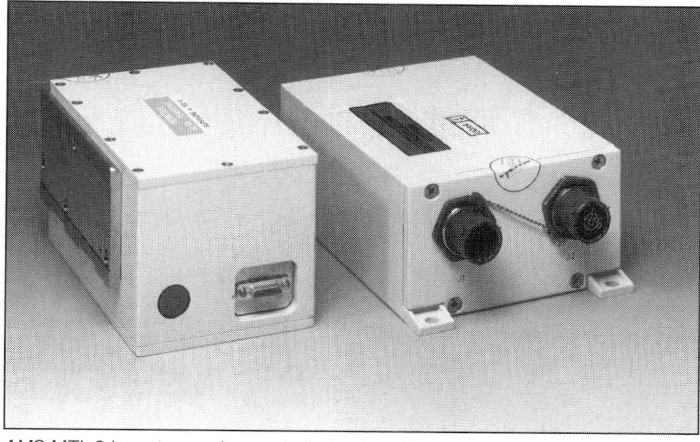

AMS MTL-8 laser transceiver and electronic unit

(a) Laser Transceiver Unit (LTU) with mini Nd:YAG laser transmitter and avalanche photodiode receiver/detector unit sharing the same aperture to minimise integration problems with the host sighting assembly
(b) Laser Electronic Unit (LEU)
(c) Appropriate sight and fire-control system interface and cable connectors.

The MTL-8 can also be integrated with many other types of existing day/night sight assemblies as required.

Operational status
Production. In service with the Italian Army (fitted to the B1 tank destroyer, C1 MBT and VCC- 80 IFV).

Specifications
Weight: approx 5.5 kg
Dimensions
LTU: 86 × 96 × 150 mm
LEU: 70 × 120 × 185 mm
Laser type: Nd:YAG
Wavelength: 1.06 μm
Operating range: 400-9,995 m
Accuracy: ± 5 m
Repetition rate: up to 1 pps
Power supply: 24 ± 5 V DC with current drawing of 1.5 A at max pulse repetition rate of 1 pps

Contractor
AMS (formerly Alenia Marconi Systems).

UPDATED

Aselsan LH-7800

Type
Land laser range-finder.

Description
The LH-7800 is a hand-held Nd:YAG laser range-finder operating in the 1.064 μm waveband and intended for use by infantry units. In addition to the laser there is a telescope and the instrument is similar in weight, size and shape to a pair of conventional 7 × 50 binoculars. It may also be used for artillery observation when mounted on a lightweight tripod with a goniometer. The instrument can measure distances to targets from 150 to 9,995 m.

Operational status
Production of these range-finders began in 1988 and 500 had been produced by the end of 1992 for Turkish forces. They are licence-made Simrad LP7s.

Specifications
Dimensions: 215 × 202 × 93 mm
Weight (excl batteries): 2 kg approx
Wavelength: 1.064 μm
Radiant energy: 8-15 mJ
Pulse length: 8 ns
Pulse rate: 12 ppm max
Transmitter beamwidth: 2 mrad
Receiver
Field of view: 1.3 mrad
Detector type: avalanche photodiode optimised for 1.06 μm
Aperture: 45 mm
Clock frequency of counter: 29.973 MHz
Range accuracy: ±5 m
Multiple targets: 1 target displayed, indication of more than 1 target registered
Min range setting: 150-4,000 m continuously

The Aselsan LH-7800 laser range-finder

Telescope
Magnification: ×7
Field of view: 125 mils
Dioptre adjustment: –4 to +3 dioptres

Contractor
Aselsan Inc, Microelectronics, Guidance and Electro-optics Division.

VERIFIED

ATCOP AR 3

Type
Laser range-finder (land).

Description
The ATCOP AR 3 laser range-finder is for use by infantry and forward observers for distance measurement and correction of gunfire. It can range targets up to 10 km with an accuracy of 5 m at all ranges. It is a hand-held system but can be quickly clamped to a goniometer (optional but recommended for fire correction).

The 7 × 50 sight is hardened against enemy lasers. A night adaptor is also available.

Operational status
Available.

Specifications
Transmitter
Laser type: Nd:YAG
Pulse energy: 12 mJ

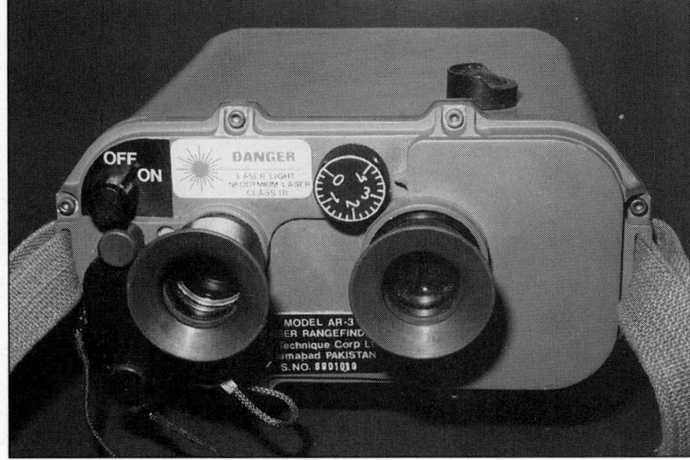

The ATCOP AR 3 laser range-finder 0004815

Pulse-width: 9 ns
Beam divergence: 1 mrad
Pulses/min: 20 (40 for short time)
Receiver
Field of view: 2 mrad
Aperture: 47 mm
Range accuracy: ±5 m
Max/min range: 9,995/200 m
Range blanking: 200-4,000 m
Sighting optics: 7 × 50 (nominal)
Dimensions: 230 × 170 × 85 mm
Weight: 2.1 kg

Contractor
Al Technique Corporation of Pakistan (Pvt) Ltd.

VERIFIED

ATCOP LDR 3

Type
Laser range-finder/designator (land).

Description
The LDR 2 laser designator/ranger is used to designate targets from the ground for close air support, thereby permitting precision bombing of strategic targets. It can also be used as a range-finder for targets up to 10 km away. An airborne version is also available for fighter aircraft and helicopters.

Operational status
Available.

Specifications
Dimensions: 6.7 × 5.5 × 4.33 cm
Weight: 10.73 kg
Pulse repetition frequency: 20 Hz max
Operating temperature: -30 to +60°C
Ranging distance: 10 km
Ranging accuracy: ±5 m

Contractor
Al Technique Corporation of Pakistan (Pvt) Ltd.

VERIFIED

The ATCOP LDR 3 laser range-finder/designator 0129211

ATCOP TR 3

Type
Laser range-finder (land).

Description
The ATCOP Model TR 3 laser range-finder is designed to retrofit to the existing day sights of T-55/T-59 tanks. It has been specially designed for the Pakistan Tank Upgrade Programme in collaboration with the Armour Corps and the Technical Development Directorate of the Pakistan Army and can be interfaced with fire-control systems.

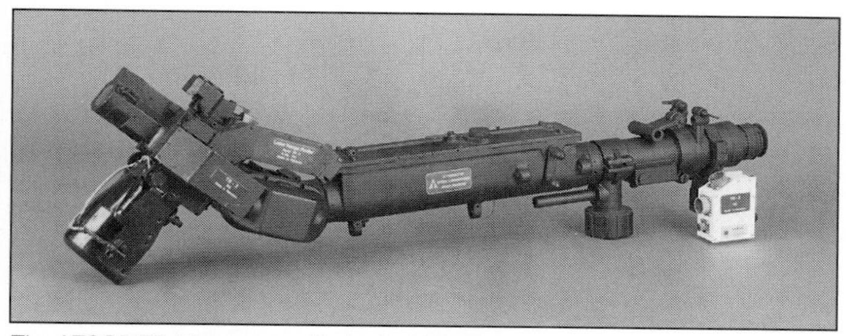

The ATCOP TR-3 laser range-finder for T-55/T-59 tanks 0053775

TR 2M can range targets up to 9 km with an accuracy of 5 m at all ranges. The sight is hardened against enemy lasers.

Operational status
Serial production.

Specifications
Transmitter
Laser type: Nd:YAG
Pulse energy: 12 mJ
Pulsewidth: 12 ns
Beam divergence: 0.7 mrad
Pulses/min: 30 (40 for short time)
Receiver
Field of view: 1.0 mrad
Aperture: 30 mm
Max/min range: 8,995/200 m
Range blanking: 200-4,000 m
Sighting optics
Boresight accuracy: 2 mrad
Field of view: same as gunner's telescope

Contractor
Al Technique Corporation of Pakistan (Pvt) Ltd.

VERIFIED

BAE Systems L20 series sight laser range-finders

Type
Land laser range-finder.

Description
The L20 series is designed to be modular periscopic sight laser range-finder units for gunners to increase their first-round hit probability, reduce the overall engagement time and maximise operational efficiency by introducing savings in ammunition expenditure, logistics, maintenance actions and overall life cycle costings.

By using a special-to-type optical head and adaptor unit, versions of the L20 can be fitted to the gunner's sight mountings in Centurion, Chieftain, Challenger 1, Vickers Mk 3 and Vijayanta MBTs as well as the M41 light tank and Scorpion tracked armoured reconnaissance vehicle.

The laser range-finder modules available are either the BAE Systems (formerly Marconi Avionics – Radar and Countermeasures Systems) Type 530 Mk 2 or Simrad LV352.

The L20 family members comprise three main modules: the optics, laser and electronics. Main operating features of the various models, which only differ in the type of laser range-finder installed and the optical head/adaptor units, can be summarised as a Nd:YAG transceiver with the range readout displayed in the gunner's eyepiece and cathode ray tube to provide a controlled aiming mark when linked to a fire-control system. In addition, the CRT displays the output from a remote thermal imager or Low-Light Level TeleVision (LLLTV) camera. The sight is provided with a ×10 magnification channel for gunnery and an integral periscopic unity surveillance channel for general observation.

Operational status
Production as required. In service with Oman and other, unspecified countries.

Specifications
Sighting telescope
Magnification: ×10
Exit pupil diameter: 4 mm
Field of view: 120 mils
Transmission
Eye response weighted: 40%
Boresight adjustment: ±10 mils
Parallax between image of target at infinity, graticule and CRT spot: 0.15 mil max
Misalignment between optical laser sight lines: 0.15 mil max
Acquisition periscope
Magnification: ×1

Field of view	**fixed eye**	**dynamic**
Horizontal	480 mils	765 mils
Vertical	125 mils	230 mils
Laser rangefinder		
Manufacturer	SIMRAD LV352	BAE Systems
Type	Nd:YAG	
Wavelength	1.064 μm	1.064 μm
Output energy	15 mJ	>20 mJ
Raw beam diameter	4 mm	4.2 mm
Clock frequency	30 MHz	30 MHz
Logic	first/last pulse	first/last pulse
Operating range	200-9,995 m	300-9,995 m
Accuracy	±5 m	±5 m
Pulse repetition rate	0.5 pps	1 pps
Duty cycle	3 pulse/15 s	continuous
Power supply	28 V DC	28 V DC

Contractor
BAE Systems Avionics Group.

UPDATED

BAE Systems Type 306 lightweight ground laser designator

Type
Land laser designator.

Description
The Type 306 is a lightweight portable laser target designator and range-finder. In addition to range-finding for conventional forward observer operations it can designate targets for all current NATO laser spot trackers and laser-guided weapons.

To allow tactical flexibility, the Type 306 allows a number of mounting systems to be used and can be co-mounted with thermal imagers and image intensifier night sights.

Options include tracking head; tripod; image intensifier; and thermal imager. With a conversion kit the Type 306 can be modified to become a navigational beacon, marked target simulator or aircraft homing device for aircraft fitted with laser spot trackers.

Operational status
In service with the UK forces and seven overseas countries.

Specifications
Dimensions: 310 × 231 × 141 mm
Weight: 9.6 kg
Laser transmitter
Type: Nd:YAG
Output energy: 80 mJ min
Pulses/s: single shot for ranging, up to 20 for designation
Beam divergence: 0.25 mrad (90% of energy)
Laser receiver
Range: 300-9,997 m (display limit)
Accuracy: ±5 m
Range discrimination: <30 m

Contractor
BAE Systems, Avionics, Tactical Systems Division, Edinburgh.

UPDATED

BAE Systems lightweight ground laser designator Type 306

BAE Systems Type 520 laser range-finder

Type
Land laser range-finder.

Description
The Type 520 third-generation Nd:YAG 1.064 μm wavelength laser range-finder is designed for integration into armoured fighting vehicle sights such as the BAE Systems L20 series (L20 for the Scorpion reconnaissance vehicle and L22 for the Chieftain MBT), M48A5 periscope retrofit kit and day/night Argus system for vehicles such as the ENGESA EE-9 Cascavel armoured car.

The Type 520 subsystems for the EE-9 comprise the following:

(a) commander's remote display
(b) laser transceiver unit with receiver, optical and power supply modules. The transmitter and receiver channels are combined to form a coaxial system. A beam expander with ×10 magnification optics can also be used with this unit
(c) gunner's sight display unit
(d) electronics unit providing the control logic for the power supply and converting the returned laser light pulses to range data for the visual display(s) and, if required, digital output for a fire-control system ballistic computer
(e) interconnecting cable set.

The other sight systems have similar sets of components but are packaged according to their individual needs.

Operational status
Production as required. In service with Oman (as part of the BAE Systems sight assembly for the Qayis Al Ardh (Chieftain) MBT and the BAE Systems L20 sight assembly for the Scorpion reconnaissance vehicle) and several other unspecified countries.

Specifications
Dimensions
Transceiver unit: 148 × 90 × 86 mm
Electronics unit: configuration to suit application
Weight
Transceiver unit: 2 kg
Electronics unit: 0.75 kg
Laser type: Nd:YAG
Wavelength: 1.064 μm
Operating range: 300-9,995 m
Accuracy: ±5 m
Resolution: 5 m
Discrimination: 30 m

Contractor
BAE Systems, Avionics, Tactical Systems Division, Edinburgh.

UPDATED

BAE Systems Type 520 laser range-finder

BAE Systems Type 629 lightweight modular laser range-finder

Type
Land laser range-finder.

Description
The BAE Systems Type 629 Nd:YAG lightweight modular laser transceiver was developed to be a fundamental building block in a number of laser applications, both for range-finding and designation. Variants of it are in service with air defence systems (as range-finder), fixed-wing air to ground applications (as range-finder) and helicopter (as designator/ranger). The common transmit/receive optical path makes it suitable, both mechanically and optically, for integration with new and existing sighting systems. In March 1987 an order for 38 Type 629G lasers was placed for use with the M113A2 mounted ADATS air defence system.

BAE Systems Type 629 laser transceiver mounted on ADATS

Type 629 lightweight modular range-finder

Operational status
The Type 629 transceiver is in service. A variant of it is in service with the ADATS system in two countries.

Specifications
Laser transmitter
Type: Nd:YAG
Output energy: >60 mJ
Pulse repetition frequency: 10 pps nominal
Beam divergence: 2 mrad max (raw beam)
Receiver
Range: 300 m-10 km
Accuracy: 2.5 m
Weight: 6 kg
Dimensions: 200 × 180 × 130 mm
Power supply: 22-30 V DC

Contractor
BAE Systems, Avionics, Tactical Systems Division, Edinburgh.

UPDATED

Brashear MLRF 100 Mini-Laser Range-Finder

Type
Land laser range-finder.

Description
The MLRF 100 is a very compact device developed for the US Army. It is intended for surveillance, surveying or measuring and its small size make it suitable for mounting on personal weapons. It is sealed to protect the internal mechanism from field and environmental conditions. The laser is eye-safe.

Operational status
Available.

Specifications
Dimensions (l × w × h): 159 × 82.5 × 44.5 mm
Weight: 567 g
Wavelength: 1.54 μm
Output energy: 5-7 mJ
Beam divergence: 1 mrad
Pulse-width: 20 ns
Repetition rate: 0.5 Hz burst mode, 5 ppm continuous
Range: 50-4,096 m
Resolution: 1 m
Accuracy: 1 m
Power requirement: 18-32 V, 0.5 A during laser charge, 50 m A standby
Power supply: external 24 V Ni:Cd battery

Contractor
Brashear LP.

VERIFIED

CEIEC Type 82 tank laser range-finder

Type
Land laser range-finder.

Description
The CEIEC Type 82 Nd:YAG 1.06 μm wavelength laser range-finder consists of a main body assembly, a power and counter unit and connecting cables. It is suitable for use with the Type 59 and Type 69 series MBTs and can be adapted for other tank types after replacing the necessary components.

It can be combined with a sight or telescope system and coupled into a fire-control system to increase the first-round hit probability.

The measured range is indicated by a three-decimal digital display on the display unit.

Operational status
Production as required. In service with the Chinese armed forces.

Specifications

Unit	Weight	Dimensions
Main body assembly		
Incl base	11.2 kg	400 × 150 × 120 mm
Power and counter unit	3.1 kg	173 × 113 × 149 mm
Range	300-3,000 m	
Accuracy	±10 m	

Contractor
China National Aero-Technology Import and Export Corporation (CATIC).

VERIFIED

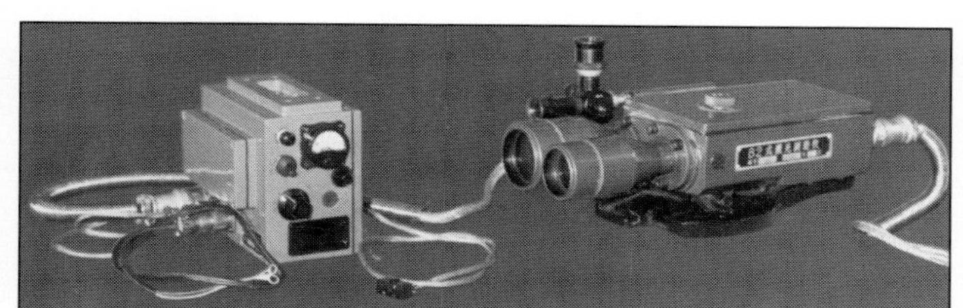

CEIEC Type 82 tank laser range-finder with main body assembly on right and counter unit on left

CILAS DHY 307 for laser-guided weapons

Type
Land laser designator.

Description
The DHY 307 is a ground-based laser target designator which is part of CILAS' TIM (Modular Illuminator Range-finder) family target markers and range-finders. It is designed to be used by small units on the ground and provides precision guidance of a laser homing missile, bomb or shell onto a tactical target. It is designed to be compatible with all laser-guided weapons conforming with STANAG No 3733 NATO standard.

DHY 307 consists of the following modules:

(a) an illuminator box consisting of modules shared by all the range-finders/illuminators of CILAS' TIM family and incorporating the elements needed for target illumination that is sighting, laser emission, user interface (ON/OFF/BITE), mechanical and electrical interfaces (power supply and protocol for dialogue with control box) and an optional range-finding function
(b) a control box connected to the illuminator by a short cable and used for selecting an illumination code among the prerecorded codes, controlling the various operating modes, initialising, triggering the Built-In Test Equipment (BITE), controlling thermal safety inhibition, checking the battery charge, triggering firing without the risk of aim-off
(c) an independent power supply guaranteeing an operating duration compatible with operational needs (battery or network on board a vehicle)
(d) a tripod with an optical angulation head supplying the target elevation and azimuth
(e) the various interconnection cables with the illuminator box, the control box and the independent power supply.

Missions in the field take place in three stages:

(a) target identification and aiming of the designator, followed by selection of one of the preprogrammed codes
(b) co-ordination with the laser-guided weapon launcher to trigger the illumination operation
(c) illumination of the target through to impact. The system may be remotely controlled and is supplied with optional night vision equipment, for example Simrad Optronics' KN200 image intensifier, SAGEM's ALIS thermal imager.

DHY 307 has been adapted for use with the Krasnopol-M laser-guided artillery projectile from KBP Instrument Design Bureau of Russia.

Operational status
This equipment has been operationally tested and validated by the French Army and Air Force and is in series production. French and foreign forces are now equipped with DHY 307. Interface with artillery computer and GPS was carried out in 1997.

Specifications
Wavelength: 1.064 μm
Magnification: ×7
Emission lens diameter: ≥25 mm
Angular field: ≥100 mrad
Laser protection: 53 dB at 1.064 μm
Dioptric adjustment: +2 to −6 dioptres
Repetition rate: according to STANAG 3733
Weight (illuminator alone): ≤8 kg
Dimensions (illuminator alone): ≤330 × 260 × 80 mm
Electrical
Average consumption at 20 Hz: ≤350 W
Max absorbed current: 20 A at 24 V
Rated voltage: 24 V DC
Operating temperature (as per MIL-STD 810 E): −40 to +50°C
PRF: 8-20 Hz

Contractor
Compagnie Industrielle des Lasers (CILAS).

VERIFIED

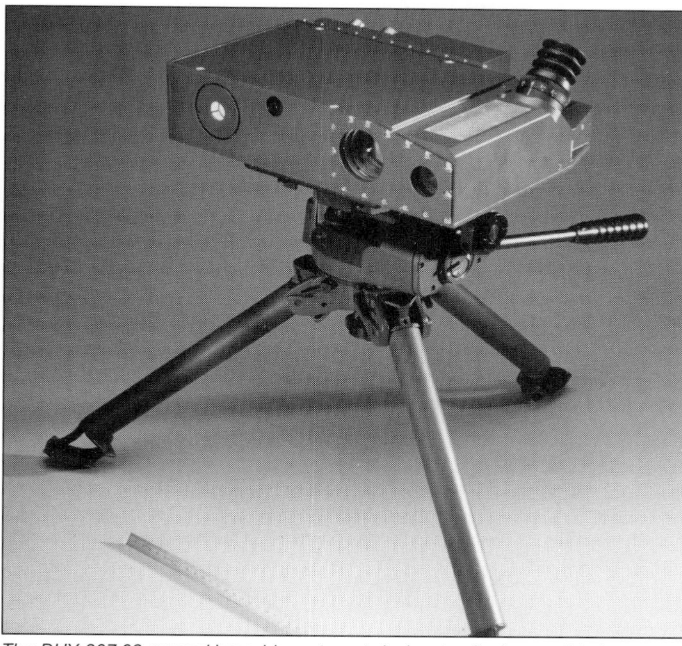

The DHY 307-03 ground-based laser target designator for laser-guided weapons
0053774

CILAS TCY 901 laser range-finder

Type
Land laser range-finder.

Description
The TCY 901 is the replacement system for the TCV 107. It is designed for use on any type of tank and comprises a compact sealed unit that contains the transmitter, receiver and power supply unit.

Its small size allows it to be installed either externally on the 90 mm Lynx 90 F1 turret or internally on the TS-90 gun and Lancelot III HOT ATGW turrets. It can be associated either with a computerised fire-control system or with the optical control box.

The three TCY 901 options available are:

(a) with the separate control box for installation where the range-finder is not associated with a fire-control system
(b) with an adjustable blinding device
(c) with an armoured box and associated port, for external mounting of the range-finder.

A dispersion meter at the front of the case allows the range-finder to be aligned with the vehicle's main gun axis. A movable adjustment lens can be provided as an accessory.

Two alignment channels (on the top and side) can be selected for use. Switching from one to the other is done through an external control and requires no disassembly. The backplate has a connector for linking to the associated system.

The target range is displayed digitally on the control box. The presence of two echoes is indicated by a pilot light. In this case the range for the furthest echo is displayed automatically and range of the closest echo can be substituted via a manual control.

Operational status
Series production. In service with undisclosed countries.

Specifications
Weight: 5.8 kg
Dimensions: 305 × 170 × 94 mm
Laser type: Nd:YAG
Wavelength: 1.064 μm
Operating range: 170-9,600 m
Resolution: 7 m
Angular discrimination: 0.5 mrad
Power supply: 20-30 V DC

Contractor
Compagnie Industrielle des Lasers (CILAS).

VERIFIED

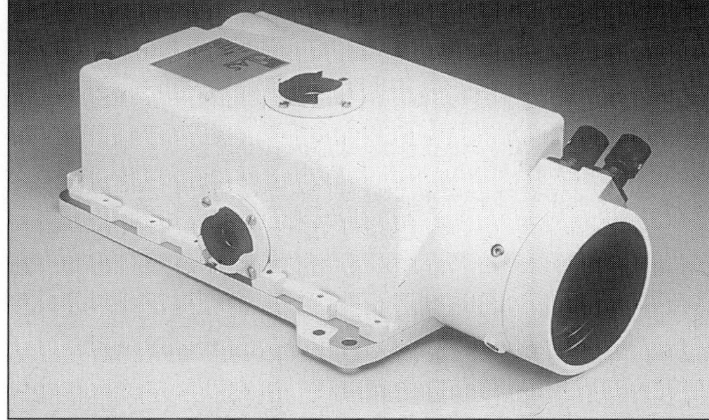
CILAS TCY 901 laser range-finder

CILAS THS 304-08 eye-safe laser range-finder for anti-aircraft fire-control systems

Type
Land laser range-finder.

Description
The THS 304-08 laser range-finder belongs to the TIM family of CILAS modular illuminators and range-finders. It is designed for continuous high pulse repetition rate operation and to improve the effectiveness of electro-optical or radar anti-aircraft fire-control systems either in the system definition phase or when already in operational service.

It consists of standard TIM family modules and sub-assemblies specifically designed for anti-aircraft fire-control systems:

(a) Q-switched 1.06 μm transmitter
(b) 1.54 μm Raman conversion cell
(c) power supply card
(d) electro-optical control card
(e) interface/control/range processing card
(f) 1.54 μm receiver
(g) low-voltage converter
(h) transmission/reception optical sub-assembly
(i) air-liquid exchanger for cooling the system
(j) de-icing device

Operational status
Production. In service with undisclosed countries.

Specifications
Weight (without options): <25 kg
Dimensions: 456 × 316 × 256 mm
Wavelength: 1.54 μm
Pulse repetition rate: 10 - 13 Hz continuous (10 min) and 20 Hz (30 s)
Operating range: 300-40,000 m
Ranging accuracy: ±4.5 m
Resolution: 2.5 m
Discrimination: ≤30 m
Echo logic: 2 echoes in one programmable window
Divergence: 1.3 mrad (typical)

Contractor
Compagnie Industrielle des Lasers (CILAS).

VERIFIED

CILAS THS 304-08 eye-safe laser range-finder for anti-aircraft fire-control systems
0004812

CILAS TM 18B/18C/18 CT

Type
Land laser range-finder.

Description
The TM 18B is a hand-held Nd:YAG laser range-finder, similar in size and appearance to a pair of field binoculars. The operator centres the cross-hairs on the target and triggers the laser by means of a push-button; the distance in metres is displayed immediately in the eyepiece. Accuracy is ±10 m whatever the target distance, within the range limits of the sight. An optical device provides protection for the operator's eyes during laser emission. Other features include a device to prevent accidental triggering of the laser, a multiple echo indicator, a blanking control adjustable between 0 and 2,300 m and an internal power supply. A tripod mounting, with elevation and azimuth control, charger for the Ni/Cd battery, remote trigger, lithium battery and a removable attenuator, is also available.

TM 18C has an RS-422 datalink for connection to goniometers. When TM 18C is combined with an electronic angulation head and a tripod, this creates the TM 18 CT artillery observation post. The tripod has a height of 340 to 630 mm and a weight of 3.2 kg while the angulation head weighs 3.4 kg.

CILAS TM 18B laser range-finder

Operational status

In large scale production for the French armed forces. TM 18C has been in production since 1992.

Specifications

Monocular sight
Field of view: 7° (+120 mrad)
Magnification: ×6
Ocular ring: 5 mm
Laser range-finder
Dimensions
TM 18B: 180 × 150 × 7.5 mm
TM 18C: 230 × 180 × 9.5 mm
Weight
TM 18B: 1.9 kg
TM 18C: 1.8 kg
Wavelength: 1.06 μm
Beam expansion ratio: 1:4
Pulse duration: 8 μs
Energy: 18 mJ
Pulse Repetition rate: 12 shots/min
Time to range: 0.1 s
Detector: PIN photodiode
Measurement range: 70-9,990 m
Range blanking: adjustable 70-2,300 m
Measurement accuracy: ±10 m
Power supply: 12 V 0.5 Ah Ni/Cd rechargeable battery giving 600 shots at 25°C (TM 18B) or 600 shots at 20°C (TM 18C)

Contractor

Compagnie Industrielle des Lasers (CILAS).

VERIFIED

CILAS TMS 303/TMY 303 eye-safe laser range-finder

Type

Laser range-finder.

Description

The TMS 303 is part of the CILAS modular family of eye-safe laser range-finder devices. The main application of the TMS 303 is as a range-finder on tanks and helicopters, integrated into a fire-control system. At present the TMS 303 optical and mechanical interfaces are compatible with those of the Leclerc MBT gunner's sight and the Viviane sight for the Eurocopter Gazelle helicopter equipped with the Euromissile HOT ATGW system. If required, other specific sight interfaces can be adapted.

A 1.06 μm version, the TMY 303, is available without the Raman conversion cell.

The TMS 303 consists of four modules:

(a) a 1.54 μm transmitter module with a Raman effect conversion cell
(b) an interface/control/range processing card with an ASIC circuit including eight programmable range-finding windows
(c) a very high-sensitivity reception module (≥10 nW)
(d) a low-voltage converter module.

These modules have been specifically developed to meet medium-range/medium-rate laser rangefinding applications.

Operational status

In production and in service with undisclosed countries.

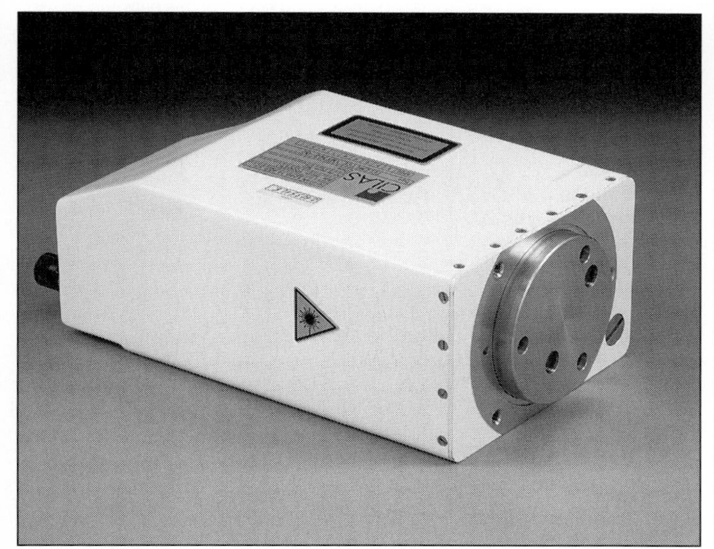

CILAS TMS 303 eye-safe laser range-finder 0022053

Specifications

Weight: 2.2 kg
Dimensions L × W × H: 190 × 119 × 78.5 mm
Transmitter module
Wavelength: 1.54 μm (TMS 303), 1.06 μm (TMY 303)
Pulse repetition rate: (2 versions) 1 Hz continuous; 3 Hz for 1 min or 8 Hz for 5 s
Interface/control/range processing
Range (min/max): 300 m to 80,000 m (depending upon visibility, type of target and the optical characteristics of the fire-control system)
Echoes logic: 2 echoes, 8 programmable windows with choice of first or last echo in each window
Range-finding window
Step: 5 m
Min range: 300 m
Min width: 50 m
Discrimination: 30 m

Contractor

Compagnie Industrielle des Lasers (CILAS).

VERIFIED

CILAS TMS 309-03 1.54 μm eye-safe laser range-finder for armoured fighting vehicles

Type

Land laser range-finder.

Description

The TMS 309-03 laser range-finder belongs to the TEMPO family of latest-generation CILAS eye-safe range-finders. It has been specially developed to meet the needs of short- and medium-range weapon systems mounted on armoured fighting vehicles such as tanks or naval turrets. The TMS 309 is externally mounted on the turret for ranging of land or airborne targets. Its repetition rate can be increased in order to meet the requirements of ground-to-air systems.

The TMS 309-03 laser range-finder is designed as a compact sealed unit. A dispersion meter at the front of the unit is used for range-finder alignment with a reference determined by the fire-control system (for example, canon axis, thermal imager or TV camera). The operation utilises a movable adjustment lens, supplied as a tool, which is mounted on the optical alignment channel.

Like TMS 303, TMS 309-03 consists of standard TEMPO family modules and specific modules and specific subassemblies. These include:

(a) 1.54 μm transmitter integrating a 1.06 μm laser
(b) Raman conversion cell and power supply
(c) interface/control/range processing card
(d) 1.54 μm receiver
(e) low voltage converter.

Other equipment include the following – transmission/reception/aiming optical head, purge screw for flushing and filling with nitrogen and dehydration cartridge. A control box is available when the range-finder is not associated with a fire-control system.

The TMS 309-03 laser range-finder was designed for any type of tank and short-range (land or naval) fire-control system. The basic version of TMS 309-03 has a repetition rate of 1 Hz. This can be optionally increased to 3 Hz and to 6 Hz in short (5 s) bursts in order to reach moving targets, for example helicopters.

Operational status

In production since 1994. In service. A new improved version is in the final stage of development.

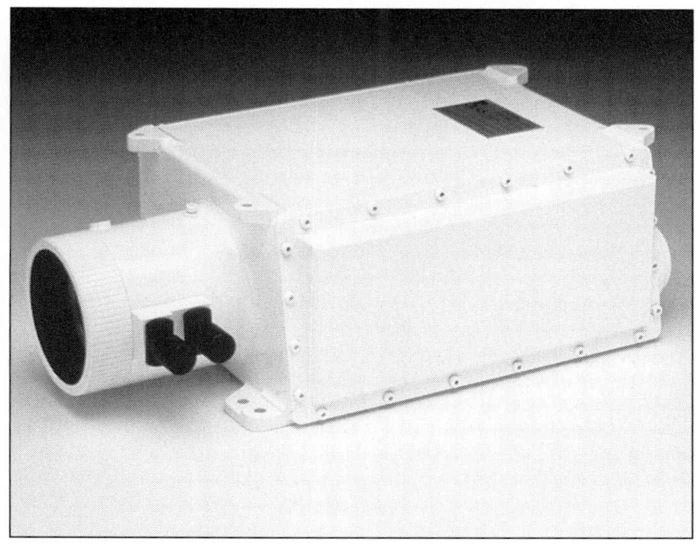

CILAS TMS 309-03 1.54 μm eye-safe laser range-finder for amoured fighting vehicles 0022054

Specifications

Measurement range: 300 - 20,000 m (depending on visibility and type of target)
Resolution: 5 m
Accuracy: ±7 m
Discrimination: ≤50 m
Echoes logic: 2 echoes and adjustable blanking range
Extinction test (Lambertian type target of 0.85 reflectivity, located at 500 m): ≥38 dB
Operating temperature: (typical performances) −33 to +55°C
Transmitter
Wavelength: 1.54 μm
Divergence: 0.5 mrad; optionally 1.5 mrad
Pulse repetition rate: 1 Hz CW; optionally 3 Hz, 6 Hz burst (5 s)
Receiver
Avalanche photodiode: InGaAs
Pupil diameter: 30 mm
Field of view: 0.5 mrad; optionally 1.5 mrad
Dimensions (l × w × h): 320 × 180 × 110 mm
Weight: ≤6 kg
Power supply: 20-30 V DC
Numerical serial link: RS-422

Contractor

Compagnie Industrielle des Lasers (CILAS).

VERIFIED

CILAS TMS 314-04

Type

Eye-safe laser range-finder.

Description

TMS 314-04 is a medium repetition rate high-performance eye-safe laser range-finder, produced with the TEMPO new-generation modules. Developed for anti-aircraft applications, it improves the effectiveness of fire-control systems in ground-to-air applications or for light and medium marine turrets.

TMS 314-04 is packaged in a single waterproof housing and features Raman-shifted Nd:YAG 1.54 μm technology and separate transmitter and receiver optical channels. The system requires no external cooling and includes a dessicator.

Operational status

Mass production started in 1998.

Specifications

Dimensions: 385 × 175 × 217 mm max
Weight: <12 kg
Rating: Class I, IIIA, IIIB (according to EN-60825) depending on repetition rate and beam divergence
Wavelength: 1.54 μm
Extinction ratio: 50 dB typical Lambertian type target of 0.85 reflectivity, located at 500 m, atmospheric visibility of 10 km, detection probability = 50%
Repetition rate: 1 Hz continuous, 3 Hz for 1 min, 8 Hz for a few s
Beam divergence: 1.4 mrad typical (0.7 mrad optional)
Measurement range: 300-20,000 m
Target discrimination: ≤50 m
Echoes logic: 2 echoes and adjustable blanking range
Resolution: 5 m
Ranging accuracy: ±7 m
Mechanical reference: within max 0.2 mrad of optical axis

Contractor

Compagnie Industrielle des Lasers (CILAS).

VERIFIED

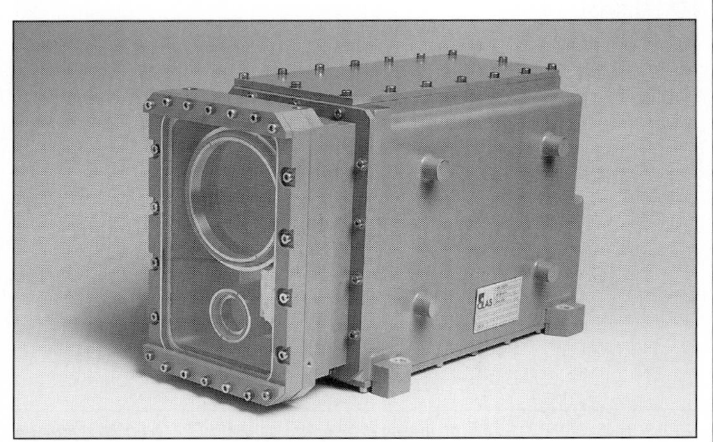

CILAS TMS 314-04 eye-safe laser range-finder 0053777

DRS AMREL advanced modular ranging eye-safe laser

Type

Laser range-finder (land).

Description

DRS Optronics Inc manufactures the AMREL family of eye-safe laser range-finders. Using a common module approach, they can be configured for use in fire-control systems on ground vehicles, in helicopters, fixed-wing and UAV aircraft, on board naval vessels and as a hand-held unit.

The Raman Shifted Nd:YAG eye-safe laser uses passive Q-switching technique, short pulse output, high repetition rate and complete absence of mechanical parts.

DRS has received a Class 1 Certification (ANSI Z136.1) from the US Army Environmental Hygiene Agency for an AMREL configured for the TOW Improved Target Acquisition System (ITAS); this being consistent with the general design of all AMREL configurations. With ranges of 20 to 20,000 m, the AMREL family of eye-safe laser range-finders provides field proven, Class 1 eye-safe ranging capability.

Operational status

In development for the US Army ITAS programme, in service with US Army ITAS and US Army Air to Ground Engagement System.

Specifications

Laser: Raman Shifted Nd:YAG
Wavelength: 1.543 μm
Pulse energy: 6 mJ
Pulsewidth: <6 ns
Beam divergence: 0.5 mrad (l/e^2)
Repetition rate: 1 Hz for 30/90 s
Boresighting: build in LED (optional)
Range display: 20 m to 19,995 m
Range accuracy: ±5 m
Range resolution: 5 m
Target discrimination: ≤20 m
Range gate setting: 20 m to 19,995 m in steps of 5 m
Self test: BIT
Receiver sensitivity: 20 nW
Operating temperature: −32 to +75°C
Storage temperature: −54 to +75°C
Size: modular design
Weight: <2.27 kg

Contractor

DRS Optronics Inc.

VERIFIED

Electromagnetica Type 1-B-36 laser range-finder

Type

Laser range-finder (land).

Description

The range-finder Type 1-B-36 is used for measuring target ranges between 200 and 9,995 m with an accuracy of ±5 m. The range-finder is made up of an optical aiming module, an electronic computer, an electronic receiving system, an accumulator, the housing and case. It may be hand-held or tripod-mounted.

Operational status

Available.

Specifications

Dimensions (l × w × h): 200 × 170 × 76 mm
Weight: 1.7 kg incl battery
Type: Nd:YAG
Power output: 2 MW
Beam divergence: 0.7 mrad
Range: 200 to 9,995 m
Range accuracy: ±5 m
Pulse repetition rate: 12/min
Endurance: 1,000 measurements per battery charge
Magnification: ×10.5
Power supply: 12 V Ni/Cd rechargeable battery

Contractor

Electromagnetica, Romania.

VERIFIED

Elop High-Repetition Laser Range-finders – HRLR and HRLR- ES

Type
Land and naval laser range-finder.

Description
Elop's High Repetition rate Laser Range-finder (HRLR) and an eye-safe version (HRLR-ES) were developed in direct response to an Israel Defence Force, Air Force combat requirement for a system that could either be used as a stand-alone unit or integrated into anti-aircraft gunfire control systems.

The HRLR is a lightweight compact system that is microprocessor controlled and fitted with a serial communications output facility. Integration into an anti-aircraft fire-control system (ground or shipborne) improves the guns' low-level accuracy and hit probability at extreme range.

The Nd:YAG laser has three modes of operation:
(a) laser On – whereby the system is operative from the moment the HRLR is connected to a power supply
(b) laser Standby – whereby the simmer power supply is operative and the laser is awaiting a fire command
(c) laser Fire – whereby the laser transmits pulses in a single shot 10 to 20 pps rate.

Operational status
Production. In service with the Israel Air Defence Force and worldwide.

Specifications

	HRLR	HRLR-ES
Weight	approx 13 kg	approx 14 kg
Dimensions	430 × 260 × 220 mm	430 × 260 × 220 mm
Laser type	Nd:YAG	Nd:YAG + OPO
Wavelength	1.064 μm	1.57 μm
Receiver operating range	250-19,995 m	250-19,995 m
Range gate	variable from 300-15,000 m	300-15,000 m
Accuracy	±5 m	±5 m
Receiver field of view	2 mrad	2 mrad
Boresight telescope		
Magnification	×6	×6
Boresight, to laser beam	>0.5 mrad	> 0.5 mrad
Power supply	28 V DC	28 V DC

Contractor
Elop Electro-Optics Industries Ltd.

VERIFIED

Elop High-Repetition Laser Range-finder (HRLR) 0059655

Elop mini-laser tank range-finder

Type
Land laser range-finder.

Description
The Elop mini-laser range-finder is designed for integration into fire-control system upgrades for main battle tanks and infantry fighting vehicles.

It provides high-accuracy range measurements using a 1.054 μm Er Glass laser and is fully compatible with the periscopes on the M32 and M36 sighting devices, replacing their existing day elbows.

Operational status
In production. In service with unspecified countries.

Specifications
Laser type: ER glass
Wavelength: 1.54 μm
Operating range: 300-9,990 m
Range accuracy: ±10 m
Resolution: 30 m
Sight: ×8 magnification with 8° field of view
Power supply: 24 V DC vehicle

Contractor
Elop Electro-Optics Industries Ltd.

VERIFIED

Eloptro LH-30 laser range-finder

Type
Land laser range-finder.

Description
The LH-30 Laser Range-finder is a compact and lightweight measuring instrument with a range capability up to 20 km. The LH-30 is ideal for a variety of hand-held and mounted applications including:
(a) artillery observation posts
(b) mortar fire-control systems
(c) ranging for anti-tank weapons
(d) support for armoured fighting vehicles
(e) remotely operated observation systems
(f) air-to-ground ranging

The LH-30 is equipped with an external RS-485 connector, providing outputs of electronic ranging information for further processing or remote display, and inputs for remote operation of the LRF. A dovetail mount allows precise connection of the unit to rigid supports.

A single Ni/Cd rechargeable battery is used to power the LH-30 but the range-finder can also operate from any external power source with a wide voltage range from 10 to 30 V DC, including vehicle batteries.

Operational status
Thousands of LH-30 laser range-finders are now in service with several armed forces around the world.

Specifications
Dimensions: 170 × 165 × 80 mm
Weight (incl battery): <1.5 kg
Magnification: ×6
Field of view: 6°
Eye relief: 18 mm
Dioptre adjustment: –4 to +2
Laser safety filter: built-in against 1,064 nm
Measuring range: 80 to 20,000 m
Range resolution: ±5 m
Standard measuring rate: 10 measurements/min
Fast measuring rate: 30 measurements/min
Multiple targets: displays first and last targets
Target discrimination: 50 m
Laser type: Nd:YAG
Reticule illumination: automatic
Power supply: 1× Ni/Cd battery
Number shots/charge: approx 700
External power: 10-30 V DC
Operating temperature: –20 to +65°C
Storage temperature: –40 to +70°C
Electronic interface: RS-485

Contractor
Eloptro, Division of Denel (Pty) Ltd.

Eloptro LH-30 laser range-finder 0110165

Eloptro LH-40 eye-safe laser range-finder

Type
Land laser range-finder.

Description
The LH-40 is a compact, high-performance eye-safe laser range-finder designed for use in hand-held or mounted applications. Range measurement is achieved using single-pulse erbium-glass laser technology operating at an eye-safe wavelength of 1.54 μm.

The LH-40 is equipped with an external communication connector, providing outputs of electronic ranging information for further processing or remote display and inputs for remote operation of the LRF. A dovetail mount allows precise connection of the unit to rigid supports. When mounted on a goniometer, range-finding data can be combined with elevation and azimuth data, providing accurate fire-control information.

Standard AA batteries are used to power the LH-40. The range-finder can also operate from an external source with a wide voltage range of 8 to 30 V DC, including vehicle batteries.

Operational status
In production.

Specifications
Dimensions: 170 × 165 × 80 mm
Weight: <1.6 kg
Measuring range: 80 to 20,000 m
Magnification: ×6
Field of view: 6°
Eye relief: 18 mm
Eyepiece dioptre adjustment: –4 to +2
Range resolution: ±5 m
Standard measuring rate: 10 measurements/min
Fast measuring rate: 20 measurements/min
Multiple targets: displays first and last targets
Target discrimination: 30 m
Laser type: Erbium glass
Safety class: Class 3A eyesafe
Laser safety filter: 1,064 nm
Reticule illumination: automatic
Power supply: 8× AA Ni/Cd batteries
Number shots/charge: approx. 400
External power: 8-30 V DC
Operating temperature: –40 to +70°C
Storage temperature: –40 to +70°C
Electronic interface: RS-232 (RS-485 optional)

Contractor
Eloptro (Pty) Ltd.

Eloptro LH-40 eye-safe laser range-finder 0041623

Eloptro LH-40C laser range-finder

Type
Land laser range-finder.

Description
The LH-40C is the latest product in Eloptro's range of compact, high performance laser range-finders. The system is based on Eloptro's proven LH-40 eyesafe laser range-finder and features an integrated digital magnetic compass.

Single-pulse erbium glass laser technology, emitting at an eye-safe wavelength of 1.54 μm, is employed to achieve measuring ranges of up to 20 km. The range counter has a resolution of 5 m with multiple target detection capability.

The integrated digital magnetic compass provides bearing and elevation data over tilt and bank ranges of up to ± 45°. Calibration is performed via an easy to follow menu guided process and the magnetic declination is user adjustable.

The LH-40C is equipped with an external communication connector, providing outputs of range, bearing and elevation data for further processing or remote display. For greater stability, the system may be mounted onto a tripod via its precise dovetail mounted interface.

The LH-40C was selected by the British Army for the Target Locating Equipment (TLE) programme. In the TLE application, the LH-40C interfaces with the in-service Rockwell Collins SPGR GPS system. The GPS system calculates and displays target co-ordinates using the range, bearing and elevation data transmitted by the LH-40C. The LH-40C incorporates a fall-of-shot function to facilitate fire correction. The system's software may be upgraded via PC download, offering long-term operational flexibility and the potential to interface with future systems.

Operational status
In production.

Specifications
Dimensions: 170 × 165 × 80 mm
Weight (inc batteries): <1.65 kg
Measuring range: 80 to 20,000 m
Range resolution: 5 m
Standard measuring rate: 10 measurements/min
Fast measuring rate: 20 measurements/min
Multiple targets: 3 targets
Target discrimination: 30 m
Laser type: erbium glass
Safety class: Class 3A eye-safe (EN 60825 (1994) and ANSI Z 136.1 (1993))
NOHD: 0 m
Magnification: ×6
Field of view: 6°
Eye relief: 18 mm
Eyepiece dioptre adjustment: –4 to +2
Laser safety filter: against 1,064 nm
Reticule illumination: automatic
Batteries: 8 × AA NiCd or Alkaline
Number ranges/charge: approx 400 (NiCd)
Interface: RS-232 (RS-485 optional)
External power: 8 to 30 V DC
Operating temperature: –30 to +55°C
Storage temperature: –40 to +70°C

Compass Module
Display standard: Mils
Azimuth accuracy: 0.5°
Elevation range: 0.2°
Elevation range: ± 45°
Bank range: ± 45°
Calibration: 4 or 12 step menu guided
Magnetic declination: user adjustable

Contractor
Eloptro (Pty) Ltd.

Eloptro LR-40 eye-safe laser range-finder

Type
Land laser range-finder.

Description
The LR-40 is a high repetition rate eye-safe laser range-finder designed for use in anti-aircraft tracking systems in land and maritime defence. It provides accurate range data at a rate up to 20 measurements per second. It has an active energy control circuit, complete built-in test of major components and is range gate width and position adjustable. The unit is mounted using a high-accuracy dovetail mount which allows units to be interchanged without the need for boresight adjustment.

Operational status
Production as required.

Specifications
Weight: 13.4 kg
Measuring range: 330 m to 20,000 m
Range resolution: ±5 m
Range rate: 0-20 Hz (user adjustable)
Emission wavelength: 1.57 μm (OPO shifted)
Laser pulse energy: 7 to 15 mJ
Beam divergence: 0.8 mrad
Receiver sensitivity: >58 dB extinction, >52 dB extinction

The LR-40 eye-safe laser range-finder from Eloptro 0041621

Range gate: user adjustable
Laser class: Class 3A - ANZI 136.1 (1993)
Power supply: 18-32 V DC, <200 W consumption at 12.5 Hz ranging
External communication: serial optocouplers
Mounting system: single dovetail, accuracy better than 1 mrad
Operating temperature: –20 to +65°C

Contractor
Eloptro (Pty) Ltd.

Fotona Laser Range-finder Module (LRM-E)

Type
Laser range-finder (land).

Development/Description
The eye-safe laser range-finder LRM-E is intended to be used in applications where there is a need for a self-contained module determining ranges to targets promptly and accurately. It is primarily designed to be used remotely as part of other systems, typically mounted on a sensor platform or on the gun mantlet of an armoured vehicle. An optional, lightly armoured protective cover shields the unit against low calibre projectiles and other particles. The module can also be used as a stand-alone instrument, provided that the laser line is aligned with an aiming device and that the operator is equipped with a simple control panel.

The erbium/glass laser transmitter enables danger-free field use. Laser transmitter maximum output energy is within the limits of Class 1 IEC specifications (Standard IEO 825 (1984), Modification 1, Amendment 1 (1990-08): Class 1 (Emax = 8 mJ): no laser irradiation hazard).

The practical ranging capability of the LRM-E is 10 km having the corresponding extinction value of about 38 dB. The LRM-E2 variant is capable of measuring ranges up to 20 km at the extinction value of about 44 dB.

The LRM-E set includes:
- LRM - Laser Range-finder Module
- RME - Removable Eyepiece for rectification.

Optionally:
- PRO – Protective Cover
- MPL – Mounting Plate
- GLE - Gunner's LRM Control Panel with ocular cable.

The LRM-E is a remotely controlled instrument. The laser is triggered and range data are obtained via the serial communication link. Data exchange protocol is according to RS-232 (any other on request). The same connector is also used for external power supply leads. The communication protocol includes remote access to every function or feature of the instrument.

Variants
LRM-E2 is a long-range version of exactly the same instrument. LRFS is a set of LRF subassembly parts, functionally complete without housing for custom completion. Intended for even more tight integration into densely packed systems like sensor heads for airborne applications or into other small size arrangements where space is critical or where there is a need to build a self-contained module comprising several opto-electronic sensors in one unit. Operation and performance of LRFS/LRFS-2 is the same as with LRM-E/LRM-E2 models respectively.

The Fotona LRM-E eye-safe laser range-finder 0129563

The interior of the Fotona LRM-E 0129564

Operational status
In production since 2001.

Specifications
Wavelength: 1.543 μm
Output beam divergence: 0.9 mrad at 90% beam intensity
Output energy: less than 8 mJ (Class 1)
Pulse duration: less than 40 nsec
Repetition rate: 20 shots per minute (a burst of 10 range measurements with 1 sec intervals permitted followed by a period of normal operation)

Practical ranging capability
LRM-E I LRFS: 10 km
LRM-E2 / LRFS-2: 20 km
Radial target resolution: <30 m
Magnification (aux eyepiece): 7 ±5%
Field of view (aux eyepiece): 7 ±5%
Range accuracy: ± 5%
Readout resolution: 2 m
Data transmission: RS-232
Power supply: 18 to 32 V DC (24 V nominal)

Contractor
Fotona dd.

Fotona Metrix laser binoculars

Type
Hand-held binoculars with laser range-finder and digital compass.

Description
Metrix is a robust, light weight, hand-held binocular viewing instrument with the capability of instantly measuring distances and oriented angles to objects in view for prompt and continuous situational awareness of an individual soldier or a small unit. Own position determination, target location and other field measurements are simplified and digitised with the possibility to transfer data to other equipment at hand. In this way, the instrument becomes part of a wider data acquisition network, while still retaining it's handiness and portability.

The erbium/glass laser transmitter technology gives eye-safety for danger-free field use. Laser transmitter maximum output energy is within the limits of Class I IEC specifications (Standard IEC 825 (1984), Modification 1, Amendment I (1990-98): Class 1 (Emax = 8 mJ): no laser irradiation hazard).

The digital compass incorporated into the instrument provides absolute angle values (measured with respect to the direction of north and from the horizontal plane) that are sufficiently accurate for most targeted applications.

Two basic modes of operation with respect to angle data treatment are available: angle read-outs at laser trigger and free running angles. The laser mode has priority over the free running angles as the co-ordinates formed at laser trigger represent the basic data group for consequent computational steps and/or data transmission. Other, less frequently used procedures are conveniently available.

Variants
Metrix Plus is the long-range version.

Operational status
Prototype evaluated and now believed to be in production.

Specifications
Wavelength: 1.543 μm
Output beam divergence: 0.9 mrad at 90% beam intensity
Output energy: less than 8 mJ (Class 1)
Pulse duration: less than 40 nsec
Repetition rate: 8 shots per minute

Practical max ranging capability
Metrix: 10 km
Metrix Plus: 20 km
Radial target resolution: < 30 in
Magnification: ×7
Field of view: 6.5
Output pupil: 5.5 mm
Dioptre adjustment: ±5
Weight: <2 kg

Digital compass
Range in azimuth: 360°
In elevation: ±20°
Resolution: 0.1

Accuracy
Azimuth: ±0.5° RMS
Elevation: ±0.20° RMS

Contractor
Fotona dd.

UPDATED

Fotona's laser binoculars 0129566

Fotona RLD-E1/RLD-E2

Type
Land laser range-finder.

Description
RLD-E1/E2 is a microprocessor-controlled hand-held laser range-finder with an eye-safe erbium:glass laser transmitter which complies with IEC 825 Class 1 requirements. Its primary application is as the ranging instrument in artillery and mortar fire control. Attached to Fotona's EMK-4P mechanical or electronic goniometer, it becomes a part of a target co-ordination acquisition system. An RS-232 data output is provided.

The operator centres the reticle upon the target through the eyepiece. By first depression of the trigger switch the power is supplied to the instrument's circuitry. By second depression of the switch the range in meters instantly appears on the display in the lower part of the eyepiece. Indications of multiple targets, of laser readiness and of battery discharge as well as some diagnostic messages are provided on the same display. The display is automatically switched off 5 seconds after the last ranging. First/last target selection logic is implemented.

Operational status
Available.

Specifications
Range
RLD-E1: 50-9,998 m
RLD-E2: 50-19,998 m

Detector
RLD-E1: InGaAs PIN diode
RLD-E2: Avalanche photodiode
Readout accuracy: ±3 m
Number of targets measured: 2 (option - 4 targets)
Number of targets displayed: 1 (first/last target selection)
Target radial resolution: 30 m
Range blocking: 200-3,000 m
Laser: Er:glass
Output energy: <8 mJ (class 1)
Repetition rate: 12 pulses/min continuous
Receiver field of view: <1.2 mrad

Telescope
Magnification: ×7
Eyepiece adjustment: ±5 dioptres
Power supply: 9 expendable AA type alkaline batteries
Autonomy: 2.5 kg
Communication interface: RS-232 (RS-422 upon request)

Contractor
Fotona d d.

UPDATED

Fotona RLD-E eye-safe laser range-finder

Hellenic Aerospace Industry Polyfimos artillery laser range-finder

Type
Laser range-finder (land).

Description
Hellenic Arms Industry's Polyfimos portable artillery laser range-finder can be used in conjunction with a goniometer and mounted on a tripod by forward observers to obtain orientation and target position. It can also be used in conjunction with a fire-control computer. A passive night vision scope compatible with the laser range-finder is offered with a second-generation image intensifier.

Operational status
Available.

Specifications
Weight: 11 kg (without battery)
Laser: Nd:YAG
Range: 300-20,000 m
Accuracy: ±5 m
Range discrimination: 10 m
Blocking: 300-5,000 m
Magnification: ×7
Field of view: 7°
Power source: Ni/Cd rechargeable (up to 600 shots)

Contractor
Hellenic Aerospace Industry SA.

Hellenic Aerospace artillery laser range-finder 0004804

Hellenic Aerospace Industry LRF II hand-held laser range-finder

Type
Laser range-finder (land).

Description
The LRF II is a hand-held binocular type laser range-finder which can measure distances of 150 to 10,000 m with an accuracy of ±5 m.

Operational status
Available.

Specifications
Dimensions: 200 × 190 × 100 mm
Weight: 2 kg
Laser: Nd:YAG
Wavelength: 1.06 μm
Beam divergence: <2 mrad
Q switch: saturable dye
Detector: avalanche photodiode
Range: 150 to10,000 m

The LRF II hand-held laser range-finder from Hellenic Aerospace 0004803

Accuracy: ±5 m
Min range blocking: 300-5,000 m
Optics
Magnification: ×7
Field of view: 7°
Aperture: 55 mm (diameter)
Power: 1.5 MW

Contractor
Hellenic Aerospace Industry SA.

Kazan 1D18

Type
Land laser range-finder.

Description
The Kazan 1D18 is a short-range hand-held laser range-finder of modular design which can be used for distances of 50 to 5,000 m. It is believed to operate in the 1.06 μm wavelength and the built-in battery allows for 1,000 shots. A tripod version with goniometer is designated 1D18-1. An illuminated graticule is an optional extra.

Operational status
Believed to be in production and in service with the armed forces of the Russian Federation and Associated States.

Specifications
Dimensions: 100 × 185 × 190 mm
Weight: 1.6 kg
Field of view: 8°
Magnification: ×8
Exit pupil diameter: 6 mm
Power supply: 11-14 V
Power consumption: 0.05 A

Contractor
Kazan Optical and Mechanical Plant.

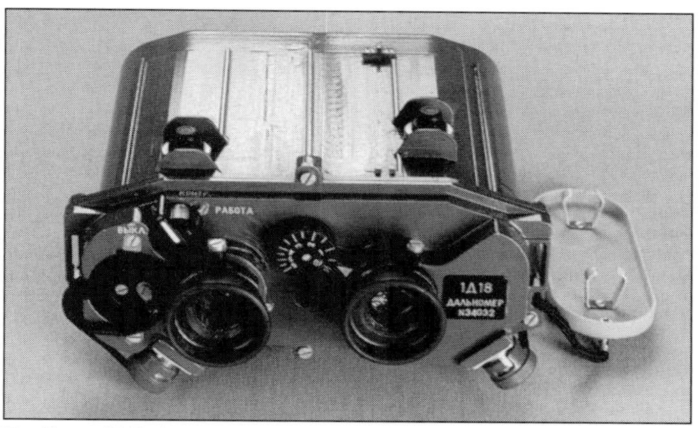

The Kazan 1D18 short-range laser range-finder

Kazan APR-1

Type
Land laser range-finder.

Description
The Kazan APR-1 is a long-range, hand-held laser range-finder with goniometer. Although designed for hand-held use it can be used with a tripod and the delivery set includes, according to the manufacturer, a charger and a 'protection device'. The laser uses a silicon avalanche photodiode detector and the display is a five-digit LED type. It is powered by a 12 V battery.

Operational status
In production and in service with the armed forces of the Russian Federation and Associated States.

Specifications
Dimensions: 110 × 215 × 225 mm
Weight: 2.5 kg
Field of view: 7°
Magnification: ×7
Wavelength: 1.06 μm
Laser detector: silicon avalanche photodiode
Pulsewidth: 6 ns
Energy: 15 mJ
Repetition rate: 20/min

The Kazan APR-1 long-range laser range-finder

Divergence: 0.6 mrad
Measurement range: 145-19,995 m
Measurement accuracy: 375 m

Contractor
Kazan Optical and Mechanical Plant.

VERIFIED

Kintex Lebed laser range-finder

Type
Land laser range-finder.

Description
The Lebed range-finder binoculars are designed for measuring distances to targets and shell bursts and for general observation purposes in the field. They can be hand-held or fixed either on a tripod or on the PAB-2 artillery gun aiming circle.

The device operates on direct line of sight to the target. The complete set consists of the laser range-finder, separate spare parts, tool kit and casing.

Operational status
Available.

Specifications
Weight: 2.8 kg
Field of view: 7°
Magnification: ×7
Pulse repetition rate: 8 ppm
Accuracy: ±5 m
Max detection range: 7,000 m
Min detection range: 250 m

Contractor
Kintex.

VERIFIED

The Lebed laser range-finder binoculars 0004809

Kintex Radian

Type
Land laser range-finder.

Description
The Radian artillery laser range-finder is designed for measuring distances and angles to objects by use of mechanical reference scales in a variety of surveillance and reconnaissance applications by tactical artillery reconnaissance units.

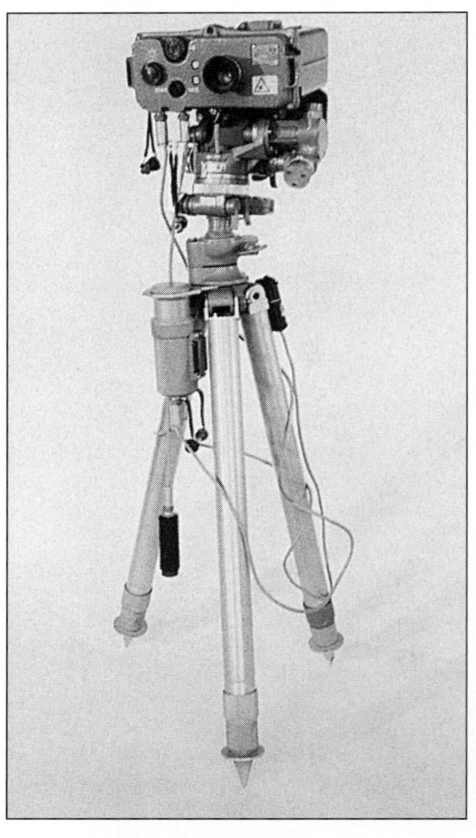

The Radian artillery laser range-finder
0004808

Operational status
Available.

Specifications
Max detection range: 9,995 m
Min detection range: 180 m
Measurement accuracy: 5 m
Output power: 1.0 mW
Divergence: 0.7 mrad
Vision magnification: ×7
Field of view: 6°
Weight: 8.2 kg

Contractor
Kintex.

VERIFIED

Norinco laser range-finders

Type
Land laser range-finder.

Description
Norinco's range of laser range-finders includes the following systems:

LR1 short-range artillery laser range-finder
This is a tripod-mounted instrument used by the artillery's reconnaissance and command detachment. It is used for observation, range-finding, angle measurement and direction. Power is supplied by a silver-zinc battery set.

Type 82 laser range-finder
This is a Nd:YAG laser range-finder for tanks with a ranging capability from 300 to 3,000 m and an accuracy of ±10 m. The measured range is indicated by a three-digit display.

Type SC-83-II hand-held laser range-finder
The hand-held laser range-finder model SC-83-II provides target observation and ranging and is intended primarily for use by artillery forward observation posts and reconnaissance groups. It has a ranging capability of greater than 10 km and an accuracy of better than ±5 m.

Norinco laser range-finder
A laser range-finder of unknown designation has been revealed by Norinco for use in artillery direction and coastal defence. The graticule is of cross-hair type with illumination. A display system can indicate three targets with the distance of one shown digitally while the other two are shown by light spots. The display also indicates the status of the laser and when the target is beyond measurement distance, which is 200 m to 6 km. However, in the coast defence role it can detect a 1,000 tonne ship at a range of 30 km.

Operational status
All these systems are believed to be in production.

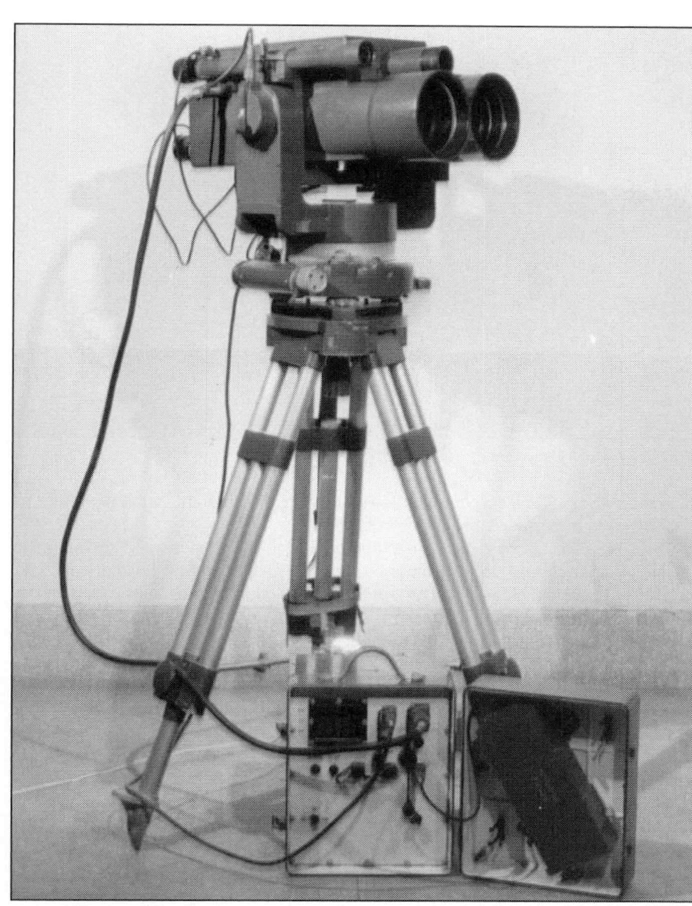

The Norinco laser range-finder

LR1 laser range-finder

Specifications

LR1
Range: 200-7,000 m
Accuracy: ±10 m
Accuracy in azimuth: ±3 mils
Weight (incl battery): 36 kg

Type 82
Range: 300-3,000 m
Accuracy: ±10 m
Power output: 15 mJ
Wavelength: 1.06 μm
Magnification: ×8

Type SC-83-II
Range: <10 km
Accuracy: <±5 m
Weight (complete unit): 2.4 kg

Norinco laser range-finder
Fields of view: 2 × 30°
Magnification: ×25
Pulse repetition rate
Intermittent: 6 pps
Continuous: 4 pps

Contractor

China North Industries Corporation (Norinco).

VERIFIED

Northrop Grumman, AN/PVS-6 Mini Eye-safe Laser IR Observation Set (MELIOS)

Type

Laser range-finder.

Description

The Mini Eye-safe Laser InfraRed Observation Set (MELIOS) is a lightweight hand-held or tripod-mounted laser range-finder. It is capable of determining ranges from 50 to 9,995 m in 5 m increments, displaying the range in the eyepiece. The range is First/Last selectable.

MELIOS can be supplied as a basic laser range-finder or with the Compass/Vertical Angle Measurement Module (C/VAM). This module provides the user with the ability to measure azimuth and vertical angle to a selected target. Target azimuth is displayed from 0 to 359°, vertical angle ±30°.

The C/VAM can operate independently or at the same time as the range-finder. MELIOS can be powered by an internal battery or from a vehicle or other 24 V-power supply. MELIOS can be configured with a GPS device to provide target location data.

MELIOS forms part of the Long Range Advanced Scout Surveillance System (LRAS3) (qv) demonstrator programme for reconnaissance vehicles being developed by the US Army Night Vision and Electronic Sensors Directorate (NVESD). It is also a candidate sensor for a mast-mounted package planned for future variants of the US Army's Stryker Reconnaissance Vehicle.

Operational status

In production.

Specifications

Laser type: Er:glass
Wavelength: 1.54 μm
Pulse energy: 16 mJ
Pulsewidth: 28 ns nominal
Pulse power: >0.5 MW
Beam divergence: <1.0 mrad
Telescope: 7 power
Field of view: 7°
Dioptre adjustment: +4 to −4 dioptres
Range: 50-9,995 m
Range accuracy: ±5 m
Fire rate: 1/6 s
Range modes: selectable first/last
Weight: 1.85 kg
Dimensions: 229 × 203 × 67 mm

Contractor

Northrop Grumman, Electronic Systems.

UPDATED

The AN/PVS-6 MELIOS mini eye-safe laser infra-red observation set

Northrop Grumman Sight Integrated Range-finder – Eye-safe (SIRE V)

Type

Laser range-finder (land).

Description

SIRE V (Sight Integrated Range-finder – Eye-safe) has been designed to fit numerous unity body type armoured vehicle periscopes, by replacing the daylight elbow. It can be retrofitted into existing M36 sighting systems in the USMC LAV-25 and AAVP7 (UGWS) turrets. This allows use with the existing image-intensified night sight. It is also compatible with thermal imaging base sights such as the Delco DIM36 and Kollsman DNRS.

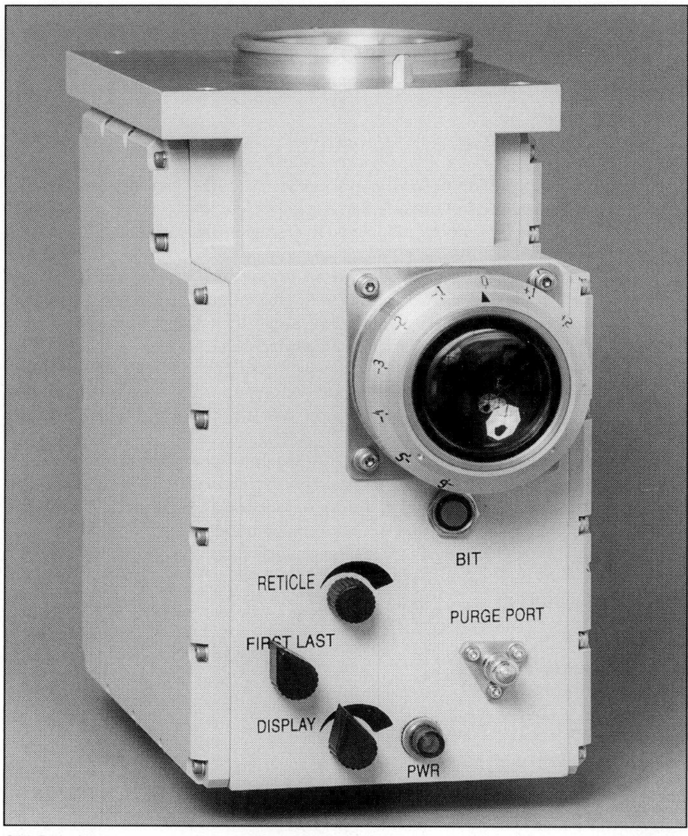

SIRE V sight integrated eye-safe laser range-finder 0098600

The KTP-OPO eyesafe laser range-finder consists of a basic unit that functions as a ×8 magnification gunsight, laser transceiver, and CCD camera for remote viewing. It contains optional controls for boresighting, eyepiece focus, brightness of reticle and display, first or last return logic and laser firing. This unit attaches to the tank periscope by four mounting bolts.

A self-contained power unit converts unfiltered vehicle power to the required level for laser operation. A built-in remote control provides range and status information to the fire-control system and remote monitor.

Operational status
Available.

Specifications
Magnification: ×8
Aperture: 48 mm
Horizontal field of view: 7.5°
Vertical field of view: 5.5°
Eye relief: 25 mm
Dioptre adjustment: +2 to −5 dioptres
Laser type: KTP-OPO
Wavelength: 1.58 μm
Beam divergence: 80% energy within 900 mrad
Receiver field of view: 1.5 mrad
Min time between rangings: 1 s
Sustained rate of ranging: 10 min 1 Hz
Laser output energy: 6 mJ
Laser pulsewidth: 10 ns
Max range: 9,995 m
Min range: 200 m
Range error: ±5 m
Range discrimination: 20 m

Contractor
Northrop Grumman, Electronic Systems, Laser Systems.

Raytheon AN/PAQ-1 laser target designator

Type
Land laser range-finder.

Description
The AN/PAQ-1 Laser Target Designator (LTD) was developed as a hand-held instrument for use by ground troops. In addition to its use for target designation to supporting ground attack aircraft and helicopters, the LTD can be used by troops who may have been cut off by opposing ground forces as a means of making their position known to rescue or supply aircraft.

The AN/PAQ-1 laser target designator

The equipment is also compatible with the AN/TVS-5 night vision sight and so may be used in darkness as well as in daylight.

The LTD consists of three easily replaceable modules designed to withstand rough field handling. The transmitter module is a sealed unit which is dry nitrogen purged to provide a clean environment for the laser and optical elements. It can be easily detached without disturbing the optical alignment or the seal. The electronics module houses the power supply and control circuitry, code-set panel, external cooling air blower and ducting, heat exchanger and trigger assembly. A quickly detachable battery module completes the basic unit. It is completely sealed and shaped to function as part of the rifle stock, forming the butt-end.

When the LTD is operated from an auxiliary power source, the power is supplied through the battery interface connectors, terminated in a module identical to the battery module so that the overall LTD configuration is maintained.

Operational status
Production of the LTD is complete with 177 sets delivered to the US armed forces. The system entered service in 1981.

Contractor
Raytheon Systems Company.

VERIFIED

Raytheon AN/PAQ-3 MULE

Type
Land laser range-finder.

Description
The AN/PAQ-3 MULE (Modular Universal Laser Equipment) integrates the AN/PAQ-1 laser target designator and AN/GVS-5 hand-held laser range-finder, plus tripod design concepts from the AN/TVQ-2 ground laser locator designator. Thus a large percentage of MULE components are interchangeable, directly, with items from the other programmes. The laser transmitter and electronics from the AN/PAQ-1 are used, with a change in output optics to meet different beam characteristic needs. Range-finder components carried over from the AN/GVS-5 include detector, video amplifier, range counter/display and low-voltage power supply.

MULE features built-in test circuits that monitor its own operation automatically. They warn the operator when the battery pack voltage is getting low, when the laser output falls below the laser threshold and certify other functions. The system consists of laser designator/range-finder and stabilised tracking tripod modules. MULE can be carried by two-person forward observer teams over rough terrain into battle areas and set up for use in less than 5 minutes. For instant target ranging, an observer trains MULE on a target manually, using its telescopic sight for precise aim. The laser beam is then fired and MULE measures the time it takes the beam to reach the target and return. The system's computer multiplies half the elapsed time by the speed of light and displays the range immediately on a digital readout. The tripod automatically determines and displays azimuth and elevation co-ordinates. This information, along with range, can be sent quickly in digital form to a fire-control centre via a digital communication terminal, or relayed to an artillery battery by a voice channel. To illuminate targets for laser-guided weapons, the observer trains his sights on the target and triggers the laser beam. Laser-homing missiles and projectiles (and aircraft with laser spot trackers) follow the reflected beam to the target.

Operational status
AN/PAQ-3 was produced for the US Marine Corps between 1983 and 1988 with some 400 delivered.

Specifications
Weight: 17.23 kg

Contractor
Raytheon Systems Company.

VERIFIED

Raytheon Avenger laser range-finder

Type
Land laser range-finder.

Description
Raytheon Systems Company has produced a CO_2 laser range-finder that provides eye-safe range and range rate data for the Boeing Avenger Pedestal-Mounted Stinger (PMS) air defence weapon system (qv). It is stated to be the first production CO_2 laser range-finder to enter the US Army inventory. The device is a stand-alone unit that can be used on multiple platforms.

Operational status
Over 750 units have been delivered under a multiyear production contract which began in 1992. In service with the US Army and US Marines on the Avenger PMS.

Specifications
Laser type: CO_2
Wavelength: 10.59 μm
Operating range: 500-9,990 m
Accuracy: ±10 m

Contractor
Raytheon Systems Company.

UPDATED

Raytheon Avenger laser range-finder

Raytheon ELITE II

Type
Land laser range-finder.

Description
The Raytheon Systems Company Eye-safe Laser with Integrated Telescope Equipment (ELITE) is designed for use on light armoured vehicles and is part of Raytheon's family of Class I eye- safe laser range-finders for military applications. The ELITE II laser contains all system elements within a single Line Replaceable Unit. The range-finders use a Raytheon-patented intracavity Raman resonator to shift the wavelength of a solid-state laser from its eye-damaging 1.06 μm wavelength to a 1.54 μm eye-safe wavelength. They use the same modular components for the laser, receiver, power supply and other electronics. The ELITE configuration includes direct view optics and built-in range display. The unit is designed to accommodate a variety of visual optical characteristics, data interface formats and is software reprogrammable for maximum user flexiblity. Other features include: laser threat protection; selectable neutral density for bright scenes; user achromatic boresight control.

In addition to the ELITE, the Raytheon Class 1 family includes: the M1 version which can directly replace the laser produced by Raytheon for the M1 Abrams, the South Korean K1 tank and the Armoured Gun System; and the Bradley Fighting Vehicle version, which Raytheon is providing for the Operation Desert Storm upgrade.

Operational status
In service in Kuwait on the Desert Warrior and on the Canadian Recce LAV-25 vehicles and has been demonstrated on other light armoured vehicles.

Specifications
Eye safety: Class 1
Range: 200 m-10 km, 50 m-20 km available
Firing rate: 1 shot/s
Resolution
All ranges: 10 m
Target discrimination: 20 m
Range logic: first/last
False return rate: 1%
BIT: fault isolatable to SRUs
Data output
DVO: 4 range digits; 5 weapon indicators; 1 user-defined indicator
TV: CCD camera option available
Remote display A: identical to DVO display
Remote display B: user configurable
Serial interface: RS-485
Telescope
Magnification: ×8 (×7 to ×10 available)
Reticle: adjustable illumination
Reticle pattern: standard NATO or user-definable system FOV (w/std eyepiece)

Contractor
Raytheon Systems Company.

UPDATED

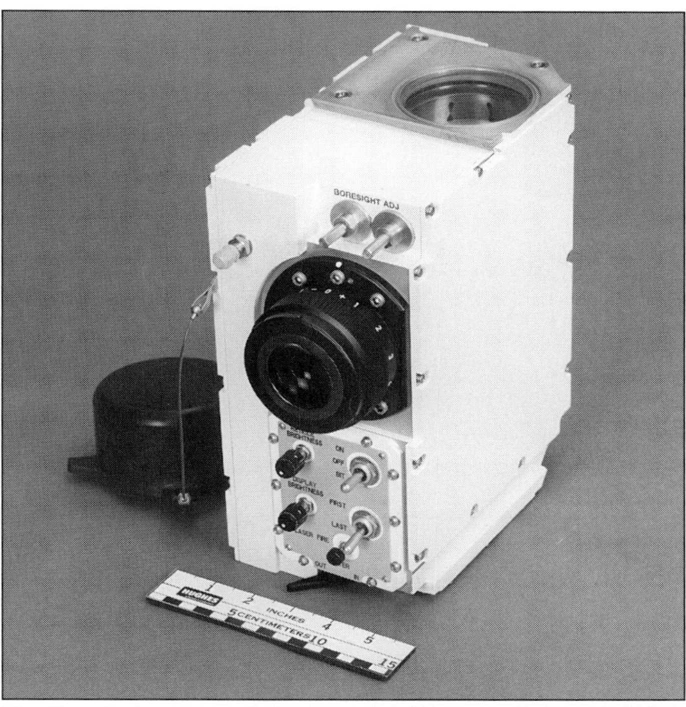

The Eye-safe Laser range-finder Integrated Telescope Equipment (ELITE), produced by Raytheon 0023151

Raytheon M1 MBT laser range-finder

Type
Land laser range-finder.

Description
The Raytheon M1 Abrams Nd:YAG laser range-finder programme started in 1978, with first deliveries being made in late 1979. The range-finder consists of a transmitter, receiver, power supply, timing and logic circuit which is integrated as a whole into the tank's fire-control system.

For the M1A2 MBT, Raytheon has upgraded the unit with a one laser pulse/s capability, which will increase the accuracy of the fire-control system when used against rapidly moving targets.

In operation the tank crew member aims at a target and triggers the laser. The beam travels to the target and is reflected back towards a receiving telescope. The elapsed time of beam travel to and from the target provides the accurate range information for the fire-control computer. This is then processed with other parameters to give the correct azimuth and elevation figures to engage the target with the main gun.

The system is capable of allowing the fire-control system to differentiate between close-up and far-off targets by allowing the gunner to select the first laser reflection signal (close target) or the last (far target).

Operational status
In production (over 11,000 delivered to date). In service with Egypt (M1A1), Kuwait (M1A2), Saudi Arabia (M1A2) and United States (M1/M1A1/M1A2), Korean K1 tank.

Specifications
Field of view: 5.0° × 2.58° narrow, 16.67° × 8.6° wide
f number: f/2.1 (4.85 in)
Wavelength: 7.5 to 12 μm
Detector: 120 × 1 (HgCdTe)
Detector element size: 40 × 58 μm
Cryocooler power requirement: 1 W
Interlace (EL): 2:1
Number of IR lines: 240
Noise equivalent temperature: ≤0.12°C
Uniformity of response: ±10%
Video format: TIS unique
Video zoom: Telescopic
Gain/level control: Manual
Power requirement: <240 W, typical
EMI: MIL-STD-461
Reliability (MTBF): 1,500 h

Contractor
Raytheon Systems Company.

VERIFIED

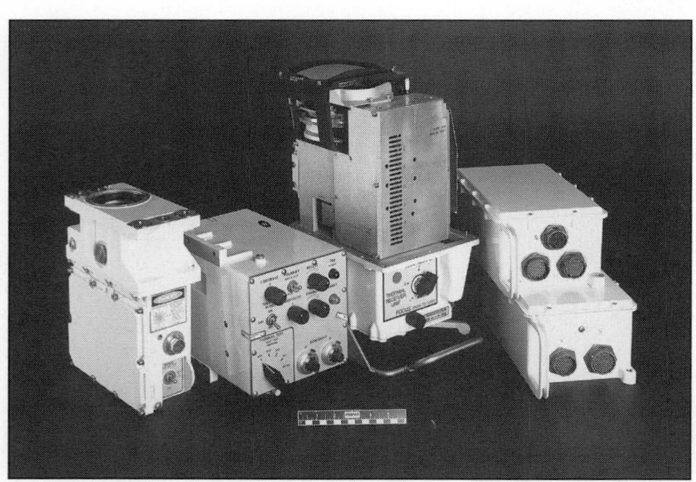

Raytheon M1BT laser range-finder (far left) and thermal imaging system

Raytheon TTS laser range-finder

Type
Land laser range-finder.

Description
Raytheon Systems Company's TTS laser range-finder has been designed to be fitted on the AN/VSG- 2 Tank Thermal Sight used in the M60A3 MBT upgrade programme. The device has a first and last target switch with the range being shown in metres on an LED display panel. Target separation capability is 20 m.

Operational status
Production as required. In service with the US Army and other countries.

Specifications
Laser type: Nd:YAG
Wavelength: 1.064 μm
Operating range: 300-7,990 m
Accuracy: ±5 m

Contractor
Raytheon Systems Company.

VERIFIED

RH-ALAN LAM-1/LAM-2

Type
Land laser charge activator.

Development
The development of the LAM series of laser charge activators was completed in 1985 with serial production starting in 1986. A number of systems have been exported to Third World countries. They are now produced for the Croatian Army.

Description
LAM-1 and LAM-2 are used to activate the electrical fuzes in explosive charges. They can be inserted into any explosive or be a part of a mine charge. Fuzes can be activated at ranges up to 1.5 km with LAM-1 and up to 5 km with LAM-2. These laser devices can be used to facilitate the destruction of stationary objects or moving targets or persons entering the preset minefield. LAM-1 is primarily used for covert or quick operations requiring fast explosive laying and activation of minefields. LAM-2 is used in tactical situations requiring the blockade of possible attack approach directions and activation of minefields.

Operational status
In production and in service with the Croatian Army.

Specifications
Charge activation range
LAM-1: up to 1.5 km
LAM-2: up to 5 km
Laser beam width (at 1,000 m): 4 m
Laser transmitter battery life: up to 100 firings
Receiver battery life: 20 h
48 Ah car battery life: 40 days
Receiving angle
LAM-1: 70°
LAM-2: 40°
Number of codes: 4
Transmitter weight: 1.4 kg
Weight of a set of 4 receivers: 8.1 kg

Contractor
RH-ALAN.

VERIFIED

LAM-2 laser charge activator

Saab Bofors Dynamics anti-aircraft laser range-finder

Type
Laser range-finder (land and naval).

Description
Saab's high repetition rate anti-aircraft laser range-finder forms part of a number of army and navy fire-control systems: Saab Bofors 9LV200 Mk 3; Thales LIOD; CS Matra Najir; Saab Bofors EOS-400; BAE Systems Sea Archer systems; and the Super-Fledermaus land system.

The system consists of a transceiver unit to be mounted on a director and a power module with range counter to be mounted in a fire-control system console. For alignment purposes, the transceiver unit is equipped with a built-in telescope, boresighted to the transceiver axis. The laser transmitter is completely sealed and is cooled by forced air which circulates through the transmitter and along the inner walls of the housing.

The system is operated by signals from the fire-control system and ranging can start 2 seconds from 'power on'. The trigger signal is a pulse train with a frequency up to 15 Hz. The target range value is given in 12 bit form. For target selection a range gate is used in combination with an adjustable minimum range. An alarm signal is generated if there is more than one target within the range gate.

Operational status
In production and in service.

Specifications
Extinction value: 53 dB
Coverage: 280-20,475 m
Accuracy: 4 m
Adjustable min range: 280-20,440 m

Range gate: length 500 m (other values optional) or at the end of the range coverage
Power consumption: 300-400 VA
Dimensions
Transceiver unit: 565 (length) × 210 (diameter) mm
Power module (example): 250 × 300 × 200 mm
Weight
Transceiver unit: 13 kg
Power module with range counter: 6-15 kg
Transmitter
Laser: Nd:YAG
Wavelength: 1.06 μm
Beamwidth: 1.2-1.5 mrad
Repetition rate: 10 pulses/s (continuous), 15 pulses/s (reduced duty cycle)
Receiver
Detector: silicon avalanche diode
Area of optics: 80 cm^2
Field of view: 1.5 mrad

Contractor
Saab Bofors Dynamics.

The Saab Bofors anti-aircraft laser range-finder

Saab Bofors Dynamics eye-safe laser range-finder

Type
Land and naval laser range-finder.

Description
Saab Bofors Dynamics has introduced a high repetition rate, eye-safe laser range-finder for integration into army and navy fire-control systems.

The laser is a Raman-shifted Nd:YAG laser emitting an eye-safe wavelength of 1.54 μm and provides accurate ranging of air and ground targets up to the instrumented maximum range of over 20 km. The system is available in either two separate units, laser transceiver unit and laser power unit, or in a single unit configuration. The transceiver unit is designed for mounting on a director and can be aligned with other sensors such as tracking radar, thermal imager or TV camera. The power unit can be mounted separately and is connected to the transceiver with a cable and if necessary over sliprings.

The laser, which is available in an army and navy version, is prepared for various primary powers and signal interfaces.

Operational status
In production and in service with several customers.

Specifications
Laser type: Raman Nd:YAG
Wavelength: 1.54 μm
Output energy: 30 mJ approx
Beamwidth (90% energy): 1 mrad
Pulse repetition rate: up to 25 Hz in bursts, 12.5 Hz nominal
Receiver
Detector: InGaAs avalanche photodiode
Area of receiver optics: 88 cm^2
Field of view: 1.5 mrad
Instrumented range: 280-20,475 m
Range resolution: 5 m
Range logic: 1st and 2nd echo registered
Range gate: 500 m basic gate, others selectable from computer
Power supply: 115-220 V AC 1,2 or 3 phase 50-400 Hz; option 28 V DC
Power consumption: C 300 W at 12.5 Hz
Data interface: serial interface RS-422 or other optional standards
Weight
Laser transceiver: 11.5 kg
Laser power unit: 5.5 kg
Dimensions
Army transceiver: 215 × 160 × 480 mm
Navy transceiver: 215 × 240 × 480 mm
Power supply: 140 × 180 × 240 mm

Contractor
Saab Bofors Dynamics.

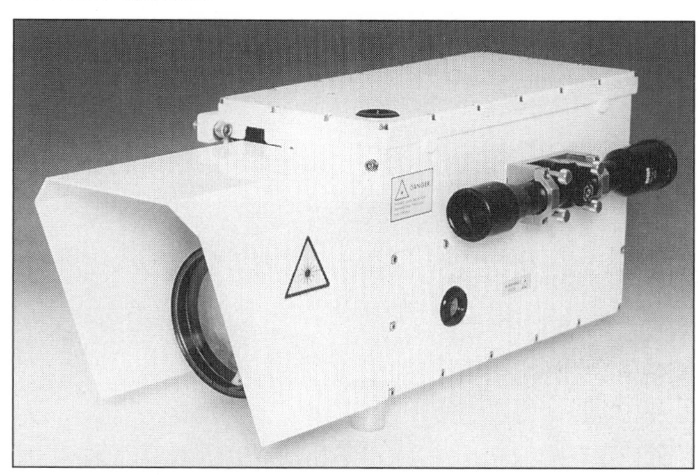

The Saab Bofors Dynamics eye-safe laser range-finder

The naval version of Saab Bofors Dynamics eye-safe laser range-finder 0004810

Saab Bofors Dynamics light anti-aircraft laser range-finder

Type
Laser range-finder.

Description
The Saab Bofors Dynamics light anti-aircraft laser range-finder is designed to measure the range of small- and high-speed targets. It forms a part of electro-optical anti-aircraft systems or provides ranging support in radar-based fire-control systems. It is used in a number of land and naval weapon systems including the Skyguard air defence system, the 9LV453 naval fire-control system, the Marksman vehicle, the LIOD II naval fire control, Seaguard naval fire control, Sting naval fire control.

The laser range-finder consists of two units, the transceiver unit and the power unit. Both units are air cooled. A number of options are available to meet different requirements regarding installation, interface and environmental conditions. Optical mounting assemblies are designed to make installation and boresighting flexible.

Operational status
In production and in service with the army and navy forces of several countries.

Specifications
Laser medium: Nd:YAG
Wavelength: 1.06 μm
Output energy: 80 mJ nominal
Laser pulse length: 20 ns

Pulse repetition rate: up to 25 Hz
Laser beamwidth: 1.5 mrad or 3 mrad
Laser output beam diameter: 30 mm
Receiver detector: silicon avalanche diode
Area of receiver optics: 88 cm^2
Field of view: 3 mrad
Extinction value: 53 dB
Instrumented range: 200-20,475 m
Range resolution: 2.5 m
Input power: 115-380 V AC, 1-3 phases, 50-400 Hz, 400 VA (Option: 28 V DC)
Dimensions
Laser transceiver unit: 150 × 220 × 360 mm
Laser power unit: 150 × 190 × 250 mm
Weight
Laser transceiver: 10 kg
Laser power unit: 5.5 kg
Interfaces: serial interface HSSL, RS-232, RS-422

Contractor
Saab Bofors Dynamics AB.

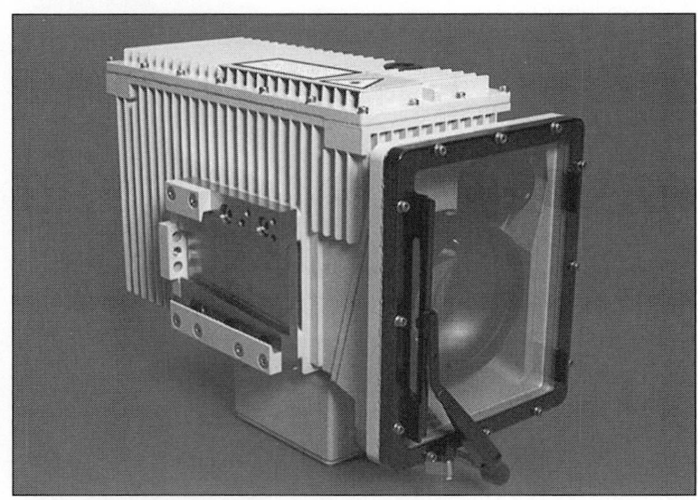

The Saab Bofors Dynamics light anti-aircraft laser range-finder

Saab Bofors Dynamics medium performance eye-safe anti-aircraft laser range-finder

Type
Laser range-finder (land and naval).

Description
Saab Bofors Dynamics has developed a compact, low-weight, medium-performance, eye-safe laser range-finder for use in compact army and naval fire-control systems.

The laser is an OPO-shifted Nd:YAG laser emitting an eye-safe wavelength of 1.57 μm with a ranging capability of around 10 km against air targets.

The single unit laser is designed for mounting on open directors or for integration into optical systems.

Saab Bofors Dynamics medium performance eye-safe anti-aircraft laser range-finder 0023152

Operational status
In evaluation and qualification.

Specifications
Laser type: OPO Nd:YAG
Wavelength: 1.57 μm
Output energy: 10 mJ
Beamwidth: 0.6-2.0 mrad
Pulse rate: <10 Hz
Instrumented range: 280-20,475 m
Processing: echo selection and range gate
Power supply: 28 V DC
Data interface: RS-422
Weight: 4 kg
Dimensions: 115 × 140 × 220 mm

Contractor
Saab Bofors Dynamics AB.

Simrad LA7

Type
Land laser range-finder.

Description
The LA7 is designed to assist fire control of a wide variety of weapons. The device is able to measure ranges up to 10 km with range data transmitted to the weapon via an RS-422 data interface. The system can be used with the Simrad KN200 image intensifier to provide night vision capability.

Operational status
In production and in service with several countries.

Specifications
Dimensions: 222 × 206 × 94 mm
Weight: approx 3.2 kg
Sighting telescope
Magnification: ×7
Field of view: 7°
Eyepiece setting: -0.75 dioptres fixed
Reticle illumination: Beta-light
Laser range-finder
Wavelength: 1.064 μm
Pulse frequency
Continuous operation: 1.5 Hz
Intermittent operation: 4 Hz
Range display limits: 150-9,995 m
Resolution: 5 m
Targets registered: 3
Range gate limits: 150-4,000 m
Data interface: full duplex RS-422

Contractor
Simrad Optronics ASA.

VERIFIED

The Simrad LA7 laser range-finder

Simrad LE7

Type
Land laser range-finder.

Description
The LE7 is an erbium:glass 1.54 μm eye-safe lightweight laser range-finder which may be hand-held (it is the same size as 7 × 50 binoculars) or mounted on a lightweight tripod support for artillery observation. The tripod has an angulation head facilitating either analogue or digital readout of azimuth and elevation.

The LE7 has adjustable minimum range, multiple target indication, digital setting of minimum range and optional remote firing. For ranging at night it may be combined with a Simrad image intensifier.

Operational status
In production.

Specifications
Dimensions: 215 × 202 × 93 mm
Weight: 2.5 kg
Laser type: Er:glass
Wavelength: 1.54 μm
Output energy (Class 1): <8 mJ
Q switch type: rotating prism
Cooling: no forced cooling
Receiver
Aperture: 45 mm
Field of view: 7°
Detector: InGaAs pin photodiode
Telescope
Field of view: 7°
Magnification: ×7
Dioptre setting: fixed, 0.75 dioptre (optional adjustable −4 to +3 dioptres)
Range counter
Min range: adjustable 100 m to 5,000 m
Max range: 9,995 m
No of targets registered: 5 (1st, 2nd, 3rd, last and last gated)
Resolution: 5 m

Contractor
Simrad Optronics ASA.

VERIFIED

The Simrad LE7 eye-safe laser range-finder

Simrad LP7

Type
Land laser range-finder.

Description
The LP7 is intended for use by infantry units in the close support role and is the size of a standard pair of 7 × 50 binoculars. It may be either hand-held or mounted on a support, and also used at night when combined with a night observation device.

Range is determined by laying the telescope graticule on to the target and pressing the fire button on top of the unit. The range is immediately displayed in the eyepiece and there is an indication if reflections from more than one target have been detected. Unwanted reflections may be gated out by a minimum range control. Indications are also given if one or more targets have been gated out. Some 600 measurements may be made on one charge of the 12 V rechargeable battery.

The transmitter is a miniaturised Q-switched Nd:YAG laser. The sighting telescope, which has a performance comparable with that of a standard observation monocular, is combined with the optical receiver by a beam-splitting technique.

The receiver uses a silicon avalanche photodiode giving a range capability of up to 10 km with a resolution of 5 m. The four-digit LED display is observed through the left eyepiece and is superimposed on the picture seen in the right eyepiece. Display intensity may be adjusted by rotating the eyepiece housing. After 3 seconds the display is automatically shut off to preserve battery power. Indication is given if the battery power is too low.

Options are digital setting of minimum range, built-in test and data output from the range counter. A rugged lightweight tripod with angulation head is available which provides azimuth and elevation readouts with a resolution of 5 mrad.

Operational status
In production both in Norway and in the United Kingdom. Licence-production of LP7 was formerly conducted by Lasergage Ltd but is now carried out by a Simrad subsidiary, Simrad Optronics Ltd of Crawley, UK.

Specifications
Dimensions: 215 × 202 × 9.3 mm
Weight (incl battery): 2.2 kg
Operating temperature: -30 to +55°C
Transmitter
Wavelength: 1.064 μm
Radiant energy: 8 mJ; 15 mJ (max)
Pulse length: 8 ns
Pulse energy: 5 mJ
Pulse frequency
Continuous: 1 every 5 s
Intermittent: 1 every 2 s
Exit pupil diameter: 18 mm
Cooling: natural convection
Beamwidth: 2 mrad
Receiver
Field of view: 1.3 mrad
Aperture: 45 mm
Time variable gain
Nominal gain change: 30 dB
Range: 0-1 km
Telescope
Field of view: 7°
Magnification: ×7
Aperture: 45 mm

Contractor
Simrad Optronics ASA.

VERIFIED

Simrad LV350 series laser range-finders

Type
Land laser range-finder.

Description
The LV350 series of laser range-finders has been designed for integration with a wide range of optical vehicle sights. The series consists of the following available models: LV352; LV353; LV354 and LV355, all of which are based on modular design parameters.

In order to simplify the optical arrangements of the sights, the transmitter and receiver channels of the LV352/353 and 354 are coaxial. The LV355 has separate transmitter and receiver channels.

The LV350 series can measure target ranges between 150 and 10,000 m with a resolution of 5 m. As an option the LV353/354/355 models can be delivered with a range counter for measuring ranges up to 20,000 m. Actual maximum range will depend upon the host sight construction.

Minimum range gate can be set from 150 to 4,000 m and a facility for testing the minimum range setting is incorporated. First or last target registration can be selected by the operator.

The LV350 series can provide its output data in serial binary form either to an external microprocessor (LV352 model) or, via opto-couplers, to a fire-control system computer.

A complete LV350 series model laser range-finder comprises two separate subsystems:

(a) a transceiver unit which is designed to be used together with an optical sight that uses a Galilean telescope with a typical magnification of ×10. The unit interfaces mechanically with the sight through dovetail or screw mounts and optically interfaces through a small window in the sight. This arrangement permits a simple installation and removal operation with no realignment required after replacement of the laser unit
(b) an electronics unit.

Operational status
Production as required. In service with several countries.

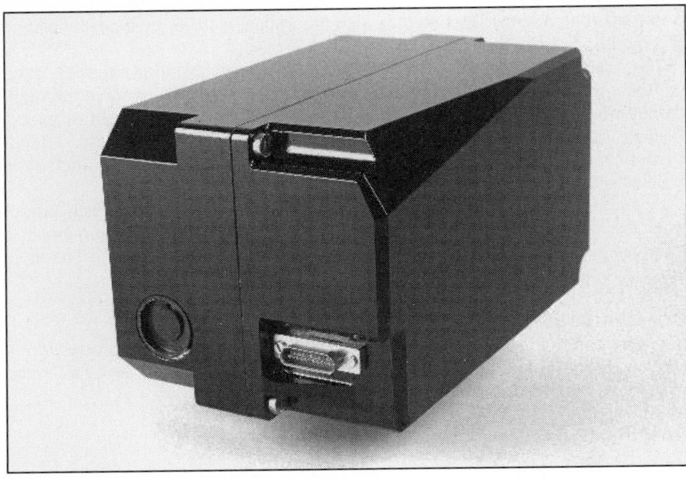

Simrad LV352 transceiver unit

Specifications

Weight
Transceiver unit: approx 2 kg
Dimensions
Transceiver unit: 150 × 92 × 90 mm
Electronic unit: 114 × 90 × 23 mm
Wavelength: 1.064 μm
Field of view
Receiver: 8 mrad
Operating range: 150-9,995 m
Resolution: 5 m
Range gate
In first target mode: 150-4,000 m
Number of targets registered: 1
Return pulse logic: registration of first or last target, selectable by operator
Power supply: 18-32 V DC (filtered input)

Contractor

Simrad Optronics ASA.

VERIFIED

Simrad LV400 series laser range-finders

Type

Land laser range-finder.

Description

The LV400 series of Nd:YAG laser range-finders has been designed for integration with a wide range of optical sighting systems.

Based on a modular concept with no forced cooling requirements, these systems can have the main module separately mounted within the sight assembly or built together as one integrated unit. If the control panel or range display is not an integral part of the sight, then a tailor-made display and/or control unit will be supplied.

The optical system transmitter/receiver channels can be coaxial or separate and the receiver field of view, variable range gate and range window requirements can be set to customer specification.

Up to six targets can be registered by the range counter with up to three target ranges displayed on the optional range display. A bidirectional opto-coupled serial datalink interface with RS-422 data format is fitted to the range counter for integrating with a fire-control system ballistic computer.

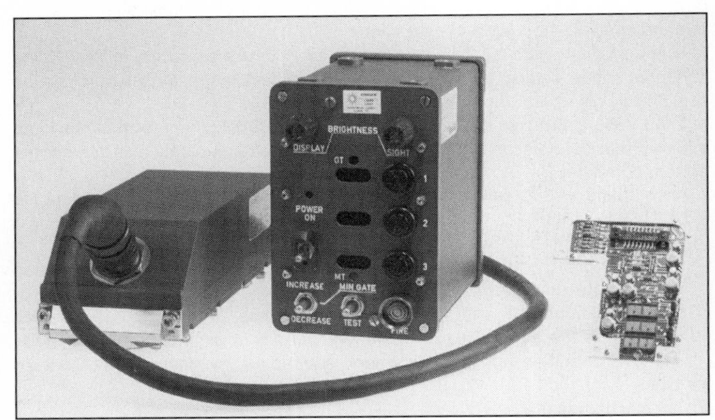

A typical configuration is the Simrad LV400 consisting of transceiver unit, transceiver cable, remote range display unit and internal range display (to be mounted inside the sight)

Operational status

Production as required. In service with several countries.

Specifications

Weight
Separate transceiver unit: 2.7 kg
Control and display unit: 2.7 kg
Dimensions
Separate transceiver unit: 242 × 112 × 70 mm
Control and display unit: 160 × 112 × 170 mm
Laser type: Nd:YAG
Operating range: 200-9,995 m (optional 200-19,995 m)
Resolution: 5 m
Power supply: 24 ± 6 V DC

Contractor

Simrad Optronics ASA.

VERIFIED

Simrad LV510

Type

Land laser range-finder.

Development

The Simrad LV510 laser range-finder is specially designed for the commander sight of the British Army's Challenger 2 main battle tank. Series production and deliveries for the Vickers Defence Challenger 2 programme started in 1993.

Description

The Simrad LV510 is supplied as a sealed and desiccated single-unit transceiver containing all the necessary optics and electronics. Operation of the LV510 is based on electrical interfacing to a fire-control computer through an RS-422 serial communication link and a turret distribution box.

Operational status

Production complete.

Specifications

Lasing medium: Nd:YAG
Wavelength: 1.064 μm
Output energy: min 5 mJ
Beam divergence: 0.7 mrad (90%) max
Repetition rate: 1 pulse every 250 ms ±25 ms, 50% duty cycle 5 s on/5 s off max
Laser aiming mark: reticle illuminated via an ON/OFF guarded switch
Max range: 9,995 m
Min range: 200 m
Range resolution: 5 m
Target selection: first or last return
Range discrimination: 30 m
Range accuracy: ±5 m
Power requirement: 28 V DC, 5 A peak current
Weight: 5.7 kg
Dimensions: 350 × 163 × 105 mm

Contractor

Simrad Optronics ASA.

UPDATED

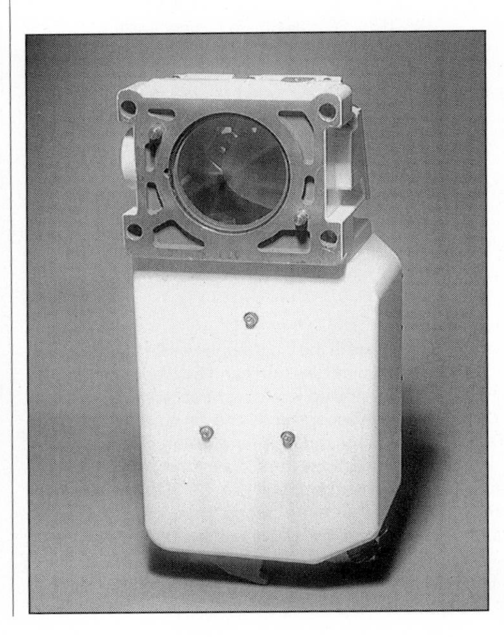

The Simrad LV510 laser range-finder

Thales Optronics HELMET

Type
Ground attack laser range-finder.

Description
HELMET (HELicopter Mounted Eye-safe Transceiver) is a military qualified, Class 1 eye-safe laser range-finder for optronic payloads mounted on helicopters and UAVs.

Operational status
In production and in service with an export customer.

Specifications
Wavelength: 1.535 μm
Laser class: Class 1
Pulse repetition rate: 0.2 Hz continuous operation 1 Hz Burst without forced cooling, 1 Hz continuous with forced cooling
Beam divergence: 0.56 mils (nominal)
Beam pointing stability: <0.3 mils
Processing performance: 20 K
Range accuracy: <±5 m
Weight: 1.8 kg
Volume: <1.6 litre
Operating temperature: –30 to +65°C
Data interface: LRF to external equipment – RS-485 bidirectional ports

Contractor
Thales Optronics (formerly Avimo Ltd).

UPDATED

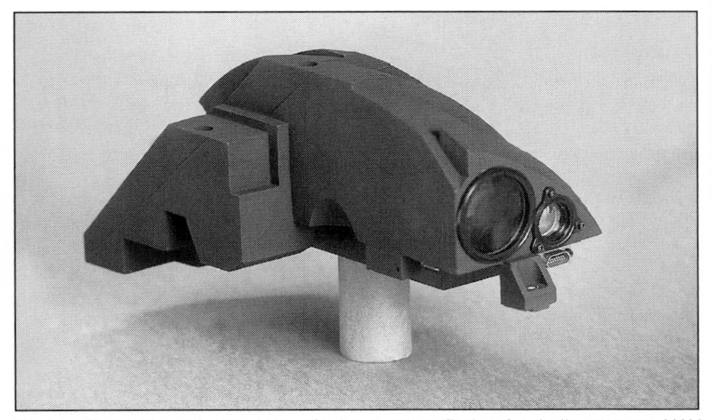

Thales Optronics airborne eye-safe laser range-finder for helicopter or UAV mounting 0053778

Thales Optronics HL58 laser range-finder

Type
Laser transceiver (land).

Description
The HL58 laser transceiver module was originally designed for the French Army's Leclerc main battle tank programme, but can be incorporated into many other systems.

Operational status
Over 1,000 units produced. In current production.

Specifications
Dimensions: 186 × 119 × 78 mm
Weight: 1.9 kg
Power supply: 18-32.5 V
Ranging performance (normal performance for temperatures –30 to +55°C, extinction ratio at 500 m)
  High gain: –43 dB min
  AGC: –29 dB
Transmitter
Laser type: Q switched Nd:YAG
Wavelength: 1.064 μm
Output energy: 10-25 mJ
Pulsewidth: 5-20 ns
Output power: 1 mW min
Beam divergence: 1.5 mrad at 50% energy
Beam size at exit aperture: 4.0 mm diameter (typical)
Pulse repetition frequency: 1 Hz max, 12 ppm average
Receiver
Field of view: ±7.5 mils to optical axis

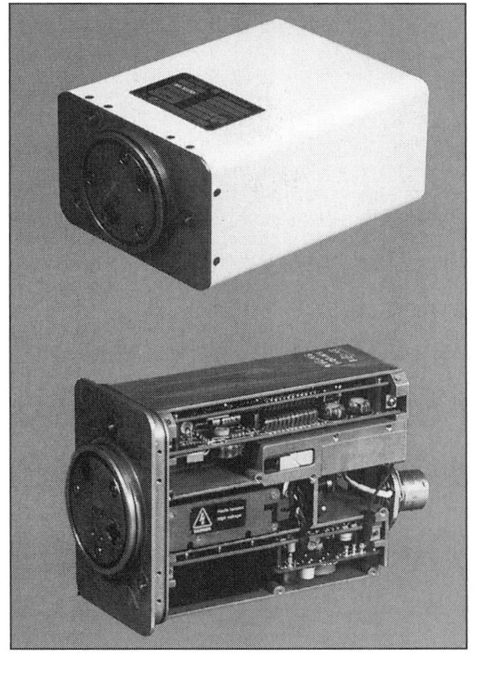

The HL58 laser transceiver chosen for the Leclerc main battle tank

Receiver sensitivity: 56 nW max
Receiver aperture: 6 mm diameter min
Min/max range: 300 to 9,995 m
Range resolution: ±5 m
Discrimination: 30 m
Number of echoes: 2
False alarm probability: <0.1%

Contractor
Thales Optronics (formerly Avimo Ltd).

UPDATED

Thales Optronics LF28A man-portable laser designator

Type
Land laser range-finder.

Description
Thales Optronics' lightweight, compact, portable laser range-finder designator, LF28A is based upon an all solid-state slab geometry laser. The slab laser does not require cooling, giving a light, quiet system. In addition the slab configuration allows the laser to change between a number of pulse repetition frequencies almost instantaneously without changing either its output or its divergence. It also reduces the necessary 'ON-time' for the laser, thus reducing power consumption.

The system is capable of designating to 10 km, using any user-defined codes at either 10 pps or 20 pps. It has an integral ×10 telescope, through which the laser output, laser receiver and sighting channels are multiplexed and includes interface for image intensifier or thermal imaging sight.

Other features of the LF28 include an advanced man/machine interface with touch panel operation, a display that allows feedback from the built-in test analysis and a clip-on battery pack.

Operational status
In operational service.

Specifications
Laser range-finder/designator
Volume: <7.5 litre
Weight: <6.5 kg
Designation range: 10 km
Ranging performance
  Max range: 9,995 m
  Min range: 300 m
  Range accuracy: ±5 m (3σ)
  Range update rate: 20 Hz max
Wavelength: 1.064 μm
Output energy: >80 mJ
Beam divergence: <300 μrad
Sighting system
Objective diameter: 60 mm
Magnification: ×10
Field of view: 3°

Contractor
Thales Optronics.

UPDATED

Thales Optronics LH90 hand-held laser range-finder

Type
Laser range-finder (land).

Description
The LH90 hand-held laser range-finder is intended primarily for daylight use to measure target ranges from any observation post, but can also be used for surveillance or with mounting points for a number of other roles. These include mortar fire-control, night ranging (with a night observation device) and target positioning (with a goniometer). There is a remote range facility with RS-485 interface. Eyepiece focus is variable for sighting and display.

Operational status
Available. Several thousand in current service.

Specifications
Dimensions: 180 × 190 × 85 mm
Weight (incl battery): 2.3 kg
Transmitter
Laser type: dye Q switched Nd:YAG
Wavelength: 1.064 μm
Output energy: 12 mJ nominal
Pulsewidth: <10.4 ns
Beam width: <1.5 mrad (90% energy)
Pulses/min: 12
Receiver
Field of view: 1.6 mrad
Effective aperture: 42 mm
Detector type: silicon avalanche photodiode
Range accuracy: ±5 m
Range discrimination: 30 m
Min/max range: 150/9,995 m (4,000 m max with training filter fitted)
Range gate setting: 150-4,000 m (continuously variable)
Optical
Magnification: ×7
Field of view: 7°
Effective aperture: 42 mm
Eyepiece focus: variable ±5 dioptres
Exit pupil diameter: 6 mm
Eye relief: 23 mm
Interocular distance: 65 mm (fixed)

Contractor
Thales Optronics (formerly Avimo Ltd).

UPDATED

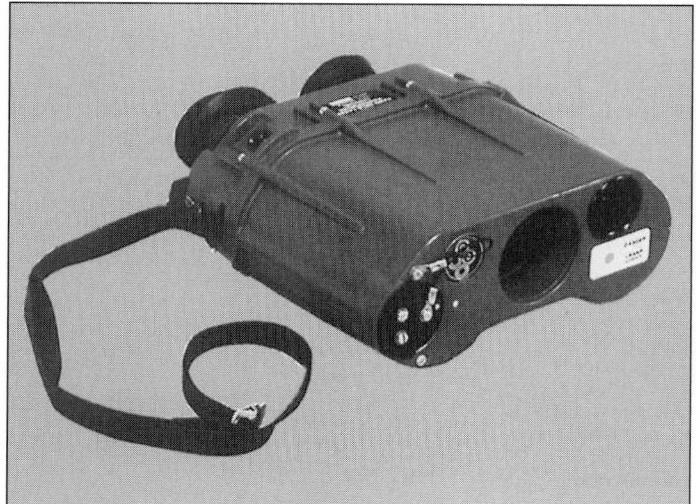

The LH90 hand-held laser range-finder

Thales Optronics MLR 30 and MLR 40 hand-held laser range-finders

Type
Laser range-finder (land).

Description
The MLR 30 and MLR 40 minilasers are suited to both hand-held and tripod-mounted operation. The MLR 30 is a Nd:YAG type and the MLR 40 is an eye-safe version. They are small and lightweight with a ranging performance out to 20 km. The MLR 40 is also available in a 7 km range version. In combination with a goniometer complete triangulation of a target is possible and the data can be externally processed if required.

Applications include:
(a) orientation of forward observers
(b) ranging for direct fire weapons
(c) provision of target information for artillery and mortar positions
(d) target ranging for armoured vehicles and helicopters.

Operational status
Available.

Specifications

	MLR30	MLR40
Wavelength	1,064 μm	1,540 μm
Laser pulse	8 mJ	8 mJ
Measuring rate	every 2 s	every 3 s
Range	80 m to 20 km	
Accuracy	±5 m	
Discrimination	50 m	
Multiple target	display first and last	
Magnification	×6	
Field of view (horizontal)	6°	
Eye relief	±5 m	
Dioptre adjustment	–4 to +2	
Display	5 digits automatic brightness control	
Power supply	8× 1.5 V AA size batteries external 10 V DC to 30 V DC	
Electronic interface	RS-422A serial link	RS-232 and -422 serial link
Weight (incl batteries)	1.6 kg	
Dimensions	170 × 165 × 80 mm	
Operating temperature	–20 to +50°C	
Storage temperature	–40 to +70°C	

Contractor
Thales Optronics (formerly OIP Sensor Systems).

UPDATED

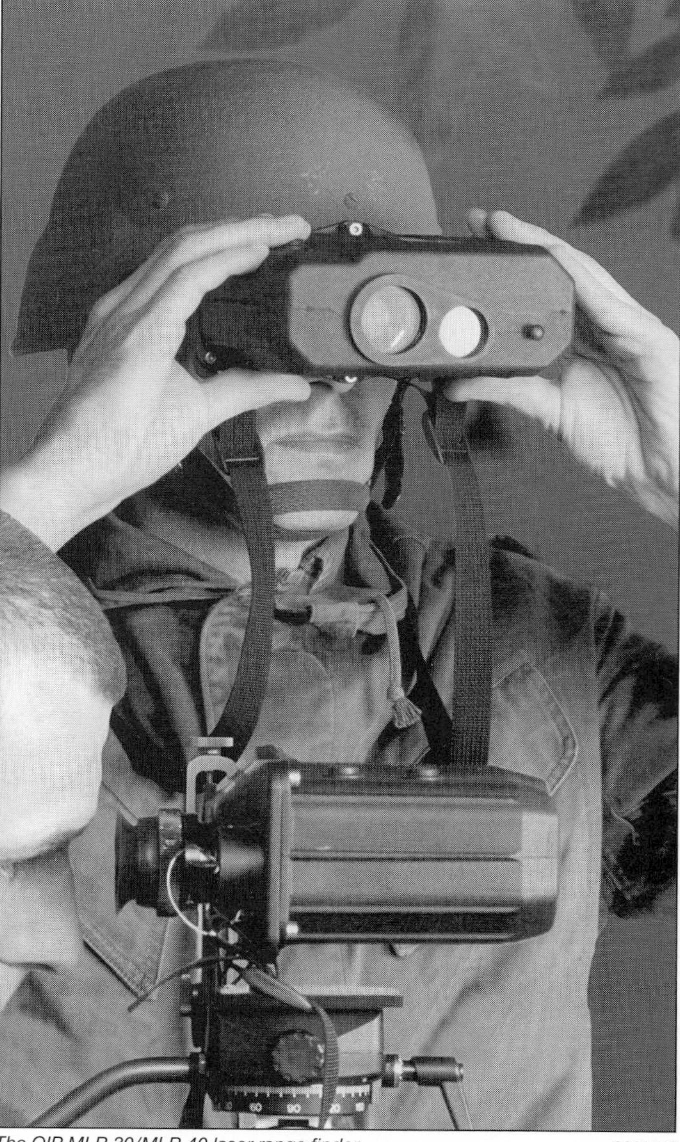

The OIP MLR 30/MLR 40 laser range-finder 0063513

Urals Optical and Mechanical Plant eye-safe laser range-finder

Type
Land laser range-finder.

Description
The UOMZ eyesafe laser range-finder is intended for measuring the distances of a selected object in the range from 100 to 10,000 m to an accuracy of 5 m. The laser range-finder can be used in any system of ground, airborne and maritime equipment in both civil and military applications.
The range-finder functions in the following main operating modes:

- built-in test mode
- single-pulse emission mode with pulse rate up to 1 Hz, continuously for up to 3 h
- pulse-train emission mode with 5 Hz repetition rate, 3 s pulse train duration and 3 min spacing

calibration/adjustment mode including generation of auxiliary pulsed emission of laser beam simulator with a wavelength of 0.82 μm in the direction of the basic laser emission propagation.
The range data as well as the input control commands are transmitted via a serial exchange channel to EIA RS-422 standard.
The range-finder is supplied from a +27 V DC power supply. Current consumption is 5 A max.

Specifications
Wavelength: 1.54 μm
Ranging limits: 100 - 10,000 m
Measurement accuracy: ± 5 m
Maximum emission pulse rate: 5 Hz
Operating temperature range: –40 to +60°C
Weight optronic module: 2.4 kg
HV power source: 0.7 kg
Dimensions
Optronic module: 94 × 275 × 112 mm
HV power source: 100 × 112 × 55 mm

Contractor
Urals Optical and Mechanical Plant (UOMZ).

VERIFIED

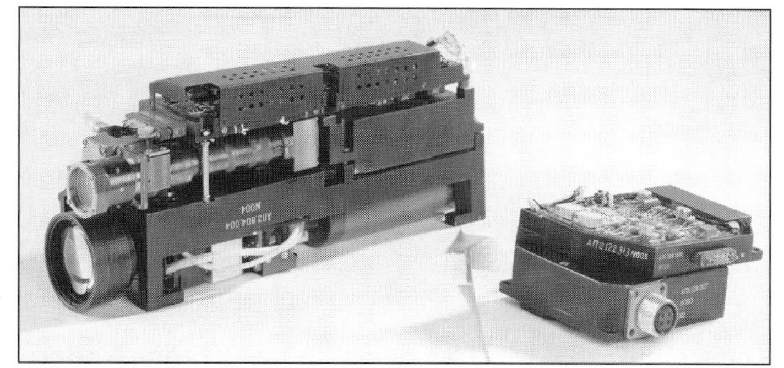
The UOMZ eye-safe laser range-finder 0089878

Vectronix Vector 1500

Type
Land laser range-finder.

Description
The Vector 1500 is a binocular eye-safe laser range-finder which can show the exact location of the target through a high-precision digital compass and inclinometer which is unaffected by tilts of up to ±35°. The Vector 1500 is powered by a 6 V lithium battery and may be mounted on a small tripod as well as being hand held. The laser is Class 1 eye-safe, in accordance with EN 60825 (1991).

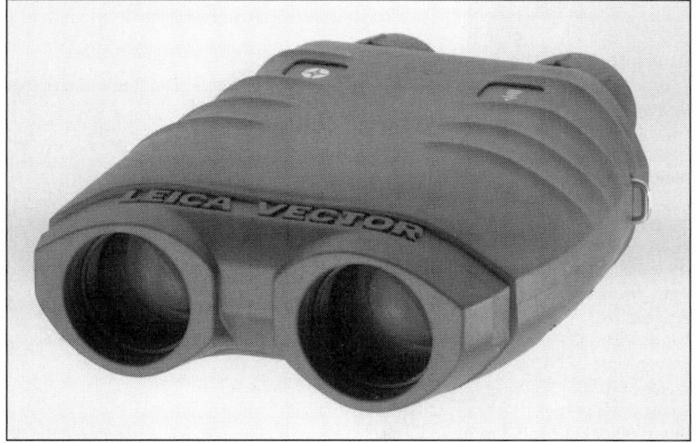

The Vectronix (formerly Leica) Vector 1500 range-finder binoculars

Operational status
In production. In service in the Netherlands, Norway, UK and elsewhere.

Specifications
Dimensions: 205 × 178 × 82 mm
Weight: 1.7 kg
Magnification: ×7
Measurement range: 5-2,500 m
Specified performance: 25 to 1,500 m at a visibility of 10 km, albedo 0.4
Accuracy: ±2 m
Resolution: 1 m
Laser type: infra-red diode
Pulse repetition rate: 12 ppm
Beam divergence: 1.5 × 0.3 mrad

Contractor
Vectronix AG.

UPDATED

Vectronix Vector IV

Type
Land laser range-finder.

Description
The Vectronix (formerly Leica) Vector IV is a compact, hand-held forward observer instrument. It combines the functions of binocular observation, distance measurement, northfinding and inclinometer. The measured target location can be output to a computer, digital message device or similar. The use of a tripod is recommended for ranges greater than 1 or 2 km. Co-mounted with a night vision device on Vectronix's (formerly Leica's) SG12 digital goniometer, the Vector IV can be used as a subsystem in a universal observation post equipment.

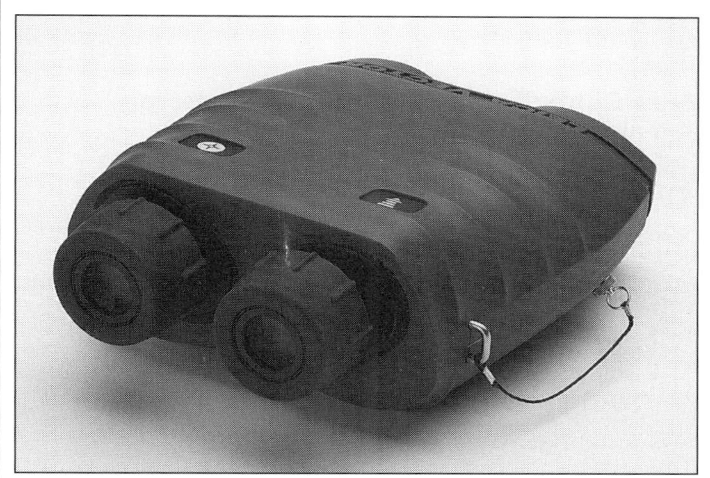
The Vectronix (formerly Leica) Vector IV range-finding binoculars 0004806

Vectronix (formerly Leica) Vector IV in use 0022055

Operational status
In series production. In service in Australia, Canada, France, the USA and elsewhere.

Specifications
Dimensions: 205 × 178 × 82 mm
Weight
Basic system: 1.59 kg
With battery, strap, eyepiece protection: 1.71 kg
Magnification: ×7
Field of view at 1,000 m: 120 m
Clear objective diameter: 42 mm
Twilight factor: 17.15
Dioptre adjustment range: ±6 dioptres
Adjustable interpupillary distance: 58.5 to 71.5 mm

Range-finder
Laser type: 1.55 μm diode laser, Class 1 eye-safe per EN 60825-1 (1994), IEC 825-1 (1993), ANSI Z 136.1 (1993)
Specified performance: 4,000 m at a visibility of 20 km, to an 8 × 8 m target with an albedo of 0.4 (1,550 nm)
Accuracy (1 σ): ±2 m
Multiple object measurement: up to 3 distances in line (gating), irrespective of distance

Digital magnetic compass (azimuth and inclination)
Azimuth measurement range: 360°
Units selectable: 6,400/6,300/6,000 mils, 360°, 400°
Azimuth accuracy: ±10 mils
Elevation accuracy: ± 3 mils
Resolution on display: 10 mils/1°/1 gon
Data interface type: RS-232 unidirectional output
Power supply: 6 V lithium battery (type 2 CR5)

Contractor
Vectronix AG.

UPDATED

Zeiss Optronik CE 658 laser range-finder for anti-aircraft applications

Type
Land laser range-finder.

Description
The eye-safe laser range-finder CE 658 was developed to be integrated into anti-aircraft fire-control systems with a fast reaction time and ranging capability against airborne targets at long distances.

The range-finder comprises an Nd:YAG laser transmitter using Raman shift technology to generate 1.5 μm laser radiation.

The unit is designed as a single-box, single-connector OEM version featuring coaxial common apertures for the transmitter and receiver channels and a self-protection mechanism to protect the receiver diode against damage caused by backscattered energy.

The laser transmitter is cooled by a closed cycle liquid cooling system including a heat-exchanger with forced air cooling.

Operational status
In production. In service with undisclosed countries.

Specifications
Dimensions: 590 × 210 × 150 mm
Weight: 18 kg
Type of LRF: OEM version
Optical conception: common apertures
Range display: no, range and status data transmission via serial interface
Data Interface: RS-422
Transmitter
Type: Nd:YAG Raman shifted
Wavelength: 1.543 μm
Pulse energy: 20 mJ
Laser class: 3B
Beam divergence: 3 mrad
Repetition rate: 12.5 Hz continuous
Receiver
Type: Avalanche diode
Self protection: >35 dB
Range: 200 - 25,000 m
Accuracy: ±5 m
Target discrimination: ≤20 m
Multiple target display: up to 3 ranges
Power supply: 18-32 V DC; 115 V AC, 400 Hz (Fan)

The CE-658 laser range-finder for anti-aircraft applications
0089920

Contractor
Zeiss Optronik GmbH.

UPDATED

Zeiss Optronik EMES 15 eye-safe laser range-finder for main battle tank applications

Type
Land laser range-finder.

Description
The EMES 15 eye-safe laser range-finder was designed to replace the non-eye-safe Nd:YAG laser range-finder currently integrated into the stabilised gunner's sight of the Leopard 2 MBT. The unit operates in a single-pulse range-finding mode providing range-finding with a pulse repetition rate up to 1 Hz.

The range-finder comprises an Nd:YAG laser transmitter using Raman shift technology to generate the eye-safe laser radiation in the 1,5 μm waveband.

Unlike the non-eye-safe version of the EMES15 LRF, the housing of the new laser range-finder includes all modules.

The unit can be easily retrofitted to replace the non-eye-safe version.

Operational status
In production. The Leopard 2 MBT is in service in Austria, Denmark, Germany, Netherlands, Spain, Sweden and Switzerland.

Specifications
Weight : 14 kg including housing
Type of LRF: OEM version; retrofit version
Optical conception: separate apertures
Power supply: 18-32 V DC
Range display: no, range and BIT data transmission via serial data interface
Data interface: RS-422
Transmitter
Type: Nd:YAG Raman shifted
Wavelength: 1.543 μm
Pulse energy: <13 mJ
Laser class: 1 H
Beam divergence: 0.5 mrad
Repetition rate: 1 Hz

Receiver
Type: Avalanche diode
Range: 200-9,990 m
Accuracy: ±5 m
Target discrimination: ≤20 m
Multiple target display: up to 6 ranges can be stored

Contractor

Zeiss Optronik GmbH.

UPDATED

Zeiss Optronik eye-safe laser range-finder for multiple applications – 6 Hz MOLEM

Type

Land laser range-finder.

Development/Description

The eye-safe laser range-finder 6 Hz MOLEM has been developed to measure the range of fast moving targets (aircraft, helicopters, fast patrol boats and so on) at medium or long distances, depending on the chosen laser divergence receiver technology.

The 6 Hz Molem is a very compact laser range-finder designed for the integration into small fire-control systems or for use as a remote-controlled stand-alone unit. It comprises a coaxial common aperture concept and a laser transmitter using a Raman shifted Nd:YAG laser technology to generate eye-safe laser radiation in the 1.5 μm waveband. The laser is cooled by a closed cycle liquid cooling system.

For boresighting the laser range-finder to other sensors two versions of the 6 Hz Molem are available, one version using projection of an aiming spot. The second version incorporates an integrated video chip which can be used for boresighting and observation.

The unit is available with different laser divergences and receiver fields of view to cover various applications.

Operational status

In production.

Specifications

Type of LRF: OEM version; stand-alone unit
Optical configuration: common coaxial apertures
Weight: 4 kg
Dimensions: 230 × 125 × 88 mm
Power supply: 18-32 V DC
Range display: no, range data via serial interface
Data interface: RS-422, RS-485, CAN-BUS
Transmitter
Type: Nd:YAG Raman shifted
Wavelength: 1.543 μm
Pulse energy: 6 mJ
Laser class: 1M
Beam divergence: max 2.5 mrad
Repetition rate: 6 Hz (average 3 Hz, burst 12 Hz)
Receiver
Type: InGaAs APD: InGaS PIN
Range: 50-40,000 m
Accuracy: ± 5 m
Target discrimination: ≤20 m
Multiple target display: up to 6 ranges are transferred via the serial interface

Contractor

Zeiss Optronik GmbH.

UPDATED

Zeiss Optronik 6 Hz eye-safe laser range-finder 0088168

Zeiss Optronik Halem 2 eye-safe hand-held range-finder

Type

Land laser range-finder.

Description

The Halem 2 eye-safe laser range-finder is a portable, hand-held device for various range-finding applications during day and night.

Using short-pulse Raman technology in conjunction with passive Q-switching, more than 2,000 range measurements are possible with one set of batteries.

Simple 'ONE BUTTON' ranging or more sophisticated range measurements (setting minimum range, multiple echo ranging, echo recall) are possible. The four control buttons allow control of all functions, without interrupting observation via the optical channel. The control buttons are ergonomically placed for the user.

Depending on the background illumination the brightness of the display is automatically controlled. Additionally the brightness can also be adjusted by the user.

The serial data interface allows complete remote control of all user functions

Interfaces for night sight equipment, standard angulation heads and target acquisition equipment are optional.

Operational status

In production. In service with undisclosed countries.

Specifications

Dimensions: 185 × 215 × 86 mm
Weight: approx 2.5 kg
Type of LRF: hand-held LRF or remote-controlled stand-alone unit
Optical conception: separate apertures for transmitter and receiver, right ocular for observation, left ocular for range display
Aiming optics: magnification ×8; field of view 110 mrad, protected against laser radiation at 1,064 nm
Range display: 2 targets with 5 digits; up to 6 targets per measurement are stored; transfer of the range data and BIT information via serial interface
User functions: FIRST/LAST ECHO selection, MIN RANGE adjust, ECHO RECALL, BRIGHTNESS CONTROL (display), LANGUAGE selection; SYSTEM TEST
Data interface: RS-422; remote control and transmission of measured data and BIT information
Transmitter
Type: Nd:YAG Raman shifted
Wavelength: 1,543 μm
Laser class: Class 1H (IEC 60825/2001)
Pulse energy: 10 mJ
Beam divergence: <1 mrad
Repetition rate: 0.5 Hz
Receiver
Type: PIN-diode
Range: 50-39,995 m
Accuracy: ±5 m
Target discrimination: ≤20 m
Power supply: 10-16 V DC; 10 batteries, size AA or rechargeable AA cells) or external supply, 2 A max
Battery life: >2,000 measurements with standard AA batteries

Contractor

Zeiss Optronik GmbH.

UPDATED

Zeiss Optronik Halem 2 0053780

Zeiss Optronik Molem

Type
Land laser range-finder.

Description
Molem is an eye-safe laser range-finder for measuring the distance of stationary and moving targets in changing combat and ambient conditions. The minimum range is 50 m and, depending on the type of target and the atmospheric conditions, distances of more than 30 km are possible. Molem is normally mounted on an existing observation system such as a TV camera or thermal sight. The system includes an RS-422 interface for remote control and data transfer from/to a separate control and computer station.

Operational status
In production. In service with undisclosed countries.

Specifications
Type of LRF: OEM version; stand-alone unit
Optical arrangement: separate apertures
Weight: 2.5 kg
Dimensions: 180 × 124 × 112 mm
Power supply: 18-32 V DC
Range display: no, range and BIT data transmission via serial interface
Data interface: RS-422
Transmitter
Type: Nd:YAG Raman shifted
Wavelength: 1,543 μm
Pulse energy: 10 mJ
Laser class: 1
Beam divergence: 0.5 mrad
Repetition rate: 1 Hz
Receiver
Type: PIN – or Avalanche diode
Range: 50-39,995 m

Zeiss Optronik Molem laser range-finder

Accuracy: ±5 m
Target discrimination: ≤20 m
Multiple target display: up to 6 ranges are transferred via the serial interface

Contractor
Zeiss Optronik GmbH.

VERIFIED

OBSERVATION AND SURVEILLANCE

IMAGE INTENSIFIER BINOCULARS

Aselsan M975/M976

Type
Image intensifier binocular.

Description
Aselan models M975 and M976 are Night Vision Binoculars (NVB) that offer image intensified night vision in a lightweight, compact, single-tube configuration. This water-resistant binocular is designed for hand-held operation and medium- to long-range night vision. For short-range use, both models can be easily converted to a ×1 binocular by a simple objective lens change. The M975 and M976 differ only in the image intensifier, M975 having a GEN II+ tube and M976 GEN III.

Operational status
In production. These products are also manufactured in the USA by Litton, the licensor.

Specifications
Weight: 1.2 kg
Magnification: ×4.2 ±5%
Field of view (min): 10°
System gain: 1,000 (M975), 1,200 (M976) typical
Resolution: 3.2 lp/mrad (M975), 3.6 lp/mrad (M976)
Focal length: 108 ±2
Focus range: 20 m to infinity
Image intensifier
M975: GEN II+, non-inverting
M976: GEN III, non-inverting
Eye relief: 27 mm
Power requirement: 2 × 1.5 V AA alkaline or 1 × 3.9 V AA lithium batteries

Contractor
Aselsan Inc, Microelectronics, Guidance and Electro-optics Division.

VERIFIED

Aselsan M975/M976 night vision binocular

Aselsan M977/M978/M978A

Type
Image intensifier binocular.

Description
Aselsan models M977 and M978/M978A are Night Vision Binoculars (NVB) that offer image intensified night vision in a lightweight, compact, single-tube configuration, for medium-range observation. For short-range use, both models can be converted to a ×1 goggle by a simple objective lens change. The M977 and M978/M978A differ only in the image intensifier, M977 using a GEN II+ tube and M978 GEN III.

Operational status
In production. These products are also manufactured in the USA by Litton, the licenser.

Specifications
Weight: 920 g
Magnification: ×2.8 ±5%
Field of view: 13.5°
System gain min: 1,600 (M977), 2,000 (M978), 2,500 (M978A)
Resolution: 1.8 lp/mrad (M977), 2.2 lp/mrad (M978), 2.6 lp/mrad (M978A)
Image intensifier
M977: GEN II+ non-inverting
M978: GEN III non-inverting
M978A: Adv. GEN II+ non-inverting
F-number: f/1.3
Eye relief: 27 mm
Power supply: 2 × 1.5 V AA alkaline or 1 × 3.9 V AA lithium batteries

Contractor
Aselsan Inc, Microelectronics, Guidance and Electro-optics Division.

VERIFIED

The Aselsan M977/M978/M987A image intensifying binocular

Aselsan M979/M980

Type
Image intensifier binoculars.

Description
Aselsan Models 979 and 980 are night vision binoculars that offer long-range high-resolution image-intensified night vision in a lightweight compact single-tube configuration. The water-resistant binocular is designed for hand-held operation at medium and long ranges. For short-range use both models can be easily converted to a ×1, ×3, or ×4 binocular by a simple objective lens change. The M979 and M980 differ only in the type of image intensifier, M979 using a GEN II+ tube and M980 using GEN III.

Operational status
In production.

Specifications
Weight (without AA batteries): 1.7 kg
Magnification: ×6
Field of view: 6.3°
Systems gain (min)
M979 (GEN II+): 900
M980 (GEN III): 1,100
Resolution, on axis (min)
M979 (GEN II+): 4.5 lp/mrad
M980 (GEN III): 5.0 lp/mrad
Focal length: 162 ±2 mm
F-number (max): f/1.7
Eye relief distance: 27 mm

Aselsan M979/M980 night vision binocular

Contractor
Aselsan Inc, Microelectronics, Guidance and Electro-optics Division.

VERIFIED

BE Delft LW 1200

Type
Image intensifier binocular.

Description
The BE Delft LW 1200 binocular image intensifiers provide a night vision system for company and platoon-level units. The LW 1200 uses 18 mm second-generation tubes powered either by two pen torch batteries or a rechargeable mercury battery. Each LW 1200 has a carrying case.

Operational status
In use with the Indian Army and paramilitary forces.

Specifications
Dimensions: 260 × 100 × 90 mm
Weight: 1.2 kg
Field of view: 9.5°
Magnification: ×4
Resolution: 0.4 mrad

Contractor
BE Delft.

VERIFIED

Econ Industries Andromeda 2

Type
Image intensifying binocular.

Description
Econ Industries' Andromeda 2 ×3.7 night vision binocular is designed for medium- and long-range surveillance. It can be configured with either second-plus or third-generation image intensifier tube, with automatic brightness cut-off. Andromeda 2 combines the US Army AN/PVS-7A night vision goggle eyepiece assembly with Econ Industries 98 mm f/1.2, t/1.7 catadioptric objective lens. The system can be hand-held or tripod mounted and can be converted to ×1 goggles.

Operational status
Available.

Specifications
Dimensions: 185 × 110 × 130 mm
Weight (with AA batteries): 1.3 kg
Magnification: ×3.7
Field of view: 10.5° (17.8 m at 100 m distance)
Resolution of axis (100 m lux): >2.3 lp/mrad
Image intensifier: 18 mm, GEN II+ or GEN III
System brightness gain: >1.300 fL/fL
Linear distortion: 4% across field of view
Focus range: 5 m to infinity
Objective lens
Lens type: catadioptric

The Andromeda 2 night vision binocular from Econ Industries

Effective focal length: 98 mm
Free aperture: 80 mm
F-number: 1.2
T-number: 1.7
Eyepiece lens
Effective focal lens: 27 mm
Eye relief: 27 mm
Dioptre adjustment range: −6 to +2 dioptres
Interpupillary adjustment range: 55-71 mm

Contractor
Econ Industries S A.

VERIFIED

Fraser-Volpe M25 Stedi-Eye day/night stabilised binoculars

Type
Image-intensified binocular.

Development
The M25 Stedi-Eye day/night stabilised binoculars were developed by Fraser-Volpe Corporation as the Stedi-Eye Patented Stabilisation Technique. The system was militarised for US government agencies including the US Army, Navy and US Coast Guard. Automatic caging and laser eye protection were added under contract from the US Army, funded by the Soldier Enhancement Program. The system is marketed commercially without the classified laser filters.

An M25 model that incorporates a laser rangefinder and compass is now being developed. This will enable an operator to pass range and bearing of targets under observation to the Command Centre by radio for instant location of the subject targets. When combined with GPS information from the observer the data will provide the command with an instant fire-control solution for those targets.

Description
M-25 Stedi-Eye is a ×14 stabilised hand-held binocular that reduces image motion caused by hand tremor and platform vibration. A touch-sensitive cage/uncage feature secures the gimbal when not in use – picking up the M-25 automatically activates the stabilisation mode and lying the unit down locks the gimbal to prevent potential damage.

The system can be configured for night use by changing the eyepieces. The night vision eyepieces use either second- or third-generation image-intensifier tubes.

M-25 has a universal camera adaptor to interface with any standard SLR camera allowing users to take photographs while continuing to scan the scene through the camera lens. It is also compatible with Fraser-Volpe's Miniature Integrated Camera Eye (MICE) system. By connecting M-25 to the MICE system transmitter/TV processor the M-25 image can be transmitted by wireless link to a command station up to 10 miles away.

Operational status
Operational and development verification tests have been completed by the US Army. The US Army is presently negotiating a multiyear procurement plan for 5,000 units. The MICE system is presently undergoing evaluation field tests by the US Army Battlelab at Fort Benning, Georgia for the Dismounted Warrior programme. Similar test efforts are scheduled for the Mounted Warrior programme and the Special Forces.

M25 Stedi-Eye day/night stabilised binoculars 0005481

Specifications

Dimensions (l × w × h)
day: 210 × 190 × 89 mm
night: 229 × 190 × 89 mm
Magnification: ×14
Objective diameter: 41 mm
Exit pupil: 3 mm
Field of view: 4.3°
Resolution
day: 4.3 s
night: >35 line pairs/mm
Stabilisation freedom: ±9°
Focus adjustment: ±5 dioptres
Max scan rate: 5-10°/s
Interpupillary adjustment: 60-70 mm
Power supply: 2 × AA cells (internal), 6-30 V DC

Contractor

Fraser-Volpe Corporation.

VERIFIED

Institute of Optronics Pakistan GP/NVB-4A and GP/NVB-5A general purpose night vision binoculars

Type

Night vision binoculars.

Description

These are lightweight, hand-held, passive night vision devices for general purpose long-range observation and surveillance. The 18 mm high performance image intensifier tube incorporated in the system creates almost 'day light' viewing conditions to facilitate the observation of a target position during night time.

The units have been engineered to minimise weight and cost without sacrificing performance and quality. They have a rugged construction and are easy to operate and maintain. Specific applications include long-range night observation and surveillance for tank commanders, forward area observers, border patrols and special forces.

Operational status

In production.

Specifications

	GP/NVB-4A	GP/NVB-5A
Length	255 mm	305 mm
Width	145 mm	155 mm
Weight	1.13 kg	3.18 kg
Magnification	×4	×6
Viewing range	450 m	1,000 m
Field of view	10.5°	6.5°
Tube	18 mm	18 mm
Objective lens focusing	10 m to infinity	20 m to infinity
Diopter adjustment	+ 2 to –6	+ 2 to –6
Interpupillary adjustment	55 to 72 mm	55 to 72 mm
Objective lens	95 mm	155 mm
Power	'AA Batteries'	'AA Batteries'
Operating temperature	–52 to + 54°C	–52 to + 54°C

General purpose night vision binoculars GP/NVB-4A and GP/NVB-5A 0089872

Contractor

Institute of Optronics Pakistan.

VERIFIED

IRT Lunatron 904

Type

Image intensifying binocular.

Description

Lunatron 904 night sight binoculars consist of two independent night sight devices which are put together to form binoculars by means of a moveable and detachable bridge. This construction means that the stereo vision is provided by two independent night sight devices with individual adjustment. Each system consists of basic body, lens image, intensifying tube/valve and eyepiece. As the bridge can be removed there is also the option of equipping two people, each with a monocular apparatus. The separated device could also be used for photography. The wide opening of the bridge gives a variety of possible eye distances and allows the attachment of a camera and/or video camera, while retaining the second picture unit with full enlargement for separate use for observation.

The binoculars are equipped with two lightweight mirror lenses with wide aperture/low obscuration (1:1.9) and which are filled with nitrogen to prevent misting/blooming. A multilayer coating prevents distracting reflections. The Lunatron 904 binoculars are equipped with two SuperGen image intensifying tubes. Automatic brightness control and protection against overloading is incorporated.

Standard equipment for each unit is a 3 V lithium battery, but 2 V batteries can also be used.

Operational status

In production and in service in many countries by the police, military and other services and in the private sector. Several thousand have so far been produced.

Specifications

Focus range: 4 m to infinity
Image intensifier: SuperGEN
Resolution: 36 lp/mm
Tube sensitivity: 600 µA/1 m

Contractor

IRT Infrarot-Technik Eiselt.

The Lunatron 904 night vision binocular from IRT

ITT F4939 night vision binocular

Type
Image intensifier binocular.

Description
ITT's F4939 night vision binocular has ×4.5 magnification and third-generation image-intensifying tubes for medium- to long-range night vision and uses technology developed for the US Army's AN/PVS-B infantry night vision goggle. It is available as the F4939M with the full military MX-10130 tube or with the ITT F4798 standard tube.

The F4939 includes an infra-red LED (Light-Emitting Diode), an infra-red-on indicator, an automatic high-light cut-off to protect the image intensifier and a low-voltage indicator. The system can also be hand-held or tripod mounted.

Operational status
Available.

Specifications

	F4939M with MX-10130 tube	F4939 with F4798 tube
Photocathode sensitivity		
at 2,856 K	800 μA/lm min	700 μA/lm min
at 880 nm	40 mA/W min	34 mA/W min
System gain, typical	1,200 fL/fL	1,200 fL/fL
Resolution		
moonlight	3.4 cy/mrad min	3.2 cy/mrad min
starlight	3.0 cy/mrad min	2.8 cy/mrad min

Collimation
Convergence: 1±1°
Dioptre adjustment: +2 to –6 dioptres
Interpupillary adjustment: 55-71 mm
Eye relief: 15 mm
Objective lens
Effective Focal Length (EFL): 116.2 mm
Eyepiece lens, EFL: 26 mm
Weight: 1.125 kg
Battery: 2 × AA alkaline or 1 × lithium (BA-5567)

Contractor
ITT Defense & Electronics, Night Vision Division.

ITT's F4939 ×4.5, GEN III night vision binocular 0005483

ITT F5050A night vision binocular

Type
Image intensifier binocular.

Description
The F5050A night vision binocular is derived from the design of the AN/AVS-6 ANVIS (Aviator's Night Vision System) but is more rugged with an aluminium main housing. One configuration eliminates the external battery pack where this is required, allowing hand-held use. The optics and image intensifier are common to the ANVIS system, but F5050A also contains an adjustable (spot/flood) infra-red illuminator and includes a new helmet mount that attaches to virtually any helmet. It has been designed for ground and sea applications that require increased depth perception.

Pirate incorporates two high-resolution third-generation image intensifier tubes using gallium arsenide technology and operating in the near-infra-red waveband. The binocular can accept ×3 magnification lenses to extend its range.

Operational status
Available for customer demonstration.

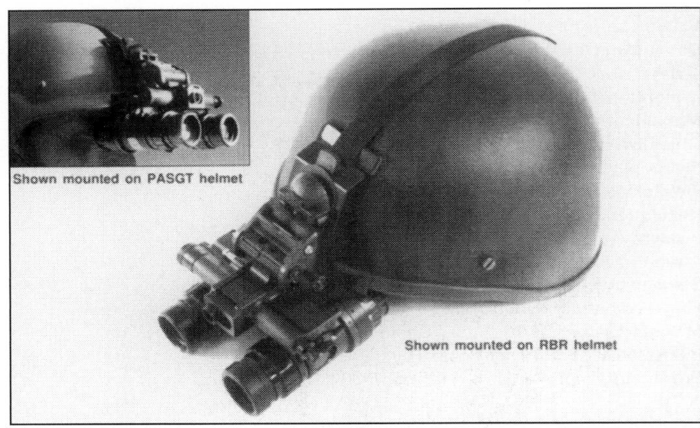

The F5050A night vision binocular shown mounted on an RBR helmet and (inset) on a PASGT helmet 0005484

A F5050A goggle assembly 0132091

Specifications
Weight (including battery): 0.65 kg
Scene illumination: 10^{-6} to 1 fc
Spectral response: visible to 0.9 μm (IR)
Field of view: 40°
Magnification: ×1 with ×3 available
Resolution: 1.3 cy/mrad min
Brightness gain: 6,000 fL/fL min
Collimation: ≤1.0° convergence, ≤0.3° dipvergence
Dioptre adjustment: +2 to –6 dioptres
Interpupillary adjustment: 52-72 mm (individual knobs)
Fore and aft adjustment: 27 mm range
Tilt adjustment: 10° min
Objective lens: EFL 27 mm F/1.23, T/1.35
Eyepiece lens: EFL 27 mm
Exit pupil/eye relief
On axis: 14 mm at 25 mm distance
Full field: 6 mm at 25 mm distance
Focus range: 41 cm to infinity
Battery: 1.5 V AA alkaline internal or 4 × AA alkaline external
Operating temperature range: –32 to +52°C

Contractor
ITT Industries, Night Vision.

Kazan 1PN33B/BN-453/Baigysh-12

Type
Image intensifier binocular.

Description
The 1PN33B are night vision binoculars using image intensifiers which may also be used for artillery observation. Features include eye flaps and a lens hood.

The BN-453 appears to be a 'commercial' version of the 1PN33B. The Baigysh-12 is similar but differs externally in the eyepieces.

Operational status
Believed to be in service with Russian armed and paramilitary forces (1PN33B) and in production.

Specifications
Dimensions: 245 × 189 × 91 mm
Weight: 1.6 kg (1.5 kg Baigysh-12)

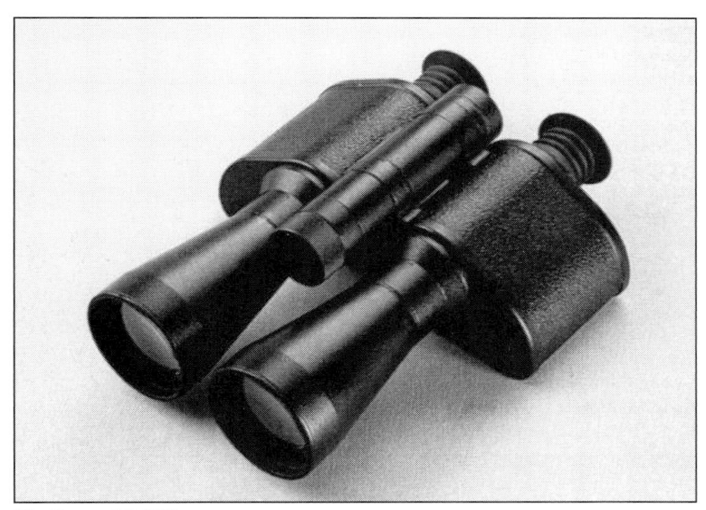

The Kazan 1PN33B

Field of view: 9° (10° Baigysh-12)
Magnification: ×3.2 (×2.8 Baigysh-12)
Exit pupil clearance: 20 mm
Power supply: 8.3-8.8 V

Contractor
Kazan Optical and Mechanical Plant (AO KOMZ).

VERIFIED

Kazan 1PN50/Baigysh-6

Type
Image intensifier binocular.

Description
The 1PN50 is an image intensifier unit for observation which uses a monocular body and wraparound rest shield. In natural light a human being may be identified at distances up to 300 m.

The 'commercial' Baigysh-6 is similar but has an 8° field of vision.

Operational status
In production and believed to be in service with the Russian armed and paramilitary forces (1PN50).

Specifications
Dimensions: 405 × 168 × 85 mm
Weight: 1.8 kg
Field of view: 11°
Magnification: ×2.4
Effective range: 350 m
Power supply: 5.73-7.3 V

Contractor
Kazan Optical and Mechanical Plant (AO KOMZ).

VERIFIED

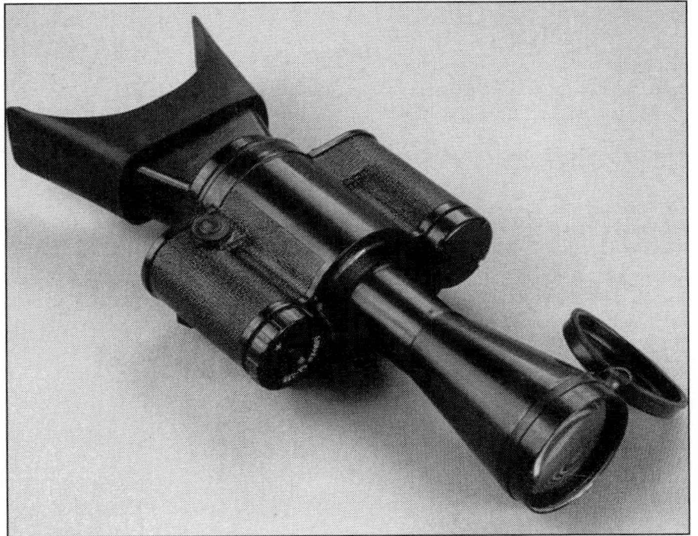

The Kazan 1PN50

LOMO RECON-1 image intensifying binoculars

Type
Image intensifier binoculars.

Description
RECON-1 is a bi-ocular pseudo binocular with a single image intensifier tube. RECON-1 is fitted with a Gen I tube which the makers state allows the viewer to see contrasting objects at 150 m at a light level equivalent to a quarter moon light. The device automatically switches off when ambient illumination is more than one lux. An infra-red emitter enables RECON-1 to operate in full darkness at short ranges.

Operational status
Available.

Specifications
Weight: 820 g
Dimensions: 110 × 70 × 60 mm
Magnification: ×2.3
Field of view: 9°
Focus range: 0.25 m to infinity
Gain: 700
Sensitivity: 250 μA/lm
Power supply: 3 V battery
Battery life: 8 h continuous use

Contractor
LOMO PLC.

VERIFIED

The LOMO RECON-1 binocular 0097951

LOMO RECON-2

Type
Image intensifier binoculars.

Description
RECON-2 is an updated version of the RECON-1 bi-ocular pseudo binocular with a single image intensifier tube. RECON-2 is fitted with a Gen II tube which the makers state allows the viewer to see contrasting objects at 400 m at a light level equivalent to a quarter moon light. The Gen II tube provides improved image quality. The device automatically switches off when ambient illumination is more than one lux. An infra-red emitter enables RECON-2 to operate in full darkness at short ranges.

The LOMO RECON-2 binocular 0097950

Operational status
Available.

Specifications
Weight: 750 g
Dimensions: 140 × 157 × 90 mm
Magnification: ×1
Field of view: 38°
Focus range: 0.25 m to infinity
Gain: 40,000
Sensitivity: 250 μA/lm
Power supply: 3 V battery
Battery life: 8 h continuous use

Contractor
LOMO PLC.

VERIFIED

Northrop Grumman M970 series night vision binocular

Type
Image intensifier binocular.

Description
The M970 series are high-resolution night vision binoculars for long- or medium-range observation. For short-range or head-mounted use, all models can be converted to ×1 goggle by an objective lens change.

Models differ in their magnification and use either second-plus or third-generation image intensifier tubes. Units may be upgraded in the field from second-plus to third-generation. They all have 27 mm eye relief and are powered by two alkaline or Ni/Cd batteries or a single mercury or lithium battery. The housing is constructed from high-tensile strength, chemical resistant polymer alloy and a purge valve is fitted for backfilling with dry nitrogen at the general support or depot maintenance level.

Operational status
In production and in service. These products are also produced in Turkey under licence by Aselsan.

Specifications

Model	Image intensifier	Magnification
M972	GEN II+	×1
M973	GEN III	×1
M977	GEN II+	×3
M978	GEN III	×3
M975	GEN II+	×4
M976	GEN III	×4
M979	GEN II+	×6
M980	GEN III	×6

Contractor
Northrop Grumman (formerly Litton) Electro-Optical Systems.

VERIFIED

PCO NPL-1 night vision binoculars

Type
Image intensifying binoculars.

Description
The NPL-1 night vision binocular is a lightweight, compact twin eyepiece electro-optic device meant for use at night time for targets and terrain observation at medium to long range. It uses a GEN II+ or SuperGEN image intensifier tube.

The PCO night vision binocular

The binocular has an interpupillary adjustment mechanism to suit individual users.

Operational status
Available.

Specifications
Dimensions: 245 × 140 × 70 mm
Weight: 1.3 kg
Magnification: ×3.4
Field of view: 12.5°
Pupil spacing: 56 to 72 mm
Dioptre adjustment: ±6
Power supply: 1 E15/51 3.6 V/1.6 Ah battery

Contractor
PCO SA, Poland.

Photonic NS-Bi

Type
Image intensifier binocular.

Description
The NS-Bi is a hand-held night observation instrument with biocular eyepiece. A switch at the top of the instrument, close to the eye shield, activates it. The instrument can take 18 mm second- or third-generation tubes and an illuminated reticle is an option. The instrument may be installed on a tripod and is powered by two 1.5 V batteries.

Operational status
In production and in service.

Specifications
Dimensions: 315 × 105 × 170 mm
Weight (with batteries): 2.2 kg
Magnification: ×2.4

Contractor
Photonic Optische Gerate GmbH.

UPDATED

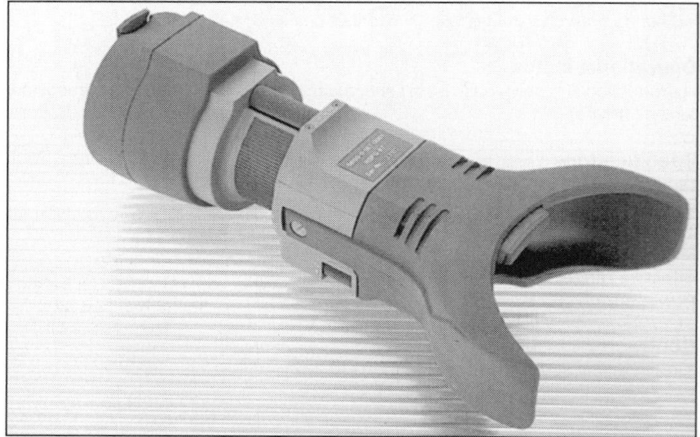

The Photonic NS-Bi night observation device with closed rubber daylight diaphram
0089881

ROMZ NZT-2MBN night vision binoculars

Type
Image intensifier binoculars.

Description
The NZT-2MBN is designed for a variety of applications requiring biocular viewing at night. It is equipped with a LED IR illuminator for use in very dark conditions. The illuminator has an indicator to make the user aware when it is switched on.

Operational status
Available.

Specifications

Model	ADSH 3.807.029	ADSH 3.807.029-01	ADSH 3.807.029-02	ADSH 3.807.029-03
Dimensions	176 × 128 × 82 mm (Interpupillary distance of 52 mm min) 176 × 162 × 57 mm (Interpupillary distance of 76 mm max)			

Model	ADSH 3.807.029	ADSH 3.807.029-01	ADSH 3.807.029-02	ADSH 3.807.029-03
Weight	<0.85 kg			
Lens-type	ADSH 5.917.010-01	ADSH 5.917.011	ADSH 5.917.010-01	ADSH 5.917.011
Objective focal length	50 mm	70 mm	50 mm	70 mm
Magnification	×3.7	×3.7	×3.3	×3.7
Field of View	≥10	≥7	≥10	≥7
Resolution (at 0.2 lx)	125 arc secs			
Dioptre adjustment	4			
Interpupillary distance (adjustable)	54-74 mm			
Operating illumination	from moonlight to starlight			
Power supply	3 V DC			
Temperature range	–15 to +40°C			

Contractor
ROMZ, Rostov Optical and Mechanical Plant (JSC).

VERIFIED

Simrad KDN250F night and day binoculars

Type
Image intensifier binocular.

Description
The KDN250F is created by combining the KN250F image intensifier with the KD binoculars to produce a day/night system for use by infantry.

The binoculars may be fitted permanently to the image intensifier or supplied with an optional mounting so that the 25 mm second- or third-generation image intensifier can be used for other tasks. A dichroic beam-splitter gives the operator the best possible view 24 hours a day without having to remove the image intensifier. If light levels become too high for the image intensifier the binoculars automatically take over.

Operational status
In production and in service with several countries.

Specifications
Dimensions (l x h x w): 230 × 140 × 158 mm
Weight: 1.5 kg
Objective lens, image intensifier: catadioptric, EFL/80 mm, f/1.1, t/1.4
Objective lens, day binoculars: doublet lens, EFL/80 mm, f/3.3
Image intensifier tube: 18 mm wafer tube, GEN II or GEN III
Magnification: ×3.5
Field of view: 12° (×3.5), 7° (×6)
Objective, image intensifier: 78 mm diameter
Objective, day binoculars: 18 mm diameter
Interocular distance: 58-73 mm
Dioptre setting: +2 to –6 dioptres
Exit pupil: 5.2 mm
Eye relief: 25 mm
Focusing range: 25 m to infinity
Power supply: 2 × 1.5 V AA batteries

Contractor
Simrad Optronics ASA.

VERIFIED

Simrad KDN250F night and day vision binoculars

Thales Bino-Kite night vision binocular

Type
Image intensifier binocular.

Description
Bino-Kite is a compact and lightweight night vision binocular. A magnification of ×4.5, makes it suitable for medium- to long-range night surveillance or fire-control observation.

Bino-Kite is supplied with either second- or third-generation image intensifier tubes. Weighing 1.1 kg, the binocular is portable and may be worn around the neck using a lanyard, or mounted to a tripod for fixed observation. It has a binocular viewer with adjustable dioptric eyepieces to suit individual user requirements.

By adding an afocal lens assembly, Bino-Kite is converted into Maxi-Bino-Kite, offering greater magnification and maximum comfort for surveillance over prolonged periods.

Operational status
In production.

Specifications
Weight: 1.1 kg
Field of view: 8.8°
Magnification: ×4.5
Focus range (to infinity): 20 m
Eye relief: 27 mm
Power supply: 2 × 1.5 V AA batteries
Eyepiece adjustment: +2 to –6 dioptres
Min range: recognition of a standing man at 400 m in starlight (10^{-3} lx) Gen 2.1.1 tube

Contractor
Thales Optics.

UPDATED

Bino-Kite night vision binoculars 0005482

Thales Maxi-Bino-Kite

Type
Image intensifier binoculars.

Description
Maxi-Bino-Kite is a compact, lightweight night vision binocular for use as a portable, long- range surveillance device. Maxi-Bino-Kite is an upgrade of Bino-Kite with an afocal lens assembly.

Maxi-Bino-Kite may be supplied with either second- or third-generation image intensified tubes without any need for modification.

Operational status
In production.

Specifications
Weight: 1.5 kg
Magnification: ×6.7
Field of view: 5.2°
Power supply: 2 × 1.5 V AA batteries
Eyepiece adjustment: +2 to –6 dioptres
Eye relief: 27 mm
Focus range: 20 m to infinity
Min range: recognition of a standing man at 580 m in starlight (10^{-3} lx) Gen 2.1.1 tube

The Thales Maxi-Bino-Kite image intensifying binocular

Contractor
Thales Optics.

UPDATED

Vectronix BIG35 night binoculars

Type
Image intensifier binocular.

Description
Vectronix's (formerly Leica's) BIG35 night binoculars are a hand-held instrument for long-term night surveillance without eye strain. They have ×3 magnification and in average conditions a man-sized target can be recognised at a range of 300 to 700 m.

BIG35 uses second-plus generation image intensifer but Super or third-generation can be fitted without modification.

Vectronix (formerly Leica) BIG35 binoculars in use

Vectronix (formerly Leica) BIG35 night binoculars

The binoculars are rubber-covered and the light-alloy housing is watertight and moistureproof.

Operational status
Production. In service with the Swiss Army and undisclosed customers.

Specifications
Dimensions: 208 × 153 × 96 mm
Weight: 1.19 kg
Magnification: ×3
Field of view: 12.5°
Limit of resolution: 3.2 lp/mrad
Objective lens: catadioptric, 75.1 mm, f/1.17
Image intensifier: GEN II, DEP XD-4 or GEN III
Power supply: 1 × 3 V lithium AA or 2 × 1.5 V alkaline

Contractor
Vectronix AG.

UPDATED

Yugoimport PN 5 × 80

Type
Image intensifier binocular.

Description
The PN 5 × 80 is a night vision system based upon an image intensifier. In low-light conditions it can be used to identify a standing human being at 600 m and a vehicle at 1.8 km. Power is by means of a Ni/Cd battery which provides 10 to 15 hours' operating time.

Operational status
In production and in service.

Specifications
Weight: 2 kg
Field of view: 10°
Magnification: ×5

Contractor
Yugoimport SDPR.

The PN 5 × 80 night vision system

Yugoimport PRD 4 × 80

Type
Image intensifier binocular.

Description
The PRD 4 × 80 is a passive, binocular-type device for night observation and is based upon an 18 mm image intensifier tube. It is powered by a Ni/Cd battery which gives 40 hours' service and in poor light conditions it permits the recognition of a standing human being at 500 m and of a vehicle at 1.5 km.

Operational status
In production and in service.

Specifications
Weight: 2.2 kg
Field of view: 10°
Magnification: ×4

Contractor
Yugoimport SDPR.

The PRD 4 × 80 observation device

OBSERVATION AND SURVEILLANCE

IMAGE INTENSIFIER CAMERAS

Allen BlackWatch 6000/4000

Type
Image intensifier camera.

Description
BlackWatch 6000 is a two stage image intensifier system, incorporating a Generation III high-performance first stage followed by a Generation I second stage. BlackWatch 4000 uses a Generation II+ first stage and Generation I second stage. The combinations offer very high system gains benefiting from high sensitivity, high signal to noise ratio, low noise first stages and the high maximum screen luminance capability of the common fit second stage. The systems are intended for very low-light level surveillance use, in conjunction with SLR, video and CCTV cameras. A range of adaptors is available, permitting connection to most current cameras. Custom made adaptors can be made for cameras not covered by the standard range.

The manufacturers state that external trials have demonstrated successful photographic and video results in light levels down to starlight (10^{-3} lx) and below for the P6000, depending on target and scene conditions and down to approximately starlight for the P4000.

The system can be used with covert IR scene illumination, where the existing illumination is inadequate, for example within darkened buildings or under heavy shade.

Operational status
In service with the UK MoD, Police and HM Customs.

Specifications
P6000 First stage
Generation III plus 18 mm wafer tube
luminous sensitivity (2856 K): 1,800 μA/lm min
radiant sensitivity 830 nm: 190 mA/W min
radiant sensitivity 880 nm: 80 mA/W min
P4000 First stage
Generation II plus 18 mm wafer tube
luminous sensitivity (2856 K): 650 μA/lm min
radiant sensitivity 800 nm: 60 mA/W min
radiant sensitivity 850 nm: 50 mA/W min
P6000 and P4000 Second stage
Generation I 18:11 m inverter tube
output phosphor: yellow-green medium persistence
P6000 Electrical
Combined intensifier gain: variable in steps; 5×10^4 to 1.5×10^6 typical
Combined intensifier resolution: 45 lp/mm typical
Signal to noise ratio: 21.0 min
EBI: 0.1 μlx typical
Max screen output
luminance: 250 cd/m^2 min
Supply voltage: 2 AAA alkaline batteries
P4000 Electrical
Combined intensifier gain: variable in steps; 5×10^4 to 1.5×10^6 typical
Combined intensifier resolution: 32 lp/mm typical
Signal to noise ratio: 18.0 min
EBI: 0.1 μlx typical
Max screen output
luminance: 250 cd/m^2 min

Allen BlackWatch system for covert night vision photography and video filming
0023154

Supply voltage: 2 AA alkaline batteries
Mechanical P6000 and 4000
Input format: C-mount (25 mm format)
Output formats: T2 to SLR film cameras (P9002BW); VL or other to video cameras (P9005BW); C-mount to CCTV cameras (P9007BW)
Dimensions
length
body: 135 mm approx
as photograph: 270 mm approx
width: 70 mm approx
height: 80 mm approx
Weight
body: 660 g approx
as photograph: 1,350 g approx

Contractor
P W Allen & Company Ltd.

Allen ICAM2-07-06E image-intensified CCD camera

Type
Image-intensified camera.

Description
The ICAM2-07-06E is a high-performance CCD camera employing a super second-generation image intensifier coupled directly, without a fibre-optic taper, to Allen's (formerly Marconi's) high-resolution 1 in image sensor, which has 1,032 pixels/line. The video output is CCIR compatible.

A separate intensifier control unit is provided to give manual gain adjustment and to switch off the intensifier automatically when the light level is excessive.

A gallium arsenide third-generation intensifier may also be used instead of second-generation for enhanced infra-red sensitivity.

Operational status
Available.

Specifications
SuperGEN II or GEN III, 18 mm multichannel plate wafer tubes
CCD type: Marconi CCD 07-06; 1 in frame transfer sensor with 1,032 pixels/line
Image format: 16 mm diagonal approx
Coupling: one-to-one fibre-optic (no taper)
Resolution: 550 TV lines typical
Signal/noise: 35 dB at 2 mlx
Blemish specifications: details on request
Genlock input: CCIR video waveform or mixed sync, input impedance 1 kΩ
Video output: 1 V peak-to-peak CCIR compatible
Intensifier gain: ×1, ×2, ×4, ×8, ×16, ×32 available settings, giving 100% video output at 2 mlx, 1 mlx and so on
Power supply: 12 V DC
Power consumption: 10 W
Lens-mount: C-mount

Contractor
P W Allen Ltd.

ICAM2-07-06E image-intensified CCD camera
0005487

Electrophysics Astroscope Model 9300 series

Type
Image intensifier camera.

Description
Electrophysics produces a range of night vision modules for cameras and camcorders.

The Astroscope 9300D2 is an image intensifying night vision module designed for use with Kodak DCS-200 digital camera systems and other Nikon-mount single-lens reflex cameras. The Astroscope incorporates Nikon hot shoe lens mount technology and turns on and off automatically with the camera.

The Astroscope 9300VL is for use with VL-mount camcorders. Full electronic control of lens iris and zoom is maintained.

The Astroscope 9323B is for use with 2/3 in bayonet-mount professional cameras and camcorders, including electronic newsgathering cameras. It is assembled by removing the camera's objective lens, attaching the 9323B module and reattaching the lens in front of the module.

Astroscope modules can have either second-plus or third-generation image intensifier tubes.

The latest version in the series is the Astroscope 9350, which is based on patent-pending Type 3 technology which permits the system's Central Intensifying Unit (CIU) to be electronically adapted so that it can be optimally configured for the camera used. In addition to the CIU, the Astroscope 9350 modules include Black Body Adapter (BBA and Front Lens Adapter (FLA). BBA/FLA pairs are available for Nikon, Canon and Minolta cameras as well as a variety of camcorders. Others are being introduced. The BBA/FLA pairs permit the electronic daylight imaging cameras to continue to communicate with their electronic lenses, leaving the original functions of the camera intact.

Operational status
Standard and custom-built instruments are currently in production.

Specifications
Image intensifier: either 18 mm GEN III GaAs photocathode, visible to 980 nm response or 18 mm GEN II+ multi-alkali photocathode, visible to 900 nm response
Dimensions
9300VL (d × l): 63.5 × 102 mm
9323B, 9350NIK-Pair: 3.25 × 5.4 in
System resolution
9300Vl: up to 450 lines
9323B: up to 450 lines
Weight
9300VL: 540 gm
9323B: 652 gm
9350NIK-Pair: 540 gm
Power: <50 mA
Housing: epoxy-coated black aluminium
System magnification (excl lens) : ×1

Contractor
Electrophysics Corporation.

The Astroscope 9323B night vision module for ⅔ in bayonet-mount cameras
0005488

IRT Lunatron 999

Type
Image intensifier camera.

Description
The Lunatron 999 Nightfoto and T Night TV system comes with a variety of parts for night vision monocular or binocular, camera or camcorder use. A subminiature CCD TV camera is also available. The system uses second- or third-generation image intensifying technology.

The Lunatron multipurpose night vison system

Operational status
Available.

Specifications
Image intensifier: GEN II or GEN III
Catadioptric lens: 1.17/68 mm
Standard TV lens: 1.3/75 mm
Subminiature CCD TV camera
Spectral response: 330-1,100 nm
Min illumination: 20 mlx

Contractor
IRT Infrarot-Technik Eiselt.

New Noga Light C8HTV ICCD series

Type
Image-intensified video camera.

Description
Noga Light's C8HTV series are image-intensified CCD cameras for low-light level applications. They are of modular design with detachable standard TV lenses and electronic control units. The systems use an 18 mm Super second-generation image intensifier tube linked to a CCD sensor via a fibre optical taper. The C8HTV is a night camera and the C8HTVG is a day/night automatically gated ICCD camera.

Operational status
In production.

Specifications
Dimensions: 112 × 54 × 54 (C8HTVG) mm
Weight: 0.5 kg
Image intensifier tube
Tube type: GEN II SUPER
Photocathode: S-25 extended
Sensitivity (at 2,856 K source): 600 μm a/Lm
Resolution: 45 lp/mm

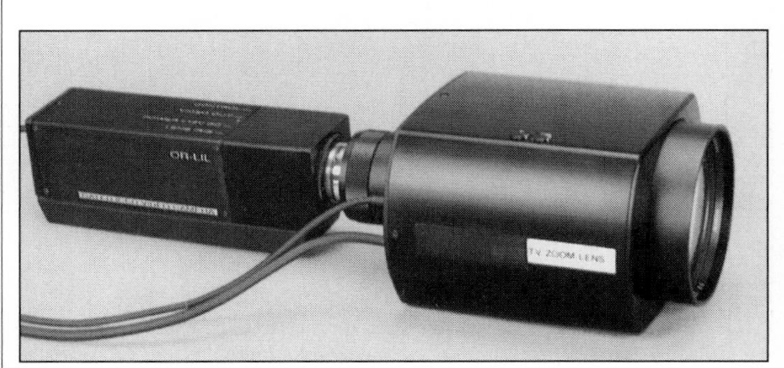

The C8HTVG day/night camera

Imager
Type: ⅔ in interline transfer CCD
Number of pixels: 756 × 581 (CCIR), 768 × 493 (EIA)
Cell size: 11 × 11 μm (CCIR), 11 × 13 μm (EIA)
Dynamic range: 67 dB, low noise, blooming suppression
System resolution (at 0.1 lx × 100% contrast): 450 (C8HTV), 400 (C8HTVG) TV lines
Illumination levels (faceplate): 10^{-6} lx min (C8HTV), 10^{-6} min to 10^{+5} lx (C8HTVG)
Video output: CCIR 625 lines/50 Hz or EIA 525 lines/60 Hz, 2:1 interlace
Power consumption: 2.5 W

Contractor
New Noga Light Ltd.

New Noga Light night video observation devices, models NL-87TV/NL-89TV

Type
Image intensifier cameras.

Description
Noga Light's NL-87TV and NL-89TV night vision TV equipment is based on CCD video cameras coupled to standard night vision equipment such as the AN/TVS-5, N.O.D, M-32/-36 'Night Elbow' and AN/VVS-2 systems.

The NL-89TV and NL87TV long-range night surveillance systems and the NL-87TV medium-range are advanced night telescopic devices. They offer improved range and image quality by means of an integrated image enhancement processor and motion detection feature. The systems use natural light, either starlight or moonlight and amplify it to the sensitivity level of the CCD imager.

Suitable for mobile, airborne or marine applications, these systems can be operated by remote control and provide optimal positioning of the camera with maximum safety for the operators. Standard video output can be recorded, displayed and output to a transmitter.

Built-in features include ambient light compensation, motion detection and real-time image enhancement, offering improved performance and a high degree of automation.

Operational status
In production.

Specifications

System	NL-87TV	NL-89TV
Optical data		
Field of view:	6.5 × 4.8°	3.8 × 2.8°
Limiting resolution		
at 0.01 ×1, 100 % contrast:	450 TVL	450 TVL
Detection range		
2.3 × 2.3 m target		
moonlight:	2,600 m	4,400 m
starlight:	1,900 m	3,200 m
Focusing range:	60 m to infinity	60 m to infinity
Objective T number:	1.6	1.7
Objective focal length:	155 m	256 mm
Environmental data		
Operating temperature:	−10 to +50°C	
Storage temperature:	−20 to +60°C	
Dimensions		
Optical sensor (ø × l):	15 × 43 cm	26 × 53 cm
Control unit (h × w × d):	25 × 33 × 28 cm	
Rechargeable battery (h × w × d):	19 × 19 × 19 cm	
Power supply:	10 × 16 × 28 cm	
Battery charger (h × w × d):	18 × 16 × 10 cm	
Weights		
Optical sensor:	3.8 kg	15 kg
Control unit:	9.5 kg	
Rechargeable battery:	9.5 kg	
Power supply:	4.5 kg	
Battery charger:	2 kg	
Power supply:	12 V DC, ±10%	
Power consumption:	30 W, max	

Major parts
ICCD Camera Unit
Image Intensifier Tube
Type: 25 mm, GEN II
Photocathode: S20, extended red
Cathode sensitivity at T=2,856 K: 400 μA/lm, min
Resolution: 36 lp/mm
CCD Camera
Horizontal resolution: 560 TVL
Photo sensitive elements (h × v): 756 × 581
Imaging area: 8.8 × 6.6 mm (⅔ in format)
Transmission standard: CCIR
Scanning: 625 lines, 50 Hz, 2:1 interlace
Control Unit
TV monitor: black/white, 7 in screen
Image Enhancement Unit
I/O video standard: CCIR
Input voltage: 1 Vpp (0.7 Vpp active video)
Sampling rate: 13.4 MHz, 8-bit resolution
Frame size: 512 lines × 512 pixels
Termination: 75 Ω
Look-up table: (LUT): 1 of 8 256 × 8 firmware programmable LUT may be selected
Rechargeable Battery
Power: 12 V DC/24 Ah.
Continuous operation, min: 7.5 h.
Power Supply
Input voltage: 210-250 V AC, 50-60 Hz
100-120 V AC, 50-60 Hz (optional)
Output voltage: 12.6 V DC regulated
Output power: 40 W
Battery Charger
Input voltage: 230 V AC ±10%
Output voltage: 12 V DC
Output current: 3 A
Indicators: 'Power On', 'Main Charge Mode', 'Constant Voltage Mode' and 'Float Mode'

Contractor
New Noga Light Ltd.

Noga Light NL-87TV/NL-89TV cameras 0053781

New Noga Light NightSPY

Type
Image intensifier camera.

Description
NightSPY from Noga Light is an image-intensified hand-held night vision camera. It has automatic brightness control and infra-red illuminator for use in total darkness.

Operational status
Available.

Specifications
Dimensions: 182 × 65 × 100 mm
Weight: 700 g
Magnification: ×1.7
Field of view: 10°

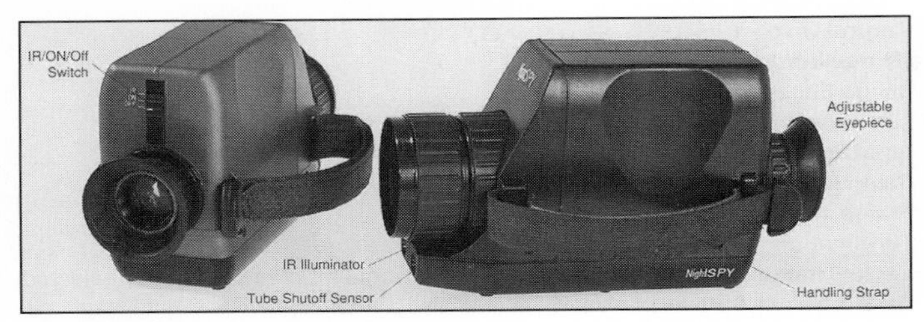

NightSPY compact nightscope

Focusing range: 250 mm to infinity
Objective focal length: 50 mm
F-number: 1.6
Dioptre range: ±3 dioptres
Power source: 3 V DC, 2 AA batteries

Contractor
New Noga Light Ltd.

ROMZ H3T-1

Type
Image intensifier camera.

Description
H3T-1 is a night vision scope for observation which can be used with any standard 'Zenit' lenses, as well as various changeable devices with thread engagement. It uses autonomous block, laser illumination.

Operational status
Available.

Specifications
Magnification (not less than): ×2.4
Field of view (not less than): 15°
Resolution, angle: >100 arcsecs
Dioptre adjustment: ≥±4
Power: 9 W
Adjustment dimensions
Lens: M42 × 1/45.5
Ocular: M32 × 0.75
Dimensions of tube (without lens and ocular): 86 × 64 × 110 mm
Mass of tube (without lens and ocular): 0.55 kg

Contractor
ROMZ, Rostov Optical-Mechanical Plant (JSC).

VERIFIED

VTUVM military CCD camera series

Type
Image intensifier and day cameras.

Description
VTUVM produces a number of CCD military cameras for day or night use including as follows:
JPJ-S-II-200 night military ICCD (image intensified) camera
JPJ-S-300 day military camera
JPJ-O-78 day military camera
They are designed for observation, detection, recognition and aiming purposes in severe environmental conditions. The cameras are built to ensure satisfactory operation when exposed to mechanical shock, vibration, dust, rain and extreme temperature ranges. They are designed to be used in reconnaisance or light combat vehicles or AFVs as well as in off-vehicle (for example, tripod-mounted) applications.

The night versions are equipped with a SuperGEN II microchannel plate image intensifier coupled via fibre optics to a CCD sensor. An electromagnetic shutter protects the night cameras from exposure to excessively bright light.

The day cameras operate in the near infra-red band (650 to 950 nm).

The cameras use the CCIR TV standard – 625 TV lines/50 Hz. The CCD sensors are 2/3″ format. All internal functions (for example aiming marks, electronic boresighting and focus correction) are controlled via RS-485 serial links.

Operational status
In service with the Czech Army.

Specifications
Model JPJ-II-200 (Night TV Camera)
Dimensions (l × w × h): 435 × 126 × 135 mm
Weight: 6.1 kg
Sensor: 2nd++ generation microchannel image tube with fibre optic-coupled CCD, frame transfer, active pixels 768 × 576
Field of view: 4 × 4°
Aiming accuracy - stability: better than 0.2 mils
Light conditions: 1,5 mlx - 1,5 lx
Target detection (tank): 1.6 km
Housing: watertight
Power supply: 21-32 V DC
Model JPJ-O-78 (day camera)
Dimensions (l × w × h): 255 × 95 × 135 mm
Weight: 3.1 kg
Field of view: 4.6 × 6.1°
Aiming accuracy: <0.3 mils
Max detection range (tank): 5 km
Power supply: 21-32 V DC

Model JPJ-S-300 (day camera)
Dimensions (l × w × h): 455 × 127 × 135 mm
Weight: 5.8 kg
Field of view: 1.2 × 1.6°
Aiming accuracy: <0.2 mils
Max detection range (tank): 10 km
Power supply: 21-32 V DC

Contractor
VTUVM Military Institute for Weapon and Ammunition Technology, Czech Republic.

UPDATED

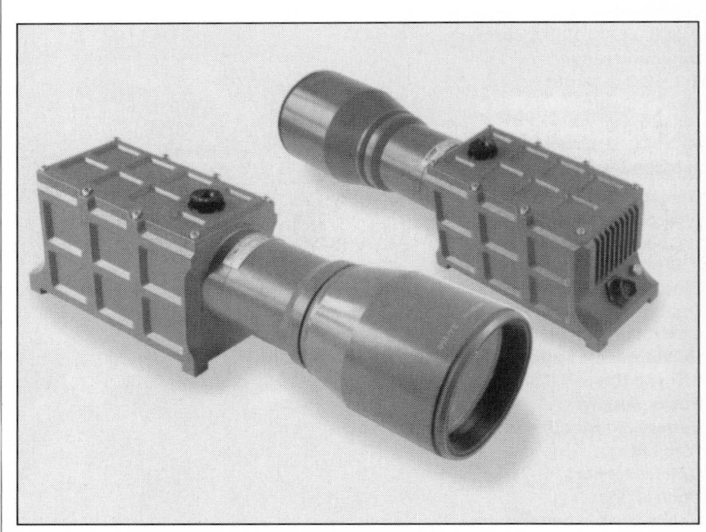
The VTUVM JPJ-II-200 night military CCD camera 0053776

OBSERVATION AND SURVEILLANCE

IMAGE INTENSIFIER GOGGLES

Angénieux Lucie

Type
Image intensifier goggles.

Description
Lucie multipurpose night vision goggles have been developed by Angénieux, a subsidiary of Thales Optronics, for the French Army.

Lucie can be fitted with a second-, Super or third-generation image intensifier with automatic brightness control and has a 50° field of view. The system also has a ×4 magnification quick-mounting magnifier lens. There is a built-in infra-red illuminator with an IR-on indicator in the left eyepiece.

Lucie can be helmet mounted with flip-up face mask or hand held.

Operational status
In mass production. In service with the French Army.

Specifications
Weight (with battery): <0.45 kg
Magnification: ×1, ×4
Field of view
×1: 50°
×4: 10°
Image intensifier: GEN II, SuperGEN or GEN III
Resolution (×1)
GEN II: 1.5 lp/mrad
SuperGEN: 1.2 lp/mrad
GEN III: 1.2 lp/mrad
Dioptre adjustment: –5 to +3 dioptres
Interpupillary distance: 56-74 mm
Eye relief: >20 mm
Focus range: 20 cm to infinity
Power supply: 1.5 V or 3.6 V lithium AA battery

Contractor
Angénieux.

UPDATED

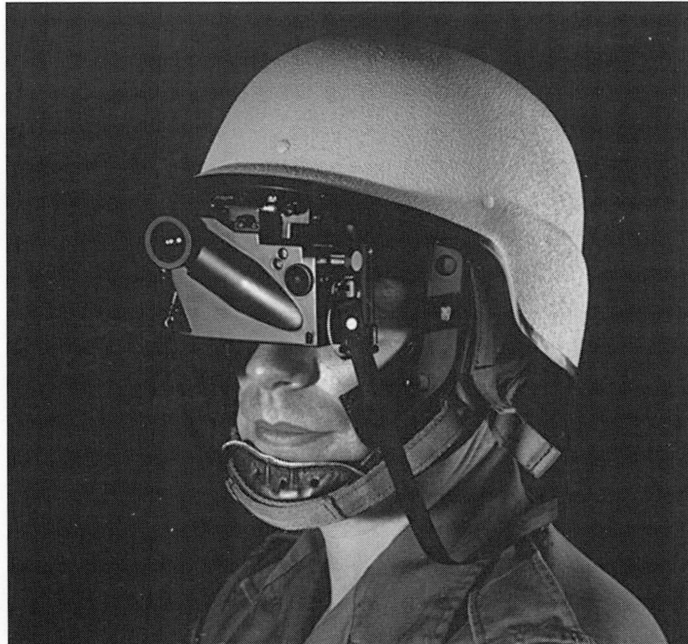

Lucie night vision goggles from Angénieux 0005459

Aselsan M972/M973

Type
Image intensifier goggles.

Description
Aselsan models M972 and M973 are Night Vision Goggles (NVG) that offer high-resolution, image-intensified night vision in a lightweight, compact, single-tube goggle configuration. The NVGs are water-resistant and can be used either with a facemask or as a hand-held viewer. The standard ×1 magnification is effective for short- to medium-range surveillance. For long-range use, both models can be field converted to a ×3, ×4 or ×6 binocular by a simple objective lens change. The M972 and M973 are identical except for the image intensifier, M972 using a GEN II+ tube and M973 a GEN III tube.

Operational status
In production. These products are also manufactured in the US by Litton, the licensor.

Specifications
Weight: 680 g
Magnification: ×1
Field of view: 40°
System gain (M972/3): 1,850/2,500 min
Resolution (M972/3): 0.76/1.01 lp/mrad min
Image intensifier
M972: GEN II+ non-inverting
M973: GEN III non-inverting
F-number: F/1.2
Eye relief: 27 mm
Power supply: 2 × 1.5 V AA alkaline or 1 × 3.9 V AA lithium battery

Contractor
Aselsan Inc, Microelectronics, Guidance and Electro-optics Division.

VERIFIED

Aselsan M972/M973 Night Vision Goggles (NVG)

Aselsan M982/M983/M983A

Type
Image intensifier goggles.

Description
Aselsan models M982 and M983/M983A are night vision monocular systems that offer high-resolution, image-intensified night vision in a lightweight single-tube and single-eyepiece configuration.

The M982 and M983/M983A models are identical except for the image intensifier, M982 using a GEN II+ tube and M983/M983A a GEN III tube. The single-eye format night vision monocular is based on independent use of each eye. One eye is equipped with a night vision device while the other eye remains free and uncovered. A major disadvantage of the standard dual-eye goggle is the one minute pupil dilation period after the goggle is removed. During this period, the user may be night-blind. With a monocular (M982/M983/M983A) goggle, the uncovered eye is already dilated and provides instant transition to ambient light viewing.

Another advantage of the monocular is the wider perceived Field Of View (FOV). The user has a 40° FOV in the right eye and 90° minimum peripheral vision with the left eye. This configuration gives the user the sense of a greater FOV.

For long-range observation, both models can be easily converted in the field to ×3 magnification and 13° FOV by an objective lens change.

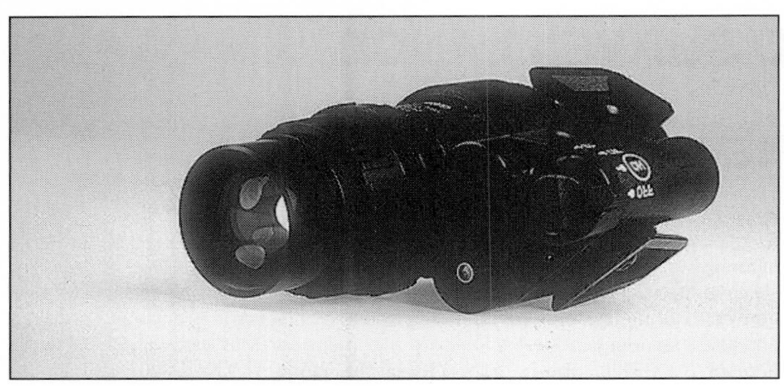

The Aselsan M982/M983/M983A night vision goggles

Operational status
In production. These products have been manufactured in the US by Litton (now part of Northrop Grumman), the licensor.

Specifications
Magnification: ×1
Field of view: 40°
Image intensifier
M982: GEN II+, inverting
M983: GEN III, inverting

M983A: Adv. GEN III, inverting
Resolution: M982 0.76 lp/mrad; M983 1.01 lp/mrad; M983A 1.28 lp/mrad
F-number: F/1.2
Eye relief: 15 mm
Weight: 350 g
Power supply: 1× 1.5 V AA alkaline or 1.5 V/3.9 V lithium battery

Contractor
Aselsan Inc, Microelectronics, Guidance and Electro-optics Division.

UPDATED

BE Delft BENG-9402 night vision goggle

Type
Image intensifier goggle.

Description
The BE Delft BENG-9402 night vision goggle is a lightweight biocular system in which both eyes share the output from one 18 mm second-generation non-inverting image intensifier tube. Both eyes see the full 40° field of view which results in less eye strain compared to the use of monocular or binoculars with partially overlapped fields.

Operational status
Available.

Specifications
Dimensions: 170 × 130 × 75 mm
Weight: 585 g
Field of view: 40°
Magnification: ×4
Resolution: 1.4 mrad

Contractor
BE-Delft Electronics Ltd, India.

VERIFIED

BENG-9402 night vision goggle 0023155

ENOSA GVN-401

Type
Image intensifying goggles.

Description
The GVN-401 night vision goggles have one objective lens, one image intensifier tube and two eyepieces. Weighing 750 g, the goggle assembly can be hand held or mounted on a facemask, freeing both hands for other tasks. The goggles have a built-in infra-red emitter.

Operational status
Production as required. In service with the Spanish Army.

Specifications
Tube: 18 mm GEN II or GEN III
Magnification: ×1
Field of view: 40°
Focus: 31 cm to infinity
Dioptric adjustment: +2 to –6 dioptres
Interpupillary adjustment: 55-71 mm
Resolution (at 10^{-3} lx): 0.66 lp/mrad
Weight: 750 g

Contractor
Empresa Nacional de Optica SA (ENOSA).

VERIFIED

GVN-401 night vision goggles from ENOSA

Institute of Optronics Pakistan AN/PVS-5A night vision goggles

Type
Image intensifier goggles.

Description
The Institute of Optronics AN/PVS-5A night vision goggles are a locally built variant of the standard US Army AN/PVS-5A night vision goggle system for general purpose work in the visible and infra-red spectral regions (up to 0.86 μm) including vehicle driving.

AN/PVS-5A night vision goggles

The goggles incorporate two second-generation, 18 mm, microchannel, wafer type, image intensifier tubes which allow the user to view at ranges from 250 mm up to infinity. A self-contained infra-red illuminator source is used for map- reading and equipment maintenance.

Operational status
Production. In service with the Pakistan Army.

Specifications
Weight: 850 g
Dimensions: 117 × 173 × 165 mm
Magnification: ×1
Brightness gain: 2,000 nominal
Resolution: 27 lp/mrad
Dioptre range: +2 to –6 dioptres
Field of view: 40°
Battery: 2 × 1.5 V AA
Focus range: 25 cm to infinity
Objective: 27 mm, f/1.0

Contractor
Institute of Optronics Pakistan.

VERIFIED

ITT AN/PVS-7D (F5001 series)

Type
Image intensifier goggles.

Description
The AN/PVS-7D is the latest development of the AN/PVS-7 night vision goggles which are in service with a number of countries. It forms part of the US Army Omnibus IV an V night vision procurement programmes and is sponsored by the Night Vision/Reconnaissance and Surveillance Target Acquisition (NV/RSTA) department of the US Army Night Vision and Electronic Sensors Directorate (NVESD). The system upgrades previous AN/PVS-7 models with improvements in the optical train and image intensifier technology that increase system resolution to 1.15 cy/mrad, boost photosensitivity and improve gain, giving increased range performance.

The top performing version of AN/PVS-7D uses ITT's third-generation MX-10130D image intensifying tube developed for Omnibus IV. The infra-red illuminator has a 'momentary' IR capability which allows the user to hold the off-on-IR switch to the IR position for temporary illumination which switches off as soon as pressure is released. This prevents accidental exposure of the IR to other night vision devices. A slip-on adjustable spot/flood lens is provided for the illuminator. The goggle can be hand-held as well as head or helmet mounted. A ×3 or ×5 magnifier and compass attachment are optional.

ITT's AN/PVS-7D (F5001) is available in the following configurations:

- F5001L – AN/PVS7D US mil version, using the MX-10130D tube (Omni V Mil Spec)
- F5001K – AN/PVS7D US mil version, using the MX-10130D tube (Omni IV Mil Spec)
- F5001P – ITT version using the F9810P tube (Omni IV performance)
- F5001N – ITT version using the F9810N tube (Omni V / Ultra performance)
- F5001J – ITT version using the F9810J tube (Omni III Plus performance)
- F5001PC – ITT version using the F9810C tube (Omni III Enhanced performance)
- F5001B – ITT version using the F9810B tube (Omni II Enhanced performance)

Operational status
Under contract for the US Army.

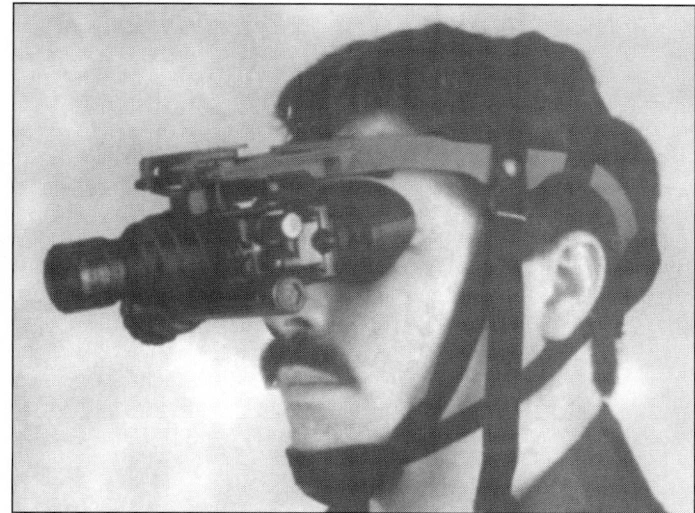
AN/PVS-7D GEN III night vision goggles developed for the US Army by ITT 0005463

Specifications
Image intensifier: GEN III 18 mm
Tube: MX-10130D
Scene illumination: 10^{-6} to 1 fc
Spectral response: visible to 0.9 μm (IR)
Magnification: ×1, ×3 or × 5 optional
Field of view: 40° circular
Resolution: 1.15 cy/mrad min
Brightness gain: 3,000 fL/fL
Collimation
Convergence: 1 ±1°
Dipvergence: ≤1/2°
Dioptre adjustment: +2 to –6 dioptres
Interpupillary adjustment: 55-71 mm
Eye relief: 15 mm
Objective lens: EFL 26 mm, F/1.2, T/1.3
Focus range: 20 cm to infinity
Weight: 0.68 kg
Power requirement: 2 × 1.5 V AA batteries or 1 × 3 V BA-5567/U lithium battery
Operating temperature range: –51 to +45°C

Contractor
ITT Defense & Electronics, Night Vision Division.

ITT F5002A – second-generation night vision goggles

Type
Image intensifier goggles.

Description
The F5002A is a second-generation version of ITT's AN/PVS-7B third-generation night vision goggles. Each unit allows objective focus, eyepiece dioptre adjustment and interpupillary distance adjustment when used in conjunction with the head mount.

The F5002A uses a second-generation image intensifier assembly and 18 mm microchannel wafer (F9806A). This image intensifier is an improved version manufactured with updated processes and techniques.

The F5002A also includes a quick-release lever, which permits one-hand mounting and unmounting, a low-voltage indicator and an automatic high-light cut-off that protects the image intensifier. The F5002A can be hand held or head mounted, making it useful in any situation requiring night vision.

Optional accessories include a slip-on or screw-in ×3 magnifier lens; slip-on compass for night-time orientation; slip-on adjustable spot/flood lens for the IR illuminator; helmet mount; C-mount adaptor that accepts any C-mount objective lens and a tripod adaptor that allows mounting on a wide variety of hardware.

Specifications
Field of view: 40°
Magnification: unity
Resolution: 0.68 cy/mrad
Brightness gain: 1,800 fL/fL, min
Collimation: 1 ±1° convergence; ≤1/2° divergence
Dioptre adjustment: +2 to –6 dioptres
Interpupillary adjustment: 55-71 mm
Eye relief: 15 mm
Objective lens: EFL 26 mm, F/1.2, T/1.3
Eyepiece lens: EFL 26 mm
Focus range: 20 cm to infinity
Voltage required: 2.7-3.0 V DC, battery (50 mA, max)
Battery type: 2 × AA size 1.5 V alkaline or 1 × 3.0 V BA 5567 lithium battery
Weight: 680 g
Operating temperature range: –51 to +45°C

Contractor
ITT Defense & Electronics, Night Vision Division.

VERIFIED

ITT Mini Night Single-Eye Acquisition Sight (Mini N/SEAS) Gen 3 or Gen 2

Type
Image intensifier monocular.

Description
ITT's Mini N/SEAS monocular is designed for the individual soldier to use in a variety of ground-based night operations, it features the superior performance of the MX-1 0160 (F9800) image intensifier tube, which is available with a wide range of performance levels to meet individual customer requirements.

Although designed as a monocular, the Mini N/SEAS may be easily configured as a dual-eyed, binocular (goggle) with the use of two monoculars attached to a

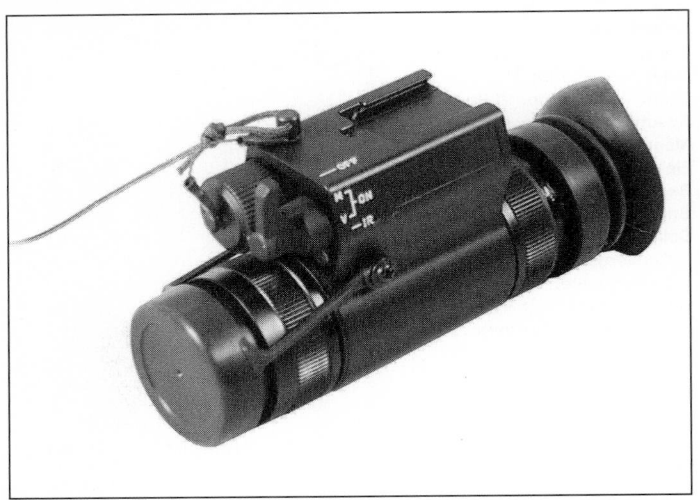

ITT's Mini N/SEAS 0132095

helmet or head mount. The manufacturer states that Mini N/SEAS is the only monocular in the world that enables collimation of better than 0.5° when combining two randomly selected units.

For weapon firing, the Mini N/SEAS mounts behind a standard collimated dot sight on the MI6/M4 rifle, and this sight provides the aim point. When adapted with a ×3 magnifier, the unit fits similarly behind the collimated dot sight. This configuration provides the soldier with a ×3 night scope, significantly increasing the range of the Mini N/SEAS for weapon firing. Designed to be used with and mounted to the MI6/M4 family of weapons, the Mini N/SEAS provides increased versatility while eliminating the need to carry multiple types of specialised equipment.

Operational status
Available and in service.

Specifications
Weight (excluding head/helmet mount): 330 g for 15 mm eye-relief eyepiece
System magnification: ×1
Field of view: 40°
On-axis resolution at optimum light level:
1.3 cycles/mrad (Gen 3)
1.1 cycles/mrad (Gen 2)
System distortion: less than 3%
Focus range: 25 cm to infinity
Dioptre adjustment: +2 to –6 dioptres
Exit pupil: 10 mm on-axis exit pupil at 25 mm or 15 mm eye relief
System brightness gain: More than 3,000
Operating temperature: –51 to +49°C
Storage temperature: –51 to +85°C

Contractor
ITT Industries, Night Vision Division.

LOMO night vision goggles

Type
Image intensifier goggles.

Description
LOMO's night vision goggles are designed to provide a night vision capability at moderate cost. The goggles have high quality optics combined with first-generation image intensifiers. An infra-red illuminator is incorporated to enhance performance in very dark conditions. The makers state that unaided performance allows contrasting objects to be observed at 100 m in quarter moonlight conditions.

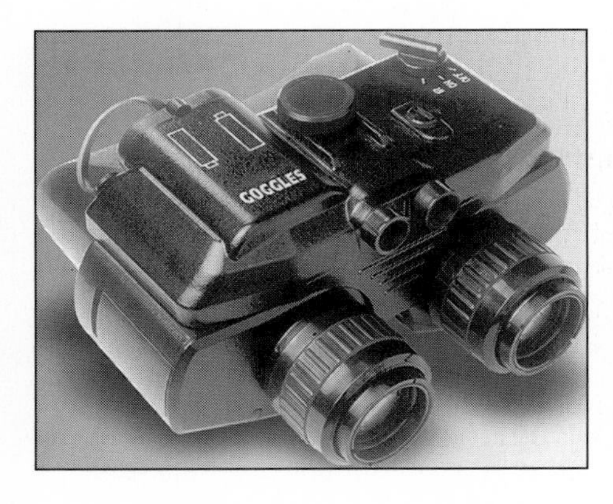

The LOMO night vision goggles
0097948

Operational status
Available.

Specifications
Weight: 730 g
Dimensions: 168 × 116 × 150 mm
Magnification: ×1
Field of view: 26°
Focus range: 0.26 m to infinity
Gain: 500
Sensitivity: 250 μA/lm
Power supply: 3 V battery
Battery life: typically 4 h continuous operation

Contractor
LOMO PLC.

UPDATED

Meopta KLARA night vision goggles

Type
Image intensifier goggles.

Description
KLARA night vision goggles are configured with a single objective lens and a GEN II Super image intensifier. The intensified image is transmitted via a beamsplitter to two eyepiece lenses to provide binocular viewing. The device may also be fitted with an afocal lens which provides magnification when required. KLARA may be hand held or head mounted.

Operational status
Available.

Specifications

Model	KLARA + afocal	KLARA
Dimensions		
length	213 mm	130 mm
width	147 mm	120 mm
height	98 mm	85 mm
Weight (inc battery)	1,025 g	450 g
Magnification	×4	×1
Field of view	10°	50°
I² Tube resolution (5 × 10⁻¹ lx)	36 l/mm	
Dioptre adjustment	–5 to +2D	–5 to +2D
Focus range	10 m to infinity	20 m to infinity
Power supply	1 AA battery	1 AA battery
Battery life	>20 h	>20 h

Contractor
Meopta Prerov, a.s.

VERIFIED

Meopta KLARA night vision goggles with afocal 0023156

New Noga Light NL-90 second-generation night vision goggles

Type
Image intensifier goggles.

Description
The Model NL-90 from Noga Light is a second-generation night vision goggle which is adjustable to fit the user's face contour. It has fixed gain, automatic brightness control and adjustable focus objective lenses and eyepiece lenses.

Operational status
In production.

Specifications
Dimensions: 120 × 171 × 171 mm
Weight: 0.96 kg
Magnification: ×1
Field of view: 40°
Focus range: 25 cm to infinity
Objective focal length: 27 mm
Objective F-number: 1.05
Eyepiece focal length: 27 mm
Eyepiece pupil diameter: 10 mm
Interpupillary distance: 60-72 mm
Dioptre adjustment: +2 to –6 dioptres
Tube type: 18 mm GEN II Super
Tube resolution (typical): 41 lp/mm
Resolution: 1.5 mrad
Brightness gain: 2,000 nominal

Contractor
New Noga Light Ltd.

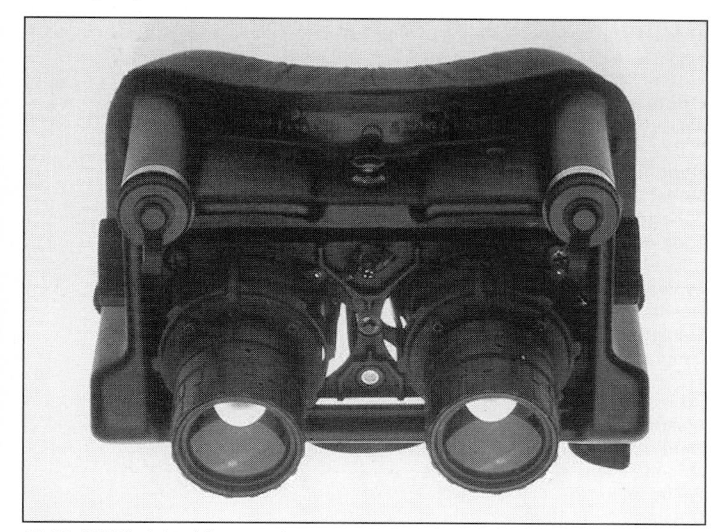

NL-90 GEN II night vision goggles

New Noga Light NL-91 night vision goggles

Type
Image intensifier goggles.

Description
The NL-91 is a lightweight, head-mounted, self-contained night vision device, equivalent to the US AN/PVS-7B. It enables the user to perform tasks such as walking, driving, weapon firing, short-range surveillance, map-reading and medical aid under both moonlight and starlight conditions.

The NL-91 has been human engineered for quick installation or removal and to be worn for long periods with comfort. It can be hand-held or head-mounted, making it usable in any situation requiring night vision.

Operational status
In production.

Specifications

Type	×1	×3	×4.5
Weight	680 g	960 g	1,125 g
Magnification	×1	×3	×4.5
Field of view	40°	14°	9°
Objective lens EFL, mm	26 mm	75 mm	116 mm
Objective lens, F number	1.2	1.5	2
Objective lens T number	1.3	1.55	2.2
Eyepiece, lens EFL, mm	26	26	26

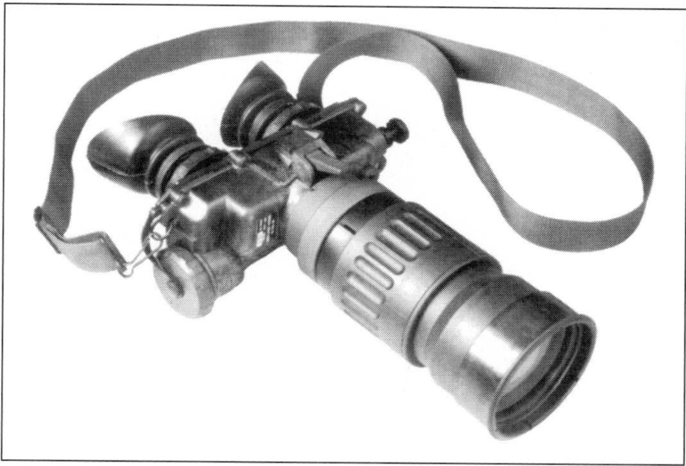

Noga Light NL-91 night vision goggles 0053782

Type	×1	×3	×4.5
Focus range	20 cm to infinity	1.52 m to infinity	35 cm to infinity
Resolution (cycles/ mrad)			
GEN II+ (with 40 lp/mm)	0.76	1.95	3
GEN II Super (with 45 lp/mm)	0.95	2.2	4.2
GEN III (with 45 lp/mm)	0.95	2.2	4.2
Brightness Gain, (fl/fl)			
GEN II+	1,800	1,400	1,000
GEN II Super	2,250	1,750	1,200
GEN III	2,250	1,750	1,200
Dioptre adjustment	–2 to –6		
Interpupillary adjustment	55 to 71 mm		
Eye relief	15 mm		
Collimation	1 ±1° convergence = 1/2° dipvergence		
Power supply	2× AA size 1.5 V alkaline or 1× lithium 3.0 V		
Operating temperature	–51 to +45°C		

Contractor
New Noga Light Ltd.

New Noga Light Personal Night Goggles (PNG2)

Type
Image intensifier goggles.

Description
Noga Light's Personal Night Goggles (PNG2) are a night-vision binocular with a head mount and a built-in infra-red illuminator.

Operational status
Available.

Specifications
Dimensions: 148 × 150 × 90 mm
Weight: 700 g
Magnification: ×1
Field of view: 20° min

PNG2 night vision goggles

Focus range: 25 cm to infinity
Objective focal length: 26 mm
Objective F-number: 1.3
Eyepiece focal length: 22 mm
Exit pupil diameter: 6 mm
Interpupillary distance: 55-75 mm
Dioptre adjustment: –6 to +3 dioptres
Tube centre resolution: 22 lp/mm

Contractor
New Noga Light Ltd.

Night Vision Equipment Company AN/PVS-5C night vision goggles NVEC Models 800/800HP/850

Type
Image intensifier goggles.

Description
The NVEC 800 series of goggles are hybrid versions of the AN/PVS-5 which incorporates either second-generation, second-generation plus or third-generation image intensifiers. Developed by the Night Vision Equipment Company, they include full MilSPEC hardware and optics with COMSPEC intensifier tubes. The tubes have automatic brightness control and high light level cut-off protection. NVEC states that the performance of the Model 800HP exceeds the US Army specification for the AN/PVS-5C and that of the Model 850 exceeds the AN/PVS/7B MilSPEC by 63 per cent.

Operational status
Available.

Specifications

Model	800	800HP	850
Tube Generation	II	II+	III
Tube class	COMSPEC	COMSPEC	COMSPEC
Typical detection range in m for still human target on green grass			
Quarter moon (10^{-2} lx)	236	253	427
Starlight (10^{-3} lx)	117	138	310
Overcast (10^{-4} lx)	95	111	163

(All models)
Weight: 960 gm
Field of view: 40°
Magnification: ×1
Focus range: 25 cm to infinity
Objective focal length: 26.8 mm
Dioptre adjustment: +2 to –6
Eye relief: 15 mm (min)
Interpupillary adjustment: 54-72 mm
Power supply: 2 × AA or 1/BA5567/U batteries

Contractor
Night Vision Equipment Company Inc.

NVEC AN/PVS-5C night vision goggles 0041409

Night Vision Equipment Company AN/PVS-7B night vision goggle NVEC Models 1500-2/1500-4/1500-5

Type
Image intensifier goggles.

Description
The AN/PVS-7B is a US Department of Defence (DoD) nomenclatured NVG design for ground forces. NVEC produces several variants of the basic design of which the Model 1500-5 is MIL-Qualified to the DoD specification. The design uses one image intensifier tube and the image from this is relayed to both eyes using beam-splitter optics. The intensifier has automatic brightness control and there is automatic protection of the tube photocathode from high brightness light sources. A built-in IR illuminator enhances viewing of details of, for example maps at close ranges. A magnifier lens (×3 or ×5) is optional.

NVEC's AN/PVS-7B night vision goggles 0041408

Operational status
Available.

Specifications

Model	1500-2	1500- 4	1500-5
Tube generation	II	II+	III
Tube class	MILSPEC	COMSPEC	MILSPEC
Typical detection range in m for still human target on green grass			
Quarter moon (10^{-2} lx)	236	256	417
Starlight (10^{-3} lx)	115	199	297
Overcast (10^{-4} lx)	90	116	163

(All models)
Weight: 680 gm
Field of view: 40°
Magnification: ×1
Focus range: 25 cm to infinity
Objective focal length: 27 mm
Dioptre adjustment: +2 to –6
Eye relief: 15 mm (min)
Interpupillary adjustment: 55-71 mm
Power supply: 2 × AA or 1/BA-5567/U batteries

Contractor
Night Vision Equipment Company Inc.

Norinco Type 1985 passive night vision goggles

Type
Image intensifier goggles.

Description
The Norinco Type 1985 passive night vision goggles is a lightweight night viewing system which uses two single-stage image intensifier tubes of the third-generation type with focusable objective lens and adjustable eyepieces and can be used with an infra-red light source as the illuminator for document or map-reading.

When in use, the goggles are attached to the user's head by chin and headstraps.

Operational status
Production as required. In service with the Chinese armed forces and other undisclosed countries.

Specifications
Weight: About 0.96 kg
Magnification: ×1
Field of view: 44°

Norinco passive night vision goggles Type 1985

Spatial resolution (target contrast = 0.85)
At 1 × 10^{-2} lx: 2.8 mrad
At 1 × 10^{-3} lx: 4 mrad
Focus range: 300 mm to infinity
Eyepiece dioptre range: +5 to −5 dioptres
Battery: Ni/Cd; 0.5 Ah capacity cell

Contractor
China North Industries Corporation (Norinco).

UPDATED

Northrop Grumman AN/PVS-5B (M912A/M915A) night vision goggles

Type
Image intensifier goggles.

Description
The M912A/M915A night vision goggles are improved versions of the US Military AN/PVS-5B. The M912A utilises two second-generation image intensifier tube assemblies, each having an 18 mm image format, which give a brightness gain of 1,000.

The M915A utilises two 18 mm GEN II+ image intensifier tubes which give the system a brightness gain of 1,700.

Operational status
Production as required. In service with the US Army.

Specifications
Weight: 960 g
Magnification: ×1
Field of view: 40°
Image intensifier
M912A: GEN II
M915A: GEN II+
Brightness gain
M912A: 1,000
M915A: 1,700
Resolution: 0.68 lp/mrad
Field of view: 40°
Battery: 2 × 1.5 V alkaline
Focus range: 25 cm to infinity

Northrop Grumman M912A/M915A night vision goggles

Objective: f/1.1
Range (1.83 m target)
M912A: 230 m
M915A: 300 m

Contractor
Northrop Grumman (formerly Litton) Electro-Optical Systems.

VERIFIED

Northrop Grumman AN/PVS-5C (Model 9876C) night vision goggles

Type
Image intensifier goggles.

Description
The Model 9876C (AN/PVS-5C) head-mounted night vision goggles are an improved version of the standard US Army AN/PVS-5A goggle system for general purpose work including vehicle driving.

The goggles incorporate two second-generation 18 mm MCP wafer image intensifier tubes with a spectral response in the visual and infra-red region (to 0.86 μm).

The user can view at ranges from 254 mm up to infinity by means of a +2 to −2 dioptre focus adjustment. A self-contained infra-red illuminator source is used for map-reading and equipment maintenance.

Operational status
Production. In service with a number of undisclosed countries.

Specifications
Weight: 0.96 kg
Magnification: ×1
Brightness gain: 2,000 nominal
Resolution: 32 Lp/mrad nominal
Field of view: 40°
Battery: 2 × 1.5 V AA or 1 × 3 V mercury/lithium
Focus range: 25.4 cm to infinity
Objective: f/1.05

Contractor
Northrop Grumman Electro-Optical Systems.

UPDATED

Northrop Grumman AN/PVS-5C night vision goggles

Northrop Grumman AN/PVS-7A (M972/M973) night vision goggles

Type
Image intensifier goggles.

Description
The M972/M973 night vision goggles are improved versions of the US Military AN/PVS-7A.

The M972 utilises a single M870, non-inverting, GEN II+ 18 mm, image intensifier tube with a brightness gain of 1,850, whereas the M973 has an M871 non-inverting GEN II+ 18 mm, image intensifier tube with a gain of 2,100.

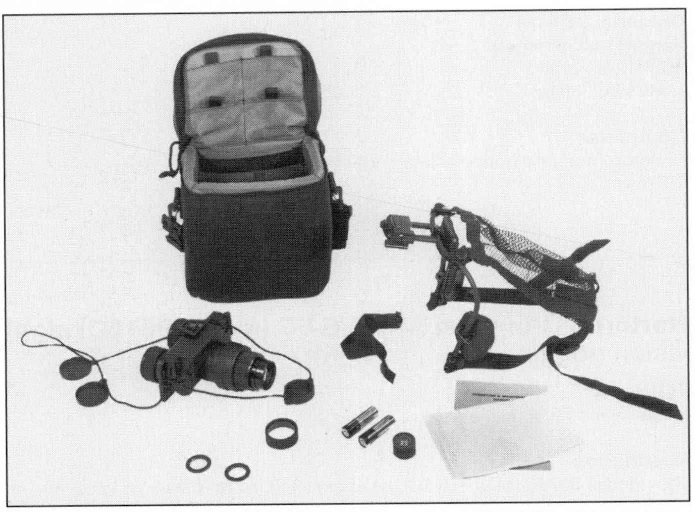

Northrop Grumman M972 night vision goggles, accessories and carrying case

If required, the 27 mm objective lens can be replaced by an optional ×4 magnification lens to convert the system into a hand-held night vision binocular.

Weight is 0.72 kg complete, magnification is ×1 and the field of view is 40° with the range on a 1.83 m target given as 250 m for the M972 and 340 m for the M973.

Operational status
Production. In service with unspecified countries.

Specifications
Weight
Complete: 720 g
Without AA batteries: 680 g
Magnification: ×1
Brightness gain
M972: 1,850
M973: 2,100
Resolution: 0.76 lp/mrad
Image intensifier
M972: GEN II+
M973: GEN III
Field of view: 40°
Battery: 1 × BA-1567/U mercury or BA-5567/U lithium or 2 × AA Ni/Cd or AA alkaline or BA-3058/U alkaline

Contractor
Northrop Grumman Electro-Optical Systems.

UPDATED

Northrop Grumman AN/PVS-7B (Model 1500) night vision goggles

Type
Image intensifier goggles.

Description
The Northrop Grumman (formerly Litton) AN/PVS-7B passive night vision goggles can be used for a wide range of night operations including vehicle driving.

It is fitted with a third-generation image intensifier tube and an optional ×3 or ×5 afocal lens can be provided. The system uses an F/1.2 objective lens and offers a 40° field of view.

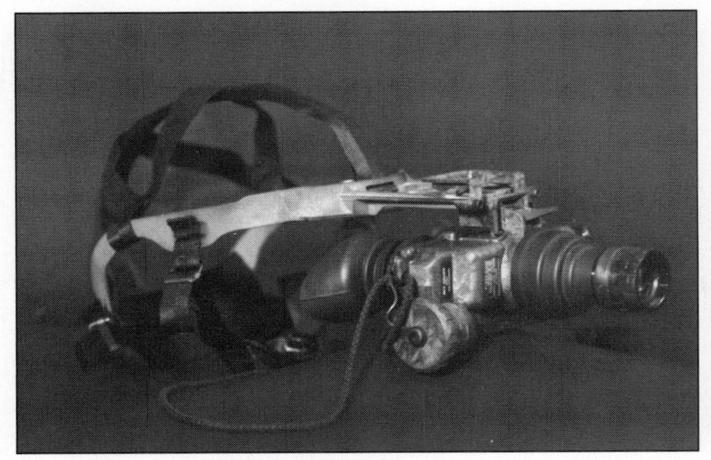

Northrop Grumman AN/PVS-7B night vision goggles

The housing is made of high tensile strength chemical resistant polymer alloy. A purge valve is provided for back-filling with dry nitrogen at the general support or depot maintenance level.

In addition to the standard head mount, Northrop Grumman also offers an optional flip-up assembly designed to be used with the US government Kevlar field helmet but which can be adapted to other helmet styles.

For map-reading or equipment maintenance a small infra-red illuminator source is provided with an eyepiece integrated on/off indicator.

Operational status
Production. In service with a number of undisclosed countries.

Specifications
Weight
Complete: 680 g
Dimensions: 152.4 × 155.6 × 101.6 mm
Magnification: ×1
Brightness gain: 1,850
Resolution: 0.81 lp/mrad (limiting)
Image intensifier: GEN III
Field of view: 40°
Battery: 2 × 1.5 V AA or 1 × 3 V mercury/lithium
Focus range: 25 cm to infinity
Helmet interface: M1 (steel), DH-145 (Kevlar) or Combat Vehicle Crewman (CVC)
Protective mask capability: M17A2, M24, M25A1, XM40, XM41, XM42 and XM43

Contractor
Northrop Grumman Electro-Optical Systems.

UPDATED

PCO MN-1 Night Vision Device (NVD)

Type
Image intensifier goggles.

Development/Description
The MN-1 Night Vision Device (NVD) is a small, compact monocular device equipped with a special harness for attaching to the user's head. It is intended for night observation and surveillance, including monitoring, during night patrols, of features including bunkers and tunnels. They may also be used for map reading at night. The MN-1 NVD uses a GEN II+ or SuperGEN Image Intensifier Tube.

Operational status
Available.

Specifications
Weight: 0.7 kg (1.1 kg ×4 magnification version)
Magnification: 1 × (optional objective – ×4)
Field of view: 40° min (magnification ×1), 10° min (magnification ×4)
Dioptre adjustment: +2 to –4
Ambient operating temperature: –35 to +45°C

Manufacturer
PCO SA, Warszawa, Poland.

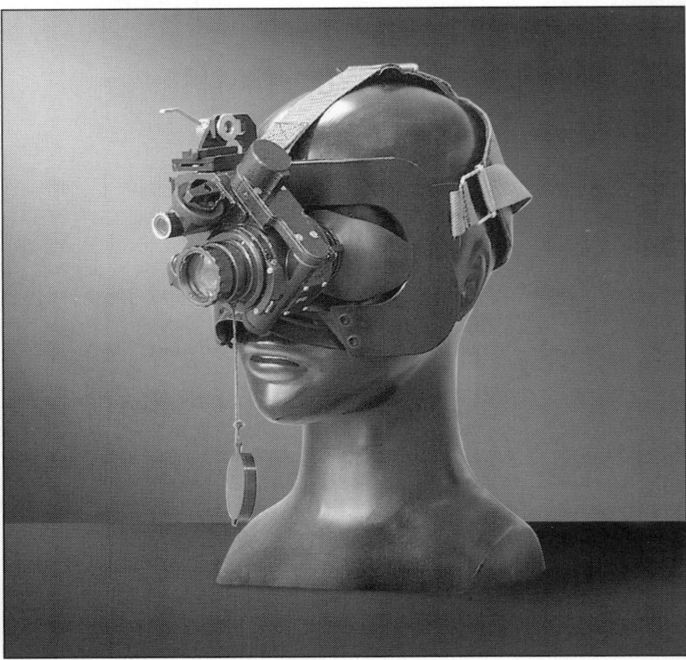

PCO MN-1 Night Vision Device (NVD) 0089916

PCO PNS-1 driver's night vision goggles

Type
Image intensifier goggles.

Description
The PCO PNS-1 night vision goggles are small and lightweight. They use two GEN II+ or GEN II Super image intensifier tubes. A user of the system can undertake tasks such as driving vehicles (including tanks), map-reading, repair work, night observation and surveillance without the need for illumination and with both hands free. An auxiliary IR light source is built-in to aid activities such as map-reading or maintenance work.

Operational status
In production. In service with Poland. Offered for export.

Specifications
Magnification: ×1
Field of view: 38° min
Focus adjustment: 0.25 m to infinity
Power supply: 1 lithium cell 3.6 V AA size
Resolution: 5.7 min
Operating temperature: −20°C to +45°C
Weight: 1 kg
Dimensions (h × l): 80 × 105 mm

Contractor
PCO SA, Warszawa, Poland.

PCO PNS-1 driver's night vision goggles 0023159

ROMZ Cyclop-22™ night vision goggles

Type
Image intensifier binoculars.

Description
The Cyclop-22™ binocular night vision goggles are designed to be head- or helmet-mounted. They are intended for use in ambient light levels down to 0.005 lx, but have an integral infra-red LED illuminator for use at lower levels. A hinge arrangement allows the goggles to be rotated up to a stowed position clear of the eyes.

Operational status
Available.

Specifications
Weight: <0.74 kg
Dimensions: 150 × 125 × 170 mm
Magnification: ×1.15 ±0.05
Field of view: >30°
Resolution: 300 arcsecs
Focusing range: 0.25 m to infinity
Dioptre adjustment: ±5
System gain: 400
Interpupillary adjustment: 54-74 mm
Voltage required: 3 V DC

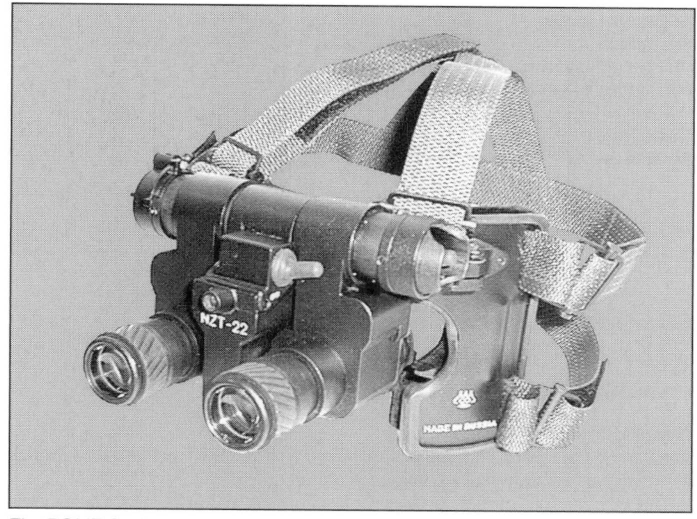

The ROMZ Cyclop-22™ night vision goggles 0088171

Contractor
ROMZ, Rostov Optical-Mechanical plant (JSC).

VERIFIED

SAGEM CLARA/CLARA MAG night vision goggles

Type
Image intensifier goggles.

Description
The CLARA night vision goggles provide a field of view of greater than 50° and are designed for general use including the driving of vehicles and short- and medium-range observation.

For tasks such as map-reading and so on, an adjustable focus is provided as well as additional infra-red lighting. For long-range observation a ×4 magnifier is available.

The CLARA goggles are compatible with SuperGen or Gen III image intensifier tubes and are powered by one AA size battery.

Operational status
In production and in service in France and abroad.

Specifications

	CLARA	CLARA with ×4 magnification
Weight	445 g	900 g
Field of view	>50°	>10°
Magnification	×1	×4

Clara night vision goggles from SAGEM 0005462

Adjustable focus	25 cm to infinity	20 m to infinity
Resolution	0.8 lp/mrad (min)	3 lp/mrad (min)

Dioptre adjustment: +2 to –5 dioptres
Image intensifier: SuperGen or Gen III

Contractor
SAGEM SA, Aerospace and Defence Division.

UPDATED

Simrad GN night vision goggles

Type
Image intensifier goggles.

Description
The Simrad GN night vision goggles are based on a patented optical design, which shortens the forward length, to improve comfort of use.

The GN controls consist of an on/off/IR switch, focus control and dioptric setting. Because of the large exit pupils, adjustment of the interocular distance is not necessary. It is worn either with an ergonomic headmount or is helmet mounted. The dovetail attachment allows adjustment for optimum viewing and enables the goggles to be flipped into a rest position when not in use.

The integral IR diode enhances performance when the goggles are used in conditions of extremely low light and provides illumination for close-range observation work such as map-reading. The GN1 can focus down to 200 mm.

Operational status
Production. In service with several undisclosed countries.

Specifications
Weight: 0.39 kg
Objective lens focal length: 26 mm
Objective lens F-number: f/1
Dimensions: 155 × 73 × 58 mm
Magnification: ×1
Field of view: 40°
Focusing range: 0.2 m to infinity
Dioptre adjustment: –6 to +2 dioptres
Image intensifier tube: 18 mm GEN II or GEN III inverting tube with glass input (ANVIS type)
Power supply: 2 × 1.5 V AA batteries
IR source: LED

Contractor
Simrad Optronics ASA.

VERIFIED

Simrad GN night vision goggles

Thales Jay

Type
Image intensifier goggles.

Description
Jay is a self-contained, head-mounted, unity magnification night vision system which is suitable for personal use when walking, driving, map-reading, or undertaking short-range surveillance and similar tasks. The goggle may be supplied with either a second- or third-generation image intensifier tube, depending upon the resolution required. The goggle may be hand held or head mounted. For close-up viewing tasks such as map-reading or vehicle maintenance, an infra-red emitting light source is incorporated into the goggle to provide the necessary illumination. Power is derived from two commercially available 1.5 V AA batteries.

Thales Jay night vision goggle 0005460

Maximum user comfort is achieved by using a binocular viewer with adjustable dioptric eyepieces to suit individual user requirements.

The Jay goggle has been designed as part of a suite of interchangeable night vision equipment. For medium- to long-range surveillance at a magnification of × 4.5, the goggle can be reconfigured as a Thales Bino-Kite by simply replacing the objective lens assembly. An even longer range upgrade can be achieved by adding the Thales Maxi-Kite adaptor.

Operational status
In production and in service.

Specifications
Weight: 680 g
Magnification: ×1
Field of view: 40°
Power supply: 2 × 1.5 V AA batteries
Eyepiece adjustment: +2 to –6 dioptres
Focus range: 25 cm to infinity
Eye relief (min): 15 mm
Range (min): recognition of standing person at 300 m in starlight conditions (10^{-3} lx)

Contractor
Thales Optics.

UPDATED

Thales Optronics AN/PVS-504

Type
Image intensifier goggles.

Description
The AN/PVS-504 and AN/PVS-504(A) goggles are a low-cost, lightweight and rugged night vision system designed to meet a range of applications under extreme environmental conditions. The goggles are ideally suited for short-range surveillance at night, patrolling and performing manual tasks such as map-reading and administering first aid.

The system is a single-tube monocular unit comprising an objective, an image intensifier tube assembly, a collimating lens assembly, two eyepiece assemblies and numerous accessories.

Operational status
Currently in production. In service in Canada and Malaysia.

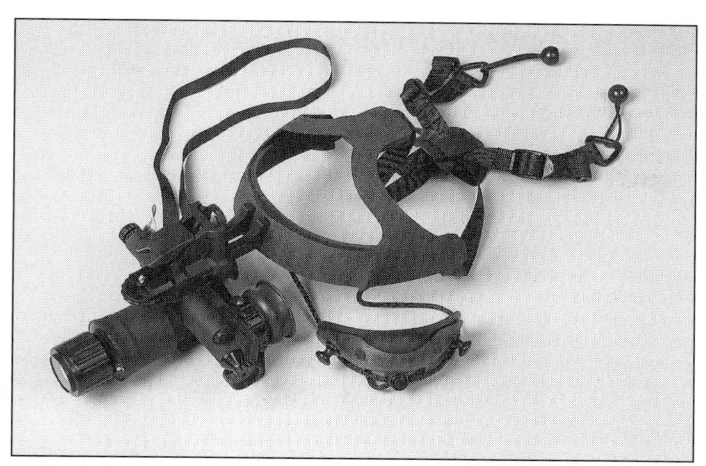

AN/PVS-504 night vision goggles from Thales Optronics

Specifications
Weight: <750 g
Magnification: ×1
Brightness gain
GEN II: 18,000-25,000
GEN III: 20,000-35,000
Resolution
GEN II: 0.68 cy/mrad
GEN III: 0.96 cy/mrad
Field of view: 40°
Batteries: 2 alkaline AA or 1 lithium AA (24 h continuous operation)
Focus ranges: 25 cm to infinity
Interpupillary adjustment: 56-72 mm

Contractor
Thales Optronics, Canada.

VERIFIED

Thales Optronics HNV-1 holographic night vision goggles

Type
Image intensifier goggles.

Description
The HNV-1 goggles use a second-generation 18 mm wafer tube unit (with optional third-generation tube) which uses Holographic Optical Element (HOE) technology to provide the user with a continuous see-through image of the real world without having to remove either the visor or goggles.

Operational status
Production as required. In service with the Belgian Army and special task forces of other NATO countries.

HNV-1 holographic night vision goggles

Specifications
Weight: approx 1 kg
Magnification: ×1
Tube: SuperGEN or GEN III (Omnibus IV spec)
Distortion: < 7%
Field of view
Night image: 30 × 40° (overlapping L+R 20°)
See-through image: unobstructed peripheral vision
Focus range: continuously adjustable from <250 mm to infinity
Interpupillary distance: 57-69 mm without adjustment
Resolution (with 90% target contrast and light level of 10 mlx): 2 mrad/lp
Battery: 2×1.5 V AA size
Centre of gravity: <50 mm from forehead

Contractor
Thales Optronics (formerly OIP Sensor Systems).

UPDATED

Thales Optronics Mono night vision goggle

Type
Image intensifier goggle.

Description
Thales (formerly Pilkington) Optronics' Mono night vision goggle is a single eye monocular which can be either hand-held, facemask mounted or helmet mounted. One eye is equipped with either second- or third-generation image intensifying night sight while the other eye retains normal vision and adaptation to surrounding light conditions. A rotating mechanism allows left or right eye operation or it can be placed in a storage position which automatically cuts off the power supply.

An infra-red illuminator is included. An indicator light in the eyepiece shows when this is activated and when battery power is low. Automatic brightness control protects the image intensifier.

An optional afocal ×3 adaptor can be attached for longer range surveillance and the goggle can be used in conjunction with a laser target marker for weapon aiming. A dual monocular configuration is available for use by drivers and pilots.

Operational status
In production and in service.

Specifications
Magnification: ×1, ×3 (with afocal adaptor)
Field of view: 40° (×1), 13.3° (×3)
Focus range: 25 cm to infinity (×1), 10 m to infinity (×3)
Dioptre adjustment: –6 to +2 dioptres
Eye relief: 15 mm
Exit pupil diameter: 10 mm
Resolution (system): 0.88 lp/mrad (GEN III)
Gain (system): 2,500 (GEN III)
Dimensions: 115 × 65 × 40 mm (mono goggle)
Weight: 330 g (incl battery), 540 g (with battery and facemask)
Power source: 1 × 1.5 V AA alkaline or lithium battery

Contractor
Thales Optronics.

VERIFIED

Thales Optronics Mono night vision goggle 0005461

Thales Optronics UGO day/night goggles

Type
Image intensifier goggle/binocular.

Description
The UGO goggle is designed for day/night observation and is suitable for short- to medium-range day and night surveillance tasks. The day/night goggles/binoculars have an 8 × 24 day binocular channel, the right hand eyepiece containing a graticule graduated horizontally and vertically in mils for range estimation. The ×1 magnification night channel is optically and mechanically integrated with the day binocular and is compatible with either second-, super- or third-generation image intensifier tubes. A ×4 magnification afocal lens adaptor, which can be fitted by the user in the field, is supplied as standard to improve target recognition.

Weighing 750 g, UGO may be hand-held or head-mounted for either wide field of view, night observation or for close-up viewing tasks.

Operational status
In production and in service with the French Army and other countries. Over 5,000 systems have been manufactured.

Specifications
Dimensions (l × h × w): 153 × 62 × 140 mm
Weight: 750 g
Day observation
Magnification: ×8
Field of view: 6°
Night vision
Magnification: ×1, ×4
Field of view: 40° (×1), 10° (×4)
Resolution: <1.5 mrad/lp (×1), <0.4 mrad/lp (×4)
Input lens diameter: 24 mm
Reticule: graduated in mils
Focus range: 25 cm to infinity
Eye relief: 14 mm min
Night channel range (min)
×1: 300 m (recognition of standing person in starlight (10_4 lx))
×4: 400 m (recognition of standing person in starlight (10_4 lx))
Power supply: 2 × 1.5 V AA or 1 × 3.5 V AA batteries
Auxiliary IR source

Contractors
Thales Optronics.

VERIFIED

The binocular mount of UGO from Thales Optronics

UGO day/night goggles 0005464

Vectronix BIG25 night vision goggles

Type
Image intensifier goggles.

Description
The BIG25 night vision goggles is one of a family of hand-held image intensifying devices that includes the BIM25 night pocketscope, the BIM35 night pocketscope and the BIG35 night binoculars.

Mounted in a watertight housing the goggles use a single image intensifier tube with binocular eyepieces. They can be strapped to the head with an individually adjustable harness.

A variety of 18 mm image intensifier tubes (improved second- or third-generation) can be fitted. With an eyebase of 56 to 72 mm and a dioptre range of –6 to +2 the BIG25 can easily be adjusted to the operator's eyes. A built-in infra-red diode provides close illumination for map-reading or for operating equipment. A luminous mark near the lower edge of the visual field shows when this facility is operating.

In average conditions a human-sized target can be recognised at a range of more than 200 m.

Operational status
Production. In service with the Swiss Army and undisclosed customers.

Specifications
Dimensions
Goggles: 143 × 153 × 71 mm
Weight
Complete with harness: 813 g
Goggles: 550 g
Harness: 290 g
Magnification: ×1
Resolution: 1.054 lp/mrad
Image intensifier: GEN II+, DEP XD-4 or GEN III
Field of view: 41°
Battery: 1 × 2.7 V lithium or 2 × 1.5 V alkaline
Focus range: 25 cm to infinity
Objective: 24.7 mm, f/1.11

Contractor
Vectronix AG.

UPDATED

BIG25 night vision goggles

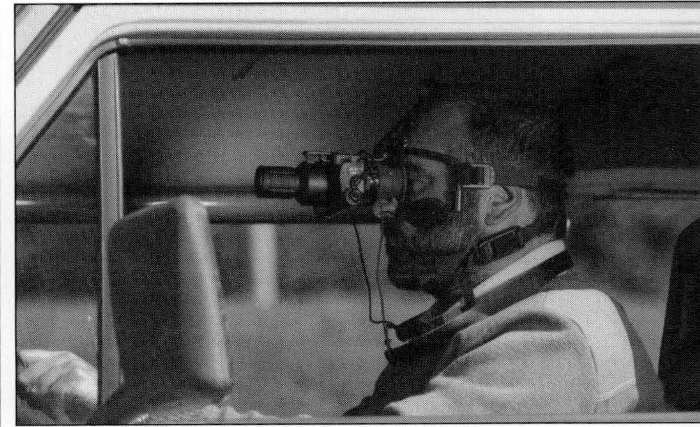
Vectronix (formerly Leica) BIG25 goggles with head harness

Yugoimport PN-2 passive night vision goggles

Type
Image intensifier goggles.

Description
The PN-2 passive night vision goggles are a lightweight night viewing system for use by individual soldiers who have to perform such duties as vehicle driving and maintenance.

The system uses two single-stage 18 mm image intensifier tubes of the second-generation type with focusable lens and adjustable eyepieces to achieve its results. It can also be used with an infra-red source to act as the illuminator for document or map-reading.

When used, the goggles are attached to the user's head by chin and headstraps.

Operational status
Production as required. In service with the former Yugoslav Army.

Specifications
Weight
With mask: 1.1 kg
Without mask: 800 g
Magnification: ×1
Resolution
1 × 10^{-3} lx: 2.7 mrad
Dioptre range: –5 to +5 dioptres
Field of view: 34°
Power supply: 3 V (battery)
Focus range: 25 cm to infinity

Contractor
Yugoimport SDPR.

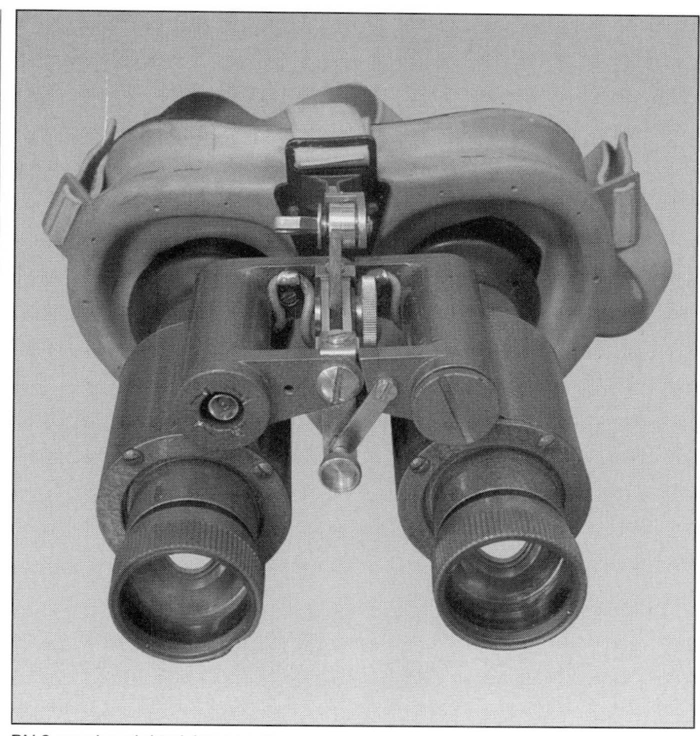

PN-2 passive night vision goggles

OBSERVATION AND SURVEILLANCE

IMAGE INTENSIFIER MONOCULARS

Allen Nite-Watch Plus

Type
Image intensifier monocular.

Development
The Nite-Watch system was developed by Marconi Applied Technology in 1990 to incorporate 18 mm second- or third-generation image intensifiers. The product line has now been acquired by P W Allen Ltd. The system has been successfully evaluated by the UK Home Office, the UK Infantry Training Development Unit and various overseas governments.

Description
The Nite-Watch Plus pocketscopes are available with either second-generation (Series 2000) or third-generation (Series 3000) image intensifying tubes. Both series have been recently extended by the addition of ruggedised versions (P2000RR/RP and P3000RR/RP). This has upgraded the systems for use in harsher environments. Adaptors are available to enable Nite-Watch to be added to video camcorders, SLR cameras and CCD TV cameras.

Operational status
In production and in service. Nite-Watch system sales have now exceeded 3,000 units for civilian and military organisations. Military units now in service include: NSN 5855-99-126- 2520 (third-generation intensifier with rotary switch) and NSN 5855-99-722-8527 (third-generation intensifier with push switch).

Specifications
Weight: 330 g incl lens and batteries
Dimensions: 46 × 120 mm
Intensifier type: GEN II or GEN III
Tube: 18 mm channel plate wafer tube
Resolution: typically 32 lp/mm
Field of view
25 mm lens: 34°
75 mm lens: 11°
Magnification
25 mm lens: ×1
75 mm lens: ×3
Objective lens (standard C mount): f/1.4 25 mm
Supply voltage: 2.5 to 3.5 V
Supply current (typically): 18 mA
Battery life: typically 30 h continuous from 3 lithium cells type DL1/3N or similar

Range performance (recognition of person with 75 mm objective lens)

Light level (lx)	GEN III (P3000)	GEN II (P2000)	GEN I
Moonlight 10^{-1}	630 m	630 m	550 m
Quarter moonlight 10^{-2}	510 m	450 m	320 m
Starlight 10^{-3}	430 m	290 m	160 m
Overcast starlight 10^{-4}	205 m	140 m	80 m

Contractor
P W Allen Company Ltd.

Allen Nite-Watch Plus 0005465

Nite-Watch Plus in a section of configurations 0023162

Corina Corporation Pantron – night vision device

Type
Image intensifier monocular.

Description
The Pantron night vision device is a second-generation monocular image intensifying device for observation and surveillance. It has a single-stage microchannel wafer tube with automatic brightness control. The eyepiece is detachable and a range of graticules is available. Tube life is stated to be approximately 2,000 hours.

Pantron can be mounted on rifles, cameras or movie cameras.

Operational status
Available.

Specifications
Length: 220 mm
Diameter: 65 mm
Weight: 1.34 kg
Magnification: ×3
Focal length: 75 mm
T/number: 1:1.3
Field of view: 13° 50′
Focal range: 2.5 m to ∞
Dioptre range: +5 to −5 dioptres
Operation: −20 to +55°C
Storage: −30 to +60°C
Image intensifier: GEN II, single-stage
Tube diameter: 18 mm
Resolution: min 30 lp/mm
Cathode sensitivity: min 340 μA/Lm
Gain: ×7,500 to ×15,000
Operating voltage: 2.4-3 V DC
Batteries per pack: 2
Type: AA/UM 3
Capacity: approx 40 h
Approx range at 30% contrast: clear (overcast)
Standing person: 400 m (300 m)
Car: 500 m (400 m)
Truck: 750 m (550 m)

Contractor
Corina Corporation AG.

Dedal Dedal-110/120/041

Type
Image intensifier monocular.

Description
The Russian joint stock company Dedal manufactures a range of night vision scopes for observation and surveillance.

The Dedal-120 is fitted with a first-generation (GEN I) image intensifier tube while the Dedal-110 is fitted with a first-plus generation tube which has a fibre-optic plate cathode. Range is stated as 150 m for Dedal-120 and 350 m for Dedal-110. Both scopes include an IR illuminator.

The Dedal-041 is equipped with a second-generation tube and is adapted for use with SLR and video cameras. It has an optional IR illuminator.

Operational status
Available.

Specifications

Dedal-110/ 120	Model A	Model B	Model C	Model D
Magnification	× 2.7	×2	×4	×2.7
Field of view	12°	17°	12°	18°
Resolution	4.3 lp/mrad	3.2 lp/mrad	3.3 lp/mrad	2.3 lp/mrad
Objective lens	85 mm	64 mm	85 mm	58 mm
Dimensions	175 × 66 × 82 mm	172 × 66 × 82 mm	175 × 66 × 82 mm	165 × 66 × 82 mm
Weight	0.76 kg	0.7 kg	0.74 kg	0.62 kg

Contractor
Dedal Joint Stock Company.

Dedal-220 night vision scope

Type
Image intensifier monocular.

Description
The baseline Dedal-220 night vision (model A) scope is equipped with a first-generation 'plus' (GEN I+) Super image intensifier tube which gives performance comparable with some GEN II devices. The tube used is of the latest technology with a fibre-optic input plate. As customer options GEN II+ (model B) or GEN III tubes (models B and C) can be supplied. Protection against excessive light exposure is provided in models B and C as is automatic brightness control and point light source protection. Dedal-220 is compatible with still video and CCTV cameras by fitting adapters.

Operational status
Available.

Specifications

Model	A	B	C
Dimensions	122 × 58 × 58 (214 × 72 × 72)* mm		
Weight	4,530 g (590 g)*		
Resolution	50 lp/mm	40 lp/mm	45 lp/mm
Magnification (standard)	×1.0 (×2.7)*	×1.4 (×3.8)*	×1.4 (×3.8)*
Field of view (100 mm f1.5 objective)	28° (10°)*		
Objective	37 mm f/1.0 (100 mm f/1.5)*		
Focus range	from 0.6 m to ∞		
Dioptre adjustment	+4 to –4		
Type of tube	GEN I+	GEN II+	GEN III
Photocathode sensitivity	280 μA/lm	350 μA/lm	1,000 μA/lm
Gain	1,000	25,000	25,000
Power supply:	1 × 3 V battery K58L (Kodak), DL1/3N, CR1/3N, 2L76 or Blik-1		
Battery life	10 h		
Operating temperature	–40 to +50°C		

*With objective lens 100 mm f/1.5

Contractor
Dedal Joint Stock Company.

The Dedal-220 night vision scope 0041425

Dedal-220 fitted to a camera 0041441

Dedal-41 night vision scope

Type
Image intensifier monocular.

Description
The Dedal-41 uses a 25 mm second-generation (GEN II) image intensifier tube. Viewing ranges of 350 to 600 m are claimed at low light levels. Effective use is claimed at light levels down to 10^{-2} lx (starlight) and 10^{-4} lx (overcast starlight). Protection against excessive light levels is provided for the image intensifier. The intensifier also has protection from point light sources and has automatic brightness control. There is a built in IR illuminator for use in very dark conditions. The device may be used with a biocular eyepiece or as a monocular. A tripod socket is built in.

Operational status
Available.

Specifications
Dimensions: 210 × 76 × 93 mm (325 × 76 × 103 mm)
Weight: 1.12 kg (1.52 kg)
Magnification monocular: ×2.7 (×3.2) ×3.8 (×4.4)
Magnification binocular: ×2.2 (×2.6) ×3.1 (×3.6)
Objective: 85 mm f/2.0 (– 100 mm f/1.5)
Field of view: 17° (14°)
Eye relief distance: 45 mm
Dioptre adjustment: +3, –4 – mono (+2, –2 – bino)
Tube type: 11, 25 mm MCP 11, 25 mm MCP
Photocathode sensitivity: min 240 μA/lm min 300 μA/lm
Gain: 50,000 35,000
Resolution: 32 lp/mm 45 lp/mm
Power supply: 2 × 3 V AA batteries
Battery life: 50 h
Operating temperature: –40 up to +50°C
Relative humidity: up to 98%

Contractor
Dedal Joint Stock Company.

Dedal-43 night vision scope

Type
Image intensifier monocular.

Description
The Dedal-43 may be used as a monocular or in a bi-ocular configuration. Viewing ranges of 450 to 700 m are claimed at low-light levels and more than 250 m in complete darkness with the Dedal IR illuminator fitted. Effective use is claimed at light levels down to 10^{-2} lx (starlight) and 10^{-4} lx (overcast starlight). Protection against excessive light levels is provided for the image intensifier. The intensifier also has protection from point light sources and has automatic brightness control.

Operational status
Available.

Specifications
Dimensions: 130 × 91 × 80 mm
Weight: 0.85 kg – biocular
0.65 kg – monocular
Objective: 85 mm f/1.4-85 mm f/1.4
Magnification: ×2.0 – biocular
×3.2 – monocular
Field of view: 12°

Dioptre adjustment: +4 to –4 (monocular)
Tube type: GEN II+ GEN II+ Super
Photocathode sensitivity: min 350 µA/lm min 600 µA/lm
Gain: 30,000 35,000
Resolution: 40 lp/mm 45 lp/mm
Power supply: 1 × 3 V type SR1123A battery
Battery life: 50 h
Operating temperature: –40 to +50°C
Relative humidity: up to 98%

Contractor
Dedal Joint Stock Company.

The Dedal-43 night vision scope 0041434

Dedal-80 night vision scope

Type
Image intensifier monocular.

Description
The Dedal-80 uses a 25 mm image intensifier tube. Viewing ranges of 1,000 to 2,000 m are claimed at low light levels. Effective use is claimed as being possible down to light levels of 10^{-2} lx (starlight) and 10^{-4} lx (overcast starlight). Protection against excessive light levels is provided for the image intensifier. The intensifier also has protection from point light sources and has automatic brightness control. A tripod socket is built in.

Operational status
Available.

Specifications

Model	A	B	C
Dimensions	370 × 250 × 240 mm	370 × 250 × 240 mm	400 × 250 × 240 mm
Weight	8.5 kg		
Magnification	×6.5	×9.2	×7.6
Field of view	7°	5.2°	5.2°
Objective	204 mm, f/1.0, T – 1.4		
Eye relief distance	45 mm	45 mm	45 mm
Dioptre setting	–5, +5 D		
Tube type	GEN II, 25 mm MCP	GEN II, 25 mm MCP	GEN II+ Super, 18 mm
Photocathode sensitivity	240 µA/lm	350 µA/lm	600 µA/lm
Gain	50,000	30,000	35,000
Resolution	32 lp/mm	45 lp/mm	45 lp/mm
Power supply	2 × 3 V AA batteries		
Battery life	50 h		
Operating temperature	–40 to +50°C		
Relative humidity	up to 98%		

Contractor
Dedal Joint Stock Company.

Econ Industries Cyclops night vision pocketscope

Type
Image intensifying monocular.

Description
Cyclops is a night vision pocketscope of modular construction which can be converted from the standard ×3 configuration to ×1 or ×4 magnification by exchanging lenses. There is an optional infra-red illuminator for map-reading, repair tasks and so on. Cyclops can be supplied with either second- or third-generation image intensifier tubes. The standard Cyclops ×3 system bears a Canon 75 mm objective lens, but optional lenses are available.

Operational status
Available.

Specifications
Magnification: ×3 (optional ×1 or ×4)
Field of view: 13.5° (3.5 m at 100 m distance)
Resolution on axis (100 mlx): >2.2 lp/mrad
Focusing distance: 0.8 m to ∞
Observation distance: >400 m (for ×3)
Image intensifier: 18 mm GEN II+ or GEN III GaAs, ANVIS type
Objective lens: 75 mm, f/1.8, T/2.0
Eyepiece lens
Equivalent focal length: 25 mm
Eye relief: 18 mm
Exit pupil: 7.5 mm
Dioptre adjustment range: +2 to –5 dioptres
Dimensions: 180 × 83 × 77 mm
Weight (with batteries): 800 g

Contractor
Econ Industries S A.

The Cyclops night vision pocketscope from Econ Industries

Hellenic Aerospace night vision monocular

Type
Image intensifier monocular.

Description
Hellenic Aerospace night vision monocular is a lightweight device for night surveillance and security. It has a single second-generation image intensifier tube and a range of lenses - standard, zoom and telephoto.

Operational status
Available.

Specifications

Objective lens	Standard	Zoom	Tele
Focal length	50 mm	70-210 mm	500 mm
Magnification	×1	×1.4-4	×10
Field of view	17°	12-4°	1.7°
Resolution	0.9 lp/mrad	1.3-3.5 lp/mrad	8 lp/mrad

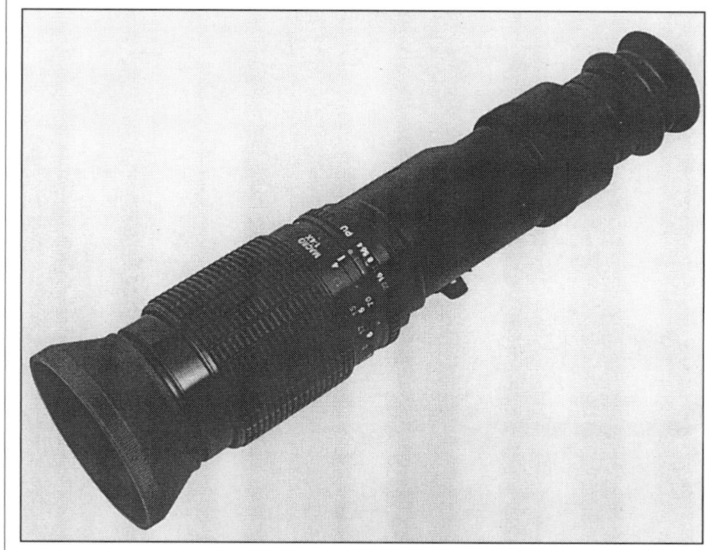
The Hellenic Aerospace night vision monocular 0005466

Weight: 650 g
Dimensions: 240 × 70 mm
Image intensifier
Photocathode sensitivity: 350 μA/lm
Tube gain: 5,500
Resolution: 30 lp/mrad
Input current: 10 mA
Voltage requirements: 3 V

Contractor
Hellenic Aerospace Industry SA.

HTS OnX2™/Super- OnX2™/ series night vision systems

Type
Image intensifier monocular.

Description
The OnX2™ and Super-OnX2™ series night vision monoculars can be used in a variety of modes including as a hand-held monocular, combined with a spotting scope or attached to a daylight rifle sight. The OnX2™ models are designed primarily for police work and similar tactical uses. The Super-OnX2™ system is military grade, small in size, lightweight, comes in submersible and non-submersible configurations and includes a short-range light emitting diode illuminator. Both series can be easily and quickly attached to daylight scopes. The Super-OnX2™ model may also be attached to a standard helmet, face mount or diver's mask. Both series are available with a choice of image intensifier standards.

Operational status
Available.

Specifications
OnX RB9712 series

Model/tube	A GEN II	B SuperGen®	C GEN III	D GEN III
Dimensions (l × w):	165 × 75 mm	165 × 75 mm	165 × 75 mm	165 × 75 mm
Weight (with batteries):	795 g	795 g	795 g	795 g
Magnification:	×1	×1	×1	×1
Field of view:	40°	40°	40°	40°
Focus range:	0.4 m to infinity	0.4 m to infinity	0.4 m to infinity	0.4 m to infinity
Dioptre adjustment:	+2 to –6	+2 to –6	+2 to –6	+2 to –6
Eye relief:	25 mm	25 mm	25 mm	25 mm
Eye relief with scope:	75 mm	75 mm	75 mm	75 mm
Photocathode sensitivity (min):				
White light:	300 μA/lm	600 μA/lm	1000 μA/lm	1600 μA/lm
At 830 nm:	25 mA/W	43 mA/W	100 mA/W	125 mA/W
At 880 nm	–	–	62 mA/W	75 mA/W
Luminance gain:	9400 fL/fc	20,000 fL/fc	20,000 fL/fc	20,000 fL/fc
Power supply:	2 × AAA batteries	2 × AAA batteries	2 × AAA batteries	2 × AAA batteries
Battery life:	>65 h	>45 h	>45 h	>45 h
Operating temperature:	–29 to +55°C	–29 to +55°C	–29 to +55°C	–29 to +55°C

The HTS OnX2™ (top) and Super-OnX2™ monoculars 0081444

Super-OnX2™ series

Model/tube	982/982S GEN II+	983/983S GEN III
Length:	148 mm	148 mm
Weight (with batteries):	568 g	568 g
Photocathode sensitivity (min):		
White light:	300 μA/lm	1,000 μA/lm
At 830 nm:	20 mA/W	100 mA/W
At 880 nm	–	62 mA/W
Luminance gain (min):	18,000 fL/fc	20,000 fL/fc
Power supply:	1 × AA battery	1 × AA battery
Battery life:	>65 h	>45 h
Operating temperature:	–51 to +55°C	–51 to +55°C

Contractor
Henry Technical Services, Inc.

VERIFIED

International Technologies (Lasers) MINI N/SEAS night single-eye acquisition sight

Type
Image intensifier monocular.

Description
The MINI N/SEAS is a night vision system in a single-eye configuration. One eye is equipped with the night vision sight while the other remains free and retains normal vision, continuously adapting to changing light conditions. The system can have either second-generation plus (GEN II+) or third-generation (GEN III) image intensifier tube with automatic brightness control.

The MINI N/SEAS can be mounted on a facemask (under or without a helmet/ helmet mounted), attached to a variety of weapons or used as a hand-held pocketscope. An optional clip-on ×3 magnifier can be attached by the user during field operations.

The system has a 'flip-off' mechanism on the facemask which shuts down the system's power automatically when the monocular is not in front of the user's eye.

The MINI N/SEAS includes infra-red (IR) illumination for close-range covert illumination. An indicator in the eyepiece lights up when the IR illuminator is activated and battery power is low. The system incorporates a high light level sensor which automatically shuts off the image intensifier tube for protection against high light environments.

A perfectly collimated binocular configuration with two randomly selected monoculars (each using a patented 'self collimation') mounted on the head or a helmet is also available.

Operational status
In production and in service with forces worldwide.

MINI N/SEAS on head mount without helmet 0023164

Specifications
Dimensions: 115 × 65 × 40 mm
Weight
With battery: 330 g
With battery and facemask: 540 g
Magnification: ×1 or ×3
Field of view: ×1, 40°; ×3, 13.3°
Focus range: ×1, 250 mm to ∞; ×3, 3 m to ∞
Dioptre adjustment: +2 to −6 dioptres
Objective lens: ×1, f/1.2; ×3, f/1.4
Resolution (system): 1.2 lp/mrad, GEN III
System gain: >3,000 (GEN III)
Power source: 1 × 1.5 V AA size alkaline or lithium battery; or 1 × 3.5 V AA size lithium battery
Battery life
Alkaline: 30 h
Lithium: 60 h

Contractor
International Technologies (Lasers) Limited.

UPDATED

MINI N/SEAS as stand-alone unit 0023165

International Technologies (Lasers) SNS-1 pocketscope

Type
Image intensifier monocular.

Description
SNS-1 is a multipurpose pocket-sized monocular night scope. It is designed to be used as a stand-alone night vision device or in combination with a variety of standard-video, still-photo or CCTV equipment. It has a standard C-mount for the objective lens, enabling standard CCTV camera lenses to be fitted.

The system employs a second- or third-generation image intensifier tube and is equipped with its own infra-red LED illuminator. In addition, a high-intensity external illuminator can be mounted on the device.

As well as being hand-held, SNS-1 can also be mounted on a facemask.

Operational status
Available.

Specifications
Dimensions: 95 × 50 × 73 mm (excl objective lens)
Weight: 0.26 kg (excl battery and objective lens)
Optical
Magnification: ×1 with 25 mm lens
Field of view: 40° with 25 mm lens
Objective lens
Mounting type: 'C' mount, 1 in 32 TPI, interchangeable
Standard lens: 25 mm, f/1.4
Lens option: range of 'C' mount lenses incl zoom, tele and wide type
Eyepiece
Focal length: 25 mm
Dioptre adjustment: +2 to −6 dioptres
Image intensifier: 18 mm, GEN II, Super or GEN III
Tube resolution: 32 lp/mm or higher on request
Photo cathode sensitivity: 300 μA/lm (min) or higher on request
IR source
Type: IR LED
Output wavelength: 840 nm
Electrical
Power supply: 1 × 1.5 V AA alkaline battery
Operating time: 50 h (10 h with IR LED illuminator)

Contractor
International Technologies (Lasers) Limited.

VERIFIED

The SNS-1 pocketscope mounted on SLR camera 0023163

ITT AN/PVS-11A – third-generation night vision pocketscope

Type
Image intensifier monocular.

Description
The AN/PVS-11A third-generation night vision pocketscope is based on the second-generation version in use by the US Navy. The AN/PVS-11A incorporates the standard MX-10160 image intensifier which provides photoresponse to the near-infra-red (IR) and meets or exceeds MIL-I-49428 (CR). Overall the system provides greater than 0.8 cy/mrad resolution and in excess of 2,000 fL/fL of brightness gain. The pocketscope has an IR illuminator which is activated by the system's three-position power switch.

The AN/PVS-11A is equipped with a light-secure eyecup that prevents system light from being detected by anyone other than the user. For hand-held use, the system comes with a wrist lanyard which screws into a standard 1/4 × 20 tripod-mount thread tapped in the housing assembly. Use of the tripod mount allows hands-free operation or adaptation to a camera/video system.

For use with a 35 mm or video camera, the eyepiece of the pocketscope can be threaded to the objective lens of a camera by adding step-down rings.

The AN/PVS-11A accepts most of the AN/PVS-7B night vision goggle accessories. These include the ×3 magnifier lens for long-range observation, the compass for use in night-time navigation or fire control, a sacrificial window to protect the optics from dirt or scratches and the protective laser filters offered in the current inventory.

Operational status
Available.

Specifications
Spectral response: Visible to 0.90 μm (IR)
Field of view: 40° ±2°
Magnification
×1 objective: 0.96 × ±0.03
×3 objective: 2.89 × ±0.26
Resolution: .80 lp/mrad

ITT AN/PVS-11 pocketscope

Brightness gain: ≥2,000 fL/fL
Dioptre adjustment: +1 to –6 dioptres
Focus range: 25 cm to ∞
Voltage required: 50 mA, max
Battery type: 2 AA size or equivalent
Weight with batteries: 26 oz
Operating temperature: –39 to +49°C

Contractor
ITT Defense & Electronics, Night Vision Division.

VERIFIED

ITT AN/PVS-14 Monocular Night Vision Device MNVD

Type
Image intensifier monocular.

Description
The AN/PVS-14 Monocular Night Vision Device (MNVD) is a multipurpose system which has been selected by the US Army as its standard MNVD. The system was developed with the NV/RSTA (Night Vision/Reconnaissance and Surveillance Target Acquisition) department of the US Army NVESD (Night Vision and Electronic Sensors Directorate).

AN/PVS-14 can be hand-held or mounted in the head harness or helmet mount of the AN/PVS-7B. It can also be mounted to the M-16A2 rifle fitted with the Picatinney rail.

The system incorporates the ITT third-generation Ultra image intensifier tube, MX-11769 (Omni V Spec), and optics used on the AN/AVS-6. The tube has a variable gain control added to its power supply and an external knob on the front of the MNVD allows the user to adjust gain. An integral infra-red LED illuminator is also fitted.

For weapon firing, the MNVD can be mounted behind a standard collimator dot sight on the M-16A2 rifle and this sight provides the aimpoint. When adapted with the AN/PVS-7 ×3 magnifier, the monocular similarly fits behind the collimated dot sight. It can be fitted with all ancillary items of the AN/PVS-7D night vision goggles.

ITT's AN/P VS-I 4 is available in the following configurations:

- F6015L - AN/PVS-1 4, US mil version, using MX-11769 tube (Omni V Spec)
- F6015K - AN/PVS-1 4, US mil version, using MX-11769 tube (Omni IV MIL Spec)
- F6015P - ITT MNVD, using the F9815P tube (Omni IV performance)
- F6015N - ITT MNVD, using the F9815N tube (Omni V / Ultra performance)
- F6015J - ITT MNVD, using the F9815J tube (Omni III Plus performance)
- F6O15C - ITT MNVD, using the F9815C tube (Omni III Enhanced performance)
- F6015B - ITT MNVD, using the F9815B tube (Omni II Enhanced performance)

Operational status
In service.

Specifications
Omni V Spec
Weight: 0.392 kg
Image intensifier: 18 mm GEN III
Tube: ITT MX-11769
Resolution (on-axis at optimum light level): 1.3 cy/mrad
Field of view: 40° ±2°

ITT's AN/PVS-14 0130952

System brightness gain: adjustable from 25 to more than 3,000 fL/fL
Magnification: ×1 ±0.03
System distortion: <3%
Range focus: 25 cm to ∞
Dioptre focus: +2 to -6 dioptres
Exit pupil and eye relief: 14 mm on axis at 25 mm eye relief
Operating Temperature: –51°C to +49°C
Storage temperature: –51°C to +85°C

Contractor
ITT Industries, Night Vision.

One of the AN/PVS-14 monoculars 0132097

ITT Night Enforcer™150 and 250

Type
Image intensifier monocular.

Description
Night Enforcer™ 150 and 250 are second-generation night vision viewers. They incorporate ITT's improved performance second-generation tube and an f/1.4 objective lens, with automatic brightness control. A variety of optional magnifier lenses can be provided: ×3; 3-in-1 (×0.42, ×2, ×3); and ×2.2 which is also usable with the 3-in-1 lens as a ×2.2 converter for ×4 and ×6 in operation. An infra-red illuminator with mount is available for Night Enforcer 150.

Operational status
Available.

Specifications
System resolution: 0.76 cy/mrad
Image intensifier: GEN II
  Photoresponse: 400 μA/lm
  Resolution: 36 lp/mm
  Gain: 20,000 min
Magnification: ×1
Field of view: 40°
Objective lens: f/1.4
Focus range: 1 ft to ∞
Dioptre range: +2 to –6 dioptres
Weight: 454 g

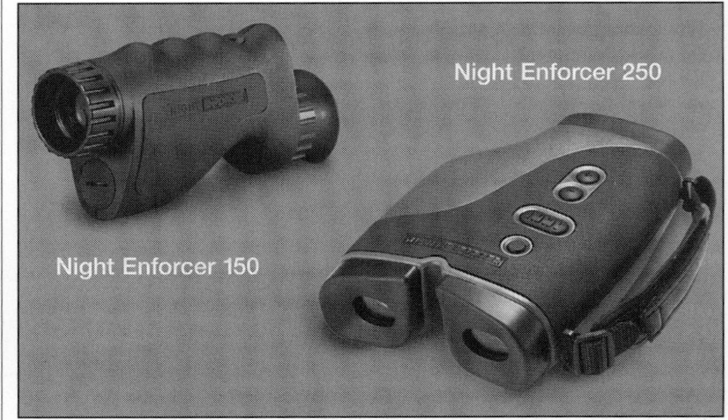

ITT's Night Enforcer™ 150 and 250 GEN II night vision viewers 0005471

Dimensions: 184 × 92 × 54 mm
Power supply: 2 × 1.5 V AA batteries

Contractor
ITT Defense & Electronics, Night Vision Division.

VERIFIED

Kazan Baigysh-3

Type
Image intensifier monocular.

Description
The Baigysh-3 is a pistol-grip, hand-held, night observation device using image intensifier technology. It features what is described as 'an afterglow effect' allowing the operator to view objects several minutes after switching off the power in order to conserve energy. The device may be used with Zenit-type cameras and it can be used to identify human beings at distances up to 200 m.

Operational status
Believed to be in service and in production.

Specifications
Dimensions: 240 × 145 × 75 mm
Weight: 0.8 kg
Field of view: 9°
Magnification: ×3
Power supply: 9 V galvanic battery

Contractor
Kazan Optical and Mechanical Plant (AO KOMZ).

UPDATED

Kazan Baigysh-7

Type
Image intensifier monocular.

Description
Baigysh-7 is a family of night observation devices with a variety of lenses. Although designed for hand-held use the Baigysh-7 may also be installed upon a tripod. In moonlight, human beings may be identified at distances between 250 and 300 m.

All versions feature automatic brightness control and are powered by a dry galvanic battery. The Baigysh-7, -7A and -7C may be fitted with the f120 Lik objective lens for panoramic observation. The Baigysh-7B, -7D and -7E may be used for the same task with the f135 mm Jupiter-37A objective lens. With special eyepieces, the Baigysh-7A, -7C, -7D and -7E may be used for detailed observation while with the use of a projection lens, the Baigysh-7C and -7D may be used with a Zenit camera.

Operational status
Believed to be in service and in production.

Specifications
Dimensions
Baigysh-7: 395 × 135 × 173 mm
Baigysh-7A: 292 × 84 × 173 mm
Baigysh-7B: 370 × 135 × 167 mm
Baigysh-7C: 320 × 84 × 173 mm
Baigysh-7D: 292 × 74 × 167 mm
Baigysh-7E: 265 × 71 × 167 mm
Weight
Baigysh-7: 1.7 kg
Baigysh-7A: 1.4 kg
Baigysh-7B: 1.5 kg
Baigysh-7C: 1.4 kg
Baigysh-7D: 1.1 kg
Baigysh-7E: 1.2 kg
Field of view
Baigysh-7, -7A, -7C: 7°
Baigysh-7B, -7D, -7E: 6°
Magnification
Biocular:
(Baigysh-7, -7A, -7C): ×2.4
(Baigysh-7B, -7D, -7E): ×2.7
Monocular:
(Baigysh-7A, -7C): ×7.8
(Baigysh-7D, -7E): ×8.7

Contractor
Kazan Optical and Mechanical Plant (AO KOMZ).

UPDATED

Lyon & Brandfield Multiscope surface and underwater night vision system

Type
Image intensifier monocular.

Description
The Multiscope is a multifunction surface and underwater night vision photographic, video and observation system. It provides hard-copy photographs, video and observation capabilities in a compact modular unit. The Multiscope is waterproof to 100 m and is suitable for use by military divers and amphibious and special forces as well as for covert surveillance and reconnaissance, police and customs. The equipment is supplied with second- or third-generation image intensifiers and is available in the following formats:

(1) waterproof with built-in photographic camera
(2) waterproof with video modification
(3) observation with photographic capability
(4) observation only, waterproof and non-waterproof
(5) observation with video modification.

Special to type configurations are provided for search and rescue operations. All multiscopes are supplied with ×7 and ×10 magnification eyepieces, lens adaptor, tripod adaptor and a choice of lenses.

Operational status
Available.

Specifications
Lenses: 25 mm, f0.85; 25 mm, f/1.4; 50 mm, f/1.4 ×2; 75 mm, f/1.4 ×3
Monocular eyepieces: ×7; ×10 magnification
Camera type: Pentax Z.20
Image intensifier: GEN II or GEN III
Weight
Lens: 505 g (25 mm, f0.85); 360 g (25 mm, f1.4); 560 g (50 mm); 790 g (75 mm)
Eyepiece: 210 g (×7), 100 g (×10)
Watertight housing and camera: 1,030 g
Lens adaptor: 55 g

Contractor
Lyon & Brandfield Ltd.

VERIFIED

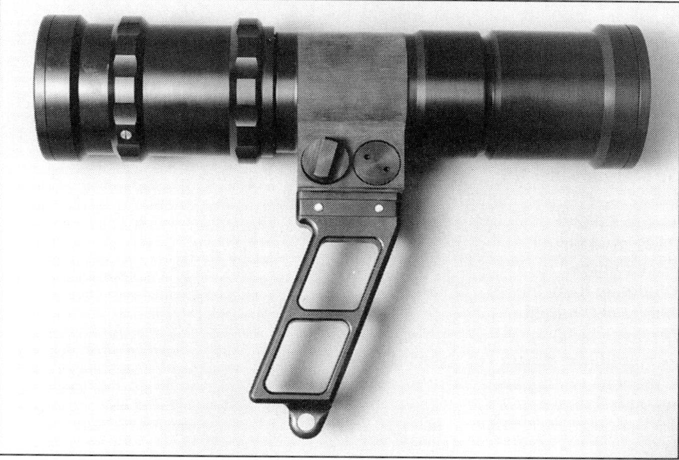

The Multiscope night vision system 0053841

MES Type VN/296 - VN396

Type
Image intensifier monocular.

Description
The MES VN/296 is an intensified monocular which can be mounted via a suspension assembly directly onto the head or alternatively with a strap arrangement over a combat helmet. When not required the monocular may be flipped up away from the face. There is a built-in infra-red illuminator to assist with map reading. There is also a low battery indicator. The VN/396 uses two VN/296 monoculars mounted on a custom bracket to form a binocular system.

Operational status
Available.

Specifications
Magnification: ×1 (×3 optional)
Objective: ×1 - 27/32 mm (optional ×3 - 75 mm)
Field of view: 40°

The MES VN/296 image intensified monocular 0023166

Dioptre adjustment: +2 to –6 dioptres
Focus range: 25 cm to ∞
Weight: 1.47 kg

Contractor
Meccanica per Elettronica e Servomeccanismi SpA (MES)

VERIFIED

New Noga Light NL-60 series mini-scopes

Type
Image intensifier monocular.

Description
The NL-60 series second-generation night vision mini-scopes are compact battery-powered sights for small weapons, but can easily be adapted for attachment to recoilless rifles for civil and military use. There are a range of models with magnifications of ×2, ×3, ×4.3, a choice of weapon sight adaptors and accessories for still and video cameras.

Operational status
In production.

Specifications
NL-62
Magnification: ×2
Field of view: 20°
Focus range: 10 m to ∞
Objective focal length: 50 mm
Objective f-number: 1.3
Dioptre range: +3 to –6 dioptres
Resolution at centre: 1.4 lp/mrad
Dimensions: 165 × 59 × 83 mm
Weight: 0.65 kg

NL-63
Magnification: ×3
Field of view: 14°

Model NL-63 GEN II mini-scope

Focus range: 10 m to ∞
Objective focal length: 75 mm
Objective f-number: 1.3
Dioptre range: +3 to –6 dioptres
Resolution at centre: 2.1 lp/mrad
Dimensions: 185 × 59 × 83 mm
Weight: 0.76 kg

NL-64
Magnification: ×4.3
Field of view: 9°
Focus range: 20 m to ∞
Objective focal length: 113 mm
Objective f-number: 1.6
Dioptre range: +3 to –6 dioptres
Resolution at centre: 3.0 lp/mrad
Dimensions: 250 × 59 × 83 mm
Weight: 0.95 kg

Contractor
New Noga Light Ltd.

VERIFIED

New Noga Light Wild Cat night vision pocketscope

Type
Image intensifier monocular.

Description
The Noga Light Wild Cat series NL-50 is a lightweight hand-held night vision pocketscope based on a second-generation image intensifier. The instrument's features include automatic brightness control, accessories for photography and video cameras, forged aluminium body and an integral infra-red illuminator. Two models are available, depending on lens, the NL-51 with 25 mm lens and the NL-52 with 50 mm lens.

Operational status
In production.

Specifications

	NL-51	NL-52
Dimensions	130 × 46 × 70 mm	156 × 46 × 70 mm
Weight	0.46 kg	0.69 kg
Magnification	×1	×2
Field of view	40°	20°
Focus range	50 cm to ∞	
Objective		
Focal length	25 mm	50 mm
F-number	1.4	1.3
Eyepiece dioptre adjustment	+6 to –6 dioptres	
Image tube		
Type	18 mm, GEN II or GEN II Super	
Resolution	GEN II 32 lp/mm, typical, GEN II Super 45 lp/mm	
System resolution	0.7 lp/mrad	1.4lp/mrad
Power	3 V DC, 2 AA batteries	

Contractor
New Noga Light Ltd.

UPDATED

The Wild Cat night vision pocketscope 0005468

Night Vision Equipment Company American Eagle Pocketscope NVEC Models 502/503/602/603

Type
Image intensifier monocular.

Description
The American Eagle pocketscope is a combination of night viewer and still or video camera night lens which incorporates automatic brightness control. Each instrument is equipped with a high quality eyepiece, carrying case, a ×1 25 mm objective lens (×2, ×3 optional) and two alkaline batteries.

Four versions of the pocketscope are available and have either commercial specification second-generation, second-generation plus or third-generation image intensifier tubes.

Operational status
Available.

Specifications

Model		**502**	**503**	**603**
Tube generation		II	II+	III
Typical detection range in m for human target on a green grass background				
Quarter moon (10^{-2} lx)	236	253	427	
Starlight (10^{-3} lx)	117	138	310	
Overcast (10^{-4} lx)	95	111	163	

(All models)
Weight: 425 gm
Field of view: 25°
Magnification: ×1
Focus range: 25 cm to ∞
Objective focal length: 25 mm
Dioptre adjustment: +2 to −6
Eye relief: 15 mm (min)
Power supply: 2 × AA batteries

Contractor
Night Vision Equipment Company Inc.

The American Eagle pocketscope 0023167

American Eagle fitted to camera 0041401

Night Vision Equipment Company MANTIS Multi-Adaptable Night Tactical Imaging System

Type
Image intensifier monocular.

Description
MANTIS is a hand-held pocket scope using second-, second plus- and third-generation tubes for a variety of purposes including direct observation and camera-mounted night surveillance.

The basic unit uses the same ×1 objective lens as the AN/PVS-7B night vision goggles, a ×10 eyepiece and Infra-Red (IR) laser. With the second plus-generation tube, a still person can be spotted at 65 m on a cloudy night and 135 m in a quarter moon, with the third-generation tube the distances are considerably increased. The objective lens accepts both ×3 and ×5 magnifiers. A MANTIS XLR version exists which includes the high-power IR laser incorporated in NVEC's GCP-1A. The XLR laser is adjustable from 0.5 mrad to 30° and has a range in excess of 5,000 m in its pinpoint mode.

With optional components it is adaptable to still and video cameras or weapons.

Operational status
Available.

Specifications

Detection range in m for still human target on meadowland

Model	**130 GEN III**	**140 OMNI-4**	**150 OMNI-5**
Quarter moon (10^{-2})	417	501	647
Starlight (10^{-3})	297	344	410
Cloudy night (10^{-4})	163	185	232

(All models)
Field of view: 40°
Magnification: unity (×1)
Dioptre adjustment: +2 to −6 dioptres
Objective lens: 27 mm f1.2, T/1.3 (Part No 300001-G2)
Focus range: 25 cm to ∞
Power source: 2 × 1.5 V DC AA cells
Weight: 450 g

Contractor
Night Vision Equipment Company Inc.

MANTIS™ Multi-Adaptable Night Tactical Imaging System 0041412

Night Vision Equipment Company NVEC Models/400/400 HP/450 Night Surveillance System (NSS)

Type
Image intensifier monocular.

Description
The 400 series NSS is a versatile night vision system, based upon an image intensifier, which may be used for photography, filming, or passive viewing. It incorporates automatic brightness control together with bright light-source protection to ensure the image remains constant.

It is fitted with an Adjust-A-Lens collar which allows the user to attach any camera lens with the appropriate T-mount adaptor. Each system has a high-quality eyepiece, a carrying case, a ×4 135 mm f/2.8 objective lens, a tripod adaptor, lens protector cap and two AA alkaline batteries. There is also a photo-adaptor ring which screws into a standard 55 mm step-up ring for attaching to any camera lens. Options include a C- mount video camera relay lens system, biocular eyepiece, and pistol-grip.

The NSS fitted to a still camera 0041410

The system is now available with a third-generation tube.

Operational status
In production.

Specifications
Dimensions: 162 × 121 × 117 mm
Weight: 0.97 kg
Field of view: 14.5°
Magnification: ×4
Dioptre adjustment: +4 to –5
Eye relief: 28 mm
Power: 2.7 to 3 V DC (2 AA alkaline batteries)

Contractor
Night Vision Equipment Company Inc.

Northrop Grumman M942/M944

Type
Image intensifier monocular.

Description
The M942/M944 modular pocketscopes use image intensifier tubes of the ANVIS type. M942 contains a second-plus generation tube and M944 third-generation. Both systems include battery compartment, selectable infra-red (IR) illuminator, integral purge valve and ON/OFF/IR switch. The modular nature of the system allows the user to select each component including ×1, ×3, ×4 and ×6 magnification lenses, 1 or⅔ in camera relay lens, 25 mm or 75 mm CCTV lens, 17.5 to 108 mm zoom lens and ×10 eyepiece. The system may be hand-held or fitted to a tripod and is adaptable to video and CCTV systems. It is powered by a single alkaline or lithium battery.

Operational status
In production.

Specifications
Weight
×1 lens: 342 g
×3 lens: 570 g
×4 lens: 870 g
×6 lens: 1.37 kg
Fields of view
×1: 40°
×3: 15°
×4: 10.2°
×6: 6.3°

Contractor
Northrop Grumman Electro-Optical Systems.

UPDATED

Novo Corporation NightMaster 2023/2033 portable night vision device

Type
Image intensifier monocular.

Description
The NightMaster 2023/2033 is a monocular multifunction night vision pocketscope device. The 2023/2033 is designed to be used both as an observation device and as a part of systems for night video/photo surveillance. The device can include an image intensifier of third-generation or GEN II Super technology. A built-in infra-red (IR) diode provides illumination of the near scene. An optional more powerful IR illuminator is available. The NightMaster 2023/2033 has an automatic shutdown system to prevent damage to the image intensifier if bright light enters the optics. There are also indicators for loss of power supply, battery discharge and IR illumination on/off.

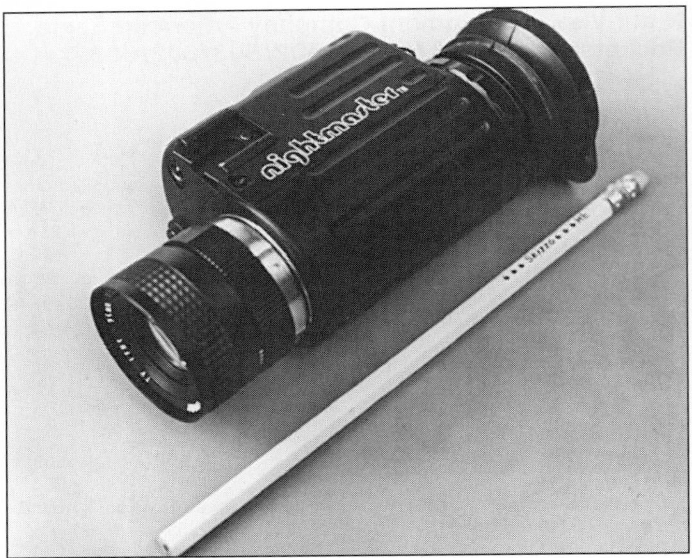

The Novo Corporation Nightmaster-2023/2033 0023168

Specifications

System			
Magnification	×1	×2	×3
Field of view (magnification ×1/2/3)	40°	18°	12°
Focusing range	0.5 m to ∞	1.0 m to ∞	1.2 m to ∞
Lens			
Focus distance	25 mm	50 mm	75 mm
Relative aperture	f/1.3	f/1.4	f/1.8
Eyepiece			
Focus distance	25 mm	25 mm	25 mm
Exit pupil distance	15 mm	15 mm	15 mm
Exit pupil dimensions	7 mm	7 mm	7 mm
Dioptre setting	+4 to –4	+4 to –4	+4 to –4
Mechanical			
Dimensions	105 × 48 × 68 mm	105 × 48 × 68 mm	105 × 48 × 68 mm
Weight	0.425 kg	0.425 kg	0.425 kg
Electrical			
Power supply source (2 × AAA batteries)	3 V	3 V	3 V
Continuous operation, hours			
using IR illumination	6	6	6
not using IR illumination	20	20	20
IR light-emitting diode illumination			
power	10 mW	10 mW	10 mW
wave length	840 nm	840 nm	840 nm
Image intensifier GEN II+ Super or GEN III			
Centre resolution limit, lp/mm	36	36	36
Gain	25,000	25,000	25,000

Operational status
In production.

Contractor
Novo Corporation.

VERIFIED

Optechs Korea Inc NT 9502 night telescope

Type
Image intensifier monocular.

Description
The NT 9502 from Optechs Korea Inc is a night vision telescope that uses a third-generation image intensifier tube and a zoom lens. The telescope can be fitted to an SLR camera.

NT 9502 night telescope from Optechs Korea Inc 0005469

Operational status
Available.

Specifications
Magnification: ×8
Field of view: 5.8°
System resolution: 5 lp/mm min
Relay lens magnification: ×1
Image intensifier
Tube type: GEN III microchannel inverter 18 mm
Tube resolution: min 36 lp/mm
Gain: 20,000 min
Dimensions
  Eye: 95 × 90 × 210 mm
  Photo: 95 × 90 × 290 mm
Weight
  Eye: 1.28 kg
  Photo: 1.95 kg

Contractor
Optechs Korea Inc.

VERIFIED

PCO PNM-1 miniaturised night vision device

Type
Image intensifier monocular.

Description
The PCO PNM-1 night vision monocular is derived from the PNS-1 driver's night vision goggles. It uses a second-generation-plus (GEN II+) or second-generation-super (GEN II Super) image intensifier tube. Users can undertake tasks such as map reading, repair work, night observation and surveillance without the need for illumination.

Operational status
In production. In service with Poland. Offered for export.

PCO PNM-1 night vision monocular 0023169

Specifications
Magnification: ×1
Field of view: 38° min
Focus adjustment: 0.25 m to ∞
Power supply: 1 lithium cell 3.6 V AA size
Resolution: 5.7′ min
Operational temperature: –20 to +45°C
Weight: 0.4 kg
Dimensions (h × l): 70 × 130 mm

Contractor
PCO SA Poland.

VERIFIED

Photonic NS-B

Type
Image intensifier monocular.

Description
The NS-B 3.3 × 80 is a lightweight and compact night observation single eyepiece device for surveillance and observation.

The instrument is used as a hand-held viewer and is provided with a lockable swivel handgrip at its bottom. This handgrip is useable with either hand and is equipped with a carrying strap.

The main switch is controlled by a top-located knob accessible with either hand.

Specifications
Dimensions: 257 × 127 × 95 mm
Weight: 1.34 kg
System magnification: ×3.3
Field of view: 187°
Objective: catadioptric type
Focussing range: 10 m to ∞
Eyepiece: monocular type
Eye relief: 35 mm
Dioptre adjustment: ±3
Image intensifier tube: 18 mm GEN II+ or GEN III
Power supply: 2 V to 3.2 V DC
Low battery warning: LED red
Batteries: 2 × 1.5 V AA, 2 × 1.2 V Accu
Environmental data
Operating temperature: –40 to +55°C
Humidity: 92% RH at 40°C/96 h
Vibration: 10 Hz to 55 Hz, max 2.5 g, 1 h, 3 axes
Shock: 10 g/6 ms
Immersion: 1 m water/2 h

Contractor
Photonic Optische Gerate GmbH.

UPDATED

The Photonic NS-B 0053842

Photonic NS-ZF

Type
Image intensifier monocular.

Description
This is a night vision system based upon second- and third-generation tubes and designed for installation through standard mounts. The instrument is powered by two 1.5 V batteries and has reticles and low battery power indicators.

Operational status
In production and in service.

Specifications
Dimensions: 150 × 120 × 95 mm
Weight (with batteries): 1.3 kg
Field of view: 10.5°
Magnification
NS-ZF3.3-80: ×3.3
NS-ZF4-80: ×4

Contractor
Photonic Optische Gerate GmbH.

VERIFIED

The NS-ZF 3.3 × 80 night vision sight 0089880

Pyser-SGI PNP2+ night vision scope

Type
Image intensifier monocular.

Description
PNP2+ is an image intensifier monocular night scope incorporating Generation2+ image intensifier technology in a compact adaptable body. Compatible with SLR and 37 mm palm-corders for evidence gathering in low-light conditions. The PNP2+ night vision scope is fully waterproof and has waterproof lens options.

Operational status
In production.

The PNP2+ night vision scope 0053832

Specifications
Dimensions: 100 × 50 × 63 mm
Weight: 395 g without batteries and objective lens
Field of view: 20°
Magnification: ×2
Dioptre adjustment: +6 to −2
F number: f/1.3
Image intensifier: 18 mm PNP-XD4
Illumination: IR LED for zero light use
Power supply: 2 × 1.5 V AA alkaline batteries
Battery Life: >100 h
Water resistance: waterproofed to IP 67 (immersion 1 m for 30 min with waterproof lens)

Contractor
Pyser-SGI Limited.

UPDATED

Pyser-SGI SNP-XD-4/SG/HG night vision scopes

Type
Image intensifier monoculars.

Description
The SNP-XD-4/SG/HG are night vision scopes incorporating the new XD-4 technology in a compact adaptable body.

The SNP-XD-4 is compatible with SLR and 37 mm palm-corders and is a versatile low-cost item for evidence gathering in low-light conditions. It is also fully waterproof and has waterproof lens options.

Operational status
In production.

Specifications
SNP-XD-4
Dimensions: 100 × 50 × 63 mm
Weight: 395 g without batteries and objective lens.
Field of view: 20°
Magnification: ×2
Dioptre adjustment: +6 to −2
F number: f/1.3
Image intensifier: 18 mm PNP-XD-4
Illumination: IR LED for zero light use
Power supply: 2 × 1 .5V AA alkaline batteries
Water resistance: waterproofed to IP67 (immersion to 1 m for 30 mins with waterproof lens)
Battery life: >100 h

Contractor
Pyser-SGI Limited.

NEW ENTRY

Romtehnica Nova-50 night vision device

Type
Image intensifier monocular.

Description
The Nova-50 monocular may be hand-held or mounted on a weapon for viewing under low-light conditions. The specification does not state the intensifier type but mentions operation down to starlight and a viewing range of 150 m at a light level of 5 mlx.

Operational status
Available.

Specifications
Weight: 500 g (without batteries)
Field of view: 26°
Magnification: ×1.5
Resolution: 24 lines/mm
Dioptre adjustment: 5±
Focus range: 6 m to ∞
Power supply: 2 × 1.5 V R6 batteries

Contractor
Romtehnica, Romania.

VERIFIED

ROMZ Cyclop-P night vision scope

Type
Image intensifier monocular.

Description
The Cyclop-P night vision monocular utilises a high-voltage power unit based on the piezoelectric effect to power the image intensifier tube. By activating a special rod gear, piezoelectric crystals are exposed to high pressure which results in a build-up of high-voltage. This voltage is quite sufficient to power the image intensifier tube. By holding the switch handle pressed, thus producing pressure on piezoelectric crystals, the necessary high-voltage power is generated. The above way of powering image intensifier tubes avoids the use of batteries and high-voltage converter units. This development enhances the reliability of night vision equipment and extends its service life. The piezoelectric night vision scope powered via crystals is simple to use and does not require any maintenance.

Operational status
Available.

Specifications
Dimensions: 160 × 60 × 65 mm
Weight: 0.4 kg
Magnification, (depends on the image tube utilised): Cyclop-P ×1.7, Cyclop-P1 ×1.9, Cyclop-P2 ×1.5
Field of view: 20°
Dioptre adjustment: ±4
Resolution
0.2 Lx: 200 arc sec.
0.01 Lx: 400 arc sec.
Guaranteed minimum number of charges using the press switch: 100,000
Observation time during one charge: 30-60 s

Contractor
ROMZ, Rostov Optical Mechanical Plant (JSC).

VERIFIED

The Romz Cyclop-P piezo-powered monocular 0088170

ROMZ night vision systems

Type
Image intensifier monocular/binoculars.

Elements of some of the ROMZ night vision systems

Description
The ROMZ (Rostovskiy Optico-Mechanicheskiy Zavod) company produces a number of night vision systems in the Cyclops and NZT range of image intensifying devices. These include first- and second-generation night scopes and pocket scopes and night vision binoculars such as CYCLOP-P (piezo-unit, no battery), NZT-2W (waterproof, floating), NZT-34 "Colibri" (miniature pocket scope with pen-like design). Most systems have an integral LED IR illuminator. Some (CYCLOP-P, CYCLOP-1) can be used with optional autonomous AP-7 or AP-12 IR LED illuminators which are easily attached to the main unit. The most advanced are the NZT-2000 (GEN 2+) and compact handheld Thermoimager.

Operational status
All these devices are believed to be in production and probably in service with the armed forces or paramilitary forces of the Russian Federation.

Contractor
ROMZ, Rostov Optical-Mechanical Plant (JSC).

VERIFIED

ROMZ NZT-20 night vision monocular

Type
Image intensifier monocular.

Description
The NZT-20 is an image intensified night vision scope. A choice of objective and eyepiece lenses is available.

Operational status
Available.

Specifications
Dimensions: 185 × 68 × 94 mm
Weight: 0.6 kg
Magnification: dependent on choice of lenses
Field of view
Objective focal length 50 mm: 10°
Objective focal length 73 mm: 7°
Resolution, at illumination of 0.2 Lx: ≤175 arc sec
Dioptre adjustment: ±4
Power: 3 V DC

Contractor
ROMZ, Rostov Optical Mechanical Plant (JSC).

VERIFIED

Senet NV-302/NV-312/Fogbuster

Type
Image intensifier monocular.

Description
These are hand-held night observation devices with second-generation tubes. The NV-302 is a passive device but the NV-312 is a version with an infra-red (IR) semiconductor laser illuminator added to the NV-302. The Fogbuster is described as 'an active-pulse type' device. The devices can be used for observation at distances up to 250 m in starlight (NV-302) or complete darkness (NV-312) and up to 1 km in full darkness and fog (Fogbuster).

Operational status
Not in service with the armed forces of the Russian Federation but possibly used by paramilitary forces. NV-302, NV-312 and Fogbuster are in production.

Specifications

	NV-302	NV- 312	Fogbuster
Dimensions	90 × 64 × 235 mm	130 × 64 × 235 mm	165 × 64 × 250 mm
Weight	1 kg	1.1 kg	1.6 kg
Magnification	×2	×2	×2-10
Field of view	15°	15°	3-11°

Contractor
Senet.

VERIFIED

Thales Optronics LORIS

Type
Image intensifier monocular.

Description
LORIS is a family of high performance monocular night vision devices to be used as hand-held observation systems or as mono or stereo goggles on a face mask or helmet mount. Complete range of observation and aiming sights. Enhanced optics permit future integration of latest technology image intensifier tubes. The concept allows LORIS to be switched within seconds from a hand-held monocular observation sight to face mask or helmet mounted mono or biocular goggles.

The device consists of an advanced low-weight housing containing:

- an ANVIS-type Gen II, SuperGen, GEN III or XD-4™ Image Intensifier Tube
- a ×1 magnifying focusable objective, with protective cover and day-training filter
- an eyepiece with dioptre setting indications, eyeguard (with demist filter) and a combined in-eyepiece IR-LED/ON and battery-low indicator
- integrated batteries (for the hand-held monocular system)
- a remote battery pack (for the face mask or helmet mount system)

Accessories and options

A variety of accessories and options for LORIS is available:

- 60° Field of View (FOV) optics
- ×4 magnifying add-on objective
- face mask — mono or stereo (with flip-up/switch off)
- helmet mount — mono or stereo (with flip-up/switch off)
- laser filter
- carry bag
- transport case.

Specifications
Dimensions: 115 × 65 × 48 mm
Weight (incl batteries): ≤320 gr
Magnification: ×1
FOV: 40° (option 60°)
Focusing range: 25 cm to ∞
Resolution (typical): =0.8 mrad/lp
Exit pupil/eye relief: 10 mm/26 mm
IR-LED: incorporated
Dioptre adjustment: +2 to –5
Battery type: 2 standard AA size
Battery life (at +20°C):
Alkaline: ≥150 h
Ni/Cd: ≥50 hours
Face mask/helmet mount sight position: movement in 3 axis
Interpupillary distance: 52 to 76 mm (bi-ocular)
Flip-up/switch off: yes
General environmental: Mil-STD-810E

Contractor
Thales Optronics (formerly OIP Sensor Systems).

UPDATED

A selection of configurations for the LORIS night vision device
0088129

Thales, LUNOS Lightweight Universal Night Observation System

Type
Image intensifier monocular (family).

Description
The LUNOS passive night vision family consists of a common body, several high-speed objectives with different magnification factors (×1, ×4 and ×6) and a number of options, such as face mask, grip, monopod and recticle for ×6.

Unlike other systems, the LUNOS night vision system allows exchange of objectives on the spot in the field. Switching from one objective lens to another is comparable with changing the lenses of a photo camera. The concept allows a change in the field, within one minute, from lightweight mask-mounted hands-free night driving goggles to medium- or long-range, tripod-mounted observation binoculars. The LUNOS night vision system type is a binocular allowing the operator to observe with both eyes (preferable over monocular). LUNOS's controls are conveniently positioned for easy operation.

The LUNOS controls are:

- Focal range.
- Eye-piece dioptre adjustment.
- Interpupilary adjustment.
- JR on/off switch.

Operational status
Full production.

Specifications

Type	LUNOS ×1	LUNOS ×4	LUNOS ×6
Magnification	×1	×4	×6
Field of view	40°	10°	6.5°
Relative aperture	f/1.1	f/1.2	f/1.5
Weight (excl batteries) approx	570 g without face mask	1,000 g	2,100 g (without handgrips)
Dimensions (L × W × H)	160 × 72 × 131 mm	230 × 104 × 136 mm	300 × 150 × 136 mm
Battery	standard AA size, 2× 1.5V	standard AA size, 2× 1.5V	standard AA size, 2× 1.5V

Contractor
Thales Optronics B.V.

UPDATED

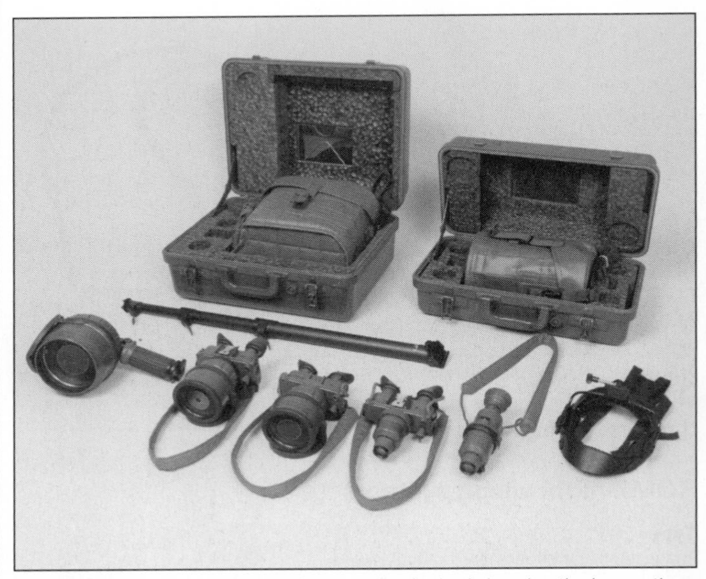

The LUNOS system with the carrying case for the body housing the lens options, the facemask assembly (for use as driving goggles) and handgrip with light source for the reticle

Transvaro TV-MON-I monocular night vision goggle

Type
Image intensifier monocular.

Description
The TV-MON-I monocular is fitted with an 18 mm third-generation image intensifier. It is equipped with an infra-red illuminator. It may be hand-held or head mounted using a custom bracket and strap headband assembly. Light weapon sight and binocular variants are also available. A ×3 magnifier lens may be fitted. The unit has low voltage and infra-red on warning indicators.

Operational status
Available.

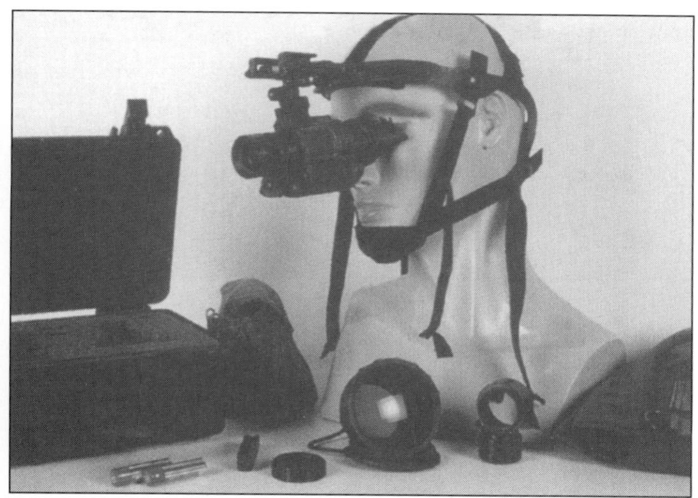

The Transvaro TV-MON-1 monocular 0023172

Contractor
Transvaro a.s, Turkey.

VERIFIED

Vectronix BIM25/35 night pocketscopes

Type
Image intensifier monocular.

Description
The Vectronix (formerly Leica) BIM25 and BIM35 pocketscopes are general purpose monocular night viewing devices, differing only in the objective lens that is used. BIM25 has ×1 magnification, while BIM35 has ×3. They are available with either second-plus, Super- or third-generation image intensifier tubes. The housing is of watertight light metal purged with nitrogen to prevent fogging of the optics.

Operational status
Production. In service with the Swiss Army and undisclosed customers.

Vectronix (formerly Leica) BIM25 pocketscope in use

Vectronix (formerly Leica) BIM25 night pocketscope mounted on a Vectronix camera

Specifications
Magnification: ×1 (BIM25), ×3 (BIM35)
Field of view: 41° (BIM25), 13° (BIM35)
Resolution: 1.035 lp/mrad (BIM25), 3.2 lp/mrad (BIM35)
Dioptre setting: +2 to –6 dioptres
Objective: 24.7 mm f/1.11 (BIM25), 75.1 mm f/1.17 (BIM35)
Image intensifier: 18 mm MCP GEN II+, DEP XD-4 or GEN III
Power supply: 2 × 1.5 V alkaline or 1 × 3 V lithium
Dimensions
BIM25: 138 × 51 × 71 mm
BIM35: 200 × 100 × 100 mm
Weight: 0.43 kg (BIM25), 1.16 kg (BIM35)

Contractor
Vectronix AG.

UPDATED

Yugoimport POD 7 × 200

Type
Image intensifier monocular.

Description
The POD 7 × 200 is a passive night observation device based upon a 20 mm image intensifier tube. It may be used for surveillance or for artillery fire control. The instrument uses a Ni/Cd battery which gives 15 to 20 hours' service and with it, a

The POD 7 × 200 night observation device

standing person may be recognised at 1 km and a vehicle at 3 km in poor light conditions.

Operational status
In production and in service.

Specifications
Weight: 13 kg
Field of view: 70°
Magnification: ×7

Contractor
Yugoimport SDPR.

VERIFIED

OBSERVATION AND SURVEILLANCE

AREA SURVEILLANCE

Alcatel BM 2000 area/border surveillance and control system

Type
Area surveillance system.

Description
The BM 2000 area/border surveillance and control system is a command and control system for users such as army, border forces, coast guards, security and police. It is a generic and open system which consists of three subsystems: an Information Processing and Presentation Subsystem (IPPS), a communication subsystem and a sensor subsystem. Scenario presentation on the screen is in a clear, easily understood form and the system is simple to operate.

The heart of BM 2000 is the IPPS. Its purpose is to process all information received from the sensor subsystem via the communication subsystem. The result is presented in real time to the operator in graphical and alphanumerical form. The operator uses this information as a basis for his decisions in the command and control process within surveillance operations.

The communication subsystem integrates the sensor subsystem and the IPPS by transmitting all sensor information to the IPPS via wire or optical fibre, HF, VHF, or UHF, or microwave and satellite communications, according to customer requirements. The sensor subsystem is, depending on the geographical and threat characteristics, a dedicated solution that can be adapted to area or border control needs by the use of one or more sensor types such as radars, optical (TV, lowlight, thermal imaging), seismic, acoustic, drone-mounted sensors, or integrated Mobile Optronic Reconnaissance System (MORS) vehicles.

Contractor
Alcatel SEL AG.

VERIFIED

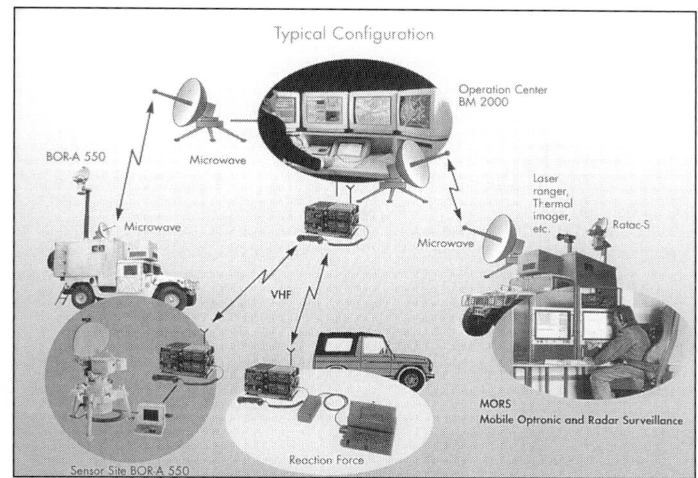

Alcatel BM 2000 typical configuration 0023174

Alcatel MORS Mobile Optronic Reconnaissance System

Type
Area surveillance system.

Description
MORS is a mobile optical, electronic and electro-optical reconnaissance system for day and night surveillance of frontier zones, coastal areas, vital and strategic sites and battlefields. In the battlefield environment, MORS serves as a surveillance vehicle for forward observers, artillery fire guidance and battlefield intelligence.

The main system components of MORS are: sensor platform with electro-optic sensors – daylight camera, latest generation thermal imager and eye-safe laser range-finder; high-precision sensor positioning system with global positioning system and north-seeking gyro; independent sensor and positioning system control units; information processing and presentation system with high-resolution single/dual monitor, keyboard and cursor control and ruggedised computer with removable disk drive; and video recording system.

Information gathered by the electro-optic sensors is processed and displayed on a digital map background of the area under surveillance. Tactical situations, hostile and friendly forces can be superimposed on the map. When combined with the RATAC-S target acquisition radar, shell impacts can be displayed.

MORS is a fully autonomous mobile sensor system. Communication interfaces allow the system to be linked with a command and control system such as the Alcatel BM 2000 area/border surveillance system. The processor uses an Open System architecture for expandability.

Operational status
Not known.

Contractor
Alcatel SEL AG.

UPDATED

MORS mobile control post container 0023175

Arkonia ARK9300 Rapid Deployment Surveillance System

Type
Area surveillance system.

Description
Traditionally the establishment of a CCTV between a target and observation point can often be a very lengthy process. The Rapid Deployment Surveillance System (RDSS) has been designed to specifically address this difficulty. Housed in a ruggedised sealed briefcase, the RDSS comprises a camera, pan and tilt device, COMMANDER 16 Remote Control System, microwave transmitter and power supplies. Customers existing microwave transmitters can be factory fitted as an option. The system can be deployed within a few seconds, and can be controlled with the COMMANDER 16 Remote Controller, to meet any application. A remotely controllable IR illuminator is available for use with the IR camera variant. Once deployed, the system is environmentally rated to 1P65, and can be safely left in the field for lengthy periods in the most demanding environmental conditions.

Operational status
Available.

Specifications
Camera: Colour supplied as standard with 18:1 zoom (72:1 digital)
Video transmission: Built in microwave transmitter
Pan & tilt: 3.5 kg load capacity
Telemetry: Via built-in COMMANDER 16 system
Operating frequency: 150 to 170 MHz
Output: 3 W
Dimensions: 46 × 36 × 15 cm
Weight: Complete system 15 kg
Power supply: Built-in 12 V re-chargeable lead acid batteries
Battery life: 36 h nominal
Environmental: IP 5

Monitor ARK9310
Display: 10 in colour 500 TV lines
Receiver: microwave c/w omnidirectional unity gain antenna
Dimensions: 50 × 29 × 29 cm
Weight: 6.2 kg

The Arkonia ARK9300 rapid deployment surveillance system 0102149

Contractor
Arkonia Systems Limited.

VERIFIED

Arkonia ARK9500 Remote Video Surveillance System

Type
Area surveillance system.

Description
The Remote Video Surveillance System (SPECTRE) has been specifically designed to fulfil stringent remote surveillance and close-support requirements in harsh environments. Ease of concealment and operation have been prime considerations in the design of the system which offers versatility for surveillance roles. SPECTRE comprises small remotely deployable RF camera sensors 'Scouts' to which are connected to Arkonia rugged PIR detectors. The scout units remain quiescent until activated by the PIR upon the occurrence of an intrusion. Upon activation, the camera automatically transmits a video signal (which also contains its sensor ident) to the Base Station. The video signal is of 15 seconds duration. The video imagery is automatically recorded and displayed at the Base Station where the date and time of the alarm are also indicated. Simultaneously, an alarm signal is also transmitted from the Base Station via a VHF radio link to miniature monitors (pagers) which are intended to be carried by the support team. The monitors indicate the ident of the sensor in alarm condition and the date and time of alarm. Body worn 'Panic buttons' operating over a UHF radio link can also be operated with the Base Station and remote pagers. Up to six remote 'scout' sensors and two 'panic buttons' can be used with each Base Station, a software package supplied with SPECTRE permits simple system configuration.

Operational status
Available.

Specifications
Optical: 25 mm lens monochrome camera (0.1 lux)
PIR: detection ranges: personnel at 30 m
Anti-tamper: Automatic activation of the camera
Dimensions: 130 × 78 × 55 mm
Weight: 1 kg (including batteries and PIR)
Power supply: 8 × AA batteries
Battery life: Nominal 7 days operation
Temperature range: –20 to +55°C (operating)

The Arkonia ARK9500 surveillance system 0100718

Base station
Indicators: Front panel LED annunciator indicates activated sensor/low battery
Pager operating frequency: 150 MHz band
Panic Button operating frequency: 400 MHz band
Display: 142 mm flat screen LCD
Video recording: Up to 3 h total elapsed recording time
Weight: 10 kg
Dimensions: 46 × 36 × 18 mm
Power supply: 2 × 7.2 Ah sealed lead acid rechargeable batteries
Battery life: 30 days nominal
Environmental: IP 65

Miniature monitor
Display: LCD
Operating frequency: 150 MHz band
Dimensions: 75 × 50 × 18 mm
Weight: 75 g

Contractor
Arkonia Systems Limited.

VERIFIED

Aselsan MARS-V Armoured Reconnaissance/Surveillance Vehicle

Type
Area surveillance/reconnaissance system.

Description
MARS-V employs a combination of advanced sensor technologies and computer-controlled infrastructure into a single platform. Mounted on 4 × 4 light armoured wheeled vehicles, this system provides multispectral surveillance using optical and radar sensors. MARS-V has elevated sensors for extended coverage and is designed for continuous monitoring of the battlefield, detecting possible targets and transmitting the collected information to command centres and reaction forces.

MARS-V uses ground surveillance radar and 2nd generation thermal imager for long-range target detection. The two independent sensor systems can perform surveillance in different sectors. When a target is detected by one of the sensors, other sensor can be automatically directed onto the target area in order to improve target recognition and classification. Target identification can be initiated using the Doppler tone generated by the radar, cueing the thermal imager and DayTV to the appropriate direction. Target information collected by the sensors and evaluated/classified in the information processing software, can be transmitted to command and monitoring centres using various communication equipments integrated in the vehicle. Target information, including co-ordinate, type, and so on, are gathered in formatted reports prepared using the software and can be transmitted securely via the frequency hopping radio, whereas full-motion target imagery can be sent through microwave video link. Still images can be incorporated in the formatted reports.

Main system functions, including control of peripheral equipment and electro-optic sensor system, is controlled by the System Control Unit as demanded by System Operator.

The MARS-V surveillance/reconnaissance vehicle 0088165

Specifications

Sensors:
- ASELSAN Ground Surveillance Radar
- 2nd Generation Thermal Imager
- DayTV
- Target co-ordinate determination system

Communication Equipment:
- 9600 VHF frequency hopping radio
- Video transmission system (MW Link)
- Field telephone

Command Control Software:
- Geographical information system
- Database
- Formatted message

Target detection range (with radar):
- 38 km (vehicle convoy)
- 15 km (personnel)

Status

The MARS-V is in service and in production.

Contractor

Aselsan Inc, Microwave and System Technologies Division, Turkey.

VERIFIED

CILAS SLD 400

Type

Sniper laser detection system.

Description

The CILAS SLD 400 is designed to detect and locate any kind of optical or electro-optic hostile sight on the battlefield or sensitive zones. It is fitted on a tripod or observation turret and can be used for standby detection of pointed optics, often associated with an immediate threat; active scanning and monitoring of specific areas; and defeating and deterring snipers. SLD 400 detects tank and armoured fighting vehicle sights, as well as snipers equipped with shooting telescopes or night vision equipment.

The SLD 400 consists three major sub-assemblies, the optronic sensor head, the pan and tilt turret and the remote control.

The optronic sensor head provides the laser illumination and observation of the scenery. It includes:

(1) A wide angle coded laser beam that illuminates the sector to be searched

(2) A special receiver featuring a low light amplifier. It receives the low-level energy reflected by the optic of the threat. It has a power supply with rechargable batteries.

The pan and tilt turret moves the optronic sensor head in azimuth and elevation for scanning the suspected area.

The remote-control unit monitors the optronic sensor and displays the image of the scenery on an observation screen. This command/control unit operates the laser and the receiver, synchronises them, digitises and processes the video signals, highlights the echoes on the observation screen and activates the warning alarm.

The basic SLD 400 system is man transportable. Its range detection is at least 500 m in the worst conditions (bright daylight) and in excess of 2 km (dark night).

It can also be coupled with other detectors and surveillance systems such as infra-red, thermal imaging cameras and acoustic detectors.

Operational status

Available.

Specifications

Dimensions
- **Sensor head:** 140 × 340 × 260 mm
- **Turret:** 285 × 300 × 240 mm
- **Remote control unit:** 220 × 600 × 500 mm

Weight
- **Sensor head:** 9.7 kg
- **Turret:** 10 kg
- **Remote control unit:** 17 kg
- **Optical head:** ≤7 kg

Laser: eye-safe, Class 1 with attenuator
Field of view: 4 to 30° adjustable, ≤6 × 4.5° (narrow), ≥35 × 26° (wide)
Azimuth setting: ±175° remotely controlled
Elevation: ±15°
Automatic scanning: 60° aperture, automatic lock on target when a relevant echo is detected
Detection range (for a typical rifle scope):
=1,000 m with a visibility of 5,000 m and an ambient light level less than 8,000 lx
≥500 m with a visibility of 5,000 m and an ambient light level between 8,000 and 20,000 lx
Power: Mains (MIL STD 1275 A)
24 V battery (option)

Consumption
Sensor head: 35 W (ambient temperature)
85 W (extreme temperature)
Remote Control Unit: 45 W average (70 W peak)
Turret: 40 W (5°/s scanning speed)
110 W (maximum speed)

Contractor

Compagnie Industrielle des Lasers (CILAS).

VERIFIED

CILAS SLD 400 sniper laser detection system 0053828

Controp CEDAR automatic intruder detection system

Type

Area surveillance system.

Description

CEDAR is a sophisticated electro-optical panoramic intruder detection system which automatically detects motion over a wide field of view. The system incorporates a 3 to 5 μm infra-red camera supplied by Controp or an alternative supplied by the customer.

There are two modes of operation:

- observation mode with live video for intruder recognition and identification
- panoramic scan mode for intruder detection.

The scan sector can be selected in both azimuth and elevation. The system is operated and controlled via a Control and Display Unit (CDU) which consists of a PC integrated with a frame grabber and a joystick. The wide azimuth and elevation range is provided by an accurate motorised pan and tilt unit controlled by the CDU. CEDAR has an option to integrate an eyesafe laser range-finder for accurate target positioning.

Specifications

Weight (pan and tilt unit): 13 kg
Dimensions (h × w × d): 400 × 380 × 250 mm
Field of regard: Horizontal ± 160°; vertical ± 30°

Flir sensor
Spectral band: 3 - 5 μm
Format: 256 × 256 pixels
Optics: dual FOV lens
- WFOV 8°, IFOV 0.6 mrad
- NFOV 1.6°, IFOV 0.12 mrad

Laser range-finder option
Wavelength: 1.54 μm
Range: 10,000 m
Range accuracy: ± 5m

Electrical interface
Power supply: 110/220 V AC or 28 V DC
Power consumption: 200 W nominal

Contractor
Controp Precision Technologies Ltd.

NEW ENTRY

Elop OPAL man-portable thermal imaging camera

Type
Thermal imaging surveillance sight.

Description
OPAL is a dual FOV, compact and ruggedised thermal imaging camera/night sight for ground applications. It utilises an advanced FPA detector sensitive in the 3 to 5 μm spectral band, coupled to a miniature closed-cycle cooler.
Main features:
- Advanced 3 – 5 μm FPA detector/cooler technology
- Dual FOV
- Superior image quality due to ELOP's proprietary ASIC
- Battery operated
- Low power consumption
- Integral CRT display and bi-ocular optics
- Ruggedised construction
- Tested and qualified to MIL-STD-810.

Applications:
- Ground to ground observation posts (fixed and deployed)
- Man-portable target acquisition systems
- Anti-tank missile launchers sight
- Laser designators
- Air defence weapon systems
- Naval and coastal observation tasks.

Specifications
Spectral band : 3 to 5 μm
Entrance aperture: 76 mm
Fields of View:
Wide: 6.0 × 4.5°
Narrow: 1.5 × 2.0°
E-Zoom: 0.75 × 1.0°

Detector
Type: InSb
No of elements: FPA, 320 × 256 elements
Cooler: Stirling closed-cycle, integral with Dewar
Video Output: CCIR, (RS-170 – optional)

Display
Miniature, high resolution internal CRT display, coupled to bi-ocular viewing optics
Hood: Rubber, light concealing hood
Communication port: RS-422, enables full remote control operation
Power: 18 to 32 V DC, Complies with MIL-STD-1275AT
Power supply:
Disposable, high-energy density $LiSO_2$ batteries
Rechargeable NiMH batteries
External 24 V DC power source
Weight: 5.8 kg
Operating temperature range: –20 to +60°C

Contractor
Elop Electro-Optics Industries Ltd.

VERIFIED

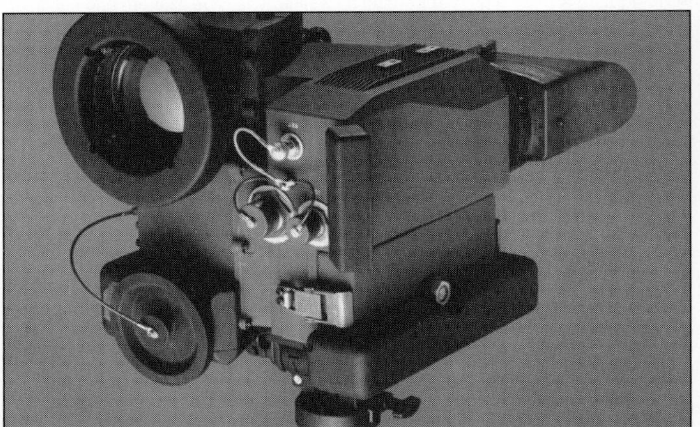

The Elop OPAL thermal imager 0109666

FLIR Systems ThermoVision Sentry POD®

Type
Thermal imager (land).

Description
The ThermoVision Sentry POD® (Personnel Observation Device) offers continuous day/night surveillance in a rugged, sealed camera module. A dual field of view infra-red imager, combined with an integrated colour CCD-TV camera, provides both situational awareness and close-up imaging. The infra-red optical system provides a 4:1 ratio between wide and narrow fields of view, while the 4:1 continuous digital electronic zoom provides for another level of magnification and fine-tuning.

The Sentry POD's® on-board processing capability includes an auto-focus function that automatically produces sharp infrared images; an InstAlert™ capability that highlights warm objects, such as people or vehicles, in red; and an auto-contrast algorithm that optimises image contrast for maximum target definition and tracking capability.

Operational status
In production.

Specifications
Dimensions: 270 × 230 × 180 mm
Weight: <6 kg
Mounting provision: 2 × 1/4 in × 20 tapped holes

Thermal imager
Detector: Uncooled microbolometer, 320 × 240
Spectral band: 7.5-13 μm

Fields of view
WFOV: 24 × 18°
NFOV: 6 × 4.5°
FOV switch time: <0.8 seconds
Electronic zoom: Continuous ×1 to ×4
Focus control: Auto or manual
Digital image resolution: 14 bit
Gain/Level adjustment: Manual or auto, with histogram equalisation

Environmental specifications
Encapsulation: Sealed enclosure, IP65, NEMA 4
Shock: 25 g, IEC 68-2-29
Vibration: 5-500 Hz, IEC 68-2-6
Operating temperature: –32 to +60°C
Solar immunity: Sun shield provided
EMI/EMC: MIL-STD-461D

Features
Imager: Auto focus, auto image optimisation, auto contrast enhancement
Target detection: InstAlert colour tagging
Image magnification: Digital zoom 1 × to 4 × continuous
On-screen symbology: Date, time, NFOV brackets, reticle, control status, momentary focus indicator
Image capture: Freeze frame
Image polarity: White hot/black hot

Contractor
FLIR Systems.

NEW ENTRY

FLIR Systems Sentry POD®
NEW/0547541

FLIR Systems ThermoVision® 2000™

Type
Thermal imager.

Description
The ThermoVision® 2000™ is a state-of-the-art, long-range, platform-mounted thermal imaging system. Its long-wave, narrow-band QWIP 320 × 240 FPA delivers high-thermal sensitivity in both warm and cold climates and makes the system resistant to existing laser aggression weapons. Three fields of view ensure mission flexibility and effectiveness. It is able to pan out for situational awareness or zoom in tight to identify targets. On-board processing features include infra-red auto focus, automatic image optimisation and network communication.

Operational status
In service.

Specifications
Dimensions (nominal): 503 × 312 × 267 mm
Weight: 18 kg (7 kg, module only)

Optics
Field of view/min switch time: <0.5 s
Lens identification: Automatic, refocused when switched
Focus: Auto or manual

Imaging performance
Detector: 320 × 240
Number of field of views: 3
Wide: 25 × 19°
Middle: 6.0 × 4.5°
Narrow: 0.99 × 0.74°
IFoV wide: 1.37 mRad
IFoV middle: 0.33 mRad
IFoV narrow: 0.054 mRad
Min focus distance: 20-100 m
Thermal sensitivity: 0.03°C
Image frequency: 50/60 Hz
Electronic zoom function: Up to 4× in real time
Detector : GaAs Quantum Well Infrared Photodetector (QWIP), 320 × 240 pixels
Spectral range: 8-9 μm
Cooling: Integrated Stirling

Image presentation
Video output: RS-170 EIA/NTSC or CCIR/PAL, composite and/or S-video
Digital image: Full dynamic, 14-bit digital image data
Vibration: Operational: 2 g IEC 68-2-6
EMI/EMC: MIL-STD 461 D, US FCC 15 J Class A, EN5008-1, EN50082-1

Interfaces
Remote control: RS-232, RS-485 and Ethernet (USB optional)

Contractor
FLIR Systems.

NEW ENTRY

ThermoVision 2000 ***NEW***/0547502

FLIR Systems ThermoVision® Ranger

Type
Thermal imager (land).

Description
Qualified to US MIL-STD-810E, the ThermoVision® Ranger is able to track vehicles in excess of 9 km, with detailed recognition at 3 km. It features a third generation Indium Antimonide (InSb) Focal Plane Array (FPA) and dual-FOV 100/500 mm telescope. The Ranger can be configured with a variety of optional high-performance sensors and has a dedicated ergonomic hand controller.

Operational status
In service.

Specifications
Dimensions (nominal): 558 (l) × 165 (d) mm (22 (l) × 6.5 (d) in)
Weight: 9.9 kg (22 lb) complete
Mounting: ¼-20 threaded base plate

Thermal imager
Detector: 256 × 256 InSb FPA
Spectral band: 3-5 μm
Optics: f/4.0
WFOV: BRS, ERS: 50 mm, 9 × 7°; BRSX, ERSX: 100 mm, 4.5 × 3.5°
NFOV: BRS, ERS: 250 mm, 1.8 × 1.6°; BRSX, ERSX: 500 mm, 0.9 × 0.8°
FOV Switching Time: <2 seconds
FPA cooling: Closed-cycle Stirling
Thermal sensitivity: 0.025°C at 23°C ambient
Camera controls: On/off/standby, auto and manual level, manual gain, uniformity correction, polarity, reticle on/off, reticle select, change FOV, focus
Video format: RS-170 or CCIR compatible
Serial interface: RS-232 (RS-422 optional)
Power requirements: 8-32 V DC
Heaters (optional): 80 W
Environmental qualifications: Fully ruggedised/MIL-STD-810E qualified for field operations
Operating temperature range: –32 to +55°C
Storage/transport case: Hard shell, lockable with fitted foam insert
AC adapter/charger: 110 V AC/220 V AC power supply
Break out box: Provides video, power and hand controller interface
Hand controller: Ergonomic design, controls all system functions
System cable: 25 ft long moulded connectors
Power supply cable: 6 ft long moulded connectors

Contractor
FLIR Systems.

NEW ENTRY

FLIR Systems ThermoVision Ranger ***NEW***/0547531

FLIR Systems ThermoVision® Ranger II™

Type
Thermal imager.

Description
Qualified to US MIL-STD 810E, the ThermoVision® Ranger II™ uses Gen III midwave (focal plane array detector) and is able to track vehicles in excess of 9 km, with detailed recognition at 3 km. It features two versions of its dual-FOV optics and 2×/4× digital zoom.

Operational status
In production.

Specifications
Dimensions (nominal): 20 (L) × 6.5 (D) in
Weight: <20 lb
Mounting: 1/4-20 threaded base plate

Thermal imager
Detector: 256 × 256 InSb FPA
Spectral band: 3-5 μm
Optics: f/4.0

WFOV:
Standard: 50 mm, 11 × 8.2°
Optional 2 × Extender: 100 mm, 5.5 × 4.1°

NFOV:
Standard: 250 mm, 2.2° × 1.6°
Optional 2 × Extender: 500 mm, 1.1 × 0.8°
FOV Switching Time: <2 seconds
FPA cooling: Closed-cycle Stirling
Thermal sensitivity: 0.025°C at 23°C
Camera controls: On/off/standby, dual-optical zoom FOV change, digital zoom 2× and 4×, front lens defroster, focus, polarity, reticule on/off, reticule select, change FOV, focus, NUC sensitivity adjustment, freeze frame, save and restore image, auto span with histogram equalisation
Video Format: S-video, RS-170 or CCIR compatible
Serial Interface: RS-422 (RS-232 optional)
Power requirements: 9-26 V DC
Heaters (optional): Lens defroster 80 W at 12 VDC
Environmental qualifications: Fully ruggedised/MIL-STD-810E qualified for field operations
Operating temperature range: –32 to + 55°C (–26 to +131°F)
Storage/Transport case: Hard shell, lockable with fitted foam insert
AC Adapter/Charger: 110 V AC/220 V AC power supply
Break out box: Provides video, power, Ethernet and hand controller interface
Hand controller: Ergonomic design, controls all system functions
System cable: 25 ft or 50 ft long moulded connectors
Power supply cable: 10 ft long moulded connectors

Contractor
FLIR Systems.

NEW ENTRY

FLIR Systems ThermoVision Ranger II on Tripod
NEW/0137864

FLIR Systems ThermoVision® Sentry™

Type
Thermal imager (land).

Description
Qualified to US MIL-STDs 810 and 461, the ThermoVision® Sentry is a completely self-contained surveillance system featuring thermal and TV imaging sensors and integrated and pan-tilt system. The thermal imager features a 320 × 240 uncooled microbolometer focal plane array detector, a dual-optical field of view for wide area surveillance and close-up inspection, an electronic zoom, freeze frame and automatic or manual level and gain control. The integrated pan and tilt drive provides fast, continuous (n × 360°) slew and lockable stow position to protect optics. The Sentry's autoscan feature enables the system to sequentially scan between multiple preset positions at varying scan rates and dwell times, reducing operator workload for monitoring of strategic areas of interest.

Operational status
In service.

Specifications
Weight: 17.3 kg
Dimensions: 38 × 28 × 23 cm

Field of regard
Azimuth: Continuous 360°
Elevation: –35 to +60° (higher elevations available)
Slew rate: 0-120°/second azimuth; 0-60°/second elevation
Pointing accuracy: ±3.0 mrad with ±0.1 mrad repeatability
Autoscan: Multiple positions, fully programmable
Park position: Lockable position protects optics

Thermal imager
Detector: Uncooled microbolometer, 320 × 240
Waveband: 7.5-13 μm
WFOV: 24 × 18°
NFOV: 6 × 4.5°
FOV switch time: <0.8 seconds
Electronic zoom: Continuous ×1 to ×4
Digital image resolution: 14 bit
Gain/Level adjustment: Auto or manual
Image processing: Histogram equalization
Palettes: Black/white, rainbow, iron

Visual Camera
Type: Colour CCD (¼ in)
Resolution: 460 television lines NTSC, 470 television lines PAL
WFOV: 48°
NFOV: 2.7°
Video format: NTSC (RS-170) or PAL (CCIR)
Serial interface: RS-485
Power: 18-32 V DC
Power consumption: 35 W (average)
Built-in-test (BIT): Self-diagnostics
Operating temperature range: –32 to +50°C

Contractor
FLIR Systems.

NEW ENTRY

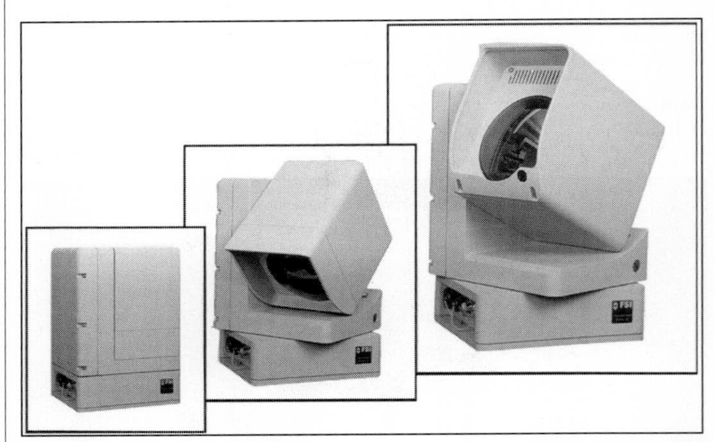

The FLIR Systems ThermoVision Sentry ***NEW***/0089906

General Dynamics Canada all-weather multispectral surveillance system

Type
Area Surveillance.

Description
The General Dynamics Canada 24-hour all-weather multispectral surveillance system is an open architecture, modular hardware and software system. The system is chassis independent and can be linked to the digital battlefield via data/image transmissions. It comprises the following subsystems:
(a) visible spectrum day/night camera
(b) thermal imager
(c) laser range-finder
(d) man-portable battery-powered Ku-band Doppler radar designed primarily for detection of personnel and vehicles.

An operator control station provides all system control and monitoring facilities. It is fitted with a high-resolution monochrome monitor and has the power conditioning and power control for the entire system. A hi-8 VCR recording system is fitted with a VME computer giving 100 per cent growth potential.

Operational status

Production as required. In service with Canadian Army (on some LAV-25 (Reconnaissance) Coyote 8 × 8 vehicles).

Specifications

Visible spectrum day/night camera
Field of view: High-resolution continuous zoom with 31 × 23 mrad NFOV and 296 × 222 mrad WFOV
Thermal imager
Spectral band: 8-12 µm
Field of view: 50 × 37 mrad NFOV and 200 × 150 mrad WFOV
Laser range-finder
Type: Eye-safe Er: glass
Range: up to 10,000 m with ±5 m accuracy
Azimuth/elevation platform
Remote controlled with pan/tilt capability and 360° traverse
Ku Doppler radar:
Scan rate: Capable of 18°/s scan rate with 180° max swath in surveillance mode,
Area coverage: 1 × 1 km area with 10 m resolution B scope (9°/s scan rate) in acquisition mode and 1 × 1 km area artillery impact (B scope, 45°/s) in fall of shot mode
Hybrid cable: Copper wire power transmission with optical fibre video transmission and built-in video/data to light converter for no-loss noise-free transmisson

Contractor

General Dynamics Canada.

VERIFIED

General Dynamics Canada all-weather multispectral surveillance system
0001630

Graflex motorised zoom lens systems

Type

Area surveillance (land).

Description

Graflex produce a range of high-performance, motorised ×27 zoom lenses suitable for surveillance applications. Five basic lenses are available, all with measured telephoto resolution exceeding 1,000 TV lines. There are also motorised extender models which double the magnification to ×54. All units feature multi-layer, high efficiency optical coatings for maximum transmission in the visible and near infra-red spectrums.

Operational status

Available.

Contractor

Graflex Inc.

VERIFIED

Indigo Systems TH-10 long-range IR surveillance camera system

Type

Long-range tracking and wide-area surveillance.

Development

Indigo Systems uses the TH-10 long-range IR surveillance camera system in its Motorized Video Surveillance Platform (MSVP - qv). The TH-10 is designed for use in harsh outdoor environments where continuous, reliable IR imaging is required. The electrical, mechanical and software interface is designed for easy integration with established tracking or surveillance systems.

Description

The system can be equipped with either the Merlin (qv) or Phoenix (qv) thermal imaging modules from Indigo Systems (both mid-wave IT 3-5µ), when it would be designated TH-10M or TH-10P as appropriate. These InSb detectors are offered with choice of dual- or triple-FOV lenses for either camera, in either mid-format (320 × 256) or large-format (640 × 512) configurations. Each E-O sensor is housed in an O-ring sealed and nitrogen-purged enclosure.

Operational status

Deployed and available.

Specifications

Weight: TH-10M - 24 kg; TH-10P - 27 kg
Dimensions: 25.4 × 76.2 × 96.5 cm
Temperature range: –20°C to +60°C
Power: single 24 V DC input (<15 amps)
Connectors: MIL-C38999 and BNC

Optics
Resolution: 320 × 256
Dual fields of view:
Narrow (250 mm): 2.2 × 1.8°
Wide (50 mm): 11 × 8°
Aperture: f/2.5
Focus range: 100 ft to ~
Triple fields of view:
Narrow (500 mm): 1.1 × 0.9°
Medium (180 mm): 3.1 × 2.4°
Wide (60 mm): 9.1 × 7.3°
Aperture: f/4.1
Focus range: 100 ft to ~
Resolution: 640 × 512
Dual fields of view:
Narrow (250 mm): 3.7 × 2.9°
Wide (50 mm): 18.2 × 14.6°
Aperture: f/2.5
Focus range: 100 ft to ~
Triple fields of view:
Narrow (500 mm): 1.8 × 1.5°
Medium (180 mm): 5.1 × 4.1°
Wide (60 mm): 15.2 × 12.2°
Aperture: f/4.1
Focus range: 100 ft to ~

Contractor

Indigo Systems Corporation (USA).

NEW ENTRY

The Indigo Systems TH-10 long-range IR surveillance camera system **NEW**/0595422

Intertechnik PASS intruder detection and surveillance system

Type
Area surveillance system.

Description
PASS is a remote ground sensor system for military and high-security area applications. It detects, classifies and remotely displays target information on personnel and vehicles.

The battery-operated system consists of remotely placed seismic and infra-red sensors connected to a sensor unit and a monitor unit. The sensor unit classifies the intrusion information using signal processing algorithms and broadcasts the alarm messages by means of a built-in VHF transmitter.

The monitor unit receives the VHF signal and presents the alarm information on a liquid crystal display. Up to ten sensor units may be used with one monitor.

Operational status
In production and in current use.

Contractor
Intertechnik Technische Productionen GmbH.

VERIFIED

ISAP Cyclops

Type
Area surveillance system.

Description
The Cyclops dual position zoom thermal imaging surveillance system is a remote-controlled or man-portable monitoring system designed for applications such as border security and monitoring of aircraft, vehicles and marine traffic. Cyclops is capable of detecting personnel at ranges in excess of eight miles. The system operates in the 3 to 5μm spectral band and is optimised for long-range surveillance in both low- and high-humidity environments.

Specifications
Spectral band: 3-5 μm
Detector type: InSb, 320 × 240 elements
NETD: 0.025°C at f/2.5
Camera Nyquist frequency: 16.7 lp/mm
Cooling: linear split-cycle Stirling
Fields of view:
Wide: 1.1°
Narrow: 0.29°
Image format: 10 mm diagonal
Resolution: ≤ 0.07 mm across field of view
Power: 6-28 V DC
Consumption: < 35 W (after cool down)

Contractor
Industrial Alliance Security Partners.

VERIFIED

Kollsman Remote FLIR system

Type
Area surveillance system.

Description
The Kollsman Remote FLIR system provides infra-red images which may be displayed on any standard TV monitor (either 525-line 60 Hz or 625-line 50 Hz) and may be used for fire and command control, monitoring high-value areas and detecting motion. The system uses US Army MCTNS thermal imaging common modules and the basic control functions include range focus, brightness, contrast and field of view selection.

The Kollsman Remote FLIR imaging head

Operational status
In production for several overseas customers.

Specifications
Weight: 18 kg
Fields of view: 3.4 × 6.8°; 1.2 × 2.2°
Resolution: 0.5 and 0.167 mrad
Power: 30 W

Contractor
Kollsman Inc.

VERIFIED

L-3 Communications REMBASS/IREMBASS

Type
Area surveillance system.

Development
Formal development by the US Defense Department of unattended ground sensors commenced in 1966. The initial goal was to develop a sensor system for use in conjunction with conventional barrier and air support systems in Southeast Asia. Development efforts were expanded in 1970 to encompass a wide range of new sensor projects. The Army decided that further tactical sensor development was to be undertaken under the REmotely Monitored BAttlefield Sensor System (REMBASS) programme.

The first full-scale development equipment was delivered to the government during the early part of FY82. In June 1984, efforts were directed to reduce the size and weight of REMBASS components to make them more manageable in use with the US Army's Special Operations Forces. REMBASS has been type-classified as standard and is fully fielded and still in use in the Army's Light Divisions.

The IREMBASS (Improved REMBASS) system fielding to the Special Forces began in January 1994 and was completed in 1995. Since that time it has also been produced for several foreign countries. L-3 Communications has continued the development of the system by including a wide range of additional sensors such as meteorological and lightweight physical security sensors.

Description
Both REMBASS and IREMBASS are designed to detect and to classify both personnel and vehicles and to provide forward units and formations with detailed information on events in the zone immediately to their front. It can also be useful in rear areas for the detection of intruders.

The system works by using remotely monitored sensors placed along likely avenues of approach. These sensors respond to a wide variety of influences; infra-red, acoustic, seismic, magnetic field changes and others. Each sensor produces a signal, indicating a detection and sends this signal by FM radio incorporated in the body of the sensor. The sensor remains passive when it is not transmitting a detection signal, so it is virtually impossible to locate. The signal may pass directly to the monitor unit, or it may be relayed by emplaced repeaters.

With the use of relays/repeaters for ground application, REMBASS is designed to provide targeting data at up to 100 km range. Both systems use digital transmissions operating at 1,200 bits/s with a burst from the sensor lasting 25 ms. Once at the monitor the signal is demodulated, decoded, displayed and recorded. The system provides real-time information over a wide area and, because of the variety of sensors, it is largely self-checking, that is, the information from one type of sensor is checked against that from another, virtually eliminating false alarms.

The prime sensors in the REMBASS system are the hand-emplaced DT-561, DT-562 and DT-565. DT-561 is a magnetic sensor powered by a lithium battery which has a minimum operating life of one month. The detector is a two-axis, brown-type ring core magnetometer with a response bandwidth of 0.04 to 1.0 Hz. It can detect tracked vehicles 25 m away, wheeled vehicles at 15 m and personnel at 4 m.

The DT-562 is the seismic-acoustic classification sensor and tests have verified a minimum of 80 per cent accuracy rate. It has the same power source and life expectancy as the DT-561 and a response bandwidth of 8 to 135 Hz. It can detect vehicles at a minimum of 350 m and personnel at 50 m. The sensor response is one report every 10 seconds.

The DT-565 is the infra-red sensor and can detect objects with only a 1.5°C temperature difference travelling at speeds of 0.1 to 40 m/s. The lithium tantalate detector operates in the 8 to 13 μm spectral band and has a field of view of 1.5 × 5.6°. It can detect vehicles at distances between 3 and 50 m and personnel at 3 to 20 m. The DT-56S, like the DT-561, will indicate the target's direction of travel and can be used to count targets and determine their speed. The response time is instantaneous.

In addition to hand-emplaced sensors, plans exist for air-delivered seismic-acoustic sensors, and for similar sensors which would be fired from 155 mm howitzers, but these have not entered production.

Data to the monitors is transmitted through the RT-1175 unmanned radio repeater, whose two lithium batteries give an operating life of at least 30 days. To

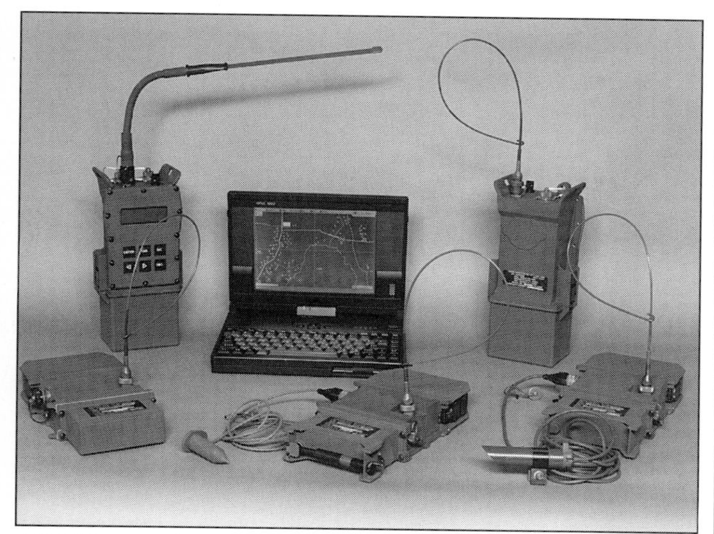
IREMBASS remotely monitored battlefield sensor system 0005455

load frequency, identity and other parameters into the sensors and the repeaters, there is the C-10434 code programmer.

IREMBASS significantly reduces the fielded REMBASS size and weight, increases functionality and simplifies logistics to enhance mission effectiveness.

For mobile sites and forward areas, the hand-held combined Monitor/Programmer Set, AN/PSQ-7, receives, decodes, displays and records data from up to 64 sensors. The AN/PSQ-7 is a dual-channel RF receiver with 599 programmed receive channels in the VHF band. The monitor/programmer presents a time-ordered readout of sensor activations identifying target classification and direction of movement. For fixed sites, the Monitor/Program is equipped with a serial communications port compatible with personal computers/workstations to provide the same information with additonal capability to display maps and details of sensor locations. The programmer is used to set the RF channel frequency, sensor identification number, armed or test functions, along with the desired mission life of the sensors and repeaters.

Event detection and classification are accomplished by battlefield proven algorithms, implemented by low-power components coupled with a target activation wake-up feature which assures a minimum battery life of 30 days based on 1,000 detected events per day.

Finally, if it is disturbed in the field, the reloadable software zeroises automatically, rendering it useless to an intruder. All IREMBASS components are fully compatible with previous REMBASS components, allowing complete interoperability.

Operational status
Production of REMBASS began in 1985 and ended in 1988. IREMBASS production started in 1991 and is in continuous production. Over 6,000 equipments have been delivered and are in use with military intelligence and special forces of the US as well as military organisations in the Middle East and Pacific regions. REMBASS/IREMBASS is deployed in Bosnia in support of the UN IFOR mission.

Contractor
L-3 Communication Systems-EAST.

VERIFIED

Octec AutoWatch

Type
Area surveillance subsystem for the automation of passive security surveillance.

Description
AutoWatch is an image processing subsystem designed to provide a significant improvement in the overall performance and effectiveness of an electro-optical surveillance system. AutoWatch accomplishes this improvement in performance by reducing operator workload, improving the quality of the video image and reducing the dependence upon operator performance.

Based upon the in-service, ADIRS integrated automatic surveillance system AutoWatch now offers these advanced capabilities to any high-performance electro-optical surveillance system.

AutoWatch, when integrated with an electro-optical surveillance system, will automatically detect and track intrusions into defined threat zones. The system can initiate alarms and provide a control centre with operational data and enhanced images of the intruder. A surveillance system fitted with AutoWatch is capable of complete autonomous operation only initiating an alarm when an intruder transgresses the predefined threat zones. The subsystem is capable of operating from images generated from TV, low-light and thermal cameras.

The AutoWatch unit is available in a rugged enclosure designed to operate in worldwide exposed environments and measuring 250 × 220 × 100 mm.

Operational Status
Available.

Contractor
Octec Ltd.

VERIFIED

Opgal reconnaissance vehicle

Type
Reconnaissance vehicle.

Description
Opgal Optronic Industries Ltd has developed a modular reconnaissance vehicle for area surveillance. The vehicle is a commercial or military vehicle equipped with a sensor head installed on a motorised pan and tilt mechanism. This is mounted on top of an hydraulic telescopic mast and controlled from inside the vehicle.

The sensor head comprises Opgal's TD92BL (or TD92CL) thermal imager, a TV camera, optional laser range-finder or other sensor. The sensor head is operated via a control panel or via a PC while all images are displayed on a TV monitor. The vehicle is equipped for autonomous operation. The sensor head can also be installed outside the vehicle on a tripod, remotely operated by cables.

Operational status
In production.

Specifications
TD92BL thermal imager
Spectral band: 8-12 μm
Cooling: split-cycle Stirling
Dimensions: 485 × 240 × 240 mm
Weight: 15 kg
Fields of view: 0.85 × 0.65°; 1.7 × 1.3°; 4.6 × 3.5°; 19.5 × 12.5°; 25 × 19° (triple field of view telescope)
Power consumption (max): 80 W

Contractor
Opgal Optronic Industries Ltd.

VERIFIED

Opgal reconnaissance vehicle with TD92BL thermal imager and a CCD camera

Opgal surveillance equipment

Type
Area surveillance system.

Description
Opgal offers a wide range of surveillance equipment, including cameras, radars, laser range-finders, towers for the mounting of sensors and fully equipped control cabins for remote sensors.

Typical building blocks of such systems include a remote-controlled thermal imager and a CCD camera with a special zoom lens, installed on a motorised pan and tilt platform. The remote-control unit is usually mounted on top of a tower or a roof of a building. It is operated via a control and display console that allows full supervision and scanning of the cameras, image display and video recording.

Opgal also offers turnkey surveillance stations, such as a tower with surveillance cameras and a fully equipped control cabin.

Opgal surveillance equipment – remote-control unit and control and display console

Operational status
Available.

Contractor
Opgal Optronic Industries Ltd.

VERIFIED

Opgal TD92BL/CL

Type
Area surveillance system.

Description
The TD92 is an upgraded version of the TD32 series using the same type of detector and it also consists of a linear drive and split Stirling closed-cycle cooling system.

The thermal imager operates in the 8 to 12 μm waveband and uses electronic multiplexing and parallel scanning to provide a sensitive, sharp picture with interchangeable fields of view. The images are then fed as a standard television signal which can be recorded, displayed or transmitted to other locations. It may be used for long-range observation in intelligence or target tracking duties. The TD92BL features semi-automatic focus and may be operated either autonomously or integrated with other equipment and systems. Dual and triple fields of view telescopes are available with the equipment, the latter offering electronic zoom, narrow, medium, wide or extra wide fields of view.

The TD92CL is based upon TD92BL but is for very long-range surveillance with a 9 in (228 mm) telescope. It is usually remotely operated and installed on an electronic pan and tilt head with all functions controlled from a remote-control unit with video monitor. There also exists an open-cycle version with Joule-Thomson cooling.

The Opgal TD92BL

The Opgal TD92CL

Operational status
In production.

Specifications
Spectral band: 8-12 μm
Cooling: split-cycle Stirling with linear drive compressor
Dimensions
TD92BL: 485 × 240 × 240 mm
TD92CL: 650 × 285 × 256 mm
Weight
TD92BL: 15 kg
TD92CL: 23-25 kg
Fields of view: 0.85 × 0.65°; 1.7 × 1.3°; 4.6 × 3.5°; 19.5 × 12.5°; 25 × 19° (TD92BL triple field of view telescope); 0.50 × 0.38°; 1.00 × 0.70°; 2.80 × 2.10°; 7.90 × 6.00°
Power consumption (max)
TD92BL: 80 W
TD92CL: 80 W

Contractor
Opgal Optronic Industries Ltd.

VERIFIED

Ortek Automatic Detection IR System – ADIR

Type
Area surveillance system.

Description
ADIR is a passive scanning infra-red system designed to detect intrusions into a 'protected' area. Applications include counter-terrorism and perimeter monitoring. The system covers up to 360° and the scanning system permits generation of fresh thermal images of the scene covered every few seconds. Single or multiple sensors my be deployed and linked to a remote control station. Detection and alarm functions are automatic. Verification of the cause of an alarm can be checked by viewing the thermal image displayed on the control unit with digital recording and playback allowing careful examination of the scene.

Operational status
Available.

Specifications
Field of regard
Horizontal: 0 to 360°
Vertical: 12° (FOV)
Scanning speed: 5 – 20/s
Image resolution
1 mrad (0 – 120° scan range)
4 mrad (>120° scan range)
Target detection: proprietary algorithm based on target temperature,
Size and motion detection
Image display format: SVGA 800 (h) × 250 (v) scrollable
Software features
Remote sensor set up and control
High resolution:
IR image display
Audio-visual alarm indication
Automatic events log
Digital image recording and playback
Sensor-control communication: IEEE 802.3 10 Base T/F-O
Power supply: 24 V DC, 110 V AC/60 Hz, or 220 V AC/50 Hz

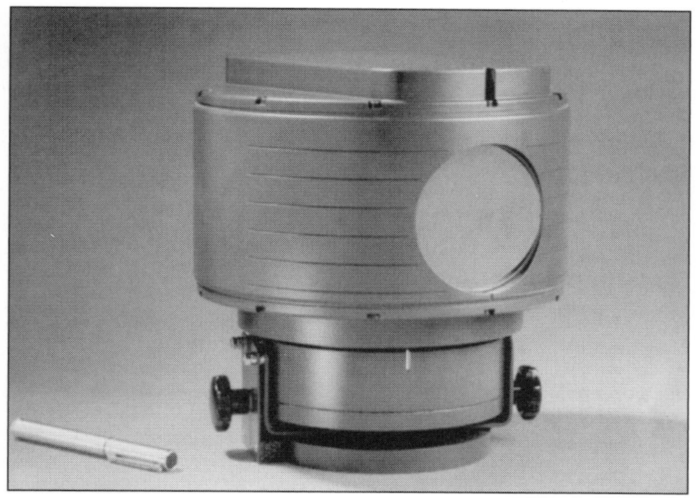

The Ortek ADIR scanning IR sensor 0089896

Contractor
Ortek Ltd Security & Surveillance Systems.

VERIFIED

Pearpoint P328 Changeover Camera

Type
Area surveillance system.

Description
Pearpoint's P328 changeover camera is a high-resolution colour CCD camera during daylight hours and an intensified monochrome camera at night. A first-generation tube is used in the intensified channel. Changeover takes less than 1.5 seconds.

Operational status
Available.

Specifications
Dimensions: 260 mm (length) × 85 mm (dia)
Weight: 2 kg
Colour channel
CCD: 1/2 in interline transfer
Pixels (h × v):
NTSC version 681 × 490
PAL version 681 × 582
Format: 6.4 × 4.8 mm
Sensitivity: 0.25 lx on sensor
Resolution: 420 lines
Monochrome intensified channel
Input window: Fibre-optic
Photocathode: S25
Gain control range: 10:1
Sensitivity: 0.2 mlx faceplate illumination (10% video)
Format: 10.8 × 8.1 mm
Resolution: 415 lines
Power (system): 11 to 30 V DC or 100 to 240 V AC, <6 W
Temperature range: −10 to +40°C

Contractor
Pearpoint Ltd.

VERIFIED

The Pearpoint P328 changeover video camera 0023177

Pyser-SGI DANOS Day and Night Observation System

Type
Area surveillance system.

Description
The Pyser-SGI DANOS surveillance observation unit is designed for a wide variety of surveillance requirements. It comprises a multisensor head combining a choice of thermal, day and image intensified night-time surveillance cameras on a pre-assembled positioner complete with tripod. Vehicle or fixed mounting options are available.

Operational status
In production.

Specifications
Camera
Thermal: TIU2 camera incorporating hybrid ferroelectric staring focal plane array
Field of view (horizontal):
choice of 6, 9 or 12° FOV fixed focal length lenses
Night vision intensified camera NIC
High-resolution monochrome CCD
Fields of view (horizontal):
41 to 7° or
21 to 2° zoom lenses
Day camera DC: ½ in high-resolution colour camera complete with zoom lens
Fields of view:
22 to 2° or
21 to 1° horizontal FOV
Control options:
Into existing CCTV system
Proprietary control equipment
Rapid deployment wired
Rapid deployment radio microwave telemetry

Display options:
CRT
Ruggedised 15 in TFT LCD flat panel
Hermetic receiving case with built-in 10.4 in TFT LCD monitor

Pan/tilt positioner
Dimensions (h × w ×d): 394 × 183 × 130 mm (depth without electrical connectors)

Contractor
Pyser-SGI Limited.

NEW ENTRY

Radamec 200 series multipurpose surveillance and tracking cameras

Type
Area surveillance cameras.

Description
The Radamec 200 series of cameras includes variants suitable for area surveillance, forward observation and target tracking in the battlefield environment. They are also suitable for naval applications. All variants use CCD technology. Image intensified cameras and high resolution colour cameras are included in the range. Options such as wash/wipe units, sunshields, electronic boresighting and remote control are available.

Operational status
Available.

Contractor
Radamec Defence Systems Ltd.

VERIFIED

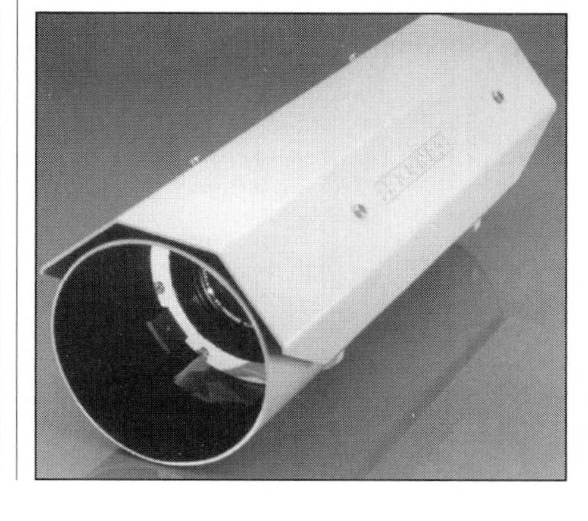

The Radamec 203-002 intensified surveillance and tracking camera 0023270

Radamec 206-000 series TV cameras

Type
Surveillance, tracking and fire-control camera.

Description
The Radamec 206-000 Series is a range of high-resolution 3-CCD colour TV cameras designed for long-range surveillance, tracking and fire control applications. They are available in standard tracking and fire control or special long-range surveillance variants, for surveillance, target tracking and weapon aiming in both land and naval environments.

The cameras employ advanced technology and feature a high-resolution format 3-CCD sensor for optimal daylight performance and stable optical boresight, coupled to a continuous zoom lens. The long-range camera is fitted with a 50:1 zoom lens with built-in ×2 optical range extender providing a focal length range from 10 to 1,000 mm, providing horizontal fields of view from 0.29° to 26°, while the standard variant utilises a more conventional 10:1 zoom lens providing horizontal fields of view from 1.7° to 15°.

Both cameras operate automatically over a wide illumination range providing high-quality images from full sunlight to twilight, using electronic shuttering, AGC and a variety of DSP based image processing features, if required.

The cameras of ruggedised COTS construction and are environmentally sealed to meet the requirements of MIL-STD-810E.

Operational status
In service on surveillance vehicles with a European customer. Selected for the UK Royal Navy Type 45 'Daring' class destroyer.

Specifications
Optical Characteristics (Standard Variant)
Lens Description: 10:1 continuous zoom
Aperture: F3 to T1500
Fields of view
Narrow: 1.3° × 1.7°
Wide: 12° × 15°
Scene Illumination: 10 lux full video
Focus Range: 2 m to 8~

Optical Characteristics (Long Range Variant)
Lens Description: 10 to 500 mm ½ in format, 50:1 motorised zoom lens with built-in ×2 range converter.
Aperture: F4 to F250
Fields of view
Narrow: 0.21° × 0.29°
Wide: 20° × 26°
Scene Illumination: 20 lux usable video
Focus Range: 5 m to 8~

Electrical Characteristics
Sensor: 3-CCD IT device
Active Elements: 752 horizontal × 582 vertical
Resolution: >500TV lines
Video Format: PAL composite and Y/C video output, 1.0V p-p into 75
Scanning: 625 lines, 25 Hz 2:1 interlace or optional 525 lines, 30 Hz.
Synchronisation: internal and external synch
S/N Ratio: 60 dB
Gain: Auto/Manual (to +30 dB + hyper)
Electronic Shutter: 8 sec to 1/100,000 sec
Remote Controls: RS-422 based serial control
Focus, field of view, auto exposure, aperture control, white balance, gain, zoom, focus, speeds and pre-sets, DSP functions, BITE, ×2 range converter (long range variant only).

Power
Camera: 28 V DC <50 W
Heater: 28 V DC 100 W

Environment
Operating Temperature: MIL-STD-810E – −25°C to +55°C
Ruggedisation: MIL-STD-810E – shock, vibration

The Radamec 206-002 colour surveillance and tracking camera
0023269

Physical Characteristics (long range variant)
Dimensions: 495 × 185 × 160 mm
Weight: <11 kg

Contractor
Radamec Defence Systems Ltd.

NEW ENTRY

Radamec 206-100 series TV cameras

Type
Surveillance camera.

Development
Introduced in 2000, the Radamec 206-100 series is a range of compact colour TV cameras designed specifically for military surveillance applications.

Description
The Radamec 206-100 Series is a family of high-resolution colour, single CCD cameras, ruggedly constructed to withstand the tough operating environments associated with battlefield surveillance (fixed or mobile), AFV indirect viewing and maritime/naval vessels.

The cameras operate automatically over a wide illumination range providing high-quality images from full sunlight to twilight and are fitted with a removable IR cut-off filter and built-in IR LEDs which allow imaging in'zero' light at close ranges.

Use of the latest COTS HAD CCD technology coupled with surface mount electronics packaged around the sensor has enabled an extremely compact camera to be produced. Additional space is provided within the camera to allow the fitting of optional extra features such as CAN-bus interface cards, external synchronisation and reticule generators.

Operational status
In service in Asia, Europe and the USA. Selected for the British Army for armoured fighting vehicles.

Specifications
Optical Characteristics
Lens Description: 18:1 (typical) continuous zoom lens

Typical fields of view (dependent on selected lens)
Narrow: 2.0° × 2.7°
Wide: 36.0° × 48.0°
Digital Zoom: ×2 and ×4
Scene Illumination: 3 lux

Electrical Characteristics
Sensor: ¼ in format CCD
Active Elements: 752 horizontal × 582 vertical
Video Format: PAL composite and Y/C video output
Scanning: 625 lines, 25 Hz 2:1 interlace or optional 525 lines, 30 Hz.
Synchronisation: internal standard (external synch available as option)
S/N Ratio: >50 dB
Gain: Auto/Manual (13 to 18 dB, 8 steps)
Electronic Shutter: 1/3 to 1/1,000 sec, 20 steps.
Remote Controls: RS-232 based serial control (CAN-bus optional)
Focus/autofocus, field of view, auto exposure, aperture control, white balance, gain, zoom, focus, speeds and pre-sets, IR filter.

Power
Camera: 28 V DC <6 W
Heater: 28 V DC 16 W

Environment
Operating Temperature: −20°C to +50°C (MIL-STD-810E)

Physical Characteristics (Typical)
Dimensions: 155 × 110 × 90 mm
Weight: <2 kg

Contractor
Radamec Defence Systems Ltd.

NEW ENTRY

Radamec 207-004 combined thermal imaging and colour TV camera

Type
Surveillance camera.

Description
The Radamec 207-004 is a lightweight, compact high-performance dual sensor package comprising a colour CCD TV camera and an uncooled staring array

thermal imager. The two sensors are combined in a rugged case containing controls, serial interfaces and power supplies. This combined camera provides a 24-hour sensor capability for high-quality images over a wide range of battlefield conditions.

The thermal imager sensor has an 8 to 14 micron Bolometric solid-state focal plane array of 320 × 240 pixels. It is combined with a switchable dual field of view lens. The imager is designed for fully automatic operation but has a range of manual controls using including gain, offset and white/black hot, accessible via an RS-422 data link.

The TV camera features a 752 × 582 pixel colour CCD coupled to an 18:1 continuous zoom lens, plus ×4 digital zoom remotely controlled in step increments to enable a maximum ×72 zoom ration to be achieved. Comprehensive controls for the camera include zoom, focus and exposure control, operated by means of a serial digital RS-232/-422 based data link.

The cameras employ advanced technology and feature a high-resolution format 3-CCD sensor for optimal daylight performance and stable optical boresight, coupled to a continuous zoom lens. The long-range camera is fitted with a 50:1 zoom lens with built-in ×2 optical range extender providing a focal length range from 10 to 1,000 mm, providing horizontal fields-of-view from 0.29° to 26°, while the standard variant utilises a more conventional 10:1 zoom lens providing horizontal fields of view from 1.7° to 15°.

Both cameras operate automatically over a wide illumination range providing high-quality images from full sunlight to twilight, using electronic shuttering, AGC and a variety of DSP based image processing features, if required.

The cameras are of ruggedised COTS construction and are environmentally sealed to meet the requirements of MIL-STD-810E.

Operational status
Pre-production trials.

Specifications
Thermal Image Characteristics
Waveband: 8 to 14 microns
Active Elements: 320 × 240 pixels
Lens Description: switchable dual magnification lens
Fields of View
Narrow: 6.0°
Wide: 18°
Focus Range: 19 m to infinity

Colour TV Camera Characteristics
Sensor: ¼ in format CCD
Active Elements: 752 × 582 pixels
Resolution: >460 TV lines
Lens description: 18:1 continuous zoom
Fields of View
Narrow: 1.3° × 1.7°
Wide: 12° × 15°
Focal Length: 4.1 to 73.8 mm
Focus Range: 1 m to infinity, manual control or auto-focus
Iris Range: F1.4 to F3.0 (W to T)
Scene Illumination: 3 lux (F1.4) (0.2 lux with IR cut-off)

Video Standard: PAL or NTSC
S/N Ratio: >50 dB
Gain: –3 dB to 18 dB, 8 steps (auto/manual)
Electronic Shutter: 1/3 to 1/10,000 sec, 20 steps

Power Requirements
Thermal Imager: 28 V DC, 20 W
CCD Camera: 28 V DC <6 W
Heater: 28 V DC 15 W

Environment
Operating Temperature: MIL-STD-810E - –40°C to +55°C with heater ()
Ruggedisation: MIL-STD-810E - shock, vibration
EMC: MIL-STD-461C (screened window only)

Physical Characteristics
Dimensions: 430 × 350 × 280 mm
Weight: 33 kg

Contractor
Radamec Defence Systems Ltd.

NEW ENTRY

Radamec System 1000L

Type
Area surveillance system.

Description
Radamec System 1000L is a lightweight electro-optical surveillance system which provides long range, covert detection and observation of potential targets 24 hours a day in all weathers. The system features a combination of high-performance thermal imagers (3-5 μm or 8-12 μm) and daylight or low-light TV cameras mounted on a precision servo head. In its basic mobile configuration it is designed to be operated and deployed from a host vehicle. The manufacturers claim a medium-sized vehicle can be detected at 12 km or more and a human being at ranges in excess of 4 km. Plans exist for seismic and radar sensors to be integrated with the system.

The system is operated from a one-person portable control unit. The operator views the video images from the sensors and controls the pointing of the head which can be located up to 30 m away. A range of automatic control facilities is available to assist operation. A single-pulse laser range-finder can be supplied as an optional extra. Other optional extras include reticle and operational data display on the video image and a video recording and playback facility.

In addition to the basic configuration, System 1000L is available in alternative arrangements. These include an elevating mast which can be attached to a vehicle and enable the director head to be raised above obstructions. Alternatively the system may be installed on a vehicle roof. A number of fixed and/or mobile systems may be linked together as part of a network controlled from a single centre.

Operational status
In service with an unidentified Middle Eastern customer.

Contractor
Radamec Defence Systems Ltd.

VERIFIED

Radamec System 1000L mounted on Land Rover Discovery

Raytheon AN/PAS-20

Type

Thermal imaging surveillance sight.

Description

The AN/PAS-20 thermal imaging system is a battery-powered portable thermal imaging system used for surveillance to assess potential hostile situations.

The system uses a linear array detector with electromechanical scanning to provide a real-time TV output. The video is displayed on an internal Cathode Ray Tube (CRT). A separate buffered output is provided for video recording and/or viewing on a remote monitor.

Operational status

Serial production. Orders for over 430 systems have been received.

Specifications

Field of view (h × v): 15.2 × 7.6°
Spectral band: 3-5 μm
Operating time internal: <5 h continuous at 25°C
Resolution: 0.83 × 1.1 mrad
Operating temperature: –32 to 49°C
Dimensions: 114.3 × 114.3 × 355.6 mm
Weight incl battery: 4.5 lb
Battery type: disposable lithium, BA 5847/U, NSN 6135-01-090-5364

Contractor

Raytheon Systems Company.

VERIFIED

Raytheon AN/PAS-20 thermal imaging surveillance sight 0005456

Raytheon Argus Falcon

Type

Area surveillance system.

Description

The Argus Falcon day/night surveillance system is a remote monitoring system designed for military and paramilitary surveillance applications such as border security, monitoring oil platforms, vessel traffic management and aircraft detection.

The Argus Falcon's infra-red (IR) imaging system incorporates Raytheon's Aurora IR camera. Aurora has a 256 × 256 indium antimonide focal plane array and has a spectral response from 3.8 to 4.2 μm Aurora is coupled to a triple field of view lens with 25, 100 and 500 mm focal lengths.

The Argus Falcon's other sensor is a high-resolution daylight monochrome CCD camera fitted with a 25 to 350 mm continuous zoom lens. The system can toggle between visible and IR images.

The system has a maximum slew rate of approximately 60°/s, taking 3 seconds to make a 180° sweep.

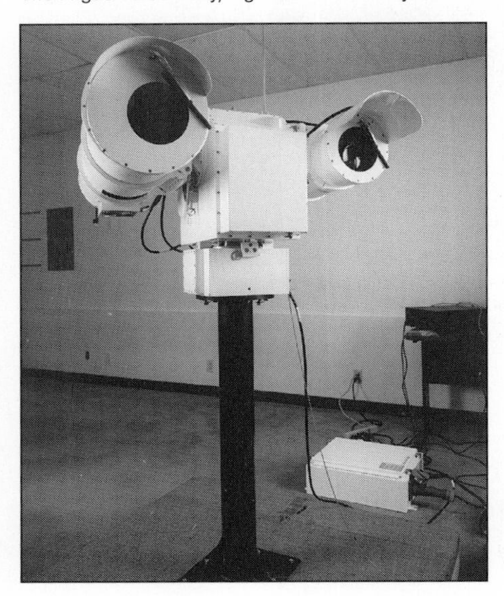

The Argus Falcon surveillance system from Raytheon Systems company 0005454

Operational status

Available.

Contractor

Raytheon Systems Company.

VERIFIED

Raytheon Nightsight 200 series mobile/stationary imager

Type

Area surveillance system.

Development

The Nightsight 200 series is a low cost short-range infra-red camera for non-battlefield surveillance applications. Like the 1000 series it uses uncooled 8 to 12 μm thermal imaging technology combined with proprietary low-cost optics. It was developed as part of a joint funding effort between Raytheon and the DARPA Technology Reinvestment Program and is being marketed for both commercial and military applications.

A hand-held version, the PalmIR 250 is also available for commercial, industrial and military applications.

Description

The 200 series can be mounted on the rooftop of patrol vehicles, on marine vessels or in a stationary setting. Mobile and marine units come equipped with an electronic 360° pan and tilt and a joystick control box.

Packaged in a weatherproof casing, the 200 series also features a thermostatically controlled anti-icing window and an optional window wiper (operator controlled).

Operational status

Available.

Specifications

Dimensions
200 camera: 19.05 × 19.69 × 22.23 cm
200 camera with pan and tilt: 30.48 × 26.04 × 24.77 cm
Weight
200 camera: 3.4 kg
200 camera with pan and tilt: 8.62 kg
Detection range (person): 732 m
Cold start to video image: ≤60 s typical
Depth of field: 22.86 m to ∞
Focus: 3.048 m to ∞, motorised adjustable (operator controlled)
Instantaneous field of view: 0.65 mrad
Standard video interface: EIA RS-170/NTSC
Video update rate: 30 Hz

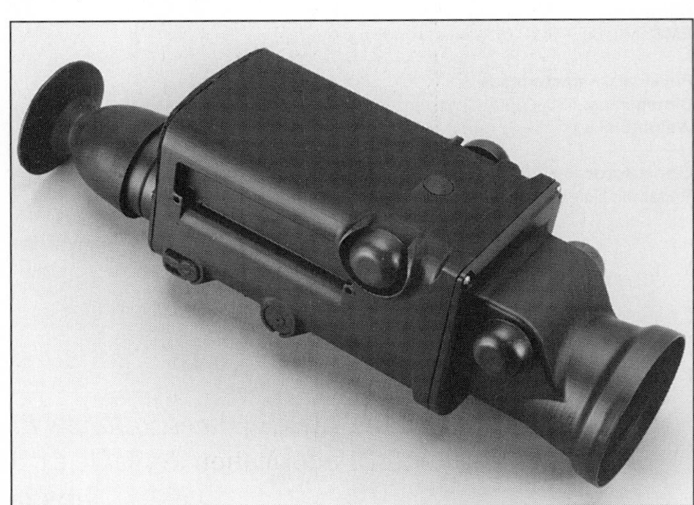

Raytheon Nightsight 200 imager 0023281

200 camera only
Field of view: 12 × 9°
Image resolution: 320 × 240 pixels

200 camera with pan and tilt
Field of view: 12 × 6°
Image resolution: 320 × 164 pixels
Pan: 360° continuous, operator controlled, ±180° auto, 0-45°/s, operator controlled
Tilt: –6 to +40°/s operator controlled

Autoscan mode
Adjustable scan range: ±2° to ±180°
Adjustable scan speed: 2°/s to 45°/s
Operating voltage: 10.5 V DC to 16 V DC, 12 V DC nominal
Power consumption: 6 W typical, 25 W max

Contractor
Raytheon Systems Company.

UPDATED

Raytheon Nightsight™ S1000 series surveillance camera

Type
Area surveillance system.

Description
The S1000 surveillance system is a part of the Nightsight™ family of thermal imaging products for military and commercial end-users. This product line currently features driver's aids (DVE) and lightweight weapon sights for targeting and acquisition (Portable Weapons Sight Model W1000). Nightsight™ products share a common, advanced, uncooled thermal imaging array and core electronics components to reduce logistics, supportability and acquisition cost.

The S1000 security system is designed for surveillance of military installations, borders, prisons, construction sights and other areas of high value. The system can be quickly deployed and can provide detection of targets at over 2 km. The system features a sensor head which can be configured from a list of optional lens sets depending on the mission need.

Operational status
Available.

Specifications
Detector: 328 × 245 uncooled staring array
Spectral band: 7.5-13 μm
Objective lens: 9°; 15°; 40°

Contractor
Raytheon Systems Company.

VERIFIED

Nightsight™ S1000 surveillance camera from Raytheon Systems Company

State Scientific Research & Engineering Institute passive IR detection

Type
Area surveillance.

Description
The State Scientific Research and Engineering Institute of Moscow has produced a passive infra-red detection sensor for detection of intruders in buildings or perimeters. The sensor's closely directional detection zone permits the sensor to be used in long passages or corridors.

The sensor is resistant to wind gusts and precipitation and produces an alarm signal if attempts are made to remove it forcibly or move an object at which it has been aimed.

Operational status
Available.

Specifications
Weight: 1.3 kg
Dimensions
Sensor body (diameter): 86 mm
Power supply unit (diameter): 95 mm
Sensor length: 240 mm
Detection range: 2 to 50 m
Detection probability: not less than 0.9
Detection zone: spatial angle of not more than 1.5°
Power supply: 1.5 V battery (built-in) or external supply of 12 to 36 V DC

Contractor
State Scientific Research and Engineering Institute.

VERIFIED

Systems & Electronics Inc (SEI), Surveillance And Battlefield Reconnaissance Equipment (SABRE)

Type
Area surveillance/reconnaissance.

Description
SEI's SABRE is a modular Reconnaissance, Surveillance and Target Acquisition (RSTA) system combining multispectrum situational awareness, precision navigation and targeting, and automation-aided digital and voice Command, Control and Communications (C^3) in a sensor- and platform-independent mission package.

The system's modular, open architecture accommodates a variety of modern sensor systems tailored to the user's mission requirements. These include:
- FLIR (thermal) imagers
- Laser range-finders/designators
- Visible spectrum (day) video cameras
- Low light level/image intensifying sensors
- Ground surveillance radar
- Combat identification systems
- Chemical biological detectors.

The sensor suite is mounted on a stabilised, reconfigurable, high-payload gimbal that provides the following system capabilities:
- Sensor stabilisation for long-range target acquisition and laser designation
- Sensor operation while on the move
- Remote, armour-protected sensor control
- Elevated, mast-mounted, or vehicle roof-mounted configuration
- Rapid, mission-based sensor reconfiguration.

On-board mission processor generates precise self- and target-location information in conjunction with a blended GPS/inertial navigation system. This data, as well as sensor imagery, can be automatically transferred to the specified battlefield C^3 system for rapid digital transmission. An ergonomic control panel and multifunction display combination provides the human-machine interface for the system. Automatic target tracking, aided target recognition, and radar/electro-optic sensor cueing are additional system functions.

The system can be integrated on multiple platforms, to include medium- and light-armoured vehicles, light wheeled vehicles, and fixed surveillance sites.

Operational status
In qualification testing.

Specifications
Sensor Gimbal
Payload:
>70 kg stabilised
>25 kg fixed mount
Stabilisation: <0.05 mrads (1σ) (stationary) (2 axis)
Gimbal weight: 70 kg

Position/navigation
Self-location: _40 m SEP (lσ) (with GPS)
Target location: <40 m CEP (lσ) @10 km

Navigation:
99 Waypoints
Position accuracy: <0.6% of distance travelled (without GPS)

Communications
Video Imagery:
Digital/analogue recording
Compressed transmission
Data Interface: MIL-STD-1553B
Serial datalink: RS-422/RS-232/RS-485

Contractor
Systems & Electronics Inc (SEI), Saint Louis, Missouri, USA.

UPDATED

Thales CLASSIC

Type
Area surveillance system.

Description
The Thales (formerly Racal) CLASSIC (Covert Local Area Sensor System for Intrusion Classification) uses covert, passive ground sensors to detect and to classify the movement of personnel and wheeled and tracked vehicles. Data from the sensors are transmitted in bursts to a hand-held monitor where they are decoded and presented on an LED display. This identifies the sensor, the type and the frequency of the intrusion. An audio tone is also generated to alert the user and a printer is an optional extra to provide a hard copy of alarms.

Seismic, passive infra-red (IR), magnetic and piezoelectric cable sensors are available and may be easily deployed up to 20 km (or more with a relay unit) from the monitor. Up to eight sensors may be used with each monitor and the sensors are designed to be easily and quickly deployed in temporary or permanent installations. CLASSIC can be computer interfaced to present deployment and alarm data in text form or overlaid onto digitised maps. The sensors can be left unattended for several months before battery changes are required. Heavy-duty battery packs and solar power units can be offered to extend this period further. Standard sensors are the TA 2781 used with the seismic transducer, the IR transducer, or the magnetic sensor. The TA 2781 has a switch to select high, medium or low seismic sensitivity and another switch to select the classification code.

The TA 2781 standard operating frequency is 148 to 162 MHz and is programmable over this range. There is a nominal transmit power of 1.5 W or 3.75 W (programmable). The seismic detector has a range of 1 to 80 m against personnel and a typical range of 2 to 7 m against vehicles. The ranges of the IR and

The monitor of the CLASSIC system is the RTA 2746

magnetic detectors are 2 to 7 m and 2 to 20 m respectively although all ranges are subject to local conditions.

The RTA 2786 monitor comprises a VHF receiver, a tone decoder and an LED display. On receipt of a transmission from a sensor unit, an audible alarm alerts the operator. The sensor's identification and alarm mode setting is displayed on a matrix of three LEDs for each of the eight possible sensors. The display is inhibited after approximately 14 seconds to avoid excessive battery drain and a push-button is provided to enable viewing on demand if required. A user's map panel provides for drawing of a diagram of a tactical deployment.

Other elements include the RTA 2785 relay unit with a range of up to 7 km the MA 2762 hard copy printer and interface, as well as battery packs, a battery charger unit and test equipment.

An RS-232 standard interface has been developed for CLASSIC to provide for the collection and presentation of data on IBM-compatible personal computers.

Each interface unit can supply data from 32 sensors and the units can be multiplexed. This makes it possible to cover larger areas than before and is particularly suited to border security requirements.

Also developed is a piezoelectric seismic sensor cable which provides surveillance over distances up to 8 km per CLASSIC system. A recent addition to the range is the camera-control unit which permits video systems to be connected to the system.

Operational status
More than 10,000 CLASSIC sensors have been produced in 1,500 systems and the system is in service in 29 countries throughout the world including nine NATO countries. It has been in service with the Australian Army since 1988 and in November 1995, Racal-Comsec (now Thales) was awarded a contract to upgrade 70 systems with the latest suite of sensors, monitors and transducers and to supply a further 28.

Specifications
Dimensions
Sensor: 95 × 250 × 42 mm
Monitor: 95 × 250 × 42 mm
Relay: 95 × 280 × 42 mm
Printer/interface: 95 × 280 × 72 mm
Weight
Sensor: 1.3 kg
Monitor: 1.3 kg
Relay: 1.4 kg
Printer/interface: 1.9 kg

Contractor
Thales Communications Systems.

VERIFIED

Thales Optronics MARGO-P

Type
Area surveillance system.

Description
The 'Multi Application Ruggedized General Optronic-Portable' (MARGO-P) is a new surveillance concept based upon two portable units. These can either be linked on a vehicle platform, linked on a remote controlled platform or used as single equipment.

These portable units are the SOPHIE new generation thermal imager and the VECTOR IV goggles which includes an eye-safe laser range-finder and a north-finder.

The pan and tilt platform is light, fully ruggedised and equipped with a quick release interface which allows dismounting of the whole sensor head from the mast of the vehicle allowing it to be put on a tripod. The sensor head can then be remotely controlled from the vehicle. Each equipment is mounted on the platform

MARGO-P from Thales Optronics 0053829

with its own quick release interface allowing hand-held operations. Boresighting can be performed in day or night conditions using the image intensifier goggles. In addition to these items, a CCTV camera can be fitted for day surveillance purposes. The operator console is based upon a rugged PC and a flat panel colour display (800 × 600, TFT). The interface is a windows-type programme entirely driven with a static joystick. The console has advanced tracking and scanning functions incorporating an automatic detection capability that generates an alarm when a new object is detected.

According to the chosen software mode, the screen features either:

(a) the real-time video from the thermal or daylight camera
(b) the self-oriented 3-D digital map, with instantaneous intervisibility field of view shown
(c) a mixture of these with a panoramic landscape at the top.

The console is connected to the on-board tactical radio, allowing automatic reporting and transmission of static pictures, pre-formatted messages and target positions.

Using the 3-D map database, automatic positioning software is able to indicate the best observation position, taking into account parameters such as meteorology and tactical stealthiness. The software also assists with target recognition.

Operational status
In production.

Specifications
Weight: <20 kg
Traverse speed: 0.1-12°/s
Localisation accuracy: 100 m
Detection range: 9 km on tanks, 4.5 km on a man sized target
Recognition range: 3.5 km on tanks, 1.6 km on a man sized target
Ranging distance: 4 km
Azimuth accuracy: 0.6°
Power supply: 24 V DC

Contractor
Thales Optronics.

VERIFIED

VistaScape Holospherix™ SM-10

Type
Area surveillance system.

Description
Holospherix SM-10 is a scanning long-range surveillance system which operates in the visual, middle and far infra-red. It provides a 360 × 320° field of regard and uses large 8 in optics. It is intended for coastal and harbour surveillance, military base and installations perimeter monitoring, border surveillance and target tracking and identification. Coverage of three parts of the spectrum improves the detection of activity and the probability of detecting hidden targets. The system employs sophisticated software to analyse the scenes in the areas covered to detect changes and inconsistencies. Constant active screening by an operator is not necessary as the software will trigger a warning when required. The system is typically mounted on a tripod.

Operational status
Believed to be at pre-production stage.

Specifications
Dimensions: 63.5 × 76.2 × 101.6 cm
Weight:
System 22.5 kg
Tripod 6.8 kg
Optics: 100 to 400 mm primary optics
Field of regard:
Azimuth 360°
Elevation +90 to −70°
Scan rate: dependant on scan pattern selected
Detectors: MCT IRCMOS FPA
Cooling: Stirling cycle microcooler with Dewar
Spectral bands:
Visible 250-1,200 nm
MWIR 3-5 μ
LWIR 8-12 μ
Resolution
Visible 0.05 mr
MWIR 0.3 mr
LWIR 0.5 mr

Contractor
VistaScape Technology Corporation.

VERIFIED

VTUVM HPU 45 and HPU 170 rugged CCD cameras

Type
Area surveillance.

Description
The HPU 45 and HPU 175 miniature cameras are designed for daylight monochrome imaging. Applications are found in small, high-performance platforms for observation, detection, recognition, identification and tracking. Most of the camera functions are controlled via an RS-232 camera control interface. A digital control algorithm for gain and electronic shutter enables a stable response (adjustable) to fast changing light conditions. The construction of the camera has been designed to minimise weight and overall dimensions and guarantee a stable line sight and remote control focusing. This product is qualified for use in combat vehicles exposed to mechanical shock, vibration, rain, dust and high or low temperatures.

Operational status
Available.

Specifications

	HPU 45	HPU 170
Weight	1,160 g	2,020 g
Dimensions (l×h×w)	185 × 104 × 84 mm	264 × 104 × 84 mm
Optics		
Field of view	6.11 × 4.57°	1.62 × 1.21°
Lens	MEOMIL F2.8/45	MEOMIL F3.5/170

Light conditions: 2 to 100,000 lx
Effective pixels: 741 (h) × 575 (v)
Pixel pitch: 6.5 × 6.25 μm
Smear: typical 0.002 %
Spectral sensitivity: maximum shifted to near infra-red
Stability: <0.3 mrad
Focusing: focus plane displacement; remotely controlled

Electrical
Sensor: 1/3 in interline HAD CCD sensor
TV standard: 625 TVL/50 Hz
Synchronization: internal, X-tal 30 ppm
External, RS-422 CS signal
Gain: 20 dB automatic
40 dB manual
Gamma: 1
Integration period: automatic or manual
10 μ sec - 20 msec
Video signal: 1 Vpp/ 75 Ω
Power supply: 12 V DC/5.5 W
Operating temperatures: −25 to +55°C full performance
−40 to +71°C reduced performance

Contractor
VTUVM Military Institute for Weapon and Ammunition Technology.

VERIFIED

The VTUVM HPU 45 and HPU 170 CCD cameras 0102147

Zeiss Optronik mobile IR monitoring system

Type
Vehicle surveillance.

Description
Zeiss Optronik has produced a mobile infra-red monitoring system for a variety of missions including border surveillance, property protection and law enforcement.

The system includes a thermal imager, monitor, video recorder, PC with own monitor for electronic map display, video display, soft control panel and tracker and pan-tilt head. A laser range-finder and a day sight CCD camera is optional. The carrier vehicle is a VW T4 synchro, long wheelbase with extended ground clearance and total weight capability of 2,890 kg, but another vehicle may be used at the customer's request. The two-person operator compartment has a control panel, monitor, PC with monitor VCR and a communications set with voice and data tranceivers.

Further sensors can be added, including global positioning system, north-seeking sensor, image intensifier and tracker.

Operational status

Available.

Specifications

Thermal sensor
Infra-red modules: advanced 2nd and 3rd generation
Spectral band: 8-12 μm
Cooling: low-noise linear Stirling cooler
Fields of view
Wide: 4.6 × 6° and °5.7 × 7°
Narrow: 1.5 × 2°
Number of lines: 480 to 576
Monitor
Screen diagonal: 17 in
Tube: monochrome or colour
Video signal: CCIR (others on request, for example PAL)
Repetition rate: 100 Hz (flicker-free)
Video recorder
Format: VHS or S-VHS
Pan-tilt head
Ranges
Azimuth: n × 360°
Elevation: ±305

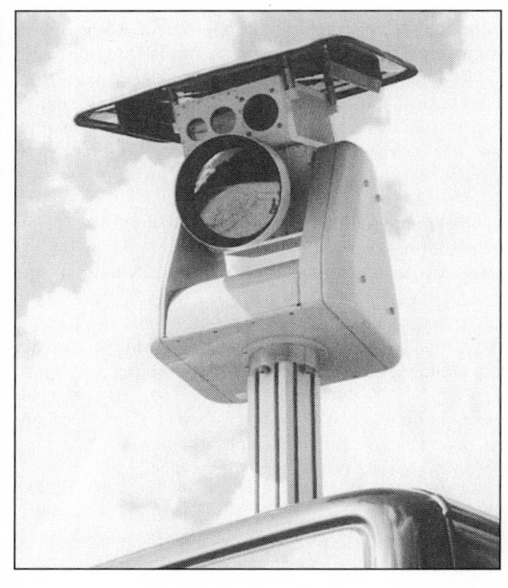

The sensor package of the Zeiss Optronik mobile IR monitoring system
0089921

Max slewing speed: 15°/s (elevation)
>30°/s (azimuth)

Contractor

Zeiss Optronik GmbH.

UPDATED

OBSERVATION AND SURVEILLANCE

THERMAL IMAGERS

Aselsan Baykus thermal camera system

Type
Thermal imager (land).

Description
The Baykus thermal camera system is a thermal surveillance system which provides the user with target detection, recognition and observation capabilities in adverse weather and obscuration conditions that reduce visibility in the combat area.

Baykus is lightweight, portable and can be operated by one person. It also has a remote-control facility. It is powered by a 12 to 24 V DC sealed lead acid battery or by 220 V for stationary applications.

The target acquisition system (HEKOS) is aimed at the target to measure its co-ordinates, range, elevation and azimuth angles. The station (own position) co-ordinates can be entered either manually or electronically through a GPS.

Operational status
In production, under licence from Raytheon, where the camera is known as the Escort.

Specifications
Thermal Camera
Dimensions: 255 × 255 × 410 mm
Weight: max 15 kg
Spectral band: 8-12 μm
Field of view
Narrow: 2.2 × 1.1°
Wide: 6.8 × 3.4°
Magnification
Narrow: ×9.5
Wide: ×3.1
Cooling system: closed-cycle linear
Video output: CCIR, 50 Hz extra video output
HEKOS
Pointing measurement accuracy: ±6 mil. azimuth; ±3 mil. elevation
Range: 10 km/±5 m

Contractor
Aselsan Inc, Microelectronics, Guidance and Electro-optics Division.

UPDATED

The Aselsan Baykus thermal camera and target acquisition system 0053833

BAE Systems 2-D staring thermal imagers

Type
Thermal imager.

Description
BAE Systems, Avionics, Tactical Systems has developed a range of thermal imager designs incorporating the latest staring focal plane array detector technologies, operating in both the long- and medium-wave infra-red bands. Imagers can be configured as calibrated radiometers. Imagers designed for airborne use with full 1553B aircraft control bus interface are available. The imager can be supplied in modular form for integration into higher level systems.

Operational status
Development/programme specific.

Specifications
Spectral band: 3-5 or 8-12 μm
Detector: 2-D focal plane array, size to suit application
NETD: 20 mK typical
Video output: compatible with STANAG 3350, Class B - 626/50 Hz, Class C
Basic field of view : determined by optics arrangement
Power consumption: 40 W typical

Contractor
BAE Systems Avionics, Sensor Systems, Basildon.

UPDATED

BAE Systems IRIS

Type
Thermal imager (land).

Description
BAE Systems (formerly Nanoquest Defence Products then BASE) produces a range of thermal imagers for a variety of applications, both military and paramilitary. These thermal imagers were previously produced by Steinheil Optronik but production was transferred to BAE Systems following a takeover.

The thermal imagers operate in the 8 to 12 μm band using eight-element cadmium mercury telluride detectors with SPRITE processing. All produce CCIR-standard TV output signals and have split-cycle Stirling cooling.

Operational status
In production. In service on Aspic VSHORAD air defence system of the French Air Force.

Specifications
Spectral band: 8-12 μm
Detector: 8-element CMT SPRITE
Cooling: Split-cycle Stirling
Video output: CCIR
Power supply: 18-32 V DC
Power consumption: 100 W (approx)

The BAE Systems IRIS 6340/30 plus 10

	IRIS 6340/2 plus 8	IRIS 6340/30 plus 10	IRIS 6340/15 plus 5	IRIS 6340/8 plus 2
Weight	28.7 kg	16.7 kg	18.4 kg	28.7 kg
Dimensions	395 × 280 × 560 mm	229 × 275 × 353 mm	229 × 275 × 503 mm	395 × 280 × 560 mm
Front lens diameter	80 mm	51 mm	111 mm	66/225 mm
Fields of view	1 × 2°; 4 × 8°	5 × 10°; 15 × 30°	2.5 × 5°; 7.5 × 15°	2 × 1°; 8 × 4°

Contractor
BAE Systems.

UPDATED

BAE Systems Lightweight Infrared Observation Night sight (LION)

Type
Thermal imager (land).

Description
BAE Systems' Lightweight Infrared Observation Night (LION) sight utilises uncooled Focal Plane Array (FPA) technology to bring the advantages of a thermal imager to ground troops.

LION is very compact, lightweight and has been ergonomically designed for ease of use under the arduous conditions of the 24-hour battlefield. The use of uncooled technology means operation of the LION is near instantaneous (<15 seconds) thereby eliminating the need for battery draining standby modes associated with cooled thermal imaging units. It is also virtually inaudible and produces no cold spot signature to other thermal imagers making it particularly suitable for covert operation.

The binocular display minimises eye strain and reduces operator fatigue allowing prolonged use of LION to gain maximum tactical benefits.

LION can be mounted onto static and mobile platforms such as jeeps and 4 × 4's, using the optional power adaptor and remote control unit. It is also fitted with a serial datalink for integration with other equipment such as laser designators.

LION comprises the main element of the Thales Optronics LION observation night sight.

Operational status
In service with UK and NATO forces.

Specifications
Weight: <2 kg
Spectral band: 8 -14 μm
Field of view: 5° vertical × 10° horizontal
Detector type: Uncooled ferro-electric micro-bolometer
Detector configuration: 256 × 128 with 2:1 microscan
Power consumption: <7 W
Battery operation: >10 hours
Temperature (operating): –30 to +50°C
NETD: <120 Mk

Contractor
BAE Systems Avionics, Sensor Systems Division.

UPDATED

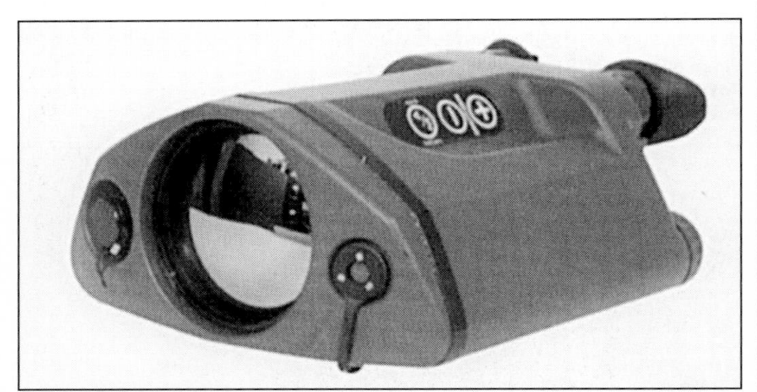

Lightweight Infrared Observation Night 0132656

BAE Systems LTC550 MICROIR™ Camera

Type
Uncooled thermal imager.

Description
The LTC55O MICROIR™ Camera is based on BAE Systems state-of-the-art MICROIR™ uncooled imaging module. At the heart of the module is a 320 × 240 uncooled microbolometer Focal Plane Array (FPA) which, when packaged with BAE Systems signal processing electronics, provides high-quality imaging performance. In addition, the MICROIR™ module provides both instant-on and silent operation.

The LTC55O MICROIR™ Camera enclosure provides a ruggedised weatherproof housing for the FPA and electronics, permitting system deployment in the harshest of environmental conditions. In addition, a custom interface bracket allows the camera to be easily integrated with off-the-shelf tripods, pan and tilt mechanisms, or other system-level equipment.

The versatile LTC550 MICROIR™ Camera can be configured with the optic required to meet application requirements, and has the ability to field swap lenses as application requirements change. BAE Systems offers a complete family of manual and motorised system optics, ranging from 18 mm single to 60 mm/180 mm dual field of view.

Operational status
Available.

Specifications
Weight: 2.5 kg (5.5 lbs)
Spectral band: 8-14 μm
Array formal: 320 × 240, 46.25 μm pixels
IFOV for 40° system: 2.1 mrad
Frame rate: 60 Hz
Video outputs: RS-170/NTSC, PAL, 8 and 16 bit digital; video
NFTT normalised to f/1.0, 30 Hz: less than 100 Mk
Available optics: 18 mm, 50 mm, 75 mm, 100 mm, 150 mm, 210 mm, 60/180 mm
Available options: Optics, controller, cabling, mechanical interface hardware
Power: <8 W

Contractor
BAE Systems North America, Information and Electronic Warfare Systems.

NEW ENTRY

BAE Systems Tank Thermal Sensor (TTS)

Type
Thermal imager (land).

Description
The modular 8 to 13 μm spectral band Tank Thermal Sensor (TTS) system is configured from the second-generation UK Thermal Imaging Common Modules (TICM II) and is a passive imaging sight providing day and night capability for target detection, recognition and tracking purposes.

It can be fitted either as a new-build item or as a retrofit vehicle installation and be integrated into a fire-control system.

The telescopes and displays can be selected by system designers to meet the operational requirements and vehicle type. TICM II modules are incorporated in the Chieftain, Challenger 1 and 2 MBTs Thermal Observation and Gunnery System (TOGS) and in the Osprey sight on the Warrior Mechanised Armoured Observation Vehicle (MAOV).

Operational status
TTS ready for production. TICM II version for British Army MAOV in production phase.

Specifications
Spectral band: 8-13 μm
Detector: 8 parallel CMT SPRITE
Detector cooling: closed-cycle split Stirling engine
Weight
Sensor head: 13.5 kg
Commander's display: 3.4 kg
Processing electronics unit: 9 kg

BAE Systems TICM II-based Tank Thermal Sensor (TTS) for the Warrior Mechanised Armoured Observation Vehicle

Dimensions
Sensor head: 355 × 230 × 325 mm
Commander's display: 130 × 175 × 290 mm
Processing electronics unit: 195 × 130 × 95 mm
Telescopes: standard range of fixed, dual or zoom types
Magnification: range ×2 to ×20 according to type
Field of view
×10 magnification type telescope: 6 × 4°
×5 magnification type telescope: 12 × 8°
Power supply: 28 V DC vehicle system
Gunner's display
Eyepiece: ×7.8 magnification
Resolution: 600 lines
Screen: 25.4 mm electrostatic CRT
Commander's direct view display
Resolution: 350 lines
Screen: 76.2 mm electromagnetic CRT

Contractor
BAE Systems Avionics, Sensor Systems, Basildon.

UPDATED

BAE Systems uncooled thermal imaging modules

Type
Thermal imager (land).

Description
BAE Systems offers the uncooled Rectangular Format Array (RFA) uncooled thermal imager modules to both prospective and existing manufacturers of dual-use thermal imager systems. The RFA module is designed for configuration into military standard equipment, offering a compact and low-cost solution for many applications.

The RFA module set incorporates a 256 × 128 element uncooled staring array detector and is provided with 2:1 microscan. Image processing is performed in a single ASIC. Output is in the form of a monochrome TV imager, the field of view being determined by the infra-red optics selected by the system configurer.

Operational status
In production. A recent example of their application is in Lightweight Infra-red Observation Nightsight (LION), a product made jointly by Thales Optronics (Nederland) and BAE Systems, Avionics, Tactical Systems.

Specifications
Spectral band: 8-14 μm
NETD: 0.17K typical with f/1 optics
Video output: compatible with STANAG 3350, CLASS B - 625/50 Hz, CLASS C - 525/60 Hz
Power consumption: 5 W typical
Weight: 0.5 kg approx

Contractor
BAE Systems Avionics, Sensor Systems, Basildon.

UPDATED

BAE Systems uncooled thermal imager modules 0005495

CEDIP Emerald MWIR

Description
The Emerald camera uses high-performance focal plane array technology to offer an infra-red image at full TV size format. Emerald uses InSb or MCT focal plane arrays of 640 × 512 pixels operating in snap shot mode and features advanced processing functions. Emerald MWIR operates in the 3 to 5 μm wavelength range and offers a typical NETD figure as low as 25 MK.

The imager is a COTS device, well suited for a variety of application such as airborne imaging or long-range surveillance. It uses the powerful CASSIOPEA imaging electronic modules which are dedicated to operate with full TV format focal plane arrays, by providing a wide variety of digital image processing features.

Emerald is fully controllable from a host computer through a communication port, for users who wish to integrate the imager into a larger system. Exposure time and external synchronisation modes as well as other key parameters are all programmable by the internal software.

Operational status
Available.

Specifications
Dimensions (l × w × h): 270 × 120 × 150 mm
Weight: 3 kg
Detector type: InSb or HgCdTe (MCT)
Number of pixels: 640 × 512
Pitch: 25 × 25 μm
Spectral range: 3-5 μm

Optics

Focal length	Field of view	IFOV
50 mm	18 × 14°	500 μrad
100 mm	9 × 7°	250 μrad
50/250 mm	18 × 14°/4 × 3°	500/100 prad

Cooling: Integral Stirling cooler
Analogue to digital: 14 bits
Remote control: RS-232 or RS-422
Analogue output: CCI R, RS-170, NTSC, PAL
Digital output: 14 bits RS-422
Integration time: 1 μs to 10 ms by steps of 1 μs
NETD: 25 MK at 300 K
External synchronisation: TTL pulse, video genlock
Power supply: 24 V DC <3 A

Contractor
CEDIP Infrared Systems.

VERIFIED

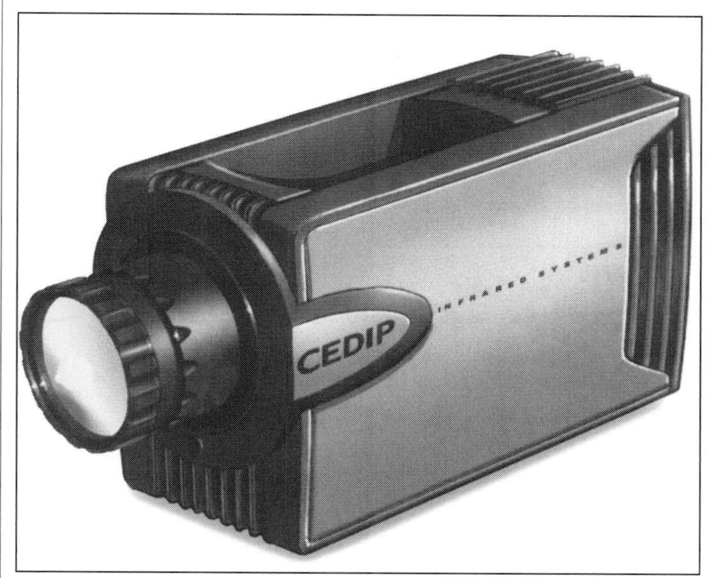

The Emerald thermal imager 0121868

CEDIP Jade

Description

The Jade fast frame-rate infra-red focal plane array cameras are designed for demanding users of IR technology, by providing high sensitivity at the highest possible frame rate. The Jade family covers the full infra-red spectrum through the three main atmospheric transmission bands as SWIR (1 to 2.5 μm), MWIR (3 to 5 μm) and LWIR (8 to 12 μm). The cameras use very high quantum efficiency materials such as InSb or MCT for the IR sensitive materials together with the latest technology in imaging and processing electronics.

The Jade variants are fully controllable from a PC through a standard RS-232 communication port and from Cirrus Win software. The frame rate, exposure time and triggering modes are software controllable as are all other key parameters of the camera. The frame rate can be changed with a step resolution of 1 Hz up to 200 Hz at full EPA size and up to 6,000 Hz in sub-window mode.

The appearance of the hardware is similar to that shown in the Emerald entry.

Operational status

Available.

Specifications

Model	Jade SWIR	Jade MWIR	Jade LWIR
Dimensions (l x w x h)	270 × 120 × 150 mm	270 × 120 × 150 mm	270 × 120 × 150 mm
Weight	2 kg	2 kg	2 kg
Detector type	MCT 320 × 256-30 μm pitch	MCT/InSb 320 × 256-30 μm	MCT 320 × 256-30 μm pitch
Spectral range	1.2-2.5 μm	3-5 μm	8-9.5 μm
Cooling	TE cooler Temperature app 200 K	Stirling engine Temperature app 80 K	Stirling engine Temperature app 80 K
Field of view			
25 mm	21 × 16°	21 × 16°	21 × 16°
50 mm	11 × 8°	11 × 8°	11 × 8°
100 mm	n/a	5.5 × 4°	5.5 × 4°
50/250 mm	n/a	11 × 8°/2.2 × 1.6°	11 × 8°/2.2 × 1.6°
Analogue to digital	14 bits	14 bits	14 bits
Remote control	RS-232-Cirrus Win	RS-232-Cirrus Win	RS-232-Cirrus Win
Analogue output	CCIR 50 Hz/RS-170 60 Hz	CCIR 50 Hz/RS-170 60 Hz	CCIR 50 Hz/RS-170 60 Hz
Digital output	14 bits RS422 Pixel clock RS-422 Frame Sync RS-422 Line Sync RS-422	14 bits RS-422 Pixel clock RS-422 Frame Sync RS-422 Line Sync RS-422	14 bits RS-422 Pixel clock RS-422 Frame Sync RS-422 Line Sync RS-422
Frame rate	Software programmable 1 Hz up to 200 Hz full frame @ Ti = 100 μs	Software programmable 1 Hz up to 200 Hz full frame @ Ti = 100 μs	Software programmable 1 Hz up to 200 Hz full frame @ Ti = 100 μs
Integration time	1 μs to 10 ms by steps of 1 μs 1 μs	1 μs to 10 ms by steps of 1 μs 1 μs	1 μs to 10 ms by steps of 1 μs 1 μs
NETD	n/a	<25 MK @ 300 K	<35 MK @ 300 K
External synch	TTL Pulse 51 W Switchable jitter < 1 μs	TTL Pulse 51 W Switchable jitter < 1 μs	TTL Pulse 51 W Switchable jitter < 1 μs
Power supply	24 V DC <3 A	24 V DC <2.5 A	24 V DC <2.5 A

Contractor

CEDIP Infrared Systems.

VERIFIED

CEO Hand-Held Thermal Imaging System (HTIS)

Type

Thermal imager (land).

Description

HTIS is a compact, lightweight hand-held thermal imager for general observation and detection. It can easily be adapted as a thermal sight for anti-aircraft and anti-tank weapon systems to enhance their day and night capability.

Operational status

Available.

Specifications

Dimensions (l × w × h): 212 × 260 × 144 mm
Weight: 3.6 kg (with lithium battery pack)
Spectral band: 8 to 10.5 μ
Detector: MCT-PV, 128 element linear array
Scan type: parallel
Objective aperture: 89 mm
Fields of view (h × v): NFOV 4.4 × 2.2°, WFOV 15 × 7.5°
Magnification: narrow × 3.56, wide × 1.05
Cooling: integral closed cycle
Video display: 1.5 in built-in CRT
Video output: CCIR, 625 lines 50 Hz
Communication port: RS-422
Power supply: lithium battery pack
rechargeable NiMH battery pack (optional)
external power supply (optional)
Power consumption: steady state ON, 22 W; standby, 10 W

Contractor

Chartered Electro-Optics Pte Ltd.

VERIFIED

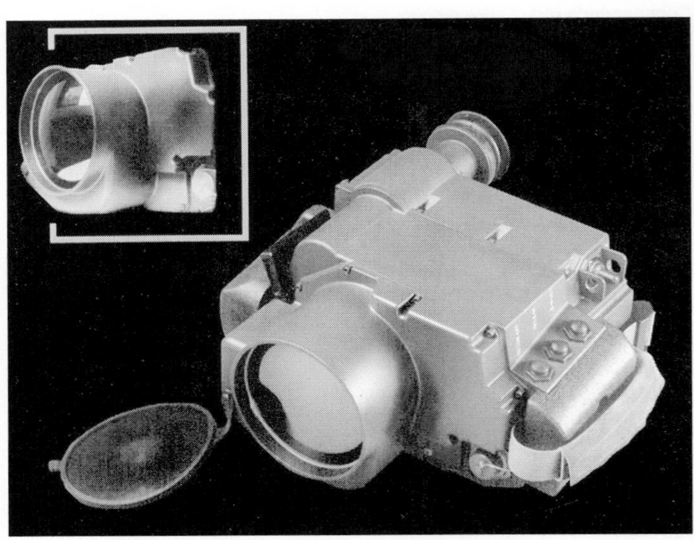

The CEO HTIS thermal imager 0089873

Chartered Industries of Singapore, Honeywell ACTIS

Type

Thermal imager (land).

Description

ACTIS is a joint development between Chartered Industries of Singapore (CIS) and Honeywell Canada Inc. ACTIS may be used as a hand-held imager but has also been adapted as a form of fire-control system (when integrated with a laser range-finder and goniometer) for crew-served weapons such as heavy machine guns and recoilless rifles, so they can engage targets at night.

When mounted on armoured fighting vehicles, for example, the AMX-13SM1 turret mantle, the ACTIS improves the crew's combat effectiveness during night engagements by allowing both passive surveillance and target acquisition. The sight picture is transferred inside to allow the crew to view the imagery whilst under armour.

ACTIS can also be used as a passive tracking system for air defence weapons when it is used in conjunction with a fire-control tracker such as the Super Fledermaus. In all applications the reticle pattern is supplied according to customer requirements.

Operational status

Production. In service with undisclosed customers.

Specifications

Weight: less than 4.6 kg
Dimensions
Without eyepiece: 247 × 190 × 135 mm
Field of view
Narrow: 3.5°
Wide: 7°
Spectral band: 8-12 μm
Detector type: CMT
Scanner type: serial scan
Cooler type: closed-cycle integral
Display type: CRT
Objective focus: 30 m to ∞
Dioptre range: +4 to −4
Reticle pattern: customised
Controls: off/standby/on/zoom – a multiple-position switch that controls the electrical power to the system and allows the observer to activate the electronic zooming polarity – black hot/white hot focus – for viewed image reticle – to adjust reticle brightness
Power supply: external power (of 18 W) or 4 × Type C lithium cells or Ni/Cd battery pack.

Contractors
Honeywell Canada Inc.
Chartered Industries of Singapore.

VERIFIED

CMC Electronics Cincinnati NightMaster™ infra-red imaging system

Type
Thermal imager (land).

Description
NightMaster™ is a high-performance infra-red imager designed to meet the US Army demand for enhanced 24-hour all-weather surveillance. It incorporates a number of advanced features to achieve low power, small size and high system magnification and builds on Cincinnati Electronics patented Indium Antimonide (InSb) reticulated focal plane array architecture. Using a staring 256 × 256 InSb IRFPA, the NightMaster™ produces excellent thermal sensitivity in a completely man-portable unit. The standard dual field of view f/4 50/250 mm telescopic lens provides a simple and effective, covert, long-range detection and recognition capability. Other built-in features include low power analogue electronics, 'on-demand' digital electronics for field calibration using an internal thermal reference and an RS-422 serial interface for full remote control capability. An optional SeeSpot and motorised 50/250 mm telescope adds to the system utility. The NightFalcon option is a high-performance long-range imager with a 400/1,500 mm motorised lens with full remote-control capability. The system is designed specifically for military surveillance, reconnaissance, forward observation, paramilitary perimeter security, counter-insurgency and force protection. It is operable and accurate in extremes of weather from desert temperatures to arctic climates from –32 to 55°C. The camera with good thermal sensitivity, quickly and clearly brings a night scene into focus and is easily configured to the majority of surveillance missions.

An important option to the NightMaster™ system is the SeeSpot imager which allows the user to view laser designation and the infra-red image simultaneously. Designed for field use and battery operated, laser imaging with this system has been demonstrated in excess of 5 km. Being able to 'see the laser spot' within the field of view of the thermal imaging system permits immediate operator feedback enabling accurate targeting. It allows the operator to view the visual scene and its possible limitations of laser illumination such as obstructions, solar saturation, battlefield obscuration, clouds and ambiguous target reflections.

Operational status
In production.

Contractor
CMC Electronics Cincinnati.

VERIFIED

The NightMaster™ infra-red imaging system with motorised lens 0023279

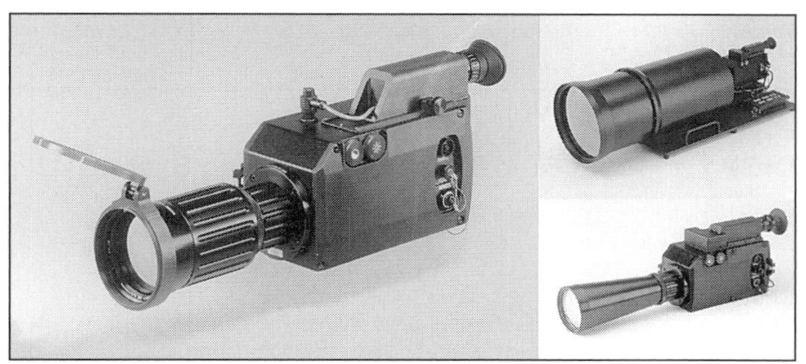

NightMaster™ variations with NightFalcon at top insert 0089900

CMC Electronics NightConqueror 640 infra-red imaging system

Type
Thermal imager (land, sea, air).

Description
NightConqueror is a high-performance infra-red thermal imager engineered for 24 hour long-range, daytime or night-time, all-weather reconnaissance, surveillance and target acquisition. Designed as a high-performance imaging module, a 640 × 512 InSb MWIR FPA produces true TV resolution for inclusion into a variety of battlefield platforms. The units combine the most current electronic technology and sensor architecture to achieve truly compact size, lighter weight and lower power consumption. The system produces exceptional thermal sensitivity and low NETD (Noise Equivalent Temperature Difference) of 20 mK at 25°C and is entirely remote controlled through a serial interface. The NightConqueror's video output may be synchronised to a laser designator for improved SeeSpot imaging or gen locked to a separate video source. A primary advantage of a SeeSpot system is that the laser spot may be viewed in the thermal image, overlaid on the image of the target, thereby increasing the operational capability of ground laser designators.

Operational status
In production.

Contractor
CMC Electronics Cincinnati.

VERIFIED

NightConquerer 640 and variants 0098189

DiOP Cadet lightweight hand-held thermal imager

Type
Thermal imager (hand-held).

Description
The Cadet handheld thermal imager incorporates the latest uncooled Focal Plane Array (FPA) technology and combines it with a unique lightweight hard carbon coated 75 mm lens. The full 320 × 240 FPA produces high-resolution surveillance type imaging, which is viewed on a 640 × 480 AMD (Active Matrix Display) internal to the system. As the FPA is thermoelectrically cooled, it is considered to be more reliable than cryogenically cooled systems. The imager can also be remotely controlled over its RS-232 port thereby rendering it as a remote surveillance platform.

Operational status
Entered service in April 2003.

The DiOP Cadet handheld thermal imager **NEW**/0594450

Specifications
Weight: < 1.4 kg
Spectral Band: 8-12μm
Field of View: 12.5° × 9.4°
Lens Type: Fixed 75 mm f/1, motorized, hard carbon exterior coating
NETD: <80 mk
LCD: 640 × 480 internal, 1× display lens with 25 mm eye relief
Controls: Polarity, Focus, AGC on/off, brightness/contrast, IR/visible channel (optional)
Eye Cup: shutter type, no light leakage
Sensor Type: Microbolometer, 320 × 240 format, 51μm × 51μm pixels
Cooling: Thermoelectric
MTBF: 10,000 hours
Video Output: RS-170, NTSC or PAL
Remote Control Link: RS-232 or -422
Power: 7.2 V DC, lithium ion, rechargeable battery
Operating Temperature: –25° to 50° C
Storage Temperature: –40° to 71°C

Contractor
Diversified Optical Products Inc (DiOP).

NEW ENTRY

DiOP ExtremeX thermal imaging system

Type
Thermal camera for long-range surveillance.

Description
The ExtremeX is a dual- or tri-field (optional) thermal camera system. The fields of view range from 0.55° × 0.41° to 2.2° × 1.65°. The system is environmentally sealed and incorporates an automatic front element heater. Operating in the 3-5μm spectral region, the system incorporates a cooled InSb focal plane array. According to the manufacturer, the sensor system is rated at 7,500 hour MTBF for most environments. The system is designed to be 'plug-and-play', permitting thermal imaging in the harshest of environments with no adjustments. Typical man-sized target detection ranges are said to be in excess of 16 km in good conditions. Detection of tank-sized targets is claimed at ranges in excess of 30 km in good conditions. All features of the camera are controllable over either an RS-232 or -422 link.

Operational status
In production since December 2002.

Specifications
Weight: <19 kg
Spectral Band: 3-5μm
Field of View
Narrow: 55° × 41°
Medium: 1.1° × .83°
Wide: 2.2° × 1.65° (optional)
Lens Type: 500/1,000 mm EFL at f/4, dual field DiOP proprietary design (tri-field optional 250/500/1,000 mm), exterior hard carbon coating, automatic front element defroster
NETD: <40 mk
Sensor Type: InSb, 320 × 256 format, 30μm × 30μm pixels
Cooling: Integral Sterling microcooler
Camera Controls: All controls available through the communications port. Focus, polarity, AGC on/off, brightness/contrast, freeze frame, region of interest, non-uniformity correction, integration time
MTBF: Camera System, 10,000 hours, Sensor Cooler, 7,500 hours
Video Format: CCIR601 NTSC or PAL-M
Remote Control Link: RS-232 or RS-422
Power: 11-16 V DC, 24 W typical

The DiOP ExtremeX thermal imaging camera system ***NEW***/0594453

Operating Temperature: –25° to 50° C
Storage Temperature: –40° to 71° C

Contractor
Diversified Optical Products Inc (DiOP).

NEW ENTRY

DiOP FieldPro 5X thermal imaging system

Type
Thermal imager.

Description
The FieldPro 5X is a continuous zoom thermal camera system. The field of view is infinitely adjustable from 1.1° × 0.83° to 5.5° × 4.12°. The system is environmentally sealed and incorporates an automatic front element heater. The system operates in the 3-5μm spectral region incorporating a cooled InSb focal plane array. The sensor system is rated at 7,500 hour MTBF for most environments. The system is designed to be plug and play, permitting thermal imaging in the harshest of environments with no adjustments. Typical man-sized target detection ranges in excess of 8 km are claimed in good conditions. Tank-sized targets have detection ranges in excess of 20 km in good conditions, the company says. All features of the camera are controllable over either an RS-232 or -422 link.

Operational status
In production since December 2002.

Specifications
Weight: <7.3 kg
Spectral Band: 3-5μm
Fields of View
Narrow: 1.1° × 0.83°
Wide: 5.5° × 4.12°
Lens Type: 100-500 mm EFL at f/4, continuous zoom of DiOP proprietary design, exterior hard carbon coating, automatic front element defroster
NETD: <40 mk
Sensor Type: InSb, 320 × 256 format, 30μm × 30μm pixels
Cooling: Integral Sterling microcooler
Camera Controls: All controls available through the communications port. Focus, polarity, AGC on/off, brightness/contrast, freeze frame, region of interest, non-uniformity correction, integration time, zoom set, preset fields of view.
MTBF: Camera System, 10,000 hours, Sensor Cooler 7,500 hours
Video Format: CCIR601 NTSC or PAL-M
Remote Control Link: RS-232 or RS-422
Power: 11-16 V DC, 24 W typical
Operating Temperature: –25 to 50° C
Storage Temperature: –40 to 71° C

Contractor
Diversified Optical Products Inc (DiOP).

NEW ENTRY

The DiOP FieldPro 5X thermal imaging system ***NEW***/0594452

DiOP LanScout 60/180 thermal imaging System

Type
Thermal imager.

Description
The LanScout 60/180 is a dual field of view, remotely controllable thermal camera system. The system is environmentally sealed and incorporates an automatic front element heater. The system operates in the 8-12m m spectral region, incorporating

The DiOP LanScout 60/180 thermal imaging system ***NEW***/0594449

an uncooled, microbolometer focal plane array. The company rates the system's MTBF at 10,000 hours, due to the use of a non-cryogenic type sensor package. The system is designed for 'plug-and-play' use, permitting thermal imaging in the harshest of environments with no adjustments. All features of the camera are controllable over either an RS-232 or -422 link.

Operational status
In production since May 2003.

Specifications
Weight: <8.1 kg
Spectral Band: 8-12μm
Field of View
Narrow: 5.1° × 3.8°
Wide: 15.1° × 11.4°
Lens Type: 60/180 mm EFL at f/1, dual field DiOP proprietary design, exterior hard carbon coating, automatic front element defroster
NETD: <80 mk
Sensor Type: Microbolometer, 320 × 240 format, 51μm × 51μm pixels
Cooling: Thermoelectric
Camera Controls: All controls available through the communications port. Focus, polarity, AGC on/off, brightness/contrast, freeze frame, region of interest.
MTBF: 10,000 hours
Video Format: CCIR601 NTSC or PAL-M
Remote Control Link: RS-232 or RS-422
Power: 4-12 V DC, <5 W typical at steady state
Operating Temperature: –25° to 50° C
Storage Temperature: –40° to 71° C

Contractor
Diversified Optical Products Inc (DiOP).

NEW ENTRY

DiOP LanScout 75 thermal imaging System

Type
Thermal imager.

Description
The LanScout 75 is a remotely controllable, environmentally sealed thermal camera system. The system operates in the 8-12μm spectral region and incorporates an uncooled, microbolometer focal plane array. The company rates the system at 10,000 hour MTBF, due to the use of a non-cryogenic type sensor package. The system is designed for 'plug-and-play' use, permitting thermal imaging in the harshest of environments with no adjustments. All features of the camera are controllable over either an RS-232 or -422 link.

The DiOP LanScout 75 thermal imaging system ***NEW***/0594448

Operational status
In production since January 2001.

Specifications
Weight: < 2.1 kg
Spectral Band: 8-12μm
Field of View: 12.5° × 9.4°
Lens Type: Fixed 75 mm f/1, motorized, hard carbon exterior coatings, automatic front element defroster
NETD: <80 mk
Sensor Type: Microbolometer, 320 × 240 format, 51μm × 51μm pixels
Cooling: Thermoelectric
Camera Controls: All controls available through the communications port. Focus, polarity, AGC on/off, brightness/contrast, freeze frame, region of interest
MTBF: 10,000 hours
Video Format: CCIR601 NTSC or PAL-M
Remote Control Link: RS-232 or RS-422
Power: 4-12 V DC, <5 W typical @ steady state
Operating Temperature: –25° to 50° C
Storage Temperature: –40° to 71° C

Contractor
Diversified Optical Products Inc (DiOP). ***NEW ENTRY***

DiOP RangePRO 50/250 thermal imaging system

Type
Thermal imager.

Description
The RangePRO 50/250 is a dual field of view, remotely controllable thermal camera system. The system is environmentally sealed and incorporates an automatic front element heater. The system operates in the 3-5μm spectral regions, incorporating a cooled InSb focal plane array. The sensor system is rated at 7,500 hour MTBF for most environments. The system is designed for 'plug-and-play' use, permitting thermal imaging in the harshest of environments with no adjustments. Typical man-sized target detection ranges in excess of 4.6 km are claimed in all conditions. Tank-sized targets have detection ranges in excess of 18 km in good conditions, the company says. All features of the camera are controllable over either an RS-232 or -422 link.

Operational status
In production since January 2002.

Specifications
Weight: <3.5 kg
Spectral Band: 3-5μm
Field of View
Narrow: 2.2° × 1.7°
Wide: 11.0° × 8.2°
Lens Type: 50/250 mm EFL at f/4,Dual field DiOP proprietary design, exterior hard carbon coating, automatic front element defroster
NETD: <40 mk
Sensor Type: InSb, 320 × 256 format, 30μm × 30μm pixels
Cooling: Integral Sterling Microcooler
Camera Controls: All controls available through the communications port. Focus, polarity, AGC on/off, brightness/contrast, freeze frame, region of interest, non-uniformity correction, integration time
MTBF: Camera System, 10,000 hours, Sensor Cooler, 7,500 hours
Video Format: CCIR601 NTSC or PAL-M
Remote Control Link: RS-232 or RS-422
Power: 11-16 V DC, 24 W typical
Operating Temperature: –25° to 50° C
Storage Temperature: –40° to 71° C

Contractor
Diversified Optical Products Inc (DiOP). ***NEW ENTRY***

The DiOP RangePRO 50/250 thermal imaging system ***NEW***/0594447

DRS Portable Forward Looking Infra-Red (FLIR)

Type
Thermal imager (land).

Description
The DRS (formerly Boeing Company) Portable Forward Looking Infra-Red (FLIR) is a lightweight thermal imager which uses cooled HgCdTe detector technology. It operates in the 3 to 5 μm spectral band. Video outputs are CCIR PAL or RS-170 NTSC composite. Portable FLIR can be supplied with a single, dual or continuous zoom lens.

Operational status
Available.

Specifications
Detector type: HgCdTe
Number of Detectors (elevation × azimuth): 256 × 256 or 320 × 240
Detector pitch: 40 or 30 μm
Spectral band: MWIR 3.8-4.8 or 3.2-4.7 μm
optional upgrade kit – LWIR (Longwave IR kit)
Type of cooling: ¼ Watt IDA (Linear Split Sterling Closed Cycle)
Cool down time: <7 min
Operating temperature: 95.0 ± 5.0 K
System MTBF (hours): >4,000 h
System Field-of-View (FOV): single FOV lens, 11.5°
Optional DFOV lens, 2.3 and 6.9°
Optional continuous zoom lens 2.3 to 6.9°
NETD 300 K background: ≤0.03°C
System dynamic range: ≤70 dB
Output video format: CCIR 'PAL' composite
RS-170 NTSC composite
12 bit digital (optional)
Size: 114 × 153 × 305 mm (with SFOV optics)
Input power: 12 V DC or 24 V DC (optional)
standard NATO 12 V DC lithium battery
110/220 V AC to 12 V DC power supply module
Power dissipation: ≤24 W nominal
Disposable Battery Power life (hours): >4 h continuous
System weight: <3.6 kg with SFOV optics

Contractor
DRS Technologies Electro-Optical Systems Group.

UPDATED

DRS Uncooled Manportable Forward Looking Infra-Red (UMFLIR)

Type
Thermal imager (land).

Description
The DRS Uncooled Manportable Forward Looking Infra-Red (UMFLIR) is a lightweight thermal imager which uses uncooled bolometer detector technology. It operates in the 8 to 12 μm spectral band. Operator controls are provided on the device but as an option it may be remotely operated. Video outputs are CCIR PAL or RS-170 NTSC composite.

Operational status
Available.

Specifications
Detector type: VO_x resistive bolometer
Detector format: 320 × 240
Detector pitch: 51 × 51 μm
Fill factor: 60%
Spectral band: 8 to 12 μm

Optics (standard)
Field of view: 12.4 × 9.3°
EFL: 75 mm
Focus: manual
Viewfinder: standard black and white
NETD: <0.1 K
System dynamic range: 12 bit
Operator controls: local (standard)
remote (optional)
Automatic gain control: standard
Automatic level control: standard
Auto start-up: <1 min
Output video format: CCIR 'PAL' composite or RS-170 NTSC composite
Dimensions: 114 × 153 × 305 mm
Input voltage: +9 to +18 V DC
Power dissipation: <15 W
External battery: BA 5598/U (8 Ahr)
System weight: <3.5 kg
Operating temperature: –30 to +54°C
Storage temperature: –40 to +80°C

Contractor
DRS Technologies, Electro-Optical Systems Group.

VERIFIED

Elettronica ELT/CAT AND ELT/IRIS

Type
Thermal imager (land).

Description
The ELT/CAT is a hand-held, binocular device which uses a Joule-Thomson cooler and staring focal plane array. It may also be installed on a tripod. Outputs to video are CCIR PAL format and with a 0.33 litre cooling gas bottle it can operate for 2½ hours.

Operational status
Not known.

Specifications
Spectral band: 3-5 μm
Field of view: 9.2 × 7°
Weight: 2.3 kg

Contractor
Elettronica.

VERIFIED

Elop ARTIM

Type
Thermal imager (land).

Description
ARTIM is a ruggedised dual field-of-view remotely controlled thermal imaging camera based on Elop's Thermal Imaging Modules (TIM) for night observation and target acquisition applications. It is designed as an upgrade (add-on) night sight for existing weapon systems.

It uses parallel scan technology and has an optional integral CRT monitor, serial communications port and remote-control panel.

Main features
- Advanced detector technology
- Excellent picture quality
- Compact and ruggedised construction to meet tracked vehicles environmental conditions
- Integral detector and Closed Cycle Cooler
- Advanced, proprietary image processing ASIC
- Flexible customer defined mechanical, electronics and control interfaces

Applications – night sight for:
- Tanks and Armoured Fighting Vehicles
- Naval/coastal Mast Mounted Applications
- Fire-Control Systems
- Air Defence Systems
- Surveillance Systems

Operational status
In production.

The Elop ARTIM thermal imaging camera 0059651

Specifications
Weight: 9.0 kg
Spectral band: 8-12 μm
Aperture: 89 mm

Fields of view
Narrow: 4.4 × 2.2°
Wide: 11.5 × 7.4°
Detector: linear array MCT-PV with on-chip pre-processing
Cooler: integral Closed Cycle
Video output: CCIR or RS-170
Monitor: integral monitor - optional
Power: 28 V DC, 44 W complies with MIL-STD-1275A
Environmental: as per MIL-STD 810
MTBF: >2,000 h

Contractor
Elop Electro-Optics Industries Ltd.

VERIFIED

Elop ARTIM-LR

Type
Thermal imager (land).

Description
ARTIM-LR thermal imaging camera/night sight is an extended-range version of Elop's ARTIM intended for fixed-based or portable observation and target acquisition missions and as an add-on night sight for anti-tank, anti-aircraft or laser designation systems.

The system is a single unit with a dual field of view, integral miniature video monitor and standard video signal.

Applications:
- Observation and surveillance tasks - fixed posts or man-portable
- TI sensor for target acquisition systems

Night sight for -
- Laser designation systems
- Anti-tank weapon systems
- Air defence weapon systems
- Paramilitary and security forces

Operational status
Development complete.

Specifications
Spectral band: 8 - 10.5 μm
Entrance aperture: 110 mm
Fields of view (CCIR)
Wide: 7.3 × 4.5°
Narrow: 2.1 × 1.3°
Detector type: CMT-PV, 240 × 1 linear array
Preprocessing: on-chip
Cooler: miniature, low-power integral with Dewar
Video standard: CCIR (RS-170 optional)
Internal display: miniature CRT with monocular eyepiece
Communication: serial, RS-422
Power: 24 V DC, nominal
Power source: Disposable LiSO2 battery

Rechargeable NiMH battery
External DC source
Power consumption: <35 W

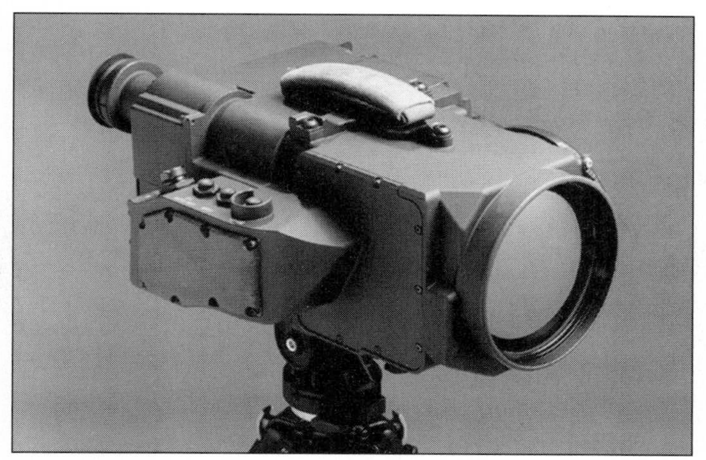

The Elop ARTIM-LR thermal imaging camera/night sight 0059650

Operating temperature range: –30 to +55°C
Weight: 6.0 kg

Contractor
Elop Electro-Optics Industries Ltd.

VERIFIED

Elop CRYSTAL-P high-resolution FPA thermal imager

Type
Thermal imager.

Description
CRYSTAL-P is a three FOV, high-resolution thermal imaging sensor utilising a Focal Plane Array (FPA) detector operating in the 3 to 5 μm spectral band, coupled to a dual-axis micro-scan mechanism, yielding high 512 × 512 sampling resolution. It is packaged in an environmentally sealed housing.

Applications are as follows:
- Long-range observation
- Night surveillance target acquisition and fire direction systems
- Add on TI sensor for ground and naval radar systems
- Night sight for air defence systems
- Maritime/coastal surveillance.

Main features:
- 3-5 μm spectral band
- InSb FPA detector 256 × 256 elements
- Micro-scan mechanism
- Miniature closed-cycle cooler, integral with the Dewar
- Three FOVs
- Non Uniformity Correction (NUC) algorithms
- Proprietary image processing ASIC
- Analogue and digital video outputs
- Single unit configuration for easy integration ruggedised and sealed housing
- Small dimensions.

Specifications
Weight: 11 kg
Spectral Band: 3-5 μm
Detector: Type In Sb Focal Plane Array (FPA) with 256 × 256 elements
Entrance Aperture (main): 100 mm

Fields of View:
Narrow 1.5 × 1.5°
Medium 5.25 × 5.25°
Wide 21 × 21°
Micro-scan: Dual axis, opto-mechanical
Sampling resolution: 512 × 512
Cooler: Closed cycle, integral with Dewar
Video output- CCIR, 625 lines, 50 Hz
Communication: Serial, synchronous, RS-422
Power requirements : 24 V DC nominal, ≤30 W
Configuration: Single unit, including electronics signal processing

Environmental
Designed and tested to MIL-STD-810
Operating temperature range: –20 to +60°C
Reliability: MTBF: > 2,000 hours

Contractor
Elop Electro-Optic Industries Ltd.

VERIFIED

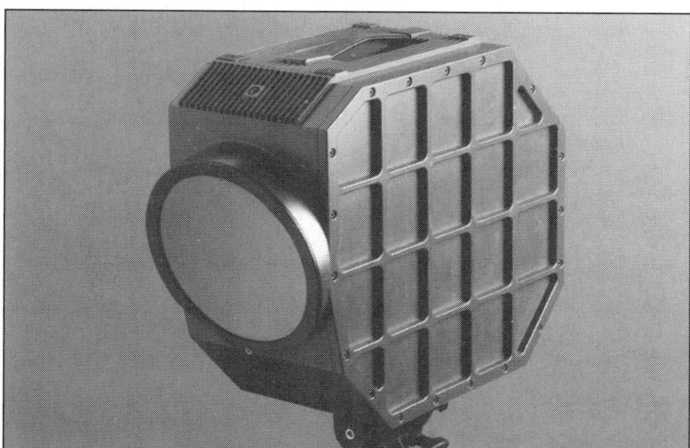

The ruggedised sealed Elop CRYSTAL-P thermal imager 0109665

Elop Head-Mounted/Helmet-Mounted Uncooled thermal imaging Viewer (HMUV)

Type
Thermal imager/viewer (land).

Description
HMUV incorporates light weight thermal imaging camera, based on a state-of-the-art uncooled microbolometer two dimension array: HMUV is a lightweight, head-mounted/helmet-mounted configuration; coupled to an integral miniature, high-resolution display, and is completely hands-free in operation.

Main features
- Lightweight
- Hands-free operation
- Right or left single-eye viewing
- Advanced Micro-Bolometric uncooled detector with 320 × 240 elements
- High image quality
- Low power consumption
- Battery powered
- Standard video output
- Integral LCD display
- Highly reliability
- Cost effective

Applications
- Infantry units
- Anti-terrorist
- Special Forces light patrol boats
- Para-military Forces Security and Perimeter Defence
- Civilian applications
- Firefighting
- Search and Rescue

Specifications
Spectral band: 8-12 μm
Entrance aperture: 16 mm
Field of view: 40 × 30°
Detector uncooled, Microbolometer: 320 × 240 pixels FPA
Integral Display: miniature monochrome LCD
Video output: CCIR or RS17ONTSC
Power: 6 V DC
Power sources: Disposable batteries belt carried power pack 2-D size batteries
Weight: less than 1 kg (head-mounted configuration)

Contractor
Elop Electro-Optic Industries Ltd.

VERIFIED

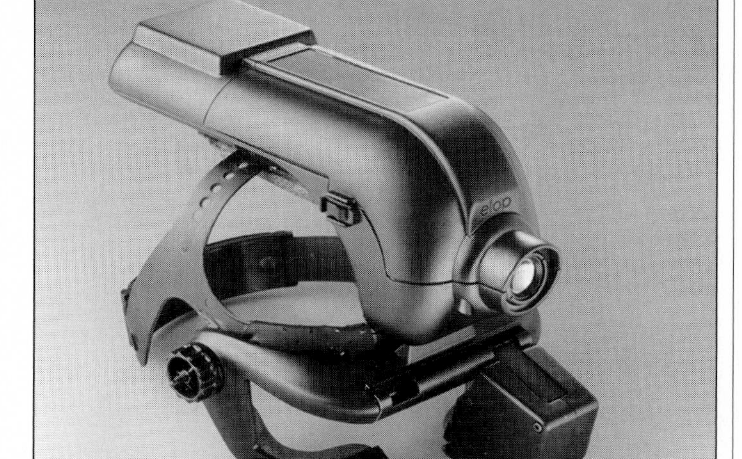

Elop HMUV thermal imager/viewer 0134532

Elop INTIM thermal imaging camera

Type
Thermal imager (land).

Description
The INTIM is a hand-held, lightweight, battery-powered thermal imaging camera, based on Elop's TIM Thermal Imaging Module and shaped and operated similarly to field binoculars. The camera is intended for use by the individual soldier in the field for infantry, scout and special forces tasks. It can also be used as a lightweight target acquisition system, as a night sight for small and medium weapon systems (anti-aircraft, anti-tank and so on), for security and perimeter defence or naval night navigation.

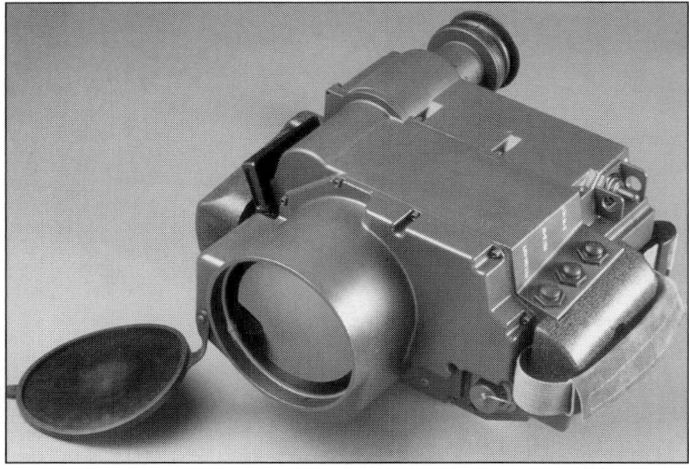
The Elop INTIM 0134533

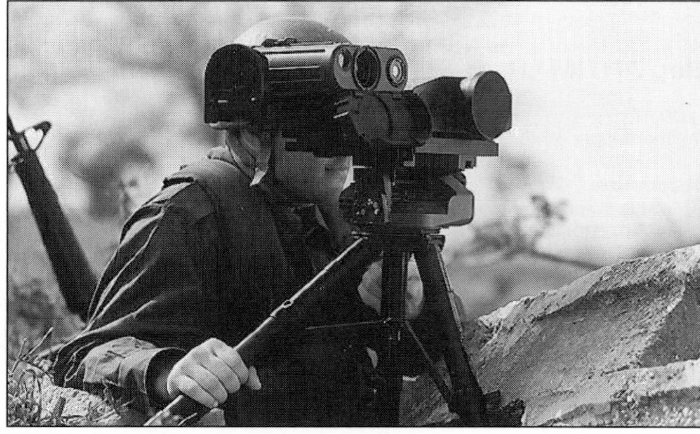
INTIM thermal camera mounted on a tripod

The thermal imager is an Advanced DDC (Detector, Dewar, Cooler) module with focal plane signal processing, integral miniature video monitor, external standard video output and miniature closed-cycle cooler.

Operational status
Development complete.

Specifications
Weight: 3.2 kg
Spectral band: 8-10.5 μm
Entrance aperture: 94 mm

Fields of view
Narrow: 4.0 × 2.5°
Wide: 13.6 × 8.5°
Detector: CMT-PV with on-chip pre-processing
Cooler: miniature integral Closed Cycle
Video standard: CCIR (RS-170 optional)
Integral monitor: miniature monochrome CRT display
Magnifying eyepiece: monocular with rubber hood
Power requirements: 12 V DC, 30 W
Battery: LiSO2
Operating temperature range: –30 to +55°C
Immersion: up to 1 m depth of water
MTBF: >1,500 h

Contractor
Elop Electro-Optics Industries Ltd.

VERIFIED

Elop TIM

Type
Thermal imager (land).

Description
The Elop TIM is a compact, lightweight thermal imaging module. It features a parallel scanning FLIR based upon an integral CMT detector, focal plane signal processing and Dewar cooling system. Consisting of a sensor and an electronics unit, it is intended for a variety of ground-based and airborne applications.

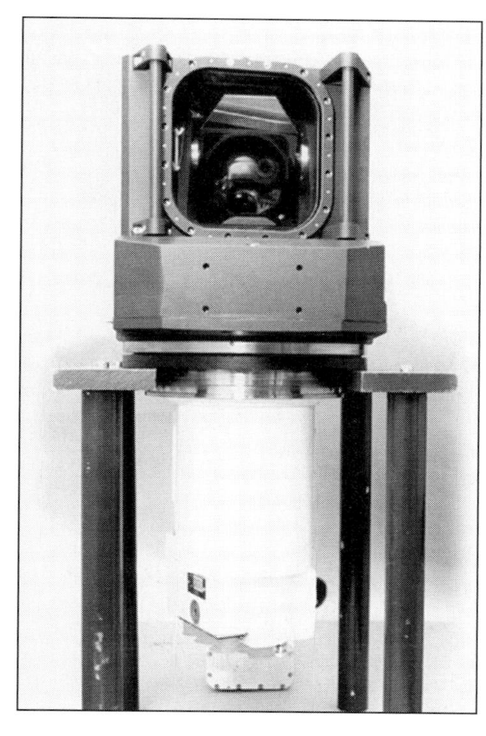

The Elop TIM sensor unit
0059667

Operational status
No longer in production.

Specifications
Detector type: CMT
Fields of view
Narrow: 4.4 × 2.2°
Wide: 15 × 7.5°

Sensor unit
Dimensions: 210 × 140 × 100 mm
Weight: 2.9 kg

Electronics unit
Dimensions: 210 × 130 × 70 mm
Weight: 1.8 kg
Power supply: 28 V DC, 40 W

Contractor
Elop Electro-Optics Industries Ltd.

UPDATED

ENOSA SVT-041 thermal camera

Type
Thermal camera.

Description
Enosa's SVT infra-red camera provides real-time television images with touchscreen display under dark and adverse light conditions.

Operational status
Production as required. In service with the Spanish Navy.

The ENOSA SVT-041 thermal camera

Specifications
Spectral bandwidth: 8-12 μm
Field of view
narrow: 6.9 × 4.3° (×5.2)
wide: 2.8 × 1.7° (×13)
Refrigeration: closed-cycle
Video output: CCIR
Performance ranges
Detection: 10 km (vehicle), 2 km (person)
Recognition: 6 km (vehicle), 1.5 km (person)
Identification: 1 km (vehicle), 0.2 km (person)
Power supply: 19-34 V DC

Contractor
Empresa Nacional de Optica SA (ENOSA).

VERIFIED

FLIR Systems MilCAM MV™

Type
Thermal imager (hand-held).

Description
The MilCAM MV™ meets MIL-STD-810E environment specs. It weighs less than 1.4 kg, complete with battery and 9° field of view lens. Standard rechargeable NiCd batteries provide more than 2.5 hours of use between charges. Unlike imager intensification or 'night vision' products, the MilCAM's sophisticated infra-red sensor does not require ambient light to view the target. Instead, it 'sees' heat emitted and converts it into a TV-quality IR image.

The MilCAM MV employs five major modules for rapid assembly. The RS-232 serial interface, autolevel and programmable reticles are standard.

Operational status
In service.

Specifications
Dimensions (l × w × h): 25 × 15.5 × 12 cm
Weight (incl battery): 1.38 kg
Detector: 256 × 256 Platinum Silicide (PtSi) FPA
Spectral band: 3-5 μm
Noise equivalent temperature: 0.10°C @ 23°C
Optics: f/1.3
Field of view: 50 mm, 9 (h) × 7° (v) (standard)
FPA cooling: Closed-cycle Stirling
Video format: RS-170 or CCIR compatible
Serial interface: RS-232 control of all camera functions
Power requirements: 6 VDC/5.5 W
Battery life: >2.5 hours
Environmental qualifications: Fully ruggedised/MIL-STD-810E qualified
Operating temperature range: –20 to +55°C
Camera controls: On/off/standby, gain, polarity, reticle on/off, reticle select, non-uniformity correction, focus
Mounting: Tripod ¼-20

Accessories
Storage/Transport case: Hard shell, lockable with fitted foam insert
Adaptor/Charger: 110 VAC/220 VAC power supply/battery charger
Batteries: 3 supplied NiCd rechargeable, COTS

Contractor
FLIR Systems.

UPDATED

FLIR Systems MilCAM MV
0137881

FLIR Systems MilCAM Recon™

Type
Thermal imager (hand-held).

Description
The MilCAM Recon is designed for military applications such as reconnaissance, surveillance and target acquisition. This unit incorporates many new advanced features such as a 320 × 240 InSb FPA detector and dual-integration time for maximum sensitivity and wide dynamic range. Vehicle targets can be detected at ranges in excess of 6,000 m and recognised out to 3,000 m in the narrow FOV. The imagery can be viewed directly through the sealed, ruggedised viewfinder or output to an external monitor through the standard BNC video output connector. The Recon utilises a unique confocal telescope design that can be operated manually or remotely through the unit's serial interface.

Operational status
In service.

Specifications
Dimensions (nominal):
355 (l) × 130 (h) × 145 (w) mm
14 (l) × 5.2 (h) × 5.7 (w) in
Weight (incl battery): 2.49 kg (<5.7 lb)

Thermal imager
Detector: 320 × 240 InSb FPA
Spectral band: 3-5 μm

Optics: f/4.0
WFOV: 50 mm, 11 (h) × 9° (v) (100 mm with lens extender)
NFOV: 250 mm, 2.2 (h) × 1.75° (v) (500 mm with lens extender)
FPA cooling: Closed-cycle Stirling
NETD: 0.025°C at 23°C ambient

Video format: RS-170 or CCIR compatible
Serial interface: RS-232 Camera and Lens
Power requirements: 7 V DC/5.5 W
Environmental qualifications: Fully ruggedised/MIL-STD-810E qualified
Operating temperature range: –32 to + 55°C
Camera controls: On/off/standby, auto and manual gain and level, polarity, reticle on/off, reticle select, non-uniformity correction, sensitivity select, change FOV lens
Mounting: Standard NATO 3-point mount

Accessories
Storage/transport case: Hard shell, lockable with fitted foam insert
Adapter/charger: 110 V AC/220 V AC power supply/battery charger
Batteries: NiMH >2.5 h (optional lithium battery >4 h)

Contractor
FLIR Systems.

UPDATED

FLIR Systems MilCAM Recon 0137880

Galileo Avionica VIRS-7 thermal imager

Type
Thermal imager (land).

Description
The VIRS-7 thermal imager system consists of the following subsystems:

(a) infra-red head - comprising a telescope module, sensor module and display and console module. The sensor module operates in the 8 to 12 μm spectral band. There are two types of telescope module available which differ in their collecting aperture sizes and hence detection and recognition ranges
(b) electronic unit - which generates the reticle and symbology and provides the power supply to the unit
(c) optional remote display and control unit.

Operational status
Out of production. In service.

Specifications
IR Head
Weight: 19 kg
Dimensions: 180 × 430 × 180 mm
Sensor module
Detector: RIR 1060
Spectral band: 8-12 μm
Detector: 60-element, linear array
Cooling: closed-cycle cooler
Cool down time: <10 min
Field of view: 48° azimuth × 20° elevation
F-number: 1.8
Focal length: 17 mm
Frame rate: 28 Hz
Lines per frame: 240

Telescope	**Type I**	**Type II**
Magnification	×12	×12
Field of view	4 × 1.7°	4 × 1.7°
IFOV		
Azimuth:	0.2 mrad	0.2 mrad
Elevation:	0.3 mrad	0.3 mrad
Collecting aperture	103 cm²	58 cm²
MRTD (3 cy/mrad)	0.4°C	0.7°C
Performance (NATO standard target, visibility 4 km, air temperature >10°C and relative humidity 80-100%)		
Detection:	6,500 m	5,000 m
Recognition:	2,500 m	2,000 m

Electronic unit
Weight: 5.4 kg
Dimensions: 190 × 210 × 140 mm
Remote display and control unit (optional)
Weight: 4.8 kg
Dimensions: 180 × 250 × 180 mm

Contractor
Galileo Avionica.

UPDATED

Galileo Avionica VTG 120

Type
Thermal imager (land).

Description
VTG 120 is a portable thermal imaging equipment for night surveillance and aiming of medium-range weapons, with particular application to the TOW anti-tank missile. It operates in the 8 to 14 μm band and is provided with an eyepiece for direct viewing. It can also be connected to an external CRT for remote monitoring of the

VTG 120 thermal imaging night sight fitted to an anti-tank missile launcher 0053839

observed scene. An electronically generated graticule is correctly positioned for aiming purposes.

The system is completely autonomous, with a battery and air bottle which allow more than 2 hours of continuous operation. A range of ancillary equipment is available including a battery charger, boresight collimator for graticule alignment, gas bottle charger and a tripod mounting interface. The VTG 120 can be supplied with a closed-cycle cooler if required. Standard performances on a tank target are 3,000 m detection and 2,000 m identification in normal meteorological conditions.

VTG 120 forms part of the Galileo TURMS forward observation system and TURBO tank fire-control system.

Operational status
In production.

Specifications
Weight: 11 kg
Field of view
Narrow: 20 × 40 mrad
Wide: 60 × 120 mrad
Magnification
Narrow FOV: ×12
Wide FOV: ×4
Spectral band: 8-14 μm
Resolution
Narrow: 0.16 mrad
Wide: 0.5 mrad
Detector: CMT 60-element array
Cooling system: Joule-Thomson minicooler

Contractor
Galileo Avionica.

UPDATED

Indigo Systems Merlin® family of cameras

Type
Thermal imaging camera.

Description
The Indigo Systems Merlin® family of cameras is designed to provide the exact solution to requirements from near IR spectroscopy to process control using long wavelength IR. Its modularity allows it to be configured to best needs of the customer, covering features such as wavelength band, imaging optics, variable integration time and data processing software.

The Merlin® is offered with a complete range of low-noise 320 × 256 FPAs covering all IR spectral bands, including Indium Gallium Arsenide (InGaAs), Indium Antimonide (InSb) and uncooled microbolometer arrays, for high-resolution imagery. These arrays are supported by the various cooling options includinging TE stabilisation (for InGaAs and microbolometers) and either liquid nitrogen or Stirling crogenic cooling (for InSb). For each FPA wavelength range, there is a spread of interchangeable IR lens from microscope to telescope optics.

All Merlin® cameras operate at a 60 Hz (50 Hz PAL) frame rate with software-adjustable integration time, or electronic iris, to accommodate a wide range of flux levels. Digital data is generated and processed at full 12-bit resolution. Both radiometric and non-radiometric models are available, with a choice of ThermaGRAM® radiometric software or Talon® digital data processing software.

While the Merlin® near-IR camera with an InGaAs FPA is essentially for scientific applications, military applications (such as long-range surveillance and use in FLIRs) are better fulfilled with the mid-IR camera, using an InSb FPA. The Merlin® Uncooled (with a microbolometer) has security applications, can penetrate smoke and haze and may be used for gas and/or mine detection.

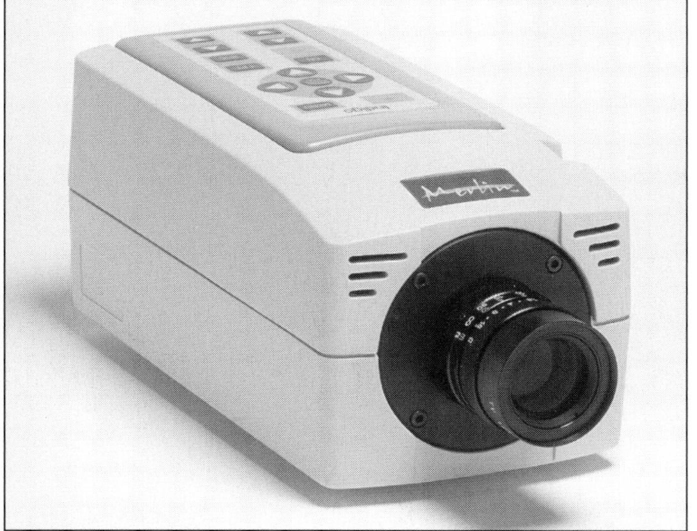
The Indigo Systems Merlin® thermal imager ***NEW***/0595415

Operational status
Available.

Specifications

Camera	Merlin NIR	Merlin Mid	Merlin Uncooled
Detector	InGaAs	InSb	Microbolometer
Spectral band	0.9 – 1.7μ	1.5-5.0μ	7.5-13.5μ
Resolution	320 × 256	320 × 256	320 × 256
Detector size	30 × 30 μ	30 × 30 μ	51 × 51 μ
Aperture	(set by lens iris)	f/2.5 or f/4.1	f/1.3
Detector cooling	TE stabilisation	Integral Stirling or LN2	TE stabilisation
NEDT	(lo-gain = ≤1E10 ph/cm2-sec)	<25 mK	<100 mK
	(hi-gain = ≤5E9 ph/cm2-sec)	(<18 mK typical)	
Dimensions (cm)	10.1 × 11.4 × 20.3	14 × 12.7 × 24.9	10.1 × 11.4 × 20.3
Weights	1.6 kg	4.1 kg	1.6 kg
Optics	NIR	MWIR	LWIR
Microscope:	×0.7 to ×22.5	×1, ×2.5 and/or ×4	n/a
13 mm lens	n/a	41 × 31° FOV	64 × 50° FOV
25 mm lens	22 × 16° FOV	22 × 16° FOV	36 × 27° FOV
50 mm lens	11 × 8° FOV	11 × 8° FOV	18 × 14° FOV
100 mm lens	n/a	5.5 × 4.1° FOV	9 × 7° FOV
200 mm lens	n/a	n/a	4.7 × 3.5° FOV
Dual FOV			
50 mm	n/a	11 × 8°	n/a
250 mm	n/a	2.2 × 1.8°	n/a
Triple FOV			
60 mm	n/a	9.1 × 7.3°	
180 mm	n/a	3.1 × 2.4°	
500 mm	n/a	1.1 × 0.9°	

NB: The NIR optics can also cover a 12.5 to 75 mm Zoom lens.

Contractor
Indigo Systems Corporation (USA).

NEW ENTRY

Indigo Systems Mid-Range Security Camera

Type
Thermal imager.

Description
The Indigo Systems Mid-Range Security Camera is sealed for outdoor applications requiring remote monitoring and security. The camera is based on 320 × 240 uncooled Vanadium Oxide microbolometer technology and includes features oriented to maximise ease of installation and operation. The camera, when mated with the sensor controller, provides remote control and video monitoring through pre-existing or easily installed standard infrastructure interfaces.

It is sealed to IP66, including a hard carbon-coated lens, and designed to endure the rigors of unattended harsh environment mobile or fixed site installations.

The standard configuration includes a 100 mm lens, with a FOV of 9° × 7°, yielding ranges in excess of 1 km. The 50 mm optional lens provides a 19° × 14° FOV.

The Indigo Systems Mid-Range Security Camera with the 50 mm optional lens ***NEW***/0595416

Operational status
Available.

Specifications
Dimensions
Without lens: 17.8 × 7.9 × 7.6 cm
With 100 mm lens: 24.8 (L) × 12.9 (Diameter)
Weight: <2.3 kg (with 100 mm lens)
Spectral Band: 8-12 μm
Detector: Vox
Array Format: 320 × 240
Lens
Standard: 100 mm f/1.4
Optional: 50 mm f/1.4
Power: 24 V DC
Control: RS-232 or RS-422
Analog Video: Analog NTSC or PAL
Digital Video: 14-bit precorrected or postcorrected
Image Manipulation: e-Zoom, freeze frame, adjustable histogram or linear AGC, gamma boost, reticle overlays, eight colour palettes, polarity, multiple manual or automatic fast-switched NUCs for changing scene dynamics.

Contractor
Indigo Systems Corporation (USA).

NEW ENTRY

Indigo Systems Omega™

Type
Thermal imaging camera.

Development
The ultra-compact Indigo Systems' Omega™ camera can be used in stand-alone applications or as an OEM module core, where space, weight and/or power are constraints and performance is critical.

This little camera, described as 'robust', delivers performance and features typically found only in larger IR systems. Typical uses might include fire-fighting, security surveillance, search-and-rescue, industrial process monitoring, or as a sensor for unmanned aerial vehicles.

Description
The Omega™ uses VOx microbolometer detectors combined with proprietary on-focal plane signal processing, DSP-based electronics and real-time algorithms that, combined with a fully optimised 160 × 128 FPA, delivers image quality and thermal resolution usually ascribed to larger arrays.

It delivers wide dynamic range (14-bit) images at real-time video rates of 30 frames per second (fps) for RS-170 or 25 fps for CCIR. An autoranging function detects very hot scenes and automatically switches into an extended temperature range mode, allowing imaging scenes up to 400°C.

Integrated with a Sony camcorder, Omega™ becomes the ThermoCorder™ (qv) handheld thermal imaging and recording system.

Operational status
Available.

Specifications
Weight: ≤120 g, lens dependent
Dimensions: 34.3 × 36.8 × 48.3 mm
Power: <1.5 W nominal
Detector: uncooled microbolometer
Spectral range: 7.5 to 13.5μ
Array format: 160 × 120 (RS-170); 160 × 128 (CCIR)

The Indigo Systems Omega™ thermal imager **NEW**/0595418

Pixel size: 51 × 51μ
NEDT: =85 Mk (equivalent to 40 mK at f/1.0)
Operating temperature range
Standard: 0 to +40°C
Optional: –40 to +55°C
Scene temperature range
Standard: up to 150°C
Optional: auto-gain modes extend to 400°C
Video output: RS-170 or CCIR
Input/output: 18-pin connector for video, power, communication, digital data
Serial commands: RS-232 interface

Contractor
Indigo Systems Corporation (USA).

NEW ENTRY

Indigo Systems Phoenix® family of cameras

Type
Thermal imaging camera.

Description
The Phoenix® family of digital IR cameras is a modular system consisting of a camera head (with a choice of near-, mid- and long-wave spectral band coverage) and a choice of two video signal processing 'back-ends' packages: Real-Time Imaging Electronics (RTIE) or Digital Acquisition System (DAS).

The IR sensors themselves use Indigo Systems' own standard CMOS ReadOut Integrated Circuits (ROICs). These offer features including snapshot (simultaneous) pixel exposure, adjustable gain variable exposure times, windowing and invert/revert facilities. The near-wave array uses an Indium Gallium Arsenide (InGaAs) detector; the mid-wave array uses an Indium Antimonide (InSb) detector; and the long-wave array uses a Gallium Arsenide (GaAs) Quantum Well IR Photodetector (QWIP).

The camera itself features 14-bit extended dynamic range, snapshot exposure mode, high frame rate capability and excellent resolution within a small rugged package. The heads are available in either 320 × 256 or 640 × 512 formats, with a range of optics from microscope level to 100 mm lens and both dual- and triple-FOV optics. The camera is designed to cope with severe environments, offering both conductive and convective cooling.

The RTIE package is an electronic subsystem that provides both analog and digital video at data rates of up to 12.2 megapixels per second. It is configured as a 'split' system being connected to the Phoenix® camera by an interface cable anything up to 15 m long. Synchronisation modes, windowing capabilities and triggering features are available. It can generate NTSC video (PAL optional) as well as S-Video. The 14-bit digital data available from the RTIE can be evaluated using the optional Talon® Ultra digital image acquisition and analysis system.

The DAS electronics package is a portable PC-based system with an interface/sync processor board capable of handling data rates of up to 40 megapixels per second. It can store at least 10 seconds of full bandwidth data from the Phoenix® camera head. DAS shares the synchronisation modes, windowing capabilities and triggering features of the RTIE but provides two additional video channels. This allows the DAS to extract maximum performance from the camera FPAs.

Operational status
In production and in service with US forces.

The Indigo Systems Phoenix® thermal imaging camera **NEW**/0595419

Specifications

Camera

	Phoenix-Near	Phoenix-Mid	Phoenix-Long
Detector	InGaAs	InSb	GaAs QWIP
Spectral band	0.9 – 1.7μ	1.5-5.0μ	8.0-9.2μ
Resolution	320 × 256 or 640 × 512	320 × 256 or 640 × 512	320 × 256 or 640 × 512
Detector size			
320 × 256	30 × 30μ	30 × 30μ	30 × 30μ
640 × 512	25 × 25μ	25 × 25μ	25 × 25μ
Detector cooling	Thermoelectric (TEC) stabilisation	Closed-cycle Stirling with LN2 optional	Closed-cycle Stirling
Power dissipation	<25 W steady (25°C)	40 W steady (25°C)	45 W steady (25°C)

Optics	320 × 256	640 ×512
Microscope	×1, ×2.5 and/or ×4	×1, ×2.5 and/or ×4
13 mm lens	40.5 × 32.9° FOV	63.2 × 52.5° FOV
25 mm lens	21.7 × 17.5° FOV	35.5 × 28.7° FOV
50 mm lens	11 × 8.8° FOV	18.2 × 14.6° FOV
100 mm lens	5.5 × 4.4° FOV	9.1 × 7.3° FOV
Dual FOV – 50 mm	11 × 8.8°	18.2 × 14.6°
-250 mm	2.2 × 1.8°	3.7 × 2.9°
Triple FOV – 60 mm	9.1 × 7.3°	15.2 × 12.2°
-180 mm	3.1 × 2.4°	5.1 × 4.1°
- 500 mm	1.1 × 0.9°	1.8 × 1.5°

Dimensions

Phoenix camera head: 19 × 11.2 × 13.2 cm
RTIE electronics: 15.2 × 15.2 × 12.7 cm
DAS electronics: 33 × 25.4 × 43.2 cm

Weights

Phoenix camera head: 3.17 kg
RTIE electronics: 2.72 kg
DAS electronics: 15.42 kg

Contractor

Indigo Systems Corporation (USA).

NEW ENTRY

ISS 300LS-3 thermal imager

Type

Thermal Imager (Land & Shipboard).

Description

The 300LS thermal imaging system is an extremely rugged device, designed to endure harsh environments, while providing a compact and lightweight form factor for reconnaissance applications requiring mast mounting. Very good standoff performance employing selectable fields of view and a low audible noise footprint make the 300LS ideal for covert operations.

The selectable fields of view are available in a two field of view and a three field of view version; with the two field of view version offering two lens options.

The 300LS is built into an environmentally sealed enclosure designed to conform with the demanding requirements imposed by MIL-STD 810, including resistance to dust, sand, snow, sleet, and rain, among others. With a heated window and optional window wiper this robust system is capable of being operated in the most challenging of military operational environments. Employing a proven 320 × 240 focal plane array sensor with a cooler offering a 5,000 hour mean time to failure (MTBF), the 300LS provides extremely high reliability and a very long service life, without the need for periodic maintenance.

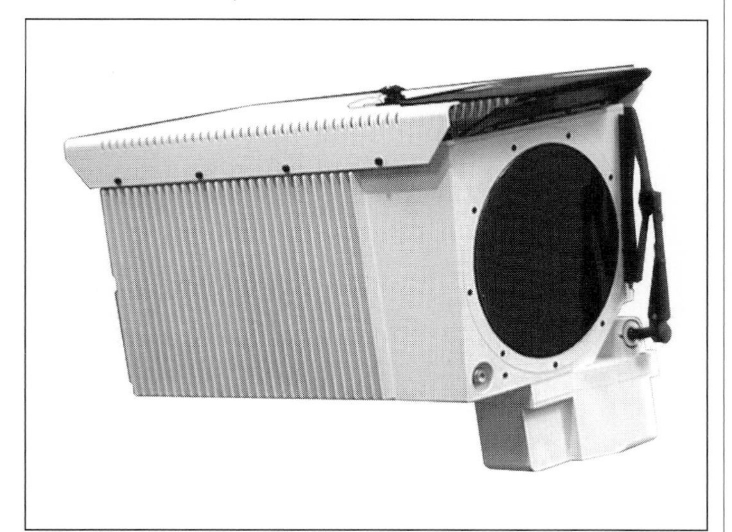

The ISS M300LS thermal imager 0101852

Broad dynamic range, coupled with 4× optical microscanning, provides substantial resolution enhancement. This high performance thermal imager is proving itself in such systems applications as perimeter security and surveillance, vehicle and man-portable reconnaissance, coastal watch, and navigation. The 300LS is also now being offered as a MWIR form/fit option to the ISS AN/TSD-501 MicroFLIR™ long wave thermal imager, which is fielded on the Canadian Army LAV-Recce Vehicle. This third generation midwave thermal imager thus provides a plug and play option for the long wave thermal imager-equipped AN/TSD-501, while maintaining the same rugged construction and reliability found in the original system.

Operational status

Available and in service.

Specifications

Size (L × W × H): 22 × 9 × 9 in (560 × 229 × 229 mm)
Weight: 34 lb (15.5 kg)
Spectral Band: 3 - 5 m m
Detector: 320 × 240 InSb FPA
Cooling: Closed cycle cryogenic
E-ZOOM: Continuously variable
Field of View Options: 2 FOV- Option (1) NFOV 2.20° H × 1.65° V
WFOV 10.97° H × 8.24° V
Option (2) NFOV 2.20° H × 1.65° V
WFOV 7.32° H × 5.50° V
3 FOV NFOV 1.20° H × 0.90° V
MFOV 5.00° H × 3.80° V
WFOV 22.00° H × 22.00° V

Power: 18 - 30 V DC (nominal), 50 W (MIL 704-D)
Video Output:NTSC or PAL
Operating temperature: −40 to +55° C
Environmental: MIL-STD-810

Contractor

Imaging Sensors and Systems, Inc.

VERIFIED

ISS AN/TSD-501 thermal imager

Type

LWIR Thermal Imager (Land and Shipboard).

Description

The AN/TSD-501 Thermal Imager Camera is a rugged device designed to endure harsh environments, while still being compact and lightweight, making it an exceptionally good fit for reconnaissance applications requiring mast mounting. Its standoff performance and low audible noise make the AN/TSD-501 suitable for covert surveillance operations. Possessing wide dynamic range and high sensitivity because of its 8-bar SPRITE detector, the AN/TSD-501 provides excellent discrimination of targets in the presence of extensive background heat sources, as well as dust, haze and smoke. Equipped with two levels of magnification and a selectable field-of-view, this camera provides clear images during rapid movement, making it highly suitable for shipboard applications.

This system has been qualified to MIL-STD-810 and MIL-STD-461 requirements for temperature, shock, vibration, sand, dust, freezing rain and EMI. Its linear resonant cooler having a Mean Time Between Failure (MTBF) of more than 6,000 hours assures a long service life without need for periodic maintenance.

The heart of the AN/TSD-501 is the Micro-FLIRÒ, a Long Wave Infra-Red (LWIR) imager, with a proven history of high performance. This mature Thermal Imager Camera has been successfully employed in systems involving helicopter fire

The ISS AN/TSD-501 imager 0101855

control, perimeter security and surveillance, coastal watch and navigation, shipboard applications. The Micro-FLIRÒ is in service with the Canadian Army on the LAV-25 Recce Vehicle, with the US Navy in the MK46 Optical Sight, and with the Belgian Army in the A-109 HeliTOW sight. It has further proven itself during extensive field trials with the Australian Army. The AN/TSD-501 is also available with a MWIR form/fit option employing the proven ISS 300LS thermal imager.

Operational status
Available and in-service in several countries.

Specifications
Dimensions (L x W x H): 17.7 × 8.6 × 7.9 in (450 × 225 × 200 mm)
Weight: 25 lb (13.3 kg)
Spectral Band: 8 – 12 μm
Detector: 8-bar SPRITE
Cooling: Closed cycle cryogenic
Field-of-View (H × V): NFOV 2.25 × 3.0°
WFOV 7.5 × 10.0°
Magnification: ×4.2 wide
×14.0 Narrow
Video Output: NTSC or PAL
Power: 75 W @ 28 V DC (12 V DC Optional)
Operating Temperature: –40 to +55°C
Qualification: MIL-STD-810, MIL-STD-461

Contractor
Imaging Sensors and Systems, Inc.

VERIFIED

Kazan TNP-1

Type
Thermal imager (land).

Description
The TNP-1 thermal imager is an artillery observation device for use in battlefield conditions. It uses bottles of compressed nitrogen to cool the detector modules and has a monocular eyepiece.

TNP-1 thermal imager from Kazan Optical and Mechanical Plant (AO KOMZ)
0005496

The TNP-1 is installed on a tripod with a goniometer and a 12 or 27 V battery. It may also be used with 1PN50 night observation binoculars. It can detect a tank at distances up to 2.5 km and can identify it at 1.5 km.

Operational status
In service with Russian armed forces and in production.

Specifications
Dimensions: 580 × 520 × 320 mm
Weight: 19 kg (operational)
Field of view: 1 × 2.5°
Resolution: 0.8 mrad
MRTD: 0.2°C
Exit pupil diameter: 5 mm
Exit pupil clearance: 22 mm
Identification: 1.5 km
Focusing limits: ±4 dioptres
Range (tank, normal conditions)
Detection: 2.5 km
Identification: 1.5 km
Power supply: 10.5-14.5 V
Power consumption: 0.8 A

Contractor
Kazan Optical and Mechanical Plant (AO KOMZ).

UPDATED

Liteye Knight-Eye

Type
Thermal imager.

Description
The Knight-Eye hand-held thermal imager combines the abilities of the Raytheon Microbolometer Camera and a Liteye Microdisplay to create a detection device that is rugged and water resistant. The Knight-Eye provides the user with a hand-held unit that is rugged enough to do the job, yet light weight enough, (only 710 g, 25 ounces), to be easy to carry or wear. Whether hand-held or helmet mounted, the microdisplay gives the user a large high-resolution picture which allows for easy target detection. With its use of AA batteries, power sources are no longer a problem, no bulky batteries, strange sizes, or rechargers are required. The Knight-Eye was designed for military and law enforcement users so it is suitable for a wide range of environments.

Operational status
Available.

Contractor
Liteye Microdisplay Systems.

Opgal Recon 1000

Type
Thermal imager (land).

Description
The Recon 1000 is a remote-controlled thermal imager from Opgal Optronic Industries Ltd. The thermal imager is the Opgal TD-92BL or TD-92CL. The system

Recon 1000 remote-controlled thermal imager from Opgal
0005498

can be integrated with a laser range-finder and CCD camera. The Remote-Control Unit (RCU) enables online operation of the thermal imager and tripod-mounted pan and tilt head from distances of up to 100 m. The RCU features a built-in video monitor with video output for external monitor, VCR or image transmission. Optional features include automatic/semi-automatic scanning by segment or specified area; millimetric wave video image transmittance; and video recording. The system is powered by a regular vehicle battery.

Operational status
In production and in service.

Contractor
Opgal Optronic Industries Ltd.

VERIFIED

Pyser-SGI TIU2 series thermal cameras

Type
Thermal imager.

Description
TIU2 is a low cost, high-quality thermal camera family operating in the 8 to 14 μm band for security applications during day and night. The models are TIU2 75/100/150.

Image resolution of 320 × 240 pixels and horizontal angle of view of 12, 9 or 6° mean that TIU2 can detect a person up to 600 m away. TIU2 75 up to 900 m, TIU2 100 up to 1.3 km and TIU2 up to 150 km away - 24 hours a day in conditions of reduced visibility due to rain, mist (dependent on water particle size) or smoke. TIU2 can be used for fire prevention by detecting heat build-up (hot spots) and creating an early warning through alarms.

TIU2 is a fully enclosed camera head and housing for integration into any new or existing surveillance system.

Operational status
Available.

Specifications
Dimensions: 250 × 150 × 170 mm
Weight: 4.5 kg (approximately)
Mounting: 4 × M6 mm on 4 in PCD
Electrical:
Input 12 V
≤3 W typical power consumption
Connector:
12 V in
contrast mode select/control
Focus adjust
Polarity change
RS-170 NTSC and PAL video out
Detector
Type: Uncooled focal plane array
Pixels: 320 × 240
Spectral band: 8-14 μm
Output: RS-170 NTSC and PAL compliant
NETD: < 0.4°C

Pyser-SGI TIU2 thermal camera 0053831

MTF: >40% at 0.38 cycles/mrad
Video update rate: 30 Hz
Operating temperatures: –40 to 55°C

Contractor
Pyser-SGI, TIU.

UPDATED

Raytheon AN/VLR-1 Avenger FLIR

Type
Thermal imager (land).

Development
The Raytheon FLIR Receiving Set (AN/VLR-1) is an integral part of the US Army Avenger air defence weapon, a component of the Army's Forward Area Air Defense System (FAADS). Raytheon is participating on the project under subcontract with Boeing Aerospace and Electronics of Huntsville, Alabama, prime contractor for Avenger.

Description
The FLIR sight is used to acquire track and identify airborne targets by day, night and adverse weather conditions. The receiving set, Infra-red (IR) AN/VLR-1 consists of a FLIR receiver and a display unit. The receiver, IR R-2448/VLR-1 is a passive, serial- scanned RS-170 compatible, IR imaging system operating in the 8 to 12 μm spectral region and using SPRITE detector technology.

The Display Unit, IP-1622/VLR-1 contains all controls required to operate the receiver and presents a real-time thermal image of the target scene to the operator. The high-resolution display is stated to be visible in direct sunlight.

The FLIR has a continuous zoom telescope and is cooled by a linear resonant Stirling cycle cooler and satisfies the environmental requirements of MIL-STD-810D.

Operational status
In service with the US Army and Marine Corps. Deliveries of the AN/VLR-1 (Avenger FLIR) started in June 1988. Over 750 systems out of a total of 1,004 have been delivered.

Specifications
Zoom telescope
Magnification: ×7.2 (max), ×1.8 (min)
Field of view
Narrow: 5.3 × 3.27°
Wide: 21 × 13.1°
Independent Field Of View (IFOV)
Narrow: 0.25 mrad
Wide: 1.0 mrad
Spectral band: 8-12 μm
MRTD: 0.35°C at 1.8 cy/mrad
Power: 120 W at 24 V DC nominal
Weight
Receiver: 20.98 kg
Display: 5.44 kg
Dimensions (w × h × l)
Receiver: 23.11 × 34.54 × 63.5 cm
Display: 19.05 × 18.54 × 34.8 cm

Contractor
Raytheon Systems Company.

VERIFIED

The FLIR receiving set of the Raytheon AN/VLR-1

Raytheon Escort-2

Type
Thermal imager (land).

Description
The Escort-2 is a continuously operating, medium-range (3,000 m), all-weather surveillance, Forward-Looking Infra-Red (FLIR) system. Escort-2 uses a fully remote, dual-field of view, portable FLIR sensor, powered either from a vehicle battery or a portable battery pack. The high-quality video output makes it suitable for a wide range of direct view or remote surveillance missions.

The Escort-2 has a full range of remote controls contained on the hand-held monitor. The sensor outputs standard NTSC or PAL video to the hand-held monitor. The front of the sensor permits the use of a ×2 extender lens assembly for doubling the range performance. Escort-2 operates on either 12 or 24 V DC input power.

Options include: remote pan and tilt; dual monitors; video recorder; tripod; high-resolution monitor. Full system/platform integration is available (that is addition of laser range-finder, daylight camera, radar, integrated platform, vehicle mount and so on).

Operational status
Available.

Specifications
Field of view (h × v)
Wide: 6.8 × 3.4°
Narrow: 2.2 × 1.1°
Range
WFOV: 1,000 m (vehicle recognition)
NFOV: 3,000 m (vehicle recognition)
Magnification/Resolution
WFOV: ×3.0/0.5 mrad
NFOV: ×9.0/0.17 mrad
MRTD: <0.3°C hand-held monitor; <0.2°C high-resolution monitor
Run time: continuous
Power: 12 or 24 V DC
Weight (sensor): <14 kg
Dimensions (sensor): 254 × 267 × 445 mm
Cooling: closed-cycle silent
Format: NSTC or PAL

Contractor
Raytheon Systems Company.

VERIFIED

Raytheon Escort-2 surveillance system

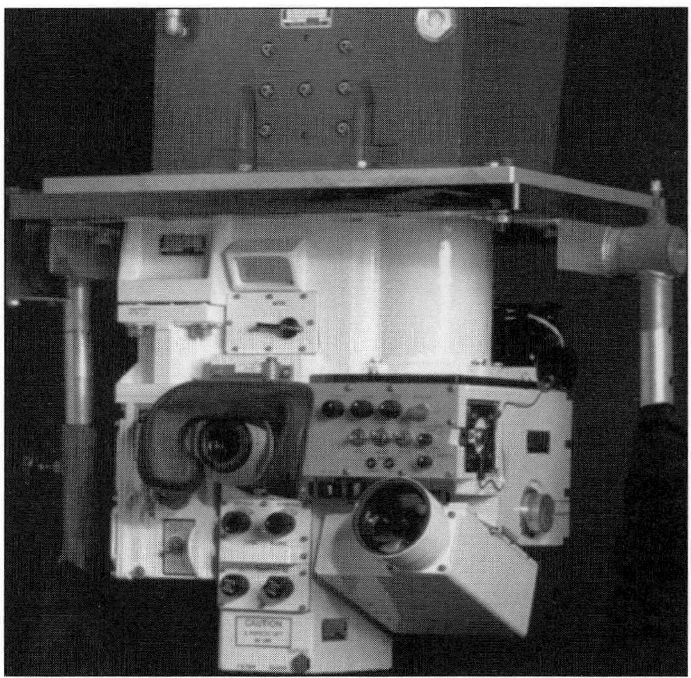

HIRE installed in a gunner's primary sight 0053830

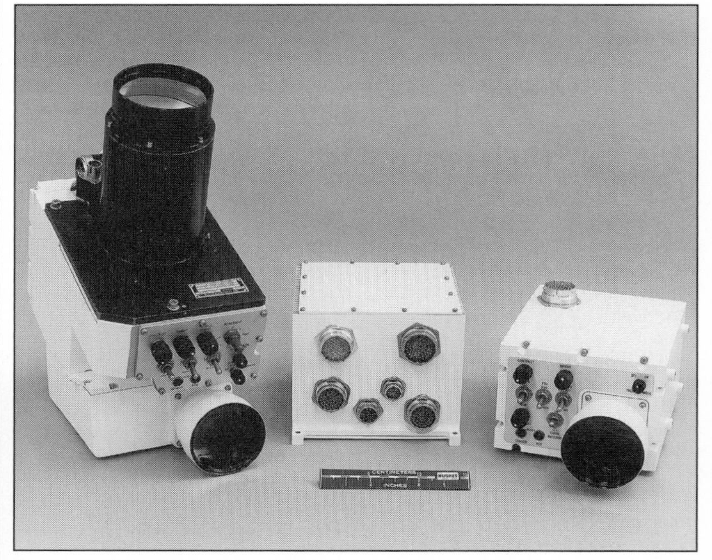

Raytheon HIRE components: Commander's and gunner's display, power supply and electronics unit

Raytheon HIRE infra-red equipment

Type
Thermal imager (land).

Description
Raytheon's HIRE infra-red system is a high-performance, lightweight, modular fire-control sight and thermal imaging system for light armoured vehicles which has been in production since 1991. In addition to providing a stand-alone thermal imaging capability, HIRE is being used in conjunction with fire-control and TOW missile systems. The HIRE sensor is installed on the GMHE Integrated Tow Sight (GITS) (qv), the Raytheon Advanced LAV sight (qv) for LAV-25 turrets and the Raytheon DNRS Day/Night Range Sight (qv). It has also been demonstrated on Desert Warrior, Piranha and M113 combat vehicles and affords commonality to the US Marine Corps in their LAV-25, LAV 105 and AAVP-7 vehicles.

HIRE also forms part of the Thales Optronics (Taunton) Ltd (formerly Avimo Ltd) SPIRE (qv) fire-control system which is to upgrade the British Army's Scimitar tracked combat reconnaissance vehicles and may also upgrade Sabre armoured combat vehicles.

HIRE is based on improved US Army Common Module components and is of modular design with a choice of image displays and control: one or two displays with monocular or biocular eyepiece. Telescope modules can be tailored for all combat vehicle applications. It can also be interfaced with the Raytheon ELITE eye-safe laser range-finder.

Operational status
In production since 1991 for several countries.

Specifications
FLIR: parallel-scanned 240 line, common module-based.
Detector: 60-element CMT
Field of view
Narrow: 4.1 azimuth × 1.7° elevation
Wide: 12.3 azimuth × 1° elevation
Spectral band: 8-12 μm
Image display: CRT
Reticles: NATO stadiametric (programmable options)
Symbology: range, ammo, fault and so on (programmable options)
Power: <110 W (meets MIL-STD-1275AT vehicle power specification)

Contractor
Raytheon Systems Company.

UPDATED

Raytheon HIRE Second-Generation

Type
Thermal imager (land).

Description
The Raytheon Second-Generation HIRE is a lightweight, modular fire-control sight and thermal imaging system for light armoured vehicles, including Piranha, Desert Warrior and LAV-25. It upgrades the HIRE system, in production since 1991, with second-generation technology that includes a Cadmium Mercury Telluride PhotoVoltaic (CMT-PV) 240 × 4 element detector focal plane array, leading to improvements in image quality and range.

Second-Generation HIRE can directly replace HIRE and does not have that system's separate electronics unit. It can be used as a stand-alone thermal imaging system and can be configured for use with TOW missile fire-control systems. It can replace most existing vehicle image intensifiers.

Specifications
Spectral band: 8-12 μm
Detector: CMT-PV 240 × 4 elements
Interlace: 2:1
Number of IR lines: 480
Field of view
Narrow: 4.1 × 1.7°
Wide: 12.3 × 5.1°
Focus: 50 m to ∞
Cool down time: <10 min
Field of view switch time: <0.5 s
Field of view alignment: <0.5 mil
Video format: RS-170 or CCIR
Zoom: ×2 electronic
Gain level/control: automatic and manual

Contractor
Raytheon Systems Company.

VERIFIED

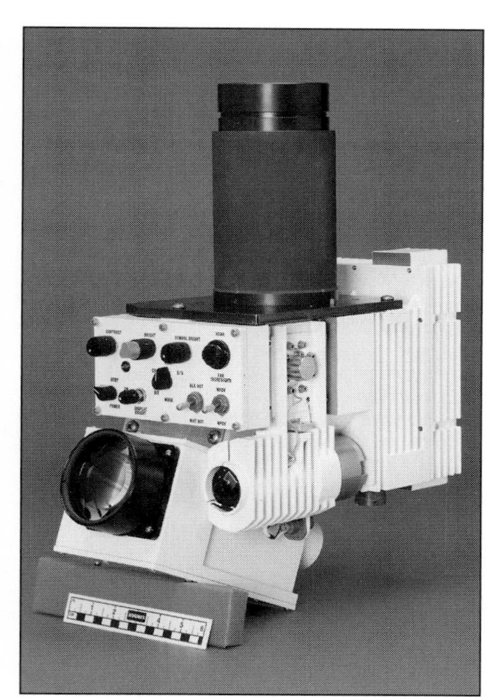

Raytheon's HIRE Second-Generation technology includes a 240 × 4 detector element focal plane array
0053840

Raytheon M1 Abrams Thermal Imaging System (TIS)

Type
Thermal imager (land).

Description
The Thermal Imaging System (TIS) was developed by Raytheon for the under-armour gunner's sight on the M1 Abrams main battle tank family. The thermal imager is based on US Army common modules and the thermal image is displayed on a CRT tube and is projected into the eyepiece of the gunner's sight.

In addition, the sight displays target range information from the Raytheon laser range-finder module and indicates the existence of more than one target.

Ready to fire indication, fire-control computer symbology and system operational status data are also provided.

The TIS consists of a thermal receiver unit, image control unit, power control unit and electronics unit. It generates a graticule pattern boresighted to the day sight graticule and laser range-finder. This allows the gunner to operate the sight just as he would the day sight and engage targets out to 2,400 m.

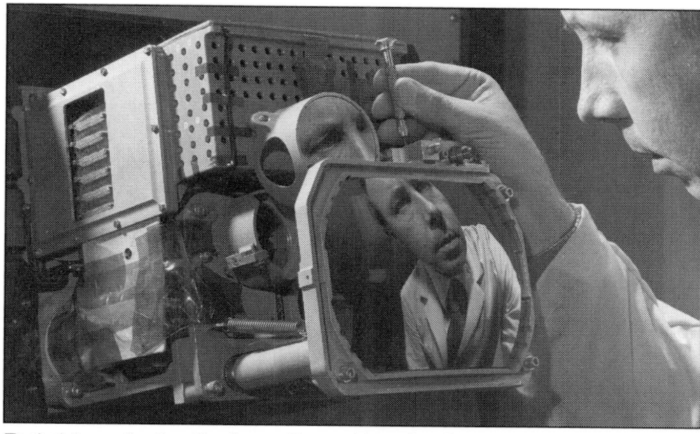

Technician adjusting Raytheon Thermal Imaging System (TIS) installed in M1/M1A1/M1A2 Abrams MBTs

If required, the TIS can be repackaged for use in lighter armoured vehicles where multiple or remote displays can be utilised.

Operational status
In production. Over 11,000 built. In service with the US Army (on M1/M1A1/M1A2 Abrams MBTs). Also fitted on export vehicles for Egypt (M1A1), Kuwait (M1A2) and Saudi Arabia (M1A2).

Contractor
Raytheon Systems Company.

VERIFIED

Raytheon MAG 2400 long range thermal imager

Type
Thermal imager (land).

Description
The Raytheon Systems Company MAG 2400 is a man-portable, high-sensitivity, staring focal plane array camera with a 3 to 5 μm, 320 × 240-element InSb detector.

The camera is designed for fire-direction team, forward observation and reconnaissance applications. Cooling is provided by a Raytheon miniature linear drive, dual-piston Stirling cryocooler. The system features an internal 5:1 continuous-zoom telescope to provide both high resolution in the narrow field of view (FOV) for long-range recognition and wide FOV for effective search.

The MAG 2400 has remote-control capability via a digital interface and incorporates both a digital image and a 30 Hz, RS-170 video output (25 Hz, CCIR video output version known as IR Lookout™ 15M). A mounting adapter for a tripod or angulation head is also supplied.

The system operates from a standard internal battery or external power. Run time on the internal rechargeable lithium battery is 2.5 hours.

Operational status
Not known.

Specifications
Size without eyecup: 194 × 152 × 260 mm
Size with eyecup: 194 × 152 × 372 mm
Weight (including battery): <4 kg including internal battery
Field of view (h × v): WFOV: 9.5 × 7.2°
(continuous zoom) (h × v): NFOV: 1.9°H × 1.4°V
Resolution (h × v): 320 × 240 × 2 samples (microscan)
IFOV, widest FOV (h × v): 0.5 mr × 0.5 mr
IFOV, narrowest FOV (h × v): 0.1 mr × 0.1 mr

The Raytheon MAG 2400 thermal imager 0053835

Detector: 320 × 240 InSb
Spectral band: 3-5 μm
Cooling: integrated detector/all-metal Dewar/dual-piston linear drive, miniature Stirling cryocooler construction
Operating temperature: –32 to +55°C (–25 to +131°F)
Storage temperature: –40 to +71°C (–40 to +160°F)
Ready time: ~5 min cooldown (at room temperature) to standby, instant from standby to on
Power required: ~10 W at room temperature
Run time: ~2.5 h at room temperature on internal rechargeable BB 2847/U lithium ion battery, NSN 6140-01-419-8194
Battery: <4 kg, including internal battery

Contractor
Raytheon Systems Company.

VERIFIED

Raytheon Radiance

Type
Thermal imager (land).

Description
The Radiance 1 is an infra-red camera from Raytheon, which operates in the medium-infra-red waveband. It is based on a 256 × 256 staring indium antimonide (InSb) focal plane array, which can discriminate temperature differences of 0.025°C.

Images are displayed on a monitor or optional viewfinder. Images can be viewed in black and white or one of many colour palettes, controlled by a press-button on the camera. To bring out the subtle details in a scene, a non-linear contrast enhancement function is included, as well as for one- or two-point normalisation of the temperature source, achieved in seconds. There is an automatic gain control button that allows hands-off operation.

Raytheon offers a version for thermographic monitoring applications, Radiance PM (Predictive Maintenance). This system is shoulder mounted with a 4 in active-matrix display attached to the front of the camera.

Operational status
In production.

Specifications
Detector: indium antimonide (InSb)
Number of elements: 256 × 256 (65,536)
Detector frame rate: 60 frames/s (50 fps w/PAL)
Spectral band: 3-5 μm (custom filter optional)
Cold filter: 3-5 μm bandpass Si
Cooling: integral direct-attach Dewar, split Stirling cryocooler
Cooler lifetime: >4,000 h
Cool-down time (300-77 K): <10 min
NEDT: <0.025 K (Q=2×10^{14} ph/cm^2/s)
Digital acquisition resolution: 12-bits
Correction method: 1- and 2-point, using internal temperature-controlled flag, or external sources

Raytheon's Radiance thermal imaging camera

Power input: +19 to +35 V DC at 55 W, regulated to 3% 95/250 V, 47/63-Hz converter supplied
Power dissipation: <40 W nominal, <45 W peak
Dimensions: 112 × 262 × 183 mm
Weight: 5.5 kg approx
Integral processor: 32-bit floating point, 16 MFLOPS, 32 MOPS
Video interfaces: RS-170, RGBS (RS-343), NTSC (PAL Optional), S-Video (Separate Y/C)
Display resolution: 640 × 482 full screen, 512 × 464 image window
Display palette: 256 colours (from palette of 16 million)
Video synchronisation: external sync input
Data interface: Raytheon High-Speed Video Bus (HSVB) (4 MHz × 16-bit RS-422)
Remote control: RS-232
On-screen symbology: 16 colours from 16M palette
Program memory: 1 Mbyte PCMCIA card; 4 Mbyte card optional
Internal frame storage: 24 16-bit frames; 48 8-bit frames, with optional PCMCIA card
Normalisation memory: 4 each 256 kbyte tables
Lens control: driver circuits to support motorised zoom and focus-compatible lenses

Contractor
Raytheon Systems Company.

VERIFIED

Raytheon Sentinel

Type
Thermal imager (land).

Description
Sentinel is a hand-held infra-red thermal camera that operates in the long waveband and uses uncooled microbolometer focal plane array technology, licensed from Honeywell Technology Center. Sentinel's imaging sensor is an integrated circuit made up of 76,800 microresistors in a 320 × 240 array, each suspended over an integrated circuit multiplexer. Each detector is essentially a small thermistor integrated onto a microbridge structure that is less than 50 μm across, providing thermal isolation and mechanical support for each detector element. The array elements change temperature because of the incident infra-red radiation from the scene. The change in corresponding resistance is measured and used to generate a video image.

It is suited to a variety of applications including surveillance, industrial condition monitoring and search and rescue.

Sentinel comes equipped with a high-resolution black and white viewfinder display and an RS-170 video interface connector for driving an external monitor or video recorder. Automatic gain control and polarity selection are standard.

A standard camcorder battery allows up to 2 hours of continuous operation, a lithium chloride battery up to 10 hours.

Raytheon produces a version for industrial predictive maintenance, the Sentinel CMS Condition Monitoring System, which is portable and includes a laser sighting radiometer and a thermographic camera, Radiance 1t, which can be mounted on a tripod or bench.

Operational status
In production.

Specifications
Detector: uncooled microbolometer
Format: 320 × 240 array (76,800 elements)
Pixel size: 50 × 50 μm
Spectral band: 8-12 μm

Raytheon's Sentinel uncooled thermal imaging system

Thermal stabilisation: thermoelectric
NEDT: <0.1°C at 25°C
Standard lens: 50 mm f/0.7
Field of view (h × v): 18 × 14°
IFOV: 1 mrad
Optional lens: 100 mm f/0.7 (future option)
Focus range: 2 ft to ∞
Time to operational: <60 s
Frame rate: 30 Hz
Scene temperature range: –20 to 150°C
Video output: RS-170 monochrome (NTSC), PAL
Remote control: RS-232
Dynamic range: 65 dB
Battery: Sony NP-80 (or equivalent)
Optional battery: lithium chloride battery (16 Ah)
Power consumption: <6 W at 25°C
Battery life: 2 h at 25°C
Dimensions: 254 × 106.7 × 95.3 mm
Weight: 4.2 lb (incl battery)
Operating temperature: –20 to 45°C
Storage temperature: –40 to 70°C
Operating humidity: 0-95% non-condensing

Contractor
Raytheon Systems Company.

VERIFIED

Raytheon Small Unit Operations (SUO) sensor system

Type
Thermal imager (land).

Development
Raytheon Systems Company has been awarded a US$4 million contract from the US Air Force, on behalf of the Defense Advanced Research Projects Agency (DARPA), to develop a man-portable long-range infra-red sensor and laser range-finder for small unit operations.

The revolutionary 8-pound system, called the Small Unit Operations (SUO) sensor system, includes a third-generation visible-through-midwave infra-red sensor and an eyesafe laser range-finder. The contract contains an option to add a Global Positioning System (GPS) receiver/digital compass and real-time video transmission. The system is designed to significantly enhance the combat capability of small units, enabling them to engage and defeat much larger enemy forces.

Raytheon will design and build an SUO sensor system under terms of the 24-month contract, which is being managed by the US Air Force Space and Missile Systems Center and sponsored by DARPA. Potential value of the contract with options is US$5.5 million.

With the addition of a GPS receiver and a digital compass, which is used with the laser range-finder, units could pinpoint the location of targets to direct artillery or other weapons.

Operational status
In development.

Contractor
Raytheon Systems Company.

VERIFIED

Raytheon, Kollsman AN/TAS-6 NODLR

Type
Thermal imager (land).

Description
The AN/TAS-6 NODLR (Night Observation Device, Long-Range) provides night observation post capability to the soldier on the battlefield. It was developed as a successor to the AN/TAS-4. When used in conjunction with the AN/GVS-5 hand-held laser rangefinder this night sight can provide information to supporting artillery in degraded battlefield conditions. The complete system is known as the AN/UAS-11.

The AN/TAS-6, in conjunction with its tripod, the AN/GVS-5 hand-held laser range-finder and a boresight mechanism, can provide a two-person team with a portable, forward observation post which allows operating capabilities in the presence of degraded conditions such as smoke, haze, darkness, camouflage and foliage.

The AN/TAS-6 is based on the basic portable infra-red receiver used in MCTNS common modules and uses the same compressed air gas bottle and rechargeable battery pack as the other portable systems. This allows the use of the same logistics supply facilities to increase support flexibility.

In the AN/UAS-11, the AN/TAS-6 and the AN/GVS-5 are mounted together, with the AN/TAS-6 at the left. The biocular eyepiece is covered with a dual-eye shroud which provides visual security for the display. This display is green in colour and is produced by an image intensifier tube.

Operational status
The US Army has bought 623 of these devices and they are no longer in production. The US Army intends to replace the NODLR with the Long Range Acquisition Scout Surveillance System (LRAS[3]), currently under development by Raytheon Systems Company and DRS Technologies.

Specifications
Spectral band: 8-12 μm
Weight: 11 kg
Fields of view: 3.4 × 6.6°; 1.1 × 2.2°
Magnification: ×3; ×9
Resolution:
  Wide: 0.5 mrad
  Narrow: 0.167 mrad
Boresight accuracy: 0.1 mrad
Power
  Closed-cycle cooling: 30 W max
  Joule-Thomson cooling: 7.5 W max

Contractor
Raytheon Systems Company.

UPDATED

Raytheon, Kollsman MCTNS

Type
Thermal imager (land).

Development
Since 1978, US manufacturers under United States Defense Department sponsorship have developed a range of basic elements for night sight systems initially used with portable weapon systems, such as TOW and Dragon, and subsequently for surveillance. This system of modular equipment is known as the portable Common Thermal Night Sights (MCTNS) family of units. Raytheon designed the basic receiver, based on common modules.

Description
Portable common modules are assembled into a chassis to become the basic portable receiver. This provides the essential FLIR functions for all the portable night sight systems.

The modules are:

Mechanical scanner: collects infra-red (IR) energy emitted from a scene or object and displays visible energy from the LED array to the operator.
IR imager (SU-108): focuses energy collected by the scanner and transfers it to the detector/Dewar module.
Detector/Dewar (DT617): detects IR energy from the imager and processes the signals into electronic energy for the amplifiers. This is a 60-element array, 120- or 180-element arrays being available for other applications. The 60-element arrays feature closed-cycle or Joule-Thomson cooling systems. The 120-element system has a split-Stirling cooling system while the 180-element type employs linear cooling.
Preamplifiers/post-amplifiers: boost the low-level electrical signals from the detector array to a level sufficient to drive the LEDs. The 120- and 180-element arrays have a field rate of 60/s and a frame rate of 30 frames/s. Digital scan conversion is to RS-170 or CCIR standards.
LED array: changes the amplified detector signals into a visible image.
Visual collimator: projects the light from the LED display to the operator.
Video auxiliary control: provides the interface function between the controls and the post-amplifiers for such functions as brightness and contrast.
30 Hz scan/interface: provides electronic control for the scanner.

The MCTNS FLIR modules

Among the non-portable sight applications is the AN/VGS-2 tank thermal sight for the M60 MBT; the AN/AAS-36 IR detecting set and the AN/AAQ-9 FLIR.

Operational status

No longer in production. The system was produced by Raytheon and Kollsman Military Systems Division. By August 1992, Raytheon had delivered more than 30,000 common module systems; by November 1988, Kollsman had delivered more than 9,500. These are used by a variety of domestic and foreign customers, frequently being employed in the AN/TAS-4, AN/TAS-5 and AN/UAS-12 anti-armour missile sights as well as the G/VLLD and AN/TAS-6 NODLR systems.

Specifications

	60 element	120 element	180 element
Field of view			
Wide	3.4 × 6.8°	11.1 × 14.7°	11.1 × 14.7°
Narrow	1.1 × 2.2°	2.7 × 3.6°	2.7 × 3.6°
Resolution			
Wide	0.5 mrad	0.8 mrad	0.8 mrad
Narrow	0.167 mrad	0.2 mrad	0.2 mrad
NET	0.2°C	0.1°C	0.1°C
Cooling time	6 min	10 min	10 min

Contractors

Raytheon Systems Company.
Kollsman Inc.

UPDATED

Raytheon/DRS Horizontal Technology Integration (HTI) FLIR

Type

Thermal imager (land).

Description

The US Army's second-generation Horizontal Technology Integration (HTI) FLIR (Forward-Looking Infra-Red) programme is designed to develop and integrate second-generation FLIR common kits into existing and future armoured combat vehicle electro-optical sights. The M1A2 Abrams Survivability Enhancement Program (SEP) and the M2A3 Bradley Fighting Vehicle already incorporate the HTI FLIR systems. The programme develops A-kits which are specific to each candidate vehicle and include the integration and installation and the B-kit which is the common FLIR sensor suite.

In April 1997, Raytheon Systems Company (RSC) and the former Hughes Aircraft Company were each awarded multi-year annually funded contracts from the US Army Communications-Electronics Command (CECOM) to produce second-generation FLIR common kits (B-kits) for integration into these electro-optic sights:

M2A3 Bradley Commander's Independent Viewer (CIV)
M2A3 Improved Bradley Acquisition system
M1A2 Abrams Commander's Independent Thermal viewer (CITV)
M1A2 Abrams Gunner's Primary Sight Thermal Imaging System
Long Range Acquisition Scout Surveillance System (LRASS3)

The B-kit FLIR is being incorporated into other Army advanced Technology Demonstrator programmes - The Hunter Sensor Suite (HSS, Future Scout Cavalry System and Targeting Acquisition.

In 1998, DRS Technologies entered into an agreement with Raytheon to purchase a portion of both its second-generation business and ground electro-optical systems (Ground E-O) and a portion of its focal plane array business. The companies are now collaborating on HTI and other programmes such as LRAS3.

HTI FLIR is stated to increase detection ranges by more than ×1.5 and identification ranges by ×2 over first-generation systems and thus be compatible with the effective range of tank gun/missile systems. The horizontal integration of the B-kit FLIR across the fleet of fighting vehicles will allow all elements to see the same battlefield, while the large quantity of common kits is intended significantly to lower the Army's life cycle costs.

The FLIR is an 8 to 12 μm scanning system which uses the Standard Advanced Dewar Assembly II (SADA II) (qv) detector which has a 480 × 4 Cadmium Mercury Telluride (CMT) focal plane array. The systems have standard RS-170 video outputs so that standard TV monitors and recording equipment can be used. The incorporation of digital video output ports allows for integration of target trackers and digital image compression hardware. Digital image compression allows FLIR images to be transmitted from the battlefield to remote-command centres.

The B-kit FLIR is being incorporated into other Army Advanced Technology Demonstration programmes - the Hunter Sensor Suite (HSS) and Targeting Acquisition.

The HTI contract was awarded by the US Army's Communications-Electronics Command (CECOM), Fort Monmouth, New Jersey. The programme is being managed by the Project Manager, Night Vision/Reconnaissance, Surveillance and Target Acquisition (NV/RSTA) at US Army NVESD, Fort Belvoir, Virginia.

Operational status

RSC delivered 242 Abrams CITV systems, 130 B-kits and portions of 378 B-kits with options for more units by February 2001.

In January 1999, DRS Sensor Systems was contracted to provide HTI systems for the sighting systems of the Abrams M1A2 enhancement package and M2A3 Bradley family of IFVs.

Specifications

FLIR
Detector: 480 × 4 CMT
Spectral range: 8-12 μm

Contractors

DRS Sensor Systems Inc.
Raytheon Systems Company.

UPDATED

ROMZ compact thermal imager

Type

Image intensifier monocular.

Description

ROMZ is developing a compact portable imager operating in the 8-14 μm waveband. It is suitable both for surveillance in total darkness or, in industrial use, to detect heat leakage from buildings or plant.

Operational status

Under development.

Specifications

Dimensions: 75 × 85 × 150 mm
Weight: 0.65 kg
Field of view (H × V): 9.5 × 4.5°
Magnification: ×2
Thermal resolving power: 0.2 K
Wavelength: 8-14 μm
Frame format: 64 × 128 elements
Frame rate: ≥30/s
Display colour: red
Power: 3 V BLIK-2 battery

Contractor

ROMZ, Rostov Optical Mechanical Plant (JSC).

VERIFIED

Saab Bofors Dynamics thermal imager

Type

Thermal imager.

Description

The Saab Bofors Dynamics thermal imager is a long-waveband system with serial-parallel scanning and advanced digital signal processing to produce real-time images. The output video signal is compatible with standard TV monitors and video recorders. Depending on the application, the thermal imager can be equipped with various telescopes.

The thermal imager can be integrated into any type of anti-aircraft and surface-to-surface fire-control system.

Operational status

In production and in service. The Saab Bofors Dynamics thermal imager equips the Swedish Army Combat Vehicle 90, the RBS 90 missile system, the Saab Bofors 9LV200 Mk 3 fire-control system for the Swedish Navy 'Göteborg' class coastal corvettes and the Saab Bofors 9LV453 fire-control system for the Australian/New Zealand 'Anzac' class frigate programme.

Specifications

Spectral band: 8-12 μm
Field of view
Narrow: 3.2 × 2°
Wide: 12 × 8°
Instantaneous field of view
Narrow: 0.12 mrad
Wide: 0.45 mrad
NETD: <0.1 K
Detector: SPRITE
Alert time: <45 s
Video output: 50 Hz 1:2 interlaced CCIR standard
Power: 28 V DC/250 W
Total weight: 30 kg
Dimensions: 607 × 346 × 237 mm

The Saab Bofors Dynamics thermal imager

Contractor
Saab Bofors Dynamics.

VERIFIED

Saab Bofors Dynamics thermal imaging modules

Type
Thermal imager (land).

Description
Saab Dynamics thermal imager modules are used in a variety of army and navy systems. The scanner module has an eight-element SPRITE detector and offers high geometric and temperature resolution. The cooling module includes a closed-cycle cooling engine and control electronics. The electronics module provides scan conversion to 50 or 60 Hz standard video, manual and automatic gain and level control, advanced image filtering, electronic reticle generation and ×2 electronic image magnification.

Operational status
In production and in service in army and navy systems, including Swedish Army: Combat Vehicle 90 (anti-aircraft version), RBS 90 missile launcher; Swedish Navy: fire-control system for 'Stockholm' and 'Göteborg' class coastal corvettes and 'Anzac' class frigates of the Australian and New Zealand navies.

Specifications
Detector: 8-element SPRITE
Field of view: 49 × 32° (scanner)
Cooling: Closed-cycle
Cooling-down time: <3 min typical
Scan conversion: to 50 or 60 Hz
Electronic image conversion: ×2
Dimensions
  Scanner module: 137 × 126 × 160 mm
  Cooling module: 150 × 110 × 145 mm
  Electronics module: 206 × 135 × 135 mm
Weight
  Scanner module: 3.8 kg
  Cooling module: 3.5 kg
  Electronics module: 2.5 kg

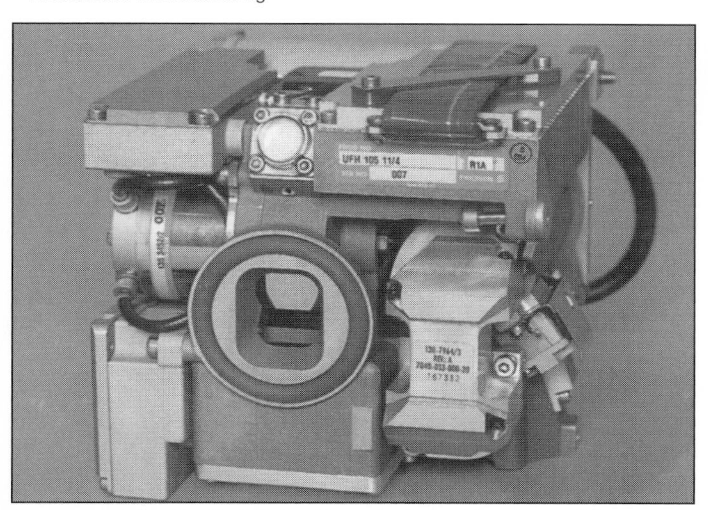

Scanner module of the Saab Bofors Dynamics thermal imaging modules 0006069

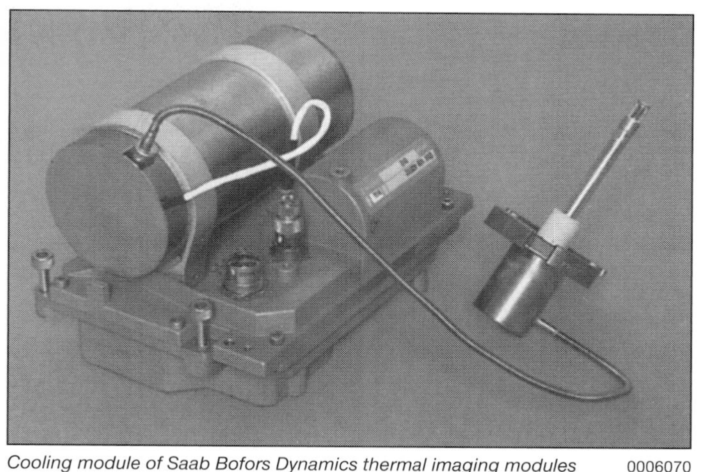

Cooling module of Saab Bofors Dynamics thermal imaging modules 0006070

Contractor
Saab Bofors Dynamics.

SAGEM Athos thermal imager

Type
Thermal imager (land).

Description
The SAGEM Athos thermal imager is based on the French SMT thermal imaging common modules. Athos can be integrated into weapon systems for observation, surveillance, detection and fire-control applications. It is fitted into the gunner's day/night sight of the Leclerc main battle tank and the Namut thermal sight (qv) for the BMP-3.

It consists of two modules, a scanning head and an electronics unit, and has CCIR compatible video output. The camera operates in the 8 to 12 μm spectral band. It has two fields of view with optional objective lenses including two zoom designs. Options include automatic controls; electronic magnification; frame integration; automatic harmonisation; electronically generated reticule and warning symbols.

Operational status
In production. In service with the French Army (Leclerc MBT) and UAE (Leclerc and BMP-3, as part of Namut thermal sight).

Specifications
Dimensions
  Scan head unit: 445 × 200 × 316 mm
  Electronics unit: 295 × 181 × 155 mm
Weight
  Scan head unit: 25 kg
  Electronics unit: 6 kg
Spectral band: 8-12 μm

The Athos thermal imaging system fitted to BMP-3 infantry combat vehicle of the United Arab Emirates Land Forces (Christopher F Foss)

Fields of view
Narrow: 1.9 × 2.9°, zoom: 0.95 × 1.45°
Wide: 5.7 × 8.6°, zoom: 2.85 × 4.3°
Number of pixels per line: 400
Pupil diameter: 150 mm
Cool down time: <10 min
Power supply: 28 V DC

Contractor
SAGEM SA, Aerospace and Defence Division.

UPDATED

SAGEM IRIS

Type
Thermal imager (land).

Development
SAGEM started autonomous development of the IRIS new-generation family of high-performance thermal imagers in 1988. The modular design has enabled IRIS to be produced in several versions for land, navy and aeronautical applications (French and foreign). Several IRIS projects are in progress, concerning tanks, anti-tank and anti-aircraft weapon systems, helicopters and aircraft, observation and surveillance systems.

Description
IRIS is an IRCMOS thermal imager which operates in the 8 to 12 μm bandwidth.

Its main characteristics are: up to three switchable fields of view; automatic gain and offset control; polarity selection; zoom ×2; athermalised focusing; extended Built-In Test Equipment (BITE); boresight alignment; digital image enhancement and hotpoint detection and tracking. The system is compatible with the CCIR standard video output.

Operational status
In production for 20 countries.

Navy: submarines (electro-optic masts and periscopes) and surface ships for Belgium, Denmark, France and Norway.
Air: Mirage 2000, Mirage III Mirage V and UAV programmes; Strix sight for Tiger HAP/HAC and Nightown sight for the Rooivalk; FLIR of the NH90-NF helicopter.
Land: Leclerc commander's and gunners sights, CRII Desert and CV90. Anti-aircraft sights: Skyshield and Skyguard.

Specifications
Spectral band: 8-12 μm
Detection module: integrated detector/Dewar microcooler device
Detector: 288 × 4 elements IRCMOS focal plane array CMT
Cooling: closed-cycle Stirling microcooler
NETD: T<0.02°C
Video output: CCIR or RS-170, digital 9-bits or 12-bits
Resolution: 576 × 768 pixels

Contractor
SAGEM SA, Aerospace and Defence Division.

UPDATED

Image taken with the SAGEM IRIS thermal imager **NEW**/0547487

The SAGEM IRIS thermal imager 0129322

SAGEM LUTIS uncooled thermal imager for observation and firing

Type
Thermal imager (land).

Description
LUTIS Mk 2 is a high-performance thermal weapon sight integrating advanced electronics coupled with the latest generation of bolometric focal plane array detector. The technical solutions implemented in LUTIS Mk 2 give this sight qualities such as: light weight, small volume, low power consumption and optronic performance.

LUTIS Mk 2 is part of a family of uncooled thermal imagers which includes: a driver-viewer, a standalone camera for surveillance and fire control, and so on.

Operational status
Available.

Specifications
Weight: 1.1 kg
Spectral range: 8 to 12 μm
Detector: micro-bolometer, 320 × 240 pixels
Optics: standard 62 mm F/1 (15° Horizontal FOV) (Other FOVs available)
Display: 640 × 480 LCD
Controls: easy operation through a 3-button menu control
Video Output: RS-170 or CCIR
Power supply: external power and standard primary/secondary batteries
Power consumption: 2 W typical, battery life up to 8 h

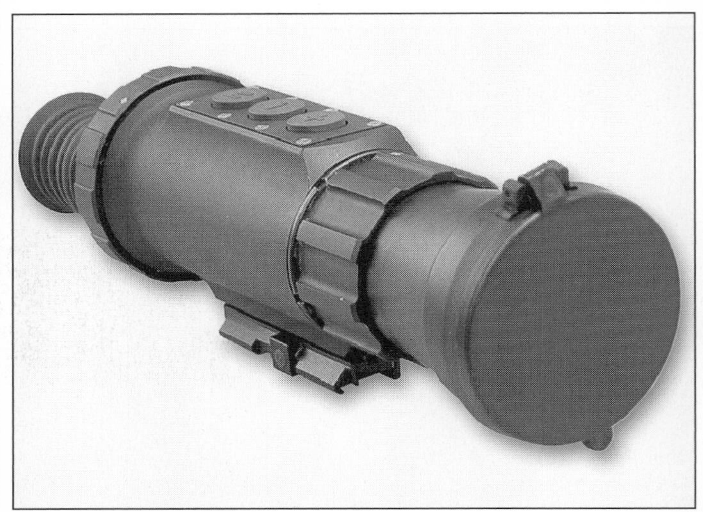

SAGEM's LUTIS thermal sight 0129321

Connector: single connector for video output, external power and remote control (RS serial link)
Environmental: fully militarised according to MIL-STD standard. Product qualified on various weapons

Contractor
SAGEM SA, Aerospace and Defence Division.

UPDATED

SAGEM MATIS hand-held third-generation thermal imager

Type
Thermal imager (land and naval).

Description
MATIS Hand Held (H-H) is a member of the SAGEM MATIS family of third generation thermal imagers, based on the latest generation of cooled FPA detectors, operating in the 3-5 μm waveband.

Extremely lightweight and compact, this hand-held thermal imager is intended for day and night short- to long-range observation and forward reconnaissance. It is particularly well adapted to highly mobile operators such as infantry and platoon leaders, and for artillery target acquisition.

The MATIS H-H is a one-piece, fully autonomous camera, integrating ergonomic biocular display, with controls and battery pack. It features low power consumption and operates from lithium or rechargeable batteries.

Operational status
In volume production for over 10 countries.

Specifications
Weight: 2.5 kg
Field-of-view
wide: 9 × 6°
narrow: 3 × 2°
electronic zoom: 1.5 × 1°
Detector: focal plane array
NEDTD: 30 mK
Spectral band: 3-5 μm
Video output: CCIR/RS-170
Reticle: aiming mark
Operating time: >5 h
NETD: 20 mK

Contractor
SAGEM SA, Aerospace and Defence Division.

UPDATED

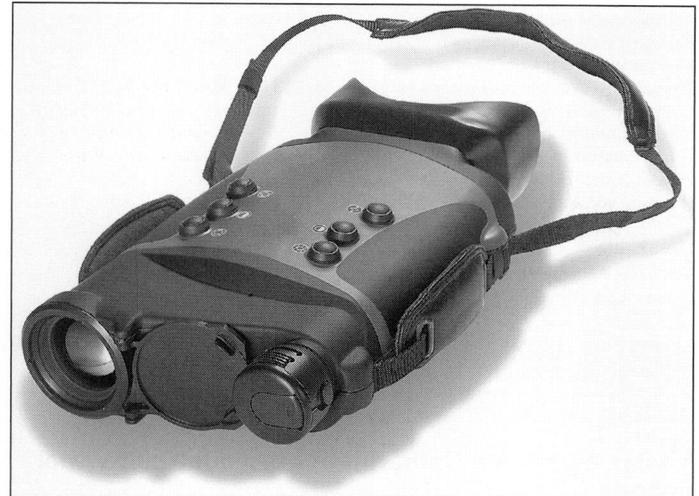

SAGEM's MATIS hand-held version 0129320

SAGEM MATIS man-portable third-generation thermal imager

Type
Thermal imager (land and naval).

Description
MATIS man-portable is a member of the SAGEM MATIS family of third-generation thermal imagers based on the latest generation of cooled FPAs and operating in the 3 to 5 μm waveband. Light-weight and compact, MATIS MP is intended for day and night short- and long-range observation posts, gun-mounted firing directors and missile manpads. Based on a dual field of view telescope, it also includes an ergonomic binocular display with integrated controls, which as a whole can be easily separated from the thermal imager.

It is powered directly from its own battery or from an external source.

Operational status
In production for several countries:
- Mistral man-portable air defence systems
- 25 mm gun turret
- long-range infantry observation post.

Specifications
Weight: <6.5 kg
Fields of view
wide: 12 × 8°
narrow: 2.5 × 1.7°
Detector: focal plane array
NETD: 30 mK
Spectral band: 3-5 μm
NETD: 20 Mk
Video output: CCIR
Reticle: aiming mark
Operating time: >7h in ambient temperature

Contractor
SAGEM SA, Aerospace and Defence Division.

UPDATED

The SAGEM MATIS man-portable version 0129317

SAGEM MATIS STD/LR third-generation thermal imager

Type
Thermal imager (land, air, naval).

Description
MATIS STD/LR is a member of the SAGEM MATIS family of third-generation thermal imagers based on the latest generation of cooled FPAs and operating in the 3 to 5 μm waveband. They are intended for day and night short- and long-range observation systems, gun-mounted firing directors and missile weapon systems. The different versions are based on dual field of view and options include: dedicated control panel, TV monitor and image stabilisation.

Operational status
In production for over 10 countries and applications include ICVs, border surveillance, anti-aircraft weapon systems, trajectography and naval fire directors.

Specifications
Detector: focal plane array
NETD: 30 Mk
Spectral band: 3-5 μm
NETD: 20 mK
Video output: CCIR and digital video
Power consumption: 28 W in ambient temperature
Power supply: external 18 to 32 V
Remote control: via RS-422 link

	MATIS STD	MATIS LR
Field of view		
wide	12 × 8°	6.2 × 4.2°
narrow	2.5 × 1.7°	1.3 × 0.9°
Electronic zoom	×2	×2
Weight	5 kG	5.5 kg
Overall dimensions	297 × 136 × 146 mm	390 × 152 × 146 mm

Contractor
SAGEM SA, Aerospace and Defence Division.

UPDATED

The SAGEM MATIS LR third-generation thermal imager 0129318

SAGEM MATIS STD thermal imager 0129316

Siemens, Vectronix FORTIS

Type
Thermal imager.

Description
The Forward Observation and Reconnaissance Thermal Imaging System (FORTIS) has been jointly developed by Siemens Switzerland Ltd and Vectronix (formerly Leica Vectronix) to allow long-term observation with minimum operator fatigue. It is designed for both military and paramilitary use with a large and selectable field of view, high target resolution and image inversion.

Operator fatigue is reduced by means of green image display and a biocular viewing lens. The portable system is ready for use within 45 seconds and has a high degree of operational autonomy. FORTIS operates in the 8 to 12 μm range and has a Joule- Thomson cooling system.

The FORTIS observation system

Targets can be detected at ranges up to 6 km, recognised at 2 km and identified at 1 km. With the long-range telescope these ranges are doubled. An external video monitor, closed-cycle cooling systems and remote control are optional extras.

Operational status
In production since 1994. In service with the Swiss Army.

Specifications
Dimensions: 460 × 240 × 300 mm
Weight: 13 kg
Spectral band: 8-12 μm
Cooling: Joule-Thomson (77 K)
Temperature resolution: 0.1 K
Fields of view: 18.5 × 9.5°; 4.5 × 2.3°
Resolution: 0.8 mrad; 0.2 mrad
Range (NATO target in standard atmosphere, 293 K):
  Detection: 6,000 m
  Recognition: 2,000 m
  Identification: ≥1,000 m
Power consumption: <4 W
Nominal voltage: 7.2 V

Contractors
Vectronix AG.
Siemens Switzerland Ltd.

UPDATED

Thales Optronics Catherine-FC fire-control thermal imager

Type
Thermal imager (land).

Description
Catherine-FC is a second-generation thermal imager developed by Thales Optronics and principally intended for AFV fire-control systems (gunnery and missile guidance) and surveillance equipment. The system provides a 24 hour passive observation capability in all light conditions, from total darkness to full sunlight, utilising the thermal radiation emitted by all objects. Catherine-FC is based upon the cooled technology 288 × 4 detector, operating in the 8-12 μm band, developed by Sofradir.

The main features compared with earlier models are:
(a) Drastic reduction of weight, volume and power consumption.
(b) Range performance improvement especially under poor weather conditions.

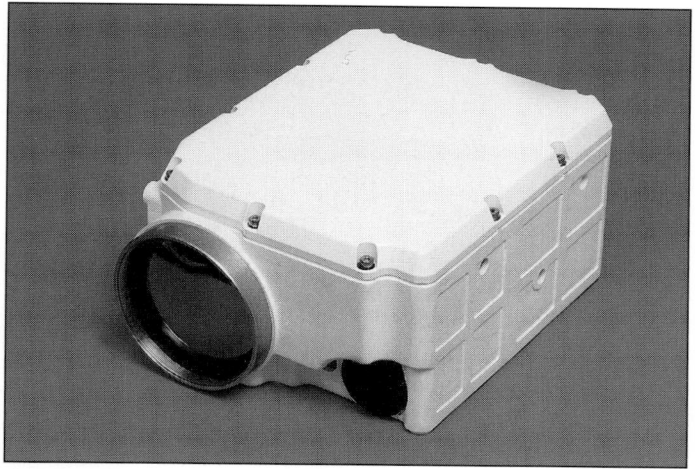

Catherine-FC fire-control thermal imager 0053930

For fire control a reticule is displayed on the video signal. The reticule can be moved for boresighting operations and according to the information delivered by the Fire-Control Computer on the serial data link.

Catherine-FC is a single LRU equipment which can be fitted on any type of land vehicle (armour fighting vehicle, light reconnaissance vehicle and so on).

Operational status

In production.

Specifications

Spectral band: 7.5-10.5 μm
Cooling: integrated microcooler, rotary stirling type
Video standard: CCIR (Europe) or SMPTE 170 M (USA)
Field of view
WFOV (H × V): 9 × 6.7° (CCIR)/7.5 × 5.62° (SMPTE 170 M)
NFOV (H × V): 3 × 2.2° (CCIR)/2.5 × 1.87° (SMPTE 170 M)
Electronic magnification (H × V): 1.5 × 1.2° (CCIR)/1.25 × 0.94° (SMPTE 170 M)
Practical range (AFV targets, standard atmospheric conditions):

	Detection	Recognition
Wide FOV	4.5 km	N/A
Narrow FOV	11.5 km	4.5 km

Weight: <5.5 kg
Power supply: 20 to 30 V DC on-board power supply
Electrical consumption: <25 W (operating)
Remote control: 2 RS-422, 1 RS-232 serial data links
Dimensions (W × H × D): 250 × 240 × 120 cm
Weight: <5.5 kg
Environment: qualification for military wheeled and tracked vehicles (vibrations) and EMI/EMC standards
Maintainability: Built-In Test Equipment (BITE)

Contractor

Thales Optronics.

VERIFIED

Thales Optronics Catherine-GP high-performance general purpose thermal imager

Type

Thermal imager (land).

Description

The Catherine-GP is a second-generation thermal imager developed by Thales Optronics, intended mainly for long-range surveillance and fire-control systems (gunnery and missile guidance). Catherine-GP provides a 24-hour passive observation capability in all light conditions, from total darkness to full sunlight and bad weather conditions and, is very effective on stealth targets. Catherine-GP is based upon the cooled technology 288 × 4 detector operating in the 8-12 μm band developed by 'Sofradir' and the high-performance SYNERGI thermal imager modules.

The main features of Catherine-GP compared with earlier models are:
(a) reduction of weight, volume and power consumption
(b) improved range performance.

For fire control, a reticule is displayed on the video signal. The reticule can be moved for boresighting operations and according to the information delivered by the fire-control computer on the serial data link.

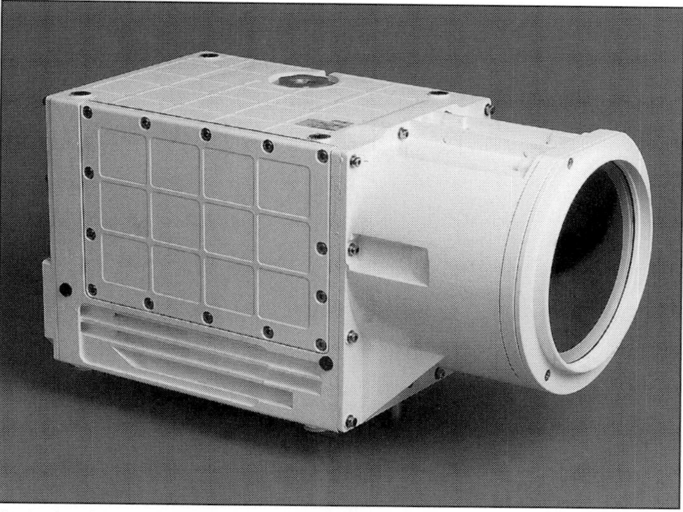

Catherine-GP high-performance thermal imager 0053931

Catherine-GP is a single LRU equipment which can be fitted to any type of land vehicle (AFVs, light reconnaissance vehicles and so on).

Operational status

In production.

Specifications

Spectral band: 7.5-10.5 μm
Cooling: linear split-stirling cooler
Analogue video standard: CCIR (Europe) or SMPTE 170 M (USA)
Digital video: 8 and 12 bits
Field of view
WFOV (H × V): 9 × 6.75 (CCIR)/7.5 × 5.62 (SMPTE 170 M)
NFOV (H × V): 3 × 2.25 (CCIE)/2.5 × 1.87 (SMPTE 170 M)
Electronic magnification (H × V): 1.5 × 1.2 (CCIR)/1.25 × 0.94 (SMPTE 170 M)
Practical range (AFV target, standard atmospheric conditions):

	Detection	Recognition
Wide FOV	6 km	n/a
Narrow FOV	15 km	6 km

Dimensions (W × H × D): 415 × 210 × 182 cm
Weight: <12.5 kg
Power supply: 20 to 30 V DC on-board power supply
Electrical consumption: <75 W (operating)
Remote control
2 serial data links RS-422
1 serial data link RS-232
Environment: qualification for military wheeled and tracked vehicles (vibrations) and EMI/EMC standards
Maintainability: Built-In Test Equipment (BITE)

Contractor

Thales Optronics.

VERIFIED

Thales Optronics Cobra

Type

Thermal imager (land).

Description

The Cobra is an autonomous, lightweight, hand-held, thermal imager operating in the 8 to 12 μm spectral band. The system incorporates compact optics, scanning system, cryogenic cooler, electronics and CRT into one package. The unit can be used for a variety of general surveillance purposes, or can be integrated with other systems such as surveillance systems, short-range radars, lightweight weapons, light turrets, laser range-finders, target designators, goniometers and fire-control systems.

The modular design allows for the replacement of faulty subassemblies. Faulty subassembly location is performed using a Second-Line Test Adapter (SLTA).

The Thales Optronics Cobra thermal imager integrated with a laser range-finder

The Cobra thermal imager has two fields of view, wide and narrow. The narrow field of view is achieved using the electro-optic zoom and allows close-up viewing without changing the telescope.

The Cobra system is currently offered in four optical configurations: Cobra ×3; Cobra ×4.9; Cobra ×7; Cobra Dual FOV (×7/×3).

Operational status

Production as required. In service with the Royal Netherlands Army and other NATO countries' armies.

Specifications

	Cobra ×3	Cobra ×4.9	Cobra ×7	Cobra Dual FOV (×7/×3)
Weight (excl battery)	3.7 kg	3.7 kg	4.1 kg	4.6 kg
Wide FOV (h × v)	9.3 × 7°	5.7 × 4.3°	4 × 3°	9.3 × 7°; 4 × 3°
Narrow FOV (h × v)	4.6 × 3.5°	2.8 × 2.1°	2 × 1.5°	4.6 × 3.5°; 2 × 1.5°
Recognition range (NFOV)	1 km	1.6 km	2.2 km	1; 2.2 km
Detection range (WFOV)	2.4 km	3.7 km	5 km	2.4; 5 km

Dimensions: 20 × 19 × 9 cm (×4.9 configuration)
Spectral range: 8-12 μm
Focus ranges: 25 m to ∞ (×4.9 configuration)
Dioptre range: +4 to –4 dioptres
Power source: 23.4 V battery pack; or 18-36 V, external vehicle power (unregulated)
Power consumption: 18 W (typical)
Battery life: up to 8 h (lithium battery)
Reticle pattern: on customer request
Video display: CRT display. A remote display adaptor is available with an STD composite video output.

Contractor

Thales Optronics, Canada.

VERIFIED

Thales Optronics LITE

Type

Thermal imager.

Description

The LITE family of hand-held thermal imagers is a development based upon stretched UK Class I Thermal Imaging Common Modules (TICM I). They operate in the 8 to 13 μm band and were designed to complement the company's existing range of imagers but have additional features that allow them to be employed in a wider variety of uses and applications such as sights for anti-tank and air defence missiles and guns.

The detector consists of an array of CMT photoconductive elements which are cooled either by a Joule-Thomson cooler or by an integrated Stirling cycle microcooler. The user has a choice of direct or indirect viewing with display on a micromonitor or TV monitor to CCIR or RS-170 standard.

A range of interchangeable telescopes can be fitted to all versions of the equipment and this includes a selection of single- and dual-field of view options. LITE imagers may be powered either by internal batteries or from a suitable remote power source.

LITE: the direct view microcooler version

Operational status

Deliveries of production units began in 1993. Customers in 26 countries have purchased LITE.

Specifications

Spectral band: 8-13 μm
Detector: CMT
Cooling: Joule-Thomson or integrated Stirling cycle microcooler
Field of view options (direct view version)
single 12.5 × 6.3°
single 6.2 × 3.2°
single 2.8 × 1.4°
dual 6.3 × 3.2° and 18.8 × 9.6°
dual 4.1 × 2.1° and 12.1 × 6.2°
Dimensions (with 12.5 × 6.3° single FOV telescope)
Microcooler version: 350 × 135 × 185 mm
Weight (with 12.5 × 6.3° single FOV telescope)
Microcooler version: 3.4 kg (excl display for IDV version)
Power consumption: microcooler version: 8 W

Contractor

Thales Optronics.

VERIFIED

Thales Optronics Sophie new generation handheld binoculars/thermal imager

Type

Thermal imager (land).

Description

SOPHIE is a hand held combination of binoculars and a thermal camera principally intended to be used for handheld or remote-controlled viewing for surveillance tasks by day, by night, and in bad weather conditions. SOPHIE is a new generation equipment based upon a cooled 288 × 4 detector developed by SOFRADIR, operating in the 8-12 μm band. The thermal image provided can be displayed either on the integrated micro monitor when SOPHIE is used as thermal binoculars or on an external monitor when used as a thermal camera. SOPHIE can be used as individual autonomous equipment or powered and remote-controlled by a distant control station. Power is supplied through batteries, rechargeable batteries, AC or DC adaptors.

Operational status

In production.

Specifications

Dimensions (W × H × D): 25 × 11 × 31 cm
Weight: 2.4 kg without battery
Spectral band: 7.5-10.5 μm
Cooling: integrated micro-cooler, rotative stirling type
Cooling down time: 5 min
Video standard: CCIR (Europe) or SMPTE 170 M (USA)
Field of view:
WFOV (H × V): 8 × 6° (CCIR)/6.7 × 5° (SMPTE 170 M)
NFOV (H × V): 4 × 3° (CCIR)/3.3 × 2.5° (SMPTE 170 M)
Electronic magnification (H × V): 2 × 1.5° (CCIR)/1.6 × 1.25° (SMPTE 170 M)
Magnification:
WFOV: ×3.2 (CCIR)/×3.8 (SMPTE 170 M)
NFOV: ×6.4 (CCIR)/×7.6 (SMPTE 170 M)
Electronic magnification: ×12.8 (CCIR)/×15 (SMPTE 170M)

The new generation Sophie binoculars/thermal camera 0053929

Typical range performance

Target	FOV	detection	recognition
man	wide FOV	2.5 km	
	narrow FOV	4.5 km	1.6 km
vehicle	wide FOV	5 km	
	narrow FOV	9 km	3.5 km

Power supply: 15 V DC
External supply: 11.5 V to 30 V DC (adaptor), 220 V AC (converter + adapter)
Electrical consumption: 8 (standby), 15 W (operating)
Endurance: 4 h with LiMn02 battery
Maintainability: Built-In Test Equipment (BITE)

Contractor
Thales Optronics.

VERIFIED

Thales Optronics STAIRS C thermal imager

Type
Thermal imager.

Development
STAIRS (Sensor Technology for Affordable IR Systems) is a UK MoD programme in collaboration with Thales Optronics and BAE Systems, to develop second-generation thermal imaging system technology to replace the UK TICM (Thermal Imaging Common Modules) used in a range of equipment. The STAIRS C detector is a cooled Cadmium Mercury Telluride (CMT) device that is intended for the PIRATE infra-red search and track system for the Eurofighter Typhoon.

The contract for production development of STAIRS C was awarded to Thales Optronics in 2000.

Description
STAIRS contains the scanning very long linear 768 × 8 element photovoltaic CMT array developed by BAE Systems for STAIRS C. The features of STAIRS C include advanced search and track processing, high-resolution SXGA and standard TV display formats, EOPM compatibility, automatic control and comprehensive graphics, zoom and freeze frame.

Operational status
In production for the Thermal Sighting System for the British Army's Starstreak Self-Propelled High Velocity Missile system.

Specifications
Weight: 20 kg
Dimensions
Sensor head: 392 × 200 × 142 mm
Electronics: 220 × 200 × 142 mm

Telescope options
Dual: 3.6 × 2.2° and 20 × 12° , aperture 180 mm
Dual: 5 × 3° and 17.5 × 10.5°, aperture 110 mm
Sightline stability: ≤0.1 mrad
Spectral band: 8-9.4 μm
Detector: 768 × 8 CMT diode array
Cooling: Integrated Detector Cooler Assembly
Displays
SXGA 1280 × 768
625 line 50 Hz (CCIR 1)
525 line 60 Hz (RS-170)
Control: automatic with manual override
Power: 120 W
Environment: specified for rotary and fixed-wing aircraft and tracked vehicles
MTBF: 5,000 h

The STAIRS C thermal imager, 110 mm aperturer 0053836

Contractor
Thales Optronics.

UPDATED

Thales Optronics, LION observation night sight

Type
Thermal imager (land).

Description
The LION is a hand-held, lightweight, uncooled thermal imaging viewer for day and night use. It has been designed as an observation sight for infantry soldiers, reconnaissance units, border surveillance, special forces and peacekeeping during day, night and conditions of poor visibility.

LION is based on a BAE Systems 256 × 128 element uncooled detector array and is fitted with a ×3 magnification optical system, providing a 10 × 5° field of view. The system is claimed to detect a vehicle at a range of more than 2 km. Imagery is presented on a binocular CRT display. As the LION is uncooled, start-up is in less than 10 seconds.

Experience from the field gives LION a good reputation for reliability.

Operational status
In production. In service with the Royal Netherlands Army, other NATO countries and Armed Forces, and law enforcement agencies worldwide.

The LION hand-held uncooled thermal imager 0005499

Specifications

Weight: 1.8 kg
Dimensions: 100 × 200 × 240 mm
Magnification: ×3, ×6
Field of view (H × V): 10 × 5° and 5 × 3.5° (electronic zoom)
Focus range: 8.5 m to ∞
NETD: 100 mK
Spectral band: 8-12 μm
Detector: 256 × 128 element lead scandium tantalate (PST), microscanned ×2
Display type: binocular, CRT
Reticle: aiming mark
Start-up time: <10 s
Acoustic noise: inaudible at 2 m
Interface options: remote control, external video, CCIR
External power: 24 or 12 V
Power consumption: 7 W
Battery type: Primary lithium, rechargeable Ni/Cd. Alkaline or Ni-Mh
Operating time: 10 h on lithium battery

Contractor

Thales Optronics, Delft, The Netherlands.

UPDATED

Zeiss Optronik Ophelios second-generation thermal imaging cameras

Type

Thermal imager.

Description

The Ophelios camera system comprises a set of basic modules that can be combined with dedicated assemblies for example, afocal, mount, control unit, monitor, to construct an infra-red imaging system.

The modules which make up the system are:

(a) scanner - a polygon scanner with six surfaces for both scanning directions and distortion-free scanning
(b) infra-red imaging and thermal reference - a thermalised infra-red imager which features a minimum number of optical components. The imager features an intermediate image plane and provides virtual diffraction-limited performance on axis (image centre). The thermal reference is used for scene related adaptation of the operating point and compensation of the variance of the detector characteristics
(c) detector/cooler assembly - this comprises 96 × 4 detector Cadmium Mercury Telluride (CMT) elements that are sensitive in the 7.5 to 10.5 μm spectral band. An integrated cooler Dewar ensures a unit which is optimised for minimum power consumption
(d) proximity electronics - these are integrated into the detector/cooler assembly and generate all clock impulses and DC voltages required for the operation of the IR multiplexer system. In addition they synchronise the IR multiplexer and the processes taking place in the electronics with the synchronising signals produced by the external system electronics
(e) MUX/ADC - this is a highly integrated electronics module which converts analogue signals to digital signals and feeds them to the system electronics
(f) system electronics - these perform image homogeneity, reformatting and geometry correction. Using a bus, they also ensure the communication with a control unit and the power supply of the entire system.

Operational status

In production and in service. Ophelios modules are used in the following systems - STN ATLAS Elektronik and Zeiss Optronik: PERI-ZTWL, PERI-R17A2 gunner's sights; STN ATLAS Elektronik: BAA observation system, BAS 2000 airborne observation system, ISOS 2000 UAV payload; Zeiss Optronik: WBG 96 × 4 thermal imager for the Brevel drone; Avimo: KITOS gunner's sight.

In Germany the platforms/applications employing Ophelios include maritime patrol aircraft, Sea King naval helicopters, patrol boats, frigate multisensor platforms and border surveillance systems.

Specifications

Spectral band: 7.5-10.5 μm
Detector: 96 × 4 CMT elements
Cooling: split-Stirling closed-cycle cooler with integrated Dewar
Fields of view (examples): 3.3 × 4.4° and 11 × 15°, 2.3 × 3.0° and 7.5 × 10°, Zoom - 3.6 × 4.8° to 29 × 38°
Number of lines: 576
Pixels per line: 768
Scan frequency: 25 Hz
Display: CCIR TV monitor
Power supply: 18-32 V DC

Contractor

Zeiss Optronik GmbH.

VERIFIED

AIRBORNE SYSTEMS

Air-launched missiles
Air-to-air missiles
Air-to-air guns
Air-to-surface missiles and munitions

Electro-optic countermeasures
Electronic countermeasures
Missile warners
Laser warners

Ground attack
Integrated systems – fixed-wing
Integrated systems – helicopter
Targeting sights
Laser range-finders

Flight aids
Laser systems
Pilot's thermal imagers
Pilot's goggles

Observation and surveillance
Air interception
Turret sensors
Maritime sensors
Unmanned aircraft sensors
Reconnaissance systems
Thermal imagers

AIRBORNE SYSTEMS – SECTION SUMMARY

Air-launched missiles/Air-to-air guns

This section includes electro-optic systems reported as deployed on air platforms or developed for airborne applications. Systems are grouped in the following sub-sections according to their type:

Air-to-air missiles
Air-launched anti-air missiles with a laser or scanning or imaging infra-red seeker in at least one variant of the missile class or with an electro-optic fire-control system or an optional electro-optic adjunct to a radar fire-control system.

Air-to-air guns
Air-to-air guns having an electro-optic adjunct.

Air-to-surface missiles and munitions
Air-launched anti-surface missiles and munitions with a laser or scanning or imaging infra-red seeker in at least one variant of the missile class or with an electro-optic fire-control system or an optional electro-optic adjunct to a radar fire-control system.

Electro-optic countermeasures

Electronic countermeasures
Airborne infra-red jammers or laser deception systems.

Missile warners
Airborne missile approach warners that include an electro-optic detection system.

Laser warners
Airborne laser warning systems or defensive aids systems which contain such systems.

Ground attack

Integrated systems – fixed wing
Ground attack systems for fixed-wing aircraft which incorporate target acquisition, laser designation and weapon delivery systems. See also targeting sights, laser range-finder/designators and pilot's thermal imagers in this section for closely related systems and components.

Integrated systems – helicopter
Ground attack systems for helicopters which incorporate target acquisition, laser designation and weapon delivery systems. See also targeting sights, laser range-finder/designators and turrets in this section for closely related systems and components.

Targeting sights
Ground attack systems for fixed-wing aircraft or helicopters that do not contain a laser designation or weapon delivery control system. See also pilot's thermal imagers and turrets for closely related systems.

Laser range-finders
Laser range-finders and or designators used on airborne platforms as part of ground attack systems.

Flight aids

Laser systems
A variety of laser systems used as altimeters, collision avoidance devices, landing guidance aids, wind shear detectors and mapping systems.

Pilot's thermal imagers
Night flying systems for use by pilots. Note that these systems normally include a ground attack capability, but because the system is used primarily as a vision system for flying the aircraft, the image field of view does not have a high magnification or zoom mode.

Pilot's goggles
Image intensifier goggles qualified for use by pilots of fixed-wing aircraft and helicopters.

Observation and surveillance

Air interception
Electro-optic sensors for providing visual means of target identification in air-to-air combat. Note that some systems also provide a limited ground attack capability.

Turret sensors
Gyrostabilised omnidirectional turrets for use on helicopters and patrol aircraft. Note some systems have a limited ground attack capability and/or may be used as a pilot's night flying aid. See also Maritime Sensors for similar systems.

Maritime sensors
Turret sensors that are used by shipborne and land-based maritime patrol aircraft or have been qualified for these roles.

Unmanned aircraft sensors
Turrets and body-fixed sensors deployed in, or developed for, unmanned air vehicles.

Reconnaissance sensors
Linescan cameras and other reconnaissance sensors predominantly for fixed-wing aircraft.

Thermal imagers
Thermal imaging systems that are deployed on, or qualified for, aircraft.

AIR-LAUNCHED MISSILES

AIR-TO-AIR MISSILES

Arsenal Central Design Bureau MK-80, UA-96 optical seekers for air-to-air missiles

Type

Air-to-air missile (seekers).

Development

Arsenal Central Design Bureau of the Ukraine has been actively engaged in the research, development, testing and production of optical seekers for close combat air-to-air missiles for 30 years. During this time an off boresight engagement capability evolved. The 60TI seeker was developed for the first AA-8 'Aphid' (R-60) missile. One of the latest developments is the MK-80 seeker for the AA-23 'Archer' R-73 missile. The UA-96 seeker has been developed for upgrading of the Aphid R-60 and R-60M missiles.

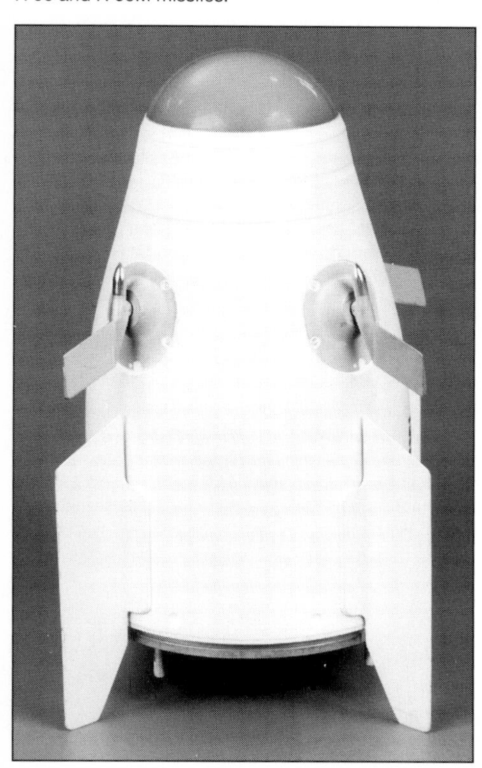

The CDO 'Arsenal' MK-80 seeker
0089889

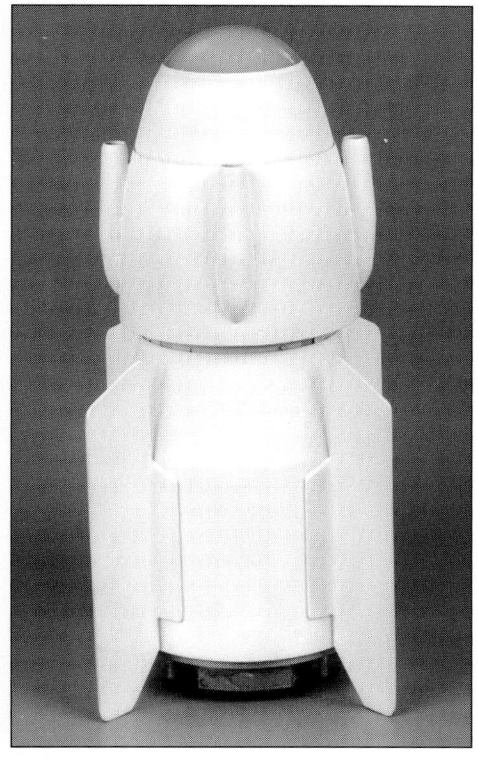

The CDO 'Arsenal' UA-96 seeker
0089888

Description

The MK-80 and UA-96 seekers operating in the mid-wave infra-red band have an all-aspect target acquisition capability. The tracking co-ordinator uses an astatic gyro with an aspheric rotor-lens. After acquisition the software can initiate an automatic lock-on and firing command if selected. The processor can also analyse the signal from the seeker to steer the missile to hit the most vulnerable area of a target. Jamming and natural interference rejection are featured.

Where the aircraft is equipped with a helmet-mounted target designation system the seeker follows the pilot's head movement. The pilot can aim the missile by turning his head and placing the aiming reticle over the target without having to turn the aircraft toward it.

Specifications

	MK-80	UA-96
Operating IR range	midwaves	
Aspect capability	All-aspects	
Operational range (in the forward hemisphere)	from 0.3 to 10- (15) km	from 0.3 to 10 km
Autotracking angular rate	60°/s	30°/s
After launch target bearing angles	±75°	±44°
Target designation angles	±45 (±60)°	±30°
Altitude	up to 25 km	up to 20 km
Calibre	170 mm	120 mm
Length	300 mm	320 mm
Mass	6 kg	3.5 kg

Contractor

Arsenal Central Design Bureau, Ukraine.

VERIFIED

BGT IRIS-T short range air-to-air missile

Type

Air-to-air missile.

Development

Canada, Germany, Greece, Italy, Norway and Sweden are jointly developing IRIS-T (Infra-Red Imaging Seeker/System - Tail/Thrust vector controlled) as a replacement for the Sidewinder missile. Based on the results of the definition phase, all six partner nations have harmonised their operational requirements for a future short-range air-to-air missile and decided to launch the full-scale development phase. The Memorandum of Understanding was signed in Spring 1998. The industrial team under the leadership of BGT started full-scale development work in January 1998. The seeker's acquisition and tracking performance under large off-boresight angles against head-on targets was already successfully demonstrated in live firing trials in July 1996, with IRIS-T seeker prototypes installed on Sidewinder aft sections with two direct hits on 25 cm diameter towed target drones. In addition, target cueing and tracking trials have been performed successfully by an RNLAF F-16 aircraft equipped with helmet-mounted sights. Seeker look angles of ±90° were demonstrated under up to 8 *g* manoeuvres and speeds of 0.9 Mach. Within the framework of the continuing development phase seeker data gathering campaigns against all common Nato fighter aircraft and Eastern targets such as MIG-29, SU-22 and Hind helicopters, all using their onboard IRCM systems, were successfully completed in 1999. A series of environmental flight tests with numerous fighter aircraft were carried out in 1999. In October 2001, Canada withdrew from the programme on cost grounds, but has been replaced by Spain.

Description

IRIS-T will be identical in weight, size and centre of gravity to the Sidewinder missile which it is designed to replace. IRIS-T is developed for digital missile interfaces of modern aircraft but can also be used on analogue Sidewinder aircraft interfaces. The IRIS-T seeker is based on BGT's IRIS infra-red imaging strap-down seeker technology, fully optimised for air-to-air applications. Its imager unit operates in the 3-5 μm band and generates IR images with 128 × 128 pixel resolution. The roll/pitch gimbal design provides ±90° look angle, thus covering the full frontal hemisphere.

The missile has a particularly stiff airframe and a combined aerodynamic and thrust vector control system to give it excellent manoeuvrability at all altitudes.

The missile will have a solid propellant motor with a thrust profile optimised for both close-in combat and intercept, a fragmentation warhead and an active radar proximity fuze activated by the imaging seeker during the final stages of intercept.

Operational status

First two demonstration firings of IRIS-T (from an F-4F) were made in May 2003 and were declared successful. Series production deliveries expected 2004-2005. The weapon will be integrated on F-16, F-18, JAS 39 Gripen and Eurofighter Typhoon.

First live firing of BGT's imaging IRIS-T seeker prototype over the North Sea in July 1996

BGT's roll-pitch 3 to 5 μm imaging infra-red seeker for IRIS-T

IRIS-T on Eurofighter wing pylon 0023268

Specifications

Seeker: strapdown imaging IR (3-5 μm scanned array)
Gimbal line of sight: ±90° pitch, roll unlimited
Length: <3.0 m
Body diameter: 127 mm
Wing span: 367 mm
Launch weight: 89 kg
Warhead: 9 kg HE fragmentation
Fuze: active radar
Guidance: Modified PN
Propulsion: solid propellant
Range: up to 25 km

Contractors

Bodenseewerk Gerätetechnik GmbH (BGT), teamed with Saab Bofors Dynamics, Alenia Marconi Systems, NAMMO, Honeywell Canada, FiatAvio, Litton Italia, Hellenic Aerospace Industry, Intracom and Pyrkal.

UPDATED

Bishnovat AA-6 'Acrid' (R-40, R-46)

Type

Air-to-air missile.

Development

Summary: There is an infra-red seeker on the R-40T/R-46TD missile variants.

AA-6 'Acrid' is the NATO designation for this former Soviet R-40 air-to-air missile developed in the late 1960s. Two versions have been reported, infra-red (R-40T) and semi-active radar-guided (R-40R). These have been updated and improved since their original design with the improved versions designated R-46TD and R-46RD. Few semi-active radar missiles have been seen and only the IR seeker version has been marketed in the 1990s. The missiles have been seen on the Su-21 'Flagon', Su-22 'Fitter', MiG-25 'Foxbat' and MiG-31 'Foxhound' aircraft. The MiG-25 'Foxbat' usually carries two IR and two radar missiles, with the radar missiles on the outboard pylons and the IR on the inboard.

Description

AA-6 'Acrid' is similar in shape to, but much larger than the AA-3 'Anab' air-to-air missile. The radar and the IR missiles are of about the same dimensions, having a length of 6.2 m, a body diameter of 355 mm and a wing span of 1.8 m. With their 70 kg HE fragmentation warhead, both missiles weigh around 475 kg at launch.

An unusual feature is, that such a large missile is rail-launched. The AA-6 'Acrid' missiles have been designed for medium-range, high-altitude air-to-air interception and it would seem most likely that there would be some form of command update and inertial mid-course guidance, followed by either semi-active radar or infra-red terminal guidance. Early infra-red seekers used T40A cooled 3 to 5 μm detectors, with later versions (the R-46TD missile) using an improved 35T2 IR seeker with improved IRCM protection. The mid-course guidance could possibly be received by a rearward-facing antenna beneath the boat tail or by the two large strip antenna on either side of the fuselage just forward of the wings.

The launch aircraft is directed towards the intercept point by the ground-to-air datalink which provides a cockpit display of the target direction and, at an appropriate range, the aircraft will switch to its own radar, with the CW illuminator used in the terminal phase of guidance for the semi-active radar missile version. It is believed that both semi-active radar and IR-guided R-40 versions have a maximum range of 30 km and that the later R-46 versions have a maximum range of 50 km.

Operational status

It is believed that the AA-6 'Acrid' (R-40) missiles entered service in the early 1970s with then Soviet Air Force and that the improved R-46 versions entered service in 1982. Missiles are being offered for export, and it is assumed that these are rebuilt missiles with an option to reopen the production line if orders are placed. Exports have been reported to Afghanistan, Algeria, Belarus, Hungary, Iraq, Kazakhstan, Libya, Poland, Syria, Ukraine and Vietnam.

Specifications

'Acrid' (R-40T/R-46TD)
Length: 6.2 m
Body diameter: 355 mm
Wing span: 1.8 m
Launch weight: 475 kg
Warhead: 70 kg HE fragmentation
Fuze: radar and active laser
Guidance: command, inertial and IR
Propulsion: solid propellant

Range
R-40: 30 km
R-46: 50 km

Contractor

Bishnovat OKB-4, Moscow.

VERIFIED

AA-6 'Acrid' (IR version) on the wing pylon of a MiG- 31, with an AA-9 'Amos' under the fuselage (Christopher F Foss)

CATIC PL-2/PL-3

Type
Air-to-air missile.

Development
Summary: the missile has an infra-red seeker.

PL-2 is a licence-built version of the AA-2 'Atoll' under an agreement reached in 1963 with the former USSR. The licence-built PL-2 entered service in 1967. In 1964 development of a new PL-2 (designated PL-2A) started and this entered service in 1970. A further development, the PL-2B, entered service in 1981.

The development of a slightly larger version of the PL-2 missile started in 1965 and this, designated the PL-3, entered service in April 1980.

PL-2/PL-3 missiles are believed to have been fitted to Chinese J-5 (MiG-17), J-6 (MiG-19) and J-7 (MiG-21) aircraft.

Description
The PL-2 is similar to AIM-9B Sidewinder with a length of 2.99 m, a body diameter of 127 mm and a wing span of 0.53 m. The missile weight is around 76 kg. The performance capability is limited to tail-aspect engagements from between 1 and 3 km range and probably boresight firings only. The improvements incorporated into the later PL-2B version included improved IR detection sensitivity, less likelihood to lock on to the sun and cloud tops, fewer fuze failures and an increased range. PL-3 had larger control surfaces, a larger warhead and an improved fuze. This resulted in a heavier launch weight, believed to be 82 kg.

Operational status
Exports of the PL-2 missile were made to Albania, Myanmar (Burma), Pakistan and Tanzania. It is believed that the missiles are no longer in production, having been replaced by the PL-5 and PL-7.

Specifications
PL-2B
Length: 2.99 m
Body diameter: 127 mm
Wing span: 0.53 m
Launch weight: 76 kg
Warhead: 11.3 kg HE blast fragmentation
Fuze: IR
Guidance: IR
Propulsion: solid propellant
Range: 3 km

PL-3
Length: 2.99 m
Body diameter: 127 mm
Wing span: 0.53 m
Launch weight: 82 kg
Warhead: 13.5 kg HE blast fragmentation
Fuze: IR
Guidance: IR
Propulsion: solid propellant
Range: 3 km

Contractor
China National Aero-Technology Import and Export Corporation (CATIC).

VERIFIED

A PL-2A air-to-air missile

CATIC PL-5

Type
Air-to-air missile.

Development
Summary: the missile has an infra-red seeker.

Development of this missile started in 1966 at the AA Missile Research Institute, but did not enter service until 1982. PL-5 is based upon the technology of the AIM-9G Sidewinder. It is believed that PL-5 is carried by the J-5, J-6, J-7, F-7M and A-5M aircraft.

A PL-5B air-to-air missile (Raymond Cheung)

Description
The PL-5 is a mixture of earlier AIM-9B Sidewinder/AA-2 'Atoll' technology and the later AIM-9G Sidewinder. The missile is 2.89 m long, with a body diameter of 127 mm, a wing span of 0.66 m and weight of 85 kg.

The improved infra-red seeker in this missile is still limited to tail-aspect engagements, but has a greater off-boresight capability than the earlier PL-2 and PL-3 systems. It must be expected that the Chinese have further modified the PL-5 series of missiles to provide an all-aspect engagement capability and photographs taken in 1987 illustrate an AIM-9L/M-shaped seeker head and fins that would presumably indicate that this has been achieved. However, published details of PL-5B, whilst introducing an improved IR fuze and new fins, do not claim an all-aspect capability.

The front control fins resemble those of the AIM-9G and it is reasonable to assume that the missile has similar solid-state electronics. There are two interchangeable fuze and warhead combinations: an IR fuze and blast fragmenting warhead or an RF fuze and continuous rod warhead.

Operational status
The PL-5 probably entered service with the People's Liberation Army Air Force in 1982 and could still be in production. It is believed that PL-5 missiles have been exported to Albania, Myanmar (Burma), Egypt, Iran, Pakistan, Tanzania and Zimbabwe.

Specifications
Length: 2.89 m
Body diameter: 127 mm
Wing span: 0.66 m
Launch weight: 85 kg
Warhead: 9 kg HE blast fragmentation or continuous rod
Fuze: IR or RF
Guidance: IR
Propulsion: solid propellant
Range: 3 km

Contractor
China National Aero-Technology Import and Export Corporation (CATIC).

VERIFIED

CATIC PL-7

Type
Air-to-air missile.

Development
Summary: the missile has an infra-red seeker.

The PL-7 missile was probably developed in the early 1980s and it is believed to be based originally upon the technology of the R550 Magic 1 missile. This missile would be carried by the J-7 (MiG-21), A-5M and F-7M, and probably by the J-8II fighter which first flew in 1984. A new version of the PL-7 missile was shown in model form in 1988, with extensions to the rotating rear wing leading edges at the root sections. This might indicate the development of a PL-7B version and perhaps the introduction of an all-aspect engagement capability.

Description
This missile is similar to the R550 Magic with a length of 2.75 m, body diameter of 158 mm and a wing span of 0.66 m. The missile weighs 90 kg and has the Magic's distinctive double fins at the nose and free rotating rear wings. The PL-7 has an IR seeker and is capable of tail-aspect engagements only up to about 3 km range. The 13 kg HE fragmentation warhead is said to have a lethal radius of 10 m against aircraft targets.

Operational status
Development started in the early 1980s and flight tests were carried out in 1985-86. The missile entered service in 1987. It is believed that an improved version, the PL-7B, has been developed and is in service. Exports of PL-7 have been reported to Iran and Zimbabwe.

PL-7 air-to-air missile under an F-7M fighter of the Air Force of the People's Liberation Army

Specifications

Length: 2.75 m
Body diameter: 158 mm
Wing span: 0.66 m
Launch weight: 90 kg
Warhead: 13 kg HE fragmentation
Fuze: IR
Guidance: IR
Propulsion: solid propellant
Range: 3 km

Contractor

China National Aero-Technology Import and Export Corporation (CATIC).

VERIFIED

CATIC PL-8

Type

Air-to-air missile.

Development

Summary: the missile has an infra-red seeker.

In March 1991 Chinese officials released details of a new missile/gun air defence system using a new IR missile designated PL-8H and the Type 715-I twin 39 mm powered anti-aircraft gun-mount. From the released information it is evident that China has acquired Israeli Python 3 missile technology, providing first direct evidence of military links between the two countries. No contract between Israel and China has ever been reported. The PL-8H may be an exported Israeli Python 3 air-to-air missile, a licence-built version or a copy.

A report in 1995 suggested that PL-8 missiles are in service, fitted to JH-7 fighter bombers. They are also carried on J-8II and Q-5 aircraft.

Description

From the released photographs the PL-8H appears to be similar to the Israeli Python 3 air-to-air missile. Assuming the missile is the same size as the Python 3, it will be 3.0 m long, have a body diameter of 160 mm, and a wing span of 0.86 m. Guidance will be by an all-aspect infra-red seeker which will give the system a head-on engagement capability against aircraft targets. If it follows the performance of Python 3 the missile will also have a countermeasure capability and be able to be slaved to the aircraft radar, or used in the scan and boresight modes. The probable launch weight of the PL-8H is 120 kg and it is expected that the missile will have an 11 kg HE fragmentation warhead with an active laser fuze. The missile range is believed to be 5 km.

A Chinese Naval Aviation J-8II armed with PL-8 missiles (US Navy) ***NEW***/0137257

Operational status

Operational with PLA Air Force and air arm of PLA Navy. There have been no reported exports of the air-to-air variant.

Specifications

Length: 3.0 m
Body diameter: 160 mm
Wing span: 0.86 m
Launch weight: 120 kg
Warhead: 11 kg HE fragmentation
Fuze: active laser
Guidance: IR
Propulsion: solid propellant
Range: 5 km

Contractor

China National Aero-Technology Import and Export Corporation (CATIC).

UPDATED

Chung Shan Institute Sky Sword 1 (Tien Chien 1)

Type

Air-to-air missile.

Development

Summary: the missile has an infra-red seeker.

This programme was first reported in development in 1986 and is now in production. The Sky Sword 1 missile is believed to have been cleared for carriage on the F-5 and the Ching-Kuo and it is expected to be cleared on F-16 and Mirage 2000 aircraft.

Description

The Sky Sword 1 is similar in appearance to the AIM-9 Sidewinder missile. The missile is 2.87 m long, has a body diameter of 127 mm and a wing span of 0.64 m. The missile is expected to weigh about 90 kg. From the few published details available, it would seem that the IR seeker has all-aspect engagement capability and that an active laser fuze is used. This indicates a performance capability similar to AIM-9L Sidewinder.

Operational status

Sky Sword 1 is believed to have entered service in 1991. The status of this programme is now unclear, as it has been reported that Taiwan has ordered both Magic 2 and AIM-9 Sidewinder missiles for use on Mirage 2000-5 and F-16 aircraft, but it is believed that Sky Sword 1 production continues.

Specifications

Length: 2.87 m
Body diameter: 127 mm
Wing span: 0.64 m
Launch weight: 90 kg
Warhead: HE blast fragmentation
Fuze: active laser
Guidance: IR
Propulsion: solid propellant
Range: 8 km

Contractor

Chung Shan Institute of Science and Technology.

VERIFIED

A view of the Ching-Kuo fighter showing Sky Sword 1 missiles on the wingtip and wing pylon, and two Sky Sword 2 missiles recessed under the fuselage.

Kentron Darter V3C/U-Darter/A-Darter

Type
Air-to-air missile.

Development
Summary: the missile has an infra-red seeker; A-Darter is to have an imaging infra-red seeker.

The V3A and B Kukri series of air-to-air missiles was developed during the late 1960s and 1970s by the Kentron Division of the Denel company. Development of an improved all-aspect version of the Kukri missile, the V3C Darter was announced in 1983. It is shorter and has a larger diameter than the V3A/B versions and includes a new two-colour seeker, automatic target acquisition and lock on and high-gain navigation with lead bias terminal guidance.

The export version of the V3C is named Darter. Darter was first displayed and offered for export in 1988 and has been designed to be compatible with Matra R550 Magic mechanical and electrical interfaces and is believed to have been cleared for carriage on the Mirage 3, Mirage F-1 and Cheetah aircraft and the Rooivalk helicopter.

In 1994 it was reported that an upgraded Darter, known as U-Darter, was in low-volume production to equip SAAF Mirage F-1s and Cheetahs. The missile is credited with an improved guidance system, larger warhead and a more powerful motor.

Kentron are developing another upgrade called A-Darter (Agile). A-Darter will have an imaging infra-red seeker and a new airframe using tail control only. A-Darter has been linked with Kentron's Guardian helmet sighting system. A version known as R-Darter with an active radar seeker is also being studied.

Description
The V3C Darter missile is a short-range, all-aspect, infra-red guided weapon powered by a solid propellant motor and armed with a fragmentation warhead. It is similar to the MBDA R550 Magic missile. The missile breaks down into two main sections. The forward section contains the electromechanical components including the IR homing head, the autopilot system and the active laser proximity fuze. The rear section contains the warhead and the propulsion unit. Guidance is by a passive infra-red seeker that has an all-aspect capability, with a two-colour detector. Infra-red filtering and signal processing is claimed to make the missile less sensitive to decoy flares once it is locked on to the target. The missile is also credited with a look-down, shoot-down ability against background clutter.

An audible target tone and a track indicator provide feedback to the pilot when the following three acquisition modes are used: CAGE mode is a reduced scan format capable of acquiring a target within a few degrees of the aircraft's boresight axis. The normal gunsight is used as an aiming device. The pilot may authorise lock on and the system will continue to track the target in TRACK mode at up to ±20° from the aircraft axis. In HELMET mode a sighting reticle is projected onto the pilot's visor, in front of his right eye and light emitting diodes on either side of his helmet are detected by the reference units in the cockpit to slave the system to his head movement. In this mode the missile's infra-red seeker is slaved to the pilot's line of sight/head movement within a cone of ±20°. Should the pilot look beyond the limits of the cone, the reticle switches off and the missile's head returns to the aircraft axis. When the pilot turns his head back within the cone, the reticle switches on again and the missile head resumes its tracking action. SCAN mode may be used without the designating helmet. The infra-red head scans a field of ±14° off axis and may be instructed to lock on to the most powerful IR source.

U-Darter version has a two-colour infra-red seeker with improved processing, a digital autopilot and a more powerful motor. The cooled indium antimonide IR seeker operates in dual bands and provides an improved level of IRCCM and background discrimination. The missile system has a choice of three acquisition modes including CAGE which limits acquisition to 3° from the boresight axis, SCAN which selects various scan patterns within a 100 × 100° field of view and 120°/s scan rate and helmet/radar, which designates a target to the missile using the helmet-mounted sight or the aircraft radar. It has been designed to use the Hermes helmet aiming system. Processing enables a high degree of seeker lead bias away from aircraft plume and into the fuselage. A-Darter will have an imaging infra-red seeker which will permit engagements at 90° off boresight angles. This has been through three design iterations and moves slowly.

Darter air-to-air missile

Operational status
Darter is believed to have entered service as the V3C missile with the South African Air Force in 1990. There are no known exports. U-Darter entered service in 1994. A-Darter moves slowly and, as of mid-2003, the first guided flight had yet to take place.

Specifications
V3C Darter
Length: 2.75 m
Body diameter: 157 mm
Wing span: 0.66 m
Launch weight: 90 kg
Warhead: 16 kg HE fragmentation
Fuze: active laser
Guidance: IR
Propulsion: solid propellant
Range: 5 km

U-Darter/A-Darter
Length: 2.75 m/2.98 m
Body diameter: 160 mm
Wing span: 0.66 m/488 mm
Launch weight: 96 kg/89 kg
Warhead: 17 kg HE fragmentation
Fuze: active laser
Guidance: IR
Propulsion: solid propellant
Range: 8 km

Contractor
Kentron Division of Denel (Pty) Ltd.

UPDATED

MBDA Air-to-Air Mistral (ATAM)

Type
Air-to-air missile.

Development
Summary: the missile has an infra-red seeker.

The Mistral short-range surface-to-air missile commenced development in the late 1980s as a multipurpose weapon system. It was primarily intended as a portable (SATCP) and small ship surface-to-air missile (SADRAL and SIMBAD) but also as a potential air-to-air armament for helicopters. In 1986 a contract was placed to develop the air-to-air adaption to helicopters. This application was originally known as *Helicoptère-Air Trés Courte Portée* (HATCP), but became Air-to-Air Mistral (ATAM) later. Development firings of ATAM were first made in 1990 from a Gazelle helicopter, using a twin launcher assembly. The ATAM has been tested on the AH-64 Apache and is now in the qualification phase for the Eurocopter Tiger. Typical helicopter loads are four missiles, two on each side.

Mistral ATAM entered service on 30 Gazelle helicopters of the French Army in June 1996. Mistral will also be fitted to French Tiger helicopters when they enter service. Mistral ATAM should also be fitted on the Rooivalk combat helicopter.

In 1996, Matra Defense and British Aerospace Dynamics Division joined to form Matra BAe Dynamics for the joint production and development of guided weapons. In 2001 Matra BAE Dynamics joined with EADS and Finmeccanica to form MBDA.

Description
The Mistral ATAM is a helicopter-launched, short-range infra-red homing missile powered by a solid propellant motor and armed with a fragmentation warhead. Guidance is by passive infra-red homing with a high sensitivity to detect low-level infra-red signals such as helicopter exhaust signatures.

ATAM can be operated in one of three modes using either a simple collimator with a slaved reticle or a gyrostabilised telescope or a helmet mounted display (TIGRE). Mistral is aimed by the Thales T.2000 sighting system on the Gazelle. Mistral ATAM has a maximum range of approximately 6 km. It can be used against either rotary- or fixed-wing aircraft.

Operational status
The Mistral ATAM completed development, with 10 missiles fired, by mid-1990. The first Gazelle helicopters equipped with Mistral ATAM entered operational service with the French Army in June 1996, although a crash programme achieved

French Gazelle-Mistral helicopter equipped with an ATAM system – 4 Mistral missiles
(F Nebinger/Matra BAe Dynamics)

an initial operational capability in time for deployment to the Gulf War in 1991. The surface-to-air variant has been exported to 24 countries. The ATAM system is being marketed to several countries.

Specifications
Length: 1.80 m
Body diameter: 90 mm
Wing span: 0.185 m
Launch weight: 19 kg
Warhead: 3 kg HE tungsten ball
Fuze: laser and impact
Guidance: passive IR fire and forget
Propulsion: solid propellant
Range: 6 km
Missile altitude: greater than 3 km
Missile speed: M 2.5

Contractor
MBDA (France). ***VERIFIED***

MBDA ASRAAM

Type
Air-to-air missile.

Development
Summary: the missile has an imaging infra-red seeker.

The Advanced Short-Range Air-to-Air Missile (ASRAAM) programme started in 1982 and was part of the Memorandum of Understanding for a family of air-to-air missiles initially signed by France, Germany, UK and US in 1980. This MoU provided for the Advanced Medium-Range Air-to-Air Missile (AMRAAM) to be developed by the USA and co-produced in Europe, while ASRAAM was developed by Europe and co- produced in the USA. ASRAAM was given the provisional US missile designator AIM-132, but this will only be used if the missile is operated by the US services. ASRAAM is designed to significantly enhance the capabilities for within visual range combat and be carried on a wide range of aircraft.

Early in 1990 the USAF indicated a reduced interest in ASRAAM and started considering alternative options, including a further Sidewinder improvement beyond AIM-9R, known as Sidewinder 2000 or AIM-9X. In 1990 the requirement was redefined and a competition opened to industry. In 1992, the UK government announced a development and production contract with British Aerospace for a revised design utilising a Raytheon imaging IR seeker, a Daimler-Benz Aerospace warhead and a Thomson-Thorn (now Thales) Missile Electronics fuze. A surface-to-air variant may also be proposed for ship defence. The UK plan is to fit ASRAAM to Harrier GR.7, Tornado F.3, Jaguar, Eurofighter Typhoon and Sea Harrier.

In 2001, BAE Systems, EADS and Finmeccanica signed a shareholders agreement to form a joint company, MBDA. In 1996, British Aerospace Dynamics Division and Matra Défense had formed a joint company, for the development and production of guided weapons known as Matra BAe Dynamics.

In 1998, ASRAAM was ordered by the Royal Australian Air Force to arm their F/A-18s.

Description
ASRAAM is an all-aspect infra-red homing missile powered by a solid propellant motor and armed with a fragmentation warhead. It is wingless, with clipped delta-tail control surfaces and lifting body aerodynamics for high manoeuvrability. The missile is of modular construction and consists of four major sections; at the front end the tapered seeker, sensor and cooling section with its seeker dome, followed by the electronics, fuze and warhead section, rocket motor section and, situated around the motor exhaust nozzle, the actuator section with its four control fins.

The ASRAAM Advanced Short-Range Air-to-Air Missile on the Tornado F.3
(Craig Hoyle/Jane's) ***NEW***/0532216

The ASRAAM imaging infra-red seeker head

ASRAAM is fitted with an HE fragmentation warhead that is being manufactured by EADS (formerly Daimler-Benz Aerospace) and this has an integral impact fuze and safety and arming unit. The active laser proximity fuze is being developed by Thomson-Thorn Missile Electronics.

Guidance is by a Raytheon Santa Barbara Research Center's imaging infra-red indium antimonide 128 × 128 staring focal plane array seeker. The IIR seeker has an all-aspect capability and target lock on before or after launch. A helmet-sighting system will be needed to take advantage of the full 90° off-boresight capability of the missile, as aircraft head-up displays have a limited field of view.

Apart from being credited with a maximum range of 15 km for a head-on engagement, ASRAAM's performance figures are classified, but the missile is said to comply with all known requirements and to be the fastest short-range missile in production or development. It is said to be effective in the presence of the most hostile countermeasures and against backgrounds which have defeated previous infra-red guidance systems. It can be fired singly or in salvo against an aggressor/aggressors anywhere in the pilot's view and, the seeker system will allow the missile to continue homing even if the target is obscured for periods after launch.

Operational status
With the selection of the British Aerospace development proposal in 1992 and a planned production order of 1,000 missiles, the ASRAAM programme will now compete with several other short-range air-to-air missile projects around the world for export orders. A second production order, believed to have been for a further 300 missiles, was placed by the UK in 1994.

In 1994 ASRAAM was selected by the USA for test and evaluation as part of the US AIM-9X programme, in addition to UK flight trials to be made at Eglin AFB from a special F-16 trials aircraft. The US did not select ASRAAM for its AIM-9X programme. Three unguided separation launches were made in 1995. The first guided firing was made in May 1996, with a total of 22 firings completed as of July 2000. Australia selected ASRAAM for use from its F/A-18 Hornet aircraft in 1998 (the first export order), and the first integration flight test was made in December 1999 using the Hornet's wing-tip pylons to carry the missiles. The missiles are expected to enter service in Australia from 2003. The RAF declared ASRAAM ready for operations in September 2002 on the Tornado F.3, and the missiles were carried operationally on missions during operation 'Telic' in March/April 2003.

Specifications
Length: 2.9 m
Body diameter: 166 mm
Finspan: 0.45 m
Launch weight: 87 kg
Warhead: HE blast fragmentation
Fuze: active laser
Guidance: imaging IR
Propulsion: solid propellant
Range: 15 km

Contractor
MBDA (UK). ***UPDATED***

MBDA MICA

Type
Air-to-air missile.

Development
Summary: the missile can be fitted with either a passive imaging infra-red or active radar seeker.

Development of the MICA (*Missile d' Interception et de Combat Aérien:* combat and air intercept missile) started in 1982 as a private venture by Matra, to provide a single missile system to replace both the short-range R550 Magic and medium-range Super 530 D/F missiles. The French Air Force and Navy agreed the requirement and interim development of MICA started in 1985. The first operational missiles were delivered in 1996. The objective of the MICA programme was to produce a single missile with interchangeable guidance systems for both interception and dogfight missions. The missiles have inertial guidance, with possible mid-course updating from the launch aircraft and, either an active radar (MICA EM) or imaging infra-red terminal seeker (MICA IR). Both radar and infra-red versions will be capable of launch at medium or short range.

Following the German decision to leave the ASRAAM programme in 1989, Matra had teamed with BGT to propose another version of MICA for Germany. This version was known as MICA-SRAAM, and would have used the MICA airframe with a BGT-developed infra-red seeker assembly in a missile with a weight in the region of 90 kg. A further variant, known as MICASRAAM, was proposed to the UK MoD as an alternative to ASRAAM by Marconi Electronic Systems and Matra. Matra had also signed teaming agreements with Alenia, Italy and Ericsson, Sweden to assist in marketing MICA in these two countries.

In 1996, Matra Défense joined with British Aerospace Dynamics Division to form a joint company for missile production, Matra BAe Dynamics. In 2001 Matra BAE Dynamics formed a joint company with EADS and Finmeccanica - MBDA.

MICA will have the benefit of commonality between two very similar missiles used for a wide multirole mission spectrum on different lightweight aircraft. It is planned for carriage by Mirage 2000 and Rafale (ACT) aircraft, with options to carry up to eight missiles per aircraft and is being considered for several other aircraft types.

Description
MICA is a medium-range infra-red or radar-guided missile powered by a solid propellant motor and armed with a fragmentation warhead. The radar seeker head has a pointed ceramic radome, housing the antenna of the AD4A radar. The infra-red seeker has a blunt tapered nose with a small optical window in the shape of a dome. Both nose sections have four long narrow cord fins that have been added.

For medium-range interceptions the missile uses inertial guidance, mid-course trajectory updates from the launch aircraft, and then an imaging infra-red seeker developed by SAGEM or a J-band (10 to 20 GHz) AD4A active pulse Doppler radar terminal guidance system developed by Dassault Electronique (now part of Thales). The infra-red seeker is dual waveband imaging, the detectors being cooled by a closed-cycle cooler.

The maximum range of MICA is claimed to be about 80 km. When the range is less or the target does not manoeuvre, the mid-course update may not be required. For short ranges the missile active radar can be locked on to the target before launch. With the infra-red-guided version the missile can be locked on to the target before or after launch and the seeker will have an all-aspect engagement capability. Complex algorithms have been developed to provide the infra-red seeker with the ability to track at longer range and to reject flare decoys.

When fitted to aircraft with track-while-scan radar the MICA weapon system is capable of attacking several targets simultaneously with individual missiles. It is presumed that the missile has common ejection and rail launcher interfaces with Magic missiles.

Operational status
The first ground firing was carried out in January 1989, and the first flight test firing of the active radar version was carried out in 1991. Flight tests of the IIR version started in 1994 with the first firing in 1995. It is believed that France has a total requirement for around 2,000 MICA missiles; the first production batch for France of 225 missiles was ordered in 1997, with deliveries starting in 1999. A second French production order was placed in January 2001 for 1,537 missiles, with deliveries starting in 2003.

MICA missile being fired from a French Mirage 2000 fighter aircraft ***NEW***/0554672

The very first production missiles were ordered by Taiwan in 1996, and deliveries of the radar version started in 1997. It is believed that Taiwan ordered 960 MICA EM missiles. Taiwan has flight tested two MICA missiles, in May 1998 and in March 2000. The MICA EM entered operational service with the French Air Force in 1999, equipping the Mirage 2000-5s of EC 1/2 'Cigones' and EC 2/2 'Cote d'Or'.

By July 2000, over 100 MICAs had been launched in trials and, by 2001, the missile had been deployed abroad for exercises in Europe, North America and the Gulf region. Qatar ordered MICA EM missiles in 1997, and deliveries started in 1998. The United Arab Emirates ordered MICA EM and IR missiles in 1998, to equip its Mirage 2000-9 aircraft, and will become the first user of both variants of the missile. Subsequently, Greece ordered 200 MICA EM and IR missiles, in August 2000, to arm its Mirage 2000-5s. French Mirage 2000-5s are currently being modified to operate the MICA IR. The full integration of the MICA with the Dassault Rafale (as part of the F1 capability standard) was completed in July 2000, following 27 successful test firings. French Navy Rafale Ms made their first operational deployment aboard the carrier *Charles De Gaulle* in early 2002, equipped with MICA missiles.

Specifications
Length: 3.10 m
Body diameter: 165 mm
Finspan: 0.56 m
Launch weight: 110 kg
Warhead: HE blast fragmentation 12 kg
Fuze: active radar
Guidance: command, inertial and active radar or imaging IR
Propulsion: solid propellant
Range: 80 km

Contractor
MBDA (France).

UPDATED

MBDA R550 Magic

Type
Air-to-air missile.

Development
Summary: the missile has an infra-red seeker.

Deliveries of the preproduction R550 Magic 1 missile to the French Air Force for evaluation trials started in 1974 and full series production began in 1975. Magic 1 was limited to mainly tail-aspect engagements by the lack of sensitivity of the infra-red detector and proximity fuze. An improved version, Magic 2, entered development in the late 1970s and was fielded in 1985. Magic 2 has many improvements over Magic 1, but principally has an all-aspect engagement capability and it takes far less time to prepare the missiles for launch. The Magic missiles have been cleared for carriage on 16 types of combat aircraft including the Mirages, Hawk, Jaguar, Super Etendard, F-5, A-4 Skyhawk, MiG- 21, MiG-23, Alpha Jet, Sea Harrier, and F-16 Fighting Falcon.

In 1996, Matra Défense and British Aerospace Dynamics Division united to form a joint company, Matra BAe Dynamics, for the production and development of guided weapons. In 2001 Matra BAE Dynamics joined with EADS and Finmeccanica to form MBDA.

Description
The R550 Magic 1 is an infra-red homing missile powered by a solid propellant motor and armed with a fragmentation warhead. A feature of the missile is its breakdown into two sections, which are packaged separately. The forward section contains all the electromechanical components including the AD550 IR homing head produced by SAGEM, the guidance and control electronics unit, the IR proximity fuze and the self-activated silver-zinc battery. The nitrogen cooling system for the IR head is installed in the launcher. The rear section contains the warhead and the butalane composite solid propellant rocket motor.

Guidance is by passive IR homing and target acquisition is visual, with the nitrogen-cooled seeker head warning the pilot when lock on has been achieved. The warhead is armed 1.8 seconds after launch, giving a minimum engagement distance of 0.3 km. Magic 1 is limited to large tail aspect engagements because of the seeker and proximity fuze designs.

Magic 2 retains the same general aerodynamic and external characteristics as Magic 1, but is fitted with a new rocket motor (increased total impulse). Magic 1 had roll control which enables the rail-launched missile to be fired up to the maximum *g* manoeuvre of the parent aircraft. Changes made to Magic 2, with regard to Magic 1, include a more sensitive multi-element seeker with head-on capability and improved infra-red counter-countermeasures. The Magic 1 infra-red proximity fuze has been replaced with an active Doppler radar fuze. A special warhead for Magic 2 was developed and tested in 1993, to provide a less sensitive munition for use on aircraft carriers. Magic 2 can be slaved to the aircraft radar or a helmet-mounted sight or it can be used in the autonomous mode to scan in either the vertical or horizontal planes and lock on to a target without help from the aircraft radar. Magic 2 is fitted with improved roll control to enhance missile manoeuvrability.

Since 1995, a new version designated Magic 2 Mk 2 has been offered to the export market. Magic 2 Mk 2 has been improved in respect to Magic 2 in four areas: more sensitive infra-red seeker while preserving clutter rejection capability;

MAGIC 2 missiles under the wings of a British Aerospace Hawk training/combat aircraft

additional flexibility in the autonomous mode of operation; reduced short firing range limits thanks to an improved digital autopilot and to new guidance and steering algorithms; and increased resistance to infra-red countermeasures.

Operational status

Magic 1 entered service in 1975 and production ceased in 1984. It has been exported to Argentina, Australia, Brazil, China, Ecuador, Egypt, Greece, India, Iraq, Kuwait, Lebanon, Libya, Morocco, Nigeria, Oman, Pakistan, Peru, South Africa, Spain, Switzerland, UAE and Venezuela. Magic 1s were used by the Argentine Air Force in the Falklands conflict in 1982, by Iraq in the Iran-Iraq War between 1981 and 1988 and by Kuwait in 1990.

Magic 2 production deliveries started in 1985 and the missiles entered service in the same year with the French Air Force and Navy. It is believed that a total of 10,000 Magic series missiles have been produced. An undisclosed Central European country placed an order for Magic 2 in January 1996 to modernise two types of aircraft of Soviet design.

Magic 2 Mk 2 has been exported to an undisclosed number of countries.

Specifications

Magic 1
Length: 2.72 m
Body diameter: 157 mm
Wing span: 0.66 m
Launch weight: 89 kg
Warhead: 12 kg HE fragmentation
Fuze: IR active
Guidance: IR single detector
Propulsion: solid propellant
Range: up to 3 km

Magic 2/Magic 2 Mk 2
Length: 2.75 m
Body diameter: 157 mm
Wing span: 0.66 m
Launch weight: 89 kg
Warhead: 12 kg HE fragmentation
Fuze: RF active
Guidance: IR multidetector
Propulsion: solid propellant
Range: up to 20 km (depending on launch altitude and velocity

Contractor

MBDA (France).

VERIFIED

Mectron MAA-1

Type

Air-to-air missile.

Development

Summary: the missile has an infra-red seeker.

Development of the MAA-1 air-to-air missile, sometimes known as Piranha or Mol, began in the mid-1970s. It was expected to enter service with the Brazilian Air Force in the early 1990s. Since 1993 the Brazilian company Mectron Engineering has been responsible for MAA-1 development.

First successful air launch of the missile took place in 1996 from an AT-26 Xavante aircraft. It is designed to be able to be launched from Mirage III, F-5 Tiger II, AT-26 Xavante and the AMX (A-1) aircraft.

Description

The MAA-1 missile is similar in appearance to the AIM-9 Sidewinder, but has a wider body diameter. The nose-mounted infra-red seeker has the capability for all-

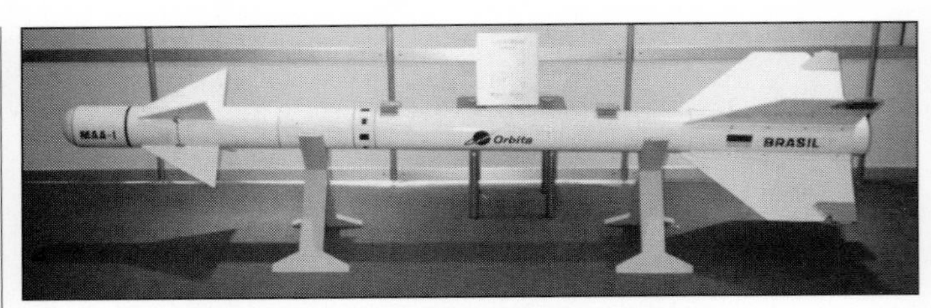

MAA-1 missile exhibited in Iraq in 1989 (Christopher F Foss)

aspect engagements and there is an active laser proximity fuze with a 12 kg HE fragmentation warhead. The seeker has a ±40° look angle and a 50°/s tracking speed is claimed. Targets can be acquired autonomously or by slaving the missile seeker to the aircraft radar. Control is by the front fins and the rear wings have rollerons like the Sidewinder. The maximum flight time is 40 seconds and the range is reported to be 5 km.

Operational status

The MAA-1 was developed for the Brazilian Air Force and was originally expected to enter service in the early 1990s. An air-launched test firing from an AT-26 Xavante aircraft was reported in mid-1996. Unconfirmed reports in 1992 suggested that Iraq had purchased MAA-1 design data and would continue with the development. Nothing further has been heard about Iraq's involvement in the programme. While there was no announced confirmation of a production order, this process may have started in 1998.

In November 2000, there was still no confirmation that the MAA-1 had entered service. In April 2001 Mectron confirmed that the MAA-1 had been cleared for wingtip carriage by the F-5 and ALX and for underwing carriage on the Mirage III and also on the ALX.

Specifications

Length: 2.82 m
Body diameter: 152 mm
Wing span: 0.65 m
Launch weight: 90 kg
Warhead: 12 kg HE fragmentation
Fuze: active laser
Guidance: IR
Propulsion: solid propellant
Range: 5 km

Contractor

Mectron Engenharia Industria e Comercio.

UPDATED

Molniya AA-8 'Aphid' (R-60)

Type

Air-to-air missile.

Development

AA-8 'Aphid' is the NATO code and designation for the Russian third-generation air-to-air missile, the R-60. 'Aphid' was developed in the late 1960s to early 1970s, as a replacement for the AA-2 'Atoll'. Although there are reports of a semi-active radar version, none have been seen and it seems likely that the IR is the only version to have entered production. An improved version, R-60M, was developed in the early 1980s. The missiles have been seen carried by MiG-21 'Fishbed', MiG-23 'Flogger', MiG-25 'Foxbat', MiG-29 'Fulcrum', MiG-31 'Foxhound', Su-17 'Fitter D', Su-22 'Fitter J', Su-24 'Fencer', Su-27 'Flanker' and reported on Su-21 'Flagon', Su-25 'Frogfoot' and Yak-38 'Forger'. The MiG-21 and MiG-31 have both been seen fitted with two missiles, mounted one above the other on underwing launcher pylons. In addition, four 'Aphid' missiles have been seen on Mi-24 'Hind D' and 'Hind E' helicopters, carried on the inner stub-wing pylons, and in 1997 fitted to an upgraded Romanian SA330L Puma helicopter.

Description

AA-8 has four rectangular fixed canards and four triangular moving control fins at the nose, with four long-chord clipped-tip delta-wings at the rear. The four fixed delta-wings have 'Sidewinder' type rollerons at the trailing edge for roll stabilisation. 'Aphid' is 2.08 m long, has a body diameter of 130 mm, a wing span of 0.43 m and a weight of 63 kg. The missile has two active radar fuze aerials located aft of the moving control fins, and a single strake running down the forward half of the body. The 3 kg HE fragmentation warhead contains 1.6 kg of uranium. The earlier 'Aphid' missiles were restricted to tail aspect engagements only and had a maximum range of 3 km. It is believed that an improved AA-8 version (Russian designator R-60M) has an active laser fuze, to match an all-aspect engagement capability; this version has an increased range to around 10 km. This version has a length of 2.09 m, a body diameter of 120 mm, and a launch weight of 43 kg. R-60M has a minimum range of 0.2 km. Intercepts can be made at altitudes between 30 m and 20 km. The missile (R-60M) is reported to be able to engage a fighter aircraft manoeuvring at up to 8 g, and to have a larger warhead with a weight of 3.5 kg. A further modification, designated R-60MK, is believed to have been introduced to enable the AA-8 missile to be designated by the pilot's helmet-mounted sight in the MiG-29 'Fulcrum'. In 1997 CDO Arsenal, from Ukraine displayed an upgraded IR seeker

Two AA-8 'Aphid' missiles on the wing pylon of a MiG-31 'Foxhound', showing the Sidewinder-type rollerons at the wing trailing edges (Duncan Lennox)

assembly with four additional IR detectors mounted externally. The additional detectors are to improve the sensitivity of the seeker in the forward hemisphere, enabling target detection out to beyond 10 km. The upgraded seeker assembly has a ±45° field of view, and can track targets crossing at up to 30°/s. The upgraded IR seeker assembly is designated UA-96, and can be retrofitted to R-60, R-60M and R-60MK standard missiles.

Operational status

It is believed that AA-8 'Aphid' entered service with the Russian Air Force and Navy in 1973 (R-60) and the improved version R-60M in 1982. There have been many export orders, including Afghanistan, Algeria, Angola, Azerbaijan, Belarus, Bulgaria, China, Croatia, Cuba, Czech Republic, Georgia, Germany, Hungary, India, Iraq, Kazakhstan, North Korea, Libya, Malaysia, Poland, Romania, Slovakia, Sudan, Syria, Ukraine, Vietnam and Yugoslavia (Serbia and Montenegro). AA-8 missiles have been built under licence in Romania, with the designator A-95, and more recently the A-960.

Specifications

R-60
Length: 2.08 m
Body diameter: 130 mm
Wing span: 0.43 m
Launch weight: 65 kg
Warhead: 3 kg HE fragmentation
Fuze: active radar
Guidance: IR
Propulsion: solid propellant
Range: 3 km

R-60M
Length: 2.09 m
Body diameter: 120 mm
Wing span: 0.43 m
Launch weight: 43 kg
Warhead: 3.5 kg HE rod
Fuze: active laser
Guidance: IR
Propulsion: solid propellant
Range: 10 km

Contractor

Molniya, OKB.

VERIFIED

Rafael Python 3/Python 4/Python 5

Type

Air-to-air missile.

Development

Summary: the missile has an infra-red seeker.

Design and development of the Python 3 missile system started in 1978 and was based upon the earlier Shafrir 1 and 2 designs. In the early days of development it was known as Shafrir 3. The major improvements were a new all-aspect infra-red seeker with increased sensitivitiy and look angle and improved countermeasures capability. Python 3 has been cleared for carriage on the Kfir, Mirage 3, Mirage F1, Mirage 2000, J-7, F-4, F-5, F-7, F-15 and F-16 aircraft.

A later version, Python 4, has been developed and is in production. It has been designed for fitment to a wide range of aircraft, optimised for close-range high off-boresight engagements and for use with helmet-mounted sights. It is believed to be the only missile in service outside Russia to have a high-boresight capability. It has been cleared for a number of aircraft including the Kfir, Mirage 3, Mirage F-1, Mirage 2000, F-15 and F-16 aircraft and is believed to have entered service with the Israel Defence Force, Air Force in 1992.

It has been known that Israel has been working on a new 'dogfight' missile since 1999 but the Python 5 did not make its public appearance until June 2003 at the Paris air show.

Two Python 3 air-to-air missiles on the wingtip and underwing pylons of an Israel Defence Force, Air Force F-16 aircraft

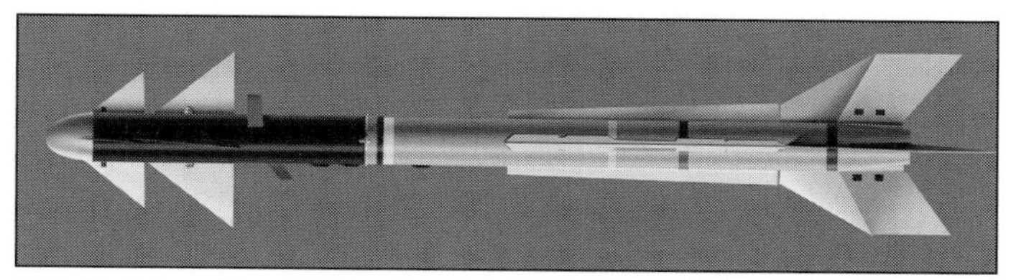

Rafael's Python 4 fourth-generation air-to-air missile 0053939

Description

Python 3 is a short-range infra-red air-to-air missile similar in appearance but slightly larger than the AIM-9 Sidewinder and Shafrir 2. The all-aspect seeker has a countermeasures capability and can be slaved to the aircraft radar or used in the scan and boresight modes. The maximum off-boresight angle is 30° prior to launch and 40° during missile flight. The missile can manoeuvre up to 40 *g*. It is believed to have a range of 5 km at low level, although reports indicate a range of up to 15 km at medium altitudes. Python 3 has a minimum range of 0.5 km.

Python 4 is a short-range, all-aspect infra-red homing missile powered by a solid propellant motor and armed with a fragmentation warhead. No official specifications have been released, but the missile is reported to be 3.0 m long, to have a body diameter of 160 mm, a wing span of 0.35 m (recent photographs suggest this may have increased to just over 0.6 m) and to weigh around 105 kg. As the body component layout and diameter is similar to that of Python 3, there is good reason to believe that the blast/fragmentation warhead and active laser fuze could well be updated versions of those used in Python 3. Guidance is by an advanced all-aspect imaging infra-red seeker with what is believed to be a ±90° look angle and a ±60° off-boresight launch capability. There are unconfirmed reports that the seeker has the ability to discriminate in three spectral bands.

In order to engage a target, even at very high off-boresight angles, the pilot simply has to look at the target through his helmet-mounted sight and fire. The solid propellant motor is said to have a long burn time, increased thrust accelerating the missile to greater than M1.0 and is credited with giving the missile a maximum range of 15 km.

Python 5 retains the airframe, rocket motor, warhead and proximity fuse used in the current Python 4 missile, but teams these with a new fully-imaging two-band seeker based on a focal-plane array. The new seeker allows lock-on at more than 100° off-boresight. It has a wider field of view than the seekers used on the ASRAAM, IRIS-T or AIM-9X, according to Rafael. This wide angle was required to ease target-acquisition in a dogfight and to allow Lock-On After Launch (LOAL) attacks. Dual-band operation, and the use of advanced guidance algorithms, are intended to provide a high degree of resistance to countermeasures. The seeker also features increased detection range, and is designed to acquire low-signature targets against adverse backgrounds, while having a low false-alarm rate.

A new missile-mounted electronics unit located just aft of the seeker incorporates an advanced computer architecture, and provides the increased computing power needed by the imaging seeker. An Inertial Navigation System (INS) mounted just ahead of the proximity fuse is used for mid-course guidance

The extreme nose of the Python 5 contains a fully imaging dual-band seeker (Doug Richardson) ***NEW***/0549686

during LOAL attacks. It allows trajectory shaping to maximise energy and improve the weapon's long-range capability, and wing-twist compensation intended to provide accurate seeker pointing. Prior to missile launch in LOAL mode, the aircraft transfers inertial data to the missile's INS, plus target data.

Once close to the target, the missile begins a search with its seeker, then begins final homing once lock-on has been obtained. The solid-propellant rocket motor has a burn time of around 7.5 seconds (unchanged from Python 4), but the guidance and control servo systems have been modified so that they remain effective for more than three times the duration achieved with the older missile. This extended control time allows long-range LOAL attacks. The maximum range of Python 5 has not been revealed, but is in the Beyond Visual Range (BVR) category.

Operational status

Python 3 entered service with the Israel Defence Force, Air Force in 1982. There have been export orders and it is believed that China, South Africa and Thailand have been amongst the purchasers.

It is reported that Python 4 is in limited production and has been in service with the Israel Defence Force, Air Force since 1992.

In October 1998, Rafael and Lockheed Martin formalised an agreement to co-operate in production and marketing of Python 4 as an extension of the PGSUS LLC joint venture established for the AGM-142 Have Nap/Popeye missile. Several reports allege that Python 4s have been delivered to Singapore, though the missile is not believed to be in service with the Republic of Singapore Air Force.

In 2000 it was reported that Ecuador had ordered the Python 4 and Rafael has released a photo of an Ecuadorian Air Force Kfir TC-2 carrying Python 4 acquisition rounds (and Ecuador has since acquired the improved Kfir 2000). The Royal Australian Air Force also evaluated the Python 4 for its AIR5400 next-generation AAM requirement (before selecting the MBDA ASRAAM). A report in 1997 suggested that China was evaluating the Python 4 against an upgraded AA-11 'Archer'.

The Israeli Air Force was carrying out evaluation firings of the Rafael Python 5 agile air-to-air missile in mid-2003, and expects to deploy the weapon in early 2005.

Specifications

	Python 3	Python 4
Length	3.0 m	3.0 m
Body diameter	160 mm	160 mm
Wing span	0.35 m	0.35 m (possibly larger)
Launch weight	120 kg	105 kg
Warhead	11 kg HE fragmentation	11 kg HE fragmentation
Fuze	active radar	active radar
Guidance	IR	IR
Propulsion	solid propellant	
Range	15 km	15 km

Contractor

Rafael Armament Development Authority.

UPDATED

Raytheon DAMASK seeker

Type

Air-to-air missile (seeker).

Development

In September 1998, Raytheon Systems Company announced that it had received a US$11 million US Navy contract to provide Uncooled Focal Plane Array technology support to the Naval Air Warfare Center Weapons Division, China Lake, California, for development of the Direct Attack Munition Affordable Seeker (DAMASK) programme. The company will apply its commercial NightSight™ uncooled infra-red technology in support of this effort.

Currently an advanced Technology Demonstration programme, DAMASK's goal is to demonstrate a low-cost, imaging upgrade kit for Global Positioning Systems/ Inertial Navigation Systems guided weapons. As an industry partner with the Naval Air Warfare Center, Raytheon will provide expertise and hardware until the end of April 2003 for a variety of precision guided weapon systems. The Joint Direct Attack Munition is scheduled to be the first candidate weapon system to incorporate Uncooled Focal Plane Array seeker technology.

"We believe imaging infra-red Uncooled Focal Plane Array camera/sensor technology will be instrumental in meeting future precision guided munitions seeker requirements," said Tony Leatham, Raytheon's Uncooled Focal Plane Array technology support program manager. "This award provides us the first opportunity to leverage Raytheon's high-quality, low-cost commercial uncooled IR products from our commercial Uncooled IR business, into a precision guided munition seeker for the US Navy. It's a 'win-win' situation."

Operational status

Development.

Contractor

Raytheon Systems Company.

VERIFIED

Raytheon FIM-92 Stinger

Type

Air-to-air missile.

Development

Summary: the missile has an Infra-Red (IR) seeker; Stinger POST (Passive Optical Seeker Technique) version has an Infra-Red/Ultra-Violet (IR/UV) seeker.

Although the portable surface-to-air FIM-92A Stinger (qv) has been in production since 1979 and trials of a helicopter-mounted air-to-air variant started in 1978, the full-scale engineering development programme for the air-to-air missile system did not start until 1984. The Air-To-Air Stinger (ATAS) programme provides a lightweight IR/UV missile for use at short range against low-flying aircraft and helicopter targets. Stinger seeker improvements incorporate a dual-colour IR and UV seeker for decoy discrimination and the latter a reprogrammable microprocessor for subsequent counter-countermeasures modifications with a rosette scan pattern for the detector. The FIM-92D Block 1 upgrade programme started in 1992 and is expected to include a new ring-laser gyro roll sensor, longer lasting lithium battery, computer memory and software upgrades.

From 1997 the Raytheon Missile Systems has been under contract for Engineering and Manufacturing Development Phase, a second upgrade programme, Block 2 (FIM-92E). This involves the development and installation of a 128 × 128 staring infra-red focal plane array seeker. The seeker will be incorporated into the standard Stinger airframe, and used for forward air defence and in ATAS missiles.

A European production capability has been set up with companies from Germany, Greece, Netherlands, and Turkey participating and being led by LFK (Daimler-Benz Aerospace). The FIM-92 Stinger has been carried by the OH-58C Kiowa, OH-58D Kiowa Warrior, UH-60A and MH-60 Black Hawk, AH-64A Apache, AH-64D Longbow Apache, AH-1S Cobra, BO 105, AH-1W SuperCobra and the A129 Mangusta helicopters. In December 2002, it emerged that had been fitted to RQ-1 Predator drones for use in armed missions over Iraq.

Description

The FIM-92 Stinger missile is a short-range missile powered by a solid propellant motor and armed with a fragmentation warhead. In the air-to-air variant (ATAS), the complete system consists of the FIM-92 Stinger missile which is stored and delivered in its launch tube, launcher with launcher electronics, launcher adaptor, coolant reservoir, interface electronics unit and the fire-control and aiming system. The most common helicopter launcher in use is a lightweight, stackable, two-round quick-reload launcher that is of modular construction and contains both coolant and electronics system and its own built-in test equipment. The launcher adaptor has an eight-missile control capability. The aiming and fire-control system varies according to launch vehicle but the ATAS system uses a simple boresight acquisition system, with the pilot uncaging the missile seeker and firing when he has a clear audio tone. The pilot's sight is normally by Head-Up Display (HUD) or

US Army OH-58C Kiowa helicopter carrying two twin FIM-92 Stinger launchers

FIM-92 Stinger shown outside its launch canister (Duncan Lennox)

reticle sight. Other launchers in use are a single missile round or a two-missile, low-drag, faired launcher for high-speed aircraft.

For the current ATAS system the early Stinger guidance seekers have been replaced by the later RMP (Reprogrammable MicroProcessor) seeker. The RMP seeker employs an advanced guidance system that uses a rosette scan as opposed to the basic Stinger which uses two rotating mirrors. The image scan method enhances target detection with the two-colour seeker providing an option to track the infra-red or ultra-violet spectral band. RMP incorporates two detector materials, one sensitive to infra-red, the other to ultra-violet energy. The advantages gained with the image scan, two-colour system are further advanced by signal processing using six microprocessors. Stinger RMP is said to be able to compare and lock on to the larger of two heat sources and to be capable of acquiring most helicopter targets from all aspects and discriminate between decoys and background clutter.

After launch the argon cooled seeker performs normally for an infra-red missile, homing on the target's exhaust plume. But shortly before impact the seeker measures the rate of change of target energy and, in the terminal phase, the target adaptive guidance circuitry veers Stinger away from the exhaust to penetrate the target airframe. Since Stinger relies upon physical impact with the target, it is normally immune to proximity deception measures.

The RMP provides the capability easily to change the missile software logic via a module external plug interface as the infra-red countermeasures threat changes.

Operational status

The air-to-air FIM-92 Stinger programme entered service with the US Army in 1988. A second source production contract was placed with Raytheon in 1987 and flight tests were completed in 1991. Delivery of FIM-92B Stinger surface-to-air missiles to Stinger RMP standard began in 1988 and these are now used in the air-to-air role. A trials programme to equip the AH-64 Apache helicopter with FIM-92 Stinger air-to-air missiles conducted firings in 1989 and 1990, with a four-missile installation. Development of Block 1 upgrades started in 1992. The Block programme was started in 2000 with operational trials 2002-2004. Although the surface-to-air missiles have been widely exported, there have been no known exports of the ATAS.

Specifications

Length: 1.52 m
Body diameter: 70 mm
Wing span: 0.14 m
Launch weight: 16 kg
Warhead: 3 kg HE blast fragmentation
Fuze: impact
Guidance: IR
Propulsion: solid propellant
Range: 3 km

Contractor

Raytheon Company.

UPDATED

Russian SA-16 'Gimlet' (9M313 Igla I) and SA-18 'Grouse' (9M39 Igla)

Type

Air-to-air missile.

Development

Summary: the missile has an infra-red seeker.

The SA-16 'Gimlet' was first confirmed in use in 1987. The missile is called Igla 1 by the Russian Federation in the surface-to-air variant, with the designator 9M313. The SA-18 'Grouse' has the indigenous designation 9M39 Igla and is now believed to have entered service before SA-16 'Gimlet', the missiles being very similar. There are reports that two further versions of SA-16 are in development, Igla 1M and Igla 2. It is believed that SA-16 and SA-18 are carried by Mi-24 'Hind' and probably Mi-28 'Havoc' helicopters.

Description

The missile has four small pop-out rectangular control fins near the nose and four rectangular fins at the tail. The improved infra-red seeker head most probably has a two-colour detection system to discriminate flares and an improved all-aspect capability. The rear body shape is similar to the US Stinger design and the seeker dome has a cone supported in front to reduce drag. Digital electronics can be expected, together with a longer standby capability. Boost and sustainer motors give the SA-16 a velocity of 570 m/s and a range of about 5 km. SA-16 has a minimum range of 500 m and can intercept targets at between 10 m and 3,000 m altitude. SA-18 'Grouse' (9M39 Igla) is similar to SA-16, except that a single spike was used in front of the seeker dome. The performance, size and weight of SA-16 and SA-18 are believed to be the same.

Operational status

It is believed that SA-18 'Grouse' entered service in Russia in 1983, followed by SA-16 'Gimlet' in 1986. It is believed that both missiles have been exported to Angola, Bulgaria, Czech Republic, Finland, Hungary, Iraq, North Korea, Poland and Slovakia. Licenced production has been carried out in Bulgaria and there are unconfirmed reports that North Korea is also building SA-16/-18 missiles. Some

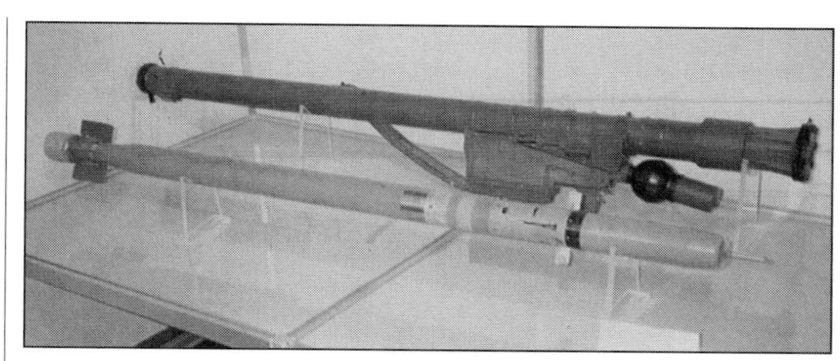

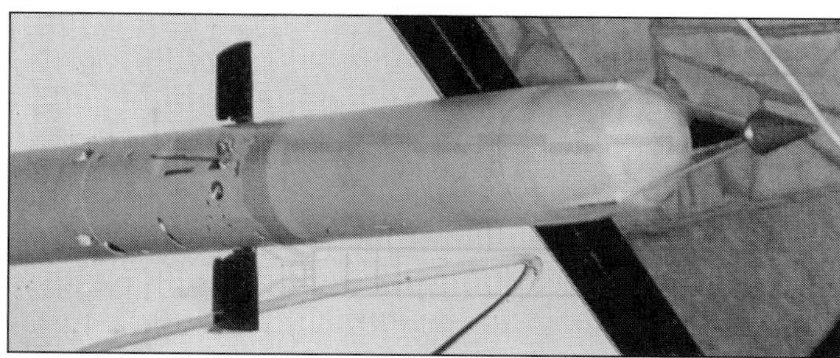

The top picture shows an SA-18 (Igla) missile out of a container, and the lower picture shows an SA-16 (Igla-1) missile nose assembly to demonstrate the different nose spike/cone arrangements (Christopher F Foss)

SA-16 missiles are believed to have been used by Iraq during the 1991 Gulf War. There have been no confirmed reports of SA-16 or SA-18 on helicopters.

Specifications

Length: 1.69 m
Body diameter: 72 mm
Wing span: 0.25 m
Launch weight: 10.8 kg
Warhead: 2.0 kg HE fragmentation
Fuze: impact
Guidance: IR
Propulsion: solid propellant
Range: 5 km

Contractor

Russian State Factories.

VERIFIED

Spetztekhnika Vympel AA-10 'Alamo' (R-27)

Type

Air-to-air missile.

Development

Summary: the R-27T and R-27ET missiles have infra-red seekers.

AA-10 'Alamo' is the NATO designation for the Russian Federation fourth-generation infra-red and radar-guided air-to-air missiles. The Russian designator is R-27 and 'Alamo' development probably started in the mid-1970s. There are six versions of the AA-10: medium-range infra-red (R-27T); medium-range semi-active radar (R-27R); extended-range infra-red (R-27ET); extended-range semi-active radar (R-27ER); long-range semi-active radar modified for low-level engagements (R-27EM); and extended-range active radar (R-27AE). It was reported in 1995 that development of the active radar version, R-27AE, had been halted. There are unconfirmed reports that a seventh version (R-27P) has been developed, with a

A medium-range IR-guided AA-10 'Alamo' (R-27T) showing behind an extended-range motor assembly indicating the different length and diameter of the rocket motors (Christopher F Foss)

Specifications

	R-27T	R-27R	R- 27ET	R-27ER	R-27AE	R- 27EM
Length	3.70 m	4.00 m	4.50 m	4.70 m	4.78 m	4.78 m
Body diameter	230 mm	230 mm	260 mm	260 mm	260 mm	260 mm
Wing span	0.77 m	0.77 m	0.8 m	0.8 m	0.8 m	0.8 m
Launch weight	254 kg	253 kg	343 kg	350 kg	350 kg	350 kg
Warhead	39 kg expanding rod	39 kg expanding rod	39 kg expanding rod	39 kg expanding rod	39 kg expanding rod	39 kg expanding rod
Fuze	Active radar	Active radar	Active radar	Active radar	Active radar	Active radar
Guidance	Inertial, command and IR	Inertial, command and SAR	Inertial, command and IR	Inertial, command and SAR	Inertial, command and active radar	Inertial, command and SAR
Propulsion	Solid propellant	Solid propellant	Solid propellant	Solid propellant	Solid propellant	Solid propellant
Range	40 km	50 km	70 km	75 km	80 km	110 km

passive radar seeker designed to intercept fighter aircraft, AEW aircraft or standoff jammers. The AA-10 is carried by the Su-27 'Flanker' and MiG-29 'Fulcrum'. It is believed that it will also be fitted to the Su-35 multirole fighter and could be retrofitted to modified MiG-21 'Fishbed', MiG-23 'Flogger' and MiG-25 'Foxbat' aircraft.

Description

AA-10 'Alamo' has four fixed chamfered rectangular canards at the nose, four large trapezoidal 'butterfly' moving control fins and four fixed clipped-tip delta-wings at the rear. The guidance system for mid-course is inertial with command updates, but alternative terminal seeker assemblies can be fitted to the six versions.

AA-10 'Alamo' has a capability from sea level to 70,000 ft with the extended-range versions possibly having some limited capability to 90,000 ft. The range against low-level targets is probably 20 km. Against medium- and high-level targets, it varies from 40 km for R-27T out to 110 km for R-27EM. Minimum firing range for all marks of AA-10 is reported to be 500 m and it is believed that the missile can follow target manoeuvres up to 8 *g*. The semi-active radar terminal seekers have a range of about 25 km and operate in J-band (10 to 20 GHz) with CW monopulse systems. The Su-27 'Flanker' can carry up to six AA-10 missiles and can carry any of the versions of the AA-10. The MiG-29 'Fulcrum' has also been seen with both IR and radar versions on its two inner underwing pylons.

Operational status

It is believed that the AA-10 'Alamo' R-27T, R-27R, R-27ET and R-27ER versions entered service in 1985 and are probably still in production. The R-27EM probably entered service in 1990, but the status of R-27AE is unclear and probably remains in development. The R-27P version remains unconfirmed, but probably in development. There are unconfirmed reports that AA-10 has been exported to Belarus, Bulgaria, China, Cuba, Czech Republic, Germany, Hungary, India, Iran, Iraq, Kazakhstan, North Korea, Malaysia, Poland, Romania, Slovakia, Syria, Ukraine and the former Yugoslavia (Serbia and Montenegro).

Contractor

Spetztekhnika Vympel NPO, Moscow.

VERIFIED

Spetztekhnika Vympel AA-11 'Archer' (R-73)

Type

Air-to-air missile.

Development

Summary: the missile has an infra-red seeker for terminal guidance.

AA-11 'Archer' is the NATO designation given to this fourth-generation short-range IR-guided air-to-air missile. This missile has the indigenous designation of R-73 and there are two versions known as R-73M1 and R-73M2. The development of 'Archer' probably started in the late 1970s and it is believed that the missile entered service around 1987. It has most probably been designed for use by fighter aircraft against agile opponent fighters, for the design emphasises manoeuvrability. The missile has been cleared for carriage on the Su-27 'Flanker', MiG-23 'Flogger', MiG-29 'Fulcrum', MiG-31 'Foxhound' and MiG-31M. It is expected to be fitted to the Su-34 and Su-35 aircraft.

Two AA-11 'Archer' missiles on the wingtip and outboard wing pylons of an Su-27 'Flanker' aircraft (Paul Beaver)

Description

AA-11 'Archer' has at the nose four small swept rectangular moving control fins or sensors, followed by four fixed rectangular fins and then four clipped-tip triangular moving fins. The missile has inertial mid-course guidance with a terminal two-colour IR seeker and is believed to have an all-aspect engagement capability as well as the ability to discriminate against flare decoys. AA-11 has the capability to be designated and to lock onto a target before launch, with designation from the aircraft radar or IRST or the pilot's helmet-mounted sight. The fuze appears to be an active radar type, similar to that used on AA-8 'Aphid', which is surprising since the West has found that active laser fuzes give a better performance for short- or medium-range all-aspect air-to-air missiles. It is therefore possible that two fuze types can be fitted. The R-73M1 version has a minimum range of 300 m and a maximum range of 20 km, with an off-boresight capability for target designation before launch out to 45°, using the helmet-mounted sight. The R-73M2 version has an off-boresight capability improved to 60° and a maximum range against a typical fighter target of 30 km. Both R-73M1 and R-73M2 missiles can track targets in flight with off-boresight angles greater than these designation limits, increasing to 60° and 80° respectively; it is believed that the missiles can follow targets manoeuvring up to 12 *g* with sightline spin rates up to 60°/s. The R-73M2 has digital control electronics and IRCCM, which presumably can be reprogrammed as decoy flares change. The greater range means that this version has a longer burning rocket motor, with suggestions that this missile can turn through 180° after launch. The R-73M2 version is believed to have a capability to attack low-flying missiles.

Operational status

It is believed that AA-11 'Archer' entered service in the former Soviet Union during 1987. There have been reports that AA-11 missiles have been exported to Belarus, Bulgaria, China, Cuba, Czech Republic, Germany, Hungary, India, Iran, Iraq, Kazakhstan, North Korea, Malaysia, Poland, Romania, Slovakia, Syria, Ukraine and the former Yugoslavia (Serbia and Montenegro).

Specifications

	R-73M1	R- 73M2
Length	2.9 m	2.9 m
Body diameter	170 mm	170 mm
Wing span	0.51 m	0.51 m
Launch weight	105 kg	110 kg
Warhead	HE 7.4 kg fragmentation	HE 7.4 kg fragmentation
Fuze	active radar	active radar
Guidance	inertial and IR	inertial and IR
Propulsion	solid propellant	solid propellant
Range	20 km	30 km

Contractor

Spetztekhnika Vympel NPO, Moscow.

VERIFIED

Thales Air-to-Air Starstreak (ATASK)

Type

Air-to-air-missile.

Development

Thales Air Defence Ltd is associated with Lockheed Martin and Boeing (formerly McDonnell Douglas Helicopter Systems) in funded trials by the US Army of an air-to-air version of their Starstreak laser beam-riding missile on the Apache helicopter. The air-to-air Starstreak (ATASK) system is designed to provide the Apache with a defensive capability to complement its air-to-ground attack role.

Phase I, valued at US$6 million and covering a technical feasibility programme including six missile firings from the AH-64D Longbow Apache during ground, hover and air tests, was successfully completed in October 1996.

Phase II, valued at US$12.9 million, was awarded in February 1997. This covers integration of the Starstreak laser guidance system into the Apache's Target Acquisition and Designation Sight (TADS) and the aircraft's advanced fire-control

Longbow Apache helicopter fitted with Starstreak ATASK missiles 0005503

Starstreak ATASK missile being fired 0005504

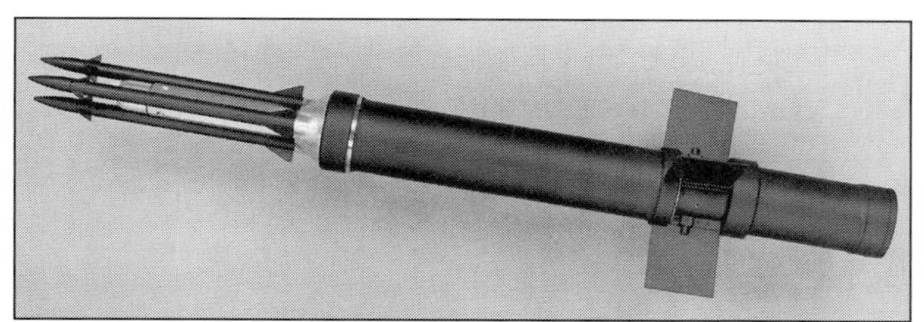

The Starstreak air-to-air missile 0053937

system, providing the pilot with a fully automatic 'hands-off' guidance capability. During a series of live-fire tests from a US Army AH-64 Apache in October and November 1998, Starstreak successfully engaged static and airborne helicopter targets six times in six attempts.

A third phase is planned as a 'shoot-off' (against Air-to-Air Stinger) and is due in the near future. It is expected to last two years and will be conducted on an AH-64D Apache Longbow.

The essentially-similar Helstreak has been considered for the AH-I Cobra, A 129 Mangusta, Tiger/Tigre and Lynx helicopters.

Description

The Starstreak ATASK missile contains three separately guided darts. Each dart contains a thermal battery, laser receiver, steering mechanisms, high-density penetrating warhead and impact fuse with delay before fragmentation and is contained in a two-tube launcher.

The missile is two-stage - in the first stage the launch motor ejects the missile from the tube. The flight motor then boosts the missile to hypervelocity and a separation system ejects the darts from the missile.

Total time-to-fire is stated as 3 seconds and the maximum range of the missile is over 5 km.

Operational status

Under development. Due for trials by the US Army on AH-64D Longbow Apache helicopter.

Specifications

Range: >5 km
Speed: M3 plus

Contractor

Thales Air Defence Ltd.

UPDATED

USAF AIM-9 Sidewinder

Type

Air-to-air missile.

Development

Summary: all versions of the missile have infra-red seekers. AIM-9X is due to have an imaging infra-red seeker.

This development programme started at what is now called the US Naval Weapons Center, China Lake, California in the late 1940s. The first prototype flew in 1953 and the first generation of Sidewinder, the AIM-9B, entered service with the US Navy and US Air Force in 1956. Designed by the US Navy staff at China Lake, the Sidewinder family is now in its third generation. The first generation of Sidewinder, AIM-9B, divided into three separate development programmes funded by the US Army, US Navy and US Air Force. The US Air Force developed the AIM-9E. These second-generation systems entered service in 1965 but were improved again and the US Air Force developed AIM-9J and AIM-9P. Export versions of AIM-9J were designated AIM-9N and AIM-9P. Finally, the US Navy and Air Force came together and moved to a joint development programme for the third generation of Sidewinder missiles in 1970, making the major performance change from earlier tail-aspect engagement only systems in the first- and second-generation Sidewinders, to an all-aspect capability with AIM-9L Sidewinder. Production of AIM-9L started in the US in 1976 and under licence in Europe and Japan in the early 1980s. Further development continued and the AIM-9M version entered production in the US in 1982 with principally an infra-red countermeasures capability to detect and reject decoy flares. AIM-9S is almost the same as AIM-9M, but with a slightly larger warhead. AIM-9R was being developed to improve further on AIM-9M and was expected in service in the early 1990s, fitted with a visual band CCD seeker, but this programme was halted in 1992.

Sidewinders have been fitted to a very large number of aircraft throughout the world and these have included the F-104, F-4, F-5, F-8, A-4, A-6, A-7, Mirage III, MiG-21, A-10, JA-37 Viggen, Kfir, F-20, OV-10, Mirage F-1, Mitsubishi F-1, Hawk, Sea Harrier, Harrier, Tornado GR1, Tornado F3, Nimrod MR2, Jaguar, F-14, Buccaneer, F-15, F-16 and F/A-18. There have been trials from helicopters over several years, including AH-64A Apache in 1987 and AH-1 Cobra in 1988. AIM-9M/S missiles have been launched from the YF-22 Lightning (ATF) prototype.

AIM-9X, or Sidewinder 2000, is under development. An 18-month demonstration/validation phase testing the new imaging infra-red seeker concluded in 1996 with Raytheon being awarded a US$169 million contract for engineering and manufacturing development of the Evolved Sidewinder. AIM-9X is a joint US Navy and Air Force programme with the US Navy serving as the executive service. US requirement is expected to be 10,000 missiles. First deliveries were

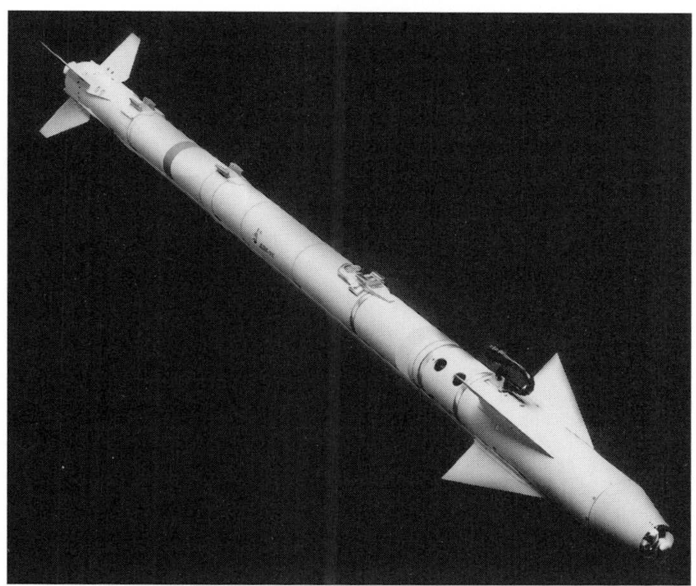

The AIM-9X missile 0053938

AIM-9L missile on the underwing pylon of EF-2000 at Paris in 1995 (Peter Humphris)

Specifications

	AIM-9B	AIM-9D	AIM- 9L/9M	AIM-9P	AIM-9S
Length	2.83 m	2.87 m	2.87 m	3.07 m	2.87 m
Body diameter	127 mm	127 mm	127 mm	127 mm	127 mm
Wing span	0.53 m	0.64 m	0.64 m	0.64 m	0.64 m
Launch weight	76 kg	90 kg	87 kg	82 kg	86 kg
Warhead	4.5 kg HE blast fragmentation	9 kg HE continuous rod	9.5 kg (11.4 kg 9M) HE blast fragmentation	12 kg HE blast fragmentation	10.15 kg HE fragmentation
Fuze	IR	RF	Active laser	RF or active laser	Active laser
Guidance	IR	IR	IR	IR	IR
Propulsion	Solid propellant	Solid propellant	Solid propellant	Solid propellant	Solid propellant
Range	2 km	3 km	8 km	8 km	8 km

made to the US Navy and US Air Force in mid-2003, with an Initial Operating Capability (IOC) on the F-15C and F/A-18C/D later in 2003.

Description

The AIM-9 Sidewinder missile has four swept front control fins and four clipped delta wings at the tail with distinctive rollerons at the trailing-edge tips. The infra-red seeker has an all-aspect engagement capability and the 9.5 kg blast fragmentation warhead has an active laser fuze. Several modes can be used depending upon the avionics fit in the carrying aircraft; primarily there is the simple boresight mode, an uncaged scan mode and a mode with the missile seeker slaved to the aircraft radar or to a helmet-mounted sight. The AIM-9M version is known as AIM-9S in USAF service and in the later 9P-3 and 9P-4 models, there is an all-aspect engagement capability; in addition, 9P-4 has an active laser fuze similar to that first introduced on AIM-9L. AIM-9P-5 has additional IRCCM capability.

The AIM-9X Evolved Sidewinder will have an advanced 128 × 128 element staring focal plane array infra-red guidance system from Raytheon, which was originally developed for the Matra BAe Dynamics ASRAAM missile. The AIM-9X missile will provide enhanced capabilites for US Navy and Air Force fighter aircraft in short-range air-to-air combat. The missile can be targeted and controlled with a helmet-mounted sight and is more manoeuvrable with a wider range of attack angles than previous versions. Alliant Techsystems will develop and produce the missile's solid propulsion rocket motor and thrust vector control system. Raytheon will produce the advanced tracker.

Operational status

AIM-9M and -9S Sidewinder are at present in production in the US and development of the AIM-9R started in 1987. Export variants and licence manufacture of AIM-9P are still in production. The total number of Sidewinder missiles built exceeds 200,000 and exports have been made to a large number of countries.

AIM-9X or Evolved Sidewinder is under development with IOC expected late-2003. A total US requirement of 10,000 missiles is projected. Apart from the USA, Poland, South Korea and Switzerland have all selected the missile.

Contractor

US Air Force.

UPDATED

AIR-LAUNCHED MISSILES

AIR-TO-AIR GUNS

Mauser short-range helicopter weapon system RMK 30

Type

Short-range weapon system for combat helicopter and other light carriers.

Description

The main assembly groups of the system are a 30 mm recoilless machine gun with a linkless ammunition feed system for caseless telescoped ammunition, a two-axis stabilised gun mount and a sensor carrier on the elevation axis of a common turret. The sensor carrier includes an electro-optical sight, comprising a FLIR camera, a high frequency laser range-finder, an intertial measurement unit and a two-axis stabilised mirror. The mirror is first stabilised and the gun follows any movement. The weapon system is optimised for air-to-air engagements and leads to very short reaction times, high hit probability and avoids dangerous recoil from gun to helicopter during fire.

In normal combat situations the target is assigned to the weapon system via the helmet mounted sight of the pilot or the gunner. Following the FLIR tracker, lock on automatic target tracking starts. From the data determined with the laser range-finder and the target tracking data in the fire-control computer, the prediction calculations are made including lead angle and elevation determination. Based on this, the gun mount is steered in the target direction. Then the combat process starts automatically after fire release.

The system is also suitable as a weapon system for light, high-mobility land vehicles which need a high fire power. In this case it will always be a remote controlled weapon system and as a minimum a CCD camera and a thermal imaging sight are planned as fire-control equipment. In addition, it is possible to adopt a multitude of variants up to the full-equipped sensor carrier to the task required.

Operational status

The recoilless gun complete with ammunition has already been tested experimentally. An advanced experimental programme for the complete weapon system was ordered by the German MoD in 1997. However, its original application on the Tiger attack helicopter has evaporated. It is now being promoted as a land vehicle-mounted system and is not expected to mature until 2006.

Mauser short-range helicopter weapon system RMK 30 0120737

Specifications

Calibre: 30 mm
Length of barrel: 1,700 mm
Muzzle velocity: 1,050 m/s
Traverse: 140°
Elevation: –45 to +20°
Speed: 175°/s
Acceleration: 260°/s^2
Mounting weight: 195 kg
Rate of fire: up to 300 rds
Ammunition: APHE, FAP, SAPHE

Contractor

Mauser-Werke Oberndorf Waffensysteme GmbH.

UPDATED

AIR-LAUNCHED MISSILES

AIR-TO-SURFACE MISSILES AND MUNITIONS

Aerospatiale AS 30

Type
Air-to-surface missile.

Development
Summary: the missile has semi-active laser terminal guidance.

The AS 30L (laser-guided) variant is a third-generation family of French air-to-surface weapons, designed to penetrate hardened targets. The first-generation AS 20 was a 132 kg radio command-guided missile that entered service in the mid-1950s. The second-generation AS 30 missile was considerably larger at 520 kg, entering service around 1960. This early version of AS 30 used the same radio command guidance system as the AS 20, but a later variant incorporated an aircraft-mounted infra-red tracker into the guidance loop, although the pilot still had to maintain the target in the centre of his sight throughout missile flight. AS 30L is the third-generation system and, with a laser target designator pod carried on the aircraft, allows the pilot to manoeuvre after missile launch to remain at a safe distance from the target area. Development of the AS 30L started in 1980 and it entered service in 1988. The missile is cleared for carriage by Jaguar, Super Etendard, F-16 Fighting Falcon, Mirage F-1 and Mirage 2000 aircraft. Adaptation to the Rafale fighter is in progress.

Description
AS 30L has four swept delta wings mid-body and four small clipped delta tail control fins. Mid-course guidance is inertial, followed by terminal phase semi-active laser homing onto a laser marked target, giving a longer range capability with lock onto target being achieved in the mid-course phase, although there is an option to lock on before launch for short-range targets. The TMV 585 Ariel laser guidance unit has been developed by Thales and is fitted in the nose of the missile. The original laser designator pod, made by Thales Optronics, is the ATLIS 2 (Automatic Tracking Laser Iluminating System). A later development, also by Thales Optronics, is called the Convertible Laser Designation Pod (CLDP). It was designed to have a rapid interchange of head, which could be fitted with either a multispectral TV camera for daytime use or an infra-red camera. Although these are the known designators in use, the AS 30L is designed to be compatible with any airborne or ground-based designating system using a 1.06 μm laser.

The warhead is believed to be based on a 250 kg GP bomb, with a weight of 240 kg and a delayed action impact fuze to ensure penetration of the target before the explosive charge is initiated. AS 30L has a maximum range of 10 km.

Operational status
In production. In service with the French Air Force on Jaguar and Mirage 2000. Deliveries to the French Navy for the Super Etendard Modernisé began in June 1996. It has been exported to Egypt, India, Iraq, Jordan, Nigeria, Oman, Pakistan and Venezuela. It is reported that orders for over 800 missiles in total have been received.

Specifications
Length: 3.65 m
Body diameter: 342 mm
Wing span: 1.0 m
Launch weight: 520 kg
Warhead: 240 kg HE
Fuze: n/k
Guidance: inertial and semi-active laser homing
Propulsion: solid propellant
Range: 10 km

Contractor
Aerospatiale (an EADS company), Missiles Division.

A Mirage 2000 aircraft carrying an AS 30L air-to- surface missile on the centre-wing pylon and a Thales Optronics CLDP on the inner-wing pylon

BAE Systems Hakim (PGM-1/2/3/4)

Type
Air-to-surface missile.

Development
Summary: Hakim A and B have semi-active laser guidance; Hakim C and D have infra-red seekers.

The development of a family of guided bombs and air-to-surface missiles by ISC in the USA is believed to have started in 1984, on behalf of the UAE. Ferranti bought ISC in 1987 and then Marconi bought the Ferranti missiles business in 1991. BAE Systems, in turn, acquired Marconi in 1999. The programme has been given several names over the years, but it is believed that the UAE knows the project as Hakim and that the designators are PGM-1 to 4 (Precision-Guided Munitions). Other names that have been ascribed to this programme include Project Alpha, GMX, Felix, Pegasus and Little Brother.

There have been reports of flight trials in France on the Mirage 2000 and in the US on F-16 aircraft. It is believed that Hakim A/B (PGM-1/2) missiles entered production in 1990 and that Hakim C/D (PGM-3A/B) entered production in 1993. Hakim missiles have been cleared for carriage on the Mirage 2000 and may have been cleared on the F-16 Fighting Falcon.

Description
Hakim A (PGM-1) has three clipped-tip delta wings just behind the nose and four nearly rectangular wings at the rear. Below the rear body is an externally mounted boost motor assembly. Hakim A has semi-active laser guidance and a range of approximately 20 km. The launch aircraft would have to carry a laser designator pod, or the missile could be guided by a ground designator illuminating the target.

Hakim B (PGM-2) is believed to have the same front end as Hakim A with a 300 mm diameter but has a longer and greater diameter rear body. The front and rear wing sets are the same shape as on Hakim A. Hakim B has two externally mounted boost motor assemblies under the rear body. Hakim B has semi-active laser guidance and a range of approximately 20 km.

Hakim C (PGM-3A) is sometimes known as Felix A and is believed to be similar to Hakim A, but with an IR guidance seeker in place of the semi-active laser seeker. This would remove the need for a separate laser designator pod, but would require a display for the pilot to select the target before launch.

Hakim D (PGM-3B) is sometimes known as Felix B and is believed to be similar to Hakim B, but with an IR seeker.

There are believed to be several warhead options for each of the Hakim missiles, including blast/fragmentation, penetration and anti-armour and minelet submunitions. The penetration warhead design is reported to be a simple steel pipe filled with HE.

PGM-4 Pegasus is believed to be a variant of the Hakim D design incorporating a turbojet engine to increase the range to over 200 km. This proposal may include inertial and GPS mid-course guidance with either an IR or a millimetric-wave radar terminal seeker.

Hakim A and B missiles shown under the wing and fuselage of a UK Tornado GR.1 aircraft; the missile under the wing is Hakim A, and the two missiles under the fuselage are Hakim B

Operational status
Hakim A and B (PGM-1/2) are believed to have entered service with the UAE in 1992 for use on Mirage 2000 aircraft. Hakim C and D (PGM-3) are believed to have started production in 1993 and to have entered service in 1995. Reports suggest that around 1,750 Hakim missiles have been ordered by the UAE.

Specifications

	Hakim A/C	Hakim B/D
Length	3.4 m	4.7 m
Body diameter	300 mm	380 mm
Wing span	1.2 m	1.3 m
Launch weight	250 kg	900 kg
Warhead	HE or submunitions	HE or submunitions
Fuze	n/k	n/k
Guidance	inertial and SAL or IR	inertial and SAL or IR
Propulsion	solid propellant	solid propellant
Range	20 km	20 km

Contractor
BAE Systems, Avionics, Radar & Countermeasures Systems, Stanmore.

VERIFIED

BGT ARMIGER air-launched anti-radiation missile

Type
Air-to-surface missile.

Development
The ARMIGER air-launched anti-radiation missile is being studied by BGT for the German Air Force. The German Air Force requirement is for a lightweight anti-radiation missile as a HARM follow-on post 2010, based upon a dual-mode seeker and ram rocket propulsion.

Captive carry flights with the 'ARAS' dual-mode seeker, replacing the Franco-German 'SPRINT', were conducted in 1997 and will continue against a variety of eastern and western air defence systems. BGT is developing the seeker in co-operation with DASA. The ram rocket has been tested at Bayern Chemie. In 1999 BGT received a contract from the German BWB for the flight demonstration of ARMIGER by 2002.

Description
ARMIGER has a dual-mode seeker with a DASA passive broadband radar sensor and a BGT long-wave imaging IR sensor. Propulsion will be provided by a throttleable Bayern Chemie boron-grain ram rocket, which can regulate the thrust within a ratio of 1:9. ARMIGER will have a 20 kg warhead and a maximum range exceeding 100 km. The German Tornados will be able to carry four ARMIGER rounds instead of two HARMs.

Operational status
Flight demonstration phase with a projected in-service date of 2008+.

Contractor
Bodenseewerk Gerätetechnik GmbH (BGT).

VERIFIED

The BGT ARMIGER air launched anti-radiation missile mounted on a Tornado
0023266

Boeing AGM-130

Type
Air-to-surface missile.

Development
Summary: the missile has an imaging infra-red or TV seeker.

A full-scale development contract was placed by the US Air Force with Boeing North American (formerly Rockwell) in 1984 for the AGM-130 powered version of the GBU-15 modular glide bomb. There were three versions being studied; the AGM-130A with the 900 kg Mk 84 general purpose warhead, the AGM-130B with the SUU-54 dispenser warhead and the AGM-130C with the BLU-109 penetration warhead.

Boeing has developed an upgrade for the AGM-130 which includes a mid-course guidance unit which has been flight tested successfully three times and, an improved infra-red seeker which was successfully flight tested by the USAF in April 1996.

The missile is certified for carriage on F-4E Phantom, F-15E Eagle and F-111F.

The USAF is considering the installation of a horizontal target mode on the missile which would allow it to attack bunkers. Other upgrades being considered include integrating a ladar (laser radar) on the nose of the missile and the addition of an automatic target recognition capability.

Description
AGM-130A/B/C has four clipped-tip fixed delta fins at the nose and four rectangular wings at the rear. In addition, a cylindrical solid propellant boost motor is attached under the missile body and is ejected during flight on completion of the 60-second boost phase. Guidance is by means of either a television or Imaging Infra-Red (IIR) seeker mounted in the missile nose, transmitting a picture back by datalink to the launch aircraft. The datalink is the AXQ-14 system developed for use with the GBU-15 and the launch aircraft carries an aircraft datalink pod. The pilot selects an aim point on a cockpit display and designates the aim point to the missile, from which point the launch aircraft is free to manoeuvre away. This is the same principle of operation as with the TV and IIR AGM-65 Maverick missiles and it is believed that AGM-130 utilises the Maverick seekers. The missile can be launched either before or after lock onto the target, or alternatively can be manually steered by the aircraft operator all the way to the target. If the launch aircraft does not have a datalink pod then the missile will be locked to the aim point before launch. AGM-130 has a radar altimeter and digital autopilot allowing the flight trajectory to be reprogrammed from the normal glide-boost-glide profile. The missile has a maximum range, when launched from a medium (30,000 ft) altitude, of 45 km.

The latest versions of the missile have a Mid-Course Guidance (MCG) unit which is based on inertial/satellite guidance. The unit includes an integrated inertial measurement unit and Global Positioning Satellite (GPS) receiver. The MCG allows pre- and post-take-off targeting by the weapon systems operator, performs autonomous guidance after launch and automatically points the seeker to assist the operator in locating the target after the weapon is launched. A new infra-red seeker, called the Improved Modular Infra-Red Seeker (IMIRS) has been flight tested and will be added to the AGM-130 upgrade programme. The IMIRS is based on focal plane array technology developed by Boeing.

Operational status
Full-scale development of the modifications and additions to the GBU-15 started in 1984. The USAF placed an early production order for 28 AGM-130A missiles in 1990, followed by a second order for 48 missiles in 1991 and 120 in 1992 for flight tests, training and production qualification. In 1994, it was reported that the USAF had about 140 AGM-130 missiles in stock, equipped with TV and IIR seekers, and that the total requirement was for 1,000 missiles. The total requirement was then

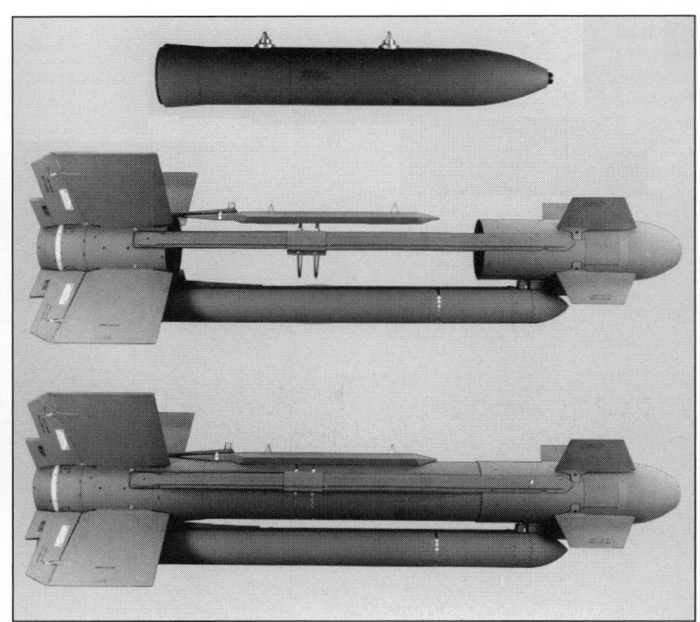

The AGM-130 missile components with the warhead (top), the airframe and motor (centre) and a complete missile assembly (bottom)

reduced to 400, which was subsequently increased to 674. In June 1999 it was believed that around 500 missiles had been delivered to the USAF. It is reported that the production line remains open, and the numbers could increase still further if it is decided to equip other aircraft. In 1996, South Korea ordered 116 AGM-130 missiles for use from its F-4E Phantom aircraft. AGM-130 missiles were used operationally in Iraq from January 1999, and against targets in Serbia and Kosovo from March to June 1999 and latest figures released by the US DoD say four were fired during Operation 'Iraqi Freedom' in March/April 2003.

Specifications

Length: 3.94 m
Body diameter: 460 mm
Wing span: 1.5 m
Launch weight: 1,323 kg
Warhead: 900 kg Mk 84 general purpose bomb
Fuze: n/k
Guidance: TV or IIR
Propulsion: solid propellant
Range: 45 km

Contractor

Boeing Integrated Defense Systems (Missile Systems and Tactical Weapons.

UPDATED

Boeing AGM-84E SLAM/AGM-84H SLAM-ER

Type

Air-to-surface missile.

Development

Summary: the missile has an imaging infra-red seeker.

In June 1971, The Boeing Company (formerly McDonnell Douglas) was selected by the US Navy as the prime contractor for the development of an anti-ship missile system called Harpoon, to include air-launched and ship-launched variants and later a submarine-launched version. All three systems entered service in 1977.

A second air- and ship-launched version known as the AGM/RGM-84E Standoff Land Attack Missile (SLAM) has been developed which utilises the Harpoon airframe with AGM-65D Maverick Imaging Infra-Red (IIR) seeker. The air-launched version of SLAM entered service in 1990.

In March 1996, there was a successful test firing of a SLAM using an automated mission planning module hosted on the Tactical Aircraft Mission Planning System (TAMPS).

AGM-84E SLAM has been cleared for carriage on the B-52, F-111C, F27 Maritime, A-6 Intruder and F/A-18 Hornet aircraft.

An extended-range version (to about 120 km) with a larger 340 kg warhead, known as AGM-84H SLAM-ER (SLAM Expanded Response), has been developed to achieve a range of about 120 km. The first flight test SLAM-ER missiles were delivered to the US Navy in December 1996 for a 13 missile flight test programme. Operational test and evaluation was completed in March 2000.

Description

The AGM-84E SLAM utilises the Harpoon airframe, engine and warhead with an AGM-65D Maverick imaging infra-red seeker. An AGM-62 Walleye video datalink enables the pilot to control the seeker and a Global Positioning System (GPS) receiver updates the mid-course inertial guidance system. SLAM weighs 628 kg at launch. The IIR seeker is activated in the terminal phase and the pilot selects his aim point on the target and locks the missile to that point. From then on the launch aircraft can manoeuvre away and the missile will be autonomous to impact. The maximum range of SLAM is believed to be 95 km.

SLAM-ER has planar wings derived from the Tomahawk missile, new warhead and an upgraded guidance system. It also has an automatic target acquisition function developed by Boeing.

An AGM-84E SLAM underneath an F/A-18 Hornet aircraft ***NEW***/0554673

Operational status

The AGM-84E SLAM entered service in 1990. Low-rate production of 60 SLAM-ER missiles began in April 1997. Full production authorised in May 2000 to meet a requirement for 700 missiles. Production is expected to continue to 2004. The majority of these missiles will be kits retrofitted to existing SLAMs. SLAM-ER was cleared for operational use in September 2002. Three SLAM-ER rounds were used in Operation 'Iraqi Freedom' in March/April 2003.

Specifications

	AGM-84E SLAM	AGM-84H SLAM- ER
Length	4.50 m	4.37 m
Body diameter	343 mm	343 mm
Wing span	0.91 m	2.43 m
Launch weight	628 kg	727 kg
Warhead	220 kg HE blast penetration	320 kg HE blast penetration
Fuze	impact	impact
Guidance	inertial with GPS and IIR	inertial with GPS and IIR
Propulsion	turbofan	turbofan
Range	95 km	190 km

Contractor

Boeing Integrated Defense Systems (Missile Systems and Tactical Weapons).

UPDATED

Chung Shan Institute Hsiung Feng 2

Type

Air-to-surface missile.

Development

Summary: the missile has a dual-mode radar/imaging infra-red terminal seeker.

The Taiwanese Ministry of Defence released details of a ship-launched anti-ship missile, Hsiung Feng 2 (Hurricane 2), in late 1988. The development of Hsiung Feng 2 probably started in the early 1980s, and the air- launched variant is expected to be carried by the Ching-kuo fighter. The Ching-kuo will carry three Hsiung Feng 2 anti-ship missiles, one under the centre fuselage and one under each wing. Reports in 1994 indicate that a Hsiung Feng 3 has been proposed, with a range increased to 200 km.

Description

Hsiung Feng 2 appears to be very similar to AGM-84 Harpoon, with the exception of the four tail fins which are triangular in shape. Guidance is inertial for mid-course with a dual-mode active radar and imaging IR terminal seeker. The active radar seeker is located in a conventional nose radome assembly, but the IIR seeker is in a 70 mm diameter dome located above and just behind the radome. The Hsiung Feng 2 has a turbojet engine and a range expected to be around 80 km. The warhead is reported to be HE semi-armour-piercing, with a weight of 225 kg.

Operational status

The Hsiung Feng II entered development in 1983. The surface-launched version completed its initial operational test and evaluation in 1990, and then entered production and deployment. The air-launched version finished its test and evaluation process in 1993 and is assumed to be available for use on the F-CK-1 Ching-Kuo. There are no known exports.

Specifications

Length: 3.9 m
Body diameter: 400 mm
Wing span: 0.9 m
Launch weight: 520 kg
Warhead: 225 kg HE semi-armour-piercing
Fuze: impact
Guidance: inertial and active radar with IIR
Propulsion: turbojet
Range: >80 km

Contractor

Chung Shan Institute of Science and Technology.

UPDATED

EADS/Saab Bofors TAURUS KEPD 350

Type

Air-to-surface missile.

Development

The MAW TAURUS KEPD 350 is a member of the TAURUS family of missiles, proposed for the German Ministry of Defence as part of the weaponry for the Tornado and in due course for the Eurofighter Typhoon 2000. Flight testing on the

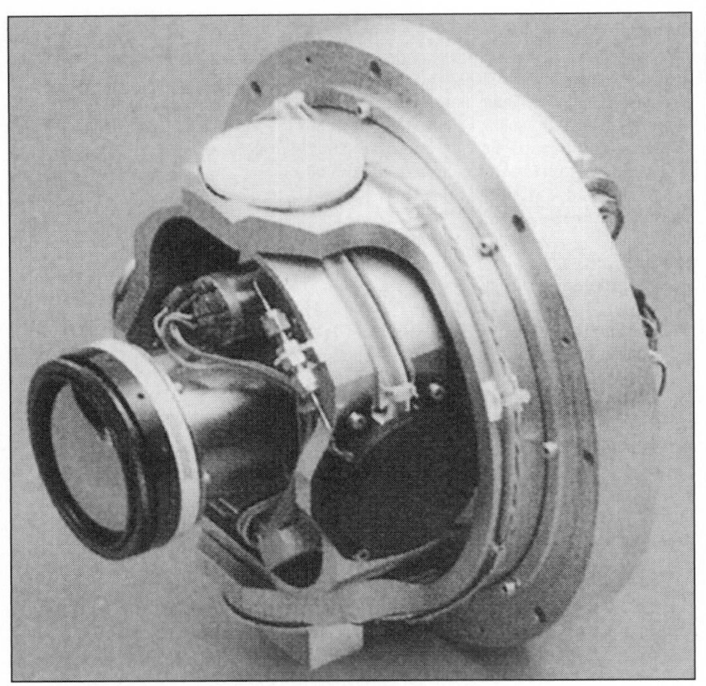

The TAURUS KEPD 350 I^3R intelligent imaging infra-red seeker

The LFK GmbH/Bofors MAW TAURUS KEPD 350 in two-store configuration on a tornado (Photo by courtesy of WTD 61) 0023376

Tornado started in 1996. Store configurations on Tornado and the Eurofighter comprise two missiles.

The Swedish Air Force studied the shorter range TAURUS KEPD 150 version but has opted for the TAURUS KEPD 350 version for the JAS 39 Gripen. Members of the Taurus missile family are also being offered to other nations including Australia, Italy and Spain.

Description

TAURUS KEPD 350 is based on expansion of the modularity concept from the DWS 39 dispenser weapon. It incorporates an indium antimonide focal plane array low- to medium-waveband infra-red seeker used for waypoint navigation updates and terminal guidance. Front end pixel enhancement electronics provides data to the processor for terrain feature correlation. 'Super Waypoint' algorithms pick out the scene features for correlation with the mission planning source data and the navigation update is derived through triangulation. The navigation system also includes other navigation features such as inertial measurements and GPS.

The warhead is a 500 kg class effector system using a chemical prefragmented precharge followed by a kinetic energy penetrator containing the main charge. The effector has intelligent fuzing controlled by the mission plan.

A high-performance turbo engine propulsion system provides the missile with the required agility and range performance.

Operational status

Key components have been tested. A German development contract was expected in 1998. The flight tests started at the end of 1997 with a safe separation release test. A final free flight test of the TAURUS KEPD 350 took place on 21 November 2002. The German Air Force will receive its first weapons in 2004, with Sweden receiving missiles after 2005. Australia is considering the weapon.

Contractors

EADS - LFK.
Saab Bofors Dynamics.

UPDATED

EMDG LR (Long-Range) TRIGAT

Type

Air-to-surface anti-armour missile.

Development

The Long Range TRIGAT (third-generation anti-tank) weapon system is being developed by Aerospatiale Matra Missiles, Matra BAe Dynamics and DASA LFK for long-range vehicle and helicopter launched applications. The requirement is to replace HOT, TOW and Swingfire systems. The MoU for development was signed in April 1988 with France, Germany and the UK participating. LR TRIGAT can be used in both the anti-armour and anti-helicopter modes. The LR TRIGAT system will be the prime armament of the Eurocopter Tiger attack helicopter. However, in 1996, the UK announced it would not adopt LR-TRIGAT for its helicopters and, in 1998, France did the same. Germany plans to introduce LR-TRIGAT on its Tiger UHT models from 2006, as does Australia.

Initial missile guidance firings were carried out in 1995 and are continuing successfully. First flight trials on the Tiger helicopter were successfully conducted in February 1996 and also are successfully continuing. First firing trials from a Panther helicopter took place in March 1998. Last of 10 qualification guided firings was of a LR-TRIGAT made from a Tiger.

Description

LR TRIGAT, in the helicopter application, uses a mast-mounted target acquisition system which employs a dual band (visible and near IR) TV and a thermal imager operating at about 10 μm.

The operator designates the target on the display screen, a fire and forget missile is selected, is guided to the target passively and automatically. Up to four missiles can be ripple-fired at different targets in under 8 seconds.

The missile has four clipped-tip folding wings at mid-body and four folding delta control fins at the rear. Guidance is by an imaging infra-red seeker, operating at around 10 μm. The operator designates the target on the display screen and the fire-and-forget missile is guided to the target automatically.

A tandem shaped charge warhead construction is combined with a terminal phase dive trajectory to maximise performance against reactive armours. The missile has a maximum range of 5,000 m and minimum range of 500 m.

Operational status

The first Long-Range TRIGAT missile guidance firings were in 1995 and are continuing successfully. The first flight trials of TRIGAT equipment on the Tiger helicopter occurred in February 1996 and are continuing successfully. LR TRIGAT is expected to enter service around 2002-05.

Specifications

Length: 1.6 m
Body diameter: 159 mm
Launch weight: 41 kg

LR TRIGAT on the Tiger Helicopter 0005505

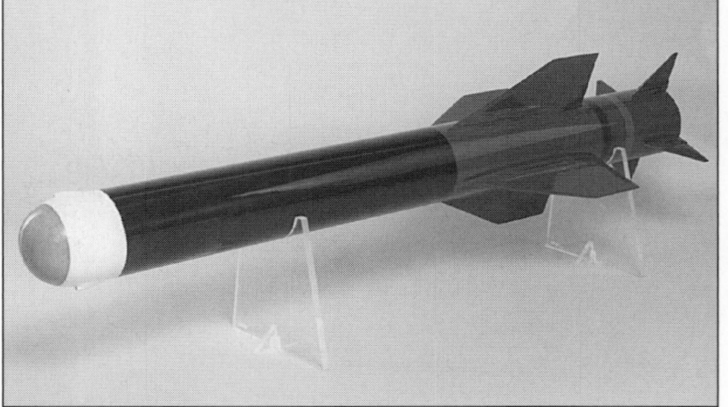

Model of the LR TRIGAT missile 0005506

Warhead: tandem shaped charge
Fuze: infra-red
Guidance: imaging IR
Propulsion: solid propellant
Speed: 290 m/s
Range: 500 m to 5 km

Contractor
Euromissile Dynamics Group (owned by MBDA and EADS).

UPDATED

IAI Nimrod

Type
Air-to-surface missile.

Development
Summary: the missile has semi-active laser terminal guidance.

Little is known of Nimrod's development and its existence was virtually unknown until it was shown in 1989. The Nimrod Advanced Laser-Guided Missile System (N/ALGMS) has been developed by Israel Aircraft Industries (IAI) as a private venture for both ground- and air-launched use. It has been tailored to the Israel Defence Force's operational needs as the next-generation anti-tank weapon. It is believed that the Nimrod missile will be carried by CH-53 and possibly Super Puma helicopters.

Description
The Nimrod missile has a reported range of 25 km and is believed to have a mid-course inertial and possibly a command update facility, in addition to the semi-active laser terminal guidance. The target is designated either by ground-based illuminators or airborne systems carried by manned or unmanned aircraft and simultaneous launches to multiple targets are achieved, using separate coded laser designators. The designation only needs to be made shortly before the missile reaches the target, probably for about the last 30 seconds of flight. Early information indicates that Nimrod approaches the target with a dive angle of around 45°, which would suggest a cruise altitude of at least 1,000 ft or higher. The missiles are stored and carried in sealed canisters, which also serve as launchers.

Operational status
Air-launched Nimrod is thought to be still in the development stage. The ground-launched version is in service and has been sold to the Israel Defence Force and export customers. The Israeli CH-53s can carry up to eight Nimrods, in place of the helicopter's standard external fuel tanks. The weapons are intended as a self-defence measure for the CH-53, to protect against ground threats, but can also be used in their own right to support air assault operations. It is not clear if the Nimrod modification has been applied to all IDF/AF CH-53s, or just to the upgraded Yas'ur 2000 variant. For some time prior to the official announcement there had already been unconfirmed reports that the Nimrod was being adopted for use on the CH-53, and it is not known exactly when this new capability was introduced.

Specifications
Length: 2.84 m
Body diameter: 210 mm
Wing span: 500 mm
Launch weight: 100 kg
Warhead: 15 kg HE
Fuze: n/k
Guidance: inertial and semi-active laser
Propulsion: solid propellant
Range: 25 km

Contractor
Israel Aircraft Industries, MBT Weapon Systems Division.

UPDATED

Nimrod semi-active laser-guided air-to-surface missile

KBP, KBM Vikhr-M (AT-16)

Type
Air-to-surface missile.

Development
A missile, believed to be the NATO-designated AT-16 air-to-surface anti-tank missile, was first seen at the 1992 Farnborough Air Show. There is some confusion in the West over the Russian name given to this missile. At Farnborough 1992, it was announced as 'Vikhr-M' (Whirlwind), a new laser beam-riding anti-tank missile. This indicates that it could be a variant of a smaller anti-tank missile displayed and offered for export at the 1991 Dubai Air Show, which was believed to have been the AT-12 (9M120 Vikhr). Nothing is known of the missile's development, but this probably started in the mid-1980s. Vikhr-M is believed to be an improved variant of the AT-12 with an extended range. The missile has been seen fitted to the Ka-50 'Hokum' (Black Shark) attack helicopter. The Ka-50 carries up to 12 Vikhr-M in two clusters of six under stub-wings. The same configuration could also be fitted to the Mi-24 'Hind', Mi-28 'Havoc' and Ka-52 'Hokum B' (Alligator) helicopters, also the Su-25 'Frogfoot' attack aircraft. There are unconfirmed reports that Vikhr-M has been tested in the air-to-air mode against other helicopters and in the air-to-surface role against small missile patrol boats. In 1999, it was reported that an improved Vikhr-S (ground-launched) missile had been developed, with an increased range. The missiles have interchangeable seeker heads; semi-active laser, infra-red, and active radar. It is not known if this version will be air launched.

Description
The AT-16 (Vikhr-M) is a tube-launched missile and, from the length and diameter of the tube, the missile is believed to be an AT-12 with a longer rocket motor. The launch tube is 2.87 m long with a body diameter of 150 mm with the missile and canister weighing 59 kg. The missile itself is 2.8 m long, has a body diameter of 130 mm, an extended wing span of 0.4 m, and is believed to weigh 72 kg at launch. The AT-16 missile has four swept-back rectangular moving control fins at the nose, probably with air-dynamic control actuators. Four fixed and hinged long chord rectangular wings are located at the rear of the missile. Guidance is by semi-active laser. To facilitate target designation by the pilot, the Ka-50 helicopter weapon system has a helmet-mounted sight and head-up display of the kind fitted to the MiG-29 fighter. However, it is also understood that the intention is to rely mainly on other aircraft or ground personnel to designate targets. The Ka-50 has two laser transmitters in the nose, one to give target range and the other to illuminate the target for the Vikhr-M missile. The laser range-finder operates at 1.54 μm and is mounted within a GOES-3 electro-optical dome assembly, which includes a FLIR and TV with both wide and narrow fields of view. The GOES-3 assembly weighs 30 kg and has a diameter of 460 mm. The warhead on the missile is believed to weigh around 8 kg, and is the tandem forward firing type with a hollow charge followed by a shaped fragmentation charge, reported capable of penetrating 1,000 mm of armour protected by ERA. The missile has both impact and proximity fuzes. The configuration carried on the Ka-50 'Werewolf' is six canisters in two groups of three, on each outer stub-wing pylon. The bottom part of the pylon is articulated to allow the missiles or other weapons it carries to be pointed downward by 10°. The solid-propellant propulsion system is thought to be of the dual boost and sustainer type and is reported to give the Vikhr-M supersonic speed (M1.8) and a range of about 10 km. At night, the maximum range is reduced to 5 km. The missile reaches 6 km in 14 seconds. The minimum firing range is 0.5 km and the missile can be launched at altitudes from 5 to 4,000 m. Two missiles can be launched in salvo to engage a single target. A longer-range version was reported in 1999, with a maximum range by day increased to 15 km and by night to 7 km.

Operational status
The Vikhr-M is believed to be in full production and is reported to have entered service around 1990. It was displayed and offered for export at the 1992 Farnborough Air Show, along with the Ka-50 helicopter. There are no known exports.

Summary: the missile has semi-active laser guidance. The Vikhr air-to-surface anti-tank missile was first seen at the 1992 Farnborough Air Show. Vikhr is jointly

A six-canister pylon load of AT-16 missiles on the outboard stub-wing of a Ka-50 'Hokum A' helicopter at Farnborough in 1996 (Duncan Lennox)

produced by the KBP Instrument Design Bureau and Krasnagorski Zavod and is intended as a multipurpose weapon system for use on both combat helicopters (for example, the Ka-50, Mi-24 and Mi-8) and fixed-wing attack aircraft (such as the Su-25T). The manufacturer claims that updating of existing aircraft with Vikhr requires only minor platform modifications.

Specifications

Length: 2.80 m
Body diameter: 130 mm
Wing span: 0.4 m
Launch weight: 72 kg
Warhead: 8 kg HE shaped charge
Fuze: n/k
Guidance: semi-active laser
Propulsion: solid propellant
Range: 10 km

Contractor

Russian State Factories.

UPDATED

Kentron Mokopa anti-armour missile

Type

Air-to-surface missile (precision guided anti-armour).

Description

Mokopa (black mamba) is a fire-and-forget anti-armour missile designed to equip the South African Rooivalk helicopter. It will complement the smaller Kentron Ingwe. It was revealed at the Paris Air Show in 1995 and is comparable to the US Hellfire missile system. Guided flight trials are believed to have been carried out on the Rooivalk in 2000.

The modular design can incorporate a variety of seeker heads as well as semi-active laser homing - imaging infra-red, laser homing and millimetre wave radar. The warhead may be of the tandem, fragmentation or penetration type. Mokopa will be capable of lock on before launch for short-range targets and lock on after launch for longer range targets. Both autonomous and remote target designation is possible.

A ground-launched version is also being developed.

Operational status

In the final stages of development.

Launch of a Kentron Mokopa from a Rooivalk ***NEW***/0533770

A vehicle launch configuration for Mokopa ***NEW***/0533771

Specifications

Length: 1.995 m
Body diameter: 178 mm
Weight: 49.8 kg
Type: semi-active laser homing
Warhead: tandem HEAT, fragmentation and penetration
Penetration: >1,350 mm RHA (with ERA)
Range: >10,000 m

Contractor

Kentron Division of Denel (Pty) Ltd.

UPDATED

Kentron MultiPurpose Stand-Off Weapon System (MUPSOW)

Type

Air-to-surface missile.

Description

It has been reported that a classified South African air-to-surface missile programme was begun in the mid-1980s to produce a low-cost missile for use in Angola. The first of these was a prototype TV-guidance system added to 450 kg bombs (probably Mk 83 type) designated H2. An upgraded version of this guided bomb, designated H3, was offered for export in 1993. In 1993, Kentron revealed details of the Multi-Purpose Stand-Off Weapon (MUPSOW). It uses standard 415 kg bombs (probably Mk 83) with guidance and propulsion assemblies added giving a range of around 50 km. An improved MUPSOW variant, with a range increased to 150 km, was announced by Kentron in 1994.

The current MUPSOW is a multi-purpose precision strike weapon designed to neutralise enemy targets such as airfields, bunkers and command-and-control centres at stand-off ranges. Accuracy is achieved by using an advanced navigation and terminal guidance technology. The missile is of modular design with a choice of seekers and payloads. It may be used in fire-and-forget or man-in-the-loop modes. Attack profiles include dive attack and lay-down missions. Seeker module type options are TV, IIR and millimetre wave.

Operational status

The latest version was planned to enter service in 2000 but has been superseded by a further development known as Torgos.

Specifications

Length: 4,920 mm
Wing span: 1,900 mm
Fuselage height: 480 mm
Fuselage width: 641 mm
All-up mass: 1,200 kg

Contractor

Kentron Division of Denel (Pty) Ltd.

UPDATED

A MUPSOW model displayed in South Africa in 1996 (Michael J Gething) 0007743

Kongsberg Penguin Mk 3 (AGM-119A) and Mk2 Mod7 (AGM-119B)

Type

Air-to-surface missile (anti-ship).

Development

Summary: Fire-and-forget, passive, IR seeker, programmable trajectory, evasive manoeuvring.

The Penguin programme started in the early 1960s with the development of the Penguin Mk1 ship-to-ship missile which entered service in 1972. The second generation – Penguin Mk2 – also developed as a ship-to-ship missile, became operational in 1980. The third generation – AGM-119A (Penguin Mk3) – an all digital version developed for carriage on fixed-wing aircraft, was fielded in 1989. Based on AGM-119A, AGM-119B (Penguin Mk2 Mod7) was developed for use from helicopters and delivered to service in 1994.

Description

Penguin Mk3 (AGM-119A), has four forward-mounted, swept canard control fins, four fixed, delta-wings just aft of the mid-body with ailerons for roll stabilisation and a single-stage motor. The Penguin Mk2 Mod7 (AGM-119B) has folding delta wings and a two-stage motor. Prior to launch the operator can choose from a range of seeker modes, trajectory options and launch modes. Both versions have similar inertial mid-course guidance systems with radio altimeter and provisions for programming the flight trajectory both in altitude and azimuth, as well as the capability to fly over land and water. The missiles can also be programmed to overfly a selected number of ship targets before making an attack. Both versions have a passive infra-red terminal seeker and a semi-armour-piercing warhead with delayed action impact fuse. Penguin is autonomous from launch to target penetration. The Penguin missile system is fully compatible with MIL-STD-1553B and is designed to be software-integrated to the aircraft's existing stores management system. Penguin requires no changes to fuselage or cockpit lay-out, or any dedicated mission equipment. The aircraft's radar or any electro-optical means that can produce bearing and range to the target, may be used for entering target and waypoint positions into the missile's navigation system.

Operational status

Penguin Mk3 (AGM-119A) entered service with the Royal Norwegian Air Force in 1989 and is carried on F-16 Fighting Falcon aircraft. The helicopter-launched Mk2

SH-60 armed with Penguin 0053933

Penguin Mk 2 Mod 7 air-to-surface missile (AGM-119B) on a Sea Hawk SH-60B 0053936

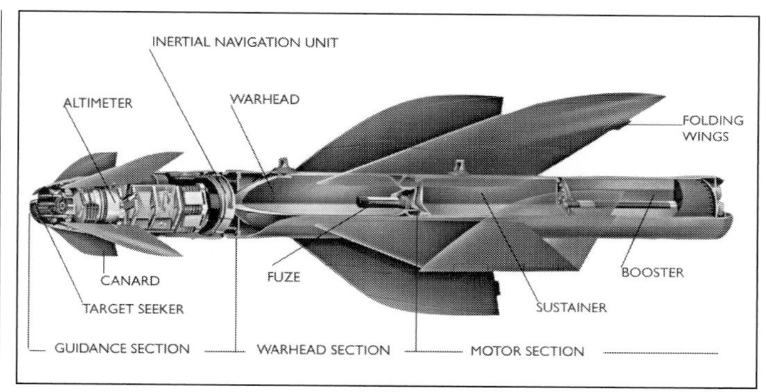

Kongsberg Penguin MK2 Mod 7N 0055094

F-16 carrying two Penguin Mk3 (AGM-119A) air-to-surface missiles 0053935

Mod7 (AGM-119B) is operational on the Sikorsky SH-60B of the USN and on the Sikorsky S-70B of the Hellenic Navy. Penguin Mk2 Mod7 has been chosen by the Royal Australian Navy to equip the SH-2G(A) Super Seasprite helicopters for the 'Anzac' class frigates, deliveries began in 2001.

Specifications

Penguin Mk 3 (AGM-119A)
Length: 3.18 m
Body diameter: 0.28 m
Wing span: 1.0 m
Launch weight: 370 kg
Warhead: 141 kg semi-armour-piercing (43 kg HE fill)
Fuze: impact, delayed action
Guidance: inertial and passive IR
Propulsion: sustainer stage, solid propellant
Range: >40 km

Penguin Mk 2 Mod7 (AGM-119B)
Length: 3.02 m
Body diameter: 0.28 m
Wing span: 1.42 m
Launch weight: 395 kg
Warhead: 123 kg semi-armour-piercing (43 kg HE fill)
Fuze: impact, delayed action
Guidance: inertial and passive IR
Propulsion: boost and sustainer stages, solid propellant
Range: >30 km

Contractor

Kongsberg Defence and Aerospace AS.

Lockheed Martin AGM-158 Joint Air-to-Surface Standoff Missile (JASSM)

Type

Air-to-surface missile.

Description

The AGM-158 Joint Air-to-Surface Standoff Missile (JASSM) is intended to enable military aircraft to defeat advanced air defence systems and well-defended, hardened, fixed or relocatable high-value ground targets from ranges outside enemy radar detection. JASSM is being designed for launch from a wide range of aircraft including the B-52, F-16, F/A-18, B-1, B-2, F-15, F-117, S-3 and P-3.

JASSM is an autonomous, long-range, conventional, air-to-ground precision cruise missile. After launch, Lockheed Martin's JASSM candidate uses an Inertial Navigation System (INS) and Global Positioning System for en route guidance. The design also features a penetrating warhead, a jamming-resistant imaging infrared seeker, an affordable mission processor and autonomous terminal guidance. Lockheed Martin's design incorporates proven technologies and subsystems in a highly survivable air vehicle intended to defeat 21st century threats.

Operational status

Project definition studies started in 1996, with a three year engineering and manufacturing development contract awarded in April 1998 calling for 40 test missiles to be built. The EMD phase was increased to four years in 1999, following

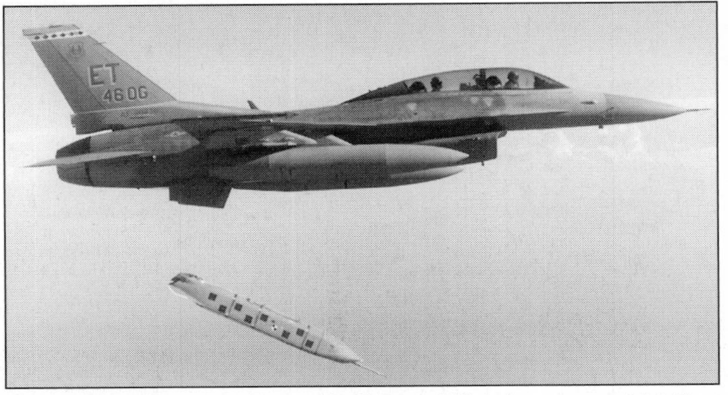

A JASSM missile is released from a US Air Force F-16 during a successful Jettison Test Vehicle launch 0053943

development problems with the composite airframe, engine and wing actuators. Initial flight tests started in 1997, with the first powered flight in November 1999. Eight Contractor Development Test & Evaluation (CDT&E) flights, to prove JASSM on the F-16 and the B-52, will complete the EMD phase of testing and allow for manufacturing process 'prove out' and validation prior to entering Low Rate Initial Production (LRIP). Full-rate production is expected to begin in early 2003. A USAF option for 1,165 missiles by 2006 is believed to have been included in the EMD proposal. By early 2001 the USAF's requirement stood at 2,400 missiles for delivery from 2003 to 2010, but by the end of the year that number has risen to 4,700. In 1999, it was reported that the US Navy was considering purchasing around 700 missiles from 2004, however, the Navy has not yet committed to a JASSM buy, though it is participating in the test programme. An export version of JASSM was reportedly offered to Australia in 1999, to replace the AGM-142 Have Nap stand-off missiles that currently equip its F-111 strike aircraft.

On 19 January 2001 a JASSM successfully performed its first CDT&E Development Test (DT-1) at White Sands Missile Range (WSMR), New Mexico. It was also the first flight using a seeker to guide to a target. For this test an F-16 launched the missile while flying at M0.80 and 15,000 ft. After weapon release, the missile navigated via pre-determined waypoints and descended to a selected altitude above ground level for target ingress, over some 70 miles. The missile performed a terminal manoeuvre that enabled it to achieve the desired 70° impact angle and accuracy. The imaging infra-red seeker and the Automatic Target Correlator (ATC) algorithm successfully reduced the target impact guidance error.

In testing DT-2, on 26 April 2001, a JASSM destroyed a target at WSMR after a low-level launch. The missile was launched from an F-16 at M0.80, and an altitude of 2,800 ft. The JASSM used its Anti-Jam Global Positioning System (AJGPS) guidance system en route. In the terminal phase, JASSM used its IIR seeker and automatic target correlation algorithms to locate the target aimpoint. The live warhead detonated upon impact, destroying the air defense target. One day earlier, on 25 April, a JASSM cleanly separated from a B-52 over WSMR and performed unpowered aerodynamic manoeuvres. In test DT-3, on 31 May 2001, a JASSM destroyed a concrete bunker at WSMR. The missile was launched from a B-52H, flying at M0.85 and an altitude of 30,000 ft. The JASSM flew for 23 minutes, navigated through 10 waypoints and traveled approximately 195 miles. Its live warhead penetrated the concrete roof of the bunker and detonated inside. This was the first launch of a full-up, live JASSM round from a bomber platform.

On 27 July 2001 a JASSM completed its fourth development flight test (DT-4) at the (WSMR), New Mexico. The missile was launched from a B-52H at an altitude of 24,000 ft at M0.85. After weapon release, the missile flew for 18 minutes, navigated through 8 waypoints and traveled approximately 158 miles. The missile used the seeker imagery to successfully guide to the desired aim point against a relocatable target. The warhead did not detonate because the fuse saved itself when communications were interrupted. In the previous live warhead flight test (DT-2), JASSM established the missile's ability to fly to, hit, and destroy surface targets when launched from an F-16 Fighting Falcon. The JASSM's seventh CDT&E launch was the final test point of 2001 and proved to be the last step before the approval of Low-Rate Initial Production (LRIP).

On 15 December 2001 a B-52 launched the seventh JASSM, clearing the way for the missile's Independent Operational Test and Evaluation (IOT&E) phase to begin in 2002 – after one more development test in early 2002. On 21 December 2001 the Pentagon approved the LRIP go-ahead for the JASSM. This decision should give the USAF a JASSM combat capability on the F-16 and the B-52 by 2003. The actual LRIP contract was placed in January 2002. It covered the delivery of 76 JASSMs and 84 Anti-Jam GPS receivers for the JASSM, by April 2004. The success of the JASSM team in achieving a unit cost of less than (the objective) US$400,000, in FY95 dollars, led to the increase in the initial USAF buy of JASSM weapons from 2,400 to 3,700 missiles. The final decision on full-rate production is expected in late 2003. A long-range version is also planned for US Air Force service as JASSM-ER (extended range) it was announced in March 2002.

Specifications
Weight: 2,250 lb
Length: 168 in
Storage life: 15 years
Range: over 100 miles

Contractor
Lockheed Martin Missiles and Fire Control.

UPDATED

Lockheed Martin LOw-Cost Autonomous Attack smart munition System (LOCAAS)

Type
Airborne anti-armour time critical targets and SEAD munition.

Development
Lockheed Martin Missiles and Fire Control, is developing the LOw-Cost Autonomous Attack System (LOCAAS) smart submunition for the US Air Force and Army. Development is focused on the powered LOCAAS. LOCAAS is one of two candidates chosen by the US Army for further evaluation after its MLRS Smart TActical Rocket (MSTAR) study which completed in December 1994. An unpowered LOCAAS has been flight-tested at Eglin Air Force Base, Florida, completing a successful airdrop and autonomously searching for a target tank. LOCAAS detected the target, correctly classified it and altered its flight path onto an attack flight profile. First system flight test, where LOCAAS autonously located and attacked a target, firing its multimode warhead to achieve target destruction, reported successful in April 2003.

Description
LOCAAS uses a low-cost, high-resolution LADAR (laser detection and ranging) seeker, also developed by Lockheed Martin Missiles and Fire Control. The solid-state system uses a direct detection, short-pulse laser system. This determines target range by measuring elapsed time from a laser transmission to the reflected return of the pulse. The LADAR system then generates high-resolution 3-D data that allows automatic target recognition for fixed targets, such as bridges, or tactical targets such as tanks. The LADAR seeker system consists of an integrated sensor head with dual-axis gimbal and optical receiver, system electronics, automatic target recognition system and associated tracking algorithms.

Both versions of the LOCAAS will carry an Alliant Techsystems multimode, explosively formed penetrator warhead. The warhead can be detonated as a long rod penetrator, an aerostable slug or as fragments, depending on the hardness of the target. Target aimpoint and warhead mode are automatically determined by the LADAR seeker.

The powered LOCAAS uses a stretched airframe to accommodate a Sundstrand TJ-50 turbojet engine, which is capable of powering the vehicle for up to 30 minutes with a maximum range of 180 km. The vehicle navigates by a GPS/INS (Global Positioning/Inertial Navigation System).

LOCAAS is expected to be dispensed from the USAF Wind Corrected Munitions Dispenser (WCMD), AGM-86C, AGM-130 and AGM-154 JSOW. They could also be dispensed from a Multiple Launch Rocket System (MLRS) or the US Army TACtical Missile System (ATACMS).

Operational status
Under development. Definition and risk reduction to 2005, with EMD following until 2008. An in-service date of 2009 is expected.

Specifications
Powered LOCAAS
Weight: 82 lb (37.2 kg)
Length: 31 in (78.7 cm)
Speed: 200 kt (370.4 kph)
Search altitude: 750 ft (230 m)
Footprint: 25 nm² (168 km²)
Range: 100 n miles (185 km)

LADAR seeker
Weight: 17 lb (7.73 kg)
Dimensions (l × d): 8 × 8 in (20.3 × 20.3 mm)

Contractor
Lockheed Martin Missiles and Fire Control.

UPDATED

The LOCAAS smart munition system 0088169

Lockheed Martin, Boeing AGM-114 Hellfire/Hellfire II

Type

Air-to-surface missile.

Development

Summary: AGM-114A/K has semi-active laser guidance; AGM-114B/C has anti-radar/infra-red or imaging infra-red seeker; AGM-114L has millimetre wave radar guidance.

The AGM-114 helicopter-launched semi-active laser fire-and-forget (Hellfire) missile was designed in the early 1970s as an anti-armour weapon and entered service in 1985. An AGM-114B version was developed for the US Marine Corps, with a low-smoke motor, safety arming device and three seeker options using semi-active laser, IIR, or a dual-mode RF/IR system. The AGM-114C version is for the US Army and is the same as AGM-114B but without the safety arming device.

A millimetre-wave seeker version (AGM-114L) has been developed for the Longbow Apache helicopter programme, which had a planned in-service date of 1997. Marconi Electronic Systems in the UK, proposed an active millimetric-wave (94 GHz) seeker as a further alternative and this system, known as Brimstone, has been chosen by the UK MoD for anti-armour weapons for both helicopters and fixed-wing aircraft. A night firing sight system, AN/UAS-12C, has been developed for use on helicopters with both BGM-71 TOW and Hellfire missiles; this system has a laser rangefinder and designator, a FLIR with ×30 magnification and a telescope for day viewing with ×13 magnification. Hellfire has been cleared for carriage on the AH-64 Apache, OH-58D Kiowa, UH-60 Black Hawk, Lynx, AH-1J SeaCobra and AH-1W SuperCobra helicopters. In addition, integration has been made with the SH-60B Seahawk, HH-60H, MD-530 and UH-1 helicopters as well as the C-130 Hercules gunship aircraft.

Hellfire II is an upgraded version which is designed to counter armoured threats well into the 21st century. Hellfire II is fully compatible with current and planned Hellfire launch platforms including the Apache, Cobra, Kiowa and Comanche helicopters. In addition, the Hellfire II warhead system, guidance and control sections are compatible with the millimetre-wave radar seeker being built by Lockheed Martin and Northrop Grumman (formerly Westinghouse Electronics Group) for the Longbow Hellfire missile, which will complement Hellfire II.

UH-60A Black Hawk helicopter carrying 16 AGM-114 Hellfire missiles

The Hellfire II missile being fired

On 26 May 1993, Electronics & Missiles received its first major production award for Hellfire II and the first production missile was completed in June 1994. Subsequent awards have brought total contract production in the US to over 9,000 missiles.

In February 1995, the US Army authorised Lockheed Martin to market the system to allied and friendly nations. International orders have been received from the UK and Netherlands. Netherlands has ordered 605 missiles to arm its Apache helicopters. These will initially equip AH-64As leased from the US Army and then be fitted on new AH-64Ds. The UK has selected Hellfire II and Longbow Hellfire as part of its Westland Apache programme.

In March 1996, the US Army exercised an option to procure an additional 750 Hellfire II missiles for Apache and SuperCobra helicopters of the US Marine Corps. In April 1996, the US Army awarded a Foreign Military Sales contract for 636 Hellfire II to supply the United Arab Emirates (UAE).

Description

The AGM-114A Hellfire missile has four clipped delta stabilising fins at the nose, with four wings at the tail, each with a moving control surface at the trailing edge. Guidance is by a semi-active laser seeker tracking a coded laser beam reflected from a designated target. The missile can also be launched from low level before lock on and will then search for the reflected beam and lock onto the target. With different designator codes it is possible to ripple-fire the missiles against different targets. The AGM-114B missile is longer and heavier, and has three seeker options. The RF/IR seeker missile is 1.73 m long and weighs 48 kg. The IIR seeker missile is 1.78 m long and weighs 48 kg. This missile also has a blast/fragmentation warhead, although the Hellfire missile family is modular with interchangeable assemblies providing a flexible set of alternatives. AGM-114F will have tandem warheads, increasing the missile length to 1.96 m and the weight to 51.2 kg.

Like Hellfire I, Hellfire II uses semi-active laser guidance technology, but the seeker has been hardened against electro-optical countermeasures and contains a digital autopilot to increase launch speeds from 300 kt to M1.1 and provide a steeper terminal dive onto armoured targets. There are no drag and weight reduction options. A pulsed rocket motor is used for increased range. The system features a more robust tandem warhead that can defeat advanced enemy reactive armour and more effective target discrimination and tracking. Range is said to be increased to 0.5 to 9 km. Hellfire II uses 100 per cent ADA software.

In May 1995, Lockheed Martin and Boeing (then Rockwell) formed a joint venture to market and produce the Hellfire and Hellfire II missiles worldwide. This does not extend to the Lockheed Martin Longbow Hellfire (AGM-114L) or the Boeing Brimstone variants.

In 1998, the US Army awarded the joint venture a US$12.4 million contract to procure 100 Hellfire IIs for the US Navy with an anti-ship warhead (blast/fragmentation). A number of allied nations have also expressed interest in the anti-ship version and in a ground launched version, which has also been demonstrated.

Operational status

AGM-114 Hellfire missiles entered service in 1985 and are in operational use with the US Army and US Marine Corps. An estimated total of 40,000 missiles will be built. Further export orders for air-launched hellfire have been proposed for Canada, Greece, Israel, South Korea and Saudi Arabia. Hellfire II is in production. Orders for Hellfire II have been received from Egypt, Netherlands and the United Arab Emirates.

Specifications

AGM-114A
Length: 1.63 m
Body diameter: 178 mm
Wing span: 0.33 m
Launch weight: 46 kg
Warhead: 8 kg HE shaped charge
Fuze: impact
Guidance: semi-active laser
Propulsion: solid propellant
Range: 8 km

AGM-114B/C
Length: 1.73 m; 1.78 m (IIR version)
Body diameter: 178 mm
Wing span: 0.33 m
Launch weight: 48 kg
Warhead: HE blast fragmentation
Guidance: semi-active laser, IIR or RF/IR
Propulsion: solid propellant
Range: 8 km

Contractors

Boeing North American Tactical Systems Division.
Lockheed Martin Missiles and Fire Control.

Lockheed Martin, Rafael AGM-142 Have Nap/Popeye medium-range standoff missile

Type
Air-to-surface missile.

Development
Summary: the missile has either a TV or an imaging infra-red seeker.

The AGM-142 Have Nap is an air-launched medium-range standoff missile system effective against high-value ground and sea targets by both day and night and in adverse weather conditions. Carried on the US Air Force B-52, the AGM-142 is a variant of the Israel Air Force 'Popeye' missile, which entered production in 1989.

Lockheed Martin and the Israel company Rafael, which developed the AGM-142, signed a teaming agreement in 1988 establishing Lockheed Martin as a co-production partner to satisfy US Air Force requirements. Subsequently, the companies formalised a joint venture for joint production and worldwide marketing. The name of the joint venture is PGSUS LLC. Following successful tests on F-4, F-111 and B-52 aircraft, the USAF adopted the AGM-142 as an interim conventional standoff weapon system for the B-52. Under an US$8.9 million contract signed with Rafael in March 1990, Lockheed Martin produced the airframes for 80 AGM-142 missiles. In August 1993, Lockheed Martin received a $1.59 million follow-on contract from Rafael to manufacture airframes for 30 AGM-142s.

Successful tests against ground targets from the B-52G and H models in September 1993, January 1995 and April 1996, further demonstrated the weapon's performance.

A smaller variant, known as 'Have Lite' or Popeye 2, has been developed which is suitable for carriage by aircraft such as the F-16. The USAF is also reported to be considering the development of a penetrator warhead, I-800, for Popeye. A further version, known as Popeye 3, has been proposed by Rafael with a turbofan engine providing a range in excess of 200 km.

Continued development has resulted in improvements under a Producibility Improvement Program (PIP) which were demonstrated during a successful test flight in June 1997 with a launch from a B52H.

Description
The AGM-142 Popeye 1 medium-range air-to-surface missile uses a zoom TV or imaging infra-red seeker for day and night operation. It employs mid-course autonomous guidance based on inertial navigation, then homes in on the target using either type of seeker, which as line replaceable units, can be selected before the mission. Guided to its target by datalink with the launch aircraft or a separate aircraft, the missile minimises aircraft exposure by permitting missile launches in various trajectories at true standoff ranges. The launch aircraft has to carry a datalink pod, which weighs around 950 kg and has separate forward- and rearward-facing aerial assemblies (AN/ASW-55 datalink pod). The TV datalink antenna is mounted below the rear motor nozzle.

Blast, fragmentation or penetrating warhead types can be used. The missile is claimed to have a hit probability of 94 per cent.

Popeye 2 is known in the US as AGM-142B 'Have Lite', and has the same external configuration as Popeye 1. However, by reducing the motor length, deploying a lightweight but stronger computer and lighter navigation and control systems the overall weight has been reduced to around 1,115 kg. It is reported that Popeye 2 operates in the same manner as Popeye 1 and can be fitted with the same choice of seeker heads and warheads, but has a slightly reduced range of 75 km.

Operational status
The AGM-142 (Popeye) entered production in 1989 and is in service with the US Air Force and the Israel Air Force. Australia announced in 1996 that it plans to buy the AGM-142 for its F-111 force. The USAF ordered 160 AGM-142 Raptor missiles in total and the last 30 missiles have imaging IR seekers replacing the earlier TV systems. Subsequent production was funded through 1997. On 1 August 1996 Lockheed Martin and Rafael set up a joint venture called PGS-US for joint production and marketing of the AGM-142/Popeye family.

In September 1998, the US Air Force awarded PGSUS a US$67.7 million contract to procure 250 additional AGM-142s. The Australian order became part of that buy in December 1998.

Two AGM-142 strike missiles under the wing of a US Air Force B-52 0053942

The Popeye missile can be fitted on B-52, F-111, F-15, F-4 and F-16 aircraft

Specifications
Popeye 1
Guidance
Mid-course: inertial
Terminal: TV or IIR
Propulsion: solid fuel
Warhead: 364 kg HE
Range: 80 km
Length: 4.8 m
Diameter: 533 mm
Horizontal span
Upper: 1.57 m
Lower: 1.73 m
Vertical span: 1.06 m
Overall weight: 1,364 kg
Warhead section weight: 455 kg

Popeye 2
Length: 4 m
Body diameter: 533 mm
Wing span: 1.73 m
Launch weight: 1,115 kg
Warhead: 340 kg HE blast or penetration
Fuze: n/k
Guidance: inertial, update and TV or IIR
Propulsion: solid propellant
Range: 75 km

Contractors
Rafael Armament Development Authority.
Lockheed Martin Missiles and Fire Control.

VERIFIED

MBDA Storm Shadow/Scalp EG

Type
Cruise missile.

Development
Storm Shadow is being developed by MBDA, a joint venture company to develop and produce guided weapons formed by Matra BAE Dynamics, EADS and Finmeccanica. Storm Shadow is identical to the Scalp EG missile being developed by Matra for the French Ministry of Defence. Scalp EG is part of the Apache family of missiles and was ordered by the French Air Force in December 1997. It was announced in July 1996 that Storm Shadow was the successful contender for the UK CASOM (Conventionally Armed StandOff Missile) programme. This contract, formally signed in February 1997, is worth over £700 million. CASOM's requirement was for a precision cruise missile for the RAF Tornado GR.4 (up to four missiles), Harrier GR.7 (two missiles) and Eurofighter.

Description
Storm Shadow will be a stealth cruise missile of approximately 1,300 kg carrying a conventional warhead. It will have the ability to destroy highly protected targets such as command bunkers and communications centres with great accuracy. Range will be over 250 km with autonomous terrain-following flight at very low altitude.

Storm Shadow's modular design is developed from the Apache family of missiles produced by Matra Défense, with a front section containing electronic equipment and the seeker, a central section with the warhead and the rear section containing a turbojet engine, tanks and control surfaces. The warhead is being produced by BAE Systems RO Defence and is a BROACH two-stage penetration type for hardened targets.

The Storm Shadow long-range air-to-surface cruise missile which won the UK MoD's CASOM competition

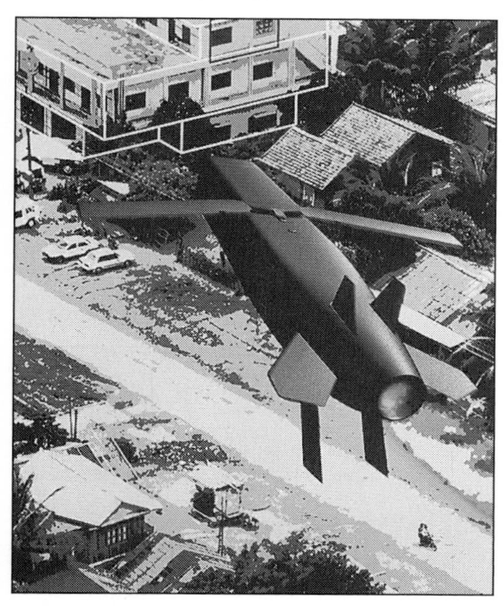

SCALP EG stealth conventional cruise missile under development for the French Ministry of Defence

The guidance system provides day/night all weather, fire-and-forget capability. There is terrain-following navigation with GPS, altimetric correlation and a laser-gyro inertial navigation unit. Terminal guidance is by imaging infra-red seeker which is being produced by BAE Systems Avionics.

Operational status

Under development. Storm Shadow has been ordered by the UK and Scalp EG by the French. The United Arab Emirates ordered an undisclosed number of Black Shaheen missiles in December 1998, and pre-series deliveries are expected to start in 2004, with full operational capability on the Mirage 2000-9 reached in 2007. The United States has expressed serious reservations at any suggestion that Black Shaheen might be integrated on the UAE's new Block 60 F-16s, now entering their initial production phase at Lockheed Martin. MBDA has described the integration of Black Shaheen to the Block 60 as 'a minor modification' and has said that the missile will be operational on that aircraft when it enters service.

In October 1999, Italy confirmed plans to acquire 200 Storm Shadows to equip its Eurofighter and Tornado aircraft. Italy's missiles are being procured through a UK contract with MBDA, under the auspices of the Defence Procurement Agency, on behalf of the Italian government. The Italian contract included industrial participation by Alenia Marconi Systems (now absorbed into MBDA), which will lead the design and development of the airborne training missile.

In August 2000 Greece selected the Scalp EG to arm its Mirage 2000-5 Mk II aircraft. Both the French and UAE missile contracts are government-to-government agreements with the French authorities. According to MBDA Missile Systems, the company holds a total of (approximately) 2,200 orders for the Scalp EG/Storm Shadow/Black Shaheen.

The UK's Royal Air Force (RAF) completed service evaluation of Storm Shadow in June 2002 and the weapon had been in operational service (although not declared) since then. Underwing carriage of Storm Shadow has yet to be cleared, so all 27 of the RAF's Storm Shadow launches during Operation 'Telic' had been from the under-fuselage weapons pylons of the Tornado GR.4.

The RAF has expressed complete satisfaction with Storm Shadow's performance during Operation 'Telic', particularly the main guidance system using a mixture of terrain-reference, inertial and GPS-guided navigation, with terminal guidance from an imaging infrared seeker. It is believed that one weapon targeted on the same bunker hit by another Storm Shadow had actually penetrated through the hole made by the first missile. The RAF is still awaiting confirmation of this claim from post-operational analysis.

Specifications

Length: 5.1 m
Wing span: 3 m
Weight (at launch): 1,300 kg approx
Propulsion: turbojet engine
Warhead: unitary penetration
Speed: 1,000 km/h approx
Range: more than 250 km
Guidance: INS/GPS with imaging infra-red seeker terminal homing

Contractor

MBDA (UK).

UPDATED

Northrop Grumman BAT Brilliant Anti-Armor Submunition

Type

Anti-tank munition.

Development

Prototype systems of the Brilliant Anti-armour Submunition (BAT) were first authorised in 1992 following a US Army critical design review of the hardware and software components. Design test flights began in 1993 and concluded in March 1996 with design verification.

A preplanned product improvement programme for the weapon began in 1995 concentrating on dual-mode seeker technologies. These include millimeter-wave radar, imaging and infra-red sensors from Alliant Techsystems Inc and Northrop Grumman's Electronic Sensors and Systems Division (formerly Westinghouse Electronic Systems Group). Northrop was scheduled to select one seeker contractor in mid-1998 to proceed to the engineering, manufacturing and development phase.

The BAT programme is managed by the Army's TACMS-BAT Project Office at Redstone Arsenal. Northrop Grumman's programme is conducted by its Combat Support Systems facility in Hawthorne, California.

From January 2003, 24 BAT sub-munitions were urgently re-packaged into a BUEV (BAT UAV Ejection Tube) for fitting on the RQ-5 Hunter Unmanned Air Vehicle (UAV). Work is also under way on a parallel development, known as 'Viper Strike', where the BAT is modified to allow use against point targets in congested urban settings.

A range camera records the BAT seconds before a direct hit on armoured vehicles during testing at White Sands Missile Range, New Mexico

The BAT UAV Ejection Tube, shown on a test helicopter. When deployed from a UAV, it is dropped whole and then ejects the BAT submunition (Northrop Grumman) ***NEW***/0532210

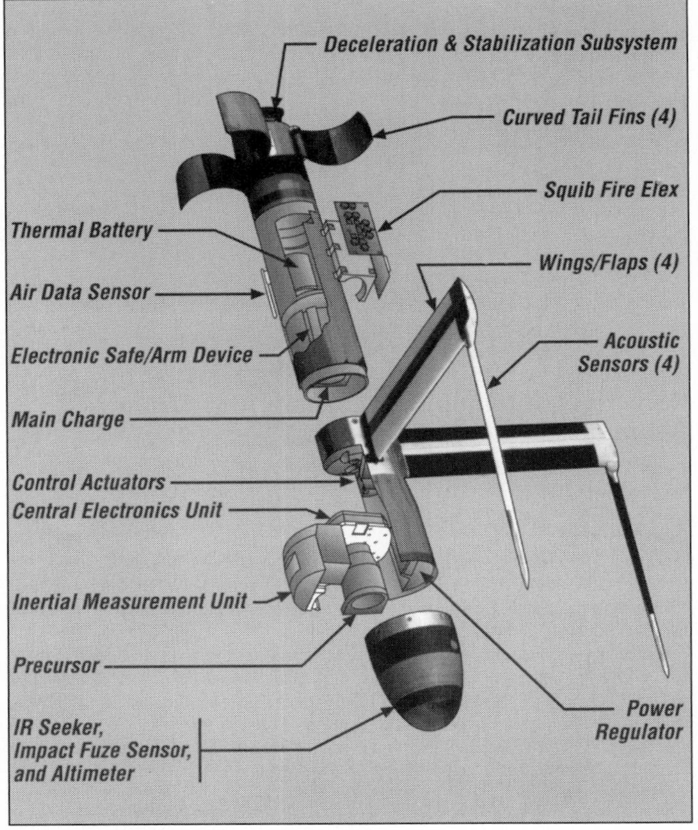

BAT submunition schematic 0053934

A BUET-armed RQ-5A Hunter UAV taking off from the White Sands Missile Range in New Mexico during trials (Northrop Grumman) ***NEW***/0529701

Description

BAT uses acoustic and infra-red sensors to locate and destroy mobile armoured vehicles. It is designed to be delivered by tactical munitions dispensers (in particular the US Army ATACMS missile) and can be used against massed mobile armour at deep strike ranges. Deployment on Tomahawk, JSOW (Joint StandOff Weapon) and SLAM (Standoff Land Attack Missile) may also be considered.

The submunition is an unpowered glider with folded wings. The current guidance system uses acoustic sensors to locate target vehicles and an infra-red sensor for terminal homing. This sensor is made by Raytheon New Hampshire, which is a member of the BAT team.

Operational status

In production. The US Army requested US$85 million in their FY98 budget for 305 BAT submunitions in the first year of procurement. In May 1998, the DoD announced a US$32 million increment of a US$80 million contract to Northrop Grumman for 89 BAT submunitions. According to the manufacturers, the US Army is buying 1,300 BATs. Under a quick-reaction programme initiated in mid-January 2003, Northrop Grumman delivered 24 re-packaged BATs and some RQ-5A Hunter UAVs modified to deliver the now-independent weapon.

Specifications

Length: 0.9 m
Diameter: 140 mm
Weight: 20 kg (wings folded)
Seeker: acoustic and infra-red

Contractor

Northrop Grumman, Electronics Systems.

UPDATED

Rafael Spice

Type

Air-to-surface electro-optic guidance kit.

Description

In June 1999 Israeli weapons manufacturer, Rafael revealed a range extension kit called SPICE (Smart Precise Impact and Cost Effective). Described as 'affordable force multiplier', SPICE converts a 895 kg Mk 84 free-fall bomb into a guided, unpowered glide weapon. The kit includes a nose section, with two swept rectangular control fins, which contains a television seeker, a global positioning navigation system and a datalink. The camera and the datalink allow the aircrew to control the weapon in the terminal phase of the attack. Four straight chord wings and four clipped-tip moving control fins are mounted behind the bomb body in a new tail assembly. SPICE is similar in some respects to the US manufactured GBU-15, but is reported to be cheaper and can be attached to any standard Mk 84 bomb. Although originally designed for use by Israeli aircraft that are unable to carry the AGM-142 Popeye air-to-surface missile, it is now being offered to air forces with limited resources that have a requirement for long-range precision-guided weapons. The SPICE kit has completed drop trials from Israeli Air Force aircraft and is offered for export but none recorded to date.

Specifications

Length: 4.2 m
Body diameter: 40 cm
Main wing span: 1.2 m
Forward canard span: 60 cm
Rear guidance fin span: 70 cm
Launch weight: 1,050 kg
SPICE kit weight: 150 kg
Warhead: 428 kg Tritonal or H-6 (Mk 84 GP bomb)
Guidance: autonomous GPS/INS, electro-optical terminal seeker
Range: over 50 km

Contractor

Rafael, Missile Division.

UPDATED

A Mk 84 bomb with Rafael's SPICE guidance kit on display at Asian Aerospace 2002 (M J Gething/Jane's) ***NEW***/0118293

Raytheon AGM-154 Joint Stand-Off Weapon (JSOW)

Type

Air-to-surface missile.

Development

In the late 1980s, the US Navy began a review of conventional weapons with the intention of reducing the number of weapon types. Three systems were selected for future development; the Advanced Bomb Family became the Joint Direct Attack Munition (JDAM) (qv), the Advanced Strike Weapon became the Tri-Service Stand-Off Attack Missile (TSSAM) and then the Joint Air-to-Surface Standoff Missile (JASSM) (qv), and the Advanced Interdiction Weapon System became the Joint Stand-Off Weapon (JSOW) when the USAF joined the programme in 1992.

The US Navy is the lead service for the JSOW, which has the designation AGM-154. It is being developed by Raytheon as a partial replacement for six existing weapons: the AGM-65 Maverick, AGM-123 Skipper, AGM-62A Walleye, Rockeye and APAM (Anti-Personnel/Anti-Material) submunition dispensers, and laser- and TV-guided bombs. The design requirements for the JSOW was to provide standoff range, with a flexible weapon that could be rapidly retargeted and deliver multiple kills per sortie. It had to be a 'joint' system, interoperable between the US Air Force, Navy and Marine Corps – and allied forces. Low cost was also a crucial consideration. The AGM-154 weapon that emerged was a 480 kg, unpowered winged glide bomb dispenser with combined GPS/INS navigation, that can be launched outside the range of enemy air defences to disperse

AGM-154 Joint StandOff weapon carried on an F-16 0053941

submunitions at a predetermined point over the target area. Three variants are being developed, each using a common 'truck' airframe. The US Navy/Marine Corps and Air Force AGM-154A, also referred to as the JSOW Baseline (BLU-97) or JSOW-A, has fully autonomous guidance and carries 145 BLU-97 combined effects bomblets (as used in the CBU-87/B). The AGM-154B (JSOW-B) for the USAF and US Navy was previously known as the JSOW/BLU-108. It is also fully autonomous and carries six BLU-108/B dispensers, each with four terminally guided Skeet anti-armour warheads. The BLU-108/B was originally developed for the CBU-97/B Sensor Fuzed Weapon system (SFW). The third JSOW version is for US Navy and Marine Corps use, and is known as the AGM-154C or JSOW/Unitary (JSOW-C). This is equipped with a 500 lb unitary warhead, and uses an imaging infra-red terminal seeker for 'man in the loop' guidance. An alternative JSOW-C configuration has also now been developed, using the BROACH staged warhead, for hardened target penetration.

Engineering and Manufacturing Development (EMD) for the AGM-154A began in 1992 and it entered Low-Rate Initial Production (LRIP) in 1997.

In 1995, approval was given for the US Navy to proceed with the development of the other two variants. EMD for the AGM-154B version also started in 1995.

In 1996, the US Navy evaluated several warhead options for AGM-154C, including an insensitive munition variant of the Mk 82 bomb (BLU-111), an Israeli option and a UK proposal (BROACH). In March 1999, Raytheon announced that the AGM-154C programme had been restructured to integrate a new terminal sensor, along with the UK-developed BROACH as the pre-planned product improvement (P3I) warhead. The BLU-111 warhead was selected as the general purpose warhead for the AGM-145C, with BROACH used for the penetration role. The seeker will be an imaging infra-red focal plane array, derived from a seeker developed by Texas Instruments (now Raytheon Defense Systems) for the cancelled AGM-137 TSSAM. The datalink will be the AN/AWW-13, which the US Navy already uses with its AGM-62 Walleye and AGM-84E SLAM weapons. In December 1999, the USAF stated that an upgrade to the AGM-88 HARM anti-radar missile targeting system (AN/ASQ-213) would allow targeting data to be passed to both HARM and JSOW.

In late 1994, it was reported that Texas Instruments (now Raytheon Defense Systems) was entering the UK CASOM competition by offering a powered version of the AGM-154 with either the BROACH penetrating warhead, the BLU-109 penetrator, or the French 'stepped-diameter' penetrator under development for APACHE. This version was further developed by the addition of a turbojet engine to produce the Powered JSOW, with greatly increased range. On 29 September 1995, a modified AGM-154 airframe powered by a Williams International turbojet made its first flight. This variant was 160 in long and weighed 1,460 lb. The prototype vehicle flew for about 75 miles, using two thirds of its available fuel supply. Maximum range for the Powered JSOW was estimated at around 120 miles. The Powered JSOW option was not adopted by the UK (the CASOM requirement was fulfilled by the MBDA Storm Shadow), or any other customer, and the programme was shelved, though the capability remains available.

In 1996, the USAF sponsored the Advanced GPS/Inertial Navigation Technology programme to develop a reliable, accurate and low-cost system for JSOW, JDAM and the Small Diameter Bomb. The USAF is also planning a LADAR programme to develop an advanced solid-state laser radar terminal seeker for the same three weapon systems. Several further JSOW variants are currently being studied. A projected preplanned product-improvement (P3I) programme would add the BAT 'smart' submunition or the 360 kg Lockheed Martin Advanced Penetrator Warhead (also known as the 1-800), and increase the total weight to approximately 680 kg. Further growth could include laser radar or millimetre-wave radar seekers, Link 16 datalink, GPS/IMU enhancements, a turbojet-powered version, and a follow-on AGM-154 that would be versatile enough to carry a variety of payloads including supplies, mines, radar jammers and other non-lethal payloads. In March 2003, the US Naval Air Systems Command announced that modifications to the control surfaces, to cure a vibration problem identified during carriage trials on an F-16, had been approved for combat use.

JSOW integration requires a standard MIL-STD-1760A/B interface and so far the weapon has been cleared on the F/A-18C/D Hornet (maximum JSOW load-out four), F-15E (five), F-16C/D Block 40/50 (four), B-1B (12), B-2 (16) and B-52H (12). Future platforms will include the Joint Strike Fighter, while possible options include the AV-8B, F-111, F-117 and Eurofighter. The JSOW has also been fit-tested on Jaguar and Tornado aircraft.

Description

The AGM-154 is an aerodynamically shaped, unpowered glide dispenser with a rectangular cross-section body shape. JSOW can be carried at supersonic speeds, but must be launched subsonically. It is made up of three major sections. A streamlined nose fairing houses the guidance and control system. A rectangular centre section payload container holds the bomblets or warhead, and this is fitted with two folding high aspect ratio wings on its upper surface, and two standard 762 mm spaced suspension lugs. The tail section has six fixed, sweptback rectangular fins positioned radially on the boat tail and contains the flight-control system. AGM-154 is 4.1 m long, has a body width of 337 mm and a depth of 442 mm, and an all-up weight of around 484 kg.

The area target version (AGM-154A) carries 145 BLU-97/B CEM bomblets. Total weight for this version is 474 kg (1,043 lb). These are cylindrical, canister-type bombs fitted with a ballute parachute retarding tail unit. An ejected front projecting tube used to sense the optimum standoff distance for activating the bomblet's shaped charge explosive. Before ejection, the CEM bomblet is 0.169 m long, has a body diameter of 64 mm and weighs 1.5 kg. Once ejected and activated, the bomblet extends to a length of 0.356 m. The warhead consists of a fragmenting case with a 287 gram shaped charge of Cyclotol and zirconium which provides an incendiary element. Further details of the BLU-97/B bomblet can be found in the CBU-87/B CEM entry.

The armoured targets version (AGM-154B, JSOW/BLU-108) carries six BLU-108/B bomblets. Total weight for this version is 470 kg (1,035 lb). This version uses the same airframe as AGM-154A but has a different dispenser system. Each BLU-108/B is a cylindrical canister with two small rectangular tailfins. It contains four 'Skeet' explosively formed projectile (EFP) warheads, carried on folding arms, while the main BLU-108 'bus' is fitted with a stabilising parachute and a rocket motor. Before ejection the bomblet is about 0.88 m long and has a body diameter of 120 mm. Each Skeet is a small canister device with a copper disc EFP and a side-mounted dual-mode laser/IR sensor. The Skeet is around 118 mm in diameter and 90 mm deep. Further details of the BLU-108/B bomblet can be found in the CBU-97/B SFW entry.

Guidance and control from the time the AGM-154 is released from the aircraft is provided by a Global Positioning System (GPS) receiver and Inertial Navigation System (INS). JSOW uses a system of pre-programmed waypoints to navigate to its target, but new targeting data can be updated from the cockpit prior to launch. An antenna on top of the AGM-154 then updates the GPS data during flight. For AGM-154A the bomb is put into a shallow dive for the terminal phase, a pyrotechnic cutter blows off the payload covers and a gas generator inflates an aluminium bladder to eject the BLU-97/B bomblets. The AGM-154B overflies the target and releases the BLU-108/B bomblets in two sticks of three from the bottom of the payload container.

The AGM-145C weighs in at 468 kg (1,030 lb). The original terminal seeker proposed for AGM-154C (JSOW/Unitary) point target version for the US Navy was an imaging infra-red system that would be equipped with a datalink to transmit seeker video to the launch aircraft, as the weapon approached the target area. It is now planned that AGM-154C will use an imaging IR seeker with Autonomous Targeting Acquisition (ATA) technology. In its hard target attack form, the AGM-145C will be fitted with the BROACH is multistage warhead, consisting of a large penetrating shaped-charge in front of a conventional follow-through bomb. The primary charge weighs 206 lb, while the secondary charge weighs 306 lb. BROACH provides blast/fragmentation effectiveness, as well as hard target penetration to a depth of over 5 ft of reinforced concrete. The secondary, augmenting charge is packaged with 3 g steel fragments for added effect. The AGM-154 can be delivered from altitudes between 200 and 40,000 ft (75 and 12,000 m) and at speeds between 250 and 650 kt. With a low altitude launch the weapon has a range of 12 nm (22 km), rising to over 40 nm (75 km) with a high-altitude launch.

Operational status

The AGM-154A entered engineering and development in 1992. Captive carry test flights on F-16C/D Fighting Falcon and F-15E Strike Eagle aircraft were started in 1994 at Eglin AFB. The first guided flight test took place in December 1994 from a F/A-18C Hornet at the Naval Air Warfare Centre, China Lake. The BLU-97/B bomblet is already in service for use with the CBU-87/B, which entered service in 1986. The USAF began development, test and evaluation flight trials of the AGM-154B in early 1996, with the first trial in July 1996 from an F-16D aircraft. In 1997, the USN placed a low-rate initial production order for 180 AGM-154A weapons for the USN and USAF, and full-rate production was authorised in January 1999. A second production contract for AGM-154A was awarded in January 2000, with 414 weapons for the USN and 74 for the USAF. In March 2003, Raytheon was awarded a US$80.8 million contract to supply the US Navy with 313 AGM-15A JSOWs, and 24 to the US Air Force. Low-rate initial production of AGM-154B was started in February 1999 and it is now in the final stages of development before operational testing. The current planned total programme buys for the JSOW family are as follows: AGM-154A 11,800 (8,800 US Navy, 3,000 USAF); AGM-154B 4,314 (1,200 US Navy, 3,114 USAF); AGM-154C (3,000 US Navy only).

The rollout ceremony for the first AGM-154A JSOW production rounds took place at NAS Fort Worth in July 1998. An F/A-18 and F-16 were on hand with a JSOW mounted on each aircraft. At the ceremony, Raytheon Systems released news that JSOW development test articles were deployed aboard the USS Nimitz in 1997 and were currently deployed on the USS Eisenhower. The first operational use of AGM-154A bombs was reported in Iraq on 25 January 1999. US Navy F/A-18 Hornets on routine 'no-fly' zone patrol over Iraq launched three JSOWs at three separate Iraqi surface-to-air sites, destroying all three. During the early days of the 1999 Kosovo crisis, the US Navy are reported to have rushed stocks of AGM-154A to Europe to improve their strike aircraft's performance in the dismal Balkan weather conditions. An undisclosed number of weapons were airlifted to Italy and then flown to the carrier USS *Theodore Roosevelt*. JSOW was used on several occasions by F/A-18C Hornet strike fighters to attack targets in conditions that would have prevented the employment of precision-guided munitions that require optical designation or guidance. By mid-2001, about 90 JSOWs had been used in Iraq and Kosovo.

The US Navy successfully launched an AGM-154C from an F/A-18 Hornet on 6 February 2003, while on 8 February, the US Air Force completed integration trials of the AGM-154A on the B-52H, firing two JSOWs at separate targets.

Specifications

Length: 4.1 m
Weight: 483 to 681 kg depending on the payload
Range
Low altitude launch: 20 km
High altitude launch: 64 km
Powered: >200 km

Contractor

Raytheon Missile Systems.

UPDATED

Raytheon AGM-65 Maverick

Type

Air-to-surface missile.

Development

Summary: AGM-65B has TV guidance; AGM-65D, AGM-65F, AGM-65G have imaging infra-red seekers; AGM-65C and AGM-65E have semi-active laser guidance.

The AGM-65 Maverick family of air-to-surface missiles has been steadily developed and improved since the mid-1960s. Designed for use against tanks and a variety of hardened targets, the later versions have a larger penetrating warhead specifically to attack ships, bunkers and hardened aircraft shelters. AGM-65A was a TV-guided missile and entered service in 1972. AGM-65B entered service in 1975. It was also TV guided but had improved optics to allow greater magnification of the target area. AGM-65C was to have been laser guided but did not enter service. AGM-65D has an Imaging Infra-Red (IIR) seeker and entered service in 1983. AGM-65E has laser guidance and entered service in 1985. AGM-65F has an IIR seeker and entered service in 1989. AGM-65G has an IIR seeker and entered service in 1990. The latest AGM-65H version has a millimetric wave active seeker with trial firings started in 1991. Different US services have sponsored the various versions of Maverick, the US Air Force being responsible for the AGM-65A/B/D/G/H, the US Marine Corps for the AGM-65E and the US Navy for the AGM-65F. Mavericks have been cleared for carriage on F-4 Phantom, F-5 Tiger, F-15E, F-16, F/A-18, F-111, A-4 Skyhawk, A-6 Intruder, A-7 Corsair, P-3 Orion, AV-8B Harrier II, A-10, AJ 37 Viggen and Hunter aircraft and the AH-64 Apache helicopter.

Description

The AGM-65 Maverick missiles all have a similar construction, with four long-chord delta wings and four moving rectangular control fins at the rear. Guidance for the A and B versions is by TV tracker and for the D, F and G versions by an IIR seeker. The AGM-65E operates on a quite different principle, with the missile homing onto coded laser energy reflected from the target. A separate laser target designator is required, either aircraft or ground mounted, to illuminate the target. However, the semi-active laser-guided E version uses the same cockpit display, with symbology to indicate the locking before launch of the missile to the selected target.

AGM-65A/B: TV-guided versions developed by USAF, with 57 kg HE shaped charge warhead. The B version has improved optics to magnify the target scene, enabling the pilot to locate the target and lock the missile to it at a greater range. Tracker logic steers the missile to the centroid of the target.

AGM-65D: IIR guidance version developed for the USAF to give day, night and bad weather capability. The range of the system was considerably improved and the IR frequency selected to cope better with battlefield smoke. The digital centroid seeker guides the missile to the centre of the target, rather than simply to the point of the greatest temperature differential.

The most widely used members of the AGM-65 Maverick missile family, from front to rear AGM-65D, -65E, -65B and -65F

AGM-65E: A semi-active laser-guided version developed for the US Marine Corps, specifically to enable Maverick to be used on targets near to friendly forces. This version was the first to introduce the new 136 kg blast penetrating warhead. With the laser guidance, the pilot does not have to identify the target visually before firing the missile.

AGM-65F: Developed for the USN, this version has the same IIR seeker as the D version but with the image processing tuned specifically for effectiveness against ship targets.

AGM-65G: Developed for the USAF to provide an improved capability over the D version for attacking a wide variety of targets. The IIR seeker is similar to that in the D version but has software changes to improve the performance against larger targets, with the pilot able to select a specific aim point within a large target complex.

AGM-65H: This active millimetric radar-guided version is in development for the USAF, with trials missiles delivered in 1991. 'Longhorn' has been proposed by Hughes (now Raytheon), to include a turbofan engine in a lengthened AGM-65F airframe. Mid-course guidance would be inertial with GPS, whilst IIR or active MMW radar terminal seekers could be used.

Operational status

AGM-65A/B/D/E/F and G Maverick are in service, whilst AGM-65H has completed development and is in early production. Over 30,000 AGM-65A/B TV-guided versions have been produced and a further 25,000 AGM-65D/E/F/G versions.

Export sales have been made to Bahrain, Belgium, Denmark, Egypt, Germany, Greece, Iran, Israel, Italy, Jordan, Kenya, South Korea, Kuwait, Malaysia, Mexico, Morocco, Netherlands, New Zealand, Norway, Pakistan, Philippines, Portugal, Saudi Arabia, Singapore, South Korea, Spain, Sudan, Sweden, Switzerland, Taiwan, Thailand, Tunisia, Turkey, UK, Venezuela, Vietnam and Yugoslavia (Serbia and Montenegro). It is reported that around 5,300 Maverick missiles were used against Iraq in the 1991 Gulf War, mostly AGM-65B (TV-guided) and AGM-65D (IIR-guided) versions. Some 810 missiles were launched in Kosovo in 1999. The UK Royal Air Force launched 30 AGM-65G2 Mavericks over Iraq in March/April 2003, while the US forces expended 918 missiles across all variants.

Contractor

Raytheon Systems Company.

UPDATED

Specifications

	A/B	D	E	F/G	H
Length	2.49 m	2.49 m	2.49 m	2.49 m	2.6 m
Body diameter	305 mm	305 mm	305 mm	305 mm	305 mm
Wing span	0.72 m	0.72 m	0.72 m	0.72 m	0.72 m
Launch weight	210 kg	220 kg	293 kg	307 kg	305 kg
Warhead	57 kg HE shaped charge	57 kg HE shaped charge	136 kg blast penetrator	136 kg blast penetrator	136 kg blast penetrator
Fuze	Impact	Impact	Impact	Impact	Impact
Guidance	TV	IIR	Semi-active laser	IIR	Active radar
Propulsion	Solid propellant	Solid propellant	Solid propellant	Solid propellant	Solid propellant
Range	3 km	20 km	20 km	25 km	25 km

Raytheon Paveway laser-guided bomb

Type
Air-to-surface bomb.

Development
The Paveway series of semi-active laser-guided bombs first saw operational service with Paveway I in Vietnam in 1968. Paveway II was developed during the late 1970s and Paveway III completed development in 1986.

The laser-guided Paveway system requires the targets to be designated/marked by either airborne, vehicle-based or hand-held designators. Early airborne designators with which Paveway could operate were Pave Knife, Pave Tack and Pave Spike. More recent target illumination has been provided by Atlis II, LTDS, TRAM, GLLD, MULE, LTM, LANTIRN, and Raytheon's own FLIR/laser designator.

In 1996, Paveway III was qualified for carriage on the F-16. The test programme included two successful GBU-22 releases.

Description
The Paveway system consists of two major bolt-on subassemblies for the conversion of conventional or specialised free-fall bombs to a laser guided bomb. All are made up of three major assemblies: the front end laser guidance and control section, the main body warhead and the stabilising tail assembly. Paveway III also has an adaptive digital autopilot.

The Paveway II front end guidance and control section contains the semi-active-laser seeker which 'weather-vanes' during bomb flight and aligns approximately along the bomb flight path. Paveway III has a new laser guidance scanning seeker so the guidance section no longer has the weather vaning nose. The new laser guidance and control section is a long light metal cylindrical tube with a clear plastic dome and has four in-line, moving clipped delta control surfaces that run from the centre to the rear of the unit. After the bomb is released the laser error detector measures the angle between the bomb's velocity vector and the line between the bomb and target. Steering corrections are made by moving the nose-mounted canard control fins to adjust the bomb's trajectory to line up with the target. The tail fins/wings are for stabilisation purposes only.

Weapons currently in use for Paveway II and III include:

1. GBU-24 general purpose and penetrator weapons are both horizontal and hard target capable and are compatible with Mk84 and BLU-109 warheads. They can be delivered from F-111, F-15E, F/A-18, A-6, F-16 and Tornado aircraft.
2. The GBU-27/B penetrator weapon, also horizontal and vertical hard-target capable, is compatible with the BLU-109 warhead and can be delivered by the F-117.
3. GBU-28A/B super penetrator weapon is a special purpose weapon developed during Desert Storm for very hard targets. It uses the BLU-113 warhead and can be delivered by the F-111 and F-15E.

Operational status
Paveway I entered service with the USAF in the late 1960s and saw operational service in Vietnam. Paveway II followed in 1976 and Paveway III entered service in 1986. The UK was the first export customer for the Paveway II and the first operational user of the LGB, when RAF Harriers dropped Paveway IIs during the Falklands conflict of 1982. The Paveway II and III are still in production and the main GBUs still in service and used by the USAF and USN during the 1991 Gulf War are: GBU-10C, D, E and F (Mk 84 warhead); GBU-10G, H and J (BLU-109/B warhead); GBU-12B, C and D (Mk 82 warhead); GBU-16A and B (Mk 83 warhead); GBU-24/A (Mk 84 warhead) and GBU-24A/B (BLU-109/B warhead). Enhanced Paveway GBU-24E/B bombs, with additional INS/GPS guidance, started testing in 1999. It was reported in 1992 that the USAF had used 8,400 LGBs during the Gulf War and it still had an inventory of 27,000. It is believed that 9,100 GBU-24s were ordered by the USAF in 1993, and some GBU-22/Bs may also have been ordered in 1998. The last Paveway II delivery to the US Armed Forces was made in 1994, but following the rapid depletion of stocks during the Kosovo conflict, Raytheon was issued with an emergency supplemental replenishment contract for 2,245 Paveway II kits, in 1999.

In addition to the USA, operators of Paveway family LGB include Australia, Canada, France, Greece, Israel, Netherlands, Saudi Arabia, South Korea, Spain, Taiwan, Thailand, Turkey and UK.

The Paveway II laser-guided bomb 0005509

Artist's impression of the Paveway III laser-guided bomb 0005510

In 1999, France ordered an undisclosed number of Paveway II and III kits, with Paveway II to be fitted to SAMP Mk 82 and Mk 83 bombs, and Paveway III to BLU-109 and Mk 84 bombs. Canada ordered some GBU-10 and GBU-12 in 2000. By 2001 over 117,000 Paveway II weapons had been delivered to the world's air forces.

In January 2001 the UK MoD signed a £42 million contract with Raytheon for the supply of Enhanced Paveway II and III LGBs, plus support equipment. The introduction of the GPS-aided Enhanced Paveway was prompted by experiences during Operation 'Allied Force' over Kosovo during 1999, when RAF aircraft armed with conventional Paveway LGBs were unable to hit their targets due to smoke and bad weather. The Enhanced Paveway contract calls for the upgrade of existing UK bombs using kits supplied and installed by Raytheon. The UK is the first customer for the Enhanced Paveway II and during March/April 2003, it was the RAF's 'weapon of choice'.

In June 2002 Raytheon delivered the 125,000th Paveway II LGB, as part of a batch of deliveries to the USAF and US Navy. The Paveway II is currently in the inventories of 32 countries. In June 2003, the UK MoD announced selection of the Paveway IV as the RAF's next-generation precision-guided bomb.

Specifications
GBU-24B
Length: 173 in
Weight: 2,315 lb
Warhead: Mk84

GBU-24A/B, GBU-24B/B
Length: 170 in
Weight: 2,350 lb
Warhead: BLU-109

GBU-27/B
Length: 167 in
Weight: 2,170 lb
Warhead: BLU-109

GBU-28A/B
Length: 230 in
Weight: 4,700 lb
Warhead: BLU-113

Contractor
Raytheon Systems Company.

UPDATED

Spetztekhnika Vympel AS-14 'Kedge' (Kh-29)

Type
Air-to-surface missile.

Development
Summary: the missile has semi-active laser or TV command-guidance.

The AS-14 'Kedge' is the NATO designation for this third-generation tactical air-to-surface missile developed in the 1970s. It is believed to have entered service around 1980. The 'Kedge' missile probably utilised the Semi-Active Laser (SAL) guidance of AS-10 'Karen' and the TV guidance of AS-13 'Kingbolt', but with a larger warhead. AS-14 has been offered for export in TV-guided X-29T and semi-active laser X-29L versions. There is an unconfirmed report that there is a third

AS-14 Kedge, semi-active laser version, on the wing pylon of an Iraqi Air Force Mirage F1 (Christopher F Foss)

version, known as Kh-29 MP, with a passive anti-radar seeker, that has been fitted to the Su-17 'Fitter'. It is possible that a later modification might have updated the TV guidance to imaging IR for day/night use. AS-14 is carried by MiG-27 'Flogger', MiG-29 'Fulcrum', Mirage F1, Su-17/22 'Fitter', Su-24 'Fencer' and Su-25 'Frogfoot' aircraft.

Description

AS-14 'Kedge' is 3.9 m long, with a body diameter of 400 mm, a wing span of 1.1 m and a weight of 660 kg for the SAL version and 680 kg for the TV. The SAL guided missile has a distinctive nose shape, where the laser guidance assembly appears to have been fitted onto a larger diameter missile body. AS-14 has a 320 kg HE general purpose bomb as the warhead. The carrying aircraft would have a laser designator pod to designate the targets for the missile. It is reported that earlier Soviet designator pods were limited in range to 10 km for accurate designation. The TV-guided version has a different nose assembly, with four rectangular strakes in front of four clipped-tip moving delta control fins and a range of 12 km. When carrying TV-guided missiles the launch aircraft would carry a TV receiver and designator pod, probably similar to that used with the AS-13 'Kingbolt' missile.

Operational status

AS-14 'Kedge' entered service around 1980 and is still in operational use. It is believed that AS-14 'Kedge' was exported to Afghanistan, Azerbaijan, Belarus, Bulgaria, Czech Republic, Georgia, Germany, Hungary, Iraq, Kazakhstan, Poland, Romania, Slovakia and Ukraine. Both TV- and SAL-guided versions were exhibited and offered for export in Dubai in 1991.

Specifications

Length: 3.9 m
Body diameter: 400 mm
Wing span: 1.1 m
Launch weight
  SAL: 660 kg
  TV: 680 kg
Warhead: 320 kg HE
Fuze: n/k
Guidance: semi-active laser or TV command
Propulsion: solid propellant
Range:
  SAL: 10 km
  TV: 12 km

Contractor

Spetztekhnika Vympel NPO, Moscow.

VERIFIED

Textron Sensor-Fuzed Weapon

Type

Air-to-surface smart munition.

Development

Textron Systems Sensor Fuzed Weapon (SFW), designated CBU-97, is the first sensor-fuzed 'smart' munition to go into production. Smart munitions have the ability to search, detect, acquire and engage targets. SFW was developed for the US Air Force and is designed to defeat multiple targets in a single pass. One weapon can neutralise moving and stationary combat vehicles within a 15 acre coverage area. It can be deployed on operational US and NATO tactical aircraft such as the F-16 and can be delivered from altitudes of 200 to 20,000 ft at speeds up to 650 kt.

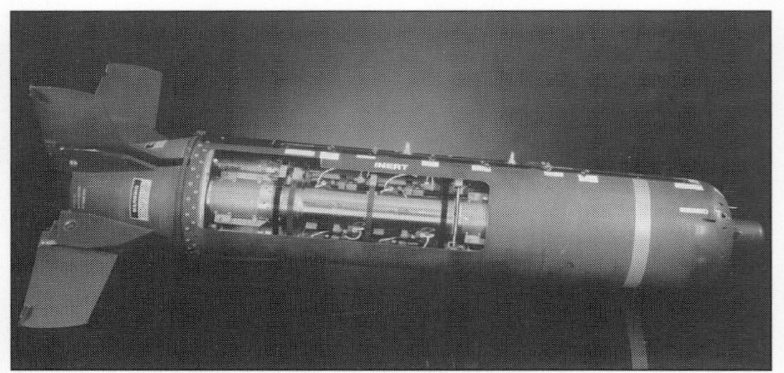

The Sensor-Fuzed Weapon showing the internal submunitions 0005507

The BLU-108 submunition of the Sensor-Fuzed Weapon 0005508

Description

The SFWs Tactical Munitions Dispenser incorporates 10 BLU-108 submunitions. Each BLU-108 has four target-sensing Skeet warheads, a total of 40 warheads per weapon. Each Skeet has a built-in infra-red sensor which searches for a target that matches a defined set of infra-red requirements and fires the Explosively Formed Projectile (EFP) into the target.

After the SFW is released from the aircraft, the 10 BLU-108 are dispensed. A main parachute is deployed, bringing the BLU-108 into a vertical position above the target area. The BLU-108's rocket motor induces upward velocity and spin and the Skeets swing out on arms and are released by centrifugal force. Each Skeet spins and wobbles forming a conical pattern while the infra-red sensor searches for the target. The Skeet will attack the first target it senses with matching infra-red signature as it sweeps an area up to 2,700 m^2. The EFP is then fired.

Operational status

The CBU-97/B entered initial low-rate production in mid-1992 and the first five complete units were delivered to the USAF in April 1993. The number of complete units to be built has been reduced from 20,000 to 5,000. A full-rate production contract was awarded in 1996, with a second contract in 1997 making a total of approximately 1,150 CBU-97/B ordered. Flight tests from a B-1B Lancer aircraft were made in May 1997, and CBU-97/B entered operational service in 1997. Alliant Techsystems will provide the Tactical Munitions Dispenser (TMD) for the weapon system.

In 1998, a third production contract was placed for a further 550 CBU-97/B, and in June 1999 a fourth contract for 300 units. The fourth contract allows for 144 BLU-108/B submunitions for the AGM-154B JSOW programme. There are unconfirmed reports that CBU-97/B were used from B-1B Lancer aircraft in Serbia and Kosovo in 1999. According to US DoD figures, 88 CBU-105 Wind Corrected Munitions Dispensers, armed with the SFW, were used in March/April 2003 during Operation 'Iraqi Freedom'.

Contractor

Textron Systems.

UPDATED

ELECTRO-OPTIC COUNTERMEASURES

ELECTRONIC COUNTERMEASURES

BAE Systems AN/ALQ-144 and 144A infra-red countermeasures system

Type
Airborne infra-red countermeasures system.

Description
The AN/ALQ-144 is an electrically powered infra-red countermeasures set which provides helicopters and small fixed-wing aircraft with protection against heat-seeking missiles. It is an omnidirectional system consisting of a cylindrical source surrounded by a modulation system to confuse the seeker of the incoming missile. A number of variants are available: the AN/ALQ-144 (V)1, AN/ALQ-144 (V)3, AN/ALQ-144A and the AN/ALQ-144 (VP) phase lock systems. The electrically heated graphite source has extremely long life and the complete system weighs less than 14 kg.

Operational status
AN/ALQ-144 (V)1/3 are reported to be in service with the US Army, US Air Force and US Marine Corps. The AN/ALQ-144 (VE) has been sold to a number of other countries. The AN/ALQ-144 (VP) has been used on the SH-60, H-3 and SH-2 aircraft of the US Navy.

Over 3,000 AN/ALQ-144 systems are in service. The US Army has equipped its helicopters with the improved AN/ALQ-144A (V) 1/3 system.

To provide expanded platform protection, the US Army recently type-classified a phaselocked pair of transmitters, the AN/ALQ A (V) 5. This variant provides locational flexibility for side-by-side or top/bottom transmitter placement based on platform mission requirements.

BAE Systems (formerly Sanders, a Lockheed Martin Company), has received contracts totalling nearly US$100 million since October 1989 for upgrades to the AN/ALQ-144 (V)1/3 configuration. In 1999, a multi-year programme to supply versions of the AN/ALQ-144 system to US and foreign military customers was awarded.

Specifications
Dimensions: (transmitter) 241 × 241 × 336 mm
Weight
Transmitter: 12.7 kg
Control unit: 0.5 kg
Power supply: 28 V DC, 1,200-2,000 W
Coverage: 360° azimuth
MTBF: 480 h

Contractor
BAE Systems, Information and Electronic Warfare Systems.

VERIFIED

The BAE Systems AN/ALQ-144 infra-red countermeasures set mounted on the upper fuselage of a helicopter

BAE Systems AN/ALQ-157 infra-red countermeasures system

Type
Air electronic jamming system.

Description
The AN/ALQ-157 provides multiple simultaneous protection for large heavy-lift helicopters and medium-size fixed-wing aircraft against SAM and AAM threats. The system employs advanced components and microprocessor technology to allow operator jamming code selection and reprogrammability for future threats. The two fuselage-mounted synchronised jammer assemblies provide continuous 360° protection against threats launched from any direction. The power module, line filter and pilot control indicator can be placed anywhere within the aircraft.

Operational status
In production. Lockheed Martin (formerly Loral) received the initial production contract in December 1983 for systems for US Marine Corps CH-46 helicopters. In service with the US Navy, US Air Force, USSOCOM, US Marine Corps, Italian Army and the Royal Air Force. US and international aircraft equipped with the AN/ALQ-157 include the SH-3, CH-46, CH-47D, H-53, Lynx, C-130 and P-3C. Nearly 1,400 systems have been sold worldwide. On 30 January 2001, BAE Systems' Pomona facility produced its 5,000th AN/ALQ-157 system.

Specifications
Weight: 99.8 kg
Power supply: 115 V AC, 3 phase, 400 Hz, 28 V DC

Contractor
BAE Systems North America - Information and Electronic Warfare Systems (formerly - Lockheed Martin Electro-Optic Systems).

UPDATED

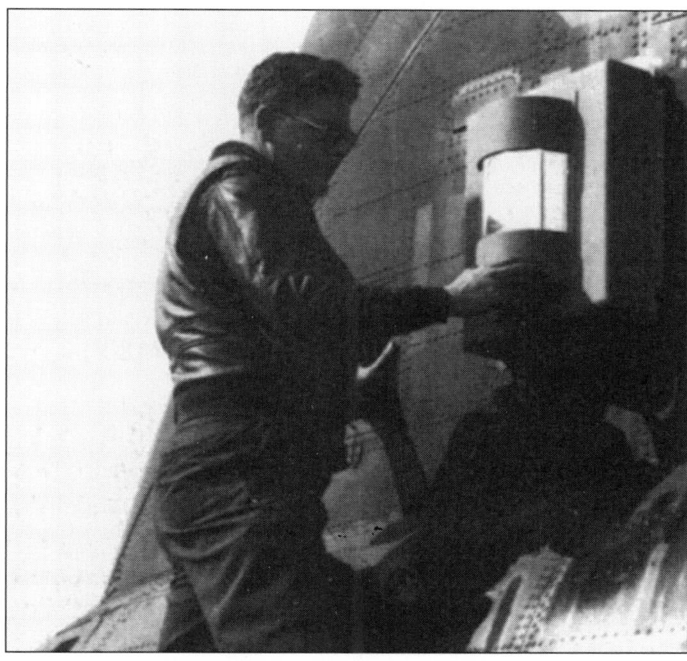

The AN/ALQ-157 infra-red jamming equipment attaches to the tail rotor pylon of the US Navy Vertol CH-46 helicopter

BAE Systems AN/ALQ-204(V) Matador IR countermeasures system

Type
Air electronic countermeasures.

Description
The AN/ALQ-204(V) Matador is a family of infra-red (IR) countermeasures systems which has 11 different configurations. The system is suitable for all types of large transport aircraft with unsuppressed engines, one transmitter per engine being recommended for maximum protection, giving 360° azimuth coverage. The basic system consists of a transmitter, a Controller Unit (CU) and an operator's controller.

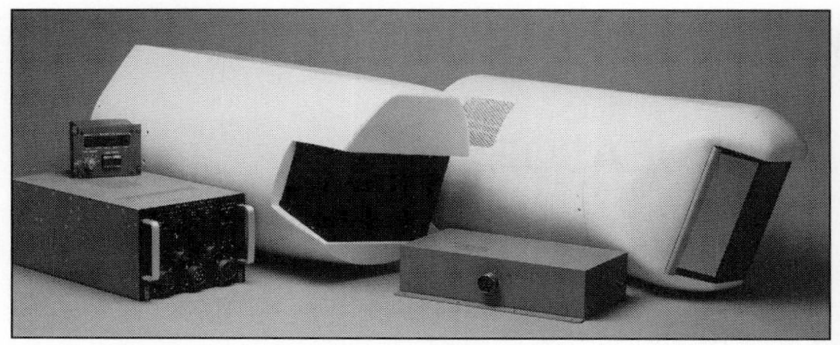

AN/ALQ-204 infra-red countermeasures system is installed in a number of large fixed-wing aircraft providing Head of State aircraft self protection

The AN/ALQ-204 mounted on the fuselage of a Royal Air Force VC10 (Paul Jackson)

Transmitters are electronically synchronised by the CU which controls and monitors one or two transmitters. The operator's controller is common to all configurations, controls from one to seven transmitters and incorporates a system status display.

Each transmitter contains an IR source which emits pulsed radiation to combat an IR missile. Preprogrammed multithreat jamming codes are selectable on the operator's control unit and all new codes can be entered as required to cope with new threats. The Matador transmitters can be aerodynamically mounted within aircraft fairings or engine pylons, or they can be pod-mounted.

Operational status

In production. In service on the Boeing 707 and 747, Lockheed L-1011 and British Aerospace VC10, BAe 146, Airbus A-340 and Gulfstream G-IV aircraft. All current installations are FAA certified for operation on Head of State, commercial and civilian aircraft.

Contractor

BAE Systems North America - Information and Electronic Warfare Systems (formerly - Lockheed Martin Electronic Defense Systems).

UPDATED

BAE Systems ATIRCM/CMWS (AN/ALQ-212(V), AN/AAR-57)

Type

Airborne infra-red countermeasures system.

Description

The AN/ALQ-212(V) Advanced Threat Infra-Red CounterMeasures (ATIRCM) system is the next generation of integrated missile warning, infra-red jamming and dispensing systems. It combines both laser and xenon lamp jamming technology, the most recent breakthrough in infra-red self-protection and an advanced dispenser of expendables, allied with other advanced technologies. The system features a co-ordinated multispectral response from its directional jammer and dispenser that is cued by the missile warner.

The effort incorporates a temporal/spatial Common Missile Warning System (CMWS) AN/AAR-57 to provide the missile launch detection role and give accurate direction of threat within a few degrees. After cueing the ATIRCM jammer will focus a beam from an optical source, modulated to provide a deceptive waveform. The system is designed to handle all currently operational infra-red guided missiles with growth to counter future missiles.

Operational status

BAE Systems (formerly the team of Sanders and Lockheed Martin IR and Imaging Systems), has developed and demonstrated a prototype ATIRCM which resulted in successful flight tests and performance against live missile firing tests during 1994. In September 1995, the US Army awarded the company an engineering and manufacturing development contract covering both ATIRCM and the tri-service CMWS missile approach warner derived from it. Additional successful live-fire missile tests were completed in 1996.

The development programme continues, with production planned for 2002. Initial fielding is scheduled for US Army special operations helicopters and the AH-64D Longbow Apache. In 1997, the former Sanders was awarded contracts for

The ATIRCM/CMWS jamming head and sensor 0055655

two key UK self-protection programmes. The AN/AAR-57 will provide missile warning for the Replacement Maritime Patrol Aircraft (RMPA) and for the UK version of the Longbow Apache, as part of the then Marconi Electronic Systems Helicopter Integrated Defensive Aids Suite (HIDAS).

Specifications

ATIRCM/CMWS four sensor suite

Weight

- **Missile warning sensor (with anti-icing) (4):** 2.6 lb ea
- **Electronic control unit:** 16.5 lb
- **Jam head control unit:** 13.0 lb
- **Jam head (2):** 31.6 lb ea
- **Laser jam source:** 22.6 lb
- **Total weight:** 125.7 lb

Power

- **Missile warning sensor (with anti-icing) (4):** 28 W ea
- **Electronics control unit:** 279 W
- **Jam head control unit:** 216 W
- **Jam head (2):** 740 W ea
- **Laser jam source:** 458 W
- **Total power:** 2,545 W

Contractor

BAE Systems, Information & Electronic Warfare Systems.

VERIFIED

BAE Systems DASS 2000 defensive aids subsystem

Type

Air electronic countermeasures.

Description

The DASS 2000 defensive aids subsystem is a turnkey, fully integrated system compatible with a wide range of aircraft. The system uses sensor fusion technology and multispectral self-protection. DASS 2000 would typically include radar warning receiver, missile warning system, laser warning system, chaff/flare dispenser, electronic countermeasures jammer, towed radar decoy and infra-red countermeasures. The particular configuration is chosen by the customer. A tactical command system interfaces DASS 2000 with other aircraft sensors, processing and communications systems.

DASS 2000 has a colour display, providing the EW operator with total tactical awareness. The display provides threat assessment as well as automatic countermeasures response by using the defensive aids resources.

Operational status

As of July 2002, the UK's Ministry of Defence was procuring DAS 2000 type EW suites for use on the Nimrod MRA Mk 4 maritime patrol and Airborne STand-Off Radar (ASTOR) aircraft. *Jane's* sources suggest that the specific Nimrod application incorporates a BAE Systems AN/ALR-56M radar warning receiver (see separate entry), a fibre optic variant of Raytheon's AN/ALE-50 active towed radar decoy (see separate entry) and Thales Shrike onboard techniques generator (see separate entry). Sources also suggest that a DASS 2000 variant has been proposed for use on C-130H and -130J transport aircraft.

Contractor

BAE Systems North America - Information and Electronic Warfare Systems (formerly Lockheed Martin Fairchild).

UPDATED

BAE Systems Helicopter Integrated Defensive Aids System (HIDAS)

Type
Air electronic countermeasures.

Description
BAE Systems, Avionics, Radar and Countermeasures Systems has developed a suite of integrated defensive aids systems suitable for installation on a variety of rotary-wing aircraft. The suite of sensors and countermeasures with each unit is optimised to defend against specific threats: radar-guided anti-aircraft artillery; vehicle-launched missiles guided by radar or laser beam; shoulder-launched heat-seeking missiles.

The complete HIDAS Helicopter Integrated Defensive Aids Suite includes the following systems:

Detection - BAE Systems Sky Guardian 2000 radar warner with HIDAS control processor. (E-J band, with C, D and K band options.); type 1220 or 1223 laser warner for detection of range-finders, designators and beam-riding missiles; missile launch detector (passive optical).

Display - dedicated unit or multifunction display via databus; audible warning by tones or synthesised speech.

Control - central control with optimised threat prioritisation and countermeasure selection.

Countermeasures - BAE Systems Apollo inboard radar jammer, pulse and CW; IR jammer for missile seduction; chaff and flare dispensers; visual, infra-red and radar stealth measures for signature reduction.

Installations can be built-in or modular units can be retrofitted. The system uses the Merlin electronic warfare threat library.

Contractor
BAE Systems, Avionics, Radar & Countermeasures Systems, Stanmore.

VERIFIED

The BAE Systems Helicopter Integrated Defensive Aids System (HIDAS) 0023377

Boeing/Lockheed Martin/Northrop Grumman YAL-1A Airborne Laser (ABL) programme

Type
High-energy laser weapon (air).

Development
The US Air Force (USAF) AirBorne Laser (ABL) programme aims to counter theatre ballistic missiles in the early boost phase using a high-energy laser beam. The Phillips Laboratory of the USAF's Matériel Command, which invented the COIL laser in the 1970s, is conducting the ABL demonstrator programme. In May 1994 it awarded 33-month contracts to two teams - Boeing/TRW/Lockheed Martin and Rockwell/Hughes/E-Systems for concept design.

On 12 November 1996, the USAF awarded Team ABL, comprising Boeing, TRW (now Northrop Grumman) and Lockheed Martin, a US$1.3 billion Project Definition and Risk Reduction (PDRR) contract to develop and flight test the ABL on board a Boeing 747-400F freighter aircraft. The PDDR phase will culminate with the destruction of a test missile by the ABL-equipped aircraft.

Boeing is responsible for overall programme management and system integration, as well as development of the ABL battle management system; modification of the Boeing 747-400F to YAL-1A configuration; and design and development of ground support subsystems. Lockheed Martin is responsible for the design, development and production of the ABL infra-red search and track target acquisition system, and beam control/fire-control hardware and software and Northrop Grumman (formerly TRW) is responsible for the design, development and production of the ABL's high-energy solid-state laser.

By January 1998, Boeing had completed a series of key wind tunnel tests that confirmed the design of two components critical to the programme, the nose turret that aims the laser and the laser exhaust system. That same month, the USAF ordered the first Boeing 747-400F aircraft for the programme. In April 1998, Lockheed Martin used a scaled laser beam control demonstrator to prove that it could meet the functional requirements of the ABL system. It demonstrated how ABL could accurately point and focus a laser at a hostile missile hundreds of miles away despite aircraft platform jitter, atmospheric turbulence, and fast engagement timelines. In June 1998, TRW (now Northrop Grumman) achieved 'first light' with the ABL flight-weighted laser module (FLM), a 'multihundred' kilowatt Chemical Oxygen Iodine Laser (COIL) that will serve as the fundamental 'building block' for the ABL megawatt-class laser. The USAF authorised Team ABL to proceed with the final design of this revolutionary weapon system on 26 June 1998.

Description
The ABL uses a Chemical Oxygen-Iodine Laser (COIL) installed aboard a modified Boeing 747-400F, designated YAL-1A by the US Air Force. An operational fleet of seven AL-1A platforms, carrying lasers rated at 2 to 3 MW is planned. The system is designed to focus a powerful laser beam to explode the fuel casing of the ballistic missile. The combat air patrol aircraft of the ABL fleet, cruising at an altitude of 40,000 to 45,000 ft, will carry sufficient fuel for 200 shots and, the support aircraft will carry fuel for a further 140 shots.

The system uses fast-steering and deformable mirrors. High-reflectivity mirror coatings are used to eliminate the need for water cooling. The mirror technology is part of the adaptive optics allowing the anti-missile systems to correct atmospheric distortion.

The Beam Control/Fire-Control (BC/FC) system ensures the weapon system's laser is accurately aligned and pointed at its target. It also performs the following functions: target acquisition, tracking and pointing; fire-control engagement sequencing and aimpoint, kill determination; high-energy and illuminator laser beam jitter control; high-energy laser beam wavefront control, including atmospheric compensation; alignment/beamwalk control and beam containment for the high-energy and illuminator laser beams; and calibration and diagnostics to provide autonomous real-time operations and post-mission analysis. The range of ABL is several hundred kilometres at altitudes of 39,000 to 70,000 ft where the ballistic missile is in the early boost phase. At altitudes less than 39,000 ft the adaptive optics would be less able to overcome atmospheric distortions and the range would be less.

The ABL system will autonomously detect a missile seconds after launch using an Infra-Red Search and Track (IRST) system. The IRST targeting information is handed off to the BC/FC system where a high-resolution infra-red sensor locates the missile plume and provides the fine track to an active illuminator laser. The illuminator laser is used to locate the missile hardbody and provide precise positioning information for the high-energy laser. The Beacon Illuminator Laser (BILL) is then used to obtain atmospheric correction signals by illuminating a small spot on the target missile and measuring the distrotion of reflected light caused by the air turbulence. This allows a deformable mirror to make compensating corrections to the COIL high-energy laser wavefront, which significantly increases the laser energy on the missile kill spot. Active and passive isolation systems are coupled with high-accuracy inertial reference units to compensate for aircraft vibration and aerodynamic motion.

Operational status
Under development. The first Boeing 747-400F to be converted to YAL-1A Attack Laser configuration made its maiden flight on 18 July 2002 and was delivered to the USAF's Flight Test Center at Edwards Air Force Base on 19 December 2002. Ground integration of equipment and systems is under way. The integration and test phases will culminate in late-2004, when the ABL destroys a theatre ballistic missile during the boost-phase of launch. Deployment is planned for 2007-2008.

Specifications
Laser type: Chemical Oxygen Iodine (COIL)
Wavelength: 1.3 μm
Laser output power: 3 MW
Delivered power: >1 MW into less than 1 μrad
Range: 450 km (at target missile altitude 39,000 to 70,000 ft)

Contractors
Boeing Integrated Defense Systems.
Lockheed Martin Space Systems.
Northrop Grumman Space Technology (formerly TRW Space and Electronics Group).

UPDATED

Lockheed Martin infra-red countermeasures testbed

Type
Infra-red countermeasures (air).

Description
The US Air Force, Air Force Research Laboratory, Wright Patterson Air Force Base has placed a contract with Lockheed Martin, formerly Loral Defense Systems, Akron, for the development of an all laser infra-red countermeasure testbed. The contract, now valued at US$30 million, was placed in 1995 and called for a system field test starting in 2001. The testbed will evaluate open- and closed-loop laser countermeasures against live infra-red air-to-air and surface-to-air missiles.

Operational status
Development programme.

Contractor
Lockheed Martin Naval Electronics & Surveillance Systems - Akron.

UPDATED

MBDA, Thales Spectra

Type
Airborne electronic countermeasures.

Development
Spectra is the electronic countermeasures suite being developed for the French Air Force's Rafale fighter aircraft. The programme began in 1990 under a French MoD contract, as a collaboration between Dassault Electronique, Matra Défense (now MBDA, France) and Thomson-CSF Detexis (now Thales Airborne Systems). The first prototype was delivered in 1993. First flights on the Rafale took place in 1994. Dassault Aviation is responsible for system integration.

Description
Spectra is designed to protect against radar, laser and infra-red threats and is stated to be able to jam multiple high-power threats simultaneously under automatic control. MBDA is responsible for the third-generation focal plane infra-red detector and decoying functions. It uses a number of advanced technologies – interferometry, digital frequency memory, electronic scanning, artificial intelligence, monolithic microwave technology and Very High-Speed Integrated Circuitry (VHSIC). Spectra also includes an active phased-array radar transmitter.

The system will be internally fitted in the Rafale and integrated through a databus and central processor.

Operational status
In service with the French Navy.

Contractors
MBDA, France.
Thales Airborne Systems.

UPDATED

The components of the Spectra EW system 0005514

Northrop Grumman AN/AAQ-24 Directional Infra-Red CounterMeasures (DIRCM) system

Type
Air electronic countermeasures.

Development
The AN/AAQ-24 Nemesis is a Directional Infra-Red CounterMeasures (DIRCM) suite that protects a wide range of fixed-wing aircraft and helicopters from infra-red missiles. Northrop Grumman Electronic Systems Integration International, Inc (NESI), a subsidiary of Northrop Grumman Corporation, leads an international team consisting of BAE Systems, Northrop Grumman Electronic Sensor and Systems Sector and the Boeing Company. DIRCM, also known as Operational Emergency Requirement (OER)3/89, is a joint UK/US programme initiated by the UK MoD in 1989 and joined by the US Special Operations Command (USSOCOM) in 1993.

In March 1995, the UK and the US Special Operations Command awarded a contract worth approximately US$450 million if all options are exercised to develop and produce the Nemesis system.

Description
Nemesis is an integrated system that provides missile warning, determines if the missile is a threat to the host aircraft, tracks its approach and activates a high-power arc-lamp-based countermeasure to defeat the missile. The four-axis turret allows easy incorporation of a laser for future growth capability. Nemesis is designed to counter a new generation of infra-red missiles which are not susceptible to existing infra-red countermeasures systems.

BAE Systems has developed the transmitter unit for both the fixed-wing and rotary-wing Nemesis suites. The transmitter unit consists of a pointing turret with payload of the Fine Track Sensor (FTS) and infra-red jammer.

Boeing is producing the FTS and this is located on the azimuth axis. The sensor uses mercury cadmium telluride mid-wave focal plane array technology with a cool

The Directional Infra-Red CounterMeasures system (DIRCM), Nemesis 0053944

down time and sensitivity that affords post-burnout tracking over an extended temperature range. The transmitter unit, when cued by the missile warning system, acquires the incoming missile, tracks it and projects a high-intensity infra-red beam at the target. The tracking system is four-axis. During a threat situation, the FTS electronically processes the image of the incoming missile. This image is then used by the Nemesis system to 'close the track loop' by locking the transmitter onto and maintaining an infra-red beam on the incoming missile until it is defeated. Each Nemesis transmitter is equipped with a laser path to enable future upgrading of the system.

The missile warning system used by Nemesis is based upon Northrop Grumman Electronic Sensors and Systems Sector (formerly Westinghouse Electronic Systems, Electro-Optic Systems Division) AN/AAR-54 PMAWS (Passive Missile Approach Warning System). This passively detects missile plume energy, tracks multiple energy sources and classifies each source as lethal missile, non-lethal missile (not intercepting the missile) or clutter. The Angle of Arrival (AOA) processing is claimed to provide detection ranges nearly double those of existing fielded passive systems and greatly reduced false alarm rates. The system uses a wide field of view sensor and compact processor. From one to six sensors can be used, providing up to full, spherical coverage.

BAE Systems is developing the system operating parameters for specific threats, the prime system AC power source and the operator/system interface control panels.

Operational status
In production for USAF' Special Operations Command (60 systems for MC-130E/H and AC-130H/U) and the UK Army Air Corps and Royal Air Force (undisclosed number covering 14 aircraft types, including WAH-64D Apache Longbow). Understood to be selected for USAF C-130 and C-17 transport aircraft. In February 2002, Northrop Grumman contracted to supply AAQ-24 with Viper™ laser source for the Royal Australian Air Force 'Wedgetail' airborne early warning platform defensive aids suite. Northrop Grumman is reported to be studying a lighter and less-expensive variant of AAQ-24 for use on unmanned aerial vehicles.

Contractor
Northrop Grumman Electronic Systems Sector.

UPDATED

Northrop Grumman AN/AAQ-8(V) (QRC 84-02) infra-red countermeasures pod

Type
Air electronic countermeasures.

Description
The AN/AAQ-8(V) is a second-generation multithreat infra-red countermeasures system capable of operating in a supersonic environment, derived from the AN/AAQ-4. The system is updated to meet new and continuing threats and has been extensively deployed on US Air Force aircraft and helicopters. The pod can be configured with a ram-air turbine, allowing protection to be independent of aircraft power and cooling resources.

Operational status
In service on fixed-wing combat and transport aircraft including the C-130, F-5E, Mirage F1-C and MC-130E Combat Talon special operations aircraft.

Specifications
Dimensions (l × d): 2,290 × 254 mm
Weight
AAQ-8(V1): 107 kg
AAQ-8(V2): 120 kg
Power supply: 115 V AC, 400 Hz, 3 phase, 4 kVA
28 V DC, 20 W

Contractor
Northrop Grumman Electronic Systems Sector - Defensive Systems Division.

UPDATED

Northrop Grumman Modularised Infra-Red Transmitting System (MIRTS)

Type
Air electronic countermeasures.

Description
The Modularised Infra-Red Transmitting System (MIRTS) is a derivative of the AN/AAQ-8(V). It is an advanced subsonic infra-red countermeasures system for deployment in a wide range of aircraft, including helicopters, which can be carried internally or pod mounted. MIRTS utilises advanced jammer technologies including a variable optics/reflector design to provide optimum aircraft infra-red signature coverage, combined with advanced digital electronics and mode-switching power supplies.

Operational status
In service. The system has been tested on the rear fuselage of a Royal Air Force VC10. Committed aircraft installations include the BAE 125, Falcon 20, Fokker 27, Boeing 707 and 747, Douglas DC-8, Puma and Raytheon Hawker 800.

The MIRTS installed on a Royal Air Force BAe 125 (Paul Jackson)

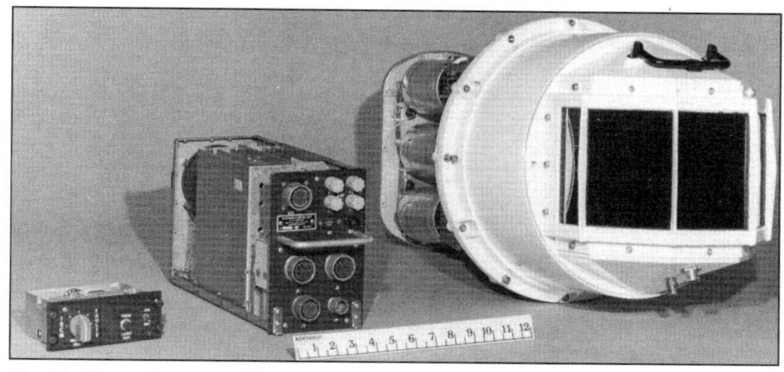
The Northrop Grumman Modular Infra-Red Transmitting System (MIRTS)

Specifications
Transmitter/receiver
Dimensions: 228 × 240 × 635 mm
Weight: 23.6 kg
Power supply: 115 V AC, 400 Hz, 3 phase, 3.3 kVA
28 V DC, 5.6 W
Operator control unit
Dimensions: 146 x 57 x 127 mm
Weight: 0.68 kg
Power supply: 28 V DC or 5 V AC, 10 VA
Electronics control unit
Dimensions: 190 × 259 × 318 mm
Weight: 6.6 kg
Power supply: 115 V AC, 400 Hz, 3 phase, 308 VA
28 V DC, 22.4 W

Contractor
Northrop Grumman Electronic Systems Sector - Defensive Systems Division.

UPDATED

Northrop Grumman Starfire self-protection suite

Type
Air electronic countermeasures.

Description
The Starfire laser-based Infra-Red CounterMeasures (IRCM) self-protection suite has been developed as the ATIRCM solution. Its jamming effectiveness is increased by the use of superior sightline stabilisation in the most demanding flight environment. The heart of the system is an advanced and combat proven pointing and tracking system, keeping Starfire's laser accurately on the target. Starfire represents an integrated and reliable IR defensive self-protection suite that can be fitted on military and commercial aircraft.

The Starfire suite provides fast accurate threat missile location in a combined missile approach and warning surveillance system. The high-power jammer acts only on demand from an alert from an approaching missile. The suite tracks missiles in all modes of operation. All-aspect self-protection is provided with a power-managed architecture. Full fitting compatibility is designed into the system for a wide range of helicopters and fixed-wing aircraft.

Contractor
Northrop Grumman, Integrated Systems Sector.

ELECTRO-OPTIC COUNTERMEASURES

MISSILE WARNERS

Alliant AN/AAR-47 missile warning set

Type
Missile warner (air).

Description
The AN/AAR-47 missile warning system is a small, lightweight, passive, electro-optic, threat warning device used to detect surface-to-air missiles fired at helicopters and low-flying fixed-wing aircraft and automatically provide countermeasures, as well as audio and visual-sector warning messages to the aircrew.

The basic system consists of multiple Optical Sensor Converter (OSC) units, a Computer Processor (CP) and a Control Indicator (CI). The set of OSC units, which normally consist of four, is mounted on the aircraft exterior to provide omnidirectional protection. The OSC detects the rocket plume of missiles and sends appropriate signals to the CP for processing. The CP analyses the data from each OSC and automatically deploys the appropriate countermeasures. The CP also contains comprehensive BIT circuitry. The CI displays the incoming direction of the threat, so that the pilot can take appropriate action.

The AN/AAR-47 basic system was first fielded in the late 1980s and is presently undergoing an extensive upgrade programme to ensure that supportability is maintained out to the 2010-2015 timeframe. The upgrade will improve the original system performance by increasing sensor sensitivity, reducing recovery time, and extending the overall temperature range. The upgrade involves an approximately 80 per cent refresh of the system: upgraded sensors, control indicator and computer processor. The laser warner is integrated into the whole system, rather than being just 'tacked on'. The laser warner capability provides detection and warning of laser-guided and laser-aided threats. The basic upgrade is designated the AN/AAR-47 (V) 1, while the laser warner capable units are designated the AN/AAR-47 (V) 2.

Operational status
In excess of 2,500 sets of the AN/AAR-47 basic system have been ordered for the US Navy, US Air Force, US Army and several foreign customers. Alliant Integrated Defense Company was appointed as second source for AN/AAR-47 basic system and the prime contractor for the AN/AAR-47 (V) 1 and AN/AAR-47 (V) 2 systems.

Contractor
Alliant Integrated Defense Company.

UPDATED

Avitronics MAW-200 missile approach warning system

Type
Missile warner (air).

Description
The MAW-200 is a passive missile launch detector and approach warner operating in the ultra-violet waveband. The system consists of a 'controller' (processor) and up to four sensors. A unique optical design, incorporating state-of-the-art filter technology, with purpose-built image-intensifier tubes and photon-counting focal-plane array processors ensures high sensitivity, which equates to long-detection range. Each sensor is served by a dedicated high-performance digital signal processor, making use of highly pipelined command execution and parallel processing.

The MAW-200 uses a distributed, hierarchical data-processing architecture to ensure optimal utilisation of information in real time. Digitisation and pre-processing functions are performed right at the detector using an advanced focal-plane processor, each sensor's data is transferred to a dedicated digital signal processor. At this stage equalisation, segmentation and feature extraction operations are executed. Each sensor processor can track and process several potential targets, passing the spatial and temporal feature data to a controller, where spatial data is integrated with real time INS information to compensate for platform movement, attitude and altitude. The controller then executes sophisticated neural net pattern recognition algorithms to ensure accurate operation with exceedingly low false alarm rates.

The MAW-200 system significantly enhances aircraft survivability in hostile air defence environments by supplying accurate and timely warnings of approaching missiles. This capability can be extended by including radar and laser warning.

Features
- Passive detection (ultraviolet)
- Long detection range
- Very low false alarm rate
- Quick response (< 1 sec) against nearby missile launches

MAW-200 sensors with Electronic Warfare Controller (EWC) and Threat Control and Display Unit (TCDU) 0097108

- Response time optimised for enhanced decoy flare effectiveness
- Accurate direction finding of approaching missiles
- Multithreat warning capability
- Neural network feature extraction and identification algorithms
- Compact, lightweight, low power, no cooling, skin mounted
- In production
- Field tested against various missiles including guided missile firings at a high-speed drone with the system installed
- Seamless tracking and handover between sensors

Operational status
In production, field tested.

Specifications
Dimensions
Sensor: 230 × 130 × 130 mm
EW Controller: 343 × 127 × 193 mm
Weight
Sensor (each): 3.2 kg
EW Controller: 8 kg
Spatial coverage: 360° azimuth (with 4 sensors)
Field of view: 94° conical per sensor
Direction finding resolution: 2°
Multithreat capability: at least 8 targets
False Alarm rate: max of 2 false alarm in 3 h in a high-clutter environment (typically better than 1 alarm in 10 operational flying hours)
Interfaces
Sensors to controller: RS 485 lines
Controller to other equipment: RS 485 or MIL-STD-1553B
Power: 28 V DC, 0.6 A per sensor

Contractor
Avitronics.

VERIFIED

BAE Systems AN/AAR-47 missile warning set

Type
Missile warner (air).

Development
The AN/AAR-47 missile warning set detects the plume of approaching surface-to-air missiles and gives the pilot an indication of range and bearing. Decoy systems, such as the AN/ALE-39 flare/chaff dispenser, can be automatically triggered and the system is also compatible with the AN/APR-39A radar warning receiver.

The feasibility of such a warning system was first demonstrated by Loral, now BAE Systems in December 1977 and, following development of the AAR-46 system in 1979, full-scale development of the AAR-47 started in March 1983.

The AN/AAR-47 missile warning set

Description

The AAR-47 is small, needs no cryogenic cooling and has an MTBF of 1,500 hours.

The system consists of six sensors, a central processor and a control indicator. The sensors are hard-mounted on the skin of the aircraft and provide full coverage, with overlap to protect against blanking. Within 30 seconds of electrical power a self-initiated BIT programme is completed and the system is operational with no in-flight down time for recalibration. The processor analyses the data from each sensor independently and as a group and, automatically deploys the appropriate countermeasures. In the case of a flare failure, the AAR-47 automatically commands a second flare.

Multidirectional threats are automatically analysed and prioritised for countermeasures sequencing. The control indicator displays the incoming direction of the highest priority threat for tactical manoeuvres.

A shorter sensor has been developed for use in locations, such as the F-16 fuselage, where depth is limited.

Lockheed Martin has developed an improved version for helicopters and transports with more advanced sensors and improved algorithms and processing.

The system has also been integrated in the Terma F-16 PIDS (Pylon Integrated Dispenser Station).

Operational status

In production since 1988. In service with the US Navy and Marine Corps, US Air Force, US Army and with other countries including the UK. The system is fitted to more than 500 aircraft and helicopters.

Specifications

Dimensions
Sensor: 120 × 200 mm
Processor: 203 × 257 × 204 mm
Weight
4 sensor system: 14 kg
Sensor: each 1.5 kg
Processor: 7.9 kg
Power supply: (4 sensor system) 28 V DC, 75 W
(4 W per sensor, 59 W for processor)
Coverage: 360° azimuth (given by 6 sensors)

Contractor

BAE Systems.

VERIFIED

CMC Electronics Cincinnati, AN/AAR-44A Missile Warning System (MWS)

Type

Missile warner (air).

Description

The AN/AAR-44A is the latest version of the passive AAR-44 airborne IR Missile Warning System (MWS). The company states that it detects missiles at the longest range of any MWS available today. They also state that it pinpoints the missile's angle of arrival more accurately than any other MWS – easily sufficient for threat hand-off to a Directional Infra-red Countermeasures (DIRCM) system. It has proven its ability to work well with Northrop Grumman's DIRCM system in USAF/Northrop DIRCM tests. It uses the threat missile's inherent infra-red energy to provide warning of missile attack. The system continuously searches to verify and track missile launches. It warns the aircrew of missile position and automatically controls countermeasures to neutralise threats and enhance survivability.

The system provides long-range, multithreat search/verification while continuously tracking each contact. It is effective over the complete flight envelope and provides hemispheric coverage with one sensor unit or spherical coverage with two sensors.

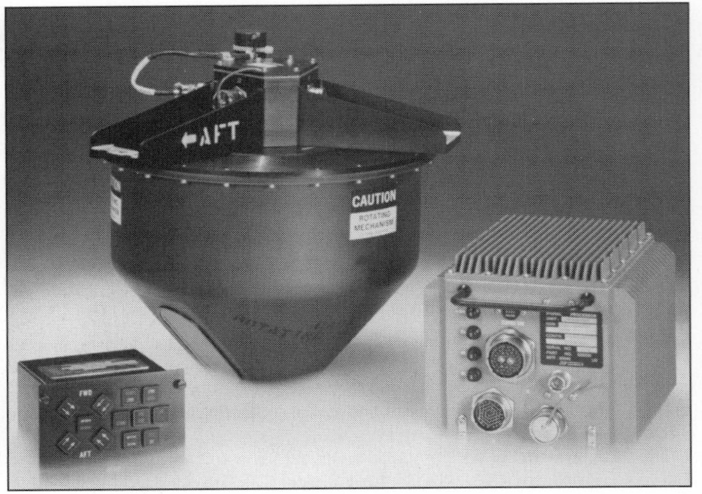

The AN/AAR-44 missile warning system with (left to right) control display unit, conical detector head and processor

Other system features include visual and audio threat warnings to aircrew, a countermeasure command ability, multispectral discrimination, full system operation with or without data-bus integration, a DIRCM interface, robust scanning technology and the ability to be reprogrammed on the flight line.

Design features include automatic warning and countermeasures command, track-while-search processing and the ability to track multiple missile threats simultaneously. Multidiscrimination modes against solar radiation, terrain and water reflection eliminate false alarms. MIL-STD- I 553B bus interface capability is incorporated.

Operational status

The AAR-44A completed USAF operational testing in 1997. It is proven, available and in service.

Specifications

Dimensions
Baseline: 101.6 × 152.4 × 330 mm
plus 152.4 mm dome
Alternative: 101.6 × 152.4 × 406.4 mm
Weight: 9.1 kg
Power supply: 115 V AC, 400 Hz, 1 A
28 V DC, 1.1 A
Coverage
Azimuth: 360°
Elevation: ±135°
Temperature range: -54 to +71°C
Altitude: up to 45,000 ft
Cueing accuracy: better than 1°
Interfaces: MIL-STD-1553B, RS-422, RS-232
MTBF: >500 h

Contractors

CMC Electronics Cincinnati.

VERIFIED

CMC Electronics, Raytheon AN/AAR-58

Type

Missile warner (air).

Description

The AN/AAR-58 passive airborne warning receiver utilises the threat missile's inherent infra-red energy to provide warning of missile attack. The system continuously searches a hemisphere to verify and track missile launches. It warns the aircrew of missile position and automatically controls countermeasures to neutralise threats and enhance survivability.

Design features include automatic warning and countermeasures command, track-while-search processing, multiple missile threat capability and countermeasures discrimination capability.

Multidiscrimination modes against solar radiation, terrain and water reflection eliminate false alarms. MIL-STD-I553B bus interface capability is incorporated.

It can detect all missile types and features hyper-hemispheric coverage; multicolour infra-red detection using color, size, intensity and trajectory discriminants; multiple simultaneous threat detection; high-speed optics for fast reaction times and greater than hemispheric coverage from a single aperture. It has an accuracy sufficient to point laser-based infra-red countermeasures systems.

It can be installed internally in high-performance aircraft or mounted in a pod. It is designed as a single LRU for ease of installation and can be integrated with

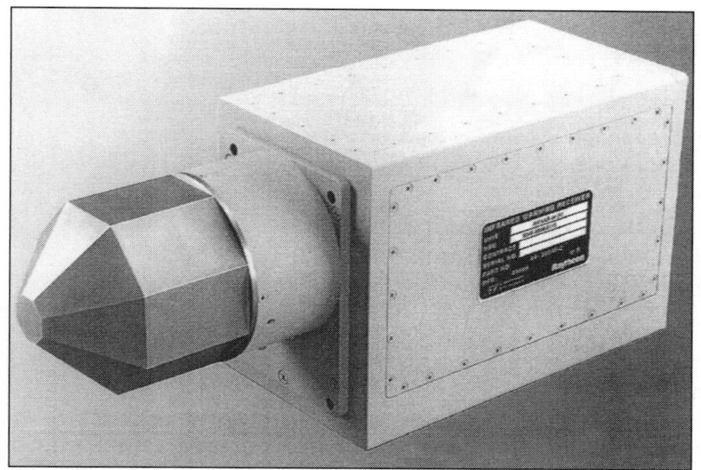

The AN/AAR-58 (previously the AN/AAR-44(V) missile warning system 0089910

directional coherent countermeasures dispensing systems. It also can be installed into the lower gondola of Raytheon's ALQ-184 ECM pod.

Specifications

Dimensions
Baseline: 101.6 × 152.4 × 330 mm plus 152.4 mm dome
Alternative: 101.6 × 152.4 × 406.4 mm
Weight: 9.1 kg
Power supply: 115 V AC, 400 Hz, I A; 28 V DC, 1.1 A
Coverage
Azimuth: 360°
Elevation: ± 135°
Temperature range: –54 to +71° C
Altitude: up to 45,000 ft
Cueing accuracy: better than 1°
Interfaces: MIL-STD-I 553B, RS-422, RS-232
MTBF: >500 h

Contractors

CMC Electronics Cincinnati.
Raytheon Systems Company.

VERIFIED

Elisra Passive Airborne Warning System (PAWS) for helicopters and transport aircraft

Type

Missile warner (air).

Description

Passive Airbone Warning System (PAWS) is a lightweight infra-red missile launch/approach warning system designed for fighter aircraft and attack helicopters. It consists of four to six IR sensors. The system may operate independently or may be integrated with a radar warning receiver.

The system detects the missile exhaust plume and tracks it even in high-clutter environments. It determines when a missile threatens the aircraft, provides an accurate readout of approach direction and an estimate of time to intercept. It can also select the appropriate narrow beam countermeasures and activate them automatically. The manufacturer states that PAWS has a very low false alarm rate, even when subject to violent manoeuvres and operating against a highly cluttered background.

PAWS consists of a high-computation power parallel processor built by Elisra and an IR sensor built by Elop Electronic Industries with advanced and mature technologies. The system provides:

(a) missile launch and approach warning
(b) discrimination between threatening and non-threatening missiles
(c) multithreat warning capability
(d) accurate approach direction
(e) time to go estimation
(f) automatic activation of countermeasures.

Operational status

Fully developed.

Specifications

Power source: 28 V DC (MIL-.STD-704D),115 V, 400 Hz
Power consumption: 200 W
Dimensions
(sensor): 120 × 120 × 230 mm
(processor): half ATR short (203 × 389 × 127 mm)

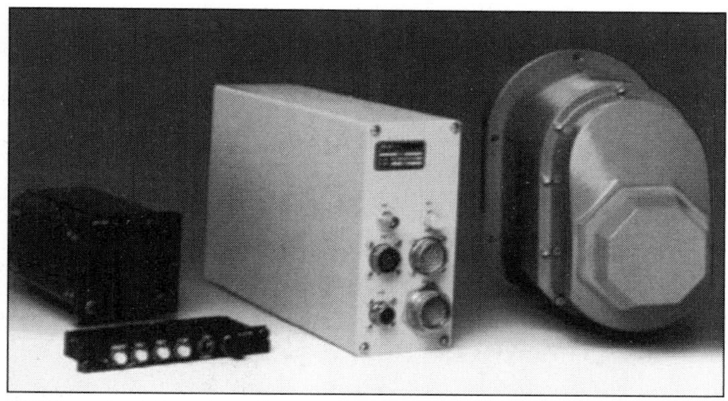

PAWS passive infra-red warning system

Weight
(sensor): 4 × 2.5 kg
(processor): 10 kg
MIL specification: MIL-STD 5400, Class 1B
Communication interfaces: RS-422, MIL-STD-1553B MUX BUS.

Contractor

Elisra Electronic Systems Ltd, Airborne Systems Division.

UPDATED

Northrop Grumman LFK MILDS, AN/AAR-60 missile launch detection system

Type

Missile warner (air).

Description

The MILDS AN/AAR-60 missile launch detection system detects and declares potential incoming missiles and indicates direction of arrival and time to go. MILDS also automatically triggers self-protection actions such as release of flares, IR jamming and flight manoeuvres. It detects and images the ultra-violet emission of missiles' plumes in the ultra-violet spectral region and uses advanced software algorithms to declare and track a threat in real time.

MILDS AN/AAR-60 is designed for four to six sensor LRUs functioning in a master/slave configuration. Each sensor contains preprocessing, signal processing and a communications processor. The master contains an additional function for sensor fusion. Inertial navigation system data are used to compensate for aircraft motion. Threat declaration includes azimuth, elevation, and prioritisation of each missile.

Operational status

Fully developed. At the end of 1996, MILDS was selected for installation on the NH-90 helicopter by Eurocopter Deutschland and NAHEMA (NATO Helicopter Design, Development and Production Agency).

Specifications

Dimensions: 120 × 120 × 120 mm
Weight: 2 kg/sensor
Field of view
Azimuth: 95° per sensor
Elevation: 95° per sensor
Response time: <0.5 s
Resolution: 1°

Contractors

LFK - Lenkflugkorpersysteme GmbH.
Northrop Grumman Applied Technology.

VERIFIED

Rafael Guitar 350 passive electro-optical warning system

Type

Missile warner.

Description

Guitar is a lightweight totally passive warning system which detects missiles by sensing the electro-optical emissions from their engines. The system provides audio and visual alarms, allowing time for the aircraft to activate its defensive systems. It has been developed for the protection of helicopters and fixed-wing aircraft.

Guitar 350 is based on a patented sensor and automatically controls the infra-red countermeasures subsystem. It emits no electromagnetic or electro-optical signals

Rafael's Guitar missile warning system is designed for both helicopter and fixed-wing applications 0009904

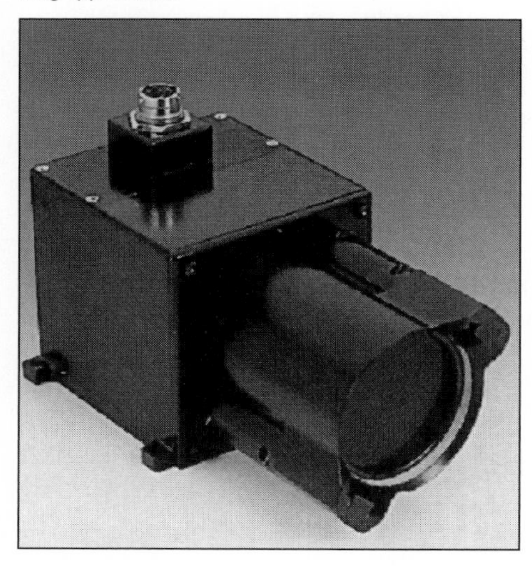

Guitar 350 passive missile warning system 0018287

which might be detected. It is designed to detect ground-to-air and air-to-air missiles and has an extremely low false alarm rate.

Guitar features coverage of all possible attack angles and high discrimination against background interference. It withstands adverse environmental conditions and is compatible with cockpit displays.

Operational status

Test flown. Suitable for helicopters, fighter and transport aircraft. Guitar 350 supercedes the earlier Guitar 300 system. A version of Guitar forms part of the ARPAM defensive aids technology demonstrator for land vehicles.

Contractor

Rafael Armament Development Authority.

VERIFIED

SAGEM SAMIR missile launch detector

Type

Missile warner.

Description

The SAMIR (*Système d'Alerte Missile InfraRouge* - also known as the DDM system) missile launch warning system provides automatic detection of a missile launch plume, locates it in flight and instantaneously feeds threat data to the aircraft ECM system. This is carried out with a high probability of detection, a low false alarm rate and in time to carry out efficient countermeasures.

SAMIR incorporates passive infra-red detector techniques to ensure covert operation in severe ECM environments and is capable of locating missiles using any type of seeker. SAMIR features advanced signal processing and an infra-red bispectral detector array. It includes an integrated cooling machine and has a multiple missile detection capability.

It can be installed in a variety of aircraft types such as combat aircraft, helicopters and transport aircraft.

Operational status

In production for France. SAMIR 2000 for the air force, Mirage 2000 and SAMIR PRIME for the naval Rafale combat aircraft. Continuing programmes for export applications. Japan has evaluated the system.

Specifications

Weight: 16 kg
Coverage : 180°/IR sensor
Accuracy: better than 2°

Contractors

MBDA Missile Systems
Vélizy-Villacoublay.
SAGEM SA, Aerospace & Defence Division.

UPDATED

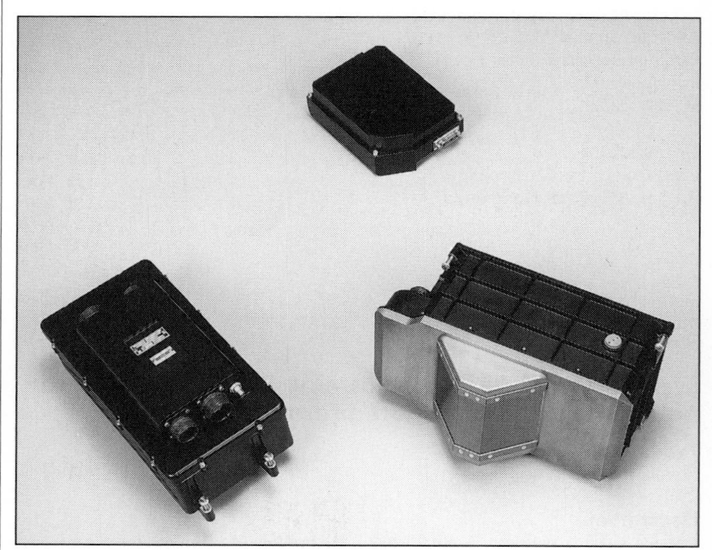

The SAMIR missile launch detector

ELECTRO-OPTIC COUNTERMEASURES

LASER WARNERS

Avitronics airborne laser warning system

Type
Air laser warners.

Description
The deployment of weapon systems utilising lasers is increasing sharply in the modern battlefield The Avitronics LWS300 System for airborne applications provides aircrew with the necessary situational awareness of air defence weapon systems using laser technology. This capability can be extended to include radar and missile approach warning.

The LWS provides threat classification and direction finding (DF) indication of laser rangefinders, designators, lasers used for missile guidance purposes and dazzler lasers. The system is designed to be stand-alone or to interface with an existing onboard RWR/ESM host-system via the EW Controller for data processing and interfacing to the host EW system. A priority interface to the countermeasures system is available for the activation of countermeasures.

On detection of a threat an audio and visual (display) alarm is generated via the host EW system or if configured as a stand-alone system, via a dedicated Threat Display & Control Unit (TDCU).

Broad coverage of the laser spectrum ensures detection of most known current threats. The sensitivity of the LWS-300 sensor has been carefully chosen to provide warning of laser threats at ranges generally 1,5 times the threats engagement range. High sensitivity with optimised installation ensures the detection of lasers targeting any part of the platform.

The system is capable of recording threat parameters encountered during a mission for post mission analyses. The threat library is user programmable and field loaded directly into the LWS or via the host EW system. The system can provide identification of detected lasers, but this is a function of the user's intelligence (threat library).

Components of the Avitronics airborne laser warning system 0041555

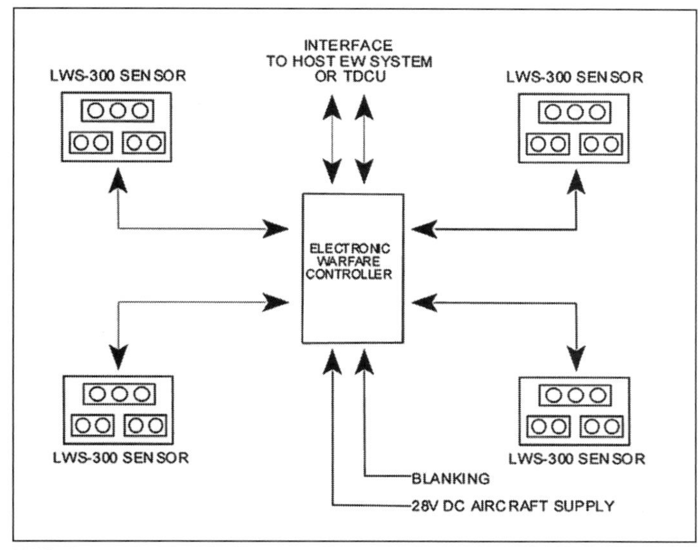

LWS block diagram – typical stand-alone configuration 0137188

Features
- Wide wavelength coverage
- High sensitivity to cater for lasers used for missile guidance purposes
- Classify range finders, designators, missile guidance and dazzler lasers
- Identify specific laser threats depending threat library availability
- Recording capability
- Direction finding ensures good situational awareness
- Very high probability of intercept

Operational status
Available.

Specifications
LWS-300
Wavelength coverage: 0.5 – 1.8 μm
Threat coverage: Doubled NdYAG, Ruby, GaAs, NdYAG, Raman Shifted, NdYAG and Erbium Glass lasers
AOA accuracy: AZ. 1SF RMS
Spatial coverage: AZ. 360E (IloEpersensor), EL. >60E
Probability of intercept: >99% for a single pulse
Dimensions: 115 × 90 × 76 mm
Mass: 1.2 kg per sensor

Electronic Warfare Controller
Dimensions: 343 × 127 × 193 mm
Mass: 8 kg

Contractor
Avitronics.

Avitronics Multi-Sensor Warning System (MSWS)

Type
Air radar/laser/missile approach warner.

Description
The fully integrated state-of-the-art Multi Sensor Warning System (MSWS) provides advanced and comprehensive multispectral situational awareness which includes radar, laser and missile approach warning functions.

Features
- Modular architecture allows flexibility to configure the system to suit user requirements with any one of the sensor types. Other sensor types can be added later as operational requirements and budgetary considerations dictate.
- Low box count allows easy installation in aircraft ranging from helicopters, transport aircraft and fighters.
- High probability of intercept.
- High sensitivity with full Pulse Doppler radar handling capability.
- User definable threat symbology.
- Pre-flight data (threat libraries) easy and quick to create and update with supplied Pre-flight Data Compiler software tool.
- Various pre-flight data sets can be loaded and selected in flight according to the mission type and/or geographical area.
- Flight-line software load/unload with Memory Loading Unit.
- Full recording and playback facilities.
- Quick analysis of recorded data with Flight Data Analyser software tool.
- Extensive Built-in-Test facilities.
- Full range of support equipment (Flight-line through to depot level Automatic Test Equipment).

Functional capabilities
- Extensive use of parallel processing.
- Internal Instantaneous Frequency Measurement Module.
- Man-Machine Interface via dedicated Multi-Function Colour Display or existing onboard colour MFD.
- Internal defensive aid computer interfaces to and controls Electronic Countermeasures systems.
- System interface via 1 553B or ARINC 429 as well as standard serial (RS-232, RS-422 and RS-485) port.
- Spherical coverage optional.

Specifications
Radar warning
Frequency coverage: 0.7-40 GHz (pulsed signals)
0.7-18 GHz (CW signals)
Direction finding: for pulsed signals in the 2-40 GHz range
Spatial coverage: 10-12° RMS 360° (azimuth or spherical)
90° (elevation or spherical)

Avitronics Multi-Sensor Warning System (MSWS) 0023380

Pulse density capability: 2.5 million pulses/s
Frequency resolution: 10 MHz
Laser warning
Wavelength coverage: 0.5-1.8 μm and 2-12 μm
AOA accuracy: AZ. 15° RMS
Spatial coverage: AZ 360° (90° per sensor)
El 60° (0.5-1.8 μm)
El 40° (2-12 μm)
Laser threat coverage: doubled NdYAG, Ruby, GaAs, NdYAG, RAMAN shifted NdYAG, Erbium glass and CO_2 lasers
Laser threat types: range-finders, designators and laser used for missile guidance (beam riders) and dazzler lasers
Probability of intercept: >99% for a single pulse
Missile approach warning: solar blind UV band
Operating frequency: Direction finding resolution: 2°
Field of view: 94° conical per sensor
Spatial coverage: AZ. 360° (with 4 sensors)
False alarm rate: max of 2 false alarms in 3 h in a high-clutter environment (typically, better than 1 alarm in 10 operational flying hours)
Probability of warning: >99%
Multi-threat capability: capable of tracking at least 8 targets simultaneously

Dimensions
EW controller: 343 × 127 × 193 mm
Threat display and control unit: 128 × 127 × 120 mm
Front end receiver: 176 × 45 × 158 mm
0.7-40 GHz spiral antenna: 110 × 110 × 67.5 mm
LWS-300 sensor: 115 × 90 × 76 mm
MAW-200 sensor: 230 × 130 × 130 mm

Weights
EW controller: 10 kg
Threat display and control unit: 2.2 kg
Front end receiver: 2 kg
0.7-40 GHz spiral antenna: 0.55 kg
LWS-400 sensor: 1.2 kg
MAW-200 sensor: 3.2 kg

Contractor
Avitronics.

BAE Systems PA7030 laser warning equipment

Type
Air laser warner.

Description
BAE Systems, Avionics, Radar and Countermeasures Systems' laser warning equipment offers instantaneous warning and bearing information on irradiation by ruby and neodymium lasers. The system consists of a direct detector head unit and up to four indirect detectors mounted externally on the aircraft. A compact display and control unit is mounted in the cockpit.

The indirect detectors are sensitive to pulsed laser radiation scattered by the aircraft and so can detect designators and range-finders even when these do not expressly illuminate the direct detector. Electronic filtering discriminates against glints and flashes to give an extremely low false alarm rate. The direct sensor gives instantaneous bearing on a radiation source to within 15°.

BAE Systems airborne laser warning system

Operational status
Development equipment has had trials in helicopters and fixed-wing aircraft in the UK, USA and Canada.

Specifications
Dimensions
PA7031 direct detector: 152 × 48 mm
PA7032 scatter detector: 105 × 80 × 130 mm
PA7033 display: ¼ ATR short
PA7034 signal conditioner: 260 × 160 × 87 mm
Weight
Direct detector: 2 kg
Scatter detector: 1.5 kg
Signal conditioner: 2 kg
Display: 2 kg
Power supply: 28 V DC, <12 W
Coverage
Direct detector: 360° (azimuth), –45 to +10° (elevation)
Scatter detector: as required

Contractors
BAE Systems, Avionics, Radar and Countermeasures Systems, Edinburgh.

VERIFIED

BAE Systems airborne laser warning system

Type
Air laser warner.

Description
BAE Systems, Avionics Sensor Systems Division has designed a family of laser warning systems for military applications. The core system comprises the sensor heads, central processor and threat warning display, with appropriate interfaces to the platform system. Using advanced scatter rejection techniques, the system can discriminate between direct and indirect hits and hence give a reliable direction of arrival of the threat. The system can analyse the incident laser source and compare it with an onboard threat library to allow identification.

Operational status
A version of this system is in advanced development for Eurofighter 2000.

Specifications
Angular coverage: complete coverage depending on platform requirement
Threat detection: simultaneous multiple sources
PRF coverage: detects from single shot to continuous wave sources
Interface: compatible with all standard databusses and defensive aids systems

Contractor
BAE Systems, Avionics, Sensor Systems Division, Edinburgh.

VERIFIED

The PA7031 direct detector and PA7032 scatter detector

BAE Systems Type 453 laser warning receiver

Type
Air laser warner.

Description
The Type 453 laser warning receiver is designed for use in fixed-wing aircraft, helicopters and armoured fighting vehicles. It provides a countermeasure to laser-guided weapon systems and provides an audible alarm and visual display showing the direction from which the threat has originated. Output from the receiver can be combined with radar warning equipment to give an integrated battlefield threat warning system or with a defensive aids system.

The system uses a number of dispersed sensors to detect incident laser radiation. This counters the problem of detecting a very narrow beam which at any instant would be illuminating only a small part of the airframe. The sensor head configuration can be tailored to provide spherical or hemispherical cover and presents the laser threat bearing to the crew as sectoral information. The number of sensors required is tailored to the aircraft type. The sensor protrudes through the aircraft skin as a 25 mm hemisphere and occupies a space of 50 mm depth behind the skin.

The sensor heads are completely passive. Laser radiation is routed to the central processing unit by fibre optic cables. This feature eliminates risks of false alarms being generated by radio frequency interference.

Operational status
In advanced development. In March 1993, Eurofighter GmbH awarded a subcontract for the Type 453 laser warning receiver for the Eurofighter 2000.

Contractor
BAE Systems, Avionics, Radar and Countermeasures Systems, Edinburgh.

VERIFIED

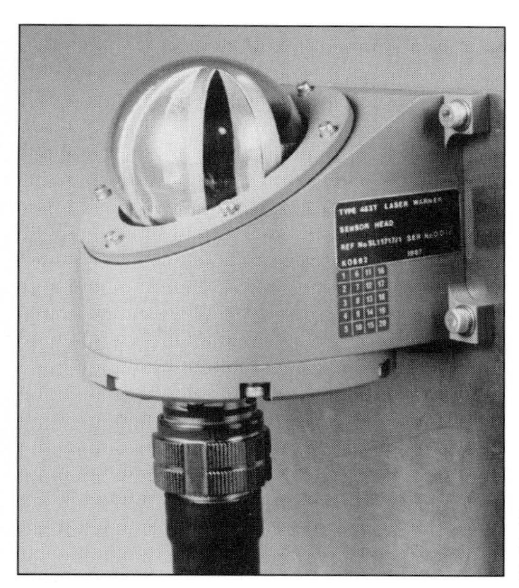

One of the sensor heads from the Type 453 laser warning receiver

BAE Systems/Elettronica Defensive Aids SubSystem

Type
Air laser warner.

Description
The Defensive Aids SubSystem (DASS) for the Eurofighter 2000 is an integrated electronic warfare suite comprising a radar warning receiver, ECM, towed offboard decoy, computer, chaff/flare dispenser, Type 453 laser warner and missile approach warner. The development of this system by a consortium, consisting of BAE Systems, Avionics in the UK and Elettronica in Italy, is expected to take about five years. Of the four nations in the Eurofighter 2000 project, Germany has opted out of DASS and Spain has reserved its position.

Operational status
The Eurodass consortium was awarded a £200 million contract in March 1992 for the development of DASS. The work-share is split approximately 60 per cent to the UK and 40 per cent to Italy. In March 1993, a contract was awarded to BAE Systems, Edinburgh for the Type 453 laser warning system for the Eurofighter 2000.

Contractors
BAE Systems, Avionics, Radar and Countermeasures Systems, Edinburgh.
Elettronica.

VERIFIED

Elisra LWS-20 laser warning system

Type
Air laser warner.

Description
LWS-20 is a laser warning system which is applicable to all airborne platforms and is in service on attack and support helicopters.

Four laser sensors receive and detect laser pulsed signals and send the detected pulses to the Laser Warning Analyser (LWA). The LWA characterises each laser pulse for its angle of arrival, relative time of arrival and amplitude. It then processes the single pulses or pulse trains and identifies the threatening emitter, based on preprogrammed library information. Identified threats are displayed on the screen of the self-protection system as alphanumeric symbols. These symbols enable the crew to establish the type, azimuth and lethality of the threats. In addition, audio warnings are sent to the crew through the aircraft intercom system.

The LWA incorporates a military 16-bit microprocessor with programmable memory for combat software, programmable tables and in-flight recording. The LWS-20 may include a 1553 multiplexer bus interface, RWR interface or other ECM interfaces.

Operational status
In production and in service.

Specifications
Dimensions
Analyser: 120 × 118 × 155 mm
Control unit: 54 × 146 × 19 mm
Display unit: 172 × 82.5 × 82.5 mm
Sensors: 64 × 110 × 150 mm
Weight
Analyser: 2.5 kg
Control unit: 0.2 kg
Display unit: 1.2 kg
Sensors: 0.6 kg each
Power supply: 28 V DC, 50 W

Contractor
Elisra Electronic Systems Ltd, Military Systems Division.

UPDATED

Elisra SPS-65 self-protection system

Type
Air laser and radar warner.

Description
The SPS-65 self-protection system is a lightweight system for helicopters which is capable of intercepting and analysing a range of emissions from the battlefield, including laser radiation sources.

The SPS-65 comprises three subsystems. The SPS-20 low-volume lightweight radar warning system detects, processes and displays radar threats operating within the 0.7 to 18 GHz frequency range. The SPS-25 high-sensitivity superheterodyne receiver subsystem detects and processes CW radar signals in the 6.5 to 18 GHz frequency range which are not recognised by conventional radar warning systems. The LWS-20 laser warning receiver detects, and identifies hostile laser-guided weapons.

A 3 in display unit provides an alphanumeric representation of the type, relative bearing, relative lethality and status up to 16 analysed threats. In addition, audio warnings are dispatched to the pilot through the intercom system. A flare/chaff dispensing system, which can be automatic or crew activated, is available for countermeasures capabilities.

The Elisra SPS-65 self-protection system for helicopters

Software and emitter tables can be loaded in the field into the emitter library giving adaptability to changing threat environments. The analyser records flight events and downloads them into a Memory Loader Verifier (MLV) for playback after flight.

Operational status

In production.

Specifications

Weight
SPS-20: 7.5 kg
SPS-25: 4 kg
LWS-20: 2.5 kg
Power supply: 28 V DC
Frequency: 0.7-18 GHz
Environmental: MIL-E-5400 Class II

Contractor

Elisra Electronic Systems Ltd, Military Systems Division.

UPDATED

Marconi Selenia Communications RALM/01 laser warning receiver

Type

Air laser warner.

Description

AMS has produced a range of laser warning systems for airborne (RALM/01) and armoured fighting vehicle (RALM/02) use, plus the RALM/I infra-red laser warning receiver for airborne, ground and naval applications.

The system typically consists of optical receivers, a processing unit and a display and control panel as well as links with countermeasures systems. Identification of the type of laser results in audio or visual warning and activation of appropriate countermeasures. Communication between system elements is via fibre optics. The system can detect and analyse multiple threats from different directions. Extensive testing has been carried out by the Italian Armed Forces and RALM has been in service since 1993.

Operational status

In production for the Italian Army's A129EES Mangusta attack helicopter and HH-3F SAR helicopter and C-130J Hercules airlifters of the Italian Air Force.

Specifications

Dimensions
Optical unit (airborne, ×2): 90 × 50 × 40 mm
Processing unit: 3/8 ATR short
Sensors
Azimuth coverage: 360°
Vertical coverage: 90°
Wavelength: 0.5-1.8 μm (optional extended band, 8-12 μm)
Operating temperature range: –40 to +60°C
Power supply: 28 V DC
Weight: 4.5 kg (electronic unit); 0.3 kg (sensor)

Contractor

Marconi Selenia Communications.

UPDATED

Raytheon AN/ALR-89(V) integrated self-protection system

Type

Air laser warner.

Description

The AN/ALR-89(V) radar and laser warning system incorporates three subsystems: the AN/ALR-90(V) RWR which detects pulsed radars in the C- to J-bands, the AN/APR-49(V) RWR which is a superhet erodyne receiver for detecting modern pulse and non-pulse radars and the AN/AVR-3(V) laser system which detects laser emissions.

Identified threats which are detected by the three subsystems are presented to the pilot on a 76 mm (3 in) display that provides an alphanumeric representation of the type, angle of arrival, relative lethality and status of the threats.

ALR-89 offers complete threat reprogrammability for new and changing environments through an extensive, updatable emitter library file. A portable loader unit enables loading of operational software and emitter tables and downloading of recorded data in the field. The integrated ALR-89 or one of its subsystems is suitable for the self-protection of helicopters or fixed-wing aircraft regardless of their size. Electrical design and mechanical configuration both adapt to new installations and provide an upgrade of AN/APR-39(V)1 installations. This upgrade can be implemented in various levels of performance using the basic APR-39 system and selected LRUs from the ALR-89 system.

Operational status

In production.

Specifications

Weight
AN/ALR-90(V): 5.76 kg
AN/APR-49(V): 4.04 kg
AN/AVR-3(V): 3.86 kg
Power supply: 28 V DC, 210 W (MIL-STD-704)
Frequency: C- to J-bands, 2 laser bands
Coverage: 360° azimuth
Environmental: MIL-E-5400, Class II

Contractor

Raytheon Systems Company.

VERIFIED

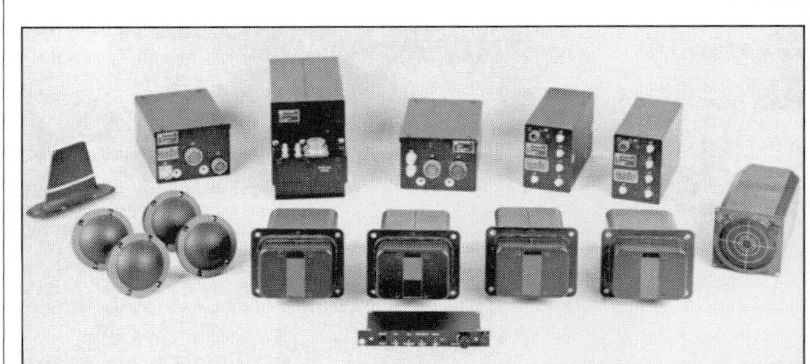

The AN/ALR-89(V) comprises three subsystems

Raytheon AN/AVR-2A(V) laser detecting set

Type

Air laser warner.

Development

The AN/AVR-2A(V) laser detecting set detects and characterises laser beam-riding missiles, laser range-finders and designators with over 360° coverage. It provides visual and audible warning of laser threats to the aircrew. It identifies the direction of threat and displays threats in priority order. It consists of four SU-130A(V) sensor

The Raytheon AN/AVR-2A laser detecting set consists of four sensor units and an interface unit comparator

units and a CM-493A interface unit comparator. AN/AVR-2A(V) interfaces with all variants of the AN/APR-39(V) series radar signal detecting set to function as an integrated radar and laser warning receiver system.

Operational status

In production for the US Army, Navy, Marine Corps and Special Forces for the AH-1F Cobra, AH-64 Apache, AH-1W Super Cobra, MH-60K Black Hawk, MH-47E Chinook, OH-58D Kiowa Warrior, HH-60H Combat Rescue and UH-1N Huey. In service with the US Army and Marine Corps. Over 1,600 systems have been ordered.

Contractor

Raytheon Systems Company.

VERIFIED

Raytheon AN/AVR-3(V) airborne laser warning system

Type

Air laser warner.

Description

AN/AVR-3(V) is an airborne microprocessor-controlled warning system for detecting laser emissions in two threat bands, with provision for a third. Identified threats, which are detected and analysed, are presented to the pilot on a 76.2 mm (3 in) CRT display that provides an alphanumeric representation of the type of laser and the angle of arrival.

AVR-3 offers complete threat reprogrammability for new and changing environments as laser weapons increase and mature. This is accomplished by an extensive and updatable emitter library file. A portable loader unit enables the loading of operational software and emitter tables and the downloading of recorded signal data in the field.

AVR-3 is suitable for the self-protection of helicopters and fixed-wing aircraft regardless of size. Electrical design and mechanical configuration provide for the use of up to eight laser sensors to ensure spatial coverage for all aircraft.

AVR-3 features detection of modern pulsed lasers, direction-finding of received signals, high-resolution emitter separation, sophisticated signal processing, alphanumeric azimuth threat display and full system BITE. The system is able to record emitter parameters during flight for post-flight playback and signal analysis. It has provision for a tie-in with flare and chaff dispensers, ECM and missile warning systems.

Operational status

In production.

Specifications

Dimensions

Laser sensor: 63.5 × 85.1 × 149.9 mm
Laser analyser: 146 × 106.2 × 78.7 mm
Control unit: 54.1 × 146 × 19 mm
Display unit: 158 × 80.5 × 80.5 mm

Weight

Laser sensor: 0.59 kg
Laser analyser: 1.5 kg
Control unit: 0.2 kg
Display unit: 1.2 kg

Power supply: 28 V DC, 90 W max (MIL-STD-704)
Frequency: 2 laser bands, provision for 4
Coverage: 360° azimuth
Environmental: MIL-E-5400

Contractor

Raytheon Systems Company.

VERIFIED

The AN/AVR-3(V) airborne laser warning system consists of a number of sensors (left and right), a laser analyser (centre left), a display unit (centre right) and a control unit (front centre)

Thales Optronics DAL laser warning receiver

Type

Air laser warner.

Description

In addition to detection, the DAL laser warning receiver also provides localisation and identification of laser threats. It has a multithreat capability and all types of threats are handled, in the 0.6 to 11 μm waveband, including monopulse and multipulse range-finders and lasers for beam-riding missiles.

The system consists of line-replaceable sensors associated with a central processor. No harmonisation on the carrier is needed.

Operational status

Currently in production as the laser warning receiver for Rafale fighter aircraft.

Contractor

Thales Optronics.

VERIFIED

Thales/Daimler Benz Threat Warning Equipment (TWE)

Type

Combined radar and laser warning system.

Description

TWE comprises radar and laser warning receivers, a Central Processing Unit (CPU), a library module, a symbol generator, a multifunction display, radar antennas, laser heads and a control box. Coverage is D- through K-band (1 to 40 GHz) in the radar domain and bands I and II for lasers. The radar warning subsystem is capable of instantaneous frequency measurement and can handle pulse, pulse Doppler and continuous wave emitters. The CPU processes received radar and laser signals, interfaces with the library module for threat identification, maintains electromagnetic compatibility and manages a self-protection suite (decoys and missile approach/launch warners) if required. The CPU is interfaced through a MIL-STD 1553B bus and is housed in a standard ARINC 600 box. The system generates a colour threat display which shows information on threat type, bearing and danger level. The visual format is backed up by an audio warning which sounds each time a radar or laser threat is detected. TWE also allows crew members to call up data lists from the library module during flight in order to change priorities for different phases of a mission. The system also incorporates a built-in test routine and because of its design, is claimed not to require a test bench for routine maintenance.

Operational status

In production for the Tiger attack helicopter and for the NH90 TTH transport helicopter. The Tiger helicopters are on order for French and German armies and the NH90 helicopters are on order for French, German, Netherlands and Italian armies.

TWE also forms the basis of the Electronic Warfare System (EWS) which is installed and qualified on the Tiger and NH90 TTH helicopters. In addition to TWE, the EWS includes the MILDS passive missile warner and the SAPHIR-M chaff and flare dispenser.

Specifications

Frequency: 1-40 GHz (radar); bands I/II (laser)
Coverage: ±45° (elevation); 360° (azimuth)
Accuracy: better than 10° RMS (DF); 20 MHz (frequency)

The TWE equipment package comprises radar warning antennas (foreground), laser warning heads (rear and left), a multifunction display (centre), a CPU (top right) and a control box 0005512

Library: 2,000 radar/laser modes (approx)
Power consumption: 200 W
Weight: <15 kg
Dimensions: 199 × 194 × 350 mm (CPU)

Contractors
EADS Germany, GmbH.
Thales Airborne Systems, Elancourt.

UPDATED

GROUND ATTACK

INTEGRATED SYSTEMS – FIXED-WING

BAE Systems TIALD ground attack pod

Type
Ground attack integrated system (fixed-wing).

Development
In the mid-1970s Ferranti Defence Systems Electro-Optics Department (now BAE Systems, Avionics, Sensors and Communications Systems) produced for the then Royal Aerospace Establishment (RAE) an airborne laser designation equipment which integrated a thermal imaging system, a laser designator and a video tracker. Flight trials of this equipment demonstrated excellent target acquisition with closed-loop video tracking and precision laser designation against small moving ground targets.

Feasibility and definition work on the next-generation laser designation equipment, TIALD (Thermal Imaging Airborne Laser Designator), began in 1983. Three years later a contract was placed by the UK Ministry of Defence (MoD) to supply TIALD for Buccaneer aircraft and systems were installed in January 1988. During the same year the MoD awarded a production contract to the company to supply TIALD for the RAF Tornado.

TIALD is designed and manufactured by BAE Systems, Avionics, Sensors and Communications Systems, Edinburgh. It was originally developed by a consortium consisting of GEC-Marconi Radar and Defence Systems, Edinburgh, GEC-Marconi Electro-Optics, Sensors Division, Basildon and British Aerospace Systems & Equipment. The thermal imaging unit is supplied by BAE Systems, Avionics, Sensors and Communications Systems and the tracker by BAE Systems, Avionics (formerly BASE).

Description
The TIALD pod is an advanced electro-optic system which enables aircraft to attack targets with precision laser-guided weapons by day, night and in poor weather conditions. It uses a combination of laser designation, thermal imaging, television and video tracking to maximise the effectiveness of both free-fall and laser-guided munitions. It is claimed to be both the only targeting pod in production which offers thermal imaging and TV sensors simultaneously and the smallest diameter targeting pod on the market at 305 mm (12 in).

The forward section of the pod contains the thermal imager and TV sensors, the telescope and the laser designator transceiver unit. The sensors have a common optical path while sightline stabilisation is provided by a gyrostabilised mirror. An important feature of the system is the combined optical path for the different wavelengths of the laser, thermal imager and TV in order to save weight and space.

The static rear sections of the pod contain the ram-air cooler, electronics units and power supplies. The automatic video tracker is included within the electronics unit.

Apart from the cockpit display and controls, the pod is self-contained, taking its power from the aircraft primary supplies. Interface with the avionics is via a MIL-STD-1553B databus.

In operation, the sightline would be directed onto the target area, either by commands from the aircraft avionics using data from the navigation system, or radar or by manual control by the crew. When the target has been selected on the video display the tracker is engaged and locked to the target. Once this has been achieved the aircraft may be manoeuvred with the pod system automatically keeping the target on boresight and designated by the laser.

The equipment is used to designate the target for laser-guided weapons delivered from either the attacking or an accompanying aircraft. Alternatively, an attack using conventional weapons could proceed by updating the aircraft's weapon aiming computer with range and bearing information from the pod. TIALD also has a passive air-to-air and a reconnaissance/surveillance capability.

Operational status
In production and in service with the RAF. Two prototype TIALD pods were used in Operation Desert Storm in 1991 by five Tornados. They were used in 49 missions, each mission consisting of two TIALD and four armed aircraft. They have since been used in the Gulf again for attack and surveillance missions.

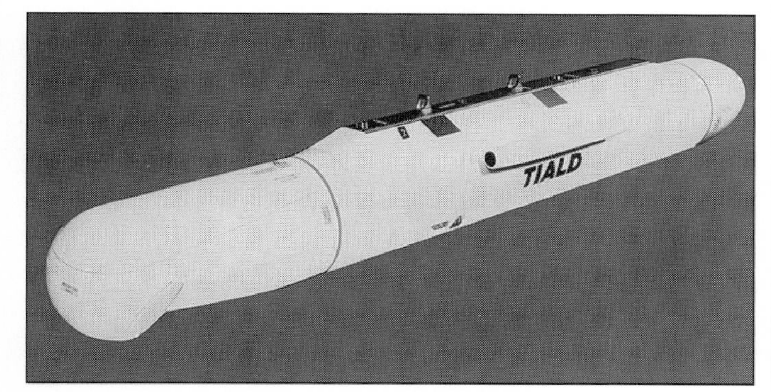

BAE Systems Thermal Imaging/Airborne Laser Designator

BAE Systems Thermal Imaging/Airborne Laser Designator TIALD on RAF Tornado

A successful Foreign Comparative Technology trial was conducted in the USA in 1994 with a pod on a US Marine Corps AV-8B.

In 1994/5, under a UK MoD Urgent Operational Requirement, TIALD was integrated on to the single-seat Jaguar GR.Mk1B and also the two seat trainer T2B and is now in service with both these aircraft. The Jaguar GR.Mk1B/TIALD combination was used to designate laser-guided bomb attacks against Bosnian Serb positions during operation Deliberate Force in the summer of 1995.

The integration of the pod to 10 Harrier GR.Mk.7 aircraft of the RAF is complete and it has been integrated into Tornado GR.Mk4 as part of the Mid-Life Upgrade programme (MLU).

Specifications
Length: 2.9 m
Diameter: 305 mm
Weight: 210 kg
Fields of view: 3.6 and 10°
Electronic zoom: ×2, ×4
Power supply: 200 V AC, 400 Hz, 3 phase, 3.3 kW max, 1.2 kW average

Contractor
BAE Systems, Avionics, Sensors and Communications Systems, Edinburgh.

UPDATED

Dassault Aviation Rubis – navigation and attack pod

Type
Ground attack integrated system (fixed-wing).

Development
Development of this system began in June 1988 under the direction of the *Services Techniques des Télécommunications et des Equipements*. The Rubis has applications as a flying or navigation aid or for target detection in air-to-surface attack for fighter aircraft. It presents a ×1 infra-red image of the landscape in the HUD.

Description
The Rubis features a Thales Optronics FLIR based upon the 8 to 12 μm SMT modules and Intertechnique processors. The front of the pod consists of two the optics with the infra-red sensor and an air-nitrogen exchanger. In the middle are another two LRUs, the electronic unit and an Intertechnique processing unit. The rear of the pod consists of the power and air conditioning systems.

Mirage 2000 fitted with Rubis navigation and attack pod 0023384

Dassault Aviation Rubis Navigation and Attack Pod

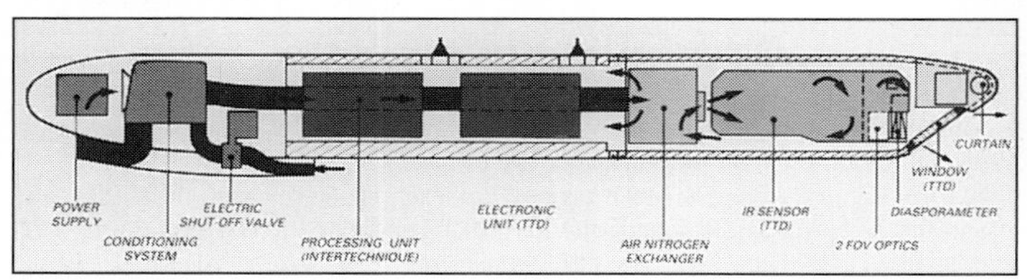

Schematic of the Rubis Pod 0005524

The wide field of view has background-overlay imagery displayed in the pilots' HUD and mission-specific symbology. The narrow field of view has gyrostabilised lock for target identification. The electro-optical zoom has dual fields of view. The pod can be flown at M1.4.

Operational status

In service with the French Air Force on the Mirage F1CR. In service with F-16 aircraft of foreign air forces. No longer offered for sale.

Specifications

Length: 2,650 mm
Diameter: 280 mm
Weight: 110 kg
Field of view
Wide: 24 × 16°
Narrow: 6 × 4°

Contractor

Dassault Aviation.

UPDATED

DRS Photronics joint service Common Boresight System (CBS)

Type

Weapon system boresighting equipment.

Development

The DRS CBS is the product of over 20 years of boresighting equipment with the Captive Boresight Harmonization Kit (CBHK) for the AH-64A Apache and the AIU36M-1 Rapid Armament Boresight Kit (RABS) for the AH-1W Cobra and the APQ-I 80 radar for the AC-130U gunship. It features the Enhanced Triaxial Measurement System (ETMS), a single-beam instrument that simultaneously measures azimuth, elevation and roll with laser accuracy. The equipment can be tailored to an extensive range of fixed and rotary wing aircraft as well as ground vehicles and naval vessels.

The ETMS provides self-contained aircraft boresighting procedures and manuals, including photographic quality pictures of procedures being carried out on the specific aircraft type.

The ETMS is common to all aircraft types, and will support aircraft as diverse as the Apache, Cobra, F-16, F-15, F-1 8, and many others.

Operational status

In service and in production.

Specifications

Accuracy: TMS ≤0.3 mrad, system ≤0.5 mrad
Operating range: 15.24 cm to 30.5 m
Laser safety: exceeds ANSIZ136.1
Interface: RS-232 and MIL-STD-1553B

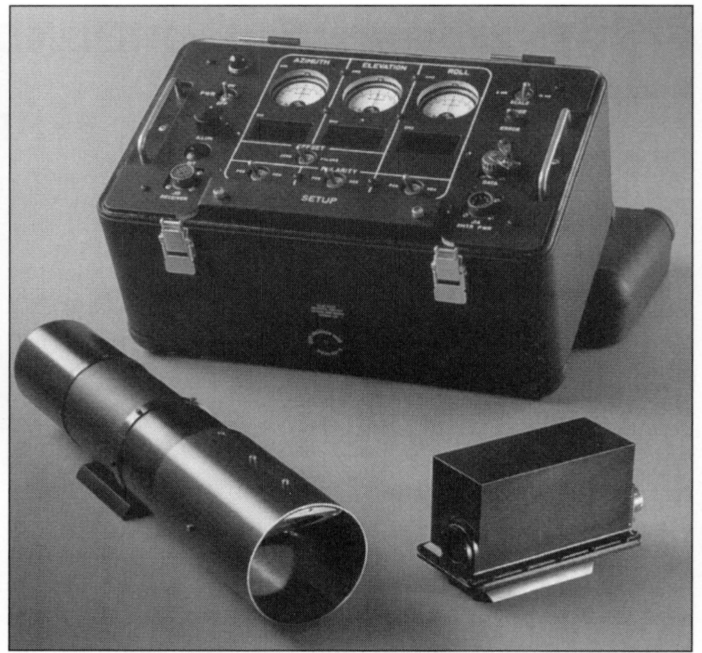

The MPBE - MultiPlatform Boresighting Equipment 0023383

Contractor

DRS Photronics, Inc.

VERIFIED

Lockheed Martin AN/AAQ-13/14 LANTIRN/Sharpshooter

Type

Ground attack integrated system (fixed-wing).

Development

Development of the Low-Altitude Navigation and Targeting Infra-Red for Night (LANTIRN) system began with the award of a US$94 million contract in 1980 to Martin Marietta (now Lockheed Martin) by the US Air Force Aeronautical Systems Division.

Initial flight testing began in July 1983 using an F-16B outfitted with a wide field of view head-up display manufactured by GEC Avionics (UK). Flight testing of operational pods began in late 1983 and continued for three years.

In February 1985, Lockheed Martin was awarded an US$87 million contract to begin production of navigation pods. The first navigation pod was delivered in April 1987 and the first targeting pod in June 1988. The system was initially designed for the F-16 Fighting Falcon but is also fitted to the F-15E Eagle and the LANTIRN targeting pod has been fitted on the US Navy F-14 Tomcat and twin-seat F111.

Approximately 20 AAQ-14 targeting pods and a larger number of AAQ-13 navigation pods were used by US Air Force F-15 aircraft during Operation Desert Storm.

In 1997, Lockheed Martin announced a company-funded upgrade, called LANTIRN 2000 which incorporates a third-generation quantum well FLIR, improved laser and computer and other features to meet the needs of F-15 and F-16 aircraft through their operational lifetimes, around 2020. LANTIRN 2000 will broaden mission capabilties to include air-to-air tracking, theatre missile defense and battle damage assessment. A version called LANTIRN 2000+ which will be available to US customers and include automatic target recognition system, laser spot tracker and digital disk recorder will provide greater targeting flexibility and a reconnaissance capability. LANTIRN 2000 is supported by several US Air Force programmes and the US Air National Guard.

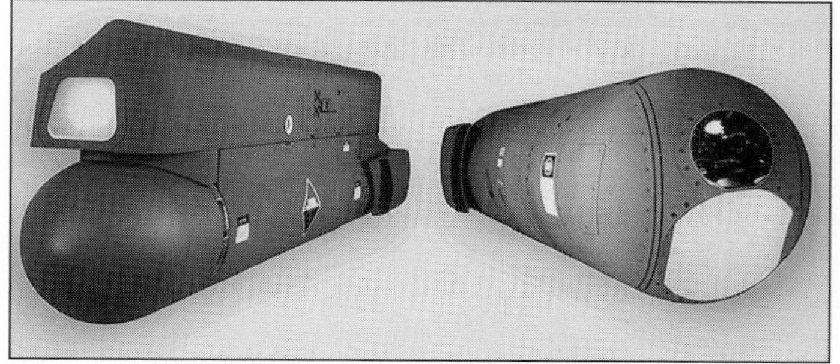

Lockheed Martin Low-Altitude Navigation and Targeting Infra-Red for Night LANTIRN

This Lockheed F-16D carries both navigation and targeting pods of the Lockheed Martin LANTIRN system under its fuselage

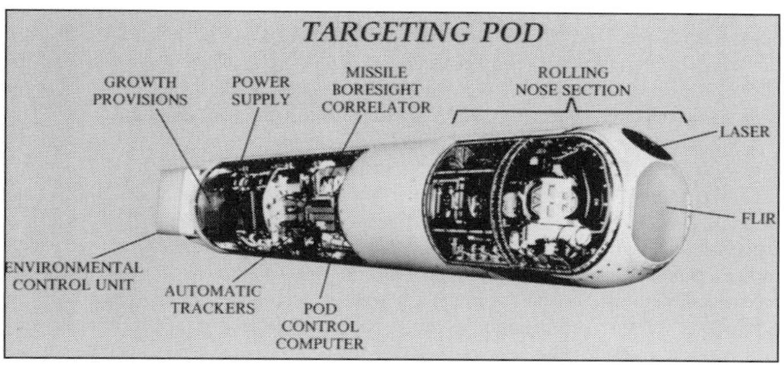

The AN/AAQ-14 targeting pod, showing the major subsystems

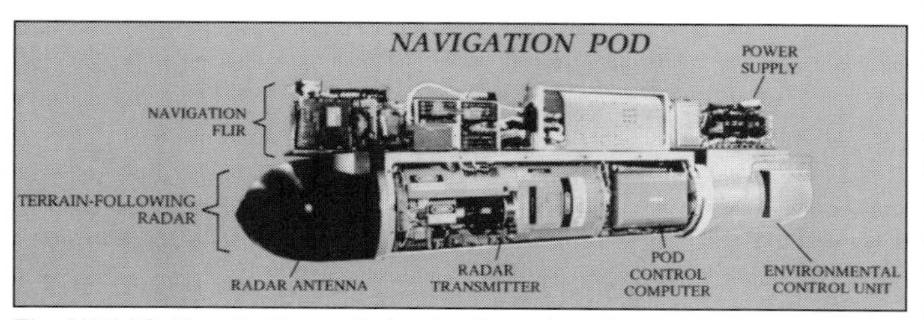

The AN/AAQ-13 navigation pod, showing the major subsystems

Description

LANTIRN is a pod-mounted system that will permit attack aircraft to acquire, track and destroy ground targets with both guided and unguided weapons at speeds exceeding 500 mph. It is carried by the single-seat F-16C and twin-seat F-16D and F-15E aircraft, but is applicable to other aircraft in the air-to-ground attack role. With the option to use either or both pods, depending on the particular mission requirement, the system provides operational flexibility and minimises system support impact.

The navigation pod (AN/AAQ-13) contains a wide field of view FLIR and a Texas Instruments Ku-band terrain-following radar. The FLIR uses a single cadmium mercury telluride array and the picture is displayed to the pilot on a wide field of view holographic head-up display developed by GEC-Marconi Avionics. This provides the pilot with night vision for safe flight at low level. The terrain-following radar enables the pilot to operate at very low altitudes.

The targeting pod (AN/AAQ-14) contains a stabilised wide and narrow fields of view targeting FLIR and a laser designator/rangefinder in the nose section, Maverick hand-off unit boresight correlator and computers in the electronics centre section and an environmental control unit in the tail section. The targeting pod interfaces with the aircraft controls and displays and with the aircraft fire-control system to permit low-level, day/night target acquisition and precision weapon delivery.

In operational use the pilot detects a target using the wide field of view FLIR displayed on a head-down display. The pilot then switches to the narrow field of view FLIR to magnify the image and engages the automatic target tracker. The tracker is dual-mode - 150° look-back angle and continuous roll. After target lock on, a weapon is selected. The line of sight is cued by the aircraft inertial navigation system. The sensor will be properly orientated for the planned attack and selected weapon delivery profile. This ensures maximum standoff and first-pass attack for guided and unguided weapon delivery. For a Maverick missile, the pod automatically hands the target off to the missile for launch with pilot consent. For a laser-guided bomb, the pilot aims the laser designator and the bomb guides to the target. For a conventional bomb, the pilot can use the laser to determine range, then the pod feeds the range data to the aircraft's fire-control system.

The Laser RangeFinder/Designator (LRF/D) has been developed by Litton's Laser Systems Division. It provides the laser designation/ranging essential to the effective delivery and guidance of laser terminally guided weapons. It operates in the 1.06 or 1.54 μm wavelength and has a maximum range readout of 24.5 km. Having a digital interface, the LRF/D provides serial or parallel commands directly to the pod control computer through the system databases.

A derivative of the AAQ-14 targeting pod, but without the infra-red missile boresight correlator used for targeting Maverick missiles, is produced for export as Sharpshooter. It can be integrated with the Pathfinder navigation pod to permit low-level flying at night.

LANTIRN 2000

LANTIRN 2000 includes a third-generation FLIR sensor using quantum well technology, an advanced diode pumped laser and an enhanced computer system. The use of quantum well technology is said to offer the following improvements: detectors that give sharp-edge imagery at significantly lower cost, target detection, recognition ranges increased by over 50 per cent; FLIR sensitivity that will allow air-to-air tracking of high-speed targets at difficult crossing angles; 50 per cent increase in standoff ranges for damage assessment.

The use of the diode-pumped laser for spotting targets and guiding munitions will result in: a narrower laser beam that extends the targeting pod's operational range while permitting operations at 40,000 ft, instead of the current 25,000 ft; fuller use of the operational envelope of the GBU-24 laser-guided weapon; and increase in reliability because of a more efficient power supply, fewer parts and cooler operating temperatures.

These developments will also lead to reduced power usage and weight of the system.

LANTIRN 2000+ which will be offered to US customers will also include: an automatic target recognition system, a laser spot tracker and a digital disk recorder.

Operational status

The system is in service on the US Air Force's F-16 C/D Fighting Falcon and F-15 C/D/E Eagle aircraft and the US Navy's F-14 and has been selected by the Air Forces of 11 allied countries.

By January 1994, Lockheed Martin had received US$3.4 billion in US Air Force contracts to produce 561 AN/AAQ-13 navigation pods, 506 AN/AAQ-14 targeting pods and 26 sets of support equipment. The last US Air Force navigation pod was delivered in March 1992 but the targeting pod continued in production until 1994 for the US Air Force. In late 1995, the US Navy purchased 75 LANTIRN systems for use on its F-14s as part of the Tomcat precision strike programmme. The first strike fighter squadron was deployed aboard the *Enterprise* aircraft carrier in June 1996.

LANTIRN is also being supplied to Turkey and South Korea. The former is receiving 40 AAQ-13 and 20 AAQ-14 while the latter is receiving 30 complete sets. Greece has also received AAQ-13 (24) and AAQ-14 (16) and Saudi Arabia has ordered 48 of each. Bahrain has ordered three AAQ-14 pods.

Israel has purchased 40 Sharpshooters and 27 AAQ-13 pods for its F-15I (first deliveries were in October 1996), and Egypt 12, together with 12 Pathfinders. Egypt has signed a Letter of Agreement to buy 18 additional Sharpshooters.

LANTIRN is being reconfigured for the Royal Netherlands Air Force F-16 Mid-Life Update (MLU) programme. The targeting pod will include electro-optic and TV sensors and a laser spot locator will be added to the navigation pod. Denmark and Taiwan became LANTIRN customers in late 1998.

The US Navy ordered a further 13 LANTIRN targeting pods for the F-14 Tomcat bringing the total fleet to 19.

LANTIRN 2000 is under development. Some of the upgrades were flight-tested by the USAF in 1997 and 1998.

Specifications

AN/AAQ-13 Navigation pod
Weight: 195 kg
Dimensions (h × w × l): 546 × 355 × 1,985 mm
Field of regard: 56 × 78°
Field of view: 21 × 28°
AN/AAQ-14 Targeting pod
Weight: 245 kg
Dimensions (w × l): 381 × 2,500 mm
Field of regard: ±150°
Fields of view: 1.7 × 1.7° (narrow) and 6 × 6° (wide)
FLIR aperture: 200 mm
Sharpshooter
Weight: 241 kg
Field of regard: ±150°
Fields of view: 1.7 × 1.7° (narrow) and 6 × 6° (wide)
FLIR aperture: 200 mm

Contractor

Lockheed Martin Missiles and Fire Control.

VERIFIED

Lockheed Martin AN/AAS-35(V) Pave Penny laser tracker

Type

Ground attack integrated system (fixed-wing).

Development

The AN/AAS-35(V) is a day and night laser-based target identification set. Used in conjunction with a laser designation system, either ground based or in a

The Lockheed Martin Pave Penny laser tracker installation on the Fairchild A-10 Thunderbolt II uses a special pylon on the front starboard fuselage

co-operating aircraft, targets can be recognised and identified rapidly and accurate steering data provided to ensure quick pilot reaction and accurate delivery of weapons. The equipment was installed initially on US Air Force A-10 close support aircraft and later on A-7D Corsair IIs. Pave Penny is also used on the Lockheed F-16 and is suitable for the McDonnell Douglas F-4 Phantom, Northrop F-5 and Dassault/Dornier Alpha Jet.

Description

A silicon pin diode detector head is used, with full lower forward hemisphere coverage plus some look-up capability. Pilot's controls permit selection of several seeker scanning patterns to improve early designator recognition. It can be used to improve the accuracy of conventional weapon delivery, or to lock up laser-guided munitions. The system is contained in a relatively small pod, which is either fuselage or pylon mounted to allow easy harmonisation with other onboard sensors. Preflight boresighting of A-10 units is claimed to allow pod attachment in a matter of minutes. An aircraft adaptor module is used to integrate the sensor data with onboard processors and a pilot's control panel provides for easy use and built-in test operations.

Operational status

In service. The first operational Pave Penny system for the A-10 was delivered in March 1977 and the US Air Force has equipped its entire fleet originally numbering 733 aircraft. Up to 380 US Air Force A-7D Corsairs also have Pave Penny installed. Pave Penny was incorporated on US Air Force Block 50D/52D F-16 aircraft in 1993.

Specifications

Dimensions (l × d): 833 × 200 mm
Weight: 14.5 kg
Power supply: 28 V DC, <10 A
Wavelength: 1.06 μm
Scan coverage
  Elevation: –90 to +15°
  Azimuth: –90 to +90°
Selectable scan patterns: Wide, narrow, depressed, offset
Output: Direction cosines of line of sight

Contractor

Lockheed Martin Missiles and Fire Control.

VERIFIED

Lockheed Martin AN/AVQ-26 Pave Tack

Type

Ground attack integrated system (fixed-wing).

Development

The need for electro-optic support in weapons delivery led the US Air Force Aeronautical Systems Laboratories to issue a request for proposals for such a system. Lockheed Martin (formerly Loral Aeronutronic) won an initial US$15 million contract in 1974 with flight testing beginning in 1976. In 1978 a US$48.5 million production contract was awarded. Production deliveries began in August 1980 with F-4 Phantom initial operational clearance achieved in late 1980 and F-111F in early 1981.

Description

The Pave Tack system provides 24-hour capability in clear and adverse weather from all altitudes, including extremely low-altitude penetration and interdiction. It provides precise target acquisition and laser target designation for a variety of

The Lockheed Martin Pave Tack installation under the fuselage, aft of the nose gear, on a General Dynamics F-111F

existing and projected laser, electro-optical and infra-red weapons, against fixed and moving targets. The aim of the Pave Tack system is to provide unvignetted full lower-hemisphere coverage for sensors and thus to permit the crew near total freedom in choice of flight path when approaching and leaving the target area. Delivery techniques can be varied to suit the weapon and the standoff range desired, the crew having the choice of dive, loft, glide, toss or level release of weapons, with virtually no manoeuvrability or flight envelope constraints.

The 1.06 μm Laser RangeFinder/Designator (LRF/D), designated AN/AVQ-25, is designed and produced by Litton Laser Systems. It provides target designation capability for the delivery of laser-guided weapons and range-to-target information for navigation and conventional munition delivery.

The AN/AAQ-9 forward-looking infra-red (FLIR) sensor is produced by Raytheon.

General Electric produces the virtual image display which is used in the rear cockpit to magnify the image presented on the display. Initially engineered for USAF F-111F aircraft, it is now being applied to expand the weapons delivery capabilities of the F-4E night attack systems and the capabilities of the RF-4C.

The Pave Tack pod is divided into two basic sections; a base section assembly and a head section assembly. The latter contains an optical bench on which are mounted the FLIR, laser, range receiver and the stabilised sight assembly. The optical bench is mounted on the turret structure, which rotates in pitch and is driven by turret servo-drive motors. The turret is mounted to the head structure, which rotates in roll with respect to the base section. Roll motion of the head and pitch motion of the turret provide complete unvignetted lower-hemisphere coverage and a highly stabilised line of sight which is fully controllable over the entire lower hemisphere of the aircraft.

The FLIR is boresighted to (and stabilised with) the laser and the entire assembly is controlled by the Pave Tack operator, aided by a pod-mounted digital aircraft interface unit. The stabilised image and rate-aided tracking mechanisation provide accurate target tracking, laser designation and weapon delivery throughout a wide spectrum of evasive aircraft manoeuvres. The FLIR has a wide field of view display for target acquisition. A narrow field of view provides high magnification to enhance target identification and allow precise target tracking.

The electronic assemblies to operate the head section are located in the base section. In addition to the power supply and control electronics, the base section contains a digital aircraft interface unit and a CRT display interface unit. The digital aircraft interface unit is integrated with the ARN-101 digital avionics on board the F-4E and RF-4C aircraft as well as with the digital avionics on the F-111 aircraft. It provides precise target location data to the aircraft weapon delivery digital computer. The computer additionally provides complete built-in testing for all components on the pod.

A videotape recorder records the video displayed in the cockpit for use in bomb damage assessment and location of targets for future missions.

Operational status

Production of these systems for the US Air Force was completed in July 1982. The US Air Force received 149 pods which were used by the F-4E and RF-4C Phantom II, together with the F-111F, in a programme which cost US$242.8 million. The system is no longer in use by the US Air Force. It is still in service with the Royal Australian Air Force who have 18 sets for use in its F-111C and the South Korean Air Force has 10 sets for the F-4E.

Specifications

Length: 4.14 m
Diameter: 0.5 m
Weight: 579 kg

Contractor

Lockheed Martin Missiles and Fire Control.

VERIFIED

Lockheed Martin NITE Hawk AN/AAS-38 targeting FLIR

Type

Ground attack integrated system (fixed-wing).

Development

Development of the NITE Hawk began in March 1978 with a requirement to provide the F/A-18 Hornet with a day/night strike capability. The first systems, designated AN/AAS-38, were delivered to the US Navy in 1983. The system became operational in February 1985 and has undergone two major operational improvements since then to become the AAS-38A and AAS-38B. Lockheed Martin

The NITE Hawk carried on the McDonnell Douglas F/A-18

The NITE Hawk SC targeting FLIR for the JAST/IHAVS flight demonstration on the AV-8B (USN)

has developed the NITE Hawk AAS-46 interface with the F/A-18E/F. A self-cooled derivative NITE Hawk SC has been developed for other tactical aircraft and has been demonstrated on the F-14, F-15, F-16 and AV-8B.

Description

The system provides day, night and bad weather attack capability by presenting the pilot with real-time passive thermal imagery in a television-formatted display to locate, identify, track and laser designate targets.

The system provides the aircraft mission computer with accurate target line of sight pointing angles and angle rates. An automatic in-flight boresight compensation is used to correct for dynamic flex and thus maintains accurate pointing angles through the aircraft's full performance envelope. Additional weapons delivery accuracy is provided in the AAS-38A through the Laser Target Designator/Ranger (LTD/R). The operational LTD/R provides precision target range data and a designation capability for laser-guided weapons. With the introduction of the AAS-38B in late 1994 the NITE Hawk also includes a laser spot tracker to search for, acquire and track targets designated by ground or airborne lasers.

The system consists of 12 Weapon-Replaceable Assemblies (WRAs)/Line-Replaceable Units (LRUs). The pod's forward section contains the laser relay optics assembly and the laser transceiver. The transceiver transmits and receives laser energy to provide the LTD/R capability. The infra-red receiver detects IR energy and converts it into a visible format which is viewed by the pilot on an 875-line television display.

The roll drive motor rotates the head section (consisting of the optic stabiliser and pod forward section) through ±540° to provide continuous line of sight tracking. The optic stabiliser is an inertial platform which provides a stable image for the infra-red receiver and is controlled by a high-speed digital servo-control. The power supply WRA receives aircraft electric power and converts it to the voltages required by the other WRAs.

The controller processor WRA is a general purpose computer which performs calculations and fault detection and isolation. The AAS-38B contains an Advanced Controller Processor (ACP) which provides improved passive air-to-air search and tracking capability at beyond-visual ranges. It also includes a multifunction tracker and processes multiple tracking algorithms and automatically selects the optimum one for target acquisition and lock on. A temperature control WRA regulates the flow of cooling air from the aircraft. Alternatively, a self-contained ram air turbine provides cooling as in the NITE Hawk SC configuration. The addition of the self-cooling feature and modified structure to contain the turbine and an aircraft interface module are all that distinguish the SC version from the basic NITE Hawk. All the other WRAs are identical.

Development of the NITE Hawk AN/AAS-46 configuration for the F/A-18E/F was completed in 1995. The self-cooled version of the system has also been developed for use on tactical aircraft other than the F/A-18.

Lockheed Martin Electronics & Missiles are now responsible for the NITE Hawk programme.

Operational status

The first production AN/AAS-38 system was delivered to the US Navy in December 1983. By 1995, a total of 524 NITE Hawk systems had been procured including 442 for the US Navy/Marine Corps, 29 for Australia, 18 for Spain, 16 for Kuwait, two for Switzerland, 13 for Canada, eight for Malaysia and one for Thailand. More than 460 of these systems had been delivered by the end of 1996.

The AAS-38 was used by US Navy and US Marine Corps aircraft during 6,000 Operation Desert Storm missions.

NITE Hawk SC has also been flight-tested on the F-14, F-15 and F-16 and AV-8B. NITE Hawk SC is the baseline laser designator pod for the Spanish Air Force's Eurofighter 2000.

Specifications

Dimensions
AN/AAS-38/38A/38B: 1,840 × 330 mm diameter
NITE Hawk (SC): 2,440 × 330 mm diameter
Weight
AN/AAS-38: 158 kg
AN/AAS-38A/38B: 168 kg
NITE Hawk SC: 195 kg
Field of view
Wide: 12 × 12°
Narrow: 3 × 3°
Field of regard
Pitch: +30 to -150°
Roll: ±540°
Accuracies
Stabilisation: 35 μrad
Tracking: 230 μrad
Pointing: 400 μrad
Tracking rate: 75°/s

Contractor

Lockheed Martin Missiles and Fire Control.

VERIFIED

Lockheed Martin Sniper/Pantera

Type

Lockheed Martin Sniper/Pantera.

Description

Sniper is a mature, flight-tested, long-range precision targeting system in a single, lightweight, affordable pod. Sniper provides positive identification, automatic tracking and laser designation of tactical size targets via real-time imagery presented on cockpit displays. The system includes an eye-safe laser range-finder, laser spot tracker, FLIR and an advanced target/scene imager tracker.

The third generation FLIR with a fast-steering mirror, microscan, 2D resolution enhancement algorithms, automatic boresighting and stabilisation techniques, delivers better performance than systems now flying.

Specifications

FLIR
Field of view
WFOV: 4°
MFOV: 1°
NFOV: 0.5°
Field of regard
Pitch: +35 to −155°
Roll: continuous
Diameter: 30 cm
Length: 230 cm
Weight: 177 kg

Contractor

Lockheed Martin Missiles and Fire Control.

VERIFIED

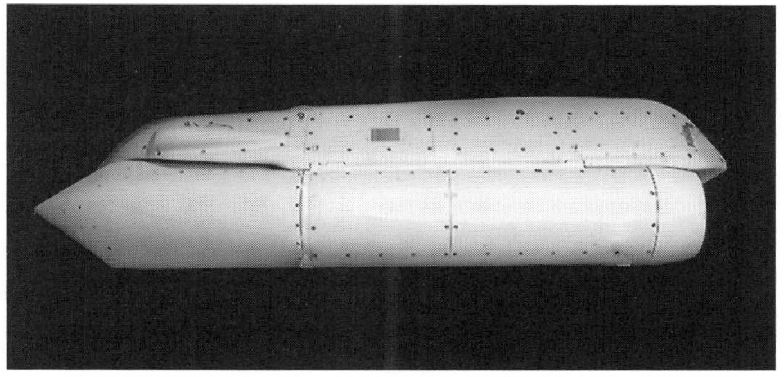

The Sniper/Pantera pod 0058708

Northrop Grumman AN/ASQ-153 and AN/AVQ-23 Pave Spike

Type
Ground attack integrated system (fixed-wing).

Description
Pave Spike is a day tracking/laser target designator system which provides the F-4D and F-4E aircraft with a self-contained laser-guided ordnance delivery capability. It was produced by Northrop Grumman (then Westinghouse) under production contracts monitored by the Aeronautical Systems Division of the USAF. The Pave Spike system (designated AN/ASQ-153(V)) is able to acquire, track and designate tactical surface targets from a manoeuvring aircraft at standoff ranges usual with laser-guided munitions.

The Pave Spike pod (designated AN/AVQ-23) contains an optical subsystem, stabilisation and beam-pointing subsystem, TV tracking sensor, laser designator/ranger subsystem, environmental control system and associated electronics. The aircraft-mounted parts of the system comprise line of sight indicator, control panel, range indicator and a modified radar control handle and weapons release computer.

The pod is divided into three sections: forward nose/roll, umbilical and aft electronics. The forward nose/roll section rotates about the pod's axis and the rotating assembly contains the electro-optical sensors and the stabilised gimbal. It is sealed and pressurised with nitrogen and temperature controlled for optimum sensor performance. The umbilical section provides physical and electrical interconnections between the rotating nose section, the aircraft and the aft electronics section. The last of these contains a thermal cold plate that runs the length of the section. Six major electronic LRUs are mounted on it.

The AN/ASQ-153(V) designates the Pave Spike system as configured for the F-4D and F-4E aircraft, the pod being mounted in the left front Sparrow missile well by means of a standard 76.2 cm (30 in) centre ECM pod adaptor. It has also been integrated with other NATO aircraft.

Operational status
No longer in production. The Pave Spike system was delivered to the USAF for use in the F-4 Phantom II between 1972 and August 1977. It remains in service with the US Air Force and the Air National Guard. Production continued for export with some 260 systems being produced for six customers including Turkey and the UK. Pave Spike was used by RAF Buccaneers in Operation Desert Storm.

Specifications
Length: 3.66 m
Diameter: 25.4 cm
Weight: 193 kg

Contractor
Northrop Grumman, Electronic Sensors and Systems Sector, Electro-optical Systems.

VERIFIED

Pave Spike laser target designator pod AN/AVQ-23 mounted in one of the Phantom's missile wells

Northrop Grumman AN/AVQ-27 LTDS

Type
Ground attack integrated system (fixed-wing).

Description
The AN/AVQ-27 Laser Target Designator Set (LTDS) is for installation on the canopy rail of the rear cockpit of the Northrop Grumman F-5F. The LTDS may be installed in most two-seat fighter or fighter-bomber aircraft with only minor wiring modification. The LTDS allows the aircraft to be equipped with target designation capability as required by the mission and have it removed when not in use.

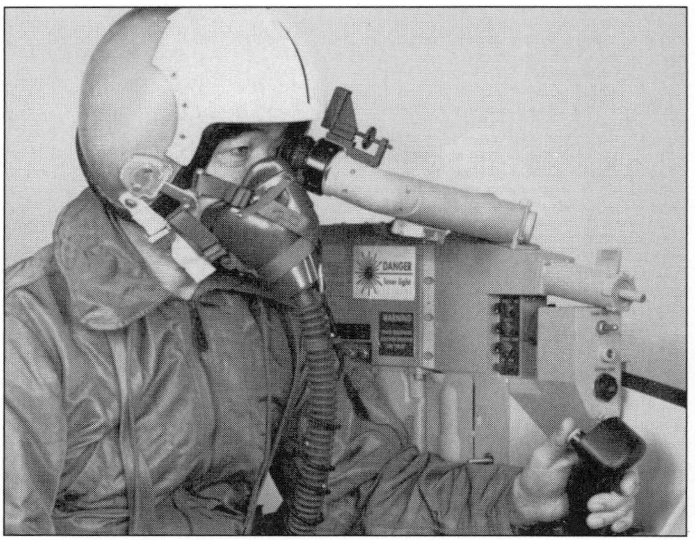

AN/AVQ-27 laser target designator set

The system's principal elements are a stabilised operator sight unit with a retractable periscopic telescope, a laser transmitter and a 16 mm mission recording camera, all of which are boresighted through common optics in a single self-contained package. The stabilised sighting assembly includes a rate-stabilised mirror which is controlled by a two-axis hand control and provides two tracking modes (rate and rate-aided), a retractable periscopic telescope and associated optics. The laser transmitter consists of a single package that houses a hermetically sealed laser and its associated optics and electronics. The 16 mm mission camera integrated into the LTDS is a version of the same KB-25A/KB-26A standard gun camera used in the Northrop Grumman F-5E/F and other aircraft.

In operation the viewfinder-sight swings across into the pilot's field of view so that the crewman can track targets. This can be conducted manually, using a two-axis hand controller and with rate or rate-aided tracking modes. The front cockpit is fitted with sight and canopy markings which allow the pilot to assist the crewman in initial target acquisition and then to maintain the target within the set's field of view. Normally one aircraft is used for designation while accompanying aircraft will deliver ordnance.

Operational status
In production and in service on foreign F-5B and F-5F aircraft.

Contractor
Northrop Grumman, Sensors and Systems Sector, Electro-Optical Systems.

VERIFIED

Northrop Grumman/Rafael Litening II

Type
Ground attack integrated system (fixed-wing).

Development
The Litening II Pod is an airborne laser target designator pod offering precision strike in day/night/under the weather conditions for fighter and tactical strike aircraft. The Litening II was chosen by the US Air Force Reserve/Air National Guard in the recent Precision Attack Targeting System competition. The Litening II evolved from the Litening which is presently fielded with a number of Air Forces worldwide, including the Israel Defence Force, Air Force.

Description
The Litening II is an externally-carried pod for high speed aircraft and contains a 256 × 256 staring FLIR with three fields of view, two visible TV cameras (both narrow and wide fields of view), a laser designator, laser marker, laser spot tracker and Inertial Navigation Sensor (INS) on the stabilised gimbal. The sensors are all fixed relative to each other, providing boresight during assembly and good boresight retention. The INS stabilises the line of sight and automatically aligns the pod boresight to the aircraft INS, making mechanical boresight unnecessary. The pod front section has an integral compressor which maintains a minimum atmospheric pressure, allowing the laser designator to operate at altitudes up to 40,000 feet. Modular design of software facilitates aircraft integration.

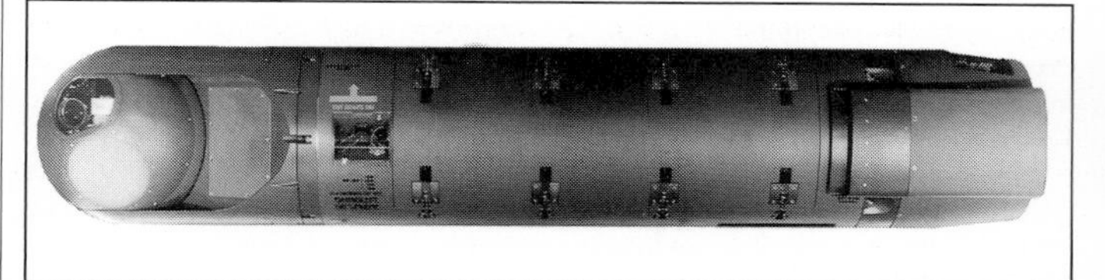

The Litening II pod 0058709

Litening II pod on an F-16 0058710

Operational status
In production for the US Air Force Reserve and Air National Guard on F-16 aircraft.

Specifications
Pod dimensions
Length: 220 cm
Depth: 40.6 cm
Weight: 200 kg
FLIR sensor
Field-of-view
Narrow: 1.0 × 1.0°
Medium: 4.0 × 4.0°
Wide: 24 × 24°
CCD camera
Field-of-view
Narrow: 1.0 × 1.0°
Wide: 3.5 × 3.5°
Elements: 768 × 494
Laser range-finder and designator
Energy: 100 mJ/pulse
Tracker: advanced area and point correlation/inertial laser spot search and track
Gimbals
Fields of regard
Pitch: +45 to –150°
Roll: ±400°
Stabilisation: 30 μrad
Flight envelope
at low level: Mach 1.2
vertical manoeuvre: 9 *g*
Cooling: self-contained environmental control unit

Contractor
Northrop Grumman, Electronic Sensors & Systems Sector, Defensive Systems Division; Rafael Missile Division.

VERIFIED

Rafael/Northrop Grumman LITE targeting and navigation pod

Type
Integrated system fixed-wing.

Description
The LITE pod is a lower cost targeting and navigation pod based on the Litening pod developed by Rafael. It is designed for light fighter aircraft, including the F-5, Mig-21, Mirage and A-4. The system is being jointly marketed by Rafael and Northrop Grumman.

The LITE pod contains a FLIR, CCD TV camera, laser designator/rangefinder and advanced correlator tracker, with laser spot detection for the delivery of laser-guided bombs as well as cluster and general purpose bombs. The FLIR has three fields of view and operates in the 8 to 12 μm waveband. The CCD TV is dual field of view.

Operational status
Available.

Specifications
Weight: 450 kg (empty)
Length: 533 cm
FLIR
Fields of view
Narrow: 1.5 × 2.0°
Medium: 4.5 × 6.0°
Wide: 18 × 24°
Spectral band: 8-12 μm
Picture element: 512 × 240
CCD camera
Fields of view
Narrow: 1.3 × 1.0°
Wide: 4.6 × 3.5°
No of elements: 768 × 494
Laser range-finder/designator
Energy: 100 MJ/pulse
Beam divergence: 0.2 mrad radius
Gimbals
Fields of regard
Pitch: +45 to –150°
Roll: 360°
Stabilisation: 40 μrad
Flight envelope (installed on F-5 with empty fuel tank)
Low level: M0.95
20,000 ft: M1.3
Vertical manoeuvre: 7 *g*

Contractors
Rafael Armament Development Authority.
Northrop Grumman, Electronics Sensors and Systems Sector, Defensive Systems Division; Precision Weapons & Electronics Systems.

VERIFIED

Rafael/Northrop Grumman Litening

Type
Ground attack integrated system (fixed-wing).

Development
Litening is an airborne infra-red pod designed to improve both day and night attack capabilities. It provides real-time FLIR and CCD imagery. It is suitable for aircraft such as the A-4, F-4, F-16, F/A-18, MiG-21 and Tornado.

Description
Litening enables the fighter crew to carry out detection, recognition, identification and laser designation of surface targets; accurate delivery of laser-guided, cluster and general purpose bombs; low-level night flights; laser spot detection; identification of aerial targets at beyond-visual ranges and surveillance, reconnaissance and battle damage assessment.

Litening is equipped with a thermal imager with three fields of view, CCD camera with two fields of view, laser designator and rangefinder, laser spot detector, digital image processing including derotation and ×2 magnification continuous zoom, inertial navigation system for accurate pointing without preflight harmonisation, area and inertial tracker and video and audio recording including telemetry data for debriefing.

All these sensors are part of the same payload and share the same optical bench. The payload is located on the stabilised platform and the sensors stare directly through the windows. The four-gimbal set provides a large area of regard with high

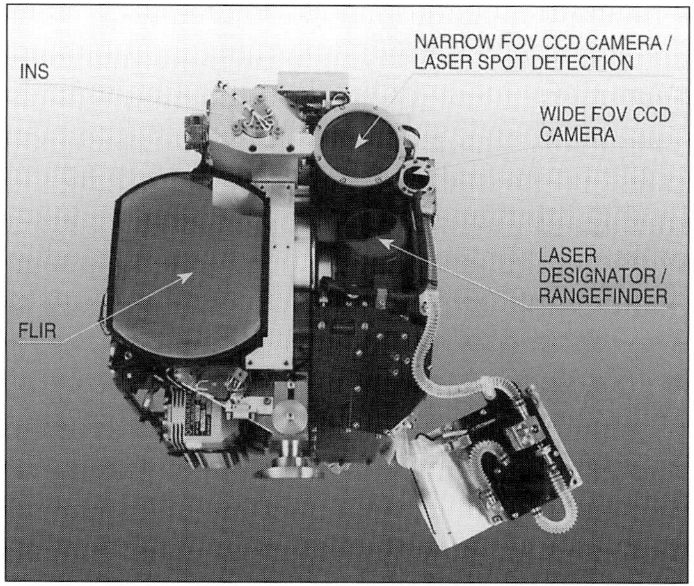

The multiple sensor arrays of the Litening pod which provide target selection, identification and verification

The Litening Pod on an F-16 0005519

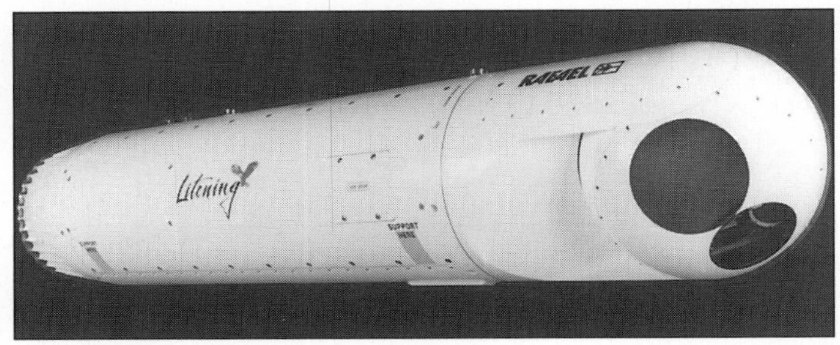

Front view of the Litening Pod 0005523

stabilisation. Shock-absorbers isolate the entire payload. This concept is stated to produce less optical loss and better line of sight retention and stabilisation.

Current plans include development of a 3 to 5 μm FLIR and replacement of the laser with a diode pumped laser.

Litening includes full BIT to detect and isolate faults to LRU and SRU levels. All LRUs are replaceable on the flight line with plug-in connectors and quick disconnect fasteners.

Operational status

In production. Litening is in service with the Israel Air Force F-16D. It has been adapted for use on the F-15, F-16, F/A-18, Tornado GR-1, AV-8B, F-4, F-5, MiG-21 and A-10. In late 1995, Rafael signed a teaming agreement with Northrop Grumman (the manufacturer of the thermal imaging system) jointly to produce the pod. Litening has been selected for German requirements (20 systems for the Luftwaffe and 16 for the Marineflieger Tornados). Zeiss Optronik supply the electro-optics systems for this requirement.

The Litening pod has also been selected by India for the Mirage 2000, Venezuela for the F-16A/B and for the Romanian modernised MiG21 Lancer fighter.

Specifications

Dimensions (l × d): 2,208 × 406.4 mm
Weight: 200 kg
Power supply: 115 V AC, 400 Hz, 3 phase, 2.5 kW
Field of regard: –150 to +45° pitch, ±400° roll
Stabilisation: 30 μrad
FLIR spectral band: 8-12 μm
FLIR field of view
  Wide: 18.4 × 24.5°
  Medium: 4.5 × 4.5°
  Narrow: 1.5 × 1.5°
FLIR pixels: 708 × 240
CCD camera field of view
  Wide: 3.5 × 3.5°
  Narrow: 1 × 1°
CCD camera pixels: 768 × 494
Laser energy: 100 mJ/pulse
Laser divergence: 0.2 mrad radius
Flight envelope: M1.2 at low level; M1.8 at 13,000 ft
9 *g* vertical manoeuvre

Contractors

Northrop Grumman Electronic Sensors & Systems Sector, Defensive Systems Division.
Rafael Missile Division.

VERIFIED

Raytheon AN/AAS-38A F/A-18 targeting pod

Type

Ground attack integrated system (fixed-wing).

Description

The AN/AAS-38A Foward-Looking Infra-red (FLIR) targeting pod enables pilots of US Navy and Marine Corps F/A-18 Hornet aircraft to attack ground targets day or night with a precision strike capability. The system provides TV-like infra-red imagery on a cockpit panel display and accommodates a laser rangefinder/designator that can pinpoint targets for both laser-guided and conventional weapon delivery.

The pod is integrated with the aircraft's avionics system through a MIL-STD-1553 databus, allowing the pod to receive command and cue signals from the onboard mission computer and provide status and targeting information to the cockpit display and weapon delivery system.

AN/AAS-38A consists of 12 Weapon Replaceable Assemblies (WRAs) including optics stabiliser, infra-red receiver, roll drive motor, controller processor, roll drive amplifier, power supply, laser transceiver. WRAs can be accessed and replaced without the need for calibration, alignment, special tools or handling equipment.

The AN/AAS-38A configuration accepts the two laser subsystem's WRAs, a laser transceiver and laser power supply to provide the aircrew with the capability for laser target designation and ranging.

The targeting FLIR, when integrated with the Raytheon-developed AN/AAR-50 Navigation FLIR, and night vision goggles, can acquire/designate targets and assess battle damage for deployment of the Pave Way/GBU-24 precision-guided weapon series on an around the clock operational cycle.

Operational status

In service. Raytheon received a contract from the US Navy in 1990 to deliver over 30 pods. Design qualification and flight test were successfully completed in August 1994 providing the US Navy with a second source for AAS-38 pods and spares.

Specifications

Length: 72 in
Diameter: 13 in
Weight: 380 lb
Field of view
  Narrow: 3 × 3°
  Wide: 12 × 12°
Field of regard
  Pitch: +30 to –150°
  Roll: ±540°
Video: RS-343, 875 lines
MTTR: O-Level 15 min
BIT fault detection: continuous 90%, initiated 95%

Contractor

Raytheon Systems Company.

VERIFIED

Raytheon AN/ASB-19 ARBS

Type

Ground attack targeting sight.

Development

The AN/ASB-19 Angle-Rate Bombing Set (ARBS) was designed for US Marine Corps aircraft to improve day and night bombing accuracy when operating in the close support role using unguided weapons.

Description

The ARBS is basically an angle-rate system. The tracker, after locking on to a target, provides the aircraft-to-target line of sight angle and angle-rate to the Weapon Delivery Computer (WDC). This information, combined with the true airspeed and altitude (from the air data computer), is processed by the WDC to provide the weapon delivery solution. Target position, weapon release and azimuth steering information are generated for display to the pilot via a head-up display (HUD). Unlike most current operational bombing systems, neither measurement of range to the target nor inertial quality platform inputs are required for ARBS.

Once the tracker is locked onto the target in either the laser or TV mode of operation, the pilot is free to manoeuvre within the gimbal limits. This freedom allows him to fly an erratic, unpredictable flight path to confuse enemy air defence tracking, while simultaneously providing target tracking information to the computer. The accuracy of ARBS is equal to the ballistic dispersion CEP of all free-fall weapons in modern inventories. A standoff, accurate delivery and identification of targets is made possible by the ×6 magnification of the TV tracker.

In the A-4M, ARBS consists of a Dual-Mode laser/TV Tracker (DMT) designed by Hughes Missile Systems (now Raytheon Systems Company), a System/4 Pi Model Sp-1 general purpose digital computer built by IBM (now Lockheed Martin) Federal Systems Division under US Navy contract and a control subsystem. The system works in conjunction with the existing cockpit controls and aircraft sensors. An AN/AVQ-24 HUD by GEC-Marconi Avionics is used to display steering information to the pilot.

The Raytheon ARBS in the nose of a US Marine Corps AV-8B

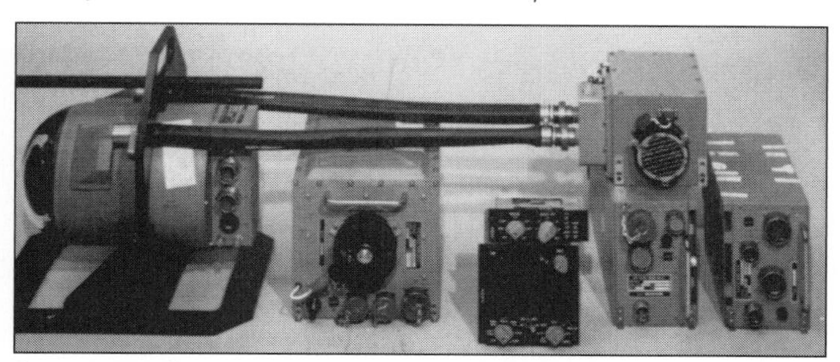

Components of the ARBS showing receiver/processor, signal data converter, bombing set control, control indicator, heat exchanger, computer and computer interface

With the AV-8B/GR. Mk 5 Harrier, Raytheon supplied the DMT to McDonnell Douglas Corporation (now Boeing) in St Louis where it is integrated into the Harrier's onboard mission computer. ARBS controls, data imagery, video imagery and steering information have been integrated into the HUD, the multipurpose display (video) and the up-front control set.

The DMT is the main element of the system. A modification of the Raytheon STALOC II TV tracker with TV camera shares a common optical element with a laser spot tracker detector on a three-gimbal stabilised platform. Tracking filter electronics, an output signal converter, power supplies and control logic complete the DMT. A dichroic filter behind the optics separates the laser energy from the visible light for sensing by a four-quadrant laser detector. The visible light is directed to form an image on the TV vidicon.

Operating in the laser mode, the sensor automatically acquires the target and, via the WDC, presents steering signals for the pilot on the HUD. After head-up TV acquisition, the pilot is shown a magnified image of the target on the cockpit TV monitor. Weapon release in both laser and optical tracking modes is either automatic or manual. A weapon data insert panel provides entry into the Weapon Delivery Computer of weapon characteristics and rack type for ordnance being carried on a particular mission.

In addition to the fuselage installation in the A-4 and the AV-8B/GR. Mk 5, a podded configuration is under consideration to give aircraft with radars or other equipment in the nose an accurate air-to-ground capability. Aircraft such as the F-5E and the F-4 are candidates for the pod.

ARBS can be employed with the complete inventory of guided and unguided weapons. Only a software update is required to adapt it to new weapons such as the laser- and infra-red-guided Maverick air-to-surface missiles. It can be given a further night attack capability by replacing the TV vidicon with a Raytheon infra-red seeker.

Operational status
In service. By October 1989 Raytheon had received contracts totalling US$196 million for the production of 613 AN/ASB-19(V)2 equipments and spares. All systems had been manufactured and delivered by mid-1990.

Specifications
Weight: 59 kg
Target acquisition range: 15 km (3 m target)
Angular coverage: ±450° (roll)
Elevation: +10 to –70°
Bearing: +37°

Contractor
Raytheon Systems Company.

VERIFIED

Raytheon AN/ASQ-228 ATFLIR

Type
Ground attack integrated system (fixed-wing).

Development
Raytheon Systems Company has been selected by The Boeing Company to develop the Advanced Targeting Forward Looking Infra-Red (ATFLIR) system for the US Navy's F/A-18 Hornet and Super Hornet. The selection followed a competition run by Boeing on behalf of the Navy to choose a contractor for the baseline system on the F/A-18E/F. The contract includes options for the first two years of low-rate initial production (24 units for the first and 33 for the second).

The Raytheon pod features targeting and navigation systems. The targeting FLIR uses the same third-generation mid-wave infra-red 640 × 480 staring focal plane array technology which Raytheon has incorporated in the US Marine Corps MV-22 Osprey. ATFLIR is being developed by Raytheon and its team-mates BAE Systems, Fairchild Controls and Advanced Conversion Technology. In addition to the targeting FLIR, ATFLIR includes a laser range-finder and target designator, laser spot tracker and navigation FLIR. BAE Systems provides the NAVFLIR as well as other important components. Flight tests indicate a considerable improvement in target detection and recognition range compared with first-generation systems.

Operational status
Low Rate Initial Production (LRIP) was authorised in March 1998, with first of 574 planned units delivered in May 2002. Entered US Navy service on F/A-18E/F in July 2002.

Contractor
Raytheon Systems Company.

UPDATED

Thales Optronics ATLIS II

Type
Ground attack integrated system (fixed-wing).

Development
In 1975, the then Thomson-CSF awarded a US$3.7 million contract to Martin Marietta Aerospace (now Lockheed Martin) for the development and test of the ATLIS pod. Thomson is responsible for manufacturing and marketing the pod. The French company CILAS provides the ITAY 139 laser illuminator/range-finder.

The prototype system, known as ATLIS I, was flight-tested during the late 1970s and the French Air Force decided to equip a number of Jaguar aircraft with ATLIS pods and cockpit viewing and control equipment.

Following completion of the ATLIS pod development, Thales and Lockheed Martin developed and flight-tested a new system, ATLIS II, for use with Jaguar and F-16 aircraft. ATLIS II is lighter than the original ATLIS and includes a computer to interface directly with a digital bus-bar.

Description
ATLIS II is a pod-mounted laser targeting system with an automatic TV tracker and CILAS ITAY 139 Nd:YAG laser designator/range-finder which includes tape recorder and interface electronics. The automatic dual-mode (visible and near infra-red) TV tracker and laser stabilisation system reduces pilot workload and permits tracking and designation by a single-seat aircraft during break and low- and medium-range standoff manoeuvres. As the system is pod mounted it may be installed on a wide variety of strike and close-support aircraft.

The stabilisation system is located in the front of the pod and provides a stabilised optical path for image data processing, laser designation/reception while also providing a mechanism for steering optical line of sight. The image data are reflected from a stabilised mirror into a fixed optical assembly which folds and focuses the image into the TV camera.

The output from the camera provides two optical fields of view to the display in the cockpit and to the automatic tracker, 6° for target acquisition and two others for identification. Laser designation energy is reflected to the dichroic portion of a combining glass located within the image path. The combining glass passes shorter wavelength image data but reflects the laser energy so that it leaves the combiner collinear with the scene data.

A steerable, stabilised mirror provides the optical line of sight and a pitch/yaw rate stabilised inertial platform provides rejection of high-frequency dynamic

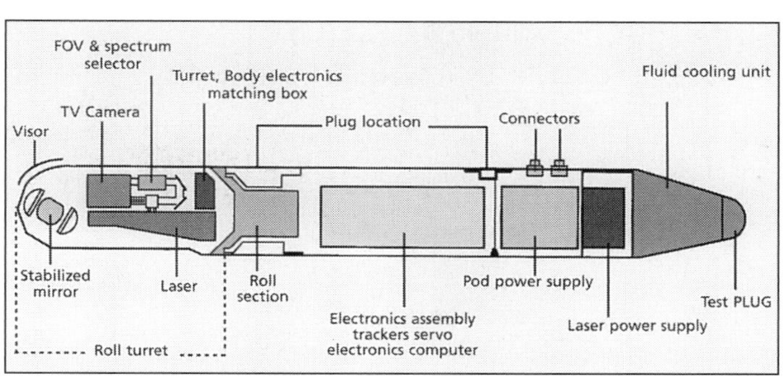

Schematic of the ATLIS II Pod 0005518

ATLIS II on the F-16 0005520

motions. A roll turret drive unit is used in conjunction with the pitch/yaw stabilisation system to provide line of sight steering. The dual-mode tracker provides both area correlation and point tracking. The area correlator mode is used to stabilise the scene and provide designation for area or low-contrast targets. The point tracking mode is used for stationary or moving point targets. The TV camera operates both in the near infra-red and visible spectrum. The image magnification and the selection of the spectral bandwidth provide the best contrast enhancement to allow greater target detection ranges.

A digital processor manages all the pod functions. It provides automatic testing, rate-aided tracking signals and fire-control computation. A closed-cycle environment control system ensures good pod performance under the most difficult conditions.

Operational status

ATLIS II has been in production for the French Air Force since 1984-85. It is in service on French Air Force Jaguars and on Mirage 2000, Upgraded Super-Etendard and F-16 aircraft in other countries. It has been delivered to several other air forces including Pakistan and Thailand.

ATLIS II was used by French Air Force Jaguars in Operation Desert Storm.

Specifications

Length: 2.54 m
Diameter: 315 mm
Weight: 180 kg
Field of regard: +5 to −160°
Accuracy: ±1 m at a firing range of 10 km
Magnification: ×20
TV tracker
Waveband: 0.5-0.7 μm and 0.7-0.9 μm
Laser
Wavelength: 1.06 μm
Type: Nd:YAG

Contractor

Thales Optronics.

VERIFIED

CLDP/TC with thermal camera on a Mirage 2000D 0005517

Thales Optronics CLDP

Type

Ground attack integrated system (fixed-wing).

Development

The Convertible Laser Designator Pod (CLDP) has been developed from Thales Optronics' ATLIS pod. Flight tests of the CLDP/TV began in mid-1986. Flight tests of the CLDP/TC version began in January 1988 using a Jaguar aircraft.

In September 1988, at the Landes flight test centre, a Jaguar successfully launched an AS 30 missile at a speed of 470 kt from an altitude of 213 m at a range of 8 km. The CLDP had acquired the target at a range of 13 km.

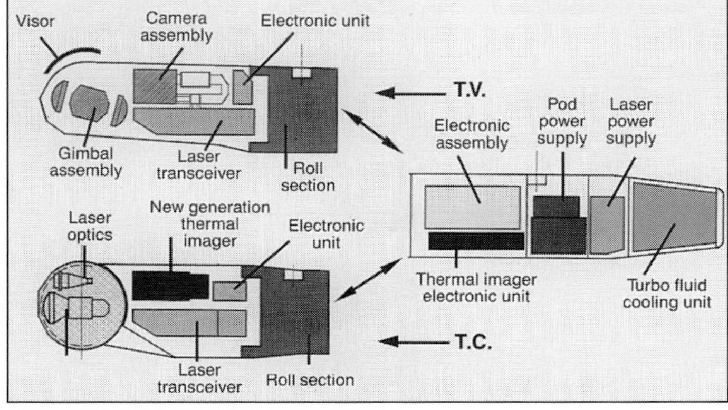

Schematic of the CLDP/TV and TC Pods 0005521

The Convertible Laser Designation Pod with thermal imager CLDP/TC

Description

In the CLDP the laser designator is supplemented by either a TV (CLDP/TV) or a thermal imager (CLDP/TC). The CLDP features a common body with laser transceiver, electronic assembly and environmental control system and two separate nose sections which can be changed in 2 hours.

The TV head features a TV camera, gimballed mirror and a roll-stabilisation device. It also has a Field Of View (FOV) selector and a visible or near-infra-red spectrum selector. This has four magnifications and corresponding fields of view.

The thermal imaging head features a gimballed optical head with a laser and thermal imager optics and roll-stabilisation device. There are four FOVs: 12/6° for navigation and 4/2° for target acquisition, reconnaissance and tracking. The thermal imager is based upon the SMT modules.

Operational status

In production. In service on French Air Force Mirage 2000D. CLDP has been ordered by Abu Dhabi for Mirage 2000 and Italy and Saudi Arabia for Tornado. It is being offered as part of the Sukhoi Su-22M5 and MiG 29 SMT upgrade.

Specifications

Weight: 290 kg
Length: 2.85 m
TV
Spectral band: 0.5-0.9 μm
Magnification: ×2.5, ×5, ×10, ×20
Fields of view: 0.75, 1.5, 3 and 6°
Thermal imager
Spectral band: 8-12 μm
Laser
Wavelength: 1.06 μm

Contractor

Thales Optonics.

VERIFIED

Thales Optronics Damocles multimode multifunction laser targeting pod

Type
Integrated system fixed-wing.

Description
Damocles is a multimode, multifunction targeting pod that incorporates a third-generation thermal imager. The thermal imager operates in the 3 to 5 μm waveband and uses a staring focal plane array. The pod also includes a laser spot tracker.

The modular pod is designed primarily for laser designation but can also be used for air-to-ground targeting, navigation, air-to-air identification and reconnaissance roles.

Operational status
Damocles is under development for the Mirage 2000-9 and Rafale aircraft. The system was due to enter flight test in mid-2000. Damocles is also marketed as a combined system with NAVFLIR when it occupies only a single weapon station. As marketed to the United Arab Emirates, this dual configuration is designated SHAHAR (Damocles) and NAHAR (NAVFLIR).

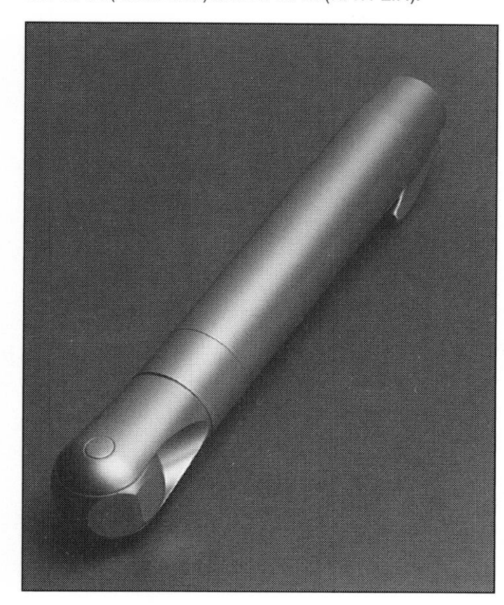

Mockup of the Damocles pod
0005516

Specifications
Thermal imager
Spectral band: 3-5 μm
Laser
Wavelength: 1.06-1.55 μm
Electronic magnification: ×2
Dimensions (l × d): 2.5 m × 370 mm
Weight: 250 kg

Contractor
Thales Optronics.

VERIFIED

Urals Optical and Mechanical Plant Sapsan

Description
The Sapsan pod is an integrated system designed to enable precision targeting of guided weapons from fast jet platforms. Standard components are as follows:
- thermal imager or thermal cueing channel
- laser designator
- laser range-finder
- FOV stabilisation
- electronics units
- thermal control system

Operational status
Not stated.

Specifications
Dimensions: 360 × 3,000 mm
Weight: 250 kg
Field of regard
  Elevation: +10 to –150°
  Azimuth: ±10°
  Roll: ±150°
Operating temperature range: –60 to + 50°C

Contractor
Production Association Urals Optical and Mechanical Plant (UOMZ).

GROUND ATTACK

INTEGRATED SYSTEMS - HELICOPTER

DRS Mast-Mounted Sight (MMS)

Type
Ground attack integrated system (helicopter).

Development
MMS was developed to meet the need for a stable image from the sighting system in the hostile environment of a helicopter mast mounting.

Description
MMS is a fully integrated stabilised electro-optical reconnaissance system capable of target acquisition, designation and handover. MMS consists of three subsystems; the turret, the control/display panel and the onboard electronics.

The turret consists of a post and a ball-shaped aerodynamic shroud, both made of graphite-epoxy composite. Patented Boeing technology, known as the 'soft mount', isolates the sensors and boresight from helicopter vibration and pitch, roll and yaw motion. The resulting stabilisation allows MMS to detect, recognise and designate targets beyond a threat weapon's range in day or night operation.

The sensor package comprises a CCD Television Camera (TV), a Thermal Imaging Sensor (TIS), a Laser Range-Finder/Designator (LRF/D) and an automatic Optical BoreSight tool (OBS).

MMS has a sophisticated video tracker which provides manual or automatic tracking of targets. There is also a digital scan converter which enhances the TIS image and provides ×2 and ×4 electronic zoom capability.

External to the turret is the Control/Display System (CDS) consisting of the master controller and is linked to the keyboard, display screen and handgrip. Additionally there is the system power supply (MCPS) and processor (MSP).

Operational status
MMS is in production with over 400 systems delivered to the US Army for the OH-58D Kiowa helicopter. MMS has been trialled on the Boeing 530 Defender helicopter. MMS was used extensively in Operation Desert Storm and was responsible for the first tank kill designation. Taiwan (Republic of China) has purchased 51 MMS systems for OH-58D helicopters.

In October 1996, the US Army ordered a further 20 MMS for the OH-58D Kiowa Warrior helicopter, which were to be delivered by February 1998. More than 500 units in total have been purchased by the US and international customers.

Specifications
System weight: 113.4 kg
Turret
Diameter: 64.77 cm
Weight: 72.57 kg
Stabilisation: 2-axis, <20 mrad jitter
Azimuth: ±190°
Elevation: ±30°
CCD television camera
Type: low-light silicon CCD
Spectral range: 0.65-0.9 μm
Field of view: 2° NFOV; 8° WFOV
Thermal imaging sensor
Detector: 120 element HgCdTe scanning
Spectral band: 8-12 μm
Aperture: 16.76 cm
Field of view: 3° NFOV; 10° WFOV

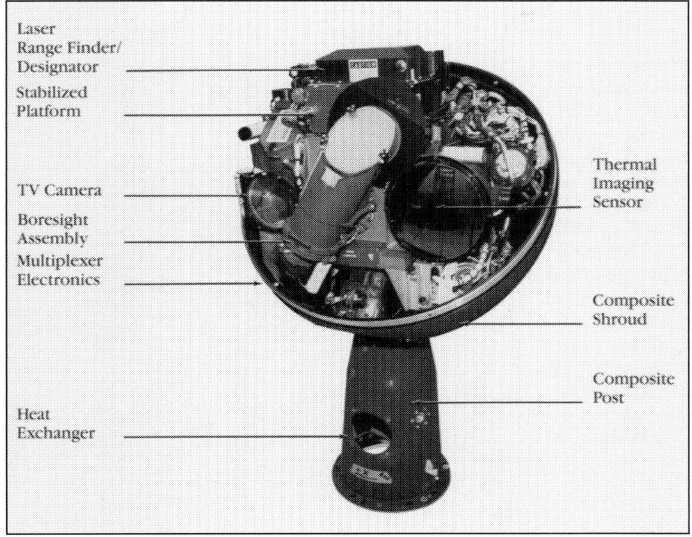

The MMS with its various sensor subsystems

The MMS installed on the OH-58D Kiowa helicopter

Laser rangefinder/designator
Type: flashlamp Nd:YAG
Wavelength: 1.06 μm

Contractor
DRS Tecchnologies, Electro-Optical Systems Group.

Kollsman Night Targeting System (NTS)

Type
Ground attack integrated system (helicopter).

Description
The Night Targeting System (NTS) was developed by Kollsman and the Tamam division of Israel Aircraft Industries under a joint US/Israel agreement and is manufactured by both companies. It is an airborne electro-optical fire-control system designed to provide a night fighting capability for the AH-1 Cobra attack helicopter. It provides the capability to detect, acquire, track, designate and attack tactical targets at night and in limited visibility or adverse weather conditions.

The NTS consists of a modified M-65 telescopic sight unit, a laser designator/range-finder, a CCD TV sensor and a FLIR sensor to provide autonomous delivery of TOW and Hellfire missiles. A video cassette recorder is also included to provide intelligence gathering and training capability. An automatic target tracker is

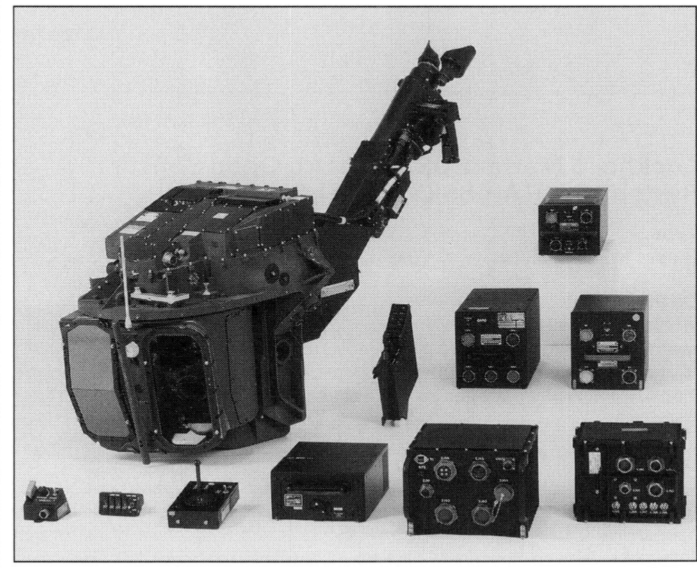

The Kollsman night targeting system equips variants of the Bell AH-1 Cobra helicopter

integrated to improve tracking accuracy for weapon delivery. A TOW-2 thermal tracker is under development.

Operational status
In production for variants of the Bell AH-1 Cobra helicopter. In service with the US Marine Corps.

Contractor
Kollsman Inc.

UPDATED

Lockheed Martin Arrowhead Modernised Target Acquisition Designation Sight/Pilot Night Vision Sensor (M-TADS/M-PNVS)

Type
Ground attack integrated system (helicopter).

Description
Targeting and Navigation System

Arrowhead is the advanced targeting and navigation system for the AH-64A/D Apache helicopter due to replace the existing TADS/PNVS, and selected in 2000. Combat-proven in design, leading-edge in technology, and greatly improved in reliability and performance, the electro-optic package will provide longer range vision and operation in adverse conditions. The manufacturer states that this major system upgrade to its AH-64 electro-optics system will improve reliability by more than 150 per cent and increase performance more than 150 per cent through the addition of Comanche forward looking infra-red (FLIR) technologies. These enhancements will reduce operation and support costs by 50 per cent. Arrowhead's modular architecture allows it to be retrofitted in the field.

Features
- Combat-proven design, upgraded with state-of-the-art reliability and performance
- Fly and fight in day, night, and adverse weather with Comanche FLIR performance
- Target cueing from Longbow fire-control radar or radar frequency interferometer for rapid, eyes-on target identification
- Digital video enhances recording capability and facilitates still-frame video imagery transmission
- Improved laser incorporates Comanche technology tactical designation, and eye-safe ranging laser with improved reliability and diagnostics
- Capability for embedded MILES training or air-to-air targeting laser on turret.

Pilotage

The upgraded Arrowhead Pilotage System (APS) provides an advanced FLIR capability with image intensification and provisions for image fusion. Coupled with the improved IHADSS, the pilot will have greatly enhanced Nap-Of-the-Earth (NOE) capability with built-in growth for use with the Comanche helmet display.

The lower turret mounts an Arrowhead Targeting System (ATS) FLIR capable of a 40 per cent increase in targeting range and an improved Charge-Coupled Device (CCD) camera for day TV viewing. Improvements to the laser adapt Comanche technology for Apache. Direct-view optics are eliminated, increasing survivable space in the crew station and enhancing the display resolution with a large flat-panel targeting display.

Contractor
Lockheed Martin Missiles and Fire Control.

UPDATED

Lockheed Martin EOSS (Electro-Optic Sensor System) for RAH-66 Comanche

Type
Ground attack integrated system (helicopter).

Development
In September 1991 Martin Marietta (now Lockheed Martin) won a US$300 million contract to build the Electro-Optical Sensor System (EOSS) for the US Army's RAH-66 Comanche advanced reconnaissance attack helicopter. Five prototype models were originally ordered but this was later reduced to two both of which have flown.

Description
The EOSS uses advanced focal plane array and digital image processing technologies coupled to the pilot's line of sight to provide targeting and navigational capabilities. The EOSS sensor suite is a component of the overall Comanche Mission Equipment package. It comprises a Solid-State TeleVision (SSTV), a two-colour laser designator/range-finder (from Elop of Israel), and two second-generation Forward-Looking Infra-Red (FLIR) sensors: the Electro-Optical Target Acquisition/Designation System (EOTADS), and the Night Vision Pilotage System (NVPS).

The Lockheed Martin Comanche nose-mounted Electro-Optical Sensor System (EOSS) mockup

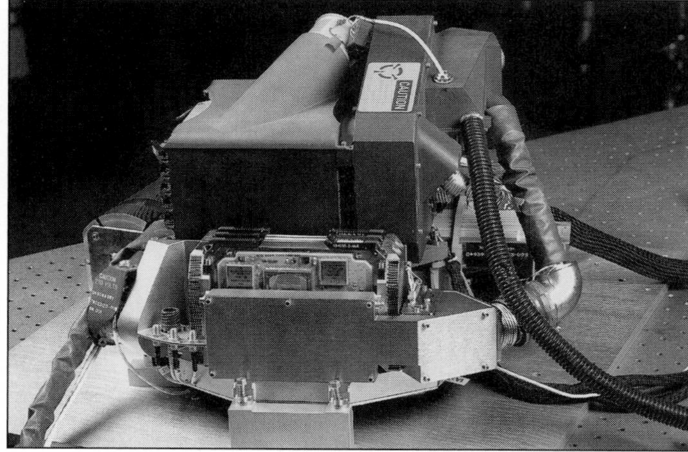

The Comanche EOSS hardware package 0058711

The EOTADS allows for long-range search and acquisition, provides laser range and weapon designation and acts as back-up for the main pilotage system. The NVPS allows the aircraft to fly at night and in adverse weather, using advanced helmet-mounted display technology. NVPS was first flown in the third quarter of 2002.

The system uses a wide area scan technique to search for airborne and ground targets. Target imagery is prioritised according to threat potential and stored in a computer memory, which permits the crew to remask while deciding combat tactics. This review can be performed manually or automatically with the Westinghouse-developed Aided Target Detection/Classification (ATD/C) system.

Operational status
Under development. In December 1997 the first flight hardware produced a laboratory image during systems integration. Five 'early operational aircraft' systems will be built and flown. Initial operating capability for the Comanche helicopter is scheduled for 2006 or later.

Specifications
EOSS System
Weight: 204.12 kg
Aircraft interface
Data: Fibre optic databus
Mechanical: Nose mount, aircraft-supplied cooling air
Power: 270 V DC, 3.4 kW (average), MIL-STD-704E
Operating envelope: –46 to +52°C; sea level to 3,000 m
NVPS Sensor
FLIR field of vision
  Azimuth: ±120°
  Elevation: ±50°
FLIR field of view: 30 × 52°
MTBF: 270 flight h
EOTADS Sensor
Field of regard
  Azimuth: ±120°
  Elevation: +25 to –50°
Fields of view
  SSTV: 0.8 × 0.6°, 2.0 × 1.5°, 8 × 6°
  FLIR: 2.0 × 1.5°, 8 × 6°
  FLIR back-up pilotage: 30 × 40°

Wavelengths
SSTV: 0.65-1.00 μm
FLIR: 8-12 μm
Laser: 1.06 μm ranging and designation, 1.54 μm ranging only
MTBF: 125 flight h

Contractor
Lockheed Martin Electronics & Missiles.

UPDATED

Lockheed Martin TADS/PNVS

Type
Ground attack integrated system (helicopter).

Development
The Target Acquisition Designation Sight and the Pilot Night Vision Sensor (TADS/PNVS) were developed during the late 1970s to provide the Boeing (formerly McDonnell Douglas) AH-64A Apache attack helicopter with the capability of operating during the day, night and in bad weather. The evaluation programme was completed during early 1980 and in early 1981 the first production orders were placed.

Description
The TADS/PNVS are two independently functioning systems which were designed for use with US Army AH-64 Apache attack helicopters. TADS provides the gunner with a search, detection and recognition capability with direct-view optics, television or Forward-Looking Infra-Red (FLIR) sighting systems. These may be used singly, or in combination, depending on tactical, weather or visibility conditions. Once acquired, targets can be tracked manually or automatically for autonomous attack with 30 mm guns, rockets or Hellfire missiles. The Northrop Grumman (was litton) laser may also be used to designate targets for remote attack by other helicopters, or by artillery units firing laser-guided munitions. The TADS or AN/ASQ-170 equipment consists of a rotating turret on the nose of the helicopter. It houses the sensor subsystems, an Optical Relay Tube (ORT) at the co-pilot/gunner's position, three electronic units in the avionics bay and cockpit-mounted controls and displays.

PNVS or AN/AAQ-11 is used by the pilot for night navigation. It consists of a FLIR sensor system packaged in a rotating turret mounted above the TADS, an electronic unit located in the avionics bay and the pilot's display and controls. The PNVS is slaved to the pilot's helmet display line of sight and provides flight information symbology which allows the helicopter to be flown at low level, contour or nap of the earth at altitudes likely to avoid, or at least minimise, enemy detection.

The Laser Range-Finder/Designator (LRF/D), provided by Litton's Laser Systems Division is integrated with the TADS and provides attack capability in day and night conditions. It consists of two line-replaceable units, a laser transceiver and an electronics unit.

TADS is designed to provide a back-up PNVS capability for the pilot in the event of the latter system failing. The pilot or the co-pilot/gunner can view on their own displays the video output from either TADS or PNVS, so increasing the possibility of mission success. While designed for combat helicopters flying nap-of-the-earth missions, PNVS may also be used as a stand-alone unit for tactical transport and cargo helicopters.

Operational status
The system became operational in 1986 and is in production. Lockheed Martin has delivered more than 900 systems to the US Army. Production contracts to date exceed US$2 billion. Further Foreign Military Sales awards are anticipated.

TADS/PNVS system on AH-64A Apache advanced attack helicopter

The system is in operation with or on order by Egypt, Greece, Israel, Netherlands, Saudi Arabia, the United Arab Emirates, the UK and the US Army. Additional orders from the US Army and allies are anticipated.

Specifications
TADS
Field of regard
Bearing: ±120°
Elevation: +30 to –60°
Fields of view
TV camera
Underscan: 0.45°
Narrow: 0.9°
Wide: 4.0°
Direct-view optics
Narrow: 3.5°
Wide: 18°
FLIR
Underscan: 1.6°
Narrow: 3.1°
Medium: 10.2°
Wide: 50°
PNVS
Field of regard
Bearing: ±90°
Elevation: +20 to –45°
Field of view: 30 × 40°

Contractor
Lockheed Martin Missiles and Fire Control.

UPDATED

MBDA TOW thermal imaging roof sight

Type
Ground attack integrated system (helicopter).

Description
MBDA (formerly Matra BAe Dynamics) has carried out a mid-life improvement programme to the optical TOW roof sight mounted on British Army Lynx helicopters. A thermal imaging system, operating in the far infra-red waveband has been incorporated to enable armoured targets to be engaged at night or in bad visibility. The system will be incorporated in all new TOW sights.

BAE Systems co-operated with the Raytheon Systems Company, manufacturer of the TOW missile and the basic LAAT sighting system, in the development of the improved sight. The thermal imaging subsystem was developed by BAE Systems Avionics. Because of the limited space available in the sight, it has been necessary to develop a special system which is significantly more compact. A full electro-optic counter-countermeasures capability has also been developed for the system.

Operational status
Operational on British Army Lynx helicopters.

Contractor
MBDA (UK).

VERIFIED

The TOW thermal imaging roof sight on a British Army Lynx helicopter 0005528

Raytheon C-NITE (Cobra-NITE)

Type
Ground attack integrated system (helicopter).

Development
Development of this system began in 1979 when Hughes (now Raytheon Systems Company) produced FACTS, a FLIR Augmented Cobra TOW Sight. It was designed to give Cobra helicopters an enhanced capability to launch, acquire and track TOW missiles in darkness, bad weather and battlefield obscurants such as smoke. The system also has an electro-optical counter-countermeasures ability and in-flight boresight capability. In the Autumn of 1984 Raytheon was awarded a contract for development of the system as Cobra-NITE (C-NITE), and in the Summer of 1987 received a US$67 million contract for production systems.

Description
The C-NITE equipment incorporates modifications to the basic M-65 LAAT which give the aircraft a FLIR capability. It is also equipped with a laser range-finder and laser training device. In addition, the sight enables the pilot to fire all versions of the TOW missile. C-NITE is designed to upgrade the two versions of the TOW fire-control system now in use, the M-65 with the standard optical sight and the M-65L with the Laser Augmented Airborne TOW (LAAT) range-finder. The system features the modified M-65L LAAT sight, a control panel, power supply and missile tracker for the FLIR.

Operational status
Raytheon has produced 100 systems for the US Army and first production deliveries were completed in December 1990. The C-NITE system is fitted to the US Army's AH-1F Cobra anti-armour helicopters.

Japanese Army Cobras have received C-NITE which is co-produced under licence by the NEC.

In November 1993, Raytheon received a US$24 million contract for 21 C-NITE systems for AH-1F Cobras of the National Guard.

In December 1994, Raytheon was awarded a US$31.9 million contract for 26 C-NITE systems, spare parts, support equipment, technical assistance and training.

In February 1995, Raytheon was awarded a US$12.3 million contract for six C-NITE and eight LAAT systems for the Bahrain Defence Force, completed in 1996.

Contractor
Raytheon Systems Company.

VERIFIED

The US Army Cobra attack helicopter is equipped with the Cobra-NITE night targeting system

Raytheon M-65L LAAT

Type
Ground attack integrated system (helicopter).

Description
An improvement to the M-65 optical sight used in the AH-1S Cobra attack helicopters is the M-65L Laser Augmented Airborne TOW (LAAT) system. It adds a laser range-finder to the basic M-65 TOW missile system sight to provide range information for the helicopter fire-control computer to direct the 20 mm cannon and 2.75 in unguided rockets. It also gives the gunner target ranges to engage the target with the TOW missile.

The laser transmitter is between the gimbal assembly and the housing plate of the turret. This unit is 13 × 13 × 4 cm in size and has a repetition rate of 4 pps for 5 seconds on and 5 seconds off. A multipulse laser range-finder is necessary because of the helicopter's forward motion during target sighting, requiring several range measurements to update the fire-control computer.

Operational status
In production since 1979. More than 2,000 M-65 and LAAT sighting systems have been delivered for use in the US Army's AH-1 Cobras since 1975 and US Marine Corps' AH-1T/AH-1W. Eight LAAT systems have been sold to the Bahrain Defence Force.

Contractor
Raytheon Systems Company.

VERIFIED

Saab Bofors Helios HeliTOW

Type
Helicopter observation and sighting system.

Description
The Helios HeliTOW system is a helicopter-mounted anit-tank system, or, if ordered without weapons provision, a scout observation system. The system includes direct view optics, TV, FLIR and laser range-finder/designator for battlefield surveillance and targeting for self-employed or co-operatively employed weapons.

Helios System
The Helios System is an observation system that can be configured as a day-only system or with FLIR added, a day/night system. Either configuration can be furnished with a laser range-finder or laser range-finder/designator. The system can stand alone or can interface with the aircraft via MIL-STD-1553B databus.

In the day only configuration, direct view optics provide the operator with a sharp, close-up, true colour image for target acquisition and following. With the FLIR added, Helios can operate at night and in adverse visibility.

The Helios HeliTOW system uses rate-integrating gyros for stabilisation. Both roof-mounted and nose-mounted configurations are available, making the system

Eurocopter BO 105 with roof-mounted HeliTOW and carrying two TOW and two Hellfire missiles 0005526

Agusta A 129 helicopter with nose-mounted HeliTOW and carrying eight TOW missiles, four each side 0005527

Agusta 109 with roof-mounted Helios and HeliTOW carrying eight TOW missiles
NEW/0125220

compatible with almost any helicopter. High-resolution optics provide both wide and narrow fields of view.

HeliTOW and HeliTOW/Hellfire
Modular design allows the user to add a variety of weapons capabilites to either the day or day/night versions of the Helios. Individual hardware and software modules interface Helios to TOW or Hellfire missiles and to helicopter guns and rockets. Interface provisions are also available to Head-Up Displays (HUD) and helmet-mounted displays.

For anti-armour, any variant of TOW or Hellfire missile can be employed, either all of the same type or in a mixed load configuration. The system recognises which missile and missile type is installed in each location so that the operator can make the most effective choice for each engagement. The system automatically implements the software controlled logic that provides the most accurate guidance for the missile selected (Basic TOW, I-TOW, TOW 2, TOW 2A or Hellfire). When Hellfire is the chosen weapon the system can operate autonomously, or as the launch platform in co-operation with some other air or ground target designator, or co-operatively as the target designator for some other launch platform.

The missile launchers can be quickly mounted and dismounted without the need to reboresight, to accommodate changes in mission requirements, such as alternating between scout and attack roles.

Operational status
In production and in service on five different helicopters (Eurocopter BO 105, Bell B406Cs Combat Scout, Agusta A 129 Mangusta, Eurocopter AS550 Fennec and the Agusta A 109) with the armies of Belgium, Denmark, Italy, Saudi Arabia, Singapore and Sweden. Over 200 systems have been produced.

Contractor
Saab Bofors Dynamics.

SAGEM APX M334 series

Type
Ground attack integrated system (helicopter).

Description
The designation APX M334 covers a family of roof-mounted day sights for helicopters. All have magnifications of ×3 and ×10. The M 334 series is designed to provide an operational system able to detect a target at a range of about 4.5 n miles (10 km) and recognise a tank at 2.3 n miles (5 km).

Several versions of the sight have been developed for observation, anti-armour missile firing or target designation for artillery.

Operational status
About 1,500 are in service with some 30 countries. The sight has been fitted on such helicopters as the Eurocopter Gazelle SA 341 and 342 and SA 321 Super Frelon; Westland Lynx; Bell 204/205 Iroquois, Bell 206/OH-58 and 212; Sikorsky SH-3D, Hughes 500 and 530, Agusta A 109, Eurocopter BO 105 and BK117.

The SFIM APX sight

Contractor
SAGEM SA, Aerospace & Defence Division.

UPDATED

SAGEM OSIRIS mast-mounted sight

Type
Integrated system helicopter.

Description
SAGEM is producing the mast-mounted OSIRIS sight as part of the LR-TRIGAT anti-armour missile system on the Tiger attack helicopter.

OSIRIS is designed to detect and acquire targets before firing long-range missiles out to the maximum range day or night and in poor atmospheric conditions. The sight is designed so that several targets can be designated to the missile seeker head via the firing station of the weapon system. It may also be used to fire the HOT anti-tank missile.

Although designed for the Franco-German Tiger helicopter, the OSIRIS sight can fit other types of helicopter able to accommodate a mast-mounted sight and also certain land vehicles.

The gyrostabilised OSIRIS sight head includes a Condor 1 thermal imager, low light TV camera and an optional laser range-finder. The system operates in three wavebands: 0.5 to 0.7 μm, 0.7 to 1.0 μm and 8 to 12 μm. The system is completely passive, allowing the helicopter to remain concealed during the reconnaissance and target-designation phases.

Operational status
OSIRIS passed its prototype acceptance trials in 1992. A small batch was produced during development to allow the operational test and evaluation of

Tiger helicopter with OSIRIS mast-mounted sight 0085466

The OSIRIS sight 0085465

LR-TRIGAT on Tiger in 2002. In production for French Tigre HAC and German Tiger UHT.

Contractor
SAGEM SA, Aerospace & Defence Division.

UPDATED

SAGEM Strix/Nightowl day/night sights

Type
Ground attack integrated system (helicopter).

Description
The Strix is a targeting sight selected by the French MoD as the main sight of the TIGRE/HAP helicopter. It is fitted with a second-generation thermal imager (SAGEM IRIS) direct view channel, TV channel, high-repetition rate eye-safe laser range-finder and micromonitor used to display the thermal and TV images and firing symbols into the sight eyepiece.

The weapon system uses the Strix sight as a data reference to determine the line of sight, sighting precision being ensured by automatic target tracking on both TV and thermal channels. The sight is designed to maintain the stabilisation and precision of the line of sight through the full operating range of the helicopter. Composite materials are used for the structure to obtain the best trade-off between the mass of the integrated sensors and its own mass.

Nightowl, a variant of the Strix sight is fitted as the nose-mounted sight of South Africa's Rooivalk attack helicopter.

Operational status
In production.

Strix roof-mounted sight on the Tiger-HAP helicopter (Eurocopter) 0129309

A Strix sight

Contractor
SAGEM SA, Aerospace & Defence Division.

UPDATED

SAGEM Viviane day/night sight

Type
Ground attack integrated system (helicopter).

Description
SFIM (now part of the SAGEM Group) has designed the Viviane day and night sight based on previous developments of electro-optic stabilised platforms flown as early as 1976. Viviane is dedicated to firing the HOT missile by day and by night. It is marketed by Euromissile as part of the HOT missile package.

Viviane's electro-optic sensors include a thermal imager and a laser range-finder.

Operational status
Viviane is in service with the French Army, equipping SA 342M Gazelle helicopters.

Contractor
SAGEM SA, Aerospace & Defence Division.

UPDATED

The SAGEM Viviane day/night sight is fitted on this Eurocopter SA 342M Gazelle

Tamam NTS/NTS-A Night Targeting System

Type
Ground attack integrated system (helicopter).

Description
Since 1982, Tamam has been involved in upgrade programmes to provide Cobra attack helicopters with laser ranging, designation and night attack capabilities. Developed by Tamam under sole-source contract from the US Marine Corps and Israeli Air Force, NTS has accumulated considerable operational use.

Night Targeting System (NTS-A) is a new version of the NTS, with enhanced performance suitable for a wide range of different attack helicopters (including the AH-1 W/S/P Cobra). In NTS-A, the original optical tube is removed, freeing space in the helicopter cockpit, for the use of flat-panel displays. This allows the system to be compatible with different types of attack helicopter and with modern glass cockpit architectures, such as the US Marine Corps AH-I W 4BW upgrade programme.

NTS-A mounted on Cobra (AH1-S) 0023385

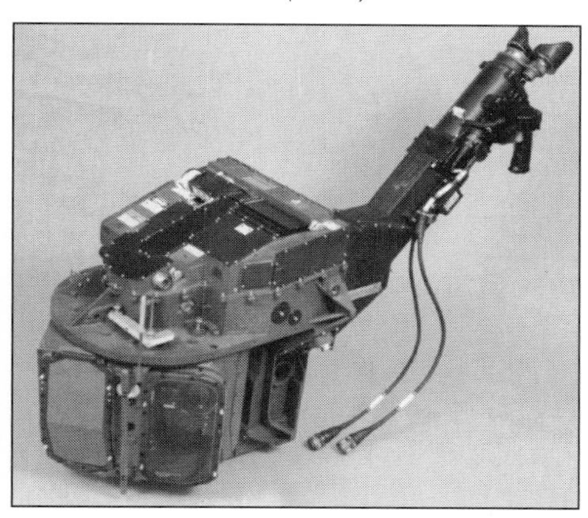

The Night Targeting System

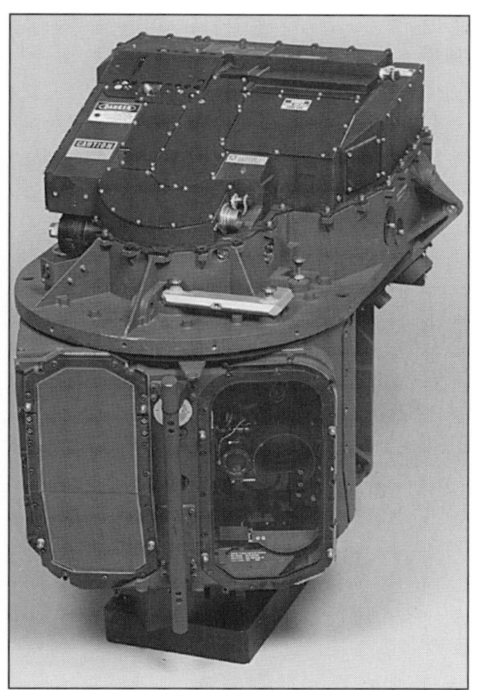

NTS-A, an enhanced version of the Night Targeting System for a range of attack helicopters 0005525

NTS-A has the following capabilities:
- Installation flexibility with TOW I, TOW II, Hellfire, and other weapons
- Target acquisition during day, night and limited visibility conditions
- Laser Designator and Range-finder System (LDRS) for laser-guided weapons and for measuring target ranges
- TV Tracker (TVI), providing target auto-tracking during day and night
- Guidance for all types of TOW missiles
- Fully-automatic in-flight boresight capability
- Navigation and tactical data display
- Display of NTS-A standard video signal (both the FLIR and TV camera pictures) on a multifunction display in both the gunner and pilot cockpits
- Built-in growth potential for future integration with other systems onboard, via (2) MIL-STD-1553B databusses.

Operational status
In production. Over 300 systems have been ordered.

Specifications
FLIR - 2nd generation

FLIR Fields of View (FOV) and magnification (relative to display viewing angle of 40 × 25°)

		FOV magnification
Wide FOV (h × v)	18 × 24°	×2
Medium FOV (h × v)	30.0 × 40°	an option for pilotage (×1)
	5.2 × 3.9°	×9
Narrow FOV (h × v)	1.47 × 1.1°	×32.1
Zoom NFOV (h × v)	0.73 × 0.55°	×64.3

Automatic TV Tracker (TVT) - with prediction, adjust and offset modes

Field of regard
Azimuth: 90° to the left, 95° to the right
Elevation: + 30 to –60°
Turret slew rate:
(low-angular velocity) 2°/s
(high-angular velocity) 90°/s
(acceleration) 600/s^2
TVC features
frame transfer Charged Couple Device (CCD), ½ in
number of pixels: 780 × 576
Missile interfaces available
Hellfire missile; Rafael NT-D missile and all types of TOW missiles (TOW, TOW-i, TOW 2A, TOW 2B)
Weight: 129 kg (excluding aircraft installation kit)

Power
average power 400 W
max power - during laser operation 650 W

Operational status
In production. Over 300 systems have been ordered.

Contractor
Israel Aircraft Industries, Electronics Group, Tamam Division.

VERIFIED

US Army AESOP

Type
Ground attack integrated system (helicopter).

Description
The Raytheon built Airborne Electro-optical Special Operations Payload (AESOP) night vision system is designed to detect and identify targets, then pinpoint them

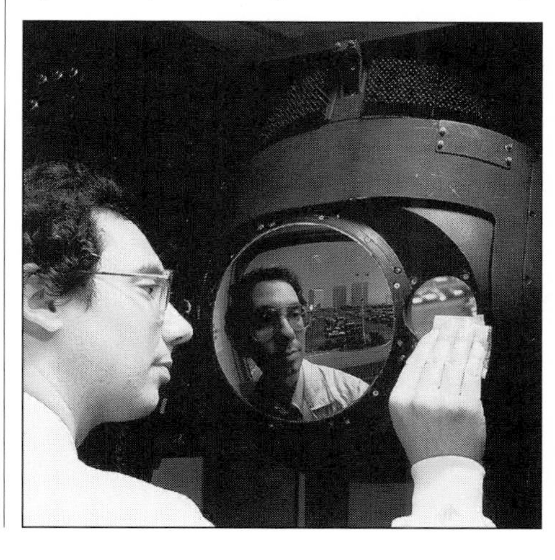

AN/AAQ-16D Airborne Electro-optical Special Operations Payload (AESOP)

with a laser designator. The system, designated AN/AAQ-16D, is intended for US Army Special Operations Force helicopters such as the MH-60L Black Hawk and the AH-6J Little Bird. It will also be installed in fixed-wing aircraft.

The system has a three fields of view infra-red telescope and a laser target designator/range-finder in a lightweight turret. The system is based upon Hi-Mag, an upgraded AN/AAQ-16 in which is provided a telescope with wide, medium and narrow fields of view, an enhanced auto-tracker as well as improved stabilisation. The system is currently being used to demonstrate the ability to detect, identify and track targets and to direct Hellfire missiles.

Operational status

In service.

Specifications

Weight (turret): <41 kg
Telescope magnifications: ×1, ×6, ×16

Contractor

US Army Night Vision and Electronic Sensors Directorate.

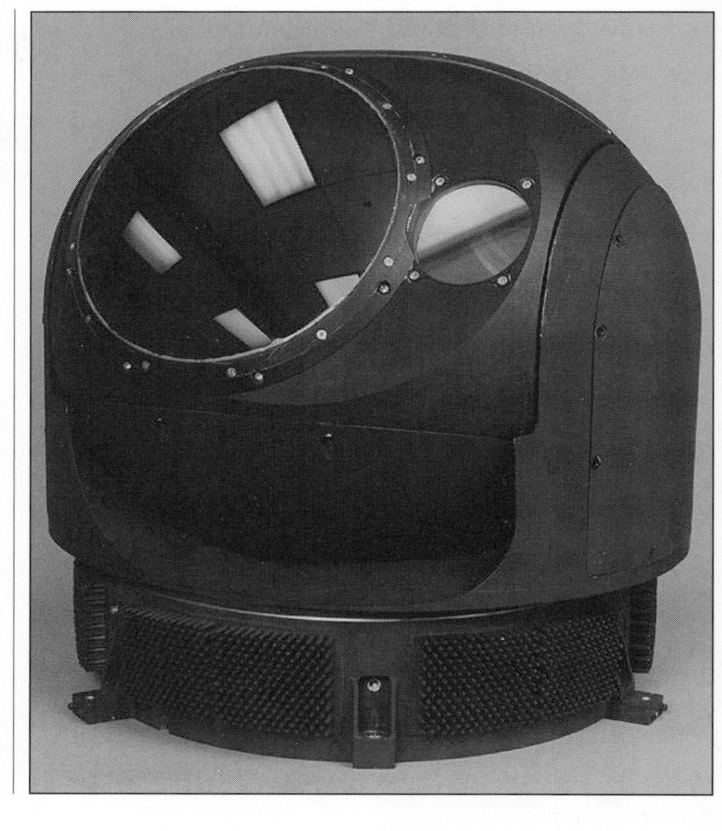

FLIR turret for Airborne Electro-optical Special Operations Payload (AESOP)

GROUND ATTACK

TARGETING SIGHTS

BAE Systems AN/AAQ-23

Type
Ground attack targeting sight.

Description
The AN/AAQ-23 Electro-Optical/Infra-red Viewing System (EOIVS) provides high-performance day or night target acquisition and navigation imagery. AN/AAQ-23 is designed to improve the reliability and maintainability of the navigation systems for the B-52 fleet. The system could also be fitted to the B-IB. A derivative of the sensor is available for other military and commercial aircraft.

The system consists of a mid-wave staring array FLIR sensor, sensor support electronics and video post-processor. Images are displayed in the cockpit on an 875-line display with TV compatibility. The AN/AAQ-23 is based on an advanced development staring infra-red focal plane that uses a platinum silicide substrate. The focal plane array, measuring 640 × 480 pixels, has two fields of view.

Operational status
Fleet-wide installation complete.

Specifications
Detector: PtSi staring FPA
Pixels: 640 × 480
Weight: 43 kg without gimbal

Contractor
BAE Systems, North America, Reconnaissance and Surveillance Systems.

UPDATED

The AN/AAQ-23 electro-optical infra-red viewing system 0079768

BAE Systems Atlantic Podded FLIR system

Type
Ground attack targeting sight.

Description
The Atlantic Navigation/Attack system incorporates the BAE Systems AN/AAR-51 modular FLIR within a low-drag supersonic pod. The system provides a high-resolution TV image overlaid on the pilot's head-up display. Automatic image optimisation enables low-level attack missions to be conducted through the night and in adverse weather. Automatic target detection options include thermal character cueing and/or a laser spot tracker.

The Atlantic system has been integrated and flown on a wide variety of ground attack aircraft including F-16, Mirage, A4, Jaguar and Tornado. The pod can be interfaced to a standard pylon or fitted to an engine nacelle's hard points as on the F-16. Control functions can be via the 1553 interface or discrete hard-wired.

Atlantic pod fitted to the nacelle of an F-16 0005529

Operational status
Atlantic is a fully developed system in service in the UK and overseas and is currently being offered for the F-16 mid-life update programme. The FLIR fitted to the pod is also in service as an internal fit on Harrier GR.Mk7/Mk10 and AV-8B aircraft. The Atlantic pod has also been selected for the RNLAF F-16 Mid-Life Update for the night attack role.

Specifications
Fields of view (selectable for aircraft HUD)
F-16/F-5: 28 × 21°
Mirage/Tornado: 25 × 16°
A4/Jaguar: 20 × 13°
Resolution CCIR system 1 (h × v): 790 × 512 pixels
Dimensions (l × diameter): 2,410 × 254 mm
Weight: 85 kg excl pylon adaptor
Power supply: 28 V DC 300 W, 115 V 400 Hz 180 W
Flight envelope: M1.6 at 72,000 ft 10.6 *g*
Temperature range: –40 to +70°C
Video output: 525-line, 60 Hz RS-170 or 625-line, 50 Hz CCIR

Contractor
BAE Systems, Avionics, Sensors and Communications Systems.

VERIFIED

BAE Systems Laser Ranger and Marked Target Seeker (LRMTS)

Type
Ground attack targeting sight.

Description
The Laser Ranger and Marked Target Seeker (LRMTS) is a dual-purpose unit which can be used as a self-contained laser ranger or as a target seeker with simultaneous range-finding. In the target seeking role it can be used to detect and attack any target designated by ground troops with a compatible laser, enhancing the effectiveness of battlefield close air support.

The LRMTS has a Nd:YAG laser mounted in a stabilised cage, which allows beam-pointing and stabilisation against aircraft movement. The seeker can detect marked targets outside the head movements' limits. It operates at a relatively high pulse repetition frequency, thus allowing continuous updating of range information during ground attack.

In a typical operational mode, a Forward Air Controller (FAC) with pulsed laser target designation equipment directs the aircraft to a location within laser detection range before switching on the ground marker equipment. Radio communications between the FAC and aircraft crew are minimised and positive identification of even small, hidden or camouflaged targets is assured.

Once the LRMTS has detected the laser energy reflected from the target, it provides steering commands to the pilot on the head-up display. Ranging data is also shown and fed directly into weapon aiming computations for the accurate and automatic release of weapons.

The LRMTS is said to be particularly effective during operations at the grazing angles used in low-level ground attack.

BAE Systems Laser Ranger and Marked Target Seeker (LRMTS)

BAE Systems Laser Ranger and Marked Target Seeker (LRMTS) on Tornado

Associated with the LRMTS head is an electronics unit which contains power supplies and ranging and seeking processing. The laser needs a transparent window and, in the Jaguar, is mounted behind a chisel-shaped nose, with two sloping panels. For the Harrier, where there is more chance of debris accumulation during VTOL operations, the optics are protected by retractable eyelid shutters.

A specially designed installation for the Royal Air Force Tornado IDS has incorporated the LRMTS into an underbelly blister.

An eye-safe Raman-shifted version of the LRMTS has been developed and successful trials have been completed. Existing LRMTS systems can be retrofitted with this modification which will significantly reduce safety restrictions for training.

Operational status

In service. Over 800 units have now been delivered for Harrier, Jaguar and Tornado aircraft. In service with the RAF and four overseas air forces. Production is still in progress for Tornado and Jaguar aircraft.

Specifications

Field of regard
Elevation: +3 to –20°
Azimuth: ±12°
Roll: ±90°
Dimensions
LRMTS head: 300 × 269 × 607 mm
Electronics unit: 330 × 127 × 432 mm
Weight
LRMTS head: 21.5 kg
Electronics unit: 14.5 kg
Power supply: 200 V AC, 400 Hz, 3 phase, 700 VA
28 V DC, 1 A
Spot Tracker
Detector type: silicon quadrant diode
Wavelength: 1.06 μm
Range-finder (Standard)
Wavelength: 1.06 μm
Max range: 10 km
Accuracy: ±6 m or 0.2% range whichever is greater
Range-finder (Eye-safe)
Wavelength: 1.54 μm
Max range: 10 km
Accuracy: ±6 m or 0.2% range whichever is greater

Contractor

BAE Systems, Avionics, Sensors and Communication Systems, Edinburgh.

VERIFIED

FLIR Systems BRITE Star™

Type

Ground attack integrated system (helicopter).

Description

The BRITE Star is a cost-effective, military-qualified, multisensor laser designation system that incorporates an advanced third-generation thermal imager, a TV camera and a laser designator/range-finder. For maximum system availability and accuracy BRITE Star incorporates an inflight boresighting capability. Derived from the well-proven SAFIRE/Star SAFIRE (AN/AAQ-21/22) family, the system is fully (backwards) compatible with all mountings of the SAFIRE family.

Options include turret invertability, an autotracker, a target accumulator, a laser spot tracker, a MIL-STD-1553B interface, an RS-232/422 interface, an ARINC interface, digital video output, and night-vision goggle compatibility.

Operational status

In service with US Marine Corps Bell UH-1N helicopters from 2001.

Specifications

Gimbal specifications: 4-axis stabilised
Turret size: 22.86 cm Diam × 30.48 cm H
Turret weight: 51.2 kg
Stabilisation: 4-axis
Azimuth: 360° continuous
Elevation: +32 to –100°
Control: HCU, serial digital
Environmental: MIL-STD-810E and MIL-STD-461D Class A1B

Thermal Imager
Detector: InSb FPA 3-5 μm response with quadrature microscan
Pixel density: 320 × 240
Sensor resolution: 640 × 480

Fields of View
Narrow: 0.8° × 0.6°
Medium: 3.4° × 2.6°
Wide: 24.9° × 18.7°

Laser Designator/Range-finder
Designator type: Nd: YAG, 1.06 μm
Classification: Class 4
Range-finder type: 1.57 μm, Class 1 (eye-safe)

CCD Sensor
Camera type: Black and white
Fields of View
Narrow: 0.8° × 0.6°
Medium: 3.4° × 2.6°
Wide: 24.90° × 18.70°
Resolution: 768 × 494

Additional capabilities
Autotracker, integral boresight module, invertible turret, MIL-STD-1553B interface, RS-232/422 interface, ARINC interface.

Contractor

FLIR Systems.

UPDATED

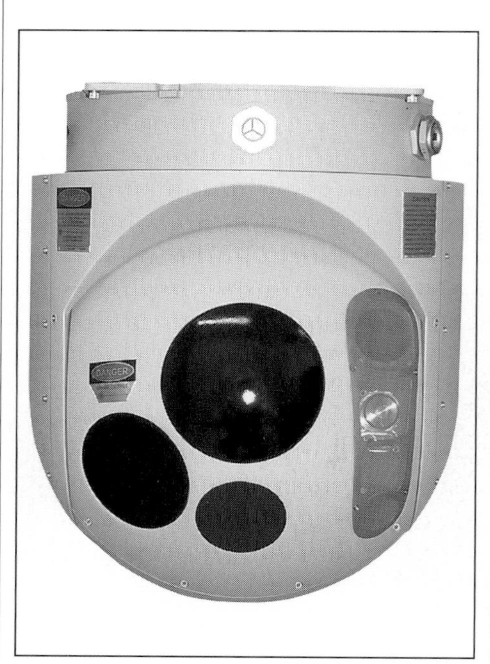

FLIR Systems BRITE Star
0137885

Lockheed Martin AN/AAQ-30 Hawkeye Target Sight System (TSS)

Type

Ground attack targeting sight

Development

The AN/AAQ-30 Hawkeye Target Sight System (TSS), part of the USMC's Bell AH-1 Upgrade Programme, facilitates target detection, recognition and identification by day or night and during adverse weather. The wide Field of View (FoV) optics provide a secondary navigation capability when light levels are low and Night Vision Goggles (NVGs) are ineffective.

Key factors which affect imaging performance include aperture size (for enhanced detection and identification ranges and poor weather performance), optics/sensitivity (for good resolution, which directly affects recognition range and thus stand-off distance according to selected magnification factor) and stabilisation (which affects resolution, designation/engagement range and third-party targeting ability).

The TSS, developed by Lockheed Martin Missiles and Fire Control, packages a variety of sensors into the same 52 cm-diameter WESCAM Model 20 turret/gimbal assembly (see separate entry) that is already operational aboard US Navy P-3C and US Coast Guard C-130 maritime patrol aircraft. The AAQ-30 also shares many key elements with Lockheed Martin's Sniper XR fixed-wing targeting system (see separate entry), recently adopted by the USAF as its Advanced Targeting Pod (ATP), and the Electro-Optical (EO) targeting system for the F-35 Joint Strike Fighter (JSF).

The TSS employs proven hardware from fielded products, with 73 per cent of its content made up of Commercial Off-The-Shelf (COTS)/non-developmental items. The system is of modular design, to permit ease of maintenance and future growth. An important part of the design is a five-axis 'soft-mount' gimbal, which provides sightline stabilisation of better than 5 picoradians. All sensors are auto-boresighted to each other, and to the helicopter's dual-EGI (Embedded GPS/INS) systems. This minimises the target-location error, which is typically 7.5 m when the helicopter's current position is known precisely but would increase to 20 m with an aircraft position error of 14 m.

The TSS accommodates a Forward Looking Infra-Red (FLIR), colour television (TV) camera, Laser Range Finder/Designator (LRFD), Inertial Measurement Unit (IMU), boresight module and Electronics Unit (EU). All units built so far also include a laser spot tracker, although the USMC has not yet decided whether this will form part of the operational fit.

The FLIR, which has a comparatively large (21.7 cm) aperture, accommodates a staring array of 640 × 512 indium antimonide (InSb) detectors, operating in the Mid-Wave Infra-Red (MWIR, 3 to 5 μm) waveband. The Sony DXC-950 TV camera, which functions in the visible and Near IR (NIR) wavebands (using a high-pass filter for operation at low light levels or in haze), includes three Charge Coupled Device (CCD) detector arrays. The camera is fitted with a Canon ×2.5 extender lens and provides continuous zoom up to a magnification of ×18. Two FoVs are matched to two of the four offered by the FLIR, permitting easy switching between the sensors.

The LRFD is of the same type as that in the LANTIRN targeting pod (see separate entry), of which approximately 1,000 units are in service, with a selectable eyesafe function.

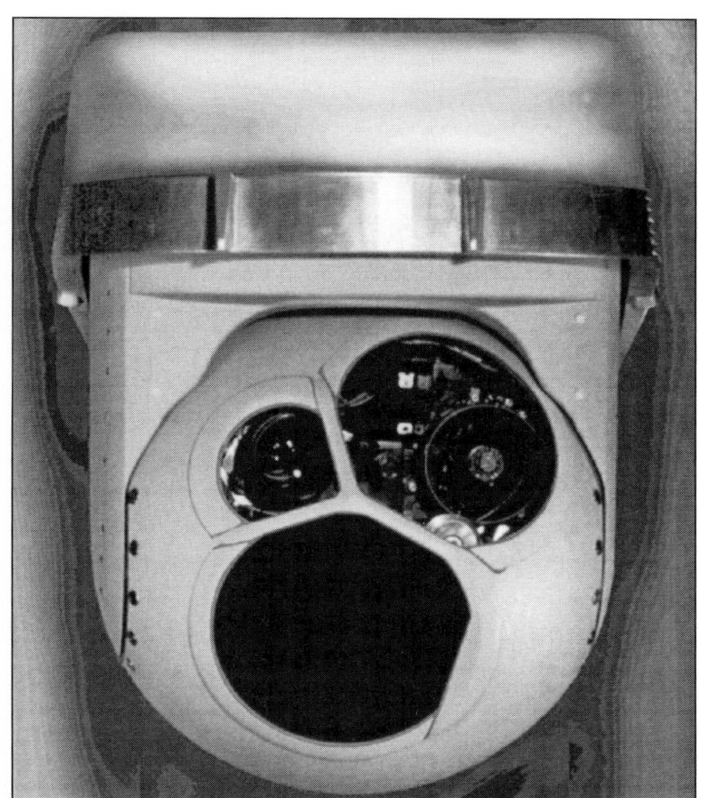

The AN/AAQ-30 TSS. The lower aperture is the FLIR sensor, while the upper left is the TV sensor and the upper right is shared by the LRFD and optional Laser Spot Tracker (LST) (Lockheed Martin) ***NEW***/0111629

The TSS supports autonomous target acquisition and re-engagement. Automatic tracking of up to three targets simultaneously, using correlation, contrast or centroid tracking modes, is available with both the FLIR and the TV camera. Each target can vary in size from a single pixel up to 80 per cent of the total FoV. The system can store track files for up to 10 additional ground targets, even after they exit the FoV, by the use of an inertially-derived 'coast' function (similar to that of the TIALD LDP - see separate entry); this function also allows track to be maintained through short-duration obscuration due to weather.

The commercially derived, open architecture of the TSS and its use of PowerPC processors, together with the comparatively large volume available in the turret, supports growth to include additional facilities. These could include sensor fusion, an NVG-compatible laser pointer, a low-light-level electron-bombarded CCD colour television camera, integration of a navigation FLIR (NAVFLIR), the adoption of a long-wave FLIR (LWIR, 8 to 12 μm) based on quantum-well devices, and other sensors. The aircraft IMU supports growth to electronic stabilisation.

Although it is not part of the TSS baseline fit, Hawkeye can also take advantage of Lockheed Martin's XR (eXtended Range) image-processing technique. This electronically enhances imagery in part of the FOV to provide a 60 per cent improvement in recognition and identification ranges. Lockheed Martin has incorporated prototype XR electronics and their associated algorithms in TSS units that it has built under the engineering and manufacturing development (EMD) programme.

Specifications

FoR: ±120° (azimuth), +45 to −120° (elevation)
Detector: staring array (640 × 512 InSb); spectral band 3 to 5 μm
FoV: 21.7 × 16.3° (wide), 4.4 × 3.3° (medium), 0.88 × 0.66° (narrow), 0.59 × 0.44° (very narrow)
System weight: 116 kg (turret 83 kg, EU 33 kg)
Tracker modes: contrast, centroid, correlation
Options: eye-safe LRFD; Laser Spot Tracker; XR image processing; 8 to 12 μm NAVFLIR

Variants

AN/AAQ-30(V) CATSeye

Lockheed Martin Missiles and Fire Control is developing a more compact version of the AN/AAQ-30 - the AAQ-30 (V) CATSeye, which employs a Wescam Model 16 (see separate entry) turret of 40 cm diameter rather than the 51 cm of the TSS, providing approximately 70 per cent of the performance available from the larger system at 70 per cent of the cost.

Potential applications include attack helicopters that are too small to accommodate the full TSS, such as the Agusta Westland A 129, together with unmanned aerial vehicles and other compact platforms. The unit can incorporate a tracker for the TOW anti-tank missile.

CATSeye weighs 61 kg, compared with 116 kg for Hawkeye. It accommodates: a MWIR TI based on a staring 640 × 480 indium antimonide (InSb) detector derived from the TSS, with a 15.2 cm aperture and three fields of view; a Sony ICX058AL intensified Charge Coupled Device (CCD) television camera; and a Laser Range Finder/Target Designator (LRF/TD) with eye-safe mode. Stabilisation of the four-axis gimbal is better than 20 μrad.

A Northrop Grumman LN-200 Inertial Measurement Unit (IMU) is mounted on the gimbal for stabilisation and aimpoint geoposition data supplied to onboard weapons employing Global Positioning System (GPS) guidance.

Lockheed Martin was reported as having completed a prototype of the AN/AAQ-30(V) CATSeye system during the first quarter of 2002.

Operational status

Forms part of the USMC upgrade for the AH-1Z attack helicopter. Five units have been manufactured under Engineering and Manufacturing Development (EMD) contract, with options covering another 21 units for Low-Rate Initial Production (LRIP). The USMC requires approximately 201 AH-1Z helicopters.

Contractor

Lockheed Martin Missiles and Fire Control.

UPDATED

Lockheed Martin AN/ASQ-173

Type

Ground attack targeting sight.

Development

The ASQ-173 Laser Spot Tracker/Strike CAMera (LST/SCAM) was designed for the F/A-18 Hornet to enhance its operational effectiveness. A laser spot tracker permits crews to identify laser-designated targets illuminated by either ground forces or supporting aircraft.

Description

Target position data from the laser detector tracker are fed into the aircraft's mission computer and are used to provide weapon-aiming and ordnance-release data to achieve superior weapon release performance. The system is based upon an integrated pod with five weapon-replaceable assemblies; the LST, an interface unit, an adaptor, the SCAM and a rotary mount.

The Lockheed Martin AN/ASQ-173 LDT/SCAM fitted to the US Navy F/A-18 Hornet in a pod

The LST consists of an inertially stabilised, gimballed optical receiver with a four-quadrant photodiode, sensors and a laser pulse decoder. They are stabilised using attitude data passed to the unit from the aircraft's inertial navigation system. The LST provides visual target cueing to the pilot through the head-up display.

The SCAM is a 35 mm variable frame rate panoramic camera housed in a rotating unit in the rear of the pod. It has a wide field of view in the lower hemisphere and can be slaved to the laser spot tracker or independently controlled by the mission computer. It provides high-resolution photographic coverage of targets before, during and after an attack, removing the need for separate damage-assessment sorties.

For strike operations the ASQ-173 will be mounted on the starboard side of the fuselage, while an AAS-38A will be installed on the port side.

Operational status

No longer in production. In service with the US Navy on F/A-18. Around 250 systems have been produced.

Specifications

Length: 2.28 m
Diameter: 20.3 cm
Weight: 73.5 kg
Scan patterns: preprogrammable/pilot selectable

Contractor

Lockheed Martin Missiles and Fire Control.

VERIFIED

Raytheon AN/AAQ-17 infra-red detecting set

Type

Ground attack targeting sight.

Description

The AN/AAQ-17 Infra-red Detecting Set (IDS) is a multipurpose thermal imaging system for navigation, search and rescue, surveillance and fire control. It is a derivative of the AN/AAQ-15 and was originally developed for the US Air Force HH-60 Nighthawk helicopter.

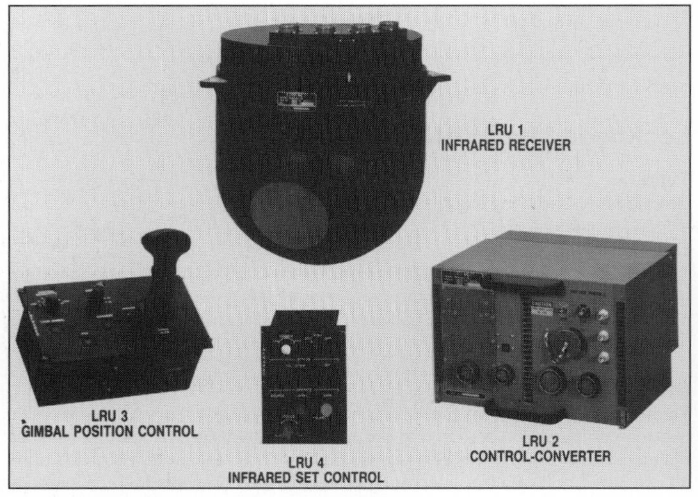

The Raytheon AN/AAQ-17 infra-red detecting set

AN/AAQ-17 consists of four LRUs: the infra-red receiver (LRU1), control-converter (LRU2), gimbal position control (LRU3) and infra-red set control (LRU4). The AN/AAQ-17 uses standard DoD FLIR common modules. A 13.7 × 18.3° wide field of view allows navigation, area search and detection of larger targets. The 3 × 4° narrow field of view allows small target detection and target recognition. A precision gimbal provides line of sight angular measurement. An adaptive gate video tracker reduces operator workload by providing hands-off automatic line of sight control. Operation of the FLIR is through manual controls or over a MIL-STD-1553B databus.

AN/AAQ-17 systems are fitted on US Air Force AC-130A, AC-130H and AC-130U gunships, HC-130P and HC-130 tankers and C-141. In the US Air Force gunship programme the AN/AAQ-17 FLIR on the AC-130H provides improved performance and reliability at lower cost.

Operational status

In production and in service on US Air Force AC-130A, AC-130H (replacing the AN/AAD-7 FLIR) and AC-130U gunships; HC-130N and HC-130P tankers; and C-141 and C-130 special operations transport aircraft. AN/AAQ-17 has also been flown on the Boeing B-52, Boeing 757 testbed, Fairchild A-10 and Sikorsky SH-60B.

Specifications

Weight
Infra-red receiver: 43.09 kg
Control converter: 21.77 kg
Gimbal position control: 2.27 kg
Infra-red set control: 1.81 kg
Field of view
Wide: 13.7 × 18.2°
Narrow: 3 × 4°
Field of regard
Azimuth: ±200°
Elevation: +15 to −105°
Power supply: 115 V AC 1,900 W max, 28 V DC, 100 W max

Contractor

Raytheon Systems Company.

VERIFIED

Ring Sights LC-40-100-NVG helicopter gun sight

Type

Ground attack targeting sight.

Development

In 1993, the RAF asked for a sight for a trial in which helicopters would fire machine guns sideways at ground targets, preferably using a laser pointer. The existing LC-40-100 sight was modified and after the trial the laser pointer was abandoned as the LC-40-100 can have a graticule which enables the gunner to aim off for helicopter movement and so on and hit a 3 m target at 500 m by day or night, using NVG. The gunner can also correct fire using bursts on target and quickly switch to successive targets as required.

Since then work has proceeded to optimise the sight and its graticule illumination and the system is now in production.

Description

The sight has a dovetail mount on top aligned with the graticule. This can accept a laser illuminator, a CCTV camera (to remote the gunner's view to the pilot or to

Ring Sights Defence Ltd LC-40-100 Night Vision Goggles NVG

record the engagement) or other device. The RAF has purchased the sight with the optional infra-red illuminator which emits a 20 mrad infra-red beam to illuminate the target at night with night vision goggles and also identifies the weapon aiming point for the commander of the aircraft. The system is used in conjunction with Ferranti's Nite-OP Night Vision Goggles.

For use in low-light conditions and with NVGs, the sight has lithium batteries, having two settings of brightness for low light and six settings for NVGs. The graticule is automatically lit by means of a dichroic light wedge, selecting one or other battery by means of a switch.

A variety of gun mounts has been developed to match the helicopter environment including M60D, M134 and GPMG 7.62 mm machine guns and the .50 in Browning HMG.

Operational status

In production. Sold to the RAF and Army Air Corps in quantity. LC-40-100-NVG is fitted on the M134 Minigun which equips the RAF CH-147D Chinook transport helicopters which have been deployed to support UN operations in the former Yugoslavia. RN procurement is in process. Fabrique Nationale has adopted it as standard equipment for its M3M airborne weapon system.

Specifications

Aperture: 40 × 30 mm
Focal length: 132 mm
Dimensions (l × h × w): 215 × 160 × 156
Weight: 2.00 kg (incl zeroing mount and laser illuminator)

Contractor

Ring Sights Holding Co Ltd.

VERIFIED

Zenit Shkval sighting system

Type

Targeting sight.

Description

The Shkval sighting system is designed as a comprehensive anti-tank electro-optical fire-control system, incorporating: TV sighting sensor, laser range-finder and target designator and laser beam-rider. It is also fitted with a three-axis field of view stabilisation system and an automatic image correlator to ensure tracking against ground, sea or sky backgrounds. Provided is ×23 magnification to extend detection/tracking ranges. Tracking angles are quoted as +15 to –80° in elevation and ±35° in azimuth. Laser guidance system accuracy is quoted as 0.6 m.

Shkval window (above YOM3 'ball') on Ka-50N Black Shark helicopter (Paul Jackson)
0018333

Operational status

Shkval is part of the weapon system on the Ka-50 Black Shark and Ka-52 Alligator helicopters and on the Su-24T and Su-39 multimission attack aircraft.

Contractor

Zenit Foreign Trade Firm, State Enterprise.

VERIFIED

GROUND ATTACK

LASER RANGE-FINDERS

BAE Systems HELD

Type
Ground attack laser designator.

Description
The HElicopter Laser Designator (HELD) is a derivative of the Type 306 laser designator which is used by the British Army. In the HELD version the Type 306 is fitted to a simple helicopter mount which requires virtually no modification to the helicopter. A gyrostabilised mirror is added to the Type 306 to compensate for the vibration and motion effects on the laser spot. The system has been successfully tested with laser-guided bombs and spot trackers.

Operational status
In production since 1980 and sold to customers in Asia, the Middle East and NATO.

Contractor
BAE Systems, Avionics, Sensors and Communications Systems, Edinburgh.

VERIFIED

A HELD laser designator being used in a Canadian Forces Kiowa helicopter

BAE Systems Type 105 airborne steerable laser range-finder

Type
Ground attack laser range-finder.

Description
The Type 105 is a high repetition rate, Nd:YAG, steerable laser range-finder developed by the former Ferranti company to provide a compact, low cost, accurate target ranging sensor for ground attack aircraft.

The device generates 1.064 μm wavelength laser pulses, receives reflection of these pulses from the target and computes slant range. The transmit and receive optical paths are common and the laser beam can be steered over a 20° full angle, conical field of view. The Type 105 is configured in two units for flexibility of interface and installation. It comes in a number of variants and has a number of options available.

One of the latest variants includes a laser spot tracker to give the aircraft the precision of laser guidance for all conventional weapons. This facility is particularly advantageous in the close air support role when the target is being designated by a co-operative laser designator. This laser spot tracker offers a much smaller package than conventional gimballed systems.

BAE Systems Type 105H airborne steerable laser range-finder 0023386

Operational status
The Type 105 has been trialled in a number of aircraft including the A-4, A-10, F-5 and Hawk 100.

Specifications

	Steerable Laser Transmitter	Electronics Unit
Weight:	11 kg	4 kg
Dimensions:	330 × 213 × 190 mm	228 × 222 × 108 mm

Emitted energy: >50 mJ
Beam divergence: <0.5 mrad
PRF: 10 pps
Field of view: 20°
Range: 10 km
Range accuracy: 3.5 m (one standard deviation)
Input power: 300 W
MTBF: 3,000 h

Contractor
BAE Systems, Avionics, Sensors and Communications Systems, Edinburgh.

UPDATED

BAE Systems Type 118 lightweight laser designator/range-finder

Type
Ground attack laser range-finder.

Description
The Type 118 laser designator/range-finder is a lightweight system intended primarily for use in helicopter applications. It was originally developed in 1980 and produced for the mast-mounted sight on the US Army OH-58D Army Helicopter Improvement Programme (AHIP). First production equipment was delivered in mid-1985.

It is of modular design, has common transmit/receive optics and can be easily integrated into a variety of sights or combined with other electro-optical sensors. The Type 118 can operate with all existing laser-guided weapons and laser spot trackers.

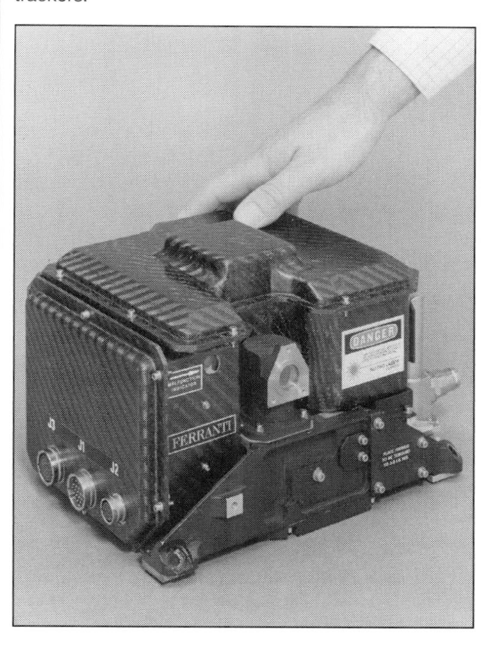

The BAE Systems Type 118 laser designator/range-finder

Operational status
In production.

Specifications
Type: Nd:YAG
Wavelength: 1.06 μm
Output energy: 110 mJ
PRF: 20 pps (max)
Beam divergence: 2 mrad (raw beam)
Resolution: 5 m
Range: 0.3-10 km

Contractor
BAE Systems, Avionics, Sensors and Communications Systems, Edinburgh.

BAE Systems Type 121 laser designator/range-finder

Type
Ground attack laser designator/range-finder.

Description
The Type 121 laser designator/range-finder is based on a standard laser transceiver in production for a number of military applications.

Operational status
The Type 121 is in service on British Army Air Corps Gazelle helicopters, where it has been retrofitted to the AF-532 roof sight.

Specifications
Transmitter type: Nd:YAG
Output energy: greater than 60 mJ
Pulse repetition frequency: up to 20 pps
Beam divergence: 1.5 mrad max (raw beam)
Receiver range capability: 300-9,900 m
Accuracy: 5 m, 1σ
Control unit
On/off switch
Laser arm switch
First/last pulse logic select
Code switch
Range display
Weight
Transceivers: 9.2 kg max
Control unit: 2 kg max
Dimensions
Transceiver: 185 × 353 × 245 mm
Control unit: 108 × 126 × 172 mm

BAE Systems Type 121 laser designator/ range-finder

BAE Systems laser designator/range-finder Type 121 on Army Air Corps Gazelle helicopter with laser-equipped AF-532 roof sight

Power requirements
Voltage: 24-29 V DC
Power firing: 410 W max
Standby: 35 W max

Contractor
BAE Systems, Avionics, Sensors and Communications Systems, Edinburgh.

BAE Systems Type 126 laser designator/range-finder

Type
Ground attack laser designator/range-finder.

Description
The Type 126 is a high-energy Nd:YAG range-finder and designator system developed for the UK's Thermal Imaging Airborne Laser Designation (TIALD) pod. The Type 126 comprises separate transmitter and power supply/control units for ease of pod or aircraft installation. The laser provides very high-output energies at pulse repetition rates of up to 20 Hz in a temperature environment ranging from –54 to +71°C.

Operational status
In production for the UK Royal Air Force in the TIALD pod. TIALD is in service with Tornado GR.4, Jaguar GR.3A and Harrier GR.7.

Specifications
Transmitter
Type: Nd:YAG
Output energy: >125 mJ
Pulse repetition frequency: up to 20 pps
Beam divergence: <2.5 mrad (raw beam)
Receiver
Range: 320 m to 25 km
Range BIT: 2.5 m
Power requirements: 115 V 400 Hz 650 VA; 28 V DC 10 W
MTBF: >1,500 h MIL-HDBK-217E
Weight
Transmitter: <6.2 kg
Power supply: <5 kg

Contractor
BAE Systems, Avionics, Sensors and Communications Systems, Edinburgh.

UPDATED

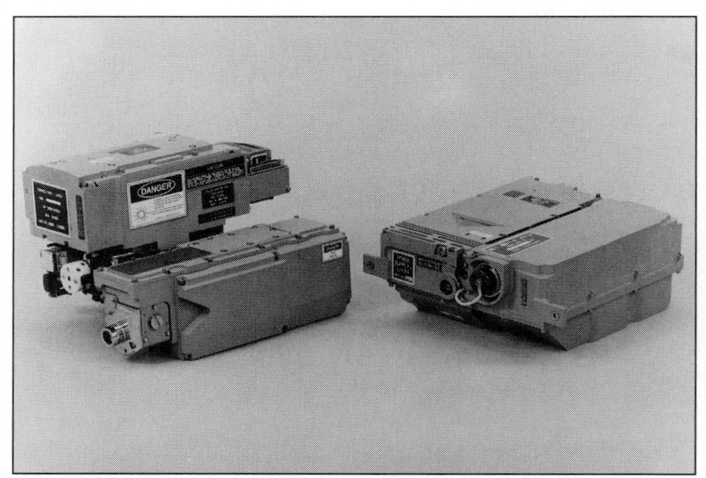

The BAE Systems Type 126 laser range-finder/designator

CILAS TIM family laser range-finders

Type
Ground attack laser range-finder.

Description
TIM laser range-finders are high repetition rate eye-safe Raman laser range-finders designed to be integrated into an airborne fire-control system to measure the distance to a ground-based, aerial or naval target.

They consist of a 1.54 μm transmitter integrating a 1.06 μm laser, a Raman conversion cell, a receiver, the power supplies, an interface chronometry/control and serial link RS-422 with the host system.

Operational status
TIM laser range-finders have been selected for the Rafale aircraft and the Tiger helicopter.

Specifications
Eye-safe laser wavelength: 1.54 μm
Dimensions: 300 × 180 × 150 mm
Weight: 8.5 kg
Power supply: 220 V AC, 3 phase or 115 V AC, 400 Hz

Contractor
Compagnie Industrielle des Lasers (CILAS).

DRS AN/AAS-32 laser designator sensor

Type
Ground attack laser designator/range-finder.

Description
The AN/AAS-32 is a laser tracker which DRS (formerly Boeing) produces for the Bell AH-1S Cobra light attack helicopter and, in derivative form, for the Target Acquisition Designation System/Pilot Night Viewing System (TADS/PNVS) in the US Army McDonnell Douglas (now Boeing) AH-64 Apache attack helicopter.

The sensor is a wide field of view unit, sensitive to 1.06 μm radiation, that can detect target designations from ground troops or co-operative aircraft. Coded pulse data is used to minimise jamming and a four-quadrant silicon detector head is used.

The gimballed optics of the receiver unit have a 127 mm aperture. Targets may be illuminated by the helicopter's own laser designator or by ground-based systems such as GLLD.

Operational status
Production of AH-1S equipment was initiated in September 1981 and first delivery was in May 1984. TADS/PNVS production began in June 1981 and deliveries began in January 1983. The system is in service with a total of 275 systems produced.

Specifications
Dimensions
Receiver unit: 214 × 188 mm
Electronics unit: 152 × 152 × 188 mm
Control panel: 144 × 66 mm
Weight
Receiver unit: 9 kg
Electronics unit: 3.4 kg
Control panel: 0.6 kg
Power supply: 115 V AC, 400 Hz 28 V DC
Field of view (scanning)
Elevation: –60 to +30°
Azimuth: –90 to +90°
Field of view (instantaneous)
Elevation: 10°
Azimuth: 20°
Optical port diameter: 127 mm (5 in)
Focal length ratio: 0.3:1

Contractor
DRS Technologies, Electro-Optical Systems Group.

Elop Comanche Laser Range-Finder/Designator (LRF/D)

Type
Ground attack laser rangefinder/designator.

Description
Elop has developed an advanced Laser Range-Finder/Designator (LRF/D) which has been selected for the RAH-66 Comanche scout/attack helicopter being developed for the US Army. The LRF/D will be integrated into the Lockheed Martin's EOSS (Electro-Optical Sensor System) for the RAH-66.

The LRF/D is capable of operating at two different spectral wavelengths within a single system, enabling target designation using the appropriate wavelength, for directing laser-guided weapons while alternatively gathering range data via different wavelengths. The system uses pumped laser diode technology and is eye-safe.

The LRF/D is also adaptable to other reconnaissance/attack platforms such as the AH-64 Apache and OH-58 helicopters.

Operational status
Selected for the US Army RAH-66 Comanche helicopter which is scheduled to enter service in 2008. Elop is under contract from Lockheed Martin to manufacture and deliver LRF/D units for the pre-production phase.

Contractor
Elop Electro-Optics Industries Ltd.

UPDATED

Elop Kiowa Warrior switchable eye-safe laser range-finder/designator

Type
Ground attack laser range-finder/designator.

Description
Elop has developed an advanced dual wavelength Switchable Eye-safe Laser Range-finder Designator (SELRD), to replace the existing single wavelength designator for the US Army OH-58D Kiowa Warrior helicopter. The SELRD enables the pilot to switch back and forth, between the designation mode at 1.06 μm and eye-safe 1.57 μm (training mission).

The new laser, which is based on developed diode pumped solid-state wavelength conversion modules, developed for the US Army Comanche helicopter will be incorporated into the OH-58D Mast-Mounted Sight (MMS), now manufactured by DRS Technologies (acquired from Boeing/McDonnell Douglas).

Operational status
A qualified SELRD unit was incorporated into an MMS and after successful live firings of laser-guided missiles, it is ready to start full-rate production.

Contractor
Elop Electro-Optics Industries Ltd.

UPDATED

Elop RFTDL laser range-finder/designator

Type
Ground attack laser range-finder/designator.

Description
Elop has developed the Range-Finder Target Designator Laser (RFTDL) as a high-repetition laser system for installation in the IAI-TAMAM/Kollsman Night Targeting System (NTS) for the US Marine Corps AH-1W Cobra helicopter. Integration of the laser designator into the NTS gives the Cobra the capability to fire Hellfire missiles.

Operational status
The system was first test flown in May 1991. Since then more than 400 units have been manufactured and delivered to the US Marine Corps.

Contractor
Elop Electro-Optics Industries Ltd.

UPDATED

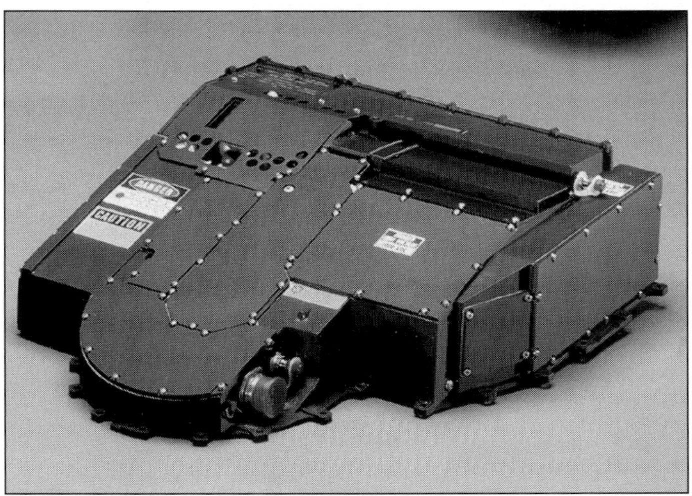

RFTDL 0059662

Elop Very Light Laser Range-finder (VLLR/D)

Type
Ground attack laser range-finder/designator.

Description
The Very Light Laser Range-finder/Designator (VLLR/D) is designed to upgrade existing platforms and pods and enhance the capabilities of new systems. It is suitable for a variety of airborne systems, including UAVs, as well as seaborne and ground platforms. The VLLR/D is compact, ruggedised and modular.

Operational status
In production for various E-O stabilised payloads.

Specifications
Weight: <3.8 kg
Power consumption: 220 W max
Wavelength: 1.064 μm
Output energy: 80 mJ
PRF: single shot or up to 20 pps
Beam divergence: less than 0.4 mrad
Pulsewidth: 15 ns max
Range: 200-9,995 m
Range accuracy: ±5 m

Contractor
Elop Electro-Optics Industries Ltd.

UPDATED

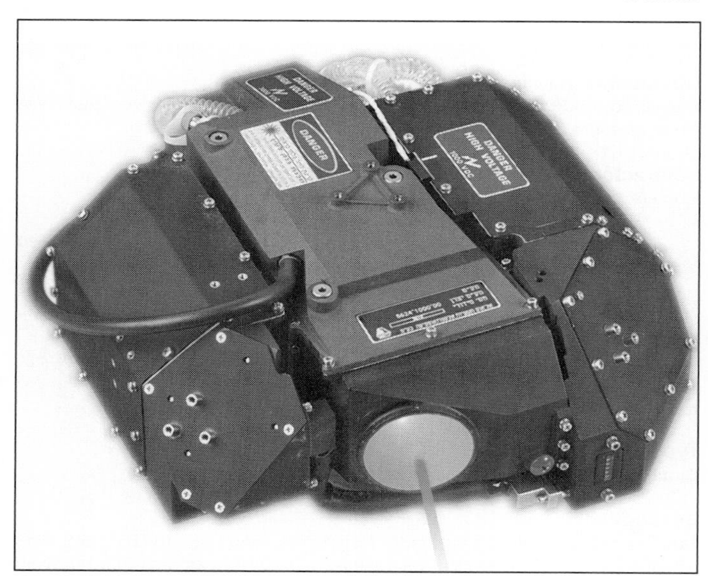

The Elop Very Light Laser Range-finder (VLLR) 0101627

Eloptro airborne laser designators

Type
Laser designator range-finder.

Description
Eloptro specialises in the development of Airborne Laser Designators to customer specifications. The company currently manufactures the NightOwl laser designator for the Rooivalk attack helicopter. This laser designator is integrated within a stabilised optronic platform and is used to guide the Mokopa Missile System manufactured by Kentron.

New developments in this field include the development of a very compact airborne laser designator with an OPO-shifted eye-safe laser range-finder function. Eloptro offers customised solutions for the integration of laser designators within the limited space envelope of in-service stabilised optronic gimbals. The designator coding is slaved to the main system control, therefore eliminating the need to transfer sensitive coding information.

Operational status
In production.

Specifications
Laser type: Nd-Yag liquid cooled
Wavelength: 1,064 nm, 1,570 nm (OPO-shifted LRF)
Output energy: >100 mJ
Designator coding: slaved to system sync pulse
Size and mechanical interface: customer specified
Electronic interface: customer specified

The Rooivalk attack helicopter fitted with the NightOwl laser designator 0110163

Contractor
Eloptro, Division of Denel (Pty) Ltd.

FIAR P0705 HELL laser range-finder

Type
Ground attack laser range-finder.

Description
The P0705 HELL is an airborne Nd:YAG laser range-finder designed for integration in helicopter stabilised electro-optic sight units with the functions of target ranging, gun and missile pointing and navigation fixing.

The fully qualified HELL system improves helicopter attack capabilities by providing extended reconnaissance and target detection ranges, increasing the probability of a kill, reducing flight times and enhancing flight safety.

A modified eye-safe version has also been developed.

Operational status
In service. The eye-safe version of HELL is in production.

Specifications
Wavelength: 1.064 μm and 1.54 μm
Pulse power: 4 MW
PRF: 2 Hz
Range: 300-10,000 m
Interface: RS-422 serial

Contractor
FIAR (Fabbrica Italiana Apparecchiature Radioelettriche) SpA.

VERIFIED

The P0705 HELL laser range-finder

Galileo Avionica P0708 laser range-finder

Type
Ground attack laser range-finder.

Description
The P0708 PULSE is an airborne steerable Nd:YAG laser range-finder which has been developed for the AMX, A-4 Skyhawk, Mirage, F-5, MiG-21 and other aircraft. It provides range data for the aircraft's main computer while also informing the pilot of the controlled aiming direction for air-to-surface attacks. The system consists of two line-replaceable units, a laser transceiver and electronics.

Operational status
In production.

Specifications
Type: Nd:YAG
Receiver field of view: 0.7 mrad
Wavelength: 1.064 μm
Pulse duration: 15 ns
Steering angle: within 20° pointing cone
PRF: 10 Hz
Output power: >8 MW
Repetition frequency: 20 Hz
Beam divergence: 1 mrad
Beam deflection
  Azimuth: +10°
  Elevation: 0-20°
Measurement range: 300-20,000 m
Range accuracy: better than 3.5 m
Interface: MIL-STD-1553

Contractor
Galileo Avionica.

UPDATED

The P0708 PULSE airborne steerable laser range-finder

IAI Cockpit Laser Designation System (CLDS)

Type
Ground attack laser designator.

Development
The Cockpit Laser Designation System (CLDS) is a lightweight small size designator designed for light attack two-seater aircraft, mounted in the rear cockpit of the aircraft which commands the tactical operation. The CLDS can designate targets for other attacking aircraft equipped with laser-guided bombs and provide the ability to deliver the laser-guided bomb in a wide variety of modes, altitudes and ranges. It is designed for operation in close air support, battlefield air interdiction, deep strike or by a forward air controller using the magnifying sight for observation only and for sea surveillance.

Description
The CLDS includes a TV camera and auto-tracker. This closed-loop hands-off tracking system reduces the pilot's workload during the critical phase of the mission and significantly improves the tracking accuracy and manoeuvrability of the designating aircraft in the target area. The video camera, interfaced to an aircraft VTR, is also utilised for debriefing and intelligence purposes.

The CLDS can be interfaced to the aircraft Weapon Delivery and Navigation System (WDNS), enabling the WDNS to direct the CLDS line of sight towards a chosen target as a highly accurate sensor for marking targets and providing position updates to the WDNS.

The MBT Weapons Systems cockpit laser designation system

Operational status
In production and sold to several countries. The CLDS can be installed in the IAI Kfir, Dassault Aviation Mirage III and 5B, Northrop F-5F and Lockheed F-16B aircraft.

Specifications
Dimensions: 560 × 420 × 360 mm
Weight: 40 kg
Field of view
  Wide ×4 magnification: 12.5°
  Narrow ×10 magnification: 5°
Field of regard
  Forward: 45°
  Backward: 25°
  Elevation: ±35°
Accuracy: 0.25 mrad
Range: 8-10 km typical

Contractor
Israel Aircraft Industries, MBT Weapon Systems Division.

VERIFIED

Northrop Grumman Dark Star/LAMPS Mk III Laser Designators

Type
Laser range-finder/designator.

Description
The Dark Star/LAMPS Mk III laser designators are designed for use with all types of laser-guided weapons.

The two-component laser designator system incorporates a newly developed High Brightness Resonator (HBR). Dark Star/LAMPS lasers feature power conversion and control electronics packaged in a High Energy Converter (HEC) assembly.

The HEC module contains power conditioning, interface and control electronics for the transmitter. Built-in test verifies critical aspects of system performance. A single large electronics array located in the HEC provides laser interface, timing and control.

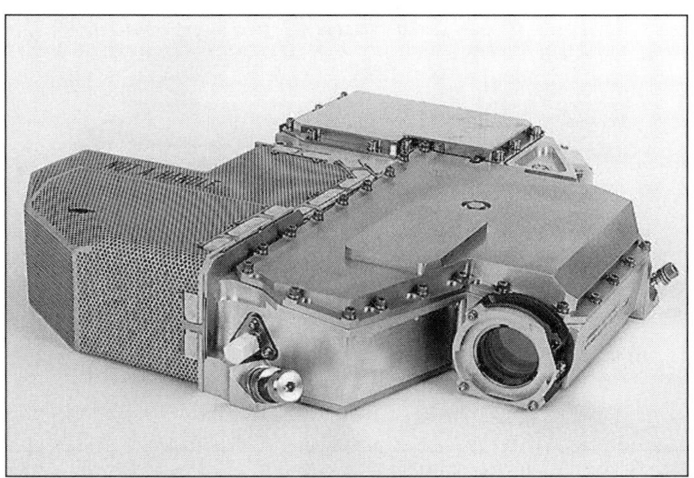

Dark Star laser designator system from Litton Laser Systems

Operational status

In production since 1990. In service on the USAF's F117A Nighthawk (Dark Star) and the US Navy's HH/SH-60 Sea Hawk (LAMPS Mk III) helicopter.

Specifications

Wavelength: 1.064 μm
Laser type: Nd:YAG
Weight: 15.4 lb (Dark Star); 11 lb (LAMPS Mk III)
Shock: 15 *g*
Vibration: 6.6 *g* RMS
Power: 115 V AC, 3 phase, 400 Hz, 250 W, ±15 V DC, 20 W
Cooling: forced ambient air
Dimensions
Transmitter: 11.5 × 11.5 × 2.4 in
HEC: 8.5 × 5.7 × 4.4 in

Contractor

Northrop Grumman, Electronic Systems.

UPDATED

Northrop Grumman MMS Laser Range-Finder/Designator (LRF/D)

Type

Ground attack laser range-finder/designator.

Description

Northrop Grumman,Electronic Systems,Laser Systems is producing the Laser Range-Finder/Designator (LRF/D) for the Boeing (formerly McDonnell Douglas) Mast-Mounted Sight (MMS) sensor suite which equips Bell OH-58D Kiowa Warrior helicopters of the US Army. Mounted above the rotor, the MMS allows the helicopter to acquire and to illuminate targets at greater standoff ranges while concealed from the battle area. Hellfire, Copperhead and other laser-guided munitions can be guided to their targets using the MMS-LRF/D.

By efficient packaging design, the MMS-LRF/D incorporates all the electronics, high-voltage power supply, cooling system, range receiver and laser transmitter in one compact lightweight line-replaceable unit. The microprocessor-controlled LRF/D features a serial asynchronous digital interface for precise communication with the MMS computer. Internally, all subassemblies interface via a common bus under control of the main processor, which maintains all system timing through software control.

Variants incorporating the ability to switch between eye-safe ranging at 1.5 μm and tactical designation at 1.06 μm have also been developed.

Operational status

In production since August 1982. In service on the US Army's OH-58D Kiowa Warrior. Over 600 systems have been delivered to the US Army. MMS is also in service with Taiwan on OH-58D helicopters. A modified MMS LRF/D has been integrated into the all-light-level television system in the Lockheed AC-130U.

Specifications

Type: Nd:YAG
Weight: 14 lb
Dimensions (l × w × d): 5.64 × 8.25 × 10.65 in
Altitude: sea level - 12,000 ft
Transmitter
Wavelength: 1.064 ±0.001 μm
Pulse repetition frequency: Band I, Band II, Tri-service codes
Q-switching: Pockel Cell
Beam diameter: 0.52 + 0.00, –0.02 in

The Northrop Grumman mast-mounted sight laser range-finder and target designator

Missed pulses: 1/10 min continuous lasing
Duty cycle: continuous
Beam alignment: ±200 μrad
Receiver
Probability of detection: 0.99
Probability of false alarm: 1%
Max range readout: 9,995 m
Supply voltage: 28 V DC, +15 V DC, 8.5 V DC

Contractor

Northrop Grumman, Electronic Systems.

UPDATED

Northrop Grumman switchable eye-safe Laser Designator Ranger (LDR)

Type

Ground attack laser designator/range-finder.

Description

The Laser Designator Ranger (LDR) is a part of the LANTIRN (Low-Altitude Navigation and Targeting Infra-Red system for Night) and TSS (Target Sighting System) being produced by Lockheed Martin. It provides tactical forces with day/night under-the-weather capability and eyesafe training.

The LDR provides the laser energy for illuminating/ranging targets for the delivery/guidance of laser terminally guided weapons. The manufacturer states that it is the only switchable eyesafe tactical laser in production.

Modular and with a digital interface, the LANTIRN LDR provides serial or parallel commands directly to the pod control computer through the system databusses. The LANTIRN LDR is equipped with self-contained, automatic and initiated BIT functions. The LANTIRN LDR consists of three replaceable units: a laser transmitter/receiver, a transmitter electronics unit and a high-voltage power supply, and TSS in an air-cooled version of the LANTIRN laser.

Operational status

In production since 1985, LANTIRN is in service with the US Air Force on F-16C/D and F-15-C/D/E aircraft and with US Navy F-14 Tomcats. Over 1,000 LANTIRN lasers have been produced in the US and Germany. TSS is currently planned for deployment on the USMC's AH-1Z Cobra helicopter upgrade as an upgrade to the existing Night Targeting Sight.

Specifications

Weight: 16.2 lb
Transmitter
Wavelength: 1.06 or 1.54 μm pulsed switchable
Pulse Repetition Frequency (PRF): Band I, Band II and PIM 1.06
Q-switching: Pockels cell
Beam diameter: 0.25 in
Beam divergance: 2.5 mrad (1.06) or 4.5 mrad (1.54)
Duty cycle: continuous –1.06 = 1-20 PPS; 1.54 = IPPS
Power requirements: ±15 V DC, +28 V DC
Receiver
Max range: 43,500 m (with single channel receiver) or 35,000 m (with old receiver)
Power requirements: +5 V DC, ±15 V DC, +28 V DC
Power supply
Requirements: 115 V 3 phase, 400 Hz AC

Contractor

Northrop Grumman, Electronic Systems.

UPDATED

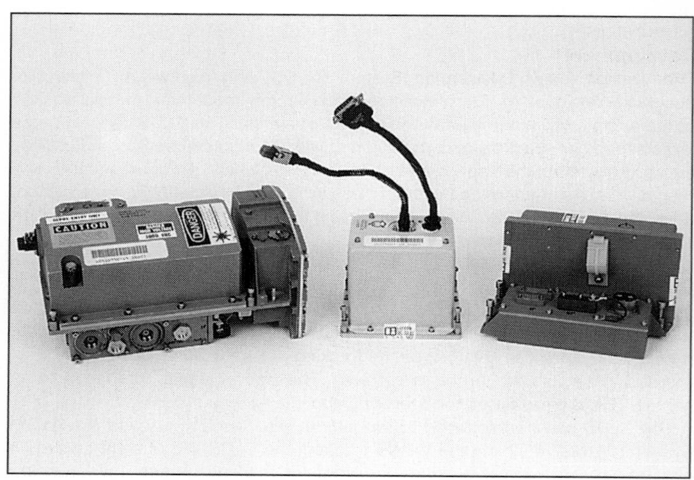

Laser components of the LANTIRN pod

0023387

Northrop Grumman, Electronic Systems, Laser Systems TADS Laser Range-Finder/Designator (LRF/D)

Type
Ground attack laser designator/range-finder.

Description
This Laser Range-Finder/Designator (LRF/D) is integrated with the Lockheed Martin Target Acquisition and Designation Sight (TADS)/Pilot Night Vision Sensor (PNVS) which is fitted to the AH-64A Apache attack helicopter. This system consists of two Line-Replaceable Units (LRUs), the laser transceiver and the electronics unit. The system includes a built-in fault location system to isolate component failures.

Operational status
In service on Boeing AH-64D Apaches of the US Army since 1982, plus more than 1,400 produced. AH-64s delivered to Egypt, Greece, Israel, Netherlands, Saudi Arabia, Singapore and the UK.

Contractor
Northrop Grumman, Electronic Systems.

UPDATED

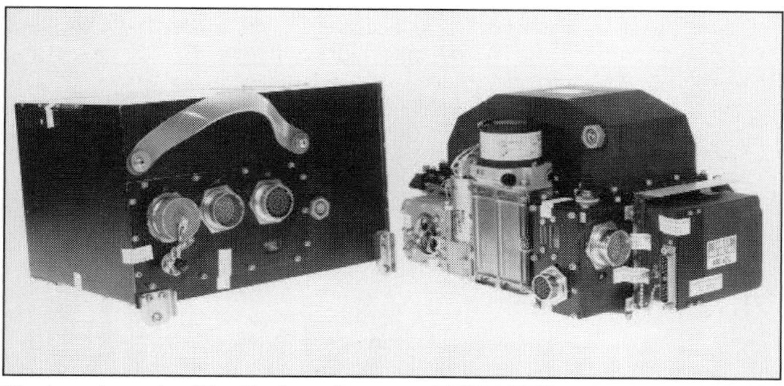

The two elements of the Northrop Grumman *TADS LRF/D*

Saab Bofors Dynamics airborne laser range-finder

Type
Ground attack laser range-finder.

Description
A compact, airborne, high repetition rate laser range-finder has been developed by Saab Bofors for upgrading of existing systems. It can be integrated into any sighting and weapon delivery system to give greater accuracy and enhanced safety in dive attacks. The Nd:YAG range-finder uses a modular design consisting of transmitter, receiver (with silicon avalanche diode), range counter and deflection unit. The laser is aimed at the target by slewing the deflection unit to the aircraft sighting system.

The range-finder has built-in safety functions and test equipment, a special training mode with increased laser safety and a flexible interface using the ARINC 429 digital bus. It can be used as the basis for an effective low-level airborne attack system with Continuously Computed Impact Point (CCIP) and Continuously Computed Release Point (CCRP) modes.

Operational status
Produced for various aircraft between 1982 and 1986 with production being resumed in 1993 for Thales.

Specifications
Laser type: Nd:YAG
Dimensions: 50 × 16 × 16 cm (approx)
Weight: 14 kg (approx)

The Saab Bofors Dynamics airborne laser range-finder

Range: 10 km (approx)
Range resolution: 5 m
Power supply: 28 V DC
Power consumption: 225 W
Transmitter
Wavelength: 1.06 μm
Pulse length: approx 10 ns
PRF: 1-10 Hz in bursts
Beam divergence: 0.7 mrad
Receiver
Field of view: 0.5 mrad
Detector: silicon avalanche diode
Deflection unit
Bearing travel: ±10°
Elevation travel: ±10°
Accuracy: <1 mrad
Slew rate: >60°/s

Contractor
Saab Bofors Dynamics.

Thales Optronics TMV 632 laser spot tracker and range-finder

Type
Ground attack laser range-finder and spot tracker.

Description
The TMV 632 ground attack laser range-finder was developed at the request of the French DGA (Delegation Generale de l'Armament) and combines eye-safe laser ranging and angular tracking functions into a single compact monobloc system, for use on tactical ground support and training aircraft. It provides the weapon systems with extremely accurate fire-control data, for both laser-guided and conventional munitions. Several variants are available, with different digital interfaces. The dual-function TMV 632 offers the same level of performance and accuracy as single-function systems. The TMV 632 is mounted in the airframe (embedded) or fitted in a mini-pod, or inside a store-carrying pylon. The TMV 632 detects and identifies the laser spot on illuminated ground-based targets. Acquisition and tracking are automatic. A sighting reticle in the head-up display allows the pilot to aim at the target. The laser beam of the range-finder is locked to the position of the tracker, making accurate measurement of the aircraft-to-target distance possible. There are two independent safety interlocks in the TMV 632. If required, cooling air is supplied by the aircraft.

Operational status
In production for the French Air Force Mirage F1 CT.

Specifications
Dimensions (l × w × h): 530 × 170 × 190 mm
Weight: 18 kg
Field of Regard: 40 × 20°
Wavelength
Laser range-finder: 1.5 μm eye-safe
Laser spot tracker: 1.06 μm
Range: up to 20 km

Contractor
Thales Optronics.

VERIFIED

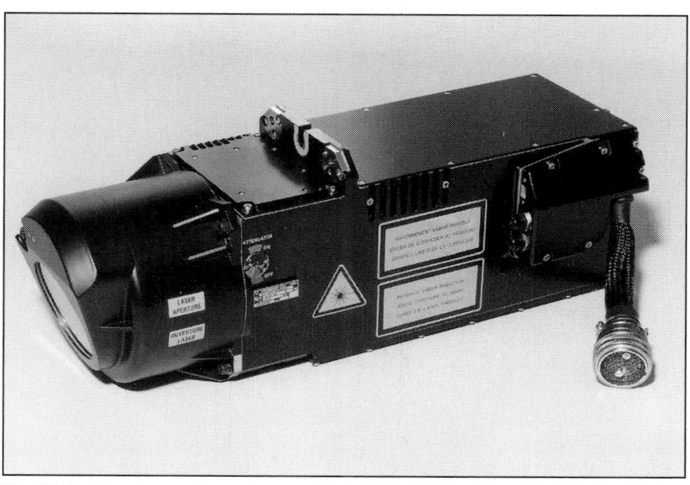

The TMV 632 LRU 0058713

Urals Optical and Mechanical Plant KLEN

Type
Laser range-finder (air).

Description
The KLEN is a laser range-finder and designator developed for the MiG-27M fighter bomber and Su-22, Su-25 attack aircraft and their derivatives. The range-finder provides measurement accuracy of range to less than 5 m at distances up to 10 km. The range of the designator against ground targets is up to 8 km. KLEN provides high efficiency of ground target damage when bombing, gunfiring and employing laser-guided missiles.

Operational status
Operational on MiG-27,Su-22,Su-25 and their variants.

Specifications
System
Weight: not more than 82 kg
Field of regard/pointing angle
Azimuth: ±12°
Elevation: +6 to −30°
Designator range: 8 km
Range-finder range: 10 km
Accuracy of range measurement: 5 m
Laser
Impulse energy: 0.07 J
Pulse repetition frequency: up to 25 Hz
Cooling: fluid cooling for all systems
Weight (incl cooling system): 2.5 kg
Overall dimensions (incl cooling system): 240 × 188 × 103 mm
Power supply: 115 V; 400 Hz, 3 phase
Power supply weight: 6.5 kg

Contractor
Production Association Urals Optical and Mechanical Plant (UOMZ).

VERIFIED

The UOMZ 'KLEN' laser designator/range-finder system suite 0089877

Urals Optical and Mechanical Plant Prichal

Type
Laser range-finder (air).

Description
The Prichal was developed in the 1980s and is a laser range-finder/designator for Su-25TK attack aircraft and Ka-50 and Mi-28 combat helicopters. Development and manufacture was achieved on the basis of experience gained on the Klyon laser range-finder/designator. The designation range and the pulse repetition frequency were increased. The Prichal provides range-finding accuracy to less than 5 m. It is claimed to have a high operating speed when updating information about the range and supplying it to the aircraft intercommunication net.

Operational status
Operational on Su-25TK, Ka-50, Mi-28.

Specifications
Mass: 46 kg
Accuracy: 5 m

Contractor
Production Association Urals Optical and Mechanical Plant (UOMZ).

VERIFIED

The Prichal laser range-finder/designator system 0090000

FLIGHT AIDS

LASER SYSTEMS

ATCOP AA 3 laser altimeter

Type
Laser altimeter.

Description
The laser altimeter measures low altitudes with an accuracy of less than 1 m. It can be retrofitted to helicopters and fighter aircraft for accurate measurement of height above ground during landing or other manoeuvres.

Operational status
In serial production.

Specifications
Max range: 80 m (extendable to 200 m)
Accuracy: ≤1.0 m
Display: 4 digit LED
Operating temperature: –30 to +60°C
Dimensions: 200 × 150 × 90 mm
Weight: 2.0 kg

Contractor
AI Technique Corporation of Pakistan (Pvt) Ltd.

VERIFIED

The AA 3 laser altimeter 0058714

BAE Systems, Thales CLARA Compact Laser Radar

Type
Laser radar.

Description
CLARA is a self-contained CO_2 laser radar system housed in an environmentally controlled pod and mounted on a helicopter or fixed-wing aircraft. Development of CLARA is as a result of an Anglo/French government-to-government initiative, resulting in a consortium comprising Marconi Avionics (now BAE Systems) and Thales Airborne Systems. Under the work-share arrangements two identical demonstrator units have been produced and tested in 1997 and 1998 on a fixed-wing aircraft in the UK and on a helicopter in France.

CLARA is designed to avoid obstacles such as cables, pylons and other hazards. It will also provide other functions such as terrain-following target ranging and designation, short-range true airspeed measurements and moving target indication.

Operational status
Flight test programme started in 1997 and is continuing.

Contractors
BAE Systems, Avionics Ltd Sensors and Systems Division.
Thales Airborne Systems.

VERIFIED

Dornier HELLAS helicopter Obstacle Warning System (OWS)

Type
Laser radar.

Description
HELLAS is an active radar-based warning system. It is designed to provide warning of all types of obstacles like towers, masts, buildings, terrain features and especially wires and cables to helicopters flying nap-of-the-earth and other low-altitude missions, at up to 1,000 m range, both for military and civil operations.

Obstacle detection is performed with an imaging eye-safe radar, which generates images of the scene in front of the helicopter, while range data processing is performed in the processing unit. Processing results can be configured for display as warning information for the pilot.

Depending on customer-specific requirements, the warning information can be indicated by means of the following display media:

- an optical warning indicator (modes: day, night, NVG).
- a helicopter display unit (MFD, HUD, HMSD).

Additionally, the system will provide a warning signal at the output of an acoustic alarm generator once a specified safety threshold is not observed by the pilot. Hellas gives a timely warning to the pilot with a high probability of detection of greater than 99.5 per cent.

Electrical and mechanical interfaces to the helicopter are standard interfaces (ARINC, STANAG MIL-1553) and will allow the fit to all types of helicopters.

Operational status
Development contract awarded by the Federal Office of Defence Technology Procurement. The system has been successfully tested since 1995 on a CH53, UH1D, Bk 17 and EC145. The German Federal Border Guard has ordered 25 units for its new EC-1 35 and EC-1 55 helicopters which are now in service. Canada's DREV , Defence Research Establishment has ordered Hellas for its research programme on a Bell 406.

Specifications
Sensor type: 3-D image - scanning laser radar
Laser transmitter: solid state at 1.55 μm, 4 kW (10 kW possible)
Laser Class III B, NOHD = 0 m; eye-safe
Receiver: In GaAs APD Hybrid (valance photo diode)
Scanning: 2 axes: horizontal fibre-optic, vertical oscillating mirror
Image Repetition Frequency: 2 - 4 Hz
Field of view: 32° v × 32° h (32° v × 40° h possible)
Range:
≥1,000 m (extended area objects, good visibility)
≥400 m (Extended objects, adverse weather)
≥600 m (wires ≥10 mm, good visibility)
≥300 m (wires ≥10 mm, adverse weather oblique incidence
Range resolution: 1 m
Angular resolution: ≤0.35° h, ≤0.2° v
Pixels: 95 h, 200 v
Volume: <36 l
Weight: <27 kg

Contractor
Dornier GmbH Friedrichshafen Germany, an EADS Company, Business Unit: Systems & Defence Electronics.

UPDATED

Hellas fully integrated into a Bk 117 of the German All-Weather Rescue Helicopter programme, AWRH. Test flights in 1997 and 2000 0113452

LaserLine laser visual guidance system

Type
Flight aid - laser system.

Description
The LaserLine system comprises two separate units: the Laser Centerline Localiser (LCL) which provides the pilot with night-time long-range visual guidance for line-up with the centreline of the runway during final approach to landing; and the Laser Glidescope Indicator (LGI) which provides similar glide path information.

Both units use a series of low-power laser beams to illuminate approach corridors. The LCL uses seven fan-shaped laser beams, red, yellow and green. The pilot approaching the runway sees a yellow light when on course, red when left of the centerline and green when right of the centerline. These lights flash at different rates depending on how far the plane is adrift. The outer LCL corridors also serve as turn indicators - the pilot initiates a turn when the rapidly flashing corridor is sighted. The steady corridor alerts the pilot that intercept is nearly complete and the appearance of the yellow light that it is complete.

Similarly, the LGI uses a system of five fan-shaped laser beams to give long-range visual glide slope information. When a pilot is well below glide slope a flashing red signal will be displayed, when slightly below a steady red signal and, when on glide slope a yellow light will be shown. If the pilot overshoots the glide slope the LGI will change to corresponding green signals.

The LCL and the LGI are positioned at different locations on the runway so that the lights will not be confused. The LCL is positioned off the approach end of the runway on the line that is an extension of the runway centerline, typically 25 to 200 ft from the end. The LGI is normally positioned off the right side of the runway so that the middle of the projected beams is parallel with the line of the approach path glide slope.

The brightness and coherence of laser light allows for long-range acquisition of signals and well-defined corridors, while its monochromatic nature means it does not change colour in haze or fog. The beams are only visible while in the approach corridor.

Operational status
In service with the US Navy on board the USS Constellation aircraft carrier since 1994 and there are plans to install it on other carriers. Currently undergoing commercial certification with the US FAA.

A system for marine use (the LSL) has also been developed where a vessel approaching either a harbour or an offshore platform would utilise the beams to ensure they are within safe water routes during their approach. It is believed the first LSL was installed atop the US Coast Guard building at Newport City Harbour, California, in 1996.

Specifications
Dimensions: 36 × 42 × 6 in (LCL); 20 × 42 × 5 in (LGI)
Power requirement: 140 W (LCL); 100 W (LGI)
Power supply: power cable or 24 V battery
Laser: Class 2
Field of projected beams
LCL: 0.5 to 5.5° elevation, ±5° azimuth
LGI: +1.5 -1.2° elevation, 5° azimuth

Contractor
LaserLine Corporation.

VERIFIED

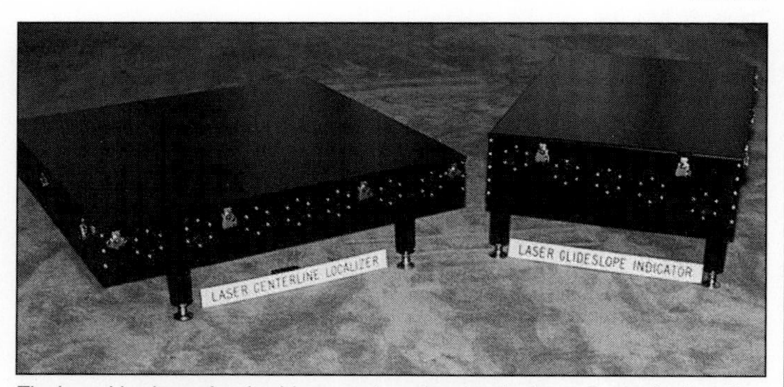

The LaserLine laser visual guidance system showing the Laser Centreline Localiser (LCL) on the left and the Laser Glideslope Indicator (LGI) on the right

Marconi Selenia Communications LOAM laser obstacle avoidance system

Type
Laser radar.

Description
The LOAM laser radar (ladar) obstacle avoidance system provides navigation capability for helicopters, particularly during low-altitude navigation. Based on eye-safe laser technology, Loam will detect any obstacle such as wires, trees or masts.

Audio and visual warnings are given whenever an obstacle is detected along the aircraft flight path. The visual warning includes display of the shape of the obstacle, position, orientation and distance. Background information can be displayed as an option.

Loam comprises three units. The scan/detection unit provides full coverage of the aircraft trajectory. The processing unit provides obstacle detection and recognition. The display unit provides visual information on obstacle position, shape, orientation and distance.

Operational status
In production for the NH 90 and EH 101 (export) helicopters.

Specifications
Horizontal field of view: 40°
Vertical field of view: 30°
Field of view steering: ±20° both in horizontal and in vertical
Max detection range: 2,000 m
Min detection range: 50 m
Obstacle detection probability: >99.5%
False alarm rate: <1 per 2 hours
Qualified according to MIL-STD 810E and MIL-STD 461C
Electrical interface: EIA RS-422 and MIL-STD-1553B
Power supply: +28 V DC, <300 W (MIL-STD-704E)
Eye-safe laser according to STANAG 3606 LAS class I
Sensor head unit
Dimensions: 320 × 239 × 419 mm
Weight: 24 kg
Control panel unit
Dimensions: 146 × 38.1 × 155 mm
Weight: 0.5 kg
Warning Unit
Dimensions: 110.2 × 25.4 × 75 mm
Weight: 0.35 kg
Display unit
Dimensions: 4 × 5 ATI
Weight: 1.7 kg

Contractor
Marconi Selenia Communications SpA.

UPDATED

Northrop Grumman OASYS Obstacle Avoidance System

Type
Laser radar.

Description
Northrop Grumman, in conjunction with the Night Vision and Electronic Sensors Directorate (NVESD) of the US Army Communications and Electronics Command (CECOM), has developed an Obstacle Avoidance SYStem (OASYS). It provides obstacle detection and situational awareness for helicopter low-altitude flight operations, enabling military helicopter pilots to fly at very low altitudes and avoid obstacles such as wires, cables, towers, trees and terrain features.

The OASYS uses a solid-state laser diode radar to detect wires, towers, poles and antennas. The laser, along with a rotating holographic scanner, generates a wide-area scan in front of the helicopter. OASYS can detect 25 mm wires at ranges up to 400 m in poor weather (2 km visibility).

The system comprises a sensor, processor and pilot control panel. Design criteria includes high detection probability, low cost, weight and volume and high reliability. Information is presented on the HUD or can be fed to a helmet-mounted display.

Operational status
Under development. OASYS has undergone successful trials on US Army JAH-1S helicopter and on the NUH-60 STAR (Systems Testbed for Avionics Research) helicopter as part of the Automated Nap Of the Earth (ANOE) programme being conducted by NASA's Ames Research Center. The UK Defence Evaluation and Research Agency (DERA) is currently conducting ground trials of the prototype system prior to a potential future flight test.

Contractor
Northrop Grumman, Electronic Sensors and Systems Sector, Defensive Systems Division.

VERIFIED

Rafael TAWS-05 altitude warning sensor

Type
Flight aid - laser system.

Description
TAWS-05 is an electro-optical sensor for monitoring the tail altitude of civilian and military helicopters. TAWS-05 uses a gallium-arsenide laser diode for high-frequency accurate measurement of the distance to the ground and warns the pilot when the limits of the safe range have been reached.

TAWS-05 represents a major improvement in safety measures for helicopters landing in poor visibility or under adverse atmospheric conditions. It interfaces with standard altitude warning systems and can be integrated into operational aircraft voice communication systems.

Operational status
TAWS-05 has been successfully tested in an Israel Air Force Sikorsky CH-53 in severe flight conditions.

Specifications
Dimensions: 64 × 150 × 34 mm
Weight: 0.5 kg
Power supply: 22-32 V DC, <1.5 VA
PRF: up to 10 kHz
Range: 0.5-15 m
Temperature range: –40 to +70°C

Contractor
Rafael Armament Development Authority.

VERIFIED

FLIGHT AIDS

COMMUNICATIONS AND BEACONS

LFD external lighting systems for NVIS operations

Type
Flight aid – anti-collision, formation, landing and identification.

Description
LFD produce a range of NVIS 'Friendly' infra-red formation lights and anti-collision beacons for both fixed-wing military aircraft and helicopters. These include bulb replacements for NVG operations, infra-red landing and anti-collision lights, and landing site marking systems. All lights are specifically designed for NVG operations and can emit conspicuity or identification codes.

Operational status
In service with UK Royal Navy, Army, Royal Air Force and several overseas customers.

Contractor
LFD Ltd.

SAGEM Telemir secure infra-red communication system

Type
Flight aid - secure, optical infra-red communication system.

Description
Telemir uses an infra-red beam for air-to-air, air-to-ground, ground-to-air and ground-to-ground communications. The airborne equipment consists of an optical head mounted on top of the tail fin and a processing unit. Jamming immunity is stated to be high.

The system is used by a carrier-based aircraft for the reception of navigational updating and reference data such as altitude, location and speed from the ship's inertial navigation system for aligning their own INS. It is also used for covert omnidirectional transmissions between aircraft in formation. A new version is now available with a MIL-bus 1553 datalink.

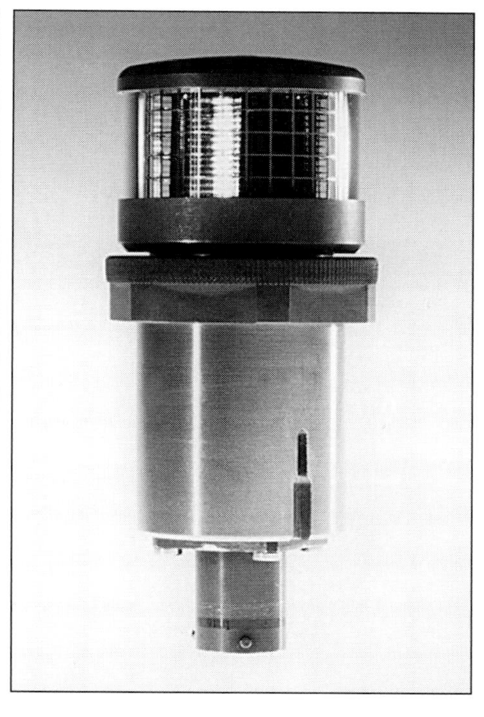

The Telemir optical head is fitted on top of the tailfin for infra-red data transfer (Rafale version)
0023388

Operational status
In service in French Navy Super Etendard and Rafale-M aircraft.

Contractor
SAGEM SA, Aerospace & Defence Division.

UPDATED

FLIGHT AIDS

PILOT'S THERMAL IMAGERS

BAE Systems Modular navigation/attack FLIR system

Type
Pilot's thermal imager.

Description
The Modular FLIR is designed to provide a passive solution to the requirements of fast jet low-level night navigation and target acquisition. The modular design enables different configurations for a variety of aircraft installations while maintaining economies of scale in production.

The equipment provides a high-resolution TV image for presentation on the pilot's head-up display. Automatic image optimisation with a ground stabilised gate area reduces pilot's workload. Optional thermal cueing provides for automatic target detection for air-to-air or air-to-ground applications.

Operational status
In production with over 550 systems delivered, equipping eight air forces. The Modular FLIR is fitted to Harrier GR.Mk 7/TMk10, Tornado and Hawk 100. The version fitted to the AV-8B Harrier Plus for the US Marine Corps, Italy and Spain is designated as AN/AAR-51.

Specifications
Field of view: selectable for aircraft HUD
Video output: 525 line 60 Hz RS-170, 625 line 50 Hz CCIR
Detector: TED
Spectral band: 8-12 μm
Detector cooling: closed-cycle Stirling
Interface: 1553 databus and/or discrete hard-wired
MTBF: >600 h

Contractor
BAE Systems, Avionics, Tactical Systems Division, Basildon.

UPDATED

The Modular FLIR in the configuration for the Harrier GR.Mk 7 and AV-8B aircraft
0005530

BAE Systems Type 239 turret

Type
Pilot's thermal imager.

Description
The Type 239 is a sensor package for surveillance by day and night and in bad weather. It has also been designed to act as a pilot's night vision system in conjunction with a helmet display system. It consists of a sensor ball with a thermal imager in the 8 to 14 μm waveband, the customer having the option of various US or UK systems.

The electrical interface features tilt and pan position demands and tilt and pan angle outputs. The video output is television compatible.

Operational status
In production since 1989 and in service on the Agusta A 129 Mangusta helicopter and on the HH-3F.

Specifications
Spectral band: 8-14 μm
Cooling: split-cycle Stirling
Pointing accuracy: 0.1°
Azimuth coverage: 260°
Elevation coverage: 135°
Slew rate: 120°/s
Slew acceleration: 1,000°/s
Dimensions: 21.6 × 27.8 cm
Weight: 10 kg
Power requirements: 28 V DC, 115 W

Contractor
BAE Systems, Avionics, Tactical Systems Division, Edinburgh.

UPDATED

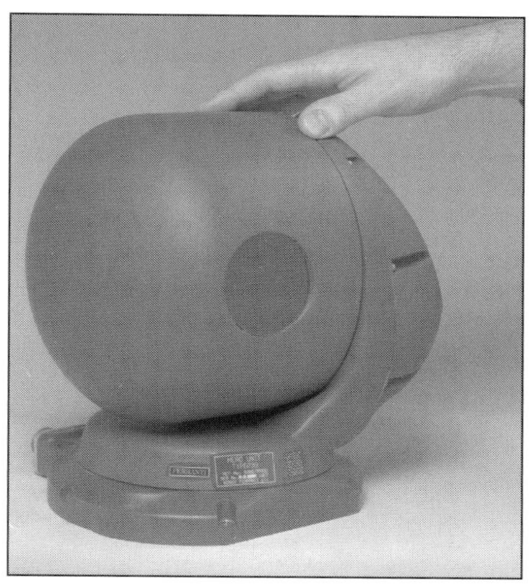

The BAE Systems Type 239 pilot's night vision system

Type 239 pilot's night vision system on Agusta A 129 helicopter
0005540

Galileo Avionica Pilot Aid and Close-In Surveillance (PACIS) FLIR

Type
Pilot's thermal imager.

Description
The Pilot Aid and Close-In Surveillance (PACIS) FLIR is a thermal imaging system in the 8 to 12 μm range, designed to be installed on helicopters and fixed-wing aircraft in order to provide them with increased capability by day, night and in adverse weather operations. It creates a TV-compatible video signal for viewing in the cockpit on a standard display.

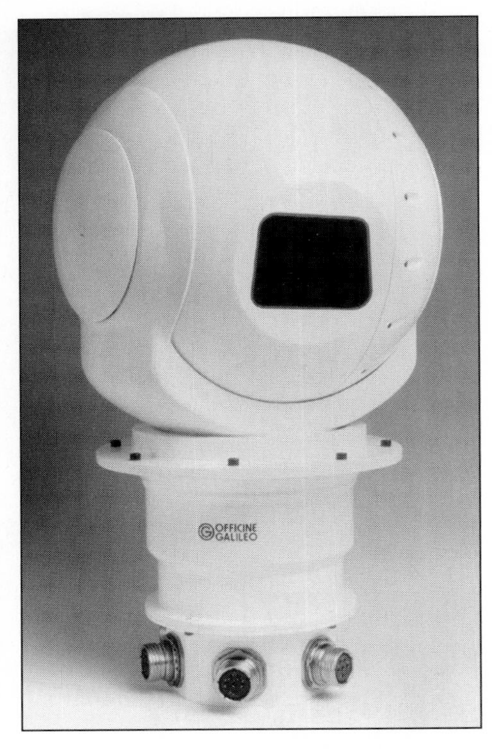

The steerable platform for the PACIS FLIR

PACIS can be used for navigation, day/night surveillance, border patrol, search and rescue, remote sensing and monitoring or as a take-off and landing aid. PACIS can be interfaced with the aircraft avionic system.

The system is composed of a steerable platform, electronic unit and FLIR control grip connected by a cable to the electronic unit. The platform aims the FLIR optical axis in azimuth and elevation. The FLIR is equipped with a two field of view telescope: the wide field of view is used for navigation and surveillance, while the other is used to identify and track targets.

Operational status
Available.

Specifications
Spectral band: 8-12 μm
Detector: 8-element CMT SPRITE
Cooling: split-cycle Stirling
Field of view
Wide: 40 × 26.7°
Magnification: ×1
Infra-red lines: 512
Frame rate: 25 Hz
Field of regard
Azimuth: ±170°
Elevation: +45 to −70°
Slew rate: 120°/s
Video format: CCIR 625 lines at 50 Hz
Interface: RS-422
Dimensions
Steerable platform: 300 × 511 × 300 mm
Electronic control unit: 170 × 225 × 390 mm
Control panel: 146 × 124 × 165 mm
Weight
Steerable platform: 23 kg
Electronic control unit: 8.5 kg
Control panel: 1.5 kg
Power supply
Average: 28 V DC, 140 W
Peak: 450 W

Contractor
Galileo Avionica.

UPDATED

Lockheed Martin Pathfinder navigation/attack system

Type
Pilot's thermal imager.

Description
Lockheed Martin has applied the technology of its LANTIRN system to produce a navigation aid for fixed-wing aircraft operating at night in conditions of poor visibility. The system is called Pathfinder (PAssive THermal Forward-looking Infra-

The Lockheed Martin Pathfinder is a FLIR sensor pod derived from the LANTIRN navigation pod

red for Navigation, Detection and Enhanced Resolution). It is essentially the AN/AAQ-13 without the terrain-following radar.

Pathfinder's FLIR presents a television-standard picture to the pilot on his head-up display.

The system consists of three line-replaceable units integrated into a pod, a pod adaptor to preserve a weapons station, or an embedded element in the aircraft structure. The LRUs are the FLIR, the power supply and the environmental control unit. The first two of these LRUs are derived directly from LANTIRN but the environmental control unit is smaller and lighter, with reduced power requirements.

Pathfinder imagery may be presented on a HUD or any other cockpit video display. Unlike the LANTIRN navigation pod FLIR, which has only a wide field of view, the Pathfinder FLIR is dual field of view and steerable; the pilot being able to slew the FLIR throughout the field of regard and zoom to a narrow field of view.

Pathfinder has been demonstrated in a pod on the Lockheed F-16, integrated into a pylon on the LTV A-7 and embedded in the Rockwell B-1B. It is suitable for installation on a wide variety of additional aircraft including the Northrop Grumman F-5, Fairchild A-10, Lockheed C-130, Dassault/Dornier Alpha Jet, Dassault Rafale, British Aerospace Hawk and Panavia Tornado.

Operational status
Flight trials were conducted in a YA-7E from October 1987 and in an F-16 from November 1987. The system is offered for the F-16 and the A/OA-10 FLIR upgrade programme. It has also been offered for a wide variety of training, transport and combat aircraft. Deliveries began in 1991. Pathfinder together with Sharpshooter is operational in Egypt and has been ordered with Sharpshooter by Taiwan.

Specifications
Dimensions
Length: 1,950 mm
Diameter: 248 mm
Weight: 90.5 kg
Power supply: 115 V AC, 400 Hz, 3 phase, 2.8 kW; 28 V DC, 2 A
Field of regard: 77 × 84°
Field of view
Wide: 21 × 28°
Narrow: 7 × 9°
Temperature range: −40 to +90°C
Reliability: 539 h MTBF

Contractor
Lockheed Martin Missiles and Fire Control.

VERIFIED

Northrop Grumman helicopter night vision system

Type
Pilot's thermal imager.

Description
The system, which started full-scale development in 1986, includes a data entry panel, master control assembly, the Honeywell Integrated Helmet And Display SubSystem (IHADSS) and control panel, video monitor/recorder and control panel, power distribution unit, heater/filter assembly, system control electronics, vapour cycle unit, symbology generator, multiplexing remote terminal, control grips and the Lockheed Martin Pilot Night Vision System (PNVS).

The PNVS is a thermal imaging system mounted as a turret on the chin of the CH-53E and is controlled by the IHADSS worn by the pilot or co-pilot. The IHADSS monitors head position and converts head movement into azimuth and elevation commands which are sent to the PNVS. Video from the turret is displayed and flight symbology is overlaid on the cockpit displays, allowing either the pilot or the co-pilot to fly the aircraft in a head- up attitude.

Operational status
In production and in service with the US Marine Corps on CH-53E transport helicopters.

Contractor
Northrop Grumman, Sensors and Systems Sector, Surveillance & Electronic Warfare Systems.

VERIFIED

Raytheon Advanced Helicopter Pilotage technology demonstrator

Type
Pilot aid.

Description
Raytheon has started development of its Advanced Helicopter Pilotage (AHP) advanced technology demonstrator programme. The technology was scheduled to be available for retrofit into AH-64 Apache and other helicopter models before the year 2000.

AHP technology is designed to prevent night-flying disorientation, caused by the separation between the pilot's head and the thermal imager. It provides helicopter pilots with a wide field of view, head trackable, fuzed view of the world combining image-intensified television and forward-looking infra-red images. Trials and development work of the AHP demonstrator were due to take place on the Boeing (formerly McDonnell Douglas Helicopters) AH-64A.

The system uses image fusion technology developed by Raytheon Instruments and will have a 30 × 50° field of view. The FLIR uses a 480 × 4 interlaced SADA 1B focal plane array and will produce a 1,000-line high-definition television output. It is installed in a common turret with the image-intensified TV sensors. The head tracker is compatible with head roll.

Operational status
In development.

Specifications
Field of view: 30 × 50°
Detector: 480 × 4 interlaced SADA 1B focal plane array

Contractor
Raytheon Systems Company.

VERIFIED

Raytheon AN/AAR-50 NAVFLIR

Type
Pilot's thermal imager.

Description
The NAVigation Forward-Looking Infra-Red (NAVFLIR), formerly Thermal Imaging Navigation Set (TINS) has been developed for the US Navy and the US Marine Corps' Boeing F/A-18D Hornet night attack aircraft and is the equivalent of the US Air Force AN/AAQ-13 navigation pod in LANTIRN.

A US Navy F/A-18 Hornet with NAVFLIR pod mounted on the starboard side near the air intake

NAVFLIR is a derivative of the Raytheon AN/AAQ-16 night vision system but is a pod-mounted system in a fixed, forward-facing position. The pod consists of four Weapon-Replaceable Assemblies (WRA): a FLIR unit, an electronics unit, thermal control unit and adaptor. It produces a TV-like image of the terrain ahead of the aircraft and projects it to the head-up display.

Operational status
In service with the US Navy and the US Marine Corps. About 200 of these systems have been delivered. It is planned that the AN/AAR-50 will also be installed on the F/A-18E2 and F/A- 18F2.

Specifications
Length: 1.98 m
Diameter: 25.4 cm
Weight: 96.61 kg; 73.48 kg (pod)
Field of view: 19.5 × 19.5°

Contractor
Raytheon Systems Company.

VERIFIED

Thales Optronics NAVFLIR

Type
Pilot's thermal imager.

Description
Thales Optronics has designed a state-of-the-art NAVFLIR for visual operation at night in a wide range of climatic conditions. Even under hot and humid conditions, the NAVFLIR's high-sensitivity 3-5 μm infra-red staring array sensor provides a night navigation and attack capability. Integrated with the inertial navigation system, the NAVFLIR significantly enhances the pilot's forward situational awareness. High-resolution infra-red imagery is presented on the head-up display on a one-to-one scale. NAVFLIR improves flight safety during all phases of a mission:

Taxiing, take-off and landing, with no airfield/runway lighting; Navigation, the pilot follows his route visually and is able to manoeuvre as in daylight, having both the horizon and terrain as visual references; Attack, the pilot detects and recognises targets as in daylight, delivering armaments with the same degree of accuracy.

The system is easily integrated onto all types of aircraft, assisted by the compact size.

Operational status
Sold to United Arab Emirates on the Mirage 2000-9.

Specifications
Dimensions: dependant upon installation
Weight: 30 kg
Detector: 3rd generation staring array
Spectral band: 3-5 μm
Field of view: 18 × 24° (or to match the HUD FOV)
Steerable line of sight
Electronic magnification: ×2
Detection range: over 50 km on 100 × 100 m target, over 10 km on 20 × 20 m target
Video standard: STANAG 3350
Databus standard: MIL-STD 1553

Contractor
Thales Optronics.

VERIFIED

FLIGHT AIDS

PILOT'S GOGGLES AND INTEGRATED HELMETS

Aselsan M929/M930 aviator's night vision goggles

Type
Pilot's goggles.

Description
Aselsan's M929/M930 aviator's goggles are designed for the comfort and effectiveness of the user, with eye relief of 25 mm, increased ranges for fore/aft and tilt controls and improved flash response when flares or bright lights enter into the field of view. The M929A and M930A are identical except for their respective mounts. The M929A features a standard mount assembly while the M930A features an offset mount for applications in the Cobra attack helicopter.

Operational status
In production by Aselsan and also by the licensor, Litton in the USA.

Specifications
Magnification: ×1
Field of view: 40°
Image intensifier
M929/M930: GEN III inverting
M929A/M930A: Advanced GEN III Omnibus-III or Omnibus-IV inverting
System gain: 2500 (Omnibus III ,Gen III)
5000 (Omnibus IV Gen III+)
Resolution, min
M929/M930: 1.01 lp/mrad
M929A/M930A: 1.28 lp/mrad
F-number: F/1.2
Eye relief: 25 mm
Weight: 590 g
Power requirement: 2 different power packs available - AA size or 3.9 V lithium

Contractor
Aselsan Inc, Microelectronics, Guidance and Electro-optics Division.

VERIFIED

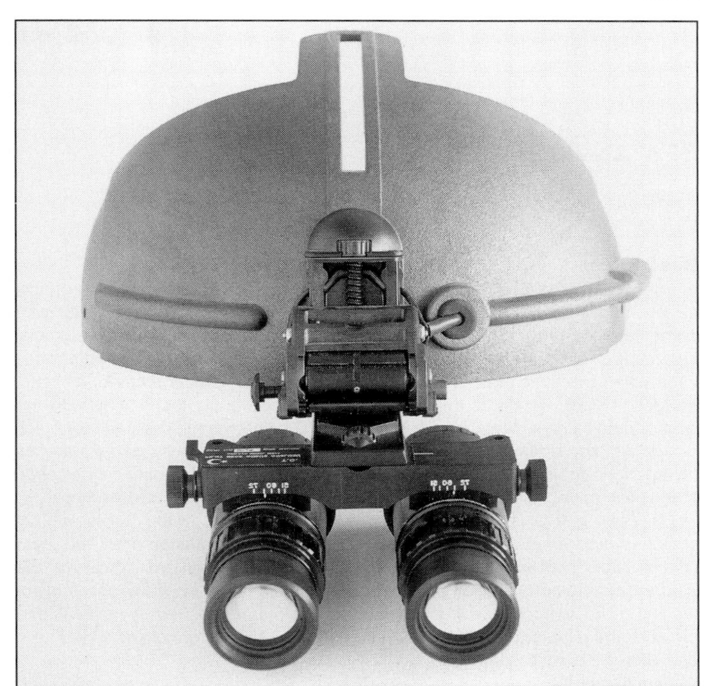

Aselsan M929/M930 pilot's night vision goggles 0099500

BAE Systems Cats Eyes night vision goggles

Type
Pilot's goggles.

Description
In the Cats Eyes night vision goggles the images from the two visual input paths, one direct and the other from the image intensifier, are registered in a 1:1 relationship and so complement one another. The advantages of this arrangement

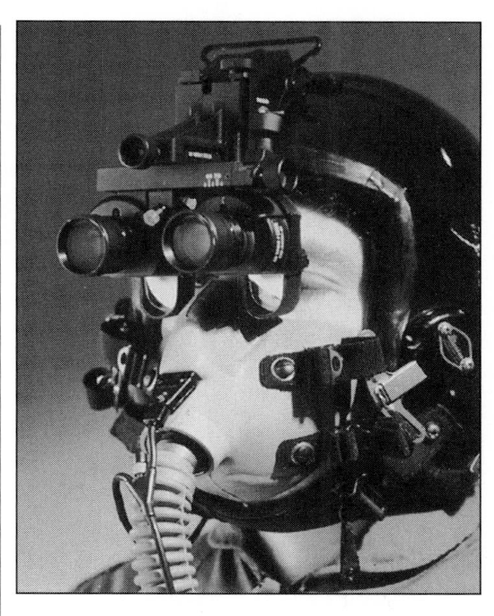

BAE Systems Cats Eyes night vision goggles

have been established during low- level night flying trials involving a fully integrated night vision cockpit with compatible lighting, conventional raster scan head-up display (portraying infra-red images of the outside world produced by a fixed forward-looking infra-red pod) and a head-down multifunction display.

The resolution and dynamic range of the head-up display seen through Cats Eyes is not impaired since it is viewed directly through the combiners as opposed to the image intensifiers. Dusk-to-dark transitions are less problematical because the image-intensifier display becomes more noticeable as the direct outside-world view becomes fainter. The shorter length of the Cats Eyes system permits the pilot a greater degree of head mobility than is possible with conventional NVGs. Cats Eyes can be configured so that an automatic separation device forms part of the overall system.

Operational status
In production. In service on F/A-18/AV-8B and other aircraft. Cats Eyes has been selected as the standard NVG for all US Navy and Marine Corps fixed-wing tactical aircraft. Over 800 Cats Eyes systems are now in service with the US Navy and they are qualified to full US standards, including Carrier EMC requirements.

Specifications
Weight: (incl mount) 803 g
Field of view: 30° circular
Magnification: unity
Eye relief: 25 mm (1 in)
Exit pupil: 10 mm (0.4 in)
Image intensifiers: GEN III

Contractor
BAE Systems, Avionics, Rochester.

VERIFIED

BAE Systems Nightbird night vision goggles

Type
Pilot's goggles.

Description
Designed specifically for fighter aircraft applications, the Nightbird NVG system has all the attributes of the NITE-OP system with the addition of full head-up display compatibility and the enhancement of pilot safety on ejection by means of automatic NVG detachment.

Utilising a special filter, the NVG optics are designed to permit viewing of HUD symbology directly, allowing the pilot to view Forward Looking Infra-Red (FLIR) raster imagery combined with HUD stroke symbology. Nightbird NVGs are designed to be employed with either.

Generation II or Generation III image intensifier tubes, permitting growth/ enhancement of the system as Gen III technology becomes more widely available.

The goggle auto-detach mechanism, activated by movement of the ejection seat and employing a small pyrotechnic charge to operate the goggle detach lever, minimises risk of injury to the pilot from the goggles on ejection by releasing them

BAE Systems Avionics Nightbird night vision goggles for fast jet aircraft

approximately 4 m/s after ejection initiation. The auto-detach system has been extensively tested and is fully qualified in all applicable RAF aircraft.

Operational status
No longer in production. In service with the UK Royal Air Force Harrier GR Mk 7, Tornado GR Mk 1/4 and Jaguar GR. Mk 1/3.

Specifications
Field of view, circular: 45° at 30 mm eye relief
Exit pupil and eye relief: 10 mm at 30 mm
Weight: 815 g
Resolution: 0.85 cycles/mm (typical)
Power source: each channel is independently powered by a 3.5 V, half AA size lithium battery
Battery endurance: 15 h (typical) above 0°C
6 h (typical) at –32°C
Adjustment ranges
Vertical: 26 mm
Fore and aft: 27 mm
Tilt: 20°
Interpupillary: 54-72 mm
Objective: fixed focus
Dioptre adjustment: +0.5 to –2.1 dioptres

Contractor
BAE Systems, Avionics, Rochester.

UPDATED

BAE Systems Nite-Op night vision goggles

Type
Pilot's goggles.

Description
The BAE Systems NITE-OP night vision goggles are specifically designed for helicopter aircrew, enabling them to fly visually at night.

NITE-OP NVGs have a fully circular 45° field of view. The optical design provides large eye relief and exit pupil which facilitates the use of an eye protection visor or NBC protective mask. The NITE-OP NVGs are lightweight and manufactured using modern composite materials. The electrical configuration increases reliability by incorporating the batteries within the NVGs. There are no external wires or connectors and each image intensifier tube is powered independently. Either second- or third-generation image intensifier tubes may be fitted.

Operational status
NITE-OP is now in service with helicopter aircrew of all UK services and has also been supplied to several overseas customers.

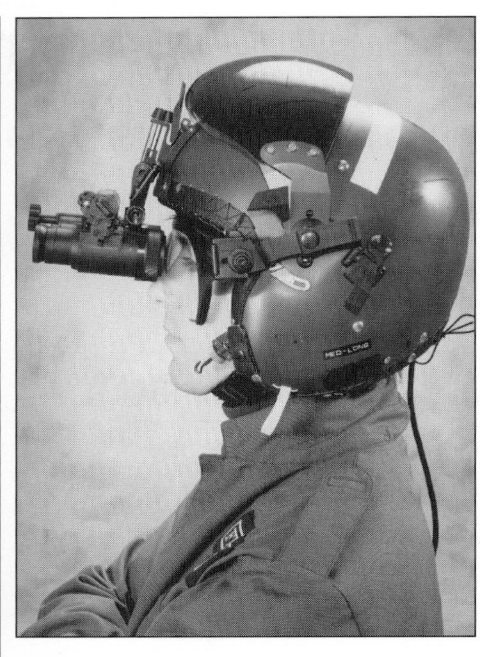

The BAE Systems NITE-OP night vision goggles for helicopters

Specifications
Weight: 0.8 kg
Power supply: 3.5 V lithium batteries (independent for each channel), 15 h endurance at 0°C
Field of view: 45° at 30 mm eye relief
Resolution: 0.85 cycles/mm (typical)
Exit pupil and eye relief: 10 mm at 30 mm
Adjustment ranges
Vertical: 26 mm
Fore and aft: 27 mm
Tilt: 20°
Interpupillary: 54-72 mm
Objective: fixed focus
Dioptre adjustment: +1.0 to –3.1 dioptres

Contractor
BAE Systems, Avionics, Rochester.

VERIFIED

EFW ANVIS/HUD

Type
Pilot's goggles.

Description
EFW's Night Vision Head Up Display (ANVIS/HUD) is an advanced electro-optical system that combines the imagery from the Night Vision Goggles (NVG) with computer-generated graphic and digital symbology. ANVIS/HUD enables crucial information to be communicated to the pilot head-up. This display integrates information the helicopter pilot needs: horizontal and vertical attitude, air data, navigation data, warnings, and more. EFW combines this technology with the comfort and safety of the crew. The ANVIS/HUD is lightweight, designed for quick disconnect and does not restrict head movements. There are no peripheral view obstructions and the helmet can be installed on either eye depending on the needs of the crew member. EFW's ANVIS/HUD contributes to mission effectiveness.

The ANVIS/HUD is a sensory gathering system. The system takes various analogue and digital rotary wing aircraft sensor information into the data accumulator, converts it into the appropriate symbology and transmits the information into an optical combiner that is attached to the objective assembly of the NVG. The symbology is then overlaid on the image provided by the NVG and provides the pilot and co-pilot with an independent display of data obtained from one central data gathering point.

System features:
- Add on existing NVG (ANVIS/HUD) second- and third-generation
- No NVG performance degradation; no NVG modification
- Can be easily installed to either monocular
- Full NVG eye relief for spectacles, NBC mask Quick disconnect for crew egress
- Lightweight
- Wide field of view
- Dual independent displays (Pilot/Co-pilot, Gunner)
- No peripheral view obstruction
- High-resolution display
- No fibre-optic cables
- No restriction to head movements
- Wide range of interfaces
- Individual fine focus adjustment
- Individual display orientation adjustment (Up/Down, Left/Right)

Four modes of operation, and four different de-cluttered modes can be programmed separately by each pilot. Changing modes while flying is made via toggle switches mounted on each collective and the control panel. Individual brightness adjustment using toggle switches mounted on each collective and the control panel.

Operational status
In service.

Specifications
Weight
E.U: 6.8 kg
OCU: 0.91 kg
HDU: 0.45 kg

Power
28 VDC: 2.5 A (MIL-STD-704 A)
26 VAC 400 Hz: 0.5 A (for reference only)
E.M.C/E.M.I: MIL-STD-46C Part 2
Environmental Conditions: MIL-E-5400 Class 1A

Display Unit
Weight (additional on the ANVIS): 120 g
Field of view: 34°
Symbology type: Stroke (X,Y)
Symbology resolution: 0.35 cycles/mrad
Symbology luminance uniformity: 1 .5:1
Symbology contrast: 2 at 0.108 lx (B.C.)
Symbology position stability: 3% of symbology FOV
Linear distortion: ±0.5%
Spatial transmission: 93% (450-1,000 nm)
Symbology focus: +0.000 to –0.050 dioptres

Contractor
EFW Inc.

Eurofighter Typhoon integrated helmet

Description
The Eurofighter Typhoon integrated helmet has been developed to meet all the requirements of the pilot for protection and life support, combined with full Helmet-Mounted Display (HMD) functions. The system is an integral part of the Eurofighter avionics suite, providing Night Vision Equipment (NVE) and Forward-Looking Infra-Red (FLIR) sensor display, combined with full navigation and weapon-aiming symbology. The helmet also provides for interface between the pilot and the Direct Voice Input (DVI) system.

The helmet is a two-part modular and re-configurable design, featuring low head-supported mass, balance and comfort. The requirement for ejection safety at up to 600 kt was found to be the major contributor to the total mass of the helmet, which, nonetheless, has been restricted to 1.9 kg for the day helmet (without NVE cameras) and 2.3 kg for the night helmet (NVE cameras fitted). This compares with approximately 1.8 kg for the day-only Guardian and JHMCS systems (see separate entries) and approximately 1.4 kg for a standard fast-jet aircrew helmet. The inner helmet is designed to provide for the high stability required to maintain the pilot within the design eye position for the optics, while remaining comfortable and providing sufficient adjustment range to fit all head sizes.

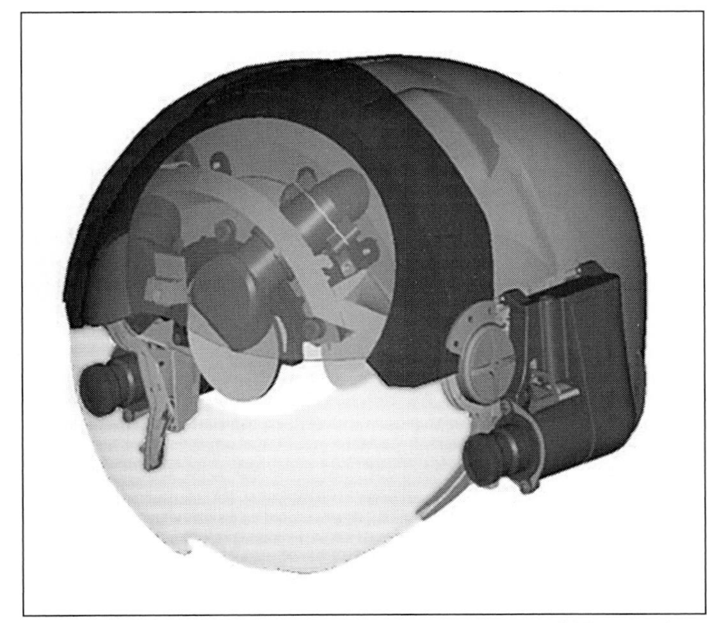

The Eurofighter outer helmet arrangement, showing the twin CRTs, visor, brow mirror, NVE cameras and blast/display visor **NEW**/0121548

The Eurofighter integrated helmet. Note the mounting position for the NVE cameras (not fitted) and outer glare visor (up position) **NEW**/0121555

The Eurofighter integrated helmet, showing mock NVE camera fitted (BAE Systems) **NEW**/0121556

The inner helmet includes a new lightweight oxygen mask, based on the MBU-20/P, which provides for pilot pressure breathing. Other features of the inner helmet are:

- Advanced suspension system;
- Forced air ventilation;
- Brow pad moulded to fit individual aircrew;
- Can be integrated with Nuclear Biological and Chemical (NBC) hood.

The outer helmet attaches to the inner helmet, combining primary aircrew protection with a lightweight and rigid platform for the optical components of the system. The outer helmet incorporates diodes for the optical helmet tracking system and all of the equipment for the projection system: twin CRTs, optical relays, a brow mirror, twin detachable NVE cameras and a blast/display visor. The optical helmet tracking system was chosen for high accuracy and low latency operation within the aircraft's performance envelope, particularly in the areas of instantaneous turn and roll rate capability. The projection system includes a fully overlapped 40° FoV (which can also be monocular if required) and features a high degree of modularity for cameras, NBC, laser protection and so on. High-performance display processing is VME-based, utilising dual-processors. The Eurofighter integrated helmet is a collaborative effort, with the Basic Mechanical Helmet (BMH) by Gentex and the display generation system by Alenia Difesa, based on its work on the Eurofighter Typhoon Enhanced Computer Signal Generator (ECSG), and the Head Equipment Assembly Processor Unit (HEAPU).

Operational status
The Eurofighter Typhoon HEA is a twin-track programme. The BMH will enter service with the first batch of Eurofighter Typhoon aircraft in 2002. The fully capable integrated helmet will enter service with the second batch of Eurofighter aircraft in 2003. All BMHs can be upgraded to full display helmets. Flight test of the system is continuing.

Specifications
Mass
2.3 kg (including NVE cameras)
1.9 kg (NVE cameras detached)

NVE cameras
Wavelength: 0.715 to 0.910 μm

Contractors
BAE Systems (Operations) Limited, leading a consortium comprising: Galileo Avionica Division and Gentex.

NEW ENTRY

Geophyzika GEO-NVG-III night vision goggles

Type
Pilot's goggles.

Description
Geophyzika claim that their GEO-NVG-III (Russian designation reported to be GEO-ONV (Ochki Nochnogo Videniya) night vision goggles employ third-generation GaAs photocathode technology to provide high responsivity in starlight/overcast conditions. The goggles are ruggedised to meet Russian military requirements for low-altitude helicopter combat, reconnaissance and search and rescue operations.

Features include: full 40° field of view, F/1.1 at 43 mm eye relief and 10 mm exit pupil; full binocular night vision; full peripheral vision; automatic brightness control; internal power supply; quick disconnect.

Operational status
Claimed to be in widespread use on Russian military helicopters.

Specifications
Illumination: 10^{-5} to 10 lx
Magnification: ×1
Field of view: 40°
Exit pupil and eye relief: 10 mm and 43 mm
Objective lens: fixed focus 25 mm, F/1.1
Focus range: 300 mm to infinity
Eyepiece lens: 25 mm
Dioptre adjustment: ± 4
Photocathode: GaAs
Luminous sensitivity 2,856 K: 1,200 uA/1m
Radiant sensitivity (830 nm): 120 mA/W
Equivalent brightness input (at 10^{-4} lx): 2.4×10^{-7} 1×
S/N (at 10^{-4} lx): 15
centre resolution: 32 mm
useful cathode diameter: 17.5 mm
Weight: 0.78 kg
Power supply: 3 V DC, 50 mA (2 × AA batteries)
Mechanical adjustment
vertical: 20 mm
fore and aft: 24 mm
tilt: 15°
interpupillary: 56-73 mm

Contractor
Geophyzika-NV.

UPDATED

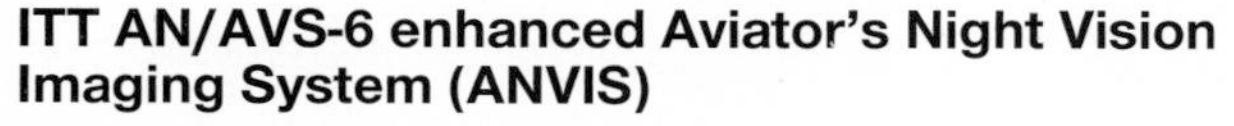

ITT AN/AVS-6 enhanced Aviator's Night Vision Imaging System (ANVIS)

Type
Pilot's goggles.

Description
ITT's enhanced Aviator's Night Vision Imaging System (ANVIS), AN/AVS-6, is a third-generation night vision goggle with new features and improved performance compared to previous versions. The new features include high-quality 25 mm eye relief eyepieces, independent eye-span adjustment for each monocular, smoother focusing, more stable mounting and increased fore-and- aft adjustment.

The enhanced ANVIS incorporates two high-resolution third-generation image intensifiers with higher gain and an increased photoresponse with the same extra-low distortion output optics. Using gallium arsenide technology, the third-generation image intensifiers operate in the near-infra-red region of the spectrum where night light is more abundant and contrasts are higher, thereby delivering improved performance over second-generation systems and a significant increase in operational tube life. The latest versions use tubes to the US Army OMNIBUS IV (OMNI IV) specification.

The 25mm eye-relief and independent eye-span adjustment accommodate a wide range of individual physical characteristics as well as eyeglasses. The system is made of Ultem engineering plastic with improved resistance to chemicals.

The lightweight binocular can be mounted to a variety of helmets. The system is powered by a dual battery compartment power pack that provides the pilot with intermediate reserve power. This power pack uses either universally available AA alkaline batteries or standard military batteries. A minus blue filter screens out the glare from the aircraft instrument lighting and automatic brightness control provides for comfortable viewing.

The enhanced AN/AVS-6 is available for SPH-4, HGU-56, Alpha and CGH helmets.

Operational status
In production and in service.

Specifications
OMNI IV version
Scene illumination: 10^{-6} to 1 fc
Spectral response: visible to 0.9 μm (IR)

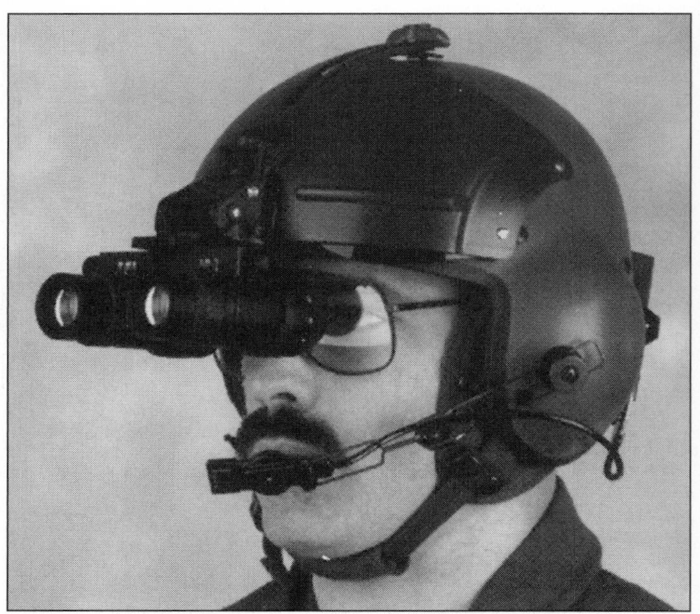

The AN/AVS-6 Aviator's Night Vision Imaging System 0005537

Field of view: 40°
Magnification: unity
Resolution: 1.3 cy/mrad, min
Brightness gain: 5,500 fL/fL, min
Collimation
Convergence: ≤1°
Dipvergence/divergence: ≤0.3°
Dioptre adjustment: +2 to –6 dioptres
Interpupillary adjustment: 51-72 mm
Fore and aft adjustment: 27 mm, range
Tilt adjustment: 10° min
Objective lens: EFL 27 mm F/1.23, T/1.35
Eyepiece lens: EFL 27 mm
Exit pupil eye relief: on-axis: 14 mm @ 25 mm distance; full-field: 6 mm @ 25 mm distance
Focus range: 25 cm to infinity
Flip up/flip down: button release
Automatic breakaway: 11-15 *g*
Voltage required: 2.7-3.0 V DC, battery (50 mA max)
Battery type: AA size alkaline or lithium (BA-5567)
Weight of binocular: 590 g Max.
Mounting
Standard: SPH-4B, APH-4AF, HGU-56P
Optional: Available for Alpha and CGF helmets

Contractor
ITT Defense & Electronics, Night Vision Division.

ITT AN/AVS-9 (F4949) Aviator's Night Vision Imaging System

Type
Pilot's goggles.

Description
ITT's AN/AVS-9 (F4949) was first developed in 1992 to meet specific requirements of the US Air National Guard and, has since evolved into a number of system variants for use with different helmets and for both helicopter and fixed-wing applications.

Versions of AN/AVS-9 that are compatible with ejection seats have a front-mounted battery pack using two AA batteries with power for about 10 hours. For extended operation, the system can be used with a rear-mounted, low-profile battery pack, which uses four AA batteries, the standard pack for helicopter aviation helmets. An adaptor is available to interface the battery pack electrically with an external power source which can extend battery life to 60 hours. An optional clip-on power source is available for both versions enabling hand-held operation.

Both AN/AVS-9 versions (fixed-wing and rotary-wing) are available with a range of image intensifier tubes, including ITT's high-performance third-generation tubes which have photoresponse ranging from 1,350 to 1,550 μA per lumen, signal-to-noise ratio greater than 18 and resolution ranging from 45 to 64 lp/mm. Both versions are also available with a minus blue filter in the objective lens and variants of the fixed-wing system also have the leaky green filter screens optimised for Head-Up Displays (HUD). These filters screen out the glare from NVG-compatible cockpit lighting while allowing the passage of enough light to view HUD symbology

The AN/AVS-9 ANVIS is compatible with the HGU-55/P helmet

The AN/AVS-9 goggles off the mounting 0132090

for both narrow band phosphor and broadband phosphor HUDs. The binocular provides interpupillary adjustment and 25 mm eye relief.

The F4949 is available in the following configurations, offering a range of image intensifier tube performance levels:

- F4949D/F/L: use the F9800J tube (high-performance Omni III)
- F4949G/H/R: use the F9800K/MX-1016 tube (Omni IV MIL Spec)
- F4949P: use the F9800P tube (Omni IV performance)

Operational status
Available.

Specifications
Omni IV
Scene illumination: 10^{-6} to 1 fc
Spectral response: visible to 0.9 μm (IR)
Field of view: 40°
Magnification: unity
Resolution: 1.3 cy/mrad, min (1.36 typical)
Brightness gain: 5,500 fL/fL, min

Collimation
Convergence: ≤ 1.0°
Dipvergence/Divergence: ≤0.3°
Dioptre adjustment: +2 to –6 dioptres
Interpupillary adjustment: 51-72 mm
Fore and aft adjustment: 27 mm
Tilt adjustment: 10° min
Objective lens: EFL 27 mm F/1.23, T/1.35
Eyepiece lens: EFL 27 mm
Exit pupil/eye relief: 14 mm @ 25 mm distance on-axis; 6 mm @ 25 mm distance full-field
Focus range: 41 cm to infinity
Flip-up/flip-down: button release
Automatic breakaway: 11-15 *g*
Voltage required: 2.7-3.0 V DC, battery (50 mA, max)
Battery type: 1/2 AA size lithium (fixed-wing), 2 AA size alkaline (rotary-wing)
Weight of binocular: 550 g max.
Weight of mount: 250 g (fixed-wing), 330 g (rotary-wing)

Contractor
ITT Defense & Electronics, Night Vision Division.

VERIFIED

ITT NW-2000 Night Vision Goggle Camera System

Type
Pilot's goggles (NVG camera).

Description
ITT Industries Night Vision, offers a way to enhance the night-time effectiveness of both new and experienced aviators. By integrating a solid-state miniature COD camera into aviators' night vision goggles, whether fixed or rotary wing versions, the pilot, mission crew, trainers and commanders will have access to the pilot's view through the goggles. The video signal can be recorded on any standard VTR/VCR or, with the proper equipment, transmitted over datalink systems.

Applications
The NW-2000 has a variety of applications. It may be used as a valuable training aid in determining the scan techniques of the pilot during any night mission. The trainer will be able not only to see how the pilot reacts during training situations, but also have a permanent recording of the flight for after-action analysis. In addition, the NW-2000 may be useful for gathering and recording intelligence information, for conducting battlefield reconnaissance and for post-conflict assessment activities including Bomb Damage Assessment (BDA).

Features
The system offers a significantly enhanced image quality as well as a 'halo' suppression mode. These features are the result of a state-of-the-art Video Control Electronics Unit (VCE-2000) that can optimise camera performance in varying light conditions. The VCE-2000 is adjustable for normal, dark, or high ambient light operation modes and is designed to be integrated into aircraft consoles.

The components of the camera system consist of a small COD camera which is integrated into one of the eyepieces of either the AN/AVS-6 or AN/AVS-9 (F4949) night vision goggles. This camera is connected to the VCE-2000 via three cables, each of which have quick disconnect connectors. The VCE-2000 accepts aircraft power for the camera and also provides a RS-170 video signal that can be sent to either a recorder or a radio transmission system. The camera may be used on either the right or left monocular and is located at the side, so as not to restrict look-under capability. The camera is switch activated and may record at any time while the night vision goggles are being used.

The ITT NVG camera 0132096

Operational status
Available.

Specifications
Camera
Black and White: Standard EIAICCIR
Detector: ½ in CCD
Scanning: 2:1 Interlaced

Resolution (NTSC)
EIA (horizontal/vertical): 570/350 TVL
CCIR (horizontal/vertical): 560/420 TVL
Synchronisation: Internal/external
Sensitivity: Down to starlight
Auto exposure: Auto to 1/100,000 sec – dual shutter
Signal-to-noise ratio: Better than 50 dB

Optical
Type: Through sight multiprisms
Field of View:
NW-2000T (narrow), 24°H × 18°V
NW-2000W (wide, 32°H × 26.5°V
Spectral response: 400-700 nm
Attenuation of light transmission: <10%
Eye Relief:
NW-2000T, 6 mm less than standard goggle
NW-2000W, 8 mm less than standard goggle
Dioptre adjustment:
Eyepiece +3 to –3
Camera +3 to –3

Electrical and mechanical
Gross weight:
NW-2000T 110 g
NW-2000W 135 g
Electronic unit dimensions (rack mount): 6 × 5.75 × 2.25 in
Power consumption: 560 mA (max) at 28 V DC (option for 12 V DC)

Environmental and reliability
Constant acceleration (g-loading): 10 g
Operating temperature: –10 to +50°C
Cyclic vibration: 7 g at 11 to 200 Hz
Random vibration: Wf = 0.025 g^2/Hz, We = 0.03 g^2/Hz
Basic shock: 15 g/11 ms
EMI: Per MIL-STD-461B
Relative humidity: 95%

Contractor
ITT Industries, Night Vision Division.

Kaiser Helmet Integrated Display Sight System (HIDSS)

Description
The HIDSS is a second-generation binocular, wide field of view, helmet-mounted display system designed for the RAH-66 Comanche helicopter. Critical for night pilotage, accurate delivery of weapons and improved situational awareness, the lightweight, high-resolution HIDSS utilises a two-piece modular helmet design, advanced optics and precision magnetic tracking to provide head-up, eyes-out operation. Driven by the high-performance SEM-E expanded display unit, HIDSS combines Gen II FLIR video, raster graphics imagery and growth to stroke symbology for a day and night, all-weather helmet-integrated display system that is adaptable to a variety of helicopter platforms and missions.

The HIDSS employs high resolution 1,280 × 1,024 AMLCDs with LED backlight. The manufacturer states that it is the only solid state digital HMD.

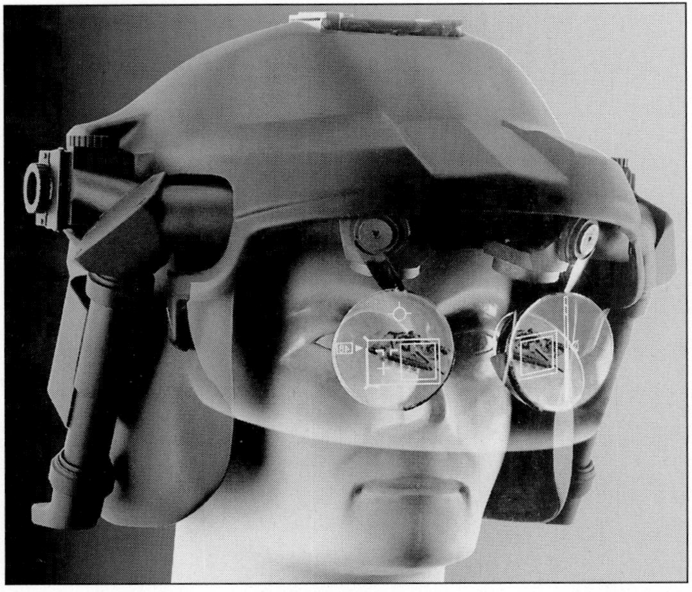

HIDSS is being developed for the RAH-66 Comanche helicopter ***NEW***/0106311

Specifications
Weight:
control panel: 0.5 kg
helmet-mounted display: 1.8 kg
expanded display electronics unit: 14.97 kg
Power supply: 270 V DC, 151 W
Field of view: 30 × 52° (18° overlap)
Field of regard:
azimuth: ±180°
elevation: ±90°
roll: ±180°
Reliability: 1,000 h MTBF

Operational status
Under flight test and qualification for the US Army RAH-66 Comanche helicopter.

Contractor
Kaiser Electronics, a Rockwell Collins business.

NEW ENTRY

New Noga Light ANVIS goggles Model NL-93

Type
Pilot's goggles.

Description
The NL-93 is a self-contained night vision device that enables improved night vision, using ambient light from the night sky (moon, stars, skyglow).

The NL-93 is a helmet-mounted passive binocular that provides the capability for pilots to fly in following contours and nap of the earth flight modes at night.

The ANVIS system consists of a light binocular, which can be mounted to a variety of helmets and powered by battery or onboard aircraft power converter.

The binocular may be flipped up and stored away from the eyes in an emergency.

Operational status
In production.

Specifications
Field of view: 40°
Magnification: ×1
Resolution (cycles/mrad)
GEN II+ (with 40 lp/mm): 0.76
GEN II Super (with 45 lp/mm): 1.0
GEN III (with 45 lp/mm): 1.0
Brightness gain (fl/fl)
GEN II+: 2,000 minimum
GEN II Super: 2,500 minimum
GEN III: 3,000 minimum
Collimation: ≤1/2° convergence/divergence
≤1/2° dipvergence
Dioptre adjustment: +0.5 to –2.1
Interpupillary adjustment: independent, 52 to 72 mm total
Vertical adjustment: 25 mm, range
Fore and aft adjustment: 27 mm, range
Tilt adjustment: 10°, minimum
Objective lens: EFL 27 mm, F/1.23, T/1.35
Eyepiece lens: EFL 27 mm
Exit pupil/eye relief
On-axis: 14 mm @ 25 mm distance
Full-field: 6 mm @ 25 mm distance
Automatic breakaway: 10 to 15 g
Power supply: 2.7-3.0 V DC, battery (50 mA, maximum)
Battery type
Fixed wing: 1/2 × AA, lithium
Rotary wing: 2 × AA, alkaline
Weight of binocular: 540 gm
Weight of mount
Fixed wing: 240 gm
Rotary wing: 330 gm
Operating temperature range: –32 to +52°C

Contractor
New Noga Lite Ltd.

Northrop Grumman Low-Profile Night Vision Goggles (LP/NVG)

Type
Pilot's goggles.

Description
The LP/NVG is a patented folded-optical design that provides a stable, low centre of gravity, self-contained goggle for multimission roles of parachuting; land/water/aircraft operation in both urban and field environments.

Its beam combiner design provides a direct vision path during transitions from low-light conditions into brightly lighted environments. This see-through capability, which provides excellent peripheral vision, offers greater flexibility and situational awareness for helicopter, cargo, and attack aircraft crew-members.

LP/NVG are powered by 1/2 or 2/3 AA batteries. The image intensifiers are turned on by an OFF/ON switch which is controlled separately from the built-in twin IR illuminators. They have their own OFF/ON/ON switch that allows the use of a two position low- and high-illumination settings.

LP/NVG is also compatible with injecting head up display information into the field of view such as symbology, alphanumerics, or FLIR imagery. Integrated GPS navigation information is readily adaptable to the HUD data entry port.

LP/NVG can be supplied with either GEN II or GEN III tubes. Each monocular has an objective lens, an image-intensifier assembly, eyepiece with HUD port input and a beamcombiner. The beam combiner folds the intensified image so that it enters the operator's eye laid on top of the scene which the operator sees directly through the beam combiner. Circuits for the left and right monocular are independent so that a failure in one will not stop operation of the other.

Operational status
Available.

Specifications
Magnification: ×1
Field of view:
40° night vision intensified
165° (H) × 90° (V) unaided eyes
Focus range: 33 cm to infinity
Dioptre adjustment: fixed at −0.25 ±0.25
Exit pupil: 10 mm on axis
Eye relief: 15 mm
Image intensifiers: 2 18 mm
IR LEDs: 2 IR LEDs
Fore and aft adjustment (vertical): 16 mm
Tilt: 14°
Interpupillary adjustment: 55 to 75 mm
Weight: 700 gms with batteries/without mount
Power supply: 3.2 volt lithium bateries
2 × 1/2 AA batteries (or)
2 × 2/3 AA batteries with cap extender
Battery life: 20 h nominal with 1/2 AA

Contractor
Northrop Grumman Electro-Optical Systems.

UPDATED

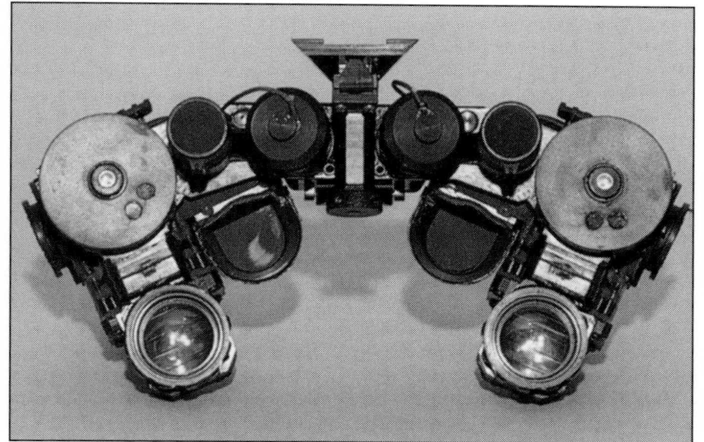

Northrop Grumman Low-Profile Night Vision Goggles (LP/NVG) 0058716

Northrop Grumman M-927/929 aviator's night vision goggles

Type
Pilot's goggles.

Description
The M-927/929 aviator's night vision goggles are improved versions of the US Military ANVIS goggle. M927 uses third-generation image intensifier tubes and M929 uses second-plus generation tubes. Both models mount to a single point on the pilot's helmet visor guard. No face mask is required, permitting nearly full peripheral vision. A flip-up feature allows the pilot to pivot the assembly up out of the way whenever he wants to use unaided vision. The weight of the binocular is counterbalanced by a dual battery pack which mounts to the rear of the helmet.

Northrop Grumman M-927/929 aviator's night vision goggles

Operational status
Production as required.

Specifications
Weight
Binocular assembly: 0.456 kg
Visor mount assembly: 0.2 kg
Magnification: unity
Field of view: 40°

Contractor
Northrop Grumman Electro-Optical Systems.

UPDATED

OIP Sensor Systems Helimun (pilot's helmet stereoscopic night vision system)

Type
Pilot's goggles.

Description
Helimun pilot's night vision system is available with both 40 and 60° field of view. Helimun is designed for pilots of helicopters or light aircraft.

Helimun incorporates SuperGen or XD-4 image intensifiers with a resolution of 0.8 lp/mrad at 1 mlux. The eyepieces have 28 mm eye relief and separate interpupillary adjustment enabling both eyes to see the full 40 or 60° field of view and reducing monocular wobble. In addition the eyepieces permit use of the Gentex inset visor developed for use with night vision goggles.

The power module is built onto the back of the helmet to increase user comfort. The mounting of Helimun can be adapted to fit a variety of pilot's helmets.

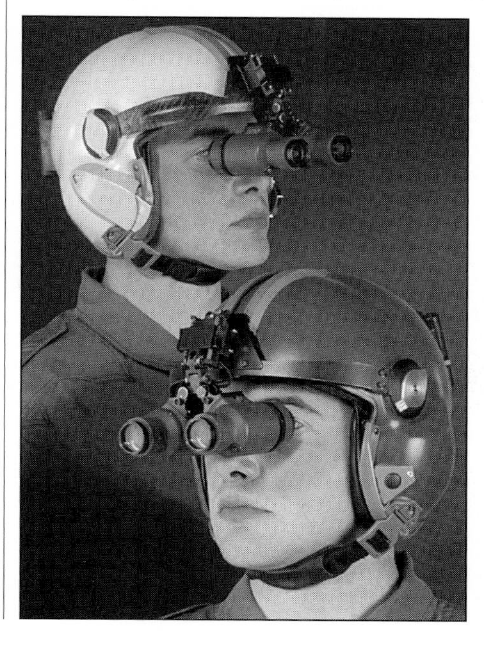

The Helimun helicopter pilot's night vision system
0023390

Operational status
Available.

Specifications

	40° FOV	60° FOV
Magnification	×1	×1
Field of view	40°	60°
Resolution (typical)	≥0.8 lp/mrad	≤ 1.1 lp/mrad
Dioptre adjustment	–5 to +2	–6 to +6
Interpupillary adjustment	52 to 76 mm	52 to 76 mm
Tilt adjustment	10° min	10° min
Exit pupil	10 mm	10 mm
Eye relief	26 mm	19 mm
Focal range	25 cm to ∞	25 cm to ∞
IR LED	incorporated	incorporated
Minus blue filter	standard	standard
Flip-up/flip down	standard	standard
Power supply	2 × AA type batteries plus back-up batteries	2 × AA type batteries plus back-up batteries
Battery life	≥150 h + 150 h	≥ 150 h + 150 h
Weight (excluding battery pack)	610 g	610 g
Environmental	MIL-STD-810E	MIL-STD-810E

Contractor
Thales Optronics.

UPDATED

PCO PNL-1 aviator's night vision goggles

Type
Pilot's goggles.

Description
The PNL-1 aviator's night vision goggles are designed for helicopter aircrew. The goggles have two independent channels using two GEN II, GEN II+ or GEN II Super intensifier tubes. They are mounted on the helmet and have adjustments to match the position of the eyepieces to the face of an individual crew member. In the event of a crash, the mounting system allows the goggles to be automatically snapped off the helmet.

Other features of the goggles are as follows:
(a) the eye relief allows viewing of the instrument panel under and around the goggles
(b) a counterweight prevents the helmet slipping forward and down due to the weight of the goggles
(c) the mounting bracket allows the goggles to be stowed upwards on the helmet, clear of the face.

Operational status
In production and in service.

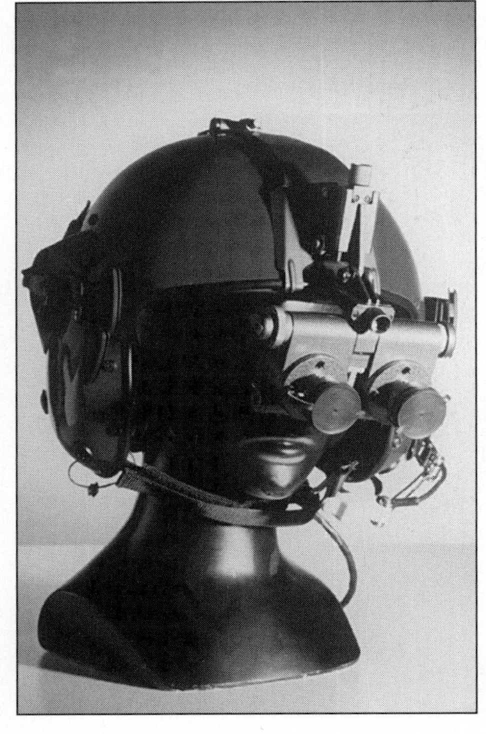

PNL-1 aviator's night vision goggles
0023389

Specifications
Magnification: ×1
Field of view: 38° min
Resolution ability: 5.7 min
Dioptre adjustment: +2 to –4
Focus range: to infinity
Power supply: 1 lithium cell 3.6 V AA size
Total weight: 1.5 kg

Contractor
PCO SA, Poland.

SAGEM CN2H night vision goggles

Type
Pilot's goggles.

Description
The CN2H night vision goggles are part of a range of night vision systems produced by SAGEM for military applications; the goggles are specifically designed for use in helicopters and fixed-wing aircraft for night piloting. They are fixed to the helmet, with a power pack on the back and a specially designed support bracket enables them to be immediately discarded in an emergency. Focusing and positional adjustments are available to suit the wearer. The goggles are compatible with most US and NATO helmets and they incorporate SuperGen or Gen III image intensifiers according to requirements.

Operational status
In production. Adopted by the French Army, Air Force and Navy and exported.

Specifications
Weight: 0.95 kg incl power pack
Power supply: 28 V DC or 3.5 V PS 31 battery or 2 × 1.5 V AA batteries
Battery life: 20 h
Field of View: 40°
Magnification: ×1

Contractor
SAGEM SA, Aerospace & Defence Division.

UPDATED

The CN2H night vision goggles

SAGEM CN2H-AA night vision goggles

Type
Pilot's goggles.

Description
The CN2H-AA Night Vision Goggles (NVGs) are a development of the CN2H. They are designed for aircrew in high-performance military aircraft and can be rapidly released with either hand before ejection. The goggles can also be used by helicopter aircrew. The NVGs are equipped with GEN II or GEN III image intensifiers, for high resolution and high sensitivity and can be mounted on any type of aircrew helmet.

Operational status
In production. Operational with French Air Force.

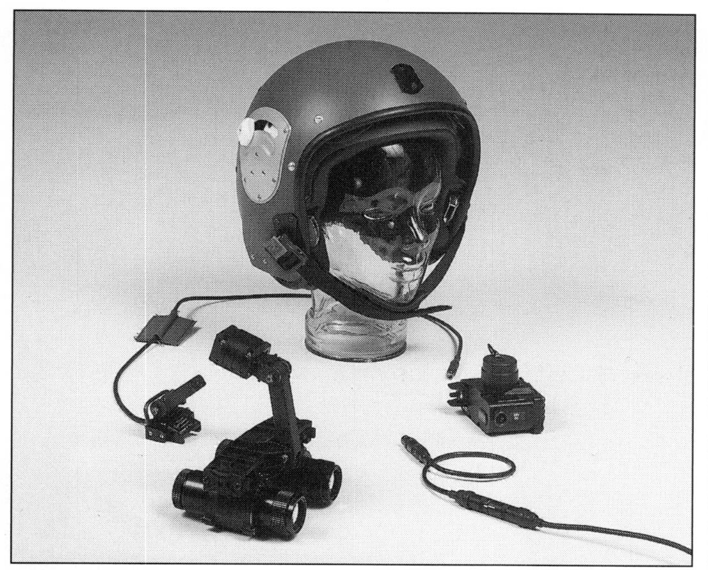

The CN2H-AA night vision goggles are equipped with GEN III image intensifier technology

Specifications
Weight: 0.59 kg
Power supply: 28 V DC or PS31 battery
Field of view: 40°
Magnification: ×1

Contractor
SAGEM SA, Aerospace & Defence Division.

VERIFIED

SAGEM JADE night vision imaging system

Type
Pilot goggles.

Description
JADE is a twin-channel NVG tailored for combat aircraft pilots. JADE is fully compatible with emergency ejection up to 600 kt. The goggles are offered with two high-resolution intensifier tubes of either GEN III or SuperGen type. Thanks to its innovative concept, unmatched by any other system, JADE is suited to operational requirements, in particular, high *g* flight conditions. The direct channel facilitates both the monitoring of the cockpit displays and instruments and viewing the external landscape. The intensified channel simultaneously superimposes the night vision image of the same landscape viewed directly with the naked eye. JADE goggles are compatible with cockpit lighting, designed in accordance with STANAG 3800.

Operational status
JADE has passed the wind blast qualification testing in the Centre d'Essais Aeronautique of Toulouse (CEAT) and flight testing in the Centre d'Essais en Vol (CEV). The equipment has successfully completed operational flight tests carried out by French Air Force at the Centre d'Expériences Aériennes Militaires (CEAM).

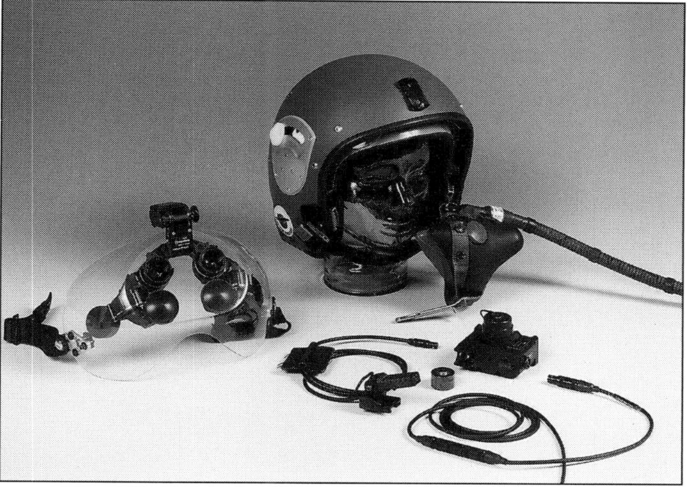

The JADE night vision system for combat aircraft is compatible with non-prepared ejection
0005536

Specifications
Intensified channel
Centre resolution
at 10^{-1} lux: ≤1.5 mrad
at 10^{-3} lux: ≤2 mrad
Gain: ≤2,700 at 10^{-2} lux
Direct vision channel
Weight
Goggles: 500 g
Equipped visor: 150 g
NVG mount: 250 g

Contractor
SAGEM SA, Defence & Security Division.

VERIFIED

Vectop Night Witness GVC-2000 W/T

Type
Pilot's goggles (NVG camera).

Description
The Vectop Night Witness CCD camera system fits to one channel of a pilot's NVGs. The camera allows recording of all that the pilot sees in the manner of a HUD camera. This enables a training or operational sortie to be analysed and debriefed. The camera interfaces with a cockpit mounted electronics unit and with a video recording system.

Operational status
Available.

Contractor
Vectop Ltd.

VERIFIED

OBSERVATION AND SURVEILLANCE

AIR INTERCEPTION

BAE Systems infra-red search and track demonstrator

Type

Air intercept sensor.

Description

BAE Systems (then Marconi) completed development of an Infra-Red Search and Track (IRST) technology demonstrator for the UK MoD in March 1995. The development of a pointing and stabilisation mechanism, a new high-performance telescope, IRST algorithm suite, target detection and tracking algorithms together with a high-performance thermal imager, completes the demonstrator equipment.

The system provides three principal modes of operation:

- Autonomous search mode – provides the capability to locate and track multiple targets over a wide field of search
- Slaved acquisition mode – accepts position demands from the mission computer, which may be sourced from the pilot's 'hands on throttle and stick', the pilot's helmet or from external sensors such as radar. Once positioned, targets within the single field of view are tracked
- Single target track mode – provides the capability to track a single designated target within the field of regard. The sensor field of view is centred on the target position, located during the autonomous search or slaved acquisition modes and uses image tracking techniques.

The system provides the functionality and performance required of an in-service equipment and can be upgraded to operate with long linear arrays and focal plane arrays.

Operational status

The equipment is installed in the UK DERA (Defence Evaluation and Research Agency) trials Tornado.

Contractor

BAE Systems, Avionics, Tactical Systems Division.

UPDATED

BAE Systems' IRST demonstrator on the nose of the DERA Tornado (Crown Copyright) 0005542

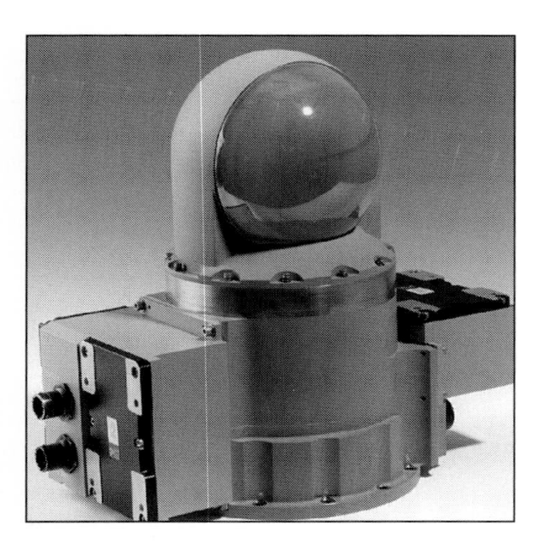

The sensor head of BAE Systems' IRST demonstrator 0005541

Galileo Avionica, Thales Optronics, Tecnobit PIRATE infra-red search and track system

Type

Air (and air-to-ground) intercept sensor.

Description

The Passive Infra-Red Airborne Tracking Equipment (PIRATE) is the FLIR/IRST for the Eurofighter Typhoon. It is in an advanced stage of development by an international consortium led by FIAR of Italy, with Thales (formerly Pilkington) Optronics of the UK and Tecnobit of Spain as major partners.

PIRATE detects the infra-red signature of aircraft at long ranges (beyond visual range), over a wide field of regard. Passive operation assists the stealthiness of an interception. It also enhances situation awareness in both air-to-air and air-to-ground missions. Main functions include:

- Automatic detection and multiple target tracking (track while scan IRST mode).
- Provision of IR imagery for cockpit displlays and helmet mounted displays (FLIR mode).

PIRATE is said to operate in cluttered scenarios in each of the following IRST modes:

- Multipe Target Tracking, where engagement (detection, tracking and prioritisation) of multiple targets in air-to-air, look up/down operation over the whole field of regard or over a selectable volume are possible. Passive ranging is also available.
- Single Target Tracking, where angle tracking of a single acquired target can be selected. Automatic target reacquisition is available.
- Slaved acquisition, where line of sight is slaved to externally defined pointing angles and automatic acquisition of a single target activated.

PIRATE also provides IR imagery to cockpit displays in the following FLIR modes:

- Flying/Landing Aid, where the IR picture is presented on the HUD, overlaid on the pilot's view of the outside world. Thermal cueing of ground targets is also provided.

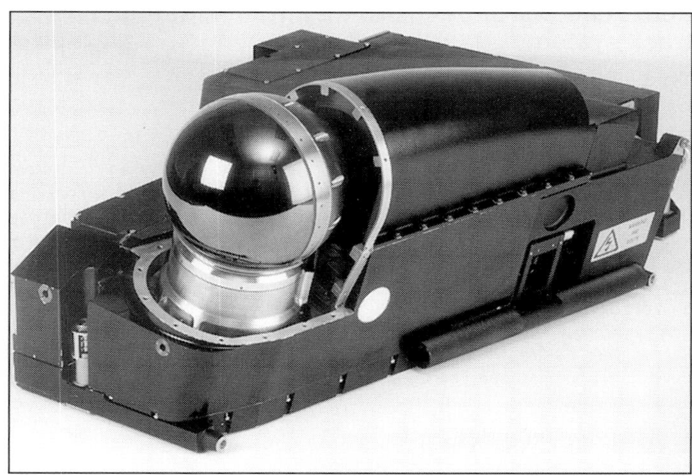

The PIRATE Passive Infra-Red Airborne Tracking Equipment being developed for Eurofighter Typhoon 0089883

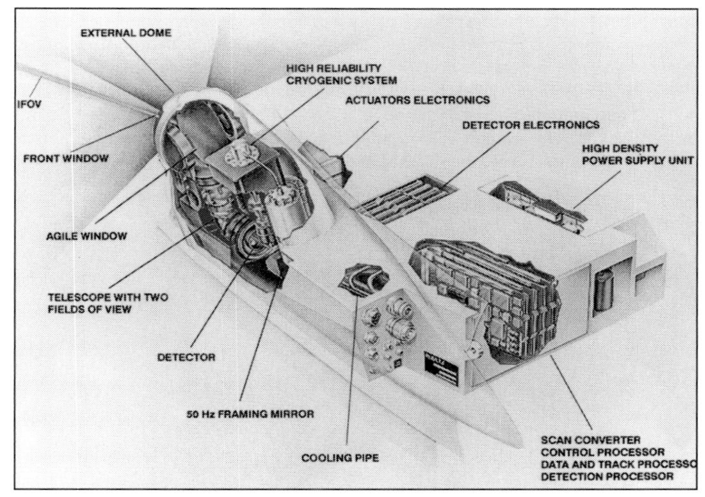

PIRATE system diagram 0058717

The Eurofighter Typhoon to be equipped with PIRATE 0089901

- Steerable IR picture on helmet, where the IR picture is presented on the HMD with the sensor pointing angle slaved to the helmet by means of a head tracker.
- Identification, when in Single Target Tracking the IR image can be presented to the pilot on the HDD and frozen to allow visual identification.

Operational status

A development contract to supply equipment for the Eurofighter Typhoon aircraft was awarded in 1992. Flight trials were conducted during 1999 using pre-production hardware, with full system testing and development of production-standard equipment during 2000/2001.

Contractors

Galileo Avionica (formerly FIAR)
Tecnobit, Spain.
Thales Optronics, United Kingdom.

UPDATED

Lockheed Martin AN/AAS-42 infra-red search and track system

Type

Air intercept sensor.

Development

The AN/AAS-42 Infra-Red Search and Track (IRST) mid-wave (MW) and long-wave (LW) systems were being produced by GE Aerospace but the programme was taken over by Lockheed Martin Electronics & Missiles in 1994. The programme was originally funded by both the US Air Force and Navy, the former planning to employ the IRST on modified F-15 aircraft. The US Air Force has now pulled out (although its Aeronautical Systems Division remains the contracting authority), citing the high cost of converting the aircraft. The US Navy is continuing with the IRST, however. The systems underwent extensive flight test evaluation at the US Navy's Pacific Missile Test Centre, Pt Mugu, California, as part of the F-14 IRST full-scale development programme. Flight test results on modified F-14A test aircraft met or exceeded performance specifications for the F-14D aircraft. Both the MW and LW IRST systems have demonstrated performance at operationally significant ranges. In addition, the IRST's imaging capability was demonstrated in flight test and further system enhancements are planned.

Description

The F-14D IRST is designed to permit the multiple tracking of thermal energy emitting targets at extremely long ranges to augment information supplied by conventional tactical radars. Using a series of sophisticated filter techniques and software algorithms, candidate detections are processed, background clutter is rejected and real targets are declared and displayed to the aircrew.

The F-14D IRST consists of two Weapon-Replaceable Assemblies (WRAs): a sensor head mounted beneath the nose of the F-14 and an electronics unit just aft of the cockpit. The system will be integrated with the F-14's central computer system and complements the AN/APG-71 radar.

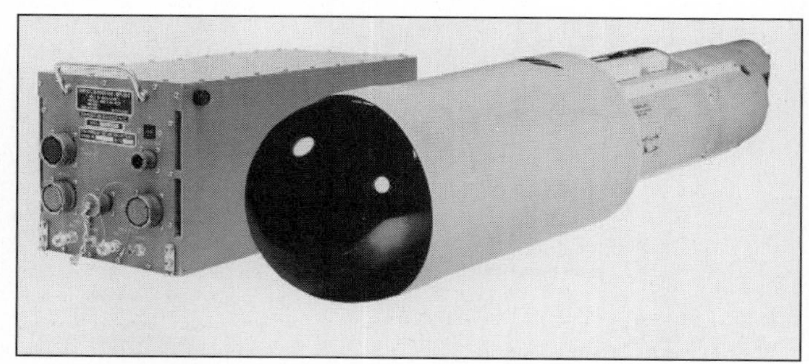

The AN/AAS-42 infra-red search and track system for the F-14D

The sensor head, mounted in a chin pod, contains the optics and infra-red detector assembly with either mid-wave (3 to 5 μm) or long-wave (8 to 12 μm) spectral band. A three axis inertially stabilised gimbal allows the system to accurately search multiple scan volumes either automatically or under manual pilot control. Required signal and data processing is performed in the air-cooled controller-processor.

The IRST functions in six separate modes similar in operation to the APG-71 tactical air intercept radar. Both azimuth and elevation scan volumes are selectable and separately controlled by the aircrew.

The F-14D is the first production US fighter in 25 years to be IRST equipped. The US Navy has approved release and demonstrations of the full performance AN/AAS-42 IRST system to several US Allied nations. The Boeing Airborne Laser (ABL) team selected the AN/AAS-42 IRST to provide early missile launch detection and tracking for the ABL programme (qv).

Operational status

Production completed for the US Navy. Operational on F-14D Tomcat fighters.

Specifications

Dimensions
Pod: 1,366.6 × 248 mm diameter
Sensor head: 914.4 × 228.6 mm diameter
Electronics unit: 190.5 × 190.5 × 482.6 mm
Weight
Pod: 86.5 kg
Sensor head: 41.28 kg
Electronics unit: 16.78 kg

Contractor

Lockheed Martin Electronics & Missiles.

UPDATED

Saab Bofors IR-OTIS – Infra-Red Search and Track (IRST) system

Type

Air intercept sensor.

Description

Saab Bofors Dynamics is currently working on design and test of a prototype of IR-OTIS, a multifunctional Infra-Red Search and Track (IRST) system, intended to provide passive situation awareness for the JAS 39 Gripen aircraft at long range, during day and night operations, against air and ground targets. The system can operate both as an IRST and a FLIR with a large scanning field. In the IRST mode, the system scans an assigned part of the airspace, detects targets automatically and tracks them while continuing to scan, thus giving the system a multifunction track-while-scan capability. In this mode, the system can scan with a narrow field of view to give long-range detection or with a wide field of view to cover a large sector in a short time. In the FLIR mode, the system generates an image covering the field of view in the assigned direction. In air-to-air combat, the IRST mode is used for automatic search, including track-while-scan, or FLIR mode in a designated direction with automatic lock-on and continuous tracking. In air-to-ground combat, the FLIR mode is used for observation and continuous tracking.

Artists impression of the IR-OTIS on the nose of the the Gripen 0005543

Operational status

In development for the Swedish Defence Material Administration for JAS 39 Gripen; the first development model was flight tested on a JAS37 Viggen aircraft. Since the sensor system is designed to be fitted on the Gripen and requires relatively little space, integration in other aircraft should be possible either in a new build arrangement or as retrofit equipment.

Specifications

Sensor wavelength: 8-12 um
Sensor elements: 1,100-1,200 elements
Field of view: several, selectable
Field of regard: more than one hemisphere, but limited by aircraft installation
Sensor unit: 30 kg and 30 litres volume
Signal processing unit: 10 kg and 10 litres volume

Contractor

Saab Bofors Dynamics AB.

Thales, SAGEM Optronique Secteur Frontal (OSF)

Type

Air intercept sensor.

Development

The Optronique Secteur Frontal (OSF) electro-optic, visual and infra-red search and tracking system is the result of a six-year collaboration by Thales Optronics and SAGEM SA under the support of the French Ministry of Defence (DGA). FOS represents one major assembly of the Dassault Aviation Rafale fighter Navigation and Attack System (SNA, Système de Navigation et d'Attaque), along with the SPECTRA countermeasures system and RBE2 radar.

Description

The FOS is a multifunction system adapted to the MICA missile fire-control system although it can also be suited to other such systems. It meets the operational requirements for both the French Air Force and the Navy, which will use it in the Rafale's weapon system.

FOS on Rafale 0058718

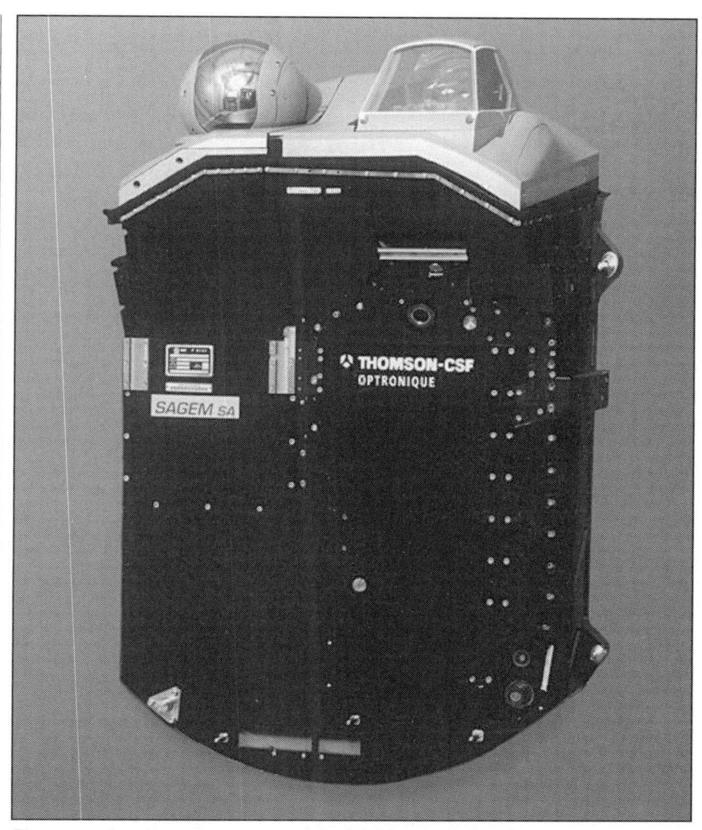

The sensor heads and processor of the FOS for the Rafale Aircraft 0005545

The Rafale's weapon system will use a variety of techniques to perform air-to-air and air-to-ground ranging, tracking and targeting functions including infra-red, television and laser rangefinding. A multispectral capability allows the FOS to perform several of these functions in parallel.

Operational status

In development for the Rafale-M of the French Navy but not yet operational on the aircraft.

Contractors

SAGEM SA Aerospace & Defence Division.
Thales Optronics.

UPDATED

Urals Optical and Mechanical Plant OEPS-27/29

Type

Targeting sight.

Description

OEPS-29 and OEPS-27 are optical-electronic laser sight systems which have been developed by the Urals Optical and Mechanical Plant (UOMZ), which has been making sights for combat aircraft of the former Soviet Army since 1941.

OEPS-29 is installed on the MiG-29 and provides the Infra-Red Search and Track (IRST) functions of search, detection and tracking of targets at all altitudes, day or night. It can range either air targets or ground targets for gunfiring. The sight system is supplied with a target acquisition device SH-3UM-1, which is fixed to the pilot's helmet. The system can track up to six targets.

OEPS-27 is installed on Su-27 fighters and derivatives. It has the same functions as OEPS-29 but has a wider field of view and longer range for air targets.

Operational status

In serial production for MiG-29, Su-27 and Su-30 fighters.

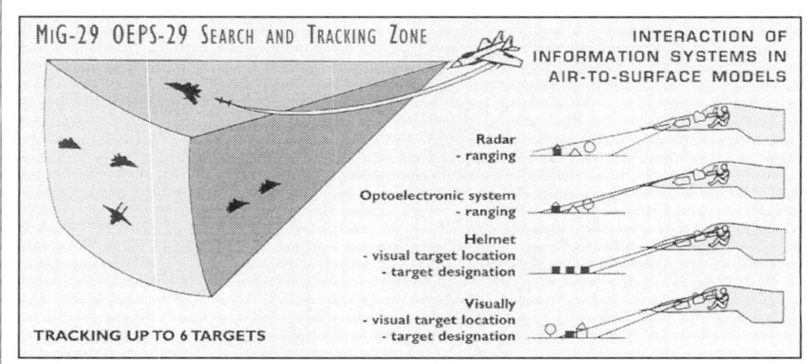

MiG-29 OEPS search and tracking zones 0003329

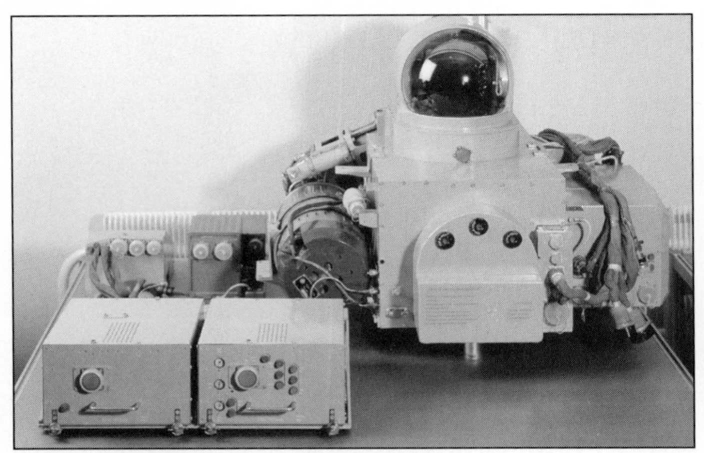

OEPS-27 optical electronic sight system 0089874

Specifications

OEPS-27

Field of regard

Azimuth: ±60°

Elevation: +60 to −15°

Weight: 174 kg

OEPS-29 optical electronic sight system 0089876

OEPS-29

Field of regard

Azimuth: ±30°

Elevation: + 30 to −15°

Weight: 78 kg

Contractor

Production Association Urals Optical and Mechanical Plant (UOMZ).

VERIFIED

OBSERVATION AND SURVEILLANCE

TURRET SENSORS

Aselsan ASELFLIR-200 airborne FLIR

Type
Turret sensor.

Description
The ASELFLIR-200 Airborne Thermal Imaging System is a light-weight, multipurpose, thermal imaging sensor for pilotage/navigation, surveillance, search-and-rescue, automatic tracking, target classification and targeting. The ASELFLIR-200 is an open architecture and hardware/software flexible unit which can be adapted to various air platforms, including rotary-wing, fixed-wing and unmanned air vehicles.

Key features of ASELFLIR-200 include Electronic Image Stabilisation (EIS), Local Area Processing (LAP) for image enhancement, MultiMode Tracking (MMT), analogue and digital video outputs for transmission and/or recording, MIL-STD 1553/ARINC and other discrete data busses to interface with onboard avionics such as radar, navigation and weapon systems. The ASELFLIR-200 has three Fields of View (FOV): narrow FOV for recognition and identification, medium FOV for detection and a unity FOV for navigation and pilotage.

The system has single-, dual- and triple-sensor configurations as following:

- FLIR only
- FLIR + colour CCD or FLIR + eye-safe laser range-finder
- FLIR + colour CCD + eye-safe laser range-finder

ASELFLIR is in full production and installed on various rotary-wing and fixed-wing platforms. It incorporates a second-generation 4 × 240 focal-plane array detector which is operating in the 8-12 μm band. The key features of the system provide greatly improved range performance over the conventional first-generation linear array detectors and improve mission capability.

There are two weapon replaceable assemblies: turret unit, WRA-1 and electronics unit, WRA-2. Options include a laser range-finder and/or a CCD day TV camera.

Operational status
In production and installed on various rotary- and fixed-wing platforms.

Specifications
Dimensions: turret unit 323.85 × 372.87 mm
electronics unit 306.3 (w) × 413.5 (l) × 199.1 mm (h)
Weight: turret unit <31.8 kg
electronics unit <22.73 kg
Fields of view: wide/medium/narrow
Parallel detector channels: 240 × 4 FPA
Spectral band: 8-12 μm
Electronic zoom: 2:1 and 4:1
Gimbal angular coverage
Azimuth: 360° continuous
Elevation: 40° up; 105° down
Gimbal slew rate: 3 rads/s

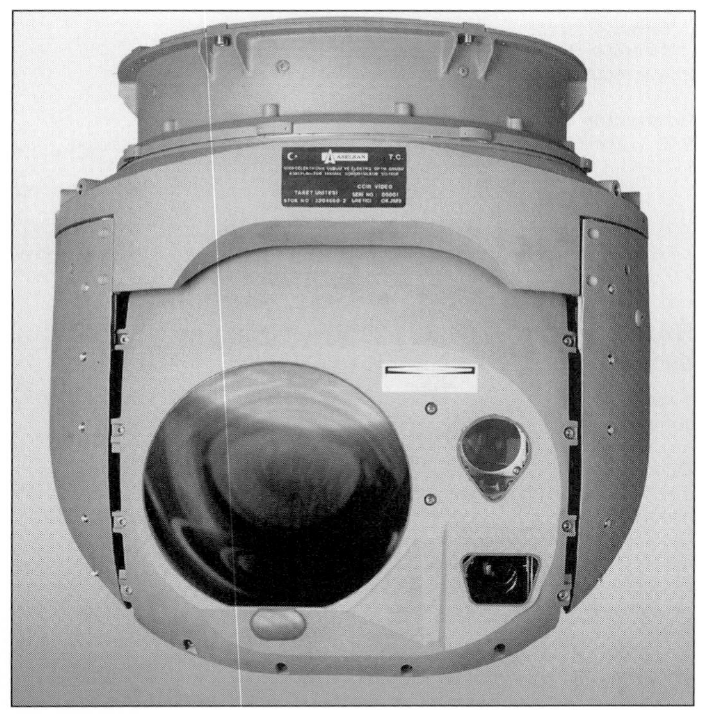

The ASELFLIR turret 0058719

Gimbal acceleration: head steering compatible at aircraft speeds
Laser range-finder: optionally provided
Day TV: optionally provided
Cooling: self-contained

Contractor
Aselsan Inc, Microelectronics, Guidance and Electro-optics Division, Turkey.

VERIFIED

BAE Systems ALLTV All-Light Level TV system

Type
Turret sensor (air).

Description
The All-Light Level TV (ALLTV) system consists of low-light level television cameras and a laser system mounted on a steerable stabilised platform based on an existing family of platforms in production for civil and military applications. The ALLTV system is designed to detect, identify and track targets which may then be engaged by the aircraft's guns. The system may also be used for damage assessment and covert surveillance.

Operational status
In production under a contract from Boeing for US Air Force AC-130U special operations aircraft.

Contractor
BAE Systems, Avionics, Sensors and Communications Systems.

UPDATED

BAE Systems Australia LRTS multirole thermal imager

Type
Thermal imager (air, land, sea).

Description
The BAE Systems Australia LRTS multirole thermal imaging system operates in the 3 to 5 μm spectral band for long-range, high-resolution performance in warm, high-humidity environments. Its small, lightweight configuration is suitable for land, sea and airborne operations and is applicable to the full range of military and commercial platforms for new or retrofit installations.

The system incorporates 3 to 5 μm staring array detector technology which simultaneously images the entire infra-red scene with over 311,000 individual detectors, without an opto-mechanical scanning system. The system offers a selection between two or three fields view providing pilot's night vision surveillance and targeting/identification capabilities. LRTS has a gimbal system with four-axis stabilisation to reduce image jitter.

Automated functions such as scan, track and flexible digital interfaces permit integration with a range of tactical platforms and systems including fixed- and rotary-wing aircraft, military vehicles and maritime forces. Interface and display facilities are provided for cueing the thermal imaging system to contacts detected by other sensors.

The Search IR thermal imaging system mounted on the side of an Australian Navy Seahawk 0005546

Integration options include sharing of display and controls with other sensors and provision of a video cassette recorder for mission reconstruction.

Operational status
In production. In service with the Royal Australian Navy and under consideration by several military forces around the world.

Specifications
Field of Regard
Azimuth: 360° continuous
Elevation: +35 to –120°
Fields of View
Narrow: 2.9 × 2.2°
Intermediate: 11.3 × 8.6°
Wide: 37.7 × 28.5°
Spectral band: 3-5 μm
Detector: platinum silicide (PtSi), 486 × 640 pixels
Video Format (frame rate): RS-170 (30 Hz) or CCIR (25 Hz) available
Dimensions
Turret (d × h): 406 × 550 mm
Control Electronics: ½ ATR 350 × 124 × 194 mm
Weight
Turret: 42 kg
Control Electronics: 9 kg
Interfaces
Power: 28 V DC, MIL-STD-704
Control/Data: dual-redundant 1553B, remote terminal unit; RS- 232 and RS-422; asynchronous serial 12-bit digital video

Contractor
BAE Systems.

VERIFIED

BAE Systems MultiSensor Turret System MST-S

Type
Turret sensor.

Description
The MultiSensor Turret System (MST-S) is a compact, lightweight thermal imaging system suitable for rotary- and subsonic fixed-wing aircraft. It combines high-resolution infra-red and CCD TV sensors for 24 hour all-weather target detection, identification and tracking.

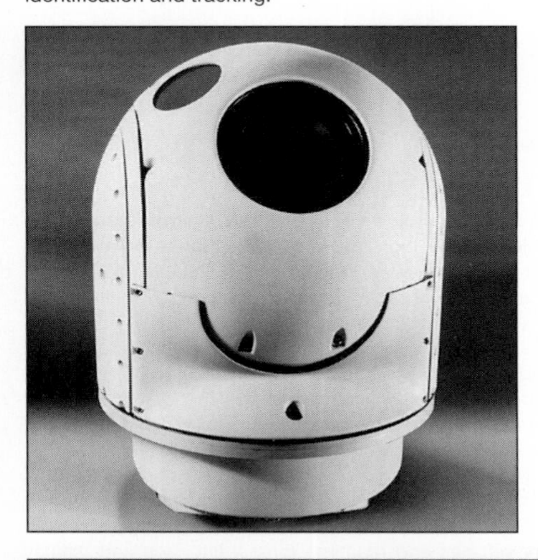

The MST-S turret
0005549

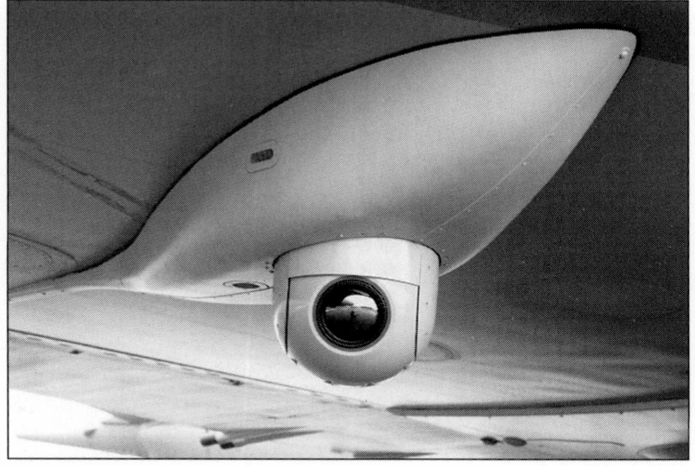

The underwing housing and turret of the MST-S installed on a UK Nimrod 0005551

MST-S installed under the starboard wing of a UK Nimrod maritime patrol aircraft
0005550

The gyrostabilised turret platform provides a 360° steerable field of regard in both azimuth and elevation and is fully operational at air speeds up to 300 kt with carriage at up to 400 kt.

The three-axis stabilised turret incorporates a modular payload with advanced scanning or staring focal plane array detectors and a continuous zoom telescope. This is combined with automatic image processing. A dual-mode video auto-tracker and MIL-STD 1553B databus are also included in the turret in a single LRU (Line-Replaceable Unit) installation.

Full control of the sensors, optics and turret functions is achieved via the databus interface, either from a dedicated, NVG-compatible controller or via soft keys on a multifunction display. Programmed search routines, handover of radar-designated targets and automatic tracking algorithms minimise operator workload. Both sensor video outputs, with or without overlaid graphics, are available simultaneously.

An extensive inventory of options enables the MST-S to be configured to suit a number of applications. The equipment is fully ruggedised, hardened and qualified to military standards.

Operational status
In production and in service in a number of countries on S-76, Sea King and Lynx helicopters, as well as Nimrod, Fokker 50, Dornier 228 and CN235 maritime patrol aircraft. BAE Systems, Avionics, Sensors and Communications Systems has received an order from GKN Westland Helicopters to supply the MST-S which will form part of the surveillance system for the EH 101 support helicopters of the Royal Air Force. BAE Systems has also received an order from Agusta to supply the MST-S to equip Agusta-Bell 412EP maritime patrol helicopters. A contract has been placed to supply MST-S for Royal Air Force Chinook helicopters.

Specifications
Spectral band: 8-13 or 3-5 μm
Detector: 8-parallel CMT/SPRITE or InSb array
Cooling: closed-cycle
Field of regard: 360° (azimuth and elevation)
Magnification: typically ×2.5 to ×20 (other ranges available)
Typical weights
Turret: 42 kg
Control switch joystick unit: 2 kg
Power requirements: 28 V DC nominal, 340 W typical

Contractor
BAE Systems, Avionics, Sensors and Communications Systems., Basildon.

UPDATED

BAE Systems Type 221 thermal imaging surveillance system

Type
Turret sensor (air).

Description
The Type 221 thermal imaging surveillance system is designed for service with military helicopters. Developed by BAE Systems in conjunction with Thales Optronics, the system incorporates an IR18 thermal imager and telescope made by the latter company with sightline stabilisation steering provided by a BAE Systems stabilised mirror. The assembly is contained in a pod, which can be mounted beneath the nose of a helicopter or can project through an aperture in the aircraft floor.

The sensor system employs SPRITE detectors cooled by a Joule-Thomson minicooler supplied with high-pressure compressed air. The air source is a bottle, mounted on the equipment and charged immediately before flight. This has a capacity of 1 litre and provides a system operation time of approximately 2.5 hours.

If greater endurance is required, other cooling options, involving the use of minicompressors permanently connected to the equipment, are available.

The BAE Systems mirror has an aluminium reflective element which is diamond-machined to a flatness of two fringes at 550 nm. This mirror has a reflectivity of greater than 97.5 per cent at 45° incidence, averaged over 8 to 12 μm. Its stabilisation system comprises a two-axis device with the integrating rate gyro as rate sensors. The mechanism is driven in each axis by a direct-drive DC torque motor while steering is obtained by torquing the integrating gyro. Angular information is derived from a resolver fitted to each axis. The mirror sightline is controlled by signals from an electronics unit which also provides power and signals to the main turret azimuth drive on the pod. System control may be exercised through a digital computer or by a hand controller unit. Infra-red spectrum vision is obtained via a germanium window on the front of the mirror turret assembly.

The display may be presented on either 525- or 625-line television monitors. The output is either in CCIR or EIA composite video formats, as required. This television-compatible output may be displayed on one or more monitors situated at various locations throughout the aircraft. It is considered desirable to have monitors at both the pilot and winch operator positions of a helicopter, in order to co-ordinate crew members for better hover control during night and bad visibility winching operations.

Operational status
In service. The BAE Systems Type 221 thermal imager has been fitted to a number of Puma helicopters operated by the Royal Air Force.

Specifications
Field of view: 38 × 25.5°
Resolution: 1.73 mrad
Pupil diameter: 14.5 mm
MRT (at 0.289 c/mrad): typically 0.17 to 0.35°C
Detector: SPRITE
Spectral band: 8-13 μm
Video Output: CCIR composite video or EIA composite video
Display: 625/525 lines/frames
Field rate: 50/60 Hz
Frame rate: 25/30 Hz
Field of regard
Elevation: +15°, −30°
Azimuth: ±178°
Telescope magnification: ×2.5/×9
Telescope field of view: 15.2 × 10.2°

BAE Systems Type 221 thermal surveillance system

Weight
Total System: 83 kg
Pod complete: 76 kg
Electronics: 8 kg
Power supply: 24 V DC nominal; 20 V min, 32 V max
Power consumption: 300 W at 24 V

Contractor
BAE Systems, Avionics, Sensors and Communications Systems, Edinburgh.

UPDATED

Controp DSP-1 day/night payload

Type
Turret sensor.

Description
The DSP-1 is a compact day/night observation system which can be used on a variety of platforms, including UAVs. It consists of a four-gimbal, gyrostabilised turret with two sensors; a third-generation indium antimonide (InSb) 3 to 5 μm focal plane array detector thermal imager (256 × 256) with ×22.5 zoom lens and a high-resolution colour CCD daylight camera with ×20 zoom. Sensor options include ICCD, a laser pointer, 8 to 12 μm first- or second-generation FLIR sensors, a video tracker, radar-designated pointing and a GPS interface. Other features include RS-422 communication with the air vehicle and PAL or NTSC video output. The sensor package can detect truck-sized targets at ranges up to 13.5 n miles (25 km; 15.5 miles) and can identify such a vehicle at 5.4 n miles (10 km; 6.2 miles) with the TV camera or 4.0 n miles (7.5 km; 4.7 miles) with the FLIR.

Operational status
The system has been in operational service with Israeli Scout and Searcher UAVs since mid-1997.

Specifications
Thermal imager
Spectral band: 3-5 μm
Detector: InSb
NETD: 0.02 mK
Cooling: closed-cycle
Lens: ×22.5 zoom
Fields of view: narrow 0.98 ×0.92°,wide 21.7 × 20.6°
Electronic zoom: 0.7° diagonal
CCD TV camera
Fields of view
Narrow: 0.92 × 0.7°
Wide: 18.6 × 13.9°
Zoom lens: ×20
Dimensions: 320 diameter × 500 mm h
Weight: 26 kg
Field of regard
Azimuth: 360° continuous
Elevation: +10 to −105°
Stabilisation: better than 25 μrad RMS

Contractor
Controp Precision Technologies Ltd.

UPDATED

Controp FSP-1 FLIR stabilised payload

Type
Turret sensor.

Description
The FSP-1 is a thermal imager payload mounted on a gyrostabilised four-axis gimbal system. The gimbal is fully stabilised in the lower hemisphere including at nadir. The FSP-1 is suitable for mounting on scout and surveillance helicopters, light aircraft, UAVs and marine patrol boats.

The FSP-1 carries a high-resolution 8 to 12 μm thermal imager with a three field of view telescope. Other sensors such as a 3 to 5 μm focal plane array thermal imager or additional daylight CCD camera are optional.

Operational status
In production and operational on various platforms.

Specifications
FLIR
Spectral band: 8-12 μm
Detector: CMT

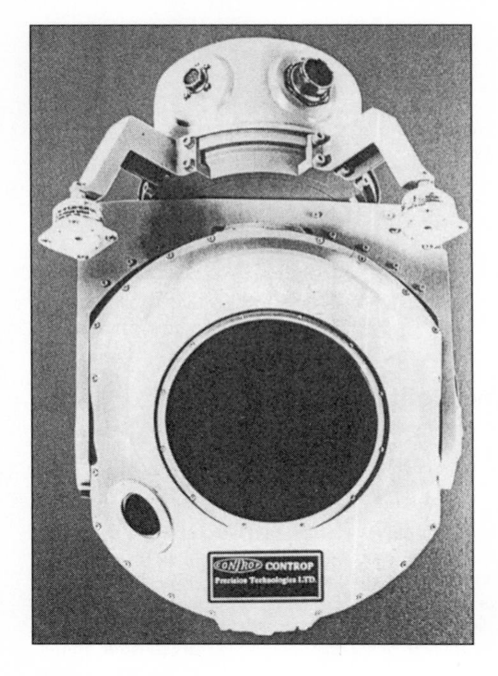

The FSP-1 FLIR stabilised payload from Controp Precision Technologies

Fields of view
Narrow: 2 × 1.5°
Medium: 7 × 5.3°
Wide: 24 × 18°
Electronic zoom: 1 × 0.75°
Resolution: 0.07 mrad × 0.11 mrad
Sensitivity: 0.8°C (MRTD at 7 cy/mrad)
Cooler: closed-cycle split-Stirling
Electromechanical
Type: 4-gimbal system
Field of regard
Azimuth: 360° continuous
Elevation: 0 to –105°
Angular velocity
Azimuth: 0 to 45°/s
Elevation: 0 to 32°/s
Stabilisation level: <25 μrad
Positioning accuracy: 0.7°
Performance
Truck recognition: 4-5 km
Truck detection: 12-20 km
Weight: 28 kg
Dimensions (d × h): 320 × 500 mm

Contractor
Controp Precision Technologies Ltd.

VERIFIED

Controp Mini-Eye

Type
Stabilised daylight electro-optical payload.

Description
Mini-Eye is a stabilised electro-optical payload designed for daylight operation and specifically for miniature UAVs with an airframe conformal design. The sensor can be used for observation, damage assessment, patrol, surveillance, search and rescue, as well as monitoring traffic. The payload features a high-resolution CCD colour TV camera with ×10 zoom lens using NTSC or PAL standard video. An option is a thermal imager, which would permit night operations.

Operational status
Believed to be available. Company publicity material suggests it has been flown in a Silver Arrow Mini-V UAV.

Specifications
Dimensions
Height: 182 mm (7.2 in)
Length: 418 mm (18.5 in)
Width: 196 mm (7.7 in)

Weights
Unit weight: 4.5 kg (9.9 lb)
Fields of view: narrow: 1.8 × 1.3°
wide: 17.6 × 13.3°
Angular velocity: 35°/s

The Controp Mini-Eye sensor payload ***NEW***/0106034

Stabilisation: 50 mrad RMS/axis
Power supply: 28 V DC or 15 V DC
Power consumption: 30 W (nominal)

Contractor
Controp Precision Technologies Ltd.

NEW ENTRY

Controp MultiSensor Stabilised Payload – MSSP-3

Description
Controp Precision Technologies' MSSP-3 is a day/night observation system especially designed for maritime patrol applications onboard aircraft, helicopters and patrol boats. In common with the MSSP-1, it is a four-gimbal system, gyrostabilised in azimuth and elevation and equipped with three sensors: a high-resolution, third-generation 3 to 5 μm InSb Focal Plane Array FLIR camera with a dual Field of View (FoV) lens; a high-performance, black and white/colour CCD camera with a ×15 zoom lens; and an optional eye-safe Laser Range-Finder (LRF).

Options include an Intensified Charge Coupled Device (ICCD), colour CCD, a ×22.5 continuous optical FLIR zoom lens for the InSb thermal imager, a second-generation 8 to 12 μm thermal imager, interface to GPS and MIL-STD-1553.

Specifications
Physical weight:
Turret: 38 kg
Control unit: 4.5 kg
Joystick: 1.5 kg
Turret Dimensions: 400 × 570 mm (D × H)

Electromechanical
Field of regard:
(azimuth) 360° continuous
(elevation) +35 to –110°

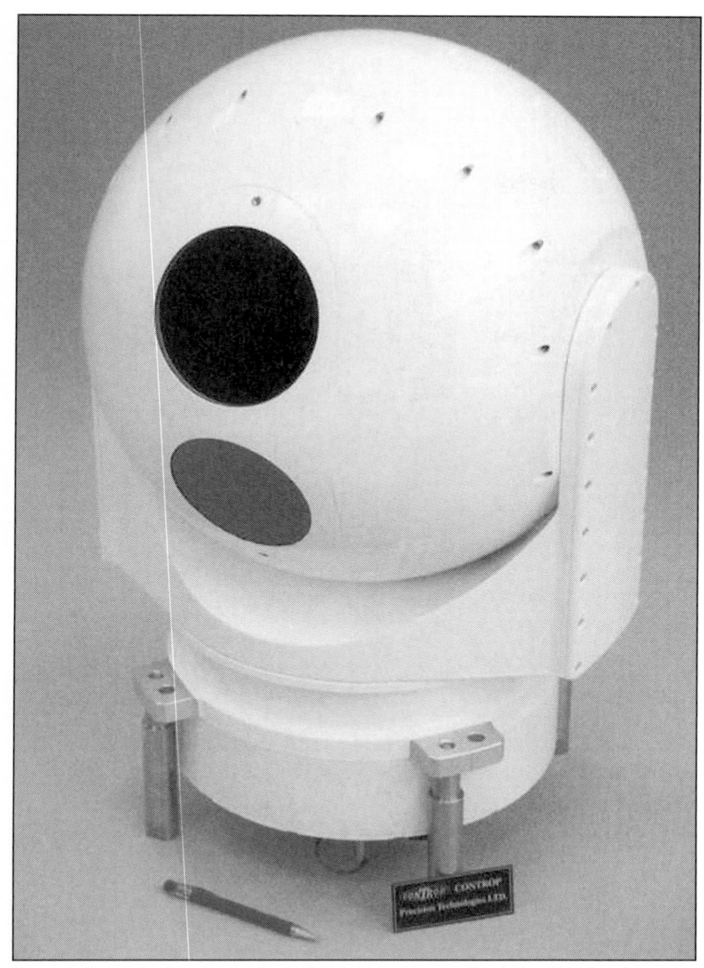

Controp Precision Technologies' MSSP-3 MultiSensor Stabilised Payload 0077827

Stabilisation: better than 25 μrad RMS
Angular velocity:
(azimuth) 60°/s (max)
(elevation) 50°/s (max)

FLIR sensor
Wavelength: 3 to 5 μm
Detector: InSb 320 × 240 FPA
Field of view:
(narrow) 2.2 × 1.65
(wide) 11.0 × 8.2
Cooler: closed-cycle cooler
Gain control: automatic/manual

Daylight camera
Camera: high-resolution black/white CCD
Resolution: 550 TV lines
Lens: ×15 zoom
Field of view:
(narrow) 1.2° × 0.9°
(wide) 18.0° × 13.7°
Gain control: automatic/manual

LRF (optional)
Wavelength: 1.54 μm
Range: 20,000 m (limited by atmospheric attenuation)
Accuracy: ±5 m
Repetition: 10 ppm (1 pps burst)

Environmental
Temperature: –20 to +50°C
Vibration: 2.5 g, 5 to 2,000 Hz
Shock: 20 g/11 ms duration
Speed: up to 300 kt

Electrical
Power: 28 V DC, 200 W
Video: CCIR or RS-170
Interface: RS-422

Contractor
Controp Precision Technologies Ltd.

NEW ENTRY

Controp Multi-Sensor Stabilised Platform MSSP-1

Description
Controp Precision Technologies' MSSP-1 is a rugged day/night surveillance system especially configured for use on attack helicopters (as well as multiwheeled terrestrial vehicles and marine patrol boats). It is a four-gimbal system, gyrostabilised in azimuth and elevation, and equipped with three sensors – a high-resolution 8 to 12 μm FLIR sensor, a high-resolution Charge Coupled Device (CCD) daylight camera and a laser range-finder.

Options include an Intensified Charge Coupled Device (ICCD), 3 to 5 μm InSb Focal Plane Array (FPA) FLIR camera, eyesafe Laser Range-Finder (LRF), extended environmental conditions capability, video tracker, radar designated pointing, MIL-STD-1553 and a GPS interface.

Specifications
FLIR sensor
Spectral range: 8-12 μm
Detector: Cadmium Mercury Telluride (CMT)
Field of view:
(narrow) 2 × 1.5°
(medium) 7 × 5.3°
(wide) 24 × 18°
Cooler: closed cycle

Daylight camera
Type: high-resolution black/white CCD
Lens: 15× zoom
Field of view:
(narrow) 1.2 × 0.9°
(wide) 18 × 13.7°

Laser range-finder
Wavelength: 1.06 μm
Range resolution: 5 m
Repetition rate: 1 pps

Electromechanical
Type: 4-gimbal
Field of regard:
(azimuth) 360° continuous
(elevation) +25 to –110°
Stabilisation: better than 25 μrad
Power: 28 V DC, 300 W
Dimensions: 400 (diameter) × 650 mm (height)
Weight: 41 kg

Contractor
Controp Precision Technologies Ltd.

NEW ENTRY

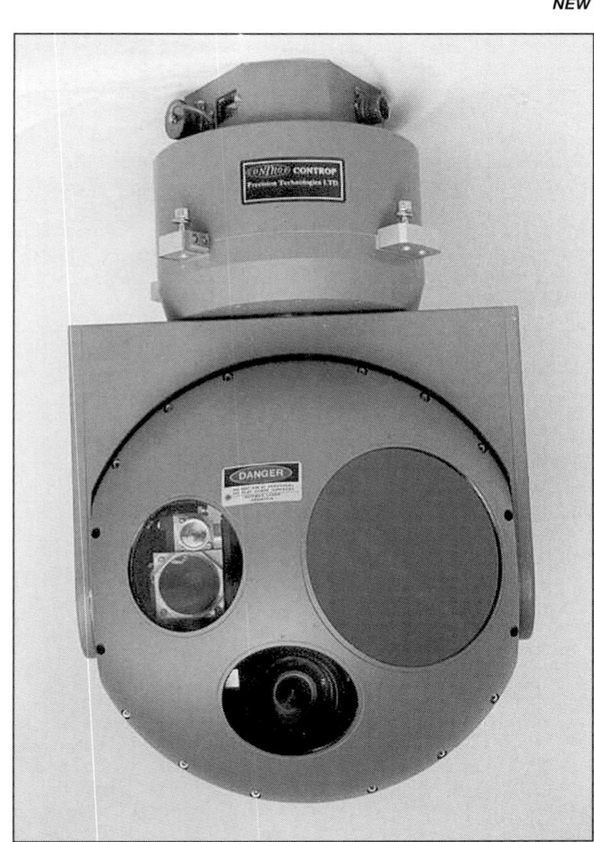

Controp Precision Technologies' MSSP-1 MultiSensor Stabilised Payload ***NEW***/0099741

CUMULUS LEO airborne surveillance systems

Type
Turret sensor.

Description
LEO*II* is a range of multisensor, gyrostabilised systems specifically designed for the paramilitary security market. Suitable for both helicopter and fixed-wing aircraft by day or night, LEO*II* is for use in applications ranging from law enforcement through search and rescue to border patrolling.

The latest model incorporates FLIR and broadcast standard TV sensors. The range also includes various single, dual and triple-sensor models fitted with a selection of FLIRs, colour TV cameras and laser range-finders.

All models are based on the same compact 400 mm platform, which features high-performance stabilisation and an open modular architecture to allow future upgradeability. The systems carry appropriate civil aviation approvals and are built to ISO 9002 and MIL-STD-10C/D/E quality standards. A comprehensive set of options allows the systems to be matched to individual requirements.

Operational status
Available. The LEO*II*-A2 variant is fitted to UK Police Air Support Unit helicopters and 80 0ther police/Search and Rescue aircraft worldwide.

Specifications
Basic system
Stabilisation: <2 µrad RMS
Dimensions : 400 mm (diameter)
Weight (basic system): 34 kg
Thermal imager: various 3-5µm and 8-12 µm FLIRs
Colour TV cameras: various 1-CCD and 3-CCD broadcast colour cameras
Options: GPS position display; search light/FLIR slaving units; auto-trackers; microwave downlinks

Contractor
CUMULUS, Business Unit of Denel (Pty) Ltd.

VERIFIED

Cumulus LEO-400-SPIR/SPTV

Type
Turret sensor.

Description
The LEO-400-SPIR/SPTV is a compact 400 mm gyrostabilised platform which combines a three-chip Sony broadcast colour TV camera and ×32 zoom lens, together with a FLIR Systems thermal imager, to give a dual day/night system with extended range TV identification capability for law enforcement, border patrol, search and rescue, reconnaissance, aerial survey, pipeline and powerline inspection.

The combination is fully compatible with existing aircraft mounting designs for LEO and current users of the proven standard one-chip LEO-400-SPIR/TV can now upgrade to SPIR/SPTV configuration with no weight, size, wiring or certification penalty.

The LEO-400-SPIR/SPTV platform is a four-axis gyrostabilised gimbal with image stability better than 20 µrad. The three-CCD broadcast colour TV with ×32 zoom lens has a narrow field of view of 0.8° for identification of persons and vehicle

The LEO-400 airborne observation system on a Bell 206 JetRanger helicopter

LEO-400-SPIR/TV day/night observation system

licence plates at ranges of 350 to 600 m, and a wide field of view of 26° for panoramic surveillance. The FLIR has dual fields of view (20 and 5°) and is optimised for medium-range surveillance, providing detection at up to 1 km of a swimmer in water, 3 km of a person on land and 10 km for a 10 m boat. It has a 12-bit dynamic range for image capture irrespective of display settings; a 32-bit microprocessor software-orientated design with image processing features; ×8 continuously variable electronic zoom function for additional magnification; freeze-frame function for detail investigation; temporal averaging function for low-contrast image enhancement; colour display options and user-set isotherm alarm levels for automatic detection.

Operational status
In production. LEO airborne observation systems are in service with police forces in Australia, Germany, Poland, South Africa, Spain and the UK.

Specifications
Stabilised platform
Platform type: full 4-axis gimbal
Stabilisation (2 axis): <20 µrad RMS
Pan look-angles: n × 360° (continuous)
Tilt look-angles: +20 to −105° (with optional image rotation at nadir)
Platform diameter: 390 mm
Thermal imager
Spectral band: 8-12 µm
Detector: 5-element SPRITE
Cooling: integral Stirling cooler (with guaranteed lifespan of 2,500 h)
Dual field of view (h × v)
  Narrow: 5 × 3.3°
  Wide: 20 × 13°
MRTD typical:
  @ 3.3 cy/mrad: 0.5° C
  @ 1.6 cy/mrad: 0.08° C
NETD: 0.18° C
FOV switchover time: <1 s
Cool down time: <5 min
Dynamic range: 12 bits
Contrast and brightness control: automatic or manual
Active IR pixels (PAL/NTSC): 800 × 450/400 (h × v)
Colour TV camera/zoom lens
Camera type (PAL/NTSC): Sony Hyper-HAD interline transfer
Sensor type: 3-CCD 0.5 in format
Resolution: 720 TV lines
Signal-to-noise ratio: 56 dB
Min sensitivity: 19 lux
Zoom ratio: ×16 (with ×2 remotely switchable extender)
Zoom focal length range: 14 to 448 mm (equivalent of 19 to 608 mm for a $\frac{2}{3}$ in sensor)
Field of view range
  ×2: 0.8 × 0.6° to 13 × 10°
  ×1: 1.6 × 1.2° to 26 × 19°
System power
Input: 22-32 V DC aircraft standard
Consumption (typical/peak): 230 W / 520 W

Weight
External platform with cameras: 33 kg
Total min system: 52 kg

Contractor
Cumulus, Business Unit of Denel (Pty) Ltd.

VERIFIED

DRS gimballed uncooled sensor

Type
Turret sensor.

Description
The DRS (formerly Boeing) gimballed uncooled sensor turret incorporates a long-wave infra-red sensor and two CCD TV sensors. The system is intended for airborne or ground vehicle use.

Operational status
Not known.

Specifications
Elevation range: 270°
Azimuth range: 360° continuous
Elevation slew rate: up to 200° continuous
Azimuth slew rate: up to 200° continuous
Elevation position resolution: 3 mrad
Azimuth position resolution: 3 mrad
Stabilisation: 2 axis
Gyrostabilisation (optional): <150 µrad
Height: 362 mm
Diameter: 254 mm
Input voltage: +12 to 28 V DC
Power dissipation (max): <200 watts
System weight: 11.4 kg
Sensors: LWIR (uncooled) and TV CCD (2)
Control: RS-232 or RS-422
Connectors: video (3), power, control
Detector type: VO_x resistive bolometer
Detector format: 320 × 240
Detector pitch: 51 × 51 µm
Fill factor: 60%
Operability: >98%
Spectral band: 8 to 12 µm
Optics (standard)
Field of view (h × v): 40 × 30°
EFL: 22 mm
System F/No: 1.0
Optics (optional)
Field of view (h × v): 12.4 × 9.3°
EFL: 75 mm
System F/No: 0.8
NETD: <0.100° K
System dynamic range: 12 bit
Automatic gain control: standard
Automatic level control: standard
Auto start-up: <1 min
Non-uniformity correction: auto calibration upon startup
Output video format: CCIR 'PAL' composite or RS-170 NTSC composite

Contractor
DRS Technologies, Electro-Optical Systems Group.

EADS Ophelia, PISA and PVS night vision systems for helicopters

Type
Turret sensor (air).

Description
Ophelia is a mast-mounted sight designed for helicopter observation in search and rescue, surveillance (for example police missions or border patrol) and military reconnaissance/combat. The system provides an unobstructed 360° view without extensive structural modifications to the fuselage (minimal modifications are required to the BO 105 rotor head) and creates no centre of gravity problems. Installation of a sensor package with a line of sight approximately 110 cm above the rotor plane allows the helicopter to maintain observation while under maximum

The BO 105 helicopter with PISA system nose mounted

cover. Inclusion of a thermal imager in the payload ensures extended operations at night and during inclement weather conditions. TV and infra-red images are displayed on either a head-up or head-down display installed in the cockpit. Images are superimposed with symbology created by a computer symbol generator. The sensor platform for Ophelia is produced by the French company SFIM.

Trials with nose-mounted sensor systems have also been carried out, including the Pilot's Vision System (PVS), a dual-sensor visually coupled system where the sensor line of sight follows exactly the pilot's head movements. Images are relayed to the pilot from electro-optical sensors mounted on a platform installed in the nose of the BO 105. These are displayed on a miniature CRT fitted to the pilot's helmet. The sensor images are superimposed with computer-generated symbology allowing the pilot to fly the aircraft without recourse to normal flight instrumentation. The system comprises a stabilised platform (by SFIM) and two electro-optical sensors: a FLIR and a low-light TV camera. The pilot is free to choose images from either sensor, depending on the flying conditions.

Another nose-mounted system flight tested on the BO 105 is the Pilot's Infrared Sighting Ability (PISA) system developed by EADS. PISA is a wide-angle sensor providing night-vision capability for orientation and observation purposes. The system consists of an infra-red thermal imager with SPRITE detectors and a single-axis steering mechanism for control in azimuth. Control in elevation is integrated in the thermal imager. It provides a 30° elevation × 60° azimuth field of view, with direct CCIR-TV output of a high-image resolution picture which is homogeneous over the total field of view. The system has a low infra-red signature on account of specially integrated cooling and can be operated either fully autonomously or integrated in a gunner's sighting system.

Operational status
Under development.

Contractor
EADS.

UPDATED

Elop COMPASS

Type
Turret sensor (air, sea, land).

Description
The Compact Multi-purpose Advanced Stabilised System (COMPASS) is designed for use in airborne, naval and ground-based applications. It can accommodate a day channel colour CCD camera with a zoom lens (interchangeable with black and white camera), a FLIR: third-generation 3 to 5 µm or second-generation TDI 8 to 12 µm and an eye-safe laser range-finder or designator. Interfacing with aircraft avionics is either RS422 or MIL-STD-1553.

Operational status
In service.

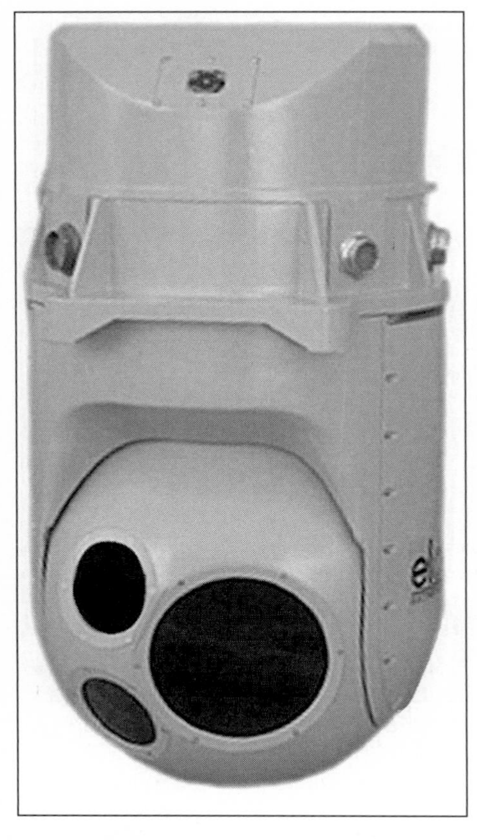

The Elop COMPASS turret
NEW/0547582

Specifications

Weight (turret and electronics): 35 kg with laser designator, 32 kg with rangefinder
Angular coverage
Azimuth: n × 360°
Elevation: +35 to –120° or –35 to +120°
Stabilisation accuracy: <20 μrad
Video output: RS170 or CCIR
Operational temperature range: –40 to +50°C

Contractor

Elop Electro-optic Industries Ltd.

UPDATED

FLIR Systems AN/AAQ-22 SAFIRE

Type

Turret sensor (air).

Description

The FLIR Systems AN/AAQ-22 SAFIRE (Shipborne/Airborne Forward-looking Infra-Red Equipment) system has been designed for surveillance, navigation and tracking. It has been fielded by all branches of the US military and is in operation with military and civil agencies worldwide.

The SAFIRE system uses an 8 to 12 μm detector array and provides real-time video output in standard TV format. Data interfaces include RS-232, RS-422, ARINC 419/429 and MIL-STD-1553 to allow seamless interfacing with aircraft systems. SAFIRE is a dual field of view system, the narrow field of view being supplemented by electronic zoom and freeze-frame features. The turret is gyrostabilised in three axes. Its modularity allows for a variety of payload configurations and provides many automated features, such as auto image adjustment and autoscan. SAFIRE is fully interchangeable for operation on ships, helicopters and fixed-wing aircraft.

Operational status

In service.

Specifications

Dimensions
Gimbal/imager: 446 × 384 × 384 mm
Central electronic unit: 320 × 260.3 × 195.6 mm
Monitor: 221.0 × 256.5 × 268.2 mm
Controller: 228.6 × 114.3 × 116.8 mm

Weight
Gimbal/imager: 42.5 kg
Central electronics unit: 10.2 kg
Monitor: 4.76 kg

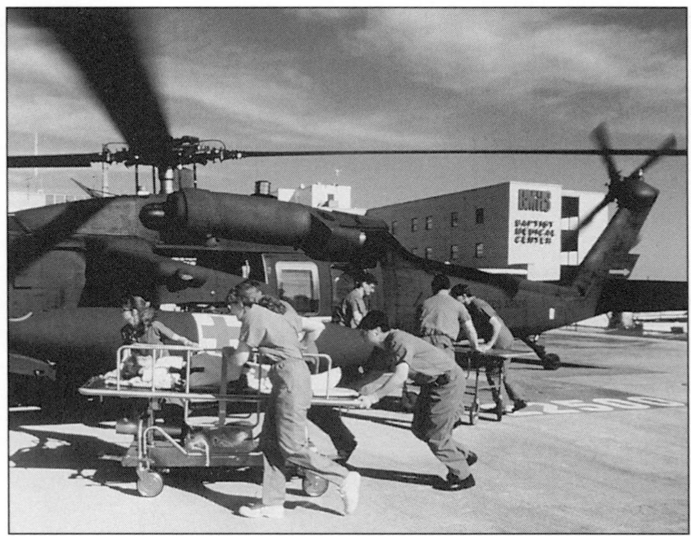

AN/AAQ-22 SAFIRE on the US Army's UH-60Q Blackhawk MEDEVAC helicopter

The SAFIRE turret 0005547

Controller: 1 kg
Cables: 2.27 kg

Thermal imager
Spectral band: 8-14 μm
Detector: 4 × 4 CMT array
Scanning: serial-parallel
Cooling: closed-cycle
Cool-down time: <6 mins
Fields of view: 28 × 16.8°; 5 × 3°
Magnification: ×1.85, ×10.5, ×21 (electronic zoom)

Field of regard
Azimuth: 360°
Elevation: +30 to –120°
Stabilisation: 3 axis
Rotation: 360°
Cooling: closed-cycle
Pixels per line: 525 × 392 active pixels

Maximum operational airspeed: >400 kt

Contractor

FLIR Systems.

VERIFIED

FLIR Systems AN/AAQ-22 Star SAFIRE

Type
Turret sensor.

Description
This military qualified, commercial-off-the-shelf (COTS), multisensor gyrostabilised airborne platform, incorporates advanced third-generation detector technology and three thermal imager fields of view in a compact system providing long-range identification and image clarity in the most adverse climates. Although compact in size, the rugged turret and modular design can incorporate up to four payloads for day or night operations, with the options to include a CCD TV camera, an eye-safe laser range-finder, NVG compatible laser illuminator, long-range spotter scope and interfacing to navigation and radar systems.

Operational status
In service.

Specifications
Gyro stabilised platform
Type: 3 axis
Slew angles
Azimuth: 360°
Elevation: +30 to −120°
Air speed: >400 kt operational
Dimensions (h × w): 446 × 383.5 mm
Weight: 43.6 kg
Thermal imager
Spectral band: 3-5 μm
Detector: 320 × 240 InSb FPA
Fields of view
Wide: 30 × 22.5°,
Medium: 5.7 × 4.3°,
Narrow: 1.4 × 1.03°
TV Sensor
Type: ⅓ in 1-CCD colour TV
Optional: Eye-safe laser range-finder; NVG-compatible laser illuminator, long-range spotter scope and interfacing to navigation and radar systems.

Contractor
FLIR Systems.

UPDATED

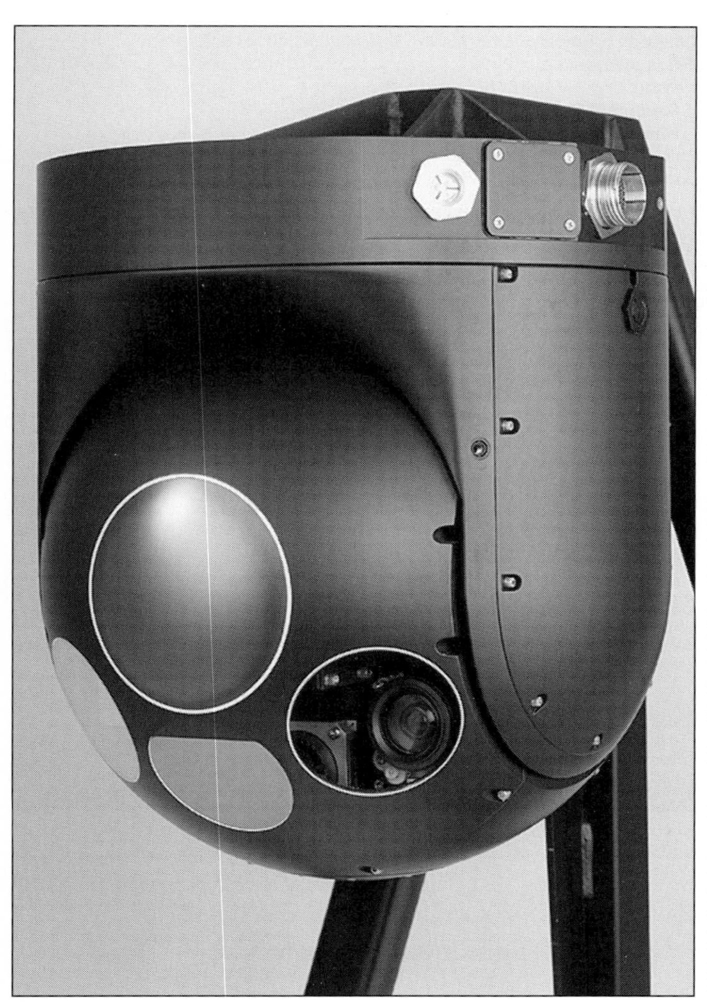

Star SAFIRE 0089852

FLIR Systems MicroSTAR™

Type
Turret sensor (air).

Description
The MicroSTAR is a dual-imaging sensor with high-resolution infra-red and boresighted CCD-TV with low-light capability. Its low-profile, fully sealed, 9 in gimbal is lightweight, compact and designed to minimise drag. This compact size translates into saved fuel, increased mission duration and improved weight and balance calculations.

The MicroSTAR provides continuous zoom optics. This allows the operator to customise the field of view to combine the right amount of target magnification with enough scene coverage to maintain situational awareness. The operator can also zoom in or out without losing sight of the target and without constant inputs. The MicroSTAR gives operators push-button access to three distinct modes of target tracking: Centroid – used to track moving targets; Scenelock – for tracking larger stationary scenes; and Correlation – which tracks small, slow-moving targets.

Operational status
In production.

Specifications
Turret size: 22.9 (d) × 34.3 (h) cm
Turret weight: 11.8 kg
Field of regard: 360° continuous in azimuth and elevation
Slew rates: 0 to 50° both axes
Stabilisation: Fiber-optic gyros, two axes

Thermal imager
Sensor: 320 × 240 InSb FPA 3-5 μm
Fields of view: 2.2-22° horizontal, 10:1 continuous zoom
Calibration: Auto calibration

Daylight imager
Pixel arrangement: 768 (h) × 494 (v) (NTSC); 752 (h) × 582 (v) (PAL)
Resolution: 470+ television lines
Telescope: 18:1 continuous zoom (×4 digital zoom)
Fields of view: 2.7 to 48° horizontal continuous zoom
CCD sensitivity: 3 lux @ f/1.4 or 0.2 lux @ f/1.4

Optional features
B/W CCD daylight camera, laser pointer ARINC and searchlight slave interface, and 30 MW laser illuminator.

Contractor
FLIR Systems.

UPDATED

FLIR Systems MicroSTAR ***NEW***/0137882

FLIR Systems Mk II™

Type
Turret sensor (air).

Description
The Mk II is a gyrostabilised, infra-red and visible light imaging system that is type-certified for many of today's most widely used helicopters, while being easily installed on fixed-wing aircraft and UAVs as well. The Mk II imaging system comes with a forward-looking infra-red imaging module and a boresighted colour CCD camera for daylight viewing. It offers two remotely controlled powers of magnification (1× and 6×), plus the added capability of instantaneous electro-optical zoom, which increases thermal sensitivity while simultaneously magnifying image size by a factor of two. The colour CCD camera contains an auto iris to maintain picture quality over varying light conditions.

Operational status
In service.

Specifications
Turret size: 9.0 (d) × 13.5 (h) in (22.9 × 34.3 cm)
Turret weight: 12.68 kg
Field of regard: 360° continuous azimuth and elevation
Slew rates: 0 to 50° both axes

Thermal imager
Four-element detector TDI: HgCDTe 8-12 μm
Magnification: ×1.5, ×3.0 (EO zoom), ×6.0, ×12 (EO zoom)
Field of view: 18.6 (h) × 14° (v); 9.3 (h) × 7° (v); 4.7 (h) × 3.5° (v); 2.3 (h) × 1.75° (v)
Field of view (CCIR): 18.6 (h) × 16° (v); 9.3 (h) × 8° (v); 4.7 (h) × 4° (v); 2.33 (h) × 2° (v)
IFOV: 1.2 mR, 1.2 mR, 0.3 mR, 0.3 mR

Colour CCD camera:
Magnification: 1.2 × to 7× continuous zoom
Field of views: 23.5 (h) × 17.8° (v) continuous to 4 (h) × 3° (v)

Daylight Imager
Pixel arrangement: 811 (h) × 508 (v) (NTSC), 795 (h) × 596 (v) (PAL)
Resolution: Four RS-170 or CCIR, 445 active IR/TV
CCD sensitivity: 5.0 lux @ f/1.2
Magnification: 1.1 × to 7× continuous zoom
Power/Max current requirements: 18 V DC to 32 V DC input/15 A

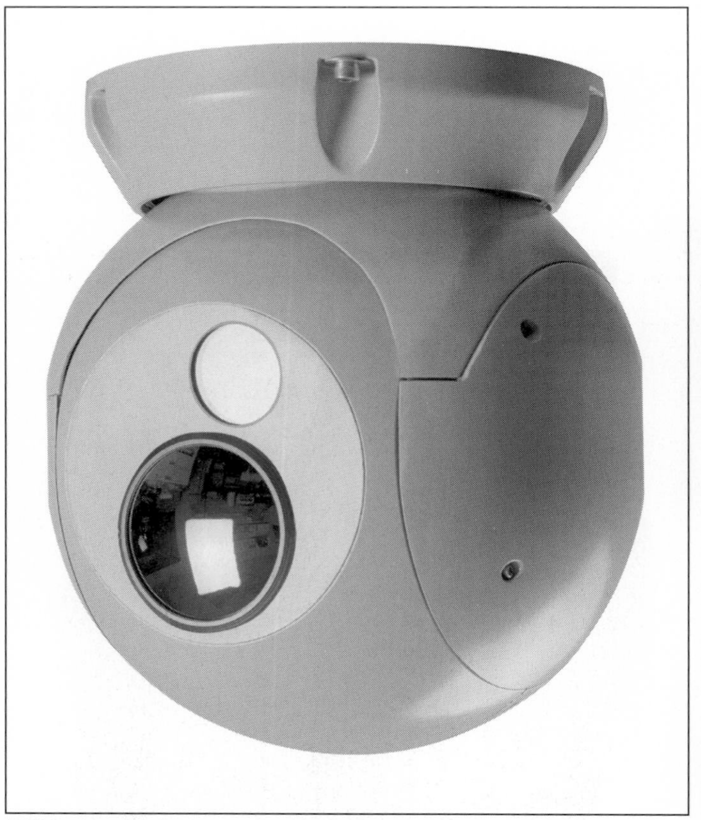

FLIR Systems Mk II 0137883

Additional information
RS-232 GPS interface; cage mode for protection, video output; on-screen display of location; azimuth and elevation position; time-date, mode.

Options
ARINC and other non-standard interfaces available, SLASS interface available.

Contractor
FLIR Systems.

UPDATED

FLIR Systems Star Q™

Type
Sensor turret (air).

Description
Star Q is a four-axis gyrostabilised multisensor platform fully ruggedised to military standards (MIL STD 810E). STAR-Q features CCD TV and a three field of view thermal imaging sensor using a third generation focal plane array detector operating in a narrow sub-band of the 8-12 μm LWIR region.

The QWIP thermal imager provides significantly enhanced platform capability, combining excellent thermal sensitivity and an optical magnification ratio of 25:1, providing target detection, recognition and identification at very high stand-off ranges.

The Star Q gimbal design features open modular architecture, allowing the addition of a variety of further payload options including: eye safe laser range-finder and very narrow field of view CCD TV spotter scope.

Operational status
In service.

Specifications
Weight: 42± kg
Stabilised Turret Unit (STU)
Turret type: full 4 axis gimbal

Thermal imager
Camera type: Triple QWIP LW with integral three FOV optics with autofocus
Spectral band: 8 – 9 μm
Detector: focal plane array (QWIP) GaAs 320 × 240° or 640 × 480°
Fields of view (320 × 240)
Wide: 25 (h) × 18.75° (v)
Medium: 6 (h) × 4.5° (v)
Narrow: 0.98 (h) × 0.74°(v)
Electric zoom: × 4 with digital interpolation
Cooling: integral Stirling cycle

FLIR Systems Star Q ***NEW***/0547501

The Star Q turret on a Swiss Air Force Super Puma ***NEW***/0547503

Colour zoom TV camera
Sensor format: 1/3 in 3-CCD sensor
Resolution: 800 TV lines (PAL)
FOV: 0.67 to 35° HFOV zoom
Focal length: 9.5 to 256.5 mm f3.5 to 512 mm f7.0
Zoom ratio: effective × 54 (×27 with ×2 switchable extender)
Format: PAL (NTSC optional)
Optional features: wide-spectrum spotter TV, laser range-finder (LP-15)

Contractor
FLIR Systems.

UPDATED

FLIR Systems Star SAFIRE II™

Type
Turret sensor (air).

Description
The AN/AAQ-22 Star SAFIRE II is a military qualified COTS multisensor gyrostabilised airborne platform featuring up to five bore-sighted payload options, including a Gen III Thermal Imager, near-IR sensitive low light CCD, high-magnification spotter scope, laser illuminators or pointers, and a laser range-finder. Star SAFIRE II offers five axis stabilisation, sophisticated automated operation and a wide array of aircraft integration features. Available with embedded moving map/ mission planning systems and is ideal for fixed or rotary-wing operations up to 405 kts.

Operational status
In service.

Specifications
Gyrostabilised platform
Turret dimensions (h × diam): 17.55 × 15.10 in (44.58 × 38.35 cm)
Turret weight: 98 lb, 44.5 kg (includes all options)

Field of regard
Azimuth: 360° continuous
Elevation: +30 to −120°
Slew rate: Variable to ≤60/sec
Control: SCU, serial digital or MIL-STD-1553B
Environmental: MIL-STD-810 and MIL-STD-461
Max air speed (flight tested): 405 kt

Thermal imager
Sensor: 320 × 240 InSb FPA 3-5 μm response
Resolution: enhanced to 640 × 480 (microscanned)

Fields of view
Narrow: 0.8 × 0.6° V
Medium: 3.4 × 2.6° V
Wide: 25.2 × 18.8° V
Optional wide field of view: 33.3 × 25°
Cooling: Stirling linear cooler
Video output: RS-170 or CCIR
Visual LL TV: colour with near-IR sensitivity (0.2 lux)
Signals format: NTSC/PAL
Sensitivity: 0.2 lux
Zoom ratio: 18× continuous zoom plus 4× electronic zoom
Fields of view: 24 to 2.1°

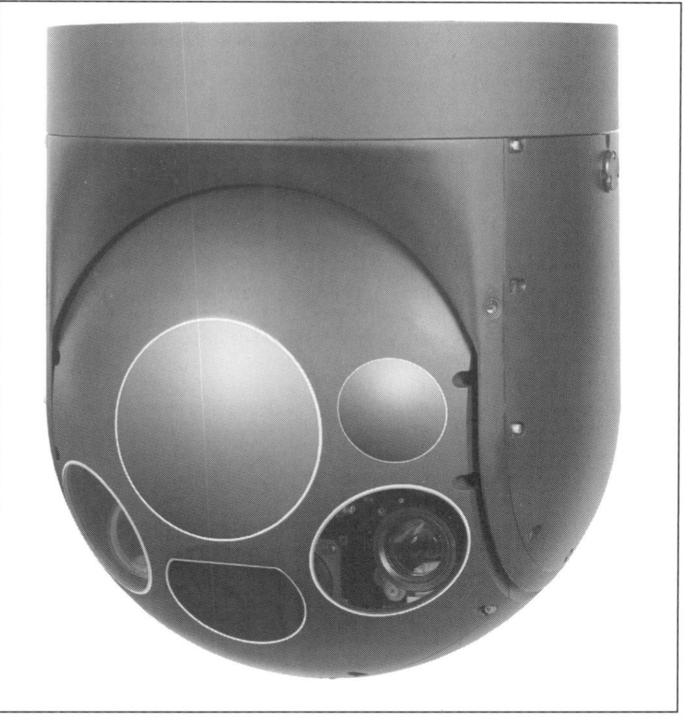

FLIR Systems Star SAFIRE II 0137861

Optional features
NTSC or PAL; multimode autotracker, high magnification spotter; laser illuminators/pointers/range-finder; haze/low light image processing, geopointing/ target positioning; MIL-STD-1553B; radar, navigation and weapon/FCS interfaces; hand held/panel mount and custom/remote system controllers; quick disconnect mounts; flat panel displays; RF downlinks and VTR accessories.

Contractor
FLIR Systems.

UPDATED

FLIR Systems Star SAFIRE™ III

Type
Turret sensor.

Description
The AN/AAQ-36 Star SAFIRE™ III is a military-qualified COTS multisensor, gyrostabilised airborne platform featuring up to six boresighted payload options,

Star SAFIRE™ III ***NEW***/0547504

including a high-resolution GEN III Thermal Imager with maximum range performance; Image Intensified Camera (ICCD); high magnification, multi-FOV spotter scope; narrow and wide area laser illuminators or pointers; and laser rangefinder. Star SAFIRE™ III is FLIR's first imaging system to offer >640 resolution, autofocus for all sensors, digital video and optional image fusion. Available with integrated Mission Management System (moving map/tactical mission planner) and is ideal for all fixed- or rotary-wing operations up to 405 kt.

Operational status
In production.

Specifications
TI: 640 × 480 InSb FPA with enhanced fidelity, 3-5 μm response; 4 FOV from 25 to 0.4° and optional wide field of views >33°
ICCD: Gen III/IV; multi-FOV matched optics for image fusion
3-CCD: Colour with >800 TVL; multi-FOV matched optics for image fusion; NFOV of 0.29°
LLTV: Colour with near-IR sensitivity; 25× continuous zoom plus 4× e-zoom; 0.2 lux sensitivity
Weight: 98 lb/44.5 kg (includes all options)
Standard: 5-axis stabilisation; autofocus; ARINC 419/429, single/dual; RS232/422/485
Optional: 320 × 240 InSb FPA (enhanced); NTSC or PAL; ICCD; high-magnification spotter; multimode autotracker, laser illuminators/pointers/rangefinder; haze/low-light image processing, geo-pointing/target positioning; MIL-STD-1553B; radar, navigation and weapon/FCS interfaces; hand-held/panel mount and custom/remote system controllers; quick disconnect mounts; flat panel displays; RF downlinks and VTR's/accessories

Contractor
FLIR Systems.

NEW ENTRY

FLIR Systems Ultra 7000™

Type
Turret sensor (air).

Description
The Ultra 7000 is an ultra-small dual-sensor system incorporating a thermal and a colour TV camera. The system includes an autotracker in the electronic control unit. The tracker has two modes of operation. For automatic tracking of moving targets it has a centroid mode of operation. This centres the target within an adjustable tracking gate shown on the display monitor. It also has a memory capability which allows a temporarily lost target to be reacquired via a coast mode which tracks on the last known movement parameters. The second mode allows locking onto a scene for detailed examination in real time. In either mode it is possible to alternate between IR and CCD sensors without loss of track or scene hold.

Operational status
In service.

Specifications
Stabilised platform
Type: dual axis
Dimensions (dia × h): 228 × 343 mm
Weight: 11.8 kg
Field of regard: 360° continuous in both axes
Slew rate: 0 to 50°/s in both axes
Maximum acceleration: 60°/s^2
Stabilisation: Fibre-optic gyros, two axes

Thermal imager
Rotation: 360°
Detector type: 256 × 256 InSb FPA
Spectral band: 3-5 μm
NETD: 0.03°C
Cooling: Stirling microcooler
Cool down time: 7-8 min
Field of view: 17.6 - 1.76° horizontal, 10:1 continuous zoom

TV sensor
Pixel layout: 811 (h) × 598 (v) (NTSC), 795 (h) × 596 (v) (PAL)
Resolution: 450 lines (NTSC), 443 lines (PAL)
CCD sensitivity: 5.0 lx @ f/1.2
Telescope: 10:1 continuous zoom
Field of view; 2.2 to 22.3° horizontal continuous zoom
Power supply: 18-32 V DC, max 15 A
Communications: RS-232/422
Video output: Four

Additional components
RS-232 serial GPS interface, cage mode for protection, on-screen display of location, azimuth and elevation position, time-date and mode.

FLR Systems Ultra 7000 0137868

Options
ARINC and other non-standard interfaces available, SLASS interface and laser point available, B/W CCD available.

Contractor
FLIR Systems.

UPDATED

FLIR Systems Ultra 7500™

Type
Turret sensor (air).

Description
The Ultra 7500's lightweight, dual-sensing gimbal with laser illuminator option is designed to provide 24-hour-a-day, higher-altitude, long-range search and surveillance from airborne platforms. Its triple payload capability features a 320 × 240 pixel InSb infra-red imager and low-light TV camera.

At just 23 cm diameter and weighing approximately 12 kg, the fully sealed gimbal is designed to minimise drag and provide more ground clearance. The low-light TV camera with ×18 optical zoom delivers clear images in daylight, while also enhancing operations during the dusk and dawn hours. The infra-red optics of the Ultra 7500 feature a continuous zoom, allowing the operator to customise the field of view. The system's autotracker feature keeps designated targets or scenes within the field of view without the need for constant operator input.

Operational status
In service.

Specifications
Turret dimensions (h × d): 34.3 × 22.9 cm
Turret weight: 11.8 kg
Field of regard: 360° continuous in azimuth and elevation
Slew rates: 0.02 to 50°/s both axes
Stabilisation: Fibre-optic gyro stabilisation

Thermal imager
Detector: 320 × 240 InSb FPA 3-5 μm
Fields of view: 2.2 × 1.65°, 22 × 16.5°, 10:1 continuous zoom
Calibration: Internal Nonuniformity Correction (NUC)

Daylight imager
Pixel arrangement: 768 (h) × 494 (v) (NTSC); 752 (h) × 582 (v) (PAL)
Resolution: 460 (PAL), 470 (NTSC) television lines
Telescope: 18:1 continuous zoom, 4× electronic zoom
Fields of view: 0.7° (e-zoom) to 48° horizontal continuous zoom, NFOV 2.7°
CCD sensitivity: 0.2 lux @ f/1.4 w/out filter, 3 lux @ f/1.4 w/filter

Electronic control unit
Power requirements: 18 V DC to 32 V DC input

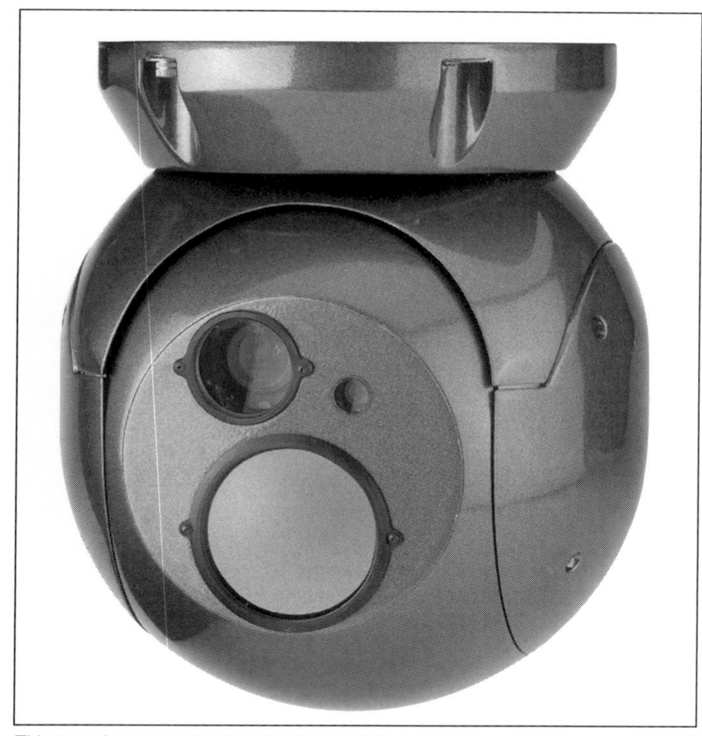

This turret is common to the Ultra 7500, 8000 and 8500 from FLIR Systems 0137867

Options
ARINC interface, RS-232 serial GPS interface, radar interface, SLASS interface, laser pointer or laser illuminator, and infra-red and CCD optical extenders.

Contractor
FLIR Systems.

UPDATED

FLIR Systems Ultra 8000™

Type
Sensor turret.

Description
The Ultra 8000™ features a compact 9 in (230 mm) stabilised turret with two boresighted sensors. Optimised for light and medium fixed- and rotary-wing aircraft operation, the system includes a high-resolution GEN III thermal imager with continuous zoom lens, color CCD TV with auto focus. Designed for law enforcement and search and rescue missions, the ULTRA 8000™ provides powerful imaging capabilities in a small and lightweight (26 lb) package.

Operational status
In production.

Specifications
TI: 256 × 256 InSb FPA 3-5 μm. IR Continuous zoom Telescope with 10× magnification – field of view range: 22 × 16.5° (WFOV) to 3.4 × 2.0° (NFOV) – PAL or NTSC Video Output. IFOV: 0.12 milliradians (NFOV)
TV: H × 494 V (NTSC); 752 H × 582 V (PAL), 460 (PAL), 470 (NTSC) television lines. Telescope: 18:1 continuous zoom, 4× electronic zoom
Field of view range: 48 × 32° (WFOV) to 3.4 × 2.0° (0.67 × 0.55° using E-Zoom) (NFOV)
Weight: 26 lb (11.8 kg)
Optional features: ARINC interface, RS-232 serial GPS interface, radar interface, SLASS interface, PIP monitor

Contractor
FLIR Systems.

NEW ENTRY

FLIR Systems Ultra 8500™

Type
Sensor turret.

Description
The Ultra 8500™ is a compact 9 in (230 mm) military-qualified, stabilised turret with up to three boresighted sensors. Optimised for light and medium fixed- and rotary-wing aircraft operation, the ULTRA 8500™ includes a high-resolution GEN III thermal imager with auto focus and continuous zoom IR lens (focal length: 25 to 250 mm), a color CCD TV with auto focus and low-light TV mode, and an optional IR laser pointer. The system provides real-time automatic image optimisation, an imbedded multimode autotracker, and an intuitive, icon-based color graphical overlay. Designed for law enforcement, search and rescue, and general surveillance missions, the Ultra 8500™ provides powerful imaging capabilities in a small and lightweight (29 lb) gimbal package.

Operational status
In production.

Specifications
TI: 320 × 240 InSb FPA 3-5 μm. IR Continuous zoom Telescope with 10× magnification – field of view range: 22 × 16.5° (WFOV) to 2.2 × 1.65° (NFOV) – PAL or NTSC video output. IFOV: 0.12 milliradians (NFOV)
TV: 752 H × 582 V (PAL) 768 × 494 (NTSC) Continuous zoom Telescope with 18× magnification and integral 4× electronic zoom – Maximum magnification: 72×
Field of view range: 48 × 32° (WFOV) to 2.7 × 2.2° (0.67 × 0.55° using E-Zoom) (NFOV)
Low-light mode: Operational down to 0.2 Lux
Resolution: 460 TV Lines (PAL), 470 TV Lines (NTSC)
Laser pointer: visible with Image Intensified Night Vision Goggles
Other available features: RS 422/232, ARINC interface, RS-232 serial GPS interface, Search Light Slave (SLASS) interface, PIP monitor, Terse Binary Protocol (TBP) for Remote and UAV control functions, compact electronics control box for UAV applications

Contractor
FLIR Systems.

NEW ENTRY

FLIR Systems UltraFORCE™ II

Type
Turret sensor (air).

Description
The UltraFORCE™ II (formerly the LEO-II-QWIP™) is a high performance multisensor airborne law enforcement system. It features three payloads: a high resolution Gen III LW thermal imager (QWIP), a 3-CCD broadcast quality high magnification TV camera and a dual camera high magnification spotter scope. The IR imager features autofocus, auto image optimisation and three field of view, high magnification lenses. The optional dual CCD spotter scope provides extreme long-range capability and the ability to have colour imaging in daylight and low light monochrome imaging at night. Optimised for airborne law enforcement, the system features numerous aircraft mounts and interfaces.

Operational status
In service.

FLIR Systems' UltraFORCE II turret sensor 0137866

Specifications

Gyrostabilised platform
Type: 4-axis gyrostabilised gimbal
Turret dimensions (h × d): 490 × 410 mm
Weight: 39.5 kg

Thermal imager
Camera type: Gen III LW QWIP FPA; three FOV optical system w/auto focus
Detector: GaAs QWIP 320 (h) × 240 (v) focal plane array
Spectral band: 8-9 μm

Fields of view
Wide: 25 (h) × 18.75° (v)
Medium: 6 (h) × 4.5° (v)
Narrow: 0.98° (h) × 0.74° (v)
Electronic zoom: ×4 with digital interpolation
Format: PAL or NTSC
Cooling: Integrated long-life Stirling cooler

Colour zoom TV camera
Sensor format: 1/3 in 3-CCD sensor
Resolution: 800 TV lines (PAL)
Active pixels: 752 (h) × 582 (v)
Field of view: 0.67° to 36° HFOV zoom
Zoom ratio: 54 × (27 × with 2× switchable extender)
Format: PAL or NTSC

Optional features
Dual sensor spotter scope, eye-safe laser range-finder, on a laser pointer. Digital autotracker, real time video and audio downlink, and a searchlight slaving capability are available as optional enhancements.

Contractor
FLIR Systems.

UPDATED

Galileo Avionica Galiflir/Astro

Type
Turret sensor (air).

Description
The Galiflir is a high-performance electro-optical sensor system for use in manned and unmanned aircraft. It consists of a stabilised or servo-driven platform with a thermal imager which operates in the 8 to 12 μm waveband.

It has eight CMT SPRITE elements and serial-parallel scanning. It may be cooled by a split-Stirling engine integrated detector-cooler assembly or a Joule-Thomson mini-cooler. The images are presented in standard CCIR 625/50 television format with 512 IR lines.

In the Galiflir Astro, the FLIR is in a two-axis stabilised platform with a two fields of view telescope. It is designed as a helicopter system for surveillance, target acquisition/tracking/designation and navigation. If a designation request is sent by an onboard radar system the operator moves the FLIR field of view window across the radar display and onto the selected target. Azimuth and elevation data are transmitted and updated through the interface with the helicopter avionics system.

Galiflir undergoing operational testing on an Agusta A 109 helicopter

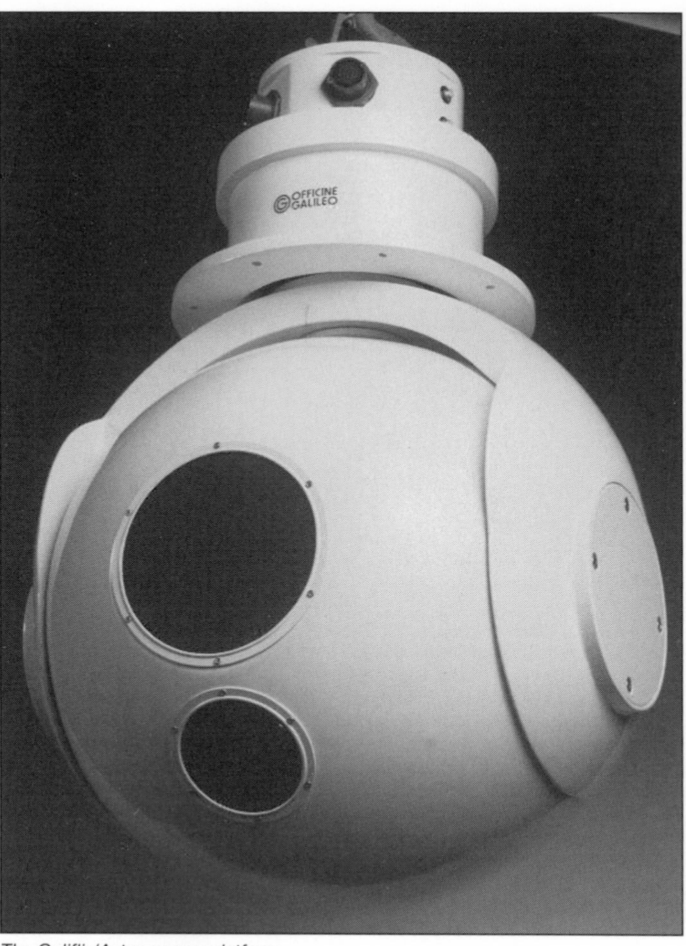

The Galiflir/Astro sensor platform

Operational status
In production. Galiflir is used by the Italian Coastguard.

Specifications
Galiflir
Spectral band: 8-12 μm
Detector: 8-element SPRITE
Cooling: Joule-Thomson
Field of regard
Azimuth: ±170°
Elevation: +30 to -75°
Fields of view
Wide: 16 × 10.7°
Narrow: 4 × 2.7°
Magnification: ×2.5, ×10
Min focus range: 50 m
Number of IR lines: 512
Slew rate: 2 rad/s
Stabilisation accuracy: 25 μrad RMS
Dimensions: 384 × 596 × 380 mm (platform)
Weight
Platform: 30 kg
Compressor: 8 kg
Control unit: 8 kg
FLIR control panel: 2 kg
Platform control panel: 1.5 kg

Contractor
Galileo Avionica.

UPDATED

ISS V14-M Multisensor airborne system

Type
Turret sensor.

Description
The V14-M Multisensor system provides fixed-wing aircraft, helicopters and Unmanned Air Vehicles (UAV's) with what the manufacturer says is an unsurpassed night/day surveillance and observation capability. The V14-M high-performance midwave thermal imager and broadcast colour TV camera, integrated

in a precision stabilisation system, provides high-resolution imagery for long-range detection, identification, and tracking of objects of interest in demanding situations both night and day.

The V14-M thermal imager includes a three field of view telescope and advanced image processing electronics that provide crisp, imagery comparable to black and white TV. The system also includes a complementary broadcast-grade three-CCD day TV sensor. Coupled to the TV sensor is a motorised lens offering precision boresight through its ×10 zoom range. Four axis gyrostabilisation gives maximum sensor performance and complete aircraft lower hemispheric coverage spanning 360° in azimuth and +35° to −110° in elevation.

An integral, eye-safe laser range-finder, aligned to the camera sensors, provides range to the object of interest. When coupled to the airborne platform's on-board navigation system, the V14-M can calculate and display the geo-location of the surveyed object in either latitude/longitude or UTM grid co-ordinates. Geo-location accuracies of less than 100 m at a range of 10 km have been routinely demonstrated operationally. Also included within the V14-M is a laser diode pointer operating in the near IR spectrum and that is compatible with NVG use. This feature permits the V14-M sensor operator to covertly highlight an object of interest, that the V14-M is focused on, to aircrew and ground personnel equipped with night vision devices.

The V14-M system can be operated directly through a laptop control console, remotely by common workstations, or in UAV applications through commonly available datalinks.

Designed with a VME chassis for expandability, the V-14M can easily incorporate optional features including trackers, symbology overlays, and other unique, user defined needs

Operational status
Available and in-service worldwide.

Specifications
Turret configuration
Dimensions: 14.5 × 17.5 × 14.5 in (37 × 44 × 37 cm)
Weight: 65 lb (30 kg)
Field of Regard: Azimuth - 360°
Continuous elevation - +35 to −110°
Field of View (IR): 3 - FOV telescope
Narrow - 1.2° H × 0.9° V
Medium - 5.0° H × 3.75° V
Wide - 22.0° H × 16.5° V
IR Spectral Band: 3-5 μm
Detector: 320 × 240 Indium Antimonide (InSb)
Cooling: Closed cycle cryogenic
TV Camera: High resolution 3-CCD Colour (NTSC, S-Video)
10× motorised zoom, 24 - 240 mm (HFOV 1.2° - 12.0°)
Laser Diode Pointer: 830 nm - 100 mW
Laser Range-finder: 20 km range
Power: 28 V DC, 350 W (maximum)

The ISS V14 turret 0101854

Contractor
Imaging Sensors and Systems, Inc.

UPDATED

L-3 WESCAM 14PS

Type
Turret sensor.

Description
Extensively deployed on unmanned aerial vehicles and high-performance aircraft for high-risk reconnaissance missions, the WESCAM™ 14PS is said to be the only actively gyro-stabilised system in its class. It has the capacity to accommodate up to five sensor payloads for total imaging and tracking capability.

Features
- Compact multisensor gyro-stabilised gimbal
- Surveillance and observation capabilities
- Available in three configurations: tri, quad or penta-sensor
- Uncompromised infra-red optics with 150 mm aperture
- DLTV spotter with full 90 mm aperture
- Optional laser employs full spotter aperture for maximum range
- Plug and play SmartLink interface.

The 14PS high-performance payloads include:
- Sensor 1 - 3-5 μm 6 field of view thermal imager
- **Sensor 2** - colour daylight CCD camera with 20× zoom lens
- **Sensor 3** - colour daylight CCD camera with 955 mm long-range spotter lens
- **Sensor 4** - eye-safe laser range-finder
- **Sensor 5** - laser illuminator.

The 14PS active gyro-stabilisation and vibration isolation allows less than 35 mrad rms line-of-sight jitter, detection and classification of persons, objects and vehicles over long distances, day or night.

Contractor
L-3 Communications, WESCAM Inc.

UPDATED

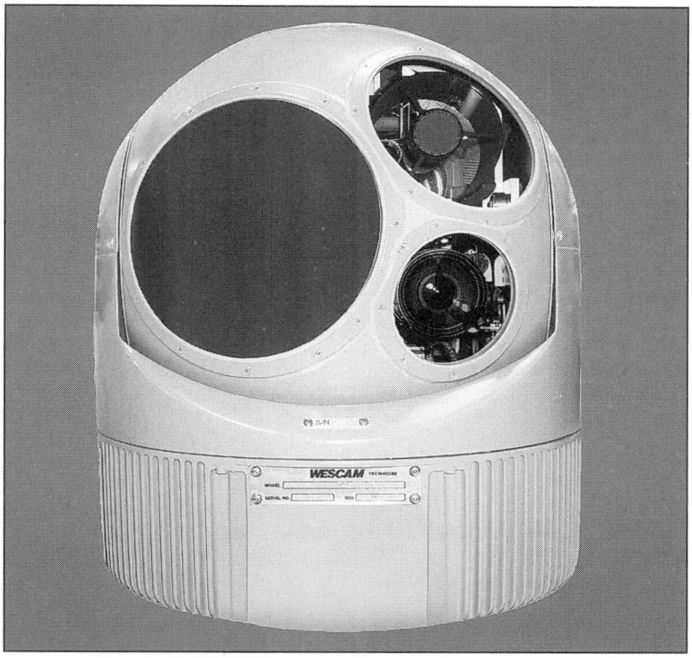

The L-3 WESCAM 14PS turret sensor 0143681

L-3 WESCAM Model 12DS

Type
Turret sensor.

Description
The WESCAM Model 12DS200 is a gyrostabilised dual-sensor camera system designed to meet the needs of the Airborne Law Enforcement (ALE) community. It is intended for the detection, recognition, identification and tracking of persons and vehicles over long distances.

Model 12DS includes an indium antimonide (InSb) staring array thermal imager with dual field of view and a colour daylight CCD TV camera with ×20 zoom lens.

Active gyrostabilisation and vibration isolation produce less than 35 µrad LOS (Line Of Sight) jitter RMS (Root Mean Square).

The small size and low weight of the Model 12DS200 allows installation on a wide variety of small rotary- and fixed-wing aircraft where weight is the primary concern. The system is designed to interface with other mission equipment such as Night Sun searchlights, moving maps, GPS, radars and navigational bearing commands.

Operational status

In June 1997, WESCAM received a contract for 22 Model 12DS systems for the Pioneer UAV. Pioneer UAV Inc is under funding from the US Joint Program Office. The system has also flown on a variety of other platforms.

Specifications

Gimbal
Active gyrostabilisation: 2 axis inner (pitch/yaw); 2 axis outer (azimuth/elevation)
Vibration isolation: 6 axis passive isolation (x/y/z/pitch/roll/yaw)
Line of sight jitter: <36 µrad RMS nominal
Azimuth/elevation slew rate: >90°/s max
Azimuth: 360° continuous
Elevation: +90 to –120°

Thermal imager
Spectral band: 3-5 µm
Detector: InSb staring focal plane array
Resolution: 256 × 256 pixels
Cooling: Stirling cycle
FOV switch over time: <0.125 s nominal
Focal length: 17 mm, 88 mm

Field of view
17 mm: 25 × 25°
88 mm: 5 × 5°
Format output: NTSC or PAL

Sony XC-999 TV camera
Pick-up device: ½ in hyper HAD interline transfer 1 CCD
Format: NTSC or PAL
TV lines: 470 NTSC/460 PAL TV lines
Min illumination: 4.5 lx at f1.2
TV zoom lens (14 × 10.5 Fujinon continuous zoom)
Focal length: 10.5-147 mm

Fields of view
Narrow: 2.5 × 1.9°
Wide: 34 × 25.7°

Dimensions
Gimbal: 305 d × 370 mm
System interface unit: 250 × 330 × 100 mm
Hand controller: 200 × 100 × 80 mm

Weight
Gimbal: 22.5 kg
System interface unit: 4.5 kg
Hand controller: 0.7 kg
Power (nominal): 28 V DC, 10 A max

Contractor

L-3 Communications, WESCAM Inc.

UPDATED

L-3 WESCAM Model 16 series including AN/AAQ-501

Type

Turret sensor.

Description

WESCAM's Model 16 thermal imaging surveillance system is mounted on a gimbal for both fixed-wing aircraft and helicopters. The systems are all gyrostabilised in four axes and vibration isolated in six axes.

Model 16DS-M is a dual-sensor system but with a medium wave (3 to 5 µm) platinum silicide thermal imager and a one-chip Sekai CCD camera and is designed for hot, humid and tropical climates.

The thermal imagers used in the Model 16DS-M have automatic gain adjustment, remote switching of field of view and electronic zoom (optional on Micro-FLIR). The Thermovision 1000 has the facility for image freezing for analysis and isotherm selection/alarm settings.

Operational status

In production. Model 16 systems have been selected for a number of defence, law enforcement and coastal surveillance applications. AN/AAQ-501 has been ordered by the Canadian Department of Defence for use in their Utility Transport Helicopter Program. Some 28 systems were ordered with deliveries completed by late 1997.

The Model 16DS-M dual-sensor system with mid-wave thermal imager

Specifications

Gimbal
Active gyrostabilisation: 2 axes inner (pitch/yaw), maintained through look-down; 2 axes outer (azimuth/elevation)
Vibration isolation: 6 axes passive isolation (x/y/z/pitch/roll/yaw)
Line of sight jitter: <35 µrad RMS (nominal)
Pan/tilt slew rate: 1 rad/s (typical)
LOS pan range: continuous 360°
LOS tilt range: +30° to –120°

Midwave thermal imager (16DS-M)
Detector: platinum silicide focal plane array
Array size: 640 × 480 pixels
Spectral band: 3-5 µm
Cooling: closed cycle
Switch over time (nominal): <1 s
Fields of view
Wide: 12.0 × 9.3° (optional 40 × 30°)
Narrow: 2.9 × 2.3°
Frame frequency (EIA/CCIR): 30/25 Hz

1CCD Sekai RSC-375 TV camera (16DS-M)
Resolution: 470 (NSTC); 460 (PAL) TV lines
Sensitivity: 2,000 lx @ f5.6
Min illumination: 4.5 lx @ f1.2
Features: automatic gain control

Dimensions
Model 16DS-M: 400 d × 560 mm
Weight (nominal)
Model 16DS-M: 44.1 kg
Power (nominal): 28 V DC, 10 A

Contractor

L-3 Communications, WESCAM Inc.

UPDATED

L-3 WESCAM MX-20

Type

Turret sensor.

Description

For demanding long-range identification requirements, the WESCAM™ MX-20 combines high-magnification sensors with high-quality gyro-stabilisation, making it suited for surveillance and intelligence applications over land and sea. The turret's 5-axis gyro-stabilisation and 6-axis vibration isolation achieve an extremely low line-of-sight jitter in both fixed- and rotary-wing aircraft. Step zoom lenses provide wider fields of view that allow the user to perform both long-range surveillance and pin-point identification without the need for the aircraft to deviate from its flight path.

The L-3 WESCAM MX-20 multisensor turret 0143684

Features

- Commonality with all MX series products
- High-performance stabilised multisensor system
- High-magnification day TV and FLIR step zoom lenses
- Geo-location capability for pin-pointing ground target location
- Flexible architecture for customisation and systems integration
- Environmentally qualified for a wide range of applications and aircraft installations.

The MX-20 is manufactured to meet demanding environmental, quality, reliability and maintainability requirements, and has been qualified to military, aeronautical and commercial aviation standards.

The entire MX series of products are based upon a common architecture. The common system interface, user interface and internal components result in logistic and service benefits on fleetwide installations, and ease of use for integrated applications that use a number of different airframes.

The AN/ASX-4 variant of the MX-20 is part of the AIMS system on board the US Navy P-3 Orion aircraft.

Contractor

L-3 Communications, WESCAM Inc.

UPDATED

Northrop Grumman LSF

Type

Turret sensor (air).

Description

The Northrop Grumman (formerly Westinghouse) Lightweight Surveillance FLIR (LSF) is a modular airborne electro-optical system for detecting targets and is designed for high-performance aircraft.

The FLIR uses a platinum silicide staring array operating in the 3.4 to 4.8 μm band that eliminates optical scanning mechanisms, thus reducing weight. It also produces a detector with improved uniformity between the elements for low noise/low NETD performance.

The small sensor head may be adapted to a number of aircraft installations and contains the system optics and IR sensors. Line of sight is controlled by stabilised gimbals and these may be adjusted by the operator to other settings. There are two fields of view, a wide one for search and a narrow one for target identification and recognition.

The processor electronics control the LSF functions and video processing to RS-170 standard. Optional extras are automatic tracking, inertial or external control of the line of sight and MIL-STD-1553 bus communications. The control unit is used to select field of view, to select from the command menu, for calibration and for built-in test functions.

Operational status

In operation. Out of production.

Specifications

Spectral band: 3.4-4.8 μm
Detector: PtSi staring array
Fields of view: 3 × 4°; 15 × 20°

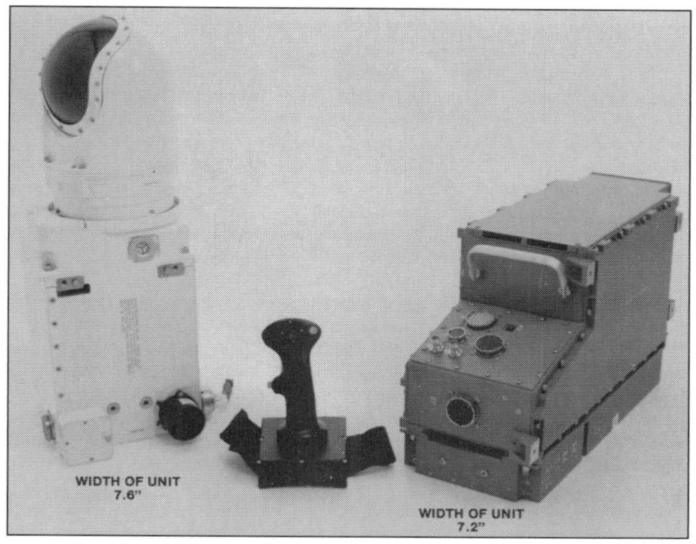

The Northrop Grumman LSF components: (from left) the sensor unit, control unit and processor electronics

Field of regard
Azimuth: 120°
Elevation: 0 to –60°
Line of sight stabilisation: <100 μrad
Dimensions: 171 mm (sensor head diameter)
Weight: 29.5 kg (total)
Power requirements: 22-29 V DC

Contractor

Northrop Grumman, Electronic Sensors and Systems Sector, Electro-optical Systems.

VERIFIED

Northrop Grumman Nightgiant long-range surveillance system

Type

Turret sensor (air).

Description

The Nightgiant long-range surveillance system was developed for the US Air Force to perform day and night electro-optical surveillance missions. Based on the Northrop Grumman (formerly Westinghouse) WF-360TL MultiSensor Surveillance and Tracking System, Nightgiant consists of a second-generation scanning Focal Plane Array (FPA) thermal imager, with three fields of view and a high-resolution television camera fitted with a zoom lens, mounted in a highly stabilised three-axis gimballed turret.

The Nightgiant FLIR utilises a 480 × 4 TDI (Time Delay Integration) mercury cadmium telluride (HgCdTe) detector array, operating in the 8 to 12 μm spectrum. The imagery generated by this system is said to be free of the linear striations and noise characteristics of earlier FLIR systems. The dynamic range allows the FLIR to discern temperature differences of less than 0.1°C and yet not be saturated by nearby objects such as jet engines with very high temperatures. Interchannel cross-talk is reduced to a minimum. The system features DC coupling of the detector,

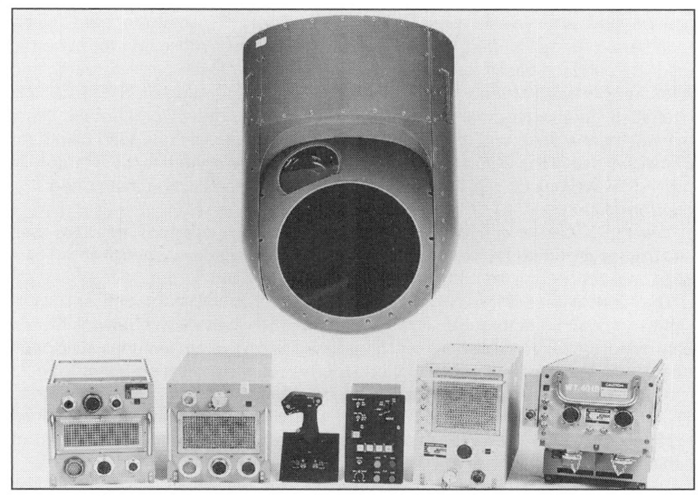

Northrop Grumman Nightgiant long-range surveillance system

together with automatic responsivity and level control (ARE/ALE). The ARE/ALE replaces the 'channel balance' associated with earlier FLIRs.

The FLIR optics provide three selectable fields of view. A large Cassegranian telescope provides the narrow field of view. Refractive afocals fold in the medium and wide fields of view optics when required. The optical system is designed so that the MFOV and WFOV elements cause no transmission obscuration of the NFOV.

The Nightgiant TV system is designed to complement the FLIR in image in terms of quality and magnification. A commercial zoom lens hardened to withstand military environments is utilised. The system can be configured with either a high-resolution CCD camera or a third-generation Low-Light Level Television (LLTV). The TV automatically zooms to match the selected FLIR field of view.

The Nightgiant incorporates two different types of automatic video trackers for greater flexibility. The operator can select the centroid tracker as necessary to maintain the system LOS pointed at a specific feature in the scene. Alternatively, the operator can select the correlation tracker to maintain the LOS pointed at a general scene. An additional tracking capability is also provided which maintains target track in the event that the line of sight is momentarily obscured by aircraft structure. Nightgiant is capable of accepting inputs from other aircraft systems such as inertial navigation systems or GPS data and directing the line of sight to designated targets or waypoints.

Symbology is software controlled and can be easily reconfigured to meet operator or mission needs. System communication can be accomplished easily by a variety of databuses, including MIL-STD-1553B, RS-422 and RS-232. Extensive built-in test (BIT) has been implemented to reduce maintenance time significantly.

Fusion of the FLIR and television images is an available option for the Nightgiant system. The Fusion subsystem provides an image that combines the best information available from each sensor at the pixel level and at multiple spatial frequencies.

Operational status

The Nightgiant system is in production and has been operationally deployed by the US Air Force on C-130 aircraft. Nightgiant has been selected to equip the UK Royal Air Force Replacement Maritime Patrol Aircraft (RMPA) Nimrod MR4. Nightgiant system deliveries for test and evaluation began in early 1999.

Specifications

Spectral band: 8-12 μm
Detector: 480 × 4 TDI CMT
Fields of View
FLIR: 0.79 × 1.0°; 3.1 × 4.2°; 12.7 × 16.9°
TV/LLTV: zoom lens matched to FLIR FOV
Field of regard
Azimuth: 360° continuous
Elevation: +15 to –100°
Line of sight stabilisation: <25 μrad
Turret size (d × h): 609.6 × 863.6 mm
System weight: 319 kg

Contractor

Northrop Grumman, Electronic Sensors and Systems, Electro-optical Systems.

UPDATED

Northrop Grumman WF-360TL

Type

Turret sensor (air).

Description

The WF-360TL multisensor surveillance system features a FLIR sensor operating in the long-wave 8 to 12 μm spectrum with options for adding a high-resolution day TV camera and an eye-safe laser range-finder. These sensors are boresighted together on an optical bed housed in a stabilised gimbal designed for airborne, maritime and land-based targets. System capabilities include passive search and track in both air-to-air and air-to-surface modes. ARINC and MIL-STD-1553 bus interfaces couple to other aircraft sensors such as the aircraft radar, flight management system and the inertial navigation system. An imbedded computer processes data from all sources to provide cueing of the system to radar targets or navigation waypoints, real-time display of the track-point co-ordinates and fire-control solutions.

The FLIR used in the WF-360TL marries US Army common modules with electronics providing DC restoration, automatic gain and level control and digital scan conversion for a US standard RS-170, 525-line video output.

The turret utilises a two-axis stabilised platform to stabilise the optical lines of sight and point them throughout the entire lower hemisphere with up-look limited only by adjacent aircraft structure. A broadband servo system featuring solid-state rate sensors and an inner acceleration loop provides stabilisation better than 35 μrad in typical aircraft environments including helicopters and fixed-wing aircraft. The turret is environmentally self-contained, employing an internal liquid-to-air heat exchanger to allow proper operation of the turret without the need for bleed or cabin air.

The WF-360TL's astronomical telescope provides two Fields Of View (FOV). The TV sensor is also dual FOV.

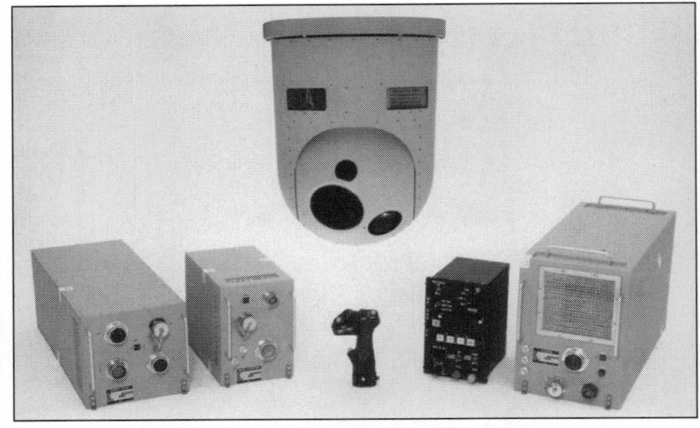

The components of the Northrop Grumman WF-360TL system

In addition to the TV and laser range-finder, other optional plug-in modules provide capabilities including FLIR-TV image fusion, wide-area correlation video tracking and laser target illumination for night TV imagery. The computer-aided track mode, in conjunction with automatic video trackers, provides a coast mode in the event of short-term target obscuration.

Operational status

The WF-360TL is in production. The system is used extensively for surveillance and drug interdiction by the US Coast Guard in the HU-25 Falcon jet and RU-38 aircraft, the US Air Force in C-26 and the US Army in the DH-7 Airborne Reconnaissance Low airplane.

Specifications

Dimensions: 406 (d) × 558 mm
Weight: 110 kg
Fields of view
FLIR: 2.7 × 3.6°, 11.1 × 14.8°
TV: 2.4 × 3.2°, 14.2 × 18.9°
Field of regard
Continuous azimuth: 360°
Elevation: +30 to –85°
Line of sight stabilisation: <50 μrad

Contractor

Northrop Grumman, Electronic Sensors and Systems, Electro-optical Systems.

UPDATED

Rafael Topaz

Type

Turret sensor.

Description

Topaz is a targeting, observation and acquisition system for UAVs, helicopters and other platforms. Sensors include thermal imager, CCD camera (colour or black and white) and optional laser range-finder or laser aiming light.

The Topaz sensors are mounted on a four gimbal system, providing a lower hemisphere field of regard and stabilisation of better than 40 μrad. Topaz also provides an automatic tracking capability. It is fully integrated with external systems

Topaz mounted on a helicopter

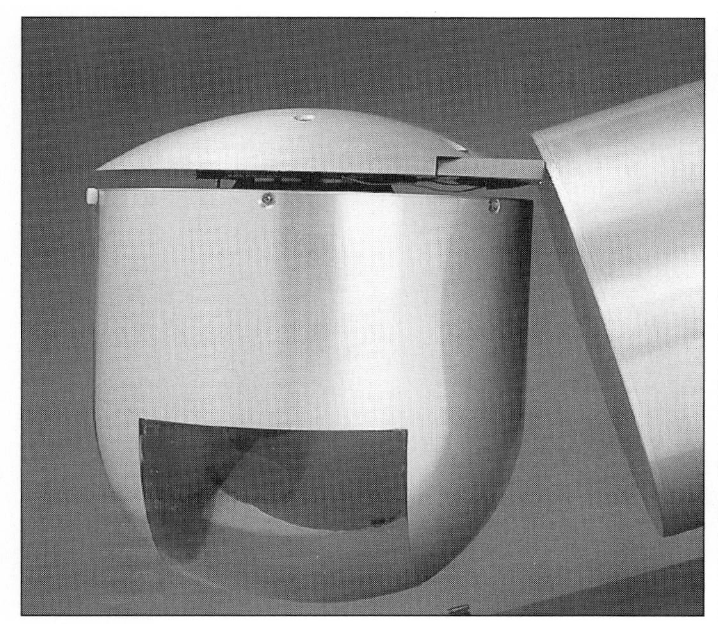

The Topaz targeting, observation and acquisition system

through its operating modes which include manual steering, scan, automatic track and slave.

Topaz has an anti-blooming capability, auto-exposure control AGC and, contrast-stretching through video processing.

Operational status
Available.

Specifications
Observation and stabilisation unit
Dimensions (d × h): 300 × 400 mm
Weight: 45 kg incl operator console, 25 kg in UAV version
Field of regard:
elevation: +15 to −110°
azimuth: 360° continuous
Stabilisation: <40 µrad RMS
FLIR
Spectral band: Long-Wave Infra-Red (LWIR)
File of view (diagonal): 2.5° narrow, 8.7° wide
Clear aperture: diameter 100 mm
Cooling: Stirling integral minicooler
CCD Sensor
Narrow FOV: 1.5 × 1.2°
Focal length: 24 to 240 mm
Elements: 768 × 576
Zoom optics: 1:10
Laser
Range: up to 10,000 m

Contractor
Rafael Armament Development Authority.

VERIFIED

Raytheon AN/AAQ-15 infra-red detection and tracking set

Type
Turret sensor (air).

Description
The AN/AAQ-15 is a lightweight infra-red tracking system which was developed in 1983 for use by the US Air Force in the HH-60 helicopter and is now in use on the MC-130H Combat Talon II aircraft. The system uses the infra-red receiver unit of the AN/AAS-36 with new electronics and lightweight gimbals. The AN/AAQ-15 has three fields of view and covers all of the lower hemisphere as well as having a 15° look-up capability. The system has a video tracker and interface for integrated helmet and display sighting system. It is used to assist search and rescue as well as special operations missions.

Operational status
In production and in service on US Air Force MC-130H Combat Talon II special operations aircraft; also installed in Japanese UH-60J search and rescue helicopters.

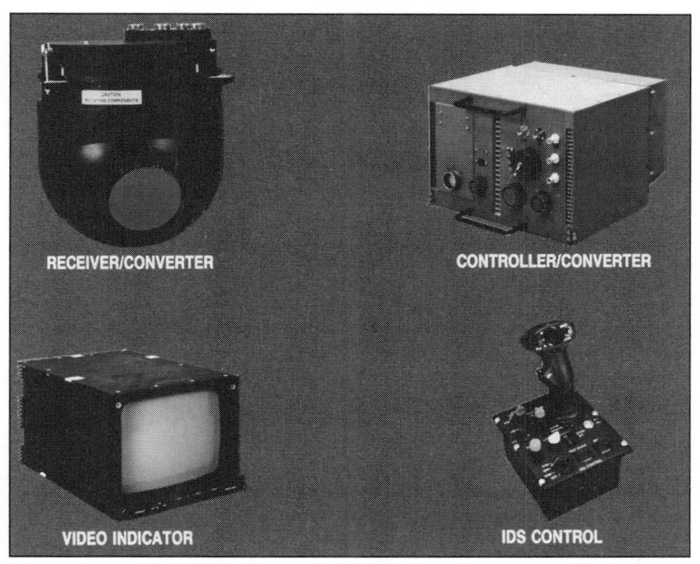

The elements of the AN/AAQ-15 system

Specifications
Fields of view
Wide: 30 × 30°
Medium: 15 × 20°
Narrow: 5 × 6.7°
Fields of regard
Elevation: +15 to −105°
Azimuth: ±200°
Weight
Infra-red receiver: 92 lb
Control converter: 49 lb

Contractor
Raytheon Systems Company.

VERIFIED

Raytheon AN/AAQ-16, AN/AAQ-27 series helicopter night vision systems

Type
Turret sensor (air).

Development
The Raytheon Systems Company (formerly Hughes) AN/AAQ-16 family of helicopter night vision systems began production in 1985 and now includes four variations: AN/AAQ-16B, AN/AAQ-16C, AN/AAQ-16D and the latest version, a medium-wave (3 to 5 µm) system now designated AN/AAQ-27 for the US Marine Corps MV-22 Osprey helicopter.

Description
The basic AN/AAQ-16B is a dual field of view long-wave FLIR system designed for night and low-visibility flying, navigation, target detection/identification and surveillance. It is microprocessor-controlled and mounted on a two-axis gimbal. The system consists of a Turret FLIR Unit (TFU) and a System Electronics Unit (SEU). Controls and displays are available and include panel- and helmet-mounted displays. The TFU houses the sensor and associated electronics which collect IR radiation and convert it to electronics signals for processing by the SEU.

The SEU houses the integral programmable symbol generator, auto-tracker, built-in test and other data processors which process the signals received from the TFU and provide an output of RS-170 (525-line) or optional RS-343 (875-line) or CCIR (625- line) video. The SEU also provides the interface between the aircraft avionics, operator controls and the AAQ-16B. The system can be installed and operated either as a stand-alone unit or integrated on a MIL-STD-1553B databus. The system design allows the AAQ-16B to be transferred quickly and easily between various types of aircraft.

In 1991 Raytheon developed the next generation known as the AN/AAQ- 16C. This is a three field of view long-wave version which includes a new telescope with three magnification levels, ×1, ×6 and ×16 as well as enhanced dual-mode auto-tracker and improved stabilisation. It is totally interchangeable with earlier models and requires no modifications to any helicopter using other configurations.

The AN/AAQ-16D is a further development which as well as having the three field of view telescope also includes a laser range-finder/designator.

The latest version now the AN/AAQ-27, which is in production for the US Marine Corps MV-22 Osprey helicopter, combines production components from the AN/AAQ-16B system using reflective optics and a 3 to 5 µm, Mid-Wavelength Infra-Red (MWIR) staring Focal Plane Array (FPA) detector assembly, produced by Raytheon.

The sensor features a MWIR indium-antimonide (InSb) staring FPA with 480 × 640 detector elements, stated to require a much smaller aperture size than long-wave infra-red systems. The total system weighs less than 93 lb, including about 50 lb for the turret.

Growth options include a video auto-tracker, third field of view and eye-safe laser range-finder.

Operational status

In production since 1985, initially for the US Army which has installed them in its Boeing CH- 47 and MH-47K Chinook and Sikorsky MH-60L and MH-60K Black Hawk helicopters.

The AAQ-16 is also installed on US Marine Corps CH-53E and US Navy SH- 60B LAMPS helicopters and the US Air Force MH-60G while a podded version is operational on the F/A-18 Hornet as AN/AAR-50.

There are more than 400 AN/AAQ-16 systems fielded or on contract worldwide.

In December 1994, Raytheon Systems Company was awarded a US$4.8 million competitive Engineering/Manufacturing Development (EMD) contract to provide advanced infra-red night vision systems for the US Marine Corps MV-22 Osprey. Raytheon was selected by the V-22 avionics integrator, Boeing, to build five EMD models of the system (AN/AAQ-27). The first EMD model was delivered in late 1996 and the system began long-lead initial production. In September 1998 Raytheon was awarded a US$8 million contract for Low Rate Initial Production (LRIP) of the AN/AAQ-27 Osprey system. Deliveries of the V-22 began in 1999 with a total of 523 destined for the US Marine Corps, US special operations units and the US Navy.

Specifications

AN/AAQ-16B
Spectral band: 8-12 μm
Field of view
×1 magnification: 30 × 40°
×6 magnification: 5 × 6.7°
Coverage
Azimuth: 210°
Elevation: +85 to −180°
Slew rate: up to 140°/s
Reliability: >450 h MTBF
Dimensions
FLIR turret (l × d): 356 × 305 mm
Electronics unit: 305 × 200 × 412 mm
Multifunction control unit: 76.2 × 76.2 × 157.5 mm
System control unit: 38.1 × 146 × 115 mm
Weights
FLIR turret: 24.54 kg
Electronics unit: 20.87 kg
Multifunction control unit: 0.59 kg
System control unit: 0.45 kg

A mock-up of the turret FLIR for the mid-wave staring infra-red system for the MV-22 Osprey

The AN/AAQ-27 is in production for the US Marine Corps MV-22 Osprey helicopter

AN/AAQ-16C
Spectral band: 8-12 μm
Field of view
×1 magnification: 30 × 40°
×6 magnification: 5 × 6.7°
×16 magnification: 1.9 × 2.5°
Electronic zoom: 2:1
Turret weight: 29.55 kg

AN/AAQ-16D
Spectral band: 8-12 μm
Turret weight: 40.9 kg

AN/AAQ-27
Spectral band: 3-5 μm
Detector: 480 × 640 InSb staring FPA
Turret weight: 22.7 kg

Contractor

Raytheon Systems Company.

VERIFIED

Raytheon AN/AAQ-18 FLIR system

Type

Turret sensor (air).

Description

The AN/AAQ-18 FLIR system was developed in 1987 for the US Air Force for use on the MH-53J helicopter, replacing the AN/AAQ-10 system. It has also been installed on the MC-130E Combat Talon aircraft.

The AN/AAQ-18 FLIR, configured for use on the HH-53 helicopter, consists of an infra-red receiver, power supply, electronic control amplifier, control indicator and mounting base. All controls are contained in the control indicator, providing the operator with facilities for one-handed control of sensor operation and servo slew commands.

AN/AAQ-18 is usually associated with the US Air Force Pave Low III system which also includes two radar systems; the AN/APN-221 and the AN/APQ-158.

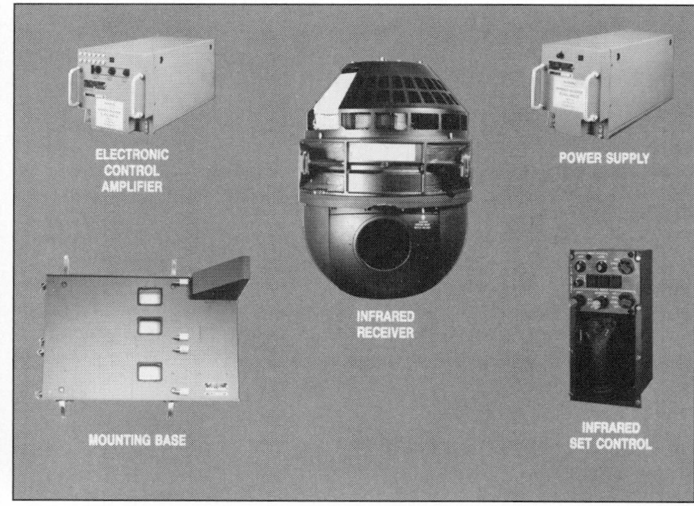

Raytheon Systems Company AN/AAQ-18 FLIR system

Operational status
Production as required. The system is used in the US Air Force HH-53 search and rescue helicopters and special forces MC-130E Combat Talon aircraft. In total, 61 systems have been delivered.

Specifications
Power supply: 2.5 kVA 115 V AC, 400 Hz, 3 phase;
28 V DC, 100 W
Field of view
Wide: 18.2 × 13.8°
Narrow: 4 × 3°
Field of regard
Azimuth: ±190°
Elevation: +15 to −105°

Contractor
Raytheon Systems Company.

VERIFIED

Raytheon AN/AAQ-27 (3 FOV)

Type
Turret sensor.

Development
Raytheon Systems Company has been awarded a contract from Kaman Aerospace International Corporation to develop and provide a turret-mounted mid-wavelength infra-red system for the Royal Australian Navy's Super Seasprite helicopters. The Seasprites will operate from the ANZAC frigates as well as long-range target detection and classification.

Subsequently in January 1998, Raytheon was awarded a multimillion dollar contract from Tenix Defence Systems to provide imaging systems for the Royal Australian Navy's Seahawk helicopters.

Description
The system is a three field of view version of the AN/AAQ-27 infra-red system that Raytheon produces for the US Marine Corps MV-22 Osprey. This features a mid-wavelength indium antimonide (InSb) staring focal plane array with 480 × 640 detector elements. A three field of view telescope is fitted.

Specifications
Dimensions
Turret (h × d): 360.4 × 202.5 mm
Electronics unit: 199.1 × 413.5 × 306.3 mm
Weight: 42.18 kg

The Raytheon AN/AAQ-27 three field of view version 0023396

Detector type: 640 × 480 InSb staring array
Spectral band: 3-5μm
Fields of view
Wide: 30 × 40° (×1)
Middle: 5 × 6.7° (×6)
Narrow: 1.3 × 1.73° (×23)

Contractor
Raytheon Systems Company.

VERIFIED

Raytheon TIFLIR™-49

Type
Turret sensor (air).

Description
The Raytheon (formerly Texas Instruments) TIFLIR™-49 system is a lightweight, multipurpose thermal imaging sensor for navigation, surveillance, maritime, search and rescue and troop transport missions. The system includes two Weapon Replaceable Assemblies (WRA): a turret unit (WRA-1) and an electronics unit (WRA-2). The system can be mounted on fixed-wing aircraft or helicopters.

The FLIR uses second-generation scanned focal plane array technology which is also used in Raytheon's AN/AAQ-17E, AN/AAS-44, AN/AAQ-26 and the Combat Vehicle Thermal Targeting System (CVTTS). The FLIR has three fields of view plus electronic zoom and electronic stabilisation of the image. It uses real-time Local-Area Processing (LAP). Also included in the system is a dual-mode video tracker along with MIL-STD-1553 databus or discrete controls. Both analogue and digital video outputs are provided.

Operational status
TIFLIR-49 has completed a successful installation and test on board a US Coast Guard HC-130 aircraft.

Specifications
Fields of view
Wide: 22.5 × 30°
Middle: 5 × 6.67°
Narrow: 1.3 × 1.7°
Electronic zoom: 2:1 and 4:1
Gimbal angular coverage:
Azimuth: 360° continuous
Elevation: 40° up; 105° down
Gimbal slew rate: 3 rad/s
Gimbal acceleration: 30 rad/s^2 at 300 kt
Max airspeed: >300 kt
Gimbal angle resolution: <100 μrad

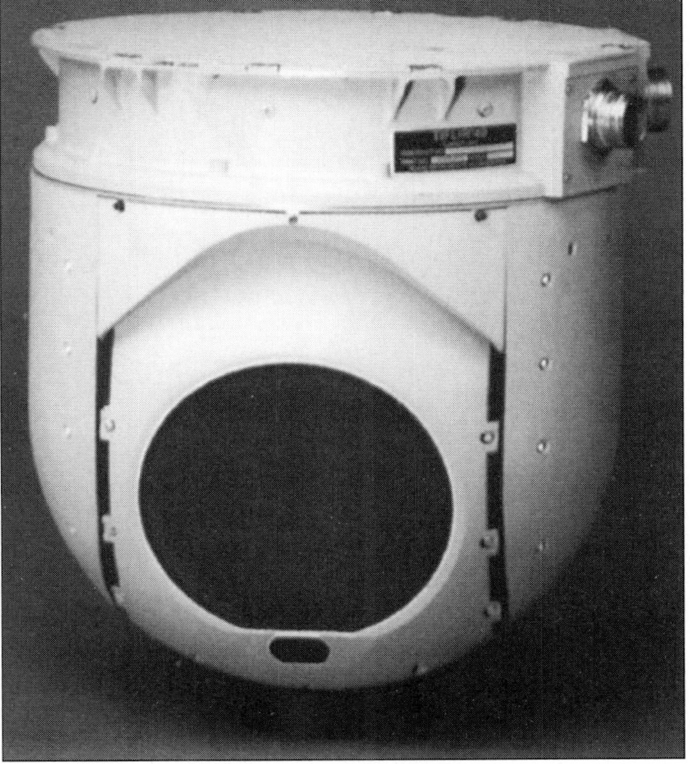

The TIFLIR™-49 thermal imaging surveillance sensor for fixed- and rotary-wing aircraft

Gimbal angular position: <2.0 mrad RMS, dynamic
Cooling: self-contained
Weight
Turret unit: 57 lb
Electronic unit: 43 lb
Dimensions
Turret unit (d × h): 12.75 × 14.68 in
Electronic unit (w × l × h): 12.06 × 12.28 × 7.84 in

Contractor
Raytheon Systems Company.

VERIFIED

Saab Bofors SEOS

Type
Turret sensor.

Description
The Saab Bofors SEOS, stabilised electro-optical multisensor system, is designated for a variety of applications. It can be mounted on a high mast for battlefield surveillance or indirect fire role. It can also be mounted on different types of helicopters or boats.

The system is designed to give high probability of detection and Line Of Sight (LOS) stabilisation sufficient for all integrated weapons out to detected targets. The intelligent man machine interface assists the system operator and reduces workload.

SEOS consists of a stabilised sensor platform, a conditioning unit for climate control, a processing unit for image enhancement and fusion of sensor signals and a control unit with a high-resolution colour monitor.

Mounted on the stabilised platform in the basic system are an 8 to 12 μm second-generation, thermal imager, laser range-finder and TV cameras. Optional sensors/designators include laser designator, missile tracker/beam rider unit, low-light level TV camera, 3 to 5 μm thermal imager.

Depending on sensor choice, SEOS can be integrated with various weapon interfaces such as optically controlled missiles, laser-guided munitions, turreted or fixed guns, rockets and artillery fire control. The sensor head can be either mast or pedestal mounted.

The image processing system features automatic target tracking and aided target detection. Image freeze/store, area tracking, thermal cueing, electronic magnification, sensor fusion, image integration and graphics generation and overlay are also included.

Operational status
Under development.

Specifications
Weight: 70 kg
Dimensions (h × d): 0.5 × 0.6 m
Azimuth: ±200° (optional 360°)
Elevation: –30 to +85°
Fields of view
Wide: 18.0°
Medium: 4.5°
Narrow: 1.5°

The Saab Bofors SEOS turret sensor 0023397

Line of sight jitter: <15 μrad RMS
Interface: 1553B, RS-422 or others

Contractor
Saab Bofors Dynamics.

SAGEM HESIS

Type
Turret sensor and surveillance system (air).

Description
HESIS is a day/night airborne surveillance system with a microwave transmission capability allowing real-time remote surveillance applications, including law enforcement, border patrol and coastguard operations. HESIS consists of an airborne gyrostabilised pod (equipped with a variety of sensors dependent on the requirement) with associated display and recorder, a fixed or mobile ground station and a microwave downlink. The system can use digital or analogue technology, according to the requirement.

The pod includes dual infra-red and daylight cameras, providing boresighted image-fused information. The field of view allows panoramic surveillance and long-range recognition/identification (3 to 5 km for a person on a landscape, 10 n miles for a surface vessel). A range of mountings allows the pod to be installed on many types of helicopters and fixed-wing aircraft.

The TV and infra-red images are continuously displayed and recorded onboard or transmitted in real time. The ground station can be mobile or fixed. For mobile applications, transmission range reaches 70 km with an auto-tracking antenna

SAGEM HESIS surveillance system

The SAGEM HESIS pod mounted on an aircraft

(one or two axis) installed on a telescopic mast. For permanent installation, HESIS can give continuous observation at a range of more than 100 km.

The HESIS system is particularly well adapted to maritime surveillance, thanks to its image processing and airborne interface unit which allows image enhancement for improved threat identification and useful image analysis functions (freeze, zoom, and so on); automatic hot point detection and auto-tracking; slaving to radar designation and overlays of flight data for threat localisation.

Operational status
In operation with the French Customs (fixed wing aircraft and helicopters) and with the security forces of several other countries. Recently chosen by the French Gendarmerie for a fleet of helicopters.

Contractor
SAGEM SA Aerospace & Defence Division.

VERIFIED

SAGEM OLOSP multi-sensor platform

Type
Turret sensor (air).

Description
OLOSP is a gyrostabilised platform allowing day and night observation and sighting independently of the movements of its carrier. This modular system allows dual or triple sensor versions to meet all mission-related requirements: observation, land or coast surveillance, long-range identification and search and rescue.

Operational status
Series production. The OLOSP systems are used in several countries, onboard a variety of carriers, such as UAVs, helicopters, observation balloons, patrol or naval surveillance aircraft, surface ships, fast patrol boats and fixed ground stations.

Specifications
Weight: 25 to 40 kg
Dimensions: 350 or 410 (diam) mm
Fields of regard: Azimuth 360°, elevation: –20 to +120°
Stabilisation: 4 axes
Payloads: dual or triple sensor configurations
High performance thermal camera: IRIS or MATIS
Number of field available: 3 (IRIS and MATIS)
Wavelengths: 8-12 μm (IRIS) or 3-5 μm (MATIS)

Day camera
Colour or black and white CCD camera
Zoom: ×10 or ×27

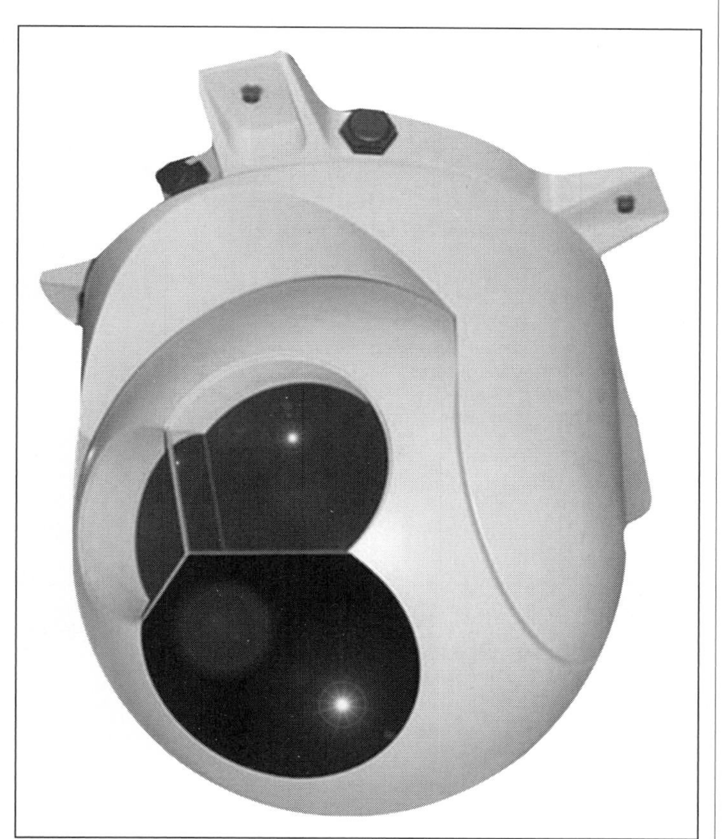

SAGEM OLOSP 410 gyrostabilised platform 0137755

The SAGEM OLOSP 350 gyrostabilised platform 0137378

Laser range-finder (eye-safe)
Automatic target tracking
Diameter: 350 mm or 410 mm
Weight: 25 to 40 kg

Control panel
Interfaces: MIL-STD-1553, ARINC 429, RS-422

Contractor
SAGEM SA Aerospace & Defence Division.

UPDATED

STN ATLAS Elektronik BAS 2000 observation and reconnaissance system

Type
Reconnaissance system.

Description
The BAS 2000 is an observation and reconnaissance system for rotary- and fixed-winged aircraft as well as UAVs. It uses a thermal imager based on the ISOS 2000 for the Brevel reconnaissance UAV which is itself based on Ophelios thermal imaging modules. The system consists of a stabilised platform, thermal imager and videotracker. The platform features two-axis coarse/fine stabilisation, rotation of the thermal imager for image derotation and stabilisation of the imager relative to

Sea King fitted with BAS 2000 0129068

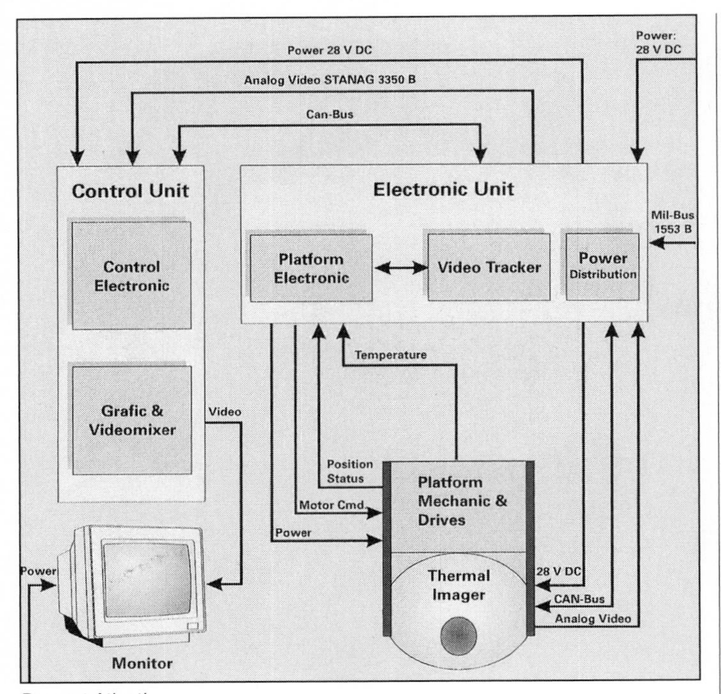

Breguet Atlantic

the air vehicle. The imager is based on a CMT IRCCD array, has a zoom telescope, remote control via a two-wire bus and integrated image derotation. The centroid and correlation tracker has automatic or manual selection of trackgate size and tracker modes.

Operational status
Series production.

Specifications
Thermal imager
Detector material: CMT PV
Detector array: 94 × 4 IRCCD
Spectral band: 7.5 to 10.5 μm
Fields of view: continuous zoom between 3.6 × 4.8° and 29 × 38°
Resolution: 576 × 756 pixels (full CCIR format)
Stabilised platform
Elevation: –80°
Azimuth: –180 to + 180°
Accuracy of stabilisation: =0.02 mrad
Tracker
Modes of operation: correlation, centroid and complementary modes
System
Weight: 39 kg excl monitor and controls
Power: 28 V DC
Power consumption: 400 W average, 550 W peak

Contractor
STN ATLAS Elektronik GmbH.

VERIFIED

Tamam Helicopter Multimission Optronic Stabilised Payload (HMOSP)

Type
Turret sensor.

Description
The Helicopter Multimission Optronic Stabilised Payload (HMOSP) is a lightweight dual- or triple-sensor day and night surveillance, target acquisition and range-finder/pointing payload version of MOSP which has been designed for installation in airborne, ground or naval platforms. It has the capability to look down through the nadir point to provide a view throughout the lower hemisphere.

The sensor package options include single monochrome or colour CCD or triple-colour CCD day channels, 8 to 12 μm first, second or third (FPA) generation FLIR, laser range-finder and laser target illuminator. The helicopter-mounted version can have either a 3 to 5 μm or 8 to 12 μm thermal imager.

Hand controllers for the operator interface via processors with the turret allowing the sensors to be pointed and all other system functions such as field of view, black hot/white hot, gain and offset to be selected.

Operational status
Over 400 systems have been supplied for rotary- and fixed-wing manned and unmanned aircraft as well as ships and mobile vehicles.

HMOSP mounted on a helicopter 0023399

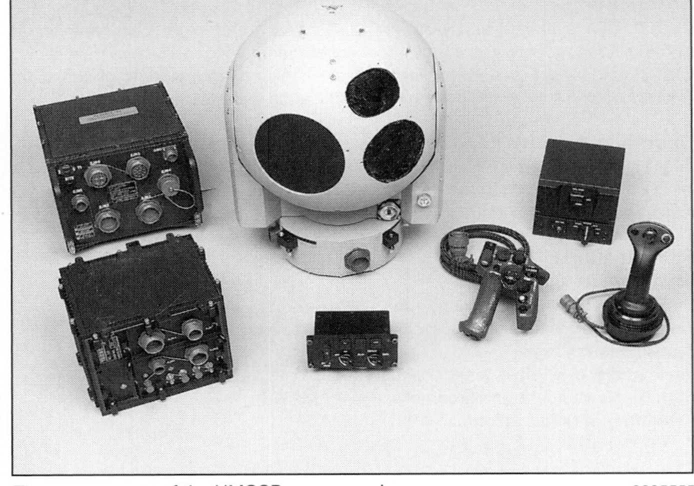
The components of the HMOSP sensor package 0005555

Specifications
Dimensions (h × d): 500 × 380 mm
Weight: <32 kg depending on sensors carried (payload)

TV fields of view
Narrow: 0.9° × 0.67°
Medium: 3.3 × 2.5°
Wide: 12 × 9°
Alternative: 20 – 770 mm zoom, zoom magnificatiom × 1 to × 35
Zoom: 1.3 to 18.2°

FLIR fields of view (3 to 5 μm, 320 × 240 pixels)
Narrow: 0.97 × 0.73°
Medium: 3.37 × 2.9°
Wide: 18 × 13.5°

Spatial coverage
Elevation: +60 to –95°
Azimuth: n × 360° (unlimited)
Power consumption: 280 W (day/night) @ 28 V DC

Contractor
Tamam Division – Israel Aircraft Industries.

VERIFIED

Tamam Plug-in Optronic Payload (POP)

Type
Turret sensor (air).

Description
The Plug-in Optronic Payload (POP) is a modular, compact, lightweight electro-optical payload designed for day/night surveillance, target acquisition, identification and location. POP is designed for light aircraft, helicopters, UAVs ground and sea applications. It consists of a 260 mm diameter stabilised platform

with a replaceable plug-in sensor unit which may be designed to meet the customer's requirements.

The sensors include an InSb focal plane array thermal imager, long-range colour CCD camera or a combination of both. The sensor unit can be replaced in the field within minutes by replacing the plug-in sensor module which slides into the gimbal envelope. The payload includes an automatic tracker and is controlled via a parallel-serial communication channel.

Operational status

In production.

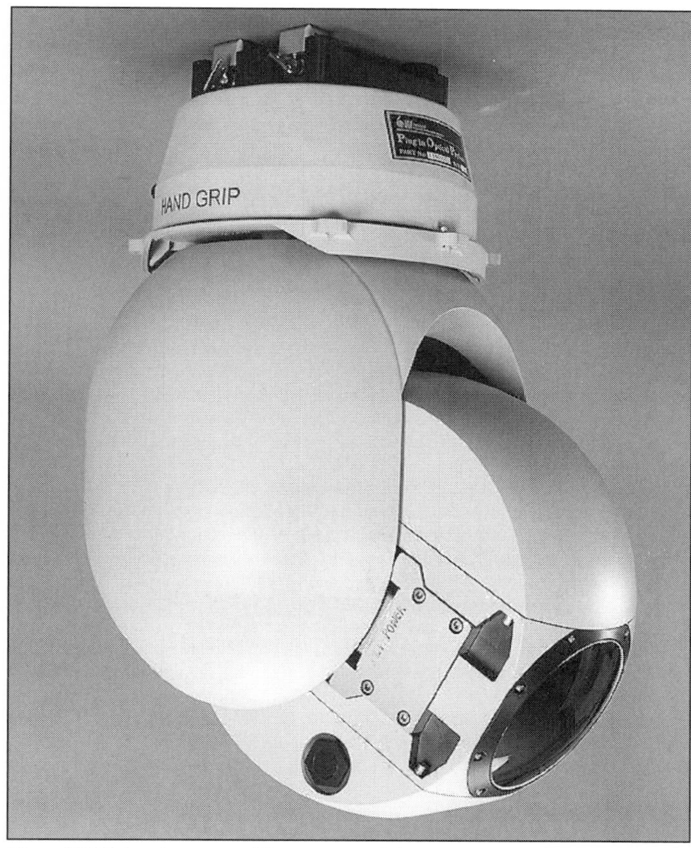

View of POP showing removable sensor module

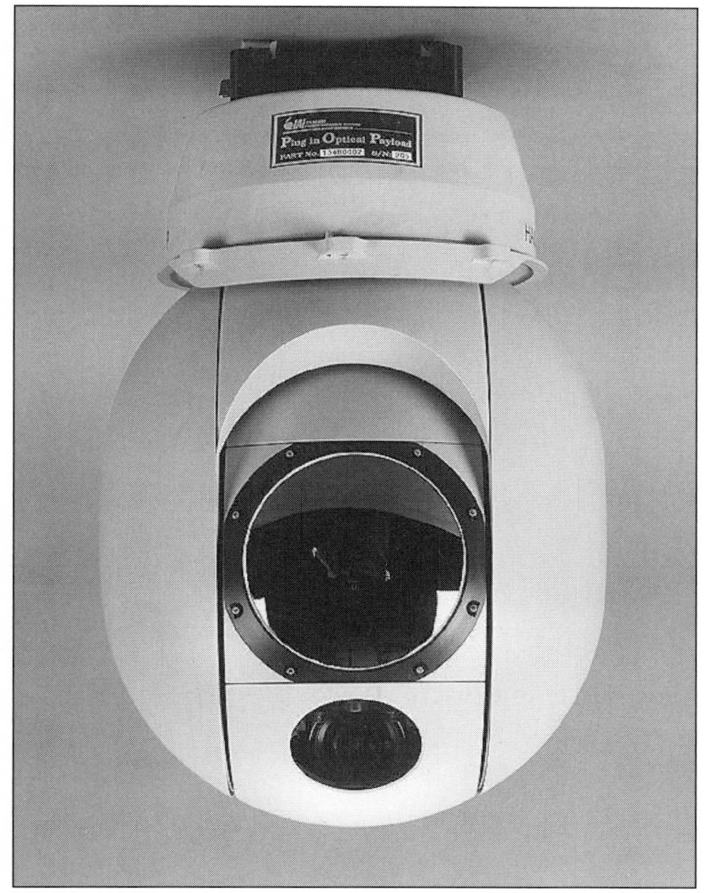

The Tamam POP

POP mounted on a helicopter 0023398

Specifications

Dimensions
Diameter: 260 mm
Height: 380 mm
Weight: 14-16 kg according to sensor type

Fields of regard
Azimuth: 360°
Elevation: +40 to −110°

Fields of view FLIR TV
NFOV: 17 × 13° zoom × 13
MFOV: 6.4 × 4.8° 1.4 to 18°
WFOV: 22 × 16°

Contractor

Tamam Division - Israel Aircraft Industries.

VERIFIED

Thales Optronics Chlio-S multisensor airborne FLIR

Type

Turret sensor.

Description

Capitalising on the Chlio FLIR, an element of the Thales Airborne Systems AMASCOS (Airborne MAritime Situation Control System), which has been in service for more than 10 years, the Chlio-S is based upon the second-generation Synergi thermal imaging modules developed by Thales Optronics and Zeiss. It is designed for long-range observation and surveillance, flying in poor visibility and for search and rescue operations.

The detector is a 288 × 4/8 to 12 μm focal plane array. The system has four fields of view and is mounted on a gyrostabilised turret which is stabilised in two axes electromechanically and two axes electro-optically.

The Chlio-S turret 0058720

Chlio-S on French Air Force RESCO Cougar Mk II (Eurocopter) 0058721

Operational status

Chlio-S is in production for C160, and French Navy Falcon 50 SURMAR (SURveillance MARitime) aircraft, Indonesian and UAE Navy CN 212 maritime patrol aircraft. It is in service on French Air Force Cougar and French Army Super Puma helicopters and with other export customers.

Specifications

Spectral band: 8-12 μm
Detector: 288 × 4 second-generation CMT
Fields of view: 24 × 18°, 12 × 9°, 3 × 2.2°, 1.5 × 1.1°

Fields of regard
Azimuth: 360°
Elevation: –97 down to +40° up
Weight: 53 kg
Option: 3-5 μm, CCD camera, laser range-finder, auto-tracking

Contractor

Thales Optronics.

UPDATED

UOMZ GOES 520 Gyrostabilised Optical Electronic Platform

Type

Turret sensor.

Description

The GOES 520 platform is intended for day and night surveillance, search and detection. Standard sensors are a thermal imager and TV camera.

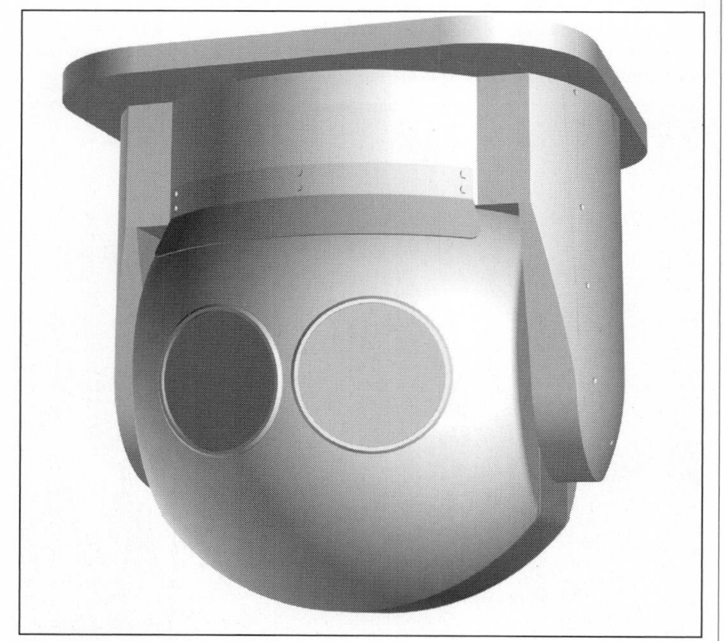

The GOES 520 turret from UOMZ 0089875

Operational status

Not stated.

Specifications

Dimensions (Diam × h): 350 × 500 mm
Weight: 45 kg
Sighting angle limits
Azimuth: ±180°
Elevation: –85 to +35°

Contractor

Production Association Urals Optical and Mechanical Plant (UOMZ).

VERIFIED

UOMZ Gyrostabilised Optical Electronic System (GOES)

Type

Turret sensor.

Description

The UOMZ GOES platforms are designed to carry thermal imagers, day and lowlight-level TV cameras, cine- and video-cameras, laser range-finders, and similar equipment. The first four models produced were designatedGOES-1/-2/-3/-4. They were all designed to carry electro-optical payloads varying from 16 to 100 kg. The systems were suitable for both civil and military use, and payload specification is by customer choice.

A new generation of GOES platforms, designated GOES-310/-320/-330 has now been developed and put into production. The designations represent single-, double- and triple-channel systems respectively. All are designed for detection and recognition of objects in a broad range of vision angles in severe rolling and vibration conditions in both civil and military environments.

Sensors available for fitment to the GOES-310/-320/-330 series include: thermal imagers, cine and video cameras, laser range-finders, and infra-red sensors in any combination of fits. System options include: compatibility with GPS, RS 232 interface, video monitor, VHS/SVHSvideotape recording, an air-ground microwave datalink.

More recent variants include the GOES 321, 342, 344, 346. The 321 includes a sighting function for unguided missiles and guns.The 344 and 346 include laser guidance for weapons such as the 'Vikhr-M' anti-tank guided missile.

Operational status

The company has stated that orders were placed for the GOES-320 system at the MAKS-97 air show. The GOES-320 unit was shown on the 'Hind' night vision proposal (Mi-24VN, also designated Mi-35O) mock-up shown at MAKS'99.

Variant systems are also fitted to the Kamov Ka-29, Ka-50N and Ka-52, and Mi-24 and -35 helicopters. The accompanying photographs taken at MAKS-97 show the YOM3 system mounted below Shkval on the Ka-50N Black Shark, although it has also been shown mounted above Shkval on the same aircraft at other times. On the Ka-52 Alligator, it is mounted above the cockpit. The company stated that the production standard electro-optical sensor fit had not yet been finalised.

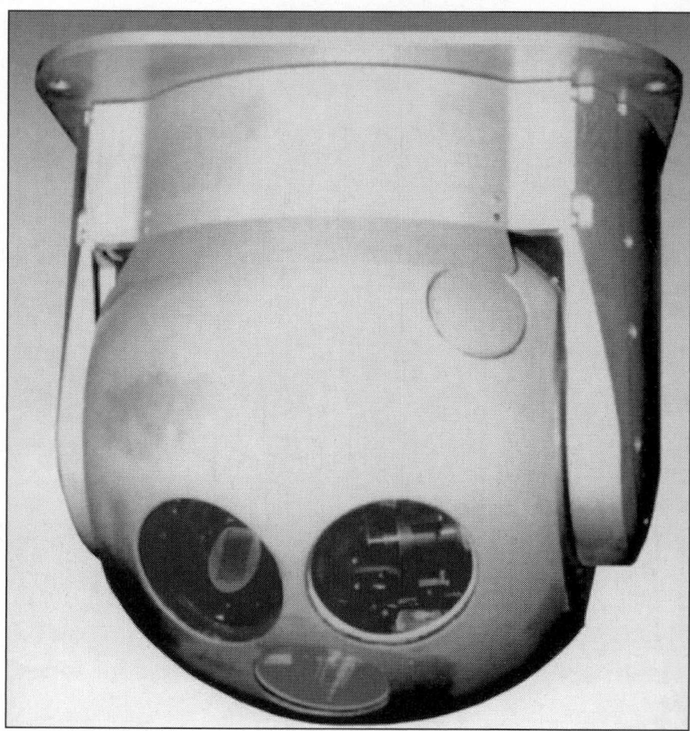

The UOMP/GOES-3 Gyrostabilised Platform (GS) 0005552

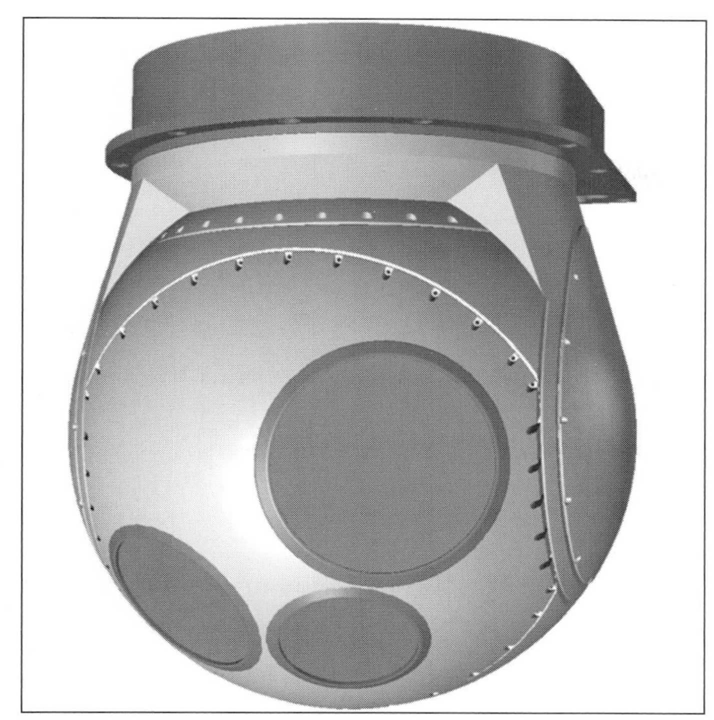

GOES 342 stabilised platforms from UOMZ 0089891

GOES 321 stabilised platforms from UOMZ 0089890

YOM3 on Ka-52 Alligator helicopter (Paul Jackson) 0018340

Specifications

Platform	**GOES-1**	**GOES-2**	**GOES-3**	**GOES-4**
Payload weight	100 kg	16 kg	30 kg	65 kg
Payload volume	85 dm^3	13.5 dm^3	20 dm^3	55 dm^3
Weight of optical turret	140 kg	25.5 kg	20 kg	85 kg
Dimensions of optical turret	0720 × 980 mm	0340 × 552 mm	0460 × 613 mm	0640 × 850 mm
Look angles				
(azimuth)	±170°	±170°	±235°	±135°
(elevation)	+80 to −40°	+10 to −30°	+45 to −115°	+10 to −30°
Stabilisation (micro radians):	50	70	50	50

GOES-310/-320/-330
Stabilisation: 5-axis
Pointing error: <50 μrad RMS

Field of regard
(azimuth): +-230°
(elevation): +30 to −110°
Maximum angular rate: 60°/s

GOES 344 stabilised platforms from UOMZ 0089892

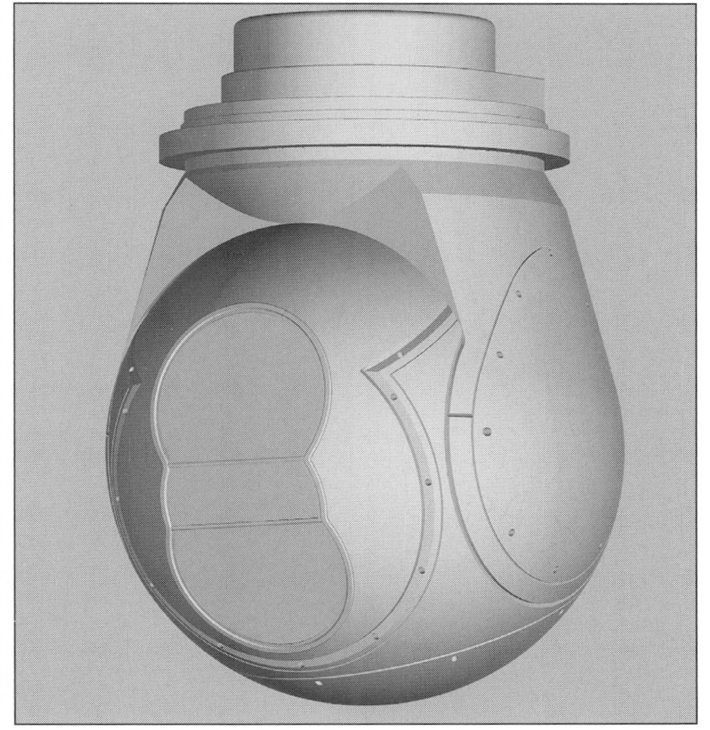

GOES 346 stabilised platforms from UOMZ 0089893

Dimensions
(opto-mechanical unit): 460 × 613 mm
(electronic unit): 330 × 485 × 225 mm
(control unit): 225 × 50 × 57 mm

Weight
(opto-mechanical unit): 55 kg
(electronic unit): 20 kg
(control unit): 0.43 kg
Power: 27 V DC, 500 W; 115/200 V AC, 3-phase, 400 Hz, 250 VA
Thermal imager: FLIR Systems ThermoVision1000
receiver: 5-bar SPRITE focal plane
spectral range: 8-12 μm
cooling: integral Stirling cooler
pixels: 580 × 386

Field of view (H&V)
(narrow): 5.0 × 3.3°
(wide): 20.0 × 13.3°
NETD: 0.18°C
TV system: CCD Sony EVI-331 colour
pixels: 752 × 582
TV lines: 480
focal range: f=5.4 to 64.8 mm with ×12 optical zoom
field of view: 48.8 × 37.6° to 4.4 × 3.3°

Laser range-finder (made by Production Association Urals Optical and Mechanical Plant (PA UOMZ):
Wavelength: 1.54 μm
Energy: 0.01-0.07 J
Beam divergence: 2-3 minutes of angle
PRF: 1 Hz
Maximum range: 10 km
Measurement error: <5 m

GOES 321/342/344/346
Standard components
GOES 321: thermal imager, laser rangefinder, stabilisation
GOES 342: thermal imager, TV channel, laser range-finder, direction finder, stabilisation
GOES 344: thermal imager, TV channel, laser range-finder, laser beam control channel, stabilisation
GOES 346: thermal imager, TV channel, laser range-finder, laser beam control channel, stabilisation

Platform	GOES 321	GOES 342	GOES 344	GOES 346
Weight of system	85 kg	185 kg	90 kg	105 kg
Dimensions of turret (diam × h)	460 × 613 mm	460 × 613 mm	460 × 613 mm	460 × 613 mm
Look angles				
(azimuth)	±230°	±230°	±150°	±230°
(elevation)	+40 to –30°	+ 25 to –115°	+85 to –20°	+30 to –115°

Contractor

Production Association Urals Optical and Mechanical Plant (PA UOMZ).

VERIFIED

OBSERVATION AND SURVEILLANCE

MARITIME SENSORS

AeroSpace Technologies marine surveillance system

Type
Maritime surveillance.

Description
The ASTA low-level airborne marine surveillance system includes a 360° Litton Canada AN/APS-504V(5) search radar; FLIR Systems 2000G thermal imaging system with recorder; a comprehensive avionics suite with DME, Omega/VLF, an optional inertial navigation system and a two-axis autopilot. These systems are integrated and operate together to detect, track and identify targets and geographically locate them with date, time and latitude and longitude co-ordinates. The system, fitted to a suitable platform, provides a surveillance package that can carry out detection and identification, often without the target being aware of surveillance.

The Litton AN/APS-504(V)5 features digital display subsystems, a track-while-scan mode capable of tracking up to 20 targets simultaneously, coherent pulse compression, frequency agility and videotape recording. Detection of a 2 m^2 target in Sea State 3 conditions at ranges up to 35 n miles is claimed.

Signals provided to the FLIR Systems 2000G by the radar and the long-range navigation system give it the capability to identify marine targets on a 24 hour basis and record these targets on video.

Operational status
The ASTA marine surveillance system was developed for the Searchmaster N22S Series 2. This aircraft is in service with the US Customs Service in its drug interdiction programme.

Contractor
AeroSpace Technologies of Australia Ltd.

VERIFIED

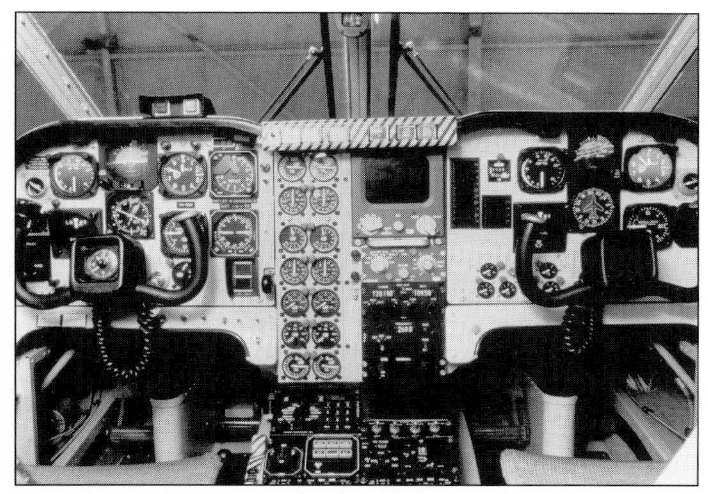

The cockpit of a Searchmaster aircraft fitted with the ASTA marine surveillance system

BAE Systems ATD-111 LIDAR

Type
Multi-mission system for mine and anti-submarine warfare.

Description
The ATD-111 is an airborne Light Detection And Ranging (LIDAR) system developed for the US Navy to detect and classify underwater threats, primarily submarines and mines. The system is a stabilised scanning laser unit mounted aboard a Navy SH-60 Seahawk helicopter. The ATD-11 employs a solid-state laser self-contained in a pod. The onboard operator control unit provides detection information, images and maps in GPS co-ordinates on a flat panel colour display.

Operational status
In development. The ATD-111 is currently undergoing upgrades for future testing to explore the potential of active laser-based technology to achieve superior mission capability in shallow waters.

Contractor
BAE Systems North America, Information and Electronic Warfare Systems.

UPDATED

BAE Systems Sea Owl passive identification device

Type
Maritime sensor (air).

Description
The Sea Owl system provides a day and night long-range target detection and identification capability for helicopters. It uses a BAE Systems thermal imaging sensor with high-magnification optics mounted in a steerable, stabilised platform sited on the nose of the helicopter. The high-resolution thermal image, automatically optimised for maximum picture quality, is displayed in the cockpit. The employment of advanced signal processing provides tracking and target cueing automatically.

The system is completely integrated with the helicopter's central tactical system and other avionics systems and offers both automatic search and acquire operation and manual control via a joystick unit.

Operational status
In production for the Royal Navy Westland Lynx HMA.8 helicopters.

Specifications
Spectral band: 8-13 μm
Detector: CMT 8-parallel TED
Cooling: minicompressor
MRTD: >0.1°C
Magnification: ×5 to ×30 switched zoom
Fields of view:
625 line 50 Hz: 12 × 8°, 2 × 1.33°
525 line 60 Hz: 9.7 × 6.5°, 1.6 × 1.08°
Field of regard
Elevation: +20 to −30°
Azimuth: +120 to −120°

The BAE Systems Sea Owl passive identification device on a Lynx helicopter

Video output: 625-line, 50 Hz or 525-line, 60 Hz
Weight
Turret: 64 kg
Signal processor: 17 kg
Tracking unit: 13 kg
Compressor: 10 kg
Power supply: 28 V DC nominal, 570 W typical
115/200 V AC, 400 Hz, 115 W typical

Contractor
BAE Systems, Avionics, Sensors and Communications Systems, Basildon.

UPDATED

FLIR Systems Sea Star SAFIRE II™

Type
Turret sensor.

Description
The Sea Star SAFIRE II™ provides range performance with 24-hour coverage. Night and day, and through a wide variety of obscurants and weather conditions, it delivers long-range imaging via a 3-5 mm InSb focal plane array with three fields of view, an 18:1 zoom CCD TV and a spotter scope housed in a gyrostabilised gimbal. The Sea Star SAFIRE II™'s modular design also accommodates a complete range of optional sensor payloads, including a laser range-finder, daylight TV, a spotter scope, a laser pointer and a CALI laser illuminator. It is designed for ship and aircraft operations and can be installed as a roll-on/roll-off asset. Fully portable, it fits with a variety of platforms and can be removed or installed in less than an hour.

Operational status
In production.

Specifications
Turret size (d × h): 44.58 × 38.35 cm (17.55 × 15.10 in)
Turret weight: 98 lb (44.5 kg)
Azimuth: 360° continuous
Elevation: +30 to −120°
Slew rate: Variable to ≤60°/sec
Control: SCU, serial digital/MIL-STD-1553B
Environmental: MIL-STD-810 and MIL-STD-461
Max air speed (flight tested): 405 kt

Thermal imager
Sensor: 320 × 240 InSb FPA 3-5 μm response
Sensor resolution: 640 × 480 (microscanned)

Fields of View
Narrow: 0.8 × 0.6°
Medium: 3.4 × 2.6°
Wide: 25.2 × 18.8°
Optional wide field of view: 33.3 × 25°

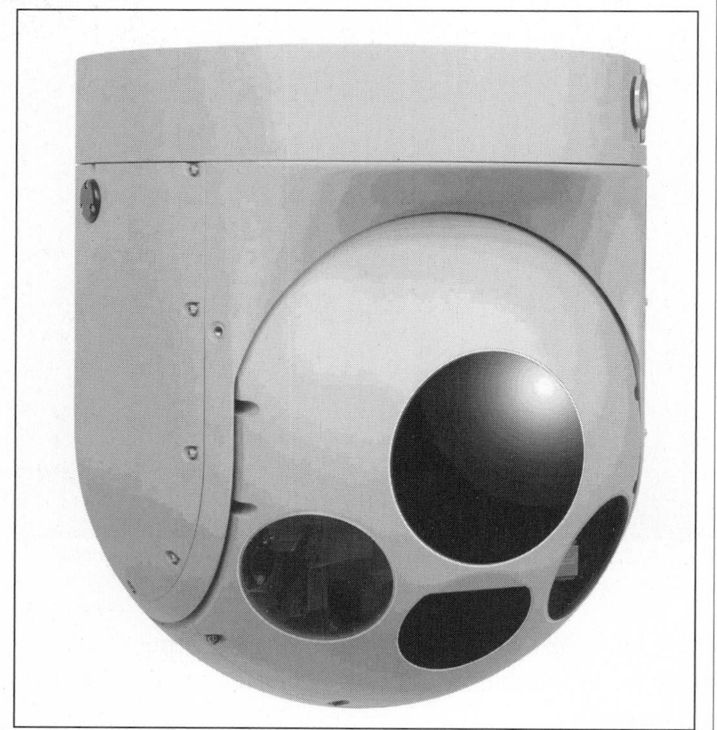

The FLIR Systems Sea Star SAFIRE II **NEW**/0547634

Cooling: Stirling linear cooler
Video output: RS-170 or CCIR

Optional payloads
Visual TV: Color CCD-TV with autofocus
Signas format: NTSC/PAL
Sensitivity: 0.2 lux
Zoom ratio: 16× plus 4× electronic zoom
Fields of view: 24° to 1.5°

Laser range-finder
Type: Erbium – glass (1.54 mm)
Classification: Class 1 (eyesafe)
Range accuracy: ±5 m
Maximum range: 20 km

CALI laser illuminator (NVG compatible – wide area illumination)
Emission wavelength: 810 nm
Classification: Class IV

Laser illuminator (NVG compatible – target pointing)
Emission wavelength: 810 nm
Classification: Class IIIb

Spotter scope
Video: PAL or NTSC – color image
Field of view: 0.4 H × 0.3° V

Optional Features: Autotracker; radar, navigation and VCR interfaces; quick disconnect mounts; Nightsun™ interface; laser illuminators for target identification; flat panel displays; moving map based Mission Management System

Contractor
FLIR Systems.

NEW ENTRY

FLIR Systems SeaFLIR™

Type
Turret sensor (maritime air).

Description
An ultra-small dual-sensor turret including a third-generation thermal imager and colour TV camera. Featuring environmental qualification to military maritime standards, the system is optimised for shipborne operation. For automatic tracking of moving targets, it has a centroid mode of operation. This centres the target within an adjustable tracking gate shown on the display monitor. It also has a memory capability which allows a temporarily lost target to be re-acquired via a coast mode, which tracks on the last known movement parameters. The turret maybe deployed in hanging or inverted mode for either ship-mounted or aircraft-mounted operation.

FLIR Systems SeaFLIR **NEW**/0137877

Operational status
In service.

Specifications
Stabilised platform
Dimensions (h × diam) 38.6 × 22.9 cm
Turret weight: 13.1 kg
Field of regard
Azimuth : 360° continuous
Elevation: 360° continuous
Slew rate: variable to ≤50°/s
Control: HCU, serial digital
Environmental: MIL-STD-810E, MIL-STD-461C, ASTM 117B
Maritime considerations: sealed housing, HCU and ECU, heated turret

Thermal imager
Spectral band: 3-5 μm
Detector: 256 × 256 InSb FPA
Resolution: IFOV (NFOV) 0.12 mRAD
Fields of view: 17.6 to 1.76° horizontal
Focal length: 25 to 250 mm continuous zoom

Optional configurations
Long-range Colour CCD Daylight Imager (SeaFLIR/SeaFLIR C)
Video format: NTSC or PAL
Imager: ⅓″ CCD 811 × 608 (PAL)
Resolution: 450 television lines
Fields of view: 48 to 2.7° horizontal continuous zoom
IR extender: 1.8 × 25-450 mm focal length (SeaFLIR C only)

Low-light monochrome CCD Daylight Imager (SeaFLIR M)
Video format: NTSC or PAL
Imager: ½ in monochrome CCD 768 × 576
Resolution: 570 television lines
Sensitivity: .0003 lux @ f-1.4
Fields of view: 29° vertical to 2.9° horizontal continuous zoom
IR extender: 1.8 × 25-450 mm focal length

Laser range-finder (SeaFLIR L)
Type: 1.5 μm, eye-safe
Maximum range: 20 km
Minimum range: 40 m
Range resolution: ±5 m
IR extender: 1.8 × 25-450 mm focal length

Contractor
FLIR Systems.

NEW ENTRY

FLIR Systems SeaFLIR™ II & SeaFLIR™ II-C

Type
Turret sensor.

Description
The SeaFLIR™ II and II-C are military-qualified, long-range, triple-sensor turrets with a third-generation thermal imager, boresighted color TV camera with low-light mode, and an optional laser pointer. The SeaFLIR™ II-C features a long-range IR continuous zoom telescope (25 to 450 mm) with an operator-selectable 1.8× optical extender. The SeaFLIR™ II's 9 in stabilised turret and all other components are sealed and hardened for maritime and over-water airborne operations. Operator-selectable turret-up/turret-down orientation allows cross-deck installations using FLIR quick-disconnect mounts. All SeaFLIR™ II's come equipped with integral turret heaters for operations in cold weather climates. Both IR and CCD-TV Autofocus, Automatic image optimisation, imbedded Autotracker and ergonomic, marinized hand controller are designed to reduce operator workload. Unique options include Radar Bearing AutoTracker Handoff, Auto Alert and Auto Detect Scanning utilities for Force Protection, as well as GPS and Searchlight Slave (SLASS) integration.

Operational status
In service.

Specifications
TI: 320 × 240 InSb FPA 3-5 μm. IR Continuous zoom Telescope with 10× magnification - Field of View range: 22 × 16.5° (WFOV) to 2.2 × 1.65° (NFOV) without extender.
Field of view range: 12.2 × 9.2° (WFOV) to 1.2 × 0.92° (NFOV) with 1.8× extender.
IFOV: 0.12 mrad (NFOV) and 0.067 mrad with 1.8× IR Extender (NFOV)
Video output: PAL, RGB or NTSC
TV: 752 H × 582 V (PAL) 768 × 494 (NTSC) Continuous zoom telescope with 18× magnification and integral 4× electronic zoom - Maximum magnification: 72×
Field of view range: 48 × 32° (WFOV) to 2.7 × 2.2° (0.67 × 0.55° using E-Zoom) (NFOV)

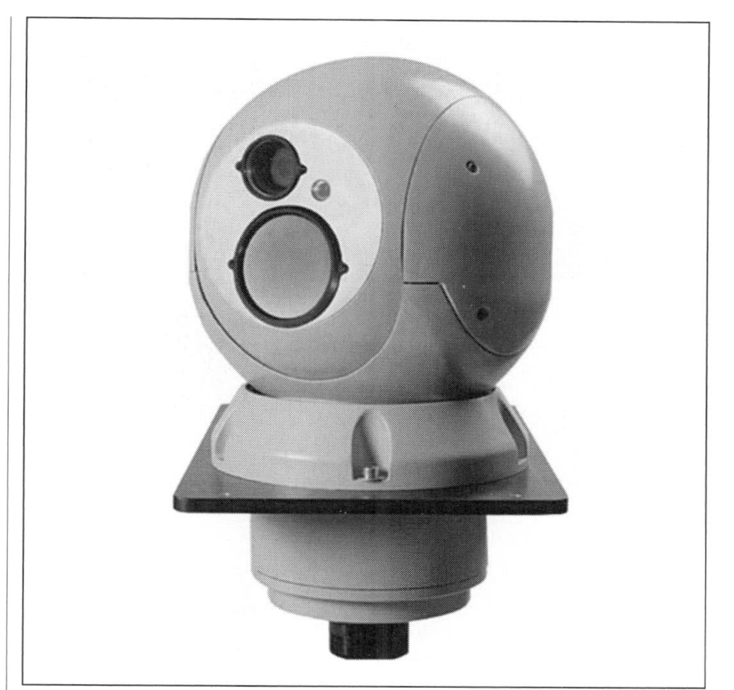

FLIR Systems SeaFLIR™ II ***NEW***/0547499

Low-light mode: Operational down to 0.02 Lux
Resolution: 460 TV Lines (PAL), 470 TV Lines (NTSC)
Laser pointer: visible with Image Intensified Night Vision Goggles
Other available features: RS 422/232, ARINC interface, RS-232 serial GPS interface, Search Light Slave (SLASS) interface, PIP monitor, Terse Binary Protocol (TBP) for Remote and UAV control functions. Integrated Eye-Safe Laser Rangefinder boresighted with IR payload only is available in other production SeaFLIR models

Contractor
FLIR Systems.

NEW ENTRY

Kaman Aerospace Magic Lantern®

Type
Laser based mine detection system (air).

Development
The laser-based Magic Lantern mine-detection systems underwent initial evaluation trials in 1988, where it was used to find, classify, and locate sea mines under real conditions. Subsequent to successful trials, an advanced version of the system was deployed aboard Navy helicopters and used during the Desert Storm Operation. Three systems have been built and are known as the Magic Lantern Deployment Contingency, ML(DC). Incorporating a GPS system, ML(DC) can detect and precisely locate sea mines. The system includes an automatic target recognition capability and multiple receivers to focus on many depth levels simultaneously. The systems are undergoing a conversion to be compatible with the MH-53E helicopter as part of the Navy's Airborne Mine Countermeasures Supporting Force concept.

The Magic Lantern® laser-based mine countermeasures system on a US Navy SH-2G helicopter

View of US Navy SH-2G helicopter equipped with Magic Lantern®

Description

Magic Lantern uses a neodymium-YAG blue-green laser linked to six intensified charge-coupled device cameras to detect mines at different depths. The system pulses the laser into the sea and pulse-timing generators electronically open the shutters of the cameras to allow them to receive a return of laser energy from preselected depths. Objects that break the laser beam at those depths appear as reflections. Energy reflected from the mine is detected by any of six highly sensitive electronic image-intensified cameras and a scanner. The scanner provides increased area coverage by moving the beam across continuous swathes of water perpendicular to the aircraft's flight path. The unique feature of the cameras is that they are able to distinguish reflected energy returned from below an object, such as a moored mine, and to produce a shadow image of that object. The received image is automatically catalogued, analysed and stored in the system's memory. It is also available for real-time analysis by an operator.

The three systems were installed on SH-2G Seasprite helicopters of USN Reserve Squadron HSL-94 and deployed in 1998.

The system is mounted in a 4 ft long pod mounted on the helicopter's fuselage. An adaptation of Magic Lantern has been demonstrated as part of a US Navy programme for mine detection from the high water mark on the beach out to 12 m water depth in advance of amphibious operations.

Operational status

The systems will be assigned to the Commander, Mine Warfare Command in Corpus Christi, Texas as contingency assets.

Contractor

Kaman Aerospace Corporation, Electro-Optics Development Center, Tucson, Arizona.

VERIFIED

L-3 WESCAM Airborne Laser-Based Enhanced Detection and Observation System (ALBEDOS)

Type

Maritime sensor.

Development

ALBEDOS is a developmental programme being funded by the Canadian National Search and Rescue Secretariat via the Department of National Defense. It uses Active Gated TeleVision (AGTV) technology and is designed to equip aircraft on search and rescue missions or maritime patrol, but may also be applicable to police and security roles.

WESCAM has built the trials unit, based on its 24D airborne TV surveillance system, which is stabilised in three axes and is marketing the system. In March 1995 the system underwent initial acceptance trials on a Bell 206 helicopter and began flight trials in October 1995 on a Bell 412.

Description

The AGTV system includes a 16.5 to 400 mm broadcast quality zoom lens with an optional 4 × 2 extender with an aperture of f/7.8 at 800 mm; a Ball Aerospace CCD Low-Light Level TV (LLLTV) camera with a third-generation image intensifier and a high-speed gated power supply; a high-power near-infra-red (810 nm) laser illuminator with variable pulsewidth and two fields of view; a switchable narrowband filter that only passes light centred about the laser-emitting wavelength; and gating control electronics. Laser illuminators by Ball Aerospace and the National Optics Institute of Canada have been tested.

ALBEDOS can operate in passive mode as an LLLTV but in active mode the laser illuminates the scene and the LLLTV is gated to suppress conventional light sources and obtain range data. The filter suppresses broadband light sources. This enables video imagery of night scenes that includes a mixture of bright lights and

ALBEDOS adds an Active Gated TeleVision (AGTV) to the standard WESCAM airborne surveillance turret

shadows, eliminating blooming. Maximum range in AGTV mode is 8 km with depths of field between 15 m and 2 km.

Operational status

Under development.

Contractor

L-3 Communications, WESCAM Inc, Canada.

UPDATED

Northrop Grumman AN/AAS-40 Seehawk FLIR

Type

Maritime sensor (air).

Development

The AN/AAS-40 Seehawk is a thermal imaging system which is designed for an aircraft or surface vessel. The system's current principal application is on board US Coast Guard Sikorsky HH-52A helicopters serving the primary role of search and rescue, law enforcement, maritime environmental control, marine and border control and navigational assistance. It is also installed on the Grumman S-2(T) ASW aircraft.

The Seehawk system has successfully completed two years service aboard a Beech 200T aircraft. The system was fully operational in conjunction with the other onboard avionics systems, demonstrating maritime patrol, surveillance and reconnaissance applications. The latest generation AN/AAS-40 Seehawk was installed aboard a modified Sikorsky S-76 helicopter used as a demonstrator in support of the US Army RAH-66 Comanche programme.

Description

The AN/AAS-40 Seehawk is a lightweight FLIR system which uses US Department of Defense common module FLIR components. It consists of a turret assembly, the control electronics unit and the power supply unit. FLIR imagery is fed to the cockpit-mounted display. An automatic scan capability provides constant search coverage in elevation and azimuth. The autosearch mode is enhanced by inclusion of automatic lock on which reacts to either large or small targets, as selected by the operator.

The AN/AAS-40 Seehawk FLIR on the US Coast Guard HH- 52A helicopter

Operational status
In production for the US Coast Guard HH-65A Dolphin helicopter and the Grumman S-2(T) ASW aircraft.

Contractor
Northrop Grumman, Electronics Sensors & Systems Sector, Surveillance & Electronic Warfare Systems.

VERIFIED

Northrop Grumman Integrated MultiSensor System (IMSS)

Type
Maritime surveillance.

Description
The IMSS was initially configured to perform day and night all-weather air interdiction and maritime patrol as part of the anti-drug efforts of the US Customs Service. By combining and integrating a multimode radar and an infra-red imaging system, IMSS is also effective in performing reconnaissance, surveillance and search and rescue missions.

The IMSS consists of the AN/APG-66 radar integrated with an infra-red detection set such as the Northrop Grumman WF-360, controls and displays and an Inertial Navigation System (INS).

The AN/APG-66 is a digital, coherent, multimode radar system developed to serve both the strike and fighter demands of the F-16 aircraft. It provides long-range detection and acquisition of targets at all altitudes and aspect angles in the presence of heavy background clutter. The current version of the APG-66 incorporates a new signal data processor which replaces several radar units and eliminates the need for special interface hardware when used in the IMSS.

The primary function of the infra-red system is short-range tracking and observation of airborne, maritime and ground targets. It can be applied independently to track a designated target by locking on to the heat differential generated by the target or it can be slaved to the radar so that the infra-red line of sight follows radar tracked airborne targets and it can be directed to acquire and track ground targets automatically, using data from the INS. Additional capabilities include a TV camera, laser range-finder and video recorder.

The IMSS controls and displays are integrated with the radar, infra-red and INS through the radar signal data processor. A hand-control unit provides the sensor operator with single-handed slew control of both sensors, including antenna elevation, radar cursor in azimuth and range, target designation and infra-red line of sight. Separate displays are provided for the radar and the infra-red and these displays can be duplicated at several stations within the aircraft.

The INS provides the inertial references required by the radar and infra-red systems, such as roll, pitch, yaw, heading and velocities in three axes.

The IMSS incorporates continuous self-testing and BIT to isolate a fault down to an easily replaceable unit.

Operational status
In operational use by US Coast Guard, US Customs, US Air National Guard aircraft. Capability improvements are planned for both the AN/APG-66 (to the AN/APG-66(V)Z mid-life update configuration) and the WF-36OTL.

Contractor
Northrop Grumman, Electronic Sensors and Systems, Electro-optical Systems.

UPDATED

Optech Airborne Laser Terrain-Mapping system (ALTM) 1020

Type
Laser radar.

Description
Optech's Airborne Laser Terrain-Mapping system (ALTM) 1020 is an all-digital, automated airborne laser scanning terrain-mapping system. It is designed remotely to collect topographic data that can then be processed, displayed and plotted using a Personal Computer (PC). Immediate post-flight playback enables air and ground data to be checked for quality and completeness. Further processing produces XYZ terrain co-ordinates and a digital terrain model.

The ALTM 1020 is suitable for applications including topographic surveys; erosion monitoring surveys; shoreline control surveys; tree-height measurements; open-cast mine operations; transmission line, road route and pipeline planning surveys; GIS base maps.

The system functions in the following sequence:

(a) a tightly focused infra-red laser, eye safe at survey altitudes, outputs very short pulses at rates up to 2,000/s. The laser's small beam divergence and spot size and its high repetition rate, are said to allow a high percentage of shots to penetrate foliage

(b) a scanner then directs laser pulses across the flight path, collects the reflected light and sends it to a receiver/discriminator

(c) the discriminator and a Time Interval Meter (TIM) measure the elapsed time between transmitting the laser pulse and receiving its reflected echo

(d) a carrier-phase differential GPS receiver determines aircraft position. An Inertial Reference System (IRS) reports the aircraft's roll, pitch and heading, as well as parameters such as position and acceleration

(e) a colour video camera records the area being scanned by the ALTM 1020. The video image is annotated with the date and time

(f) a VME airborne computer time-stamps the data and manages the recording of all data onto a high-density digital data tape

(g) an operator interface enables the operator to control the airborne module through a touchscreen display keeping the operator informed of the status of the ALTM 1020 during flight

(h) after the survey flight, the data tapes are transferred to the ground-based PC. A display of the recorded data as a function of time is immediately available in graphical form.

The ALTM 1020 software processes the data for each flightline and for the entire survey area in steps. Operator displays and data qualify verification are available at each step. The data can be displayed in several modes by using AutoCAD. Printer and plotter options are also available to make hard copies of the data displays. All processed data sets are stored and easily archived and can be exported to GIS using standard formats such as DXF.

The ALTM 1020 is portable, the sensor module fits standard aerial camera mounts and can be installed by two people.

Specifications
Airborne module

Operating altitude: 330-1,000 m nominal (assuming 20% of terrain reflectance)
Scan angle: variable from 0 to ±0.20°
Swath width: variable from 0 to 0.73 × altitude
Laser classification: Class IV laser product (FDA CFR 21)
Eye-safe range: 308 m (single shot)
Power requirements: 28 V DC @ 15 A
Sensor: fits all existing camera mounts or is directly mounted to aircraft floor
Dimensions: 290 × 250 × 430 mm
Weight: 11.4 kg
Control racks: 1 stackable vibration-isolated portable case
Dimensions: 600 × 600 × 650 mm, excl GPS
Weight: 45.4 kg, incl shipping covers and cables
Video format: NTSC or PAL
Data storage: 12 h capacity (8 mm digital data tape)

Ground-based module

XYZ module
Vegetation module
Powerline module

Contractor
Optech Inc.

UPDATED

Optech SHOALS-Hawkeye Scanning Lidar Bathymeter (SLB)

Type
Laser radar.

Development
Optech's laser-based Scanning Lidar Bathymeter (SLB) SHOALS-Hawkeye is a system that allows hydrographers a means of surveying shallow coastal and inland waters, with depth and position accuracy to International Hydrographic Organisation (IHO) standards. It is most suitable for shoal areas (1 to 50 m) where sonar is least cost-effective. Optech's first-generation SLB, the LARSEN 500, has been operating since 1985 for the Canadian Hydrographic Service. In 1987 Optech built a prototype system for DARPA in the USA for the detection of submerged mines. Optech has signed a collaborative agreement for the production and marketing of SHOALS-Hawkeye with Saab Bofors Dynamics.

Description
Designed to operate from a small- to medium-sized helicopter or fixed-wing aircraft, the SLB operates at aircraft speeds and scans across the flight path, thus generating a wide swath of soundings. This results in an improved rate of area coverage and nearly uniform sounding density.

The SLB provides soundings to IHO standards, with depth measurements accurate to 15 cm and horizontal positioning accuracy better than 4 m using differential GPS.

A short pulse of infra-red and a co-linear short pulse of green laser radiation are simultaneously transmitted toward the water at an off-nadir angle. The infra-red pulse is scattered from the water surface, while the green pulse penetrates the water and is scattered from the bottom as well. Scattering at both wavelengths is detected by a receiver on the aircraft and the elapsed time between the two is used to determine the water depth.

The scanning laser system consists of four major subsystems: transceiver; positioning; data acquisition, control and display; and ground-based data processing. The first three acquire depth and positional data while the fourth processes the raw data to provide an XYZ database of soundings.

The transceiver subsystem generates the laser pulses, scans the beam, performs the detection and preprocessing of the optical return signals and provides a high-resolution video image of the sounded area.

The positioning subsystem receives aircraft position data from the global positioning system. An inertial reference system gives the angular orientation of the transceiver which, together with the scan angle data, is used to determine the co-ordinates of each sounding.

The data acquisition, control and display system provides the system interface for the operator, captures and stores all data and displays the system status and quality control parameters. It also produces information and guidance displays for flight-line management by the pilot.

The ground-based data processing subsystem analyses the acquired data to produce corrected depths, assigns confidence values to each depth and merges it with positional data. It enables data examination and editing, generates the desired XYZ databases and provides hard-copy (sounding plot) data output.

A SHOALS-Hawkeye electronics rack

The SHOALS-Hawkeye pod shown mounted on a Bell 212 helicopter

Operational status
Systems are operational in the United States, Indonesia and Norway. In addition, the system has been evaluated for underwater mine detection and ASW, mounted on a HKP-4 helicopter, by the Royal Swedish Navy.

Specifications
Water depth penetration: Kd = 4 day; Kd = 5 night (K is the water diffuse attenuation coefficient, d is the depth penetration for example K = 0.1 m^{-1}, d = 40 m max in daylight; for K = 0.5 m^{-1}, d = 8 m max in daylight)
Depth accuracy: 15 cm
Horizontal accuracy: 4 m
Operating altitude: 100 to 1,000 m, 300 m typical
Ground speed: 0 to 100 m/s
Swath width: up to 85 % of operating altitude
Sounding rate: 200 Hz
Area coverage rate: 3 to 80 km^2/h
Safety: eye safe from operating altitude
Aircraft positioning: GPS
Dimensions
Airborne transceiver pod: 3 × 0.6 m
Airborne operator consoles (×2): 1 × 0.5 × 0.6 m
Weight
Airborne: 350 kg
Ground-based: 80 kg

Contractor
Optech Inc.

UPDATED

Raytheon AN/AAS-36 infra-red detection set

Type
Maritime sensor (air).

Development
The AN/AAS-36 FLIR system was developed in 1975 for US Navy P-3C maritime patrol aircraft. The primary missions for the system are Anti-Submarine Warfare (ASW), maritime surveillance and air-to-air interdiction. The system was initially designed and developed to meet a P-3C update programme requirement, but the equipment has also been fitted to earlier P-3C and P-3B aircraft.

Production of the system commenced in 1977 following a testing, evaluation and demonstration programme which used 10 preproduction systems to assure the US Navy that design specifications were either met or exceeded. The service's Initial Operational Capability (IOC) was realised in 1979.

Description
Based on Raytheon's US Department of Defense common modules, the AN/AAS-36 is a stand-alone system requiring only electrical power for operation. The common-module infra-red receiver is mounted in an azimuth over elevation stabilised gimbal. Additional weapon replaceable assemblies provide system power, servo-control, FLIR system control, slew commands and a real-time video display. The display presentation is on an 875-line RS-343 composite television monitor which permits the operator to identify, as well as observe, vessels.

Features include automatic optical temperature compensation, gimbal pointing outputs for servo platform slaving, self-contained stabilisation and a two field of view optical system. A digital computer interface is available for online gimbal control. The system contains built-in self-test facilities which permit checkout down to weapon replaceable assembly level and these themselves are compatible with automatic test equipment.

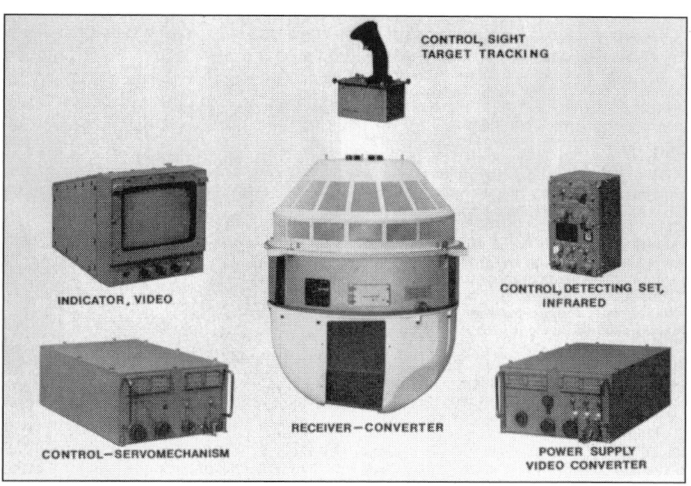

The AN/AAS-36 FLIR system is used in US Navy P-3C aircraft

The system is also being supplied to many non-US operators of the P-3 for upgrading to US Navy standards. The receiver-converter weapon replaceable assembly of the AN/AAS-36 has been fitted to Sikorsky CH-53 helicopters and Cessna Citation, Beechcraft E-90, King Air 200 and other unspecified aircraft.

Operational status
In service.

Specifications
Spectral band: 8-12 μm
Detector: CMT
Coverage
Azimuth: ±200°
Elevation: +16 to -82°
Fields of view: 15 × 20°, 5 × 6.7°
Weight: 136.36 kg
Power supply: 115 V AC, 400 Hz, 3-phase, 2.5 kVA
28 V DC, 100 W

Contractor
Raytheon Systems Company.

UPDATED

Raytheon OR-5008/AA FLIR

Type
Maritime sensor (air).

Description
The Raytheon OR-5008 FLIR system is a derivative of the company's OR-89/AA, adapted for use on the CP-140 Aurora ASW aircraft operated by the Canadian Forces. It was designed to support a variety of mission requirements in the maritime patrol field including search and rescue, shipping and fisheries surveillance, mapping, ice reconnaissance and defence surveillance.

The system is mounted in the lower part of the CP-140 radome in a similar manner to that of the AN/AAS-36 FLIR system used in the US Navy's Lockheed P-3 aircraft. The specification is similar to that of the OR-89/AA except that it uses the US Navy P-3 interface casting, provides for a 5° look-up capability, uses offline control for the extend-retract function and has additional composite video outputs.

Operational status
In service in Canadian Forces CP-140 Aurora aircraft.

Contractor
Raytheon Systems Company.

VERIFIED

Raytheon Rapid Airborne Mine Clearance System (RAMICS)

Type
Mine clearance system (air).

Description
Raytheon Rapid Airborne Mine Clearance System (RAMICS) is a helicopter-based naval Mine CounterMeasures (MCM) system adaptable to dedicated mine clearance as well as an organic battle group MCM capability.

RAMICS targeting is performed by a blue-green laser that penetrates water to locate and target a shallow mine. The laser, aboard the hovering or circling helicopter targets the mine from more than one direction to reduce the possibility of false alarms. The laser-derived co-ordinates are provided to the RAMICS controller.

The controller automatically directs and holds the stabilised rapid-fire gun on target, firing a burst of 20 to 50 rounds of the RAMICS super cavitating projectile. Spin-stabilised in air, the projectile is designed to enter the water at oblique angles to the surface. Upon water entry, its shape and speed produce a cavitation envelope in which the projectile rides at very low drag. The projectile strikes the mine at high velocity and penetrates the mine case and begins to break up as it enters the explosive. As the projectile breaks up, a reactive material within it is then released. Energetic destruction of the mine results from the combined effects of the projectile's reactive material and its kinetic energy. Positive evidence of the reaction is visible at the surface, providing assurance that the mine has been destroyed.

Contractor
Raytheon Systems Company, Naval & Maritime Systems, Fullerton.

VERIFIED

Sensytech, Inc Imaging AADS1221 IR/UV maritime surveillance scanner

Type
Maritime surveillance sensor.

Description
The AADS1221 IR/UV scanner is a sensor subsystem of the Swedish Space Corporation Maritime Surveillance System. It is specifically designed for use in maritime surveillance applications.

The IR/UV scanner, operating in the 8.5 to 12.5 μm range for infra-red and the 0.32 to 0.38 μm range for ultra-violet, is used at low altitude to obtain high-resolution imagery of oil spills. IR data can be obtained by both day and night, providing information on the spreading of oil and also indicating the relative oil thickness within the oil slick. Usually 90 per cent of the oil is concentrated within less than 10 per cent of the visual slick. By using IR information, clean-up operations can be directed for maximum efficiency.

UV data is obtained during daylight and maps the entire extent of the slick, irrespective of thickness. The UV data adds confidence to the IR registration by distinguishing natural thermal phenomena, such as cold upwelling water, from suspected oil pollution.

In the Maritime Surveillance System, the IR/UV video outputs are presented in real time on a split-screen format colour TV monitor, with IR on the left and UV on the right, with false colour coding for image enhancement.

Operational status
In production and in service with the coastguard organisations of a number of countries as part of the Swedish Space Corporation Maritime Surveillance System.

Specifications
Dimensions
Scan head: 380 × 380 × 380 mm
Operator console: 180 × 490 × 340 mm
Weight
Scan head: 17 kg
Operator console: 12 kg
Power supply: 28 V DC, 15 A
Field of view: 5 mrad instantaneous, 87° total
Scan rate: 160 scans/s

Contractor
Sensytech, Inc Imaging Group.

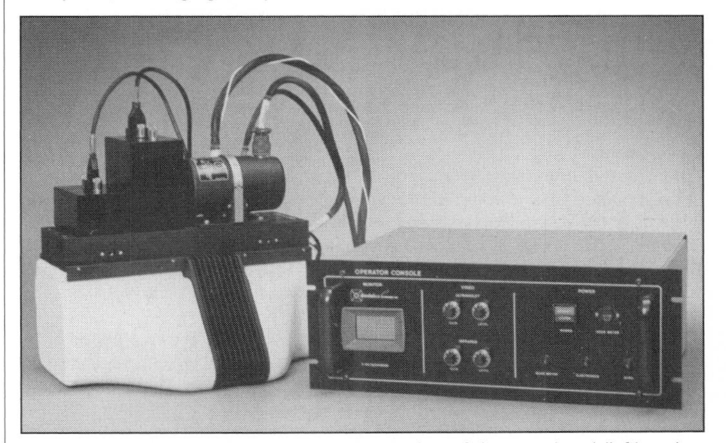

The IR/UV maritime surveillance scanner consists of the scan head (left) and an operator's console (right)

Swedish Space Corporation Maritime Surveillance System MSS 5000

Type
Maritime surveillance system.

Description
The MSS was originally developed for the Swedish Coast Guard and has been in operation for more than 25 years. The newest version, the MSS 5000, was taken into operation by the Norwegian Pollution Control Authority in 1998 and by the Swedish, Polish, Greek and US Coast Guards in 2000 and 2001. In 2003 the MSS 5000 will be replaced by the MSS 6000 with incorporation of ship transponder data and search radar target tracks in the on-board map presentation for enhanced situation overview and with improved air to ground connection for exchange of reports and instructions with surface units. The MSS 5000/6000 is used for maritime surveillance tasks including Exclusive Economic Zone (EEZ) protection, oil pollution detection and assessment, fishing activities monitoring and border patrol.

The MSS consists of:

(a) sensor package – a Side Looking Airborne Radar (SLAR), Infra-Red/Ultra-Violet (IR/UV) scanner, camera(s), video camera and optional sensors such as FLIR, microwave radiometer, interface to aircraft's search radar, AIS (automatic identification system) transponder, other user defined sensors

(b) computer network for sensor data processing, presentation, storage and data link for transmission of images and reports on target observations.

The MSS 5000/6000 has been improved over previous versions of the MSS in a number of ways. The most important improvement is the integration of the sensor presentation with the digital map database, so that radar and infrared images can be presented as geographically referenced map overlays in real time. In the MSS 6000 ship transponder data and other background information is also incorporated in the map database to further enhance the situation overview in the system.

The SLAR is an imaging radar that gives a high sensitivity map-like presentation of the sea surface over a 160 km wide swath. It is designed exclusively for maritime surveillance and is used primarily for detecting small variations in the sea clutter such as those caused by oil spills or small solid targets. Its main applications are surveillance of sea traffic, oil spill detection, fishery protection, search and rescue and sea ice mapping.

The IR/UV scanner is used primarily to obtain high-resolution imagery of oil spills in ship wakes and at accident sites.

The cameras are used for high-quality, high-resolution photographic evidence. Each frame is automatically annotated with time, position, mission number, and so on by the system.

All digital images and target observations are automatically correlated by the system using the time and position of acquisition and presented together with the electronic map.

With the MSS 5000/MSS 6000 the system operator can view several sensor images simultaneously in different windows in traditional format or geographically corrected and superimposed on the digital map. All images and observations are available on line to facilitate real time production of a mission report and/or a situation overview for transmission to a command centre or to co-operating surface units.

Operational status
MSS systems or subsystems have been delivered to Coast Guard organisations in China, Germany, Greece, India, Netherlands, Norway, Poland, Portugal, Sweden, UK and USA. They have been installed on aircraft such as the C-130 Hercules, CASA 212, Cessna 402, 404 and 406, Dornier Do-28, Fairchild Merlin, PZL M28, Turbolet L410 and Y-12. In 2003 MSS 5000 systems are being installed also on the Bombardier 415MP amphibian and on the EH101 helicopter.

Contractor
Swedish Space Corporation.

UPDATED

Tenix LADS Corporation Laser Airborne Depth Sounder (LADS)

Type
LIDAR (Light Detection and Ranging) bathymetric survey system.

Development
The Laser Airborne Depth Sounder (LADS) is a laser bathymetry system designed to be mounted in a fixed-wing aircraft. The original LADS was built by Vision Systems under contract to the Australian government to a design specified by the Australian government's Defence Science and Technology Organisation (DSTO). This original system took approximately four years to build, test and trial and has been in routine operational service with the Royal Australian Navy (RAN) since February 1993.

The LADS system was developed to prototype by the DSTO over a period of 20 years. Vision Systems manufactured the operational LADS which was accepted into naval service in 1993 in a Fokker F-27 fixed-wing platform. The original LADS used by the RAN is based on a flash-pumped Nd:YAG laser firing infra-red pulses of wavelength 1,064 nm at 168 Hz. At the Fokker F-27's nominal operating velocity of 75 m/s at the nominal operating altitude of 500 m, a sounding density of 10 × 10 m is produced.

The laser platform delivered to the United States Navy (USN) Naval Research Laboratories (NRL) in January 1996 is based on a more efficient diode-pumped laser firing infra-red pulses at 264 Hz. In addition to improved power efficiency, the higher firing rate provides a closer sounding density and accommodates a faster aircraft platform. The USN NRL laser platform is installed in an Orion P-3B fixed-wing platform.

In 1998, LADS Corporation completed development and construction of the second-generation LADS Mk II, incorporating improved laser, computer and navigation technology mounted in a faster aircraft to provide closer sounding densities (5 × 5 m nominal), faster survey rate (up to 65 km^2/sec), increased depth capability (0.5 to 70 m) and more efficient use of aircraft power and payload. LADS Corporation uses LADS Mk II to provide a contract survey service worldwide.

Description
The Laser Airborne Depth Sounder comprises three basic components:

(1) Stabilised laser platform

The purpose-designed solid-state infra-red laser, operating at 900 Hz is mounted in a platform that is stabilised for aircraft roll, pitch and yaw. Vertical reference is provided by a strapdown inertial system. The laser's output is frequency doubled to produce visible green light of wavelength 532 nm, in 1 MW, 5 ns duration pulses. An optical coupler splits the output into infra-red and green components.

Infra-red pulses are emitted vertically from the aircraft and reflect from the sea surface to provide an initial reference. A scanning mirror directs the green pulses in a rectilinear pattern across the survey track. Returning pulses are collected by the scanning mirror and are directed to the green receiving telescope which contains spectral, spacial and polarising filters. The pulses are detected by a sensitive photomultiplier with controlled gain and propagation characteristics.

(2) Airborne laser control and data collection software and equipment

Following amplification, the pulses are digitised and combined with positional data and recorded on tape for subsequent analysis by the ground equipment. Accurate ground position is determined by a satellite Global Positioning System (GPS).

(3) Data processing software and equipment

The data is converted into discrete depth soundings for each sounding pulse of the laser using analysis software which corrects for system geometry, surface datum, refraction, depth bias, tides and position. The data is then processed to select automatically and record hydrographically significant shoal depths.

The post-processing system automatically produces sortie plans which drive future surveys through the aircraft's automatic pilot.

LADS is capable of measuring depths from 0.5 to 70 m to the standard required by the International Hydrographic Organisation (IHO). The laser fires at 900 Hz and scans 18 times/s across a 240 m swath. Operating at 90 m/s (175 kt) airspeed and 500 m altitude, the aircraft can gather 3.24 million soundings in a 5 × 5 m grid pattern in 1 hour of operation. The system can survey 65 km^2/h. Survey data is fully processed at a 1:1 time ratio with data collection, namely for every hour of data collected, 1 hour of processing time is required.

The LADS Laser Airborne Depth Sounder is a laser bathymetry system which is mounted on a fixed-wing aircraft

The maximum sortie duration is 8 hours. The time on survey task depends on the distance between the operating base and the survey area and is generally between 3 and 7 hours. The maximum area that can be surveyed by LADS Mk II in a single sortie is 420 km^2.

The laser is eye-safe when viewed from the ground when the aircraft is at its operating height. Automatic cut-off inhibits prevent the laser operating at other times.

Operational status

The original LADS is in routine service with the Royal Australian Navy (RAN) Hydrographic Service. It and has surveyed over 60,000 km^2 for the production of nautical charts since February 1993.

The United States Navy (USN) Naval Research Laboratories (NRL) took delivery of a LADS laser platform in January 1996.

LADS Mk II made its first flight in February 1998 and commenced commercial survey operations in July 1998.

LADS Mk II has now executed commercial contracts for nautical charting in Finland, Norway, New Zealand and the sub-Antarctic; for oil and gas exploration in Australia to demonstrate the use of laser bathymetry in strategic defence applications and for the assessment of water characteristics in the UK.

Specifications

LADS Mk II
Laser: Solid state
Wavelength (IR): 1,064 nm
Wavelength (green): 532 nm
Laser tuning: automatic
Beam geometry: vertical 1,064 nm and scanned 532 pulses
Laser safety: eye-safe to IEL 825-1: 1993 and AS/N25 2211.1 1997 at survey height
Depth sounding accuracy: IHO Order 1
Positional accuracy: 3 m
Depth sounding rate: 900 Hz
Swath width: 240 m
Scan pattern: orthogonal to planned track
Sounding density: 5 × 5 m (2 × 2 m, 3 × 3 m, 4 × 4 m capable)
Soundings per km^2: 40,000
Soundings per h: 3,240,000
Survey speed: 90 m/s
Survey altitude: 500 m (360 m capable)
Depth range: 0 to 70 m
Area coverage: 65 km^2/h
Survey endurance: 8 h
Area coverage: 100%
Data processing time: 1:1 with survey time

Contractor

Tenix LADS Corporation Limited.

Terma Elektronik airborne surveillance system

Type

Maritime surveillance system.

Description

The Terma surveillance system is designed for detection of oil spills, identification and documentation of fishing violations, performance of search and rescue missions and ice-mapping by day and night and during periods of poor visibility. It integrates a wide variety of surveillance sensors, navigation equipment and video systems.

A Side-Looking Airborne Radar (SLAR) has become the primary long-range sensor for oil pollution surveillance, typically covering a 37 km (20 n miles) swath from preferred search altitudes. An oil slick is detected by variation in reflected radar signals between oil covered water and normal seawater. In applications like ice-mapping and ship surveillance the SLAR covers a 74 km (40 n mile) swath.

An Infra-Red/Ultra-Violet (IR/UV) scanner is provided for close-range imagery and allows a rough area estimation to be made, as the aircraft passes overhead, of the oil slick detected by the SLAR. The IR system can be operated by both day and night. It provides information on the spreading oil and indicates the relative thickness within the oil slick. The ultra-violet sensor is only used during daylight. It maps the complete area covered with oil, irrespective of thickness.

A scanning radiometer system is provided for oil thickness measurements and quantification, enabling clean-up operations to attack the worst part of the spill first. The MicroWave Radiometer (MWR) measures microwaves originating from the sea surface at X- and Ka-band wavelengths.

Video cameras are used to secure evidence of oil pollution, fishery violation and so on. Information can be recorded on videotape or stored as still photographs in the computer. Real-time navigation data is integrated into the picture. The video can be normal colour, low-light level TV or IR. A hand-held camera, with a real-time data annotation capability, can be integrated into the system.

Data downlink equipment is used for transmission of real-time or stored data to a ground- or ship-based station.

Information from microwave and optical sensors can be recorded either on a standard VTR or on a high-resolution digital tape recorder.

The operator console for the Terma surveillance system

The 356 mm Sensor Image Display (SID) provides the operator with sensor information. The SID presents the current sensor image whether it is the SLAR image, the IR/UV scanner image or the radiometer image that is selected. Real-time navigational data is integrated into the bottom of the SID format. Information on aircraft position, heading, speed and altitude, as well as date and time is presented to the operator. By means of the trackerball, the operator can move a cursor on the SID and the target position is then annotated with real time.

The 254 mm colour map display provides the operator with an outlined map of the area under surveillance. The map display is integrated with the video system. The operator can select map information, video information or both simultaneously. Map information is available from customised map data. The map can be zoomed in close steps and the operator can insert symbols at any position.

The 254 mm control panel display facilitates the operation of all surveillance sensors, back-up stores and video systems.

Specifications

Dimensions
Operator console: 1,150 × 575 × 1,400 mm
Observer console: 1,150 × 775 × 1,190 mm
Weight
Operator console: 145 kg
Observer console: 150 kg
Power supply: 28 V DC, 900 VA max

Contractor

Terma Elektronik A/S.

VERIFIED

Thales AMASCOS multisensor system

Type

Maritime sensor (air).

Description

The Airborne MAritime Situation COntrol System (AMASCOS) is designed for real-time tactical situation build-up and update, as well as a decision aid to operators. It is a family of maritime patrol systems using a modular approach to system design and makes it possible to integrate AMASCOS systems on board any type of fixed- or rotary-wing aircraft.

There are three versions of AMASCOS – AMASCOS 100, 200 and 300 which correspond to the three broad categories of mission requirements from simple maritime surveillance to anti-surface and anti-submarine warfare. The typical AMASCOS configuration integrates Thales Optronics equipment (radar, Forward-Looking Infra-Red (FLIR), Electronic Support Measures (ESM), sonics, Magnetic Anomaly Detector (MAD) and communications) but its modular architecture makes it possible to tailor each system to specific requirements. The radar in this typical configuration is the Ocean Master equipment developed jointly by Thales and EADS.

AMASCOS 100
AMASCOS 100 is the lightweight configuration weighing less than 250 kg. It includes radar and FLIR plus a tactical computer and data link. It is suited for carrying out a wide range of missions, such as Economic Exclusion Zone (EEZ) surveillance, search and rescue and law enforcement.

AMASCOS 200
AMASCOS 200 adds ESM equipment to the AMASCOS 100 to offer additional functions. This version is suitable for anti-surface missions and can be extended to provide an Anti-Submarine Warfare (ASW) capability.

AMASCOS 300
In addition to basic maritime patrol and surveillance functions, AMASCOS 300 offers both anti-surface and anti-submarine warfare capabilities. This version is suitable for naval operations command and control assignments.

The heart of the system is an open processing architecture which collates, processes and fuses data from the different sensors or received through a tactical link, to provide for a comprehensive tactical situation assessment. The use of international standards and open architecture maximises interfacing capabilities and provides virtually unlimited growth potential.

The concept of the operator stations is based on reconfigurable multifunctions displays with windowing capabilities. AMASCOS 300 is offered as a baseline with the following sensors:
Radar— Ocean Master with ISAR-SAR (Thales Airborne Systems)
FLIR — Chlio (Thales Optronics)
ESM — DR3000 (Thales Airborne Systems)
Acoustic — SADANG MK II (Thomson Marconi Sonar)
FMS with TOTEM 3000 IRS-GPS — (Sextant Avionique)
Data Link — (Thales)
Other selected customer equipment can also be integrated to the system.

Operational status
In 1996, Indonesia signed a contract for the supply of AMASCOS 100 (including Ocean Master radar, FLIR Chlio and TOTEM 3000-IRS-GPS) to be fitted on NC212, plus Ocean Master radars for NB105 helicopters.

In 1998, the first French Navy Falcon 50 equipped with AMASCOS 100 including Ocean Master radar and Chlio FLIR was delivered for I.O.C.

In 1998, F.O.C. was declared for the Pakistan modernised ATLANTIC and F27 aircraft fitted with AMASCOS systems.

In 1998, UAE selected AMASCOS 300 in its most complete version to equip its CN235 Maritime Patrol Aircraft.

Contractor
Thales Airborne Systems.

VERIFIED

Thales Optronics Electro-Optic Search and Detection System (EOSDS)

Type
Maritime sensor.

Description
The EOSDS is for maritime patrol aircraft. The system combines a second-generation cooled thermal imager, EPIC (developed jointly with the UK Defence Research Agency as part of the STAIRS C, Sensor Technology for Affordable IR Systems programme) fitted on a fully steerable platform and a modified infra-red search and track processor. The EOSDS also has a TV system to record incidents required by the mission controller. A major feature of the system is its ability to track multiple targets automatically and simultaneously. The manufacturer reports that the system will cover a search area 20 times larger than that covered by conventional systems, thereby increasing the detection probability while at the same time reducing search time and operator fatigue.

Operational status
In development.

Specifications
EPIC thermal imager
Spectral band: 8-9.4 μm
Detector: 768 × 6 CMT diode array
Cooling: long-life linear Stirling engine
Fields of view
Single: 5 × 3°
Dual: 5 × 3°, 17.5 × 10.5°

Contractor
Thales Optronics.

VERIFIED

Thales Optronics Tango thermal imager

Type
Maritime sensor (air).

Description
The Tango is a modular thermal imaging system for long-range maritime patrol aircraft. It is an 8 to 12 μm thermal imager with three fields of view. The system was developed for the Dassault Aviation Atlantique 2 maritime surveillance aircraft. Key features include: a large aperture for very long range imaging; high resolution; fine image stabilisation; automatic aiming towards designated targets; aircraft databus coupling; and line-to-line integration. It is incorporated in a gyrostabilised platform fixed under the nose of the Dassault Aviation Atlantique 2. Day and night missions include passive detection of ships and snorkels, long-range identification of surface vessels, reconnaissance, and search and rescue.

Tango 2G is a second-generation thermal imager based on the Synergi thermal imaging modules developed by Thales Optronics and Zeiss Optronik. Synergi uses a 288 × 4 IRCCD detector. Tango 2G has four fields of view. It is a multisensor integration and incorporates a CCD detector for the visible and near-infra-red channel and has an auto-search mode. It is being developed for the Atlantic third-generation, the aircraft ATL3G.

Operational status
Tango is in service on the French Navy Atlantique 2.

Specifications
Tango modular thermal imaging system
Dimensions: 600 mm turret diameter
Turret weight: 87 kg
System weight: 120 kg
Gyrostabilised field of view
Azimuth: ±110°
Elevation: +15 to –60°
Tracking speed with speed/accuracy optimisation: 1 rad/s
Infra-red channel
Cooling: Stirling engine
Spectral band: 8-12 μm
Fields of view
Wide: 6.45 × 4.30°
Medium: 2.15 × 1.43°
Narrow: 1.07 × 0.7°
Tango 2G
Dimensions: 600 mm turret diameter
Turret weight: 75 kg
System weight: 98 kg
Gyrostabilised field of regard
Azimuth: ±360°
Elevation: +15 to –93°
Tracking speed with speed/accuracy optimisation, auto-search mode: 1 rad/s
Infra-red channel
Detector: 288 × 4 IRCCD focal plane array
Spectral band: 8-12 μm

The Tango 2G modular thermal imaging system for long-range maritime patrol aircraft

Fields of view: 15 × 11.2°
7.5 × 5.6°
1.5 × 1.1°
0.75 × 0.6°
Visible and near infra-red channel
Detector: CCD
Field of view: 1.5 × 1.1°

Contractor
Thales Optronics.

UPDATED

UOMZ Optical-TV 24-hour sight (OTV-24)

Description
The optical-TV 24-hour sight is intended to provide detection, identification and tracking of ground and surface targets 24 hours per day. It has a high-resolution optical channel, with magnification, to aid identification of surface targets in the role of border control and maritime exclusion zone policing.

Operational status
The optical-TV 24-hour sight is produced for An-72P coastguard aircraft.

Optical-TV 24-hour sight 0003332

Contractor
Urals Optical and Mechanical Plant (UOMZ).

VERIFIED

OBSERVATION AND SURVEILLANCE

UNMANNED AIRCRAFT SENSORS

BAE Systems Phoenix UAV payload

Type
UAV sensor.

Development
The Phoenix battlefield target acquisition and surveillance UAV system was designed as the British Army's first fully equipped unmanned aircraft system for real-time surveillance and target acquisition, by day and night, in all weather conditions.

First flight took place in May 1986; the 100th flight was made in early 1991. Phoenix's primary role is to locate and designate targets for the British Army's MLRS (Multiple Launch Rocket System) and other artillery. Phoenix is also to be integrated with both the Ptarmigan C^2 system and BATES (Battlefield Artillery Targeting Engagement System).

Description
The airborne component of Phoenix is totally sensor-orientated, to the extent that the air vehicle is an aerial 'taxi' for a detachable underfuselage pod containing the imaging sensor and the airborne portion of the datalink. Subsystems on board the 'taxi' include those for flight control, navigation, parachute recovery and pod stabilisation.

In the target acquisition and surveillance role, the payload comprises an infra-red sensor based on the UK Thermal Imaging Common Modules (TICM II), mounted in a two-axis stabilised turret. The third axis, in roll, is stabilised by the stabilisation of the complete mission pod. This has the advantage of being 'horizon up' for maximum ease of interpretation and correlation with the surrounding terrain. Also in the payload pod are the processing electronics and the airborne end of the datalink, which utilises two fully steerable antennas, one at each end of the pod to provide full directional coverage. The datalink provides two-way communications between the air vehicle and the ground data terminal, a mobile, mast-mounted tracker assembly containing processing electronics and a transmitter/receiver unit linked to the GCS.

The IR payload has a 360° scan in azimuth, a line of sight that can be steered through more than 70° in elevation and a zoom lens providing continuous magnifications from ×2.5 to ×10. It can be locked fore and aft at a preset elevation during the cruise phase of a mission. Sector scanning can be selected for area search and the imager's sightline can be steered automatically to remain aligned with a target being orbited by the UAV.

Operational status
Following an initial design certification of the Phoenix system in September 1993, an Agreed Programme of Work (APW) was initiated to incorporate improvements which included an airbag recovery system. This system is now in full production. Regimental conversion training started in November 1997 and Phoenix entered service in 1998. Limited operational missions were flown over Kosovo in June 1999.

Specifications
Coverage
Azimuth: 360°
Elevation min: 70°
Magnification: continuous from ×2.5 to ×10
Payload weight: 50 kg

Contractor
BAE Systems, Rochester.

VERIFIED

The complete British Army Phoenix system

Controp ESP-1H

Type
UAV sensor.

Description
The ESP-1H is a lightweight observation system designed for UAVs, light aircraft and observation balloons. It is capable of detecting a truck-size target at ranges up to 20 km (10.8 n miles; 12.4 miles) and recognising it at 7 km (3.8 n miles; 4.3 miles).

The three-gimbal mount is gyrostabilised in azimuth and elevation and the ESP-1H carries two camera channels: a high-resolution black and white or colour CCD TV camera with a ×6 (50 to 300 mm) zoom lens and a wide-angle camera with 16 mm fixed focal length lens.

Operational status
Operational on various (unspecified) platforms.

Specifications
Dimensions (d × h): 283 × 415 mm
Weight (system): 8 kg
Field of regard:
Azimuth: 360° continuous
Elevation: +10 to −110°
Field of view: 1.5 × 26°
Stabilisation: >50 mrad
Angular velocity: 0-50°/s
Pointing accuracy: 0.7°

Contractor
Controp Precision Technologies Ltd.

VERIFIED

Controp ESP-600C

Type
UAV sensor.

Description
The ESP-600C is a high-resolution, lightweight stabilised daylight observation system which has been designed for use in light fixed-wing, rotary-wing and unmanned aircraft as well as land vehicles. It is a three-gimbal system gyrostabilised in both azimuth and elevation. It is protected against wind loads, humidity and dust by a sealed rotating dome with a built-in optical window. All the host electronics are contained within the gimbal assembly. It carries a high-resolution (450 lines) colour CCD camera with ×15 (40 to 600 mm) zoom lens (20 to 300 mm zoom optional). The sensor can detect a truck-size target at 30 km (16.2 n miles; 18.6 miles) and recognise it at 12 km (6.5 n miles; 7.5 miles).

The Controp ESP-600C payload

Operational status
In service in manned and unmanned applications. The latter includes the IAI Scout/or Searcher UAVs and the EES Dogan.

Specifications
Dimensions (d × h): 300 × 435 mm
Weight: 12.3 kg
Field of regard:
Azimuth: 360° continuous
Elevation: +10 to −105°
Field of view:
Standard: 0.75 × 11.5°
Optional: 1.5 × 22.9°
Stabilisation: better than 10 mrad
Angular velocity: 0-40°/s
Pointing accuracy: 0.7°
Power requirement: (28 V DC) 40 W
Power consumption (nominal): 40 W

Contractor
Controp Precision Technologies Ltd.

UPDATED

Controp FSP-1

Type
UAV sensor.

Description
The FSP-1 (Flir Stabilised Payload) is a sophisticated, high-resolution, day and night FLIR observation system, designed primarily for scout helicopters, reconnaissance light aircraft, UAVs and marine patrol boats. It is capable of very long target acquisition ranges for various surveillance applications, as well as very detailed fine resolution at closer ranges for police or similar use.

The FSP-1 features a four-gimbal system, gyrostabilised in azimuth and elevation and is protected against wind, humidity and dust by a rotating dome with a built-in optical window. The standard FLIR sensor is the El-Op MLFS (qv), but other sensors such as a 3 to 5 μm focal plane array camera, additional daylight CCD camera or LLLTV camera, are optional. Autotracking is also available.

Remote controls include azimuth and elevation; FoV selection; focus; power on/off; FLIR gain; FLIR polarity; and FLIR reticle on/off.

Operational status
In service in several applications, including IAI Searcher UAV.

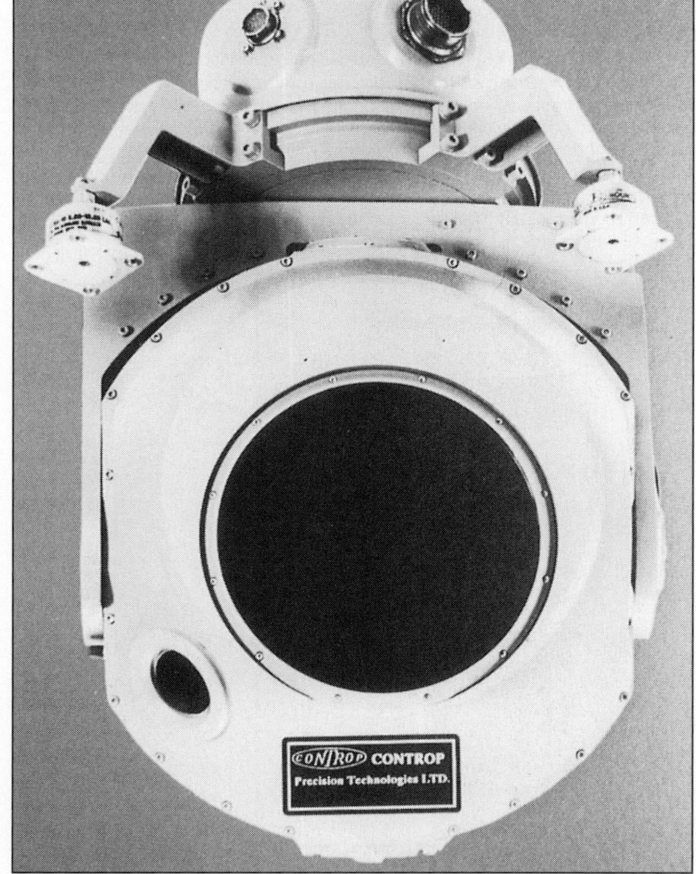

Controp FSP-1

Specifications
Dimensions
Gimbal assembly (d × h): 320 × 500 mm
FLIR electric box (l × w × h): 210 × 225 × 100 mm
Payload interface (l × w × h): 225 × 200 × 100 mm
Weights
Gimbal assembly: 21 kg
FLIR electric box: 6 kg
Payload interface: 1 kg
Total weight: 28 kg
Field of regard
Azimuth: 360° continuous
Elevation: +10 to −105°
Gimbal angular rate:
Azimuth: up to 45°/s
Elevation: up to 32°/s
Stabilisation: >25 mrad
Pointing accuracy: 0.7°
Target acquisition range (narrow FoV)
truck detection: 12-20 km (6.5-10.8 n miles; 7.5-12.4 miles)
truck recognition: 4-5 km (2.2-2.7 n miles; 2.5-3.1 miles)
FoV change time (max): 1 s
Altitude: up to 6,100 m (20,000 ft)
Acceleration: up to 5g
Temperature
Operational: −20 to +50°C
Storage: −32 to +71°C
Humidity: up to 95%
Serial communication (bidirectional): RS-422A
Nominal power supply: 28 V DC
Power consumption: 125 W

Contractor
Controp Precision Technologies Ltd.

VERIFIED

Cumulus Goshawk 350 and 400

Type
UAV sensor.

Development
The Cumulus (previously Kentron) Goshawk 350 was developed for the Kentron Seeker UAV but was designed also to be operated by small manned aircraft with a payload capability of about 25 kg (55 lb).

Description
The **Goshawk 350** is one of a range of airborne observation products manufactured by this company, and can be installed in RPVs, UAVs, light fixed-wing aircraft or helicopters. It can undertake surveillance of objects on land or water, enabling detection, recognition and identification of objects in day and night conditions. The payload turret is four-gimbal, two-axis gyrostabilised to reject platform motion and vibration, permitting vibration-free video imagery at long ranges. Options include automatic video tracking. The surveillance operator controls the payload by means of a control panel, a video monitor and a video recorder located in the GCS. Features include window de-icing and a built-in measuring system that can report to the UAV the angular displacement of the rotating platform, in both azimuth and elevation, in relation to the air vehicle's frame of reference.

The Goshawk 350 has a CCD colour TV camera with several options, an optional 1.54 μm eyesafe laser, and an IR sensor which can be in the 3 to 5 μm or 8 to 12 μm waveband, with electronic zoom of 2:1 of narrow FoV and a UAV interface to RS-422 standard. The 12.7 mm (½ in) TV cameras have a PAL CCIR, 625 lines at 50 Hz output. Three versions are available: a 1-CCD colour version with 450 × 450 pixel resolution and focal length of 16 to 244 mm; a 3-CCD colour version with 560 × 560 pixel resolution and 14 to 448 mm focal length; and a monochrome camera with 580 × 580 pixel resolution and 1,000 mm (celestron) focal length.

The laser range-finder has a range of up to 10.8 n miles (20 km; 12.4 miles) with an accuracy of ±5 m. The 8 to 12 μm sensor has a second-generation 288 × 4 CMT CCD focal plane array detector; the 3 to 5 μm sensor has either a 384 × 288 CMT or 256 × 256 indium antimonide (InSb) focal plane array detector.

The Goshawk 350 family was formerly marketed in MSP (Multi-Sensor Payload), SCP (Super Colour Payload) and UCP (Universal Camera Payload) versions, but these have now been replaced by dedicated packages:

Goshawk 350-100: 8 to 12 μm IR, 1-CCD TV camera
Goshawk 350-200: 8 to 12 μm IR, 1-CCD TV camera, laser
Goshawk 350-300: 3 to 5 μm CMT IR, 3-CCD TV camera
Goshawk 350-400: 3 to 5 μm CMT IR, 1-CCD TV camera, spotter TV camera
Goshawk 350-500: 3 to 5 μm CMT IR, 1-CCD TV camera, spotter TV camera
Goshawk 350-600: 3 to 5 μm InSb IR, 3-CCD TV camera
Goshawk 350-700: 3 to 5 μm InSb IR, 3-CCD TV camera, laser
Goshawk 350-800: 3 to 5 μm InSb IR, 1-CCD TV camera, spotter TV camera
Goshawk 350-900: 3-CCD TV camera, spotter TV camera, laser

The **Goshawk 400** is similar, but with a larger turret, and is intended for medium- or high-altitude UAVs. It uses similar sensors to the Goshawk 350, but does not have a 1-CCD colour TV camera. The dedicated sensor package versions are:

Goshawk 400-100: 8 to 12 μm IR, 3-CCD TV camera
Goshawk 400-200: 8 to 12 μm IR, 3-CCD TV camera, laser
Goshawk 400-300: 8 to 12 μm IR, 3-CCD TV camera, spotter TV camera
Goshawk 400-400: 3 to 5 μm CMT IR, 3-CCD TV camera
Goshawk 400-500: 3 to 5 μm CMT IR, 3-CCD TV camera, laser
Goshawk 400-600: 3 to 5 μm CMT IR, 3-CCD TV camera, spotter TV camera
Goshawk 400-700: 3 to 5 μm InSb IR, 3-CCD TV camera
Goshawk 400-800: 3 to 5 μm InSb IR, 3-CCD TV camera, laser
Goshawk 400-900: 3 to 5 μm InSb IR, 3-CCD TV camera, spotter TV camera.

Operational status
In production and service; UAV applications include the Kentron Seeker (South Africa and Abu Dhabi) and SAGEM Sperwer.

Specifications

Diameter:	350 350 mm (13.8 in), 400 410 mm (16.1 in)
Height:	350 490 mm (19.3 in), 400 500 mm (19.7 in)
System weight	
350:	25-32 kg (55.1-70.5 lb)
400:	35-42 kg (77.2-92.6 lb)
System field of regard	
azimuth:	360°
elevation:	350 +15/−120°, 400 +20/−105°
Line of sight stabilisation:	<33 mrad
Stabilisation error:	350 <25 mrad RMS, 400 <20 mrad RMS
Sensor field of view: CCD TV (narrow)	
1-CCD (colour):	1.5°
3-CCD (colour):	0.8°
Monochrome:	0.3°
CCD TV (wide)	
1-CCD (colour):	22.6°
3-CCD (colour):	26.0°
IR (narrow)	
3-5 μm InSb:	1.8°
3-5 μm CMT:	3.0°
8-12 μm:	4.0°
IR (wide):	18.0°
Positional accuracy:	<3 mrad
Slew rate:	>1 rad/s
Target detection range:	
CCD TV, IR:	>4.3 n miles (8 km; 5 miles)
Environmental temperature:	+50/−20°C
Power supply:	28 (±10) V DC

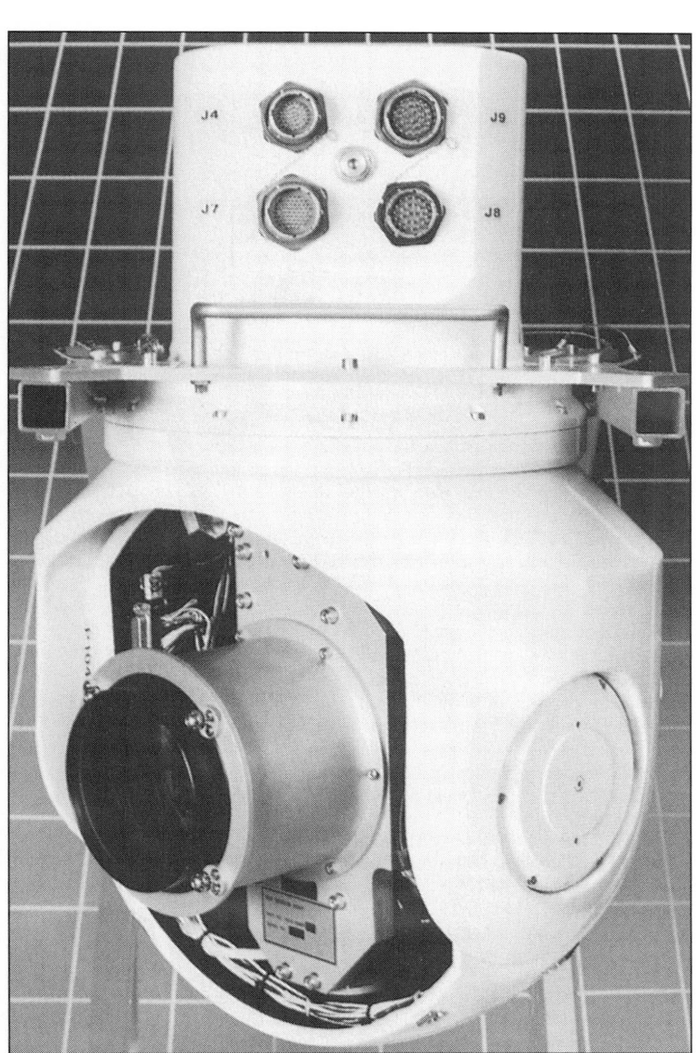

Goshawk two-axis stabilised E-O payload turret

Prime contractor
Cumulus Business Unit of Denel (Pty) Ltd
Centurion.

VERIFIED

DRS SPIRI^{2}T

Type
UAV sensor.

Description
DRS (formerly Boeing) with Tadiran and S-TRON has developed a multisensor payload for UAVs called the Stabilised Payload Infra-red Reconnaissance Image Intensifier Turret (SPIRI^{2}T).

It features a FLIR based upon Boeing's Producible Alternative to Cadmium telluride for Epitaxy (PACE-1) cadmium mercury telluride staring focal plane array with 256 × 256 detectors which operates in the 3.8 to 4.8 μm waveband. The FLIR is fitted with an S-TRON 6.25:1 continuous zoom lens. This features a 100 mm aperture, f/2.5 reimaging design which uses silicon lens material. The zoom actions are controlled by a microprocessor. The other sensor is a third-generation tube image intensifier CCD television, with 1,134 × 486 detectors, from which video records may be made. Closed-cycle cooling is used based upon a linear-drive, reactional-design 1 W cryogenic cooler.

The gimbal is produced by Tadiran and is a derivative of the Moked 400 which provides line of sight coverage. The design comprises two axis and three gimbals.

Operational status
Not known. Ground tests were conducted by the UAV Joint Project Office in January 1992. Captive flight tests were undertaken in July 1992.

Specifications
FLIR
Detector: 256 × 256 CMT staring focal plane array
Spectral band: 3.8-4.8 μm
Cooling: closed-cycle

Low-light level CCD camera
Detector: 1,134 × 486

Fields of view
FLIR: 2.4 × 2.4°; 15 × 15°
CCD: 6.5 × 6.5°
Slew rate: 40°/s

The SPIRI^{2}T UAV package

Coverage
Azimuth: 360°
Elevation: +15 to –85°
Pointing accuracy: <5 μrad (both axes)
Power requirements: <170 W
Weight: <22.5 kg

Contractor
DRS Technologies, Electro-Optical Systems Group.

L-3 WESCAM Model 14TS/14QS

Type
UAV sensor.

Description
The Model 14QS Quad-Sensor and Model 14TS Tri-Sensor gyrostabilised gimbals are surveillance systems for Unmanned Aerial Vehicles (UAV). The 14QS Quad-Sensor can accommodate up to four sensor payloads while the Model 14TS Tri-Sensor can accommodate up to three. The range of sensor systems available for the Model 14 series includes a high-resolution TV camera with zoom, a 955 mm long-range spotter scope with a colour camera, a 3 to 5 μm platinum silicide or indium antinomide multiple field of view thermal imager and an optional eye-safe laser range-finder.

The Model 14 systems can be installed on a wide range of aircraft. The active gyrostabilisation and vibration isolation allow less than 35 μrad RMS line of sight jitter, enabling detection and classification of personnel, objects and vehicles over long distances by day or night. Options include an auto-tracker, GPS display and information overlay, LCD/CRT display monitors and videotape recorders.

Operational status
Operational on Predator.

Specifications
Gimbal
Active gyrostabilisation: 4-axis
Vibration stabilisation: 3-axis passive
Slew rate: max 90°/s
Azimuth range: 360° continuous
Elevation range: +90 to –120°
TV camera – two 1CCD Sony XC-999 (one dedicated to zoom lens, one dedicated to spotter scope)
Resolution: 470 (NTSC)/469 (PAL) lines
Sensitivity: 2,000 lx @ f5.6
Min illumination: 4.5 lx @ f1.2
TV zoom lens
Focal length: 16-160 mm
Narrow FOV (h × v): 2.3° × 1.7°
Wide FOV (h × v): 23° × 17°
TV spotter scope
Focal length: 955 mm
Narrow FOV (h × v): 0.38° × 0.29°

L-3 WESCAM's Model 14TS/14QS can carry up to four sensor payloads 0023401

Platinum silicide thermal imager
Detector: PtSi staring array
Resolution: 512 × 512
Spectral range: 3-5 μm
Mean time between failure: 4,000 h
Cooling: Stirling cycle cooler
Nominal switchover time: <1 s
Fields of view (h × v)
Focal length 19 mm: 40.9 × 31.3°
Focal length 70 mm: 10.9 × 8.4°
Focal length 280 mm: 2.7 × 2.1°
Focal length 560 mm: 1.4 × 1.0°
Focal length extenders: ×1, ×2
Gain adjustment: 5-position selectable
Indium antinomide thermal imager
Detector: InSb staring array
Resolution: 256 × 256
Spectral range: 3-5 μm
Mean time between failure: 4,000 h
Cooling: Stirling cycle cooler
Nominal switchover time: <1 s
Fields of view (h × v)
Focal length 11 mm: 40.9 × 40.9°
Focal length 70 mm: 6.3 × 6.3°
Focal length 280 mm: 1.6 × 1.6°
Focal length 560 mm: 0.8 × 0.8°
Focal length extenders: ×1, ×2
Gain adjustment: 5-position selectable
Zoom: electronic zoom
Freeze: image freeze for analysis
Eye-safe laser rangefinder
Laser type: Er:glass
Wavelength: 1.54 μm
Pulse rate at 100% duty: 60 ppm
Receive area: 50 cm^2
Max range of counter: 49,995 m
Accuracy: ±5m

Contractor
L-3 Communications, WESCAM Inc, Canada.

UPDATED

L-3 WESCAM SST

Type
UAV sensor.

Description
The WESCAM™ Step-Stare Turret (SST) is the first completely digital tri-sensor gimbal developed by WESCAM that has high-speed step-stare capability. Introduced into the market primarily for tactical unmanned aerial vehicle programmes, this high-performance system is intended for small aircraft requiring very long-range reconnaissance and surveillance.

Offering a third field of view long-range optic, the system captures 7 video frames per second, which are then tiled together to create a hi-resolution digital image. Offering unprecedented coverage and resolution with 4,000 m standoff, the SST covers up to 1,200 sq km per hour, to give complete and accurate coverage of the target area while increasing the effectiveness of the UAV.

Features
- High-speed step-stare capability
- All digital video collection, display and storage (IEEE 1394/firewire)
- High-quality gyro-stabilisation performance in a small compact turret
- Tri-sensor payload (IR/Daylight, LRF)
- Digital, large format 3-5 μm InSb thermal imager and 3 field of view lens.

The SST tri-sensor payloads include:
- **Sensor 1** - 3 to 5 μm 3 field of view thermal imager, with Indium Antimonide staring array.
- **Sensor 2** - colour daylight CCD camera with ×14 zoom lens
- **Sensor 3** - eye-safe laser range-finder.

Contractor
L-3 Communications, WESCAM Inc.

UPDATED

Northrop Grumman Reconnaissance, Infra-red Surveillance, Target Acquisition (RISTA)

Type
UAV sensor.

Description
Northrop Grumman, under contract to the US Army's Joint Precision Strike Demonstration Office, has developed an advanced multimode/multimission electro-optical reconnaissance system that detects, identifies and pinpoints threat targets from an airborne platform.

The system consists of a second-generation infra-red sensor located in the airborne platform, a ground-based processor for Aided Target Recognition (AiTR) and an image processing facility (IPF) which houses the AiTR and processes detected target images for dissemination to using organisations. This advanced concept reconnaissance system provides the army with the capability to search over a wide area and locate threats in real-time to provide critical intelligence to battlefield commanders.

The approach to RISTA builds upon advanced image processing algorithms developed on the Comanche program and represents the culmination of more than two decades of advanced image processing research.

The RISTA sensor is derived from the army's Airborne Standoff Minefield Detection System (ASTAMIDS) program. That programme's second-generation sensor has been upgraded to include a forward looking infra-red (FLIR), while retaining the line scanning capability inherent in ASTAMIDS. Similarly, the AiTR processor uses a design employed on the army's MSAT-AIR programme.

RISTA's IPF is derived from the army's ETRAC programme. Configured for use in a HMMWV, the workstations provide mission planning and sensor control, real-time image display and exploitation and reporting capability.

The system has two modes of operation: spotlight and line scanning with either mode selectable during flight from the IPF. Each mode is designed to support a particular portion of the reconnaissance mission.

Spotlight mode
The spotlight mode (FLIR) is used to perform battle damage assessment or provide longer viewing of a suspected target to enhance recognition. In this mode, the line-of-sight of the RISTA sensor is geographically stabilised (geo track) to a set of UTM co-ordinates commanded by the sensor control function within the IPF. Imagery is transmitted to the IPF at a 30 Hz rate.

Line scanning mode
The line scanning mode is used to perform reconnaissance over a broad area. In this mode, the sensor swaths out an eight-mile wide track on the ground and the AiTR screens the incoming data for likely targets. Image target chips centred about the AiTR target detections are presented to the image analyst for confirmation and dissemination.

Operational status
Development.

Contractor
Northrop Grumman Electronic Sensors and Systems Division, Electro-optical Systems

VERIFIED

Raytheon Airborne Standoff Minefield Detection System (ASTAMIDS)

Development
The Airborne Standoff Minefield Detection System (ASTAMIDS) is intended to be a highly reliable airborne minefield detection system which will provide a commander with the accurate location of minefields. The system will be required to detect and identify the boundaries of buried and surface patterned minefields plus scatterable mines and detect individual buried nuisance mines on unpaved roads. The minefield location, combined with other intelligence, surveillance and target acquisition operations, will provide alternatives for command decisions.

Following a development phase during which Northrop Grumman and Raytheon were contenders, in early 1994 Raytheon, Missile Systems Division, combined with Cambridge Parallel Processing (CPP) of Irvine, California, were awarded a US$1.52 million increment as part of a US$22,261,448 cost plus incentive fee contract. This was to develop ASTAMIDS further over a 32 month demonstration and validation period, during which ASTAMIDS was demonstrated in 1997 with a surrogate air vehicle and a mobile display station.

The contracting activity was the Countermine Division of the Night Vision and Electronic Sensors Directorate (NVESD), Fort Belvoir, Virginia. Project Manager is the Office of the Project Manager for Mines, Countermine and Demolitions, Picatinny Arsenal, New Jersey.

Description
ASTAMIDS is an airborne mine detection system consisting of an imaging sensor mounted on an Unmanned Aerial Vehicle (UAV) and a processor/algorithm into the UAV Ground Control Station (GCS). The system allows the user to obtain information for formulating an operational plan and gather further information to modify that plan still more.

The ASTAMIDS sensor will be mounted on the UAV and controlled by the UAV GCS, transmitting minefield imagery via a datalink to the GCS for processing in near-realtime by a very high-speed parallel processor from CPP. The latter involves a Distributed Array Processor (DAP) with a growth capability of up to 450 per cent. The resultant information will be displayed and disseminated to users in a manner similar to other reconnaissance, intelligence, surveillance and target acquisition data.

The Raytheon-developed sensor involves laser diode technology, including a laser receiver and a UAV-mounted passive longwave infra-red and active laser sensor detection system. The ground-based portion of the ASTAMIDS includes the Minefield Detection Algorithm and Processor (MIDAP).

Operational status
Under development.

Contractor
Raytheon Systems Company.

VERIFIED

Raytheon Global Hawk integrated sensor suite

Type
UAV sensor.

Development
The Global Hawk UAV began development in 1995 and is part of the Tier II Plus programme sponsored by the US Defense Research Projects Agency (DARPA) acting for the Defense Airborne Reconnaissance Office. Teledyne Ryan Aeronautical heads the team selected for the advanced development and flight test of the system, with Raytheon responsible for the Global Hawk integrated sensor suite. The first prototype Global Hawk was rolled out in February 1997.

Description
The Global Hawk High-Altitude Endurance Unmanned Aerial Reconnaissance System includes Synthetic Aperture Radar (SAR), electro-optical and infra-red sensors. The SAR can operate simultaneously with either the EO or the IR sensor, enabling wide area coverage for situational awareness and/or the ability to focus on specific targets for threat assessment, targeting or bomb damage assessment. It can operate for more than 40 hours from an altitude of over 21,000 m.

The EO system incorporates a third-generation 3.6 to 5 μm infra-red sensor based on an indium antimonide staring focal plane array and Kodak digital CCD visible camera with a 0.4-0.8 μm silicon focal plane array. The mid-wavelength infra-red sensor technology is based on that used in the system Raytheon is producing for the MV-22 Osprey helicopter (AN/AAQ-27) and for the F/A 18 ATFLIR (Advanced Targeting FLIR).

SAR imagery and a Moving Target Indicator (MTI) capability are provided by COTS (Commercial-Off-The-Shelf) hardware and software. The gimballed radar antenna scans from either side of the vehicle to obtain 0.3 m imagery in Spot mode, 1 m imagery in wide area search mode and a 4 kt minimum detectable velocity in MTI mode. This imagery can be transmitted via datalink in near-realtime over a number of communications paths.

Operational status
First prototype rolled out on 20 February 1997 and engine runs under way by June. The first flight eventually took place from Edwards AFB on 28 February 1998. During the 56 minute flight the aircraft reached 9,750 m (32,000 ft). The second

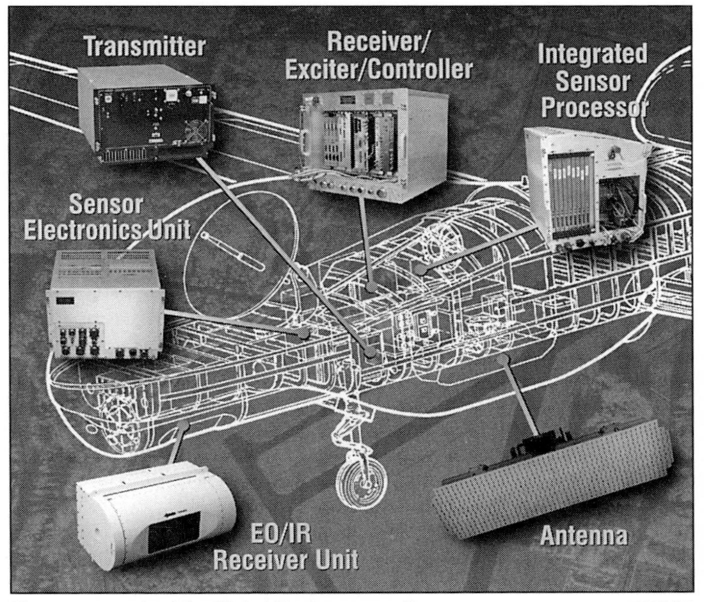

The components of the Global Hawk sensor suite 0005557

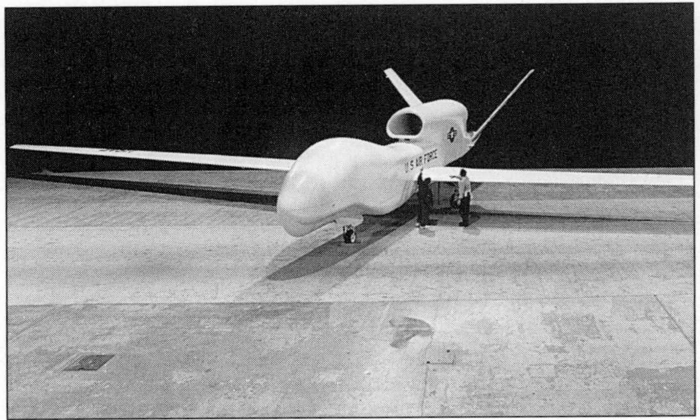
The Global Hawk unmanned air vehicle 0005556

test flight, on 10 May 1998, lasted 2 hours 24 minutes, reached 12,500 m (41,000 ft), and demonstrated two-way control handover via satcom command link. The third flight on 30 May lasted 5 hours 22 minutes, reached 15,600 m (51,200 ft), and covered 1,918 n miles (3,540 km; 2,200 miles).

Second prototype flew for the first time in November 1998, with a second flight on 4 December. On 22 January 1999, it employed all three sensors during a single 6 hour 24 minute sortie, capturing 21 E-O, three SAR (mapping mode) and two IR scenes. It was lost in a crash on 29 March 1999 after inadvertently receiving a flight termination signal from another UAV some distance away. Global Hawk No 3 suffered minor damage (collapsed nose gear) on 6 December 1999 when a software problem affecting its taxying speed caused it to veer off the Edwards runway; flying was resumed on 11 March 2000.

At least two Global Hawks were deployed in Operation 'Enduring Freedom', for surveillance over Afghanistan, following the terrorist attacks on New York and Washington of 11 September 2001. First imagery was relayed on 27 November 2001; one air vehicle was lost due to a control surface malfunction on 30 December, while returning to its base at Al Dhafra, Abu Dhabi.

Specifications
Total weight: 402 kg
EO/IR system
Focal length: 1.75 m
Aperture: 0.28 m
Infra-red sensor
Spectral band: 3.6-5 μm
Detector: InSb 480 × 640 staring focal plane array
Array field of view: 5.5 × 7.3 mrad
Pixel independent field of view: 11.4 μrad
CCD camera
Spectral band: 0.4-0.8 μm
Array elements: 1,024 × 1,024
Array field of view: 5.1 × 5.2 mrad
Pixel independent field of view: 5.1 μrad
Performance
Wide area search mode: 138,000 km²/day
Spotlight mode: 1,900 spots/day
Radar sensor
Frequency: X band
Bandwidth: 600 MHz
Peak power: 3.5 kW
Antenna field of regard: ±45°
Performance
Wide area search mode: 138,000 km²/day, 200 km range
Spotlight mode: 1,900 images/day, 200 km range, squint to ±45°
MTI mode: 15,000 km²/min, min detectable velocity 4 kt

Contractor
Raytheon Systems Company.

UPDATED

Recon/Optical CA-236 LOROP digital camera

Type
UAV sensor.

Development
CA-236 is an electro-optical panoramic scan reconnaissance sensor which has been developed for the DarkStar High Altitude Endurance UAV, developed by Boeing and Lockheed Martin for the Tier III Minus programme, sponsored by the US Defense Advanced Research Projects Agency (DARPA).

Description
CA-236 is a medium/high-altitude panoramic scan reconnaissance sensor with Forward Motion Compensation (FMC) which has a 36 in focal length and medium field of view. The Image Sensor Unit (ISU) contains a high-resolution 12,064 × 32 pixel silicon CCD focal plane array. The DarkStar application employs either 4,020 or 6,024 pixels depending on operating mode.

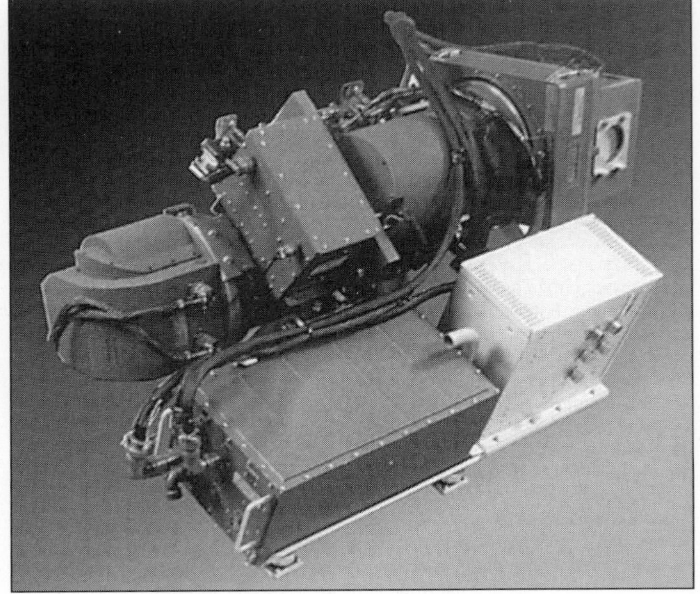
The CA-236 sensor for the DarkStar Tier III Minus programme 0005563

CA-236 demonstration imagery 0023179

The system has built-in electronic haze penetration (contrast enhancement) and the resulting imagery can be recorded directly onto an airborne digital tape recorder and/or transmitted to a gound station via a digital datalink.

Operational status
The CA-236 DarkStar sensor was flown on a US Air National Guard C-130 Pacer Coin surrogate platform in April 1997 and flight testing was in 1998 for final aircraft integration. The DarkStar programme itself was cancelled in the second quarter of 1999.

Contractor
Recon/Optical Inc, CAI Division.

SAGEM Corsaire

Type
UAV sensor.

Description
The Corsaire infra-red reconnaissance system is a derivative of Super Cyclope for low-altitude reconnaissance UAVs and features a wider field of view. It uses a CMT detector, a catadioptric optical system and a four-faced rotating prism driven by a DC motor. In addition to the detector and a photographic unit recording on 70 mm film, the system consists of an electronic line-replaceable unit.

The SAGEM Corsaire infra-red reconnaissance system

The Corsaire system is a single analyser in which the terrain is scanned along a line in a vertical plane. When the analysis head is installed in a UAV, the analysis of the terrain comprises juxtaposed lines at right angles to the flight path. During the process of scanning along one of these lines, the cell generates electric current and the variations from this reproduce the variations of energy received from the terrain to form the video signal. To enable the image to be reconstituted from the individual lines, pulses are recorded at the start and finish of each line to constitute the synchronisation signal.

Operational status

The system has been in production since 1988 for the Canadair/EADS (Dornier) CL-289 and is in operational service in the French and German armies.

Specifications

Spectral band: 8-12 μm
Weight: 12 kg
Field of view: 120°
Film speed: 5.4-87 mm/s
Power supply: 28 V DC
Cooling: closed-cycle Stirling or liquid nitrogen

Contractor

SAGEM SA, Aerospace and Defence Division.

UPDATED

Tamam Multimission Optronic Stabilised Payload (MOSP)

Type

UAV sensor.

Description

The Tamam Multimission Optronic Stabilised Payload (MOSP) is a lightweight dual or triple sensor day and night surveillance, target acquisition and range-finder/pointing payload which has been designed for installation in airborne, ground or naval platforms. It has the capability to look down through the nadir point to provide a view throughout the lower hemisphere.

MOSP mounted on a 'Searcher' UAV 0023400

The MOSP Multimission Optronic Stabilised Payload

The sensor package options include single monochrome or colour CCD or triple-colour CCD day channels, 8 to 12 μm first-, second- or third- (FPA) generation FLIR, laser range-finder of up to 6 pps, or laser target illuminator. A helicopter-mounted version is available called HMOSP.

Operational status

Over 400 systems have been supplied for fixed, manned and unmanned aircraft and rotary-wing platforms as well as ships and mobile ground vehicles.

Specifications

Dimensions: 548 × 354 mm
Weight: 26-36 kg depending on sensors carried
TV fields of view
Narrow: 0.37° (for the long range MOSP)
Zoom: 1.3 to 18.2°
FLIR fields of view
Narrow: 2.4°
Medium: 8.2°
Wide: 29.2°
Spatial coverage
Elevation: +15 to –105°
Azimuth: n × 360° (unlimited)
Power consumption: 280 W (day/night) @ 28 V DC

Contractor

Tamam Division - Israel Aircraft Industries.

VERIFIED

Zeiss Optronik WBG 96 × 4 airborne thermal imager

Type

UAV sensor.

Description

The WBG 96 × 4 has been developed and selected for the Brevel Franco-German drone system. It forms part of the ISOS 2000 (qv) system being developed by STN ATLAS Elektronik and Zeiss Optronik. WBG 96 × 4 is based upon the German OPHELIOS thermal imaging common modules, it is an IRCCD detector using 96 × 4 elements and operating in the 7.5 to 10.5 μm waveband. It includes a high-performance Zeiss ×8 zoom lens. Features include image derotation; active athermalisation; automatic, scene-dependent restoration; automatic gain and control.

Operational status

In production.

Specifications

Dimensions (l × d): 330 × 180 mm
Weight: 10.7 kg
Spectral band: 7.5-10.5 μm
Detector: 96 × 4 IRCCD array
Zoom lens: ×8
Fields of view: 28.8 × 38.4° to 3.6 × 4.8°
Power: 28 V DC, 85 W

Contractor

Zeiss Optronik GmbH.

VERIFIED

Zeiss Optronik airborne thermal imager WBG 96 × 4

OBSERVATION AND SURVEILLANCE

RECONNAISSANCE SYSTEMS

BAE Systems Advanced Tactical Airborne Reconnaissance System

Description

The Advanced Tactical Airborne Reconnaissance System, ATARS, is a high-performance reconnaissance system offering a high level of aircraft integration. The system is installed in the nose bay of the USMC F/A-18D. The low altitude sensor (LAEO) and the medium altitude sensor (MAEO) combine with the infra-red line scanner to provide high-resolution digital reconnaissance imagery both day and night. In addition, the system also interfaces with the aircraft's radar, processing Synthetic Aperture Radar (SAR) imagery expanding ATARS capabilities to 'all weather'. The imagery is recorded on board two digital tape recorders each capable of recording over 40 minutes of imagery. Two recorders give the system versatility to dual-record critical targets, or to simultaneously record the data from two sensors, or to double the record time available to a single sensor. The operators can review tapes while in the air and re-task operations based on data gathered as well as review these tapes on the ground. The system is controlled by a Reconnaissance Management System (RMS). The RMS is the central computer that controls all the system sensors and recorders whilst interfacing to the aircraft mission computer. The RMS software is programmable to meet user needs. A centreline-mounted pod is used to transmit the imagery by datalink down to the ground station in near-realtime.

Operational status

Originally configured for an internal fit for the F/A and RF-4C aircraft, ATARS can be configured for a pod configuration. ATARS has completed Milestone III and has been successfully deployed.

Specifications

Low Altitude Electro-Optical (LAEO) sensor
Type: High resolution, dawn to dusk below weather, high-speed sensor
Operating height: 200-3000 ft above ground level
Field of view: 140°
Viewing angle: Vertical or forward oblique

Medium Altitude Electro-Optical (MAEO) sensor
Type: High resolution, medium-range standoff sensor for daylight operation in high threat environments.
High resolution, low- to medium-altitude day/night reconnaissance capability
Operating height: 2,000- 25,000 ft above ground level
Field of view: 22°
Field of regard: Horizon to horizon

Infra-Red Line Scanner (IRLS) sensor
Operating height: 200-10,000 ft above ground level
Field of view:140° or 70°
Spectral band: 8-12 μm

Contractor

BAE Systems, North America, Reconnaissance and Surveillance Systems.

UPDATED

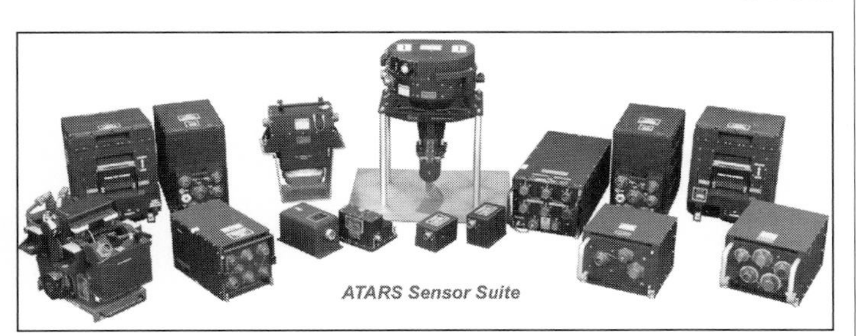

The BAE ATARS reconnaissance system 0079784

BAE Systems AN/AAD-5 infra-red reconnaissance set

Type

Infra-red linescanner.

Description

The AN/AAD-5 is a dual-field of view high-performance infra-red reconnaissance system which scans the terrain beneath an aircraft's flight path. The system consists of seven Line-Replaceable Units (LRUs): the receiver, recorder, film magazine, control indicator, infra-red performance analyser, cooler and power supply.

The long-wave infra-red receiver includes the scanning optics, cooler cryostat, two 12-element detector arrays and preamplifiers and the associated buffer electronics. The detector arrays are Cadmium Mercury Telluride (CMT) photoconductors, one for the wide field of view, one for narrow field of view. The receiver also contains the spin motor regulator and an encoder to provide system timing pulses to the recorder.

Received infra-red energy is reflected through an optical scanner system and focused on two detector arrays enclosed in a vacuum-sealed Dewar chamber. These arrays are maintained at a temperature of approximately 80K by a cryostat which is part of a closed-cycle nitrogen system inserted into the Dewar. The different-sized 12-element detector arrays convert received infra-red energy into electrical signals.

The analogue voltage representing the velocity/height (V/H) ratio determines the number and type of detectors to be used. At a low V/H ratio the small area array yields better resolution which improves high-altitude sensor performance and provides a better scale for image interpretation. At high V/H ratios the large area array is used to meet the higher data rate. The V/H ratio also determines the number of detectors used in each array, from a minimum of one to a maximum of 12, and establishes the optimum number of overlaps.

Detector signals are processed into a video format which is displayed as an intensity modulated trace on a 127 mm (5 in) multibeam CRT in the system's recorder. The CRT image is optically coupled to the film magazine and recorded on moving film at a speed which is also controlled by the V/H ratio analogue voltage. The system compensates for aircraft roll by using inputs from the INS.

The high-resolution beam CRT provides spatial correspondence between the readout and the detector array. It also affords flexibility in the selection of field of view options. The CRT is of a ruggedised type suitable for airborne applications.

An autofocus opto-electronic subsystem is used to maintain optimum focus in the electronic optics of the multibeam CRT. It maintains correct spot brilliance on the faceplate and provides a sensor for the built-in checkout system which monitors performance of the display. A recorder head assembly for the auxiliary data annotation set is also provided.

The cryogenic refrigerator compressor receives gaseous nitrogen from the receiver cryostat, compresses it to approximately 2,200 lb/sq in and returns it to the cryostat for expansion to a liquid state. The compressor contains a motor, turn-on relays, fans, adsorber and safety thermostats.

An improved version, the AN/AAD-5(RC), is derived from the AN/AAD-5. The AN/AAD-5(RC) imaging process is identical to that of the AN/AAD-5, but the receiver is two-thirds the size of the AN/AAD-5 receiver and uses an integral 0.25 W split-Stirling miniature cooler. This eliminates the separate cryogenic cooler LRU, as well as the need for the nitrogen supply tanks carried on board the aircraft.

While the effective receiver aperture has been reduced, the imaging performance has been kept equal to that of the AN/AAD-5 receiver through improvements in the infra-red detector performance. The same materials have been used in the scaled receiver. The spin mirror shaft encoder has not been scaled because of the need to maintain scan position accuracy. The scaling has resulted in a smaller, lighter and more reliable receiver.

The AN/AAD-5(RC) preamplifier assembly has also been upgraded. Previously, preamplifiers occupied a significant portion of the receiver envelope. The AN/AAD-5(RC) hybrid monolithic preamplifier assembly occupies a smaller volume than the existing AN/AAD-5 set.

Various electro-optical output options are available for both AN/AAD-5- and AN/AAD-5(RC)-equipped systems. Upgrades using the Lockheed Martin (formerly Loral) IR linescanner real-time display system can add a cockpit real-time display with onboard video recording, with an optional datalink to ground stations, while maintaining the current film system operation.

Operational status

In service with the US Air National Guard in the RF-4C and the US Navy in the F-14 TARPS. Also used by a number of foreign air forces including Australia for the RF-111 and Germany, Greece, Korea, Spain and Turkey for the RF-4. Over 600 AN/AAD-5 systems and 168 AN/AAD-5(RC) systems have been delivered. The AN/AAD-5(RC) is still in production.

Specifications

Spectral band: 8-14 μm
Detectors: 2 12-element CMT arrays
Cooling: closed-cycle nitrogen (AN/AAD-5); integral 0.25 W split Stirling minicooler
Dimensions
Receiver: 460 × 380 × 330 mm (AN/AAD-5), 430 × 345 × 279 mm (AN/AAD-5(RC))
Recorder: 410 × 580 × 230 mm
Magazine: 380 × 180 × 250 mm
Control indicator: 150 × 100 × 80 mm
Cooler: 410 × 20 × 300 mm
Analyser: 150 × 50 × 300 mm
Power supply: 250 × 430 × 280 mm

The AN/AAD-5

Weight
AN/AAD-5: 130 kg
AN/AAD-5(RC): 63.2 kg
Power supply: 115 V AC, 400 Hz, 725 VA nominal
28 V DC, 118 W nominal
Magazine capacity
Conventional film: 106 m
Thin-based film: 213 m

Contractor
BAE Systems (formerly Lockheed Martin IR Imaging Systems).

VERIFIED

BAE Systems D-500 infra-red linescanner

Type
Infra-red linescanner.

Description
The D-500 is an advanced version of the AN/AAD-5 but with identical performance, which is designed for installation in pods, small aircraft and UAVs.

The system consists of four basic Line-Replaceable Units (LRUs); the receiver in which the IR image is formed in a CMT detector array, the film recorder in which a visible image is formed through a CRT trace from the detector signal, the film magazine and a power supply. The intensity modulated CRT trace is coupled by a fibre optic faceplate to the recording film upon which the image is composed on a line-by-line basis. Their detector is composed of two linear arrays of 12 detectors, each aligned in the direction of the flight.

Between one and 12 detectors are used for recording, depending upon the velocity/height profile of the aircraft. The signal from the receiver is used to drive a 12-beam CRT which sweeps the signal onto film, mimicking the fashion in which the signals are received through the scan mirror at the detector. The scene sensed by the instrument can have a large intrinsic dynamic range which is first limited by the detectors to 12 bits. The range of levels displayed on the film is reduced below this by both the limited dynamic range of the CRT and by the film itself. The range is typically in the order of 5 bits or 32 grey levels in the film image.

The D-500 imaging process is similar to that of the AAD-5, the most significant differences being the CRT/film coupling and the video electronics packaging. Both optical efficiency and bulk have been improved by replacing the AAD-5 CRT relay lens with the fibre optic faceplate CRT. Options for the D-500 include videotape recording, onboard real-time display and a datalink capability.

The system is electronically roll corrected. Up to ±20° of roll excursions can be corrected in both the wide and narrow modes. The automatic roll correction is accomplished by shifting a display unblank pulse within the 180° timing window provided by each facet of the scan mirror. The display unblank pulse gates out the proper 'section' of data from the number of degrees of video collected during a scan line.

Operational status
In production and in service with the Egyptian Air Force on Beech 1900 light aircraft and the Teledyne Ryan Model 324 Scarab and Lear Astronics R4E SkyEye UAVs. It is also being produced for the reconnaissance pod of Japanese FJ-4Es.

The D-500 reconnaissance system

Specifications
Fields of view: 120°; 60°
Focal length: f7.62
Aperture: 165 mm
Film length: 106-213 m

Contractor
BAE Systems North America (formerly Lockheed Martin IR Imaging Systems).

UPDATED

BAE Systems D-500A infra-red linescanner

Type
Infra-red linescanner.

Description
The D-500A Infra-Red LineScanner (IRLS) is a high-performance reconnaissance sensor developed for next-generation reconnaissance aircraft. The D500A has evolved from the BAE Systems (formerly Lockheed Martin IR Imaging Systems) AN/AAD-5 family of products currently deployed on tactical aircraft worldwide. It is a flexible system that can be used for digital real-time datalinking and with video cassette or solid-state recorders. The D500A IRLS is a derivative of the linescanner being produced for the US Marine Corps' F/A-18 ATARS programme. For this programme, the digital version is being provided to BAE Systems for integration into the ATARS sensor suite. The compact design permits use on new lightweight fighters or future unmanned air vehicles.

The system consists of two Line-Replaceable Units (LRUs): the infra-red receiver which is upgraded from that used in the D-500 system for the next-generation manned or unmanned tactical reconnaissance aircraft; and the Signal Electronics Unit (SEU), which processes the raw imagery data and provides either an analogue or a digital output signal for real-time display, datalink and recording. The system has two fields of view with selectable left- or right-oblique viewing modes in the narrow FOV. A step zoom feature allows imaging at full resolution.

Operational status
Operational.

Specifications
Spectral band: 8-12 μm
Standoff capability: electronic oblique
Cooling: sealed, linear split-cycle Stirling
Fields of view
Wide: 140°
Narrow: 70°
Field of regard
WFOV: 140°
NFOV: 140°
Focal length: 7.3 in
CMT detector array: 12 × 1 (NFOV); 11 × 1 (WFOV)
Detector size: 45 μm^2 (NFOV); 91.4 μm^2 (WFOV)
NETD typical
NFOV: 0.5-0.6°C
WFOV: 0.2-0.3°C
Weight: <72 lb
Electronic roll stabilisation: ±30°
Power (average): 298 W
Interface: MIL-STD-1553

Contractor
BAE Systems, North America, Reconnaissance and Surveillance Systems.

UPDATED

BAE Systems F-9120 Advanced Airborne Reconnaissance System (AARS)

Type
Reconnaissance system.

Description
The F-9120 Electro-Optical, Dual-Band (EO/IR), Long-Range Oblique Photographic System (EO/IR LOROPS) was designed and developed as an IR&D programme but with the demanding requirements of the US Navy's and foreign military programme requirements. The system provides simultaneous imagery in the visible and infra-red spectra. The visible focal plane operates in the silicon spectrum from .450 μm to .950 μm and the infra-red uses an Indium Antimonide (InSb) focal plane operating in the 3.4 to 5 μm spectrum range. Within one sensor are 3 different fields of view each with focal lengths matched to produce the same size/scale of image in the ground station. The low level FOV provides contiguous imagery at 550 kt from about 1,000 ft above ground level, and stereo from about 2,500 ft up to a slant range of approximately 11,000 ft when automatic switchover to the medium FOV takes place. This FOV operates out to a slant range of approximately 15 miles when automatic switching to the long-range/high-level FOV takes place. The high-level FOV will give NIRRS rating of 5 out to over 75 miles in the visible mode. Control of the sensor is from a reconnaissance management unit and data is stored in a solid state recorder. These two functions are combined within one box, the RMU/SSR. The system is enclosed in a BAE Systems-developed Advanced Recce Pod suitable for carriage on many types of tactical fighter and larger commercial type aircraft. The pod contains an environmental control system at the rear and a datalink can be fitted in the front.

Operational status
BAE Systems AARS is currently undergoing flight trials on an F-16 aircraft.

Specifications
Spectral band: Simultaneous visible and infra-red imaging (Vis – .450 - .950 μm; IR – 3.4 to 5 μm)
Optical system: 13 in aperture
Field of regard: Horizon to horizon field of regard
Field of view: 3 field of view optical system
Weights: Sensor < 300 lbs
Complete pod plus datalink < 1,600 lb
Pod: Length 165 in/4.20 m
Diameter = 25 in/635 mm

Contractor
BAE Systems, Reconnaissance and Surveillance Systems.

UPDATED

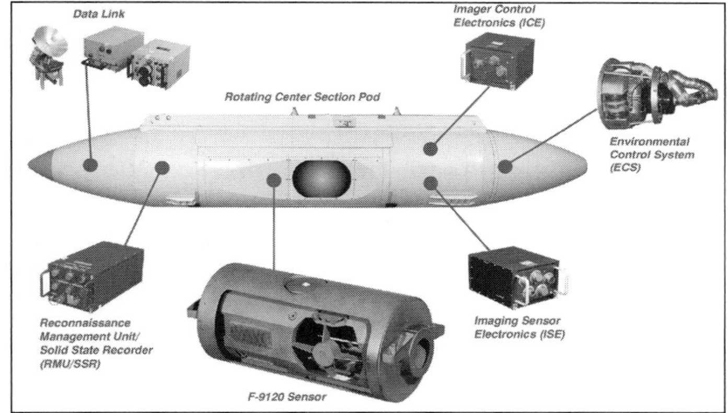

The BAE Advanced Airborne Reconnaissance System 0079783

BAE Systems IIS Infra-red Imaging System (IIS)

Type
Infra-red linescanner.

Description
The Infra-red Imaging System (IIS) is a near real-time, 180° field of view, high-resolution IR linescanner system developed for the German Air Force Tornado ECR aircraft. The system consists of five interactive units. Two of these units, the Recorder/Film Processor Unit (R/FPU) and the Scanner Receiver Unit (SRU) are produced by Lockheed Martin. These two units provide the infra-red detection and recording functions of the system.

The SRU, used to detect horizon-to-horizon infra-red radiation emanating from the terrain, senses medium band infra-red radiation and converts it to 12 parallel video signal channels. SRU electronics provide a constant resolution footprint of the target, regardless of scan angle, throughout the widest practical ground coverage. These electrical signals are then sent to the R/FPU. The SRU is controlled by digital commands from the control unit, which are sent via the aircraft databus.

The BAE Systems IIS is a major part of the reconnaissance system in the German Tornado ECR

The SRU can be broken down into eight functional groups. The spin mirror group performs optical scanning of the IR scene. The IR sensor and amplifier group collect the IR energy of the scanned scene. The channel processor group performs the initial amplification and processing of the 34 preamplified detector signals. The mapping group processes signals from the channel processor group and combines them into 12 parallel video signals. The delay buffer and timing group provides proper timing for the 12 video channels and generates the file video output. The gyro and electronics group provide the roll compensation for maintaining the stability of the scanned imagery. The BIT and control group provide the token ring bus interface, central processing for the SRU and BIT circuitry. The video group provides the video output.

Only 12 infra-red signal channels, video syncronised signals and roll sensor data are passed directly between the SRU and the R/FPU. Modular design concepts are used within each unit in order to simplify maintenance and logistics.

The function of the R/FPU is to receive and process the video signals from the SRU, provide the aircraft crew with the ability to view the imagery in near-realtime and produce a permanent film record of the IR imagery scanned by the SRU.

A precision CRT converts the amplitude modulated video signals to intensity modulation of the CRT beam as it sweeps across the tube face in synchronism with the rotation of the spin mirror in the SRU. An autofocus detector and focusing circuit measure the intensity of the beam and automatically corrects the focus. The film is exposed by the variable intensity CRT beam as it moves in front of the faceplate. Annotation data, obtained from the aircraft via the databus, is applied to the film at proper time intervals.

The dry process film is developed within the R/FPU for onboard viewing. The film moves, at constant speed, between a rotating drum and a heated shoe. A servo-controlled heater inside the shoe maintains the temperature at 130°C, the level required for the dry process chemicals on the film to be activated by the heat, thus developing the film on board the aircraft.

Processed film is temporarily stored on the film manipulator. The film storage mechanism consists of two concentric powered drums on which the film is wound or unwound as required, to place the desired portion of the film in the viewing area. Each drum is controlled separately so that newly processed film can be wound on to the manipulator as it accumulates, while previously stored film is being wound back on to the manipulator as the operator commands viewing of earlier images.

As the film leaves the manipulator, it passes over a light table. This electroluminescent panel causes the image on the film directly above it to be projected on to the lens of a TV camera. The camera output is displayed on the operator's console. The operator can view any part of the exposed film. When previously exposed film is to be viewed, it is rerolled from its storage spool, moved back across the light table for viewing and temporarily stored on one of the two film manipulator drums. This action does not prevent newly exposed film from being stored on the other drum.

Selectable lenses give the TV camera three fields of view, effectively a zoom capability as the film is being viewed. Areas of interest on the film may be marked during viewing so that they may be quickly located at a later time.

A 107 mm reel of unexposed dry process film is available to the R/FPU. The film is provided at the correct time and speed for exposure by the CRT. The exposed film is stored on a separately controlled take-up reel.

Operational status
In production and in service with the German Air Force which has a requirement for 51 systems for Tornado ECR aircraft.

Specifications
Dimensions
SRU: 520.7 × 381 × 509.8 mm
R/FPU: 600 × 640 × 476 mm
Weight
SRU: 53 ±2 kg
R/FPU: 87 kg

Contractor
BAE Systems (formerly Lockheed Martin IR Imaging Systems).

UPDATED

BAE Systems SU-172/ZSD-1(V) Medium-Altitude Electro-Optical sensor (MAEO)

Type
Reconnaissance system.

Description
The Medium-Altitude Electro-Optical (MAEO) sensor operates from 2,000 to 25,000 ft with a 22° narrow field of view, 304.8 mm (12 in) focal length, f/5.6 lens for daylight operations. It is designed for medium-altitude tactical airborne reconnaissance penetration and standoff missions. High-resolution imagery is captured on a 12,000, 10 × 10 µm pixel CCD.

The SU-172/ZSD-1(V) is designed with emphasis on an integrated system approach to reduce life cycle costs. The sensor is automated to reduce aircrew workload and for use in UAVs. Stabilisation allows for target tracking in the presence of aircraft manoeuvres.

Specifications
Dimensions
Imaging unit: 511 × 381 × 330.2 mm
Weight: <62.14 kg
Power supply: 28 V DC, 440 W
Altitude: 2,000-25,000 ft

Operational status
In production for the US Marine Corps.

Contractor
BAE Systems, North America, Reconnaissance and Surveillance Systems.

UPDATED

BAE Systems SU-173/ZSD-1(V) Low-Altitude Electro-Optical sensor (LAEO)

Type
Reconnaissance system.

Description
The SU-173/ZSD-1(V) Low-Altitude Electro-Optical (LAEO) sensor operates from 200 to 3,000 ft with a low-distortion 140° wide field of view and vertical or forward oblique fixed focus. The sensor provides high-resolution visible spectrum imagery for high-speed, low-altitude area coverage on tactical reconnaissance missions.

The SU-173/ZSD-1(V) consists of a sensor and electronics unit that collect imagery, provide roll correction of imagery, perform preliminary image processing and output digital image data. The low-altitude sensor system has been designed to be flexible and responsive to both mission and platform tailoring. The system may be internally mounted in manned aircraft such as the RF-4C/E or F/A-18(RC) or pod-mounted in the F-14, F-16, Tornado, Mirage, Jaguar and JAS 39, in addition to being configured for unmanned aerial vehicles.

Specifications
Dimensions:
(imaging unit) 284.5 × 121.9 × 363.2 mm
Weight: <24.5 kg
Power supply: 28 V DC, 231 W
Altitude: 200-3,000 ft

Operational status
In production for the US Navy and US Marine Corps.

Contractor
BAE Systems, North America, Reconnaissance and Surveillance Systems.

UPDATED

BAE Systems TARS Theater Airborne Reconnaissance System

Type
Electro-optical reconnaissance system.

Description
The Theater Airborne Reconnaissance System (TARS), designed and manufactured by BAE Systems Reconnaissance & Surveillance Systems, provides the USAF with the ability to capture reconnaissance data in adverse, under-the-weather daylight conditions. The Air National Guard operates the system on the F-16 C Block 25-32, Multirole aircraft. The TARS suite consists of a forward facing electro-optical framing sensor, CA-. 260/25A and the BAE Systems Medium Altitude Electro-Optical (MAEO) sensor installed in a Modular Reconnaissance Pod (MRP). Command and control of the TARS system

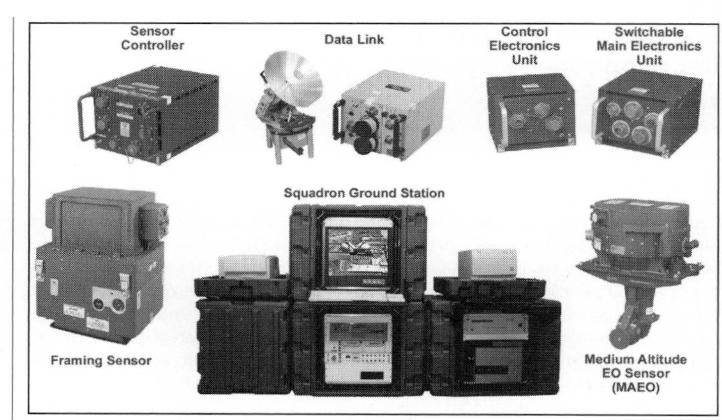

The TARS system including the ground-based elements 0079767

is integrated through the F-16 main computer using the AN-ALQ-213 Counter Measures System Control Panel and the F-16 MFD hardware. A BAE Systems Squadron Ground System (SGS) provides a five transit box, field deployable viewing, screening and exploitation capability to give the squadron commander the near-realtime reconnaissance data critical to making timely decisions on the battlefield. The USANG is presently implementing Pre-Planned Product Improvements (P^31) to the TARS system by implementing a STANAG 4575 compliant Solid State Recorder and an Airborne Information Transmission (AB IT) Common Data Link.

Operational status
BAE Systems has delivered the 20 required airborne systems and 5 Squadron Ground Stations. In February 2001 the USAF officially declared Initial Operational Capability (IOC). The systems are presently deployed at two USANG bases.

Specifications
CA-260/25A: 5 × 5K pixel frame format
3,000-10,000 altitude operation
NIJRS 5 imagery at 30,000 ft slant range (12 in lens/15° depression)
MAEO: 2,000 - 25,000 ft altitude operation
22° field of view
horizon to horizon field of regard

Contractor
BAE Systems, North America, Reconnaissance and Surveillance Systems.

UPDATED

Computing Devices Ltd, Thales Optronics (Vinten) Tornado infra-red linescanning reconnaissance system

Type
Reconnaissance system.

Description
The Tornado reconnaissance system is fitted to Royal Air Force Tornado GR.Mk 1A aircraft. Computing Devices Ltd is the prime contractor and the system design authority. Computing Devices Ltd also makes the recorders and control panel. The sensor system consists of a WVL Type 4000 infra-red linescan system in a blister under the fuselage and two Side-Looking Infra-Red (SLIR) sensors from TICM II modules, made by BAE Systems and configured by Thales Optronics (formerly W Vinten Ltd). These are roll stabilised up to 30° and cover a normal vertical displacement of 10° to the horizon in the fuselage to provide extra resolution on the edge of the linescan's field of view. The linescan covers 180° across the track, but rectilinearisation algorithms in the image processing unit compensate for extreme angular distortions and onboard scan conversion of the sensor signals allows imagery to be displayed in the rear cockpit. Various upgrades are already under consideration. A higher resolution onboard monitor may be included.

Operational status
Development of the system began in 1982 and the first operational aircraft were delivered in January 1989. The system was used during Operation Desert Storm and, among other tasks, located 'Scud' missile launch sites as well as surveying the route to be taken by the British ground forces. In service with the Royal Air Force and Saudi Royal Air Force. Computing Devices is at present working on the next-generation system for the UK MoD.

Contractors
Computing Devices Company Ltd.
Thales Optronics (Vinten) Ltd.

VERIFIED

Dassault reconnaissance pods

Type
Reconnaissance pod.

Description
Dassault makes 10 types of optical and infra-red reconnaissance pod for high-performance combat aircraft. Two of these are:

- the COR 2 which has focal lengths up to 600 mm and produces 114 × 114 format pictures and contains the SAGEM Super Cyclope infra-red sensor. It is used for high-speed, low-altitude reconaissance
- the HAROLD which has a 1,700 mm focal length and is a long-range oblique view strategic reconnaissance pod. The environment of the photographic equipment and its extremely high quality gives exceptional performance. Its integration into the aircraft weapon system allows automatic filming in prepared and planned mission.

Operational status
The COR 2 pod is no longer in production. COR 2 and Harold (no longer in production) pods are used with Mirage F1 and Mirage 2000 aircraft. The company has produced 11 Harold and 19 COR 2 pods.

Contractor
Dassault Aviation.

VERIFIED

Mirage F1 fitted with COR 2 pod 0022254

EADS France Aerospatiale reconnaissance pod

Type
Reconnaissance pod.

Description
The Aerospatiale reconnaissance pod was developed to assist deployment of the Canadair/DASA CL-289 UAV, known to the French Army as Piver 289. The pod was designed for training by providing calibration for the ground station but has since become a surveillance product in its own right for use in helicopters and light aircraft.

The pod houses the same sensors as the CL-289, that is, a three lens Zeiss Optronik KRb 8/24 photographic camera for stereoscopic terrain coverage and a SAGEM Corsaire infra-red linescanner operating in the 8 to 12 μm waveband.

Operational status
Available.

The Aerospatiale recon pod attached to a Gazelle helicopter

Specifications
Length: 1.40 m
Diameter: 38 cm
Weight: 78 kg

Contractor
EADS France (formerly Aerospatiale Missiles Division).

UPDATED

EADS reconnaissance pod

Type
Reconnaissance pod.

Description
This is a general purpose reconnaissance system in which the sensors are housed in the pod's central module. The sensors consist of the Zeiss TRb 60/24 LHOV and Zeiss KRb 6/24 LLDC cameras and the Raytheon (formerly Texas Instruments) RS-710 infra-red linescanner. The rear module features the Litef reconnaissance interface system which controls the system and co-ordinates both sensor and aircraft data.

The LHOV camera can either be focused from 600 m to infinity during flight, or be operated with an auto focus and with different horizontal, oblique or vertical modes relative to the direction of the flight. In addition, the camera also supplies stereoscopic pictures for direct interpretation. The LLDC produces distortion-free imagery of the terrain from horizon-to-horizon during low-level missions during day or night.

The RS-710 scans thermal sources under all light conditions and records them on film. It has a terrain coverage similar to that of the LLDC system which assists comparison during post-mission interpretation.

The Litef system records digitally the latitude, longitude, altitude, velocity, heading and time of targets overflown. The pod also includes built-in test equipment and an environmental control unit.

Operational status
In service with the German Navy and the Italian Air Force since 1985, both of which use the pods with the Tornado. EADS (formerly DASA) has produced 50 of these pods. German Air Force Tornados participated in UN/NATO peacekeeping missions in Bosnia and Kosovo using this type of pod.

Specifications
Length: 410 cm
Diameter: 58 cm
Weight: 380 kg

Contractor
EADS, Military Aircraft.

VERIFIED

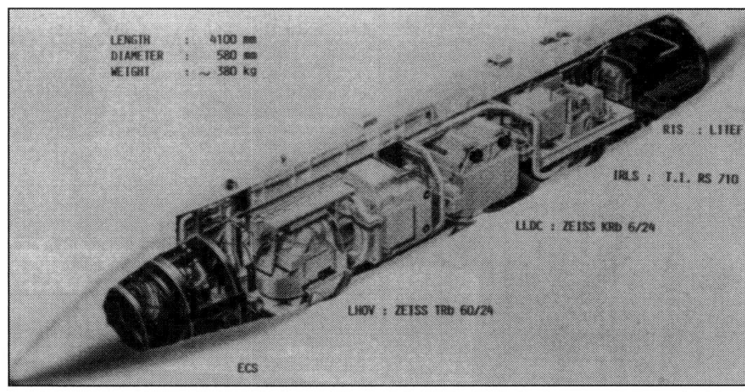

EADS reconnaissance pod

EADS tactical reconnaissance pod system

Type
Reconnaissance pod.

Development
In 1994, DASA, now EADS, started development of a new-generation reconnaissance pod for the German Air Force. In 1997, an additional reconnaissance pod solution was contracted by the German Air Force.

Description
The sensors within the pod are: an optical camera with a Zeiss KS-153 Trilens which covers the forward oblique area and records on film; an optical camera using a Zeiss KS-153 Pentalens which is directed vertically downwards and records on

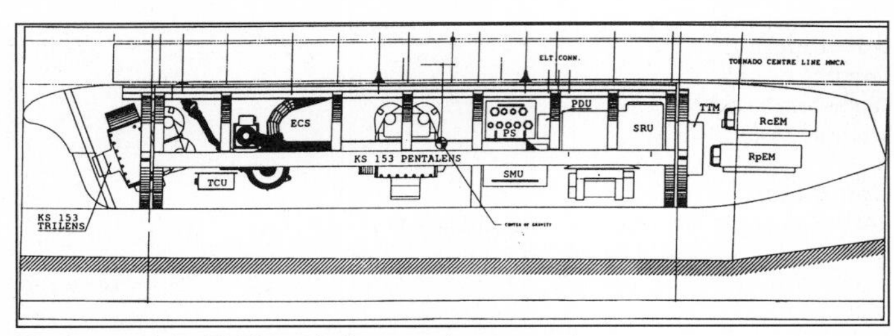

The layout of the EADS tactical reconnaissance pod

film; and an Infra-Red LineScanner (IRLS). The IRLS sensor front end is the Honeywell Scanner Receiver Unit (SRU) as used in the Electronic Combat Recce (ECR) on the Panavia Tornado.

Further components are: a Recce Management Unit (RMU) as the interface between the aircraft avionic system and the reconnaissance pod; and an Ampex DCRSi 107 digital tape recorder consisting of tape transport module, recording electronic module and reproduce electronic module for IRLS image recording and replay.

The pod is carried on the centreline pylon of the Tornado. The pod receives aircraft data via a serial datalink (Panavia Link). In addition, a Recce Control Panel for sensor operation is installed in the Tornado rear cockpit. The IRLS imagery is displayed in the rear cockpit via a video link for onboard image evaluation. The German Air Force reconnaissance programme includes the development of the pod system, integration into the Tornado aircraft avionics and the development of a ground exploitation system.

The present layout is shown in the diagram. The following abbreviations for the subsystems are used:

ECS Environmental Control System
PDU Power Distribution Unit
PS Power Supply
RcEM Recording Electronic Module
RpEM Reproduce Electronic Module
RMU Recce Management Unit
SRU Scanner Receiver Unit
TCU Temperature Control Unit
TTM Tape Transport Module
DTR Digital tape Recorder

In the additional version, the Zeiss KS-153 Trilens camera is replaced by a Zeiss KS-153 Telelens camera with a focal length of 610 mm. The infra-red linescanner is modified to increase the operational altitude.

Operational status

In development since 1994. The first series production system was delivered in April 1999, and now in service with the German Air Force on 37 Tornado aircraft.

Contractor

EADS, Military Aircraft.

UPDATED

EADS Tornado ECR

Type

Reconnaissance pod.

Development

In 1978, the German MoD began to consider some means of overcoming its shortfall in tactical reconnaissance and electronic warfare capabilities from the late 1980s. The requirement was defined by 1984 as the Tornado ECR (Electronic Combat and Reconnaissance). The first prototype flew in August 1988.

Description

The Tornado ECR was developed to provide all-weather, day and night tactical reconnaissance, the suppression of enemy air defence systems, neutralising enemy command, control and communication systems, determining low-risk penetration routes for other attack aircraft, together with reconnaissance and targeting for follow-on attack aircraft using near-realtime datalink transmissions.

The airframe follows the Tornado IDS configuration, but includes a number of subsystems to meet the operational requirement. These include an infra-red reconnaissance system, an Emitter Location System (ELS), a digital databus system and advanced displays and computers. A Zeiss Optronik FLIR is included for operations in poor visibility. This generates video images which are projected onto the pilot's HUD (Head-Up Display) as well as to TV tabulator displays in the weapon systems officer's position.

Meanwhile, the infra-red reconnaissance system has been taken out and is no longer used in the ECR Tornado.

For reconnaissance, Honeywell Sondertechnik was the prime contractor for the linescanner system, which was mounted nearly flush with the fuselage. This has a wide hemispherical field of view of the area below the aircraft. The system was the Loral IIS but has been adapted to use dry silver film which can be developed in flight without chemicals. The exposed film was developed by a heat processor then passes through a film manipulator. It then passes over a light table above which

The Tornado ECR is in service with the German Air Force

was a television camera and from there the film moves onto the take-up spool. With the aid of the television camera, the developed film was displayed in the cockpit for rapid evaluation by the weapon system operator in the rear cockpit, through the use of the MIL-STD-1553B digital databus system and the Litef ODIN (Operational Data INterface) system. This can superimpose navigation information and event markers on the reconnaissance data which are presented on a display using a VDO symbol generator.

The ELS is designed to detect, identify and locate hostile emitters passively by reference to an onboard emitter library. It features multi-octave RF coverage, phase interferometric antenna arrays for precision direction-finding, passive ranging channelised receivers and a multiple MIL-STD-1750A digital processor. Location is achieved using triangular methods and the databus not only distributes this and linescanner data to the aircraft's tactical displays, but allows them to be transmitted to ground command posts and follow-on attack forces.

For active electronic warfare roles the aircraft carries an airborne ECM pod and AGM-88 HARM missiles. Missiles are cued using co-ordinates identified by the ELS.

Operational status

In production, with 35 aircraft delivered to the German Air Force between 1990 and 1992, although problems with the ELS delayed inertial operational capability until April 1993. The Italian Air Force has 16 ECR variants, converted from IDS airframes. GAF ECR and Reconnaissance Tornadoes participated in UN/NATO peacekeeping missions in Bosnia and Kosovo.

Contractor

EADS, Miltary Aircraft.

UPDATED

Elop E-O/IR LOROP airborne reconnaissance system

Type

Reconnaissance pod.

Description

The Electro-Optic Long-Range Oblique Photograph System is an airborne standoff reconnaissance system with real-time data transmission or real-time data record/transmit options, in pod configurations. The system provides imagery in the visible and IR spectra simultaneously. The reconnaissance pod houses the camera, containing a Cassegrain Ritchey-Chretien mirror telescope with a linear array of butted CCD detectors and IR detectors in the focal planes, scanning-mirror video processing unit.

The peripheral units, such as the data link, digital solid state recorder, air conditioning unit, power supply and reconnaissance management unit, also reside in the pod.

Applications include combat aircraft, business jets and maritime patrol aircraft.

System control modes include pre-loaded mission file, remote control from the ground station and cockpit control. The collected imagery can be recorded in flight and/or transmitted in real time to the ground station. The ground station incorporates the tracking antenna, datalink transmitter and receiver, advanced intelligence capabilities, image enhancement, archiving capability and hard copy and soft copy displays.

Major System Characteristics include:

- full turnkey system from mission planning to data dissemination
- day and night standoff missions

Sample imagery from the E-O/IR LOROP pod taken at 25,000 ft **NEW**/0547530

Elop's E-O/IR LOROP pod **NEW**/0547527

- autonomous navigation
- pilot override for targets of opportunity
- on-board mission data recording
- wide band data link for real time image transmission
- self-contained pod configuration
- simultaneous visible and IR photography
- fully complies with the F-16 and RF-4 flight envelope
- MultiMission Modularity (M3) pod with interchangeable centre section for tactical and strategic mission
- simple interface with existing aircraft avionics
- fully certified for F-16 and RF-4 aircraft
- minimal certification effort for other aircraft.

Operational status

In operation. The system is also available for non-pod installations in various types of air vehicle.

Contractor

Elop Electro-Optics Industries Ltd.

UPDATED

Goodrich DB-110 dual-band reconnaissance system

Type

Reconnaissance system.

Description

The Goodrich (formerly Hughes, then Raytheon) DB-110 dual-band reconnaissance system combines electro-optical and infra-red imaging capabilities in a compact lightweight modular design which is available as a dual-band, day only or night only sensor. The day sensor has a silicon CCD array and the infra-red sensor is mid-waveband with a choice of indium antimonide arrays. NIIRS 5 quality imagery is obtained by both the visible and infra-red sensor.

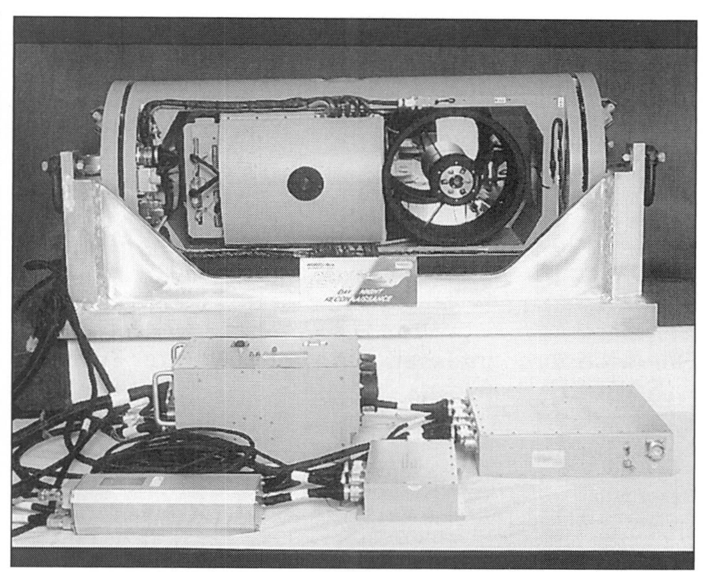

The DB-110 dual-band reconnaissance system 0005558

The DB-110 can interleave wide area search, spot collection and target tracking/stereo modes so as to allow imaging of widely dispersed targets. This minimises flight time and maximises crew safety. Depending on the mission-planning scenario, in excess of 200 targets can be acquired when in spot collection mode. In target-tracking mode, up to 10 scans of the same sector can be performed. In wide area search mode, ground coverage can be up to 90,000 n miles^2/h. Continuous operation is possible using a datalink, limited only by the range of data transmission and aircraft endurance. Record time is determined by data rate and tape capacity when using an onboard recorder.

The system is compatible with nearly all reconnaissance platforms including UAVs, whether inboard or podded configurations are required. If pod mounted, DB-110 is adaptable to stores stations of tactical aircraft such as F-4, F-14, F-16, F/A-18, Mirage, Tornado, AM-X. With the pod the aircraft has a flight envelope equivalent to that carrying a 275 gallon fuel tank. For inboard installation, the small size of the system allows internal nose mounting using existing windows and access hatches.

Operational status

DB-110 made its first flights in a flight qualified pod for the Tornado aircraft in February 1997. IOC on RAF Tornado GR.4A in September 2002.

Specifications

Sensor
Dimensions (l × d): 127 × 47 cm
Weight: 140 kg
Power: 350 W, 115 V AC, 400 Hz
Electronics
Dimensions (5 units)
PPDV: 457 × 406 × 102 mm
SCU: 165 × 198 × 418 mm
RMS: 257 × 410 × 236 mm
PAA: 72 × 255 × 175 mm
INS: 178 × 119 × 303 mm
Weight: 50 kg total
Power: 1,300 W total, 115 V AC, 400 Hz
Optical
Type: cassegrain reflector
Focal length (adjustable to fit application)
Visible: 280 cm nominal
Infra-red: 140 cm nominal
Aperture
Visible: f/10
Infra-red: f/5
Focal planes
Visible
Spectral band: 0.5-1.0 μm
Type: silicon CCD 5,120 × 64 line array
Infra-red
Spectral band: 3-5 μm
Type: 512 × 484 InSb array (640 × 480 available)
Max output data rate: 333 Mbits/s compressed to 180 Mbits/s
Data recording rate (digital tape recorder): up to 240 Mbits/s continuously variable
Max data rate (digital datalink): 150 Mbits/s, J-band, line of sight
Altitude: 10,000-80,000 ft
Ground speed: M0.1-1.6
Field of regard: 180° across line of flight, ±20° along line of flight
Panoramic/sector scan: 4 to 28°
Overlap: variable from 1 to 100%

Performance
Visible: up to NIIRS 6
Infra-red: up to NIIRS 5

Contractor
Goodrich Surveillance and Reconnaissance.

UPDATED

Goodrich Reconnaissance Airborne Pod for Tornado - (RAPTOR)

Type
Reconnaissance pod.

Description
Raytheon Systems Company was selected in September 1997 to provide an end to end tactical reconnaissance system for the Royal Air Force's two Tornado reconnaissance squadrons. The programme SR(A)(OE)1368 is called Reconnaissance Airborne Pod for Tornado (RAPTOR). The original requirement specified a dual-waveband day/night long-range system able to operate simultaneously in the video and thermal spectrums. Target recognition at 72 km (visual) and 36 km (infra-red) was also specified. The chosen system is based on the DB-110 electro-optical/infra-red dual-band sensor (qv) which is designed for both tactical and high-altitude stand-off applications. Goodrich acquired the RAPTOR business from Raytheon in 2001. The DB-110 system is intended to enable high-quality reconnaissance imagery to be collected at stand-off target to sensor ranges of 5 km to more than 45 km. The system will include pod, datalink and ground exploitation stations. The contract, estimated at £55 million, calls for the delivery of eight airborne pods and two ground stations. The pod could also be deployed on Eurofighter 2000.

Operational status
In September 2002 the UK Tornado GR4A received Initial Operating Capability (IOC) with RAPTOR. The Japanese Maritime Self Defence Force will also take delivery of DB-110 systems to equip an unknown number of the service's P-3C multisensor ocean surveillance aircraft.

Contractor
Goodrich Surveillance and Reconnaissance.

UPDATED

The Goodrich RAPTOR reconnaissance pod on port shoulder pylon of an RAF Tornado GR.4. Note the large square optical panel is closed by a sliding cover (RAF/Crown copyright) ***NEW***/0547529

An RAF Tornado GR.4 with the Goodrich RAPTOR reconnaissance pod on its port shoulder pylon. Note the large square optical panel is open (RAF/Crown copyright) ***NEW***/0547528

Graflex ultra-compact ×13 zoom lens colour camera

Type
Reconnaissance camera.

Description
Graflex produce a compact zoom lens colour camera, suitable for a variety of military applications where small size and high reliability is required, including UAVs and helicopter turrets. The camera is of rugged construction and it produces high-resolution imagery. There is a choice of lenses, giving fields of view of 1.87 to 24° and 1.39 to 18°.

Operational status
Available.

Specifications
Zoom lens
Fields of view: 1.87 to 24° or 1.39 to 18°
Free aperture: 40 mm
Focus range: 15 ft to ¥
Resolution: 230 lp/mm (112 mm lens) or 190 lp/mm (150 mm lens)
Dimensions (l × h × w):
97 × 68 × 55 mm (112 mm lens)
98.5 × 68 × 55 mm (150 mm lens)

Colour camera
Type: ¼ in transfer CCD NTSC (PAL available)
Image area: 3.65 × 2.75 mm
Resolution: 480 TV lines
Minimum illuminance: 2.5 lx
Power supply: 12 V DC ± 1 V

Contractor
Graflex Inc.

VERIFIED

Intertechnique Camelia infra-red linescan camera

Type
Reconnaissance system.

Description
The Camelia infra-red linescan camera from Intertechnique has been designed to provide drones, light aircraft or helicopters with an observation and surveillance capability.

Camelia consists of a linear infra-red detector with thermoelectric cooling, a single-axis scanning system with interface and image display electronics. It features a large field of view (up to 60°) coupled with a spatial resolution of 1 mrad (typical) in the infra-red band II (3.5 μm). Image resolutoin is 2,048 dots per line. The image is coded so that a standard analogue video transmitter can be used. The transmitted coded video signal is then decoded in the ground station by means of a simple PC board. The coding/decoding process is achieved without data compression which means full integrity of the transmitted image compared to the original onboard image. Data can then be displayed on a high-resolution display system and, if needed, stored on the ground station disk system.

The Camelia IR linescan camera 0022255

Camelia imagery from 1,700 ft altitude 0022256

Operational status
In production, deliveries in progress.

Specifications
Weight: 5.2 kg
Dimensions: 222 × 160 × 170 mm
Roll compensation: ±15°
Field of view: 60°
Line resolution: 2,048 dots
Infrared detector: 3.5 μm
Power supply: 20-30 V DC
Power consumption: 60 W

Contractor
Intertechnique, Optronics & Image Processing Department.

VERIFIED

Lockheed Martin Special Avionics Mission Strap-On-Now (SAMSON) pod

Type
Reconnaissance pod.

Description
The Lockheed-Martin Special Avionics Mission Strap-On-Now (SAMSON) is a modular pod-mounted surveillance system designed to be installed on the C-130 Hercules transport to provide enhanced capability without degrading the aircraft's primary transport role.

The system consists of a wing-mounted pod housing a variety of sensors and a palletised control, display and data processing system. Control and power interface is provided by a set of quickly installed electrical cables.

The pod, converted from an external fuel tank, can be installed or removed within one working shift by a team of four.

For reconnaissance, the pod may be supplemented by a terrain-following radar while for border patrol the radar may be a Moving Target Indicator (MTI) system supplemented by a photographic camera and a low-light level television. A search and rescue version would have a sea-search radar, a navigation system, an infra-red linescanner and a searchlight.

Lockheed Martin Special Avionics Mission Strap-On-Now (SAMSON) pod

The SAMSON pod on a Royal Norwegian Air Force C-130 0058723

The SAMSON pod for the Open Skies Treaty carries three Recon-Optical KS-87B framing film cameras, one Recon-Optical KS-116A panoramic film camera and two video cameras. Space has been reserved for future additions of a synthetic aperture radar and an infra-red linescanner.

Operational status
The first SAMSON Open Skies Pod system was delivered to the Belgian Air Force in October 1996. It has been used for training missions by the pod group consisting of Belgium, Canada, France, Greece, Italy, Luxemburg, Netherlands, Norway, Portugal and Spain and is probably now in service. Norway has used the pod to overfly Russia.

Specifications
Dimensions (l × d): 8.7 × 1.2 m

Contractor
Lockheed Martin Aeronautics Company.

VERIFIED

Raytheon ARL airborne minefield detection and reconnaissance system

Type
Reconnaissance system.

Description
The ARL (Air Reconnaissance Low) airborne minefield detection and reconnaissance system can be mounted on manned and unmanned, fixed- and rotary-wing platforms for both day and night missions.

The sensor is a high-resolution infra-red linescanner employing an eight-channel SPRITE detector and split-cycle Stirling linear resonant cooler with an MTBF stated to be greater than 2,500 hours. The optical system includes an oblique scanner which is not dependent upon critical alignment requirements.

The linescanner is coupled to a three-axis stabilisation system. The pitch and yaw axes are stabilised electromechanically using low-drift gyroscopes and high-torque DC motors. The roll axis is stabilised electronically by employing processing techniques that compensate for both roll rate and roll position effects. In addition, scanline gap and overlap are removed by a processor-controlled electro-optical technique. All the stabilisation components perform together to achieve a stabilisation down to the subpixel level.

The stabilisation system also provides for independent pointing of the linescanner in any of the three axes.

The detector outputs of the scanner are connected to a scene-dependent video processing system. The video processing electronics include high-frequency peaking, Automatic Low-Frequency Gain Limiting (ALFGL) automatic gain and level compensation and dynamic gain matching; all under processor control. The system utilises the actual digitised imagery, together with an optimising processor algorithm to dynamically adjust the system parameters.

The display/control system allows the operator complete control of the sensor system while displaying the imagery on a waterfall display subsystem. The display/control system can be interfaced either directly to the sensor system (manned systems) or through a digital datalink (unmanned systems). Some of the functions that can be controlled by the operator include FOV of the sensor, pointing position of the sensor in a roll, pitch and yaw and focus. Sensor and mission status, along with the imagery can be displayed on the control subsystem and recorded on a variety of digital storing devices.

The IR Reconnaissance System is modular and has BIT and integrated diagnostics.

Operational status

Initial delivery was made in April 1987. Various iterations have been developed and a total of seven systems was scheduled for delivery.

Specifications

Sensor
Dimensions: 18 × 22 × 16 in
Weight: 64.9 lb
Power: 28 V DC; 600 W nominal
Scan mechanism: oblique scanner
Detector: 8-parallel CMT SPRITE
Spectral band: 8-12 μm
Scan speed: 5,144 rpm
f/no: 3.26
Altitude range: 300-10,000 ft AGL
Pointing range
Roll: ±40°
Pitch: ±30°
Yaw: ±10°
Field of view: 0-120°
Focus (manual): 300 ft to ∞
Display/Control
Dimensions: 19 × 12 × 12 in
Weight: 35 lb
Power: 5 V DC ± 12 V DC 170 W nominal
Display zoom: ×16 to ×1
Display mode waterfall/freeze
Video output: 2 channels, RS-170 video
Display format: 512 × 480 pixels each channel

Contractor

Raytheon Systems Company.

UPDATED

Raytheon ILR 100 Imaging Laser Radar

Type

Reconnaissance system.

Description

The 3-D ILR 100 imaging laser radar is an active gallium arsenide laser linescan sensing system capable of simultaneously detecting contrast variations in the fore-and-aft (X) and left/right (Y) dimensions of the ground, as well as instantaneous range to the X-Y pixel. The system can be used in both manned and unmanned high-performance aircraft for military reconnaissance purposes. The system provides outputs for real-time viewing on a CRT display, suitable hard-copy output and transmission to the ground by datalink.

Operational status

Flight testing completed in 1993.

Specifications

Detector: silicon avalanche photodiode
Altitude: 400-1,500 ft
Coverage: 1,036 m at 300 m (4,000 ft at 1,000 ft)

Contractor

Raytheon Systems Company.

VERIFIED

Raytheon RS-700 series infra-red linescanner

Type

Infra-red linescanner.

Description

The Raytheon RS-700 airborne infra-red linescanner operates in the 8 to 14 μm band and uses a common module closed-cycle cooling subsystem. The optical system focuses radiation on to the detectors to produce video electrical signals that correspond with the picture formed by the radiation pattern scanned. The system converts the video signals to visible wavelengths by light-emitting diodes for recording on film.

The RS-700 comprises three subassemblies mechanically mounted to form a single assembly for aircraft installation. Features of the equipment include: manual or automatic gain selection, manual or automatic level control, video compression of unusually hot or cold objects, continuous scanning over the whole velocity/height range, adjustable hot- spot marker, event marker and built-in test equipment.

Data annotation is an optional feature of the system. Basic numerical data annotation in the film margin is used for mission identification. With this annotation, mission date, time, heading and other information can be determined. MIL-STD-782 C code matrix data annotation is also available.

Another option is roll stabilisation. This option allows the RS-700 to be attitude stabilised up to angles of ±30° from the nadir. Oblique slew, when added to the roll-stabilised system, permits offsets of up to +15° from the nadir. This allows hostile targets to be overflown while providing profile data from targets of special interest. The RS-700 is also available mounted in a reconnaissance pod.

Operational status

In production. Some 180 RS-700 series linescan systems have been supplied to the air forces of Denmark, Germany, Italy, Malaysia, Saudi Arabia, Singapore, Sweden, Switzerland and the USA.

Specifications

Spectral band: 8-14 μm
Detector: CMT
Cooling: closed-cycle Stirling 77 K
Field of view: 120°
Scan mirror facets: 4
Optical aperture: 38.4 cm^2
Detector type: CMT

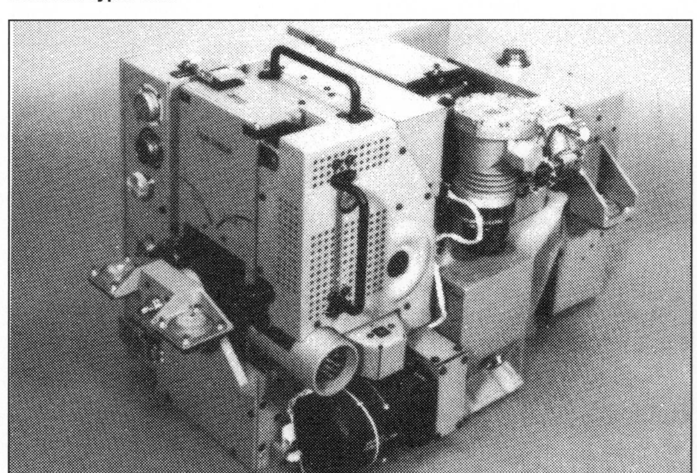

Raytheon RS-700 infra-red linescan equipment for the Panavia Tornado

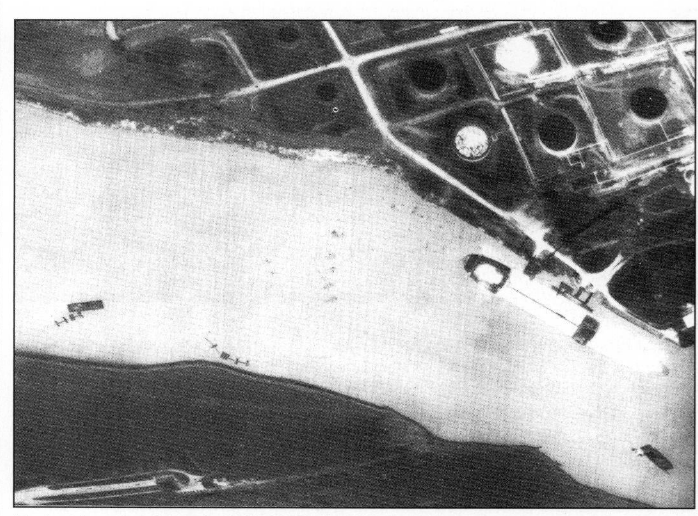

Typical imaging from the RS-700 Series infra-red linescanner shows hot details as white in the area surrounding a port

Recording light source: GaAs phosphor diodes
Film width: 70 mm
Film capacity: 46 m, 70 m
Velocity/height range: 0.2-5
Thermal resolution: 0.2°C
Spatial resolution: 0.5-1.5 mrad
Weight
  Without roll stabilisation: 32 kg
  With roll stabilisation: 42 kg

Contractor
Raytheon Systems Company.

VERIFIED

Recon/Optical CA-260® digital framing camera

Type
Reconnaissance camera.

Description
The CA-260® advanced development Electro-Optical (E-O) digital framing camera is designed specifically to provide near-photographic quality images, while enhancing the survivability of the tactical reconnaissance platform at low- to medium-altitudes. Electro-optic framing technology is stated to reduce the time required to cover a target area when compared to linescan or pushbroom sensors. Patented on-chip graded image motion compensation improves resolution.

The CA-260® is configured for external podded or internal mounting on a wide variety of reconnaissance platforms. The mounting configuration and physical envelope is identical to the Recon/Optical KS-87 film camera which allows virtual drop-in replacement. At 2.5 frames/s, the CA-260® digital output directly connects to an airborne digital video cassette recorder. Camera control is selectable through an RS-232, RS-422 or optional Mil-STD-1553 interface.

The CA-260® digital framing camera can provide stereo coverage of targets at 56 per cent overlap. When used with a gimbal or turret, the sensor's multi-aspect imaging capability allows viewing from multiple angles.

Operational status
CA-260®/4 and CA-260®/25 are in production. CA-260®/25 production deliveries for the USAF TARS programme are completed.

Specifications
Lens
37.5 mm (1.5 in) fl, f/4.5
75.0 mm (3 in) fl, f/4.5
150.0 mm (6 in) fl, f/4.0
300.0 mm (12 in) fl, f/4.0
1,676.4 mm (66 in) fl, f/5.6 and f/8

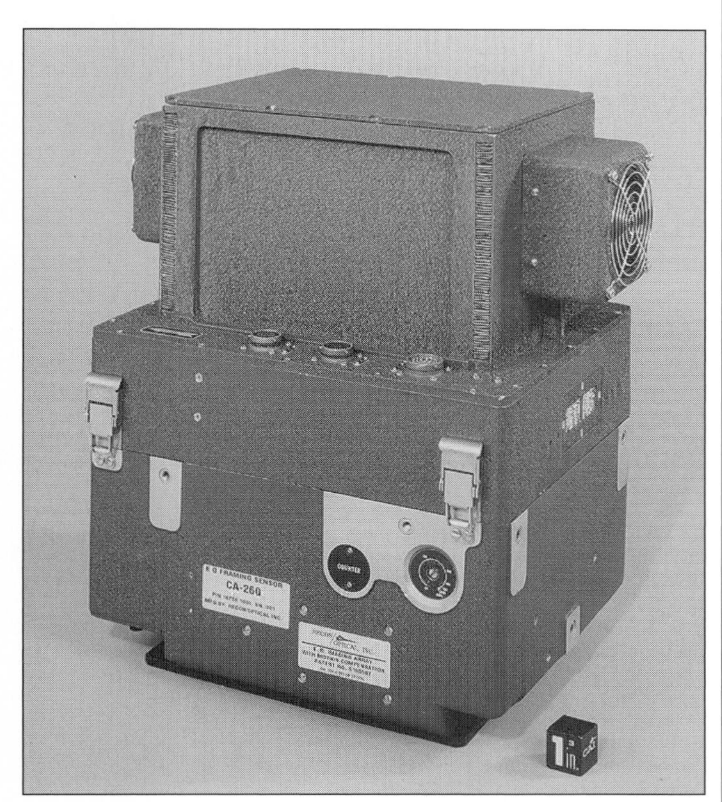

CA-260 digital framing camera

An image from the CA-260® digital framing camera taken 11 minutes after sunset from a TARPS-equipped F-14 0005562

CCD array size
  CA-260/4: 2,048 × 2,048 pixels
  CA-260/25: 5,040 × 5,040 pixels
Pixel size: 0.12 × 0.12 mm
Frame rate: 2.5 frames/s
Weight
  Camera: 33.6 kg (1.5 in lens)
  Power supply: 8 kg
Dimensions
  Camera (h × w × d): "A" × 262.16 × 41.3 mm. "A" varies with lens
  Power supply (h × w × d): 134.6 × 274.3 × 464.8 mm

Contractor
Recon/Optical Inc, CAI Division.

Recon/Optical CA-261® digital step framing camera

Type
Reconnaissance system.

Description
The CA-261® electro-optical (digital) step framing camera provides photographic quality images while enhancing the survivability of the tactical reconnaissance platform at medium altitudes. It is configured for external pod or internal aircraft mounting on a variety of reconnaissance platforms.

CA-261® combines the proven performance of the CA-260® 25 Mpixel digital framing camera with the stepping capability of a proven two axis stabilised step head with derotation prism. The combination produces a system that captures a series of 25 Mpixel images through a 12 in focal length lens in the cross-track direction. These images allow for wide area coverage of up to 180° in an E-O framing format.

The captured imagery can be displayed in mosaic form to provide a bird's eye view of the entire area of interest. The display retains the capability to manipulate individual images in the same manner as the current CA-260®. Derotation optics are employed to eliminate image rotation. The high resolution of the CA-260® is

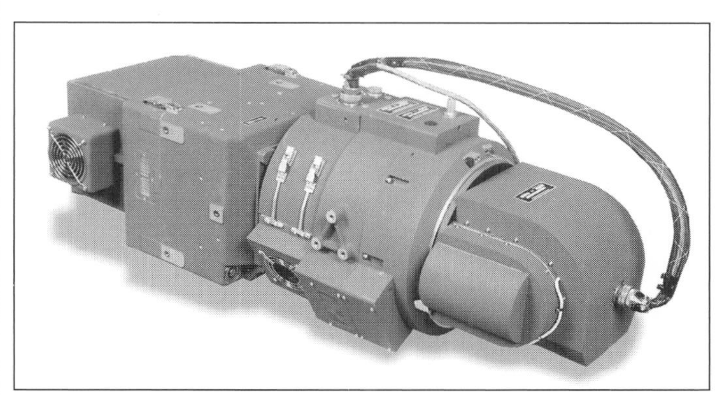

The CA-261® E-O step framing camera 0125120

CA-261® step framing imagery using 7:1 data compression 0023180

CA-261® flight test imagery (CA261/25®) lens 12 in f1, altitude 20,425 ft, slant range 44,990 ft 0023181

maintained by incorporating Recon-Optical developed stabilisation electronics and software. This combination yields residual rates <0.001 rad/s even at disturbance input rates of 10°/s.

Advanced digital step framing technology reduces the amount of time required to cover a target area compared to conventional E-O linescan sensors. Recon/Optical employed wafer-scale processing to develop the 25 Mpixel array CCDs needed for wide field of view image from the CA-260®. On-chip motion compensation eliminates the image blur normally associated with a wide field of view framing camera.

Operational status
In production and in service with several countries.

Specifications
Dimensions: 35.6 × 40.6 × 96.58 cm
Weight: 76.4 kg
Lens: 30.48 cm f/1, f/6 or 40.57 cm f1, f/8
Field of view: 11.3 × 11.3° each frame (30.48 cm lens); 7.6 × 7.6° (40.57 cm lens)
Field of regard: 180° in cross-track × 11.3° in line of flight
Frame rate: 2.5 frames/s
Angular resolution: 39.4 μrad/pixel (30.48 cm lens); 26.2 μrad/pixel 40.57 mm lens
Max V/R: >0.44 rad/s 30.48 cm lens; > 0.30 rad/s 40.57 cm lens
Example of coverage: 4,000 n miles^2/h at 20,000 ft, 300 km centered at Nadir, resolution >IIRS 5
Power supply: 110 W, 28 V DC; 600 VA or 115 V AC, 400 Hz, 3 phase

Contractor
Recon/Optical Inc, CAI Division.

Recon/Optical CA-265™ infra-red digital framing camera

Type
Reconnaissanse camera.

Description
The CA-265 infra-red digital framing camera is designed to provide state-of-the-art performance for low- to medium-altitude missions in high-speed tactical aircraft. The camera uses Mid-Wave IR focal plane array technology. A 1968 × 1968 PtSi detector is used. Patented on-chip image motion compensation eliminates image blur. Near realtime digital output is provided. The camera may be fitted in a vertical, forward-oblique or side-oblique mounting. The manufacturer states that the CA-265 IR is the world's largest framing IR camera.

Specifications
Dimensions (l × w × d):
Image sensing unit: 12.3 × 12.8 × 14.5 in
Video data processing unit: 22 × 10.2 × 11.5 in
Power supply: 13.6 × 11.9 × 3.6 in
Spectral band: 3 to 5 μm
Detector: 1968 x 1968 PtSi
Lens focal length/f number: 12 in (304.8 mm), f/2, option 6 in (152.4 mm) f/2
Field of view/frame: 11.6 × 11.6 (12 in f/2), 21.9 × 21.9(6 in f/2)
Exposure time: variable
Aircraft speed: 100 – 580 kt
Operating altitude: 500 to 50,000 ft
Control interface: RS-232
Video output: RS-170
Operating temperature range: 0 to 46°C

Contractor
Recon/Optical Inc, CAI Division.

Recon/Optical CA-265 IR framing camera 0089914

Recon/Optical CA-270 dual-band digital framing camera

Type
Reconnaissance camera.

Description
The Recon/Optical Inc CA-270 dual-spectral band (visible/infra-red) digital step framing camera was designed to meet the needs of the next generation of tactical reconnaissance. This capability combines high performance in both visible spectrum digital imaging and infra-red imaging. The CA-270 is intended to provide a capability for both day and night reconnaissance.

The CA-270 builds upon ROI's experience in visible-spectrum framing cameras by adding an infra-red imaging channel to provide 24 hour operation. Both channels share a common aperture and have identical fields of view. Step framing operation provides wide across-track coverage simultaneously in both spectral bands.

Using a large format, mid-wave infra-red staring array, the IR channel provides the benefits of high-resolution electro-optical framing operation in the 3 to 5 μm spectrum. ROI states that in both spectra, their patented, graded, on-focal-plane image motion compensation ensures high-performance imaging.

Specifications
Dimensions (L × D): 109.3 × 44.4 mm (43 × 17.5 in)
Spectral band: Visible/near IR, 515-900 nm
Infra-red, 3 to 5 μm

Recon/Optical CA-270 dual-band camera 0125118

Lens focal Length/f number
Visible spectrum: 12.0 in (304.8 mm) fl, f/6 or 18 in (457.2 mm), f/8
Infra-red spectrum: 12.0 in (304.8 mm), f/3.5 or 18 in (457.2 mm), f/5

Detector
Visible spectrum: 25 Mpixel silicon CCD, square format, 10 μm pixel pitch
Infra-red spectrum: 3.8 Mpixel array, MWIR, square format, 25 μm pixel pitch
Field of view/frame: visible, 9.5 × 9.5°, infra-red 6.3 × 6.3°
Frame rates: visible up to 4 frames/s, infra-red 2.5 frames/s
Operating Modes: visible alone, infra-red alone and VIS-IR simultaneously
Forward motion compensation: both Spectra, on chip, graded
Stabilization: two-axis stabilised, roll and azimuth. Roll axis <1.0 mrad/s residual against a 10°/s roll

Contractor
Recon/Optical, Inc, CAI Division.

Recon/Optical CA-295 dual-band digital framing camera

Type
Reconnaissance camera.

Description
Recon/Optical, Inc's CA-295™ is a dual spectral band (visible and infra-red), long-range oblique, step framing, digital reconnaissance camera. It captures medium- to high-altitude, high-resolution digital imagery both day and night. The CA-295 provides simultaneous dual-band coverage, or coverage in either band individually and stereo imagery for enhanced photo interpretation.

The camera's actively stabilized optical system yields excellent resolution in a variety of airborne applications. The CA-295's common aperture allows both the visible and IR channels to use the same primary optical element, allowing precise registration between the visible and IR imagery. A choice of relay optics in the visible channel provides a selection of camera focal lengths, giving the CA-295 the ability to meet a variety of mission requirements with a standard system design.

The CA-295 can provide precision pointing and target location using an integrated inertial navigation system/global positioning system (INS/GPS). This capability permits highly accurate georegistration of the images and generates three-dimensional data for moving target indication and precision location.

Operational status
Flight tested and available.

Specifications
Dimensions (L × W × H): 1,244.6 × 508 × 508 mm
Weight: 181.4 kg

	Visible	Infra-red
Spectral band	500-900 μm	3.0-5.0 μm
Focal plane	25 Mpixel	4 Mpixel
Detector format	5,040 × 5,040	2,016 × 2,016
Detector pitch	10 μm	25 μm
Frame rate (nominal)	4.0 frames/s	4.0 frames/s
Angular resolution	4.7 μrad/pixel	19.7 μrad/pixel
Lenses	1,270 mm (50 in) fl, f/4.0 1,829 mm (72 in) fl, f/5.8 2,134 mm (84 in) f1, f6.7	1,270 mm (50 in), f/4.0
Field of view (per frame)	2.27 × 2.27° (50 in fl) 1.58 × 1.58° (72 in f1) 1.35 × 1.35° (84 in f1)	2.27 × 2.27° (50 in fl)

Field of regard: Horizon to horizon (window dependent) both bands
Operational modes: visible only, IR only, simultaneous visible/IR
Coverage modes: wide area, stereo area, multi-aspect spot; each band or independent operation.

Recon/Optical CA-295 dual-band framing camera 0125117

Output compatibility: ampex DCRsi DTR, MIL-STD-2179 DIR, SSR, fibre channel, NITF 2.1, STANAG 7023, STANAG 7085, others available.

Contractor
Recon/Optical, Inc, CAI Division.

Recon/Optical CA-880 pod-mounted long-range oblique photographic system

Type
Reconnaissance system.

Description
The CA-880 performs tactical and strategic LOng-Range Oblique Photographic (LOROP) reconnaissance missions. The oblique-looking KS-146B camera is installed in a modified fuel tank which has been certified for centreline carriage on the RF-4, F-4E, Mirage III, and Mirage V. The system also includes cockpit control and status panels, left and right oblique sights for camera pointing in multiple modes and a master power distribution unit.

Key system features include the 1,676 mm focal length, f/5.6 lens, both passive and active stabilisation, forward motion compensation, automatic exposure control, autocollimation for focus optimisation, manual and automatic pointing control, pod and camera thermal control systems and built-in test functions. Either Electro-Optical (E-O) framing or E-O scanning Image Sensor Units (ISUs) can be used. Aircraft annotation data are provided via the MIL-STD-1553D databus with camera parameters electronically integrated before to imaging.

Operational status
In production and service.

Specifications
Fields of view: 3.9° in line of flight; 3.9, 7.5, 11, 14.5 or 21.6° across line of flight
Film length: 305 m
Cycle rate: 0.45 cycle/s max
Exposure times: 1/30-1/1,500 s
Overlap: 12 or 56%
Weight: 666 kg
Power: >3 kVA of 115 V AC, 400 Hz, 3 phase; 300 W of 28 V DC

Contractor
Recon/Optical Inc, CAI Division.

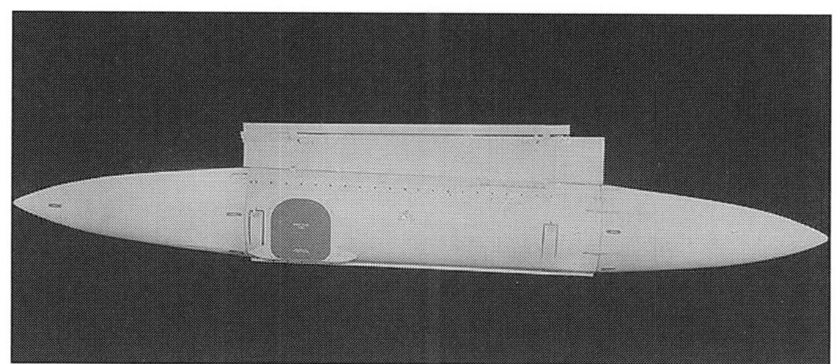

Recon/Optical Inc CA-880 Long-Range Oblique Photographic system (LOROP)

Recon/Optical CA-890 tactical reconnaissance pod

Type

Reconnaissance pod.

Description

The CA-890 tactical reconnaissance pod performs reconnaissance missions at altitudes of 60 to 9,144 m and employs one to three sensors singly or simultaneously. The nose section houses a KS-153A Tri-lens camera on a forward-oblique or vertical in-flight rotatable mount. The centre section contains a KA-95B panoramic camera configured for 40, 90 and 140° vertical scans and 40° left and right oblique scans; all are in-flight selectable. The tail section sensor is an infra-red linescanner, configured for in-flight selectable vertical fields of view.

The sensors are controlled by the CAI Reconnaissance Interface Unit (RIU). The RIU is microprocessor based and houses three identical sensor interface CCAs, data annotation functions, the MIL-STD-1553 interface bus function and power supply functions.

The pod is a modified 1,400 litre (370 gallon) fuel tank that has been flight certified for centreline carriage on the F-4E. The environmental control system is integral to the pod and controls the nose and centre section. The tail section is open to atmosphere for the IRLS receiver. The system also includes cockpit control and status panels and a master power distribution unit.

Operational status

In production and operational.

Specifications

Weight: 544 kg
Power: <6 k VA, 115 V AC, 400 Hz 3 phase; 300 W, 28 V DC

Contractor

Recon/Optical Inc, CAI Division.

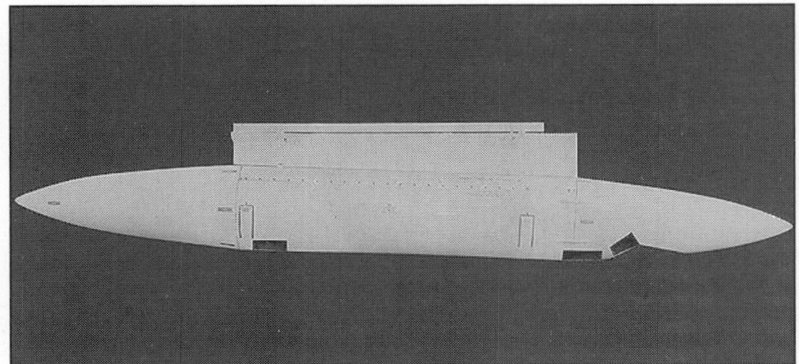

The Recon/Optical Inc CA-890 Tactical Reconnaissance Pod

KS-127B imagery from 26,000 ft, slant range 15.5 n miles 0023184

A ×21 enlargement of detail from KS-127-B imagery 0023185

Recon/Optical KS-127B long-range oblique film camera

Type

Reconnaissance camera.

Description

The KS-127B is an optical LOng-Range Oblique Photographic (LOROP) camera designed to fit into the nose sensor compartment of the RF-4 Phantom II aircraft. The camera provides tactical and strategic reconnaissance from safe, long-range standoff distances. Typical missions are 10,000 m altitude, with a range to the target of 55 km or greater. The camera has a 1,676.4 mm focal length, f/8.0 lens and contains active and passive stabilisation, autofocus and thermal stabilisation.

Recon/Optical Inc KS-127B LOng-Range Oblique Photographic (LOROP) camera 0018497

The KS-127B is a stepping frame camera that has two modes of operation: automatic and manual. The manual mode provides 3.9 × 3.9° coverage in a single-step mode of operation. The automatic mode offers two-step vertical, or three- or four-step oblique coverage. The two-step mode provides 3.9 × 7.5° of coverage; the three-step mode provides 3.9 × 11° of coverage; and the four-step mode provides 3.9 × 14.5° of coverage. Manual mode is controlled by the sights in the rear cockpit that are used to point the camera at selected targets.

Operational status

In production and service.

Specifications

Weight: 272 kg
Film length: 305 m
Lens: 1,676.4 mm, f/8.0
Power: 28 V DC, 280 W; 115/208 V RMS, 400 Hz, 700 VA
Shutter Speed: 1/300-1/1,500 s
Overlap: 12 or 56%

Contractor

Recon/Optical Inc, CAI Division.

Recon/Optical KS-146A long-range oblique film camera

Type

Reconnaissance camera.

Description

The KS-146A is an optical LOng-Range Oblique Photographic (LOROP) camera designed for both tactical and strategic reconnaissance from a safe, long-range, standoff distance.

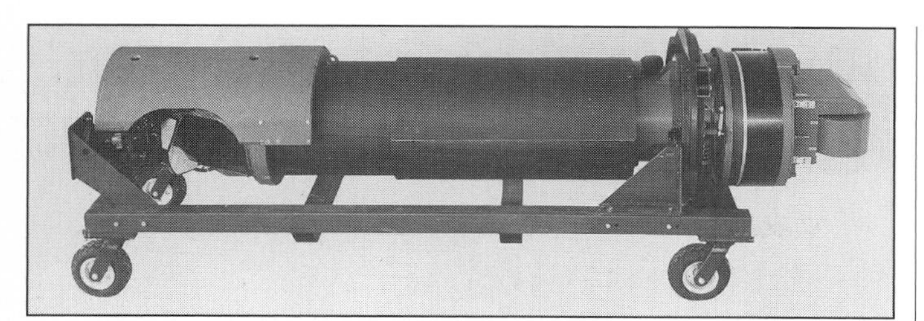

Recon/Optical Inc KS-146A Long-Range Oblique Photographic (LOROP) camera

Key features include a 1,676 mm focal length, f/5.6 lens, a two-axis, gyrostabilised scan head, a passive isolation mount, closed-loop autocollimation and a self-contained thermal system. The system is divided into two subsystems: the camera assembly and the electronics unit.

The camera assembly contains the scan head, lens, autofocus drive, thermal system, roll drive, shutter, optical filter and film magazine. The electronics unit contains power supplies, microprocessor-based control electronics and, servo controls for camera operation including built-in test, cycle rate, exposure, focus, stabilisation, roll drive and thermal control.

The scan head is a two-axis gimbal providing scan mirror mounting. It provides pointing, stepping and active stabilisation for roll and azimuth axes. The optical system uses a nearly diffraction limited lens yielding high modulation even at low contrast levels.

Operational status

In production and in service with several countries.

Specifications

Weight
System: 385 kg
Camera: 317 kg
Fields of view: 3.9-14.5°
Film length: 305 m
Shutter speed: 1/30-1/1,500 s
Power: 300 W, 28 V DC, 1,000 VA, 115 V AC, 400 Hz

Contractor

Recon/Optical Inc, CAI Division.

Recon/Optical KS-147A long-range oblique film camera

Type

Reconnaissance camera.

Description

The KS-147 is a LOng-Range Oblique Photographic (LOROP) camera designed to provide tactical and strategic reconnaissance from long ranges. It includes a 1,676 mm focal length, f/5.6 lens, a two-axis, gyrostabilised scan head, a passive isolation mount, closed-loop autocollimation autofocus and a self-contained thermal system.

To meet space and weight constraints, the camera is subdivided into the camera assembly and the electronics assembly. The former contains the scan head, lens, fold mirrors, autofocus drive thermal system, roll drive, shutter, optical filter and film magazine. The latter has the power supply, control electronics and servo-controls. In the RF-5E, the camera assembly is on a removable pallet which becomes part of the aircraft's nose structure.

Operational status

In production and in service in the RF-5E.

Specifications

Weight
System: 270 kg
Camera: 245 kg

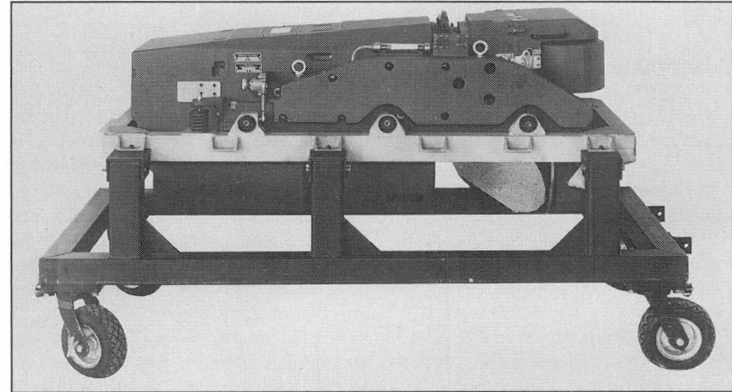

Recon/Optical Inc KS-147A Long-Range Oblique Photographic (LOROP) camera

Film length: 305 m
Angular coverage: 3.9 × 3.9 × 7.4°, 3.9 × 10.9 × 14.4°
Depression angle: 4-39°
Cycle rate: 0.375-1.5 cycles/s
Overlap: 12 or 56%
Power requirements: 28 V DC, 300 W; 115/208 V AC, 400 Hz, 700 VA

Contractor

Recon/Optical Inc, CAI Division.

Recon/Optical KS-157A long-range oblique film camera

Type

Reconnaissance camera.

Description

The KS-157A is a compact, folded LOng-Range Oblique Photographic (LOROP) camera designed for installation in business jet aircraft. It performs both tactical and strategic reconnaissance from safe, long-range, standoff distances of 55.5 km or beyond. The camera uses a 1,676 mm focal length, f/5.6 optical system to provide high-resolution frame imagery suitable for the detection and identification of military targets. The system offers in-flight selection of photographing either the right or left oblique positions and can record targets at depression angles from the horizon to 30° below.

The KS-157A features high-resolution performance even under low-light conditions. Controls in the cabin select variable depression angles and frame coverages which feature a wide range of available altitudes, velocities and standoff distances for mission planning. The controls also provide for repeated frames on a specific target.

The stabilisation system uses gas-bearing gyroscopes, specially developed hybrid electronic circuitry and a custom-designed passive isolation system.

Operational status

In production and in service with several countries.

Specifications

Weight
System: 385 kg
Camera: 317 kg
Fields of view: 3.9-14.5°
Film length: 305 m
Shutter speed: 1/30-1/1,500 s
Power: 300 W, 28 V DC, 1,000 VA, 115 V AC, 400 Hz

Contractor

Recon/Optical Inc, CAI Division.

Recon/Optical Inc KS-157A Long-Range Oblique Photographic (LOROP) camera

SAGEM Cyclope 2000 infra-red linescan sensor

Type

Infra-red linescanner.

Description

The Cyclope 2000 is a reconnaissance system designed for use in small, manned reconnaissance aircraft and UAVs. It features an infra-red linescanner with high performance in thermal and spatial resolution, V/H range, data processing/recording and display.

The basic system consists of an infra-red linescanner sensor which may be expanded by the addition of electronic cards to meet various mission and airframe

The SAGEM Cyclope 2000 infra-red linescanner

requirements. These processing elements are linked to the sensor to provide, for example, a bus interface for navigation updates. The system can use either closed-cycle Stirling or open- cycle Joule-Thomson cooling. The Cyclope 2000 is designed to store images on videotape but it may also be used purely for monitoring.

A real-time air-to-ground datalink may be used with the system. The basic modular configuration can be adapted to various aircraft. Improved versions are offered with a different spectral bandwidth, multispectral detection or stereoscopy.

Operational status

Development is complete and Cyclope 2000 is in operational service in the Crecerelle UAV for the French Army French Customs Service Rheims F-406 Caravan II (for the POLMAR mission) and foreign coastguards.

Specifications

Spectral band: 8-12 μm
Sensitivity: 0.1 or 0.2°C
Roll angle compensation: ±15°
Field of view: 90-120°
Angular resolution: 1 mrad
Dimensions: 170 × 170 × 200 mm
Weight: <5 kg
Power supply: 115 V AC, 400 Hz, 3 phase, 70 W or 28 V DC, 70 W

Contractor

SAGEM SA, Aerospace and Defence Division.

UPDATED

SAGEM Super Cyclope

Type

Reconnaissance system.

Description

Super Cyclope is a successor to the Cyclope system. In addition to the detector and a photographic unit recording on 30 or 70 mm film, the system consists of an electronic line-replaceable unit.

The video and synchronisation signals are recorded by an onboard photographic unit consisting of a high-resolution cathode ray tube. The data can be either recorded on tape, or transmitted in real time by a radio transmitter to a ground station equipped with a photographic recorder. For the latter, the data passes through a processing unit to the 126 mm high-resolution photographic unit. These ground stations have been designed to be carried in an air conditioned shelter on a truck.

Super Cyclope can operate in three ways. In the first, the analysis head and photographic recorders are placed on the carrier and thermal images are directly recorded on film; when the aircraft returns the film is processed. In the second, the analysis head and its recorders are linked to an onboard radio transmitter; the data are transmitted during the mission and when the aircraft returns, are compared with the data recorded by the aircraft system. Thirdly, the analysis head and a video recorder are installed and upon the aircraft's return the tape is processed.

Operational status

Super Cyclope SCM 2450 has been installed in the Mirage IVP, Mirage F1-CR and Mirage 2000. It is in operational service with the French Air Force and, since 1976, with foreign air forces.

The SAGEM Super Cyclope

Specifications

Spectral band: 8-12 μm
Detector: CMT
Cooling: either closed-cycle Stirling or liquid nitrogen
Angular resolution: ±1.5 mrad
Sweep angle: ±60°
Roll compensation: ±20°
Film speed: 5.4-87 mm/s
Weight: 72 kg
Power supply: 208 V, 3 phase 400 Hz, and 27 V DC

Contractor

SAGEM SA, Aerospace and Defence Division.

UPDATED

Terma Industries Modular Reconnaissance Pod (MRP)

Type

Reconnaissance pod.

Development

The Modular Reconnaissance Pod (MRP) is a family of reconnaissance pods for fighter aircraft including the F-16, Tornado, F/A-18, F-15 and others. It accommodates wet film systems, infra-red line-scanners and a range of electro-optical sensors. Terma Industries Grenaa (formerly Per Udsen) started a development and test programme in 1994 under contract to the Royal Danish Air Force. Flight tests began in the first quarter of 1996 and major certification was completed in 1996. The MRP has been adopted for use with the Belgian and Danish air forces' F-16s, which have selected Vinten 70 mm framing cameras and controls configured to allow the pods to be subsequently upgraded with the Type 8010 EO day camera.

In March 1996, Terma signed a memorandum of understanding with Lockheed Martin Tactical Aircraft Systems to collaborate on systems for the US Air National Guard Theater Airborne Reconnaissance System (TARS), for which both companies have offered solutions.

Description

The MRP incorporates a flexible, modular-mounting system which will accommodate most current and future sensor suites. A rail system at the top of the upper part of the pod provides interfaces for mounting of equipment. The pod is designed in three sections, a platform-specific section, a common pod section and a sensor-specific section. The common section is the pod body which includes the structural strongback where the rail is located and to where all loads are transferred. The sensor-specific section includes the set of fittings which holds all the sensors, recorders and various LRUs and also the access doors and windows. The platform- specific section is the attachment fitting which provides the mechanical and electrical interface to the aircraft fuselage or pylon rack.

The MRP can accommodate different sensor suites including existing wet film systems with large magazines, infra-red linescanning systems, low- and medium-altitude electro-optic sensor suites such as the Recon/Optical CA-260/4(25) and the BAE Systems (formerly Lockheed Martin Fairchild Systems) ATARS and large

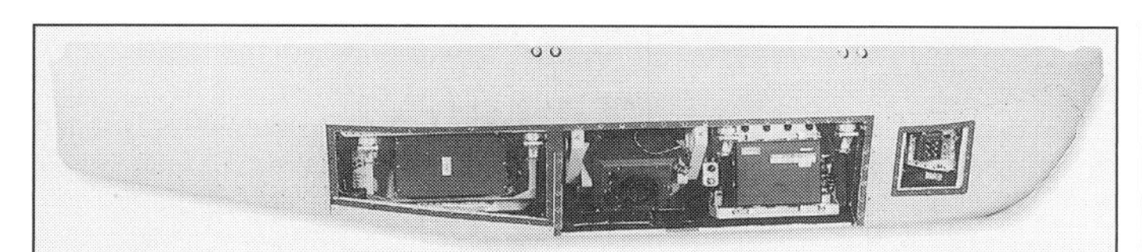

The Terma Modular Reconnaissance Pod can accommodate a range of sensor suites

The Terma Modular Reconnaissance Pod on an F-16

E-O-LOROPS sensor suites. The pod delivered to the USANG was equipped with Recon/Optical CA-260 sensors. Pods for the Danish Air Force have W Vinten Ltd wet film sensors.

The pod can be operated through various control systems including the Terma EWMS which is compatible with the F-16 MLU.

Operational status

Flight tests began in early 1996 and all certification is now complete. In production and delivered to the Belgian Air Force, Royal Netherlands Air Force, Royal Danish Air Force and the USAF Air National Guard.

Specifications

Dimensions

Overall length: 4,496 mm
Height: 762 mm
Width: 610 mm

Weight

Max weight: 544 kg
Empty weight: 227 kg
Max payload: 318 kg

Contractor

Terma Industries.

Thales Infra-Red LineScanner (IRLS)

Type

Infra-red linescanner.

Description

Thales Optronics (formerly Signaal USFA), is developing the Infra-Red LineScanner (IRLS), which will equip the next-generation tactical reconnaissance pod for the Royal Netherlands Air Force F-16s. The IRLS development is funded by Netherlands government's development committee (CoDeMa).

Operational status

In development.

Contractors

Thales Optronics B.V.

UPDATED

Thales Optronics (Vinten) Ltd Type 8010/Type 8011/Type 8012 electro-optical sensor

Type

Reconnaissance sensor.

Description

The Type 8010 pushbroom electro-optical sensor uses a Charged Coupled Device (CCD) with either a 4,096 or a 6,144 element linear array. The sensor may be fitted internally in a wide range of aircraft types, RPVs/drones and UMAs or in podded systems to give the following surveillance roles:

(a) Day low- and medium-altitude tactical reconnaissance operations.

(b) Day low- and medium-altitude surveillance operations to aid various government agencies, including police, coastguard, drug enforcement and fishery protection.

The Electro-Optical (E-O) sensor offers improvements in contrast enhancement and haze penetration. Haze or low-contrast scenes present what is effectively a spatially varying offset signal to the video output of the Focal Plane Array (FPA). The E-O sensor's internal central processing unit performs real-time monitoring of the overall FPA video signal and extracts the haze characteristics from it. A compensating signal is then fed back to the signal processing circuitry which effectively subtracts the haze content from the imagery before further signal manipulation. Consequently, high-contrast image signals result even when imaging under very low contrast conditions. The sensor also employs an anti-blooming facility.

E-O imagery can be replayed and exploited in real time onboard the aircraft on an airborne display system. Imagery is also recorded on an airborne video cassette recorder or digital tape recorder and exploited on a Ground Exploitation System (GES).

The Type 8011 is the digital variant of the Type 8010 and has been in use for several years including for Open Skies methodology trials. The Type 8012 (6,144 element linear CCD array) has been selected for inclusion in the Gripen Reconnaissance Pod.

Operational status

In production and in current use with European customers.

Specifications

CCD: 4,096 or 6,144 element linear CCD array
Pixel size: 12 μm (4,096 element); 8 μm (6,144 element)
Scan speed: 1,900 cycles/s (12 μm); 1,200 cycles/s (8 μm)
Dimensions: 226 × 180 × 181 mm
Weight: 7.8 kg plus lens
Lens options: 38; 76; 152 mm
Fields of view: 18.3° (152 mm); 35.7° (76 mm); 65.6° (38 mm)
Power: +28 V DC, 50 W

Contractor

Thales Optronics (Vinten) Ltd.

VERIFIED

Thales Optronics (Vinten) Ltd Type 8010 Electro-Optical Sensor

Thales Optronics (Vinten) Ltd Ground Imagery Exploitation System 0089907

Thales Optronics (Vinten) Ltd Type 8040 B electro-optical sensor

Type
Reconnaissance sensor.

Description
The Type 8040 B Electro-Optical (E-O) pushbroom sensor incorporates a 450 mm focal length lens, with a CCD linear Focal Plane Array (FPA) of 12,288 active pixels on a single chip. This device is manufactured using advanced techniques which allow such long arrays to be produced with excellent sensitivity and uniformity, as well as removing the alignment errors and discontinuities associated with mechanically abutted arrays.

The pixel size within the focal plane array is 8 μm; which gives 'photographic quality' resolution in the imagery. The lenses offer a corresponding high-modulation transfer function.

Imagery is recorded on a digital tape recorder or, depending upon the application, an airborne video cassette recorder.

Operational status
In production and in service with the Belgian Air Force (F-16) and UK Royal Air Force (Jaguar).

Contractor
Thales Optronics (Vinten) Ltd.

VERIFIED

The Type 8040 B electro-optic sensor 0089908

Thales Optronics (Vinten) Ltd Vicon 18 series reconnaissance pods

Type
Reconnaissance pod.

Description
The 18 in diameter Vicon 18 pod is one of an extremely versatile series of pods designed to provide supersonic attack or strike aircraft with a reconnaissance capability. Typically, the pods can be carried at M1.1 at sea level and at M1.8 at 11,000 m.

The following variations are available:

Series 300
Variants of this pod are available for tactical low-level reconnaissance, tactical standoff reconnaissance, medium-altitude reconnaissance and LOng Range Oblique Photography (LOROP). The equipment fit can be varied to suit the customer requirements. For example, the series 327A is an advanced tactical low-altitude day/night reconnaissance system with a FLIR sensor in the nose recording on to videotape and is capable of transmitting data to the ground. Also included are two Type 900 panoramic cameras and a Vicon 2000 reconnaissance management system.

The dual-role Series 329 is a LOROP and medium-altitude reconnaissance pod. It is fitted with the Type 690 126 mm LOROP framing camera in the nose and the Type 950 panoramic camera with the Vicon 2000 reconnaissance management system. The Series 329B is a purely LOROP system featuring the Type 8040 LOROP camera with electro-optical air-to-ground transmission features and the Type 900B panoramic tracking camera as well as the Vicon 2000 reconnaissance management system.

Series 400
This pod is designed to meet the mechanical interface requirements for aircraft types which cannot accept the longer Series 300 pod, the depth over the saddle being 482 to 508 mm and the length of the pod being 2,450 mm. Several have been designed to meet the requirements of the V/STOL Harrier, being hung from centreline pylons where there is limited clearance to the aircraft control surfaces and high-lift devices.

Two Harrier pods are typical of the range. The Series 401 is for tactical low-altitude day and night reconnaissance with the Thales (formerly W Vinten Ltd) Type 512 forward facing framing camera, the Thales Type 753 panoramic camera and the Type 401 infra-red linescan system, together with the Vicon 2000 reconnaissance management system. All these sensors record onto film.

The Series 403 is an advanced tactical low-altitude day and night electro-optical reconnaissance pod system based upon the Type 4000 infra-red linescan system and with Type 8010 electro-optical sensors as obliques. These sensors record onto videotape, have an air-to-ground transmission capability and are managed by a Vicon 2000 reconnaissance management system.

Series 600
These pods are designed for LOROP and tactical standoff missions as well as low-, medium- and high-altitude missions. They are designed for fitting on either aircraft centreline or underwing pylons.

The Series 601 has two variants, the Stand-Off E-O Reconnaissance Pod and the E-O Day/Night Tactical Reconnaissance pod.

The StandOff pod is equipped with two sensors, the Type 8040B day sensor and the VIGIL IRLS. The Type 8040B incorporates a state-of-the-art 8 μm 12,288 pixel CCD linear array and an Ultra High Resolution (UHR) 450 mm focal length lens which automatically focuses from 150 m to infinity. The sensor with a field of view of 12.4°, is fitted in the rotating nose of the pod giving a field of regard of 180° from horizon to horizon. The IRLS fitted in the rear of the pod provides high-resolution day and night horizon to horizon imagery at low and medium altitude.

The Day/Night Tactical pod is equipped with two Type 8010A sensors and the VIGIL IRLS. The Types 8010A are each fitted with a 12 μm 4,096 pixel CCD linear array and a range of lenses may be fitted. The sensors may be configured in a number of oblique or vertical split pair configurations which, in combination with the IR sensor, meet the requirements of a wide range of tactical reconnaissance roles.

The IRLS sensor is fitted in the rear of the pod and provides high-resolution day and night horizon to horizon reconnaissance imagery at low and medium altitude.

The Series 603 has the Type 690 126 mm format LOROP framing camera complete with a 900 mm focal length lens. The lens has a mirror attached and is fitted in a rotating nosecone. Under control of the aircraft operator, the nose can be rotated to selected depression angles either port or starboard. The Type 8040 electro-optical sensor is an alternative fit that interfaces with a video cassette recorder, a datalink transmitter and/or a Thales (Vinten) image display processor. Further upgrades could include the video-based recording system being replaced by a solid state video recorder which would allow a greater volume of data to be recorded, while image quality would be enhanced by operating at higher data rates. If used in conjunction with a datalink real-time reconnaissance will be possible.

Operational status
Many of the Series 300, 400 and 600 have been produced for installation on aircraft such as the A-4 Skyhawk, F-4 Phantom II, F-5 Freedom Fighter, F-16 Fighting Falcon, F-104 Starfighter, BAE Systems Hawk Series 60/100/200, Jaguar,

The Vicon 18 series 300 reconnaissance pod

Vicon 18 series 400 pod for the Harrier

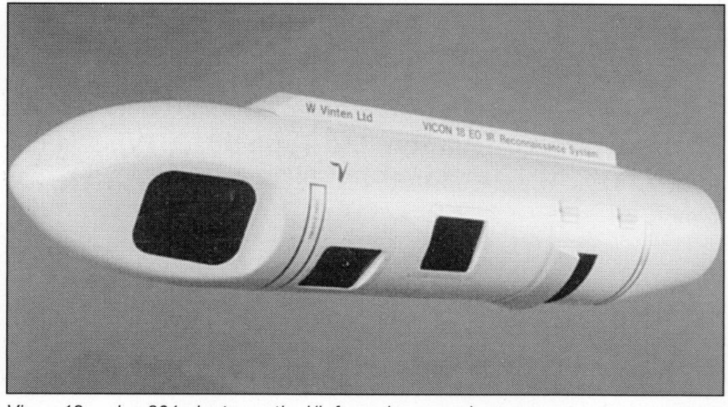

Vicon 18 series 601 electro-optical/infra-red reconnaissance system 0089909

Tornado, MiG-21 and Mirage III/5. The Series 401 is used by the Sea Harrier; the Series 403 can be used by the Harrier GR.Mk 5/7 and the AV-8B; and the 601 on the Harrier GR.Mk 5/7 as well as Tornado and Jaguar. The Vicon 18 is used by numerous air and naval forces.

Specifications

Model	Series 300	Series 400	Series 600
Length	3,420 mm	2,450 mm	2,260 mm
Diameter	457 mm	457 mm	457 mm
Depth over saddle	508 mm	482 mm	508 mm

Contractor
Thales Optronics (Vinten) Ltd.

UPDATED

Thales Optronics (Vinten) Ltd Vicon 70

Type
Reconnaissance pod.

Description
The Vicon 70 is a modular general purpose pod designed largely for reconnaissance and surveillance. It is designed for use by high-performance aircraft and is cleared for speeds of up to M2.2 but may also be used by helicopters and light aircraft.

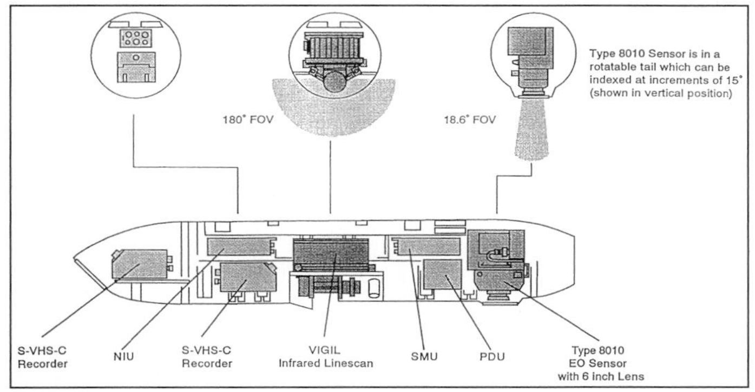

A diagram of the Vicon 70 series 5C equipment installation

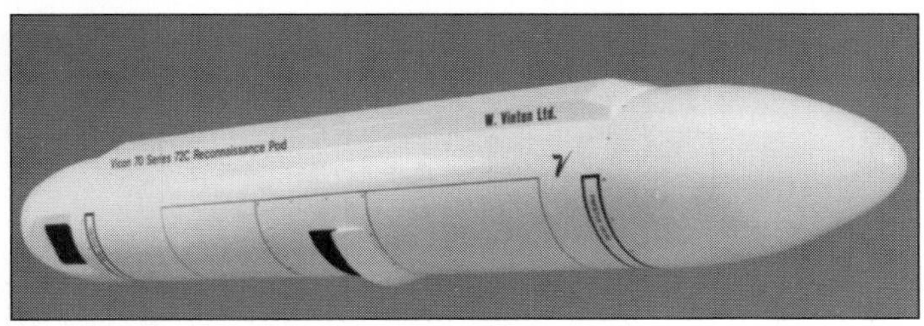

The Vicon 70 series 72C reconnaissance pod 0058726

Vicon 70 series 72C pod on Gripen aircraft 0058727

It will accept a wide range of sensors including film framing and panoramic cameras recording on 70 mm film, infra-red linescanners recording either on 70 mm film or on videotape and night illumination systems operating either with white light or IR flash. In addition, a number of FLIRs have been integrated into the Vicon 70 pods for various applications including navigation and attack.

The Vicon 70 series 72C is a daylight low and medium level EO/IR tactical reconnaissance version designed for use on the shoulder or centreline stations of the Gripen aircraft.

Operational status
In production and in service. More than 50 variants have been designed or built. The Vicon 70 series 72C has been selected by BAE Systems/Saab for EBS Gripen.

Specifications
Length: 1.44-2.30 m
Diameter: 35.5 cm

Contractor
W Vinten Ltd.

VERIFIED

Thales Optronics (Vinten) Ltd VIGIL infra-red linescan sensor

Type
Infra-red linescanner.

Description
The Thales Optronics (formerly W Vinten Ltd) InteGrated Infra-red Linescan sensor (VIGIL) is a single Line-Replaceable Unit (LRU) sensor which incorporates all the electronics and the control interface which may be either discrete signals or RS-422 serial command link for direct interface into the Reconnaissance Management Systems. It is designed for fixed- or rotary-wing aircraft and for UAVs. The sensor provides horizon to horizon coverage in the 8 to 14 μm wavelength from a single-element SPRITE detector. The detector is cooled by a continuous rated Stirling closed-cycle cryogenic cooling engine and the lightweight sensor is operated from 28 V DC power. VIGIL produces an analogue signal containing both line synchronisation and video which may be interfaced with a video cassette recorder, a datalink transmitter and/or a W Vinten display processor.

Operational status
In production and in service, VIGIL replaces the company's Type 200, 1000 and 2000 series infra-red linescanners.

Specifications
Spectral band; 8-14 μm
Detector: single-element SPRITE CMT
Cooling: closed-cycle Stirling
Transverse field of view
Scanned: 190°
Displayed: 180°
Angular resolution: >0.67 mrad
Thermal sensitivity: >0.16° NET
Velocity/height range: up to 2.5 rads/s
Dimensions: 309 × 254 × 247 mm
Weight: 10.5 kg

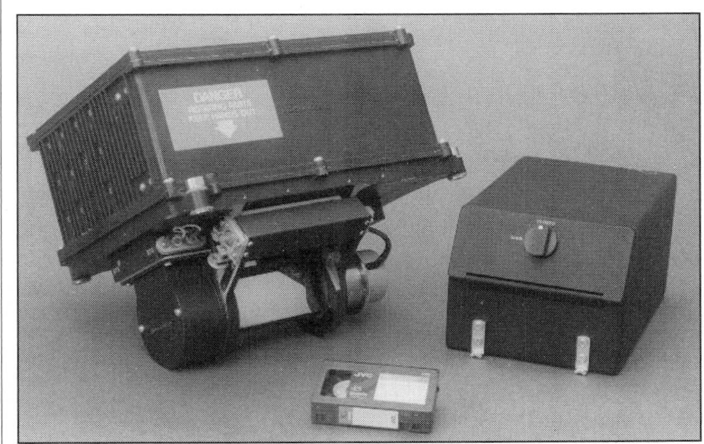

The VIGIL infra-red linescan sensor

Contractor
Thales Optronics (Vinten) Ltd.

VERIFIED

Thales Optronics (Vinten) Type 4000 infra-red linescan sensor

Type
Infra-red linescanner.

Description
This is a small, high-performance sensor designed for high-speed, low-level tactical reconnaissance aircraft. It may be installed in an external underwing pod, under or within the aircraft's fuselage. The Type 4000 sensor, which has a very wide field of view, collects infra-red radiation from the terrain being overflown and, by means of highly advanced conversion and processing electronics, provides a video output signal. This allows the scene below the aircraft to be presented in real time on a television monitor and recorded on a videotape recorder and/or the data may be transmitted to a ground receiving station. Such a ground station has a similar display and recording equipment, together with a hard-copy imagery printer.

The sensor features an aluminium mainframe on top of which are the detector assembly, the closed-cycle helium cooling engine and the housing for the electronic circuitry, scanner pick-offs and control system. The optical mirror elements are on the underside of the framework where the scanner, with its rotational axis parallel to the direction of the flight, collects radiation from the ground and transmits it to a pair of parabolic mirrors which focus, via a combining ridge mirror, on to the detector. In the Type 4000 the scanner is of triangular section.

Operational status
In production and in service for the Royal Air Force for the Tornado GR.Mk1A, the Royal Saudi Air Force and for foreign customers.

Specifications
Spectral band: 8-14 μm
Detector: multiple-element CMT
Cooling: closed-cycle Stirling
Angular resolution: <1 mrad
Thermal sensitivity: <0.2° NET
Transverse field of view: 190° scanned, 180° displayed
Roll stabilisation (2 modes): ±30° with signal input from 5 wire 26 V, 400 Hz; synchro to ARINC 407; locked to airframe
Dimensions
Sensor unit: 309 × 254 × 247 mm
Electronics unit: 302 × 337 × 202 mm
Weight
Sensor unit: 10.5 kg
Electronics unit: 12.5 kg
Power requirement: 56 W at 28 V DC, 110 VA at 115 V 1 phase 400 Hz, 90 VA at 200 V 3 phase 400 Hz
Outputs: video, line synchronisation, sample clock, BITE, system ready

Contractor
Thales Optronics (Vinten) Ltd.

VERIFIED

The IRLS 4000 infra-red linescan sensor from Thales Optronics (Vinten) Ltd

Thales Optronics (Vinten) Type 401 infra-red linescan

Type
Infra-red linescanner.

Description
The Type 401 infra-red linescan system is designed for low-level, high-speed tactical airborne reconnaissance at speeds up to 600 kt and altitudes down to 60 m. The system operates in the 8 to 14 μm waveband and employs a five-element cadmium mercury telluride detector array, which is cooled to liquid-air temperature by a Joule-Thomson high-pressure air minicooler. Typical endurance times from a standard air cylinder are 120 minutes at 20°C and 45 minutes at 70°C soak temperatures. Detector cool-down time is less than 1 minute.

Imagery is recorded on standard 70 mm photographic film which is contained in a large-capacity (92 m) quick-change magazine. Automatic velocity/height ratio control is provided and the flight navigation data are recorded on the film edge, reducing both aircrew and photo-interpreter workload.

The transverse field of view covers 120° and can be offset left or right to provide horizon-to-horizon coverage. Roll stabilisation to +55° is provided and allows high angles of aircraft bank without loss of continuity of the scanned area or the resulting imagery.

Operational status
In service with seven air forces. No longer in production.

Specifications
Spectral band: 8-14 μm
Detector: 5-element CMT array
Cooling: Joule-Thomson high-pressure air minicooler
Resolution: >1 mrad
Sensitivity: >0.2° NET
Velocity/height range: 0.025-5 rad/s
Field of view: 120° (30° left/right offset)
Reliability: >200 h MTBF
Roll stabilisation (3 modes): stabilised +30° from vertical; stabilised port or starboard oblique; slaved to airframe vertical or port or starboard oblique
Dimensions
Linescan: 604 × 320 × 280 mm
Cooling pack: 390 × 230 × 120 mm
Weight: 43 kg (total)

Contractor
Thales Optronics (Vinten) Ltd.

VERIFIED

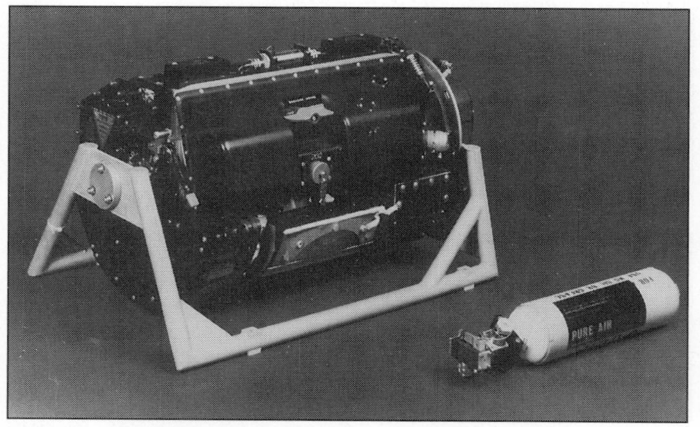

Thales Optronics (Vinten) Ltd Type 401 infra-red linescan sensor

Thales Optronics MDS 610 (MultiDistance Sensor)

Type
Reconnaissance sensor.

Description
The MDS 610 is a passive electro-optical airborne reconnaissance sensor designed for the following daytime intelligence gathering missions: low-, medium- and high-altitude tactical reconnaissance; standoff oblique and vertical reconnaissance; fixed- or moving-target localisation and identification.

The system is gyrostabilised in two axes and can be pod or fuselage-mounted on combat, reconnaissance and surveillance aircraft. The sensor has both pushbroom and panoramic scanning modes. It can be manually or automatically controlled with real-time image display in the cockpit and is compatible with real-time transmission via datalink. Automatic control is achieved by preprogramming according to the mission.

MDS 610 is of modular design with an electro-optical CCD detector, giving unlimited mission duration, stereo viewing of small surface areas and in-flight recording and replay capability.

MDS 610 also exists as a film version.

Operational status
MDS 610 has been fitted in a Thales reconnaissance pod demonstrator which has been tested on a Mirage F1 CR aircraft. It is carried in the PRESTO pod, based on the demonstrator. Seven PRESTO pods were ordered by the French Air Force and they were scheduled to equip Mirage F1 CR aircraft from 1999.

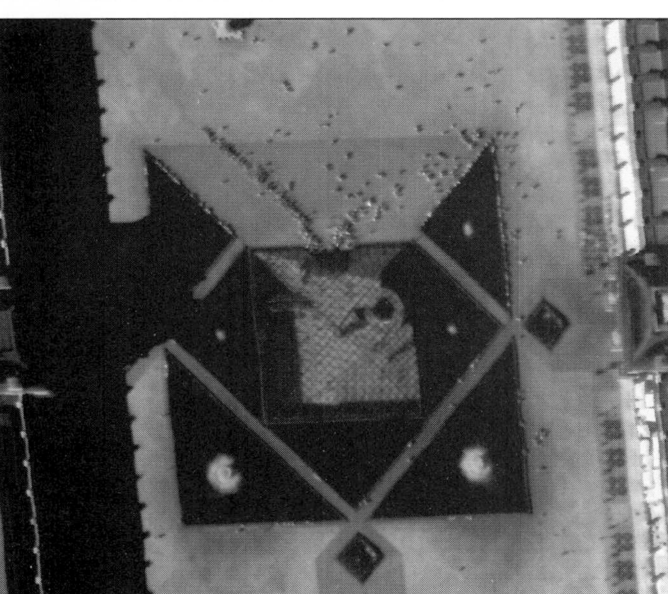

Typical images taken with the MDS 610. The top image is taken using a ×4 zoom from 31,000 ft. The bottom image is part of the same scene using a ×8 zoom 0058729/0058728

A French Air Force F1 CR made the first successful medium-range reconnaissance flight carrying the pod-mounted MDS 610 electro-optical sensor (PRESTO demonstrator)

Specifications

Weight: 120 kg
Dimensions (d × l): 350 × 1,500 mm
Lens focal length: 610 mm - f/4
CCD detector: 10,000 pixels (0.4-1.1 μm)
Field of view: 11°

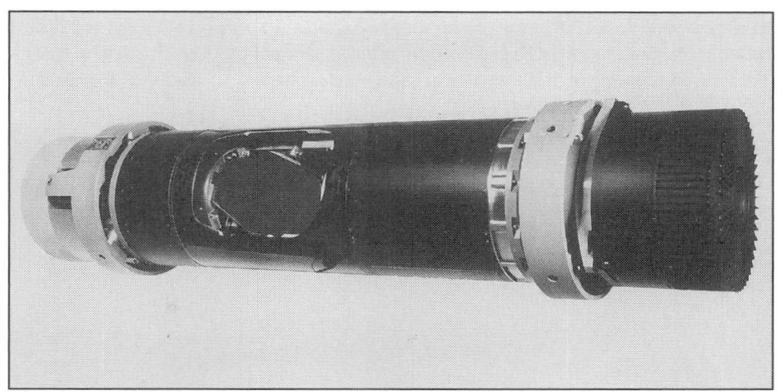

Thales Optronics MultiDistance Sensor (MDS) 610

Lateral coverage: ±110°
Longitudinal coverage: ±20°

Contractor

Thales Optronics.

VERIFIED

Thales Optronics PRESTO reconnaissance pod

Type

Reconnaissance system.

Description

The PRESTO pod has been developed by Thales Optronics under a French Ministry of Defence programme. It is designed for the Mirage F1 CR and Mirage 2000 for stand-off or high-speed penetration missions. The system includes the Thales Optronics MDS 610 Multi-Distance Sensor, a reconnaissance management

The PRESTO demonstrator reconnaissance pod under Mirage F1 CR 0005561

PRESTO on Mirage 2000 0058724

system, a high-speed digital recorder and an associated ground station. PRESTO is capable of operating in both pushbroom and panoramic modes. The pod provides real-time imaging, digital recording and transmission to ground processing stations.

Operational status
The French Air Force placed an order for seven PRESTO pods. Believed to be in service.

Contractor
Thales Optronics.

VERIFIED

Thales Optronics SDS 250 electro-optical reconnaissance sensor

Type
Reconnaissance sensor.

Description
The SDS 250 is a compact passive electro-optical airborne reconnaissance sensor designed to perform the following daytime intelligence gathering missions: low- and medium-altitude tactical reconnaissance, vertical reconnaissance and up to 20 km standoff oblique reconnaissance.

The sensor has pushbroom scanning mode, electronic roll stabilisation and selectable operating spectral band - either visible or near infra-red or both. The system provides real-time image display in the cockpit and line of sight position is selected from the cockpit. The system is compatible with real-time image transmission via datalink. The compactness of the sensor allows fuselage installation.

Operational status
SDS 250 has been selected by the French Air Force for Mirage F1 CR and by the French Aeronavale for Upgraded Super Etendard aircraft. In production.

Specifications
Lens focal length: 250 mm - f/5.6
CCD detector: 6,000 pixels (0.4-1.1 μm)
Field of view: 14°
Lateral coverage: ±85°
Weight: 30 kg
Dimensions: 410 × 400 × 230 mm

Contractor
Thales Optronics.

VERIFIED

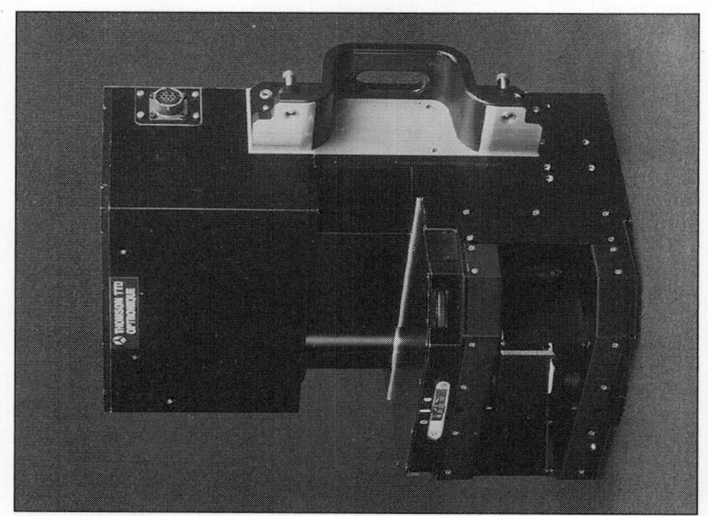

Thales Optronics SDS 250 medium-/low-altitude electro-optic sensor 0063743

Z/I Imaging KRb8/24f drone camera

Type
Reconnaissance system.

Description
The lightweight KRb/24F is a state of the art camera system with high resolution optics that achieves wide-angle panoramic coverage with undistorted framing camera geometry. The camera is suitable for low-to-medium altitude reconnaissance missions and is the main sensor payload in the mobile CL-289 Drone System.

Operational status
In sevice and available.

Contractor
Z/I Imaging Corporation.

Z/I Imaging KS-153 camera family

Type
Reconnaissance system.

Description
Z/I Imaging Corporation produces a number of camera systems. The modular KS-153 camera family offers a high cycle rate, pulse-operated framing camera with modular design and proven performance. The camera consists of three (80 mm, 57 mm, and 610 mm) interchangeable focal length lenses that can be used for a variety of reconnaissance and remote-sensing applications. The KS-153 is built and tested to exceed reliability and maintainabilty criteria for use in modern aircraft, pods and drones.

A digital module with a solid-state detector is currently in preparation. It will be possible to interchange the existing film module with the digital module as required.

Operational status
In service and available.

Contractor
Z/I Imaging Corporation.

Zeiss Optronik VOS 60 airborne electro-optical sensor

Type
Linescan imager.

Description
The VOS 60 sensor was designed for low- to medium-altitude reconnaissance missions. It is a high-performance electro-optical colour system and consists of the sensor head, the electronic control unit and the control panel. The image will be acquired with a 3 × 6,000 pixel colour detector in pushbroom technique. The digital output of the sensor can be recorded on a digital recorder or viewed on board on a high-resolution monitor. Additional mission-related data can be added. The system comprises a variety of high-performance Zeiss-Optronik lenses for different applications and image scale requirements.

Operational status
In production. In service for the Open Skies Treaty programme.

The VOS 60 air reconnaissance camera

Specifications
CCD sensor: 3 × 6,000 pixel colour
Pixel size: 12 × 12 μm
Video channels: red, green, blue (RGB)
Line rate: 1.6 kHz max
Lens: Carl Zeiss 60 mm, f/3.5 Distagon or others
Field of view: 60° across flight path (for 60 mm lens)
Dimensions
VOS 60 sensor: 154 × 154 × 221 mm
Electronic control unit: 133 × 483 × 340 mm
Control panel: 213 × 129 × 107 mm
Weight
Sensor: 6 kg
Electronic control unit: 11 kg
Control panel: 1.5 kg
Power: 28 V DC, 200 W

Contractor
Zeiss Optronik GmbH.

VERIFIED

Zenit A-84 panoramic aerial camera

Type
Reconnaissance camera.

Description
The A-84 panoramic aerial camera is designed to provide wide-area photographic coverage of the earth's surface during daylight conditions, from medium and high altitudes. It can be set to operate automatically by onboard control system, or be controlled manually from its control panel.

The A-84 camera is equipped with image motion compensation and automatic exposure control. The camera film records navigation data from its airport navigation system.

Operational status
Fitted to Tu-22 medium bomber (presumably the Tu-22MR variant) and to the Tu-154 medium transport aircraft for 'Open Skies' operations. Also reportedly fitted to the M-17 high-altitude reconnaissance and research aircraft.

Specifications
Focal length: 300 mm
Aperture: f/4.5
Frame size: 118 × 748 mm
Film size: 480 × 130 mm
Nominal overlap in centre frame: 25%
Along track filming distance: 160 × aircraft altitude
Linear resolution:
0.4 m @ range = 2 × altitude
0.8 m @ range = 6 × altitude

A-84 panoramic aerial camera 0018339

Power
27 V DC: less than 300 W
115 V AC, 400 Hz: less than 900 V A
Weight: 160 kg

Contractor
Zenit Foreign Trade Firm.

VERIFIED

Zenit AC-707 spectrozonal aerial camera

Type
Reconnaissance camera.

Description
The AC-707 aerial camera has four separate photographic channels, four separate lenses, four filters and four automatic exposure control systems to provide photography in four predetermined spectral bands. Each lens is corrected for focal length and distortion. The camera contains image motion compensation, vacuum back film platen, range focus and temperature focus compensation. It records navigation information obtained from the aircraft navigation system and it can be fitted with gyrostabilisation. It can be controlled automatically or manually from its control panel.

Operational status
Installed on Mi-8 helicopters.

Specification
Focal length: 140 mm
Aperture: f/2.8-f/22
Spectral bands:
Channel 1 (blue): 400-500 nm
Channel 2 (green): 480-600 nm
Channel 3 (red): 580-700 nm
Channel 4 (infra-red): 700-860 nm
Frame size: 180 × 180 mm
Sub picture size: 70 × 70 mm
Film type (l × w): MLU-4, 240 m × 190 mm
Operating altitude: over 50 m
Exposure time: 1/20-1/300 s
Along track filming distance: 400 × altitude
Across track coverage: 0.5 × altitude
Power:
27 V DC: less than 30 W
115 V AC, 400 Hz: less than 5 VA
36 V AC, 400 Hz: less than 6 VA
Weight: 212.9 kg

Contractor
Zenit Foreign Trade Firm.

VERIFIED

AC-707 spectrozonal aerial camera 0018338

OBSERVATION AND SURVEILLANCE

THERMAL IMAGERS

Boeing, Raytheon Airborne Surveillance Testbed (AST)

Type
Thermal imager.

Development
The Airborne Surveillance Testbed (AST) project is a key element of the US Department of Defense Ballistic Missile Defense Organisation (BMDO). The AST collects infra-red optical data on theatre and intercontinental ballistic missiles, thereby helping in the development of effective defence systems. Additionally, the AST provides an airborne testbed platform for developing and demonstrating anti-ballistic missile sensors and systems.

Boeing was awarded the initial contract in July 1984. Formerly called the Airborne Optical Adjunct (AOA), the system initially was developed as a technology demonstrator programme to determine whether airborne optical sensors could provide early warning and tracking of enemy intercontinental ballistic missiles and their warheads. The AOA programme was later renamed AST and the programme focus changed to data-gathering and issue resolution for ballistic missile defence.

Under contract with the US Army's Space and Strategic Defense Command, Boeing Defense and Space Group has modified a 767 model commercial jet aircraft to carry a long-wave infra-red sensor housed in a cupola atop the Model 767's fuselage. It is hoped that the infra-red sensors will be able to detect a wide range of objects as they generate heat entering the earth's atmosphere and be able to discriminate warheads from other debris that may re-enter with them. Subcontractors to Boeing are Raytheon, who are the sensor developer and Harris Corp for the Nighthawk onboard data processor.

The 767 AST is being used in a series of experimental system test and evaluation flights designed to collect infra-red data needed by developers of missile defence systems. These data collection missions also serve to validate and demonstrate the functional performance of the sensor on an airborne platform. The aircraft was modified at Boeing's Seattle, Washington facilities. It was recertified in 1987 and flew its first mission with the sensor on board in May 1990.

The flights also validate the onboard software, computers and other electronic equipment necessary for instantly processing the tracking data and transmitting that data to ground-based radar.

Description
The cupola housing the infra-red sensor is 7.7 ft high and has a 3.9 ft wide, 6.3 ft high viewing port which has to be open while the sensor is scanning. This means that the sensor is exposed to air turbulence and so Boeing has designed an air-flow equalisation scheme that creates an 'invisible curtain' of almost zero turbulence across the port. Raytheon has designed a pointing/stabilisation system composed of rails, gimbals, ring- laser gyros and servo-electronic controls. The sensor is mounted on the rails and can move laterally, fore and aft in the cupola. Gimbal design permits four-axis stabilisation.

An optical telescope is mounted on the pointing/stabilisation assembly. It integrates three fused-quartz mirrors that focus the scene onto the focal plane assembly through a zinc selenide window.

Raytheon's sensor is based on focal plane array technology. The assembly accommodates 44 modules, each module containing four sensor chip arrays and each sensor chip array has 640 detectors. The 38,400 gallium-doped silicon detector elements are precisely aligned and indium-bonded to 38,400 MOSFET preamplifiers and associated readout electronics. The readout chip performs first-level noise reduction and multiplexes the detector output signals. Cooling is provided by a closed-cycle helium cryogenic system.

The modified Boeing 767 with Raytheon long-wave infra-red sensor which forms the Airborne Surveillance Testbed programme

The Raytheon AST infra-red sensor subassembly

Three levels of signal processing are incorporated into the subsystem. Parts are mounted on the telescope, the rest in the 767 cabin areas. At the first level, detector output signals are conditioned, amplified, buffered and gain-normalised and then the signal is converted from analogue to digital. For the second level, a time-dependent processor further improves signal-to-noise ratio and makes the first attempt at bulk-filtering in the digital mode. The third level, the object dependent processor, analyses relevant objects' characteristics (position, velocity and amplitude) to weed out false alarms. This is accomplished by parallel processing which permits analysis in near-realtime. Finally sensor subsystem output is handed-off to the onboard data processing subsystem for high-speed refinement and analysis.

The aircraft is positioned safely to the side of the missile trajectory, gathering data as the dummy warheads or TBM targets pass through the sensor's field of view. Onboard computers process the sensor data for handover to ground-based radar and record information for post-mission analysis. Based in Seattle, the AST carries flight and test crews of 10 to 15 members in missions typically lasting 4 to 6 hours. The aircraft flies at altitudes somewhat higher than those flown in normal commercial flights.

Operational status
Experimental programme.

Contractors
The Boeing Company.
Raytheon Systems Company.

VERIFIED

DRS Multi-Mission Infra-red Surveillance and Targeting Imager (MISTI)

Type
Thermal imager

Description
The DRS MISTI thermal imager is suited to a variety of applications including airborne and naval systems. It operates in the mid-wave band and uses a HgCdTe detector. There are two fields of view with the option of a third.

Operational status
Available

Specifications
Detector type: HgCdTe
Elements (elevation × azimuth): 320 × 240 or 640 × 480
Detector Pitch: .30 or .27 μm
Spectral band: MWIR (3.8-4.8) μm
Type of cooling: 1 W IDA (Linear Split Sterling Closed Cycle)

Cool down time: <7 min
Operating temperature: 95.0±5.0 K
System MTBF: >3000 h

Field-of-view:
Narrow: 2.0°
Wide: 8.0°
Super wide: 20° (Optional)
NETD, 300K background: <0.025 K
System dynamic range: >70 dB
Data resolution/camera output: 12 bits
Adjustable integration time: 0.2 to 15 ms
Output video format: CCIR 'PAL' composite, RS-170 NTSC composite, 12-bit digital (optional)
Power supply: 17-32 V DC
Power dissipation: 35 W nominal
System weight: <11 kg

Contractor
DRS Technologies, Electro-Optical Systems Group.

Elop CRYSTAL high resolution FPA thermal imager

Type
Thermal imager.

Description
CRYSTAL is a three FOV, high resolution thermal imaging sensor utilising a Focal Plane Array (FPA) detector sensitive in the 3 to 5 μm spectral band, coupled to a dual-axis micro-scan mechanism, yielding high 512 × 512 sampling resolution.

CRYSTAL is built in a single unit, open frame, lightweight configuration, especially designed for easy on-gimbals installation and integration, in stabilised electro-optical payloads.

Applications are as follows:
- Stabilised gimballed E-O payloads – airborne, heliborne or naval
- Stabilised and non-stabilised mast mounted E-O sensor clusters
- Target acquisition systems
- Night sight systems.

Main features:
- 3-5 μm spectral band
- InSb FPA detector
- Micro-scan mechanism
- Miniature closed-cycle cooler, integral with the Dewar
- Three FOVs
- Non Uniformity Correction (NUC) algorithms
- Proprietary image processing ASIC
- Single unit configuration for easy integration
- Lightweight
- Small dimensions.

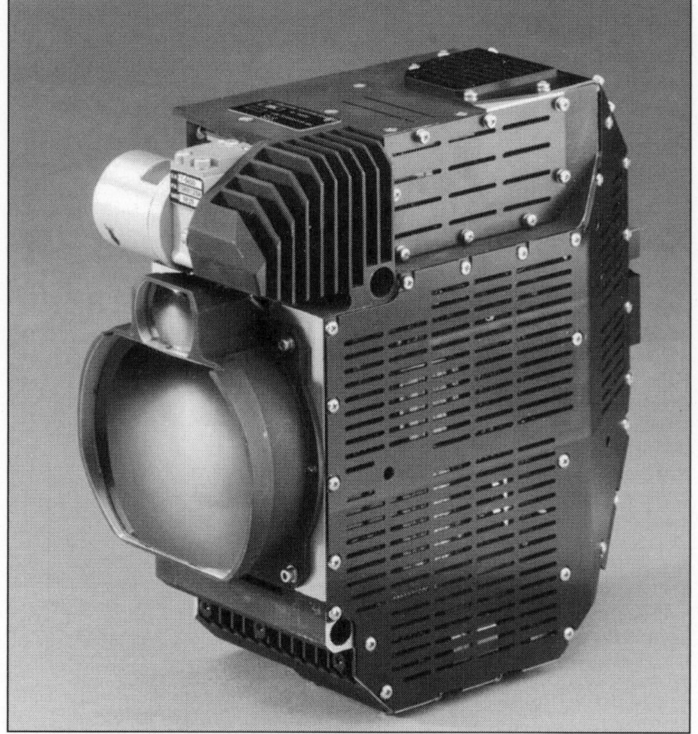
The Elop CRYSTAL thermal imager 0109664

Specifications
Detector: Type In Sb Focal Plane Array (FPA) with 256 × 256 elements
Spectral Band: 3-5 μm
Entrance aperture (main): 100 mm

Fields of View:
Narrow 1.5 × 1.5°
Medium 5.25 × 5.25°
Wide 21 × 21°
Micro-scan: Dual axis, opto-mechanical
Sampling Resolution: 512 × 512
Cooler: Closed-cycle, integral with Dewar
Video output- CCIR, 625 lines, 50 Hz
Communication: Serial, synchronous, RS-422
Power requirements : 24 V DC nominal, 30 W
Configuration: Single unit, including electronics signal processing
Weight: 5.0 kg
Environmental: Designed and tested to MIL-STD-810
Operating temperature range: 20°C to +60°C
Reliability: MTBF: > 2,000 h

Contractor
Elop Electro-Optic Industries Ltd.

VERIFIED

Elop TADIR lightweight thermal imager

Type
Thermal imager.

Description
TADIR is an advanced second-generation lightweight thermal imager which utilises a 2nd generation 480 × 4 element Time Delay and Integration (TDI) Detector.

It is described as using a high-reliability closed-cycle cooler. The optics have a wide range (20:1) continuous optical zoom with automatic athermalisation. It has a combination of high resolution with a continuously variable FOV. Powerful signal processing is based on proprietary ASIC with real-time algorithms for image enhancement. The image output uses CCIR/RS-170 composite or 8/12 bit digital video output.

TADIR has a single unit configuration. No separate electronics box is needed. A minimum number of host platform system slip-rings are used to maximise reliability.

Applications are as follows:
- Stabilised airborne payloads
- Attack helicopters targeting systems
- Helicopter night piloting systems
- Long-range air defence fire-control systems
- Stabilised naval payloads.

The Elop TADIR lightweight thermal imager 0109667

Specifications
Dimensions: 250 × 250 × 150 mm
Weight: 7 kg
Detector: MCT-PV, TDI, 480 × 4 elements
Spectral Band: 8-12 μm
Entrance aperture: 150 mm
Optical zoom: (20:1) continuous optical zoom from FOV of 2.0 × 1.5° to 40 × 30°
Electronic zoom: × 2
Cooler: High-reliability, closed-cycle, split onfiguration
Video output: RS170 or CCIR or 8/12-bit digital video
Serial communication interface: RS-422
Power consumption: 70 W @ 28 V DC, conforms with MIL STD 704
Options: HDTV video output

Contractor
Elop Electro-Optic Industries Ltd.

VERIFIED

Indigo Systems Phoenix®

Type
Thermal imager module.

Description
The Indigo Systems Phoenix® thermal imager module (OEM version) is based on Indigo's proven 640 × 512 staring InSb FPA, integrated in an all-metal, sealed Dewar cooled with split Stirling linear cooler.

The OEM version is especially suited for gimbal applications in a military environment. It is a compact, high performance module transmitting 40 MHz digital data (Hotlink) off the gimbal to user-selectable video processing electronics embeddable in a small rack or PC-based electronics.

Operational status
In service with US forces on the Northrop Grumman/Rafael Litening ER and AT targeting pods.

Specifications
Dimensions: 12.7 × 7 × 11.4 cm
Weight: 2 kg
Spectral range: 3-5μ (1-5μ custom)
Detector material: InSb
Array format: 640 × 512
Detector size: 25μ
Readout: ISC9803
Detector cooling: closed-cycle Stirling
Cooldown time: 5 minutes at 22°C and under 10 minutes at 71°C
Optical interface: custom
Video output: RS-170 and 8-bit digital
Power consumption (SS): 25 W typical

Contractor
Indigo Systems Corporation (USA).

NEW ENTRY

The Indigo Systems Phoenix® thermal imaging module ***NEW***/0595420

ISS M320i-3 MWIR camera

Type
Thermal imager.

Description
The M320i-3 MWIR thermal sensor is a high-performance sensor, based on an Indium Antimonide (InSb) 320 × 240 element Focal Plane Array (FPA). This sensor consists of an integrated detector-dewar-cooler assembly, 3 field of view lens, detector interface electronics, and signal processing electronics. Features of this camera include broad dynamic range, auto gain/level, integral digital processing electronics and optional microscanner. These features provide a high-resolution output that approaches the performance of the much higher priced, larger arrays. It is available in an environmentally sealed enclosure.

The M320i-3 electronics provide desirable features including: image sharpening 2× electronic zoom, dynamic range enhancement, and a symbology overlay capability. An RS-232/422 serial datalink provides control of the M320i-3 functions, as well as feedback of Built-In-Test (BIT) status information.

Responsivity in the 3 to 5 μm waveband, coupled with the wide dynamic range and high sensitivity found in the 320i-3, allow for discrimination of targets in high humidity environments. Its multiple FOV optics provides operators with very good target acquisition and identification capabilities. The wide field of view is ideal for navigation and situational awareness.

The M320i-3 is small and lightweight, which makes it ideal for use in the limited space of airborne gimbal systems. It is the baseline thermal imager used in the ISS V14-M Multisensor system. It also fits most applications where the ISS MicroFLIR™ thermal imaging camera is now used.

Operational status
Available and in service.

Specifications
Dimensions: 11 × 8 × 8 in (280 × 203 × 203 mm)
Weight: <12 lb (<5.5 kg)
Spectral Band: 3 - 5 μm
Detector: 320 × 240 InSb FPA
Cooling: Sterling linear cryogenic
Field of View: 0.9 × 1.2°
MFOV 3.8 × 5.0°
WFOV 16.5 × 22.0°
E-Zoom: ×2
Video Output: NTSC or PAL
Power: 18 - 30 V DC (nominal), 50 W (MIL-407-D)
Operating temperature: –40 to +55°C

Contractor
Imaging Sensors and Systems, Inc.

VERIFIED

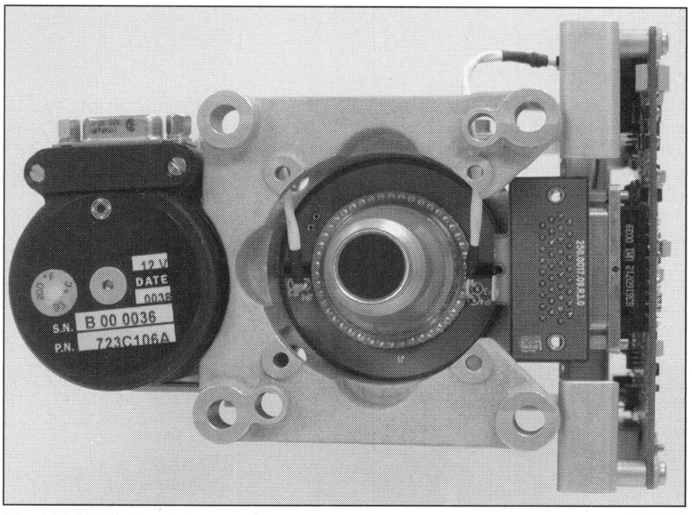

The ISS M320i imager 0101853

Kentron KENIS high-resolution imaging infra-red camera

Type
Thermal imager.

Description
The Kentron KENIS imaging infra-red (IIR) camera is a high-performance, third-generation thermal imager, using a 384 × 288 CMT focal plane array. The IIR camera employs microscanning to enhance the resolution fourfold. The stand-

alone infra-red camera provides high-quality thermal images in the 3-5 μm spectral band and is ideal for long-range imaging in hot and humid conditions.

Passive athermalisation, an FOV switching mechanism and onboard fully integrated electronics are employed to achieve a very compact and lightweight design. High sensitivity, combined with a dual field of view, results in improved acquisition and tracking capabilities. The IIR camera is contained in a single integrated unit and implementation of most of the system functionality is software-based. This software orientation allows rapid incorporation of customer requirements and can be easily upgraded. This IIR camera has been developed for the target acquisition and accurate terminal guidance requirements of stand-off weapons and similar observation applications. It is ideal for a wide range of applications, from surveillance to naval applications in poor visibility conditions.

Operational status
Available.

Specifications
Dimensions (h × l × w): 135 × 260 × 150 mm
Weight: 3.5 kg
Thermal Imager
Detector: 384 × 288
Spectral band: 3-5 μm
Cooling: integral Stirling cooler
Switch-over time: <1 s
Field of view:
Narrow: 2.25 × 3° (f/2)
Wide: 13 × 18° (f/2)
Focus: (100 m to ∞)
Focus control: fixed (athermalised)
Electrical zoom: ×2
MRTD Typical
@ 1.1 cycles/mrad: 0.05°C
@ 3.0 cycles/mrad: 0.09°C
@ 5.2 cycles/mrad: 0.13°C
NETD (300K): 0.04-0.08°C
Dynamic range: 12 bits
Power supply: 20-35 V DC, 40 W (max)
Shock: 20 *g* 11 ms (half-sine)
Vibration: 0.002 *g*/Hz bandwidth, 20-2,000 Hz
Operating temperature: −15 to +55°C
Storage temperature: −40 to +70°C

Contractor
Kentron Division of Denel (Pty) Ltd.

Opgal M2TIS

Type
Thermal imager.

Description
M2TIS is a modular thermal imaging system operating in the 8 to 12 μm spectral range with Stirling closed-cycle cooling. It consists of a sensor and an electronic unit and is available as an independent unit for integration in a customer's system or as part of a stabilised turret with other sensors. In such a system it can be used for fire control.

The system is designed for use in manned fixed- and rotary-wing aircraft as well as UAVs. However, it may also be used in vehicles and ships.

Operational status
In service with the Israel Air Force and US Marine Corps, integrated in the Night Targeting System of the M-65 turret of AH-1W and AH-1S Cobra helicopters.

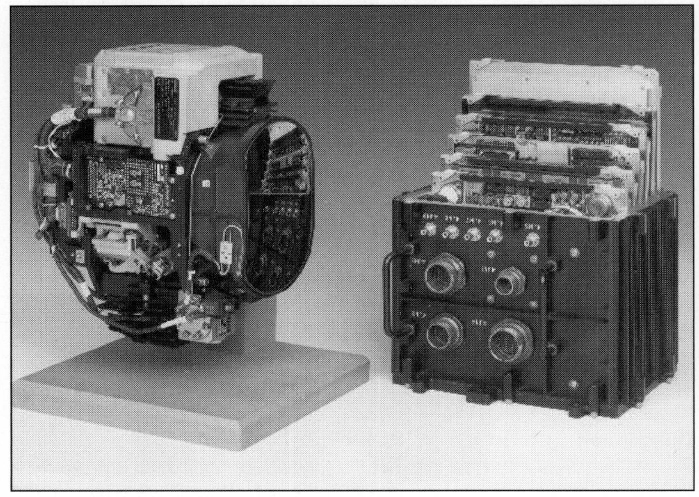

The M2TIS sensor and electronic unit

Specifications
Spectral band: 8-12 μm
Cooling: closed-cycle Stirling
Fields of view
Wide: 24.5 × 18.4°
Medium: 7 × 5.2°
Narrow: 2 × 1.5°
Aperture: 148 mm
Focal length: 180 mm
Weight
Sensor: 6 kg
Electronic unit: 7 kg

Contractor
Opgal Optronic Industries Ltd.

VERIFIED

Raytheon F/A-18 FLIR infra-red imaging subsystem

Type
Thermal imager.

Description
The AN/AAS-38 was developed by Ford Aeronutronic (now Lockheed Martin) and Raytheon in 1979 for the US Navy and Marine Corps for use on the F/A-18 Hornet aircraft. The Infra-Red Imaging Subsystem (IRIS) is part of the AN/AAS-38 FLIR imaging system. This is a self-contained pod for target acquisition and recognition and weapon delivery. It also allows reconnaissance under day, night and adverse weather conditions. AN/AAS-38A includes laser target designator and range-finder.

Major subassemblies of the IRIS include the infra-red receiver, controller-processor, power supply and the infra-red afocal optics group. Lockheed Martin Aeronutronic manufactures the pod and gimbal.

The IRIS includes dual fields of view with automatic thermal focus. Built-in image derotation provides a natural horizon display to the pilot. Microprocessor-controlled BIT circuits provide 98 per cent fault detection.

A key feature of IRIS is the automatic video tracker contained in the controller-processor. This tracker provides automatic target acquisition, line of sight control and offset designation for accurate weapon delivery.

The controller-processor provides video processing necessary to produce 875-line RS-343 television video. The processor adds track and field of view reticles for cockpit displays. A MIL-STD-1553 digital databus provides communications with the aircraft AN/AYK-14 mission computer.

The infra-red receiver uses US Department of Defense FLIR common modules produced by Raytheon. The IRIS is packaged similarly to the US Air Force AN/AAQ-9 infra-red receiver.

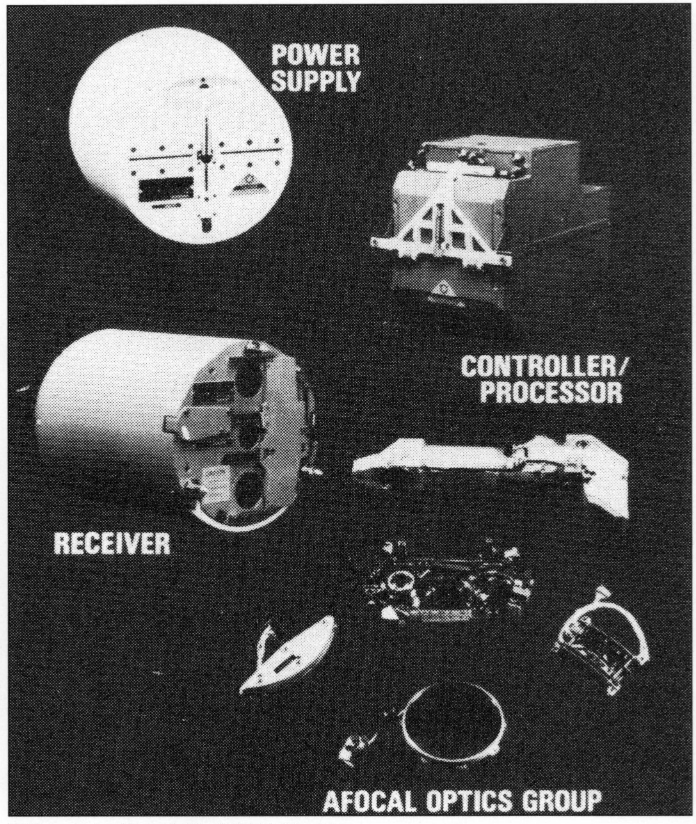

IRIS is part of the AN/AAS-38 FLIR system

Operational status
In service on the US Navy McDonnell Douglas (now Boeing) F/A-18 aircraft.

Specifications
Fields of view: 12 × 12° and 3 × 3°
Fields of regard
Pitch: +30 to −150°
Roll: ±540°
Weight: 370 lb
Power: 115 V AC, 400 Hz, 3 phase, 2,700 VA max; 28 V DC 150 W max

Contractor
Raytheon Systems Company.

UPDATED

Raytheon Radiance HS

Type
Thermal imager.

Development
Raytheon announced the Radiance HS high-performance infra-red camera in March 1998. It has been developed for various aerial imaging, surveillance and non-destructive testing applications.

Description
Radiance HS incorporates Raytheon's AE173 256 × 256 indium antimonide (InSb) snapshot mode focal plane array featuring 30 μm pitch pixels and a nominal spectral response from 3-5 μm. Custom cold filters and cold shields are available as options. The camera operates at up to 140 frames per second in full frame mode and allows microscanning to boost resolution.

Engineered to meet payload limitations of likely applications Radiance is of light weight and small. It is designed to work with a range of optics from f/2.5 to f4.1. It is possible to package long focal length optics into smaller envelopes than is possible with scanning and platinum silicide imagers. The electronics can be located on or off the mounting gimbal.

Operational status
Available.

Specifications
Dimensions
145 × 107 × 171.5 mm (electronics on gimbal)
145 × 107 × 107 mm (electronics off gimbal)
Weight
4.1 kg (electronics on gimbal)
3.2 kg (electronics off gimbal)

Contractor
Raytheon Systems Company (Goleta).

VERIFIED

Thales Optronics Victor thermal camera

Type
Thermal imager.

Description
The Victor thermal camera is designed to be integrated into Viviane gyrostabilised sights for SA342 Gazelle HOT armed escort helicopters. It can also be fitted to AS365M Panther, BO 105 or other types of helicopters.

Victor is suitable for air-to-ground and air-to-air gun or missile firing, rocket firing and as a flying aid by day and night and under adverse weather conditions.

The system displays symbols in the eyepiece of the sight or on a TV monitor and has a magnifying function of ×2 to enlarge the image. Initial sight stabilisation is by the platform and electronic fine stabilisation is by the imager. There is a specific video output for tracking.

Operational status
Victor entered service in 1988.

Specifications
Spectral band: 8-13 μm
Trifocal lens: 30 × 20°, 6 × 4°, 2.4 × 1.6°
Range: up to 4,000 m
Weight
Total: 25 kg
On roof: 17 kg
Power supply: 20-32 V DC, 140 W

The Victor thermal camera, integrated into SFIM's Viviane sight, is shown above the cabin of a French Army Gazelle helicopter

Contractor
Thales Optronics.

UPDATED

US Air Force Space-Based Infra-Red System (SBIRS)

Type
Space-based ballistic missile warning system.

Development
The Space-Based Infra-Red System (SBIRS) is a US Air Force Space and Missile Systems Center programme to provide warning of ballistic missile launches and cue US missile defence systems. It is to replace the Defense Support Program which has been the US ballistic missile warning system for the last 20 years. It will also provide data on infra-red target signatures. It will consist of four satellites in geosynchronous earth orbit, two in highly elliptical orbits (SBIRS High) and a number of satellites in low earth orbit (SBIRS Low or the Space and Missile Tracking System - SMTS). SBIRS Low is the only programme that will provide mid-course tracking of ballistic missiles.

The Space and Missile Systems Center awarded Lockheed Martin Missiles and Space a contract for Engineering and Manufacturing Development (EMD) of the SBIRS High component in 1996. It is to deliver five geosynchronous satellites (based on the A2100 bus), a ground system and two High Elliptical Orbit Payloads to be installed on other satellites. Aerojet and Northrop Grumman will supply the primary infra-red sensor payload. SBIRS High will provide global and theatre infra-red data and processed messages concerning launch, flight and impact of strategic and theatre missiles.

SMTS, the Low component, will concentrate on mid-course tracking of targets. Two contractor teams have been given pre-Engineering and Manufacturing Development (EMD) contracts: Raytheon/TRW with a two satellite flight demonstration system (Low Altitude Demonstration System - LADS) and Boeing North American/Lockheed Martin with a single satellite system.

Description
SMTS satellites, flying in multiple rings, will operate in pairs, communicating via 60 GHz crosslinks, to provide stereoscopic viewing. Each will carry a scanning short waveband infra-red sensor with wide field of view for acquisition and a staring multiband (mid-, mid/long- and long-wave infra-red, together with visible light) tracking sensor with narrow field of view. These will operate in sequence, below and then above the horizon to acquire and track the missile plume, post-boost vehicles and cold re-entry vehicles. Mid-course tracking and discrimination between warheads and other objects will provide cueing information to defensive systems.

Operational status
Under development. The SBIRS High EMD contract was scheduled to end in September 1997. The SBIRS High Critical Design Review was due in 1999. Both SMTS flight-experiment satellites were scheduled to be launched in 1999 with engineering and manufacturing satellites to be launched in 2006. However this may be accelerated to 2004 if recommendations to the Pentagon are accepted.

Contractor
US Air Force, Space and Missile Systems Center.

VERIFIED

KEY TECHNOLOGIES FOR ELECTRO-OPTIC SYSTEMS

KEY TECHNOLOGIES FOR ELECTRO-OPTIC SYSTEMS – SECTION SUMMARY

This section includes key system building blocks which do not readily fit into either the Naval, Land or Airborne systems section. System adjuncts with significant operational benefits are also described.

Infra-red detectors and coolers

Detector devices which are state of the art or widely used in a variety of applications. Cryogenic cooling systems which may be applied to multiple types of infra-red sensors.

Thermal imager modules

Thermal imager modules which may be supplied to form key components of imagers produced by multiple manufacturers.

Video trackers

Electronic video trackers which are an essential part of fire-control, sighting and surveillance systems.

Anti-detection systems

Devices which provide a significant operational enhancement by preventing the detection of electro-optic systems via glint from optics or the 'cold window' of thermal imagers.

INFRA-RED DETECTORS AND COOLERS

BAE Systems 256 × 128 pyroelectric un-cooled IR detector array

Type
Infra-red detector (uncooled).

Description
The BAE systems Merlin un-cooled pyroelectric infra-red detector array is designed for imaging and instrumentation applications in the 8 - 12 μm waveband. It is intended to be simple to operate with high electro-optic performance at ambient temperatures. The detector uses BAE's Lead Scandium Tantalate (PST) technology which they believe is the highest performance pyroelectric material for infra-red imaging purposes.

Status
Available.

Specifications
Array format: 256 × 128 pixels
Element size: 56 μm
Element pitch: 56 m × 56 μm
Active area: 14.3 × 7.2 mm
Spectral band: 8 to 12 μm
Technology: CMOS
Window: Anti-reflective coated Ge (8 to 12 μm)
NETD (mean): 0.11 K @ 50 Hz frame rate, f1, typical
Defective elements, typical: 0.1%
Response: 1.5 mV/K
Modulation Transfer Function: 40%
Electrical characteristics
Frame rate: 10 to 75 Hz
Power consumption: <150 mW typical
Output impedance:< 500 Ω typical
Environmental
Operating temperature: -30 to +50°C
Vibration survivability: 20 Hz to 2 kHz
Shock survivability: 500 g, 1 ms half sine pulse
Acceleration survivability: 100 ms^{-2}

Contractor
BAE Systems, Avionics, Tactical Systems Division.

UPDATED

The BAE Systems Merlin 256 × 128 pyroelectric uncooled infra-red detector array
0089864

BAE Systems Condor LW 320 × 256 multiple quantum well integrated detector cooler assembly

Type
Integrated detector cooler assembly.

Description
The BAE Systems Condor LW detector/cooler uses the company's multiple quantum well process and is designed for high-performance imaging in the long (8 - 10 μm) waveband. It operates as a 2-D 'blinking mode' infra-red detector. It is described as giving high electro-optic performance with low crosstalk and automatic anti-blooming at the pixel level. In outward appearance the assembly looks similar to the BAE Systems Osprey assembly.

Status
Available.

Specifications
Array
Array format: 320 × 256 pixels
Element size: 30 μ
Active area: 9.6 × 7.68 mm
Spectral band: 8 to 10 μm
Technology: CMOS
NETD (median): 20 mK
Defective elements: <1%
Temperature: 70 K
Cooling: Rotary Stirling engine
Array/ multiplexer
Intrinsic MUX noise: 100 μV rms
Scan format: blinking
Electrical characteristics
Pixel rate: up to 5 MHz
Max operating voltage: 7 V
Power consumption: 50 mW
Output impedance: <80 Ω typical
Assembly power consumption: <10 W steady state
Environmental
Operating temperature: –45 to + 70°C
Mechanical
Weight: ≈600 g

Contractor
BAE Systems, Avionics Tactical Systems Division.

UPDATED

BAE Systems Condor MW 320 × 256 Indium Antimonide integrated detector cooler assembly

Type
Integrated detector cooler assembly.

Description
The BAE Systems Condor MW detector/cooler uses the company's Indium Antimonide (InSb) process and is designed for high-performance low-cost imaging in the medium (3-5 μm) waveband. It operates as a 2-D 'blinking mode' infra-red detector. It is described as giving high electro-optic performance with low crosstalk and automatic anti-blooming at the pixel level. In outward appearance the assembly looks similar to the BAE Systems Osprey assembly.

Status
Available.

Specifications
Array
Array format: 320 × 256 pixels
Element size: 30 μ
Active area: 9.6 × 7.68 mm
Spectral band: 3 to 5 μm
Technology: CMOS
NETD (median): 10 Mk
Defective elements: <1%
Temperature (nominal): 80 K
Cooling: Rotary Stirling engine
Array/ multiplexer
Intrinsic MUX noise: <200 μV
Scan format: blinking
Electrical characteristics
Pixel rate: up to 5 MHz
Max operating voltage: 7 V
Power consumption: 50 mW
Output impedance: <80 Ω typical
Assembly power consumption: <8 W steady state

Environmental
Operating temperature: –45 to + 70°C

Mechanical
Weight: ≈600 g

Contractor
BAE Systems, Avionics, Tactical Systems Division.

UPDATED

BAE Systems Falcon 576 channel CMT integrated detector/cooler assembly

Type
Integrated detector cooler assembly.

Description
The BAE Systems Falcon detector/cooler uses the company's CMT process to provide high environmental integrity in a focal plane detector. It is intended to provide high performance at low cost. The detector operates in the long waveband (8 - 10 μm). Falcon uses Time Delay Integration (TDI) and a memory deselection function.

Status
Available.

Specifications
Array
Array format: 576 channels arranged as 2 blocks of 288 channels
Element size: 30 μ
Element stagger: 15 μm line to line

The BAE Falcon 576 channel CMT integrated detector/cooler assembly 0089866

The 2 block detector used in the Falcon assembly 0089911

Active length: 14.3 × 7.2 mm
Spectral band: 8 to 10 μm
Detector: CMT
Technology: CMOS
NETD (mean): 25 mK @ 50 μs integration, f1.5
Dead channels: Zero
Temperature (nominal): 75 K
Cooling: rotary Stirling engine
Array/ multiplexer
Intrinsic MUX noise: <200 μV
Electrical characteristics
Pixel rate: up to 5 MHz per output
Power consumption: <150 mW
Output impedance: <80 Ω typical
Assembly power consumption: <10 W steady state

Environmental
Operating temperature: –45 to + 70°C

Mechanical
Weight: ≈600 g

Contractor
BAE Systems, Avionics, Tactical Systems Division.

UPDATED

BAE Systems Infra-Red detector technology

Type
Infra-red detectors.

Description
BAE Systems, Avionics, Infra-Red is Europe's largest supplier of infra-red detectors. The company has more than 40 years' experience in the design and manufacture of infra-red detectors for military, industrial and space applications.

The company manufactures photoconductive and photovoltaic infra-red detectors in Cadmium Mercury Telluride, including Focal Plane Arrays (FPAs) in Dewars for Joule-Thompson or engine cooling and thermoelectrically cooled infra-red CMT detectors. Pyroelectric detectors are manufactured in Lithium Tantalate, Deuterated L-Alanine-doped Tryglycene Sulphate and ceramic uncooled FPAs are also available.

Contractor
BAE Systems, Avionics, Tactical Systems Division.

UPDATED

BAE Systems North America 1st Generation HgCdTe Infra-red Imaging Systems

Type
Infra-red detectors.

Description
BAE Systems North America has developed a Long-Wave Infra-Red (LWIR) 128 × 128 staring two-dimensional focal plane array for system applications in ground-based interceptors, anti-tank and air-to-air missiles.

Operational status
Available. BAE Systems North America has developed Dewar packages for the US Common Modules and Imaging Infra-Red (IIR) Maverick missile system as well as chemical agent detector units. Over 40,000 units have been produced in support of these programmes.

Specifications
Detector array: 128 × 128 HgCdTe staring array
Detector element size: 45 × 45 μm
Number of outputs: 4
Output data rate: 2 MHz
Channel operability: >95% typical
Frame rate: <480 Hz
Cold shield f/number: 2.4
1/f noise knee: <20 Hz
Crosstalk: <5%
Detector electronics power: <0.5 W
MRTD: 0.02°C
Scene dynamic range: –23 to +87°C
Ambient operating temperature: –55 to +81°C
FPA operating temperature: –193°C (80 K)

Contractor
BAE Systems North America.

UPDATED

BAE Systems Osprey 384 × 288 CMT integrated detector cooler assembly

Type
Integrated detector cooler assembly.

Description
The BAE Systems detector/cooler assembly uses the company's CMT process to provide high environmental integrity in a focal plane detector. It is intended to provide high performance at low cost. The detector operates in the mid-waveband (3 - 5 μm). The detector array is very small to allow for miniaturisation of the encapsulation and optics. It operates in a 2-D 'blinking mode'. The detector operating temperature (120 K) is higher than for earlier cooled devices.

Status
Available.

Specifications
Array
Array format: 384 × 288 pixels
Element size: 20 μ
Active area: 7.68 × 5.76 mm
Spectral band: 3 to 5 μm
Technology: CMOS
NETD (median): 14 mK
Defective elements: <1%
Temperature: <120 K
Cooling: rotary Stirling engine
Array/ multiplexer
Scan format: blinking
Intrinsic MUX noise:100 μV rms
Electrical characteristics
Pixel rate: up to 5 MHz per output
Max operating voltage: 7 V
Power consumption: 60 mW
Output impedance: < 80 Ω typical
Assembly power consumption: <5W steady state
Environmental
Operating temperature: –45 to +70°C
Mechanical
Weight: ≈600 g

Contractor
BAE Systems, Avionics, Tactical Systems Division.

UPDATED

BAE Systems 384 × 288 CMT detector used in the Osprey detector cooler assembly 0089918

BAE Systems SPRITE/TED CMT integrated detector cooler assembly

Type
Integrated detector cooler assembly.

Description
The BAE Systems LW detector/cooler with a CMT SPRITE (Signal Processing In The Element) uses the company's multiple quantum well process and is designed for high-quality imaging in the long (8 - 12 μm) waveband.

Status
Available. Used in IR imagers by defence forces worldwide.

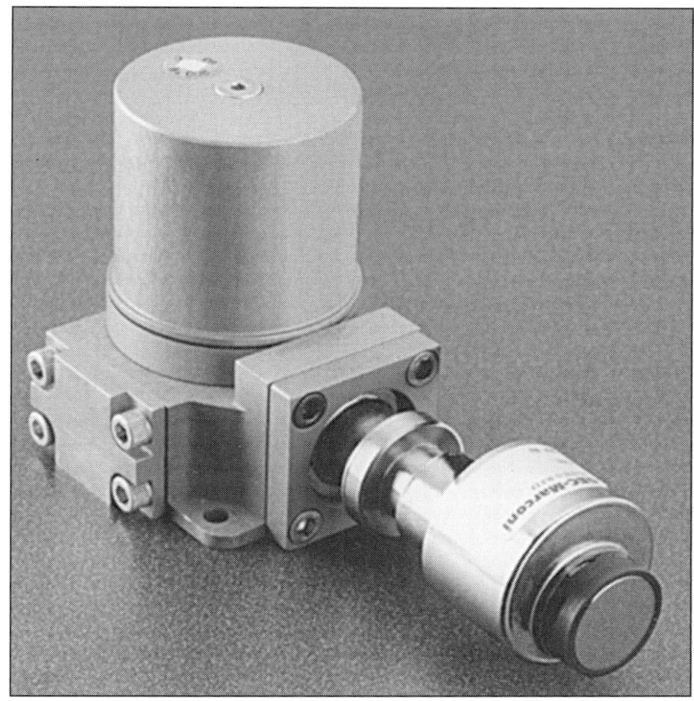

BAE Systems Integrated detector cooler assembly with SPRITE CMT array 0089868

Specifications
Array
Array format: 8 parallel in-line
Element sensitive area: 700 × 62.5 μm
Element readout area: 62.5 × 62.5 μm
Element pitch: 75 μm
Spectral band: 8 to 12 μm
Temperature: 80 K
Operating voltage: 2.1 V typical
Power consumption: 75 mW typical
Cooling: rotary Stirling engine
Cool-down time: 5 min
Assembly power consumption: during cool down 24 W, steady state <10 W typical
Environmental
Operating temperature: –45 to + 70 °C
Mechanical Weight: ≈600 g

Contractor
BAE Systems, Avionics, Tactical Systems Division.

UPDATED

DRS 640 × 480 infra-red focal plane array, Integrated Dewar Cooling Assembly (IDCA)

Type
Infra-red detector.

Description
DRS Infrared Technologies produce a range of scanning and staring infra-red Focal Plane Arrays (FPAs). Typical of these is a 640 × 480 element device, which is available in both Long-Wave Infra-Red (LWIR) and Medium-Wave Infra-Red (MWIR) variants. The FPAs use HgCdTe infra-red sensitive material, tuned to the desired cut-off wavelengths.

Operational status
In service and in production.

Specifications
Spectral band: LWIR 10-10.5 μm, MWIR 5-5.5 μm
Pixel pitch: 25 μm
Outputs: 4
Normalised detectivity:
(LWIR) D*>5 × 10^9 Jones
(MWIR) D*>1.5 × 10^{10} Jones

Contractor
DRS Infrared Technologies.

VERIFIED

DRS TCM2000 mid-wave infra-red focal plane array

Type
Infra-red dectector.

Description
DRS is a supplier of Infra-Red (IR) focal planes for strategic defence, tactical systems and commercial, scientific and astronomy applications.

The TCM2000 is a 256 × 256 (65,536 active pixels) Mercury Cadmium Telluride Focal-Plance Array (FPA), mated to a single-output CMOS multiplexer that operates in a ripple readout mode (30 × 10^6 electron well-capacity). This array is designed for high-background (10^{12} to 10^{15} photon/cm^2/s) military and commercial imaging applications such as surveillance cameras, fire-control FLIRs, navigation, missile warning, predictive maintenance and medical imaging.

Operational status
Available.

Specifications
Detector type: photovoltaic MCT
Array format: 256 × 256
Detector size: 40 × 40 μm
Fill factor: =85%
Detector pitch: 40 μm
Spectral response: 1-4.85 μm
Cutoff wavelength: 4.85 μm
Quantum efficiency: >60%
D* (Qb=2 × 10^{14} ph/cm^2/s): 4.0 × 10^{11} cm-√Hz/W

Contractor
DRS Technologies, Electro-Optical Systems Group.

DRS U3000 uncooled microbolometer infra-red sensor

Type
Infra-red detector.

Description
The U3000 is a 320 × 240 (76,800 active pixels) uncooled focal-plane that is based on Vo_x resistive technology. The focal plane provides a single-channel signal output at up to 60 Hz frame rates. Both horizontal and vertical blanking can be externally controlled by the user to achieve signal formats compatible with either NTSC or PAL TV standards. The U3000 is packaged in a rugged, miniaturised assembly that incorporates an LWIR infra-red window, a solid-state temperature sensor and a refireable vacuum getter. Room-temperature operation eliminates the need for cyrocooling, which improves sensor reliability while reducing power consumption and cost. The U3000 is designed to operate under the full range of terrestrial imaging conditions for both military and commercial LWIR imaging and remote-sensing applications.

The uncooled FPAs have a wide variety of applications, ranging from military night sights and driving aids to area surveillance, firefighting, industrial radiometry, search and rescue, border patrol and vehicle collision-avoidance.

Operational status
Available.

Specifications
Array format: 320 × 240
Multiplexes: CMOS ripple integration
Detector type: Vo_x resistive bolometer
Detector pitch: 51 × 51 μm
Video output: Standard TV format, CMOS/PAL, NTSC Apache
Area fill-factor: ~60 %
Operability: >98%
Spectral response: 8-14 μm
NETD: ≤100 mK
Nominal data rate: 5 MHz
Nominal frame rate: 60 Hz

Contractor
DRS technologies, Electro-Optical Systems Group.

UPDATED

Indigo Systems' IR focal plane arrays

Type
Infra-red detector.

Description
Indigo Systems is a premier supplier of Indium Antimonide (InSb) focal plane arrays in 320 x 256 and 640 × 512 pixel formats. These arrays are based on Indigo's

The Indigo Systems ISC 9803 InSb focal-plane array ***NEW**/0595414*

standard Read-Out Integrated Circuits (ROICs), providing high performance combined with a flexible architecture. The ISC9705 based arrays are the industry standard for 320 × 256 format on 30μ pitch. Indigo's large format arrays – the ISC9803 with 25μ pitch and the ISC9901 with 20μ pitch are the core of today's high-resolution IR systems.

Operational status
Available, in production.

Specifications
Type: ISC9705 InSb FPA
Array format: 320 × 256
Array size: 11.35 × 10.65 mm
Element sensitive area: 30μ
Element readout area: 30μ
Element pitch: 30μ
Spectral band: 3-5μ
Frame Rate: 110 Hz, 202 Hz, 346 Hz for 1, 2 and 4 outputs
Operability: >99.9% typical
Data rate: 10 MHz
Integration time: >5 u sec
Temperature: 80 K
Cooling: LN2 or closed-cycle
Electrical Characteristics: 19 interface pads for Default mode
Power Consumption: <30 mW (one output), <120 mW (four outputs)
Output: 1, 2 or 4 outputs
Well Capacity: 18Me-

Type: ISC9803 InSb FPA
Array format: 640 × 512
Array size: 17 × 17.5 mm
Element sensitive area: 25μ
Element readout area: 25μ
Element pitch: 25μ
Spectral band: 3-5μ
Frame Rate: 30 Hz, 58 Hz, 107 Hz for 1, 2 and 4 outputs
Operability: >99.9% typical
Data rate: 10 MHz
Integration time: >9.6 u sec
Temperature: 80 K
Cooling: LN2 or closed-cycle
Electrical Characteristics: 19 interface pads for Default Mode
Power Consumption: <90 mW (one output), <180 mW (four outputs)
Output: 1, 2 or 4 outputs
Well Capacity: 11Me-

Type: ISC9901 InSb FPA
Array format: 640 × 512
Array size: 14.65 × 14.71 mm
Element sensitive area: 20μ
Element readout area: 20μ
Element pitch: 20μ
Spectral band: 3-5μ
Frame Rate: 30 Hz, 55 Hz, 97 Hz for 1, 2 and 4 outputs
Operability: >99.9% typical
Data rate: 10 MHz
Integration time: >100 u sec
Temperature: 80 K

Specifications Raytheon focal plane arrays:

Modes	AE159, AE188	AE166	AE173	AE187	AE194	AE195	AE197
Array format	128 × 128	256 × 256	256 × 256	512 × 512	256 × 256 (320 × 240)	384 × 256	640 × 512
Integration mode	Sequential	Sequential	Snapshot	Snapshot	Snapshot	Snapshot	Snapshot
Detector type	InSb, CMT, Si:Ga	InSb	InSb	InSb	InSb	InSb	InSb
Junction size	43 μm	31 μm	27 μm	19 μm			
Pixel pitch	50 μm	38 μm	30 μm	25 μm	30 μm	25 μm	25 μm
Array size (side)	6.40 mm	9.73 mm	7.67 mm	2.80 mm	7.67 mm	9.60 × 6.40 mm	18 × 16.3 mm
Array size diagonal	9.04 mm	13.77 mm	10.84 mm	18.11 mm	10.85 mm	11.53 mm	
Integration time	3-97% T frame	3-97% T frame	2 μs-99% T frame	2-98% T frame	2 μs-99% T frame	2 μs-99% T frame	2 μs-99% T frame
Frame rate	≤1,000 Hz	=240 Hz	120-6,600 Hz	68-19,000 Hz	140-6,600 Hz	150-15,000 Hz	120-50,000 Hz

Cooling: LN2 or closed-cycle
Electrical Characteristics: 19 interface pads for Default Mode
Power Consumption: <90mW (one output), <180 mW (four outputs)
Output: 1, 2 or 4 outputs
Well Capacity: 7Me-

Contractor
Indigo Systems Corporation (USA)

NEW ENTRY

Raytheon 7060 Cryocooler

Type
Cooler assembly.

Description
The 7060 linear-drive split-Stirling cryocooler has been developed to meet stringent performance and environmental specifications for most airborne and ground-based systems. Three types of coldfingers are available to accomodate first-generation, SADA IIIA, and SADA IIIB standard Army C/F configurations. Its versatile, compact, light-weight design is suited to a variety of commercial, industrial, and military applications.

Features

- cooling capacity up to 0.5 W @ 77 K, 23°C ambient
- production proven
- performance tested to US military specifications and environments
- linear resonant design
- maintenance free
- system compatibility: first-generation slip-on and SADA III coldfingers
- coldfinger size options: 5.067 and 6.6 mm (.196 and .260 in) in diameter
- slip-on or IDA compatible design
- low acoustic noise
- low output vibration
- high vibration applications
- lightweight and compact
- temperature control electronics available
- custom transfer tube length and formations are available.

Specifications

Parameter	7060-196	7060-260S	7060-260L
Refrigeration capacity @ 77 K:			
(in 23°C ambient, W)	0.45	0.50	0.50
(in 50°C ambient, W)	0.25	0.35	0.35
Cooldown time to 77°C:			
(in 23°C ambient, min)	≤6.0	<4.5	<5.0
(in 50°C ambient, min)	≤8.0	<6.0	<7.0
Thermal mass:			
(296 K-77 K, Joules)	250	250	250
Max AC input power:			
(@ 23°C ambient, W)	≤18	=27	≤27
AC RMS input voltage:			
(@ 23°C ambient)	11.0 ±0.5	11.0 ±0.5	11.0 ±0.5
(@ 50°C ambient)	12.0 ±0.4	12.0 ±0.4	12.0 ±0.4
Frequency (Hz)	60	60	60
Operating temperature range (°C)	−40 to +60	−40 to +60	−40 to +60
Generated vibration output:			
(lb/rms)	≤0.3	=0.3	≤0.3
MTTF (predicted life in hrs):	>4,000	>4,000	>4,000
Weight (lb):	1.2	1.25	1.3

Notes:
1. Specifications subject to change.
2. Performance data assumes transfer tube length of 6 in or less for 196, 260 (S,260)L.
3. Special transfer tube lengths/configurations available upon request.

Contractor
Raytheon Systems Company (Raytheon Infrared Operations).

UPDATED

Raytheon focal plane arrays

Type
Infra-red detectors.

Description
Raytheon produces mid-waveband indium antimonide (InSb) focal plane arrays as well as mid- and long-wave focal plane arrays using Cadmium Mercury Telluride (CMT/HgCdTe), or gallium-doped silicon (Si:Ga), for integration into dewar packages or to support turnkey imaging needs. These arrays use either sequential readout mode or simultaneous integration ('snapshot' mode).

Specifications
See above.

Operational status
In production.

Contractor
Raytheon Systems Company.

VERIFIED

Raytheon focal plane arrays

Type
Infra-red detectors.

Description
Raytheon manufactures a broad range of infra-red detectors/Dewar assemblies, both long and mid-wave, incorporating linear and two-dimensional focal plane arrays for scanning and staring applications. Detectors are produced using various materials including Cadmium Mercury Telluride (CMT or HgCdTe) and indium antimonide (InSb), platinum silicide (PtSi) and arsenic-doped silicon (Si:As IBC).

SBRC began production of second-generation InSb focal plane arrays in 1982 and in 1989 developed VISMIR technology which improves spectral sensitivity in the visible as well as the infra-red spectral range (0.5 to 5.5 μm).

Raytheon Santa Barbara also makes a 240 × 320 silicon microbolometer array for the new generation of uncooled thermal imaging technology. Potential applications include night vision, process controls, infra-red spectroscopy and medical imaging.

Operational status
In production. Applications for SBRC's focal plane arrays include the Raytheon Thermal Weapon Sight AN/PAS-13 (thermo-electrically cooled HgCdTe 40 × 16), ASRAAM missile seeker (InSb 128 × 128 staring FPA) and TOW LightWeight Launcher (LWL) (HgCdTe 240/288 × 4 FPA). FLIR Systems and Inframetrics use SBRC platinum silicide detectors for their lightweight, portable infra-red cameras.

Specifications
InSb focal plane arrays
128 × 128 InSb MWIR VISMIR hybrid detector array
Element size: 40 × 40 μm
Overall chip size: 9.2 × 7.3 mm
Spectral response Σ≥90%: 0.5-5.4 μm
Dynamic range: ≥70 dB
Operability: ≥90%
Frame rate: Variable to 120 Hz
Integration times: 10 μs to 8.6 ms (selectable, dependent on frame/output rate)
Data rate: ≤5 MHz (per channel)
Power dissipation: ≤70 mW
Package: sealed metal Dewar

SBRC's 128 × 128 indium antimonide medium-wave infra-red hybrid detector array

The 256 × 256 indium antimonide medium-wave infra-red hybrid detector array from SBRC

256 × 256 InSb MWIR VISMIR hybrid detector array
Element size: 30 × 30 μm
Overall chip size: 10.5 × 9.4 mm
Spectral response $\Sigma \geq 90\%$: 0.5-5.4 μm
Dynamic range: ≥70 dB
Operability: ≥90%
Frame rate: Variable to 120 Hz
Integration times: 10 μs to 8.6 ms (selectable, dependent on frame/output rate
Data rate: ≤5 MHz (per channel)
Power dissipation: ≤100 mW

480 × 640 InSb MWIR VISMIR focal plane array
Element size: 20 × 20 μm
Overall chip size: 14.2 × 15.5 mm
Spectral response $\Sigma \geq 90\%$: 0.5-5.1 μm Σ_c=5.45 ±0.1 μm (also available with IR response only, 3.0-5.1 μm)
Dynamic range: ≥70 dB
Operability: ≥90%
Frame rate: variable to 60 Hz
Integration times: >10 μs to 7 ms
Data rate: ≤6 MHz, (per channel)
Power dissipation: ≤75 mW
Package: sealed metal Dewar

HgCdTe focal plane arrays
240/288 × 4 LWIR HgCdTe focal plane array
Element size: 28 μm in scan, 38 μm cross scan; or 22 μm in scan, 25 μm cross scan
Element pitch: 44 μm in scan × 51 μm cross scan
Sample spacing: 22 μm in scan, 25.5 μm cross scan
Overall chip size: 93.5 × 110.9 mm
Spectral response: 7.65 ±0.25 μm to 10.0 + 0.5-0.0 μm
NEDT: 0.018K at f/2.31 and 34.7 kHz
Dynamic rating: >72 dB

The 480 × 640 indium antimonide VISMIR (visible through medium-wave infra-red) focal plane array from SBRC

Operability: >97%
Data rate: 5.0 MHz max
Package: standard Dewar

256 × 256 LWIR HgCdTe focal plane array
Element size: 30 × 30 μm
Overall chip size: 10.5 × 9.4 mm
Spectral response (80K): 8.5-11.0 μm
Dynamic range: ≥66 dB
Operability: ≥99.0%
Frame rate: variable to 120 Hz
Integration times: 10 μs to 8.6 ms (selectable, dependent on frame/output rate)
Data rate: ≤5 MHz (per channel)
Power dissipation: ≤ 100 mW
Package: sealed metal Dewar

A 240 × 320 silicon microbolometer uncooled focal plane array, with spectral response 8-14 μm

Thermo-electric Cooled Mid-wave HgCdTe focal plane array/ Dewar assembly

Model	**TE12840**	**TE12850**	**TE25630**
HgCdTe detector array	128^2 pixels	128^2 pixels	256^2 pixels
Pixel size and centre to centre spacing	40 × 40 μm	50 × 50 μm	30 × 30 μm
Spectral 50% response cutoff available	>4.5 μm	>4.5 μm	>4.5 μm
Operability	99%	99%	99%
Dynamic range at 2×10^{14} photons/cm^2- s	>3,000	>3,000	>3,000
NEDT at 25°C ambient, f/1.6 optics & 1 ms integration	<.050°C*	<.050°C*	<.065°C*
Frame rate	120 Hz max	120 Hz max	100 Hz max
TE cooler power with 35°C base temperature	6/3 V/A	6/3 V/A	6/3 V/A

*Preliminary

SBRC's 256 × 256 cadmium mercury telluride long-wave infra-red detector

240 × 320 Silicon microbolometer uncooled focal plane array
Element size: 48 × 48 µm
Overall chip size: 17.4 × 14.6 mm
Spectral response: 8-14 µm
Signal response: >6.5 MV/°C
NEDT (in 7-13.5 µm spectral band): <50 mK at f/1.0, 300 K
Output swing: 2.0 V
Offset non-uniformity: <100 mV p-p
Output noisefloor: 350 µV RMS
Dynamic range: 70 dB
Frame rate: 60 Hz or less
Data rate: <6 MHz

Contractor
Raytheon Systems Company.

UPDATED

Raytheon Santa Barbara Standard Advanced Dewar Assembly (SADA)

Type
Infra-red detector.

Description
The Standard Advanced Dewar Assembly (SADA) is an important feature of Raytheon second-generation infra-red sensor technology. SADA II has multimode output, 480 lines, 480 lines with 2:1 electronic interlace or 240 lines. It can have up to six photovoltaic detectors in time delay integration for increased sensitivity and range. Cooling is split Stirling with twin-piston linear drive for low vibration.

It is used on the improved target acquisition system for the TOW anti-tank missile launcher, Hughes Second-Generation Tank Sight and Horizontal Technology Integration (HTI) FLIR.

SBRC's SADA IIIB detector/Dewar assembly

SADA IIIB has four photovoltaic detectors in time delay integration mode and output is 240 or 280 lines programmable for either PAL or NTSC operation. It is used on the Spanish Lightweight TOW Launcher thermal targeting system and others. SADA detectors are suitable for a wide range of thermal imaging applications including night vision, infra-red search and track and thermal targeting sights.

Operational status
In production.

Specifications
SADA II
Detector: 480 × 4 photovoltaic CMT
Min resolvable temperature: 0.02°C
Scene dynamic range: –33 to +87°C
Sample rate: 20-160 kHz
Video output rate: 0.5-5.0 MHz
Coldshield f-number: 2.5 standard, custom available
1/f noise knee: 30 Hz
Crosstalk: 0.1%
Cooler input power: 20 W nominal, 60 W max
Detector electronics power: 4 W
Weight: 0.635 kg (Dewar); 1.905 kg (cooler)
Dimensions (Dewar assembly): 8.687 × 10.617 × 10.668 cm

SADA IIIB
Detector pitch (azimuth × elevation): 44 × 51 µm
Detector optical area (azimuth × elevation): 28 × 38 µm
NEDT: 0.020°C
Scene dynamic range: –33 to +87°
Sample rate: 20-83 kHz
Video output rate: 1.0-5.0 MHz
Coldshield f/number: 2.31 standard, custom available
Cooler input power: 12 W nominal, 24 W max
Detector electronics power: 1.5 W
Weight: 0.295 kg (Dewar assembly); 0.454 kg (cooler)
Dimensions: 11.532 × 6.045 cm

Contractor
Raytheon Systems Company.

UPDATED

SCD Blue Fairy 320 × 256 Advanced InSb FPA

Type
Infra-red detector.

Description
Blue Fairy is an advanced InSb Focal Plane Array.

Features
- Based on SCD's BF I readout
- Integrate-while-read, Integrate-then-read, and a new combined mode
- Outstanding linearity
- Negligible blinking pixels
- Very stable residual non-uniformity

Options
- Stand alone FPA or integrated in a long vacuum-life tactical dewar
- Integral or split-Stirling cooler
- Proximity electronics board supporting 1,2 or 4 video lines
- Supporting video electronics on request

Operational status
Available.

Specifications
(at 80 K operating temperature)
Detector Type: InSb Focal Plane Array
Format: 320 × 256 pixels
Readout architecture: Si-CMOS, 0.5 µm, DPTM
Integration mode: Snapshot, IWR/ITR/proprietary combined mode
FPA size: 11.4 × 10.7 mm^2
Spectral response: 1 – 5 µm
Optical fill factor: 90%
Quantum Efficiency: 75%
Pixel dark current (78 K): 10 pA approx
Detector input current: 1 pA 1.5 µA
Integration Capacity: 10 Me- @ IWR
20 Me- @ ITR
20 Me- @ combined mode
Operability: 99.50%
Noise floor (nominal): 1,000 electrons*
Non-uniformity (uncorrected): 5% (σ/mean)

Integration time: >1μs to 99% frame time @ IWR mode
NETD (60% well fill): 17 mK
Outputs: 1, 2 or 4 selectable
* at max gain setting

Contractor
SCD SemiConductor Devices.

VERIFIED

SCD Flamingo InSb 640 × 512 IDCA (20 μm)

Type
Integrated Dewar cooler assembly.

Description
The Flamingo Integrated Dewar cooler Assembly (IDCA) combines the custom ISC9901 readout and SCD's knowledge of detector technology.

Features
- Based on Indigo Systems 990 custom ROIC (20 μm pitch)
- Snapshot Integrate-then-read operation mode
- Negligible blinking pixels
- Very stable residual non-uniformity

Options
- Long vacuum-life tactical dewar
- Integral or split-Stirling cooler
- Proximity electronics board supporting 1, 2 or 4 video lines
- Supporting video electronics on request

Operational status
Available.

Specifications
(at 80K operating temperature)
Format: 640 × 512 pixels
Spectral response: 3-5 μm
Readout: Indigo Systems 9901
Pixel pitch: 20 μm
Storage capacity:
7e[6] electrons (350 F Cflt)
3e[6] electrons (150 W Cint)
Operability: 99.5%
Dynamic range: > 72 Db
Readout noise: 300 electrons *
Non-linearity: <0.5%
Cross talk: <0.1%
Input polarity: p on n
Integration time: >5 μsec, adjustable
Outputs: 1, 2 or 4 selectable
Output signal swing: 2.5 V
Power consumption:
<90 mW at 30 Hz single output
<180 mW at 107 Hz, 4 outputs
Max frame rates: 1 output, 30 Hz
(640 × 480) 2 outputs, 58
4 outputs, 107 Hz
Video output: NTSC or PAL
*at max gain setting

Contractor
SCD SemiConductor Devices.

SCD PV 256 256 ×1 MCT LWIR IDCA

Type
Infra-red detector cooler assembly.

Description
The PV 256 IDCA is a high-performance PhotoVoltaic Mercury Cadmium Telluride (PV-MCT) array, designed for thermal imaging applications. The compactness and light weight of the integral detector-dewar-cooler, make it an ideal choice for applications such as drivers viewers, personal weapon sights and hand-held thermal imagers.

Features
- State-of-the-art CMOS readout architecture
- 256 scan lines
- Four video output lines and simple system interface
- Advanced dewar technology with RICOR's K508 cooler

Operational status
Available.

Specifications
Detector size: (cross-scan/inscan) 24 × 24 μm²
Detector pitch: 20 μrn
Output dynamic range: 3 V
Sample rate: up to 50 KHz (4 video outputs)
Video output range: up to 7 MHz per output
Typical integration time: 50 μsec
IDDCA weight: < 500 gm
Operating temperature: –40°C/70°C
Vacuum integrity: > 5 years (at room temp)

	Figure of Merit		Typical value (Remarks)
Waveband	7.8-10.2 μm		
Cold shield F#	Standard f/I .5		Upon customer request
Cool down time	<4.5 min @ 23°C		
Steady-state cooler Input power	<5.5 W		300 K background uncorrected
D*(peak) @ Tint	20 μs		7 × 10[10] Jones
Signal Responsivity		3% RMS	
Uniformity after NUC	<0.5% RMS		After 2 Point Correction
Crosstalk to Nearest		Combined optical	
Neighbour	<2%		Electrical
Array Operability	<99%		D* > 0.5D*(avg)

Contractor
SCD SemiConductor Devices.

UPDATED

SCD TDI 480 LWIR Integrated Detector-Cooler Assembly

Type
Integrated detector cooler.

Description
A high sensitivity MCT-LWIR Integrated Detector-Cooler Assembly, having 480 scan lines with Time Delay Integration (TDI) on six lines and advanced CMOS readout architecture. The TDI 480 is intended for high-end LWIR applications, where high resolution and sensitivity are necessary.

Features include:
- Programmable integration time
- Adjustable gain
- Bi-directional scan
- Bad pixel deselection
- Good linearity and stability.

Operational status
Available.

Specifications
Elements size (cross-scan/inscan): 35 × 28 μm²
Output dynamic range: 3 V
Spectral response: LWIR
Sample rate: up to 50 KHz @ 4 video outputs
Video output frequency rate: up to 7 MHz per output
Typical integration time: 20 s selectable
IDDCA weight: 0.5 kg
Operating temperature: –40 to +70°C
Vacuum integrity: >5 years without activation
F No: F/3
Steady-state power consumption: 20 W
Signal responsivity uniformity: 3% RMS typical uncorrected
Cross talk to nearest neighbour: <2% combined optical and electrical
Array operability: < 99%, D* >0.5 × D* (avg)

Contractor
SCD SemiConductor Devices.

UPDATED

Sofradir infra-red detector technology

Type
Infra-red detectors.

Description
Sofradir produces a range of second- and third-generation infra-red detectors using Mercury Cadmium Telluride (MCT). Detectors are delivered as Detector Dewar Assemblies (DDA) or Integrated Detector Dewar Cooler Assemblies (IDDCA). These detectors are adapted to thermal imager, seeker, IRST, POD, space and other applications.

Sofradir's technology has matured significantly since the early 1990's with detector prices falling over the period 1991 to 1996 by 60 per cent. In addition, in 1996 they received Texas Instruments' Supplier Recognition Award for the timely and compliant deliveries of over 500 detector assemblies.

Current capability provides photovoltaic detectors that may be tuned to 1 to 3 µm, 3 to 5 µm or 8 to 12 µm bands. A common planar process is used which creates homojunctions which interconnect with dry etching and ion implantation throughout. Hybridisation is via a fully automated indium bump reflow process.

Research is carried out by LIR (Infra-red Research Laboratory) near the Sofradir plant in Grenoble. Here basic research into photovoltaic devices, hybridisation techniques and focal plane electronics is underway. Transitioning to production in Sofradir has been smooth and Sofradir is now producing more then 200 detectors a month. In addition, standard hybridisation techniques have been applied to QWIP's detectors. Bolometers arrays operating at room temperature are now under manufacture.

MCT linear CCD arrays are produced with 32 to 1,500 pixels. All have CMOS readout techology. Linear arrays with in-built time-delay and integration are available in 4 × 48 to 6 × 480 element arrays. 3.5 µm, 5 µm and 10.5 µm cut-off wavelengths are available throughout most of the range.

Area arrays are available with 128 × 128, 256 × 256 and 320 × 256 elements. Again 3.5 µm, 5 µm and 10.5 µm are available. All have CMOS readout. Larger arrays of 640 × 480 elements are now being delivered by Sofradir.

Operational status
Sofradir's 288 × 4 detector arrays are used in many production infra-red systems such as Sophie, IRIS, Synergi, DNTSS and the AN/AAS-17. Their custom devices are also in development and production for a wide range of programmes including: TRIGAT (LR) missiles, TRIGAT (LR) sight, MICA missile, the OSF IRST for the RAFALE fighter, VAMPIR B naval IRST, HELIOS II and US programmes. The 480 × 6 (SADA II) is qualified and production deliveries are being made for US Army orders. For the SADA II detector, Sofradir entered the Low-Rate Initial Production (LRIP) phase at the beginning of 1999. At the end of 1997 Sofradir joined the biggest European stand-off missile seeker programme which represents more then 1,500 DDAs. LRIP started in early 1999.

Over 6,000 detector Dewar assemblies have been delivered in recent years, with some 30 per cent of these for export markets.

The Sofradir ID TL008 480 × 6 long wave infra-red focal plane array 0023273

Specifications
IDTL060 288 × 4 LWIR FPA/IDCA (long-wave Infrared focal plane array/integrated detector/cooler assembly)
Array type: MCT 8-12 µm TDI (Time Delay and Integration)/multiplexed focal plane array
Detector: 288 × 4 MCT PV diodes
TDI: available on 4 elements
Number of signal outputs: 16
Output dynamic range: 2.5 V (>80 dB)
Element sensitive size: (in scan) 25 µm × (cross scan) 28 µm
NETD (average): 20 mK
Pixel rate: 2 MHz max per output

IDML018 128 × 128 LWIR FPA/IDCA
Array type: 8-12 µm staring/snapshot focal plane array
Number of elements: 16,384 MCT PV diodes
Array format: 128 × 128
Dynamic range: 2.8 V (81 dB)
NETD (average): 10 mK
Frame rate: up to 300 Hz

The Sofradir 288 × 4 focal plane array 0005501

The Sofradir 320 × 256 medium-wave infra-red focal plane array/integrated detector Dewar cooler assembly 0023272

IDTL056 480 × 6 LWIR FPA/IDCA (long wave infra-red focal plane array)
Array type: MCT 8-12 μm TDI (Time Delay and Integration) multiplexed focal plane array
Detector: 480 × 6 MCT PV diodes
TDI: available on 6 elements
Number of signal outputs: 16
Output dynamic range: 3 V (74 dB)
NETD (average): 20 mK
Pixel rate: 5 MHz max per output

IDMM016 320 × 256 MWIR FPA/IDDCA
Array type: MCT 3-5 μm staring-snapshot focal plane array
Number of elements: 81,920 MCT PV diodes
Array format: 320 × 256
NETD (average): 88 mK
Frame rate: variable to 400 Hz
Number of signal output: 1-4 selectable

Contractor

Sofradir.

UPDATED

THERMAL IMAGER MODULES

BAE Systems TICM II Thermal Imaging Common Modules

Type
Thermal imager modules.

Development
The Thermal Imaging Common Module (TICM) programme was sponsored by the UK Ministry of Defence. BAE Systems (then Marconi) was contracted to manufacture TICM II (the indirect view system) which began to enter service in the early 1980s. Development of the system has continued with a range of product enhancements.

Description
TICM II thermal imaging modules are built to full military standard and use advanced optical and electronic components, including the BAE Systems Infra-Red Ltd SPRITE detector, to give high resolution and sensitivity even at long range. Optical and electronic processing modules are based on the coaxial scanning polygon which requires only one drive motor for both azimuth and elevation scans. The modular basis gives flexibility in system design. Telescopes and displays can be selected by system designers to meet precise operational requirements.

Images from these indirect view systems can be displayed on television monitors or head-up and head-down displays. They are fully automatic in operation and allow module replacement without any adjustment or set up time.

The systems are suited to applications demanding the highest resolution and reliability such as pilot night vision, target detection and tracking, weapon aiming and surveillance.

Operational status
TICM modules are in large-scale production for a variety of purposes and are in service with a number of countries.

Specifications
Spectral band: 8-13 μm
Field of view
625-line 50 Hz: 60 × 40°
525-line 60 Hz: 48.4 × 32.5°
Resolution: 2.27 mrad (60° field of view)
Pupil diameter: 10 mm
Min resolvable temperature difference: typically better than 0.1°C
Detector: 8 parallel CMT SPRITE
Video output: CCIR Systems I/EIA-RS-170, compatible 625/525-line 50/60 Hz
Weights (modules in enclosures)
Scanner head: 9.5 kg
Processing electronics: 6.5 kg
System control panel: 1.03 kg
×20 telescope: 10 kg

Contractor
BAE Systems, Avionics, Tactical Systems Division.

UPDATED

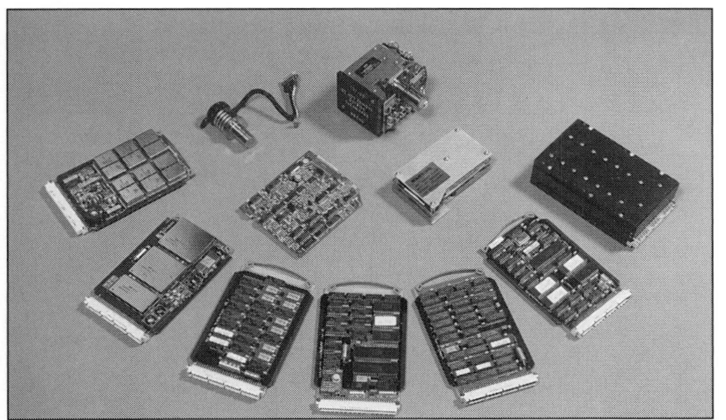

BAE Systems TICM II thermal imaging common modules 0005494

BAE Systems Wizard uncooled thermal imaging module set

Type
Thermal imager module.

Description
The Wizard thermal imaging module uses the Merlin 256 × 128 uncooled detector operating in the 8 to 12 μm band. The design includes a number of signal processing functions including image difference processing, automatic dead element detection and concealment, full dynamic range digital output and responsivity normalisation with in-the-field calibration. Through field sequential scanning using a microscan chopper assembly the effective resolution is increased to 256 × 512 equivalent elements. Image break-up during motion is also eliminated. The module may be applied to hand-held or remote imagers.

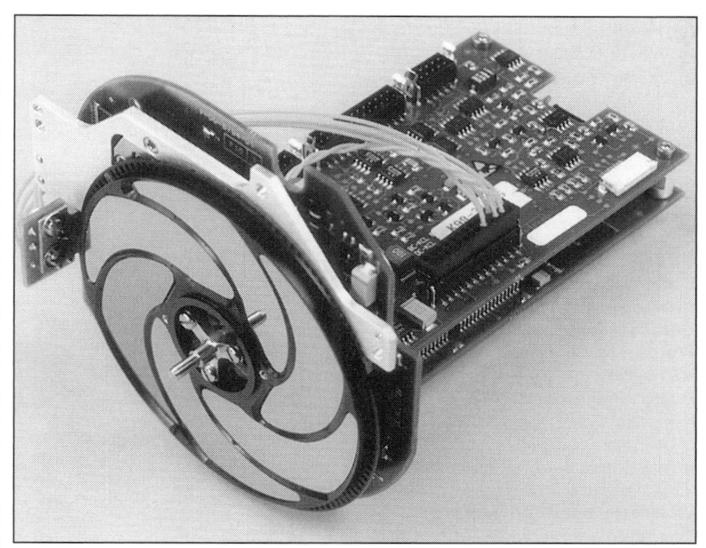

BAE Systems Wizard thermal imaging module 0089865

Operational status
Available.

Specifications
Spectral band: 8 to 12 μm
Detector: 256 × 128 element Lead Scandium Tantalate (PST) ferroelectric focal plane array
Element pitch: 56 μm
Scanning: field sequential microscan
Video output: 8 bit digital
Video update: 50/60 Hz
Electronic zoom: × 2 magnification

Contractor
BAE Systems, Tactical Systems Division.

UPDATED

DRS SE-U20 uncooled IR imager module

Type
Thermal imager modules.

Description
The DRS (formerly Boeing) SE-U20 uncooled IR imager module is a high-performance long-wave imager electronics module, designed to interface with Boeing U3000 series 320 × 240 uncooled Focal Plane Arrays (FPAs). The imaging electronics produce an NTSC/PAL video image and 12-bit digital data from a single 6 V DC supply provided by a camcorder battery or wall-mounted AC/DC Adaptor.

The SE-U20 uncooled IR imager module is designed to allow customers to evaluate the performance of the DRS U3000 series uncooled microbolometer FPAs for a variety of applications. The electronics can also be integrated directly into an OEM's product. The user can change integration time and voltage to vary the dynamic range and optimise the detector's performance for the desired application. The imager electronics also perform a 10-bit analog offset correction, digital gain and offset correction, bad pixel arrangement, brightness and contrast adjustment and video display. All camera functions can be controlled through an RS-232 serial port or a manual port button panel.

The uncooled IR imager module demonstrates the performance of the DRS U3000 series uncooled microbolometer detectors. The high-sensitivity of the uncooled imager module is suitable for commercial and military imaging applications such as surveillance, machine vision, predictive maintenance, firefighter vision, traffic monitoring, process monitoring, thermography, medical imaging, rifle sights and other applications.

The module features:
(a) clock and bias generation
(b) NTSC/PAL, composite and S-Video, 12-bit digital output

(c) non-uniformity correction
(d) time and date display
(e) automatic gain/level control
(f) gamma correction
(g) monochrome and pseudocolour display
(h) thermal electric cooler controller
(i) cursor display
(j) one or two point calibration
(k) reverse colour
(l) symbology overlay
(m) external control through RS-232 port or button panel.

Operational status
Available.

Specifications
Detector type: VO_x resistive bolometer
Array size: 320 × 240
Spectral response: 8-14 μm
Frame rate: 60 Hz
NETD (f/1): <80 mK
Board dimensions:
(control board) 3.3 × 6.5 in
(support board) 3.3 × 6.5 in
(power supply) 3.3 × 3.0 in
(button panel) 2.3 × 1.9 in
Weight (no housing/lens): 405 g
Operating temperature: −20 to +60°C
Video interfaces: 4-pin Mini-DIN, BNC
Serial port interface: DB-9 (RS-232)
Power supply: AC/DC adaptor or battery
Power consumption: 12 W

Contractor
DRS Electro-Optical Systems.

UPDATED

FLIR Systems FTI (hand-held thermal imager)

Type
Thermal imager module.

Description
FTI is a third-generation thermal imager based on Quantum Well Infra-red Photodetector (QWIP) technology. The detector is a 320 × 240 element focal plane array. Sensitive in the Long-Wave Infra-Red (LWIR) (for example, the 8-12 μm band), it is cooled to its operating temperature by an integrated Stirling rotary microcooler. FTI has two fields of view to enable surveillance and target acquisition in the wide field of view and long-range target identification in the narrow field of view. The imager also has a biocular display. It is powered by replaceable, rechargeable batteries, but can also be powered from external sources.

Operational status
Under development.

Specifications
Weight: <3.5 kg
Dimensions: 260 × 180 × 120 mm

FLIR Systems FTI Thermal Imager **NEW**/0137884

Detector
Type: 320 × 240 QWIP
Spectral band: 7.5-9.3 μm
Cooling: Integrated Stirling rotary microcooler
NETD: <50 mK

Fields of View
Wide: 13 × 10°
Narrow: 3.5 × 2.6°

Optics
Magnification: ×2 (WFOV), ×7.4 (NFOV)
Aperture: 78 mm
Field of View change: <0.5 s
Focus: Manual from 50 m to infinity; automatic parfocus 500 m to infinity; autofocus operating on image content
Interfaces: Remote control: RS-422; Video: standard CCIR Monochrome Pal/NTSC
Power supply: 8-16 V Rechargeable Li-Ion battery or external power via power adapter
Operating temperature: −30 to +60°C
Storage temperature: −40 to +70°C
Features: Biocular display; automatic/manual gain and offset; stand-by mode; electronic zoom 2×/4×; user-defined reticles; BITE

Contractor
FLIR Systems.

NEW ENTRY

FLIR Systems LIRC compact thermal imager

Type
Thermal imager (land).

Description
LIRC is a third-generation thermal imager based on Quantum Well Infrared Photodetector (QWIP) technology. The detector is a 320 × 240 element Focal Plane Array (FPA). The detector is sensitive in the Long Wave InfraRed (LWIR), that is the 8-12 μm band, and is cooled to its operating temperature by an integrated Stirling rotary micro-cooler. A 640 × 480 element version will be available later. LIRC is designed for integration into an AFV or MBT sight or on its own on a vehicle or observation platform. The system has two Fields Of View (FOV) to enable easy surveillance and target acquisition in the Wide Field Of View (WFOV) and accurate long-range target identification in the Narrow Field Of View (NFOV).

The imager is completely remote controlled by an RS-422 interface enabling control either from a computer or a remotely located control box. The output of the imager is a standard CCIR/PAL analogue video signal. LIRC has been designed to withstand the stringent environmental requirements of military use and to survive the vibrations and shocks experienced in a tracked combat vehicle, retaining boresight accuracy during gun firing.

FLIR Systems LIRC **NEW**/0137859

Operational status
In production for Swedish Army's CV9040C upgrade; being integrated into the vehicle systems by Bofors Defence.

Specifications
Weight: 4 kg
Dimensions: 240 × 160 × 110 mm

Detector
Type: QWIP
Spectral band: 7.5-9.3 μm
Elements: 320 × 240; 640 × 480
Cooling: integrated Stirling rotary micro-cooler

Fields of View
Wide: 10 × 7.5° (320 × 240), 13 × 10° (640 × 480)
Narrow: 3.5 × 2.6° (320 × 240), 4.6 × 3.5° (640 × 480)
Frame rate: 50/60 Hz

Optics
Aperture: 78 mm
Field of View change: <0.5 s
Focus: Manual from 50 m to infinity
Automatic parfocus 500 m to infinity
Auto-focus operating on image content

Features
Upgradeable from 320 to 640 detector
Automatic/manual gain and offset
Stand-by mode
Electronic Zoom ×2, ×4
User defined reticles
BITE

Contractor
FLIR Systems.

NEW ENTRY

FLIR Systems ThermoVision® 6-24 UC II

Type
Thermal imager module.

Description
The ThermoVision 6-24 UC II is an uncooled imaging detector. Utilising a 320 × 240 uncooled microbolometer focal plane array, it provides reliable and fast start-up over conventional cryogenically cooled infra-red cameras. The focal plane array is environmentally housed, containing the detector and electronic components with dual field of view and electronic zoom. The ThermoVision 6-24 UC II is an OEM/integrator module and has been widely sold into both military and civil applications.

Operational status
In production.

Specifications
Detector: Uncooled microbolometer, 320 × 240

FLIR Systems ThermoVision® 6-24 UC II **NEW**/0137886

Fields of view
Narrow: 6 × 4.5°
Wide: 24 × 18°
Spectral band: 7.5-13 microns
Electronic zoom: Continuous 1× to 4×
FOC switch over time: <0.7 s
Digital image resolution: 14 bit
Gain/Level adjustment: Auto or manual
Image processing: Histogram equalisation
Palettes: Black/White, rainbow, iron
Video format: NTSC (RS-170) or PAL (CCIR)
Serial interface: RS-422/485
Control: Remote control and/or host computer, such as a PC
Power: 11 V-16 V
Power consumption: 12 W typical
Built-In-Test (BIT): Self-diagnostics
Weight: 3.2 kg
Dimentions: 201 × 159 × 225 mm
Operating temperature range: –32 to +50°C
Environmental qualifications: MIL-STD-810 and MIL-STD-461 (EMI)
Start-up time: <45 sec
Encapsulation: IP 65/NEMA 4

Contractor
FLIR Systems.

NEW ENTRY

FLIR Systems Triple QWIP Imaging Module

Type
Thermal imager module.

Description
The Triple QWIP (Quantum Well Infra-red Photodetector) Imaging Module is a Gen III focal plane array based system offering long-wave performance in the 8-9 μm spectral range. It is designed for airborne and ground sensor integration. Its electronics and optics are designed for full compatibility with large format (640 × 480) QWIP arrays. Designed for ruggedness and reliability in a military environment, the WFOV provides air and/or ground crews situational awareness, while the MFOV is for scene tracking. The NFOV mode delivers target detection and recognition capabilities from long stand-off distances.

Operational status
In service.

Specifications
Dimensions: 303 × 240 × 188 mm
Weight: 7.1 kg

Detector
Type: GaAs, Quantum Well Infrared Photodetector (QWIP), 320 × 240 pixels
Spectral band: 8-9 μm
Cooling: Integrated long-life Stirling cycle
Start-up time: <6 min
Number of field of views: 3
Wide: 25 × 19°
Middle: 6.0 × 4.5°
Narrow: 0.99 × 0.74°
IfoV wide: 1.37 mRad
IfoV middle: 0.33 mRad
IfoV narrow: 0.054 mRad
Minimum focus distance: 20-100 m
Thermal sensitivity: 0.03°C
Image frequency: 50/60 Hz (rolling or snap-shot mode)
Electronic zoom function: 4:1 continuous digital zoom w/pixel interpolation, live or frozen image
Image freeze: 14 bit digital freeze/changeable parameter and zoom

Image Presentation
Video output: RS-170 EIA/NTSC or CCIR/PAL, composite and S-video 14-bit digital serial link
Presentation: 2 colour palettes /B&W or W&B selectable
Range performance: Nato Tank, STANAG 4349
Detection: >17 km

Digital Image
Type: Full dynamic, 14-bit digital image data

Optics
Field of view/min switch time: <0.5 sec
Lens identification: Automatic, refocused when switched
Coating: High efficiency. Optional, hard carbon
Focus: Auto or manual

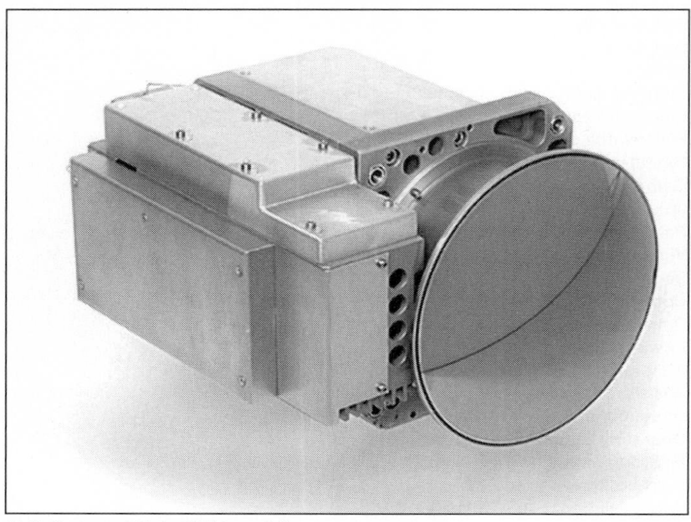

FLIR Systems Triple QWIP module **NEW**/0137869

Power Input
Voltage: 12-35 V DC
Power consumption: 20 W
Environmental specification
Operating temperature range: –32 to +55°C
Storage temperature range: –40 to +70°C
Humidity: Operating and storage: 10% to 95%, non-condensing
Bump: Operational: 25 g IEC 68-2-29
Vibration: Operational: 2 g IEC 68-2-6
EMI/EMC: MIL-STD 461 D, US FCC 15 J Class A, EN5008-1, EN50082-1

Physical Characteristics
Weight: 7.1 kg
Size: 303 × 240 × 188 mm

Interfaces
Remote control: RS-232, RS-485, Ethernet, built in
Ext. Sync: Gen. lock input (optional)

Contractor
FLIR Systems.

UPDATED

Indigo Systems Phoenix® thermal imager module (OEM version)

Type
Thermal imager module.

Description
The Indigo Systems Phoenix® thermal imager module (OEM version) is based on Indigo's proven 640 × 512 staring InSb FPA, integrated in an all-metal, sealed Dewar cooled with split Stirling linear cooler.

The Indigo Systems Phoenix® thermal imaging module **NEW**/0595420

The OEM version is especially suited for gimbal applications in a military environment. It is a compact, high performance module transmitting 40 MHz digital data (Hotlink) off the gimbal to user-selectable video processing electronics embeddable in a small rack or PC-based electronics.

Operational status
In service with US forces on the Northrop Grumman/Rafael Litening ER and AT targeting pods.

Specifications
Dimensions: 12.7 × 7 × 11.4 cm
Weight: 2 kg
Spectral range: 3-5µ (1-5µ custom)
Detector material: InSb
Array format: 640 × 512
Detector size: 25µ
Readout: ISC9803
Detector cooling: closed-cycle Stirling
Cooldown time: 5 minutes at 22°C and under 10 minutes at 71°C
Optical interface: custom
Video output: RS-170 and 8-bit digital
Power consumption (SS): 25 W typical

Contractor
Indigo Systems Corporation (USA).

NEW ENTRY

Indigo Systems Thermal Imaging Sensor Assembly (TISA)

Type
Thermal imager module.

Description
Indigo's Thermal Imaging Sensor Assembly (TISA), is based on Indigo's proven 640 × 512 InSb staring FPA, integrated in an all-metal, sealed Dewar cooled with either a rotary-integral or split Stirling linear cooler.

As a modular thermal imager, TISA is easily customised and packaged for various applications with specific mechanical, optical and electrical requirements and interfaces. This includes hardening for harsh vibration, temperature, marine, EMI and space environments. Its small size, weight and power requirements make it the ideal module for system integrators needing a high-performance camera for lightweight or man-portable and unmanned aerial vehicles applications

Operational status
In service with the US Army's Lightweight Laser Designator Receiver (LLDR).

Specifications
Dimensions: 15.2 × 7.6 × 7.6 cm
Weight
Rotary: <0.91 kg
Linear: <1.82 kg
Spectral range: 3-5µ (1-5µ custom)
Detector material: InSb
Array formats: 640 × 512
Detector size: 25µ
Readout: ISC9803

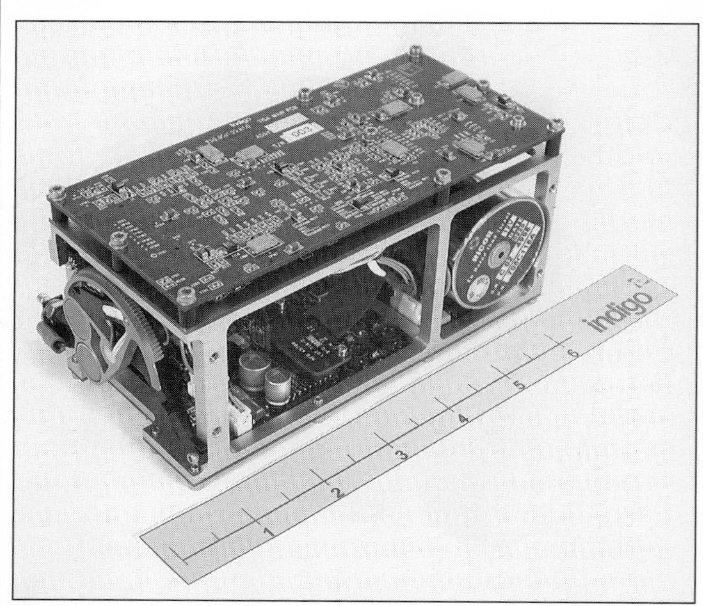

The Indigo Systems Thermal Imaging Sensor Assembly **NEW**/0595413

Detector cooling: Closed-cycle Stirling (rotary or linear)
Cooldown time: five minutes at 22°C and under 10 minutes at 71°C
Optical interface: Custom
Video output: RS-170 and 8-bit digital
Power consumption (SS)
Rotary: 14 W typical
Linear: 20 W typical

Contractor
Indigo Systems Corporation (USA).

NEW ENTRY

Qwiptech QWIP Chip™ focal plane arrays

Type
Infra-red detectors.

Description
Qwiptech produces a range of focal plane array infra-red detectors including mid large and extra large format devices of 320 × 256, 640 × 512 and 1,024 × 1,024 layouts respectively. These have a mean temporal NETD (@ f/2 optics, 300 K background temperature, 15 ms integration time) of 354 MK when cooled to 75 K and 20 MK at 60 K. Custom designs are available.

Operational Status
Available.

Contractor
Qwip Technologies.

UPDATED

SCD Falcon InSb 640 × 512 IDCA (25 µm)

Type
Integrated Detector-Cooler Assembly.

Description
Combining the standard ISC9803 readout with SCD's InSb technology, Falcon is intended for high end thermal imaging applications.

Features
- Based on Indigo Systems 9803 standard ROIC
- Excellent response from 1 µm to 5 µm
- Negligible blinking pixels
- Very stable residual non-uniformity.

Options
- Stand alone FPA or integrated in a long vacuum-life tactical dewar
- Split-Stirling cooler
- Proximity electronics board supporting 1,2 or 4 video lines
- Supporting video electronics on request.

Operational status
Available.

Specifications
(at 80K operating temperature)
Format: 640 × 512 pixels
Spectral response: 15.45 µm
Readout: Indigo Systems 9803
Pixel pitch: 25 µm
Storage capacity:
11e[6] electrons (350 fF Cint)
3e[6] electrons (100 fF Cint)
Operability: 99.5%
Dynamic Range: >72 dB
Readout noise: 345 electrons*
Non-linearity: <0.5%
Crosstalk: <0.1%
Input polarity: p on n
Integration time: > 5 µsec, adjustable
Outputs: 1, 2 or 4 selectable
Output signal swing: 2.5 V
Power consumption:
<90 mW at 30 Hz single output
<180 mW 107 Hz, 4 outputs
Max frame rates: 1 output, 30 Hz
(640 × 480) 2 outputs, 58 Hz
4 outputs, 107 Hz
Video Output: NTSC or PAL
* at max gain setting

Notes:
1. Specifications subject to change without notice.

Contractor
SCD SemiConductor Devices.

VERIFIED

SCD Gemini 320 × 256 InSb Focal Plane array

Type
Infra-red detector.

Description
Gemini is the first of SCD's mid-format INSb (Indium Antimonide) 2-D arrays and is manufactured in quantities of several thousands. It is fielded worldwide in a variety of naval, land forces and airborne platforms. Gemini has been recognised as a first class InSb FPA by users and research institutes.

Features:
- Advanced snapshot Si-CMOS readout architecture
- Excellent response from 1 µm to 5 µm
- Low power dissipation
- Good flux range capability (~1 pA to 15 ~µA per cell)
- Very stable Residual Non-Uniformity after 2-point correction
- Negligible blinking pixels
- A variety of a Frame-to-Frame user selectable operating options, controlled by a 12 bit serial command
- Low frame-to-frame lag, blooming and crosstalk

Options:
- Stand-alone chip or integrated in a long vacuum-life operational dewar
- Proximity board supporting 1, 2, 4 or 8 video output lines
- Joule-Thomson or Stirling coolers
- Germanium or Sapphire window
- Supporting video electronics on request

Operational status
Available and in service worldwide.

Specifications
(typical at 80 K operating temperature)
Detector type: Photo-Voltaic InSb 2-D array
Spectral response: 1 to 5.45 µm
Array format: 320 × 256 elements
FPA dimensions (side): 13.3 × 13.3 mm
FPA dimensions (diagonal): 17 mm
Readout architecture: Si-CMOS, 1 µm SPDM
Integration mode: Snapshot, integrate-then-read
Optical (effective) fill factor: ~90%
Quantum efficiency (with AR Coating): ≥75%
Pixel Dark Current (@ 78K): ~5 Pa
Detector input current: ~1pA ~ 5 µA
Integration capacity: 7.5 to 11 × 10[6] electrons
Operability: 99.5% (99.9% typical)
Non-uniformity (uncorrected): <5% (σ/mean)
Pixel read rate: ≤5 MHz/output line
Integration time (tunable): 0.2 µsec (^{T}frame – ^{T}read)
Number of output lines: 1, 2, 4 or 8
Frame rate (via all outputs):
≤440 Hz (320 × 256 – full-frame)
≤2,000 Hz (128 × 128 – window)
Sub-array access (windowing): 25 row/column combinations
Output voltage swing: 1.9 V (–0.5 to 4.0 V)
Read Noise Floor (nominal): <700 electrons
Power dissipation @ full-frame rate 120 Hz: ≤30 mW
Number of clocks required: 5
Number of biases required: 6 V DC +3 IDC

Contractor
SCD SemiConductor Devices.

UPDATED

Thales Cryogenic SA cryogenic coolers

Type

Cryogenic coolers.

Description

Cryotechnologies, in co-operation with the Advanced Cryogenic Technologies operation of Thomson-CSF Signaal (now Thales Nederland), produces a range of Stirling coolers (linear or rotary), pulse tube coolers, Joule-Thomson mini-coolers and thermoelectric coolers for high-temperature superconductivity or for the cooling of other types of components in commercial, tactical or space applications.

Operational status

Available.

Contractor

Thales Cryogenic SA.

UPDATED

Thales Optronics, Zeiss Optronik Synergi

Type

Thermal imager module.

Description

Synergi is a family of second-generation thermal modules for infra-red cameras capable of being deployed in a wide range of airborne, naval and land-based roles. The Synergi programme to develop a European second-generation thermal imager began in 1992 and has been jointly developed by Thales Optronics (UK and France), and Zeiss-Optronik (Germany). The system uses a second-generation 288 × 4 IRCCD detector developed by Sofradir of France.

The camera features a highly modular design concept based on a long linear array detector operating in the 8 to 12 μm waveband. Closed-cycle cooling, direct analogue and digital video outputs operate from a single 28 V display. This accommodates individual customer configurations which can be supplied with user-defined telescope options.

The modules include: scanner; detector optics; proximity electronics; control board; digital scan converter board; power supply and cooler.

Synergi cameras are compact, lightweight units and integration with future systems is facilitated by simple interfaces and smaller optics. Other features include large-scale electronic integration, fully digital architecture and control; manual and automatic brightness and contrast; edge enhancement; electronic and underscan zoom modes; opto-mechanical fine stabilisation; and BITE.

Synergi modules are being used in the Thales Catherine thermal camera, Rafael/Northrop Grumman Litening pod and the Swedish 'Bause' air defence system.

Operational status

Serial production.

Specifications

Spectral band: 8-12 μm
Detector: 288 × 4 IRCCD
Analogue video output: CCIR or RS-170
Digital video output: 8-bits
Full dynamic range digital video output: 12-bits
Typical configuration
Weight: 11 kg
Volume: 12 litres
Power consumption: 90 W
NETD: 0.045 K

Contractors

Thales Optronics.
Zeiss Optronik GmbH.

VERIFIED

VIDEO TRACKERS FOR MILITARY APPLICATIONS

Octec ADEPT 30 + automatic video tracker

Type
Turret sensor tracker.

Description
Octec has produced a range of video trackers to enhance the capability of electro-optic systems such as turret sensor systems. The ADEPT 30 tracker is the core video tracker in the range. It comprises an integrated software design on a single VME card designed for use in a wide range of systems, including airborne sensor platforms, fire-control systems and weapon test ranges. It has been designed for easy integration into other manufacturers VME electronics.

The ADEPT 30 unit has a great deal of flexibility in its design, allowing closed loop control of a platform and easy integration to control panels and other man machine interfaces. An intelligent breaklock and reacquisition facility is provided, as is video output with symbology overlay.

A range of daughter boards give greater flexibility and performance, allowing input of high-speed digital data, further signal processing, picture stabilisation, improved detection capability or additional tracking channels.

A range of enclosures is available from a commerical 19 in rack unit to ruggedised enclosure.

Octec have integrated their trackers with over 50 different platforms, including most widely used helicopter sensor platforms.

Operational status
Over 600 ADEPT 30 trackers have been shipped to customers worldwide and the system continues in full production, with enhanced facilities.

Specifications
Mechanical: VMEbus; forced air or conduction cooled variants available
Dimensions: 233.4 × 160 mm (Double Euro)
Video Input: Composite video 625/525 line CCIR or RS-170
No of video inputs: 2
Track modes: Centroid, Correlation, Edge, Scenelock
Video output: 1 with symbology overlay
Interfaces available: VME, RS-232/422, analogue, discretes
Automatic video detection: variable from 2-90% of FoV
Power: +5 V 3 A, +12 V 0.2 A, -12 V 0.2 A

Contractor
Octec Ltd.

VERIFIED

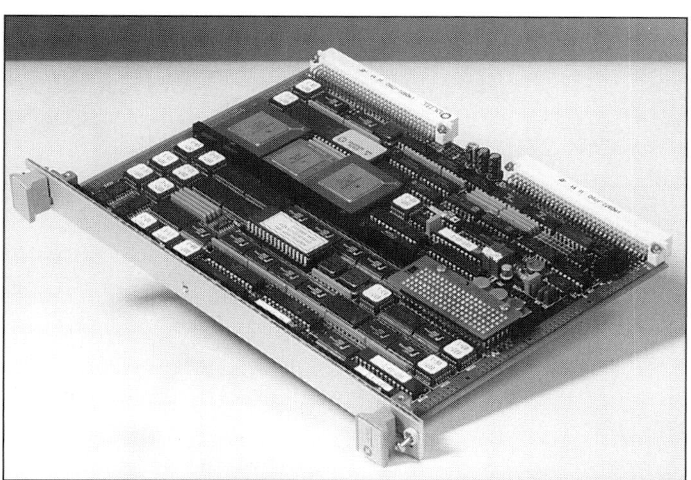

ADEPT 30+ video tracker VMEbus card 0023395

Octec ADEPT 33 + automatic video tracker

Type
Turret sensor tracker.

Description
The ADEPT 33 tracker is one of the range of Octec video trackers and has a smaller form factor than previous models, such as the ADEPT 30, for requirements where size and weight are more critical. It comprises an integrated software design on a single Eurocard for easy integration into other manufacturer's electronics.

The ADEPT 33 unit has a great deal of flexibility in its design, allowing closed loop control of a platform and easy integration to control panels. Built-in interfaces include RS-232/422 and analogue input from a joystick and analogue output to the sensor platform.

Operational status
In service.

Specifications
Mechanical: single height VME board; forces air or conduction cooled variants available
Dimensions: 100 × 160 mm single Euro (3U) format
Video Input: composite video 625/525 line CCIR or RS-170
No of Video Inputs: 2
Track modes: Centroid, Scenelock
Video output: 2
Interfaces available: RS-232/422, analogue, discretes
Automatic video detection: variable from 2 to 90% of FoV
Power: +5 V 2.5 A, +12 V 0.2 A, –12 V 0.2 A

Contractor
Octec Ltd.

VERIFIED

Octec CAATS 2 – Compact Airborne Automatic Video Tracker

Description
The CAATS 2 unit provides the same, or even better, tracking performance than the earlier CAATS unit, but is lighter, less than half the size, and uses less power than the original design. It is a sealed unit using conduction cooling. Several versions are available to meet different environmental scenarios. CAATS 2 also has the capability, using daughter boards, to enhance performance including electronic image stabilisation and input of digital video.

The CAATS 'Scene Lock' tracking feature is retained and performance improved. This is a particularly important feature for the helicopter fit. Centroid tracking is also provided together with a range of submodes, including automatic cueing, which allows optimisation for particular target scenarios. A video symbology generator is integrated to allow on-screen data to be shown and recorded.

Specifications
Weight: 2.8 kg
Dimensions: 250 (w) × 210 (d) × 50 mm (h)
Power: 28 V DC, 1.0 A mean or 115/220V AC using adaptor
Video input: RS-170 or CCIR video input 50/60 Hz
Number of video inputs: 2
Tracking modes: centroid, scenelock (other modes optional)
Video output: 1 with symbology overlay.
Automatic video detection: variable from 2-90% of FoV

Operational status
Selected by UK government agencies and US DoD.

Contractor
Octec Ltd.

UPDATED

ANTI-DETECTION DEVICES

Tenebræx KillFlash™ Anti-Reflection Devices (ARDs)

Type
Anti-reflection devices.

Description
Reflections or glint from objective lenses or the laser filters of optical and electro-optical devices such as sights, laser range-finders and thermal imagers can often reveal the position of the user to the enemy. Thermal imagers can also be detected due to the 'cold window' produced by the cooled detector which looks black or white dependant on the polarity setting when viewed via another thermal imager. Tenebræx has developed the KillFlash™ family of Anti-Reflection Devices (ARDs) to minimise the risk of detection. In most cases the ARD can be easily fitted over the objective lens. The result is the equivalent of fitting a lens hood many times the length of a conventional hood. The ARD is constructed using a Nomex honeycomb with a special matt black coating. The specification of an example device are given below.

Operational status
Available for a broad range of field optics including M-22 binoculars, Leica VECTOR laser range-finder Leupold Mk IV rifle scope and others.

Specifications
Example M144 spotting scope ARD
Mechanical
Dimensions (l × d): 114 × 89 mm
Weight: 270 gm
Housing material: anodised aluminium
Filter material: resin reinforced Nomex honeycomb with matt black coating
Temperature range: +71 to −51°C
Optical
light attenuation: 15% approx

Contractor
Tenebræx Corporation.

VERIFIED

View of glint from binoculars without ARD and lack of glint with KillFlash™ filters 0054782

Close-up of binoculars with one lens fitted with KillFlash™ showing filter's honeycomb structure 0054781

CONTRACTORS

Contractors

ADI Ltd (plant)
Communications and Surveillance Systems
4 Talavera Road
North Ryde
New South Wales 2113
Australia
Tel: (+61 2) 93 25 15 00
Fax: (+61 2) 93 25 16 00

ADI Ltd, Systems Group
Locked Bag 14
Post Office
Lidcombe
New South Wales 2141
Australia
Tel: (+61 2) 99 11 86 66
Fax: (+61 2) 99 11 86 99

AeroSpace Technologies of Australia Ltd
226 Lorimer Street
Port Melbourne
Victoria 3207
Australia
Tel: (+61 3) 96 47 31 11
Fax: (+61 3) 96 46 22 53
Web: http://www.asta.bnd.boeing.com

Aimpoint AB
Jägershillgatan 15
SE-213 75 Malmö
Sweden
Tel: (+46 40) 671 50 20
Fax: (+46 40) 21 92 38
e-mail: info@aimpoint.se
Web: http://www.aimpoint.se

Aimpoint Inc
7702 Leesburg Pike,
Falls Church
Virginia 22043
USA
Tel: (+1 703) 749 23 20
Fax: (+1 703) 749 23 23
Web: http://www.aimpoint.com

Aims Optronics NV SA
Rue F Kinnenstraat 30
B-1950 Kraainem
Belgium
Tel: (+32 27) 85 00 25
Fax: (+32 27) 31 89 18
Web: http://www.aimsoptronics.be

Alcatel Deutschland GmbH
Lorenzstrasse 10
D-70435 Stuttgart
Germany
Tel: (+49 711) 82 10
Fax: (+49 711) 821 11 11

Alliant Integrated Defense Company
13133 N 34th Street
PO Box 4648
Clearwater
Florida 33762
USA
Tel: (+1 813) 572 19 00
Fax: (+1 813) 572 21 80

Alliant Techsystems Inc
600 Second Street NE
Hopkins
Minnesota 55343-8384
USA
Tel: (+1 612) 931 60 00
Fax: (+1 612) 931 59 20
e-mail: public_affairs@atk.com
Web: http://www.atk.com

Allied Ordnance of Singapore (Pte) Ltd
2D Ayer Rajah Crescent #08-01
Singapore 139938
Singapore
Tel: (+65) 779 79 55
Fax: (+65) 779 02 84

ALST (Advanced Laser Systems Technology) Inc
6860 Edgewater Commerce Parkway
Suite 500
Orlando
Florida 32810
USA
Tel: (+1 407) 295 58 78
Fax: (+1 407) 298 60 12
e-mail: sales@alst.com
Web: http://www.alst.com

AL-Technique Corporation of Pakistan (Pvt) Ltd
PO No 1878
Islamabad
Pakistan
Tel: (+92 51) 920 65 58
Fax: (+92 51) 922 10 15

Alvis Hägglunds AB
Head Office
SE-891 82 Örnsköldsvik
Sweden
Tel: (+46 660) 800 00
Fax: (+46 660) 809 95
Web: http://www.alvishagglunds.se

Alvis Vickers Defence Limited
Armstrong Works
Scotswood Road
Newcastle upon Tyne NE99 1BX
UK
Tel: (+44 191) 273 88 88
Fax: (+44 191) 273 23 24
Web: http://www.alvisvickers.co.uk

Alvis Vickers Ltd
Hadley Castle Works
PO Box 106
Telford
Shropshire TF1 6QW
UK
Tel: (+44 1952) 22 45 00
Fax: (+44 1952) 24 39 10
e-mail: name@alvisvehicles.co.uk

Amcoram Ltd
7 Haplada Street
PO Box 696
IL-60218 Or-Yehuda
Israel
Tel: (+9 723) 538 86 66
Fax: (+9 723) 533 92 23

Angenieux
Boulevard Ravel de Malval
F-42570 Saint-Heand
France
Tel: (+33 4) 77 90 78 00
Fax: (+33 4) 77 90 78 02
e-mail: angenieux@calva.net

Applied Laser Systems
2160 NW Vine Street
Unit A
Grants Pass
Oregon 97526
USA
Tel: (+1 503) 479 04 84
Fax: (+1 503) 476 51 05

Armscor (Armaments Corporation of South Africa Ltd)
Private Bag X337
0001 Pretoria
South Africa
Tel: (+27 12) 428 19 11
Fax: (+27 12) 428 56 35
e-mail: info@armscor.co.za
Web: http://www.armscor.co.za

Arsenal Central Design Bureau
8 Moscovskaya Street
Kiev
101010, Ukraine
Tel: (+38 044) 293 00 62
Fax: (+38 044) 293 15 09
e-mail: cdoars@gu.kiev.ua

Aselsan
Microwave and System Technologies Division
PO Box 101
TR-06172 Yenimahalle
Ankara
Turkey
Tel: (+90 312) 385 19 00
Fax: (+90 312) 354 19 46
e-mail: marketing@mst.aselsan.com.tr
Web: http://www.aselsan.com.tr

Aselsan Inc
Microelectronics, Guidance and Electro-optics Division
PO Box 30
Etlik
TR-06011 Ankara
Turkey
Tel: (+90 312) 847 53 00
Fax: (+90 312) 847 53 20

Avitronics
PO Box 8492
0046 Centurion
South Africa
Tel: (+27 12) 672 60 00
Fax: (+27 12) 672 62 22
e-mail: defense@grintek.com
Web: http://www.grintek.com

AV Technology Corporation
50405 Patricia Drive
Chesterfield
Michigan 48051
USA
Tel: (+1 313) 949 58 50
Fax: (+1 313) 949 87 60

Azimuth Technologies Ltd
15 Hataasiah Street
PO Box 2497
IL-43654 Raanana
Israel
Tel: (+972 9) 761 25 00
Fax: (+972 9) 742 87 67
e-mail: mkt@azimuth.co.il
Web: http://www.azimuth.co.il

BAE Systems
Corporate Headquarters
Warwick House
Farnborough Aerospace Centre
Farnborough
Hampshire GU14 6YU
UK
Tel: (+1252) 37 32 32
Fax: (+1252) 38 30 00
Web: http://www.bae.co.uk

BAE Systems, Avionics
Christopher Martin Road
Basildon
Essex SS14 3EL
UK
Tel: (+44 1268) 52 28 22
Fax: (+44 1268) 88 31 40

BAE Systems, Avionics
Clittaford Road
Southway
Plymouth
Devon PL6 6DE
UK
Tel: (+44 1752) 69 56 95
Fax: (+44 1752) 69 55 00
Web: http://www.bae.co.uk

BAE Systems, Avionics, Infra-Red
Millbrook Industrial Estate
PO Box 217
Southampton
Hampshire SO15 0EG
UK
Tel: (+44 23) 80 70 23 00
Fax: (+44 23) 80 31 67 77
e-mail: Infrared.sales@gecm.com

BAE Systems, Avionics, Milton Keynes
Foxhunter Drive
Linford Wood
Milton Keynes
Buckinghamshire MK14 6LA
UK
Tel: (+44 1908) 22 00 44
Fax: (+44 1908) 31 71 37

BAE Systems, Avionics, Radar & Countermeasures Systems, Edinburgh
Silverknowes
Ferry Road
Edinburgh EH4 4AD
UK
Tel: (+44 131) 332 24 11
Fax: (+44 131) 343 57 29

BAE Systems, Avionics, Stanmore
The Grove
Warren Lane
Stanmore
Middlesex HA7 4LY
UK
Tel: (+44 20) 89 54 23 11
Fax: (+44 20) 84 20 39 90
Web: http://www.baesystems.com

BAE Systems, C4ISR - Lane Systems, Leicester
New Parks
Leicester LE3 1UK
UK
Tel: (+44 116) 256 10 12
Fax: (+44 116) 256 10 50
Web: http://www.bae.cu.uk

BAE Systems, C4ISR, Rochester
Airport Works
Rochester
Kent ME1 2XX
UK
Tel: (+44 1634) 84 44 00
Fax: (+44 1634) 82 73 32
Web: http://www.baesystems.com

BAE Systems, Information and Electronic Systems Integration
Business Development Group, IR Imaging Systems
2 Forbes Road M/S LEX 112
Lexington
Massachusetts 02421 - 7306
USA
Tel: (+1 781) 863 36 84
Fax: (+1 781) 863 41 93
Web: http://www.lews.na.baesystems.com

BAE Systems, North America
Information & Electronic Systems Integration
2 Forbes Road
Lexington
Massachusetts 02421 - 7306
USA
Tel: (+1 603) 863 36 84
Fax: (+1 603) 863 41 93
Web: http://www.iews.na.baesystems.com

BAE Systems, North America
6500 Tracor Lane
Austin
Texas 78725-2070
USA
Tel: (+1 512) 929 28 00
Fax: (+1 512) 929 23 80
Web: http://www.baesystems.com

BAE Systems, North America, Reconnaissance and Surveillance Systems
300 Robbins Lane
Syosset
New York 11791-6012
USA
Tel: (+1 516) 349 22 00
Fax: (+1 516) 931 40 37
Web: http://www.baesystems.com

BAE Systems North America, Threat Warning and Defensive Systems
1 Ridge Hill
Yonkers
New York 10710 5598
USA
Tel: (+1 914) 964 25 07
Fax: (+1 914) 964 07 49
Web: http://www.baesystems.com

Ball Aerospace & Technologies Corporation, Telecommunications Product Division
9675 W 108th Circle
Broomfield
Colorado 80021
USA
Tel: (+1 303) 939 40 00
Fax: (+1 303) 533 70 10

BE Delft Electronics Ltd
EL-30 J Block
Bhosari Industrial Area
Pune 411026
India
Tel: (+91 20) 712 29 81
Fax: (+91 20) 712 05 89
e-mail: bedelft@vsnl.com

Bharat Electronics
Trade Centre
116/2 Race Course Road
Bangalore 560 001
India
Tel: (+91 80) 228 76 71
Fax: (+91 80) 225 84 10
e-mail: dgmmcco@vsnl.net
Web: http://www.bel-india.com

Bodenseewerk Geraetetechnik GmbH (BGT)
PO Box 101155
D-88641 Uberlingen
Germany
Tel: (+49 7551) 89 01
Fax: (+49 7551) 89 28 22
e-mail: pr@bgt.de

Boeing Company, Defense and Space Group, Electronic Systems Division
PO Box 3999
Seattle
Washington 98124-2499
USA
Tel: (+1 206) 773 30 45
Fax: (+1 206) 657 30 33

Boeing Defense & Space Group, Missiles & Space Division
PO Box 24002
Huntsville
Alabama 35824
USA
Tel: (+1 205) 461 21 21
Fax: (+1 205) 561 22 52
Web: http://www.boeing.com

Brashear LP
615 Epsilon Drive
Pittsburgh
Pennsylvania 15238
USA
Tel: (+1 412) 967 77 00
Fax: (+1 412) 967 79 73
e-mail: sales@brashearlp.com
Web: http://www.brashearlp.com

Calzoni S.r.L
Via Bargellino 25/a
40012 Calderara Di Reno
Bologna
Italy
Tel: (+39 051) 413 77
Fax: (+39 051) 413 75 55
e-mail: calzoni@calzonispa.com
Web: http://www.calzonispa.com

CEDIP Infrared Industries
19, bd Bidault
F-77183
Croissy-Beaubourg
France
Tel: (+33 1) 60 37 01 00
Fax: (+33 1) 64 11 37 55
e-mail: info@cedip-infrared.com
Web: http://www.cedip-infrared.com

China National Aero-Technology Import and Export Corporation (CATIC)
CATIC Plaza
18 Beichen Dong Street
Chaoyang District
100101 Beijing
China
Tel: (+86 10) 64 94 22 55
Fax: (+86 10) 64 94 10 88
e-mail: webmaster@catic.com.cn
Web: http://www.catic.com.cn

China National Electronics Import & Export Corporation (CEIEC)
Electronics Building A23
PO Box 140
100036 Beijing
China
Tel: (+86 10) 68 29 62 13
Fax: (+86 10) 68 21 23 61
e-mail: ceiec@ceiec.com.cn
Web: http://www.ceic.com.cn

China National Precision Machinery Import & Export Corporation
30 Haidian Nanlu
Haidan Qu
100080 Beijing
China
Tel: (+86 10) 68 74 88 77
Fax: (+86 10) 68 74 88 44
Web: http://www.cpmiec.com.cn

China North Industries Corporation (Norinco)
PO Box 2137
100823 Beijing
China
Tel: (+86 10) 68 51 22 44
Fax: (+86 10) 68 53 32 36; 68 03 32 36

Chung Shan Institute of Science and Technology
PO Box 90008-1
Lung-Tan
Tai Chung 407
Taiwan
Tel: (+886 347) 122 01
Fax: (+886 347) 110 57
e-mail: rdr115@eagle.seed.net.tw

CILAS (Compagnie Industrielle des Lasers)
route de Nozay
PO Box 27
F-91460 Marcoussis
France
Tel: (+33 1) 64 54 48 00
Fax: (+33 1) 69 01 37 39
Web: http://www.cilas.com

CMC Electonics Cincinnati
7500 Innovation Way
Mason
Ohio 45040-9699
USA
Tel: (+1 513) 573 61 00
Fax: (+1 513) 573 62 90
Web: http://www.cmccinci.com

CMI, Cockerill Mechanical Industries SA
Cockerill Mechanical Industries SA
Defence Division
134 Voie de l'Ardenne
B-4053 Embourg
Belgium
Tel: (+32 4) 330 25 39
Fax: (+32 4) 361 14 57

Controp Precision Technologies Ltd
PO Box 611
Hod Hasharon
IL-45105 Israel
Tel: (+972 9) 744 06 61
Fax: (+972 9) 744 06 62
e-mail: cntrpnir@netvision.net.il
Web: http://www.controp.co.il

Corina Corporation AG
Unter Allmend 16
CH-8702 Zollikon
Switzerland
Tel: (+41 1) 391 96 30
Fax: (+41 1) 391 96 26

Crimson Trace Corporation
8089 SW Cirrus Drive
Beaverton
Oregon 97008
USA
Tel: (+1 800) 442 24 06
Fax: (+1 503) 627 01 66
e-mail: customer@crimsontrace.com
Web: http://www.crimsontrace.com

Cumulus, Business Unit of Denel (Pty) Ltd
PO Box 8859
Centurion 0046
South Africa
Tel: (+27 12) 674 00 00
Fax: (+27 12) 674 01 98; 01 99

Dassault Aviation
78 Quai Marcel Dassault-cedex 300
St Cloud F-92552
France
Tel: (+33 1) 47 11 40 00
Fax: (+33 1) 47 11 49 01

DCN International
Direction des Constructions Navales
10 rue Sextius Michel
F-75015 Paris Armées
France
Tel: (+33 1) 40 59 50 00
Fax: (+33 1) 40 59 54 08
e-mail; dcncomm@wanadoo.fr
Web: http://www.dcnintl.com

Dedal (JSC)
PO Box 109
Moscow 107014
Russian Federation
Tel: (+7 095) 269 96 65
Fax: (+7 11) 269 96 74
e-mail: info@nightvision.ru

Defence Research and Development Organisation
South Block
Ministry of Defence
New Delhi 110 001
India
Tel: (+91 11) 301 09 10
Fax: (+91 11) 379 30 08

Delegation Generale pour l'Armement (DGA)
14 rue Saint Dominique
F-00457 Paris Armées
France
Tel: (+33 1) 42 19 45 46
Fax: (+33 1) 42 19 47 95

Diehl Stiftung & Co
Stephanstrasse 49
D-90478 Nurnberg 30
Germany
Tel: (+49 911) 947 24 92
Fax: (+49 911) 947 36 43
Web: http://www.diehl.com

Diel (Pty) Ltd
PO Box 228
7800 Plumstead
Capetown
South Africa
Tel: (+27 21) 715 11 20
Fax: (+27 21) 715 11 23
e-mail: diel@global.co.za
Web: http://www.diel.co.za

Diversified Optical Products Inc (DIOP)
27, Poulcott
Wraysbury
Staines
Middlesex
TW19 5EN
United Kingdom
Tel: (+44 1784) 48 18 75
Fax: (+44 1784) 48 12 54
Web: http://www.diop.com

Diversified Optical Products Inc (DIOP)
282 Main Street
Salem
New Hampshire
03079
United States
Tel: (+1 603) 898 18 80
Fax: (+1 603) 893 43 59
e-mail: info@diop.com
Web: http://www.diop.com

Dornier GmbH (an EADS company)
D-88039 Friedrichshafen
Germany
Tel: (+49) 754 58 00
Fax (+49) 754 58 44 11
e-mail: simulation.training@dornier.eads.net

DRS Infrared Technologies
13532 N Central Expressway
Dallas
Texas 75243
Tel: (+1 877) 377 47 83
Fax: (+1 972) 560 60 49

DRS Optronics Inc
2330 Commerce Park Drive NE
Palm Bay
Florida 32905
USA
Tel: (+1 407) 984 90 30
Fax: (+1 321) 984 87 46
Web: http://www.drs.com

DRS Photronics, Inc
138 Bauer Drive
Oakland
New Jersey 07436
USA
Tel: (+1 201) 337 38 00
Fax: (+1 201) 337 47 75

DRS Sensor & Targeting Systems Inc
2500 Torrance Boulevard
Torrance
California 90503
USA
Tel: (+1 310) 750 32 00
Fax: (+1 310) 750 31 97

Dynamit Nobel GmbH Explosivstoff- und Systemtechnik
Dr-Hermanfleck Allee 8
D-57299 Burbach
Germany
Tel: (+49 27) 36 46 20 00
Fax: (+49 27) 36 46 21 02
e-mail: infogm@dynamit-nobel.com
Web: http://www.dynamit-nobel.de

E2V Technologies
106 Waterhouse Lane
Chelmsford
Essex CM1 2QU
UK
Tel: (+44 1245) 49 34 93
Fax: (+44 1245) 49 24 92
e-mail: enquiries@e2vtechnologies.com
Web: http://www.e2vtechnologies.com

EADS Deutschland GmbH
PO Box 801109
D-81663 Munich
Germany
Tel: (+49 89) 60 70
Fax: (+49 89) 60 72 64 81
Web: http://www.eads.net

EADS France (was Aerospatiale, Missiles Division)
Chatillon Plant
2 rue Beranger
PO Box 84
F-92322 Chatillon Cedex
France
Tel: (+33 1) 47 46 21 21
Fax: (+33 1) 47 46 27 77

Econ Industries SA
32, Kifissias Avenue
Atrina Tower
GR-15125 Maroussi
Athens
Greece
Tel: (+30 1) 682 86 01
Fax: (+30 1) 684 15 24

EFW Inc.
4700 Marine Creek Parkway
PO Box 136969
Fort Worth
Texas 76179-6969
USA
Tel: (+1 817) 234 66 00
Fax: (+1 817) 234 67 92
e-mail: efw_inc@efw.com
Web: http://www.efw.com

Elbit Systems Ltd
Advanced Technology Center
PO Box 539
IL-31053 Haifa
Israel
Tel: (+972 4) 831 53 15
Fax: (+972 4) 855 00 02
e-mail: marcom@elbit.co.il
Web: http://www.elbit.co.il

Electro Optic Systems Pty Ltd
Locked Bag 2 Post Office
Queanbeyan
New South Wales 2620
Australia
Tel: (+61 2) 62 98 80 00
Fax: (+61 2) 62 99 76 87
e-mail: eos@dynamite.com.au
Web: http://www.eos-aus.com

Electromagnetica
266-268 Calea Rachovei
Sector 5
76402 Bucharest
Romania
Tel: (+40 1) 423 20 23

Electron National Research Institute
68 Prospect M.Toreza
194223 St Petersburg
Russian Federation
Tel: (+7 812) 552 36 00; 552 22 76
Fax: (+7 812) 552 31 97
e-mail: electron@nevsky.net
Web: http://www.electron.spb.ru

Electronintorg Ltd
Usievicha 24/2
125315 Moscow
Russian Federation
Tel: (+7 095) 155 40 26
Fax: (+7 502) 224 57 67
e-mail: elers@redline.ru

Electrophysics Corporation
373 Route 46 West
Fairfield
New Jersey 07004-2442
USA
Tel: (+1 973) 882 02 11
Fax: (+1 973) 882 09 97
e-mail: info@electrophysics.com
Web: http://www.electrophysics.com

Elettronica SpA
Via Tiburtina Valeria - Km 13,7
Loc Settecamini
I-00131 Rome
Italy
Tel: (+39 06) 415 41
Fax: (+39 06) 415 49 24

Elisra Electronic Systems Ltd
48 Mivtza Kadesh Street
IL-51203 Bene Barak
Israel
Tel: (+972 3) 617 51 11; 55 22
Fax: (+972 3) 617 58 50
e-mail: marketing@elisra.com
Web: http://www.elisra.com

ELOP Electro-Optics Industries Ltd, a wholly owned subsidiary of Elbit Systems
Advanced Technology Park
Kiryat Weizman
PO Box 1165
IL-76111 Rehovot
Israel
Tel: (+972 8) 938 62 11
Fax: (+972 8) 938 60 10
e-mail: marketing@elop.co.il
Web: http://www.elop.co.il

Eloptro Denel (Pty) Ltd
Atlas and de Havilland Road
PO Box 869
1620 Kempton Park
South Africa
Tel: (+27 11) 921 41 17
Fax: (+27 11) 973 24 47
e-mail: marketing@eloptro.co.za
Web: http://www.eloptro.com

Euroatlas GmbH
Zum Panrepel 2
D-28307 Bremen
Germany
Tel: (+49 421) 48 69 30
Fax: (+49 421) 486 93 77
e-mail: info@euroatlas.de
Web: http://www.euroatlas.de

Euromissile Dynamics Group
28 rue de la Redoute
F-92260 Fontenay aux Roses
France
Tel: (+33 1) 41 13 27 20
Fax: (+33 1) 46 60 27 21

Fabbrica Italiana Apparecchiature Radioelettriche (FIAR) SpA
Avionic Systems and Equipment Division
Via G.B. Grassi 93
I-20157 Milan
Italy
Tel: (+39 02) 35 79 01
Fax: (+39 02) 356 73 25

Fakel Engineering Design Bureau
33 Academician Grushin Street
Khimki
141400 Moscow Region
Russian Federation
Tel: (+7 095) 571 44 94
Fax: (+7 095) 573 51 85

Fakel Experimental Design Bureau
181 Moscovoskiy Prospekt
236001 Kaliningrad
Russian Federation
Tel: (+7 0112) 45 19 64
Fax: (+7 0112) 46 17 62

Federal Stake Unitary Enterprise Production Association - Urals Optical and Mechanical plant
33-b Vostchnaya Strasse
620100 Ekaterinburg
Russian Federation
Tel: (+7 3432) 24 17 01
Fax: (+7 3432) 24 18 44
e-mail: trank@gin.global-one.ru
Web: http://www.uralregion.ru

FLIR Systems AB - Sweden
European Operations
Worldwide Thermography Centre
Rinkebyvagen 19
PO Box 3
SE-182 11
Danderyd
Sweden
Tel: (+46 8) 753 25 00
Fax: (+46 8) 753 26 64
Web: http://www.flir.com

FLIR Systems Boston, Inc.
16 Esquire Road
North Billerica
Massachusetts 01862
USA
Tel: (+1 978) 901 80 00
Fax: (+1 978) 901 88 85
Free Phone: 800-GO INFRA
e-mail: sales@flir.com
Web: http://www.flir.com

FLIR Systems Inc. (HQ)
16505 SW 72nd Avenue
Portland
Oregon 97224
USA
Tel: (+1 503) 684 37 31
Tel: (+1 800) 322 37 31
Fax: (+1 503) 684 32 07
e-mail: sales@flir.com
Web: http://www.flir.com

FLIR Systems Ltd. - Canada
5230 South Service Road
Suite 125
Burlington, Ontario L7L 4K2
Canada
Tel: (+1 800) 613 05 07
Tel: (+1 905) 637 56 96
Fax: (+1 905) 639 54 88
Web: http://www.flir.com

FLIR Systems Ltd. - Europe
2 Kings Hill Avenue
West Malling, Kent
ME19 4AQ
United Kingdom
Tel: (+44 1732) 22 00 11
Fax: (+44 1732) 22 00 14
e-mail: marketing@flir.uk.com
Web: http://www.flir.com

FLIR Systems Ltd. - Middle East
C/O Middle East Optronics, FZCO
Dubai Airport Free Zone, Unit C-13
PO Box 54262
Dubai
United Arab Emirates
Tel: (+971 4) 299 68 98
Fax: (+971 4) 299 68 95
Web: http://www.flir.com

Fotona DD
Stegne 7
1210 Ljubljana
Slovenia
Tel: (+386 1) 500 93 28
Fax: (+386 1) 500 92 04
e-mail: info@fotona.si
Web: http://www.fotona.si

Fraser-Volpe Corporation (FVC)
1025 Thomas Drive
Warminster
Pennsylvania 18974
USA
Tel: (+1 215) 443 42 50
Fax: (+1 215) 443 09 66
e-mail: info@fraser-volpe.com
Web: http://www.fraser-volpe.com

Galileo Avionica S.p.A - a Finmeccanica Company
Surface Systems Business Unit
Via Albert Einstein
50013 Campi Bissenzio (Fi)
Italy
Tel: (+39) 05 58 95 01
Fax: (+39) 05 58 95 06 00
e-mail: stc.galileo@pn.itnet.it
Web: http://www.galileoavionica.it

Galileo Avionica SpA
Roma Division
PO Box 7083
I-0131 Rome
Italy
Tel: (+39 06) 41 88 31
Fax: (+39 06) 41 88 32 98

General Dynamics
Ordnance and Tactical Systems
10101 9th Street North
St Petersburg, Florida 33716
USA
Tel: (+1 727) 578 81 00
Fax: (+1 727) 578 81 19
Web: http://www.gd-ots.com

General Dynamics Canada
3785 Richmond Road
Ottawa
Ontario K2H 5B7
Canada
Tel: (+1 613) 896 77 86
Fax: (+1 613) 820 50 81
e-mail: wwwinfo@gdcanada.com
Web: http://www.gdcanada.com

General Dynamics Corp, Land Systems
38500 Mound Road
Sterling Heights
Michigan 48310-3200
USA
Tel: (+1 810) 825 40 00
Fax: (+1 810) 825 40 13
Web: http://www.gdls.com

General Dynamics Group
Armament and Technical Products
128 Lakeside Avenue
Burlington
05401-4985 Vermont
USA
Tel: (+1 802) 657 64 86
Fax: (+1 802) 657 61 04
Web: http://www.lockheedmartin.com

General Dynamics United Kingdom Ltd
Castleham Road
St Leonards on Sea
East Sussex TN38 9NJ
UK
Tel: (+44 1424) 85 34 81
Fax: (+44 1424) 85 15 20
Web: http://www.generaldynamics.uk.com

Gessellschaft für Intelligente Wirksysteme GmbH (GIWS)
Kupferstrasse 4
Postfach 2926
D-90013 Nürnberg
Germany
Tel: (+49 911) 462 62 60
Fax: (+49 911) 46 26 26 40
e-mail: info@giws.de
Web: http://www.giws.de

Giat Industries
13 route de la Miniere
F-78034 Versailles Cedex
France
Tel: (+33 1) 30 97 39 91
Fax: (+33 1) 30 97 39 67
e-mail: marketing@giat-industries.fr
Web: http://www.giat-industries.fr

GM Defense - Delco Defense Systems
7410 Hollister Avenue
Goleta
California 93117
USA
Tel: (+1 805) 961 74 77
Fax: (+1 805) 961 50 14
e-mail: info@delcodefense.com
Web: http://www.delcosystems.com

Goodrich Corporation
4 Hartwell Place
Lexington
Massachusetts 02421-3122
USA
Tel: (+1 781) 402 21 90
Fax: (+1 781) 402 22 16
Web: http://www.bfgoodrich.com

Graflex, Inc
1281 North Ocean Drive
Suite 201
Riviera Beach
Florida 33404
USA
Tel: (+1 561) 842 26 00
Fax: (+1 561) 842 30 20
e-mail: sales@graflex.com
Web: http://www.graflex.com

Hall and Watts Defence Optics Limited
266 Hatfield Road
St Albans
Hertfordshire AL1 4UN
UK
Tel: (+44 1727) 85 92 88
Fax: (+44 1727) 83 56 83
e-mail: managing@hallwatts.co.uk
Web: http://www.hallwatts.co.uk

Hellenic Aerospace Industry SA
Electronics Business Unit
PO Box 23
GR-320 09 Schimatari
Greece
Tel: (+30 22) 62 05 20 00
Fax: (+30 22) 62 05 21 70
e-mail: abourtog@haicorp.com
Web: http://www.haicorp.com

Hensoldt Systemtecknik GmbH
Gloelstrasse 3-5
D-35539 Wetzlar
Germany
Tel: (+49 6441) 40 43 80
Fax: (+49 6441) 40 43 22
e-mail: infohst@zeiss.de
Web: http://www.hensoldt.de

HKV
1325 Airmotive Way
Suite 175-N
Reno
Nevada 85902
USA
Tel: (+1 702) 786 26 55

Imaging Sensors and Systems Inc
925 South Samoran Boulevard
Suite 114
Winter Park
Florida 32792
USA
Tel: (+1 407) 673 83 33
Fax: (+1 407) 673 82 28
e-mail: iss@imagesensors.com
Web: http://www.imagesensors.com

Imatronic Limited
9A Chester Road
Borehamwood
Hertfordshire WD6 1LP
UK
Tel: (+44 20) 89 53 90 55
Fax: (+44 20) 89 53 62 37

Indra Group
Ctra. De Loeches 9
E-28850 Torrejon de Ardoz
Madrid
Spain
Tel: (+34 91) 626 81 97
Fax: (+34 91) 626 80 02
Web: http://www.indra.es

Indra Sistemas SA
Joaquin Rodrigo 11
E-28300 Aranjuez
Madrid
Spain
Tel: (+34 1) 894 88 00
Fax: (+34 1) 891 80 56
Web: http://www.indra.es

Industrial Security Alliance Partners
450 A Street
Suite 501
San Diego
California 92101-4217
Tel: (+1 619) 232 70 41
Fax: (+1 619) 232 70 58
e-mail: Isap@ispausa.com

Insight Technology Inc
3 Technology Drive
Londonderry
New Hampshire 03053
USA
Tel: (+1 603) 626 48 00
Fax: (+1 603) 626 48 88

Instalaza SA
Montreal 27
E-50002 Zaragoza
Spain
Tel: (+34) 976 29 34 23
Fax: (+34) 976 29 93 31
e-mail: instalaza@instalaza.es
Web: http://www.instalaza.es

Institute of Industrial Control Systems (formerly Khan, Dr A Q, Laboratories)
PO Box 1398
Dhoke Nusah, Dakhli Gangal, Nr Chatri Chowk
Rawalpindi 46000
Pakistan
Tel: (+92 51) 928 05 41
Fax: (+92 51) 928 05 42
e-mail: iics@mail.comsats.net.pk

Institute of Optoelectronics
PO Box MG-22
RO-76900 Bucharest
Romania
Tel: (+40 1) 780 66 40
Fax: (+40 1) 423 25 32

Institute of Optronics Pakistan
Ministry of Defence
PO Box 1596 Chaklala
Rawalpindi
Pakistan
Tel: (+92 80) 05 22; 3
Fax: (+92 80) 05 21

Instrument Design Bureau KBP Tula
Ulitsa Scheglovskaya Zaseka
300001 Tula
Russian Federation
Tel: (+7 0872) 41 07 50
Fax: (+7 0872) 42 61 39
e-mail: kbkedr@tula.net

International Technologies (Lasers) Limited
12 Hachoma Street
PO Box 4099
IL-75140 Rishon-Lezion
Israel
Tel: (+972 3) 961 65 67
Fax: (+972 3) 961 65 63
e-mail: itl@lasers.com
Web: http://www.itlasers.com

Intertechnik Technische Produktionen GmbH
PO Box 100
Industriezeile 56
A-4040 Linz
Austria
Tel: (+43 70) 732 78 92
Fax: (+43 70) 732 78 92 13
e-mail: info@intertechnik.at

Intertechnique, Optronics & Image Processing Department
BP 1
61 rue Pierre Curie
F-78373 Plaisir Cedex
France
Tel: (+33 1) 30 54 82 00
Fax: (+33 1) 30 55 71 61
e-mail: arriviere@intertechnique.com
Web: http://www.intertechnique.fr

IRT Infrarot-Technik Eiselt
PO Box 500 323
D-70333 Stuttgart
Germany
Tel: (+49 7031) 22 22 12
Fax: (+49 7031) 22 22 28

Israel Aircraft Industries Ltd
Tamam Division
PO Box 75
Yahud Industrial Zone
IL-56100 Yahud
Israel
Tel: (+972 3) 531 50 03
Fax: (+972 3) 531 51 40
e-mail: infotmm@tamam.iai.co.il
Web: http://www.iai.co.il

Israel Aircraft Industries Ltd
MBT Weapon Systems Division
Yehud Industrial Zone
PO Box 105
IL-56000 Yehud
Israel
Tel: (+972 3) 531 55 55
Fax: (+972 3) 531 52 05; 33 76

Israel Military Industries
64 Bialik Boulevard
PO Box 1044
IL-47100 Ramat Hasharon
Israel
Tel: (+972 3) 598 56 19
Fax: (+972 3) 540 69 08

ITT Defense & Electronics
Night Vision Division
7635 Plantation Road
Roanoke
Virginia 24019
USA
Tel: (+1 540) 362 80 00
Fax: (+1 703) 366 90 15
Web: http://www.ittnv.com

IZAR
Velazquez Street 132
E-28006 Madrid
Spain
Tel: (+34 91) 335 84 00
Fax: (+34 91) 335 86 52
e-mail: izar@izar.es
Web: http://www.enbazan.es

JAI AS
Produktionsvej 1
DK-2600
Copenhagen
Denmark
Tel: (+45) 44 57 88 88
Fax: (+45) 44 91 32 52
e-mail: mail@jai.com
Web: http://www.jai.com

Joint Stock Company Research and Production Association Molniya
6 Novoposelkovaya Street
123459 Moskva
Russian Federation
Tel: (+7 095) 493 50 53
Fax: (+7 095) 497 47 23
e-mail: molniya@dol.ru
Web: http://www.buran.ru

Kaman Aerospace Corporation, Electro-Optics Development Center
Old Windsor Road
PO Box 2
Bloomfield
Connecticut 06002-0002
USA
Tel: (+1 860) 243 73 36
Fax: (+1 860) 293 70 43
e-mail: dml-corpakaman.com
Web: http://www.kamanaero.com

Kawasaki Heavy Industries Ltd
4-1 Hamamatsu-Cho 2
World Trade Center Building
Minato-ku
Tokyo 105-6116
Japan
Tel: (+81 3) 34 35 21 11
Fax: (+81 3) 34 36 30 37

Kazan Optical and Mechanical Plant (AO KOMZ)
2 Stantsionnaya Street
420018 Kazan
Tatarstan
Russian Federation
Tel: (+7 8432) 71 17 43
Fax: (+7 8432) 76 60 94

Kearfott Guidance and Navigation Corporation
150 Totowa Rd
Wayne
New Jersey 07474-0946
USA
Tel: (+1 201) 785 60 00
Fax: (+1 201) 785 60 25
Web: http://www.kearfott.com

Kentron Division of Denel (Pty) Ltd
PO Box 7412
0046 Hennopsmeer
South Africa
Tel: (+27 12) 671 12 39
Fax: (+27 12) 671 17 79

Kintex
66 James Baucher Street
PO Box 209
1407 Sofia
Bulgaria
Tel: (+359 2) 66 23 11
Fax: (+359 2) 65 81 91

Kollmorgen Corporation
Electro-Optical Division
347 King Street
Northampton
Massachusetts 01060
USA
Tel: (+1 413) 586 23 30
Fax: (+1 413) 586 13 24
e-mail: sales@kollmorgen.com or info@adg.kollmorgen.com

Kollsman Inc
220 Daniel Webster Highway
Merrimack
New Hampshire 03054-4844
USA
Tel: (+1 603) 886 22 73
Fax: (+1 603) 595 60 80
e-mail: marcom@kollsman.com
Web: http://www.kollsman.com

Kongsberg Gruppen ASA
Headquarters
Kirkegårdsveien 45
PO Box 1003
N-3601 Kongsberg
Norway
Tel: (+47) 32 73 82 00
Fax: (+47) 32 73 82 48; 85 86
e-mail: office@kongsberg.com
Web: http://www.kongsberg.com

Korea Aerospace Industries Ltd (KAI)
Aerospace Division Seoul Office
22nd Floor Daewoo Center Building
541 5-ga Namdaemun-ro Jung-gu
Seoul
Republic of Korea
Tel: (+82 2) 727 32 84
Fax: (+82 2) 726 32 80

Kurganmashzavod JSC
17 Machinostroitelei Avenue
640631 Kurgan
Russian Federation
Tel: (+7 35) 22 23 22 44
Fax: (+7 35) 22 23 39 96
e-mail: adm@kmz.kurgan.su

Kvaerner Eureka AS
Defence Division
Joseph Kellers Vei 20
N-3401 Lier
Norway
Tel: (+47) 32 85 90 00
Fax: (+47) 32 85 22 55
e-mail: kvaerner.eureka@kvaerner.com
Web: http://www.kvaerner.com/oilgas/eureka/defence/defence.stm

L-3 Communications
Communication Systems East
1 Federal Street
Camden
New Jersey 08102
USA
Tel: (+1 609) 338 30 00
Fax: (+1 609) 338 60 14

L-3 Communications WESCAM
649 North Service Rd W
Burlington
Ontario L7P 5B9
Canada
Tel: (+1 905) 633 40 00
Fax: (+1 905) 633 41 00
Web: http://www.wescam.com

Laserline Corporation
225 South Lake Avenue
Suite 601
Pasadena
California 91101
Tel: (+1 818) 564 49 02

LFD Ltd
White Hart Road
Portsmouth
PO12 2JE
UK
Tel: (+44 23 92) 78 23 66
Fax: (+44 23 92) 78 23 77
e-mail: lfd1@compuserve.com
Web: http://www.lfd.ltd.uk

LFK - Lenkflugkorpersysteme GmbH
PO Box 1661
D-85705 Unterschleissheim
Germany
Tel: (+49 89) 31 79 25 49
Fax: (+49 89) 31 79 38 71
e-mail: rainer.ackerman@lfk.dasa.de
Web: http://www.eads-nv.com

LFK Lenkflugkorpersysteme GmbH, Friedrichshafen
Plant Frierichshafen-Lowental
D-88039 Friedrichshafen
Germany
Tel: (+49 7545) 860 05
Fax: (+49 7545) 862 93

Liteye Microdisplay Systems
12415 Dumont Way Unit 103
Littleton
Colorado 80215
USA
Tel: (+1 303) 470 80 49
Fax: (+1 303) 470 81 53
e-mail: info@liteye.com
Web: http://www.liteye.com

LIW
PO Box 7710
0001 Pretoria
South Africa
Tel: (+27 12) 620 34 05
Fax: (+27 12) 620 53 06; 24 31
e-mail: liwcorporate@liw.denel.co.za
Web: http://www.denel.co.za

Lockheed Martin Aeronautics Company
Lockheed Blvd.
Fort Worth
Texas 76108
Tel: (+1 817) 777 20 00
Fax: (+1 817) 777 21 10
Web: http://www.lockheedmartin.com

Lockheed Martin Electronic Defense Systems
600 East Bonita
Pomona
California 91769
USA
Tel: (+1 909) 624 80 21
Fax: (+1 909) 447 22 30
Web: http://www.lockheedmartin.com

Lockheed Martin Naval Electronics & Surveillance Systems - Akron
1210 Massillon Road
Akron
Ohio 44315-0001
USA
Tel: (+1 330) 796 21 22
Fax: (+1 330) 796 32 74
e-mail: cary.j.dell@imco.com
Web: http://www.lockheedmartin.com

LOMO PLC
20 Chungunnaya St
St Petersburg 194044
Russian Federation
Tel: (+7 812) 248 50 09
Fax: (+7 812) 542 18 39

Lyon & Brandfield Ltd
4-5 Inverness Mews
London W2 3JQ
UK
Tel: (+44 20) 72 29 91 25
Fax: (+44 20) 72 29 92 17

Marconi Selenia Communications Do Brasil
Brasilia, SHIS Q126 Conj 5 Casa 4 Bario
Lago sul Brasilia, Distrito Federal, 71670-050
Brasil
Tel: (+55 61) 367 35 30
Fax: (+55 61) 367 44 12
Web: http://www.marconiselenia.com

Marconi Selenia Communications GmbH
Gartenstrasse 106, Postfach 1980
D-71509, Backnang
Germany
Tel: (+49 71) 911 30
Fax: (+49 71) 91 13 38 21
Web: http://www.marconiselenia.com

Marconi Selenia Communications Ltd
Marconi House, New Street
Chelmsford, CM1 1PL
United Kingdom
Tel: (+44 1245) 35 32 21
Fax: (+44 1245) 28 71 25
Web: http://www.marconiselenia.com

Marconi Selenia Communications Romania Srl
8 Dr Louis Pasteur
R-76206, Bucharest
Romania
Tel: (+40 1) 410 95 30
Fax: (+40 1) 410 95 50
Web: http://www.marconiselenia.com

Marconi Selenia Communications SpA
Viale dell'Industria, 4
I-00040 Pomezia, Roma
Italy
Tel: (+39 06) 91 09 11
Fax: (+39 06) 910 93 39
e-mail: marketing@marconiselenia.com
Web: http://www.marconiselenia.com

Marconi Selenia Communications SpA
Via A, Negrone 1/A
I-16153 Cornigliano, Genova
Italy
Tel: (+39 010) 660 21
Fax: (+39 010) 650 18 97
Web: http://www.marconiselenia.com

Marconi Selenia Kominikasyon AS
Konya Yolu Km 25
TR-06830, Gölbasi, Ankara
Turkey
Tel: (+90 312) 484 51 81
Fax: (+90 312) 484 43 32
Web: http://www.marconiselenia.com

Matra Défense Equipements & Systèmes
Orsay Centre
6 avenue des Tropiques
ZA de Courtaboeuf 2 - BP 80
F-91943 Les Ulis Cedex
France
Tel: (+33 1) 34 63 23 00
Fax: (+33 1) 34 63 23 23

Mauser-Werke Oberndorf Waffensysteme GmbH
PO Box 1349
D-78722 Oberndorf a.N
Germany
Tel: (+49 7423) 700
Fax: (+49 7423) 706 70
e-mail: info@mauser.de
Web: http://www.mauser-online.com

MBDA (Euromissile Dynamics Group)
12 rue de la Redoute
F-92260 Fontenay aux Roses
France
Tel: (+33 1) 41 13 27 20
Fax: (+33 1) 41 13 27 21

MBDA (France)
20-22 rue Grange Dame Rose
F-78141 Velizy
France
Tel: (+33 1) 34 88 30 00
Fax: (+33 1) 34 88 14 24

MBDA (UK) Ltd
Headquarters
11 The Strand
London
WC2N 5RJ
UK
Tel: (+44 20) 74 51 60 00
Fax: (+44 20) 74 51 60 89

Meccanica per l'Elettronica e Servomeccanismi SpA (MES)
Via Tiburtina 1292
I-00131 Rome
Italy
Tel: (+39 06) 41 62 71
Fax: (+39 06) 41 62 72 18, 41 62 72 20

MEO Products Ltd
Special Equipment Division
Suite 2V7, Cooper House
2 Michael Road
London SW6 2AD
UK
Tel: (+44 20) 77 31 88 20
Fax: (+44 20) 77 31 49 90
e-mail: sales@meo.co.uk
Web: http://www.meo.co.uk/sed

Meopta Prerov, a.s.
Kabelikova 1
CZ-75002 Prerov
Czech Republic
Tel: (+420 641) 24 11 11
Fax: (+420 641) 24 22 22
e-mail: meopta@meopta.com
Web: http://www.meopta.com

Mitsubishi Heavy Industries
5-1 Marunouchi 2-chome
Tokyo 100
PO Box 645
Japan
Tel: (+81 3) 32 18 31 11
Fax: (+81 3) 34 53 64 34

Moked Engineering (1969) Limited
Merkaz Shattner
5 Givat Shaul
PO Box 3454
IL-91034 Jerusalem
Israel
Tel: (+972 2) 652 71 41
Fax: (+972 2) 651 11 33
e-mail: moked-en@moked-en.com

MSI-Defence Systems Ltd
Salhouse Road
Norwich
Norfolk NR7 9AY
UK
Tel: (+44 1603) 48 40 65
Fax: (+44 1603) 41 56 49
e-mail: contact@msi-dsl.com
Web: http://www.msi-dsl.com

M-Tek
PO Box 10239
Hennopsmeer 0046
South Africa
Tel: (+27 12) 660 22 30

Naval Research Laboratory
4555 Overlook Avenue SW
Code 1030
Washington DC 20375-5320
USA
Tel: (+1 202) 767 25 41
Fax: (+1 202) 767 69 91
e-mail: nrl1030@ccf.nrl.navy.mil
Web: http://www.nrl.navy.mil

New Noga Light Ltd
Jewish Agency Building
Zerifin
PO Box 760
IL-72100 Ramla
Israel
Tel: (+972 8) 920 08 01
Fax: (+972 8) 920 08 14
e-mail: nogalight@nogalight.com

Night Vision & Electronic Sensors Directorate (NVESD)
10221 Burbeck Road
Fort Belvoir
Virginia 22060-5806
USA
Tel: (+1 703) 704 11 58
Fax: (+1 703) 704 12 15
e-mail: martha.mccaslin@nvl.army.mil

Night Vision Equipment Company Inc
PO Box 266
Emmaus
Pennsylvania 18049-0266
USA
Tel: (+1 610) 391 91 01
Fax: (+1 610) 391 92 20
e-mail: sales@nvec-night-vision.com
Web: http://www.nvec-night-vision.com

Night Vision Technology Corporation
1620 Oakland Road
Suite D-101
San Jose
California 95131
USA
Tel: (+1 408) 526 01 16
Fax: (+1 408) 436 28 30

Northrop Grumman (Litton Electro-Optical Systems)
1215 South 52nd Street
Tempe
Arizona 85281
USA
Tel: (+1 602) 968 44 71
Fax: (+1 602) 966 90 55

Northrop Grumman Advanced Systems
5225 Hellyer Avenue
Suite 100
San Jose, California 95138
USA
Tel: (+1 408) 365 47 47
Fax: (+1 408) 365 40 40
e-mail: info@littonas.com
Web: http://www.littonas.com

Northrop Grumman Corporation
Electronic Systems, Electro-Optical Systems
3414 Herrmann Avenue
Garland
Texas 75041
USA
Tel: (+1 972) 840 56 00
Fax: (+1 972) 271 21 95
e-mail: sales@litton-eos.com
Web: http://www.littoneos.com

Northrop Grumman Corporation
Electronic Systems
1580-A West Nursery Road
MS A255
Linthicum
Baltimore
Maryland 21090
USA
Tel: (+1 410) 993 87 71
Web: http://www.northropgrumman.com

Northrop Grumman Corporation, Defensive Systems Division
600 Hicks Road
Rolling Meadows
Illinois 60008-1098
USA
Tel: (+1 708) 259 96 00
Fax: (+1 708) 870 57 05
Web: http://www.northgrum.com

Northrthrop Grumman Corporation (Systems Division)
2787 South Orange Blossom Trail
Apopka
Florida 32703
USA
Tel: (+1 407) 297 40 10
Fax: (+1 407) 297 46 40
Web: http://www.littonlaser.com

Northrup Grumman Space and Electronics Group
Space and Technology Division
One Space Park
5/2090
Redondo Beach
California 90278-1001
USA
Tel: (+1 310) 813 43 21
Fax: (+1 310) 813 33 31
Web: http://www.trw.com

NOVO Corporation
PO Box 34
6 Dubki Street
127434 Moscow
Russian Federation
Tel: (+7 095) 210 27 60
Fax: (+7 095) 210 27 71
e-mail: novo@novo.com.ru

Novosibirsk Instrument Making Plant
63049 Novosibirsk 49
Russian Federation

Octec Ltd
12 The Western Centre
Western Road
Bracknell
Berkshire RG12 1RW
UK
Tel: (+44 1) 344 46 52 00
Fax: (+44 1) 344 46 52 01
e-mail: octec@octec.co.uk
Web: http://www.octec.co.uk

Oerlikon Contraves AG
Birchstrasse 155
CH-8050 Zurich
Switzerland
Tel: (+41 1) 316 22 11
Fax: (+41 1) 311 31 54
Web: http://www.oerlikoncontraves.com

Oerlikon Contraves Inc
225 boulevard du Seminaire Sud
Saint Jean sur Richelieu
Quebec J3B 8E9
Canada
Tel: (+1 450) 358 20 00
Fax: (+1 450) 358 17 44
e-mail: info@oerlikon.ca
Web: http://www.oerlikon.ca

OIP Sensor Systems
Westerring 21
B-9700 Oudenaarde
Belgium
Tel: (+32 55) 33 38 11
Fax: (+32 55) 31 68 95
e-mail: sales@oip.be
Web: http://www.oip.be

OnX2 - Henry Technical Services Inc
4037 Miranda Court
Springfield
Missouri 65807
USA
Tel: (+1 417) 877 17 33
Fax: (+1 417) 883 11 87
e-mail: susangilmore@prodigy.net

Open Joint-Stock Company VA Degtyarev Plant
Truda Street
601900 Kovrov
Vademir Region
Russian Federation
Tel: (+7 09232) 326 91
Fax: (+7 09232) 535 64
Web: http://www.ZID.ru

Opgal Optronic Industries Ltd
Industrial Area
PO Box 462
IL-20101 Karmiel
Israel
Tel: (+972 4) 995 39 07
Fax: (+972 4) 995 39 00

Optech Systems Inc
100 Wildcat Road
North York
Ontario M3J 2Z9
Canada
Tel: (+1 416) 661 59 04
Fax: (+1 416) 661 41 68
e-mail: general@optech.on.ca
Web: http://www.optech.on.ca

Optechs Korea Inc
291-1 Kaya-dong
Seo-ku
Inchon 404-250
South Korea
Tel: (+82 32) 578 14 41
Fax: (+82 32) 578 14 42
Web: http://www.koptictech.koreasme.com

Optical Systems Technologies Inc
110 Kountz Lane
Freeport
Pennsylvania 16229
USA
Tel: (+1 412) 295 28 80
Fax: (+1 412) 295 33 66

ORTEK Limited, a wholly owned subsidiary of Elbit Systems
PO Box 388
IL-80100 Sederot
Israel
Tel: (+972 8) 869 16 91
Fax: (+972 8) 869 16 95
e-mail: marketing@orteklt d.com
Web: http://www.ortekltd.com

Oto Melara S.p.A, a Finmeccanica Company
Via Valdilocchi 15
Casilla Postalle 337
I-19136 La Spezia
Italy
Tel: (+39 0187) 58 11 11
Fax: (+39 0187) 58 26 69
Web: http://www.otomelara.it

P W Allen & Company Ltd
Allen House
Alexandra Way
Ashchurch Business Centre
Tewkesbury
GL20 8TD
UK
Tel: (+44 1684) 85 11 00
Fax: (+44 1684) 85 11 01
e-mail: sales@pwallen.co.uk
Web: http://www.security.pwallen.co.uk

PCO Przemyslowe Centrum Optyki SA
Ulitsa Ostrobramska 75
PL-04-175 Warsaw
Poland
Tel: (+48 22) 13 70 21
Fax: (+48 22) 13 92 15
e-mail: pco@ikp.atm.com.pl
Web: http://www.wat.waw.pl

Pearpoint Ltd
Pearpoint House
47 Woolmer Way
Woolmer Trading Estate
Bordon
Hampshire GU35 9QF
UK
Tel: (+44 1420) 48 99 01
Fax: (+44 1420) 47 75 97

Peleng
23 Makayonok Street
Minsk 220023
Belarus
Tel: (+375 17) 263 82 04
Fax: (+375 17) 263 65 42

Photonic Optische Gerate GmbH
Seebockgasse 59
A-1160 Vienna 1160
Austria
Tel: (+43 1) 486 56 91-0
Fax: (+43 1) 486 56 91 33
e-mail: kittagaphotonic.at
Web: http://www.photonic.at

Profesionalna Elektronika Rudi Cajavec
Brace Pavlica 23a
7800 Banja Luka
Bosnia-Herzegovina
Tel: (+387 78) 467 07
Fax: (+387 78) 334 82

Pyser-SGI Ltd - Security Products Division
Fircroft Way
Edenbridge
Kent TN8 6HA
UK
Tel: (+44 1732) 86 41 11
Fax: (+44 1732) 86 55 44
e-mail: sales@pyser-sgi.com
Web: http://www.pyser-sgi.com

PZO Warszawa (Polskie Zaklady Opyczne)
04-839 Warazawaul.
Grochowska 316/320
NIP 525-00-00-788
Poland
Tel: (+48 22) 813 20 21; 20 11
Fax: (+48 22) 813 66 51
e-mail: info@pzo.com.pl

Qwip Technologies
2400 Lincoln Avenue, 239
Altadena
California 91001
USA
Tel: (+1 626) 296 64 32
Fax: (+2 626) 296 64 56
Web: http://www.qwip.com

Radamec Defence Systems Ltd
Bridge Road
Chertsey
Surrey KT16 8LJ
UK
Tel: (+44 1932) 56 11 81
Fax: (+44 1932) 56 87 75
e-mail: rdsl@radamec.co.uk
Web: http://www.radamec.co.uk

Rafael Armament Development Authority
Corporate Headquarters
PO Box 2082
IL-31021 Haifa
Israel
Tel: (+972 4) 877 40 57
Fax: (+972 4) 879 24 13
e-mail: intl-mkt@rafael.co.il
Web: http://www.rafael.co.il

RAM Systems GmbH
Daimlerstrasse 11
D-85521 Ottobrun
Germany
Tel: (+49 89) 60 80 03-0
Fax: (+49 89) 60 80 03 16

Raytheon Company
Tucson Operations
1151 East Hermans Road
Tucson
Arizona 85706
USA
Tel: (+1 520) 794 30 00
Fax: (+1 520) 794 13 15
Web: http://www.raytheon.com

Raytheon Company, Bedford
180 Hartwell Road
Bedford
Massachusetts 01730
USA
Tel: (+1 617) 274 31 55
Fax: (+1 617) 274 32 30
Web: http://www.raytheon.com

Raytheon Company, Command Control and Communications Systems
1001 Boston Post Road
Marlborough
Massachusetts 01752-3789
USA
Tel: (+1 508) 490 10 00
Web: http://www.raytheon.com

Raytheon Company, Dallas
13532 N Dallas Expressway MS 37
Dallas
Texas 75234
USA
Tel: (+1 972) 927 37 56
Web: http://www.raytheon.com

Raytheon Company, Garland
1200 South Jupiter Road
PO Box 660023
Garland
Texas 75266-0023
USA
Fax: (+1 972) 205 70 52

Raytheon Company, Goleta
Raytheon Infrared Center of Excellence (RICOE)
75 Coromar Drive
Goleta
California 93117
Tel: (+1 805) 968 35 11
Fax: (+1 805) 685 82 27
Web: http://www.rsc.raytheon.com/es

Raytheon Company (TI Systems), Lewisville
2501 South Highway 121
PO Box 405
Lewisville
Texas 75067-8122
USA
Tel: (+1 972) 462 65 01
Fax: (+1 972) 462 65 08
Web: http://www.raytheon.com

RDM Technology b.v.
PO Box 1039
NL-3000 BA Rotterdam
Netherlands
Tel: (+31 10) 487 27 53
Fax: (+31 10) 487 22 99

Recon/Optical Inc
550 West Northwest Highway
Barrington
Illinois 60010-3094
USA
Tel: (+1 847) 381 24 00
Fax: (+1 847) 381 13 90
Web: http://www.roi.bournes.com

Reutech Systems Pty Ltd
PO Box 35
1685 Halfway House
South Africa
Tel: (+27 11) 652 55 55
Fax: (+27 11) 652 54 71
e-mail: info@reutech.co.za
Web: http://www.reutech.co.za

RH-ALAN
Bosanska 26
10000 Zagreb
Croatia
Tel: (+385 1) 378 08 00
Fax: (+385 1) 378 08 38
e-mail: antun.persin@rh-alan.tel.hr

Rheinmetall Industrie AG
PO Box 1663
D-40836 Ratingen
Germany
Tel: (+49 821) 02 90 01
Fax: (+49 821) 02 47 35 53
Web: http://www.rheinmetall.com

Rheinmetall Landsysteme GmbH (includes KUKA)
PO Box 43 13 69
D-86073 Augsburg
Germany
Tel: (+49 821) 79 70
Fax: (+49 821) 797 12 07

Ring Sights Ltd
PO Box 22
Bordon
Hampshire GU35 9PD
UK
Tel: (+44 1420) 47 22 60
Fax: (+44 1420) 47 83 59
e-mail: ringshights@easynet.co.uk

Romtehnica SA
5C Timis oara Boulevard
Nr. 77311
Bucharest, 6
Romania
Tel: (+40 21) 410 34 05
Fax: (+40 21) 410 14 67
e-mail: romtechnica@rdsnet.ro
Web: http://www.romtehnica.com.ro

Rosoboronexport Federal State Unitary Enterprise
21 Gogolevsky Boulevard
119865 Moscow
Russian Federation
Tel: (+7 095) 202 66 03
Fax: (+7 095) 202 45 94
Web: http://www.rusarm.ru

Rostov Optical-Mechanical Plant (ROMZ)
Savinskoye Schosse 1
Yaroslav Region
Rostov-the-Great 152150
Russia
Tel: (+7 08536) 307 46
Fax: (+7 08536) 338 18
e-mail: info@romz.ru
Web: http://www.romz.ru

Rudjer Boskovic Institute
Division of Laser and Atomic Research and Development
10001 Zagreb
Bijenicka c54
Croatia
Tel: (+385 1) 468 01 10
Fax: (+385 1) 468 01 04

Saab Bofors Dynamics
SE-69180 Karlskoga
Sweden
Tel: (+46 586) 810 00
Fax: (+46 586) 853 90
Web: http://www.saab.se

Saab Bofors Dynamics AB
SE-581 88 Linkoping
Sweden
Tel: (+46 13) 18 60 00
Fax: (+46 13) 18 60 06
Web: http://www.saab.se/dynamics

SaabTech Electronics AB
SE-175 88 Järfälla
Sweden
Tel: (+46 8) 58 08 40 00
Fax: (+46 8) 58 03 22 44
e-mail: gyhoe@informatics.saab.sc
Web: http://www.saab.se

SaabTech Systems AB
SE-175 85 Jarfalla
Sweden
Tel: (+46 8) 58 08 40 00
Fax: (+46 8) 58 03 22 44
e-mail: guhoe@systems.saab.se
Web: http://www.saab.se

SABCA (Société Anonyme Belge de Constructions Aeronautiques)
1470 Chaussée de Haecht/Haachtesteenweg 1470
B-1130 Brussels
Belgium
Tel: (+32 2) 729 55 11
Fax: (+32 2) 216 15 70
e-mail: sabca.secr@sabca.be
Web: http://www.sabca.com

SAGEM SA
Head Office
27 Rue Leblanc
F-75512 Paris Cedex 15
France
Tel: (+33 1) 40 70 63 63
Fax: (+33 1) 40 70 66 00
Web: http://www.sagem.com

SAGEM SA Aerospace & Defence Division
Le Ponant de Paris
27 rue LeBlanc
F-75512 Paris Cedex
France
Tel: (+33 1) 40 70 63 63
Fax: (+33 1) 40 70 66 00
Web: http://www.sagem.com

Sakr Factory for Developed Industries A.O.I
Suez Road
PO Box 33
Heliopolis
Cairo
Egypt
Tel: (+20 2) 415 02 30
Fax: (+20 2) 290 12 10

Samsung Thales Defence
Advanced Electronic Systems Division
17th Floor
Daechi Building 889-11
Daechi 4-Dong
Kangnam-Ku
Seoul 135-284
Korea, South
Tel: (+82 2) 39 58 11 81

Samsung Thales Defence
18th Floor
Daechi Bldg 889-11
Daechi 4-Dong
Kangnam-Ku
Seoul
Korea, South
Tel: (+8 22) 34 58 11 83
Fax: (+8 22) 34 58 11 39
Web: http://www.sec.co.kr

Seiler Instrument & Manufacturing Company Inc
170E Kirkham Avenue
St Louis
Missouri 63119
Tel: (+1 314) 968 22 82
Fax: (+1 314) 968 26 37
e-mail: mfg@seilerinst.com

Selectron Management Corporation
4711 Falcon Drive
Suite 355
Mesa
Arizona 85215
USA
Tel: (+1 602) 654 00 73
Fax: (+1 602) 654 00 68
Web: http://www.tronmcorp.com

Senet
16-1-19 Kravchenko Street
Moscow
Russian Federation
Tel: (+7 095) 211 87 83
Fax: (+7 095) 956 33 75

Sensytech Inc (Corporate HQ)
8419 Terminal Road
PO Box 1430
Newington Virginia 22122-1430
USA
Tel: (+1 703) 550 70 00
Fax: (+1 703) 550 74 70
Web: http://www.sensytech.com

Sensytech, Inc
Imaging Group
300 Parkland Plaza
PO Box 1869
Ann Arbor
Michigan 48106-1869
USA
Tel: (+1 313) 769 56 49
Fax: (+1 313) 769 04 29
e-mail: enginer@mich.com
Web: http://www.sensytech.com

Siemens Switzerland Ltd
Civil and National Security
Freilagerstrasse 40
CH-8047 Zurich
Switzerland
Tel: (+41 58) 558 48 49
Fax: (+41 58) 558 38 16
Web: http://www.siemens.ch/ics

Simrad Optronics ASA
Ensjøveien 236
PO Box 6114
Etterstad
N-0602 Oslo
Norway
Tel: (+47) 22 67 04 90
Fax: (+47) 22 19 29 91
Web: http://www.simrad-optronics.com

Sofradir
43/47 rue Camille Pelletan
F-92290 Chatenay-Malabry
France
Tel: (+33 1) 41 13 45 30
Fax: (+33 1) 46 61 58 84
e-mail: sofradir@sofradir.com
Web: http://www.sofradir.com

Sperry Marine
1070 Seminole Trail
Charlottesville
Virginia 22901-2891
USA
Tel: (+1 434) 974 20 00
Fax: (+1 434) 974 22 59
Web: http://www.sperry-marine.com

Spetztekhnika Vympel NPO, Moscow
90 Volocolamskoe Schosse
123424 Moscow
Russian Federation
Fax: (+7 095) 491 87 83

State Scientific Research and Engineering Institute
6 Shosse Entusiastov Street
143900 Balashikha
Moscow Region
Russian Federation
Tel: (+7 095) 521 71 71

Steyr-Daimler-Puch Spezialfahrzeug AG
2-Haidequerstrasse 3
PO Box 100
A-1111 Vienna
Austria
Tel: (+43 1) 760 64
Fax: (+43 1) 769 81 49

STN ATLAS Elektronik GmbH
Sebaldsbrucker Heerstr 235
D-28305 Bremen
Germany
Tel: (+49 421) 45 70
Fax: (+49 421) 457 29 00
e-mail: systemtechnik@stn-atlas.de
Web: http://www.stn-atlas.de

Swedish Space Corporation
PO Box 4207
Solna Strandvag 86
SE-171 04 Solna
Sweden
Tel: (+46 8) 627 62 00
Fax: (+46 8) 98 70 69
Web: http://www.ssc.sc

Systems & Electronics Inc (SEI)
201 Evans Lane
St Louis
Missouri 63121-1126
USA
Tel: (+1 314) 553 40 00
Fax: (+1 314) 553 49 49
e-mail: webmaster@seistl.com
Web: http://www.seistl.com

Tenebræx Corporation
326 A Street
Boston
Massachussets 02210
USA
Tel: (+1 617) 514 99 00
Fax: (+1 617) 514 99 98
e-mail: sales@camouflage.com
Web: http://www.camouflage.com

Tenix LADS Corporation Pty Ltd
Second Avenue
Technology Park
Mawson Lakes
South Australia 5095
Australia
Tel: (+61 8) 83 00 44 47
Fax: (+618) 83 49 75 28
e-mail: lads2@tenix.com
Web: http://www.tenix.com

Terma A/S
Headquarters
Hovmarken 4
DK-8520 Lystrup
Denmark
Tel: (+45) 87 43 60 00
Fax: (+45) 87 43 60 01
e-mail: terma.hq@terma.com
Web: http://www.terma.com

Terma A/S
Grenaa Office
Fabrikvej 1
DK-8500 Grenaa
Denmark
Tel: (+45) 86 32 19 88
Fax: (+45) 86 32 14 48
e-mail: terma.ind@terma.com
Web: http://www.terma.com

Terma A/S
Radar Systems Division
Hovmarken 4
DK-8520 Lystrup
Denmark
Tel: (+45) 87 43 60 00
Fax: (+45) 87 43 60 01
e-mail: terma.rsy@terma.com
Web: http://www.terma.com/radar

Textron Marine & Land Systems
19401 Chef Menteur Highway
New Orleans
Louisiana 70129
USA
Tel: (+1 504) 245 66 00
Fax: (+1 504) 254 80 01
Web: http://www.systems.textron.com

Thales AFV Systems Ltd (formerly Helio-Ltd)
Crabtree Manorway North
Belvedere
Kent DA17 6AY
UK
Tel: (+44 20) 83 19 75 00
Fax: (+44 20) 83 19 75 25
e-mail: info@helio-ltd.com
Web: http//www.helio-ltd.com

Thales Air Defence
Alanbrooke Road
Castlereagh
Belfast BT6 9HB
UK
Tel: (+44 28) 90 45 84 44
Fax: (+44 28) 90 73 38 72
e-mail: info.tad@fr.thalesgroup.com
Web: http://www.shortsmissiles.co.uk

Thales Air Defence SA
9 rue des Mathurins
PO Box 150
F-92221 Bagneux Cedex
France
Tel: (+33 1) 40 84 40 00
Fax: (+33 1) 40 84 33 81
Web: http://www.thalesgroup.com

Thales Air Defence Systems
7-9 rue des Mathurins
F-92223 Bagneux Cedex
France
Tel: (+33 1) 40 84 40 00
Fax: (+33 1) 40 84 33 81
e-mail: info.tad@fr.thalesgroup.com
Web: http://www.thales-airdefence.com

Thales Communications
Milford Industrial Estate
Tollgate Road
Salisbury
Wiltshire SP1 2JG
UK
Tel: (+44 1722) 32 39 11
Fax: (+44 1722) 33 00 16

Thales Cryogenie SA
4 rue Marcel Doret - BP 22
F-31701 Blagnac Cedex
France
Tel: (+33 5) 62 74 58 00
Fax: (+33 5) 62 74 58 58
e-mail: cryotechnologies@wanadoo.fr

Thales Nederland
Zuidelijke Havenweg 40
PO Box 42
NL-7550 GD Hengelo Ov
Netherlands
Tel: (+31 742) 48 81 11
Fax: (+31 742) 42 59 36
e-mail: info@nl.thalesgroup.com
Web: http://www.thales-nederland.com

Thales Nederland
Meerenakkerweg 1
PO Box 6034
NL-5600 HA Eindhoven
Netherlands
Tel: (+31 40) 250 36 03
Fax: (+31 40) 250 37 77
e-mail: info@nl.thalesgroup.com
Web: http://www.thales-nederland.nl

Thales Optics
Glascoed Road
St Asaph
Clwyd
LL17 0LL
UK
Tel: (+44 1745) 58 80 00
Fax: (+44 1745) 58 42 58
Web: http://www.thales-optics.co.uk

Thales Optronics
1 Linthouse Road
Govan
Glasgow G51 4BZ
UK
Tel: (+44 141) 440 40 00
Fax: (+44 141) 440 40 01
Web: http://www.thalesgroup-optronics.com

Thales Optronics (Taunton) Ltd
Lisieux Way
Taunton
Somerset TA1 2JZ
UK
Tel: (+44 1823) 33 10 71
Fax: (+44 1823) 27 44 13
e-mail: sales@avimo.co.uk
Web: http://www.thalesgroup.com

Thales Optronics (Vinten) Ltd
Vicon House
Western Way
Bury St Edmunds
Suffolk IP33 3SP
UK
Tel: (+44 1284) 75 05 99
Fax: (+44 1284) 75 05 98
e-mail: sales@wvintenltd.com
Web: http://www.thalesgroup-optronics.com

Thales Optronics Canada Inc
4868 Levy Street
Ville St Laurent
Montréal
Quebec H4R 2P1
Canada
Tel: (+1 514) 337 78 78
Fax: (+1 514) 337 11 07

Thales Optronique
rue Guynemer
BP 55
F-78283 Guyancourt Cedex
France
Tel: (+33 1) 30 96 70 00
Fax: (+33 1) 30 96 75 50
Web: http://www.thalesgroup.com

Thales Systemes Aeroportes SA
La Clef de St Pierre
1 boulevard Jean Moulin
BP 7
F-78852 Elancourt Cedex
France
Tel: (+33 1) 34 59 60 00
Fax: (+33 1) 34 59 62 36
e-mail: info@detexis.thomson-csf.com
Web: http://www.thalesgroup.com

Thalis Sensors S.A.
7 Stratigi Street
GR-154 51 Neo Psychiko
Greece
Tel: (+30 10) 672 86 10
Fax: (+30 10) 672 86 24
Web: http://www.thalis.gr

The Boeing Company
Integrated Defense Systems, Duluth
1800 Satellite Boulevard
Duluth
Georgia 30136
USA
Tel: (+1 770) 476 63 00
Fax: (+1 770) 476 65 78
Web: http://www.boeing.com

The Boeing Company
Autonetics and Missile Systems Division
3370 Miraloma Avenue
PO Box 3105
Anaheim
California 92803-3105
USA
Tel: (+1 714) 762 04 38
Fax: (+1 714) 762 12 43

Toshiba (Tokyo Shibaura Electric) Company Limited
1-6 Uchisaiwacho 1-chome
Chiyoda-ku
Tokyo 100
Japan
Fax: (+81 3) 34 56 16 31

Transvaro A.S.
Fatih Cadesi
Dereboyu SOLAK No 12
TR-34660 Halkali
Istanbul
Turkey
Tel: (+90 212) 473 01 00; 473 01 53
Fax: (+90 212) 548 52 84; 473 01 55
e-mail: mail@transvaro.com.tr
Web: http://www@transvaro.com.tr

Ulyanovsk Mechanical Plant
90 Moskovskoye Shosse
432008 Ulyanovsk
Russian Federation
Tel: (+7 8422) 31 75 58
Fax: (+7 8422) 32 61 63

United Defense Limited Partnership (UDLP)
Ground Systems Division, Pennysylvania
PO Box 15512
York
Pennsylvania 17405
USA
Tel: (+1 717) 225 80 04
Fax: (+1 717) 225 80 03

USAF Materiel Command
Kirtland AFB
3550 Aberdeen Avenue SE
New Mexico 87117-5776
USA
Tel: (+1 505) 846 19 11
Fax: (+1 505) 846 04 23
Web: http://www.de.afrl.af.mil

Vectop Ltd
PO Box 11388
IL-48901 Rosh Ha'ayin
Israel
Tel: (+972 3) 938 69 20
Fax: (+972 3) 902 40 54
e-mail: vectop@vectop.co.il
Web: http://www.vectop.co.il

Vectronix AG
Max-Schmidherny-Strasse
CH-9435 Heerbrugg
Switzerland
Tel: (+41 71) 727 47 47
Fax: (+41 71) 727 46 79

Vistar Night Vision Ltd
24 Doman Road
Camberley
Surrey GU15 3DF
UK
Tel: (+44 1276) 70 88 00
Fax: (+44 1276) 70 88 07
e-mail info@vistar.co.uk

VistaScape Technology Corporation
300 Galleria Parkway
Suite 690
Atlanta
Georgia 30339
USA
Tel: (+1 678) 919 11 30
Fax: (+1 678) 919 11 42
e-mail: info@vistascape.com
Web: http://www.vistascape.com

VTUVM - Military Institute for Weapon and Ammunition Technology
Dlouha 300
CZ-763 21 Slavicin
Czech Republic
Tel: (+42 577) 34 12 53
Fax: (+42 577) 34 12 52
e-mail: director@vtuvm.cz
Web: http://www.vtuvm.cz

Yugoimport SDPR
2 Bulevar Umetnosti
PO Box 89
11000 Novi Beograd
Serbia
Federal Republic of Yugoslavia
Tel: (+381 11) 14 39 33
Fax: (+381 11) 13 02 63
e-mail: sdpr@yugoimport.co.yu
Web: http://www.yugoimport.co.yu

Z/I Imaging Corporation
Worldwide Headquarters
301 Cochran Road
Suite 9
Huntsville
Alabama 35824
USA
Tel: (+1 888) 538 07 13
Fax: (+1 256) 730 15 90
Web: http://www.ziimaging.com

Zeiss Optronik GmbH
Carl Zeiss Strasse 22
D-73447 Oberkochen
Germany
Tel: (+49 7364) 20 65 30
Fax: (+49 7364) 20 36 97
e-mail: marketing-zeo@zeiss.de
Web: http://www.zeiss-optronik.de

Zenit Foreign Trade Firm, State Enterprise P/C S.A Zverev Krasnogorsky Zavod
8 Rechnaya Street
Krasnogorsk
143400 Moscow
Russian Federation
Tel: (+7 095) 561 33 77
Fax: (+7 095) 562 82 75

INDEXES

Manufacturers' Index

Alphabetical index

To help users of this title evaluate the published data, *Jane's Information Group* has divided entries into three categories.
[N] NEW ENTRY Information on new equipment and/or systems appears for the first time in the title.
[V] VERIFIED The editor has made a detailed examination of the entry's content and checked its relevancy and accuracy for publication in the new edition to the best of his ability.
[U] UPDATED During the verification process, significant changes to content have been made to reflect the latest position known to *Jane's* at the time of publication.
Items in italics refer to entries which have been deleted from this edition. These entries are still available in the online version of this product–jeos.janes.com.

N

O

P

Q

R

S

T

U

V

W

X

Z

NOTES

NOTES

NOTES

NOTES